AA002405

Proceedings of ASME 2023 International Design Engineering Technical Conferences and Computers and Information in Engineering Conference

(IDETC-CIE2023)

Volume 3A

49th Design Automation Conference (DAC)

August 20-23, 2023
Boston, Massachusetts

Conference Sponsors
Design Engineering Division

Computers and Information
in Engineering Division

THE AMERICAN SOCIETY OF MECHANICAL ENGINEERS
Two Park Avenue * New York, N.Y. 10016

© 2023 The American Society of Mechanical Engineers, 2 Park Avenue, New York, NY 10016, USA
(www.asme.org)

All rights reserved. Printed in the United States of America. Except as permitted under the United States Copyright Act of 1976, no part of this publication may be reproduced or distributed in any form or by any means, or stored in a database or retrieval system, without the prior written permission of the publisher.

INFORMATION CONTAINED IN THIS WORK HAS BEEN OBTAINED BY THE AMERICAN SOCIETY OF MECHANICAL ENGINEERS FROM SOURCES BELIEVED TO BE RELIABLE. HOWEVER, NEITHER ASME NOR ITS AUTHORS OR EDITORS GUARANTEE THE ACCURACY OR COMPLETENESS OF ANY INFORMATION PUBLISHED IN THIS WORK. NEITHER ASME NOR ITS AUTHORS AND EDITORS SHALL BE RESPONSIBLE FOR ANY ERRORS, OMISSIONS, OR DAMAGES ARISING OUT OF THE USE OF THIS INFORMATION. THE WORK IS PUBLISHED WITH THE UNDERSTANDING THAT ASME AND ITS AUTHORS AND EDITORS ARE SUPPLYING INFORMATION BUT ARE NOT ATTEMPTING TO RENDER ENGINEERING OR OTHER PROFESSIONAL SERVICES. IF SUCH ENGINEERING OR PROFESSIONAL SERVICES ARE REQUIRED, THE ASSISTANCE OF AN APPROPRIATE PROFESSIONAL SHOULD BE SOUGHT.

ASME shall not be responsible for statements or opinions advanced in papers or . . . printed in its publications (B7.1.3). Statement from the Bylaws.

For authorization to photocopy material for internal or personal use under those circumstances not falling within the fair use provisions of the Copyright Act, contact the Copyright Clearance Center (CCC), 222 Rosewood Drive, Danvers, MA 01923, tel: 978-750-8400, www.copyright.com.

Requests for special permission or bulk reproduction should be addressed to the ASME Publishing Department, or submitted online at: https://www.asme.org/publications-submissions/journals/information-for-authors/journalguidelines/rights-and-permissions

ISBN: 978-0-7918-8730-1

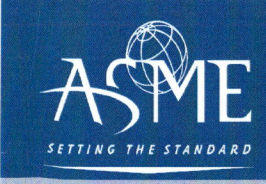

2023

PROCEEDINGS OF THE ASME

INTERNATIONAL DESIGN ENGINEERING TECHNICAL CONFERENCES AND COMPUTERS AND
INFORMATION IN ENGINEERING CONFERENCE

IDETC-CIE2023

VOLUME 3A

49TH DESIGN AUTOMATION CONFERENCE

(DAC 2023)

THE AMERICAN SOCIETY OF MECHANICAL ENGINEERS® (ASME)®

WELCOME TO ASME IDETC-CIE 2023!

On behalf of the ASME IDETC-CIE 2023 Conference Organizing Committee, it is our great pleasure to welcome you to Boston Park Plaza to participate in this year's conference! The International Design Engineering Technical Conferences & Computers and Information in Engineering Conference (IDETC-CIE) stands proudly as one of the premier ASME conferences. The 2023 ASME-IDETC-CIE marks a significant milestone as the second in-person conference after the challenges of the COVID-19 pandemic.

This year, the response to our call for presentations was beyond our expectations! We appreciate the excitement and commitment shown by our community, and we can't wait to experience the breadth of innovative ideas in the presentations. A special thanks goes to all the sub-conference organizing committees and the technical committees. Without your dedication and hard work, the success of this conference would not have been possible.

This year, we have dedicated efforts to enhance industry participation as we recognize the importance of collaboration between industry and academia. Recognizing the significance of diversity and inclusivity, we have curated activities that support our underserved and underrepresented community members.

Located in the vibrant city of Boston, a central hub for global engagement, this conference is very accessible to participants around the world. We invite you to take full advantage of this opportunity to share knowledge, collaborate, and forge new connections within our community at ASME IDETC-CIE 2023!!

Sincerely,

Sachin Goyal
Associate Professor, Department of Mechanical Engineering
University of California, Merced
Chair, ASME IDETC-CIE 2023 Conference Organizing Committee

Andreas Müller
Professor, Institute of Robotics
Johannes Kepler University Linz
Chair, ASME IDETC-CIE 2023 Conference Organizing Committee

Faez Ahmed
Assistant Professor, Department of Mechanical Engineering
Massachusetts Institute of Technology
Local Chair, ASME IDETC-CIE 2023 Conference Organizing Committee

Beshoy Morkos
Associate Professor, College of Engineering
University of Georgia
Student Activities Chair, ASME IDETC-CIE 2023 Conference Organizing Committee

ASME 25TH INTERNATIONAL CONFERENCE ON ADVANCED VEHICLE TECHNOLOGIES (AVT)

Ole Balling *Conference Chair*	**Angelo Bonfitto** *Program Chair*

The Vehicle Design Committee (VDC) promotes innovative analytical, computational, and experimental investigations in the dynamics, control, and design of full vehicle systems, subsystems, and components. With the increasing demands on driving safety and autonomy, the human-vehicle interaction, advanced driver assistance systems, and connected vehicles as well as sustainable propulsion systems and their coupling with the driver/vehicle system are included in the spectrum of topics addressed by VDC. Our members perform fundamental and applied research, and they implement technology for light/heavy vehicle design, modeling, and validation. The VDC is pleased to welcome you to the 24th International Conference on Advanced Vehicle Technologies, held as a part of the 2022 ASME IDETC-CIE. This year the AVT conference will consist of eight symposia for a total of eight sessions in the areas of: Ground Vehicles Dynamics and Controls; Modeling and Testing Tire-Terrain Interaction; Methods for Ground Vehicle Systems Design; Vehicle Electrification and Powertrain Design; Light Vehicles Design; Military and Commercial Ground Vehicle Design; and Intelligent Vehicles. We sincerely appreciate the time and services of these symposium organizers.

This Year the VDC is especially honored to host Dr. Bo Persson from the Peter Grünberg Institute, Jülich, Germany, for the William Milliken Lecture, which is entitled "Rubber friction, tire dynamics and ABS braking simulations". In addition, VDC with support from ASME-DED organizes a Workshop on Autonomous and Connected Vehicles that is open to any ASME member or IDETC attendee.

A Best Paper and a Student Best Paper (for papers authored and submitted by a student as the primary author) are awarded for conference papers that best exemplify the research advances in ground vehicle engineering based on peer reviews and the award committee's ranking.
We truly hope that this year's AVT Conference will provide you with an exciting, enriching, and rewarding experience!

43ᴿᴰ COMPUTERS AND INFORMATION
IN ENGINEERING DIVISION CONFERENCE (CIE)

Greetings All Attendees!

The Computers and Information in Engineering Division of ASME welcomes all IDETC-CIE 2023 Conference participants to the 43rd Annual Computers and Information in Engineering Conference (CIE) in Boston, MA (USA).

The CIE conference is a premier venue for the international exchange of technical, scientific, and application knowledge related to the theory and practice of computing to support engineering activities. It provides a forum for researchers, practitioners, educators, and students from academia, industry, and government research labs to share their latest findings and challenges with the broader research community, foster collaborations, and build a sustainable research and education community.

This year we are pleased to report that there will be over 140 technical presentations in the following technical and special topic sessions, organized around the four Technical Committees of the CIE Division, namely: Advanced Modeling and Simulation, Computer-Aided Product and Process Design, Systems Engineering and Information Knowledge Management, and Virtual Environments and Systems.

Advanced Modeling and Simulation (AMS):
- Inverse Problems in Science and Engineering
- Computational Multiphysics Applications
- Uncertainty Quantification in Simulation and Model Verification & Validation
- Simulation in Advanced Manufacturing
- Material Characterization Methods and Applications

Computer-Aided Product and Process Development (CAPPD):
- Human-In-the Loop for Product Design and Automation
- Digital Human Modeling for Design and Manufacturing
- Product and Process Design Automation for Industry 4.0
- Data-Driven Product Design and Fabrication

Systems Engineering Information Knowledge Management (SEIKM):
- Design Informatics
- Systems Engineering and Complex Systems
- Knowledge Capture, Reuse, and Management
- Smart Manufacturing Informatics
- Advanced Manufacturing for Bioeconomy and Circular Economy

Virtual Environments and Systems (VES):
- Designing User Experiences for Virtual Environments
- Virtual Systems for Engineering Applications
- VES Show-and-Tell

AI + ML Approaches for Engineering (General)

Joint Sessions:
- Digital Twin: Advanced Human Modeling and Simulation in Engineering
- Digital Twin Modeling and Analytics for Advanced Manufacturing
- Physics-Informed Machine Learning for Design and Advanced Manufacturing
- Artificial Intelligence and Machine Learning in Design and Manufacturing
- Design, Simulation and Optimization for Additive Manufacturing

In addition to the technical presentations, we will host several specialized events. Accompanying a CIE Keynote Talk, four panels of leading experts from industry, government, and academia will convene to discuss topics related to the future of Computers and Information in Engineering. The Journal of Computing and Information Science in Engineering (JCISE) Spotlight panel session will highlight top articles published over the past year. At the graduate student poster session, select graduate students, each the recipient of an award stipend, will showcase their excellent works.

In addition, we will use the CIE Luncheon to recognize conference best paper awards and the CIE Division awards. We invite you all to join us at the CIE Awards Ceremony Luncheon on Tuesday August 22nd to recognize some of the outstanding research being conducted by peers, colleagues, and students alike. As always, this year's conference would not be possible without the outstanding efforts andcontributions from ASME volunteers. This year's CIE Technical Committee meetings and Division meeting will be held on the evening of Tuesday, August 22nd. It is at these meetings where we acknowledge contributors from the past year while setting the stage for the upcoming year's activities. Please plan to attend and/or join one of these meetings to become further involved in CIE activities.

We would like to thank and recognize the Technical Committee leadership this year for their hard work and contributions:
Advanced Modeling and Simulation (AMS)
- **Piyush Pandita**, Chair
- **Ahn Tran**, Vice Chair
Computer Aided Product and Process Design (CAPPD)
- **Anand Balu Nellippallil**, Chair
- **Jida Huang**, Vice Chair
Systems Engineering and Information Knowledge Management (SEIKM)
- **Douglas Van Bossuyt**, Chair
- **Dazhong Wu**, Vice Chair
Virtual Environments and Systems (VES)
- **Vinayak Krishnamurthy**, Chair
- **Yunbo "WILL" Zhang**, Vice Chair

We would like to use this opportunity to thank our symposium organizers, including Seung-Kyum Choi, Piyush Pandita, Ahn Tran, James Yang, Ashish M. Chaudhari, John Michopoulos, John Steuben, Brian Dennis, Athanasios Iliopoulos, Guanglu Zhang, Zhimin Xi, Chao Hu, Yan Wang, Gaurav Ameta, Bjorn Johansson, Chiradeep Sen, Ehsan Esfahani, Anand Balu Nellippallil, Jida Huang, Tsz Ho Kwok, Giorgio Colombo, Daniele Regazzoni, Satchit Ramnath, Marco Rossoni, Anand Balu Nellippallil, Giovanni Berselli, Weiss Cohen, Jida Huang, Jun Wang, Luis Segura, Yan Lu, Zhuo Yang, Dazgong Wu, Douglas Van Bossuyt, Yaoyao Fiona Zhao, Ying Liu, Zhenghui Sha, Farhad Ameri, Chris Hoyle, Mutahar Safdar, Hyunwoong Ko, Boonserm Kulvatunyou, Evan

Wallace, Vincenzo Ferrero, Senthil Chandrasegaran, Rebecca Friesen, Ronak Mohanty, Vinayak Krishnamurthy, Junfeng Ma, Jinjuan She, Yunbo "WILL" Zhang, Yujiang Xiang, Xianlian Alex Zhou, Dehao Liu, Sheng Yang, Yanglong Lu, Jiarui Xie, Yaoyao Fiona Zhao, Jaehyuk Kim, Fahad Milaat, Jun Wang, Chih-Hsing Chu, Dehao Liu for their efforts and hard work in paper review coordination and recommendation. We would like to thank all reviewers for their time to provide valuable feedback and help maintain high standards and improve the quality of the conference. Last but not the least, we thank all authors for submitting and sharing their latest work to shape the research directions in this community.

Moreover, we thank you for your participation in the various activities of our CIE community. We look forward to seeing you all again next year!

49TH ASME DESIGN
AUTOMATION CONFERENCE (DAC)

Dear Colleagues,

On behalf of the DAC Executive Committee, welcome to the **49th ASME Design Automation Conference (DAC)**!

Following a rigorous review process, this year's DAC technical program consists of 124 accepted papers in 25 active research areas (corresponding approximately to an acceptance rate of 92%). For the first time this year, we also solicited and accepted 23 presentation-only submissions. The technical program will be presented from Monday, August 21 to Wednesday, August 23.

Complementing our technical sessions, we will host a signature event on "**Design for Safe and Reliable Autonomous Systems**", consisting of a panel of top experts from AI modeling, autonomous vehicles, and additive manufacturing, including:
- *Dr. Qi Hommes*, Senior Director, ZooX
- *Mr. Chris Robinson*, Senior Product Manager, Ansys
- *Dr. Heng Huang*, Professor, University of Maryland – College Park
- *Dr. Rajiv Malhotra*, Associate Professor, Rutgers University – New Brunswick

Please join us for the DAC committee meeting on the evening of Tuesday, August 22. During that meeting, we will also present the DAC Young Investigator Award winner and the DAC Best Paper Award winner. We look forward to having our community come together, meet old friends, and make new ones.

From the accepted papers, ten were identified as "Papers of Distinction". These papers are listed below (ordered by paper number and including the assigned session):

- DETC2023-109380: "*Model Consistency for Mechanical Design: Bridging Lumped and Distributed Parameter Models With a Priori Guarantees*", by Randi Wang and Morad Behandish

- DETC2023-110756: "*Mixed-Variable Global Sensitiviy Analysis With Applications to Data-Driven Combinatorial Materials Design*", by Yigitcan Comlek, Liwei Wang, and Wei Chen
- DETC2023-114999: "*Machine Learning-Based Model Bias Correction by Fusing Cae Data With Test Data for Vehicle Crashworthiness*", by Yang Li, Saeed Barbat, Zhenyan Gao, Guosong Li, Ying Zhao, Jice Zeng, and Zhen Hu.
- DETC2023-116586: "*A Reliability-Based Optimization Framework for Planning Operational Profiles for Unmanned Systems*", by Indranil Hazra, Joseph Southgate, Arko Chatterjee, Shapour Azarm, Katrina M. Groth, and Matthew Weiner.
- DETC2023-116622: "*Accounting for Model and Data Uncertainty in Machine Learning Assisted Mechanical Design*", by Xiaoping Du
- DETC2023-116743: "*Characterizing Designs via Isometric Embeddings: Applications to Airfoil Inverse Design*", by Qiuyi Chen and Mark Fuge
- DETC2023-116896: "*Concurrent Probabilistic Control Co-Design and Layout Optimization of Wave Energy Converter Farms Using Surrogate Modeling*", by Saeed Azad and Daniel R. Herber
- DETC2023-116962: "*Advise: AI-Accelerated Design of Evidence Synthesis for Global Development*", by Kristen Edwards, Binyang Song, Jaron Porciello, Carolyn Huang, Faez Ahmed, and Mark Engelbert.
- DETC2023-117013: "*Integrated Sustainable Product Design With Warranty and End-of-Use Considerations*", by Xinyang Liu and Pingfeng Wang
- DETC2023-117400: "*Car Drag Coefficient Prediction With Depth and Normal Renderings*", by Binyang Song, Chenyang Yuan, Faez Ahmed, Nikos Arechiga, and Frank Permenter

Authors from our community will present these and many other excellent papers throughout the conference. We encourage you to support your colleagues by attending their presentations and participating in the discussions.

Finally, organizing the conference requires the generous effort of many individuals. We are particularly grateful to all session organizers and paper review coordinators:

Faez Ahmed, Janet K. Allen, Jesse Austin-Breneman, A. Emrah Bayrak, Morad Behandish, Bill Bernstein, Ramin Bostanabad, Amy Bilton, Wei (Wayne) Chen, Souma Chowdhury, Daniel Cooper, Xiaoping Du, Bryony DuPont, Paul Egan, Ehsan Esfahani, Cong Feng, Yan Fu, Payam Ghassemi, Joshua Hamel, Daniel Herber, Zhen Hu, Horea Ilies, Namwoo Kang, Leifur Leifsson, Mian Li, Xingchen Liu, Yuanzhi Liu, Nordica MacCarty, Ali Mehmani, Nicholas Meisel, Zhenjun Ming, Farrokh Mistree, Seung Ki Moon, Beshoy Morkos, Venkat Nemani, Saigopal Nelaturi, Julián Norato, Philip Odonkor, Herschel Pangborn, Rahul Renu, Daniel Selva, Ada-Rhodes Short, Binyang Song, Eun Suk Suh, Ahn Tran, Zequn Wang, Kate Whitefoot, Natasha Wright, Hongyi Xu, Nita Yodo, Jie Zhang, Zhibo Zhang, Fiona Zhao, Yuqing Zhou

On behalf of the entire DAC community, we welcome you to another enjoyable and thought-provoking Design Automation Conference.

We look forward to seeing you in Boston!

Christopher McComb
Conference Chair

Chao Hu
Program Chair

20TH INTERNATIONAL CONFERENCE ON DESIGN EDUCATION (DEC)

On behalf of the Design Education Committee, we welcome you to the 20th annual International Conference on Design Education. The focus of this conference is on design education among educators, practitioners, and researchers.

This year's DEC Program consists of four technical symposia – (DEC-1) *Implementation, Assessment and Research Methods Across the Curriculum* (DEC-1)*, Diversity and Inclusion in Design Education* (DEC-2)*, Innovative Practices in Design Education* (DEC-3), and *Demos and Presentation Only* (DEC-4). The Demos and Presentation Only session will include presentations and provide ample opportunity for discussion with the presenters to give feedback on emerging design education research. Refer to the conference Technical Program for the times and locations of the technical sessions. In addition to our technical symposia, we will be continuing our mentorship program for graduate students.

The DEC Best Paper for the 2023 Conference is:

IDETC2023-116688, "Nature Versus Nurture: The Influence of Classroom Creative Climate on Risk-Taking Preferences of Engineering Students," Authors: Aoran Peng, Jessica Menold, Scarlett Miller

We extend special appreciation to our technical session Review Coordinators: Mohammad Fazelpour, Elizabeth Starkey, and Charlotte de Vries. We also give our sincerest thanks to all the reviewers of technical papers; they have ensured the quality of this year's conference.

The DEC technical committee meeting will be posted in the Technical Program. At the meeting we present many of the DEC Awards and plan for next year's conference, which includes the election of new committee leadership members. Everyone is welcome to attend, including new attendees and graduate students. Our meeting is streamlined to respect members' participation in other committees.

Nicholas Meisel
Conference Chair

Rahul Renu
Conference Program Chair

28TH DESIGN FOR MANUFACTURING AND THE LIFE CYCLE CONFERENCE (DFMLC)

The ASME Design for Manufacturing and the Life Cycle Committee welcomes participants to the 28th Annual Design for Manufacturing and the Life Cycle Conference. The ASME Design for Manufacturing and the Life Cycle Conference is the main international forum for the exchange of technical and scientific information on the theory and practice of Integrated Product and Process Development, Sustainable Design and Manufacturing, Product Lifecycle Management (PLM), and Design for X (DFX) Methods. This conference provides a forum for researchers, practitioners, and educators from academia, government organizations, and industry to share their latest results and challenges with the research community.

We are happy to report that this year's conference continues to feature many new and exciting results and methods to be presented as part of the conference's technical sessions. This year's DFMLC conference includes 21 technical papers and 22 technical presentations across 8 sessions, as follows:

- Session 1: Life Cycle & Human Factors Decision Making
- Session 2: Modeling and Optimization for Sustainable Design and Manufacturing
- Session 3: Design for Supply Chain, End of Life Recovery, and Large Systems
- Session 4: Design for Manufacturing and Assembly
- Session 5: Design for Additive Manufacturing 1
- Session 6: Design for Additive Manufacturing 2
- Session 7: Design of Product-Service and Energy Systems
- Session 8: Special Session: Design Tool & Commercialization Showcase

We would like to thank all the authors for submitting papers, the paper reviewers for sharing their time and expertise, and the session chairs/co-chairs for their participation. Special thanks go to the DFMLC Special Session Chair, Albert Patterson, and the paper review coordinators/co-coordinators for managing the papers through the review process: Hao Zhang, Vincenzo Ferrero, William Bernstein, Bryony Dupont, Yong Hoon Lee, Sara Behdad, Yongxian Zhu, Soonjo Kwon, Satya Peddada, Yaoyao Zhao, Xinyi (Serena) Xiao, Albert Patterson, Amin Mirkouei, Abigail Clarke-Sather, Paul Egan. Your participation and hard work have been vital for the success of the DFMLC conference!

This year, Dr. Gul Kremer, Dean for the University of Dayton School of Engineering, will present the DFMLC keynote lecture. Professor Kremer's research accomplishments focus on applied decision sciences and operations research for product and design systems, and other research interests include sustainability, system complexity, design creativity, and engineering education. There will be a presentation of the 2023 DFMLC Conference Kos Ishii-Toshiba Award for sustained and meritorious contributions to design for manufacturing and the life cycle at the DED luncheon on Monday, August 21st.

The 2023 DFMLC Conference also features a special presentation session. The "Design Tool & Commercialization Showcase" highlights new design tools developed by the members of the ASME Design community in both digital and physical forms.

The DFMLC technical committee meeting will include a review of DFMLC activities during the 2022-2023 cycle. The DFMLC Awards, including the Best Paper Award for the 2023 DFMLC conference, will also be presented in this meeting, and the technical committee will plan for next year's conference. Everyone is welcome to attend.

On behalf of the entire DFMLC community, we welcome you to the 28th Design for Manufacturing and the Life Cycle virtual conference!

Paul Egan
Conference Program Chair

Daniel Cooper
Conference Chair

35TH INTERNATIONAL CONFERENCE ON DESIGN THEORY AND METHODOLOGY (DTM)

On behalf of the ASME Design Theory and Methodology Committee, we would like to welcome you to the 35th International Conference on Design Theory and Methodology (DTM). Our conference focuses on fundamental design theory and methodologies, and their application in engineering contexts, with contributions provided by both researchers and practitioners.

This 2023 DTM conference includes 54 technical paper presentations and 10 lightning talks. Thematically, the conference includes contributions associated with our four broad foci: Design Theory, Design Methods, Design People, and Design Practice. In addition, this year's conference features a joint session between the Design Education Committee, the Design for Manufacture and Lifecycle Committee, and DTM titled *DfAM Principles and their Education*. This year's conference also features a student poster session where selected Ph.D. students showcase their dissertation proposals.

There were 72 papers submitted and reviewed by an incredible cohort of review coordinators and reviewers. A total of 239 reviews were completed by 155 reviewers. The review coordinators for this year's conference include: Ambrosio Valencia Romero, Astrid Layton, Christine Toh, Hyeonik Song, James Righter, Jinjuan She, Joshua Summers, Kelley Dugan, Kosa Goucher-Lambert, Maha Haji, Mansur M. Arief, Paul Grogan, Rohan Prabhu, Srinivasan Venkataraman, Vivek Rao, Vrushank Phadnis, Youyi Bi, and Zhenghui Sha. It is through the service of these individuals that we are able to maintain the high-quality expectations of the DTM conference.

We are excited to welcome you to this year's conference and hope that you find it engaging, informative, and beneficial.

Conference Chair
Dr. Vimal K. Viswanathan
San Jose State University

Program Chair
Dr. Rahul S. Renu
Francis Marion University

19TH IEEE/ASME INTERNATIONAL CONFERENCE ON MECHATRONICS AND EMBEDDED SYSTEMS AND APPLICATIONS (MESA 2023)

We are pleased to welcome everyone to the 19th IEEE/ASME International Conference on Mechatronics and Embedded Systems and Applications (MESA 2023). The goal of the MESA 2023 is to bring together experts from the fields of mechatronic and embedded systems, disseminate the recent advances in the area, discuss future research directions, and exchange application experience. MESA 2023 will especially bring out and highlight the latest research results and developments in Industry 4.0 and Artificial Intelligence (AI) in the fields of mechatronics and embedded systems. The success of MESA 2023 would be impossible without the tireless effort and dedicated work of the Members of the Organizing Committees. We would like to express our sincere thanks to Symposium Chairs for their wisdom and hard work in coordinating the review of all submitted papers. We are grateful for Members of the International Program Committee and reviewers for their thorough review of the papers. This year the program committee selected about 30 technical presentations following a review process by two or more expert reviewers for each proposed paper. We sincerely hope that MESA 2023 will be a place for excellent discussions that will put forward new ideas advance educational endeavors and promote active research collaborations.

Conference Chair,
Prof. Adriano Mancini
Università Politecnica delle Marche, Ancona, IT

Program Chair,
Prof. Matteo Claudio Palpacelli
Università Politecnica delle Marche, Ancona, IT

17TH INTERNATIONAL CONFERENCE ON MICRO- AND NANOSYSTEMS (MNS)

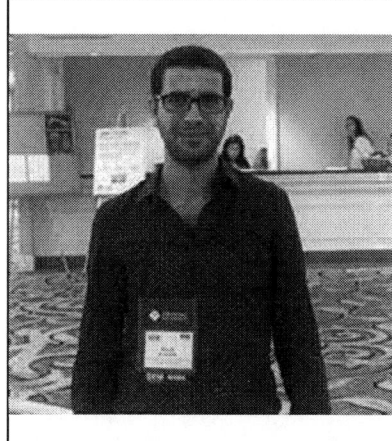	**Conference Chair:** Prof. Najib Kacem, Ph.D., FEMTO-ST Institute, France, najib.kacem@femto-st.fr		**Program Chair:** Prof. Jian Zhao, Ph.D., Dalian University of Technology, China, jzhao@dlut.edu.cn

Welcome to the17th International Conference on Micro- and Nano-systems (MNS) with the topic of "The Next Advances in MEMS", we would like to welcome you and thank you for participating. This conference, sponsored by the Technical Committee of Micro and Nano-systems, an integral part of the ASME Design Engineering Division, will provide researchers in industry, academia, and government a forum to exchange scientific and technical information related to recent developments and emerging issues in the design, mechanics, dynamics, control, and fabrication of micro- (MEMS) and nano-scale (NEMS) systems.

This conference is organized around 5 technical sessions, one of which is jointly offered with the 19th International Conference on Multibody Systems, Nonlinear Dynamics, and Control and the 35th Conference on Mechanical Vibration and Noise:

- Keynote Lecture: Professor Ashwin A. Seshia
- MNS-1: Nonlinear Dynamics and Vibrations of MEMS and NEMS (joint session with MSNDC and VIB)
    - Organizers: Najib Kacem (najib.kacem@femto-st.fr), Hanna Cho (cho.867@osu.edu), Jian Zhao (jzhao@dlut.edu.cn)
- MNS-2: Micro/Nano Bioengineering
    - Organizers: Dumitru Caruntu (Dumitru.Caruntu@utrgv.edu), Brian Jensen (bdjensen@byu.edu), Chu-Yu Huang (tomhuang@nchu.edu.tw)
- MNS-3: Micro/Nano Robotics and Functional Materials
    - Organizers: Irene Fassi (Irene.Fassi@stiima.cnr.it), Yu Liu (yu.liu@vip.163.com), Hoe Joon Kim (joonkim@dgist.ac.kr), Mohammad H. Hasan (hhasan_mohammad@columbusstate.edu), Longquiu Li (longqiuli@hit.edu.cn)
- MNS-4: Micro/Nano IoT, Sensors and Computing
    - Organizers: Muhammad Raziuddin A. Khan (muhammad.khan@navy.mil), Fadi Alsaleem (falsaleem2@unl.edu), Pourkamali Anaraki Siavash (Siavash.Pourkamali@utdallas.edu)
- MNS-5: Micro/Nano Power Sources and Storage
    - Organizers: Oliver M. Barham (oliver@olivermbarham.com), Muhammad Raziuddin A. Khan (muhammad.khan@navy.mil), Marc Litz (marc.s.litz.civ@mail.mil)

This conference provides a forum for researchers, practitioners, educators, and students from industry, academia, and government research labs to share their latest findings and challenges with the broader research community, foster collaborations, and build a sustainable research community.

We are pleased to offer Pr Ashwin A. Seshia as the MNS keynote speaker. Ashwin A. Seshia is a Professor of Microsystems Technology in the Department of Engineering at Cambridge University and a Fellow of Queens' College, Cambridge. He has acted in numerous service and leadership roles for the MEMS, sensors, and frequency control technical communities, and is currently an Editor of the IEEE Journal of Microelectromechanical Systems and a member of the executive committee of the European Frequency and Time Forum.

We would like to thank all the authors for submitting papers and talks and sharing their work in our conference. We would also like to thank the reviewers for providing valuable feedback to help improve the reporting and the quality of the conference, and finally the session chairs and co-chairs that worked on coordinating the paper review process.

We welcome conference participants to become involved with our technical committee. If you are interested in becoming involved in helping to organize our conference, please contact a conference organizer to inquire, and feel free to attend the technical committee meeting which will be held on Tuesday evening, Aug 22nd, from 6-7pm. This meeting is open to all. Room locations are announced in the program. Our community will continue to grow and flourish with your active participation as we work to define our vision for future events.

We welcome you to the17th International Conference on Micro- and Nanosystems (MNS)!
Sincerely,
Najib, Jian, and the entire 2023 MNS Conference team.

47TH MECHANISMS AND ROBOTICS CONFERENCE (MR)

The Mechanisms and Robotics Technical Committee of the ASME Design Engineering Division would like to warmly welcome you to the 47th Mechanisms and Robotics Conference, the premier international forum for the exchange of technical and scientific information on the theory and application of mechanical systems, mechanisms, and robotics.
The first conference, as The Conference on Mechanisms, was held at Purdue University, West Lafayette, Indiana, in 1953. ASME took over the conference and formed the ASME Biennial Mechanisms Conference in 1964. The conference was renamed the ASME Biennial Mechanisms and Robotics Conference in 2000. Starting in 2005, the conference became an annual conference, the ASME Mechanisms and Robotics Conference. Nowadays, the Mechanisms and Robotics Conference is held annually as a part of the ASME International Design Engineering Technical Conferences & Computers and Information in Engineering Conference.

This year we have assembled an exciting conference program and a slate of activities for the attendees, with 100 peer-reviewed technical papers and six technical presentations/posters organized into 9 technical symposia, a keynote speech, an early career invited talk session, a symposium keynote speech, and the Student Mechanisms and Robot Design Competition. Paper topics range throughout areas central to the design of mechanical, mechatronic, and robotic systems, including kinematics, dynamics, design, analysis and validation, compliant mechanisms, origami-based design, metamaterials for mechanisms, novel mechanisms and robots, mobile robots, and various applications. Our Keynote Speech will be given by Prof. Jian Dai, Director of Institute of Robotics at Southern University of Science and Technology and Honorary Chair Professor of King's College London, with his speech entitled: "Reconfiguration that Evolves into Robotics in Arts, Healthcare, and Production".

Submitted papers were eligible for several awards, including the Mechanisms and Robotics Best Paper award, A.T. Yang Memorial award, and Compliant Mechanisms award. The authors of selected papers of the Mechanisms and Robotics Conference are invited to submit enhanced archival versions of their papers to an IDETC Special Issue of the *ASME Journal of Mechanisms and Robotics*. We would like to thank Jian Dai, Chair of the Awards Committee, for coordinating the selection of the awards. Please attend our award session preceding the MR Keynote speech for the presentation of these awards.

The conference and program chairs would like to extend special thanks to all the volunteers who participated in the peer-review process to produce this high-quality program, especially the symposium organizers who coordinated the process:

MR-1: Mechanisms Synthesis & Analysis: Latifah Nurahmi, Kuan-Lun Hsu, Jieyu Wang
MR-2: Theoretical & Computational Kinematics (A.T. Yang Symposium): Nina Robson, Hongliang Shi, Haohan Zhang
MR-3: Compliant Mechanisms: Hongzhe Zhao, Jovana Jovanova, Giovanni Berselli
MR-4: Origami-Based Engineering Design: Shikui Chen, Suyi Li, Jared Butler
MR-5: Motion Planning, Dynamics, and Control of Robots: Damien Chablat, Joo Kim, Andreas Mueller, Jeffrey W. Herrmann
MR-6: Medical and Rehabilitation Robotics: Carl Nelson, Abbas Fattah
MR-7: Novel Mechanisms, Robots, and Applications: Guowu Wei, Reza Fotouhi, Salih Abdelaziz
MR-8: Soft and Continuum Mechanisms: Girish Krishnan, Sree Kalyan Patiballa, Vishesh Vikas
MR-9: Design, Analysis and Fabrication of Architected Materials and Structures: Nilesh D. Mankame, Pablo D. Zavattieri, David Restrepo, Tian "Tim" Chen
MR-10: Student Mechanism and Robot Design Competition: Long Wang, Gaurav Singh, Haiyang Li, Huijuan Feng, Colette Abah
MR-11: Special Early Career Session of Invited Presentations: Mark Plecnik

We extend special thanks to all authors, reviewers, presenters, symposium organizers, session chairs, and other volunteers who have contributed to the overall success of the conference. We trust that you will enjoy the conference and look forward to your continued support to our future Mechanisms and Robotics Conferences.

Conference Chair and Co-Chair:
Dongming Gan, Purdue University
Ketao Zhang, Queens Mary University of London

Program Chair and co-Chairs:
Guangbo Hao, University College Cork
Yu She, Purdue University
Mark Plecnik, University of Notre Dame

19TH INTERNATIONAL CONFERENCE ON MULTIBODY SYSTEMS, NONLINEAR DYNAMICS, AND CONTROL (MSNDC)

On behalf of the ASME Technical Committee on Multibody Systems and Nonlinear Dynamics, we extend a wholehearted welcome to the attendees of the 19th International Conference on Multibody Systems, Nonlinear Dynamics, and Control (MSNDC). Consisting of 14 symposia, the conference features nearly 70 presentations covering traditional and emerging topics in the broad areas of multibody systems and nonlinear dynamics. This event presents a unique opportunity for researchers, practitioners, educators, and students to report their accomplishments, exchange ideas, and become familiar with emerging trends in the field. The conference is organizing the MSNDC Best Paper and Best Student Paper competitions.

This year, we are honored to recognize Professor Gábor Stépán as the recipient of the Lyapunov Award for his seminal contributions in numerical methods and cutting-edge applications such as machine tool vibrations, balancing, wheel shimmy, vehicle traction, breaking, stability of robot control, human balancing and traffic control. Established in 2003, the Lyapunov Award recognizes lifelong contributions to the field of nonlinear dynamics.

We are honored to host two keynote lectures by Professor Arvind Raman and Professor Aki Mikkola. Professor Arvind Raman is the Robert V. Adams Professor in Mechanical Engineering at Purdue University. His research focuses on exploiting nonlinear dynamics for innovations in diverse interdisciplinary areas such as nanotechnology, biomechanics, and appropriate technologies for sustainable development. He is an ASME fellow, an ASME Gustus Larson Memorial Award recipient, Keeley fellow (Oxford), College of Engineering outstanding young investigator awardee, and NSF CAREER awardee.

Professor Aki Mikkola is a Professor in the Department of Mechanical Engineering at LUT University, Lappeenranta, Finland. Currently, he leads the research team of the Laboratory of Machine Design. He has been awarded five patents, has contributed to more than 150 peer-reviewed journal papers, and has presented more than 100 conference articles. His major research activities are related to flexible multibody dynamics, rotating structures, and biomechanics. He is currently Editor-in-Chief of the Journal of Multibody System Dynamics (Springer).

Last but not least, we would like to acknowledge the all-important effort and contribution made by the symposium organizers as well as manuscript reviewers – thank you very much. Your help has been essential. We would also like to thank all contributors for choosing this conference as the venue for sharing the outcomes of their intellectual pursuits.

Conference Co-Chairs:
Hiroyuki Sugiyama
University of Iowa
Pierpaolo Belardinelli
University of Marche

Program Co-Chairs:
Kiran X. D'Souza
Ohio State University
Grzegorz Orzechowski
LUT University

2023 POWER TRANSMISSION AND GEARING CONFERENCE (PTG)

On behalf of the ASME Technical Committee on Power Transmission and Gearing (PTG), we would like to extend a wholehearted welcome to the attendees of the 2023 International Conference on Power Transmission and Gearing. We thank all of the authors for choosing this forum to share their latest research findings, and all of those who chose to attend this conference. The contributions of the leading researchers and practitioners from around the world make this conference an ideal forum and opportunity for enhancing the technology of power transmission and gearing and exchanging ideas. We hope you will take advantage of this unique opportunity to learn about the latest research work, become familiar with emerging trends in the field and network as well.

PTG 2023 features outstanding full research papers and presentations covering a wide range of topics on power transmission and gearing, which include:

- Gear Geometry
- Gear Analysis
- Materials, Fatigue
- Gear Dynamics and Noise
- Gearbox Design, Reliability, and Diagnostics
- Gear Manufacturing
- Lubrication and Efficiency
- Bearings, Clutches, Couplings, and Splines
-

We acknowledge and thank all the reviewers for their support and assistance and all the following members of the PTG Committee for their dedicated service and efforts in organizing this conference:

Paris Altidis, Flexco Steel Lacing Company
Christopher Cooley, Oakland University
Richard Dippery, Kettering University
Brian Dykas, US Army Research Laboratory
Qi Fan, The Gleason Works
Alfonso Fuentes, Rochester Institute of Technology
Robert Giachetti, Exponent
Robert Handschuh, NASA Glenn Research Center
Adrian Hood, US Army Research Laboratory
Mohammad Hotait, General Motors
Don Houser, The Ohio State University
Murat Inalpolat, University of Massachusetts Lowell
Ahmet Kahraman, The Ohio State University
Mark Klein, Honda Motor Company
Mohsen Kolivand, Meritor
Timothy Krantz, NASA Glenn Research Center
Sheng Li, Wright State University
Teik C. Lim, New Jersey Institute of Technology
Kenneth Nowaczyk, Ford Motor Company
Robert Parker, University of Utah
Alfred Pettinger, Engineering Systems Inc.
Steve Siegert, Borg Warner
Avinash Singh, General Motors
Jeremy Wagner, John Deere Product Engineering Center
Yawen Wang, University of Texas, Arlington
Jon Williams, Hilliard Corporation
Brian Wilson, Advanced Drivetrain Engineering
Carlos Wink, Eaton Vehicle Group

We look forward to a successful conference and hope that you enjoy all the events and your stay in Boston as well.

David Talbot
Ohio State University
Conference Chair

Hai Xu
General Motors
Program Chair

35TH CONFERENCE ON MECHANICAL VIBRATION AND NOISE (VIB)

On behalf of the Technical Committee on Vibration and Sound (TCVS), we cordially welcome you to the 35th Conference on Vibration and Noise (VIB). This conference covers a broad spectrum of topics related to vibratory systems including those at emerging frontiers of science and engineering as well as traditional fields where mechanical vibrations are essential. VIB provides a setting for dissemination and discussion of the state of the art of modeling, analysis, and experimentation in all aspects of vibration and noise research. This year's conference includes close collaborations with other IDETC tracks to bring together researchers with similar interests, enhance the technical program, and improve the attendee experience. The following symposia make up this year's VIB:

VIB-1	Dynamics and Waves in Structures and Metamaterials
VIB-2	Vibration and Stability of Mechanical Systems
VIB-3	Energy Harvesting
VIB-4 and MR-4	Origami-Inspired Engineering: Design, Dynamics, and Everything in Between
VIB-5, MNS-1 and MSNDC-13	Nonlinear Dynamics and Vibrations of MEMS and NEMS
VIB-6 and MSNDC-3	Contact Dynamics and Jointed Structures
VIB-7 and MSNDC-14	Industrial Applications of Vibration, Acoustics and Dynamics
VIB-8 and MSNDC-12	Rotating Systems and Rotor Dynamics
VIB-9 and MSNDC-4	Nonlinear Systems & Phenomena
VIB-10 and MSNDC-6	Machine Learning Applications in Vibrations and Dynamics
VIB-11 and MSNDC-7	Time-delay, Time-varying and Discontinuous Dynamical Systems
VIB-12 and MSNDC-11	Dynamics and Control of Smart Structures and Systems
VIB-13	Dynamics of Biological, Bio-Inspired and Biomimetic Systems
VIB-14 and PTG-3	Gear Dynamics and Noise
VIB-15	Vibration Measurement, Signal Processing, and Structural Damage Detection

VIB is highlighted by keynote lectures celebrating top honors given by TCVS. Professor Steven Shaw from the Florida Institute of Technology is the recipient of the J. P. Den Hartog Award for lifetime contributions to the teaching and practice of vibration engineering. His keynote talk is titled "Centrifugal Pendulum Vibration Absorbers – from Den Hartog to Now". Professor Bogdan Epureanu from the University of Michigan will present a keynote talk titled "Physics-Informed Data-Driven Methods for Reduced Order Modeling in Dynamics." He is the recipient of the N. O. Myklestad Award in recognition of his innovate contributions to vibration engineering in the area of multi-physical systems. Professor Kathryn Matlack from the University of Illinois Urbana-Champaign is the recipient of the C. D. Mote, Jr. Early Career Award for research excellence in the field of vibration and acoustics. Her keynote talk is titled "Manipulating Vibrations with Phononic Materials".

As part of VIB, TCVS graciously sponsors a student paper competition and student travel support program. This year's conference includes a new undergraduate research symposium on dynamics, vibration and acoustics.

We gratefully acknowledge the efforts of the VIB symposium organizers, reviewers, and authors. It is your efforts that make this conference vibrant.

Christopher G. Cooley	Mark Jankauski
Oakland University	Montana State University
Conference Chair	Technical Program Chair

2023 IDETC-CIE CONFERENCE ORGANIZING COMMITTEE

Sachin Goyal
University of California
Conference Co-Chair

Andreas Müller
Johannes Kepler University Linz
Conference Co-Chair

Beshoy Morkos
University of Georgia
Student Activities Chair

Faez Ahmed
Massachusetts Institute of Technology
Local Chair

Computers & Information in Engineering Division Executive Committee (CIE)
Past Chair/Awards Chair: Mahesh Mani
Chair: Paul Witherell
Vice Chair & Conference Chair: Caterina Rizzi
Program Chair: Robert Wendrich
Secretary: Krishanand Kaipa
Member at Large: John Steuben
Member at Large: Daniela Faas
Liaison: Marc Halpern

Design Engineering Division Executive Committee (DED)
Past Chair: Dane Quinn
Chair/Vice Chair: Micky Caruntu
Technical Committee Liaison: Scott Ferguson
Conference Executive: Scarlet R. Miller
Secretary: Mary Frecker
Honors & Awards: Stefano Lenci
Publications: Michael Kokkolaras

CONFERENCE ORGANIZERS

25th International Conference on Advanced Vehicle Technologies (AVT)

Costin Untaroiu
Virginia Tech University
Conference Chair

Ole Balling
Aarhus University, Denmark
Program Chair

43rd Computers and Information in Engineering Division Conference (CIE)

 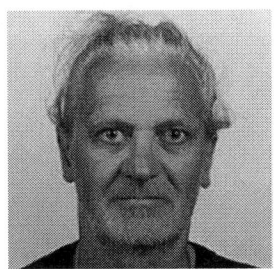

Caterina Rizzi
University of Bergamo
Conference Chair

Robert E. Wendrich
Rawshaping Technology RBSO
Program Chair

49th ASME Design Automation Conference (DAC)

Chris McComb
Penn State University
Conference Chair

Chao Hu
University of Connecticut
Program Chair

20th International Conference on Design Education (DEC)

Nicholas Alexander Meisel
Penn State University
Conference Co-Chair

Rahul Renu
Francis Marion University
Conference Co-Chair

28th Design for Manufacturing and the Life Cycle Conference (DFMLC)

Daniel Cooper
University of Michigan
Conference Chair

Paul Egan
University of Michigan
Texas Tech University

35th International Conference on Design Theory and Methodology (DTM)

Vimal K. Viswanathan
San Jose State University
Conference Chair

Rahul Renu
Francis Marion University
Program Chair

19th IEEE/ASME International Conference on Mechatronic and Embedded Systems and Applications (MESA)

Adriano Mancini
Università Politecnica delle Marche
Conference Chair

Matteo-Claudio Palpacelli
Università Politecnica delle Marche
Program Chair

47th Mechanisms and Robotics Conference (MR)

Dongming Gan
Purdue University
Conference Chair

Ketao Zhang
*Queen Mary
University of London*
Conference Co-Chair

Dongming Gan
Guangbo Hao
Program Chair

Yu She
Purdue University
Program Co-Chair

Mark Plecnik
*University of Notre
Dame*
Program Co-Chair

17th International Conference on Micro- and Nano-Systems (MNS)

Najib Kasem
*Univ. Bourgogne
Franche-Comté,*
Program Co-Chair

Jian Zhao
*Dalian University of
Technology*
Program Co-Chair

19th International Conference on Multibody Systems, Nonlinear Dynamics, and Control (MSNDC)

Pierpaolo Belardinelli
Polytechnic University of Marche
Conference Chair

Hiroyuki Sugiyama
University of Iowa
Conference Co-Chair

Kiran X. D'Souza
Ohio State University
Program Co-Chair

Grzegorz Orzechowski
LUT University
Program Co-Chair

2023 International Power Transmission and Gearing Conference (PTG)

David Talbot
The Ohio State University
Program Chair

Hai Xu
Oakland University
Program Co-Chair

35th Conference on Mechanical Vibration and Noise (VIB)

Chris Cooley
Oakland University
Conference Chair

Mark Jankauski
Montana State University
Program Chair

REVIEWERS

Mohammad A. Al-Shudeifat

Colette Abah

Salih Abdelaziz

Abdessattar Abdelkefi

Lanie Abi

Gizem Acar

Gabriele Maria Achilli

Hamid Afshari

Fatemeh Afzali

Alice Agogino

Milton Aguirre

Malena Agyemang

Faez Ahmed

Muhammad Hassaan Ahmed

Christopher Aksland

Louay Al Roomi

Oreoluwa Alabi

Michael Alexander

Khaled Alhazza

Hessein Ali

Sami Alkharabsheh

Douglas Allaire

Janet K. Allen

James Allison

Ali Almandeel

Mohammad Alsager Alzayed

Fadi Alsaleem

Abdulrahman Al-Shanoon

Farhad Ameri

Gaurav Ameta

Ali Amoozandeh

Di An

Luis Angel

Anusha Anisetti

Joel Anstrom

Nicole Apetre

K. Arabian

Vigen Arakelyan

Andrea Arena

Mojtaba Arezoomand

Mansur Arief

Keisuke Arikawa

Ryan Arlitt

Jessica Armstrong

Mauricio Arredondo

Alessio Artoni

Manan Arya

Keivan Asadi

Alireza Asadpoure

Doris Aschenbrenner

Simone Asci

Omar Ashour

Vanessa Audrey

Daniel Aukes

Jesse Austin-Breneman

Alkim Avsar

Saeed Azad

Enrico Babilio

Erin E. Bachynski-Polić

Noah Bagazinski

Duygu Bagci Das

XianXu Bai

Youdun Bai

Nikhil Bajaj

Albin Bajrami

Firooz Bakhtiari-Nejad

Ole Balling

Federico Ballo

Hyunseung Bang

Qifang Bao

Oumar Barry

Vito Basile

Alain Batailly

Mattia Battarra

Alparslan Emrah Bayrak

Sara Behdad

Amir Behjat

MohammadMahdi Behzadi

Pierpaolo Belardinelli

Pinhas Ben-Tzvi

William Bernstein

Giovanni Berselli

Harish Bezawada

Anindya Bhaduri

Kiran Bhole

Anudeep Bhoopalam

Pranav Bhounsule

Youyi Bi

Michele Bici

Dustin Bielecki

Amy Bilton

Andrew Birnbaum

Thijs Blad

Lucienne Blessing

Angelo Bonfitto

Joran Booth

Monica Bordegoni

Yuri Borgianni

Ramin Bostanabad

Armin Bosten

Joe Bradley

Adam R. Brink

Levent Burak Kara

Ravi Burla

Grace Burleson

Alex Burnap

Jared Butler

Asad Butt

Srikumar C. Gopalakrishnan

Jonathan Cagan

Runze Cai

Massimo Callegari

Benjamin Calmé

Chris Cameron

Jake Campbell

Stephen Canfield

Antonio Caputi

Luca Carbonari

Biagio Carboni

Stephane Caro

Julia Carroll

Marina Carulli

Dumitru Caruntu

Giandomenico Caruso

Arnaldo Casalotti

Damien Chablat

James Chagdes

Surya Chakrabarti

Nilanjan Chakraborty

Joel Chan

Kuei-Yuan Chan

Senthil Chandrasegaran

Ching-Yuan Chang

Tse-Shao Chang

Abheek Chatterjee

Ashish Chaudhari

Prathamesh Chaudhari

Brian Chell

Shikui Chen

Yu-Hsun Chen

Liuqing Chen

Jiangce Chen

Shuping Chen

Wei Chen

Cheng Chen

Genliang Chen

Zhen Chen

Hongrui Chen

Chong Chen

Zheyuan Chen

Yangquan Chen

Yao Cheng

Christine Chevallereau

Heng Chi

Shital Chiddarwar

Hyunkyoo Cho

Hanna Cho

Seung-Kyum Choi

Leah Chong

Souma Chowdhury

Sanjib Chowdhury

Chih-Hsing Chu

Ching-Wei Chuang

Erik Chumacero-Polanco

Hyun-Joon Chung

Wu Chunnong

Ender Cigeroglu

Abigail Clarke-Sather

Timothy Cleary

François Cluzel

Bryan Cochran

Peter Coffin

David Cohen

Courtney Cole

Arianne Collopy

Giorgio Colombo

Justin Conzola

Christopher Cooley

Daniele Costa

Ronald Couch

Tonghui Cui

Yaxin Cui

Xu Cui

Ranting Cui

Daicong Da

Xiang Dai

Shanna Daly	Kiran D'souza	Vincenzo Ferrero
Nicole Damen	Xiaoping Du	Francesco Ferrise
Revanth Damerla	Xianping Du	Brett Fiedler
Karthik Dantu	Ping Du	Matteo Forlini
Francesco Danzi	Daniel Duecker	Reza Fotouhi
Oguzhan Das	Kelley Dugan	Mary Frecker
Madhurima Das	Shammo Dutta	Rebecca Friesen
Tim Davenport	Salvador Echeveste	Matthew Fronk
Michael Dawson	Kristen Edwards	Katherine Fu
Jan De Jong	Paul Egan	Jiaming Fu
Charlotte De Vries	Mohammed El Kihal	Yan Fu
Ebru Demir	Richard Ellingham	Kazuko Fuchi
Onan Demirel	Aliakbar Eranpurwala	Alfonso Fuentes Aznar
Shiguang Deng	Jose Escalona	Mark Fuge
Kshiteej Deshmukh	Ehsan T Esfahani	Lawrence Funke
Shrinath Deshpande	Lorenzo Failla	Alessandro Galdelli
Harish Devaraj	Huashuai Fan	Juan Galvis
Gaurang Dharap	Qi Fan	Dongming Gan
Somayajulu L. Dhulipala	Hongbin Fang	Anthony Garland
Ahmet Dindar	Lezheng Fang	Diego Garzon-Alvarado
Joseph Distefano	Irene Fassi	Joseph Gattas
Donald Docimo	Abbas Fattah	John Gero
Zoltan Dombovari	Claudio Favi	Johannes Gerstmayr
Bin Dong	Mohammad Fazelpour	Masood Ghasemi
Jiayuan Dong	Brian Feeny	Payam Ghassemi
Guoying Dong	Zhang Feihong	Bogdan-George Gherman
Andy Dong	Cong Feng	Anna Ghidotti
Ata Donmez	Huijuan Feng	Seyede Fatemeh Ghoreishi
Daniel Dopico	Zhipeng Feng	Sayan Ghosh
Greg Dorgant	Yanbiao Feng	Amaninder Gill
Alberto Doria	Scott Ferguson	Andrew Gillman
Arinan Dourado	Javier Fernández Aceituno	Daniel Giraldo-Guzman

Massimiliano Gobbi

Francisco Gonzalez

Alex Gorodetsky

Kosa Goucher-Lambert

Marc Gouttefarde

Benjamin Graber

Paul Graham

Daniele Grandi

Paul Grogan

Magdalena Grohman

James Guest

David Guirguis

Yi Guo

Tinghao Guo

Xin Guo

Aakash Gupta

Joshua Gyory

Karl Haapala

Mahdi Haghshenas-Jaryani

Ichiro Hagiwara

David Hajdu

Maha Haji

Amal Z. Hajjaj

John Hall

Luke Hallum

Denise Halverson

Josh Hamel

Ji Han

Buddhika Hapuwatte

Brianne Hargrove

Megan Harris

Md Nahid Hasan

Mohammad Hasan

George Hazelrigg

Haiyang He

Shawky Hegazy

Seyed Mohammadreza Heidari

Daniel Herber

Just Herder

Jeffrey Herrmann

Nathan Hertlein

Amin Heyrani Nobari

Ethan Hilton

Dion Hogervorst

Derek Hollenbeck

Katja Holtta-Otto

Takatoshi Hondo

Yao Hong

Isaac Hong

Jonathan Hopkins

Imre Horvath

Yehia Hossam El Din

Mohammad Hotait

Larry Howell

Christopher Hoyle

Kuan-Lun Hsu

Chao Hu

Lingnan Hu

Weiwei Hu

Zhen Hu

Fu Hu

Chien-Ming Huang

Jida Huang

Chu-Yu Huang

Shih-Chun Huang

Aihua Huang

Shyh-Chour Huang

Jianzhe Huang

Guanyu Huang

Shen Huipijng

Daniel Hulse

Christine Human

Dongwook Hwang

Edoardo Ida

Horea Ilies

Athanasos Iliopoulos

Farhad Imani

Murat Inalpolat

Giovanni Incerti

Nizar Jaber

Prakhar Jaiswal

Ankur Jaiswal

Sagil James

Anand Jammulamadaka

Laurent Jay

Paramsothy Jayakumar

Brian Jensen

David Jensen

Nand Jha

Hao Ji

Chen Jiang

Long Jiang

Zhujin Jiang

Jiefeng Jiang

Yi Jin

Jian Jin

Yan Jin

Enze Jin

Bjorn Johansson

Albert Jones

Cole Joslyn

Jovana Jovanova

Sri Sadhan Jujjavarapu

Rama Krishna K.

Najib Kacem

Ibrahim Falih Kadhim

Ahmet Kahraman

Vinayak J Kalas

Amir Mohamad Kamalirad

Elisabeth Kames

Namwoo Kang

Sean Kelly

Pramod Khadilkar

Qasim Khadim

Mostafa Khalil

Muhammad Khan

Firas Khasawneh

Raj Pradip Khawale

Jivtesh Khurana

Namhun Kim

Hoe Joon Kim

Harrison Kim

Joo H. Kim

Jaehyuk Kim

Euiyoung Kim

Yves Klett

Adam Klodowski

Hyunwoong Ko

Eric Kolb

Ali Kolivand

Xianwen Kong

Lingyu Kong

Rasool Koosha

Stijn Koppen

Maulik C. Kotecha

Jozsef Kovecses

Julia Kramer

Timothy Krantz

Vinayak Raman Krishnamurthy

Prajit Krisshnakumar

Sai Aditya Raman Kuchibhatla

Boonserm Kulvatunyou

Ajeet Kumar

Rajesh Kumar

Shivesh Kumar

Jyoti Kumar

Chandan Kumar Sahu

Chin-Hsing Kuo

Eric Kurstak

Michael Kutzer

Tsz Ho Kwok

Elisa Kwon

Soonjo Kwon

Kai Lan

Brandon Lane

Daniel Lanzoni

Carlye Lauff

Astrid Layton

Michael Leamy

Yong Hoon Lee

Ikjin Lee

Jyh-jone Lee

Mathias Legrand

David Lehotzky

Leifur Leifsson

Matthew Leineweber

Stefano Lenci

Nikita Letov

Honglin Li

Suyi Li

Sheng Li

Jinghao Li

Hewenxuan LI

Haiyang Li

Yan Li

Xinyu Li

Zhuoxuan Li

Mian Li

Bowen Li

Longqiu Li

shiyao Li

Xiaofan Li

Yuxiang Li

Yuhua Li

Wei Li

Yan Liang

Haiguang Liao

Hao-Yu Liao

Jiankan Liao

Ting Liao

Po Ting Lin

Jianing Lin

Julie S. Linsey

Marc Litz

Xin Liu

Dehao Liu

Zuolin Liu

Zhao Liu

Yu Liu

Tianchen Liu

Yuanzhi Liu

Chang Liu

Xingchen Liu

Chao Liu

Feng Liu

Alvaro Lopez- Varela

Mostaan Lotfalian Saremi

Karen Lozano

Yanglong Lu

Yaodong Lu

Yan Lu

Lele Luan

Jose Lugo

Urbano Lugris

Chuan Luo

Jianxi Luo

Craig Lusk

Liye Lv

Matthew Lynch

Shengnan Lyu

Junfeng Ma

Nordica Maccarty

Ameneh Maghsoodi

Spencer Magleby

Gargi Majumder

Richard Malak Jr.

Adriano Mancini

Charles Manion

Hemanth Manjunatha

Peter Manzl

Jessica Gissella Maradey Lazaro

Edoardo Marconi

Marco Marconi

Carianne Martinez

Valeriy Martynyuk

Matteo Massaro

Ion Matei

Pawandeep Matharu

Jayant Mathur

Christopher Mattson

Christopher Mccomb

Lachlan McKenzie

John Mcphee

Lior Medina

Ali Mehmani

Nicholas Meisel

Giovanni Meneghetti

Jessica Menold

Diwesh Meshram

John Michopoulos

Aki Mikkola

Fahad Milaat

Jelena Milisavljevic-Syed

Scarlett R. Miller

Zhenjun Ming

Amin Mirkouei

Farrokh Mistree

Kentaro Miura

Ronak Mohanty

Tushar Mollik

Tamas Molnar

Ryan Monroe

Youngjin Moon

Marco Morandini

Beshoy Morkos

Federico Morosi

Daniel Morris

Simir Moschini

Zissimos Mourelatos

Huda Mousavi

Chandan Mozumder

Andreas Mueller

Andreas Muller

Luis Munoz

Alexander Murphy

Andrew Murray

Rosemarie Murray

David Myszka

Frank Naets

Jacquelyn Nagel

David Najera-Flores

Joel Najmon

Ananya Nandy

Austin Nash

Paromita Nath	Jitesh Panchal	Rohan Prabhu
Anand Balu Nellippallil	Piyush Pandita	Phanisri Pratapa
Todd Nelson	Herschel Pangborn	Christopher G. Pretty
Carl Nelson	Meghashyam Panyam	Anurag Purwar
Venkat Nemani	Dimitrios Papadimitriou	Feng Qian
Federico Neri	Kijung Park	Xiaoping Qian
Paul Nesline	Hyongju Park	Jian Qin
Rodrigo Nicoletti	Yeun Park	Lifang Qiu
Dean Nieusma	Robert G. Parker	Justin Quan
Joep Nijssen	Matthew Parkinson	D. Dane Quinn
Yutaka Nomaguchi	Victor Parque	Eliott Radcliffe
Julian Norato	Apurva Patel	Sajjad Raeisi
Mostafa Nouh	Parth Patel	Madhu Raghavan
Kenneth Nowaczyk	Sree Kalyan Patiballa	Hossam Ragheb
Konstantina Ntarladima	Albert Patterson	Jubeyer Rahman
Latifah Nurahmi	Steve Paul	Ayush Raina
Zachary Ochitwa	Satya Peddada	Vishal Ramadoss
Manaswin Oddiraju	William Peng	Venkat Ramakrishnan
Philip Odonkor	Tao Peng	Adhiti Raman
Eduardo Okabe	Giuseppe Pennisi	Devarajan Ramanujan
Alison Olechowski	Alfred Pettinger	Satchit Ramnath
Andrew Olewnik	James Pflumm	Vivek Rao
Grzegorz Orzechowski	Vrushank Phadnis	Sandipp Krishnan Ravi
Kevin Otto	Cyril Picard	Antonio Recuero
Emilio Ottonello	Bharath Pidaparthi	Daniele Regazzoni
Hassen Ouakad	Rocco Pietrini	Lyle Regenwetter
Enes Timur Ozdemir	Cecil Piya	William Regli
Alberto Padovan	Mark Plecnik	Tahira Reid
Safvan Palathingal	Alex Pletta	Yi Ren
Giacomo Palmieri	Jonisha Pollard	Rahul Sharan Renu
Matteo C. Palpacelli	Mirco Polonara	James Righter
Tan Pan	Siavash Pourkamali	David Rios-Zapata

Caterina Rizzi

Nina Robson

Steven Rodriguez

Bao Rong

David Rosen

Marco Rossoni

Clark Roubicek

Sipu Ruan

Davide Russo

Lokaditya Ryali

Jana Saadi

Mostafa Sabbaghi

Omid Saber

Mutahar Safdar

Alex Sahar

Tarik Sahin

Anshuman Kumar Sahu

Michael Saidani

Akira Saito

Alejandro Salado

Andrea Salvatore

Corina Sandu

Brandon Sargent

Bahadir Sarikaya

Soumalya Sarkar

Prabir Sarkar

Sree Shankar Satheesh Babu

Anupam Saxena

Mark Schenk

Ryan Schkoda

James Schmiedeler

Adam Schroeder

Christopher Sebastian

Carolyn Seepersad

Hullas Sehgal

Robert Seifried

Daniel Selva

Chiradeep Sen

Radu Serban

Valeria Settimi

Thurston Sexton

Zhenghui Sha

Divya Shah

Devanshi Shah

V.R. Shanmukhasundaram

Ashu Sharma

Saurav Sharma

Shashank Sharma

Mohammad Shavezipur

Jinjuan She

Yu She

Tripp Shealy

Gulai Shen

i.y. Shen

Chengzhi Shi

Hongliang Shi

Zhenghong Shi

Yong Shi

Yi-Pei Shih

Shailesh Shirguppikar

Geng Shixiong

Ada-Rhodes Short

Timothy Simpson

Siddharth Singh

Yogesh Singh

Vishal Singh

Shubhendu Kumar Singh

Alok Sinha

Kevin Skenes

Brian Slaboch

Gim Song Soh

Hyeonik Song

Xueguan Song

Binyang Song

Rujun Song

Carl Sorensen

Nicolas F. Soria Zurita

Christian Spreafico

Saketh Sridhara

Tino Stankovic

Elizabeth Starkey

Giulia Stefani

John Steuben

Daniel Suarez

Hiroyuki Sugiyama

Joshua Summers

Xiaoguang Sun

Sijie Sun

Krishnan Suresh

Sita Syal

Henrik Tamás Sykora

Brian Sylcott

Wei-Che Tai

Shun Takai

David Talbot

Tsz Ling Elaine Tang

Yunlong Tang

Alessandro Tasora

Ayse Tekes

Yishen Tian

Meng-Hsuan Tien

Christine Toh

Serife Tol

Daniel Torres

Andres Tovar

Shahrzad Towfighian

Alex Towse

Anh Tran

Cameron Turner

Pedro Urda

Kedar Vaidya

Nima Valadbeigi

Ambrosio Valencia-Romero

Homero Valladares

Anton van Beek

Douglas Van Bossuyt

Noe Vargas Hernandez

Soumya Vasisht

Pavan Tejaswi Velivela

Srinivasan Venkataraman

Swaminath Venkataswaran

Christopher Vermillion

Matteo Verotti

Adwait Verulkar

Vishesh Vikas

Jairo Viola

Antoni Viros I Martin

Vimal Viswanathan

Andrea Vitali

Nguyen Vu Linh

Evan Wallace

Shu Wan

Zequn Wang

Pingfeng Wang

Jing Wang

Pai Wang

Ruyue Wang

Fengxia Wang

Tingwei Wang

Haoqi Wang

Jieyu Wang

Jun Wang

Randi Wang

Kai Wang

Yan Wang

Wei Wang

Zihan Wang

Liwei Wang

Shoufei Wang

Gou-jen Wang

Wang Wang

Sibao Wang

Jonathan Weaver-Rosen

Guowu Wei

Xinqi Wei

Kate Whitefoot

Richard Wiebe

Gloria Wiens

Justin Wilbanks

Jon Williams

Matthew Williams

Carlos H. Wink

Amos Winter

Paul Witherell

Andrew Wodehouse

Kristin Wood

Natasha Wright

Dazhong Wu

Guangqiang Wu

Hao Wu

Di Wu

Zhimin Xi

Guanghui Xia

Songtao Xia

Yiwei Xia

Yujiang Xiang

Jiarui Xie

Siyuan Xing

Cenbo Xiong

Yi Xiong

Haohua Xiu

Hongyi Xu

Hai Xu

Qingsong Xu

Leidong Xu

Dong Xu

Mostafa Yacoub

Darshan Yadav

Hiroki Yamashita

Peng Yan

Zhipei Yan

Wei Yan

Haosen Yang

Zhuo Yang

Sheng Yang

Jingyi Yang

Hang Yang

Xiaonan Yang

Wenhao Yang

James Yang

Ruoyu Yang

Maria C. Yang

Xiaoou Yang

Chao-Lung Yang

Yuan Yao

T.-J. Yeh

Karthik Yerrapragada

Juan Yi

Zhong You

Behrooz Yousefzadeh

Liangyao Yu

Xinxin Yu

Xiangping Yu

Yongguang Yu

Sichen Yuan

Andrea Zanoni

Chen Zeng

Jice Zeng

Yuxin Zhai

Jie Zhang

Yunbo Zhang

Zhen Zhang

Haifeng Zhang

Ying Zhang

Peng Zhang

Hao Zhang

Dong Zhang

Qi Zhang

Siqi Zhang

Zhibo Zhang

Xiaoxu Zhang

Yang Zhang

Shengli Zhang

Guanglu Zhang

Xiaojia Shelly Zhang

Siyuan Zhang

Haohan Zhang

He Zhang

Shanglong Zhang

Yaoyao Zhao

Donghua Zhao

Hongzhe Zhao

Ping Zhao

Jian Zhao

Liqian Zhao

Yinjun Zhao

Linchuan Zhao

Zhen Zhao

Bujingda Zheng

Yi Zheng

Kevin Zheng

Kai Zhou

Yuqing Zhou

Hong Zhou

Xianlian Zhou

Jianhua Zhou

Xiang Zhou

Shengxi Zhou

Weidong Zhu

Jiaxiang Zhu

Damijan Zorko

Hongxiang Zou

Chengzhe Zou

Jianyong Zuo

Andreas Zwölfer

Marcel Zydeck

Wei Yan

Haosen Yang

Zhuo Yang

Sheng Yang

Jingyi Yang

Hang Yang

Xiaonan Yang

Wenhao Yang

James Yang

Ruoyu Yang

Maria C. Yang

Xiaoou Yang

Chao Lung Yang

Yuan Yao

T.-J. Yeh

Karthik Yerrapragada

Juan Yi

Zhong You

Behrooz Yousefzadeh

Liangyao Yu

Xinxin Yu

Xiangping Yu

Yongguang Yu

Sichen Yuan

Andrea Zanoni

Chen Zeng

Jice Zeng

Yuxin Zhai

Jie Zhang

Yunbo Zhang

Zhen Zhang

Haifeng Zhang

Ying Zhang

Peng Zhang

Hao Zhang

Dong Zhang

Qi Zhang

Siqi Zhang

Zhibo Zhang

Xiaoxu Zhang

Yang Zhang

Shengli Zhang

Guanglu Zhang

Xiaojia Shelly Zhang

Siyuan Zhang

Haohan Zhang

He Zhang

Shanglong Zhang

Yaoyao Zhao

Donghua Zhao

Hongzhe Zhao

Ping Zhao

Jian Zhao

Liqian Zhao

Yinjun Zhao

Linchuan Zhao

Zhen Zhao

Bujingda Zheng

Yi Zheng

Kevin Zheng

Kai Zhou

Yuqing Zhou

Hong Zhou

Xianlian Zhou

Jianhua Zhou

Xiang Zhou

Shengxi Zhou

Weidong Zhu

Jiaxiang Zhu

Damijan Zorko

Hongxiang Zou

Chengzhe Zou

Jianyong Zuo

Andreas Zwölfer

Marcel Zydeck

PROCEEDINGS OF ASME 2023 INTERNATIONAL DESIGN ENGINEERING TECHNICAL CONFERENCES AND COMPUTERS AND INFORMATION IN ENGINEERING CONFERENCE (IDETC-CIE2023)

49th Design Automation Conference (DAC) Part One
Table of Contents

Control Co-Design

Control Co-Design With Approximate Explicit Model Predictive Controllers   **DETC2023-109551**
Ying-Kuan Tsai and Richard J. Malak, Jr.

Control Co-Design With Varying Available Information Applied to Vehicle Suspensions   **DETC2023-114690**
Saeid Bayat and James T. Allison

A Two-Timescale Reinforcement Learning Approach for Control Co-Design Problems   **DETC2023-116567**
Eddieb Sadat, Mostaan Lotfalian Saremi, and Alparslan Emrah Bayrak

Platform Hydrodynamic and Structural Control Co-Optimization for the Floating Offshore Wind Turbines   **DETC2023-117541**
Jinbin Liang, Xianping Du, Jianbo Yi, Guowei Qian, Peng Xie, and Hongyi Xu

Artificial Intelligence and Machine Learning for Challenging Real-World Problems in Design Automation

A Heuristic Approach to Classify Geometrically Defective Bead Segments Based on Range of Curvature, Range of Sound Power and Maximum Height   **DETC2023-114741**
Nowrin Akter Surovi and Gim Song Soh

Automating Style Analysis and Visualization With Explainable AI - Case Studies on Brand Recognition   **DETC2023-115150**
Yu-Hsuan Chen, Levant Burak Kara, and Jonathan Cagan

Characterizing Designs via Isometric Embeddings: Applications to Airfoil Inverse Design   **DETC2023-116743**
Qiuyi Chen and Mark D. Fuge

Counterfactuals for Design: A Model-Agnostic Method for Design Recommendations   **DETC2023-117216**
Lyle Regenwetter, Yazan Abu Obaideh, and Faez Ahmed

Novel AI or ML Frameworks for Design or Systems Science

Bayesian Mesh Optimization for Graph Neural Networks to Enhance Engineering Performance Prediction
Jangseop Park and Namwoo Kang

DETC2023-113308

On the Use of Geometric Deep Learning Towards the Evaluation of Graph-Centric Engineering Systems
Anthony Sirico, Jr. and Daniel R. Herber

DETC2023-114592

AutoSurf: Automated Expert-Guided Meshing With Graph Neural Networks and Conformal Predictions
Amin Heyrani Nobari, Justin Rey, Suhas Kodali, Matthew Jones, and Faez Ahmed

DETC2023-115065

Diffusing the Optimal Topology: A Generative Optimization Approach
Giorgio Giannone and Faez Ahmed

DETC2023-116595

Document Understanding-Based Design Support: Language Model Based Design Knowledge Extraction
Yunjian Qiu and Yan Jin

DETC2023-116746

Topology Optimization Using Neural Networks With Conditioning Field Initialization for Improved Efficiency
Hongrui Chen, Aditya Joglekar, and Levent Burak Kara

DETC2023-116937

Data-Driven Design

Dated: Guidelines for Creating Synthetic Datasets for Engineering Design Applications
Cyril Picard, Jürg Schiffmann, and Faez Ahmed

DETC2023-111609

Deep Generative Model-Based Synthesis of Four-Bar Linkage Mechanisms Considering Both Kinematic and Dynamic Conditions
Sumin Lee, Jihoon Kim, and Namwoo Kang

DETC2023-114464

GradeS: An AI-Driven Graphic Design Support System for Design Style Analysis
Jinyu Song, Weitao You, Shuhui Shi, Ziwei Tu, Juntao Ji, Kaixin Han, and Lingyun Sun

DETC2023-114660

On-the-Fly Dual Reduction Method on Transient Fluid Topology Optimization
Tianye Wang and Xiaoping Qian

DETC2023-114806

Investigating the Effect of a Brand Factor in Product Design Based on a Data-Driven Approach Using Online Reviews
Seyoung Park and Harrison Kim

DETC2023-114966

A Visual Representation of Engineering Catalogs Using Variational Autoencoders
Saketh Sridhara and Krishnan Suresh

DETC2023-115029

Interpretable Neural Network Analyses for Understanding Complex Physical Interactions in Engineering Design
Tuba Dolar, Doksoo Lee, and Wei Chen

DETC2023-115103

Investigate Customer Preferences Using Online Video Reviews - Preliminary Results   DETC2023-115206
Kangcheng Lin and Harrison Kim

Heat Sink Design Optimization via GAN-CNN Combined Deep-Learning   DETC2023-116429
Nathan Flynn and Xiaoping Qian

On the Connectedness of the Topology Optimization Predictors   DETC2023-116574
Mohammad Mahdi Behzadi and Horea T. Ilies

When Is it Actually Worth Learning Inverse Design?   DETC2023-116678
Milad Habibi, Jun Wang, and Mark Fuge

Design of Self-Organizing Systems Using Multi-Agent Reinforcement Learning and the Compromise Decision Support Problem Construct   DETC2023-116703
Mingfei Jiang, Zhenjun Ming, Chuanhao Li, Farrokh Mistree, and Janet K. Allen

Using Machine Learning to Predict the Adoption of Building Electrification Technologies in US Households   DETC2023-116751
Andrew Majowicz and Philip Odonkor

Ship-D: Ship Hull Dataset for Design Optimization Using Machine Learning   DETC2023-117003
Noah J. Bagazinski and Faez Ahmed

Surrogate Modeling of Car Drag Coefficient With Depth and Normal Renderings   DETC2023-117400
Binyang Song, Chenyang Yuan, Frank Permenter, Nikos Arechiga, and Faez Ahmed

Decision Making in Engineering Design

Modeling the Dynamics of Customer Demand to Determine the Optimal Time to Release Product Updates: A Cognitive Approach   DETC2023-115259
Ian Walter, Philip E. Paré, and Jitesh H. Panchal

A Framework to Support Multilevel Robust Co-Design of Manufacturing Supply Networks   DETC2023-117145
Mathew Baby, Akshay Guptan, Jacob Broussard, Janet K. Allen, Farrokh Mistree, and Anand Balu Nellippallil

Design and Optimization of Energy Systems

Improving Nonuniform Utilization of Li-Ion Pouch Cells Using Tapered Electrodes Through Calendering   DETC2023-111360
Changik Cho, Seth Kelley, Joseph G. Tylka, Miao He, Naresh N. Nandola, and Christopher D. Rahn

System Level Techno-Economic and Environmental Design Optimization for Ocean Wave Energy   DETC2023-114607
Rebecca McCabe, Madison Dietrich, Alan Liu, and Maha Haji

A Multi-Fidelity Gaussian Process Regression Method for Probabilistic Wind Farm Power Curve Estimation
Honglin Li, Cong Feng, and Jie Zhang

DETC2023-114762

Exploration of Building Clustering Potential With Energy Storage in New York City
Gregory Kaminski and Philip Odonkor

DETC2023-115046

Multi-Fidelity Modeling for Dynamic Power Control and Optimization of Nuclear-Renewable Hybrid Energy Systems
In-Bum Chung and Pingfeng Wang

DETC2023-116914

Multiphysics-Informed Machine Learning for Battery Design and Health Monitoring
Parth Bansal and Yumeng Li

DETC2023-117113

Modeling a Concentrated Solar Collector (CSC) - Convection-Enhanced Evaporation (CEE) System for Small-Scale Brine Management
Nallely Guillen Rodriguez, Mustafa F. Kaddoura, and Natasha C. Wright

DETC2023-117236

Design for Additive Manufacturing

Empirically Tuned Mechanical Simulation Model of 3D-Printed Biaxial Weaves
Marc Wirth and Kristina Shea

DETC2023-111278

Build Orientation Optimization for Five-Axis 3D Printing
Ghazi Alonayni and Matthew I. Campbell

DETC2023-111726

Process-Aware Prediction of Geometric Accuracy for Additive Manufacturing via Transfer Learning
Daphne Lin and Carolyn Seepersad

DETC2023-111826

Topology Optimization in Consideration of Overhang Constraint for Additive Manufacturing Based on Coupled Fictitious Physical Model in Thermal Design Problem
Mikihiro Tajima and Takayuki Yamada

DETC2023-112499

Application of a Continuous Variable Density Infill: Manipulating Center of Gravity
Patrick N. Murphy and Bashir Khoda

DETC2023-116644

A Framework Establishing the Bounds of Small Angle Assumptions in Multi-Material Additively Manufactured Compliant Mechanisms
Evelyn Thomas, Nicholas Meisel, and Jared Butler

DETC2023-116865

Design for Market Systems

Evolutionary Co-Mention Network Analysis via Social Media Mining
Phillip A. O. Gavino, Yinshuang Xiao, Yaxin Cui, Wei Chen, and Zhenghui Sha

DETC2023-115114

Exploring How the Design Hierarchy of Needs Explains Correlations Between Design Decisions and Online Customer Review Ratings
Lisa Retzlaff and Scott Ferguson

DETC2023-117051

Design for Resilience and Failure Recover

A Reliability-Based Optimization Framework for Planning Operational Profiles for Unmanned Systems
    Indranil Hazra, Arko Chatterjee, Joseph Southgate, Matthew J. Weiner, Katrina M. Groth, and Shapour Azarm

DETC2023-116586

Data-Driven Control Co-Design for Indirect Liquid Cooling Plate With Microchannels for Battery Thermal Management
    Zheng Liu, Yanwen Xu, Hao Wu, Pingfeng Wang, and Yumeng Li

DETC2023-116921

Physics-Informed Neural Networks for Degradation Diagnostics of Lithium-Ion Batteries
    Sina Navidi, Adam Thelen, Tingkai Li, and Chao Hu

DETC2023-116940

Sensor Network Design for Permanent Magnet Synchronous Motor Fault Diagnosis
    Sara Kohtz, Junhan Zhao, Anabel Renteria, Anand Vikas Lalwani, Xiaolong Zhang, Kiruba Sivasubramaniam Haran, Debbie Senesky, and Pingfeng Wang

DETC2023-116972

Mean Time to Failure Prediction for Complex Systems With Adaptive Surrogate Modeling
    Hao Wu, Yanwen Xu, Zheng Liu, and Pingfeng Wang

DETC2023-117177

TABLE OF CONTENTS

Control Co-Design With Approximate Explicit Model Predictive Controllers 1
Ying-Kuan Tsai, Richard J. Malak Jr.

Control Co-Design With Varying Available Information Applied to Vehicle Suspensions 15
Saeid Bayat, James T. Allison

A Two-Timescale Reinforcement Learning Approach for Control Co-Design Problems 29
Eddieb Sadat, Mostaan Lotfalian Saremi, Alparslan Emrah Bayrak

Platform Hydrodynamic and Structural Control Co-Optimization for the Floating Offshore Wind
Turbines 38
Jinbin Liang, Xianping Du, Jianbo Yi, Guowei Qian, Peng Xie, Hongyi Xu

A Heuristic Approach to Classify Geometrically Defective Bead Segments Based on Range of
Curvature, Range of Sound Power and Maximum Height 47
Nowrin Akter Surovi, Gim Song Soh

Automating Style Analysis and Visualization With Explainable AI - Case Studies on Brand
Recognition 57
Yu-Hsuan Chen, Levant Burak Kara, Jonathan Cagan

Characterizing Designs via Isometric Embeddings: Applications to Airfoil Inverse Design 71
Qiuyi Chen, Mark D. Fuge

Counterfactuals for Design: A Model-Agnostic Method for Design Recommendations 85
Lyle Regenwetter, Yazan Abu Obaideh, Faez Ahmed

Bayesian Mesh Optimization for Graph Neural Networks to Enhance Engineering Performance
Prediction 98
Jangseop Park, Namwoo Kang

On the Use of Geometric Deep Learning Towards the Evaluation of Graph-Centric Engineering
Systems 108
Anthony Sirico Jr., Daniel R. Herber

AutoSurf: Automated Expert-Guided Meshing With Graph Neural Networks and Conformal
Predictions 121
Amin Heyrani Nobari, Justin Rey, Suhas Kodali, Matthew Jones, Faez Ahmed

Diffusing the Optimal Topology: A Generative Optimization Approach 137
Giorgio Giannone, Faez Ahmed

Document Understanding-Based Design Support: Language Model Based Design Knowledge
Extraction 153
Yunjian Qiu, Yan Jin

Topology Optimization Using Neural Networks With Conditioning Field Initialization for
Improved Efficiency 166
Hongrui Chen, Aditya Joglekar, Levent Burak Kara

Dated: Guidelines for Creating Synthetic Datasets for Engineering Design Applications 177
Cyril Picard, Jurg Schiffmann, Faez Ahmed

Deep Generative Model-Based Synthesis of Four-Bar Linkage Mechanisms Considering Both Kinematic and Dynamic Conditions... 190
Sumin Lee, Jihoon Kim, Namwoo Kang

GradeS: An AI-Driven Graphic Design Support System for Design Style Analysis 198
Jinyu Song, Weitao You, Shuhui Shi, Ziwei Tu, Juntao Ji, Kaixin Han, Lingyun Sun

On-the-Fly Dual Reduction Method on Transient Fluid Topology Optimization.............................. 208
Tianye Wang, Xiaoping Qian

Investigating the Effect of a Brand Factor in Product Design Based on a Data-Driven Approach Using Online Reviews... 218
Seyoung Park, Harrison Kim

A Visual Representation of Engineering Catalogs Using Variational Autoencoders 228
Saketh Sridhara, Krishnan Suresh

Interpretable Neural Network Analyses for Understanding Complex Physical Interactions in Engineering Design 238
Tuba Dolar, Doksoo Lee, Wei Chen

Investigate Customer Preferences Using Online Video Reviews - Preliminary Results 250
Kangcheng Lin, Harrison Kim

Heat Sink Design Optimization via GAN-CNN Combined Deep-Learning 259
Nathan Flynn, Xiaoping Qian

On the Connectedness of the Topology Optimization Predictors 268
Mohammad Mahdi Behzadi, Horea T. Ilies

When Is it Actually Worth Learning Inverse Design? 276
Milad Habibi, Jun Wang, Mark Fuge

Design of Self-Organizing Systems Using Multi-Agent Reinforcement Learning and the Compromise Decision Support Problem Construct.. 289
Mingfei Jiang, Zhenjun Ming, Chuanhao Li, Farrokh Mistree, Janet K. Allen

Using Machine Learning to Predict the Adoption of Building Electrification Technologies in US Households 302
Andrew Majowicz, Philip Odonkor

Ship-D: Ship Hull Dataset for Design Optimization Using Machine Learning................................. 311
Noah J. Bagazinski, Faez Ahmed

Surrogate Modeling of Car Drag Coefficient With Depth and Normal Renderings 326
Binyang Song, Chenyang Yuan, Frank Permenter, Nikos Arechiga, Faez Ahmed

Modeling the Dynamics of Customer Demand to Determine the Optimal Time to Release Product Updates: A Cognitive Approach 339
Ian Walter, Philip E. Pare, Jitesh H. Panchal

A Framework to Support Multilevel Robust Co-Design of Manufacturing Supply Networks........................ 354
Mathew Baby, Akshay Guptan, Jacob Broussard, Janet K. Allen, Farrokh Mistree, Anand Balu Nellippallil

Improving Nonuniform Utilization of Li-Ion Pouch Cells Using Tapered Electrodes Through Calendering 371

Changik Cho, Seth Kelley, Joseph G. Tylka, Miao He, Naresh N. Nandola, Christopher D. Rahn

System Level Techno-Economic and Environmental Design Optimization for Ocean Wave Energy 381

Rebecca McCabe, Madison Dietrich, Alan Liu, Maha Haji

A Multi-Fidelity Gaussian Process Regression Method for Probabilistic Wind Farm Power Curve Estimation 391

Honglin Li, Cong Feng, Jie Zhang

Exploration of Building Clustering Potential With Energy Storage in New York City 401

Gregory Kaminski, Philip Odonkor

Multi-Fidelity Modeling for Dynamic Power Control and Optimization of Nuclear-Renewable Hybrid Energy Systems 411

In-Bum Chung, Pingfeng Wang

Multiphysics-Informed Machine Learning for Battery Design and Health Monitoring 421

Parth Bansal, Yumeng Li

Modeling a Concentrated Solar Collector (CSC) - Convection-Enhanced Evaporation (CEE) System for Small-Scale Brine Management 431

Nallely Guillen Rodriguez, Mustafa F. Kaddoura, Natasha C. Wright

Empirically Tuned Mechanical Simulation Model of 3D-Printed Biaxial Weaves 442

Marc Wirth, Kristina Shea

Build Orientation Optimization for Five-Axis 3D Printing 455

Ghazi Alonayni, Matthew I. Campbell

Process-Aware Prediction of Geometric Accuracy for Additive Manufacturing via Transfer Learning 464

Daphne Lin, Carolyn Seepersad

Topology Optimization in Consideration of Overhang Constraint for Additive Manufacturing Based on Coupled Fictitious Physical Model in Thermal Design Problem 481

Mikihiro Tajima, Takayuki Yamada

Application of a Continuous Variable Density Infill: Manipulating Center of Gravity 488

Patrick N. Murphy, Bashir Khoda

A Framework Establishing the Bounds of Small Angle Assumptions in Multi-Material Additively Manufactured Compliant Mechanisms 495

Evelyn Thomas, Nicholas Meisel, Jared Butler

Evolutionary Co-Mention Network Analysis via Social Media Mining 507

Phillip A. O. Gavino, Yinshuang Xiao, Yaxin Cui, Wei Chen, Zhenghui Sha

Exploring How the Design Hierarchy of Needs Explains Correlations Between Design Decisions and Online Customer Review Ratings 519

Lisa Retzlaff, Scott Ferguson

A Reliability-Based Optimization Framework for Planning Operational Profiles for Unmanned Systems..532
   Indranil Hazra, Arko Chatterjee, Joseph Southgate, Matthew J. Weiner, Katrina M. Groth, Shapour Azarm

Data-Driven Control Co-Design for Indirect Liquid Cooling Plate With Microchannels for Battery Thermal Management..544
   Zheng Liu, Yanwen Xu, Hao Wu, Pingfeng Wang, Yumeng Li

Physics-Informed Neural Networks for Degradation Diagnostics of Lithium-Ion Batteries............551
   Sina Navidi, Adam Thelen, Tingkai Li, Chao Hu

Sensor Network Design for Permanent Magnet Synchronous Motor Fault Diagnosis.....................559
   Sara Kohtz, Junhan Zhao, Anabel Renteria, Anand Vikas Lalwani, Xiaolong Zhang, Kiruba Sivasubramaniam Haran, Debbie Senesky, Pingfeng Wang

Mean Time to Failure Prediction for Complex Systems With Adaptive Surrogate Modeling............567
   Hao Wu, Yanwen Xu, Zheng Liu, Pingfeng Wang

Author Index

Proceedings of the ASME 2023
International Design Engineering Technical Conferences and
Computers and Information in Engineering Conference
IDETC-CIE2023
August 20-23, 2023, Boston, Massachusetts

DETC2023-109551

CONTROL CO-DESIGN WITH APPROXIMATE EXPLICIT MODEL PREDICTIVE CONTROLLERS

Ying-Kuan Tsai
Design Systems Laboratory
Department of Mechanical Engineering
Texas A&M University
College Station, Texas 77843
Email: yktsai0121@tamu.edu

Richard J. Malak Jr.
Design Systems Laboratory
Department of Mechanical Engineering
Texas A&M University
College Station, Texas 77843
Email: rmalak@tamu.edu

ABSTRACT

This paper presents a novel methodology for control co-design (CCD) with approximate explicit model predictive controllers. This paper also investigates how approximation errors impact the decision-making for design of dynamic systems. The explicit nonlinear control policies are approximated and represented as surrogate models by a function of states and plant parameters, which allows for offline solving control problems and reducing the computational time required for CCD optimization. Set invariance theory and Lyapunov stability theory are employed to ensure the stability of designed systems. To facilitate design exploration, a multi-objective optimization problem is formulated to trade-off between performance and probability of state-constraint satisfaction. A probability assessment for system stability with the sampled initial states and multiple precomputed controllers is included in the optimization constraints. The approach is benchmarked using a numerical example and an engineering case study of a satellite attitude control system. Some of the non-dominated designs are selected to visualize their state and control trajectories for comparison. The visualization shows the successful implementation of explicit model predictive controllers in CCD problems while ensuring stability. The results of this study highlight the importance of including plant design and performing CCD to achieve system-optimal solutions and better understand dynamic characteristics.

NOMENCLATURE

$\mathbf{F}$	Function of dynamics
$\mathbf{F_c}$	Function of closed-loop dynamics
$\mathbf{G}$	Vector of constraint functions
$\mathbb{P}$	Optimal control problem
$\mathbf{P, Q, R}$	Weighting matrices
$\mathbf{u}$	Vector of control inputs
V_N	Cost function with N-horizon
k	Time index
$\mathbf{K}$	Vector of control gains
κ	Feedback control law
$\mathfrak{R}$	Family of feedback control laws
$\boldsymbol{\theta}_p$	Vector of plant parameters
$\mathbf{x_c}$	Vector of control design variables
$\mathbf{x_p}$	Vector of plant design variables
$\boldsymbol{\xi}$	Vector of state variables
$\mathbb{X}, \mathbb{U}$	State and control constraint sets
CCD	Control co-design
MPC	Model predictive control
MPI	Maximal positively invariant
mp-MPC	Multi-parametric model predictive control
P3GA	Predicted parameterized Pareto genetic algorithm
sp-NLPC	State-parameterized nonlinear programming control

Copyright © 2023 by ASME

1 INTRODUCTION

Integrating physical and control systems design (known as *control co-design, CCD*) has received much attention because it achieves superior system performance compared to sequentially optimizing both systems [1–3]. However, control design is challenging due to the involvement of time-dependent variables and responses. While open-loop control methods that find the optimal sequence of discretized control actions can give insights of how the best performance can be achieved without restricting the types of controllers [4–13], such control schemes may present challenges in real-world applications, particularly with respect to robustness and disturbance control.

Model predictive control (MPC) is a powerful feedback control technique for many engineering applications since it can find the optimal actions in real-time and allows constraints and nonlinear dynamic models with a finite horizon in the formulation [14–16]. Examples of CCD using MPC include power and energy systems [17] and aircraft thermal management systems [18]. Nevertheless, it can suffer from real-time delays when the computation is expensive or the dynamics are fast. This can significantly impact the performance or even lead to instability [19, 20]. Although an explicit MPC technique called Multi-Parametric MPC (mp-MPC) offers an effective way to avoid the circumstance by solving the explicit feedback control policies *offline* [21–23], it is limited by a single-objective formulation with linear constraints and dynamics and a quadratic cost function.

Due to practical engineering considerations (e.g., the complexity of dynamic models and constraints), many other explicit MPC techniques aim to *approximate* control policies using deep learning and other advanced computation methods [19, 20, 24–27]. However, one should ensure the designed controllers are capable of stabilizing the systems along with satisfying the state and control constraints under the presence of approximation errors. The literature indicates that their results provide evidence for the guarantee of system stability and constraint satisfaction. Despite the development, there is a lack of attention on taking into account the design or redesign of the plants, hindering the possibility of achieving system-optimal performance. Enhancing design flexibility through the inclusion of plant design by using CCD formulations and methods can enable designers to develop more robust and higher-performance systems [28].

Although significant advancements have been made recently in CCD and explicit MPC separately, limited work has been done to incorporate explicit MPC methods into CCD formulations. There are several research challenges to performing CCD with explicit MPC:

1. Explicit control policies may differ significantly when the plant is modified or redesigned. This necessitates repetitive processes of generating data on optimal points and training surrogate models every time the plant is changed, which can be both time-consuming and expensive.

2. In addition to the control policies, many properties in the MPC formulation vary across different plants. For example, the criteria for ensuring stability and constraint satisfaction become more complicated when the plant is changed. Furthermore, some plants may be designed such that they cannot be stabilized by any controller [29]. In such cases, additional efforts are needed to verify the feasibility of the dynamics for a specific plant and controller combination.

3. Explicit MPC solutions using approximation techniques always involve some level of approximation errors, which can have a significant impact on system performance and stability.

This article aims to address these gaps through the following contributions. First, the state-dependent control solution is extended to a CCD version by adding *plant parameters*. Among existing explicit MPC techniques, the *State-Parameterized Nonlinear Programming Control (sp-NLPC)* is chosen because it is data-efficient for training an explicit control policy while ensuring stability [20]. Second, set invariance theory and Lyapunov stability theory are used to ensure stability. Third, since the use of sp-NLPC significantly reduces the time spent on function evaluations, probabilistic assessment of stability by running many dynamic simulations is incorporated into optimization constraints. Lastly, approximate Pareto fronts are generated using multi-objective optimization to assist designers in conducting a trade-off study between performance and other metrics. This paper also investigates how approximation errors impact system performance and CCD solutions.

The remainder of this article is structured as follows: Section 2 presents the background and preliminaries, Section 3 presents the proposed CCD method, optimization formulation, and algorithm, Section 4 implements and demonstrates the proposed approach using a numerical example and an engineering case study, and Section 5 summarizes the conclusion and highlights future research directions.

2 BACKGROUND

Notation: The sets of non-negative and positive integers and non-negative reals are denoted by $\mathbb{N}$, $\mathbb{N}_+$, and $\mathbb{R}_+$. Given $a, b \in \mathbb{N}$ such that $a < b$, we denote $\mathbb{N}_{[a,b]} := \{a, a+1, ..., b\}$. A function belongs to class $\mathcal{K}$ if it is continuous, zero at zero, and strictly increasing; a function belongs to class $\mathcal{K}_\infty$ if it is in class $\mathcal{K}$ and unbounded. A function $\beta(\cdot)$ belongs to class $\mathcal{KL}$ if it is continuous and if, for each $k \geq 0$, $\beta(\cdot, k)$ is a class $\mathcal{K}$ function and for each $s \geq 0$, $\beta(s, \cdot)$ is a class $\mathcal{L}$ function, which is $\beta(s, \cdot) > 0$, nonincreasing and $\beta(s, i)$ converges to 0 as $i \to \infty$.

Copyright © 2023 by ASME

2.1 System Description

Consider a discrete-time and linear time-invariant system:

$$\boldsymbol{\xi}(k+1) = \mathbf{F}(\boldsymbol{\xi}(k), \mathbf{u}(k)) := \mathbf{A}\boldsymbol{\xi}(k) + \mathbf{B}\mathbf{u}(k), \tag{1}$$
$$\forall k \in \mathbb{N}_{[0,N-1]},$$

where $\boldsymbol{\xi}(k) \in \mathbb{R}^{n_\xi}$ is the state vector, $\mathbf{u}(k) \in \mathbb{R}^{n_u}$ is the control input, at time k, $\mathbf{F} : \mathbb{R}^{n_\xi} \times \mathbb{R}^{n_u} \mapsto \mathbb{R}^{n_\xi}$ is the dynamic function assumed to be continuous with the system matrices $\mathbf{A} \in \mathbb{R}^{n_\xi \times n_\xi}$ and $\mathbf{B} \in \mathbb{R}^{n_\xi \times n_u}$, and $N \in \mathbb{N}$ is the horizon. Suppose the origin $\boldsymbol{\xi} = \mathbf{0}$ is the equilibrium point of system (1), then $\mathbf{F}(\mathbf{0}, \mathbf{0}) = \mathbf{0}$ is satisfied.

Assumption 1 (State and input constraints). *System (1) is subject to the following constraints:*

$$\boldsymbol{\xi}(k) \in \mathbb{X} \subset \mathbb{R}^{n_\xi}, \tag{2a}$$
$$\mathbf{u}(k) \in \mathbb{U} \subset \mathbb{R}^{n_u}, \tag{2b}$$

where $\mathbb{X}$ and $\mathbb{U}$ are assumed to be polyhedrons.

Assumption 2 (Stabilizability and measurability). *The pair of $(\mathbf{A}, \mathbf{B})$ from system (1) is stabilizable and the states of system (1) can be measured such that $\boldsymbol{\xi}(k)$ is known for all $k \in \mathbb{N}_{[0,N]}$.*

Given a feedback control law $\mathbf{u}(k) := \kappa(\boldsymbol{\xi}(k))$, system (1) becomes closed-loop and can be formulated as:

$$\boldsymbol{\xi}(k+1) = \mathbf{F_c}(\boldsymbol{\xi}(k)) := \mathbf{A}\boldsymbol{\xi}(k) + \mathbf{B}\kappa(\boldsymbol{\xi}(k)), \ \forall k \in \mathbb{N}_{[0,N-1]}. \tag{3}$$

The following definitions from set invariance theory [30] will be important to implementing control design.

Definition 1 (Control invariant set [31]). *A set $\mathbb{X}_c \subseteq \mathbb{X}$ is control invariant for system (1) and constraint set $(\mathbb{X}, \mathbb{U})$ if $\mathbb{X}_c \subseteq \mathbb{X}$ and if, for all $\boldsymbol{\xi}(k) \in \mathbb{X}_c$, there exists a $\mathbf{u}(k) \in \mathbb{U}$ such that $\mathbf{F}(\boldsymbol{\xi}(k), \mathbf{u}(k)) \in \mathbb{X}_c$.*

Definition 2 (Positively invariant set [31]). *A set $\mathbb{X}_p \subseteq \mathbb{X}$ is positively invariant for the closed-loop system (3) and the constraint set $\mathbb{X}$ if $\mathbb{X}_p \subseteq \mathbb{X}$ and if, for all $\boldsymbol{\xi}(k) \in \mathbb{X}_p$, $\mathbf{F_c}(\boldsymbol{\xi}(k)) \in \mathbb{X}_p$.*

Remark: A control invariant set is a set of states where, for every state in the set, there exists at least one feasible control input that will keep the state evolution of the system within the set for all future times. In other words, a control invariant set is a set of states that can be maintained under control. A positively invariant set, on the other hand, is a set of states where the evolved state of the closed-loop system will stay within the set for all future times.

Definition 3 (Maximal positively invariant [32]). *The maximal positively invariant (MPI) set for system (3) and constraints in Eq. (2a) is a positively invariant set for system (3) and constraints in Eq. (2a) that contains all other positively invariant sets for system (3) and constraints in Eq. (2a).*

To ensure system stability, it is critical to define stability by Lyapunov definitions and theorems. Let $\phi(k; \boldsymbol{\xi})$ denote the solution to Eq. (3), i.e., the evolved state of system (3), at time k for a given state $\boldsymbol{\xi}$ at time 0 [16].

Definition 4 (Asymptotically stable [16]). *Suppose $\mathbb{X}_p$ is positive invariant for system (3). The origin is asymptotically stable for system (3) in $\mathbb{X}_p$ if there exists a $\mathcal{KL}$ function $\beta(\cdot)$ such that, for each $\mathbb{X}_p$,*

$$|\phi(k; \boldsymbol{\xi})| \le \beta(|\boldsymbol{\xi}|, k), \ \forall k \in \mathbb{N}. \tag{4}$$

Definition 5 (Lyapunov function [16]). *Suppose $\mathbb{X}_p$ is positive invariant for system (3). A function $V : \mathbb{R}^{n_\xi} \mapsto \mathbb{R}_+$ is said to be a Lyapunov function in $\mathbb{X}_p$ for system (3) if there exist functions $\alpha_1, \alpha_2 \in \mathcal{K}_\infty$ and a continuous, positive definite function α_3 such that for any $\boldsymbol{\xi} \in \mathbb{X}_p$,*

$$V(\boldsymbol{\xi}) \ge \alpha_1(|\boldsymbol{\xi}|), \tag{5a}$$
$$V(\boldsymbol{\xi}) \le \alpha_2(|\boldsymbol{\xi}|), \tag{5b}$$
$$V(\mathbf{F_c}(\boldsymbol{\xi})) - V(\boldsymbol{\xi}) \le -\alpha_3(|\boldsymbol{\xi}|). \tag{5c}$$

Theorem 1 (Lyapunov stability theorem [16]). *Suppose $\mathbb{X}_p \subset \mathbb{R}^{n_\xi}$ is positive invariant for system (3). If there exists a Lyapunov function in $\mathbb{X}_p$ for system (3), then the origin is asymptotically stable in $\mathbb{X}_p$ for system (3).*

Generally speaking, given the control law $\kappa(\cdot)$ for system (1), the origin of the closed-loop system (3) can be said asymptotically stable if Eq. (4) and Eq. (5) are satisfied.

2.2 Problem Statement of Control Design

MPC is a model-based technique that aims to frequently solve finite-horizon optimal control problems by predicting the system behaviors at each sampling instant [16]. An N-horizon optimal control problem given the initial state $\boldsymbol{\xi}_0$ can be defined as $\mathbb{P}_N(\boldsymbol{\xi}_0)$:

$$\min_{\mathbf{U}} V_N(\mathbf{U}, \boldsymbol{\xi}_0) = \sum_{k=0}^{N-1} \left[||\boldsymbol{\xi}(k)||_{\mathbf{Q}}^2 + ||\mathbf{u}(k)||_{\mathbf{R}}^2 \right] + ||\boldsymbol{\xi}(N)||_{\mathbf{P}}^2, \tag{6}$$

Copyright © 2023 by ASME

subject to:

$$\xi(k+1) - \mathbf{F}(\xi(k), \mathbf{u}(k)) = \mathbf{0}, \ \forall k \in \mathbb{N}_{[0,N-1]}, \quad (7a)$$

$$\xi(k) \in \mathbb{X}, \ \mathbf{u}(k) \in \mathbb{U}, \ \forall k \in \mathbb{N}_{[0,N-1]}, \quad (7b)$$

$$\xi(N) \in \mathbb{X}_f, \ \xi(0) := \xi_0 \text{ is given}, \quad (7c)$$

where $\|\xi\|_{\mathbf{Q}}^2 = \xi^{\top} \mathbf{Q} \xi$ represents the quadratic operation of vector ξ, the weighting matrices $\mathbf{Q} > 0$ and $\mathbf{R} > 0$ are symmetric, and $\mathbf{P}$ is the terminal cost matrix. Equation (7a) is from system (1), and Eq. (7b) is from the state and input constraint sets Eq. (2). The set $\mathbb{X}_f \subset \mathbb{X}$ in (7c) is the terminal constraint set that satisfies Definition 3, which is defined to guarantee stability [14] and known as the region of attraction [16]. We denote Ξ and $\mathbf{U}$ by the matrices for the sequences of states and control inputs, respectively:

$$\Xi = [\xi(0), \xi(1), ..., \xi(N)] \in \mathbb{R}^{n_\xi \times N}, \quad (8a)$$

$$\mathbf{U} = [\mathbf{u}(0), \mathbf{u}(1), ..., \mathbf{u}(N-1)] \in \mathbb{R}^{n_u \times (N-1)}. \quad (8b)$$

Definition 6 (Feasible set [33]). *A set $\mathbb{X}_{fea}^{(N)}$ is said to be the feasible set if and only if, for all $\xi_0 \in \mathbb{X}_{fea}^{(N)}$, the N-horizon optimal control problem $\mathbb{P}_N(\xi_0)$ from Eq. (6) and (7) has feasible solutions, i.e.,*

$$\mathbb{X}_{fea}^{(N)} := \{\xi_0 \in \mathbb{X} \mid \exists \ (\mathbf{U}, \Xi) \text{ satisfying } (7)\} \quad (9)$$

Remark: $\mathbb{X}_{fea}^{(N)}$ apparently satisfies Definition 1 and can be also interpreted as the *maximal control invariant set* by means of the MPC with the prediction horizon N and terminal set $\mathbb{X}_f$ [33]. Because the feasible set is related to the optimal control problem $\mathbb{P}_N$ which depends on the prediction horizon considered, the superscript (N) is used for the feasible set. The feasible set can be explicitly identified using the algorithm from Ref. [33] for linear systems.

The optimal control problem $\mathbb{P}_N(\xi_0)$ from Eq. (6) and (7) is a *parametric* optimization problem because both the cost and the constraints depend on the *parameter* ξ_0. Parameters are the variable affecting the optimal solution but are beyond the direct control of a designer [34–36]. The traditional techniques generate the samples of initial states, repeatedly solve $\mathbb{P}_N(\xi_0)$ with different ξ_0's, and train a surrogate model for the explicit nonlinear control policy using the data points. In contrast, our previous work, state-parameterized nonlinear programming control (sp-NLPC), can solve $\mathbb{P}_N(\xi_0)$ *once* without additional sampling processes. In Ref. [20], the comparative study is shown using the benchmarking problems of inverted pendulum systems.

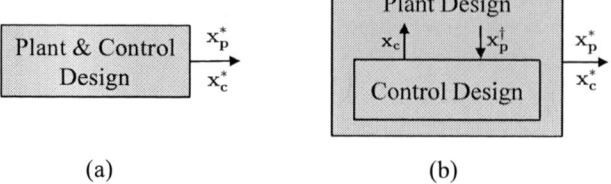

(a)   (b)

FIGURE 1: Two existing design architectures for CCD: (a) simultaneous (b) nested, where $\mathbf{x}_\mathbf{p}^{\dagger}$ denotes a candidate plant design and $\mathbf{x}_\mathbf{p}^{*}$ and $\mathbf{x}_\mathbf{c}^{*}$ are the optimal designs of plant and control, respectively.

2.3 Control Co-Design (CCD)

In this section, two most common CCD strategies are discussed: (1) simultaneous, and (2) nested. A simultaneous CCD strategy is to co-optimize the design variables for plants and controllers by integrating both design problems into a single formulation. For a nested CCD strategy, a bi-level design framework including an outer loop and an inner loop is formulated. The outer loop optimizes the plant design, whereas the inner loop is to find the optimal control parameters given the candidate plant design $\mathbf{x}_\mathbf{p}^{\dagger}$ from the outer loop. Figure 1 shows their architectures.

Both strategies possess different features, requiring designers to assess the problems and check if they can combine both formulations into one. Also, they need to consider their potential trade-offs between solution quality and computational expense. More details of the comparative study between these two formulations can be referred to Ref. [37] and Ref. [38]. However, either real-time MPC or training explicit MPC for the control design can make the CCD optimization too slow because the optimal control solution needs to be resolved whenever a plant design is re-determined. To avoid the computationally demanding task for solving CCD problems, a novel CCD formulation and method is needed and will be explained in the following section.

3 PROPOSED METHOD

This work aims to solve CCD problems with approximate explicit model predictive controllers. To avoid the time-consuming process during optimizations, pre-computing the controllers is preferred. A straightforward way is to parameterize all the plant design variables into the parametric optimization problem from Eq. (6) and (7) and find the surrogate model that is a function of $\mathbf{x_p}$ and ξ. However, as the dimensionality of the parametric optimization problem increases, finding the optima becomes more challenging. Instead of parameterizing all plant design variables, let us define $\theta_p = \theta_p(\mathbf{x_p})$ as a vector of *dependent* variables (i.e., a function of $\mathbf{x_p}$) to minimize the dimensionality but still can represent the physical system. That is, θ_p denotes the parameters that we really care about when doing the

Copyright © 2023 by ASME

control design and analyzing the dynamics. For example, springs and dampers have several parameters to be designed, such as diameters for wires, helix, valves, pistons, spring pitch, and number of coils [4]. However, their dynamic properties, such as the coefficients of the springs and dampers, are of greater interest at the level of system dynamics. In this case, the coefficients of the springs and dampers are the intermediate parameters for physical systems, defined by θ_p, whereas the other independent parameters to design the springs and dampers are treated as the plant design variables $\mathbf{x_p}$.

3.1 Problem Formulation

One of the main ideas for this paper is to compute and train the approximate explicit controllers offline. Therefore, in the optimization, we can directly implement the controllers and simulate the dynamic performance with the cheap function evaluations. In this way, the outer-loop optimization problem is finding optimal plant variables using the plant-parameterized controllers. Figure 2 conceptually shows the design architecture of the proposed approach with the detailed procedure in Fig. 3. In this part, two optimization formulations will be presented: (1) problem formulation for explicit controller design, and (2) problem formulation for multi-objective optimization with the pre-trained controllers.

3.1.1 Design of Approximate Explicit Controllers

In Ref. [20], the state-dependent optimal control solution can be approximated using parametric optimization and metamodeling techniques with the consideration of system stability. In this paper, the plant parameters θ_p are included, thus the optimal control variables $\mathbf{x_c}$ with the corresponding parameters θ_p and ξ_0 can be obtained. The parametric optimization can be formulated as:

$$\mathbf{x_c^*}(\theta_p, \xi_0) = \arg\min_{\mathbf{x_c}} V_N(\mathbf{x_c}, \theta_p, \xi_0), \qquad (10)$$

subject to:

$$\mathbf{u}(k) = \pi(\mathbf{x_c}(\theta_p, \xi_0), \xi(k)) \in \mathbb{U}, \ \forall k \in \mathbb{N}_{[0, N-1]}, \qquad (11a)$$

$$\xi(k+1) - \mathbf{F}(\theta_p, \xi(k), \mathbf{u}(k)) = \mathbf{0}, \ \forall k \in \mathbb{N}_{[0, N-1]}, \qquad (11b)$$

$$\xi(k) \in \mathbb{X}, \ \forall k \in \mathbb{N}_{[0, N-1]}, \qquad (11c)$$

$$\xi(N) \in \Omega, \ \xi(0) := \xi_0 \in \mathbb{X}_{fea}^{(N)}(\theta_p), \qquad (11d)$$

$$\theta_p \in \Theta_p, \qquad (11e)$$

where Θ_p is the feasible set of θ_p and $\pi(\cdot)$ denotes the prescribed control architecture. For simplicity, a state-dependent feedback controller is used so that Eq. (11a) becomes $\mathbf{u}(k) =$

$-\mathbf{K}(\xi_0)\xi(k)$ with the state-dependent vector of control gains $\mathbf{K}(\xi_0) = [K_1(\xi_0), ..., K_{n_\xi}(\xi_0)]$ as the control variables [39]. In Eq. (11d), Ω is the terminal set of the stability constraint defined by:

$$\Omega = \left\{ \xi \in \mathbb{R}^{n_\xi} \ \middle| \ \frac{||\xi||}{||\xi_0||} \leq \epsilon \right\}, \qquad (12)$$

representing the set that the terminal state is supposed to be closer to the origin (equilibrium point) than the initial point, where ϵ is the tolerance whose value is within $[0,1]$. Using the normalized $\left\{ \left(\theta_p^{*(j)}, \xi_0^{*(j)} \right), \mathbf{u}^{*(j)} \right\}, \ \forall j = 1, ..., n_{sol}$ as the training data, where $\left(\theta_p^{*(j)}, \xi_0^{*(j)} \right)$ is input data, $\mathbf{u}^{*(j)} := -\mathbf{K}^{*(j)}\xi_0^{*(j)}$ is output data, and n_{sol} is the number of the data sets, the surrogate model of optimal control input can be created using radial basis function approximations:

$$\mathbf{u} = \kappa(\theta_p, \xi), \qquad (13)$$

which means the control input can be obtained given any plant θ_p and state ξ. Because the dynamics discussed in this paper are time-invariant systems which have fixed dynamics that do not change over time, their behavior is only determined by the current state and the control input. It is not necessary to distinguish between initial and current states, thus the initial state ξ_0 can be replaced by any current state ξ.

3.1.2 Multi-Objective Optimization for Plant Design

In addition to evaluating the cost of dynamic performance over the simulation time interval using Eq. (6), there are still other metrics needed to be considered by taking into account the inherent approximation errors from the designed controllers. One is the *reliability* of the dynamic responses, which can be quantified by the probability that a system performs its intended function in a period of time without failures [40]. Due to uncertain errors from controllers, the dynamic responses can be treated as failures if the system violates the constraints. To avoid the violation, one can tighten the constraints from Eq. (11a) and Eq. (11c) in the stage of designing controllers. The amount of tightening can be chosen by designers based on their level of conservatism [41]. Ref. [28] uses set invariance theory to define the tightened constraints for the MPC formulation. However, one of the aims of this paper is to examine how controller errors can negatively impact dynamic behaviors, such as system performance and reliability. Therefore, the original constraints are used and the probability of violating the state constraint is evaluated to assess the reliability of the controlled systems. Reliability-based design optimization (RBDO) is to formulate constraints as

Copyright © 2023 by ASME

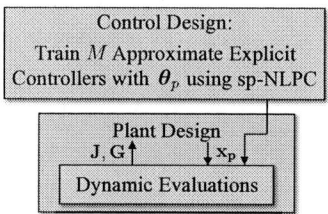

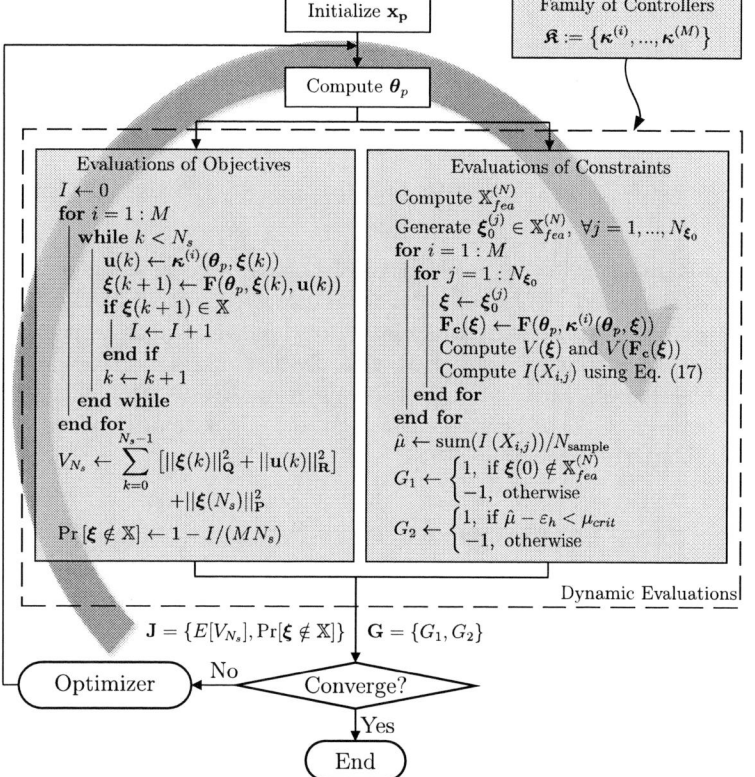

FIGURE 2: Design architecture of the proposed method: CCD using approximate explicit controllers, where M denotes the number of trained controllers, $\boldsymbol{\theta}_p$ denotes the vector of plant parameters, $\mathbf{J}$ is the vector of optimization objectives, and $\mathbf{G}$ represents the set of optimization constraints.

FIGURE 3: Flow chart of the proposed approach, detailing the formulations and quantities from Fig. 2.

inequality equations for the probability of failure [42–44]. Nevertheless, to facilitate design exploration and make a trade-off between performance and reliability, a multi-objective optimization for the CCD problem using the offline designed explicit controllers can be formulated as:

$$\min_{\mathbf{x_p}} \mathbf{J}(\mathbf{x_p}) = \left\{ E[V_{N_s}], \Pr[\boldsymbol{\xi} \notin \mathbb{X}] \right\}, \qquad (14)$$

subject to:

$$\mathbf{g_p}(\mathbf{x_p}) \leq \mathbf{0}, \qquad (15a)$$

$$\mathbf{G}\left(\mathbf{x_p}, \mathfrak{K}, \mathbb{X}, \mathbb{U}\right) \leq \mathbf{0}, \qquad (15b)$$

where $E[V_{N_s}]$ denotes the expected value of the costs with the N_s-horizon evaluations, $\Pr[\boldsymbol{\xi} \notin \mathbb{X}]$ denotes the probability of the state-constraint violation, $\mathbf{g_p}(\cdot)$ represents the constraints that depend on the plant design variables only (including the bounds of $\mathbf{x_p}$), $\mathfrak{K} := \left\{ \boldsymbol{\kappa}^{(i)}, ..., \boldsymbol{\kappa}^{(M)} \right\}$ denotes the family of M trained explicit controllers, and $\mathbf{G}\left(\mathbf{x_p}, \mathfrak{K}, \mathbb{X}, \mathbb{U}\right)$ in Eq. (15b) is the set of the functions for inequality constraints, including:

- $G_1\left(\mathbf{x_p}, \mathfrak{K}, \mathbb{X}, \mathbb{U}\right)$: identify if the initial state is in the feasible set $\mathbb{X}_{fea}^{(N)}$ from Definition 6; if not, then **return** 1.
- $G_2\left(\mathbf{x_p}, \mathfrak{K}, \mathbb{X}, \mathbb{U}\right)$: identify if the criteria of stability are satisfied via the probabilistic assessment (with the sampled initial conditions and trained controllers); if not, then **return** 1.

To evaluate the second constraint G_2, the probabilistic assessment method to validate system stability of closed-loop systems proposed by Ref. [25] is adopted in the paper. However, unlike Ref. [25], in addition to the sampled initial conditions, the uncertainty of developing approximate explicit controllers is also considered. Therefore, for all $\boldsymbol{\theta}_p \in \Theta_p$, we define:

$$X_{i,j} := \left\{ \mathbf{F_c}(\boldsymbol{\xi}) := \mathbf{F}(\boldsymbol{\theta}_p, \boldsymbol{\xi}, \boldsymbol{\kappa}^{(i)}(\boldsymbol{\xi})) \text{ with } \boldsymbol{\xi} := \boldsymbol{\xi}_0^{(j)} \in \mathbb{X}_{fea}^{(N)} \right\}, \qquad (16)$$

which denotes a sample of the successor state for the closed-loop system using the ith controller with the jth initial state. Then, let

$$I(X_{i,j}) := \begin{cases} 1, & \text{if (Definitions 4 \& 5 and Theorem 1} \\ & \text{are satisfied)} \cap \left(\boldsymbol{\xi}_0^{(j)} \in \mathbb{X}_p\right) \\ 0, & \text{otherwise} \end{cases} \qquad (17)$$

be an indicator function, which specifies whether the stability criteria are satisfied and the initial state is positively invariant given the sample $X_{i,j}$.

Scenario-based Monte Carlo Simulation (MCS) is used to evaluate the probabilistic constraint to validate stability using the N_{sample} dynamic simulations based on the M trained controllers and the N_{ξ_0} generated samples of the initial state, where $N_{\text{sample}} = M \times N_{\xi_0}$. While there are some efficient techniques for reliability analysis, like the first-order reliability method (FORM) and the second-order reliability method (SORM) [45], they require the probability information of uncertain variables, such as parameters for probabilistic distributions. Nevertheless, it is difficult to fit a probabilistic distribution for the approximation errors due to the unpredictable nature of approximating optimal explicit controllers.

Copyright © 2023 by ASME

For the N_{sample} samples, we define the empirical risk as

$$\hat{\mu} := \frac{1}{N_{\text{sample}}} \sum_{l=1}^{N_{\text{sample}}} I(X_l). \qquad (18)$$

Since $I(X_l) = 1$ is what we expect to ensure stability, we define the probability for $I(X_l) = 1$ as $\mu := \Pr[I(X_l) = 1]$ for X_l with independent, identically distributed (iid) samples of initial states and trained controllers. Hoeffding's inequality is used because it characterizes how well the empirical mean approximates the expected value by providing an upper bound on the probability of the deviation by more than a certain amount.

Lemma 1 (Hoeffding's inequality [46]). *Let* $I(X_l)$, $\forall l = 1, ..., N_{\text{sample}}$ *be* N_{sample} *iid random variables with* $0 \leq I(X_l) \leq 1$. *Then,*

$$\Pr[|\hat{\mu} - \mu| \geq \varepsilon_h] \leq \delta_h, \qquad (19)$$

where $\delta_h := 2\exp(-2N_{\text{sample}}\varepsilon_h^2)$ *denotes the confidence level.*

Remark: Note that the higher N_{sample} is, the smaller δ_h becomes, meaning we are more confident about the satisfaction of the inequality. For convenience, we express ε_h as:

$$\varepsilon_h = \sqrt{-\frac{\ln(\delta_h/2)}{2N_{\text{sample}}}}. \qquad (20)$$

Based on Lemma 1, it can be also said that with the confidence of at least $1 - \delta_h$,

$$\Pr[I(X_l = 1)] = \mu \geq \hat{\mu} - \varepsilon_h \geq \mu_{crit}, \qquad (21)$$

with the chosen critical bound μ_{crit}. If Eq. (21) holds, it implies that with $1 - \delta_h$ confidence the probability that system stability is guaranteed with a random initial condition from $\mathbb{X}_{fea}^{(N)}$ using a random controller from $\Re$.

3.2 Algorithm & Procedure

Figure 3 shows the detailed procedure of the proposed approach based on the design architecture from Fig. 2. Before starting the optimization, the family of explicit controllers $\Re$, which is the set of surrogate models $\kappa^{(i)}(\theta_p, \xi)$, needed to be provided using Section 3.1.1.

For the outer-loop CCD to optimize the plant design variables, with the given design $\mathbf{x_p}$, the values of θ_p are computed to define the dynamics. Then, in the process of dynamic evaluations, both objectives and constraints are evaluated. For the

simulation in objective evaluations, the M controllers are used to generate the state and control trajectories given a specific initial condition. The cost functions from Eq. (6) and the probability of the state-constraint violation based on the M trajectories are computed with the simulation period N_s. For evaluating the constraints, two criteria for ensuring that the given initial condition is in the feasible set and the closed-loop systems can be asymptotically stabilized are considered. The first constraint is violated if $\xi(0) \notin \mathbb{X}_{fea}^{(N)}$. To assess the second constraint, in addition to $\Re$, the N_{ξ_0} initial conditions are generated from the feasible set $\mathbb{X}_{fea}^{(N)}$ to examine if the designed plant can satisfy the stability condition based on Eq. (17) using all the pre-trained controllers from $\Re$. The empirical risk is then estimated using the indicator $I(X_{i,j})$ and Eq. (18). The Hoeffding's inequality from Lemma 1 and Eq. (19)-(21) is used to identify if the constraint G_2 is satisfied.

Finally, the optimizer will produce a new set of population members based on the current population members until the generation reaches the maximum generation. P3GA [47–50], a genetic algorithm-based optimizer, is used to solve the multi-objective CCD optimization problem and the non-dominated solutions can be obtained.

4 ILLUSTRATIVE EXAMPLES

The proposed methodology and sp-NLPC can be used to solve CCD problems for both linear and nonlinear dynamics. However, to effectively validate the system stability and identify the sets of $\mathbb{X}_{fea}^{(N)}$ and $\mathbb{X}_f$, two linear systems (a numerical example and an engineering case study of satellite systems) are chosen to demonstrate the proposed method without loss of generality. Since the development of set invariance for nonlinear systems is relatively premature compared to linear cases, the authors do not claim to provide a complicated and time-consuming implementation due to the nonlinearity of dynamics.

In addition, true solutions to the control problems can be obtained when the cost functions are quadratic and the dynamic equations and all other constraints are linear. The section will visualize the approximate trajectories by comparing them with the true ones.

4.1 Numerical Example

An inherently unstable double integrator from Ref. [51] is selected as the numerical example to benchmark the proposed method. The dynamic equation is defined as:

$$\xi(k+1) = \begin{bmatrix} 1 & 1 \\ 0 & 1 \end{bmatrix} \xi(k) + \begin{bmatrix} 0.5 \\ \theta_p \end{bmatrix} u(k), \ \forall k \in \mathbb{N} \qquad (22)$$

Copyright © 2023 by ASME

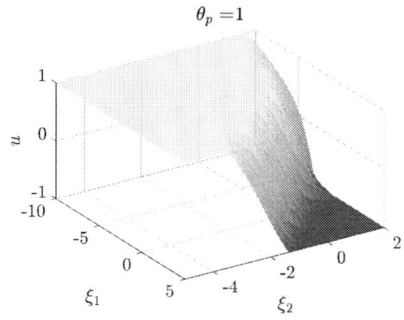

FIGURE 4: An example of the visualized explicit control policy for the numerical example with $\theta_p = 1$.

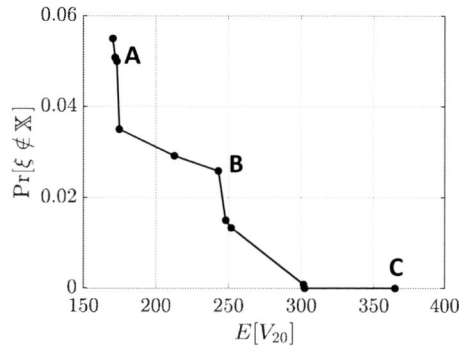

FIGURE 5: Approximate Pareto front of the solutions for the numerical example using the proposed method.

where $\theta_p = \frac{a \sin(b\pi)}{2} + 1.0 \in [0.5, 1.5]$ is the dependent plant variable, and $a \in [-1, 1]$ and $b \in [-1, 1]$ are the independent variables for designing the plant (i.e., $\mathbf{x_p} = [a, b]^\top$). The trade-off between high or low θ_p plays a key role in this design problem because a high θ_p improves the controllability of the system, but it also leads to a greater susceptibility to dynamic behaviors resulting from control or actuation uncertainty. The state is constrained by $\boldsymbol{\xi}(k) \in \mathbb{X} := \{\boldsymbol{\xi} \mid \xi_1 \in [-10, 5], \xi_2 \in [-5, 2]\}$; the control constraint is $u(k) \in \mathbb{U} := \{u \mid u \in [-1, 1]\}$. The simulation period $N_s = 20$ is used for evaluating the cost of dynamic performance. The initial condition is given by $\boldsymbol{\xi}_0 = [-5, -2]^\top$. The functions α_1, α_2, and α_3 from Eq. (5) are defined by $0.01|\boldsymbol{\xi}|$, $1000|\boldsymbol{\xi}|$, and $0.01|\boldsymbol{\xi}|$. For the probabilistic assessment, the values for δ_h and μ_{crit} from Eq. (20) and (21) are 0.01 and 0.9, respectively.

Before solving the outer-loop CCD optimization problem, sixty explicit controllers ($M = 60$) are approximated using Section 3.1.1 with the parameter setting: $\epsilon = 1$, $\Theta_p = [0.5, 1.5]$, $\mathbf{Q} = \mathbf{I}_{2 \times 2}$, $R = 0.1$, and $\mathbf{P}$ is the terminal weighting matrix for the cost function Eq. (10) and obtained from solving the discrete-time Riccati equation, and the horizon $N = 10$. One example of the explicit feedback control policy with $\theta_p = 1$ is visualized in

TABLE 1. Optimal plant design variables and objectives for Solutions A, B, and C for the numerical example.

Solution	Design variables			Objectives	
	a	b	θ_p	$E[V_{20}]$	$\Pr[\boldsymbol{\xi} \notin \mathbb{X}]$
A	0.7187	0.4450	1.3540	172.98	0.0500
B	-0.4043	0.2662	0.8500	243.03	0.0258
C	-0.8878	0.4762	0.5573	364.71	0

Fig. 4, where the control input is saturated by ± 1. The higher M is, the more confidence we have in evaluating both objectives and constraints in the sense of probabilistic assessment. A main advantage of sp-NLPC is its efficiency of computational time and data size, making the synthesis of many explicit controllers more computationally tractable. Solving each controller only took about 2 minutes using the machine: an Intel(R) Core(TM) i7-9700 CPU @ 3.00 GHz, 32 GB RAM, WINDOWS 10 64-bit, and MATLAB 2020a with parallel computing. It is noted that the processes were completed *offline* and each function evaluation is extremely cheap, reducing the computational time in the optimization significantly.

Figure 5 shows the approximate Pareto front by the proposed approach using 20 generations and 30 populations. It is noted that these non-dominated solutions are all feasible, which means the constraint G_2 satisfies and thus the systems are asymptotically stable with statistical significance after the probabilistic assessment using Eq. (16) to Eq. (21), where $N_{sample} = 3000$. The result demonstrates the trade-off between having less expected cost for performance and having less probability of state-constraint violations. To have a better comparison of these non-dominated solutions, three of them (Solutions A, B, and C) are selected. Their values of design variables, plant parameters, and objectives are shown in Table 1. Their state and control trajectories are visualized in Fig. 6. The corresponding true solutions are also shown in Fig. 6 for reference. As we see in Table 1, since Solution A has the highest θ_p, it can be said that the plant is more "sensitive" to the control actuation, meaning its behaviors are more easily altered by control inputs. Therefore, the states and control inputs of Solution A converge faster to the equilibrium (which is the origin in this case) and zero, respectively, as shown in Fig.6a-6c and Fig.6d-6f, in comparison to Solutions B and C. This is a great advantage for performance. It is worth noting that Solution A has a larger size of $\mathbb{X}_f$ (is also the MPI set satisfying Definition 3), which is the gray region shown in Fig. 6a. This feature fundamentally explains that Solution A possesses good properties of stability and convergence. In Fig. 7, Solution A also has the largest size of feasible set $\mathbb{X}_{fea}^{(10)}$ covering the entire region of $\mathbb{X}$, allowing for a greater range of starting conditions that can result in feasible solutions. However, in the perspective of reliability, the state trajectories of Solution A have a higher

Copyright © 2023 by ASME

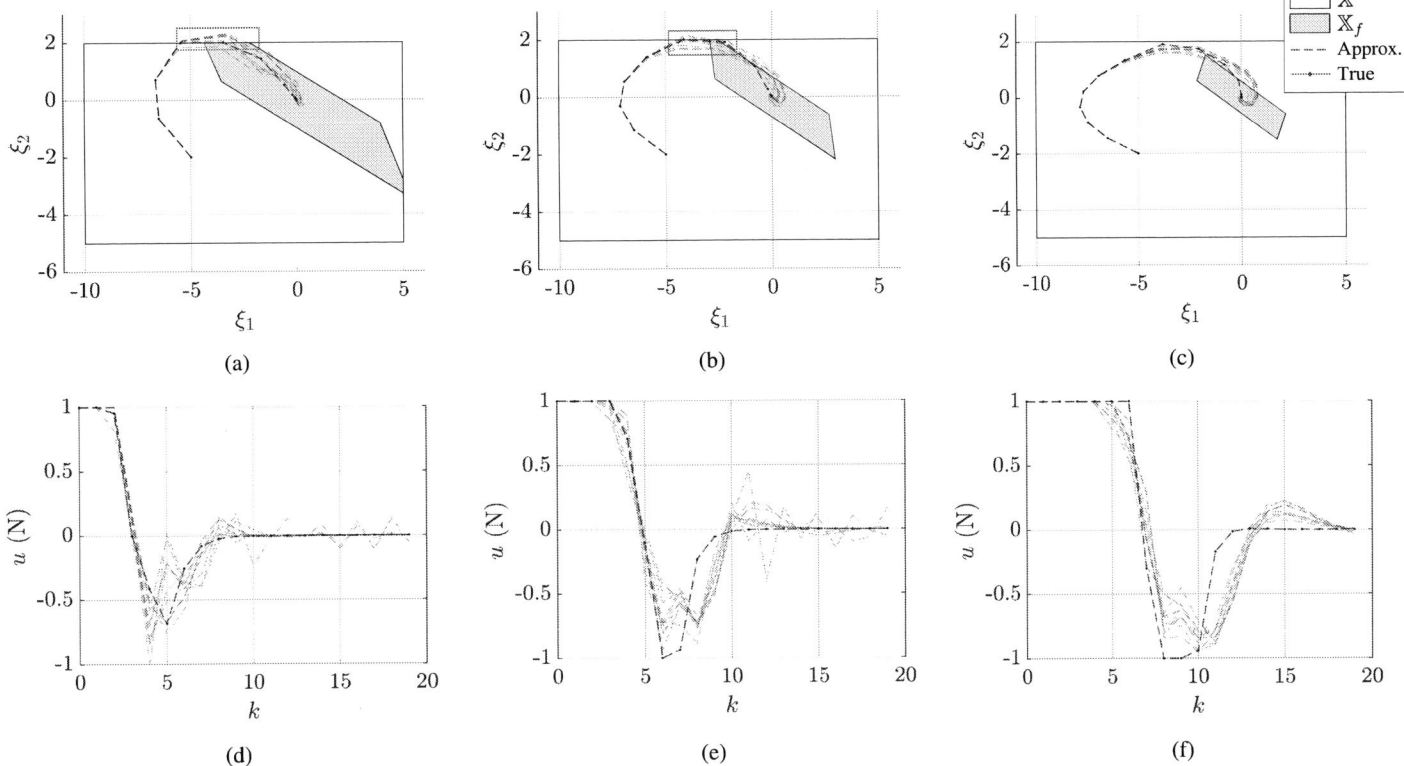

FIGURE 6: State trajectories (a-c) and control trajectories (d-f) of Solutions A, B, and C for the numerical example, respectively. The dashed lines denote the approximate explicit controllers (ten out of the sixty controllers are selected randomly for visualization), whereas the solid lines are the trajectories for the true solutions. The terminal sets $\mathbb{X}_f$ satisfy Definition 3 and are identified via Multi-Parametric Toolbox 3 [52]. (Online version in color.)

chance of going out of the bound (highlighted in Fig. 6a) and violating the state constraint. On the other hand, Solutions B and C have less or even no violations of the state constraint. However, they need more control efforts than Solution A does.

4.2 Case Study: Satellite Attitude Control

Satellites typically need attitude control to make antennas point toward a particular location on Earth or to orient the panel toward the direction of the sun by changing the satellite orientation [53]. The continuous-time and the discrete-time models of the one-dimensional (1-D) satellite attitude control system are derived in Fig. 8. The plant design variables are: the distance between the propulsion and the center of mass, $d \in [5, 10]$ m, and the angle, $\alpha \in [10°, 45°]$, whereas the plant parameter is $\theta_p = \frac{d \cos \alpha}{I}$. The control force comes from the reaction jet $u(k)$ bounded by $[-3000, 3000]$ N. The state is constrained by $\boldsymbol{\xi}(k) \in \mathbb{X} := \{\boldsymbol{\xi} \mid \psi \in [-10°, 30°], \ \dot{\psi} \in [-0.5, 1] \text{ rad/s}\}$. For the CCD optimization, the cost function is defined by (6) with $\mathbf{Q} = \text{diag}\left(\begin{bmatrix} 10^5, 10^4 \end{bmatrix}\right)$, $R = 10^{-4}$, $\mathbf{P}$ from the solution of the Ric-

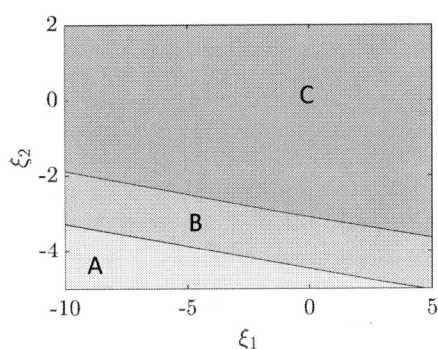

FIGURE 7: Feasible sets $\mathbb{X}_{fea}^{(10)}$ of Solutions A, B, and C for the numerical example via Multi-Parametric Toolbox 3 [52], where Solution A covers the entire region of $\mathbb{X}$.

cati equation, and $N_s = 100$. The initial condition is given by $\boldsymbol{\xi}(0) = [25°, 0]^\top$. The control task is to force the state back to

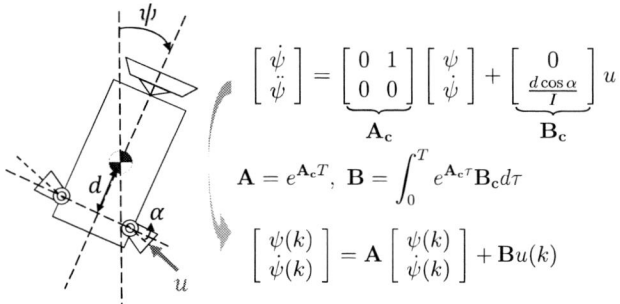

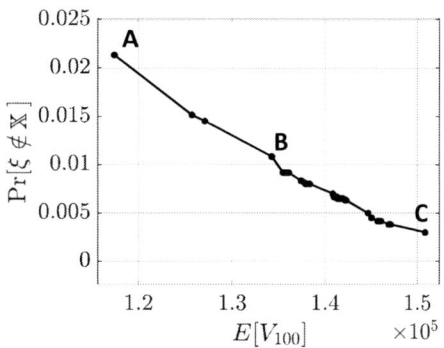

FIGURE 8: 1-D satellite model with the continuous-time and discrete-time equations, where $T = 0.1$ sec is the sampling time, d denotes the distance between propulsion and the center of mass, α denotes the angle of propulsion, and $I = 1.1472 \times 10^5$ kg·m^2 is the moment of inertia whose parameter value is from Ref. [54].

FIGURE 9: Approximate Pareto front of the solutions for the satellite attitude control system using the proposed method.

the origin ($\psi = 0$, $\dot{\psi} = 0$). For the control design, sixty controllers ($M = 60$) are developed with the cost function that uses the same parameters above but with $N = 10$ as the prediction horizon. The tolerance in Eq. (12) is $\epsilon = 1$ and the bound for the plant parameter is $\Theta = [2.03, 5.65] \times 10^{-4}$. The functions α_1, α_2, and α_3 from Eq. (5) are defined by $0.001|\xi|$, $10^6|\xi|$, and $0.001|\xi|$. For the probabilistic assessment, the values for δ_h and μ_{crit} from Eq. (20) and (21) are 0.01 and 0.9, respectively.

The approximate Pareto front that consists of several non-dominated solutions by 20 generations and 30 populations is shown in Fig. 9. These non-dominated solutions are feasible by ensuring the stability using Eq. (16) to Eq. (21), where $N_{sample} = 3000$. Similar to the numerical example, we compare the selected solutions (A, B, and C) and visualize their trajectories in Fig. 10. As we see in Table 2, the value of $\theta_p := \frac{d\cos\alpha}{I}$ in matrix $\mathbf{B_c}$ has a significant impact on the controllability. Like the results from the numerical example, the system of Solution A can

TABLE 2. Optimal plant design variables and objectives for Solutions A, B, and C for the case study of satellite attitude control.

Solution	Design variables			Objectives	
	d	α	$\theta_p := \frac{d\cos\alpha}{I}$	$E[V_{100}]$	$\Pr[\xi \notin \mathbb{X}]$
A	9.7998	21.44°	5.23×10^{-4}	1.17×10^5	0.0213
B	7.5087	37.43°	3.42×10^{-4}	1.34×10^5	0.0108
C	6.1808	45.00°	3.18×10^{-4}	1.51×10^5	0.0030

be easily controlled because of having a higher θ_p. Based on the comparison, Solution A has the lowest cost but the highest probability of violating the state constraint. In Fig. 10d, the state goes to the origin with fewer control efforts compared to Solutions B and C in Fig.10e and Fig. 10f. Especially for Solution C, the control is sometimes saturated at the top limit in Fig. 10f, while it does not happen in Solutions A and B. On the other hand, the state trajectories of Solution A go out of the window more often than Solutions B and C shown in Fig. 10a - 10c.

The examples discussed above illustrate the substantial effect that changes in plant design can have on the dynamic characteristics of a system, even when the controllers are optimized. This highlights the importance of considering both plant and controller design for the purpose of achieving system-optimal solutions. To facilitate design exploration and avoid overlooking any potentially viable solutions, Pareto fronts based on bi-objective optimization are identified. Since no designs outperform the others, system designers need to conduct a trade-off study and choose the one that best fits their requirements and preference. For example, if the system designer is willing to tolerate a higher amount of violations for the sake of better overall performance, then Solution A might be a suitable option. On the other hand, if the system is anticipated to be more conservative to the violation, then the system designer can choose Solution C to guarantee satisfaction, though sacrificing some levels of performance. If no specific preference between both objectives, Solution B appears to be an excellent solution that strikes a good balance between both metrics. Ultimately, the choice of solution will depend on the specific requirements and priorities of the design problem.

5 CONCLUSIONS AND FUTURE WORK

This work is to propose a novel approach for Control Co-Design (CCD) with approximate explicit model predictive controllers. An important aspect of understanding how approximation errors from designed control policies impact the performance and decision-making of designing the plants is also investigated. The explicit MPC technique, state-parameterized nonlinear programming control (sp-NLPC), is extended to the CCD by including plant parameters in the surrogate models. This en-

Copyright © 2023 by ASME

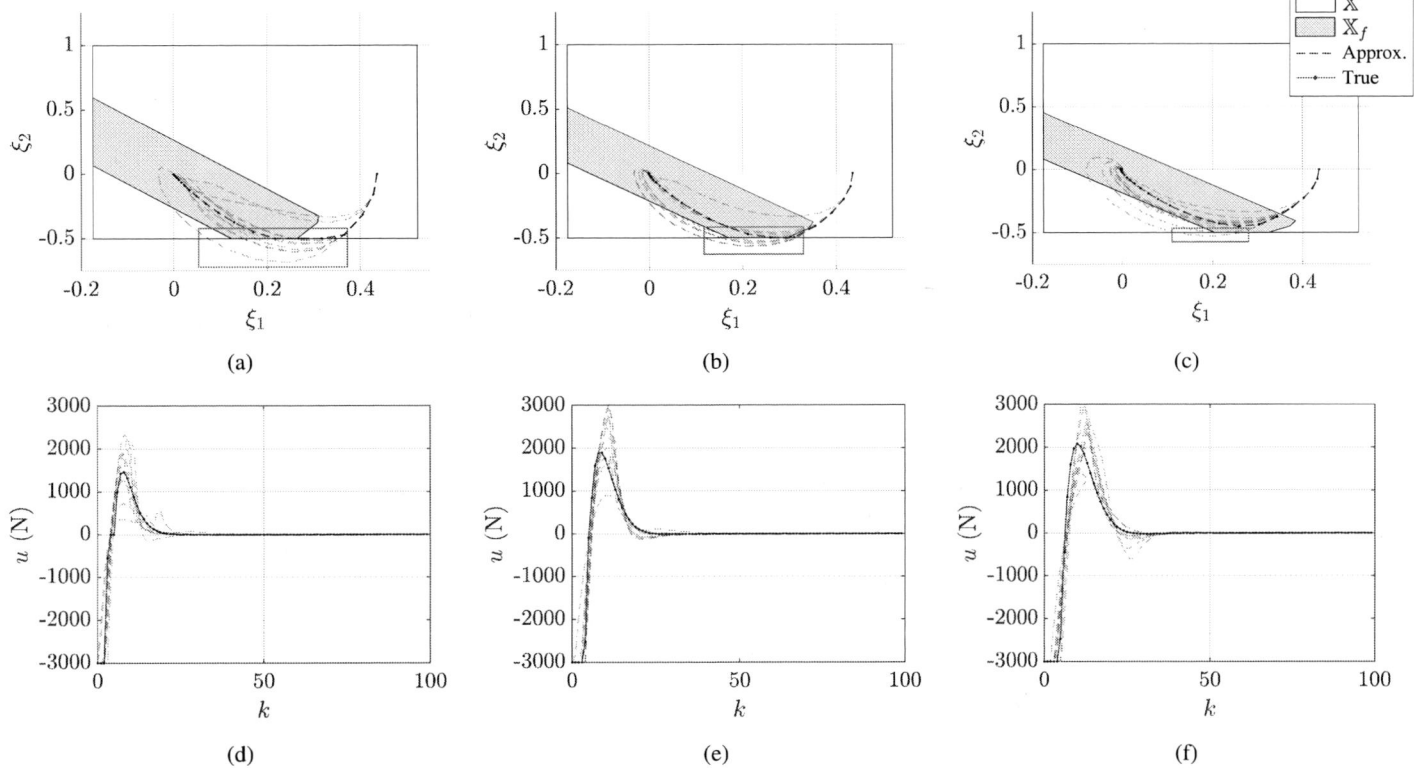

FIGURE 10: State trajectories (a-c) and control trajectories (d-f) of Solutions A, B, and C for the satellite attitude control system, respectively. The dashed lines denote the approximate explicit controllers (ten out of the sixty controllers are selected randomly for visualization), whereas the solid lines are the trajectories for the true solutions. The terminal sets $\mathbb{X}_f$ satisfy Definition 3 and are identified via Multi-Parametric Toolbox 3 [52]. Note that the sampling time $T = 0.1$ sec so that $k = 10$ means $t = 1$ sec and the simulation period is 10 sec. (Online version in color.)

ables solving the control problem offline and avoids considerable computational time in the CCD optimization. Even with the presence of approximation errors in explicit feedback controllers, the closed-loop system given the plant design can be guaranteed to be asymptotically stable using set invariance theory and Lyapunov stability theory. In the optimization constraints, the Hoeffding's inequality is adopted with the probabilistic assessments. The Pareto fronts are approximated with the identified non-dominated solutions using the multi-objective optimization formulation and the solver of P3GA. Therefore, system designers can conduct trade-off studies and select the design that best fits their requirements and scenarios. The illustrative examples include a numerical example and a case study of co-designing the satellite attitude control system. The results of this study demonstrate that the feedback controllers synthesized by sp-NLPC are effective in steering the system states toward equilibrium while satisfying control constraints. The visualized trajectories provide clear evidence of this. Given that the cost functions are quadratic and all the constraints (state and control bounds, as well as lin-

ear dynamics) are linear, the true solutions to the control problems can be obtained through quadratic programming. The trajectories for the true solutions are also visualized for comparing the approximate results and giving insights into approximation errors from the explicit controllers. Additionally, these results show that changing or redesigning the physical systems significantly impacts the dynamic characteristics and performance, which highlights the importance of performing CCD.

Despite the contributions of the proposed method and the investigation in this paper, the authors would also like to point out the limitations and the potential future work. Although the control design method, sp-NLPC, is capable of solving nonlinear control problems and stabilizing highly nonlinear and unstable systems, there are many theorems and definitions to guarantee stability and identify the control properties (e.g., $\mathbb{X}_f$ and $\mathbb{X}_{fea}^{(N)}$) for linear systems so that both the numerical example and the case study of engineering systems are linear. However, for the need of designing and controlling nonlinear systems in practice, it requires nonlinear MPC techniques to theoretically guarantee

Copyright © 2023 by ASME

stability.

The paper uses model-based control techniques and assumes the errors only come from approximate optimal explicit controllers. Nevertheless, apart from the approximation errors, there are still many other possible uncertainties from estimation errors, unknown/uncertain system parameters, and external disturbances that we need to think of during the design stage [55–57]. An alternative for future work is to develop the methods of CCD under uncertainties using robust and stochastic model predictive controllers [28, 58, 59].

REFERENCES

[1] Garcia-Sanz, M., 2019. "Control co-design: an engineering game changer". *Advanced Control for Applications: Engineering and Industrial Systems, 1*(1), p. e18.

[2] Allison, J. T., and Herber, D. R., 2014. "Special section on multidisciplinary design optimization: multidisciplinary design optimization of dynamic engineering systems". *AIAA journal, 52*(4), pp. 691–710.

[3] Fathy, H. K., Reyer, J. A., Papalambros, P. Y., and Ulsov, A., 2001. "On the coupling between the plant and controller optimization problems". In Proceedings of the 2001 American Control Conference.(Cat. No. 01CH37148), Vol. 3, IEEE, pp. 1864–1869.

[4] Allison, J. T., Guo, T., and Han, Z., 2014. "Co-design of an active suspension using simultaneous dynamic optimization". *Journal of Mechanical Design, 136*(8).

[5] Deshmukh, A. P., Herber, D. R., and Allison, J. T., 2015. "Bridging the gap between open-loop and closed-loop control in co-design: A framework for complete optimal plant and control architecture design". In 2015 American Control Conference (ACC), IEEE, pp. 4916–4922.

[6] Deshmukh, A. P., and Allison, J. T., 2016. "Multidisciplinary dynamic optimization of horizontal axis wind turbine design". *Structural and Multidisciplinary Optimization, 53*(1), pp. 15–27.

[7] Chilan, C. M., Herber, D. R., Nakka, Y. K., Chung, S.-J., Allison, J. T., Aldrich, J. B., and Alvarez-Salazar, O. S., 2017. "Co-design of strain-actuated solar arrays for spacecraft precision pointing and jitter reduction". *AIAA journal, 55*(9), pp. 3180–3195.

[8] Azad, S., Behtash, M., Houshmand, A., and Alexander-Ramos, M. J., 2019. "Phev powertrain co-design with vehicle performance considerations using mdsdo". *Structural and Multidisciplinary Optimization, 60*(3), pp. 1155–1169.

[9] Cui, T., Zheng, Z., and Wang, P., 2022. "Control co-design of lithium-ion batteries for enhanced fast-charging and cycle life performances". *Journal of Electrochemical Energy Conversion and Storage, 19*(3).

[10] Sundarrajan, A. K., Lee, Y. H., Allison, J. T., and Herber, D. R., 2021. "Open-loop control co-design of floating off-shore wind turbines using linear parameter-varying models". In International Design Engineering Technical Conferences and Computers and Information in Engineering Conference, Vol. 85383, American Society of Mechanical Engineers, p. V03AT03A010.

[11] Jiang, B., Amini, M. R., Liao, Y., Martins, J. R., and Sun, J., 2022. "Control co-design of a hydrokinetic turbine with open-loop optimal control". In International Conference on Offshore Mechanics and Arctic Engineering, Vol. 85932, American Society of Mechanical Engineers, p. V008T09A006.

[12] Vercellino, R., Markey, E., Limb, B. J., Pisciotta, M., Huyett, J., Garland, S., Bandhauer, T., Quinn, J. C., Psarras, P., and Herber, D. R., 2022. "Control co-design optimization of natural gas power plants with carbon capture and thermal storage". In International Design Engineering Technical Conferences and Computers and Information in Engineering Conference, Vol. 86229, American Society of Mechanical Engineers, p. V03AT03A010.

[13] Zhang, T., Fu, T., and Song, X., 2022. "Design of unmanned cable shovel based on multiobjective co-design optimization of structural and control parameters". *Journal of Mechanical Design, 144*(9), p. 091708.

[14] Mayne, D. Q., Rawlings, J. B., Rao, C. V., and Scokaert, P. O., 2000. "Constrained model predictive control: Stability and optimality". *Automatica, 36*(6), pp. 789–814.

[15] Mayne, D. Q., 2014. "Model predictive control: Recent developments and future promise". *Automatica, 50*(12), pp. 2967–2986.

[16] Rawlings, J. B., Mayne, D. Q., and Diehl, M., 2017. *Model predictive control: theory, computation, and design*, Vol. 2. Nob Hill Publishing Madison, WI.

[17] Docimo, D. J., Kang, Z., James, K. A., and Alleyne, A. G., 2021. "Plant and controller optimization for power and energy systems with model predictive control". *Journal of Dynamic Systems, Measurement, and Control, 143*(8).

[18] Nash, A. L., Pangborn, H. C., and Jain, N., 2021. "Robust control co-design with receding-horizon mpc". In 2021 American Control Conference (ACC), IEEE, pp. 373–379.

[19] Tsai, Y.-K., and Malak Jr, R. J., 2021. "A methodology for designing a nonlinear feedback controller via parametric optimization: State-parameterized nonlinear programming control". In International Design Engineering Technical Conferences and Computers and Information in Engineering Conference, Vol. 85383, American Society of Mechanical Engineers, p. V03AT03A011.

[20] Tsai, Y.-K., and Malak Jr, R. J., 2022. "Design of approximate explicit model predictive controller using parametric optimization". *Journal of Mechanical Design, 144*(12), p. 124501.

[21] Bemporad, A., Morari, M., Dua, V., and Pistikopoulos, E. N., 2002. "The explicit linear quadratic regulator for

Copyright © 2023 by ASME

constrained systems". *Automatica, 38*(1), pp. 3–20.

[22] Domínguez, L. F., and Pistikopoulos, E. N., 2010. "A novel mp-nlp algorithm for explicit/multi-parametric nmpc". *IFAC Proceedings Volumes, 43*(14), pp. 539–544.

[23] Pistikopoulos, E. N., 2009. "Perspectives in multiparametric programming and explicit model predictive control". *AIChE journal, 55*(8), pp. 1918–1925.

[24] Paulson, J. A., and Mesbah, A., 2020. "Approximate closed-loop robust model predictive control with guaranteed stability and constraint satisfaction". *IEEE Control Systems Letters, 4*(3), pp. 719–724.

[25] Hertneck, M., Köhler, J., Trimpe, S., and Allgöwer, F., 2018. "Learning an approximate model predictive controller with guarantees". *IEEE Control Systems Letters, 2*(3), pp. 543–548.

[26] Drgoňa, J., Picard, D., Kvasnica, M., and Helsen, L., 2018. "Approximate model predictive building control via machine learning". *Applied Energy, 218*, pp. 199–216.

[27] Drgoňa, J., Mukherjee, S., Tuor, A., Halappanavar, M., and Vrabie, D., 2022. "Learning stochastic parametric diferentiable predictive control policies". *IFAC-PapersOnLine, 55*(25), pp. 121–126.

[28] Tsai, Y.-K., and Malak Jr., R. J., 2023. "Robust control co-design using tube-based model predictive control". In 2023 American Control Conference (ACC), IEEE.

[29] Tsai, Y.-K., and Malak Jr, R. J., 2022. "A constraint-handling technique for parametric optimization and control co-design". In International Design Engineering Technical Conferences and Computers and Information in Engineering Conference, Vol. 86229, American Society of Mechanical Engineers, p. V03AT03A009.

[30] Raković, S., Kerrigan, E., Kouramas, K., and Mayne, D., 2004. *Invariant approximations of robustly positively invariant sets for constrained linear discrete-time systems subject to bounded disturbances.* University of Cambridge, Department of Engineering Cambridge.

[31] Blanchini, F., 1999. "Set invariance in control". *Automatica, 35*(11), pp. 1747–1767.

[32] Raković, S. V., and Zhang, S., 2022. "The implicit maximal positively invariant set". *IEEE Transactions on Automatic Control.*

[33] Scibilia, F., Olaru, S., and Hovd, M., 2011. "On feasible sets for mpc and their approximations". *Automatica, 47*(1), pp. 133–139.

[34] Weaver-Rosen, J., Tsai, Y.-K., Schoppe, J., Terada, Y., Malak, R., Cizmas, P. G., and Lazzara, D. S., 2022. "Surrogate modeling and parametric optimization strategy for minimizing sonic boom in a morphing aircraft". In AIAA SCITECH 2022 Forum, p. 0097.

[35] Weaver-Rosen, J. M., 2021. "Multi-objective efficient parametric optimization". PhD thesis, Texas A&M Univeristy.

[36] Galvan, E., 2016. "Parametric optimization: Application in

systems design". PhD thesis, Texas A&M Univeristy.

[37] Herber, D. R., and Allison, J. T., 2019. "Nested and simultaneous solution strategies for general combined plant and control design problems". *Journal of Mechanical Design, 141*(1).

[38] Sundarrajan, A. K., and Herber, D. R., 2021. "Towards a fair comparison between the nested and simultaneous control co-design methods using an active suspension case study". In 2021 American Control Conference (ACC), IEEE, pp. 358–365.

[39] Çimen, T., 2010. "Systematic and effective design of nonlinear feedback controllers via the state-dependent riccati equation (sdre) method". *Annual Reviews in control, 34*(1), pp. 32–51.

[40] Wu, H., Hu, Z., and Du, X., 2021. "Time-dependent system reliability analysis with second-order reliability method". *Journal of Mechanical Design, 143*(3).

[41] Azad, S., and Alexander-Ramos, M. J., 2020. "Robust mdsdo for co-design of stochastic dynamic systems". *Journal of Mechanical design, 142*(1).

[42] Agarwal, H., 2004. "Reliability based design optimization: formulations and methodologies". PhD thesis, University of Notre Dame.

[43] Chiralaksanakul, A., and Mahadevan, S., 2004. "First-Order Approximation Methods in Reliability-Based Design Optimization". *Journal of Mechanical Design, 127*(5), 10, pp. 851–857.

[44] Azad, S., and Alexander-Ramos, M. J., 2020. "A single-loop reliability-based mdsdo formulation for combined design and control optimization of stochastic dynamic systems". *Journal of Mechanical Design, 143*(2), p. 021703.

[45] Haldar, A., and Mahadevan, S., 1995. "First-order and second-order reliability methods". *Probabilistic Structural Mechanics Handbook: theory and industrial applications*, pp. 27–52.

[46] Von Luxburg, U., and Schölkopf, B., 2011. "Statistical learning theory: Models, concepts, and results". In *Handbook of the History of Logic*, Vol. 10. Elsevier, pp. 651–706.

[47] Galvan, E., and Malak, R. J., 2015. "P3ga: An algorithm for technology characterization". *Journal of Mechanical Design, 137*(1).

[48] Hartl, D. J., Galvan, E., Malak, R. J., and Baur, J. W., 2016. "Parameterized design optimization of a magnetohydrodynamic liquid metal active cooling concept". *Journal of Mechanical Design, 138*(3).

[49] Weaver-Rosen, J. M., Leal, P. B., Hartl, D. J., and Malak, R. J., 2020. "Parametric optimization for morphing structures design: application to morphing wings adapting to changing flight conditions". *Structural and Multidisciplinary Optimization, 62*(6), pp. 2995–3007.

[50] Galvan, E., Malak, R. J., Hartl, D. J., and Baur, J. W., 2018. "Performance assessment of a multi-objective parametric

Copyright © 2023 by ASME

optimization algorithm with application to a multi-physical engineering system". *Structural and Multidisciplinary Optimization,* **58**(2), pp. 489–509.

[51] Mayne, D. Q., Seron, M. M., and Raković, S., 2005. "Robust model predictive control of constrained linear systems with bounded disturbances". *Automatica,* **41**(2), pp. 219–224.

[52] Herceg, M., Kvasnica, M., Jones, C., and Morari, M., 2013. "Multi-Parametric Toolbox 3.0". In Proc. of the European Control Conference, pp. 502–510. `http://control.ee.ethz.ch/~mpt`.

[53] Franklin, G. F., Powell, J. D., Emami-Naeini, A., and Powell, J. D., 2002. *Feedback control of dynamic systems,* Vol. 4. Prentice hall Upper Saddle River.

[54] Sabatini, M., Pisculli, A., Polomini, A., Monti, R., Gasbarri, P., Palmerini, G., Baldesi, G., and Dumontel, M., 2012. "Control parameters transition during deploying operations of a space flexible structure via multi-body approach". In Proceedings of the 63th International Astronautical Congress, IAC, pp. 6339–6349.

[55] Karg, B., Alamo, T., and Lucia, S., 2021. "Probabilistic performance validation of deep learning-based robust nmpc controllers". *International Journal of Robust and Nonlinear Control,* **31**(18), pp. 8855–8876.

[56] Cui, T., Allison, J. T., and Wang, P., 2020. "Reliability-based co-design of state-constrained stochastic dynamical systems". In AIAA Scitech 2020 Forum, p. 0413.

[57] Cui, T., Allison, J. T., and Wang, P., 2021. "Reliability-based control co-design of horizontal axis wind turbines". *Structural and Multidisciplinary Optimization,* **64**, pp. 3653–3679.

[58] Mayne, D., 2016. "Robust and stochastic model predictive control: Are we going in the right direction?". *Annual Reviews in Control,* **41**, pp. 184–192.

[59] Azad, S., and Herber, D. R., 2022. "Control co-design under uncertainties: Formulations". In International Design Engineering Technical Conferences and Computers and Information in Engineering Conference, Vol. 86229, American Society of Mechanical Engineers, p. V03AT03A008.

Copyright © 2023 by ASME

Proceedings of the ASME 2023
International Design Engineering Technical Conferences and
Computers and Information in Engineering Conference
IDETC-CIE2023
August 20-23, 2023, Boston, Massachusetts

DETC2023-114690

CONTROL CO-DESIGN WITH VARYING AVAILABLE INFORMATION APPLIED TO VEHICLE SUSPENSIONS

Saeid Bayat[1],*, James T Allison[1]

[1]Department of Industrial and Enterprise Systems Engineering, University of Illinois at Urbana-Champaign, Urbana, IL, USA

ABSTRACT

Recent optimization strategies for Control Co-Design (CCD) often utilize open-loop optimal control (OLOC) to explore the physical performance limits of actively controlled engineering systems. For most real systems, however, closed-loop control (CLC) is required for implementation. OLOC methods incorporate the use of present, past, and future information in making control decisions at each point in time. For systems with any uncertainty, CLC is needed for stability and robustness. The physical (plant) design generated by an OLOC CCD method will normally not interact optimally with CLC, producing results that are not system optimal. The ideal outcome of CCD optimization is the combined physical and control system design that produces maximum system utility, while accounting for the realities of implementable control systems, such as causality and other limits on information available to inform real-time control decisions. In this article an intuitive strategy is presented for investigating empirically the impact of information availability on CCD optimization results. Model Predictive Control (MPC) provides a flexible means to vary what information is used in making real-time control decisions. This is used as a proxy for the vast space of potential controllers, from simple to sophisticated. This method for studying information-based characteristics of CCD problems is demonstrated using a canonical CCD problem based on an active automotive suspension problem. Different plant architectures with various plant design variables are considered. Results show that varying the amount of information in the control design yields different plant designs and different objective values, and has the potential to yield insights into promising CLC architectures (beyond MPC), fruitful directions to head for plant design, and a deeper understanding of the interface between physical and control system design. It is also observed in the studies here that by using more advanced system architectures, the MPC prediction horizon can be reduced and still produce superior system performance compared to cases with simpler architectures and a

longer prediction horizon, or, even when using an open-loop controller. Comparison studies are performed that provide in-depth knowledge of the effect of control sampling time and plant design on states and control. Furthermore, a hybrid Kalman filter is designed that couples with the MPC controller to provide state estimation in the presence of measurement noise and process noise. This article introduces the concept of information-based studies in CCD, but utilizes an applied approach based on MPC to generate insights. A more theoretical approach could be taken in the future that yields more generalizable understanding of how information limitations influence CCD optimization outcomes.

Keywords: Control Co-Design, Optimal Control, Model Predictive Control, Vehicle Suspension, Dynamic Optimization

1. INTRODUCTION

Conventionally, actively controlled engineering systems are designed sequentially, where the physical aspects of the system (i.e., the plant) are designed first, and then control experts design the controller for the plant to meet desired dynamic system performance properties. This approach does not use the synergy between plant and control design decisions and yields a sub-optimal solution. In Control Co-Design (CCD), the plant and controller are designed together, resulting in a higher-performance design compared to the traditional sequential approach. This improvement is possible due to coupling between physical and control system design decisions, specifically, changing plant design influences how control systems should be designed for enhanced behavior, and similarly, how differences in control design influence what plant design decisions may be best. This unified approach can be implemented via careful coordination between human design decision makers in a system development process; CCD can also be implemented using formal mathematical optimization strategies.

The benefit of CCD over sequential engineering design methods is often more pronounced in cases with rich dynamics and high levels of coupling between the plant and control aspects of a system. Dynamic and physical system coupling typically pro-

*Corresponding author: bayat2@illinois.edu
Documentation for `asmeconf.cls`: Version 1.34, May 12, 2023.

Copyright © 2023 by ASME

duces *design decision coupling*, i.e., changes in one set of design decisions influence how choices in other decision sets should be made to improve results. When mathematical optimization is used, design decision coupling strength can be quantified, and CCD optimization methods can more comprehensively leverage design coupling relationships to improve system performance compared to traditional plant then control sequential design [1, 2].

The first step to constructing a CCD optimization study is to evaluate which of the problem formulation strategies is most appropriate. A discussion of these formulation methods is presented in Sec. 1.1. A CCD problem formulation requires definition of a control design problem, and important control system types are described in Sec. 1.2. Most control systems require a state estimation strategy; this topic is reviewed in Sec. 1.3. In the studies presented here, Model Predictive Control (MPC) is used because it provides a convenient means to adjust closed-loop control system complexity and information availability; MPC is discussed in Sec. 1.4. Additionally, a complete overview of the method used in this paper to analyze the impact of information limitation on CCD results is presented in Sec. 1.5.

1.1 CCD Methods

There are two main approaches to solve control co-design problems: nested CCD and simultaneous CCD. In the nested approach, the dynamic optimization process is divided into two stages: the outer-loop and the inner-loop. In the outer-loop, the plant design is modified, whereas in the inner-loop the optimal control for each candidate plant design is determined. In the simultaneous approach, the plant and the controller are optimized together [1, 3]. It can be shown, with mild assumptions, that the nested approach is equivalent to the simultaneous CCD approach [4]. A graphical depiction of these two methods, along with the iterated sequential approach, is shown in Fig. 1, where x_p indicates plant design variables, and $u(t)$ represents control signals.

1.2 Open- and Closed-Loop Control in CCD

In recent CCD research studies, open-loop optimal control (OLOC) is very commonly used. This involves an underlying assumption that control decisions may be made using complete information over the entire time horizon. OLOC CCD studies provide insight into the best possible physical performance of active systems, but do not account for the limitations of realistic implementable control systems. In reality, most implemented controllers are closed-loop and must determine control signal values with incomplete information. In very limited cases a closed-loop controller (CLC) might be identifiable that is capable of generating a control trajectory that is identical to the OLOC trajectory $u^*(t)$ based on information that is available in practice to the controller, but in general this is not possible. If a large difference exists between $u^*(t)$ and the closest possible CLC, the optimal plant design from the OLOC CCD study will in general not be system optimal when a CLC must be used. This is shown to be the case in the studies presented in this article.

The optimal CCD problem with CLC can be posed by assuming a CLC type, such as full-state feedback or Proportional-Integral-Derivative (PID) control, and then optimizing the objective function with respect to x_p and x_c (instead of $u(t)$), where x_c is a finite-dimensional vector of control design variables, such as control gains, that govern the behavior of the selected CLC. Such an approach can often work, but is highly limited in control behavior. In contrast, OLOC CCD provides tremendous flexibility in control system design, providing the potential to discover non-obvious novel control strategies. How might one explore control design flexibility in CCD, while accounting for the information-based limitations of real control systems?

One approach might be to propose a set of control architectures (i.e., ways of determining $u(t)$ based only on information actually available to the controller), and then solve the optimal CLC CCD problem for each control architecture. This enumerative approach may work well if it is possible to propose a set of control architectures that includes one or more candidates that happen to perform well in the context of a holistically optimized system. In some cases, perhaps expert intuition will yield a successful set of candidate architectures. But what about more complex cases where intuition is insufficient, or where systems have the potential for novel high-performance active dynamic behavior? The space of distinct candidate control architectures is a vast, infinite combinatorial space that is profoundly difficult to navigate. Naïve enumerative strategies break down quickly. A more targeted approach is needed. The efficient enumeration strategies of Herber and Allison [5] may be applied with some practical success to expand the scope of control architectures that can be explored in a CLC CCD optimization problem, but these will also reach combinatorial limits quickly. Does a more elegant strategy exist that capitalizes upon the unique properties of active dynamics systems?

An information-theoretic approach may yield generalizable insights, but this is a topic left for future work. Here we propose an initial empirical strategy for investigating the impact of varying information quality on CCD optimization results. So far we have discussed two CCD approaches with starkly distinct information properties: 1) OLOC CCD (perfect information) and 2) standard feedback control architectures with highly limited information. Perhaps the ability to more gradually vary the amount of information utilized by a controller would yield insight into how physical systems and their controllers should be designed as we transition from OLOC CCD to options with implementable control systems. Here we propose the use of Model Predictive Control (MPC) as a tool for varying the degree of information availability to be used in studying the relationship between OLOC and CLC in CCD optimization.

MPC is a class of closed-loop control that utilizes an online (real time) model of the system to predict at regular intervals how the system will respond to different control signals, and to determine the optimal control signal at each time interval based on a relatively low-fidelity optimal control problem. Several aspects of MPC provide a means to adjust very finely the quality of information used by a controller. The amount of time in the future used in real-time optimization, the number of time steps, and the fidelity of the real-time model all can be adjusted. Increasing the prediction horizon, using smaller time steps, and using a higher-order model all can improve information quality. A simple model with a very short time horizon is close in nature

Copyright © 2023 by ASME

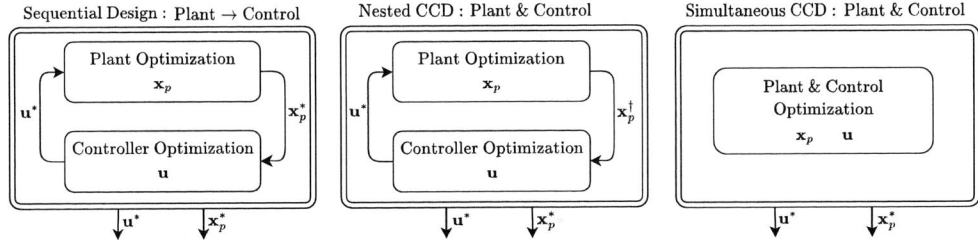

FIGURE 1: THREE KEY CLASSES OF METHODS FOR SOLVING AN ACTIVELY CONTROLLED ENGINEERING SYSTEM DESIGN PROBLEM

to basic CLC architectures, whereas in the limit, as we improve model fidelity and lengthen time horizon, MPC approaches the accuracy of OLOC (albeit without explicit incorporation of past information).

While open-loop control or parameterized closed loop controllers are commonly employed in CCD within the existing literature, there have been instances where MPC has been applied. For example, Nash et al. [6] proposed a nested control co-design approach utilizing robust model predictive control to effectively address disturbances or uncertainties within a closed-loop system. Additionally, Docimo et al. [7] employed MPC for CCD of power and energy systems. However, despite these applications of MPC in CCD, to the authors' knowledge, no study has yet investigated the impact of varying available information used in MPC on the design solution.

This article presents a study of CCD optimization where information quality is varied by modifying time horizon, and the trends in optimization results are investigated. This is an initial empirical study that reveals some insight into the relationship between OLOC and CLC in CCD, and may inspire further analysis strategies that also focus on how CCD behaves with varying degrees of information quality. The rest of this section details several topics, including state estimation, that are central to the MPC-based CCD method analysis strategy that is introduced in this article.

1.3 State Estimation

Closed-loop control systems typically determine an appropriate control response at each point in time as a function of system state variable quantities. In real systems it is usually impractical to sense in real time the value of each state variable. Many cannot be measured directly, and often the use of sensors needs to be limited due to cost, space, or other factors. This limitation requires that a strategy for estimating system state as a function of measurable outputs is implemented to provide input to the controller [8].

Furthermore, systems usually experience unexpected disturbances and sensor outputs exhibit noise in their signals. Therefore, a state estimation system that can mitigate measurement noise and disturbances is needed. Kalman filters are powerful state estimators that perform well in the presence of measurement noise and process noise (disturbance) [9]. Kalman equations can be presented in discrete, continuous, and hybrid formats. Engineering systems are often modeled using continuous-time differential equations, but controllers usually rely upon discretely

sampled data obtained using digital hardware. This motivates the use of a hybrid Kalman filter in this article.

The model and system measurement in the hybrid format are shown in Eq. (1):

$$
\begin{aligned}
\dot{x} &= Ax(t) + Bu(t) + w(t), & w(t) &\sim N(0, Q) \\
y_k &= Cx_k + v_k, & v_k &\sim N(0, R_k),
\end{aligned}
\tag{1}
$$

where $\mathbf{A}$ is the state transition matrix, $\mathbf{B}$ is input matrix, x is the state signal, u is control signal, w is process noise which is assumed to be a normal random variable with covariance $\mathbf{Q}$, y_k is the measurement at time t_k, $\mathbf{C}$, is system output matrix, and v_k is measurement noise generated by a normal distribution with covariance equal to $\mathbf{R}_k$. The Kalman filter goal is to minimize the variance of estimation error, and according to Ref. [10], if w_k and v_k are Gaussian, zero-mean, uncorrelated, and white, the Kalman filter is the optimal solution for this problem. If they are not Gaussian, Kalman is still the best linear solution. If the covariance of the measurement and process noise are not accessible, or if the goal is to minimize the maximum absolute estimation error instead of variance, then an estimation strategy known as H_∞ can be employed [11].

1.4 Model Predictive Control

MPC is a powerful control technique used in the studies presented here; the value of MPC in these studies is the ability to flexibly adjust information quality used in deciding control actions. MPC uses a simplified model of the plant to predict how the system will behave dynamically over a desired time horizon, and uses this predictive capability to find optimal control signal values based on this prediction and upon state estimates for the previous time step. This optimization problem is solved every T_s seconds, which is known as the control sampling time and indicates how fast the control signals can change. The prediction horizon is defined as p, which is an integer and specifies the number of time steps into the future for which the system dynamics are predicted, and the length of the prediction horizon in seconds is pT_s. To reduce the dimension and computational expense of the MPC optimization problem, it can be helpful to hold the optimized control signal constant toward the end of the prediction horizon instead of allowing it to vary across all p time steps. The control horizon $m \le p$ is the number of steps that the control signal is allowed to vary in the MPC optimization problem. This article assumes that the control horizon is equal to the prediction horizon (i.e., $m = p$). The rationale behind this is that certain

Copyright © 2023 by ASME

studies mentioned in this article involve extending the prediction horizon, which is generally anticipated to yield a better design solution. However, when $m < p$, an increase in p necessitates a corresponding increase in m to ensure a fair comparison. Nevertheless, the specific magnitude of the increase in m remains unclear. To ensure an equitable comparison, this article adopts the assumption that m is equal to p.

MPC may be used in conjunction with a Kalman filter employed as a state estimator. Figure 2 illustrates how dynamic plant outputs (y_k) are used by the Kalman filter to estimate states ($\hat{x}_{k-1|k-1}$, i.e., the state estimate at time t_{k-1} given information up to time t_{k-1}); these state estimates are then provided to an MPC controller that computes real-time control signals for the plant actuators ($u(t)$). Please note that output measurements at time t_k is not used by the MPC controller to determine the control signal. In the MPC plant model, the plant dynamics are represented as a continuous state-space model: $\dot{x} = Ax(t) + Bu(t)$. This continuous model was used instead of a discrete dynamic model to enhance accuracy of the results. This implementation eliminates dependence of plant simulation results on discrete time choice. Also, in these studies we make the simplifying assumption that the plant model used within MPC is the same as the real plant model. The term $w(t)$ is process noise that is included to account for some discrepancy between real plant dynamics and the dynamics predicted within the MPC controller. The term $v(t)$ is measurement noise that helps account for discrepancy between sensor output and actual plant dynamics. In real implementations, the MPC plant model is typically a reduced-order model and could then have notably different dynamics compared to the real plant. Adjusting MPC plant model order is one option for tuning information quality by improving predictions of future information. MPC plant model adjustments are not used in the studies here, but could be an interesting topic for future work.

In the depiction of the Kalman filter in Fig. 2 (the element that performs state estimation), K is Kalman gain, and $\mathbf{P}$ is the covariance of the estimation error. $\mathbf{C}$ (the same as $\mathbf{H}$) is a matrix that maps states to outputs. The algorithm is initialized with $\hat{x}_{0|0}$, $x_{0|0}$, and $\mathbf{P}_{0|0}$, which correspond to the initial value of estimated state, true state, and error covariance, respectively.

The MPC optimization problem is solved every T_s seconds to compute a control signal value u_k that is applied to the real plant. The plant output obtained by sensors, y_k, is sent to the Kalman filter. The Kalman filter executes two phases: prediction and update. In the prediction phase, it uses the assumed model and computes the state trajectory and error covariance. Here we use the same assumed model in the Kalman filter as the real plant model and the MPC model; in general there will be a difference between the assumed and real plant model. After computing $x(t_k)$ and $\mathbf{P}(t_k)$, the Kalman filter updates the state based on the received measurement, y_k. These updated states, $\hat{x}_{k|k}$, and error covariance $\mathbf{P}_{k|k}$, are used as the initial values for the next iteration. It should be noted that in the linear case, the Kalman gain and Kalman estimation error are not a function of measured data, so they can be computed offline, reducing Kalman filter computational time substantially. However, storing all values of $\mathbf{K}$ and $\mathbf{P}$ for each iteration is memory-intensive. To tackle this issue, a constant gain can be used instead of a time varying $\mathbf{K}$.

In a Kalman filter with time varying K, the gain will converge to some value after several iterations; therefore, this steady state value can be used from the beginning [12]. In this paper, it is shown that the result is not significantly different from the time varying case.

One benefit of using an estimator is that it can use the assumed model to help predict all states even if not all states can be sensed directly. In the real plant model, the output equation $y = Cx(t) + v(t)$ captures this situation. Specifically, the output matrix C maps true state values to sensed outputs, and often the number of sensors is less than the number of states. This is the case in the suspension study presented later in this article. As described above, $v(t)$ represents measurement noise.

A comparison between the information used in MPC and open-loop optimal control (OLOC) is shown in Fig. 3. Here the prediction horizon is assumed to be $p = 3T_s$. At the initial time (k_0), the open-loop controller has full information of the future, and MPC has estimated future information up to the length of the prediction horizon. In the next step (k_1), the same thing happens as at k_0; however, the prediction horizon for MPC is moved forward by one step and its length is constant. In this paper, CCD is implemented using both MPC and OLOC, each using the nested CCD strategy. Each case has plant design variables that are updated in the outer loop while the corresponding controller is used in the inner-loop to solve the control-only optimization problem. MPC utilizes limited information according to its prediction horizon, whereas OLOC utilizes complete information across the entire time horizon.

MPC is an implementable feedback control system for active dynamic systems, but is constrained by real-time computational resources (motivating simpler plant models and shorter time horizons), and produces a control signal that is only optimal with respect to the simplified plant model and the limited prediction horizon. As prediction horizon is enlarged and plant model fidelity is increased, MPC control performance can improve, but requires more sophisticated control hardware. If we remove the constraint of real-time computation, control optimization can be performed as an offline activity using OLOC. This can make the use of high-fidelity plant models and long time horizons possible, and, while not practical for real-time control, supports discovery of ultimate physical system performance limits.

1.5 Overview of Computational Experiment Strategy

The process used here to test CCD results with MPC implementations that have varying levels of information quality is illustrated in Fig. 4. The first step is to select the system architecture, which is a discrete decision. In this article, architecture selection is not optimized, but rather six unique representative architectures selected by the authors are studied. The second step is to solve the nested CCD problem for each architecture. Two main hyper parameters at this stage are prediction horizon (p) and control sampling time (T_s). This produces an actively controlled system design that is optimal with respect to the limitations of the MPC implementation, and not accounting explicitly for process and measurement noise. The third step then evaluates the impact of noise on the resulting designs by including a Kalman filter, as well the effect of estimating state from limited measurements.

Copyright © 2023 by ASME

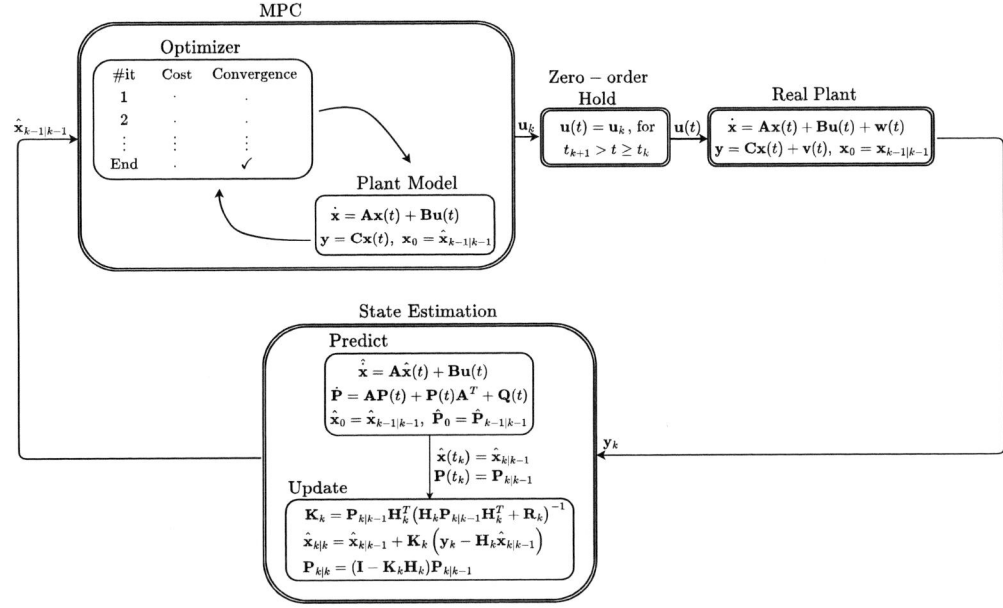

FIGURE 2: A REPRESENTATION OF AN ACTIVE DYNAMIC SYSTEM USING BOTH STATE ESTIMATION AND MODEL PREDICTIVE CONTROL

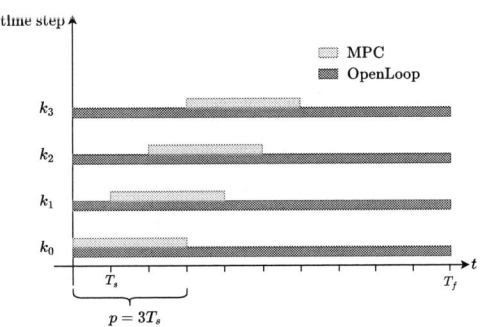

FIGURE 3: AVAILABLE FUTURE INFORMATION FOR EACH CONTROL ARCHITECTURE AT EACH TIME STEP (k_i)

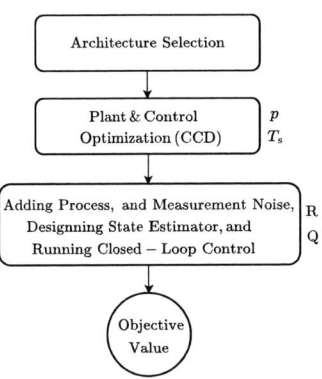

FIGURE 4: PROCESS USED TO PERFORM COMPUTATIONAL EXPERIMENTS TO STUDY THE IMPACT OF SYSTEM ARCHITECTURE, PREDICTION HORIZON, CONTROL SAMPLING TIME, MEASUREMENT NOISE, AND PROCESS NOISE ON SYSTEM PERFORMANCE AND RESULTING OPTIMAL PLANT DESIGNS.

Details of this step are shown in Fig. 2. The two main hyper parameters here are measurement noise covariance ($\mathbf{R}$) and process noise covariance $\mathbf{Q}$. In the third step we perform simulation with the Kalman filter (no optimization is performed). To evaluate the utility of the system while accounting for noise, in the last step we take the simulation results from step three and compute the objective function value based on dynamic behavior across the entire time span used for simulation, i.e., $t_0 \leq t \leq t_F$. The specific objective function used for the studies here is defined in Eq. (2).

For each architecture, this process can be performed for different hyper-parameter values: p, T_s, $\mathbf{R}$, and $\mathbf{Q}$. These results then can provide insights into the the tradeoffs between the quality of information provided to the controller, system performance, and difference in plant design under different scenarios.

Figure 5 illustrates a conceptualization of the relationships between the associated hyper-parameters and other considerations that are important for the designer when choosing a final system design. This decision involves many tradeoffs. For example, sensor unit price, p, T_s, and $\mathbf{R}$ all have an important effect on the sensor selection. If a long prediction horizon with low noise measurement and fast update rate is needed, costly sensors will be required. Control sampling time, actuator unit price, and hardware unit price also affect hardware and actuator selection. If an actuator with a fast update rate is required, then more advanced actuators capable of providing fast responses must be used. Also, a large p and a small T_s will increase the computational expense of solving the MPC optimization problem in real time, requiring use of more advanced computing hardware. Enhanced sensing and control fidelity will improve system performance, but comes as a cost. Selecting the plant design involves consideration of other factors, such as plant manufacturability, plant complexity, the number/type of required sensors and actuators, and possible disturbance on the system. The combination of all these results can help support the designer in making a final specification of the system. In addition, for the purposes of the studies here, these data provide insights into how optimal plant and control designs change as information available for control decisions becomes more limited.

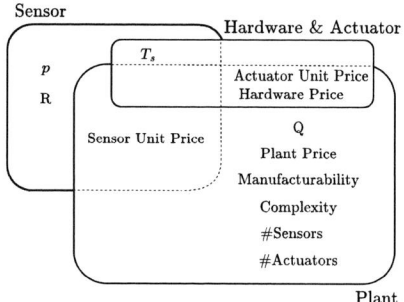

FIGURE 5: IMPORTANT PARAMETERS AND FACTORS THAT SHOULD BE CONSIDERED IN THE SELECTION OF EACH ARCHITECTURE COMPONENT.

To compare fairly the design obtained by different controllers, we need to ensure that the optimal global solutions are achieved for each control assumption. To reach this aim, a global optimization method should be used for the outer loop. Here, we use the Covariance Matrix Adaptation Evolution Strategy (CMA-ES). It is an evolutionary algorithm that is stochastic, gradient-free, and is a global optimization method that works for non-convex, nonlinear optimization problems It should be noted that while a global optimal solution is not guaranteed, if the population size is sufficiently large and the algorithm is run for many iterations, the results are likely to converge to the global optimum. In this method, the covariance matrix of design variables is updated in each iteration. Adaptation of the covariance matrix is similar to inverse Hessian approximation in the quasi-newton method in that it attempts to learn a second-order model of the objective function. According to Ref. [13], CMA-ES has demonstrated successful performance in optimizing non-separable, non-convex, ill-conditioned, multi-modal, or noisy objective functions.

This article aims to investigate the impact of controllers with varying amounts of information and sampling time on CCD outcomes. The study shows that increasing the complexity of the plant alone may not lead to better design solutions unless control and sensor characteristics such as sampling time and prediction horizon are also considered. A case study is presented to illustrate these concepts, and opportunities for future research are identified that involve theoretical development. Section 2 introduces a case study involving an active automotive suspension system and problem formulation elements. The results of this case study under different scenarios are presented in Sec. 3. Finally, Sec. 4 summarizes the findings and offers concluding remarks.

2. SUSPENSION CASE STUDY

In this article we have selected a well-established automotive suspension CCD problem as a basis for our information-based studies. In this section we provide a brief overview of this case study, but refer readers to relevant references for complete details.

Automotive suspensions serve multiple functions, including providing a smooth ride for passengers by isolating them from road disturbances using components such as springs, dampers, linkages, and in some cases even active or semi-active actuators. A fundamental tradeoff exists between passenger comfort and handling performance. Incorporating active components can help simultaneously improve comfort and handling characteristics. An interesting system configuration design problem for this application is determining what components should exist in this vibration isolation system, how they should be configured, sized, and controlled [5]. Active suspension CCD problems can be formulated at many different levels of fidelity. Here we have chosen to use the linear quarter-car models introduced in Ref. [5] for our case study. Reference [5] developed a comprehensive approach to suspension architecture design, including system configuration, plant, and open-loop control design; the investigation generated a rich set of different optimal system designs for different levels of complexity (i.e., number of system components).

This article presents an investigation into the design of plant and control systems for six different vehicle suspension architectures, as illustrated in Fig. 6. Important quantities in this problem involve $\delta(t)$, U, S, $z_U(t)$, and $z_S(t)$ represent the road profile, unsprung mass, sprung mass, position of the unsprung mass, and the position of the sprung mass, respectively. A varying number of components are utilized in each system configuration to connect the sprung and unsprung masses. $F(t)$, m_i, k_i, and b_i denote control force, value of the the ith auxiliary mass, ith spring constant, and ith damper coefficient, respectively. The component index i can take on the values of 1 or 2 in the configurations presented. In the context of the CCD problem, $F(t)$ is the control signal for the actuator (active configurations include just one force actuator), whereas m_i, k_i, and b_i are the plant design variables. The CCD problem is solved in a nested manner [4] with either open-loop optimal control or MPC with different prediction horizons within the inner-loop problem.

To obtain the dynamics of the six vehicle suspension architectures illustrated in Fig. 6, Newton's laws are used, although the details are not included in this paper for brevity. The objective function is defined following the approach of Ref. [5] and is presented in Eq. (2). The function consists of three terms, where the time integral of $w_1(z_U - \delta)^2$ represents handling performance, $w_2\ddot{z}_S^2$ denotes passenger comfort, and w_3F^2 serves as a penalty function for control effort. Using the same objective function as previous studies supports comparison of results.

$$\Psi_d = \int_{t_0}^{t_F} \left(w_1(z_U - \delta)^2 + w_2\ddot{z}_S^2 + w_3F^2 \right) dt \qquad (2)$$

The initial state values are set to zero, while the displacements between sprung and unsprung mass are limited by r_{max}, and all displacements between two consecutive masses are restricted by s_{max} (geometric interference constraints). We use the same system parameters as defined in Ref. [5], where k_t and b_t denote tire equivalent stiffness and damping, respectively. Table 1 presents these system parameters and the lower and upper

Copyright © 2023 by ASME

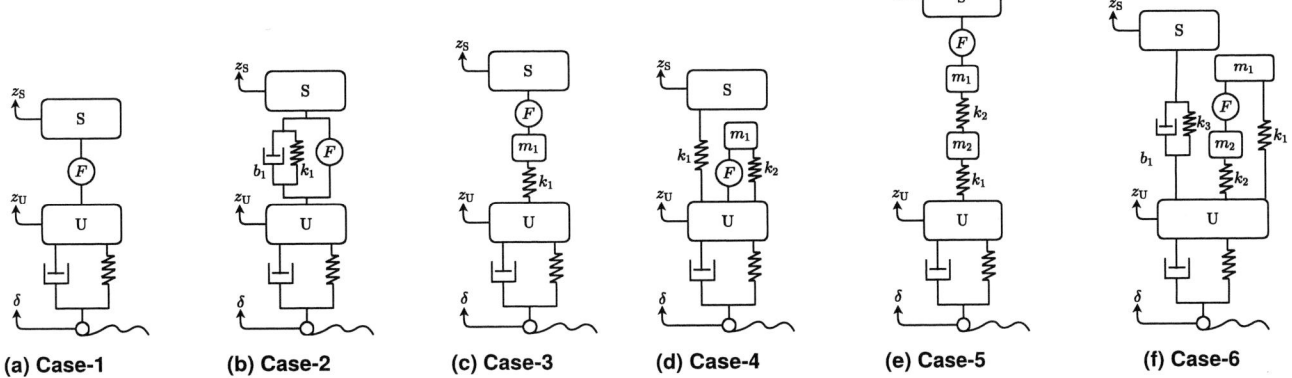

(a) Case-1 **(b) Case-2** **(c) Case-3** **(d) Case-4** **(e) Case-5** **(f) Case-6**

FIGURE 6: SIX UNIQUE VEHICLE SUSPENSION ARCHITECTURES USED IN THE CASE STUDY. PLANT DESIGN VARIABLES INCLUDE m_i **(MASS),** b_i **(DAMPER), AND** k_i **(SPRING) COMPONENT VALUES.** $F(t)$ **IS THE CONTROL SIGNAL FOR THE TRANSLATIONAL FORCE ACTUATOR (ONE EXISTS IN EACH CONFIGURATION).**

TABLE 1: CO-DESIGN PROBLEM PARAMETERS

Parameter	Value	Parameter	Value
t_0	0 s	t_f	3 s
m_{min}	10^{-2} kg	m_{max}	10^1 kg
b_{min}	10^2 Ns/m	b_{max}	10^5 Ns/m
k_{min}	10^2 N/m	k_{max}	10^6 N/m
r_{max}	0.04 m	s_{max}	0.04 m
m_U	65 kg	m_S	325 kg
w_1	$10^5\ s^{-1}m^{-2}$	k_t	232.5×10^3 N/m
w_2	$0.5\ s^3m^{-2}$	b_t	0 Ns/m
w_3	$10^{-5}\ s^{-1}N^{-2}$		

bounds of plant design variables considered in the CCD problem.

It will be demonstrated that for simpler suspension structures, optimization results produce a more sophisticated controller design, and a larger prediction horizon is necessary to achieve an objective function value that is comparable to the open-loop control case. On the other hand, for those architectures that feature additional plant design variables, proper plant design can ease the controller's workload and allow for smaller prediction horizons. With this outcome, the designer has greater flexibility in choosing an appropriate architecture according to the required prediction horizon and the complexity of the plant. Furthermore, the results of MPC with varying characteristics are also useful in making informed decisions regarding sensor, actuator, and hardware selection.

3. RESULTS

This section presents the optimization results obtained for the suspension example under various scenarios. Two symbols, namely x_{OL}^* and x_{CCD}^*, are used repeatedly throughout this section. Here, x^* represents the optimal plant, while x_{OL}^* represents the optimal plant obtained using CCD with an open-loop controller (OLC) optimized in the inner loop. This optimal plant may subsequently be used to evaluate the system's objective value when an MPC controller is utilized. The reason for this is that in the recent CCD literature, open-loop controllers are typically used to obtain optimal plants. However, in practice, closed-loop

controllers are used, which can lead to suboptimal solutions and underutilization of the coupling between the plant and controller design.

In some comparisons, as described above, we will be interested in how well the plant design x_{OL}^* performs when OLC is swapped with MPC. It is expected that performance degradation will be larger for some systems than others. In addition, we are interested in how well a system using MPC can perform if the plant is optimized in conjunction with an MPC controller. In this case, plant optimization decisions are aligned with the controller actually used for objective function evaluation. When the optimal plant design used is based on the control strategy used for quantitative evaluation (i.e., plant optimization is aligned with control strategy), we use x_{CCD}^* to denote optimal plant design (whether the control strategy is OLC or MPC). For example, if MPC is applied in the actual system to evaluate the system's objective value, the same controller is used in CCD. It will be demonstrated that x_{CCD}^* yields better design solutions.

This section illustrates several different scenarios used to derive insights regarding the impact of information quality on optimal plant design and related questions. The following subsection present the details of each study. A summary of each study is listed in the outline below, including study titles, goal, and main findings. For a detailed explanation of the results, please refer to the corresponding subsection.

3.1 Study-1: MPC vs. OLOC CCD with $T_s = 0.01$

- **Goal:** Comparison of CCD results with OLOC and MPC, using a small T_s value, while varying the prediction horizon p.

- **Main Findings:**
  - x_{CCD}^* yields better objective function value compared to x_{OL}^*
  - Increasing prediction horizon brings MPC system performance closer to OLOC, resulting in x_{CCD}^* being closer to x_{OL}^*. As a result, If p is large enough, the optimal plant can be obtained using

Copyright © 2023 by ASME

the OL controller in CCD, and then coupled with MPC in the actual implementation

- Increasing the plant complexity enhances design space richness, and consistently leads to better system performance (but may cost more)

- Plant complexity and control sampling time trade-offs exist. Systems with more plant design variables and larger T_s may yield better solutions compared to simpler models with smaller T_s or even open-loop controllers.

3.2 Study-2: MPC vs OLOC CCD with $T_s = 0.1$

- **Goal:** Increase T_s, observe its impact on optimal solution, and compare with Case Study 1.

- **Main Findings:**

  - Increasing T_s changes the optimal plant and objective value

  - The difference between x^*_{CCD} and x^*_{OL} gets bigger as T_s increases

  - Increasing plant complexity does not always lead to better solutions as certain plants without dampers may require a small T_s

3.3 Study-3: CCD vs. Plant-Only Optimization

- **Goal:** Compare passive and active systems

- **Main Findings:**

  - The objective function value of passive systems is an upper bound for active systems, and this gap reduces with increasing T_s.

  - For some system the result of plant only optimization is the same as CCD having large T_s. In these cases, the control authority is minimal (near zero), so removing the controller does not change the optimal solution.

  - Increasing plant complexity does not always lead to better solutions as certain plants without dampers may require a small T_s.

  - In systems where there is no damper between sprung and unsprung mass, plant-only optimization is not possible. These systems require control to maintain system stability and meet path constraints.

3.4 Study-4: Effect of external input on CCD result

- **Goal:** See the effect of external input (road profile) on design solution

- **Main Findings:**

  - Reducing the external input frequency leads to a controller with a lower frequency

  - Decreasing the control frequency reduces the gap between MPC and OLOC. Therefore, as the control frequency decreases, x^*_{OL} becomes closer to x^*_{CCD}

  - Knowing the road profile frequency can aid in selecting appropriate sensors and actuators. If the frequency is low, a cost-effective sensor with a small prediction horizon can suffice.

3.5 Study-5: State Estimation using Kalman Filter

- **Goal:** Kalman filter for state estimation and disturbance rejection

- **Main Findings:**

  - Estimated states are close to the real states

  - Estimation error is lower than measurement error

  - Constant-gain Kalman is close to variable-gain Kalman.

3.6 Study-6: CMAES Results

- **Goal:** Demonstrate CMAES results

- **Main Findings:**

  - CMAES is capable of generating design solutions that are highly probable to represent the global optimum

  - New populations are generated around the best population obtained from the previous iteration

3.1 Study-1: MPC vs OLOC CCD with $T_s = 0.01$

This section presents a comparison between MPC and OLOC. Objective function values are displayed in Table 2 and MPC controllers with corresponding optimized plant design variables are shown in Table 3. In this table, each row represents the controller used in the CCD, and each column represents the plant design variable used in the inner loop. In the x^*_{OL} columns, the objective value of MPC with $p = 40$ is close to the open-loop objective value, which is the lower bound for the MPC case. Therefore, the plant design for MPC with $p = 40$ should be close to the open-loop plant design.

In Case-1, depicted in Fig. 6, there are no plant design variable, and only the control problem is solved with MPC and open-loop controllers. As the prediction horizon is decreased, the objective value increases. In Case-2, there are two plant design variables: b_1 and k_1, as shown in Table 3. The use of different controllers in the inner loop of CCD results in different optimized plant design variables. As shown in Table. 2, the CCD objective value for Case-2 under the open-loop control assumption is 1.99, which is lower than the values of 2.32 and 2.80 obtained for the MPC cases with different prediction horizons. In the x^*_{OL} column, the plant design variables obtained by open-loop CCD are used. It is shown that when the plant is optimized by the open-loop assumption and then MPC is applied, the objective value is 3.96. On the other hand, when MPC is assumed in the inner loop of CCD, the objective value is 2.80. Furthermore, as shown in Table 3, the plant design variables obtained by these approaches are different, which explains why the objective values of the x^*_{CCD} column are different from those of the x^*_{OL} column.

The same concept is shown in Cases 3-6, where the plant design variables are gradually increased, resulting in a more complex architecture. As a consequence, the optimizer has more

Copyright © 2023 by ASME

TABLE 2: VEHICLE SUSPENSION OBJECTIVE VALUE, T_s = 0.01

		Case-1	Case-2		Case-3		Case-4		Case-5		Case-6	
		-	x^*_{CCD}	x^*_{OL}	x^*_{CCD}	x^*_{OL}	x^*_{CCD}	x^*_{OL}	x^*_{CCD}	x^*_{OL}	x^*_{CCD}	x^*_{OL}
Ψ_d	Open-Loop	2.24	1.99	1.99	0.84	0.84	0.64	0.64	0.47	0.47	0.39	0.39
	MPC$_{(p=40, T_p=0.4)}$	2.28	2.02	2.04	0.88	0.9	0.65	0.65	0.47	0.47	0.4	0.4
	MPC$_{(p=10, T_p=0.1)}$	2.97	2.32	2.73	1.2	1.57	0.95	1.00	1.08	3.72	0.54	0.85
	MPC$_{(p=5, T_p=0.05)}$	4.85	2.80	3.96	2.94	6.48	2.64	4.33	3.02	8.93	1.97	6.84

TABLE 3: VEHICLE SUSPENSION PLANT DESIGN VARIABLES, T_s = 0.01. ALL k HAVE UNITS OF KN/M, b HAVE UNITS OF NS/M, AND m HAVE UNITS OF KG

Architecture	Design variable	Open-Loop	MPC$_{(p=10, T_p=0.1)}$	MPC$_{(p=5, T_p=0.05)}$
Case-1	-	-	-	-
Case-2	b_1	100.8	947	1332.3
	k_1	21.5	16.08	17.0
Case-3	m_1	3.8	3.6	2.3
	k_1	14.0	21.6	15.8
Case-4	m_1	5.8	6.3	7.9
	k_1	3.6	4.2	10.2
	k_2	13.1	12.0	7.0
Case-5	m_1	2.8	3.7	0.8
	m_2	10.0	10.0	0.9
	k_1	38.1	68.6	13.8
	k_2	11.7	25.5	503.1
Case-6	m_1	2.2	2.7	3.1
	m_2	9.0	8.8	7.8
	b_1	100.0	157.0	877.8
	k_1	15.7	17.2	313.4
	k_2	14.0	8.0	7.1
	k_3	3.8	2.7	11.3

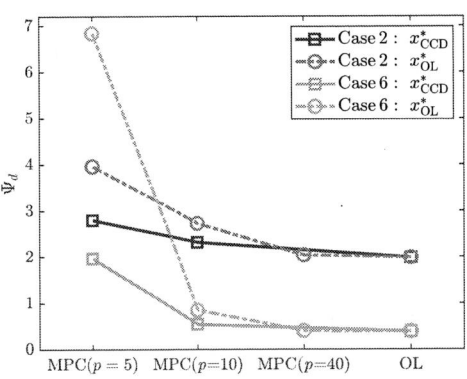

FIGURE 7: INVESTIGATING THE EFFECT OF MODEL COMPLEXITY AND PREDICTION HORIZON ON THE OBJECTIVE VALUE

flexibility to find an improved solution, leading to a lower objective value. An interesting finding is that even with a very small prediction horizon of 0.05 seconds, the objective value of Case-6 is lower (1.97) than that of Case-2 with an open-loop controller (1.99), thanks to the more elaborate plant model with more design variables. In Fig. 7, the objective values for different controllers and optimized plant design variables are plotted against each other, showing that increasing the prediction horizon leads to results that are closer to open loop. Additionally, the objective values obtained by x^*_{CCD} are lower than those of x^*_{OL}. Furthermore, the objective value of Case-6 with x^*_{CCD} and a small prediction horizon is lower than that of Case-2 with an open-loop controller. These findings provide useful insights for the designer who can choose between designing a complex plant with many design variables or a simple plant with a more sophisticated control system (e.g., high-accuracy sensors, more capable microcontrollers) that can provide larger prediction horizons.

Exploring the presence of a pattern in the objective value gap between MPC and OLOC poses an intriguing question. However, in order to provide a conclusive answer, it is necessary to consider a broader range of architectures. As of now, due to this limitation, we cannot make any definitive statements. In the future, we plan to undertake further analysis by incorporating multiple architectures and extracting valuable insights from the optimization data

3.2 Study-2: MPC vs OLOC CCD with T_s = 0.1

A new study is conducted in this section with a control sampling time of T_s = 0.1, which was ten times greater than the previous study. The results of this study are shown in Table 4, and the optimized plant design variables are shown in Table 5. In general, the objective values obtained in this study are higher than those of the previous study, which is expected given that the control sampling time has been increased by a factor of 10. This could be attributed to the fact that, with the longer sampling time, the system's response is slower and therefore may miss opportunities to optimize the system quickly. Moreover, changing the control sampling time affects the plant design, as the optimized plant design variables are different from the previous study. The difference between the objective values of x^*_{OL} and x^*_{CCD} is greater than in the previous case due to the increased control sampling time, which makes the difference between open loop and MPC higher. In this study, increasing model complexity does not necessarily decrease the objective value because some complex architectures require small control sampling times to handle all constraints and converge to the optimal point. For instance, the CCD objective function values in Cases 4 and 5 are greater than in Case-2, while in the previous case where the control sampling time was smaller, these objective values were smaller than the objective function value of Case-2. Case-2 and Case-6 have a damper between the sprung and unsprung masses, which helps damp out the oscillations induced by road disturbance and increases the system stability. As a result, the system remains stable even when the control sampling time is large. However, in other cases, there is no damping between the sprung and unsprung masses, so the system heavily depends on the control action. Thus, when the control sampling time is increased, the controller does not have enough freedom to result in a low objective value. By conducting these studies, the designer can understand the dependence of each design configuration on variables such as control sampling time, which determines the proper processors, actuators, and sensors

Copyright © 2023 by ASME

to be employed.

Figure 8 is shown to aid better understanding of the effect of control sampling time on states and controllers. Here, Case-2 with $T_p = 0.1s$ is used with two different control sampling times: 0.1 and 0.01 s. The plant design variables are also obtained via CCD optimization, as shown in Tables 3 and 5. It is generally observed that the MPC with the smaller sampling time is closer to the open-loop case. This difference may be higher for other architectures because the objective function values of Case-2 under different control architectures are not significantly different. However, as shown in Tables 3 and 5, this difference is much more significant for other architectures.

Figure 9 shows how optimal control trajectories and corresponding states are different for two distinct plant designs: x^*_{OL}, and x^*_{MPC}. As shown in Table 3, the objective function value of the x^*_{MPC} case is smaller than for the x^*_{OL} case. Additionally, by looking at the signals in this figure, it can be observed that the states and controls have smaller values in the x^*_{CCD} case, resulting in a smaller objective function value, defined by Eq. (2), compared to the x^*_{OL} case.

3.3 Study-3: CCD vs. Plant-Only Optimization

In this section, the focus is on optimizing the passive suspension system by removing the actuator and designing only the plant. Cases 2 and 6, which have a damper between the sprung and unsprung mass, are selected for analysis as they exhibit reasonable stability even without an actuator. The results of plant-only optimization are compared with those obtained using CCD. Table 6 shows that the objective function values of plant-only optimization are at the upper bound compared to CCD, as the presence of a control signal provides more freedom to the NLP to reach a lower value. The results indicate that for Case-6, the objective value of plant-only optimization is the same as that of CCD with a larger control sampling time. Control signals for Cases 2, 4, and 6 are plotted in Fig. 10, and the control mean, standard deviation, and maximum value are shown in Table 7. In Case-6, the control signal drops almost to zero when the control sampling time is increased from 0.01 to 0.1, and the result of plant-only optimization is the same as this case where the force is close to zero. For Case-2, the control value decreases with increasing control sampling time, but not as much as in Case-6. Hence, the objective function value of Case-2 with $T_s = 0.1$ is not the same as that of the plant-only optimization case. In contrast, for Case-4, where there is no damper between sprung and unsprung mass, the control signal for both MPC cases with different T_s is large, indicating that plant-only optimization cannot be performed for this case as the system response is heavily dependent on the control signal. Table 6 also shows the optimal plant design variables. The optimal plant design variables obtained through plant-only optimization are similar to those of the MPC case with a larger control sampling time and minimal control authority. For Case-6, the plant-only optimization values are identical to those of MPC with $p = 1$.

3.4 Study-4: Effect of external input on CCD result

As previously demonstrated, there are differences in the objective function values obtained by MPC and open-loop control.

However, these differences are smaller for Case-2 and Case-6, which feature a damper between sprung and unsprung mass, as the damper can aid the active element in providing desirable system dynamics. Generally, if the required optimal control signal does not involve high frequency content, MPC control is closer to open-loop control. To demonstrate this, a simpler road profile is used. Figure 11 shows the original road profile velocity ($\dot{\delta}_1$) and the simpler road profile ($\dot{\delta}_2$). As the simpler road profile has lower complexity, we expect the open-loop control signal to have lower frequency, and therefore, the MPC control signal to be closer to the open-loop control signal. Figure 12 shows the open-loop and MPC control signals for these two road profiles. For the simpler road profile, the MPC control signal is closer to the open-loop control signal, leading to an objective function value that is expected to be close to that of the open-loop. Table 8 presents the objective value and plant design variables for the case where $\dot{\delta}_2$ is used for the road profile. The results show that the objective function value of MPC with $p = 10$ (0.104) is almost the same as that of the open-loop control (0.100), whereas in Table 2, where $\dot{\delta}_1$ is used, the MPC objective value (2.32) is not as close to the open-loop case (1.99). This study highlights the importance of having information on the frequency of the road profile (as a disturbance), as it can help engineers determine the necessary sensors and actuators. If the road profile frequency is not high, an economical sensor that can provide a sufficiently small prediction horizon may suffice, which can impact the cost and the Pareto optimal solution.

3.5 Study-5: State Estimation using Kalman Filter

In the previous analysis, it was assumed that all states were available, and there was no measurement or disturbance noise. However, in practical situations, state estimation is necessary to provide feedback for the controller and to minimize the effect of process and measurement noise. To address this, the present study uses a Kalman filter. The magnitude of the measurement and process noise is set to 10% of the true value. Figure 13 illustrates the Kalman filter's output for Case-6, assuming that only the displacement between sprung and unsprung mass is measured. By obtaining the measurement, the Kalman filter reduces the estimation error and, over time, converges to a value. Figure 13 (a) displays the estimation error variance for the state that shows displacement between unsprung mass and absorber mass m_2 as a function of time. The σ^{2-} and σ^{2+} terms indicate the variance before and after the measurement, respectively. Since the Kalman filter incorporates the information contained in the measurement, σ^{2+} is always lower than σ^{2-}, resulting in a lower estimation error variance. Figure 13 (b) indicates that all innovation values lie between -2σ and 2σ, indicating that the Kalman filter is operating correctly. Figure 13 (c) shows the Probability Density Function (PDF) of estimation error and measurement error. Since the Kalman filter utilizes both system dynamics and measurement, the estimation error is lower than the measurement error. Figure 13 (d) illustrates the true estimation and measurement signal. As observed, the estimation signal lies between the measurement and true value. Figure 13 (e) displays one of the states, which is the displacement between unsprung mass and absorber mass m_2. The estimated signal is sufficiently close to the

Copyright © 2023 by ASME

TABLE 4: VEHICLE SUSPENSION OBJECTIVE VALUE, $T_s = 0.1$

		Case-1	Case-2		Case-3		Case-4		Case-5		Case-6	
		x^*_{OL}	x^*_{CCD}	x^*_{OL}	x^*_{CCD}	x^*_{OL}	x^*_{CCD}	x^*_{OL}	x^*_{CCD}	x^*_{OL}	x^*_{CCD}	x^*_{OL}
Ψ_d	Open-Loop	2.24	1.99	1.99	0.84	0.84	0.64	0.64	0.47	0.47	0.39	0.39
	MPC$_{(p=4\,T_p=0.4)}$	49.27	3.00	14.55	8.74	23.17	8.81	38.99	5.53	25.11	2.21	15.80
	MPC$_{(p=3,T_p=0.3)}$	50.60	3.02	14.7	10.45	21.93	9.49	41.8	6.36	25.30	2.25	16.8
	MPC$_{(p=2,T_p=0.2)}$	54.20	3.12	15.07	10.95	24.70	9.65	39.02	8.54	28.44	2.50	22.07
	MPC$_{(p=1,T_p=0.1)}$	75.00	3.26	15.25	11.63	33.9	9.89	39.7	12.82	33.75	2.71	78.1

TABLE 5: VEHICLE SUSPENSION PLANT DESIGN VARIABLES, $T_s = 0.1$. ALL k HAVE UNITS OF KN/M, b HAVE UNITS OF NS/M, AND m HAVE UNITS OF KG

Figure	Design Variable	OpenLoop	MPC$_{(p=4,T_p=0.4)}$	MPC$_{(p=3,T_p=0.3)}$	MPC$_{(p=2,T_p=0.2)}$	MPC$_{(p=1,T_p=0.1)}$
Case-1	-	-	-	-		
Case-2	b_1	100.8	1423.9	1431.6	1525.1	1580.6
	k_1	21.5	36.4	36.8	29.6	21.3
Case-3	m_1	3.8	8.2	9.8	0.9	1.2
	k_1	14.0	25.9	29.8	24.9	3.1
	m_1	5.8	10.0	6.8	10.0	7.2
Case-4	k_1	3.6	87.9	88.9	88.4	88.9
	k_2	13.1	26.1	19.2	26.9	21.2
	m_1	2.8	9.9	10.0	6.8	4.7
Case-5	m_2	10.0	10.0	10.0	6.9	9.5
	k_1	38.1	72.0	64.9	37.5	55.4
	k_2	11.7	117.5	1000	208.0	28.7
	m_1	2.2	0.1	0.01	0.01	0.03
	m_2	9.0	9.4	10.0	9.61	0.19
Case-6	b_1	100.0	1223.6	1232.1	1232.9	1506.4
	k_1	15.7	3.7	65.6	46.1	0.1
	k_2	14.0	19.8	21.0	19.8	0.1
	k_3	3.8	11.0	11.9	12.5	18.2

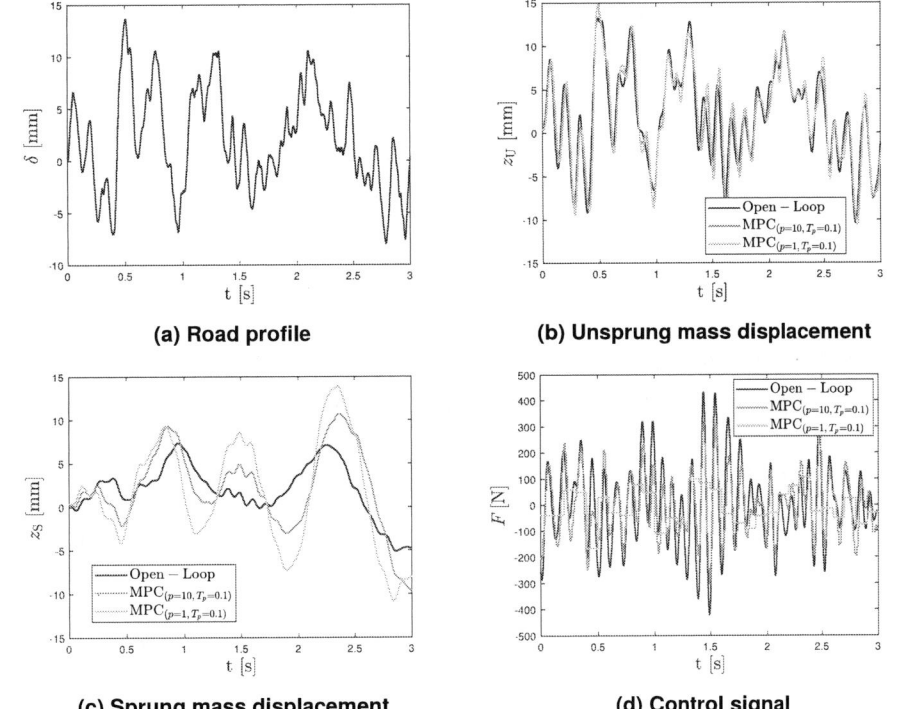

(a) Road profile   (b) Unsprung mass displacement

(c) Sprung mass displacement   (d) Control signal

FIGURE 8: COMPARISON STUDY FOR CASE-2 TO REVEAL THE EFFECT OF CONTROL SAMPLING TIME ON STATES AND CONTROL. THE IMPLEMENTED CONTROLLERS ARE: OPEN-LOOP; MPC WITH $T_p = 0.1$, $p = 10$ ($T_s = 0.01$); MPC WITH $T_p = 0.1$, $p = 1$ ($T_s = 0.1$)

true value. To reduce the computation cost of the Kalman filter, a constant gain can be used. Figure 13 (f) compares the result of using a constant Kalman gain to a variable gain and demonstrates that the constant gain produces a result close to the variable gain.

Copyright © 2023 by ASME

TABLE 6: COMPARISON BETWEEN CCD AND PLANT OPTIMIZATION

Case-study	CCD/Plant	objective-value	m1	m2	b1	k1	k2	k3
Case-2	Open-Loop	1.99	-	-	100.8	21.5	-	-
	MPC ($p = 10$)	2.32	-	-	947.0	16.1	-	-
	MPC ($p = 1$)	3.26	-	-	1580.6	21.3	-	-
	Just plant optimization	3.45	-	-	1496.0	18.9	-	-
Case-6	Open-Loop	0.39	2.2	9.0	100.0	15.7	14.0	3.8
	MPC ($p = 10$)	0.54	2.7	8.8	157.0	17.2	8.0	2.7
	MPC ($p = 1$)	2.71	10.0	3.1	1216.8	16.2	4.5	14.4
	Just plant optimization	2.71	10.0	3.1	1216.8	16.2	4.5	14.4

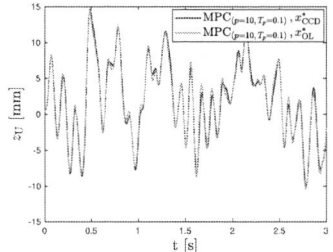

(a) Unsprung mass displacement

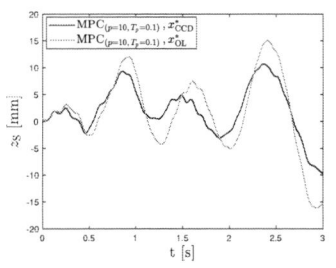

(b) Sprung mass displacement

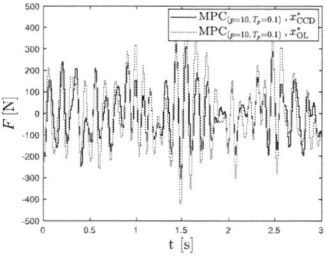

(c) Control signal

FIGURE 9: COMPARISON STUDY FOR CASE-2 TO SEE THE EFFECT OF PLANT DESIGN VARIABLES ON STATES AND CONTROL. THE IMPLEMENTED CONTROLLER IS MPC WITH $T_p = 0.1$, $p = 10$ ($T_s = 0.01$), AND THE PLANT DESIGN VARIABLES ARE: x^*_{CCD}, AND x^*_{OL}

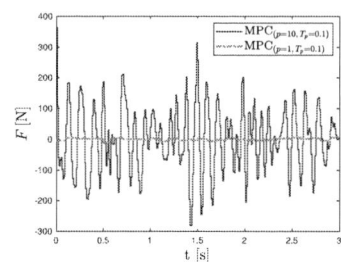

(a) Case-6

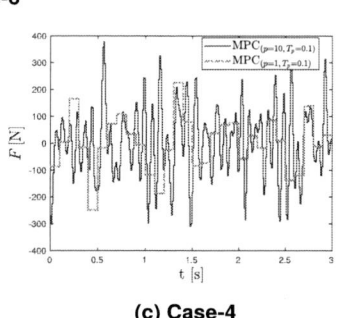

(b) Case-2   (c) Case-4

FIGURE 10: COMPARISON BETWEEN CCD AND PLANT OPTIMIZATION (CONTROL SIGNALS)

TABLE 7: COMPARISON BETWEEN CCD AND PLANT OPTIMIZATION (MEAN, STD, AND MAX OF CONTROLLER)

Case-study	Case-2	Case-6	Case-4
MPC ($p = 10$)	0, 117, 320	0, 107, 362	2, 132, 375
MPC ($p = 1$)	-8, 66, 106	0, 4, 5	-2, 102, 225

3.6 Study-6: CMAES Results

The outer loop of the optimization process utilizes the CMA-ES algorithm, which is a gradient-free method. This algorithm generates a population of potential solutions, evaluates their objectives, and generates a new population based on the performance of the previous generation. The CMA-ES algorithm is utilized to explore the design space and identify plant design variables that minimize the objective function value. To demonstrate an example of the population's evolution and convergence, Fig. 14 depicts the explored designs in the two-dimensional space of the objective value (Ψ_d) and k_1 for Case-2, with a control sampling time of 0.01 s and a prediction horizon of 0.1 s. It is noteworthy that Case-2 has two plant design variables: b_1 and

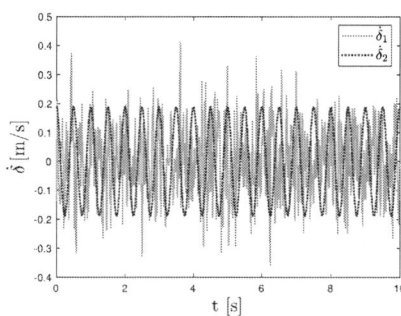

FIGURE 11: ROAD PROFILE VELOCITY

k_1, but only k_1 is displayed in the figure.

4. CONCLUSION

In this paper, a CCD method with varying amounts of available information in the controller design was presented. It was noted that CCD assumes an open-loop controller with complete information at the start of optimization, which is not the case in reality, resulting in suboptimal designs. To address this, it was demonstrated that defining controllers with different available in-

Copyright © 2023 by ASME

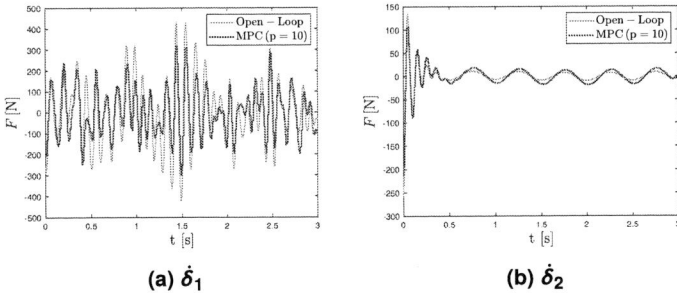

(a) $\dot{\delta}_1$ **(b) $\dot{\delta}_2$**

FIGURE 12: OPTIMAL CONTROL SIGNAL OBTAINED BY OPEN-LOOP AND MPC ($p = 10$) UNDER TWO DIFFERENT ROAD PROFILES, $\dot{\delta}_1$ AND $\dot{\delta}_2$ FOR CASE-2.

TABLE 8: OPTIMAL DESIGN OF CASE-2 WITH $\dot{\delta}_2$ AS ROAD PROFILE UNDER DIFFERENT CONTROL AUTHORITIES

Case-study	CCD/Plant	objective-value	b_1	k_1
	Open-Loop	0.100	100.0	2.85
Case-2	MPC ($p = 10$)	0.104	100.0	2.79
	MPC ($p = 1$)	0.190	380.5	2.76
	Just plant optimization	0.209	327.4	2.74

formation in CCD, such as MPC, can lead to an optimal solution. Additionally, state estimation can be achieved in the presence of process and measurement noise by designing a Kalman filter with constant or varying gain.

The case study focused on six different architectures of vehicle suspensions, but the concept of varying amounts of information in CCD can be applied to other dynamical systems. This approach can yield better designs since available information during optimization is the same as that in reality. However, conventional CCD uses an open-loop controller with unlimited information during optimization and a closed-loop control with much less information during implementation.

To design a dynamical system, the designer must consider various system characteristics, such as system complexity, control hardware equipment, and sensor capabilities. One simple measure of system complexity is the number of plant design variables that can be optimized to handle some of the design objectives. The limitations of control hardware equipment dictate the minimum sampling time the controller can have, and sensor capabilities influence possible prediction horizon and associated sensor noise. If sensors provide only a small prediction horizon, a more complex system may be needed. However, if sensors provide a long enough prediction horizon, a simpler plant design with fewer design variables can be considered since the controller can produce a reasonable objective value. The nature of the system disturbance, such as high-frequency content, was also observed to have a significant impact on optimal system design decisions. Such disturbances can be mitigated using some combination of passive dynamic elements and faster control.

Comparing the dependence of each design on the control sampling time can help identify a design from the Pareto set that is suitable for a particular design scenario. For example, Case-2 and Case-6 are not heavily dependent on control sampling time (even the actuator can be removed). This reduces system cost and complexity. By using these different knobs to design systems, the designer can obtain a series of optimal designs with the same or close objective values, but different characteristics. If the designer is more interested in simple architectures, they can utilize more advanced sensors and control hardware, which may be more expensive. However, if physical system complexity is not an issue, or if the cost of advanced sensors is too high, a more complex plant configuration can be selected that does not require advanced sensors or control hardware.

REFERENCES

[1] Allison, James T and Herber, Daniel R. "Multidisciplinary Design Optimization of Dynamic Engineering Systems." *AIAA journal* Vol. 52 No. 4 (2014): pp. 691–710.

[2] Allison, James T., Guo, Tinghao and Han, Zhi. "Co-design of an Active Suspension using Simultaneous Dynamic Optimization." *Journal of Mechanical Design* Vol. 136 No. 8 (2014): p. 081003. DOI 10.1115/1.4027335.

[3] Herber, Daniel Ronald. "Advances in Combined Architecture, plant, and Control Design." Ph.D. Thesis, University of Illinois at Urbana-Champaign. 2017.

[4] Herber, Daniel R. and Allison, James T. "Nested and Simultaneous Solution Strategies for General Combined Plant and Control Design Problems." *Journal of Mechanical Design* Vol. 141 No. 1 (2019): p. 011402. DOI 10.1115/1.4040705.

[5] Herber, Daniel R and Allison, James T. "A Problem Class with Combined Architecture, Plant, and Control Design Applied to Vehicle Suspensions." *Journal of Mechanical Design* Vol. 141 No. 10 (2019).

[6] Nash, Austin L, Pangborn, Herschel C and Jain, Neera. "Robust Control Co-Design with Receding-horizon MPC." *2021 American Control Conference (ACC)*: pp. 373–379. 2021. IEEE.

[7] Docimo, Donald J, Kang, Ziliang, James, Kai A and Alleyne, Andrew G. "Plant and Controller Optimization for Power and Energy Systems with Model Predictive Control." *Journal of Dynamic Systems, Measurement, and Control* Vol. 143 No. 8 (2021).

[8] Bayat, Saeid, Pishkenari, Hossein Nejat and Salarieh, Hassan. "Observer Design for a Nano-positioning System using Neural, Fuzzy and ANFIS Networks." *Mechatronics* Vol. 59 (2019): pp. 10–24.

[9] Bayat, S, Nejat Pishkenari, H and Salarieh, H. "Observation of Stage Position in a 2-Axis Nano-positioner using Hybrid Kalman Filter." *Scientia Iranica* Vol. 28 No. 5 (2021): pp. 2628–2638.

[10] Simon, Dan. *Optimal State Estimation: Kalman, H infinity, and Nonlinear Approaches*. John Wiley & Sons (2006).

[11] Zhao, Junbo and Mili, Lamine. "A Theoretical Framework of Robust H-infinity Unscented Kalman Filter and its Application to Power System Dynamic State Estimation." *IEEE Transactions on Signal Processing* Vol. 67 No. 10 (2019): pp. 2734–2746.

[12] Golestan, Saeed, Guerrero, Josep M and Vasquez, Juan C. "Steady-state Linear Kalman Filter-based PLLs for Power Applications: A second Look." *IEEE Transactions on Industrial Electronics* Vol. 65 No. 12 (2018): pp. 9795–9800.

Copyright © 2023 by ASME

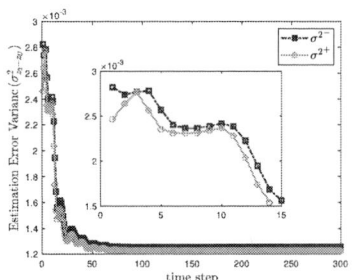
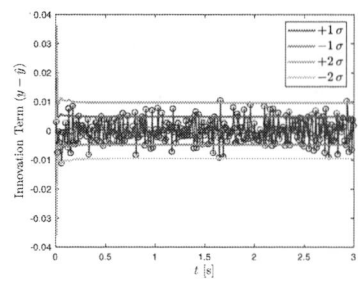
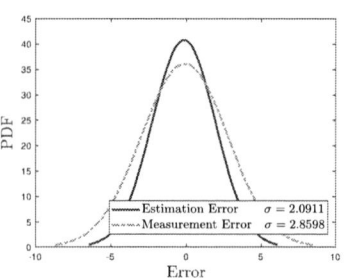

(a) Estimation error variance before receiving measurement (σ^{2^-}) and after receiving measurement (σ^{2^+})

(b) Comparison of Innovation term points with $\pm 1\sigma$ and $\pm 2\sigma$, where $\sigma = \sqrt{H_k P_k^- H_k^T + R_k}$

(c) Probability density function of estimation error and measurement error

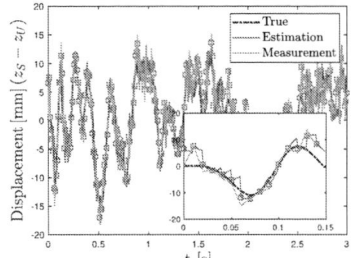
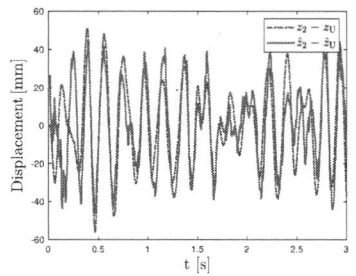
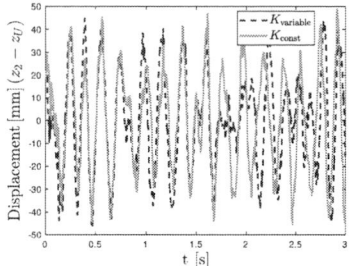

(d) Comparison of true, estimated, and measured displacement

(e) Comparison of true state and estimated state

(f) Time varying and constant Kalman gain

FIGURE 13: KALMAN FILTER RESULTS

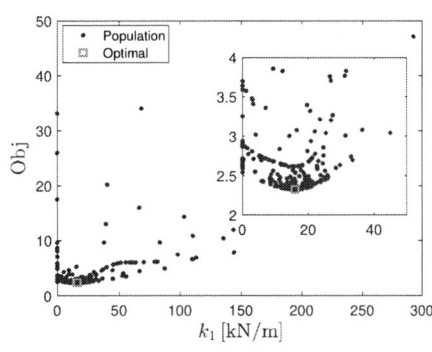

FIGURE 14: EXPLORED DESIGNS IN THE TWO-DIMENSIONAL SPACE OF OBJECTIVE (Ψ_d) AND k_1 FOR CASE-2 WITH MPC CONTROLLER HAVING $T_p = 0.1$, AND $p = 10$

[13] Ryan, Pole In C et al. "References to CMA-ES Applications." *Strategies* Vol. 4527 No. 467 (2007).

Copyright © 2023 by ASME

Proceedings of the ASME 2023
International Design Engineering Technical Conferences and
Computers and Information in Engineering Conference
IDETC-CIE2023
August 20-23, 2023, Boston, Massachusetts

DETC2023-116567

A TWO-TIMESCALE REINFORCEMENT LEARNING APPROACH FOR CONTROL CO-DESIGN PROBLEMS

Eddieb Sadat[1], Mostaan Lotfalian Saremi[1], Alparslan Emrah Bayrak[1,*]

[1]Stevens Institute of Technology, School of Systems and Enterprises, Hoboken, NJ

ABSTRACT

Design of smart (or active) systems that perform automated tasks intelligently based on the interaction with their environments requires a collective solution of the physical and control system design problems together. In this paper, we present a model-free on-policy reinforcement learning approach to solve control co-design problems for such smart systems. This approach uses a discrete two timescale reinforcement learning that addresses the control system design in an inner loop with a fast time scale and the physical system design in an outer loop with a slower time scale. Both design problems use the same temporal difference-based Q-learning formulation. We apply this two-time-scale reinforcement approach to the online video game EcoRacer where the physical system involves the design of a gear ratio for an electric vehicle and the control system involves acceleration and braking decisions over time to finish a track with minimum energy consumption within a limited time. The results show the ability of the proposed approach to find the system optimal solution for the EcoRacer case study within a reasonable computation time without requiring any knowledge of the physics governing the system. The proposed method is generalizable and has the potential to take advantage of the ongoing developments in the field of reinforcement learning.

Keywords: Control co-design, model-free learning, reinforcement learning, video games

1. INTRODUCTION

As the practical applications of smart systems grow in number, the need for methods to effectively design such systems becomes more critical. In addition to the physical system (also known as the embodiment or plant), a control system that enables the smart system to intelligently perform the required tasks based on the interactions with its environment must be designed concurrently [1]. It is well established in the literature that these two design problems are coupled, i.e., the design decisions in one system affect the decisions in the other one [2]. Histori-

*Corresponding author: ebayrak@stevens.edu

cally, the typical method for system design optimization has been to sequentially develop models by independently looking at the physical system design first, then the control system later [3]. Though this method has, and will continue, to work for many applications of design optimization, it does not always guarantee an optimal outcome for the combined physical and control system [4].

Control co-design (CCD) is an emerging field that aims to address this issue by concurrently optimizing the physical and control systems. Multiple CCD methods have been successfully applied to many application domains including automotive [5, 6] and energy systems [7, 8]. Almost all these methods rely on a model-based optimization approach where design decisions are based on the knowledge of the governing physical system model. Developing such models linking physical and control design decisions to system objectives and constraints with sufficient fidelity may not always be possible particularly for complex systems.

As an alternative, in this paper we explore whether a model-free approach to a control co-design problem is a viable method for design optimization. Here, the term "model-free" does not refer to the absence of a model in the study. A model-free approach does not rely on the predictions of the response from a system to the design decisions when searching for a solution. Instead, these approaches "learn" what design decisions are feasible and improve the objective through trial and error [9]. A model-free method uses direct responses from the system, i.e., rewards, to the design actions as a basis for the trial and error process. These responses may come from a physical system setup or a simulation engine used for model training. The success of model-free reinforcement learning (RL) methods in control [10, 11] and more recently in design [12–14] domains separately makes these methods appealing for CCD problems. The proposed approach in this paper makes it possible to integrate existing successful RL methods from the literature to address CCD problems.

The proposed approach borrows a two-time-scale RL approach introduced in the operations control literature [15–17]. Conceptually, two-time-scale RL formulations separate decisions that need to made very frequently (referred to as fast time scale)

Copyright © 2023 by ASME

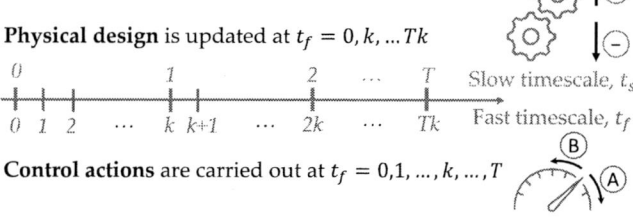

Physical design is updated at $t_f = 0, k, ... Tk$

Slow timescale, t_s

Fast timescale, t_f

Control actions are carried out at $t_f = 0, 1, ..., k, ..., T$

FIGURE 1: DESIGN DECISIONS IN TWO TIMESCALES

and made less often (referred to as slow time scale) in a bi-level problem representation. As seen in Fig. 1, we adopt this idea to represent the control system design problem with a fast time scale formulation and the physical system problem with a slow time scale formulation.

We illustrate the application of the proposed approach using the existing online video game EcoRacer where the physical system involves the design of a gear ratio of an electric vehicle and the control system involves acceleration and braking decisions to finish a track with minimum energy consumption within a limited time [18].

The rest of this paper is organized as follows. Section 2 briefly reviews relevant literature, Section 3 describes the application platform EcoRacer, Section 4 presents the mathematical details of the proposed method, Section 5 shows the results followed by conclusions in Section 6.

2. BACKGROUND

Here we provide a brief background for the techniques used in this paper. The literature review provided in this section is partial and compiled to highlight a knowledge gap along with our contributions to address this gap.

2.1 Control Co-Design

CCD is an emerging field in recognition of the coupling between physical and control system design problems in smart systems. Fathy et al. groups the existing methods to solve the combined physical and control system design problems into four categories: (i) sequential, (ii) iterative, (iii) bi-level (or nested) and (iv) simultaneous formulation [4]. The first two approaches are known to provide suboptimal solutions [4] where the last two are able to achieve system optimality under certain assumptions [2]. The bi-level formulation nests the control problem inside the physical design problem, allowing it to leverage the existing control methods in CCD problems. As an alternative, the simultaneous formulations solve both problems at the same level. The most commonly used method in simultaneous CCD approaches is direct transcription [5]. A comparison of these two approaches is provided in [2, 19] including pros and cons of each approach. Existing approaches from both categories are model-based and hence rely on a mathematical model of the system that links the control and physical system design decisions to system objectives and constraints.

2.2 Reinforcement Learning

RL is a field of machine learning that builds intelligent policies for making decisions based on interactions with the environment. This method does not rely on prior data but rather learns rewarding decisions through trial and error. One particular RL technique is Q-learning, which is a model-free algorithm, meaning it does not require the knowledge of the system model and only relies on the direct responses from the system to assess the value of a state-action pair when training an optimal policy [9]. This optimal policy, either a network-type mathematical function or a look-up table, determines what actions to take for an observed or estimated system state. There are several types of RL formulations in the literature most of which are developed for control problems [10, 11]. The recent success of the RL methods in several challenging problems, where building a physics-based model is not possible, including Atari games [20], Chess and Go [21], and even Starcarft [22], shows the promise of this approach to address complex problems. Further, the application of RL to centralized and decentralized coordinatation of multiple agents in team of autonomous systems further amplifies the potential that RL offers in complex controls problems [23–25]. More recently, there are a few successful applications of RL methods in physical system design as well [12–14]. While RL has been applied to physical and control system design problems separately, the application to CCD problems has been missing.

Two time-scale RL is an application of RL on problems that operate on different time scales in operations control domain [15–17, 26–28]. This approach employs a fast timescale for decisions made frequently in the short term and a slow timescale for decisions made less often in the longer term. For instance, to regulate voltage in energy distribution grids, utility-owned load-tap-changing transformers are configured on an hourly or daily basis whereas smart inverters are controlled on 30-second basis to eliminate instantaneous voltage fluctuations [16, 17]. This approach treats both problems as separate RL models that exchange data to update their respective parameters simultaneously. Due to the complexity of operating in multiple timescales and the nonlinearity of these problems, a problem formulation at a single level is computationally challenging to solve.

2.3 Control Co-Design with Reinforcement Learning

Robotics literature is a particular domain where CCD problems are commonly seen. Historically, this body of literature has focused more on the control problems with the physical design of robotic systems given a priori. RL is a common tool used for controllers in the robotics literature [10, 11]. There is a small body of literature that includes robot manipulator design on top of the RL-based control. The common principle is to maintain a population of potential physical design candidates and develop a control policy for each candidate using a common state and action space. Then, the fitness values of the physical system population are updated using gradient-based approaches in each control iteration [29, 30]. Alternatively, an actor-critic formulation where control problem is solved with RL and the physical system design problem is solved with particle swarm optimization can be found in the literature [31].

Copyright © 2023 by ASME

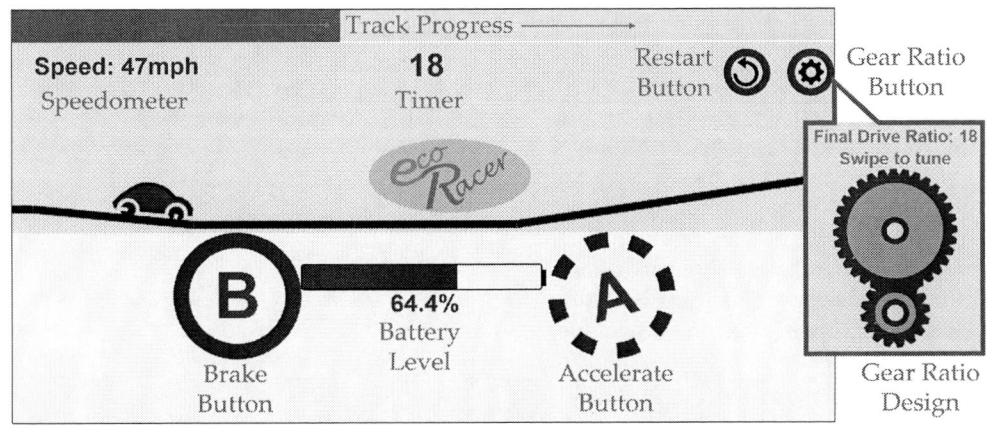

FIGURE 2: ECORACER GAME INTERFACE

2.4 Contributions

To the best of our knowledge, the application of a model-free RL approach to concurrently solve physical and control system design problems is missing in the literature. We introduce a two-time-scale RL formulation based on Q-learning with temporal difference to solve physical and control system design problems. The key novelty in this approach is to use the same RL-based mathematical formulation in both the physical and control system design problems, albeit with different model hyper-parameters. This bi-level (nested) approach serves as a model-free alternative to the existing CCD formulations. We show the application of the method to the existing online game EcoRacer [18] and demonstrate the convergence of the present method to the system optimal solution. The proposed approach is generalizable and has the potential to take advantage of the ongoing developments in the field of reinforcement learning.

3. APPLICATION PLATFORM: ECORACER

EcoRacer is an online game originally designed to analyze human computation process based on the data collected from game players for challenging design and control problems [18]. As shown in Fig. 2, the game models, in a 2-D plane, an electric vehicle that follows a pre-generated track of about 910 distance units. The vehicle is powered by an electric motor connected to a battery with limited capacity. The electric motor is connected to the vehicle through a single gear. Players are allowed to select the gear ratio as a design variable from a list of options. The objective of the game is to successfully complete the track with the highest remaining battery possible. To successfully complete the track, players must reach the finish line before the timer runs out (about 36 in-game seconds). After starting the game, players are given two input options over time for control, a brake button and an accelerate button. As vehicle accelerates the battery level drops, while braking regenerates some battery energy.

The original EcoRacer study finds the optimum battery level to be 43.8% with a gear ratio of 18 using dynamic programming for the control problem and enumeration of all design alternatives for the physical system design problem [18].

4. MODELING

4.1 Control System Design (Fast Timescale)

We use Q-learning with temporal difference [32] to formulate the control system design problem in EcoRacer. Q-learning represents the control problem as a Markov decision process where the goal is to determine the best policy π, i.e., the best action to take at a given state based on a reward function. In EcoRacer, we represent the system state by the position of the car in the track. We discretize the position in the track into a finite number of equal steps. For each state s_k, we define three commands (accelerate, brake, or do nothing) as possible actions a_k. We then create a Q-table $Q^{(c)}$ that represents the value of each possible action at each possible control system state. The values in the Q-table are updated during the training episodes based on the rewards received from the game for the actions taken in each episode.

The reward function that represents the incentive for good actions significantly affects the performance of reinforcement learning [33] and is usually developed specific to the application. For EcoRacer, we develop a reward function $R^{(c)}$ that balances finishing the track in time and conserving battery as follows:

$$R_v(s_k, a_k) = \Delta v(s_k, a_k) \cdot e^{-\lambda \cdot x(s_k)}, \tag{1}$$

$$R_b(s_k, a_k) = \Delta b(s_k, a_k) \cdot (1 - e^{-\lambda \cdot x(s_k)}), \tag{2}$$

$$R^{(c)}(s_k, a_k, w) = w \cdot R_v(s_k, a_k) + (1 - w) \cdot R_b(s_k, a_k),$$
$$R^{(c)}(s_K) = R_{win}. \tag{3}$$

In Eqn. (3), R_v is the reward for increasing the velocity which incentivizes finishing the track in time, R_b is the reward to conserve battery energy. Here, Δv is the change in velocity and the change in battery state of charge between two game states. In this reward, Δv is scaled by an exponential term that diminishes as the vehicles progresses in the track, emphasizing speeding up early in the game. On the other hand, Δb is scaled by another exponential term that emphasizes conserving battery later in the track. The parameter λ in the reward function determines how fast the speed reward diminishes and the battery reward grows with the vehicle position x. The two pieces of the reward are weighted by $0 \le w \le 1$ to control their importance in the reward function. An additional reward R_{win} for the final state s_K is used to incentivize

Copyright © 2023 by ASME

completing the track. A control policy that cannot reach the end of the track cannot receive this reward. For EcoRacer, the completion reward is set to final battery percentage as a way to favor completing the track with more remaining battery.

Note that originally, finishing the track in time is a constraint from an optimization point of view. However, since RL does not have an explicit constraint handling mechanism, this constraint has to be integrated into the objective, which is manifested by the reward model in RL. We choose to use a weighted sum to balance completing the game in time and conserving battery but other formulations to study the trade-offs between the two are possible.

Using this reward, we update $Q_j^{(c)}$ at the fast timescale episode j in RL training using temporal difference as follows:

$$Q_{j+1}^{(c)}(s_k, a_k) = Q_j^{(c)}(s_k, a_k) + \alpha \cdot \varepsilon_{TD}, \qquad (4)$$

$$\varepsilon_{TD} = R_j^{(c)}(s_k, a_k, w) + \gamma \cdot \max\{Q_j^{(c)}(s_{k+1})\} - Q_j^{(c)}(s_k, a_k). \quad (5)$$

For the next fast timescale episode $j + 1$, the Q-table $Q_{j+1}^{(c)}$ is calculated by updating Q-table $Q_j^{(c)}$ based on the reward R_j^c from the current episode and the maximum predicted Q value for all actions in the next state $\max\{Q_j^{(c)}(s_{k+1})\}$. For each state s_k in $Q_j^{(c)}$, only the Q value corresponding to the action taken is updated. In Eqn. (5), ε_{TD} is referred to as the temporal difference error. The importance of Q values for future states in this temporal difference equation is controlled by the discount factor $0 \leq \gamma \leq 1$. The importance of the temporal difference in Q update is controlled by the learning rate $0 \leq \alpha \leq 1$. As $\alpha \to 1$, the importance of the reward obtained from the current set of actions in the Q update increases compared to the existing values in the current Q-table.

The overall control system design problem is solved as follows. The process starts with a random Q-table. In the first five episodes, the policy $\pi_j(s_k)$ defined by the Q-table is ignored and the control actions are taken at random. In the later episodes, a policy of taking the action with the highest Q-value at each game state is calculated. Note that this is an on-policy approach in RL, where the best action (also referred to as target action) is followed during the training process. The game is played with the selected control policy $\pi_j(s_k)$ and returns a reward defined by Eqn. (3). Then, the Q-table is updated following Eqn. (4). We adopt a commonly used ϵ-greedy algorithm [32] to improve the exploration capability of Q-learning [34], which introduces random actions determined by a parameter $0 \leq \epsilon \leq 1$. In each episode at a given state, there is an ϵ chance that the policy takes a random action instead of the action specified by the Q-table.

4.2 Physical System Design (Slow Timescale)

We use the same Q-learning algorithm presented in the preceding section to optimize the physical system design ξ, the gear ratio of the vehicle in case of the EcoRacer game. For the Q-learning formulation, we discretize the domain of feasible gear ratios and use the discrete gear ratio values to represent the physical system state. For each physical system state at z_i, we set increasing, decreasing, and maintaining the gear ratio as three

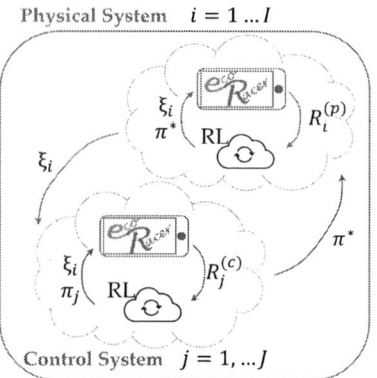

FIGURE 3: CCD PROCESS INFORMATION FLOW

possible physical system actions (u_i). Therefore, for a given slow timescale episode i, the following state transition holds:

$$z_{i+1} = z_i + u_i. \qquad (6)$$

Note that in our particular case the physical system design variable is the same as the system state, i.e., $\xi_i = z_i$, but in a general case, ξ_i can be a function of the system state. Using these states and actions, we define a Q-table $Q^{(p)}$ for the physical system design problem. We compute the reward for the physical system problem $R^{(p)}(z_i, u_i, w)$ as the total reward from the control system:

$$R_i^{(p)}(z_i, u_i, w) = \sum_{k=1}^{K} R_i^{(c)}(s_k, a_k, w), \qquad (7)$$

where and K denotes the number of discrete states in the control system problem. Note that in the notation the dependency of the control reward $R_i^{(c)}$ on the physical system state and actions z_i and u_i is not shown explicitly in this equation for brevity. The control reward depends on the physical system design due to couplings. We use the same update scheme from Eqn.s (4) and (5) but with $R^{(p)}$ instead of $R^{(c)}$ for updating the $Q^{(p)}$. At each episode i, we define the policy of taking the action with the highest Q-value to update the physical system state. Note that the hyperparameters of the physical system design problem (w, λ, α, γ and ϵ) must be tuned differently from the control system design problem to achieve convergence.

4.3 Control Co-Design

Here, we present the proposed model-free approach for solving the CCD problem using a two timescale RL formulation, where the physical and control system designs concurrently evolve through time. Our approach relies on the formulation presented in Sections 4.1 and 4.2. As Fig. 3 shows, our model involves training the control system design policy (fast timescale RL) for a specific number of iterations, denoted by J in an inner loop, and then passing the best policy π^* to the physical system design problem (slow timescale RL) in an outer loop. The physical system then performs one slow timescale iteration to update the physical system state and passes the current physical system design variable values ξ_i down the control system design problem. We summarize the CCD process in Algorithm 1. We

Copyright © 2023 by ASME

32

assume that the state space for the control system remains the same for all feasible physical design candidates to apply the same control system formulation throughout the CCD solution process.

Note that unlike the existing bi-level approaches in CCD literature, we do not wait until the inner control problem converges. Instead, we pass a pre-mature control policy π after J iterations in the inner loop to the physical system to allow both design solutions to concurrently evolve throughout the process. This approach also remedies the computation burden of the nested CCD solutions.

Due to inherent randomness in the RL process, for a given slow timescale iteration i, the final fast timescale iteration J does not necessarily yield the best control policy but the best policy might be achieved earlier than J iterations. Therefore, there is a need to choose the best control policy from RL training episodes and pass it to the physical system design problem. Our model chooses the best fast timescale iteration in the control system design as follows:

$$R_i^{(p)}(z_i, u_i, w) = \max\{\sum_{k=1}^{K} R_j^{(c)}(s_k, a_k, w) | \forall j \in \{1, \ldots, J\}\}. \tag{8}$$

In the first slow timescale episode $i = 1$, we allow the model to run with completely random actions for the first five fast timescale iterations (just as described in Section 4.1) and then start updating the control system design actions (a_k) using $Q^{(c)}$ in the control section. After the first slow timescale iteration, we maintain $Q^{(c)}(s_k, a_k)$ from the previous slow timescale iterations to preserve the knowledge learned throughout the present. In this CCD model, we follow the rewards weights and hyperparameters set separately in the control system and physical system design problems.

Algorithm 1 CONTROL CO-DESIGN ALGORITHM

Initialize $\xi_0, Q_0^{(c)}, Q_0^{(p)}$
Set the values of $w, \lambda, \alpha, \gamma$ and ϵ
for $i = 1 \rightarrow I$ **do**
    for $j = 1 \rightarrow J$ **do**
        if $i \leq 5$ **then**
            Set $a_k, \forall k \in \{1, \ldots, K\}$ to random
        end if
        Run the simulation (Game)
        Calculate the $R_j^{(c)}$
        Update the $Q_j^{(c)}$ using Eqns. (4,5)
        Update $a_k, \forall k \in \{1, \ldots, K\}$ using $Q_{j+1}^{(c)}$
    end for
    Calculate $R_i^{(p)}$ using Eqn. (8)
    Update $Q_i^{(p)}$
    Update u_i
    Update ξ_i
end for

---

5. RESULTS

In this section we present the results from the EcoRacer problem using our methods presented in Section 4. We first present our results for the control system design problem with physical design fixed at the system optimum and the results for the physical system design problem with control policy fixed at the system optimum. These results will serve as a sanity check to present the capabilities of the individual design methods. Finally, we present the solution of the CCD problem with the proposed bi-level approach. The hyperparameter values used in the solution of the physical system and control system design problems are shown in Tbl. 1.

TABLE 1: HYPERPARAMETER VALUES

Parameter	Control Design	Physical Design
w	0.75	0.75
λ	2/850	2/850
α	0.3	0.1
γ	0.4	0.9
ϵ	0.02	0.3

5.1 Control System with Fixed Physical Design

This section shows the solution of the optimal control problem with the gear ratio set to 18, which is the true system optimal design reported in [18] with the remaining battery of 43.8%. We tuned the parameters in Eqns. (1–5) to achieve convergence through trial-and-error.

We discretize the track into 180 states s_k, of approximately five distance units each. The finer the discretization is, the more precise the control policy becomes at the expense of increased computational cost. We set the first 16 control actions to acceleration in all control episodes to provide the initial excitation that the car needs. Starting from a random Q-table, we train the Q-learning model for 150 episodes with the hyper-parameters shown in Tbl. 1. In Eqns. (1) and (2), $\lambda = 2/850$ makes the contribution of speed disappear at the final phases of the track. In Eqn. (3), the weights were set to $w_1 = 0.75$ and $w_2 = 0.25$, emphasizing R_v or finishing the track more than the battery. Since the most challenging part in EcoRacer is finishing the track in time, it is more important to learn how to finish the track and then to adjust the strategy to fine tune for energy conservation. In Eqns. (4) and (5) $\alpha = 0.3$ and $\gamma = 0.4$ primarily tuned to improve the solution as much as possible. Finally, $\epsilon = 0.02$ leaves a 2% chance to select a random action at a given state during the game. Due to the randomness in the initialization of $Q^{(c)}$ and from ϵ, each trial run yields different results. We present the best policy after running the model multiple times that results in a final battery of 43.2%, which is only 0.6% lower than that of reported in [18].

Fig. 4 shows the convergence of the learning process over 150 iterations (referred to as fast timescale episodes). The model initially learns very quickly where the final position of the vehicle in the track drastically increases within the first 25 episodes. The model eventually learns how to complete the track after Episode 55 and then converges to a solution around Episode 70, after which there are no major improvements. The best policy that completes the track with the highest remaining battery is achieved in Episode 80.

The control policy to achieve the best result is shown in Fig. 5. In this policy, the vehicle accelerates up to the most

Copyright © 2023 by ASME

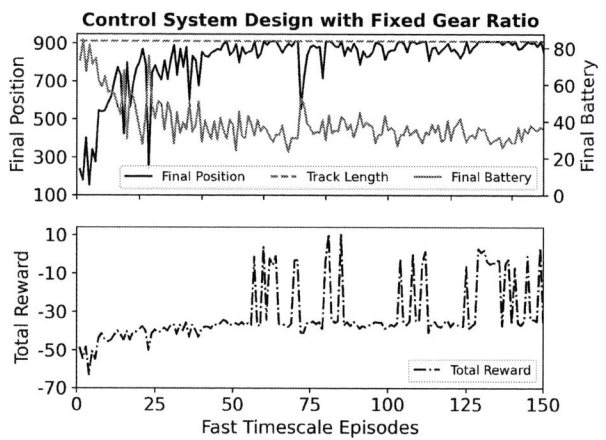

FIGURE 4: CONTROL SYSTEM DESIGN WITH FIXED GEAR RATIO

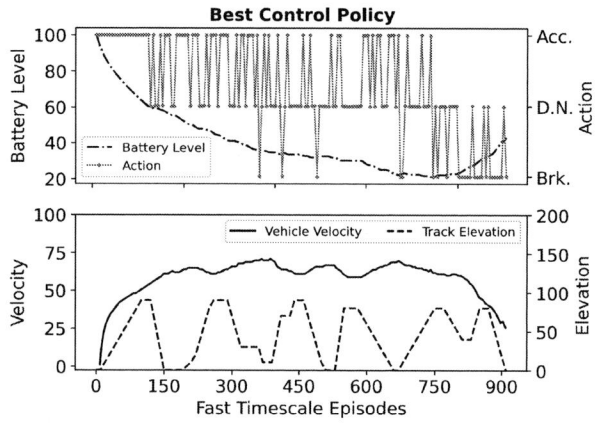

FIGURE 5: BEST CONTROL POLICY

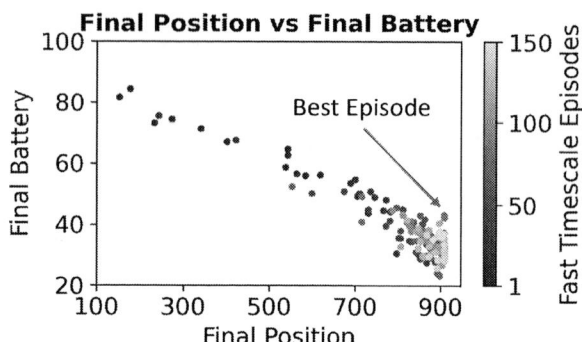

FIGURE 6: FINAL POSITION VS BATTERY DURING CONTROL POLICY TRAINING

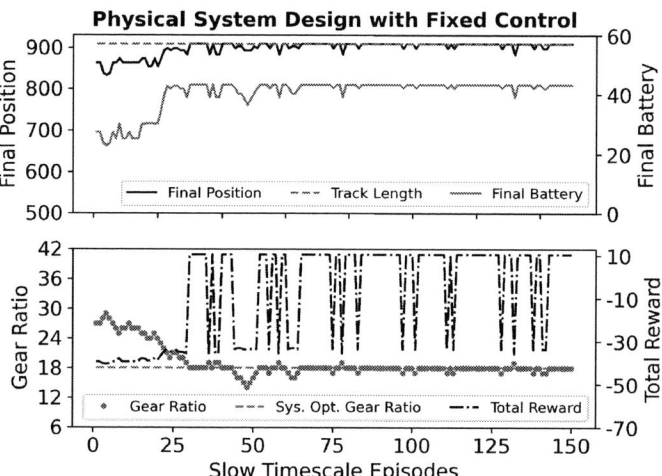

FIGURE 7: PHYSICAL SYSTEM DESIGN WITH FIXED CONTROL

efficient speed around 70mph (in-game units) and stays close to that speed up until around three quarters of the track. This speed also allows the vehicle to finish the track in time. Then, in the last quarter of the track, when λ diminishes the influence of speed, the control policy heavily favors battery regeneration by braking while finishing the track before the time runs out. Note the this policy is similar to that of reported in [18] providing additional confidence in the ability of the proposed RL method to find the optimal policy. The policy we report is more noisy compared to the solution obtained by dynamic programming in the original EcoRacer paper.

Fig. 6 provides additional insight by plotting the relationship between the final position and the remaining battery from all 150 fast timescale episodes. The episodes that have a remaining battery larger than 43.2% fail to reach the final position, either due to running out of time or getting locked into a state (position) by fully braking. It is also evident that the model learns throughout each episode as the later episodes (marked with yellow color) more frequently appear towards the final position.

5.2 Physical System with Fixed Control Design

In this section, we show the results of the physical system design with the control policy fixed at the optimal policy (π^*) from Section 5.1. In EcoRacer, the gear ratio, is the sole physical

system design variable. Here, we initialize the Q-table values with all zeroes and the gear ratio with an arbitrary value $\xi_0 = 27$. Then, we train the model with the hyperparameters from Tbl. 1 over $I = 150$ slow timescale episodes. The parameter values in both problems are different (as expected) since mathematical nature of the physical system and control system design problems are different.

In this model, we set the physical system actions as $u_i \in \{-1, 0, 1\}$ for each state $10 \leq z_i \leq 40$. In other words, the RL method is allowed to increase or decrease the gear ratio by 1 or keep it the same for gear ratio values between 10 and 40. These bounds are set to reflect the bounds set in the original EcoRacer game. Fig. 7 illustrates the convergence of the RL to an optimal gear ratio, where the x-axis denotes the slow timescale episodes, the dots indicate the gear ratio in each episode, the black dash line represents the reward $R_i^{(p)}$, and the red dash line denotes the optimal gear ratio reported in [18]. The same figure also shows the final position and final battery associated with this run. Similar to control system design, the method can learn how to finish the track very quickly within the first 25 slow timescale episodes and fine tunes the gear ratio to optimize the final battery. The results demonstrate that the model consistently converged to the system optimal gear ratio $\xi^* = 18$ in 60 episodes. Running this model multiple runs consistently gives the same result, albeit

Copyright © 2023 by ASME

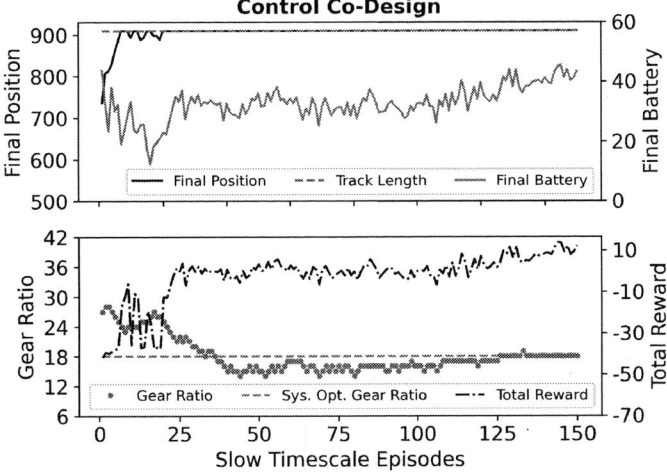

FIGURE 8: CONTROL CO DESIGN

with varying rate of convergence.

Note the fluctuations in reward and final batteries for different physical system designs under the same control policy. The optimal control policy for $\xi^* = 18$ provides a final battery of 43.2% while that value is lower than 30% for the initial design $\xi_0 = 27$. This is an evidence for the coupling between physical and control system design problems and an argument for solving both problems concurrently.

5.3 Control Co-Design

Here, we present the results of the concurrent solution of the physical and control system design problems. For the design of each system, we used the hyper-parameters and the reward weights from Sections 5.1 and 5.2 when training the CCD model. Note that in this particular problem, we use the same weight for both physical and control system rewards. However, in other applications, different weights may be necessary for convergence. For a general application, an appropriate reward function that is tailored for the particular problem must be developed.

We performed $I = 150$ slow timescale episodes in the physical system design problem and $J = 20$ fast timescale episodes in the control system design problem. As in the previous sections, we use an initial random $Q_0^{(c)}$ for the control system problem and an initial all-zero $Q_0^{(p)}$ for the physical system design problem.

Fig. 8 shows the evolution of design over slow timescale episodes. Note that within each slow time episode, we have $J = 20$ fast timescale episodes that is not shown in these results for brevity. The results show that the model eventually learns the system optimal gear ratio though slower than the results from Section 5.2. It is expected because in the CCD problem, both system designs co-evolve over those slow timescale episodes. The best physical and control system design is achieved in Episode 143 with the final battery of 44.9%. This is even higher than what was reported in [18] due to the difference in discretizations in both studies. In this study, we use a much finer discretization of states compared to the original EcoRacer paper. This result is also better than that of the control system design problem in Section 5.1. Considering that the CCD problem iterates over

143×20 fast timescale episodes until this optimal result, there is a higher chance that a better control policy will be found.

In this architecture, the RL model for CCD problem evaluates the physical system designs with the current control policy which is only optimal at the end of the process. Therefore, even though the RL process evaluates the physical system design $\xi = 18$ around Episode 40, that evaluation does not provide an accurate assessment of the performance of that design. It is merely an approximation since the control policy is still evolving with the physical design. The overall behavior of the process highly depends on the choice of the reward function since it is the only driver of design decisions as in any other RL approach. Particularly in constrained optimization problems (like EcoRacer), the balance between satisfying the constraints and improving the objective function must be carefully designed to find system optimal designs.

6. CONCLUSION

In this paper, we presented an RL-based model-free approach to concurrently design a physical and control system using a bi-level formulation. We used a discrete two-timescale on-policy RL method with temporal difference to iteratively update physical and control design decisions until convergence. We showed the application of our approach on the existing online video game EcoRacer. The results showed that the present approach can identify the system optimal solutions in a reasonable amount of computation time in the selected application problem. The proposed method is generalizable to other CCD problems with a possibility to take advantage of the ongoing developments in the area of RL. For instance, it can be applied to the re-design of existing systems where a control policy has already been trained with RL.

Limitations of the present approach include the following. First, we have not assessed the scalability of the two timescale RL approach. As the complexity of physical system design problem increase, computation time may be intractable with the given problem discretization. Deep reinforcement learning approaches may address this scalability issue as they operate on a Q-network rather than a discrete Q-table, which has much better scalability. Second, a comparison of the present approach with the existing approaches in the literature have not been made. Such an analysis may identify strengths and weaknesses of model-based and model-free approaches and shed light on when to use which one. It is expected that the model-based approaches may be computationally more efficient than the model-free ones since they take advantage of the knowledge of physics behind the system. However, as the systems become more complex, physics models may also become more inaccurate, the performance of the model-based approaches may be questionable. Finally, a formal study on the coordination of physical and control design problems with reference to the convergence characteristics is left to a future study. We have used only one coordination strategy where we pass a premature control policy a small number of iterations to the physical design problem. A parametric study on the number of iterations to run for the inner loop problem may provide useful insights about the convergence characteristics of the present CCD approach.

Copyright © 2023 by ASME

REFERENCES

[1] Allison, James T. and Herber, Daniel R. "Multidisciplinary Design Optimization of Dynamic Engineering Systems." *AIAA Journal* Vol. 52 No. 4 (2014): pp. 691–710.

[2] Herber, Daniel R and Allison, James T. "Nested and simultaneous solution strategies for general combined plant and control design problems." *Journal of Mechanical Design* Vol. 141 No. 1 (2019).

[3] Papalambros, Panos Y and Wilde, Douglass J. "Systems Design." *Principles of Optimal Design: Modeling and Computation, 3rd Ed.* Cambridge University Press, New York, NY (2017): Chap. 8, pp. 355–420.

[4] Fathy, Hosam K, Reyer, Julie A, Papalambros, Panos Y and Ulsov, AG. "On the coupling between the plant and controller optimization problems." *Proceedings of the 2001 American Control Conference.(Cat. No. 01CH37148)*, Vol. 3: pp. 1864–1869. 2001. IEEE.

[5] Allison, James T, Guo, Tinghao and Han, Zhi. "Co-design of an active suspension using simultaneous dynamic optimization." *Journal of Mechanical Design* Vol. 136 No. 8 (2014).

[6] Bayrak, Alparslan E, Kang, Namwoo and Papalambros, Panos Y. "Decomposition-based design optimization of hybrid electric powertrain architectures: Simultaneous configuration and sizing design." *Journal of Mechanical Design* Vol. 138 No. 7 (2016): p. 071405.

[7] Deese, Joe, Tkacik, Peter and Vermillion, Chris. "Gaussian Process-Driven, Nested Experimental Co-Design: Theoretical Framework and Application to an Airborne Wind Energy System." *Journal of Dynamic Systems, Measurement, and Control* Vol. 143 No. 5 (2021).

[8] Cui, Tonghui, Allison, James T and Wang, Pingfeng. "Reliability-based control co-design of horizontal axis wind turbines." *Structural and Multidisciplinary Optimization* Vol. 64 (2021): pp. 3653–3679.

[9] Sutton, Richard S and Barto, Andrew G. *Reinforcement learning: An introduction.* MIT Press, Cambridge, MA (2018).

[10] Pierson, Harry A and Gashler, Michael S. "Deep learning in robotics: a review of recent research." *Advanced Robotics* Vol. 31 No. 16 (2017): pp. 821–835.

[11] Arulkumaran, Kai, Deisenroth, Marc Peter, Brundage, Miles and Bharath, Anil Anthony. "Deep reinforcement learning: A brief survey." *IEEE Signal Processing Magazine* Vol. 34 No. 6 (2017): pp. 26–38.

[12] Vermeer, Kaz, Kuppens, Reinier and Herder, Justus. "Kinematic synthesis using reinforcement learning." *International Design Engineering Technical Conferences and Computers and Information in Engineering Conference*, Vol. 51753: p. V02AT03A009. 2018. American Society of Mechanical Engineers.

[13] Ororbia, Maximilian E and Warn, Gordon P. "Design synthesis of structural systems as a Markov decision process solved with deep reinforcement learning." *Journal of Mechanical Design* Vol. 145 No. 6 (2023): p. 061701.

[14] Rahman, Molla H, Bayrak, Alparslan E and Sha, Zhenghui. "A Reinforcement Learning Approach to Predicting Human Design Actions Using a Data-Driven Reward Formulation." *Proceedings of the Design Society* Vol. 2 (2022): pp. 1709–1718.

[15] Xue, Wenqian, Fan, Jialu, Lopez, Victor G, Li, Jinna, Jiang, Yi, Chai, Tianyou and Lewis, Frank L. "New methods for optimal operational control of industrial processes using reinforcement learning on two time scales." *IEEE Transactions on Industrial Informatics* Vol. 16 No. 5 (2019): pp. 3085–3099.

[16] Yang, Qiuling, Wang, Gang, Sadeghi, Alireza, Giannakis, Georgios B and Sun, Jian. "Two-timescale voltage control in distribution grids using deep reinforcement learning." *IEEE Transactions on Smart Grid* Vol. 11 No. 3 (2019): pp. 2313–2323.

[17] Liu, Haotian, Wu, Wenchuan and Wang, Yao. "Bi-level Off-policy Reinforcement Learning for Two-Timescale Volt/VAR Control in Active Distribution Networks." *IEEE Transactions on Power Systems* (2022).

[18] Ren, Yi, Bayrak, Alparslan E and Papalambros, Panos Y. "EcoRacer: game-based optimal electric vehicle design and driver control using human players." *Journal of Mechanical Design* Vol. 138 No. 6 (2016).

[19] Sundarrajan, Athul K and Herber, Daniel R. "Towards a fair comparison between the nested and simultaneous control co-design methods using an active suspension case study." *2021 American Control Conference (ACC)*: pp. 358–365. 2021. IEEE.

[20] Mnih, Volodymyr, Kavukcuoglu, Koray, Silver, David, Graves, Alex, Antonoglou, Ioannis, Wierstra, Daan and Riedmiller, Martin. "Playing atari with deep reinforcement learning." *arXiv preprint arXiv:1312.5602* (2013).

[21] Silver, David, Hubert, Thomas, Schrittwieser, Julian, Antonoglou, Ioannis, Lai, Matthew, Guez, Arthur, Lanctot, Marc, Sifre, Laurent, Kumaran, Dharshan, Graepel, Thore et al. "A general reinforcement learning algorithm that masters chess, shogi, and Go through self-play." *Science* Vol. 362 No. 6419 (2018): pp. 1140–1144.

[22] Vinyals, Oriol, Babuschkin, Igor, Czarnecki, Wojciech M, Mathieu, Michaël, Dudzik, Andrew, Chung, Junyoung, Choi, David H, Powell, Richard, Ewalds, Timo, Georgiev, Petko et al. "Grandmaster level in StarCraft II using multi-agent reinforcement learning." *Nature* Vol. 575 No. 7782 (2019): pp. 350–354.

[23] Gronauer, Sven and Diepold, Klaus. "Multi-agent deep reinforcement learning: a survey." *Artificial Intelligence Review* (2022): pp. 1–49.

[24] Wu, Haochen, Ghadami, Amin, Bayrak, Alparslan E, Smereka, Jonathon M and Epureanu, Bogdan I. "Impact of heterogeneity and risk aversion on task allocation in multi-agent teams." *IEEE Robotics and Automation Letters* Vol. 6 No. 4 (2021): pp. 7065–7072.

[25] Wu, Haochen, Ghadami, Amin, Bayrak, Alparslan E, Smereka, Jonathon M and Epureanu, Bogdan I. "Evaluating Emergent Coordination in Multi-Agent Task Allocation through Causal Inference and Sub-Team Identification." *IEEE Robotics and Automation Letters* (2022).

[26] Li, Jinna, Chai, Tianyou, Lewis, Frank L, Fan, Jialu, Ding, Zhengtao and Ding, Jinliang. "Off-policy Q-learning: Setpoint design for optimizing dual-rate rougher flotation operational processes." *IEEE Transactions on Industrial Electronics* Vol. 65 No. 5 (2017): pp. 4092–4102.

[27] Liu, Fangzhou, Gao, Huijun, Qiu, Jianbin, Yin, Shen, Fan, Jialu and Chai, Tianyou. "Networked multirate output feedback control for setpoints compensation and its application to rougher flotation process." *IEEE Transactions on Industrial Electronics* Vol. 61 No. 1 (2013): pp. 460–468.

[28] Li, Jinna, Kiumarsi, Bahare, Chai, Tianyou, Lewis, Frank L and Fan, Jialu. "Off-policy reinforcement learning: Optimal operational control for two-time-scale industrial processes." *IEEE Transactions on Cybernetics* Vol. 47 No. 12 (2017): pp. 4547–4558.

[29] Schaff, Charles, Yunis, David, Chakrabarti, Ayan and Walter, Matthew R. "Jointly optimizing placement and inference for beacon-based localization." *2017 IEEE/RSJ International Conference on Intelligent Robots and Systems (IROS)*: pp. 6609–6616. 2017. IEEE.

[30] Ha, David. "Reinforcement learning for improving agent design." *Artificial life* Vol. 25 No. 4 (2019): pp. 352–365.

[31] Luck, Kevin Sebastian, Amor, Heni Ben and Calandra, Roberto. "Data-efficient co-adaptation of morphology and behaviour with deep reinforcement learning." *Conference on Robot Learning*: pp. 854–869. 2020. PMLR.

[32] Watkins, Christopher J C H. "Learning from Delayed Rewards." Ph.D. Thesis, King's College, Cambridge, UK. 1989.

[33] Ng, Andrew Y., Harada, Daishi and Russell, Stuart J. "Policy Invariance Under Reward Transformations: Theory and Application to Reward Shaping." *Proceedings of the Sixteenth International Conference on Machine Learning*: p. 278–287. 1999. Morgan Kaufmann Publishers Inc., San Francisco, CA, USA.

[34] Vermorel, Joannes and Mohri, Mehryar. "Multi-armed bandit algorithms and empirical evaluation." *Machine Learning: ECML 2005: 16th European Conference on Machine Learning, Porto, Portugal, October 3-7, 2005. Proceedings 16*: pp. 437–448. 2005. Springer.

Proceedings of the ASME 2023
International Design Engineering Technical Conferences and
Computers and Information in Engineering Conference
IDETC-CIE2023
August 20-23, 2023, Boston, Massachusetts

DETC2023-117541

PLATFORM HYDRODYNAMIC AND STRUCTURAL CONTROL CO-OPTIMIZATION FOR THE FLOATING OFFSHORE WIND TURBINES

Jinbin Liang, Xianping Du[1], Jianbo Yi, Guowei qian, Peng Xie

School of Marine Engineering and Technology, Sun Yat-Sen University, & Southern Marine Science and Engineering Guangdong Laboratory (Zhuhai), Zhuhai, China

Hongyi Xu

Department of Mechanical Engineering, University of Connecticut, Storrs, CT 06269, USA

ABSTRACT

The design of floating offshore wind turbines requires a rigorous method for handling the complex multi-systems and multi-physics interactions involved. The traditional method optimizes subsystems of different disciplines sequentially and independently and the control is always implemented at the last step. This neglects the coupling between sub-systems and may miss the global optimum. Control co-design could formulate the plant and control in a single formula for reaching the global optimum. The floating offshore wind turbine is a complex system with the interaction of aerodynamics, structural, hydrodynamics, and control, which causes the difficulty of co-design problem formulation. In this study, a framework for the co-design of platform hydrodynamics and structural control is established using the coupled time-domain simulation. A case study is conducted with the DTU 10 MW reference wind turbine and the NAUTILUS-10 semi-submersible platform for demonstration. The platform shapes and tuned mass dampers are co-optimized by establishing and solving a formula with eight design variables under the constraints of power generation and platform stability. An efficient Kriging model-enhanced genetic algorithm solves problems for minimizing platform mass. Compared with the baseline model, the platform mass is reduced with the lowered platform motion, which increased the annual energy production by 2.96%. The time history analysis on rated condition suggests
lowered blade tip deflection and rotor thrust. This demonstrates the improvement made by the co-design method under limited numbers of constraints and loading conditions.

Keyword: Wind energy; floating offshore wind turbine; control co-design; platform hydrodynamics; tuned mass damper; frequency domain; time domain

1. INTRODUCTION

With the rapid development of wind energy, it was seen as one of the most hopeful alternatives to traditional fossil fuels [1]. In the past decades, inshore wind energy has increased greatly while offshore wind is emerging as the new trend due to the advantages of the deep sea with the rich wind, no land occupation, and being far from residential areas [2]. The floating offshore wind turbine (FOWT) is a technique to extract wind energy in the deep sea, which is studied widely and seen as a potential for the next generation of wind turbines [3].

The FOWT stabilizes the system by a giant floating platform, which is expensive and increases the Levelized Cost Of Energy (LCOE) of wind energy. Also, the current floating platform follows the design rules for the oil platforms, which can meet the technical requirements of stabilization but may not be the optimal choice for trading off the cost and structural stability [4]. Optimization is a way to balance the different aspects of requirements.

---

[1] Corresponding: duxp3@mail.sysu.edu.cn

Copyright © 2023 by ASME

These were some optimization studies for stability or mass reduction purposes. Clauss and Birk proposed a hydrodynamic shape optimization method for improving the seakeeping qualities of offshore structures [5]. Hall et al optimized the platform structures using frequency domain analysis with different types of platforms evaluated [6]. Karimi et al modeled the FOWT in the frequency domain and conducted the optimization for the spar and semi-submersible platforms under different mooring methods [7]. Dou et al optimized the platform under the wind and wave excitations by developing a four-degree-of-freedom model in the frequency domain [8]. However, most of these studies focus on the platform frequency domain response without the time-domain responses of the FOWT. The frequency domain method is suitable for solving the steady-state problem but not the transient response, which may also affect the structural loads or stability. Compared with the frequency domain method, the time domain method is suggested to be better for structural fatigue load calculation [9]. In this way, the time domain method is a better option for optimizing the platform under the integrated FOWT simulation.

Based on a specific system design, structural control is often used to stabilize the platform by, for example, the tuned mass damper (TMD). The TMD could mitigate the motion by matching motion frequency and dissipating the kinematic energy as the heat of the damper [10]. However, in tradition, the control is often designed for a specific system [11]. The TMD control is always implemented after completing the design of the platform and mooring systems. In this sequential process, the bilateral coupling between the platform plant and control is neglected, which may result in the global optimum not being achieved.

The control co-design (CCD) method involves formulating the plan and controller together and solving for the global optimum. The CCD was widely used in different engineering fields after being proposed at the end of the last century, for example, vehicle active suspension [12], electric vehicle powertrain [13], aircraft vertical tail [14], and wave energy converter [15], wind turbines [16-18] and hydrokinetic turbines [19, 20]. The hull shape of a semi-submersible platform is optimized with a nested framework with wind turbine controller tunned for each hull shape in [21], which is similar with the work done in [22] with different wind turbine control strategies considered for a spar type platform. As compared with the conventional sequential method, CCD could result in a significant improvement in optimal performance at an acceptable cost increase [23]. Nevertheless, the current studies do not take into account the co-design of the floating platform and its passive structural control for motion stability improvement.

This paper presents a framework for the CCD of the floating platform and TMD control using the integrated dynamics of the FOWT. The geometric sizes of platforms are parameterized and evaluated in the frequency domain. By using the frequency domain characteristics, e.g., added mass and radiation damping, the parameterized platform is coupled with the time domain simulation tool of the wind turbine. TMDs are installed in the platform for the integrated simulations. Then, the time domain responses, e.g., the annual energy production (AEP) and platform pitch angle, under the wind and wave interactions are calculated. CCD problem can be formulated and efficiently solved by a Kriging model enhanced Genetic algorithm.

Based on these works, the contributions are summarized as follows:

- A framework is developed for coupling the parametrized floating platform with the wind turbine for time domain simulations;
- Under the constraints of energy production and motion stability of FOWT, CCD is implemented for the platform hydrodynamics and TMD control for reducing platform mass;
- A surrogate model-enhanced genetic algorithm is developed for solving the CCD problem efficiently.

The remainder of this manuscript is organized as follows. The established co-optimization framework is introduced in Section 2. Using this framework, a case study is conducted with the reference model introduced and parameterized in Section 3. The co-optimization problem is formulated and solved in Section 4 and the results are analyzed in Section 5. Conclusions are drawn in Section 6.

2. CO-DESIGN FRAMEWORK FOR HYDRODYNAMIC AND TMD SYSTEM

The CCD framework for co-optimizing FOWT platform hydrodynamics and TMD control is shown in FIGURE 1. It is composed of three sub-modules, that is, the platform hydrodynamics and wind turbine coupling (SM#1), surrogate modeling (SM#2), and optimization (SM#3) sub-modules.

SM#1 couples the platform hydrodynamics in Ansys AQWA with a dynamical wind turbine model in OpenFAST [24]. In AQWA, the platform shape is parameterized by geometric sizes. For different platforms, the mass, mass center, inertial properties, and fairlead positions are calculated by MATLAB code to update the model parameters automatically. The dynamical characteristics, that is, the added mass, radiation damping, and force RAO (response amplitude operator) are calculated and sent to the OpenFAST for synthesizing time domain response. Meanwhile, the platform mass and inertia parameters in OpenFAST are updated. The structural control can be implemented based on the geometric features of each platform. The coupled time domain simulations for FOWT are carried out using OpenFAST under the complex interaction between wind and wave.

In SM#2, a CCD formula has to be defined first for generating the design space. In the design space, a set of points can be sampled. Based on SM#1, these samples can be evaluated under specific wind and wave conditions to form the initial design dataset. The data will be partitioned into the training and validation subsets with a specific ratio of sample numbers (9:1 in this study). The Kriging model is trained and tested by the training and validation subsets, respectively [25]. The tested model could predict the response of new design alternatives efficiently and accurately.

Copyright © 2023 by ASME

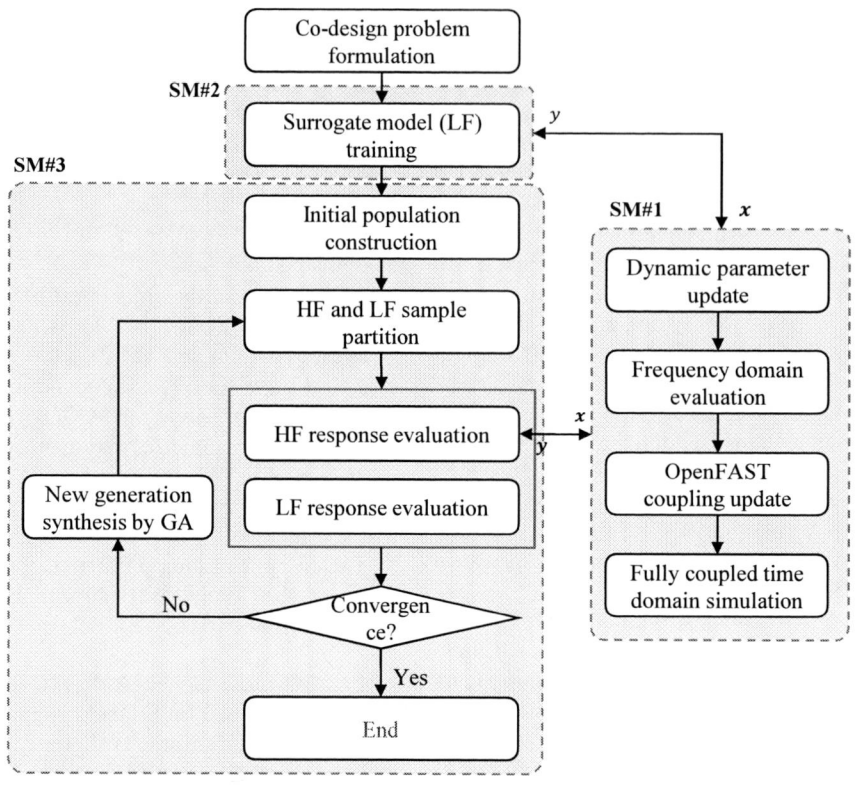

FIGURE 1: CCD FRAMEWORK FOR PLATFORM HYDRODYNAMICS AND CONTROL

In SM#3, the Kriging surrogate model could accelerate the Genetic algorithm process [26]. The initial population is generated and split into two subsets for evaluation by the surrogate and AQWA-OpenFAST models. The surrogate model predicts the responses quite efficiently and the AQWA-OpenFAST evaluates the samples in parallel for improving computational efficiency. The optimizer determines the next generation by crossover and mutation. The optimization is terminated after reaching a specific number of generations.

3. CASE STUDY： NATILUS-10 PLATFORM AND TMD CO-OPTIMIZATION FOR DTU 10MW MODEL

To demonstrate the CCD framework, a case study is conducted using the DTU (Technical University of Denmark) 10 MW reference wind turbine (RWT) with the NATILUS-10 semi-submersible floating platform [27, 28].

3.1 Model introduction

The DTU 10 MW RWT was developed by the Technical University of Denmark by size scaling and sequential design of different disciplines. It is widely used as the reference model. The basic configurations are shown in TABLE 1. The DTU 10 RWT semi-submersible platform can be integrated with the NAUTILUS-10, which was developed in the LIFE S50+ project and well-validated by model tests. The basic dynamic parameters are listed in TABLE 2. The integrated geometry model of the DTU 10 MW RWT and the NAUTILUS-10 platform is generated and presented in FIGURE 2.

The TMD system can be installed on the platform, nacelle, blade, etc. by specifying its installation locations in OpenFAST. In this study, four TMDs are installed in the four columns of platform after removing the ballast water. The mass, spring stiffness and damping ratio can be changed for tuning its dynamics.

TABLE 1: THE VALUES OF THE DTU 10 MW RWT SYSTEM PARAMETERS

Parameter	Value
Rated power (MW)	10
Rotor orientation, configuration	Upwind, 3 blades
Wind turbine control	Variable speed, collective blade pitch
Rotor, Hub diameter (m)	178, 5.6
Hub height (m)	119
Cut-in, Rated, Cut-out speeds (m/s)	4, 11.4, 25
Cut-in, Rated rotor speed (rpm)	6, 9.6
Rated tip speed (m/s)	90

For evaluating the hydrodynamics of different platforms, an AQWA model is developed as shown in FIGURE 2 since the OpenFAST cannot change its geometric structures. Coupling the AQWA and OpenFAST can achieve the parameterized platform and TMD systems for co-optimization.

Copyright © 2023 by ASME

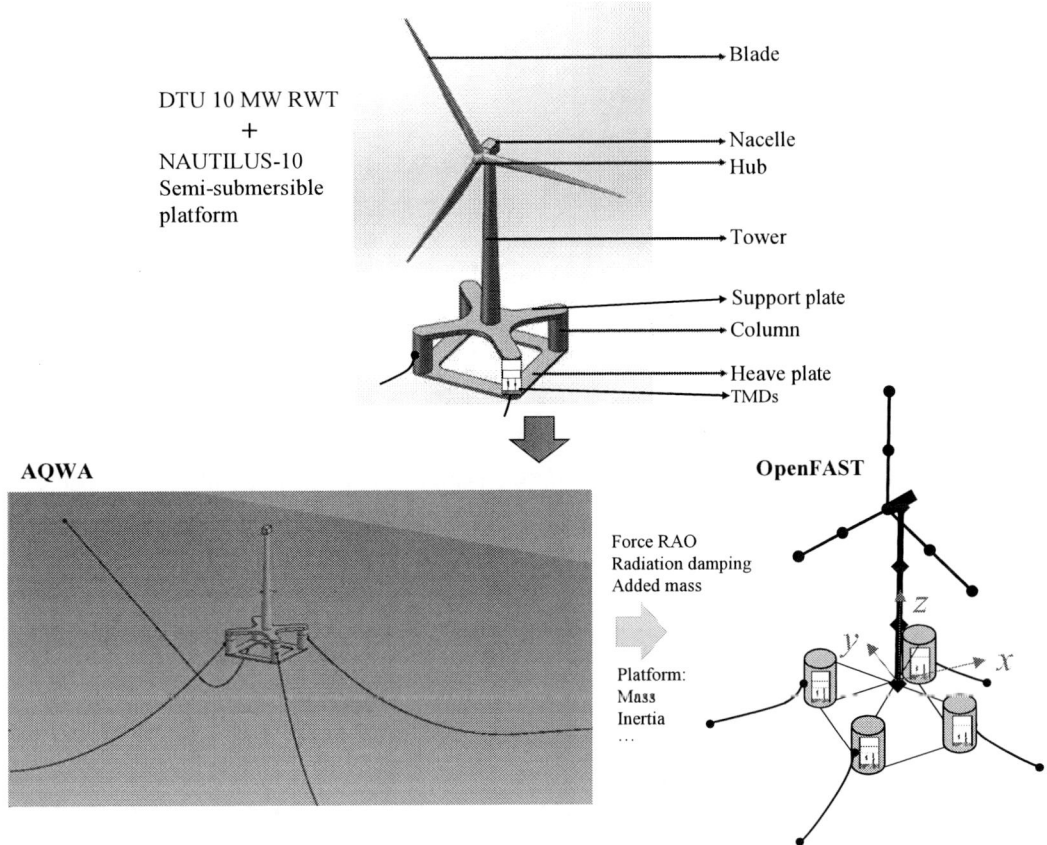

FIGURE 2: GEOMETRIC MODEL OF THE DTU 10 MW RWT WITH THE NAUTILUS-10 SEMI-SUBMERSIBLE

TABLE 2: PARAMETERS OF THE NAUTILUS-10 PLATFORM [28]

Parameter	Value
Overall substructure mass (kg)	7.781E6
Centre of Mass (CM) below MSL (m)	14.283
Substructure roll inertia about CM (kg·m²)	4.829E9
Substructure pitch inertia about CM (kg·m²)	4.829E9
Substructure yaw inertia about CM (kg·m²)	7.451E9

3.2 Model parametrization

In this study, we focus on the co-optimization of platform hydrodynamic and TMD control parameters. The former is greatly impacted by geometric shapes. Thus, five shape parameters are used as the plant design variables as denoted in FIGURE 3. They are the column diameter (D), heave plate length (L), heave plate width (W), column gap (L_B), column draft (D_B). Their ranges are determined in TABLE 3.

Based on the five parameters, the platform is parameterized in the ANSYS Workbench. It could update the model parameters and change the platform geometry. The new geometric model is processed with the boundary element model generated for the hydrodynamic calculation.

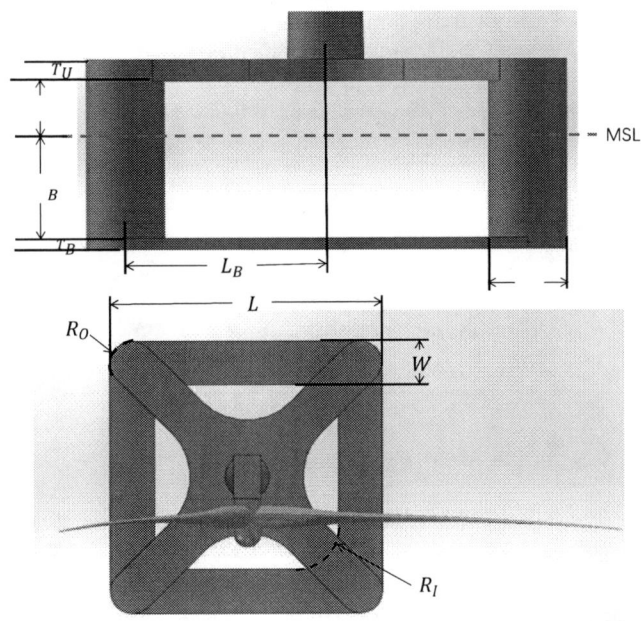

FIGURE 3: DEFINITION OF PLATFORM SHAPE PARAMETERS

Copyright © 2023 by ASME

TABLE 3: THE RANGES OF THE PLATFORM SHAPE VARIABLES

Variable	Min	Max	Baseline model
D (m)	8	14	10.5
L (m)	64	70	65.25
W (m)	8.5	15	10.5
L_B (m)	25	35	27.375
D_B (m)	12	20	15.5

Due to the minor effect on the hydrodynamics, the other parameters in FIGURE 3 are set as the constants. The thickness of the support plates (T_U) and heavy plate (T_B), the remaining height of the column (D_D), and the outer (R_O) and inner fillet radius (R_I) of the heave plate are assigned with the values of the baseline model which are 3 m, 1.5 m, 6 m, 5.25 m, and 5.25 m, respectively. In addition, with the constant Z coordinate, the fairlead position in x and y directions are calculated based on geometric sizes by,

$$\left.\begin{matrix} x \\ y \end{matrix}\right\} = L_B + \sin\left(\frac{\pi}{4}\right) * \frac{1}{2}D. \tag{1}$$

In this study, we focus on the co-optimization of platform hydrodynamic and TMD parameters. The former is greatly impacted by geometric shapes. Thus, five shape parameters are used as the plant design variables as denoted in FIGURE 3. They are the column diameter (D), heave plate length (L), heave plate width (W), column gap (L_B), column draft (D_B). Their ranges are determined in TABLE 3.

The TMD is installed in the four columns of the platform after removing the active water ballast of the baseline NAUTILUS-10 model as the appearance in FIGURE 2. To adapt with the physical limitation, the mass block size, which is assumed the steel material, is determined considered the inner space of four columns. As the prior knowledge, the weight of the TMD mass block (M_T) is often defined as 0.2% ~2% of the whole system [29]. The spring stiffness (K_T) and damping coefficient

(B_T) impact the response of the platform motion under stochastic wave so they are set as the design variables for reaching the optimal under irregular wave conditions. Their ranges are determined as shown in TABLE 4.

TABLE 4: RANGES OF THE TMD DESIGN VARIABLES

Variable	Min	Max
M_T (kg)	10904	18076
B_T (N/(m/s))	8000	15000
K_T (N/m)	200	2000

3.3 Co-design for platform hydrodynamic and TMD

Using the established parameterized model, the CCD problem is formulated before optimization. For evaluating the wind turbine performance, environmental conditions are defined for the coupled simulations. The design load case (DLC) 1.2 of inflow wind is used to evaluate the power generation and fatigue responses as requested by the standard of IEC 61400-3 [30]. The Weibull distribution formulates the probability ($p(v)$) of different wind speeds at a specific location by,

$$p(v) = \frac{k}{\lambda}\left(\frac{v}{\lambda}\right)^{k-1} e^{\left(-\frac{v}{\lambda}\right)^k}. \tag{2}$$

where the coefficients k, and λ control the distribution shape, and they are assigned as 2, and 9.59, respectively, for demonstration. The continuous distribution is discretized into 12 bins from the cut-in (4m/s) to cut-out (25 m/s) speeds.

Referring to [31], twelve combinations of the mean wind speed (WS) of each bin, significant wave height (H_s), and wave period (T_p) are generated for a single evaluation as shown in TABLE 5. The normal turbulent wind profiles are generated by the TurbSim with the A-class Kaimal spectrum. The irregular wave condition is modeled by the Jonswap spectrum. Every simulation of the twelve combinations is running for 650 s with the first 50 s removed for eliminating the transient response.

TABLE 5: ENVIRONMENTAL CONDITIONS WITH THE WS, H_s AND T_p FOR CO-OPTIMIZATION

Case	1	2	3	4	5	6	7	8	9	10	11	12
WS (m/s)	5.0	7.0	9.0	11.0	11.4	13.0	15.0	17.0	19.0	21.0	23.0	25.0
H_s (m)	1.38	1.66	1.98	2.36	2.46	2.83	3.38	4.01	4.79	5.70	6.85	8.31
T_p (s)	7.00	9.95	8.00	8.29	8.46	9.13	9.64	9.89	10.65	11.85	12.34	12.00

(1) Formulation of the CCD problem

The eight design variables for the co-optimization can be denoted as $x = [x_p, x_c] = [[D, L, W, L_B, D_B], [M_T, B_T, K_T]]$, where the x_p and x_c represent the plant and control variables, respectively. Our design objective is to reduce the cost of the floating platform without loss of the platform motion stability and wind turbine power generation. In [23], the cost of the platform can be approximated by the multiplication of weight and material unit price. In this study, we just use the mass ($M(x)$) of the platform steel (M_{steel}) and ballast ($ballast$) as the

objective for simplification, which is defined by,

$$M(x) = M_{ballast}(x_p) + M_{steel}(x). \tag{3}$$

In this study, three types of constraints are defined, namely geometric, kinematic, and power generation constraints. Geometric constraints could prevent physical interactions between geometric features. In one respect, changing the L and W may result in the contact of two opposite heave plates. To maintain a constant radius of the inner fillet of the adjacent two plates, as shown in FIGURE 4, the distance between the two fillets is defined and constrained as,

Copyright © 2023 by ASME

$$S_1 = \frac{1}{2}L - W - R_I \geq 0. \tag{4}$$

On another aspect, to avoid the four columns exceeding the heave plates, the distance between the outer surfaces of a column and the corresponding outer fillet of heave plates on diagonal direction, in FIGURE 4, is calculated and constrained as,

$$S_2 = \left[\frac{L}{\sin\left(\frac{\pi}{4}\right)} - 2 \cdot R_O\left(\frac{1}{\sin\left(\frac{\pi}{4}\right)} - 1\right)\right] - \left[2 \cdot \frac{L_B}{\sin\left(\frac{\pi}{4}\right)} + D\right] \geq 0. \tag{5}$$

For evaluating motion stability, the platform pitch motion (α) is constrained and calculated by,

$$\alpha = \bar{\alpha} + 3\bar{\sigma}, \tag{6}$$

where $\bar{\alpha}$ and $\bar{\sigma}$ are the means of the average and standard deviation of each simulation's platform pitch motion. $3\bar{\sigma}$ guarantees the 99.7% probability of better pitch motion stability than the reference model.

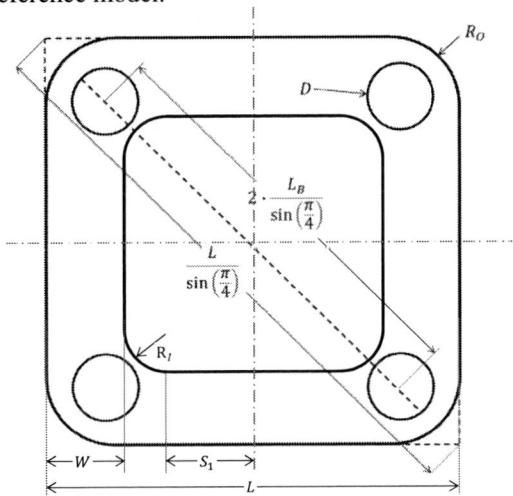

FIGURE 4: GEOMETRIC CHARACTERISTICS OF PLATFORM FOR CONSTRAINTS

Power generation is also maintained by considering the annual energy production (AEP). The AEP is calculated by,

$$AEP = N_k \sum_{i=1}^{N} [F(V_i) - F(V_{i-1})] \cdot \frac{P_{i-1} + P_i}{2}, \tag{7}$$

where N_k is the number of hours per year (24*365h in this study) and N is the number of bins. $F(V_i)$ is the integration of wind distribution function between 0 and V_i and P_i is the mean power of the ith bin. To keep the energy production, the AEP is contained no less than the reference model (AEP_{ref}).

Finally, the formulation of the CCD problem can be expressed by,

$$\begin{aligned}
\text{Min} \quad & M(x), \\
\text{s.t.} \quad & \\
& x_L \leq x \leq x_U, \\
& S_1 \geq 0, \\
& S_2 \geq 0, \\
& \alpha \leq \alpha_{ref}, \\
& AEP \geq AEP_{ref},
\end{aligned} \tag{8}$$

where x_U and x_L are the upper and lower limits of the plant and control design variables in TABLE 3 and TABLE 4. An efficient Kriging model-enhanced Genetic algorithm is constructed and solves this formula.

(2) Kriging model enhanced Genetic algorithm for solving the CCD problem

The Kriging model, which is an equivalent of the Gaussian process regression, is widely used as the surrogate model for expensive engineering problems [32, 33]. For exploring the design space defined in Eq.(8), 200 points are sampled randomly.

The 200 samples are subjected to the process of AQWA-OpeFAST coupling evaluation with the responses in Eq.(8) calculated. This forms the initial design dataset. Ninety percent of the dataset trains three Kriging models for the $M(x)$, AEP, and α, which is tested by the remaining 10% of design data. Two measures, that is, the mean absolute (E_a) and relative (E_r) errors are used to account for the accuracy of each Kriging model as expressed by,

$$E_a = \frac{1}{n}\sum|Y_i - Y_i^r|, \tag{9}$$

$$E_r = \frac{1}{n}\sum_{i=1}^{n}\frac{|Y_i - Y_i^r|}{Y_i^r}. \tag{10}$$

where n is the number of test data. Y_i and Y_i^r represent the predicted and real values of ith samples. The calculated error values are listed in TABLE 6. All relative errors are less than 5%, which shows the high prediction accuracy of the Kriging model.

TABLE 6: TEST ERRORS OF THE KRIGING MODEL

Response	E_a	E_r
$M(x)$	2.18e4 kg	0.234 %
AEP	5.07e4 kWh	0.093 %
α	0.238 °	4.92 %

Based on the Kriging model, the optimization process solves Eq.(8) with 30 populations and 50 generations for configuring the Genetic algorithm. The Kriging model enhances the optimization process by evaluating 80% of the 30 populations. The remaining 6 samples are evaluated in parallel by coupled simulations.

3.4 RESULTS

(1) Analysis of the optimal design

By the CCD scheme, the co-optimization of the platform hydrodynamics and TMD control are solved. The convergence history is shown in FIGURE 5. The objective value is unchanged after the fifth generation, which shows the convergence of the CCD optimization process.

The optimal design after 50 generations is extracted as manifested in TABLE 7. The generated CCD solution reduced the sizes of all geometric variables of the platform. Meanwhile, the M_T, B_T and C_T are moving close to their corresponding lower limits. This is caused by the single objective of the CCD problem, which may force the optimum to the bound of design space under the constraints. Meanwhile, the natural frequencies in heave (f_z) and pitch (f_{Ry}) DOF of the CCD solution model is calculated and compared with the baseline.

Copyright © 2023 by ASME

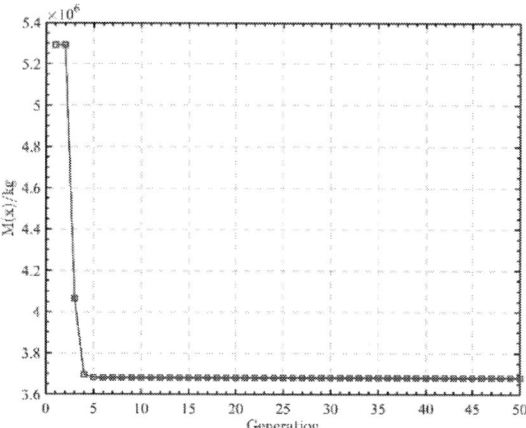

FIGURE 5: CONVERGENCE OF THE CCD OPTIMIZATION PROCESS

TABLE 7: COMPARISON OF THE OPTIMAL DESIGN AND BASELINE MODEL

Variable	Baseline model	CCD solution	Difference (%)
D (m)	10.5	8	-23.8
L (m)	65.25	64	-1.9
W (m)	10.5	8.5	-19.0
L_B (m)	27.375	26.2	-4.3
D_B (m)	15.5	12	-22.6
M_T (kg)	\	1.09E5	\
B_T (N/(m/s))	\	8050	\
K_T (N/m)	\	440	\

Quantitative comparisons of the responses between the CCD solution and the baseline model are made in TABLE 8. The platform mass (M) of the CCD solution shows a 44.1% reduction from the baseline model, which shows the great improvement made by the co-optimization. Meanwhile, regarding the two constraints, AEP, and α are also decreased by 2.96% and 49.9%, respectively. These significant improvements suggest the capability of the CCD method to improve the power generation capability and structural stability.

TABLE 8: COMPARISON OF RESPONSE BETWEEN THE CCD SOLUTION MODEL AND BASELINE MODEL

Response	Baseline model	CCD solution	Difference (%)
M (kg)	6.58E6	3.68E6	-44.1
AEP (kW·h)	4.06E7	4.18E7	2.96
α (°)	9.47	4.73	-49.9
f_z	0.053	0.172	224.5%
f_{Ry}	0.033	0.067	103.0%

In addition, the natural frequencies of the CCD solution model are also compared with the baseline. The platform resonant motion frequencies in heave and pitch degree of freedoms are increased from the baseline model. These impact the dynamical response of the platform under turbulent wind and irregular wave conditons.

In addition, FIGURE 6 compares the platform geometric features of the CCD solution and the baseline model. Due to the reduction of the column diameter, the width of the support plates is reduced, which may cause weakened structural strength. Also, the width of heave plates is decreased, which may be compensated by the damping of the TMD system but it may also undermine the structural strength. These may require additional structural strength analysis that is not included in this study and has to be done in the future.

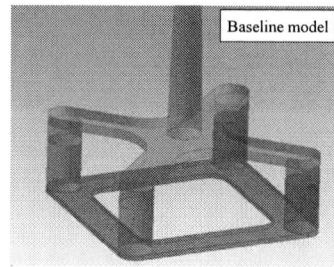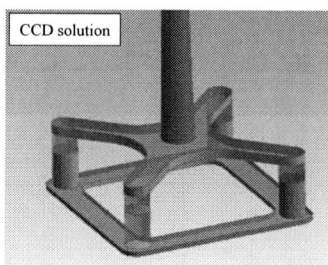

FIGURE 6: COMPARISON OF GEOMETRY BETWEEN THE BASELINE MODEL AND THE MODEL OF CCD SOLUTION

(2) Analysis of integrated time-domain simulations

In addition, the time history responses are analyzed. The case of $WS = 11.4\,m/s$ that is, Case 5 in TABLE 5, is taken as the example as shown in FIGURE 7.

In FIGURE 7 (a), after the CCD process, the platform pitch motion was stabilized as suggested by the lowered magnitude. This increases the power generation with significantly lowered perturbation in FIGURE 7 (b), which increases the AEP in TABLE 8. Meanwhile, the blade deflection and rotor thrust of the optimal design manifest a lowered degree of perturbation in FIGURE 7 (c) and (d).

These demonstrate that compared with the reference model, the CCD method reduces the platform mass significantly by the co-optimization of platform hydrodynamics and TMD control system. Also, the platform motion stability and wind turbine power generation capability are raised greatly by the lowered platform pitch angles and AEP. The time-domain simulation also revealed the stabilized platform pitch motion and power generation. These suggest the high capability of the CCD method for the integrated optimization of platform hydrodynamic and structural control.

Copyright © 2023 by ASME

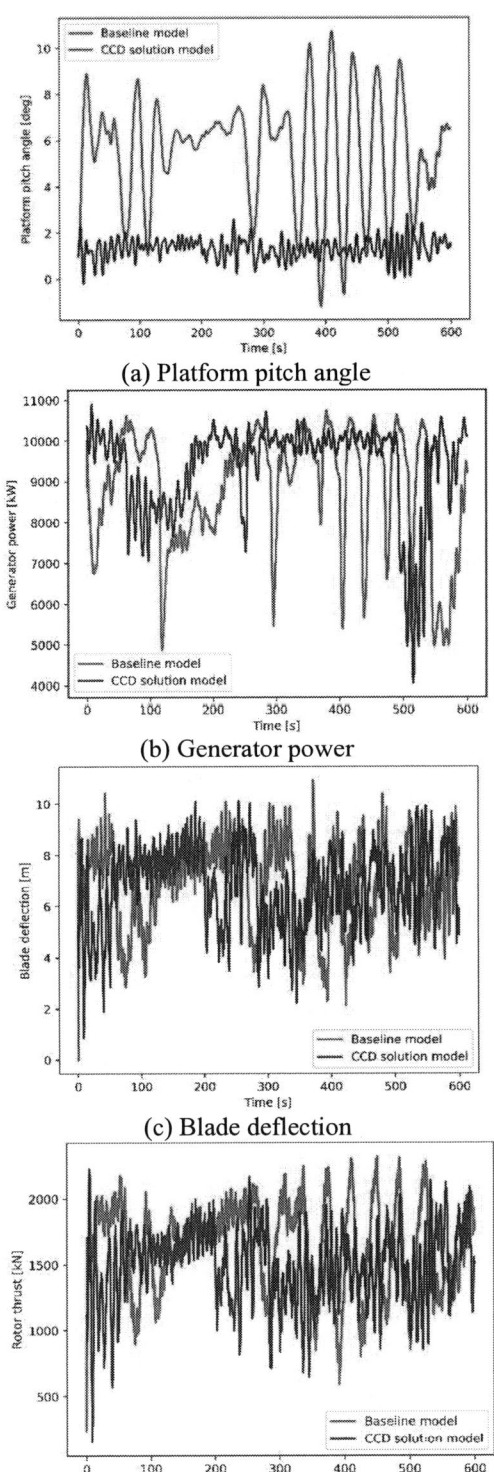

(a) Platform pitch angle

(b) Generator power

(c) Blade deflection

(d) Rotor thrust

FIGURE 7: COMPARISON BETWEEN THE BASELINE AND CCD SOLUTION MODELS WITH RESPECT TO THE RESPONSES OF THE PLATFORM PITCH, GENERATOR POWER, BLADE DEFLECTION, AND ROTOR THRUST.

4. CONCLUSIONS

To overcome the performance limitations of traditional sequential design, this study proposes a CCD framework for the co-optimization of platform hydrodynamics and TMD controllers. A coupling scheme is developed to realize the coupled time domain simulation of the parameterized platform and wind turbines with a TMD system. The platform hydrodynamics is evaluated by AQWA in the frequency domain, which is sent to the OpenFAST for synthesizing the hydrodynamic force in the time domain. A TMD system is implemented and parameterized by OpenFAST. A Kriging model enhanced- Genetic algorithm is implemented outside the coupling scheme to realize the simultaneous optimization of platform geometry and TMD controller.

To demonstrate the functionality of the CCD framework, a case study is conducted using the DTU 10 MW RWT with the NAUTILUS-10 semisubmersible platform. The platform is parameterized by five geometric design variables while another three design variables represent the TMD controller. Using the DLC1.2 and Jonswap wave spectrum with specific wave heights and periods for each DLC case, the CCD is solved by a Kriging surrogate model enhanced-Genetic algorithm (KS-GA). Thirty populations and fifty generations are defined for the optimization, where 80% and 20% of samples are evaluated by the surrogate model and AQWA-OpenFAST model respectively. Based on the results of the CCD process, the platform mass is reduced by 44.1% while the constraint values of AEP and α are raised by 2.96% and decreased by 49.9%, respectively. The blade tip deflection and rotor thrust force are also reduced by the CCD process by time history analysis. These demonstrate the improvement made by the CCD method to reduce the platform mass and stabilize the platform, which, in turn, benefits the power generation quantity and quality.

However, limitations of this preliminary study need to be fixed in the future. Firstly, only DLC 1.2 is used and more design load cases have to be involved for CCD optimization with more practical constraints. Secondly, the function of platform cost should be refined but not only the material weight. In addition, structural flexibility has to be considered for considering hydroelectricity and structural loads. These will be done in the future.

ACKNOWLEDGEMENTS

The J. Liang and X. Du would like to thank the financial support from the National Natural Science Foundation of China (No. 52205294), the Innovation Group Project of Southern Marine Science and Engineering Guangdong Laboratory (Zhuhai), and the start-up funding of Sun Yat-Sen University in China. The P. Xie acknowledge the support from the Marine Economy Development (Six Marine Industries) Special Foundation of Department of Natural Resources of Guangdong Province (GDOE [2021]43).

REFERENCES

[1] Leung, D.Y.C., and Y. Yang, Wind energy development and its environmental impact: A review. Renewable and Sustainable Energy

Copyright © 2023 by ASME

Reviews, 2012. 16(1): p. 1031-1039.

[2] Breton, S.-P., and G. Moe, Status, plans and technologies for offshore wind turbines in Europe and North America. Renewable energy, 2009. 34(3): p. 646-654.

[3] Schwartz, M., D. Heimiller, S. Haymes, et al., Assessment of offshore wind energy resources for the United States. 2010, National Renewable Energy Lab.(NREL), Golden, CO (United States).

[4] Wang, C., T. Utsunomiya, S. Wee, et al., Research on floating wind turbines: a literature survey. The IES Journal Part A: Civil & Structural Engineering, 2010. 3(4): p. 267-277.

[5] Clauss, G.F., and L. Birk, Hydrodynamic shape optimization of large offshore structures. Applied Ocean Research, 1996. 18(4): p. 157-171.

[6] Hall, M., B. Buckham, and C. Crawford, Evaluating the importance of mooring line model fidelity in floating offshore wind turbine simulations. Wind energy, 2014. 17(12): p. 1835-1853.

[7] Karimi, M., M. Hall, B. Buckham, et al., A multi-objective design optimization approach for floating offshore wind turbine support structures. Journal of Ocean Engineering and Marine Energy, 2017. 3(1): p. 69-87.

[8] Dou, S., A. Pegalajar-Jurado, S. Wang, et al., Optimization of floating wind turbine support structures using frequency-domain analysis and analytical gradients. Journal of physics. Conference series, 2020. 1618(4): p. 42028.

[9] Pillai, A.C., P.R. Thies, and L. Johanning. Comparing frequency and time domain simulations for geometry optimization of a floating offshore wind turbine mooring system. in International Conference on Offshore Mechanics and Arctic Engineering. 2018. American Society of Mechanical Engineers.

[10] He, J., X. Jin, S. Xie, et al., Multi-body dynamics modeling and TMD optimization based on the improved AFSA for floating wind turbines. Renewable Energy, 2019. 141: p. 305-321.

[11] Chen, L., X. Du, B. Hu, et al., Drivetrain Oscillation Analysis of Grid Forming Type-IV Wind Turbine. IEEE Transactions on Energy Conversion, 2022. 37(4): p. 2321-2337.

[12] Haemers, M., C.-M. Ionescu, K. Stockman, et al., Optimal Hardware and Control Co-Design Applied to an Active Car Suspension Setup. Machines, 2021. 9(3): p. 55.

[13] Azad, S., M. Behtash, A. Houshmand, et al., PHEV powertrain co-design with vehicle performance considerations using MDSDO. Structural and Multidisciplinary Optimization, 2019. 60(3): p. 1155-1169.

[14] Nguyen Van, E., D. Alazard, C. Döll, et al., Co-design of aircraft vertical tail and control laws with distributed electric propulsion and flight envelop constraints. CEAS Aeronautical Journal, 2021. 12(1): p. 101-113.

[15] O'Sullivan, A.C., and G. Lightbody, Co-design of a wave energy converter using constrained predictive control. Renewable Energy, 2017. 102: p. 142-156.

[16] Du, X., L. Burlion, and O. Bilgen. Control Co-Design for Rotor Blades of Floating Offshore Wind Turbines. in ASME 2020 International Mechanical Engineering Congress and Exposition. 2020. Virtual conference: American Society of Mechanical Engineers.

[17] Bilgen, O., L. Burlion, D.-G. Caprace, et al. Multidisciplinary Control Co-Design Optimization (MCCDO) Framework with Mixed-Fidelity Fluid and Structure Analysis. in Wind Energy Science Conference. 2021.

[18] López Muro, J., X. Du, J.-P. Condomines, et al., Wind Turbine Tower Thickness and Blade Pitch Control Co-Design Optimization, in AIAA SCITECH 2022 Forum. 2021, American Institute of Aeronautics and Astronautics.

[19] Jiang, B., M.R. Amini, Y. Liao, et al., Control Co-design of a Hydrokinetic Turbine with Open-loop Optimal Control. arXiv preprint arXiv:2204.01134, 2022.

[20] Ross, H., M. Hall, D.R. Herber, et al., Development of a Control Co-Design Modeling Tool for Marine Hydrokinetic Turbines. 2022, National Renewable Energy Lab.(NREL), Golden, CO (United States).

[21] Lemmer, F., K. Müller, W. Yu, et al. Optimization of floating offshore wind turbine platforms with a self-tuning controller. in International Conference on Offshore Mechanics and Arctic Engineering. 2017. American Society of Mechanical Engineers.

[22] Hegseth, J.M., E.E. Bachynski, and J.R. Martins, Design Optimization of Spar Floating Wind Turbines Considering Different Control Strategies. Journal of Physics: Conference Series, 2020. 1669(1).

[23] Deshmukh, A.P., and J.T. Allison, Multidisciplinary dynamic optimization of horizontal axis wind turbine design. Structural and Multidisciplinary Optimization, 2016. 53(1): p. 15-27.

[24] Jonkman, J.M., and M.L. Buhl, Development and Verification of a Fully Coupled Simulator for Offshore Wind Turbines. 2007.

[25] Du, X., H. Xu, and F. Zhu, Understanding the effect of hyperparameter optimization on machine learning models for structure design problems. Computer-Aided Design, 2021. 135: p. 103013.

[26] Du, X., and F. Zhu, A new data-driven design methodology for mechanical systems with high dimensional design variables. Advances in Engineering Software, 2018. 117(2017): p. 18-28.

[27] Bak, C., F. Zahle, R. Bitsche, et al. The DTU 10-MW reference wind turbine. 2013.

[28] Ustutt, L.B., Qualification of innovative floating substructures for 10MW wind turbines and water depths greater than 50m. 2018.

[29] Lackner, M.A., and M.A. Rotea, Structural control of floating wind turbines. Mechatronics, 2011. 21(4): p. 704-719.

[30] Commission, I.E., IEC 61400-3 wind turbines Part3: design requirements for offshore wind turbines. International Electrotechnical Commission: Geneva, Switzerland, 2009.

[31] Pegalajar-Jurado, A., H. Bredmose, M. Borg, et al. State-of-the-art model for the LIFES50+ OO-Star Wind Floater Semi 10MW floating wind turbine. 2018. IOP Publishing.

[32] Du, X., H. Xu, and F. Zhu, A data mining method for structure design with uncertainty in design variables. Computers & Structures, 2021. 244.

[33] Du, X., O. Bilgen, and H. Xu. Generating Pseudo-Data to Enhance the Performance of Classification-Based Engineering Design: A Preliminary Investigation. in ASME 2020 International Mechanical Engineering Congress and Exposition. 2020. Virtual conference: American Society of Mechanical Engineers Digital Collection.

Copyright © 2023 by ASME

Proceedings of the ASME 2023
International Design Engineering Technical Conferences and
Computers and Information in Engineering Conference
IDETC-CIE2023
August 20-23, 2023, Boston, Massachusetts

DETC2023-114741

A HEURISTIC APPROACH TO CLASSIFY GEOMETRICALLY DEFECTIVE BEAD SEGMENTS BASED ON RANGE OF CURVATURE, RANGE OF SOUND POWER AND MAXIMUM HEIGHT

Nowrin Akter Surovi
Engineering Product Development
Singapore University of
Technology and Design
Singapore 487372
Email: surovi_akter@mymail.sutd.edu.sg

Gim Song Soh
Engineering Product Development
Singapore University of
Technology and Design
Singapore 487372
Email: sohgimsong@sutd.edu.sg

ABSTRACT

The paper presents a classification methodology for determining geometrically good and defective bead segments for the Wire Arc Additive Manufacturing (WAAM) process based on optimally determined thresholds of three parameter metrics: namely, the range of curvature, range of sound power, and maximum height. We show how a heuristic search can be performed to systematically identify the suitable thresholds for the various parameter metrics to classify bead segments. The heuristic search for each parameter metric is performed by finding a threshold value that minimizes overlapping area between the Kernel Density Estimation (KDE)s formed by the distribution of the good and defective bead segments. We combine these thresholds to yield a three-dimensional threshold window for identifying geometrically defective bead segments. This approach is helpful for dataset labeling that overcomes the visual inspection labeling used in the literature. Experiments are conducted on the Inconel 718 material, with 660 bead segment data. We verify the results of our threshold selection using real bead scans and evaluate its mean squared error (MSE) with the parabolic bead model commonly used in literature to model bead profiles. We find that our proposed approach can separate the starting and ending segments of the beads well and identify geometrically defective bead segments between them. We also observe that the MSE of geometrically defective bead segments' is much higher than the good

segments, demonstrating the validity of our approach to classify geometrically defective bead segments.

1 INTRODUCTION

Wire Arc Additive Manufacturing (WAAM) is a droplet-based direct energy deposition additive manufacturing process that is very promising for the direct energy deposition of complex 3D parts [1]. It employs an electric arc as the heat source and metal wires as a material feedstock to fabricate parts. During the fabrication process, an electric arc melts the metal wire within a shielded environment provided by gas such as argon or helium. Using a shielding gas reduces the potential for contamination, helps with the material transfer, and protects the molten pool and its surroundings from oxidation [2]. WAAM features a high deposition rate, which is suitable for fabricating moderate to large-scale components. Furthermore, it is low-cost and has a high material utilization ratio. This makes such an approach advantageous over alternative additive manufacturing processes in the industry now-a-days, especially in the aerospace, oil, gas, and marine offshore industries [3]. One of the most promising WAAM approaches uses the cold metal transfer (CMT) process, providing material deposition at low thermal input. During the CMT process, the metal wire tip is heated, melted, and transferred into the melt pool, and then the wire is separated from the melt pool by pulling it back [4]. Subsequently, the melted

Copyright © 2023 by ASME

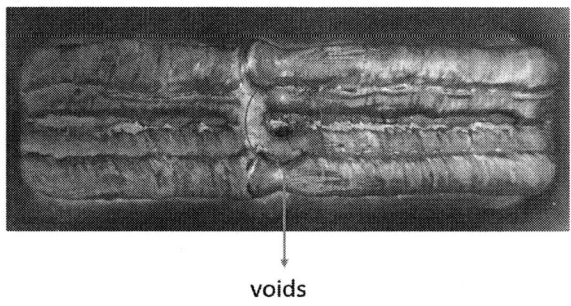

voids

FIGURE 1: Voids created by geometric defects

wire solidifies at the melt pool boundary to form the deposited material.

During the WAAM process, the stability of arc ignition and extinction, and the choice of wrong process parameters like torch speed (TS) and wire feed rate (WFR) can cause geometrical fluctuations in weld beads, which affects its uniformity and therefore produces geometrically defective beads, which we term geometric defects [5,6]. Geometrically defective beads often result in lack of fusion between segments of successive beads, due to the variation of a segment geometry from the required cross-sectional overlapping distance, leading to voids within a final printed product as shown in Figure 1. Hence, it is important to identify the defective print segments early to reduce the amount of post-processing repair and material wastage [7].

Numerous types of research have looked into defects identification for WAAM, and these include distortion [8], porosity [9], crack [10], oxidation [11] etc. Among these, a common approach is to use supervised machine learning (ML) models to monitor the weld bead deposition process and use it to identify defective print segments. However, the accuracy of a supervised ML model largely depends on how well a dataset is labeled for training. The current approaches to labeling are based on visual inspection, which is rather subjective and prone to error. For instance, Cho et al. [12] classified their training and testing melt pool image dataset into three categories: normal, abnormal, and unknown, based on two experts' opinions. Drissi-Daoudi et al. [13] labeled their acoustic dataset manually to associate them with fusion pores, conduction mode, and keyhole pores defects for the laser powder-bed fusion (LPBF) deposition process based on the x-ray tomography images captured. Chen et al. [14] manually tagged the microscope images of the resulting laser-directed energy deposition (LDED) process to identify the location of cracks and keyhole pores defects. Tempelman et al. [15] used X-ray radiography images to identify the spatial coordinates of keyhole pores for the LPBF deposition process and label their dataset manually. Okaro et al. [16] labeled the photo-diode data as faulty or acceptable by using a specific value of ultimate tensile strength (UTS) for the LPBF deposition process.

In all the research mentioned above, researchers used visual inspection approaches for labeling datasets which is error-prone and time-consuming when the dataset is large. On the other hand, in our paper, we propose a heuristic approach to label datasets of geometrically defective bead segments based on three different parameter metrics, namely "range curvature", "range power" and "maximum height". We chose the three-parameter metrics because these parameters have a relation with geometric defects. For example, Li et al. [17] introduced local curvatures into the ML model to detect geometric defects in additive manufacturing (AM). They found from the correlation matrix that mean curvature has a strong correlation with defect labels and concluded that adding this feature improves the performance of the ML model. Since curvature is a sensitive parameter, its value changes significantly, even if there is a slight change in bead geometry. Polajnar et al. [18] showed that irregularities in the bead geometry are reflected in the intensity of acoustic signals. Since power transfer during the material deposition process is related to sound intensity, it is also related to the geometric defect. Furthermore, maximum bead height at the start and ending section is often abnormal compared with the middle region due to the arc ignition and extinction process [19]. This also forms the basis for us to use this as a parameter metric to pick out a print's start and end segments.

2 GEOMETRIC DEFECT CLASSIFICATION

In this section, we explain our methodology of geometric defect classification and the heuristic search performed on three of our proposed parameter metrics to separate the bad bead segments from good bead segments. This approach is inspired by Alewijn al. [20] where they showed that some overlap between the Kernel Density Estimate (KDE) of the organic and conventional samples is observed, even though they are from two distinct distributions. This means that to associate between good and bad bead segments, we could apply the concept of minimizing the overlapping area of the KDEs distribution to search for optimal thresholds to separate the various bead segments.

2.1 Overview

Figure 2 gives an overview of our classification approach for identifying geometrically defective bead segments. First, we collect point cloud scan data and acoustic signals for all bead segments. Next, for each bead segment, we extract the various key parameter metrics used for classifying the bead segments. From the point cloud, i) we compute the mean curvature at each point in that segment, followed by the range of all these mean curvatures (*range curvature* RC); and ii) we compute the bead segment maximum height (*maximum height* MH). Similarly, from the acoustic signal, we compute the instantaneous power of its sound intensity, followed by the range of its instantaneous power (*range power* RP). Then, we perform a heuristic search to determine the optimal thresholds to separate the bead segments based

Copyright © 2023 by ASME

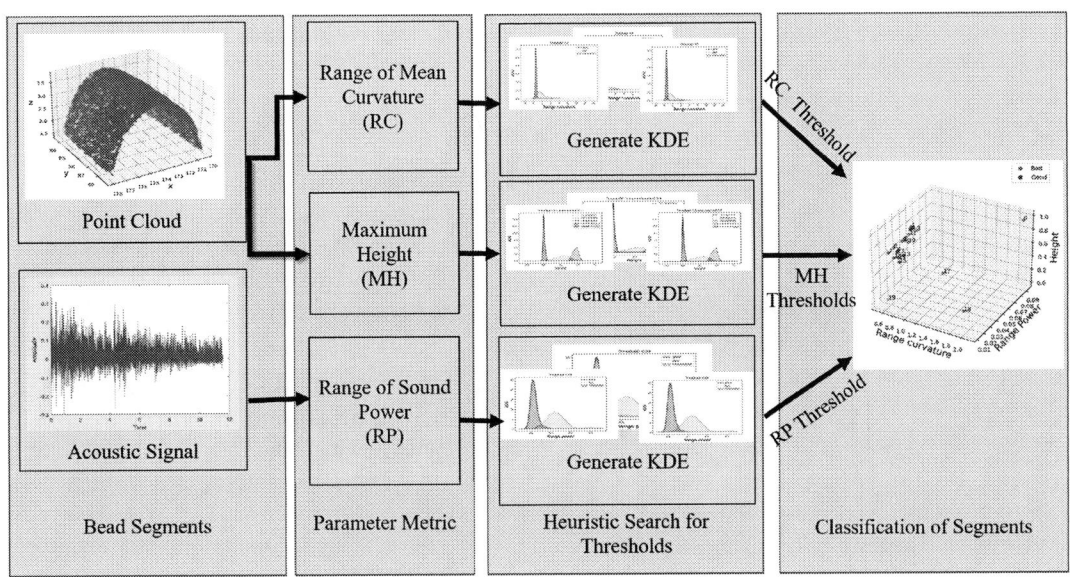

FIGURE 2: Overview of the workflow of our classification approach for identifying Geometrically Defective Bead Segments

on the proposed parameter metrics. This is a two-step procedure where we search through various threshold values that separate the bead segments and determine a candidate threshold that minimizes the overlapping area between their associated Kernel Density Estimation (KDE). An example of a heuristic search method (RC threshold) is shown in Algorithm 1. Finally, we combine the various optimal parameter metrics thresholds to classify geometrically good and defective segments. We consider all bead segments are bad above these thresholds. Otherwise, they are good.

Note that, for the RC and RP, we compute a set of two KDEs based on a chosen threshold to separate geometrically good and defective segments. The threshold that minimizes the overlapping area between the two KDEs is used as the optimal threshold for both parameter metrics. Similarly, for the MH, we compute a set of three KDEs based on a pair of chosen thresholds to separate the start, middle, and end segments of a bead. The reason for using two thresholds to separate the bead segments into three types is due to the abnormal heights observed at the start and end segments of the bead due to arc ignition, and extinction [19]. Thus the two thresholds that minimize the overlapping area between the start, middle, and end segments of the bead are used as the optimal thresholds.

2.2 Good and Bad Bead Segment Classification based on Range of Mean Curvature

The curvature of a bead segment surface gives an indication of how much it curves. The higher the curvature at a point, the sharper the curve "turns" at that point [21,22]. The range of bead curvature indicates how much the curvature change in a bead segment. For a smooth or uniform bead surface, the range curvature

---

Algorithm 1 Heuristic search algorithm for RC threshold

Input: $RT \in (RT_1, RT_2, \ldots, RT_m) \leftarrow$ Range curvature thresholds
$RC \in (RC_1, RC_2, \ldots, RC_n) \leftarrow$ Range curvature values for n bead segments
$A \leftarrow$ list of overlapping areas of all bead segments.

1: **for** RT **do**
2:   $G \leftarrow$ list of good segments.
3:   $B \leftarrow$ list of bad segments.
4:   **for** RC **do**
5:     **if** $RT \geq RC$ **then**
6:       $G \leftarrow RC$
7:     **else**
8:       **if** $RT < RC$ **then**
9:         $B \leftarrow RC$
10:       **end if**
11:     **end if**
12:   **end for**
13:   Estimate 2 KDEs for G and B
14:   Estimate overlapping area between 2 KDEs, A_i
15:   $A \leftarrow A_i$
16: **end for**
   return Optimal threshold for RC, $RT_{i*} = \underset{i}{\arg\min} \|A\|$

---

is small, and for a non-uniform bead segment surface, the range curvature is high.

In our approach, we compute the mean curvature at each point of a bead segment because we know the value of mean curvature is more stable and describes the defect feature more

Copyright © 2023 by ASME

effectively than Gaussian curvature [17]. In order to explain the relationship between the range of mean curvature and bead segment geometry, we take two bead segments with their associated range of mean curvature value as shown in Fig. 3. Notice how

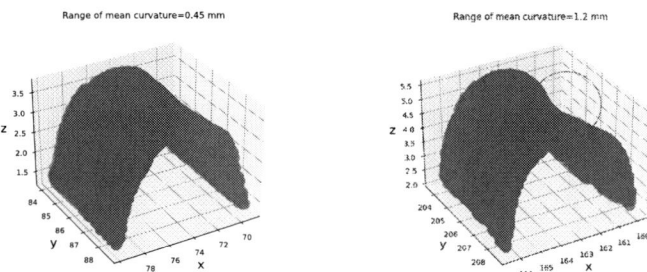

FIGURE 3: Range of mean curvature of geometrically good (left) and bad (right) bead segment. The red circle indicates a geometric defect.

the range curvature is related to the uniformity of a bead segment. From this example, we can see that the higher the RC value, the more non-uniform it looks, indicating that the bead segment has a geometric defect and vice versa [23].

2.2.1 Heuristic-based search of range curvature threshold

Let $S = (s_1, s_2, \ldots, s_i, \ldots, s_n)$ be the sequence of segments, and $p_{i1}, p_{i2}, \ldots, p_{ij}, \ldots, p_{im}$ be the set of points on a particular segment s_i. We approximate the parametric spherical surface at each point on segment s_i using the moving least squares (MLS) method [24]. To compute the mean curvature, κ, at each point of this parametric surface, let κ_{ij} be the mean-curvature at j-th point p_{ij} on segment s_i. The mean curvature [25] at this point can be found as the average of the principal curvatures, κ_1 and κ_2.

$$\kappa_{ij}(p_{ij}) = \frac{1}{2}(\kappa_1(p_{ij}) + \kappa_2(p_{ij})) \qquad (1)$$

Such computation can be easily performed using MeshLab [26]. In our case, we apply to colorize curvature (APSS) filter on the point cloud of a bead segment and get mean curvature from it. Next, we perform the following steps to determine the optimal range curvature threshold.

1. **Range Curvature Parameter Metric Computation.** Let RC_i be the range of mean-curvatures of segment s_i. Then, we can compute range of mean-curvatures $(RC_1, RC_2, \ldots, RC_n)$ for n segments using

$$RC_i = \max_j(\kappa_{ij}) - \min_j(\kappa_{ij}) \quad i = 1, \ldots, n \qquad (2)$$

2. **Density Estimation.** Let $G_i = \{s_j \in S : RC_j \leq RC_i\}$ be the set of good bead segments and $B_i = \{s_j \in S : RC_j > RC_i\}$ as the set of bad segments for range curvature RC_i. Next, we estimate the distribution of feasibly good segments $\hat{f}(G_i)$ and bad segments $\hat{f}(B_i)$ using the kernel density estimates function provided in Python.

3. **Heuristic Search.** To find the optimal RC_{i*}, we heuristically search through various RC_i thresholds. We find the optimal RC_{i*} as the threshold that minimizes the overlapping area between two KDE distributions of $\hat{f}(G_i)$ and $\hat{f}(B_i)$.

2.3 Good and Bad Bead Segment Classification based on Range of Sound Power

Sound power is the rate at which the sound wave transfers energy. When sound propagates through a medium, acoustic sound power is transferred [27]. Sound power is transferred uniformly through the medium if the printing process is stable. Having a high range of instantaneous power means that power transfer during the deposition process is interrupted at that segment. Such interruption may indicate the existence of a defect.

2.3.1 Heuristic-based search of range power threshold

Let $S = (s_1, s_2, \ldots, s_i, \ldots, s_n)$ be the sequence of bead segments and $x_i^1, x_i^2, \ldots, x_i^t, \ldots, x_i^m$ be the set of temporal sound amplitudes associated with a particular segment s_i. Let P_i^t be the instantaneous sound power [28] of x_i^t at time t and it can be computed using

$$P_i^t = |x_i^t|^2 \qquad (3)$$

Next, we perform the following steps to determine the optimal range power threshold.

1. **Range Power Parameter Metric Computation.** Let PR_i be the range of sound power of segment s_i. Then, we can compute range of instantaneous sound powers $(PR_1, PR_2, \ldots, PR_n)$ for n segments using

$$PR_i = \max_j(P_i^t) - \min_j(P_i^t) \quad i = 1, \ldots, n \qquad (4)$$

2. **Density Estimation.** Let, $G_i = \{s_j \in S : PR_j \leq PR_i\}$ be the set of good bead segments and the set $B_i = \{s_j \in S : PR_j > PR_i\}$ be the set of bad bead segments for PR_i. Next, we estimate the distribution of feasibly good segments $\hat{f}(G_i)$ and bad segments $\hat{f}(B_i)$ using the kernel density estimates function provided in Python.

3. **Heuristic Search.** To find the optimal PR_{i*}, we heuristically search through the various PR_i thresholds. We find the optimal PR_{i*} threshold such that it minimizes the overlapping

Copyright © 2023 by ASME

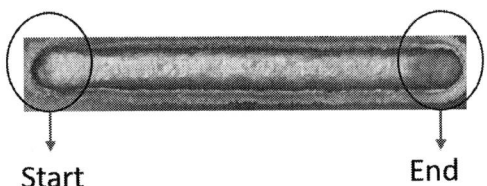

Start   End

FIGURE 4: The height of the middle bead segments is lower than the starting segments but higher than the ending segments.

area between their associated two KDE distributions $\hat{f}(G_i)$ and $\hat{f}(B_i)$.

2.4 Start, Middle, and End Bead Segment Classification based on Maximum Bead Segment Height

Due to arc ignition and extinction of the welding process, the height of the middle bead segments is lower than the starting segments but higher than the ending segments, as shown in Fig. 4. The starting height of a bead is high and bulkier because when the arc strikes, the weld pool experiences an initial non-steady state at a shorter length, and the equivalent wire feeding volume would generate a bulkier bead geometry as compared to the middle part. On the other hand, the height at the end of the bead is low and slanted because of the sudden shutdown of heat input and the rapid solidification process [19]. The above-mentioned two ends of the bead segments are also considered geometrically defective and need to be classified appropriately.

2.4.1 Heuristic-based search of maximum height thresholds
Let $S = (s_1, s_2, \ldots, s_i, \ldots, s_n)$ be the sequence of bead segments. Then, we perform the following steps to determine the optimal maximum height threshold.

1. **Maximum Height Parameter Metric Computation.** Let MH_i be the maximum height on segment s_i. Again, let $T_1, T_2 \in \{MH_1, MH_2, \ldots, MH_n\}$ as the maximum height-pairs such that $T_1 > T_2$. T_1 separates the start and middle segments and T_2 separates the end and middle segments.
2. **Density Estimation.** Let $G_{T_1, T_2} = \{s_j \in S : T_2 \leq MH_j \leq T_1\}$ be the set of good bead segments and both $B_{T_1} = \{s_j \in S : MH_j > T_1\}$ and $B_{T_2} = \{s_j \in S : T_2 > MH_j\}$ as the set of bad segments. Therefore, we combine all bad segments as $B_{T_1, T_2} = S \setminus G_{T_1, T_2}$ where $B_{T_1, T_2} = B_{t_1} \cup B_{t_2}$. Next, we estimate the distribution of feasibly good segments $\hat{f}(G_{T_1, T_2})$ and bad segments $\hat{f}(B_{T_1, T_2})$ using kernel density estimates function provided in python.
3. **Heuristic Search.** To find the optimal (T_{1*}, T_{2*}), we heuristically search a pair of maximum heights in $(MH_1, MH_2, \ldots, MH_n)$. Finally, we select the (T_{1*}, T_{2*}) that minimizes the overlapping area between three estimates

$\hat{f}(G_{T_1, T_2})$ and $\hat{f}(B_{T_1, T_2})$.

3 EXPERIMENTAL SETUP AND DATA COLLECTION

In this section, we explain our experimental setup, data collection process, and acoustic signal segmentation approach.

3.1 Experimental Setup

The experiments for bead printing and acoustic data collection were conducted on our robotic WAAM system as shown in Figure 5 at the Singapore University of Technology and Design (SUTD).

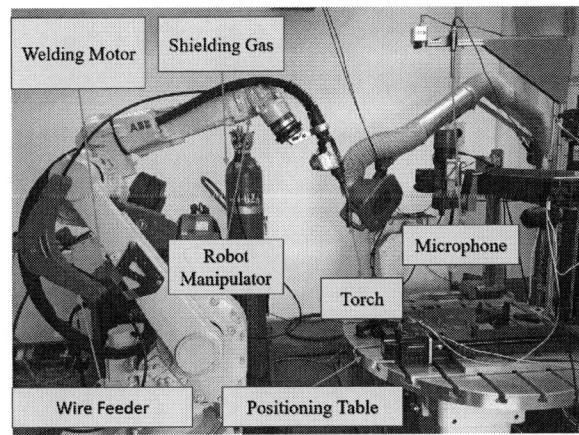

FIGURE 5: Experimental setup of SUTD Robotic WAAM for Bead Printing and Acoustic Data Collection

The Robotic WAAM system consists of a robot manipulator (ABB IRB1660ID), a welding power source (Fronius TPS 400i) equipped with a welding torch (Fronius WF 25i Robacta Drive), a cartesian coordinate robot made up of three linear rails (PMI KM4510) powered by three servos (SmartMotor SM34165DT) and a microphone (UMIK-1 miniDSP) installed at around 80 cm above the substrate in order to minimize environmental noise.

3.2 Data Collection and Segmentation

We printed 33 weld beads of lengths 100 mm using the Inconel 718 (BÖHLER 3D print AM 718) wires using a different combination of torch speed and wire feed rate to obtain different weld bead geometry. The torch speed and wire feed rate used to span the entire material process map and were in the range of [1,12] mm/s and [2,8] m/min, respectively. We used around 70% Ar and 30% He as our shielding gas with a constant gas flow rate of around 25 L/min. We set the nozzle to deposit the material at around 15 mm above the substrate surface.

Copyright © 2023 by ASME

We collected point cloud data of the printed beads using the GOM ATOS III Triple scanner. Examples of the measured bead point cloud are shown in Fig. 6. Next, we separated all the beads

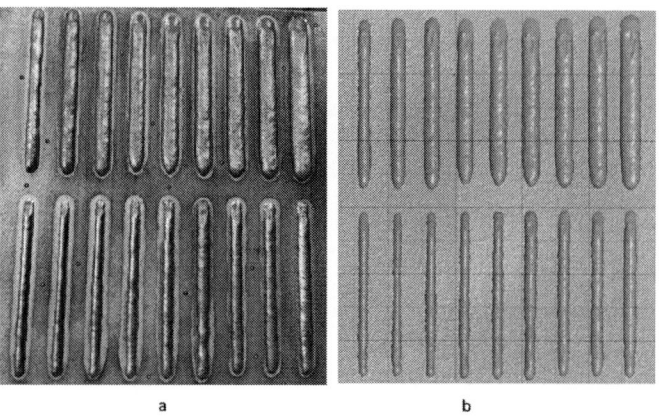

FIGURE 6: (a) Original beads with plate (b) Mesh representation of beads with plate.

from the substrate using RANSAC plane segmentation. Then, we divided each bead into $N = 20$ segments, with each segment measuring about 5 mm.

Similarly, we collected acoustic signals at 44 kHz during the printing process and segmented each bead signals into 20 acoustic signal segments, with each segment containing the acoustic signal for a 5 mm bead segment. Thus, with 33 beads, we get a total of 660 signal segments.

4 EXPERIMENTAL RESULTS

In this section, we show the results of the heuristic threshold search performed on all the printed bead segments to determine the appropriate classification threshold for the range of curvature (Section 4.1.1), the range of sound power (Section 4.1.2), and bead segment maximum height (Section 4.1.3). Then we show the bead classification performance (Section 4.2) for a variety of printed beads based on the combination of these three thresholds. Subsequently, we verify the results of the identified geometrically defective bead segments and compare its Mean Square Error (MSE) with a parabolic bead profile, a bead profile commonly used in literature to model the geometry of a printed bead, to check for the accuracy of our proposed classifier.

4.1 Thresholds Selection

In the following, we discuss the results of the optimal threshold selection based on our three proposed parameter metrics.

4.1.1 Threshold selection based on the range curvature
The distribution plots based on different candidate thresholds of range curvature are as shown in Figure 7. Due to

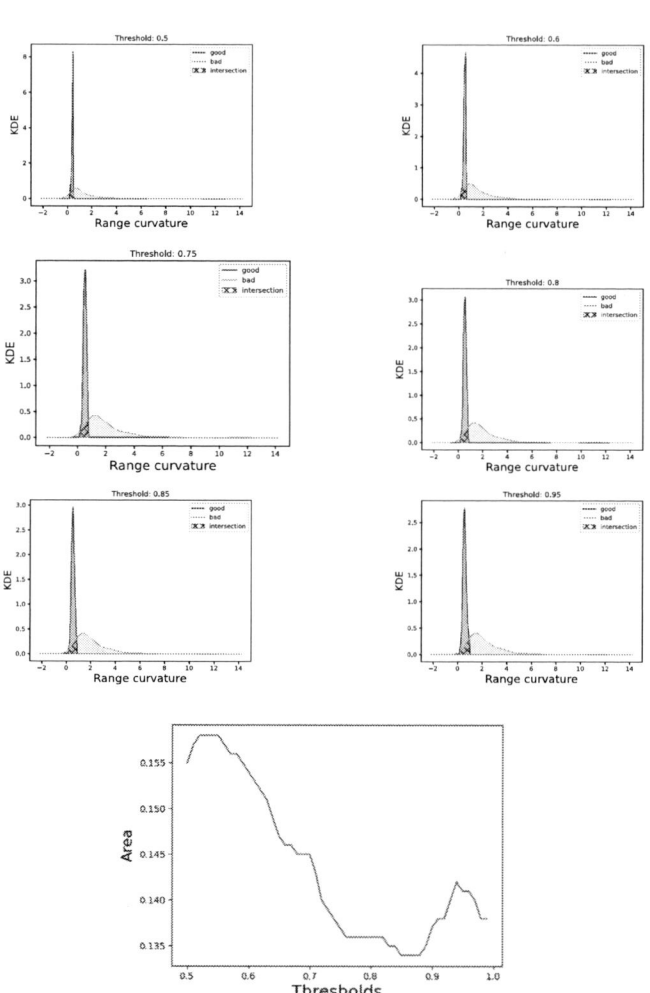

FIGURE 7: Examples of Distribution plots for good and bad segments for different candidate thresholds based on range curvature (First six plots). Different candidate thresholds vs overlapping area (last plot).

page limitations, we only include six example plots with curvature threshold ranges of 0.5, 0.6, 0.75, 0.8, 0.85, and 0.95. Notice that different range curvature candidate thresholds yield different overlapping areas. Based on a heuristic search, we find that the overlapping area between two distributions is minimum when the candidate threshold is 0.85 mm as shown at the bottom of Figure 7. Therefore, we consider any bead segments above this threshold to be bad. Otherwise, they are good.

Copyright © 2023 by ASME

4.1.2 Threshold selection based on the range of sound power

The distribution plots based on different candidate thresholds of range power are as shown in Figure 8. Due

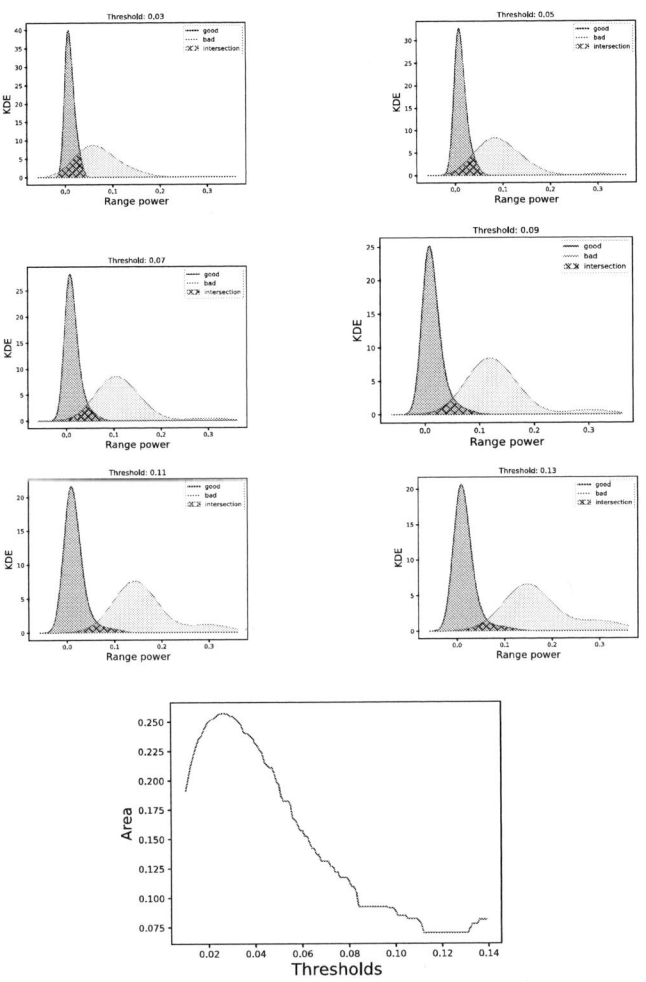

FIGURE 8: Distribution plots for good and bad segments of beads for different candidate thresholds based on range power (First six plots). Different candidate thresholds vs overlapping area (last plot).

to page limitations, we only include six example plots with a power threshold range of 0.03, 0.05, 0.07, 0.09, 0.11, and 0.13. Similarly, we observe from the six example plots that different range power candidate thresholds yield different overlapping areas. Based on a heuristic search, we find that the overlapping area between two distributions is minimum when the candidate threshold is 0.11 mm as shown at the bottom of Figure 8. Therefore, we consider any bead segments that are above this threshold

as bad. Otherwise, they are good.

4.1.3 Threshold selection based on maximum bead segment height

The distribution plots based on different candidate thresholds of bead segment maximum height is as shown in Figure 9. Note that the height value ranges from

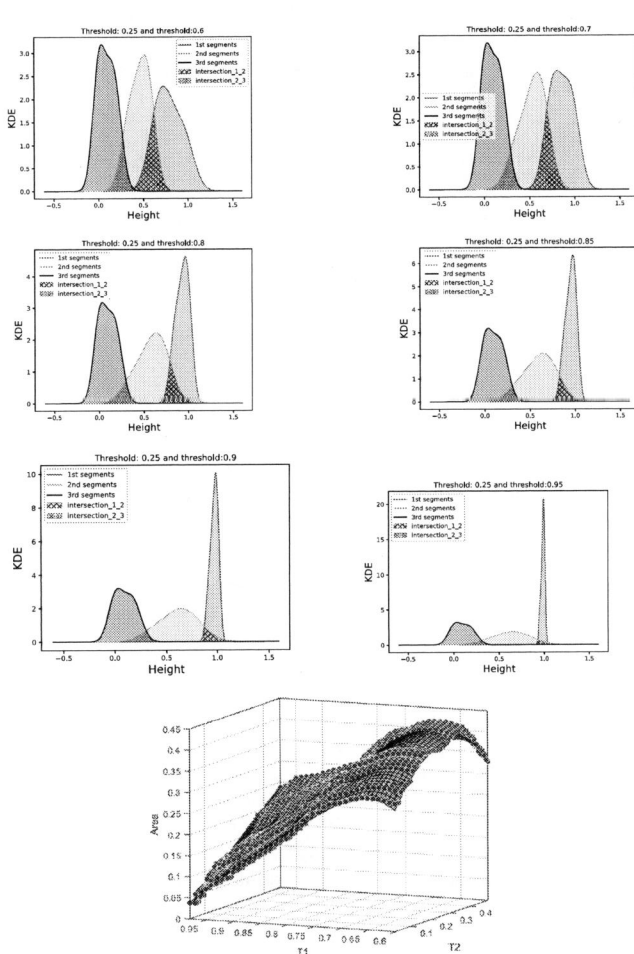

FIGURE 9: Distribution plots for 1st, 2nd, and 3rd segments for different combinations of candidate thresholds based on normalized maximum height (First six plots). Different candidate thresholds vs overlapping area (last plot).

0 to 1 here due to height normalization. This is because different process parameters produce beads of different heights, and to compare effectively, we need to normalize the maximum bead segment height with respect to the absolute maximum of the bead so that the KDE can be constructed. For this parameter metric, we heuristically take different candidate thresholds T_1 and

Copyright © 2023 by ASME

FIGURE 10: Original beads images. The white color rectangle indicates bad segments

T_2 to categorize all the bead segments into three sub-segments: start, middle, and end bead segments and find their corresponding overlapping areas. Again, due to page limitations, we include example plots of height threshold (T_1, T_2) for six example plots of value (0.25, 0.6), (0.25, 0.7), (0.25, 0.8), (0.25, 0.9) and (0.25, 0.95). Based on a heuristic search, we find that the overlapping area between the three distributions is minimum when the candidate has a max height threshold of $T_1 = 0.98$ and $T_2 = 0.02$ as shown at the bottom of Figure 8. Therefore, a bead segment with $0.02 \leq$ normalized maximum height ≤ 0.98 is considered a good segment. Otherwise, it is considered a bad segment.

4.2 Bead Segment Classification Performance based on the Three Parameter Metric Thresholds

Figure 11 shows the performance of our approach to separate the geometrically defective bead segments for a variety of beads, printed using different process parameters. For each bead, its segments are labeled from 0 to 19 on the graph, with 0 denoted as the starting of the bead segment and 19 as the ending bead segment. The identified geometrically defective bead segments are labeled in red. From the plots, we observe that our classification approach is able to separate the starting and ending segments of the beads well and identify geometrically defective bead segments that lie along the middle portion of the bead. These are shown in Figure 10. Notice that segment 3 is flagged out as a geometrically defective segment for beads printed using process parameters of TS= 10 mm/s, WFR=6.5 m/min, and TS=7 mm/s, WFR= 8.5 m/min.

To verify if the identified segments are indeed defective, we use the actual bead segments and fitted segments generated using a parabolic bead profile model [29]. The parabolic bead profile model approximates cross-sectional profiles of single beads based on a bead segment height and width and can be expressed

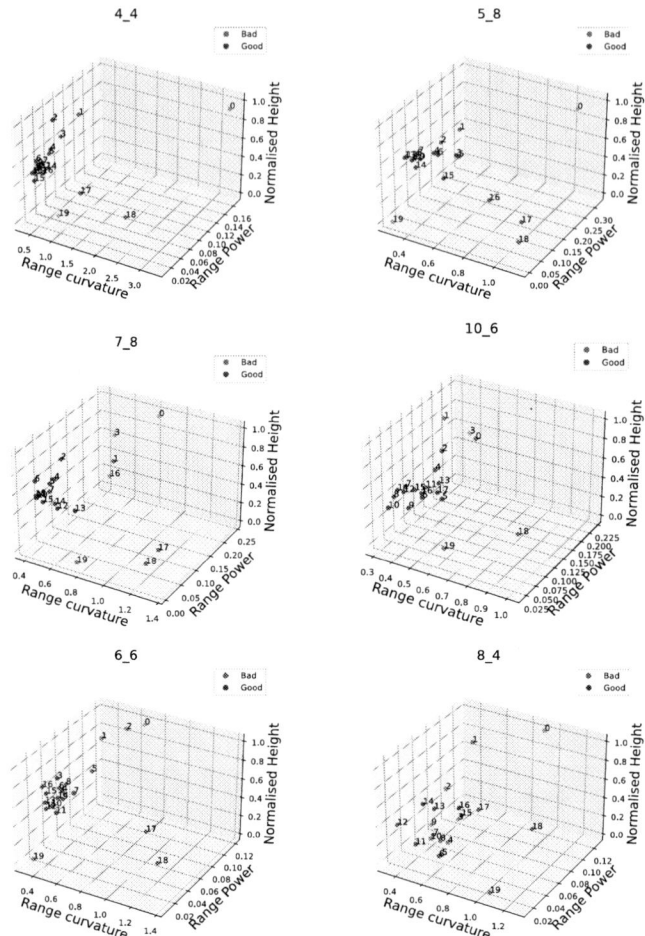

FIGURE 11: 3D plots separating good and bad (geometrically defective) segments for different process parameters. Top left: TS=4 mm/s and WFR=4.5 m/min, Top right: TS=5 mm/s and WFR=8.5 m/min, Middle left: TS=7 mm/s and WFR=8.5 m/min, Middle right: TS=10 mm/s and WFR=6.5 m/min, Bottom left: TS=6 mm/s and WFR=6.5 m/min, Bottom right: TS=8 mm/s and WFR=4.5 m/min.

as:

$$z = ax^2 + c \qquad (5)$$

Here, c is the bead segment height and $2\sqrt{-\frac{c}{a}}$ is the bead segment width [30]. Then we measure the Mean Square Error (MSE) of the various bead segments with the parabolic fitted bead segment. The MSE between the actual and fitted profiles of two beads' segments are shown in Figure 12 (a) and (b). These contain a defective segment 3. From Figure 12 (c), we observe that the bad segments identified based on our classification

Copyright © 2023 by ASME

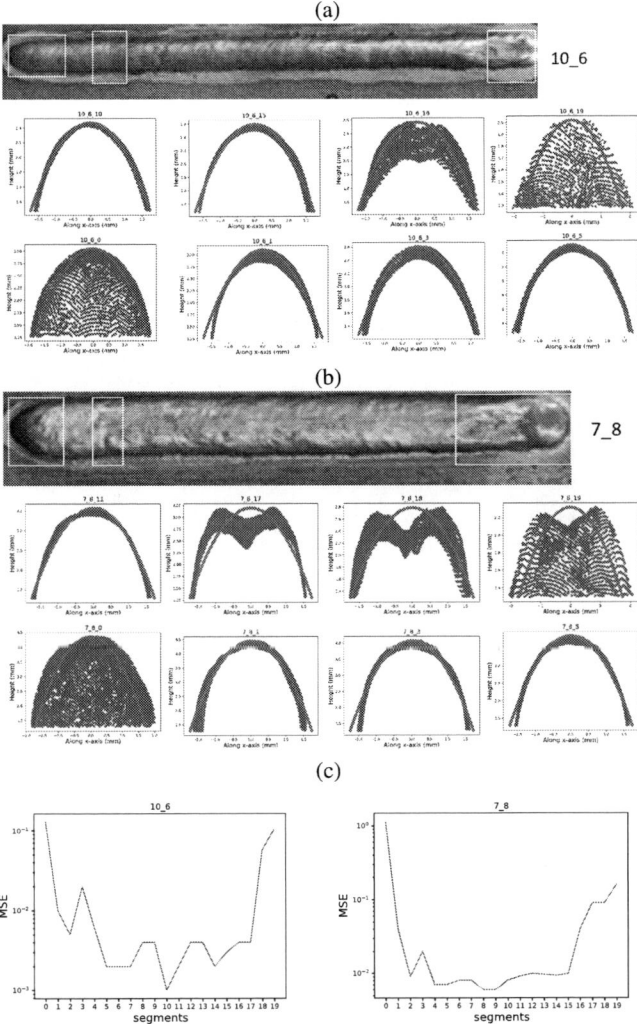

FIGURE 12: (a) (Top) TS=10 mm/s and WFR= 6.5 m/min original bead image (Bottom) Comparison of fitted and measured bead segments profiles. Green lines represent the actual bead segment while red lines represent the literature bead model. (b) (Top) TS=7 mm/s and WFR=8.5 m/min original bead image (Bottom) Comparison of fitted and measured bead segments profiles (c) segments vs MSE value of actual and literature model of the two beads (in log scale).

method have a much higher MSE than good segments. Therefore, we can conclude that our method can effectively identify the geometrically defective bead segments.

5 CONCLUSION

In this paper, we have explored three different parameter metrics, namely the range curvature, range power, and maximum height of bead segments to separate geometrically defective bead segments for the WAAM process based on Inconel 718 material.

We show how heuristic search can be performed on the various parameter metrics to find suitable thresholds for classification. The heuristic search is done by minimizing the overlapping area between the KDEs, made up of different threshold value choices. We combine these thresholds to make a threshold window for identifying geometrically defective segments. We verify our method using real bead scans and evaluate its mean squared error (MSE) with bead models found in the literature. We find that the MSE of bad segments' is much higher than the good segments, demonstrating the feasibility of such an approach. This approach is helpful for dataset labeling for a supervised ML classification model that overcomes the error-prone manual labeling in the literature. For future works, we intend to use the proposed classification method to label our training datasets for training with our ML models and subsequently use them for identifying geometric defects in real time on our WAAM system.

6 ACKNOWLEDGEMENT

The authors gratefully acknowledge the support of the Growth Plan Grant for Aviation at the Singapore University of Technology and Design.

REFERENCES

[1] Yi, H., Qi, L., Luo, J., Zhang, D., Li, H., and Hou, X., 2018. "Effect of the surface morphology of solidified droplet on remelting between neighboring aluminum droplets". *International Journal of Machine Tools and Manufacture, 130*, pp. 1–11.

[2] Kah, P., and Martikainen, J., 2013. "Influence of shielding gases in the welding of metals". *The International Journal of Advanced Manufacturing Technology, 64*, pp. 1411–1421.

[3] Xia, C., Pan, Z., Polden, J., Li, H., Xu, Y., Chen, S., and Zhang, Y., 2020. "A review on wire arc additive manufacturing: Monitoring, control and a framework of automated system". *Journal of Manufacturing Systems, 57*, pp. 31–45.

[4] Selvi, S., Vishvaksenan, A., and Rajasekar, E., 2018. "Cold metal transfer (cmt) technology-an overview". *Defence technology, 14*(1), pp. 28–44.

[5] Surovi, N. A., Dharmawan, A. G., and Soh, G. S., 2021. "A study on the acoustic signal based frameworks for the real-time identification of geometrically defective wire arc bead". In International Design Engineering Technical Conferences and Computers and Information in Engineering Conference, Vol. 85383, American Society of Mechanical Engineers, p. V03AT03A003.

[6] Surovi, N. A., and Soh, G. S., 2022. "Process map generation of geometrically uniform beads using support vector machine". *Materials Today: Proceedings, 70*, pp. 113–118.

[7] Surovi, N. A., Hussain, S., and Soh, G. S., 2022. "A study

of machine learning framework for enabling early defect detection in wire arc additive manufacturing processes". In International Design Engineering Technical Conferences and Computers and Information in Engineering Conference, Vol. 86229, American Society of Mechanical Engineers, p. V03AT03A002.

[8] Montevecchi, F., Venturini, G., Grossi, N., Scippa, A., and Campatelli, G., 2017. "Finite element mesh coarsening for effective distortion prediction in wire arc additive manufacturing". *Additive Manufacturing, 18*, pp. 145–155.

[9] Cheepu, M., 2022. "Machine learning approach for the prediction of defect characteristics in wire arc additive manufacturing". *Transactions of the Indian Institute of Metals*, pp. 1–9.

[10] Seow, C. E., Zhang, J., Coules, H. E., Wu, G., Jones, C., Ding, J., and Williams, S., 2020. "Effect of crack-like defects on the fracture behaviour of wire+ arc additively manufactured nickel-base alloy 718". *Additive Manufacturing, 36*, p. 101578.

[11] Caballero, A., Ding, J., Bandari, Y., and Williams, S., 2019. "Oxidation of ti-6al-4v during wire and arc additive manufacture". *3D Printing and Additive Manufacturing, 6*(2), pp. 91–98.

[12] Cho, H.-W., Shin, S.-J., Seo, G.-J., Kim, D. B., and Lee, D.-H., 2022. "Real-time anomaly detection using convolutional neural network in wire arc additive manufacturing: molybdenum material". *Journal of Materials Processing Technology, 302*, p. 117495.

[13] Drissi-Daoudi, R., Pandiyan, V., Logé, R., Shevchik, S., Masinelli, G., Ghasemi-Tabasi, H., Parrilli, A., and Wasmer, K., 2022. "Differentiation of materials and laser powder bed fusion processing regimes from airborne acoustic emission combined with machine learning". *Virtual and Physical Prototyping, 17*(2), pp. 181–204.

[14] Chen, L., Yao, X., Tan, C., He, W., Su, J., Weng, F., Chew, Y., Ng, N. P. H., and Moon, S. K. "In-situ crack and keyhole pore detection in laser directed energy deposition through acoustic signal and deep learning". *Available at SSRN 4308023*.

[15] Tempelman, J. R., Wachtor, A. J., Flynn, E. B., Depond, P. J., Forien, J.-B., Guss, G. M., Calta, N. P., and Matthews, M. J., 2022. "Detection of keyhole pore formations in laser powder-bed fusion using acoustic process monitoring measurements". *Additive Manufacturing, 55*, p. 102735.

[16] Okaro, I. A., Jayasinghe, S., Sutcliffe, C., Black, K., Paoletti, P., and Green, P. L., 2019. "Automatic fault detection for laser powder-bed fusion using semi-supervised machine learning". *Additive Manufacturing, 27*, pp. 42–53.

[17] Li, R., Jin, M., Pei, Z., and Wang, D., 2022. "Geometrical defect detection on additive manufacturing parts with curvature feature and machine learning". *The International Journal of Advanced Manufacturing Technology, 120*(5-6), pp. 3719–3729.

[18] Polajnar, I., Bergant, Z., and Grum, J., 2013. "Arc welding process monitoring by audible sound". In 12th International Conference of the Slovenian Society for Non-Destructive Testing: Application of Contemporary Non-Destructive Testing in Engineering, ICNDT 2013-Conference Proceedings, pp. 613–20.

[19] Hu, Z., Qin, X., Shao, T., and Liu, H., 2018. "Understanding and overcoming of abnormity at start and end of the weld bead in additive manufacturing with gmaw". *The International Journal of Advanced Manufacturing Technology, 95*, pp. 2357–2368.

[20] Alewijn, M., van der Voet, H., and van Ruth, S., 2016. "Validation of multivariate classification methods using analytical fingerprints–concept and case study on organic feed for laying hens". *Journal of Food Composition and Analysis, 51*, pp. 15–23.

[21] Do Carmo, M. P., 2016. *Differential geometry of curves and surfaces: revised and updated second edition*. Courier Dover Publications.

[22] Devaraj, A., 2020. "An overview of curvature". *Retrieved May, 5*, p. 2021.

[23] Surovi, N. A., and Soh, G. S., 2023. "Acoustic feature based geometric defect identification in wire arc additive manufacturing". *Virtual and Physical Prototyping*.

[24] Guennebaud, G., and Gross, M., 2007. "Algebraic point set surfaces". In *ACM SIGGRAPH 2007 papers*. pp. 23–es.

[25] Jia, Y.-B., 2020. "Gaussian and mean curvatures". *Com S, 477*(577), pp. 1–7.

[26] Cignoni, P., Callieri, M., Corsini, M., Dellepiane, M., Ganovelli, F., and Ranzuglia, G., 2008. "MeshLab: an Open-Source Mesh Processing Tool". In Eurographics Italian Chapter Conference, V. Scarano, R. D. Chiara, and U. Erra, eds., The Eurographics Association.

[27] Baken, R. J., and Orlikoff, R. F., 2000. *Clinical measurement of speech and voice*. Cengage Learning.

[28] Page, C. H., 1952. "Instantaneous power spectra". *Journal of Applied Physics, 23*(1), pp. 103–106.

[29] Xiong, J., Zhang, G., Gao, H., and Wu, L., 2013. "Modeling of bead section profile and overlapping beads with experimental validation for robotic gmaw-based rapid manufacturing". *Robotics and Computer-Integrated Manufacturing, 29*(2), pp. 417–423.

[30] Oh, X. Y., and Soh, G. S., 2020. "A study on the machine learning framework for the geometric modelling of wire arc bead profile". In International Design Engineering Technical Conferences and Computers and Information in Engineering Conference, Vol. 83952, American Society of Mechanical Engineers, p. V006T06A001.

Proceedings of the ASME 2023
International Design Engineering Technical Conferences and
Computers and Information in Engineering Conference
IDETC-CIE2023
August 20-23, 2023, Boston, Massachusetts

DETC2023-115150

AUTOMATING STYLE ANALYSIS AND VISUALIZATION WITH EXPLAINABLE AI - CASE STUDIES ON BRAND RECOGNITION

Yu–hsuan Chen, Levent Burak Kara, Jonathan Cagan

Department of Mechanical Engineering
Carnegie Mellon University
Pittsburgh, PA, 15213, USA

ABSTRACT

Incorporating style-related objectives into shape design has been centrally important to maximize product appeal. However, stylistic features such as aesthetics and semantic attributes are hard to codify even for experts. As such, algorithmic style capture and reuse have not fully benefited from automated data-driven methodologies due to the challenging nature of design describability. This paper proposes an AI-driven method to fully automate the discovery of brand-related features. Our approach introduces BIGNet, a two-tier Brand Identification Graph Neural Network (GNN) to classify and analyze scalar vector graphics (SVG). First, to tackle the scarcity of vectorized product images, this research proposes two data acquisition workflows: parametric modeling from small curve-based datasets, and vectorization from large pixel-based datasets. Secondly, this study constructs a novel hierarchical GNN architecture to learn from both SVG's curve-level and chunk-level parameters. In the first case study, BIGNet not only classifies phone brands but also captures brand-related features across multiple scales, such as the location of the lens, the height-width ratio, and the screen-frame gap, as confirmed by AI evaluation. In the second study, this paper showcases the generalizability of BIGNet learning from a vectorized car image dataset and validates the consistency and robustness of its predictions given four scenarios. The results match the difference commonly observed in luxury vs. economy brands in the automobile market. Finally, this paper also visualizes the activation maps generated from a convolutional neural network and shows BIGNet's advantage of being a more human-friendly, explainable, and explicit style-capturing agent.

Keywords: graph neural network, scalar vector graphics, explainable AI, feature recognition, design automation, deep learning, signal processing

1. INTRODUCTION

Recognizing, codifying and incorporating desired stylistic objectives into shape design has long been a focus of product development [1,2]. While market appeal is important, conveying specific aesthetic styles through design can be challenging and unpredictable due to the need for differentiation from previous designs and the time-variant nature of style [3,4].

Attempting to relate aesthetic attributes to consumer response and market success, Liu et al., [5] studied three aspects of car aesthetics impact on the market: segment prototypicality (SP), brand consistency (BC), and cross-segment mimicking (CSM). Among all, BC has shown to have the most consistent effect that positively relates to profit, indicating that BC maintenance is one of the key factors of a successful design. To maintain BC, brand feature encoding is crucial for designers to manage the brands' essence and to produce consistent and competitive designs. However, codifying and modeling BC-related features is challenging and subjective, because they are articulated primarily by humans. Because brand features are often subtle and difficult to systematically quantify, designers often have to go through a laborious process to master the brand features of a product.

Previously, research has shown the possibility to construct shape grammars – a sequential and systematic shape description system for product design [6] – to capture product brand. While previous research showed that shape grammars can describe a variety of products' brand features or semantic languages [7–14], the process of finding shape grammars was mostly achieved through human perception, which is a time-consuming and hard to transfer process. Despite the difficulty of automatic shape grammars induction, studies in this field showed the feasibility of constructing describable and quantifiable systems for brand consistency. During such a process, human designers learn the unique features shared among products in each brand. As a realization process resembles supervised learning, a natural

Copyright © 2023 by ASME

question arises: can an AI agent learn brand consistency through a fully automated process and free humans from laborious shape-to-shape comparison? This paper, therefore, models the brand recognition process as a data-driven fine-grained classification task. By examining the trained neural network classifier, humans are expected to gain brand-related feature knowledge from the AI's attention.

There are multiple challenges. First is data scarcity. As most brands only annually release several products that share similar functionality and have distinct exteriors, a class would have only on order of tens of data samples. Since deep learning models' performance hugely relies on a large number of samples to learn meaningful content [15], augmentation techniques to expand the dataset are necessary to study. Second, ensuring the interpretability of the constructed AI imposes great challenges. Since AlexNet [16], convolutional neural networks (CNN) have seen a dramatic accuracy improvement on classifying pixel images. However, extracting key primitives in a parametric fashion from such CNN remains difficult. While class activation mapping (CAM) [17] techniques attempted to interpret CNN pictorially, the resulting heat maps are often fuzzy and lack of precision. Since humans tend to learn, reason, and design based on curves and shapes, this type of architecture and workflow may not effectively capture intuitive brand features. Much less can it be expected to quantify or edit these features using this approach. This paper proposes Brand Identification Graph Neural Network (BIGNet), a curve-based AI that can capture and visualize explicit features. Using the proposed approach, humans can utilize AI as a communicative and explainable style discovery agent and accelerate the design process.

This research's main contributions are:

1. Two vectorized data acquisition approaches for style recognition: parametric modeling from small curve-based datasets, and vectorization from large pixel-based datasets.
2. BIGNet, a novel hierarchical Graph neural network (GNN) that can learn from both scalar vector graphics' (SVG) curve-level and chunk-level features.
3. Evaluation study to produce design insight and feature visualization, which shows BIGNet's capability of perceiving explicit and explainable brand-related features.

2. RELATED WORK

This research aims to accelerate product style design, based on deep learning. This section will first review how previous research attempted to construct systematic approaches to convey stylized ideation. Second, this section will review the progress and limitations in fine-grained image classification and curve-based deep learning methods.

Shape Grammars. Shape grammars have been used as a computational tool for explicit feature representation and generation for over five decades. A shape grammar consists of a set of shape rules that sequentially eliminate, edit, or generate design primitives [18]. Because of a shape grammar's explicit expression to describe stylized concepts, it later became a feasible method for product designers to capture brand-related features of exterior design [5], which is a crucial subset of

branding [19]. Agarwal and Cagan [7] brought shape grammars into industrial products, and they used it to successfully describe coffee machines' shape generation rules and find brands' discriminative features. It was further shown that shape grammars could describe a variety of products' brand features or semantic languages [8–14]. However, the process of defining shape grammars was mostly done by human perception, which is still a time-consuming process and difficult to transfer from one product to another. This research takes a different approach on identifying differentiating geometric features as a means to automate the feature perception task in a data-driven method, thus accelerating the design cycle.

Fine-grained classification and attention visualization. Detecting style-oriented features for better object recognition accuracy has been a challenge. However. explicit detectors [20,21] and descriptor design [22–24] paved the way for deep CNNs [25] to reach high accuracy by learning complex and transformation-invariant features [16,26]. To generalize CNN's application to a variety of tasks, finetuning on pre-trained networks by only training one fully connected layer from scratch was studied and found to result in much faster convergence and better accuracy [27–29]. These advancements, however, couldn't promise CNN to have full interpretability of the describable features. While it showed possibility to visualize localized regions on fine-grained classification [17,30–33], because CNN is learning from pixelated information, it is still unclear which shapes or curves are important, therefore diminishing the usage for designers. Therefore, the attempt to train a deep learning model on curve-based image representation is proposed in this work to visualize human-readable features.

Curve-based recognition methods. As representing images in curves is much closer to how humans see an image, research on curve-based recognition models focuses on building AI that can learn descriptive features from human sketches for classification. In the field of sketch recognition, early studies [34–36] use Support Vector Machine (SVM) as the classifier to differentiate rasterized sketch images. To visualize stroke importance, Schneider and Tuytelaars [35] also adopted a leave-one-feature-out (LOFO) technique to remove one stroke at a time and see how it affects the classification score. More recent studies [37–39] took advantage of CNN and achieved better performance and robustness on the recognition task. To recognize curve-based images with multiple abstract levels of features, Yu et al., [37] proposed a multi-scale, multi-channel CNN architecture to learn from partial images segmented from stroke order. However, it still has not incorporated grouping information of curve-based images, which is useful for learning more descriptive features [40]. To address this limitation, Li et al., [38] then proposed sketch-R2CNN to classify sequentially rendered images and paired with a recurrent neural network (RNN) attention mechanism that enabled better accuracy and feature visualization. Although sketch recognition has made significant progress, most studies focused on recognizing simple human sketches comprising only a few tens of strokes. These strokes are typically drawn with straight lines rather than curves, and the sketches are often low resolution, which is a simple

Copyright © 2023 by ASME

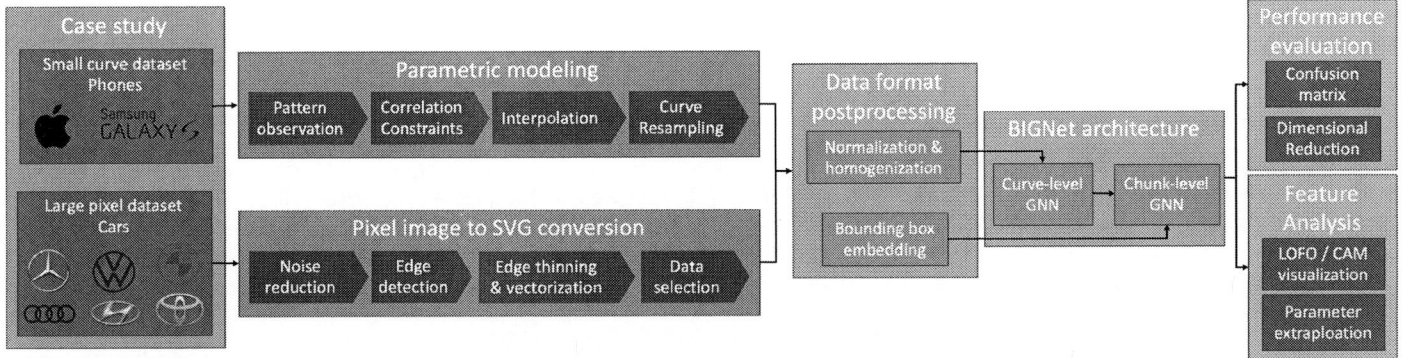

Figure 1: Workflow of this research, illustrating the key stages of product, brand, and model selection (red), data acquisition (blue), neural network's message passing design (green), and AI analysis (yellow). The logos of phones and cars are shown to represent the brands selected for identification, but the classification task is based on product shapes rather than logos. Two case studies demonstrate the framework's adaptability to different product domains and scales on the design spectrum.

representation and limits the complexity that can be achieved with the task. As scaling the recognition onto industrial products like cars can easily have thousands of curves, previous studies would not be applicable due to expensive rendering computation, and vanishing gradient problem for RNN-based architectures. Furthermore, industrial products typically consist of dozens to hundreds of groups of curves (chunks) that may contain higher-level brand-related features. There has been little research on analyzing the inter-chunk relationships at a scale of 10,000 or more. The above studies conclude that since CNN is restricted to work on pixelated images, it is not possible to analyze SVGs without first rasterizing, which results in a sparse image and loses grouping information.

Spatial Graph Neural network. GNN refers to the domain of deep learning methods designed to deduce information from general non-Euclidean graphs. A graph $\mathcal{G}$ is defined as $\mathcal{G}(\mathcal{V}, \mathcal{E})$ where $\mathcal{V}$ is the set of vertices or nodes and $\mathcal{E}$ is the set of edges [41]. Among the branches, spatial-based convolution GNNs (convGNN) are found to be most analogous to conventional CNNs, while allowing the nodes to have an arbitrary number of neighbors and offering flexibility on connectivity strength as well as the aggregation process. Neural Network for Graphs (NN4G) [42] was the first work towards spatial ConvGNNs that performed aggregation by summing up each node's neighborhood information directly. After that, multiple architecture improvements were proposed around flexible aggregation [43–45] and sampling [46] strategies. Among all, Graph Attention Networks (GAT) [45] adopted attention mechanisms to learn the relative energy (weights) between two connected nodes. By enabling specification of different weights to different nodes, it has achieved state-of-the-art prediction results in Cora, Citeseer, and Pubmed benchmarks. From an engineering perspective [47], GNN has shown its capability of tackling various problems including physical modeling [48], chemical reaction prediction [49], traffic state prediction [50], and engineering drawings' segmentation [51,52]. Inspired by the recent successful applications, this research models each SVG as a two-tier graph and builds a spatial GNN with learnable chunk-level attention mechanisms to perform graph-level classification.

3. METHODOLOGY

To enhance designers' ability to edit parametric curves in real time and evaluate the impact on brand consistency, the goal of this research is to build a curve-based AI-driven feature retrieval surrogate that is both explainable and describable. As brand consistency is shown to be important yet abstract for humans to easily identify, the case studies of the proposed methodology are applied to industrial products' brand recognition. The research workflow is shown in Figure 1.

3.1 Data representation and acquisition

This research focuses on the front view of product models because just as human beings are more recognizable by their faces, designers tend to place the most recognizable features in products' front view as well [53]. SVG format is chosen to represent the objective products, as it composes an image of chunks of geometry defined explicitly by parametric control points. To maintain data homogeneity, all curves are converted to cubic Bezier curves, with each image represented by an arbitrary number of chunks of curves, while each curve is parameterized by four control points (eight scalar values).

Facing the scarcity of product data, this research first synthesizes intermediate designs as a data augmentation method. This is achieved by creating and interpolating unified design rules for each brand of products through human observation. For the simplicity of geometry and planar design, a case study is run to identify mobile phones' exterior features. Second, this research attempts to generate SVGs from vectorizing a generic pixel dataset. As images are taken from the same car at slightly different post angles, this perspective difference contributes to part of the data augmentation. This is implemented with an image processing pipeline including background removal, noise reduction, edge detection and vectorization. A case study on cars' recognition is run for the second approach.

3.2 AI architecture

To learn the discrepancy among curve-based images in terms of brand-related styles, this paper proposes Brand Identification Graph Neural Network (BIGNet), a two-tier spatial GNN that can learn from the SVG format dataset (Figure

Copyright © 2023 by ASME

2). In the first layer, a chunk of curves is represented as a graph, while each node is a curve, and connectivity is determined by its neighbor curves. More precisely, the model first samples and aggregates the neighborhood of each node, feeds each node into fully connected layers (FCs), and then reads out the response by average pooling. In the second layer, an SVG picture is represented as a graph, while each node is a chunk of curves, and connectivity strength is determined via the weights learned from the bounding box parameters. After aggregation, the hidden layers are concatenated and passed to the last fully connected layer to get the prediction. BIGNet's forward propagation is summarized in Algorithm 1. The parameters used in the two case studies are slightly different to adjust for the images' complexity level and are listed in Table 1.

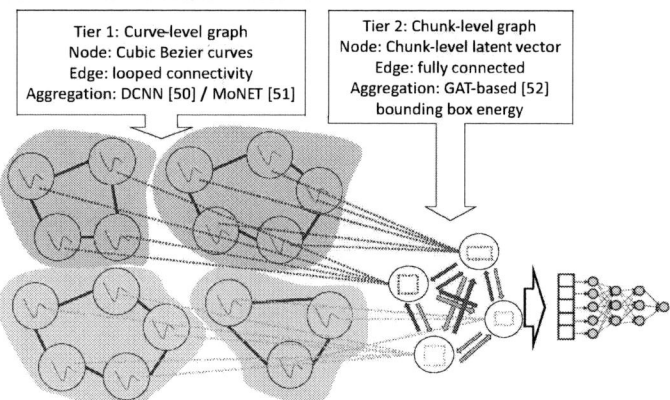

Figure 2: Schematic diagram of the two-tier BIGNet structure.

3.3 AI evaluation overview

After successfully training the network, a series of evaluation criteria is studied to deduce explainable and quantifiable results.

The evaluation of the synthetic phone dataset aims to determine whether BIGNet can accurately perceive intricate details, and whether these details can also be transferred to humans. As the dataset is synthesized using shape rules created by human through observation from a reference source dataset, the accuracy of the reference source dataset is first examined to verify the manually constructed shape rules' validity. After that, a dimensional reduction is performed on the networks' latent vector to see if the brands are well separated. To further investigate the network's attention, an ablation study using leave-one-feature-out (LOFO) is performed both at the curve and chunk level. This test removes a chunk or a curve from the picture at a time. Those curves that result in a prediction performance drop when removed are considered to be important features. Finally, based on the localized features observed from the ablation study, parameter extrapolation (Partial Dependence Plot [54]) is implemented on the original shape rules to visualize the confidence change. This step further checks the importance of highlighted curves to the discrimination task to understand if the products' brand features are successfully extracted by the AI.

The evaluation of BIGNet on the vectorized car dataset assesses its ability to recognize brand-related styles in complex, automatically generated data. This includes testing the model's

---

Algorithm 1: BIGNet forward propagation algorithm

Input: SVG graph $\mathcal{G}(\mathcal{V}_1, \mathcal{E}_1)$; graph-level FCs f_1; aggregated chunk-level FCs f_2; primitive chunk-level FCs f_3; curve-level FCs f_4; chunk attention matrix FC f^*; chunk-level depth D_1; curve-level depth D_2; chunk attention matrix W_1; curve diffusion weight W_2; chunk aggregating function A_1; curve aggregating function A_2; chunk graphs $\{\mathcal{G}(\mathcal{V}_2(v_1), \mathcal{E}_2(v_1)), \forall v_1 \in \mathcal{V}_1\}$; curve features $\{\{x_{v_2}, \forall v_2 \in \mathcal{V}_2(v_1)\}, \forall v_1 \in \mathcal{V}_1\}$

Output: Vector representation $z_{\mathcal{G}}$

for $v_1 \in \mathcal{V}_1$ do
    for $v_2 \in \mathcal{V}_2(v_1)$ do
        $h_{v_2}^0 \leftarrow x_{v_2}$
        $e_{v_2}^0 \leftarrow x_{v_2}$
        for $d_2 = 1 \ldots D_2$ do
            $e_{v_2}^{d_2} \leftarrow A_2(e_{v_2}^{d_2-1}, \mathcal{E}_2(v_2), W_2)$
            $h_{v_2}^{d_2} \leftarrow concat(h_{v_2}^{d_2-1}, e_{v_2}^{d_2})$
        end
    end
    $H_{v_1}^* \leftarrow pool(\{f_4(h_{v_2}^{D_2}), \forall v_2 \in \mathcal{V}_2(v_1)\})$
    $H_{v_1}^0 \leftarrow f_3(H_{v_1}^*)$
    $E_{v_1}^0 \leftarrow H_{v_1}^0$
    for $d_1 = 1 \ldots D_1$ do
        $E_{v_1}^{d_1} \leftarrow A_1(E_{v_1}^{d_1-1}, \mathcal{E}_1, f^*(W_1))$
        $H_{v_1}^{d_1} \leftarrow concat(H_{v_1}^{d_1-1}, E_{v_1}^{d_1})$
    end
end
$H_{\mathcal{G}}^* \leftarrow pool(\{f_2(H_{v_1}^{d_1}), \forall v_1 \in \mathcal{V}_1\})$
$z_{\mathcal{G}} \leftarrow f_1(H_{\mathcal{G}}^*)$

Table 1: BIGNet's parameters for the two case studies.

Case	Phone	Cars
Activation	LeakyReLU	
Optimizer	Adam	
Pooling	Average pooling	
Loss	Binary cross entropy	Categorical cross entropy
$\mathcal{E}_2$	Loop graph (each node has 2 neighbors)	
D_2	Bidirectional, depth: 2	Bidirectional, depth: 2
A_2	$e_{v_2}^{d_2} = (1-W_2)e_{v_2}^{d_2-1} + W_2 \times \mathcal{E}_2 e_{v_2}^{d_2-1}$	$e_{v_2}^{d_2} = (1-W_2)e_{v_2}^{d_2-1} + W_2 \times Linear(8 \to 8)(\mathcal{E}_2 e_{v_2}^{d_2-1})$
W_2	1	0.5
f_4	$Linear(32 \to 24 \to 12)$	$Linear(24 \to 32 \to 24)$
f_3	pass	$Linear(24 \to 24 \to 24)$
$\mathcal{E}_1$	Fully connected	
D_1	2	2
W_1	$N \times 5$	$N^2 \times 5$
f^*	$Linear(5 \to 12)$	$Linear(5 \to 24)$
A_1	$E_{v_1}^{d_1} = f^*(W_1) \times E_{v_1}^{d_1-1}$	
f_2	pass	$Linear(72 \to 24 \to 24 \to 24)$
f_1	$Linear(36 \to 18 \to 8 \to 2)$	$Linear(24 \to 18 \to 12 \to brand\#)$
Learnable parameters	$2000 \sim 2076$	$6716 \sim 6812$

robustness and consistency across different tasks and scenarios. Firstly, Confusion matrix and dimension-reduced latent vectors plots are calculated to learn the brands' differences in distinguishability. Secondly, AI's chunk-level attention is visualized using a class activation mapping (CAM) [17]– inspired algorithm. By highlighting the chunk that contributes more to correct identification, CAM is shown to have much more robustness than LOFO on SVGs with higher order of chunks. Finally, to compare the curve-based approach to the pixel-based approach, a CNN is finetuned using ResNet-50 that was pre-trained with simCLR [28]. The attention of this CNN is then visualized using Grad-CAM [33]. As CNN is expected to reach a better accuracy level due to its quantitively much larger model size, this study focuses on examining whether BIGNet conveys more explicit and describable design features than CNN's class attention visualization.

4. CASE STUDY - PHONES
4.1 Brand and model selection
This study compares and differentiates the front views of the products of the two most popular cellphone brands – Apple and Samsung [55,56]. Among the many lines of Samsung phones, the Samsung Galaxy S series has the most similar functionality and price range as Apple's iPhone and therefore is chosen to be the competitor of Apple's iPhone. To preserve a reasonable degree of homogeneity, all the phone models chosen are without home buttons, which are also more contemporary designs.

4.2 Parametric modeling from small curve datasets
4.2.1 Synthetic Dataset Generation
Challenges exist in finding an abundant and well-measured dataset. After realizing the need of increasing the sample size of an existing dataset collected from Dimensions.com (shown in Table 2a), this study then observes the patterns for the selected models, and by using parameter interpolation on the manually established shape rules (number and types of parameters are listed in Table 2b, and shape rules example of Apple is shown in Figure 3a), a synthetic SVG dataset with 20,000 synthetic phones, 10,000 for both Apple and Samsung is successfully created (some results are shown in Figure 3b).

4.2.1 Synthetic image preprocessing
After generating the synthetic SVG dataset, the next important step is to adapt it to a more homogenous format, so that the AI can learn the difference between brands' shapes instead of the difference between the brands' creative processes. Therefore, this study then rasterizes the images and vectorizes each into a cubic Bezier SVG using Potrace [57]. All the phones' heights are then normalized to 1 since the synthetic dataset has relatively larger Samsung phones than Apple phones. This will enable the AI to learn meaningful design languages.

4.3 Results and discussion of phone case study
4.3.1 Model's training process and performance
Using BIGNet's architecture and parameters from Table 1 column 1, after 105 epochs, the model was able to reach over

Table 2a: The phone models of an existing dataset.

Apple iPhone	Submodel Names	Total Number of Submodels
X	XR, X, XS, XS Max	
11	11, 11 Pro, 11 Pro Max	15
12	12 mini, 12, 12 Pro, 12 Pro Max	
13	13 mini, 13, 13 Pro, 13 Pro Max	

Samsung Galaxy S	Submodel Names	Total Number of Submodels
S10	S10e, S10, S10+, S10 5G	7
S20	S20, S20+, S20 Ultra	

Table 2b: Number and types of parameters used in shape rules to make the synthetic SVG dataset.

Shape Rules Parameters	Continuous (ex: height, width, fillet)	Discrete (ex: lens position)	Regulation (ex: height-width ratio)
Apple	28	5	6
Samsung	25	1	12

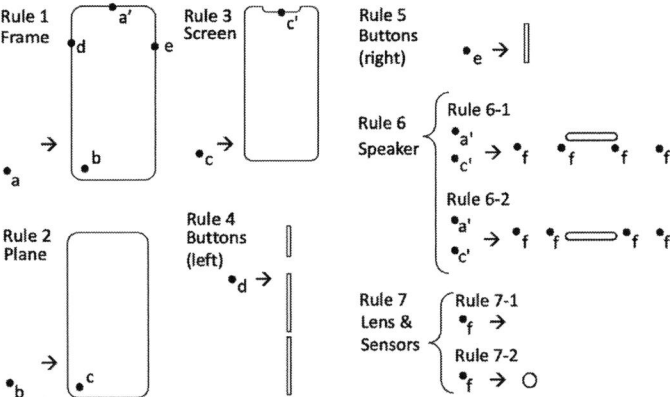

Figure 3a: Shape rules used to generate Apple phones. By sequentially applying rule 1 to rule 7 with parameters of Apple in table 2b, a synthetic Apple phone is made.

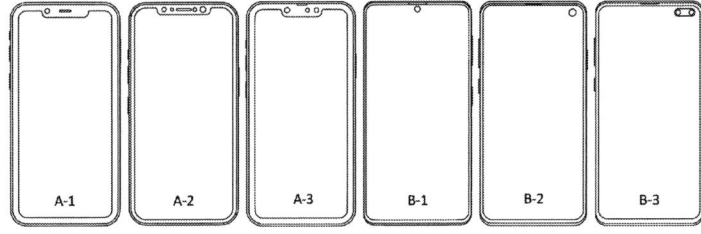

Figure 3b: Synthetic examples of Apple (A-1~A-3) and Samsung (B-1~B-3). For Apple, there are two distinct types of speaker position: at the middle (A-1, A-2) or at the top (A-3) of the notch. The number of circles representing lens and sensors range from zero to four. For Samsung, there are three distinct types of lens: one at the middle (B-1), one at the upper right corner (B-2) and two at the upper right corner (B-3).

Copyright © 2023 by ASME

99.8% for both training and testing accuracy (Figure 4) and can 100% predict the reference sources' brands as well (Table 3a). The small difference between train and test accuracy could be attributed to the fact that both sets are generated using identical interpolated shape rules. Moreover, the high accuracy result is likely due to the same reason which gives the two classes significant brand consistency, and therefore makes the problem highly manageable for BIGNet. Table 3b shows the confusion matrices on the three datasets.

4.3.2 Dimensional reduction

By visualizing the last hidden layer with t-distributed Stochastic Neighbor Embedding (t-SNE) and Principal Component Analysis (PCA) in Figure 5, Apple phones (red) and Samsung phones (blue) are clustered and well-separated from each other, demonstrating that the network can very clearly discriminate the two phone brands.

4.3.3 LOFO visualization study

To visualize localized features, important chunks are colored red, and curves are colored blue for each picture (some results are shown in Figure 6). Discriminative features of the two brands are then observed and summarized in Table 4. Since both brands highlight lens, fillet, width and the gaps between screen and the frame, the study then examines partial dependence plots created by parameter extrapolation on these features in the following section to further understand the network's attention.

4.3.4 Partial dependence plot

(1) Lens horizontal position

The goal of this experiment is to see if prediction confidence drops while extrapolating Apple's lens horizontal position. Since the trained BIGNet is robust by looking at multiple features, the result in Figure 7 is plotted when feature i-2, i-4 and i-6 are all shifted to Samsung's dimensions range. The confidence curve drops when the lens is at the middle and to the right of the notch, which is because those are both Samsung's possible lens locations (see b-1, b-2 in Figure 6).

(2) Fillet Radius and height-width ratio

Similar results from (1) are found while extrapolating Apple's width. The interesting thing is although Samsung is shorter in dimension of both width and normalized width, the model considers wider phones as Samsung. This is the result from the model looking at the length of the frame's segment instead of the whole width, which also takes fillet radius into consideration (Table 5). In Figure 8, this explanation is verified since the crossover width lays between the boundary of the two brands' normalized segment length ranges.

(3) Screen-frame gaps

Since both brands also highlight the gaps among screen, frames' inner width (plane) and outer width (edge) in 4.3.3, a 2-D extrapolation experiment on the two gap parameters is done on Samsung's shape rules. In Figure 9, results show that phones with smaller gaps between the screen and the frame are more likely to be predicted as Samsung, which matches the interpolation range of the two brands.

Table 3a: Trained model's loss and accuracy on the 3 datasets.

Dataset	number of samples	loss	accuracy
train	18000	0.0053	99.83%
test	2000	0.0038	99.85%
reference	28	0.0046	100%

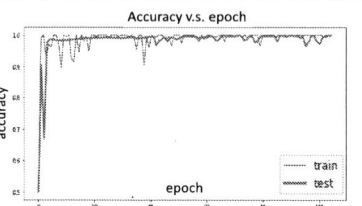

Figure 4: Accuracy converges to 98% within 20 epochs.

Table 3b: Confusion matrices (not normalized)

Train Set Confusion Matrix		
prediction / truth	Apple	Samsung
Apple	9000	0
samsung	30	8970

Test Set Confusion Matrix		
prediction / truth	Apple	Samsung
Apple	999	1
samsung	2	998

Reference Set Confusion Matrix		
prediction / truth	Apple	Samsung
Apple	15	0
samsung	0	13

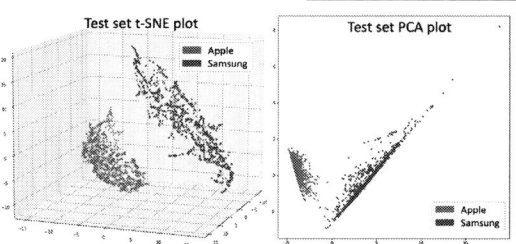

Figure 5: t-SNE and PCA plots from test set's latent vectors.

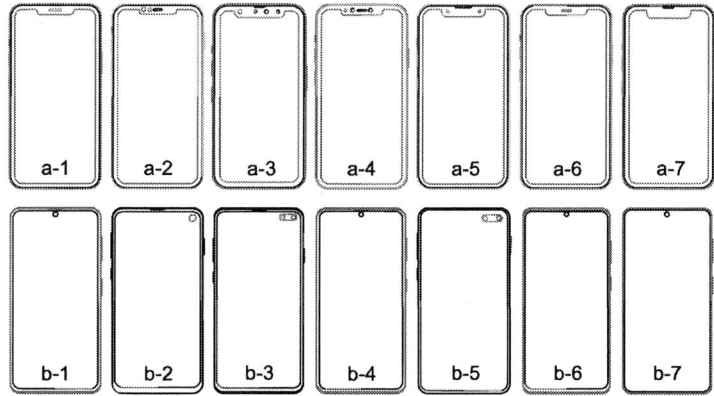

Figure 6: LOFO results from Apple (a-1~7) and Samsung (b-1~7).

Table 4: A summary of the model's attention by observing the LOFO visualization results. One thing to notice is i-1~4 are similar parameters to s-1~4, therefore this study continues to experiment on their parameter extrapolations.

Brand	index	Observed Features	Figure
Apple	i-1	lens	a-2, a-3, a-4, a-5
	i-2	corner's fillet	a-1, a-3, a-5, a-7
	i-3	width	a-1, a-2, a-3, a-4, a-6, a-7
	i-4	screen-frame gap	a-1, a-2, a-3, a-4, a-5, a-6, a-7
	i-5	speaker when at the middle	a-1, a-2, a-4, a-6
	i-6	notch related features	a-1, a-2, a-4, a-5, a-6, a-7
	i-7	mute button	a-1, a-2, a-4, a-5, a-6
Samsung	s-1	lens when at the right corner	b-2, b-3, b-5
	s-2	corner's fillet	b-1, b-2, b-3, b-5, b-7
	s-3	width	b-1, b-2, b-3, b-4, b-5, b-6, b-7
	s-4	screen-frame gap	b-1, b-2, b-4, b-6, b-7

Copyright © 2023 by ASME

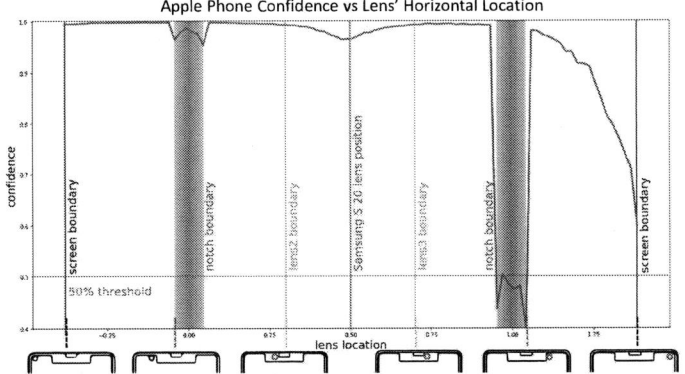

Figure 7: Confidence change while extrapolating Apple's lens horizontal position.

Table 5: Although Samsung has a shorter normalized width, it also has a relatively smaller fillet radius, therefore its segment length that the model perceives is longer (last column).

Dimension / Brand	Width (mm)	Fillet (mm)	Height (mm)	Normalized width	Normalized segment length
Apple	71.15	10.75	146.15	0.49	0.34
Samsung	73.1	7.415	154.55	0.47	0.38

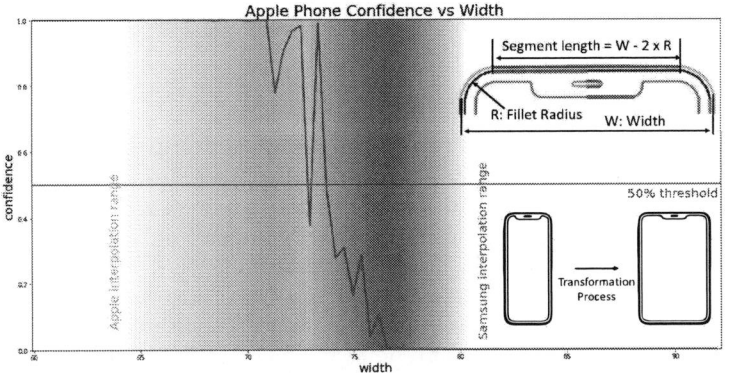

Figure 8: Confidence change while extrapolating Apple phone's width. The orange and red regions represent the range of Apple and Samsung's normalized segment length.

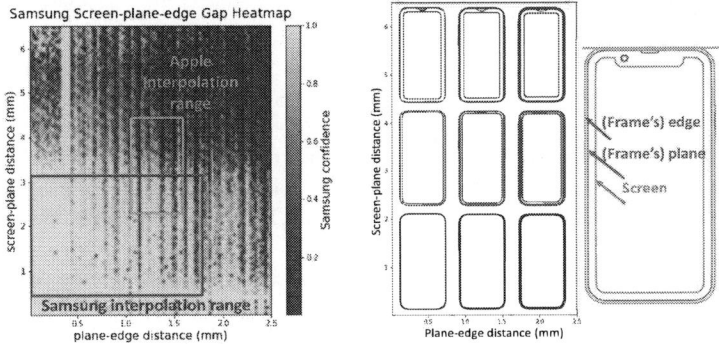

Figure 9: Since Samsung has shorter distances between both screen to plane and plane to edge, the heatmap shows a greater prediction confidence at the lower left corner, meaning this is also a discriminative feature to the model.

5. CASE STUDY - CARS

5.1 Background

While the phone case study has shown the viability of GNN learning from curve-based representation, laborious work has to be done on observing and parameterizing to synthesize an augmented dataset. It is doable on two phone brands promptly, but as unified, interpolatable parametric expressions have to be established for every studied model in every brand; generalizing without human attention is challenging. In addition, for products like cars with more complex shapes and more variety of models, although it will be difficult to construct unified shape grammars, there exist pixel datasets that have thousands of images for each brand. If SVGs can be acquired from such resources, BIGNet can be applied to learn from these large datasets with complex product geometries, and the workflow of extracting brand-related features can be even further automated.

This case study, therefore, aims to explore the feasibility of converting pixel images to curve images to create a data-driven, hands-free recognition system. Since cars have distinctive functionality and design criteria that differ significantly from those of phones, this study not only demonstrates the potential of fully automated SVG retrieval but also attempts to showcase the adaptability and generalizability of BIGNet across different product domains and design scales. Therefore, the following distinct yet comparable training scenarios are run to examine the model's flexibility, robustness, consistency and explainability:

Classifying different number of brands. All vectorized images in this case study are generated from the same automated pipeline without the need of parametric modeling. As a result, expanding the number of models and brands for a more comprehensive style classification is made possible with little human effort. Yet, one of the counters of deep learning methods is their lack of reproducibility. This is caused by having redundant freedom of parameters that would lead to suboptimal convergence. However, a style perception agent is expected to consistently exhibit the same features regardless of the training scheme, or data processing nuances, to enable designers ability to reason and make decisions from its inference. To investigate the generalizability of BIGNet and showcase the ease of dataset regeneration, this study conducts both six- and ten-brand classifications. Although a decrease in overall confidence when moving from six to ten brands is expected, it is also anticipated that BIGNet will still exhibit similar patterns in terms of which brands are easily identifiable and which are not.

Logo removal. While logos aid in brand recognition, the geometry of logo is not necessarily the brand-related features design engineers are attempting to extract. Therefore, identifying logos as part of the learned brand features is not the primary objective, as this may lead to overfitting on the logo and hinder the attention given to other important design features. To assess the effect of logos on brand classification, separate models are trained on cars with logos and without logos.

Comparison to CNN trained on pixel images. BIGNet trained on SVGs is claimed to offer more explainability, but CNN trained on pixel images is widely used and offers high identification accuracy. To compare the two approaches, this

Copyright © 2023 by ASME

study finetunes a simCLR-pretrained ResNet-50 using the exact train-test split of pixel images before vectorization. Since ResNet-50 has 23 million learnable parameters and pixel images contain richer information, CNN is expected to have better accuracy than BIGNet. Despite that, as the goal of this research is to extract explicit and usable attention features, this comparison will focus on comparing feature visualization mapping and validate whether BIGNet provides more explicit and parametric results.

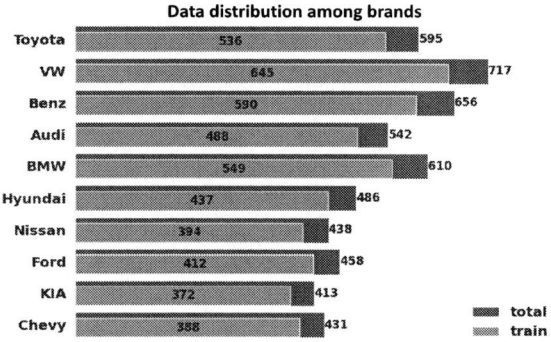

Figure 10: Data distribution among brands shows a nonnegligible imbalance. The largest class Volkswagen (VW) has 73.6% more data than the smallest class KIA. Additionally, with only several hundreds of samples per brand, this study performs a 9:1 stratified train-test split.

Table 6: Data distribution among segments and years. Among all brands, major segments are sedan, SUV, and hatchback, while most images come from 2009-2015. This demonstrates not only the balanced segment and year ratio in each brand, but also validates the concurrent brand competency during 2009-2015.

brand / segment	Toyota	VW	Benz	Audi	BMW	Hyundai	Nissan	Ford	KIA	Chevy	total	ratio
sedan	114	190	119	127	231	171	72	76	153	94	1347	38%
SUV	108	23	151	59	117	87	57	24	35	32	693	20%
hatchback	74	80	27	29	28	40	50	88	63	74	553	16%
MPV	34	55	50	0	0	10	9	35	65	0	258	7%
others	21	166	110	117	97	0	60	38	13	36	658	19%
labeled %	59%	72%	70%	61%	78%	63%	57%	57%	63%	55%	64%	

brand / year	Toyota	VW	Benz	Audi	BMW	Hyundai	Nissan	Ford	KIA	Chevy	total	ratio
<2008	22	21	8	6	8	29	28	8	11	0	141	3%
2008	6	27	6	4	31	22	36	15	7	7	176	3%
2009	53	72	34	36	50	61	17	40	25	26	414	8%
2010	58	64	81	102	61	37	51	16	31	70	571	11%
2011	61	105	75	33	108	78	54	47	67	60	688	13%
2012	111	122	80	74	63	44	61	83	64	48	750	14%
2013	110	112	116	129	123	81	91	151	56	86	1055	20%
2014	127	121	155	107	129	78	87	31	111	100	1046	20%
2015	39	72	96	46	31	51	21	42	33	32	463	9%
>2015	4	1	2	5	6	5	1	3	0	2	29	1%
labeled %	99.3%	100.0%	99.5%	100.0%	100.0%	100.0%	98.9%	99.8%	100.0%	100.0%	99.8%	

5.2 Data selection and preprocessing

HK Comp cars [58], the largest brand- and orientation-labeled car dataset, is chosen to be the source of pixel images. To ensure a fair representation of the market, this research selects the top six and top ten most abundant car brands as the dataset, including both luxurious and affordable brands with an international distribution in Asia, Europe, and America. While the resulting selection is expected to be non-biased and diverse, the total number of images per brand is relatively small, with

only a few hundred images per brand. Therefore, to provide enough training data for BIGNet, a train-test ratio of 9:1 is split. Nonetheless, the dataset has an imbalanced number of images among brands (Figure 10). To address this issue, two steps are taken to ensure no falsely high accuracy will occur due to bias towards brands with more data. First, a stratified split is performed to preserve the same ratio in both sets. Second, the algorithm randomly oversamples minority classes during training to ensure each brand has equal representation. More statistics on data distribution, including year and car segments, are provided in Table 6 to ensure the variance of each brand.

To achieve successful vectorization, an image preprocessing pipeline is proposed. First of all, background noise is removed by applying detectron2 [59], a maskRCNN-based AI to detect and apply the mask on the original image. Second, Google cloud vision API is used to detect and remove the logo for a comparison dataset. Edge detection is then implemented before curve-fitting. Through comparing state-of-the-art edge detection methods, the transformer-based EDTER [60] is found to achieve the best performance at preserving object curves and eliminating reflections on the cars' glossy surfaces. After applying edge thinning on EDTER's response, Potrace [57] is used to vectorize the edges into SVG. To maintain a reasonable degree of homogeneity, all vectorized results are converted to cubic Bezier curves. At last, each SVG has its height normalized to 1, and the bounding box information of center coordinate, width, height, and area are pre-computed to enhance chunk-level aggregation.

5.3 AI modification for increased data complexity

Since BIGNet is trained with the same type of image format, the message passing flow in this case study shares a lot of common blocks in the phone study's architecture. However, cars' exterior shape has many more components than phones, and the vectorization from generic images also unavoidably leads to redundant curves. The two factors result in around 20 times more chunks and 10 times more curves than phones. To cope with the data's increased scale of complexity, BIGNet architecture is modified from the phone case study's parameters (Table 1 column 2), and can be summarized into three aspects:

(1) *Increase the size of GNN.* First, a more flexible curve-level aggregation policy with one fully connected layer (FC) is enhanced. Second, chunk-level FCs are added for better digestion both before and after the chunk-level aggregation. Lastly, the widths of most FCs are doubled to allow GNN's bandwidth of carrying more features per node. This is a reasonable modification because the geometry of car parts is much more complicated than that of a phone's, with a lot more organic shapes. Overall, number of learnable parameters is increased from 2000 to 6716.

(2) *Better use of bounding box information.* One of BIGNet's key components is the chunk-level connectivity strength learned from bounding box attention. In the phone study, as all data share the common largest chunk being the phone's outer frame, the connectivity strength matrix is derived from the bounding box relationship normalized by the maximum bounding box. Such homogeneity doesn't exist on vectorized

Copyright © 2023 by ASME

SVGs of cars. Furthermore, inter-chunk relationship, which is the square of number of chunks, becomes roughly 100 times larger than synthetic phones, and has a much larger variance. All these factors impose great challenges to the previous parameters. To tackle the increased complexity, the ratio values in the correlation matrix, namely area, width and height, are first normalized to between 0 and 1 by taking logarithmic values. After that, an FC is applied to adapt the shape to desired features. The pseudo-code is showed in Algorithm 2.

(3) *Augmentation*. First, horizontal flip is done because cars' front views are symmetric. Second, another augmentation is applied by running two distinct EDTER models to retrieve slightly different edge detection response. The two combined techniques enlarge the dataset four times. Lastly, in this dataset, multiple images are often taken on the exact make and model at slightly different perspectives of front view. As humans can identify car parts from slightly off perspectives, this research also treats this as a natural perspective augmentation. It is expected to prevent GNN from overfitting and therefore achieve a more robust model.

Algorithm 2: chunk level aggregation in car case study

Input: graph with N nodes $\mathcal{G}(\mathcal{V} = \{v_1 \dots v_N\}, \mathcal{E})$;
Node features: $E_{N \times m} = \{e_{1 \times m}(v_1) \dots e_{1 \times m}(v_N)\}$;
Nodes' bounding box features (horizontal location, vertical location, width, height, area):
$\beta_{1 \times 5}(v) = \{x(v), y(v), w(v), h(v), a(v)\}, \forall v \in \mathcal{V}_N$;
Linear layer $f^* = f^*(5 \to m)$

Output: Node features after one aggregation:
$E'_{N \times m} = \{e'_{1 \times m}(v_1) \dots e'_{1 \times m}(v_N)\}$

Init $B_{primitve} = \mathbf{0}_{N \times N \times 5}$
Init $B_{adapted} = \mathbf{0}_{N \times N \times m}$
Init $E'_{N \times m} = \{e'_{1 \times m}(v_1) \dots e'_{1 \times m}(v_N)\} = \mathbf{0}_{N \times m}$
Init $\mathbf{E}' = \{e'_{N \times m}(v_1) \dots e'_{N \times m}(v_N)\} = \mathbf{0}_{N \times N \times m}$
for i in 1…N **do**
    for j in 1…N **do**
        $B_{primitve}[i, j, 0] \leftarrow x(v_i) - x(v_j)$
        $B_{primitve}[i, j, 1] \leftarrow y(v_i) - y(v_j)$
        $B_{primitve}[i, j, 2] \leftarrow \log\left(\frac{w(v_i)}{w(v_j)}\right)$
        $B_{primitve}[i, j, 3] \leftarrow \log\left(\frac{h(v_i)}{h(v_j)}\right)$
        $B_{primitve}[i, j, 4] \leftarrow \log\left(\frac{a(v_i)}{a(v_j)}\right)$
        $B_{adapted}[i, j, :] \leftarrow f^*(B_{primitve}[i, j])$
    end
    $e_{N \times m}(v_i) \leftarrow stack\left(e_{1 \times m}(v_i)\right) N \ times$
    $e'_{N \times m}(v_i) \leftarrow B_{adapted}[i, :, :] \cdot e_{N \times m}(v_i)$
    $e'_{1 \times m}(v_i) \leftarrow pool(e'_{N \times m}(v_i), axis = 0)$
end

5.4 Results and discussion of car case study

This section will first focus on the results of BIGNet's six-brand classification without logo removal, and then have an extensive comparison with other training scenarios.

5.4.1 Training

During training, batch size is chosen to be 100, and learning rate is initialized to be 0.001. After ~27000 iterations, the learning rate is decreased to 0.0001 for fine-tuning since both train and test accuracy are stagnated (Figure 11). As the end of the model starts overfitting, maximized test accuracy, which is at the 723rd epoch, is selected for evaluation. It is able to reach 89.3% training accuracy and 80.6% test accuracy.

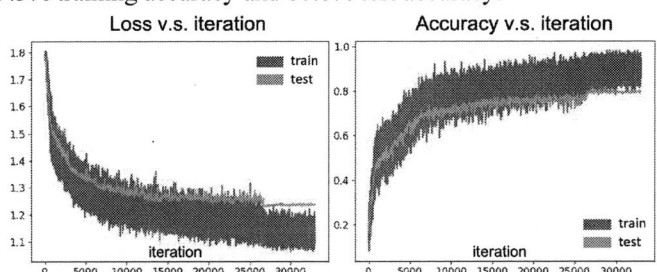

Figure 11: Accuracy and loss during the training process.

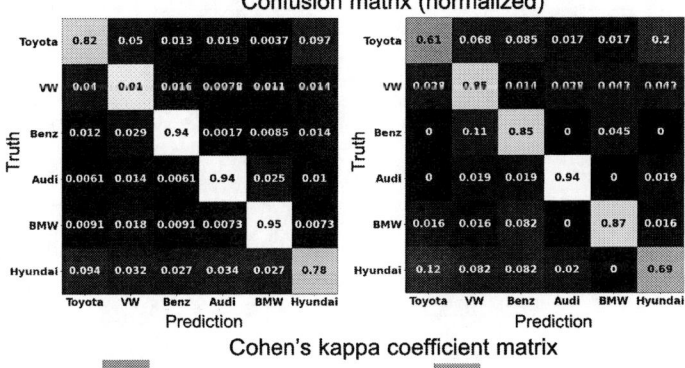

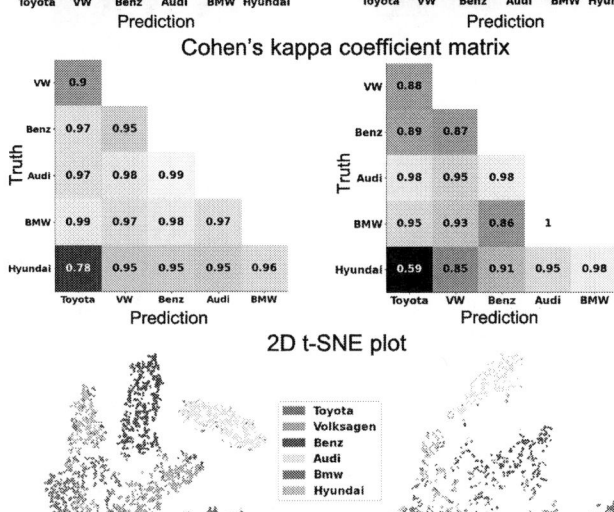

Figure 12: BIGNet evaluation on train set (column 1) and test set (column 2) in terms of: Confusion matrix (row 1), Cohen's kappa coefficient matrix (row 2) and 2D t-SNE plot of data's latent vectors (row 3). Both train and test sets show Benz, Audi and BMW having better recognition rate, and that Toyota and Hyundai get confused with each other more often.

Copyright © 2023 by ASME

5.4.2 Performance evaluation

First, confusion matrices are examined. In Figure 12, both train and test sets show a consistent trend of being able to predict Audi, BMW and Benz most correctly. The result also shows that Toyota and Hyundai, while having the least accuracy among the six, also have a relatively higher chance to confuse with each other. Cohen's kappa coefficient matrix is then calculated to validate this finding, as lower values between 0 and 1 indicate more inconsistency between ground truth and model prediction. Dimensional reduction on the last hidden layer using 2D t-SNE also shows evidence that Audi, BMW and Benz are the three distinct clusters in latent space, while Hyundai and Toyota are much more entangled with each other. In other words, Hyundai

and Toyota have better segment prototypicality (SP) than other brands, making them harder to differentiate by BIGNet. These findings are consistent with the conclusion of Liu et al. [5], who found that luxurious brands yield higher brand consistency (BC) and that SP has a stronger effect on economy cars than BC.

5.4.3 Feature analysis

As mentioned in Section 5.3, the increased data complexity also negatively impacts attempts to visualize the attention of BIGNet using LOFO. LOFO requires to run each image as many times as the number of curves and chunks, leading to huge computational cost. Further, the confidence score change from LOFO often doesn't reflect desirable car features either. The ablation of individual shapes is either too subtle due to the robustness of larger graphs, or is too biased to chunks with large bounding boxes, making the contour always highlighted. This case study, therefore, implements a CAM[17]-inspired algorithm that efficiently runs only one inference per image. In the algorithm, it looks at the contributions of each chunk's latent vector and visualizes the chunks that contribute the most to a correct prediction.

Table 7: Most highlighted features by BIGNet.

brand	number of data observed in test set	1st most highlighted feature	%	2nd most highlighted feature	%
Toyota	54	fog lights	92.6%	logo	79.6%
VW	70	logo	68.6%	grille	64.3%
Benz	65	grille	89.2%	headlights	76.9%
Audi	63	grille	77.8%	headlights	33.3%
Bmw	65	headlights	84.6%	grille	72.3%
Hyundai	49	headlights	69.4%	fog lights	24.5%

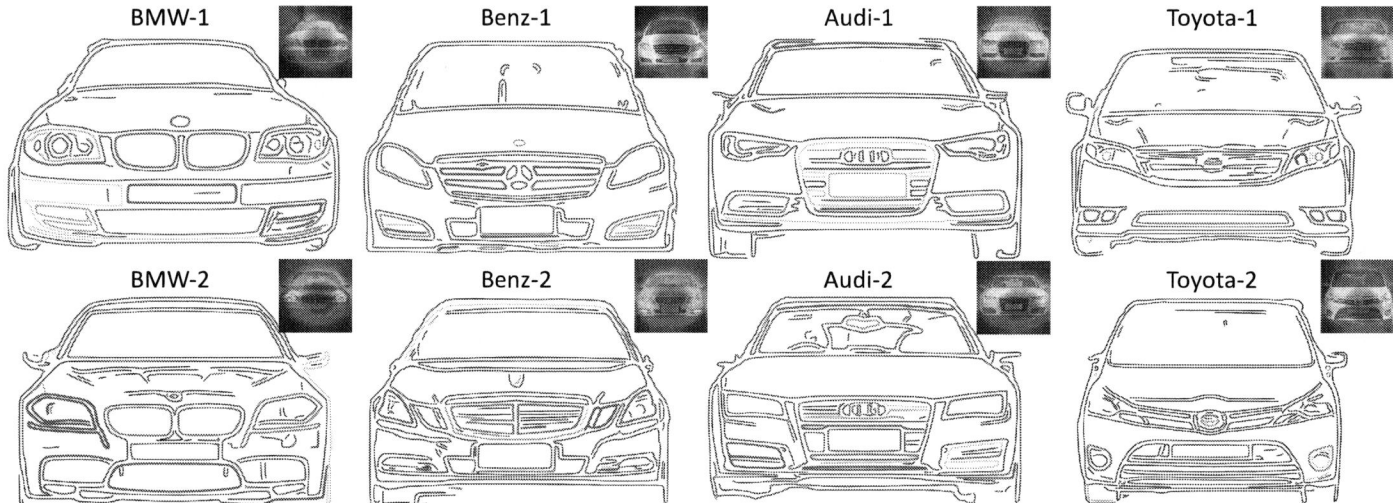

Figure 13: CAM-based BIGNet brand-related features' visualization on test set. The Grad-CAM visualizations on CNN are located in the upper right corner of each image. It is obvious that BIGNet captures luxury segments' well-distinguishable car parts including grille, headlights and fog lights, while there are much fewer geometric clues on affordable cars (Toyota) that it has to rely on logo detection.

Table 8: Comparison of the four training scenarios. All cases share consistency of having better recognition of luxurious brands and more entanglement between affordable brands.

brands	has logo	accuracy	Well clustered classes (t-SNE)	Well recognizable (Confusion matrix accuracy)	Badly recognizable (Confusion matrix accuracy)	Entanglement pairs (Cohen's kappa coefficient matrix)
6	yes	train: 89.4% test: 80.6%	BMW>Audi>Benz>VW	train: BMW>Audi=Benz>VW test: Audi>BMW>Benz=VW	train: Hyundai<Toyota test: Toyota<Hyundai	Toyota-Hyundai
	no	train: 87.1% test: 77.3%		train: BMW>Audi>Benz test: BMW>VW>Benz	train: Toyota<Hyundai test: Hyundai=Toyota	Toyota-Hyundai Toyota-VW
10	yes	train: 64.9% test: 63.4%	BMW>Audi>Benz	train: BMW>Audi>VW test: Audi>VW>BMW	train: Hyundai<Nissan<KIA test: Ford<Hyundai<Nissan	Toyota-Hyundai Toyota-Nissan Ford-Hyundai Ford-KIA
	no	train: 65.8% test: 59.1%		train: BMW>Audi>Benz>VW test: BMW>Audi>VW>Benz	train: Nissan<Ford<Hyundai test: Nissan<Hyundai<Ford	

Copyright © 2023 by ASME

The visualization results (Figure 13) show BIGNet's consistent attention to certain parts of each brand. Table 7 summarizes the most frequently highlighted attention in test set, and their percentage of being visualized. Among the six brands, the luxury brands (BMW, Benz, Audi) exhibit more explainable and intuitive attention, while all three tending to highlight the curves related to the grille and headlights. This suggests that luxury brands prioritize preserving brand consistency in the same car parts while incorporating different geometries. Finally, ResNet-50 is finetuned as the CNN to compare with BIGNet. Although CNN reaches almost 100% accuracy for both train and test sets, it fails to show the explicit features of a brand using Grad-CAM. BIGNet's results, on the other hand, are not only explainable, but also editable as each curve is parameterized by control points. This makes it a much more useful tool as a surrogate for analyzing brand-related features.

5.4.4 Generalizability study

In this section, BIGNet's reliability and consistency are examined by further increasing the classification difficulty, with the effects of logo removal and classifying ten brands (original six brands plus KIA, Chevy, Ford, and Nissan). For each experiment, the accuracy, recognizability ranking, and entanglement pairs are summarized in Table 8. As BMW, Audi, and Benz are frequently getting higher accuracy across all four scenarios, economy cars are also frequently getting lower accuracy and more entanglements, which substantiates the findings in Section 5.4.2. It is also found that adding more brands to the classification problem has a bigger effect than removing the logo, which can be because of all four additional brands are in the economy segments, which have lower brand consistency. It is also notable that logo removal only slightly decreases the accuracy, showing BIGNet's capability of recognizing higher level features. Lastly, for fair comparison, the trained ten-brand classifiers share the same BIGNet architecture that was designed for the six-brand classification task, except for the final linear layer. Therefore, the low accuracy result is very likely because of underfitting, and is possible to be improved by increasing the number of BIGNet's learnable parameters.

6 LIMITATIONS AND FUTURE WORK

Although BIGNet's classification accuracy is lower than pixel CNN, there are currently only less than 7000 learnable parameters in BIGNet. This is much fewer than Resnet-50 having 23 million parameters. On one hand, there is a huge space to improve accuracy from hyperparameter tuning and data processing. While the current BIGNet can already demonstrate explicit and explainable visualization results on brand-related features, with increased accuracy, visualization results would be expected to yield even more explainability. On the other, as curve-based images have more condensed information than pixel images, both dataset and AI architecture require smaller storage, which may have applications on lightweight AI design.

This research has shown it possible to extract explicit and editable features by a deep network agent. To help actual designers identify and quantify features in an automatic way, studies are planned to examine how humans can interact and collaborate with such a surrogate system and achieve design objectives. Aside from recognition, there is potential for actively generating or transferring stylized content building upon BIGNet's framework. This may open up a new avenue for future research in data-driven, explainable generative models.

As this research proposes a general stylization workflow, it has wide-ranging applications including image segmentation, engineering design, market positioning and technical appraisement. Aside from classifications to recognize and preserve brand consistency, safety, semantics and ergonomics, with subtle modifications it also has the potential of learning regression problems to recognize or predict contents' labels like year or price. With the strong explainability of learning from curve-based representations, this research opens an avenue of deducing information from curve representations and is expected to outperform pixel base approaches in domains that values interpretability more than accuracy.

7 CONCLUSIONS

This research proposes an automatic workflow to analyze and visualize style content explicitly and performs case studies on products' brand classification to recognize and preserve brand consistency. To mimic a human designer's thought process, data is constructed as SVG and is classified using BIGNet, a two-tier spatial GNN. In the phone study, it shows the model able to learn on parametrically synthesized SVG data. By visualizing attention using LOFO, BIGNet demonstrates capability of capturing brand-related features at intra-curve, inter-curve, intra-chunk, inter-chunk levels. Partial dependence plot on model's confidence variation during parameter extrapolation further substantiates that BIGNet learned continuous and meaningful features, including lens' location, height-width ratio, and screen-frame gap. The car study further explores the potential of a fully automated recognition system, and investigates the generalizability of the workflow. With some architecture modifications from the phone study, BIGNet can learn from generic vectorized car images and reach 80.6% test accuracy on a six-brand classification task. During the evaluation process, BMW, Benz, and Audi are found to achieve higher recognition rates compared to other brands. This finding matches the optimized marketing strategy that luxurious cars value brand consistency more than economy cars. CAM visualization further shows that BIGNet has consistent attention on luxury brands' grille and headlights. Finally, comparison of BIGNet with a CNN baseline demonstrates that a curve-based deep learning model produces more interpretable visualizations, while the image format is also more editable. Therefore, BIGNet, as a deep learning model, can identify brand-related features and can be applied to various product categories with distinguishing geometries, enabling humans to finally utilize it as a communicative and explainable style discovery agent, which significantly accelerates the aesthetic design process. Future research will explore a wider range of potential applications in other stylization domains.

ACKNOWLEDGEMENT

This work was partially funded by the National Science Foundation under grant Award CMMI-2113301.

Copyright © 2023 by ASME

REFERENCES

[1] Orbay, G., Fu, L., and Kara, B., 2015, "Deciphering the Influence of Product Shape on Consumer Judgments Through Geometric Abstraction," Journal of Mechanical Design, 137(8), p. 081103.

[2] Ersin Yumer, M., Chaudhuri, S., Hodgins, J. K., and Burak Kara, L., 2015, "Semantic Shape Editing Using Deformation Handles," *ACM Transactions on Graphics (TOG)*, pp. 1–12.

[3] Ravasi, D., and Stigliani, I., 2012, "Product Design: A Review and Research Agenda for Management Studies," International Journal of Management Reviews, 14(4), pp. 464–488.

[4] Bloch, P. H., 2011, "Product Design and Marketing: Reflections after Fifteen Years," Journal of Product Innovation Management, 28(3), pp. 378–380.

[5] Liu, Y., Li, K. J., Chen, H. A., and Balachander, S., 2017, "The Effects of Products' Aesthetic Design on Demand and Marketing-Mix Effectiveness: The Role of Segment Prototypicality and Brand Consistency," Journal of Marketing, 81(1), pp. 83–102.

[6] Stiny, G., 1991, "The Algebras of Design," Research in Engineering Design, 2, pp. 171–181.

[7] Agawal M and Cagan J, 1997, "A Blend of Different Tastes- the Language of Coffeemakers," Environment and Planning B, 25, pp. 205–226.

[8] Ang, M. C., Ng, K. W., and Pham, D. T., 2013, "Combining the Bees Algorithm and Shape Grammar to Generate Branded Product Concepts," Proceedings of the Institution of Mechanical Engineers, Part B: Journal of Engineering Manufacture, 227(12), pp. 1860–1873.

[9] Chau, H. H., Chen, X., McKay, A., and de Pennington, A., 2004, "Evaluation of a 3D Shape Grammar Implementation," *Design Computing and Cognition '04*, Springer Netherlands, pp. 357–376.

[10] Michael J. Pugliese and Jonathan Cagan, 2002, "Capturing a Rebel- Modeling the Harley-Davidson Brand through a Motorcycle Shape Grammar," Research in Engineering Design, pp. 139–156.

[11] Jay P. McCormack and Jonethan Cagan, 2004, "Speaking the Buick Language- Capturing, Understanding, and Exploring Brand Identity with Shape Grammars," Design Studies, 25(1), pp. 1–29.

[12] Aqeel, A. B., 2015, "Development of Visual Aspect of Porsche Brand Using CAD Technology," Procedia Technology, 20, pp. 170–177.

[13] Hsiao, S.-W., and Huang, H. C., 2002, "A Neural Network Based Approach for Product Form Design," Design studies, 23(1), pp. 67–84.

[14] Chung-Chih Lin and Shih-Wen Hsiao, 2003, *A Study on Applying Feature-Based Modeling and Neural Network to Shape Generation.*

[15] Goodfellow, I., Bengio, Y., and Courville, A., 2016, *Deep Learning*, MIT press.

[16] Krizhevsky, A., Sutskever, I., and Hinton, G. E., 2017, "ImageNet Classification with Deep Convolutional Neural Networks," Communications of the ACM, 60(6), pp. 84–90.

[17] Zhou, B., Khosla, A., Lapedriza, A., Oliva, A., and Torralba, A., 2016, "Learning Deep Features for Discriminative Localization," *Proceedings of the IEEE Conference on Computer Vision and Pattern Recognition*, pp. 2921–2929.

[18] Stiny, G., and Gips, J., 1971, "Shape Grammars and the Generative Specification of Painting and Sculpture," *IFIP Congress*, p. 128.

[19] Boatwright, P., Cagan, J., Kapur, D., and Saltiel, A., 2009, "A Step-by-Step Process to Build Valued Brands," Journal of Product and Brand Management, 18(1), pp. 38–49.

[20] Harris, C., and Stephens, M., 1988, "A Combined Corner and Edge Detector," British Machine Vision Association and Society for Pattern Recognition, p. 23.1-23.6.

[21] Carlo Tmosi and Takeo Kanade, 1992, "Shape and Motion from Image Streams: A Factorization Method," International journal of computer vision, 9(2), pp. 137–154.

[22] Lowe, D. G., 2004, "Distinctive Image Features from Scale-Invariant Keypoints," International Journal of Computer Vision, 60, pp. 91–110.

[23] Dalal, N., and Triggs, B., 2005, "Histograms of Oriented Gradients for Human Detection," *IEEE Computer Society Conference on Computer Vision and Pattern Recognition*, pp. 886–893.

[24] Calonder, M., Lepetit, V., Strecha, C., and Fua, P., 2010, "BRIEF: Binary Robust Independent Elementary Features ⋆," SpringerGradient-Based Learning Applied to Document Recognition, pp. 778–792.

[25] Lecun, Y., Bottou, L., Bengio, Y., and Haffner, P., 1998, "Gradient-Based Learning Applied to Document Recognition," Proc. IEEE, 86(11), pp. 2278–2324.

[26] Simonyan, K., and Zisserman, A., 2015, "Very Deep Convolutional Networks for Large-Scale Image Recognition."

[27] Iandola, F. N., Han, S., Moskewicz, M. W., Ashraf, K., Dally, W. J., and Keutzer, K., 2017, "SqueezeNet: AlexNet-Level Accuracy with 50x Fewer Parameters and <0.5MB Model Size," *International Conference on Learning Representations*.

[28] Chen, T., Kornblith, S., Norouzi, M., and Hinton, G., 2020, "A Simple Framework for Contrastive Learning of Visual Representations," *PMLR*.

[29] Chen, T., Kornblith, S., Swersky, K., Norouzi, M., and Hinton, G., 2020, "Big Self-Supervised Models Are Strong Semi-Supervised Learners," Advances in neural information processing systems, 33, pp. 22243–22255.

[30] Chabot, F., Chaouch, M., Rabarisoa, J., Teuliere, C., and Chateau, T., 2017, "Deep Edge-Color Invariant Features for 2D/3D Car Fine-Grained Classification," *IEEE Intelligent Vehicles Symposium, Proceedings*, Institute of Electrical and Electronics Engineers Inc., pp. 733–738.

[31] Zhang, Q., Wu, Y. N., and Zhu, S.-C., 2018, "Interpretable Convolutional Neural Networks," *Proceedings of the*

Copyright © 2023 by ASME

IEEE Conference on Computer Vision and Pattern Recognition, pp. 8827–8836.

[32] Chang, D., Ding, Y., Xie, J., Bhunia, A. K., Li, X., Ma, Z., Wu, M., Guo, J., and Song, Y.-Z., 2020, "The Devil Is in the Channels: Mutual-Channel Loss for Fine-Grained Image Classification," *IEEE Transactions on Image Processing*, pp. 4683–4695.

[33] Selvaraju, R. R., Cogswell, M., Das, A., Vedantam, R., Parikh, D., and Batra, D., 2017, "Grad-CAM: Visual Explanations from Deep Networks via Gradient-Based Localization," *Proceedings of the IEEE International Conference on Computer Vision*, pp. 618–626.

[34] Eitz, M., Hays, J., and Alexa, M., 2012, "How Do Humans Sketch Objects?," ACM Transactions on graphics (TOG), **31**(4), pp. 1–10.

[35] Schneider, R. G., and Tuytelaars, T., 2014, "Sketch Classification and Classification-Driven Analysis Using Fisher Vectors," ACM Trans. Graph., **33**(6), pp. 1–9.

[36] Li, Y., Hospedales, T. M., Song, Y.-Z., and Gong, S., 2015, "Free-Hand Sketch Recognition by Multi-Kernel Feature Learning," Computer Vision and Image Understanding, **137**, pp. 1–11.

[37] Yu, Q., Yang, Y., Song, Y.-Z., Xiang, T., and Hospedales, T., 2017, "Sketch-a-Net: A Deep Neural Network That Beats Humans," International journal of computer vision, **122**, pp. 411–425.

[38] Li, L., Zou, C., Zheng, Y., Su, Q., Fu, H., and Tai, C.-L., 2020, "Sketch-R2CNN: An RNN-Rasterization-CNN Architecture for Vector Sketch Recognition," IEEE transactions on visualization and computer graphics, **27**(9), pp. 3745–3754.

[39] Hu, C., Li, D., Song, Y.-Z., Xiang, T., and Hospedales, T. M., 2018, "Sketch-a-Classifier: Sketch-Based Photo Classifier Generation," *Proceedings of the IEEE Conference on Computer Vision and Pattern Recognition*, pp. 9136–9144.

[40] Xu, P., Huang, Y., Yuan, T., Pang, K., Song, Y.-Z., Xiang, T., Hospedales, T. M., Ma, Z., and Guo, J., 2018, "SketchMate: Deep Hashing for Million-Scale Human Sketch Retrieval," *Proceedings of the IEEE Conference on Computer Vision and Pattern Recognition*, arXiv, pp. 8090–8098.

[41] Wu, Z., Pan, S., Chen, F., Long, G., Zhang, C., and Yu, P. S., 2020, "A Comprehensive Survey on Graph Neural Networks," IEEE Transactions on Neural Networks and Learning Systems, **32**(1), pp. 4–24.

[42] Micheli, A., 2009, "Neural Network for Graphs: A Contextual Constructive Approach," IEEE Transactions on Neural Networks, **20**(3), pp. 498–511.

[43] Atwood, J., and Towsley, D., 2016, "Diffusion-Convolutional Neural Networks," *Advances in Neural Information Processing Systems*.

[44] Monti, F., Boscaini, D., Masci, J., Rodolà, E., Svoboda, J., and Bronstein, M. M., 2017, "Geometric Deep Learning on Graphs and Manifolds Using Mixture Model CNNs,"

Proceedings of the IEEE Conference on Computer Vision and Pattern Recognition, pp. 5115–5124.

[45] Veličkovi´veličkovi´c, P., Cucurull, G., Casanova, A., Romero, A., Lì, P., and Bengio, Y., 2017, "Graph Attention Networks," stat, **1050**(20), pp. 10–48550.

[46] Hamilton, W. L., Ying, R., and Leskovec, J., 2017, "Inductive Representation Learning on Large Graphs," *Advances in Neural Information Processing Systems*.

[47] Zhou, J., Cui, G., Hu, S., Zhang, Z., Yang, C., Liu, Z., Wang, L., Li, C., and Sun, M., 2020, "Graph Neural Networks: A Review of Methods and Applications," AI Open, **1**, pp. 57–81.

[48] Sanchez-Gonzalez, A., Heess, N., Springenberg, J. T., Merel, J., Riedmiller, M., Hadsell, R., and Battaglia, P., 2018, "Graph Networks as Learnable Physics Engines for Inference and Control," *International Conference on Machine Learning*, PMLR, pp. 4470–4479.

[49] Do, K., Tran, T., and Venkatesh, S., 2019, "Graph Transformation Policy Network for Chemical Reaction Prediction," *Proceedings of the 25th ACM SIGKDD International Conference on Knowledge Discovery & Data Mining*, pp. 750–760.

[50] Guo, S., Lin, Y., Feng, N., Song, C., and Wan, H., 2019, "Attention Based Spatial-Temporal Graph Convolutional Networks for Traffic Flow Forecasting," AAAI, **33**(01), pp. 922–929.

[51] Xie, L., Lu, Y., Furuhata, T., Yamakawa, S., Zhang, W., Regmi, A., Kara, L., and Shimada, K., 2022, "Graph Neural Network-Enabled Manufacturing Method Classification from Engineering Drawings," Computers in Industry, **142**, p. 103967.

[52] Zhang, W., Joseph, J., Yin, Y., Xie, L., Furuhata, T., Yamakawa, S., Shimada, K., and Kara, L. B., 2023, "Component Segmentation of Engineering Drawings Using Graph Convolutional Networks," Computers in Industry, **147**, p. 103885.

[53] Ranscombe, C., Hicks, B., Mullineux, G., and Singh, B., 2012, "Visually Decomposing Vehicle Images: Exploring the Influence of Different Aesthetic Features on Consumer Perception of Brand," Design Studies, **33**(4), pp. 319–341.

[54] Friedman, J. H., 2001, "Greedy Function Approximation: A Gradient Boosting Machine," The Annals of Statistics, **29**(5), pp. 1189–1232.

[55] Akkucuk, U., and Esmaeili, J., 2016, "The Impact of Brands on Consumer Buying Behavior," International Journal of Research in Business and Social Science (2147-4478), **5**(4), pp. 1–16.

[56] Hussain Shaheed Zulfikar Ali Bhutto, S., and Raheem Ahmed, R., 2020, "Smartphone Buying Behaviors in a Framework of Brand Experience and Brand Equity," Transformations in Business & Economics, **19**(2).

[57] Selinger, P., 2003, *Potrace: A Polygon-Based Tracing Algorithm*.

[58] Yang, L., Luo, P., Loy, C. C., and Tang, X., 2015, "A Large-Scale Car Dataset for Fine-Grained Categorization and Verification," *Proceedings of the IEEE Conference on*

Copyright © 2023 by ASME

Computer Vision and Pattern Recognition, pp. 3973–3981.

[59] Wu, Y., Kirillov, A., Massa, F., Lo, W.-Y., and Girshick, R., 2019, "Detectron2."

[60] Pu, M., Huang, Y., Liu, Y., Guan, Q., and Ling, H., 2022, "EDTER: Edge Detection with Transformer," *Proceedings of the IEEE/CVF Conference on Computer Vision and Pattern Recognition*, pp. 1402–1412.

Proceedings of the ASME 2023
International Design Engineering Technical Conferences and
Computers and Information in Engineering Conference
IDETC-CIE2023
August 20-23, 2023, Boston, Massachusetts

DETC2023-116743

CHARACTERIZING DESIGNS VIA ISOMETRIC EMBEDDINGS: APPLICATIONS TO AIRFOIL INVERSE DESIGN

Qiuyi Chen[1], Mark D. Fuge[1,*]

[1] University of Maryland, College Park, Maryland, 20720

ABSTRACT

Many data analysis and design problems involve reasoning about points in high-dimensional space. A common strategy is to embed points from this high-dimensional space into a low-dimensional one. As we will show in this paper, a critical property of good embeddings is that they preserve isometry—i.e., preserving the geodesic distance between points on the original data manifold within their embedded locations in the latent space. However, enforcing isometry is non-trivial for common Neural embedding models, such as autoencoders and generative models. Moreover, while theoretically appealing, it is not clear to what extent enforcing isometry is really necessary for a given design or analysis task. This paper answers these questions by constructing an isometric embedding via an isometric autoencoder, which we employ to analyze an inverse airfoil design problem. Specifically, the paper describes how to train an isometric autoencoder and demonstrates its usefulness compared to non-isometric autoencoders on both simple pedagogical examples and for airfoil embeddings using the UIUC airfoil dataset.

Our ablation study illustrates that enforcing isometry is necessary to accurately discover latent space clusters—a common analysis method researchers typically perform on low-dimensional embeddings. We also show how isometric autoencoders can uncover pathologies in typical gradient-based Shape Optimization solvers through an analysis on the SU2-optimized airfoil dataset, wherein we find an over-reliance of the gradient solver on angle of attack. Overall, this paper motivates the use of isometry constraints in Neural embedding models, particularly in cases where researchers or designer intend to use distance-based analysis measures (such as clustering, k-Nearest Neighbors methods, etc.) to analyze designs within the latent space. While this work focuses on airfoil design as an illustrative example, it applies to any domain where analyzing isometric design or data embeddings would be useful.

1. INTRODUCTION

Analyzing past design data via Machine Learning has opened up new avenues for accelerating both human- and computer-

generate designs in several ways [1]. For instance, in works like [2–7], researchers developed conditional inverse design models that can generate new designs satisfying the performance requirements, without going through time-consuming optimization. Other researchers [8–12] have focused on unconditional generation of designs, to either create more efficient shape parameterization functions or to augment an existing dataset with high-quality designs. Lastly, surrogate modeling methods have a long history using data-driven models to predict and optimize a design's performance so as to limit the need for computationally intensive first-principles solvers, such as Finite Element or Finite Volume solvers [13–17].

However, for all data-driven models researchers often need to ensure a dataset's quality by analyzing: how non-uniform it is; whether the data are concentrated, multimodal, or biased over regions of data space; or whether the data points are noisy. Data analysis tools such as clustering or computing density or topological properties are typically used to characterize some of these factors, yet the curse of dimensionality [18] undermines their use in high dimensional data space. Thus, practitioners usually first embed the data into some low dimensional latent space via a dimension reduction method, and then conduct the analyses there instead [19–21]. But, given the large number of existing embedding methods, what properties do we need from the embedding to make such latent analyses reasonable? Chen et al. [22] proposed that for design data, we should care about the embedding's preservation of the geodesic distance, but that paper did not address the question of how to actually construct such an embedding.

This paper answers that question by proposing the recently developed *isometric autoencoder* [23–25] based on Riemann geometry to embed designs in a bidirectional and distance-preserving manner. We demonstrate the isometric embedding's necessity and practicality via a latent space analysis of the airfoil inverse design problem. Specifically, the paper provides the following contributions:

1. We describe how to produce a bidirectional isometric representation with the isometric autoencoder. We apply this architecture to both some pedogogical toy problems and

---
*Corresponding author: fuge@umd.edu

Copyright © 2023 by ASME

the real-world UIUC airfoil dataset. These representations preserves the geodesic distance on the UIUC airfoil manifold in the form of the Euclidean distance in the latent space. We show how to use this isometric low dimensional embedding as a proxy to investigate the quality of the UIUC dataset—*i.e.*, its non-uniformity and multimodality—through the lens of HDBSCAN clustering.

2. We illustrate why preserving the geodesic distance when learning the design representation is necessary through a pedagogical counterexample incorporating flow-based models and optimal transport. We show that the lack of isometricity can lead to ambiguity when analyzing the UIUC airfoils' latent embedding.

3. We use the isometric airfoil representation to study the quality and properties of the SU2 dataset of optimized airfoils. This dataset was produced by a gradient-based (i.e., adjoint) SU2 CFD solver and was used to train an CEBGAN for inverse airfoil design. We unearth a pathology of the SU2 adjoint optimizer that it favors optimizing the angle of attack more than the shape, and provide insights for future improvement. In addition, the isometric embedding sheds light on the robustness of different airfoils' to various flow conditions, which has implications for work in robust design optimization.

2. BACKGROUND AND RELATED WORK

Before introducing the isometric autoencoder and how we use it for latent space cluster analysis, this section briefly reviews background and related work in clustering methods, how we define metric functions (in both the Euclidean and Geodesic sense), common latent space dimension reduction methods, and lastly basic definitions of isometry that we use through the rest of the paper.

2.1 Clustering

A design dataset can be regarded as a collection of designs sampled from a probability distribution supported by the set of valid designs. Among all the techniques that analyze the data samples to characterize their underlying distributions, clustering is a fundamental and popular one. It aims to assign the samples to different clusters by a certain algorithm on an unsupervised basis, insofar as the samples in the same cluster are more *similar* to each other than to those in other clusters. Owing to its ability to taxonomically describe the data distribution, clustering can highlight a distribution's non-uniformity and multi-modality, which makes it especially suitable to design datasets, as they are in general non-uniform and may contain several exemplary and timeless groups of designs that engineers desire to extract and imitate.

Out of all the existing clustering schemes, density-based clustering excels when the dataset consists of an unknown number of clusters of arbitrary shapes. Empirically [26, 27], it is an ideal choice when the data distribution is supported by a set comprised of several separated-by-closed-neighbourhoods components and the dataset is rich enough to delineate each, such that every disconnected component can be accurately identified as a cluster.

Because of these properties, density-based clustering is ideal for cases where topological separation in data space among designs can indicate crucial design variation. There exist many density-based clustering models, among which DBSCAN [26] is probably the most celebrated one thanks to its efficacy withstanding the test of time [28, 29], yet it is still not perfect [29, 30] as a matter of course. Several successors like OPTICS [31], DENCLUE [32] and HDBSCAN [33] have since then being proposed to refine it. HDBSCAN [33] is a recent density-based hierarchical innovation with many great improvements, among which it in particular discards the annoying length scale hyperparameter ϵ of DBSCAN, hence we shall use this model in the later experiments for our convenience.

Despite clustering's usefulness, there is one critical aspect upon which any method's success hinges: the chosen distance function that describes similiarity between points. Indeed, for clustering schemes either connectivity-based like hierarchical clustering [34], or centroid-based like k-means [35], or distribution-based like Gaussian mixture [36], or density-based like DBSCAN [26], selecting the distance function to quantify the resemblance between data samples is inevitably the first step and the foundation of the remaining process. As such, the distance function has outsized influence on the final result. Beyond clustering, this distance function is also important to any analyses wherein the data samples need to be juxtaposed to perceive their difference, such as in nearest neighbor search [37], DPP diversity quantification [38, 39], etc., therefore it is worth being discussed in depth next.

2.2 Distance Functions

The canonical distance function (or metric) for an Euclidean space is the Euclidean distance. Despite being the most intuitive metric for low dimensional spaces—thus usually being the default choice for clustering models—the Euclidean distance is frequently challenged in high dimensional spaces. One well-known curse of dimensionality is the diminishing contrast between the maximum and the minimum L_p distance from a random query point to a series of random data points as the space dimensionality increases [40, 41]. This cripples the effectiveness of distance-based algorithms like nearest neighbors or clustering on high-dimensional data. Apart from suffering the contrast-loss, Euclidean distance is also not even aware of the semantics of many high dimensional data (*i.e.*, what object(s) each data represents). For example, an image of object A may appear closer in Euclidean distance to an image of object B than to another image representing the same object A but in a different pose [42]. One plausible interpretation of this is that most high dimensional data only reside on low dimensional manifolds [43], yet an Euclidean distances defined over the ambient space (*i.e.*, the Euclidean space containing that manifold) cannot take the shape and curvature of the embedded data manifold into account, as illustrated in Fig. 1.

Many alternative metrics have been proposed to overcome these issues. As an example, it is suggested that employing L_p distance functions of smaller (even fractional) p can mitigate the contrast-loss in high dimensional spaces [41]. Nevertheless, just like their L_2 counterpart, these candidates are still unaware of the geometry of the data manifold, not to mention it is dubious to

Copyright © 2023 by ASME

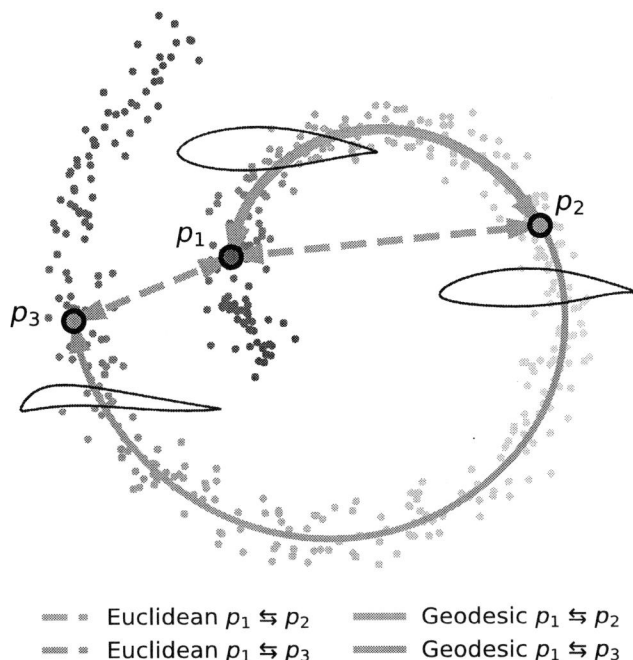

$\cdots$ Euclidean $p_1 \leftrightarrows p_2$ ▬ Geodesic $p_1 \leftrightarrows p_2$

$\cdots$ Euclidean $p_1 \leftrightarrows p_3$ ▬ Geodesic $p_1 \leftrightarrows p_3$

FIGURE 1: DIFFERENCE BETWEEN EUCLIDEAN DISTANCE AND GEODESIC DISTANCE WHEN QUANTIFYING DESIGN DIFFERENCES.

adopt them simply because of their better contrast, as L_p distance functions with $p < 1$ are not even well-defined, violating the triangular inequality that a legitimate metric should obey. Mahalanobis distance is another popular option [42, 44, 45], which is equivalent to performing a linear transformation over the entire ambient space prior to measuring distances with the Euclidean distance, so that when comparing points, different directions can be either emphasized or downplayed depending on their contribution to the semantics. However, such linear transformation needs to be trained beforehand by leveraging additional knowledge about the data's semantics, which is not always available. On top of that, linear transformation could be too primitive to simplify complicated nonlinear data manifold structure.

A probably more proper and natural choice of distance function for high dimensional data is the geodesic distance (or Riemannian distance) [46, 47] induced by the Euclidean metric over the data manifold, provided that the data manifold resides on a connected Riemann manifold—which can be intuitively called *extended* data manifold—such that the geodesics on it are well defined. This distance function adapts to the geometry of the data manifold because it is defined as the length of the shortest path on it connecting two given points, thereby it is designed exclusively for that particular manifold and makes more sense both intuitively and empirically [48–52]. The greatest downside of this metric is its general lack of closed-form expression, since the evaluation of the geodesic length involves integration over the irregular manifold. Luckily, we can to some extent circumvent this issue by preserving this geodesic distance in a low dimensional latent

Euclidean space after establishing an *isometry* between the data manifold and a latent set in the latent space, so that the latent set serves as a proxy for that data manifold equipped with geodesic distance and we can equivalently perform geodesic-based analyses on it instead. To initiate this latent analysis, we need to start with dimension reduction to construct a map between the data space and the latent space at first.

2.3 Dimension Reduction

Dimension reduction aims to map data points in the high dimensional space into a low dimensional latent space while preserving some necessary information, such that the validity of certain analyses performed in the latent space can be ensured. For instance, t-SNE [20] retains the disconnectedness of the dataset [53], so that the disconnected subsets remain disconnected in the latent space. Isomap [54] preserves the graph distance on the neighborhood graph as an approximation of the true geodesic distance.

Generally, the data manifold is nonlinear, which often renders linear dimension reduction methods like PCA and NMF futile [54, 55]. Out of all the nonlinear dimension reduction models, the autoencoder [56] is special for its simple reconstruction-loss-based formulation and the ability to not only map forward but also backward from the latent space to the data space. This ability to transform data bidirectionally is ideal for designs, as it can not only help us analyze designs conveniently in a low dimensional setting—which is the focus of this paper—but also enable us to use its backward mode as a design parameterization to synthesize novel designs after we efficiently perform optimization or construct conditional generative models in the low dimensional latent space. In addition, being a parameteric model, once trained the autoencoder can be immediately applied to unseen new designs to derive the corresponding latent codes without starting from scratch. This advantage should compound as the dimensionality and scale of the design problem grows. It is therefore worthwhile to model that isometry for designs with the autoencoder rather than the other unidirectional, non-parametric methods like Isomap.

Compared with its counterparts like Isomap [54], LLE [55], UMAP [21], or diffusion map [57] that need to scrutinize each data point's neighborhood to infer the local manifold structure, the autoencoder has a more straightforward formulation which only needs us to globally minimize the reconstruction loss. However, it is still a dimension reduction method that can preserve a dataset's topological properties. This is because it approximately learns a *topological embedding*—which is a homeomorphism—between the dataset in the high dimensional data space and latent set in the low dimensional latent space, such that the original dataset's crucial topological properties like connectedness—which are invariant under homeomorphism—are preserved on the latent set [58]. Our claim about the autoencoder's homeomorphicity is bolstered by the rationale that both the encoder e and the decoder g are modeled by *continuous* neural networks, and the minimization of the reconstruction loss $\mathbb{E}_{x \sim \mathcal{X}} \| g \circ e(x) - x \|$ encourages the composition $g \circ e$ to be an identity function over the dataset $\mathcal{X}$ and drives both g and e to be *bijective* between the dataset $\mathcal{X}$ and the latent set $e(\mathcal{X})$ [59], which is the very definition of a topological

Copyright © 2023 by ASME

embedding [58].

However, preserving the topological properties alone is not enough for clustering on the latent set, since there exists an infinite number of latent sets that each has different pairwise distances but is still homeomorphic to the same dataset (this non-uniqueness of homeomorphism is why flow-based models [60–62] can approximate different distributions with a fixed latent distribution, as we will see in §4.1.3). This may in consequence induce an infinite number of results for latent clustering and lead to ambiguity. Therefore, preserving a selected distance function in the data space is necessary, and this brings us to the preservation of the aforementioned geodesic distance on the data manifold.

2.4 Isometry

An *isometry* between two Riemann manifolds is a *locally isometric* diffeomorphism (*i.e.*, a smooth homeomorphism), which preserves the geodesic distances between points. For simplicity, if we restrict this generic definition to our special case where this map is modeled by an autoencoder, then based on the discussions in [23–25], we say the autoencoder establishes an isometry between a Riemann data manifold in the high dimensional data space and a Riemann latent manifold in the low dimensional latent space—provided that they are *connected* sets and inherit their Riemann metrics from their Euclidean ambient spaces respectively—if the autoencoder is:

1. A diffeomorphism: Both the encoder and the decoder are modelled by *smooth* neural networks, and the autoencoder's reconstruction loss is also minimized to near zero over the Riemann data manifold, creating a *homeomorphism*.

2. Locally isometric: All singular values of the Jacobians of both the decoder and the encoder need to be 1 at every point over the Riemann manifolds. Intuitively, this means the local linear transformation (*i.e.*, the differential) does not stretch or compress the input along any direction tangent to the manifold.

On top of that, ideally we hope the latent space's dimension is equal to the latent manifold's dimension and the latent manifold is a convex set—which aligns its geodesic distance with the Euclidean distance—such that the geodesic distance between any pair of points on the data manifold equals the Euclidean distance between their corresponding latent codes. However, this ideal setting is not always encountered for the following reasons:

1. The dataset (and thus the latent set) may not be connected. Remember this is what motivates us to use HDBSCAN to investigate the dataset's topology. Intuitively, this means there exist some intervals between these components that are not regularized for the autoencoder, such that while the distance within each component is preserved, these components can be arbitrarily close to each other in the latent space.

2. The latent space's dimension could be larger than the dataset's dimension. There are many causes for this issue. For example, the dataset may not be a manifold, as it could be not locally Euclidean somewhere. The dataset could also be a manifold unable to be embedded in a space of the same

dimension, as will be discussed later. Moreover, the dimension of the latent space needs to be determined beforehand, but we may not be able to estimate it accurately.

3. The latent set is not necessarily convex, as its shape is determined by the shape of the dataset through the isometry. In addition, if the above dimension misalignment exists, the nonlinearity of the latent set may also destroy its convexity.

All these issues suggest that we need to somehow extend the dataset into a well-behaved connected Riemann manifold to establish that isometry robustly. We introduce methods to do this next.

3. METHODOLOGY

This section covers the two primary methods that we use in this paper's later experiments: (1) the Isometric Autoencoder and (2) a method for estimating the intrinsic dimension of the data manifold.

3.1 Isometric Autoencoder

Since the reconstruction loss enforces homeomorphicity, and both the decoder and the encoder are smooth models, to enforce isometry within an autoencoder we only need to make the encoder and the decoder locally isometric. We can accomplish this by having their Jacobians' singular values all equal to one over the dataset (or equivalently the latent set). This is easier said than done, due to the substantial computational cost entailed in explicitly deriving the Jacobian's singular values, particularly in high dimensional cases. Thus, instead of regularizing those singular values explicitly, researchers typically impose the isometry constraint implicitly via random vectors sampled uniformly from a unit sphere $\mathbb{S}^{m-1}$ embedded in the m-D latent space, such that $\|J_g v\| = 1$ and $\|v^\top J_e\| = 1$ for every $v \in \mathbb{S}^{m-1}$, where J_g and J_e are the Jacobians of the decoder and the encoder respectively. It can be verified that this leads to unitary singular values [24]. More importantly, we can compute the Jacobian-Vector Product (JVP) or Vector-Jacobian Product (VJP) far more efficiently than the full Jacobian via modern packages like functorch and JAX, which can also invoke the optimal auto-differentiation mode (forward-mode for JVP and backward-mode for VJP) providing efficient backpropagation.

Enforcing isometry only over the dataset may not be enough, due to the three problems mentioned in §2.4. To overcome these issues, we employ the latent interpolation or mixup method [23, 63] to additionally enforce isometry over some interpolated latent points. Specifically, these interpolated points are sampled uniformly from the lines connecting some random pairs of real latent points (*i.e.*, the corresponding latent points of real data samples). This latent interpolation approximately samples points from the *convex hull* of all real latent points, so that by enforcing isometry over this extended convex latent set, we equivalently enforce isometry over its image in the data space, which is a connected Riemann manifold covering the dataset. This connected manifold can thus be regarded as an extended data manifold that has a well-defined geodesic distance and a

Copyright © 2023 by ASME

convex latent set. To make sure the extended latent set is homeomorphic to the extended data manifold, we additionally minimize the cycle consistency loss [64] over the interpolated latent points.

Altogether, the loss function of the isometric autoencoder consists of four square terms:

$$
L = \mathbb{E}_{x \sim \mathcal{X},\, z \sim \overline{e(\mathcal{X})},\, v \sim \mathbb{S}^{m-1}} \{ [g \circ e(x) - x]^2 + \beta [e \circ g(z) - z]^2
$$
$$
+ \lambda [\|J_g(z) \cdot v\| - 1]^2 + \lambda [\|v^\top \cdot J_e(x)\| - 1]^2 \} \quad (1)
$$

where β and λ are weight coefficients for cycle consistency loss and isometry regularization respectively, $\overline{e(\mathcal{X})}$ is the extended latent set consisting of the interpolated points, and the random vectors (x & v) are sampled uniformly from their respective sets. To sample v uniformly from $\mathbb{S}^{m-1}$, we can take advantage of the radial symmetry of the unit Gaussian distribution by simply normalizing the vectors sampled from it.

3.2 Intrinsic Dimension Estimation

To construct the autoencoder, we must first determine the latent space dimension (m). This dimension is vital for two reasons. On one hand, if the dataset constitutes a manifold—which is locally Euclidean of certain dimension—and we attempt to fit an autoencoder whose latent space is lower than this intrinsic dimension, the autoencoder will never establish a homeomorphism between the dataset and the latent set due to the topological invariance of dimension [65]. This would cause the autoencoder to map different data points to the same latent point, making the encoder no longer injective. This "data collapse" is detrimental to latent clustering for the following reasons. First, the autoencoder is thereby no longer isometric, which makes the intrinsic latent space metric unreliable for clustering. In addition, although the connectedness of the dataset is still preserved on the latent set thanks to the encoder's continuity, the disconnectedness is not guaranteed to remain; this means that multiple disconnected clusters in the data space may be merged into a single connected one in the latent space. This merging would clearly mislead any clustering algorithm. While some may argue that this collapse is not always unfortunate since some trivial data dimensions might be eliminated in our favor in the latent space, this removal is out of our control and it is unwise to rely on luck. On the other hand, avoiding this collapse problem by setting the latent space dimension higher than absolutely necessary creates its own problems: our whole reason for conducting dimension reduction in the first place is to reduce clustering problems that occur in high dimensional spaces and make isometry regularization more efficient.

To this end, we employ the Maximum Likelihood Estimation (MLE) [66] with bias correction [67] as the intrinsic dimension estimator, which performed well in our previous work [68]. In brief, MLE presumes locally constant data density and Poisson-distributed number of neighbors around each point. Under that model, the likelihood maximization leads to the local estimate

$$
\hat{m}_k(x) = \left[\frac{1}{k-1} \sum_{j=1}^{k-1} \log \frac{T_k(x)}{T_j(x)} \right]^{-1} \quad (2)
$$

where k is the pre-selected number of neighbors for each evaluation and $T_j(x)$ is the Euclidean distance between x and its jth

nearest neighbor. The debiased global estimator [67] summarizing these local results is

$$
\bar{m}_k = \left[\frac{1}{n} \sum_{i=1}^{n} \hat{m}_k(x_i)^{-1} \right]^{-1} = \left[\frac{1}{n(k-1)} \sum_{i=1}^{n} \sum_{j=1}^{k-1} \log \frac{T_k(x_i)}{T_j(x_i)} \right]^{-1} \quad (3)
$$

More details about improving its precision can be found in our previous work [68].

Even when we estimate the data manifold's dimension d accurately, it is not necessarily the proper number for the latent dimension, because we cannot always embed some manifolds of a given dimension into an ambient Euclidean space of the same dimension. The Klein bottle is a well-known example—this 2-manifold can only embed successfully in at-least-4D Euclidean spaces. Luckily, the Whitney embedding theorem [65] suggests that to obtain a homeomorphic autoencoder we can upper bound the latent dimension by $2d$. Therefore, a practical way to attain the optimal latent dimension is to incrementally increase it from d all the way to $2d$ until the final reconstruction loss becomes marginal. For data manifolds that do not have complexities or pathologies similar to the Klein bottle example, we would expect the result to be much closer to d than $2d$, and thus attainable in a few trial-and-error shots.

4. EXPERIMENTS AND DISCUSSION

In this section, we use the airfoil designs as a concrete example to demonstrate how the isometric autoencoder can help engineers analyze high dimensional designs intuitively in a low dimensional setting for different purposes. Overall, we first apply the isometric autoencoder to the UIUC airfoil dataset to obtain a low dimensional isometric latent representation of the historical airfoil designs. Then, we perform latent HDBSCAN to locate common past airfoil designs—for pedagogical effect, we will disrupt the isometricity of this autoencoder in one experiment to highlight the importance of preserving distance for latent clustering. In the second half of our experiments, we introduce and employ the resulting isometric autoencoder to derive the latent representation of a subset of airfoils optimized by the SU2 suite under a large variety of boundary conditions. We then use this isometric embedding to investigate how the airfoil shape and angle of attack are related to different input conditions.

4.1 UIUC Airfoil Latent Clustering

4.1.1 Isometric Representation of UIUC Airfoils.
To help generate smooth airfoil curves, we integrate the Bézier layer used in our previous works [11] into the decoder as its output layer. Our decoder and encoder also inherit their general architectures respectively from the generator and discriminator in [11], since their complexity was previously sufficient for this dataset. We set both β and λ to 0.01 for the isometry regularization. We set the latent space dimension to 3, based on prior published experiments with the UIUC dataset [68]. After constructing the autoencoder, we then train it with an Adam optimizer of learning rate 0.0001 for 6000 epochs with batch size 32 over the UIUC dataset consisting of 1528 airfoils. During each epoch, the autoencoder is trained 48 times over different shuffled mini-batches.

Copyright © 2023 by ASME

After training, the autoencoder achieves a reconstruction error of approximately 5×10^{-5} and a cycle consistency loss around 9×10^{-4}, while the isometry regularization errors of the decoder and the encoder are around 8×10^{-5} and 4.5×10^{-3} respectively. These results indicate that the isometric autoencoder does not sacrifice reconstruction error relative to the unregularized version. We then use the isometric encoder to map all of the UIUC airfoils into the latent space, and we use these encoded coordinates for the following latent analyses. Figure 2b shows the 2D principal projection of this (3D) latent set.

4.1.2 Latent Clustering with HDBSCAN using the Isometric Autoencoder.

At first, we perform HDBSCAN on the latent set with both m_{pts} and m_{clSize} set to 5—namely the minimum number of neighbors a given point should have to qualify as a core and the minimum number of cores that a cluster should have. This HDBSCAN configuration classifies $< 10\%$ of the latent points as noise and does not produce as many trivial tiny clusters, thus reasonably grasping the overall structure of the dataset. Fig. 2 shows that over 90% of airfoils in the UIUC dataset are considered density-connected by HDBSCAN under this hyperparameter setting and assigned to cluster #3, while there are only a few (139) outliers left surrounding it (Fig. 2b). In other words, in UIUC dataset there exists a dominant connected subset in which the airfoils are densely distributed everywhere (otherwise they would not become cores in HDBSCAN). This cluster can be regarded as the group of 'canonical' airfoil designs. The illustrations in Fig. 2c and Fig. 2d seem to testify to this claim, as subjectively those noise-airfoils have a higher variety of unusual shapes than those in cluster #3, although this is a qualitative conjecture on our part.

Despite its mechanical and systematic procedure, HDBSCAN is still highly sensitive to varying m_{pts} and m_{clSize}, which can significantly alter each point's cluster assignments. For instance, when we increase m_{pts} and m_{clSize} to 10 and 50 respectively to sift out low density regions and small clusters more radically, we can get a different result as shown in Fig. 3. Due to the current higher standard for what constitutes a cluster, only a few (4) regions of higher density in the previous cluster #3 now qualify as clusters. In that sense, we can regard these four as the most typical groups of airfoils among all canonical airfoil designs. We can plot the mass centers of these clusters as the representatives of them, as shown in Fig. 3c.

4.1.3 What happens to Latent Clustering when we destroy Isometry?.

It may not be easy to appreciate the importance of isometry regularization without a contrast, so here we purposefully sabotage the airfoil autoencoder's isometricity to demonstrate how its absence may lead to distortion of the latent set and hence to a misleading clustering result. Specifically, our overall 'trick' below will be to enforce the exactly same autoencoder reconstruction loss, but selectively destroy the latent space's isometricity using a tunable bijective distortion that allows us to increasingly "break" only the isometricity between the design and latent spaces.

To do this in practice, we first unlock the autoencoder's isometricity by attaching a flow-based model like RealNVP [61] to its latent space. Specifically, let g and e be the decoder and

encoder of the isometric airfoil autoencoder and f be the RealNVP, then construct a new latent set with $f \circ e(\mathcal{X})$, where $\mathcal{X}$ denotes the UIUC dataset. We can thereby regard $e' := f \circ e$ and $g' := g \circ f^{-1}$ as the new pair of encoder and decoder between the UIUC dataset and the new latent set, where f^{-1} can be readily retrieved given that f is a diffeomorphism between the old and new latent spaces by construction. Therefore, the reconstruction loss does not change at all for any flow f, as $\|g' \circ e'(x) - x\| = \|g \circ f^{-1} \circ f \circ e - x\| = \|g \circ e(x) - x\|$. In other words, through this new autoencoder we get a new latent set that is also homeomorphic to the dataset (and thus to the old isometric latent set), but we can train f while fixing e and g to tamper with its isometricity.

Next we show this unlocked autoencoder can obtain a latent set (as shown in Fig. 4) dramatically different from the isometric one in Fig. 2b. We start by constructing a 3D Gaussian mixture target distribution $p_t(z)$ of three components centered at $[1, 0, 0]$, $[0, 1, 0]$ and $[0, 0, 1]$ respectively, each having an isotropic standard deviation equal to 0.1. Then we drive the empirical distribution $p_f(z)$ of the new latent codes $e'(\mathcal{X})$ to p_t by only training f and leaving g and e fixed. Since we do not know the probability density function over the isometric latent set $e(\mathcal{X})$, there is no way to train f via log likelihood maximization, so we achieve this instead by minimizing the Sinkhorn divergence between p_t and p_f. More information about this method can be found in [69]. This encourages f to transform the old isometric latent set into a new one with most of its points concentrating around $[1, 0, 0]$, $[0, 1, 0]$ and $[0, 0, 1]$, and thus may form three big clusters instead of one. If our claim is true that the lack of isometricity can lead to distortion, this 3-cluster latent set should be attainable as long as f has enough complexity.

Figure 4 shows the clustering result of the non-isometric latent set after training. It now consists of three major clusters (#7, #8, #13) corresponding to the three Gaussian components of p_t. This stands in stark contrast to the single giant cluster in Fig. 2b, despite the two models having identical reconstruction error, and exemplifies the ambiguity problem mentioned in §2.3. That is, we cannot use reconstruction error alone to know that the resulting latent space distances preserve isometry, and the isometric autoencoder provides a mean to regularize this directly. While in real applications an autoencoder may not distort the dataset's distance and topology quite as severely as in our extreme example, naïve models have no protection against it. As such, it is therefore always worthwhile to switch on the isometry regularization when doing latent analyses relying on distance functions.

4.2 Latent Shape Analysis of SU2 Airfoil Optimization

As mentioned earlier in §2.3, one great advantage of the autoencoder over many other non-parameterized dimension reduction methods is its ability to perform *amortized* inference, *i.e.*, it can immediately process unseen data samples without retraining. In the following experiments, we exploit this advantage to analyze the SU2 airfoil dataset [5] consisting of 1245 airfoil-AoA (angle of attack) pairs that are optimized under a variety of boundary conditions by the SU2 CFD toolset. Specifically, we postulate that these optimized airfoil shapes still reside on the UIUC airfoil manifold, and apply the pre-trained UIUC airfoil autoencoder di-

Copyright © 2023 by ASME

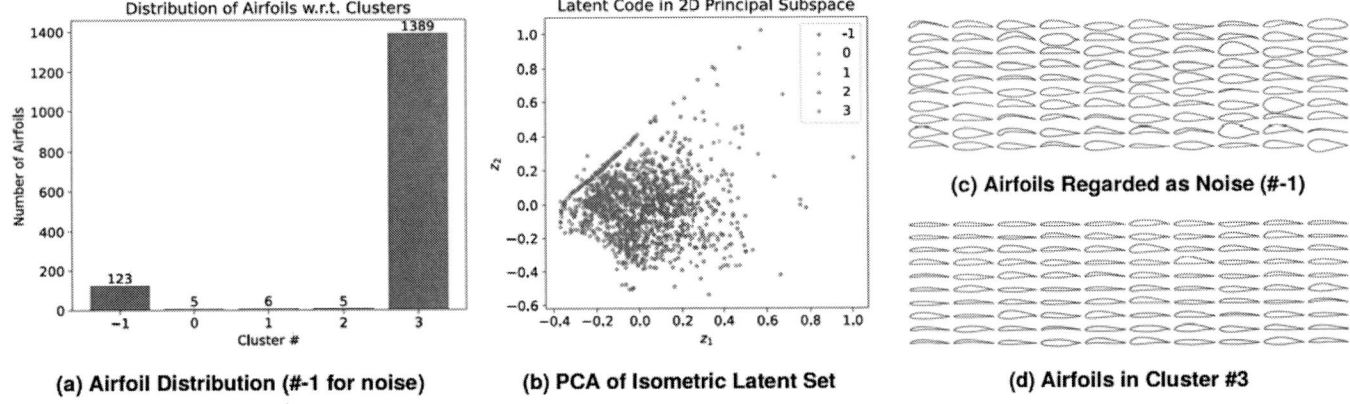

(a) Airfoil Distribution (#-1 for noise)   **(b) PCA of Isometric Latent Set**   **(c) Airfoils Regarded as Noise (#-1)**

(d) Airfoils in Cluster #3

FIGURE 2: RESULT OF LATENT HDBSCAN ON UIUC AIRFOIL DATASET, WITH $m_{pts} = m_{clSize} = 5$.

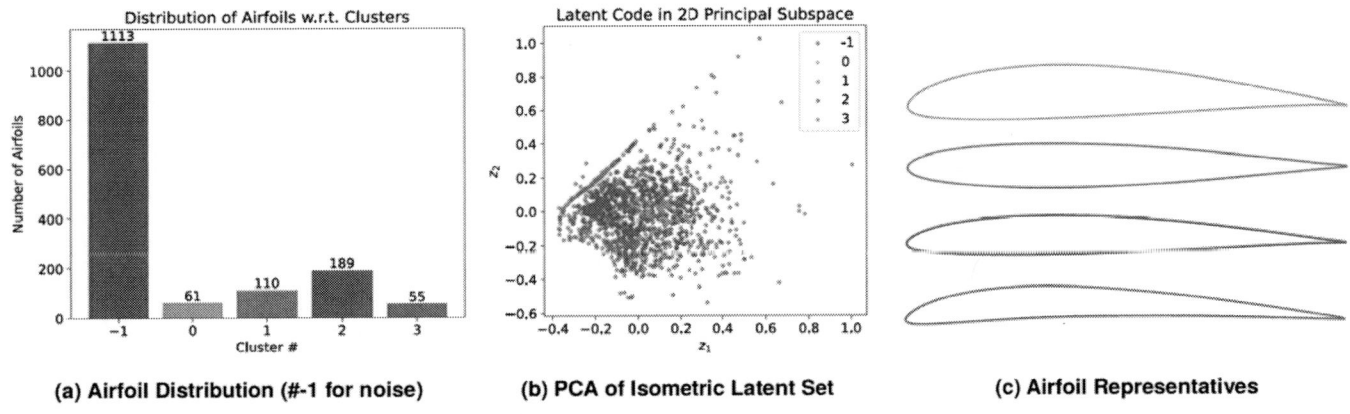

(a) Airfoil Distribution (#-1 for noise)   **(b) PCA of Isometric Latent Set**   **(c) Airfoil Representatives**

FIGURE 3: RESULT OF LATENT HDBSCAN ON UIUC AIRFOIL DATASET, WITH $m_{pts} = 10$ AND $m_{clSize} = 50$.

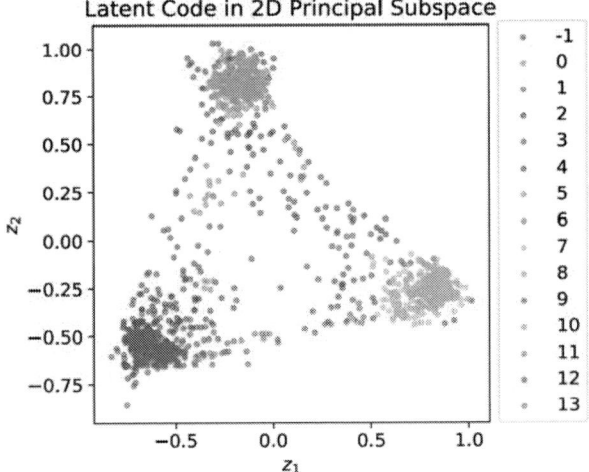

FIGURE 4: PCA OF AND HDBSCAN ON NON-ISOMETRIC LATENT SET, WITH $m_{pts} = m_{clSize} = 5$.

rectly on the SU2 dataset (shapes only, without AoAs) to derive its isometric latent representation for the following latent analyses.

4.2.1 Latent Clustering of Optimized Airfoils. We perform HDBSCAN with $m_{pts} = 10$ and $m_{clSize} = 30$ on the latent

set of SU2 airfoils first and illustrate its result in Fig. 5. Compared with the 1528 UIUC airfoils that cover a large area and comprise a giant cluster, the 1245 SU2 airfoils only occupy a few small regions and form 5 clusters, as we can see in Fig. 5a. This higher regional density is the prime reason why in this case we do not reuse the previous $m_{pts} = m_{clSize} = 5$ setting for Fig. 2, as otherwise it will produce over 30 tiny clusters and leave about 25% airfoils categorized as noise, which is not reasonable.

The SU2 airfoils' distinctive regional concentration probably stems from the way they were created. Chen et al. [5] observed that the gradient-based airfoil adjoint optimization often arrived at sub-optimal local optima. Consequently, for each input condition they performed eight restarts of the adjoint optimization by selecting a diverse set of starting airfoils sampled from Bézier-GAN approximating the UIUC airfoil distribution. In [5], the final training set included only the highest efficiency final design from these eight optimization trials. In practice, in many cases the SU2 optimizer found that altering only the AoA was sufficient to find the most efficiency design, compared to modifying the airfoil shape. As such, we would expect to find dense clusters of initial shapes (in the latent space) for the cases where the optimized design only modified the AoA. We shall verify if this is true next.

Copyright © 2023 by ASME

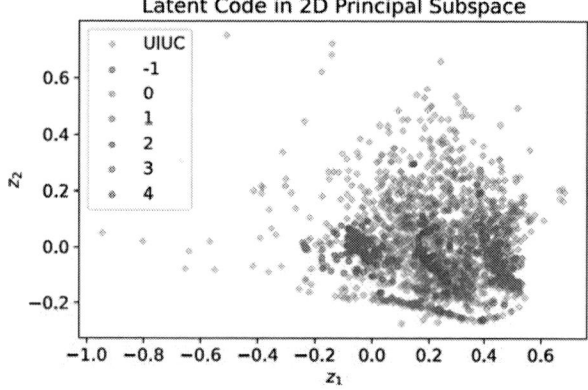

(a) PCA of SU2 Dataset's Isometric Latent Set

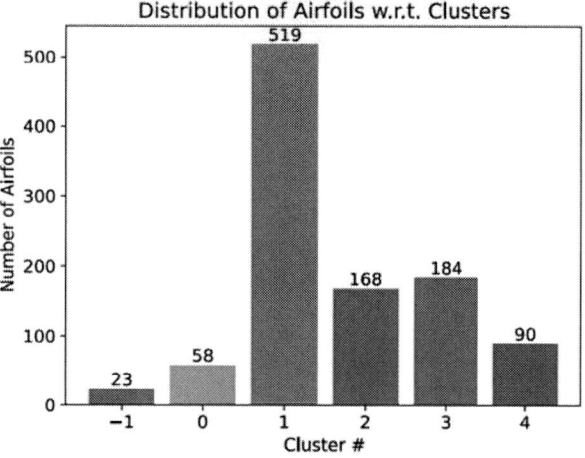

(b) SU2 Airfoil Distribution (#-1 for noise)

FIGURE 5: RESULT OF LATENT HDBSCAN ON SU2 AIRFOIL DATASET, WITH $m_{pts} = 10$ AND $m_{clSize} = 30$.

4.2.2 How much do airfoils morph in optimization?. To investigate the airfoil shape's degree of variation during optimization, especially in comparison to that of the AoA, we perform PCA on the Cartesian product of the normalized 3D latent code and normalized AoA—namely their 4D concatenation. To avoid distortion of the latent code, when normalizing/standardizing the code we use the standard deviation across all three latent dimensions as the scaling factor, instead of scaling it dimension-wisely after mean centering. By doing this, we can make sure the normalized latent set is still isometric to the dataset *up to a scale factor* [23], so that some potentially trivial latent dimensions will not be emphasized relative to the principal ones after normalization.

The new PCA result is plotted in Fig. 6, where each point is colored according to the cluster found earlier plotted in Fig. 5. In contrast to Fig. 5a, the introduction of AoA in PCA induces a reorientation that reveals each cluster's linear pattern along the AoA direction (which is illustrated by the dashed line in Fig. 6). There are two naïve takeaways from this graph, if we (for now) ignore some technical caveats and presume that smaller variation in the direction perpendicular to the AoA direction indicates

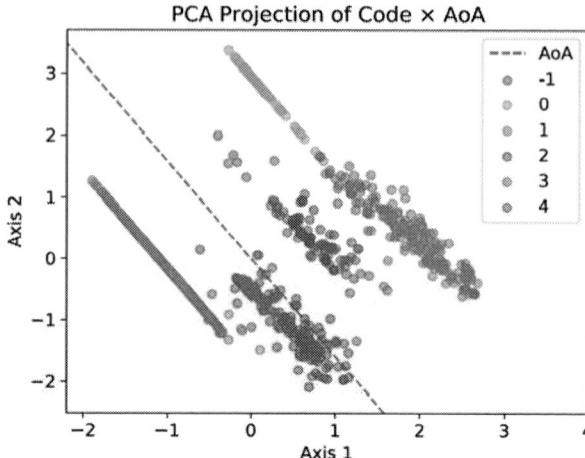

FIGURE 6: 2D PRINCIPAL PROJECTION OF THE CARTESIAN PRODUCT OF AIRFOIL LATENT CODE × AOA.

smaller change in shape:

1. In general, the SU2 adjoint optimizer tends to optimize the airfoil's lift-drag efficiency for different input conditions—*Re*, *Ma*, & Lift Coefficient—more by adjusting its AoA than by morphing its shape, given that each cluster's variation along the AoA direction is more substantial than that along the perpendicular direction.

2. Not all airfoils are created equal. We can notice that the airfoil shapes in cluster #0 and #1 have much smaller variation compared with the ones in cluster #2, #3 and #4. If it is true that almost all airfoils in each cluster are optimized from the same initial design from among the provided eight restarts,[1] then this suggests some of the eight initial airfoils are better starting shapes (*i.e.*, lie close to the basin of the optima) under certain ranges of boundary conditions compared to others, leading the SU2 optimizer to only need to adjust the AoA to improve efficiency. The overwhelming number of airfoils in cluster #1 (Fig. 5b) also reflects this inequality, suggesting that the initial design that, when optimized, yielded the shapes in cluster #1 is not just optimal for some conditions, but rather broadly optimal under a wide range of boundary conditions, compared with the others.

Despite seeming plausible, these two takeaways need to be taken with a grain of salt, as there are several caveats that may undermine their validity:

1. The high variation along the AoA direction in Fig. 6 may not be attributed to the AoA alone, as some variations in the latent code may also have been projected along it under PCA. However, we think it is still safe to regard AoA as the dominant contributor to this direction's variation and ignore the airfoil shape's influence, as otherwise we should have already seen a similar linear pattern in Fig. 5a, *i.e.*, the PCA without AoA.

---

[1]This is likely considering how separated these clusters are away from one another in the latent space. We will also see this visually later in Fig. 7

2. The distance variation among the latent codes shown in Fig. 6 may not be directly comparable to variations in AoA. For instance, when comparing data A and data B, we may come up with a Mahalanobis distance function for A that scales up A's space arbitrarily large, such that the variation of data A measured by it is also arbitrarily large. It is then pointless to compare this variation with a normal Euclidean-based one of data B. In our case, the variations of shape and AoA are based on the Euclidean distances defined on the normalized latent space and normalized AoA space respectively. Because the latent set is isometric to the dataset, the variation of shape is also equivalently based on the geodesic distance on the shape manifold, up to the shape code's normalization factor.

3. The previous caveat seems to not affect our second take-away, because for that claim we only compare each cluster's shape variation (not including the AoA). The variation of shape, as mentioned above, is based on the geodesic distance between shapes. However, despite its awareness of the data manifold's geometry, whether or not the geodesic distance is reasonable for comparing shapes is still an open question. For instance, it might actually not be aligned with a human's "perceptual metric" [70], such that a large difference between airfoils in terms of the geodesic distance only corresponds to a small visual difference, and *vice versa*. In addition, sometimes a subtle visual change in shape may lead to a huge shift in the design's performance. Since the design's performance is what we care about ultimately, it might be more reasonable to compare shapes by measuring their difference in performance.

These concerns, together with our above takeaways, require future research. Nonetheless, it could be informative to investigate the third caveat tentatively by illustrating all airfoil shapes in each cluster. This allows us to see how varied they are visually and whether the degrees of visual variation agree with the Euclidean/geodesic based variations in Fig. 6. We can then assess, albeit qualitatively, if the geodesic distance agrees with our "perceptual metric."

To do this, for each cluster we superpose all its airfoils and plot them altogether on the left of Fig. 7. Visually, we see that cluster #0 and #1 (Fig. 7a, 7c) have much lower variation in their airfoil shapes compared to cluster #2 to #4 (Fig. 7e, 7g, 7i). This agrees well with the variation difference shown in Fig. 6. This suggests the geodesic distance is at least a reasonable choice for evaluating the visual difference between shapes. On the right side of Fig. 7, we also plot the five initial airfoil designs—handpicked out of the eight restart candidates—that look the most similar to the airfoils in different clusters. We can see that for all clusters (maybe except #3), the optimized airfoils look almost identical to their corresponding initial designs, which testifies to both of our takeaways.

4.2.3 Relationships between Airfoil Shape, AoA and Boundary Condition.
So far we have only analyzed the shape-AoA configurations of SU2 airfoils, without taking any boundary conditions into account. One potential hypothesis for the clustering of shapes in the latent space could be that each cluster

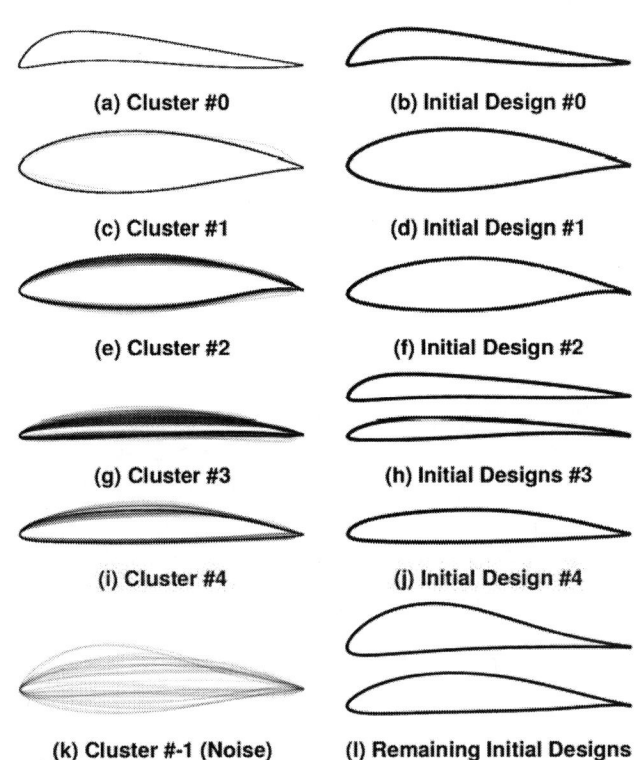

FIGURE 7: AIRFOILS IN DIFFERENT CLUSTERS AND THEIR INITIAL DESIGNS

corresponds to some identifiable change in the boundary conditions—*i.e.*, that certain clusters arise naturally when optimizing a design within a range of conditions, and that the optimal cluster switches at some flow regime.

To investigate this hypothesis, we demonstrate the conditional distribution of airfoil shapes w.r.t. to the three boundary condition parameters—Reynolds number, Mach number and lift coefficient. Figure 8 scatter plots all the boundary conditions in the SU2 dataset, with each point colored according to its corresponding shape's cluster. We observe several features from this plot:

1. There is a conspicuous laminated pattern along the Mach number dimension (bottom left and bottom right), which divides this dimension into five distinctive segments. Each segment is prominently occupied by the airfoil shapes in a single cluster. As the Mach number incrementally increases from ~0.2 to ~0.8, the optimal airfoil shape morphs from cluster to cluster following the order #0 → #1 → #2 → #4 → #3. Not only that, this morphing is also in general *monotonic* or *injective*, namely no airfoil cluster appears *dominantly* more than once in different Mach number segments.

2. In contrast, the optimal shape is in general independent of Reynolds number and lift coefficient, as least within the SU2 dataset's cubic boundary condition regime. This is reflected in both how uniformly the shape clusters distribute in the Re-Lift subspace (top right), and how perpendicular

Copyright © 2023 by ASME

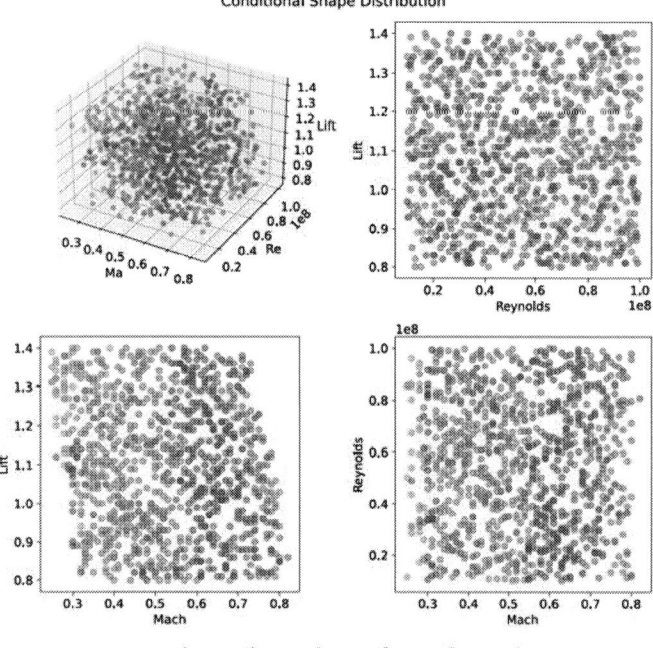

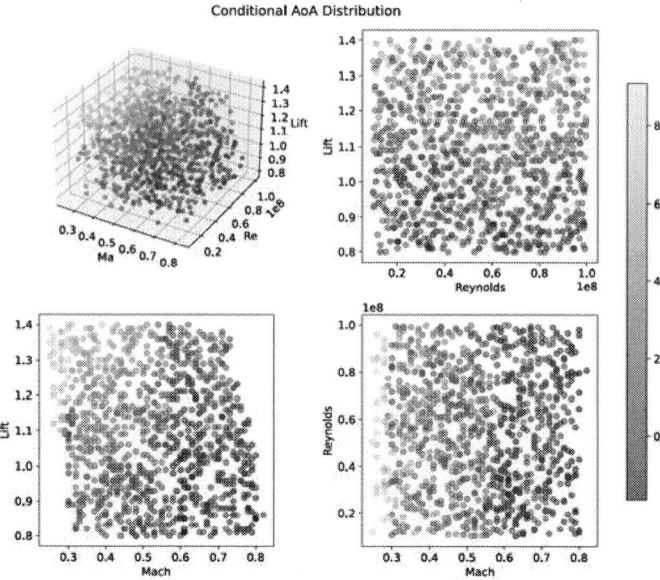

FIGURE 9: DISTRIBUTION OF AOA W.R.T. MACH NUMBER, REYNOLDS NUMBER AND LIFT COEFFICIENT

FIGURE 8: DISTRIBUTION OF SHAPE CLUSTERS W.R.T. MACH NUMBER, REYNOLDS NUMBER AND LIFT COEFFICIENT

the lamination boundaries are to the Ma dimension (bottom left and bottom right).

3. Again, not all airfoils are created equal. For example, cluster #1 dominates the wide flow velocity regime roughly between Mach 0.3 to 0.55, whereas cluster #4 concentrates around Ma 0.7, at the boundary between cluster #2 and #3.

Overall, in aerodynamic design optimization, the airfoil shape morphs primarily to adapt to its working velocity. This is consistent with existing design practice of customizing aerodynamic surfaces to different speed regimes (*e.g.*, Boeing 737 vs Concorde). Not only that, as expected, an airfoil shape that is optimal within a given speed regime is not likely to be optimal within another. Moreover, different airfoil shapes also have different sensitivities to velocity, as some are adaptable to a wider range of speeds than the others. It is intriguing that Fig. 8 found this known behavior in an unsupervised manner only through analyzing the learned latent space rather than being directly trained on the boundary conditions.

Figure 9 demonstrates whether this same pattern exists for changes in the AoA, with each point colorized according to its AoA value. This graph shows that, in general, the optimal AoA is correlated not only with lift coefficient but also with Mach number. The latter, however, might be considerably affected by the shape lamination along the Ma dimension, as each cluster of shapes may have a distinctive optimal range of AoAs that couples with it, such that the AoA-Ma correlation mainly results from this cluster-wise coupling. Indeed, we can notice this coupling between shape cluster and AoA in Fig. 10.

To avoid the influence of this shape lamination, we instead study how AoA varies w.r.t Ma, Re, and Lift with the airfoil

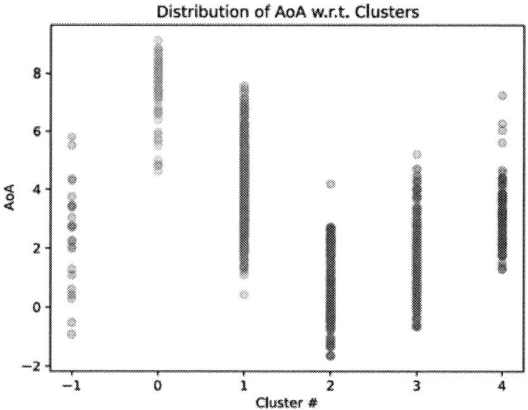

FIGURE 10: DISTRIBUTION OF AOA W.R.T. CLUSTERS

shape *fixed*. Specifically, we evaluate the Pearson correlation coefficients (PCC) between AoA and the three conditions in each cluster (recall that the shapes in each cluster look very similar, thus may be roughly regarded as fixed). The results are demonstrated in Fig. 11, which reveals AoA's moderate negative correlation with Mach number (PCC on average -0.57, weighted by cluster size), high positive correlation with lift coefficient (PCC on average 0.83, likewise) and negligible correlation with Reynolds number (PCC on average 0.03, likewise).

5. CONCLUSIONS

In this paper, we employed the isometric autoencoder to learn an isometric representation of the airfoil designs that preserves the geodesic distance. Then we performed distance-based analyses such as clustering in the isometric latent space to study different airfoil dataset's characteristics and complexity, while investigating the necessity and validity of preserving the geodesic distance

Copyright © 2023 by ASME

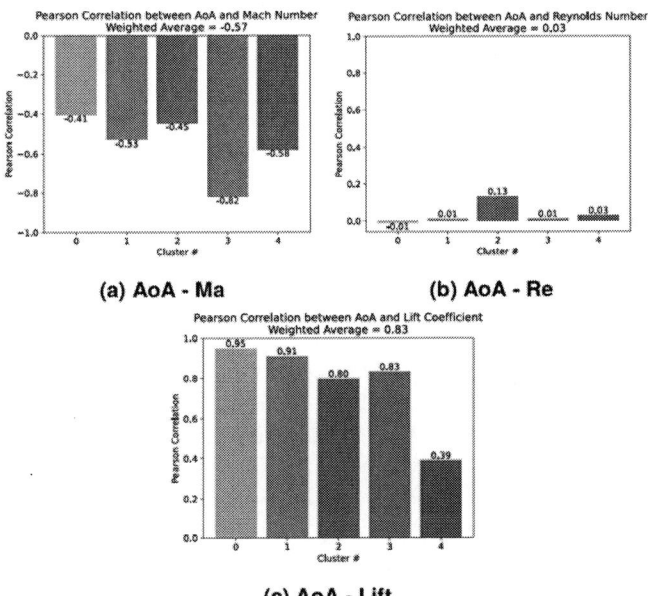

(a) AoA - Ma   (b) AoA - Re

(c) AoA - Lift

FIGURE 11: PEARSON CORRELATIONS BETWEEN AOA AND BOUNDARY CONDITIONS IN DIFFERENT CLUSTERS

in the latent space.

As to the preservation of the geodesic distance, we found in §4.1.3 that without it the latent set produced by the autoencoder is at risk of comprising an arbitrary number of clusters, which may mislead engineers' interpretation of the design distribution. It is therefore necessary to impart isometricity to the autoencoder when analyzing data in its latent space. We also verified that the geodesic distance agrees well with our visual perception when it comes to detecting shape variation (at least for the airfoils, §4.2.2), hence it is reasonable to preserve it for shape analysis.

Through the lens of the isometric autoencoder and HDB-SCAN clustering, we found that compared to the high-quality and diverse UIUC dataset, the SU2 dataset of optimized airfoils has far less variation in its airfoil shapes. This may be blamed on two culprits. First, when optimizing the airfoil's L/D efficiency, the SU2 adjoint optimizer prefers adjusting the angle of attack (AoA) to morphing the airfoil shape, probably because this is more effective in increasing the L/D ratio in terms of the adjoint gradient (specifically, the L/D objective's gradient w.r.t. the AoA may have much larger norm than that w.r.t. the airfoil spline's control parameters). Second, when creating the SU2 dataset with the SU2 optimizer, for each boundary condition we only performed eight restarts from the eight candidate airfoils and kept only the best final design. This, paired with the first issue, leads to the lack of diversity in the SU2 dataset. A future way to improve the quality of the SU2 dataset would be to either introduce more candidate airfoils or replace the current adjoint-based optimizer.

Despite the SU2 dataset's diversity issue, analyzing it in the isometric latent space still provides many insights into the conditional distribution of optimized airfoils. It shows that the cruising speed is the primary factor in the design of airfoil shapes, and not all airfoil shapes are created equal, as some can work optimally in a broader range of speed. In addition, an airfoil shape that

works optimally in one speed regime is not likely to do so in another—in other words, it is unlikely to find an airfoil that is universally optimal at every speed. Moreover, when the airfoil shape is determined, if we want to increase or maintain its lift coefficient while the speed goes down, the airfoil should pitch up. In conjunction, these two results suggests the condition distribution $p(\text{shape}, \text{AoA} \mid \text{Ma}, \text{Re}, \text{Lift})$ that the inverse airfoil design models in [5] tried to capture might be factorized and simplified into $p(\text{AoA} \mid \text{Lift}, \text{Ma}, \text{shape}) \cdot p(\text{shape} \mid \text{Ma})$. Although the insights on this specific domain are already known from past human efforts in airfoil design, what is unique is that the proposed Isometric AE uncovered these without explicitly being trained to do so, and that this technique can be applied to many other domains. This demonstrates the value of constructing isometry via Isometric AEs to new, more complex problems. We are investigating more complicated, high-dimensional design problems as one avenue of future work, and expecting more meticulous research on the sensitivity of the autoencoder and clustering method's hyperparameters.

ACKNOWLEDGEMENTS

We acknowledge the support from the National Science Foundation through award #1943699 as well as ARPA-E award DE-AR0001216.

REFERENCES

[1] Panchal, Jitesh H, Fuge, Mark, Liu, Ying, Missoum, Samy and Tucker, Conrad. "Machine learning for engineering design." *Journal of Mechanical Design* Vol. 141 No. 11 (2019).

[2] *Topology Design With Conditional Generative Adversarial Networks*, Vol. Volume 2A: 45th Design Automation Conference of *International Design Engineering Technical Conferences and Computers and Information in Engineering Conference* (2019). DOI 10.1115/DETC2019-97833.

[3] Yu, Yonggyun, Hur, Taeil, Jung, Jaeho and Jang, In Gwun. "Deep learning for determining a near-optimal topological design without any iteration." *Structural and Multidisciplinary Optimization* Vol. 59 No. 3 (2019): pp. 787–799. DOI 10.1007/s00158-018-2101-5.

[4] Yilmaz, Emre and German, Brian. "Conditional Generative Adversarial Network Framework for Airfoil Inverse Design." *AIAA AVIATION 2020 FORUM*. 2020. American Institute of Aeronautics and Astronautics, VIRTUAL EVENT. DOI 10.2514/6.2020-3185.

[5] Chen, Qiuyi, Wang, Jun, Pope, Phillip, Chen, Wei and Fuge, Mark. "Inverse design of two-dimensional airfoils using conditional generative models and surrogate log-likelihoods." *Journal of Mechanical Design* Vol. 144 No. 2 (2022): p. 021712.

[6] Wang, Jun, Chen, Wei (Wayne), Da, Daicong, Fuge, Mark and Rai, Rahul. "IH-GAN: A conditional generative model for implicit surface-based inverse design of cellular structures." *Computer Methods in Applied Mechanics and Engineering* Vol. 396 (2022): p. 115060. DOI 10.1016/j.cma.2022.115060.

Copyright © 2023 by ASME

[7] Nobari, Amin Heyrani, Chen, Wei and Ahmed, Faez. "Range-GAN: Range-Constrained Generative Adversarial Network for Conditioned Design Synthesis." (2021). DOI 10.48550/ARXIV.2103.06230.

[8] Yang, Zijiang, Li, Xiaolin, Catherine Brinson, L., Choudhary, Alok N., Chen, Wei and Agrawal, Ankit. "Microstructural Materials Design Via Deep Adversarial Learning Methodology." *Journal of Mechanical Design* Vol. 140 No. 11 (2018). DOI 10.1115/1.4041371.

[9] Mosser, Lukas, Dubrule, Olivier and Blunt, Martin J. "Reconstruction of three-dimensional porous media using generative adversarial neural networks." *Phys. Rev. E* Vol. 96 (2017): p. 043309. DOI 10.1103/PhysRevE.96.043309.

[10] Xue, Tianju, Wallin, Thomas J., Menguc, Yigit, Adriaenssens, Sigrid and Chiaramonte, Maurizio. "Machine learning generative models for automatic design of multi-material 3D printed composite solids." *Extreme Mechanics Letters* Vol. 41 (2020): p. 100992. DOI 10.1016/j.eml.2020.100992.

[11] Chen, Wei, Chiu, Kevin and Fuge, Mark. "Aerodynamic design optimization and shape exploration using generative adversarial networks." *AIAA Scitech 2019 Forum*: p. 2351. 2019.

[12] Chen, Wei and Ahmed, Faez. "PaDGAN: Learning to Generate High-Quality Novel Designs." *Journal of Mechanical Design* Vol. 143 No. 3 (2020). DOI 10.1115/1.4048626.

[13] Liang, Liang, Liu, Minliang, Martin, Caitlin and Sun, Wei. "A deep learning approach to estimate stress distribution: a fast and accurate surrogate of finite-element analysis." *Journal of the Royal Society, Interface* Vol. 15 No. 138 (2018): p. 20170844. DOI 10.1098/rsif.2017.0844.

[14] Jiang, Haoliang, Nie, Zhenguo, Yeo, Roselyn, Farimani, Amir Barati and Kara, Levent Burak. "StressGAN: A Generative Deep Learning Model for Two-Dimensional Stress Distribution Prediction." *Journal of Applied Mechanics* Vol. 88 No. 5 (2021). DOI 10.1115/1.4049805.

[15] Nie, Zhenguo, Jiang, Haoliang and Kara, Levent Burak. "Stress Field Prediction in Cantilevered Structures Using Convolutional Neural Networks." *Journal of Computing and Information Science in Engineering* Vol. 20 No. 1 (2019). DOI 10.1115/1.4044097.

[16] Kochkov, Dmitrii, Smith, Jamie A., Alieva, Ayya, Wang, Qing, Brenner, Michael P. and Hoyer, Stephan. "Machine learning–accelerated computational fluid dynamics." *Proceedings of the National Academy of Sciences* Vol. 118 No. 21 (2021): p. e2101784118. DOI 10.1073/pnas.2101784118.

[17] Kim, Byungsoo, Azevedo, Vinicius C., Thuerey, Nils, Kim, Theodore, Gross, Markus and Solenthaler, Barbara. "Deep Fluids: A Generative Network for Parameterized Fluid Simulations." *Computer Graphics Forum* Vol. 38 No. 2 (2019): pp. 59–70. DOI 10.1111/cgf.13619.

[18] Bellman, R. *Dynamic Programming*. Princeton Landmarks in Mathematics and Physics, Princeton University Press (2010).

[19] Belkin, Mikhail and Niyogi, Partha. "Laplacian Eigenmaps and Spectral Techniques for Embedding and Clustering." Dietterich, T., Becker, S. and Ghahramani, Z. (eds.). *Advances in Neural Information Processing Systems*, Vol. 14. 2001. MIT Press.

[20] Van der Maaten, Laurens and Hinton, Geoffrey. "Visualizing data using t-SNE." *Journal of machine learning research* Vol. 9 No. 11 (2008).

[21] McInnes, Leland, Healy, John and Melville, James. "Umap: Uniform manifold approximation and projection for dimension reduction." *arXiv preprint arXiv:1802.03426* (2018).

[22] Chen, Wei, Fuge, Mark and Chazan, Jonah. "Design Manifolds Capture the Intrinsic Complexity and Dimension of Design Spaces." *Journal of Mechanical Design* Vol. 139 No. 5 (2017). DOI 10.1115/1.4036134.

[23] Chen, Nutan, Klushyn, Alexej, Ferroni, Francesco, Bayer, Justin and Van Der Smagt, Patrick. "Learning flat latent manifolds with vaes." *arXiv preprint arXiv:2002.04881* (2020).

[24] Gropp, Amos, Atzmon, Matan and Lipman, Yaron. "Isometric autoencoders." *arXiv preprint arXiv:2006.09289* (2020).

[25] Yonghyeon, LEE, Yoon, Sangwoong, Son, MinJun and Park, Frank C. "Regularized autoencoders for isometric representation learning." *International Conference on Learning Representations*. 2022.

[26] Ester, Martin, Kriegel, Hans-Peter, Sander, Jörg, Xu, Xiaowei et al. "A density-based algorithm for discovering clusters in large spatial databases with noise." *kdd*, Vol. 96. 34: pp. 226–231. 1996.

[27] Campello, Ricardo JGB, Kröger, Peer, Sander, Jörg and Zimek, Arthur. "Density-based clustering." *Wiley Interdisciplinary Reviews: Data Mining and Knowledge Discovery* Vol. 10 No. 2 (2020): p. e1343.

[28] SIGKDD, A. "sigkdd test of time award,"." (2014).

[29] Schubert, Erich, Sander, Jörg, Ester, Martin, Kriegel, Hans Peter and Xu, Xiaowei. "DBSCAN revisited, revisited: why and how you should (still) use DBSCAN." *ACM Transactions on Database Systems (TODS)* Vol. 42 No. 3 (2017): pp. 1–21.

[30] Gan, Junhao and Tao, Yufei. "DBSCAN revisited: Misclaim, un-fixability, and approximation." *Proceedings of the 2015 ACM SIGMOD international conference on management of data*: pp. 519–530. 2015.

[31] Ankerst, Mihael, Breunig, Markus M, Kriegel, Hans-Peter and Sander, Jörg. "OPTICS: Ordering points to identify the clustering structure." *ACM Sigmod record* Vol. 28 No. 2 (1999): pp. 49–60.

[32] Hinneburg, Alexander, Keim, Daniel A et al. *An efficient approach to clustering in large multimedia databases with noise*. Vol. 98. Bibliothek der Universität Konstanz (1998).

[33] McInnes, Leland, Healy, John and Astels, Steve. "hdbscan: Hierarchical density based clustering." *J. Open Source Softw.* Vol. 2 No. 11 (2017): p. 205.

[34] Johnson, Stephen C. "Hierarchical clustering schemes." *Psychometrika* Vol. 32 No. 3 (1967): pp. 241–254.

[35] MacQueen, J. "Classification and analysis of multivariate observations." *5th Berkeley Symp. Math. Statist. Proba-*

bility: pp. 281–297. 1967. University of California Los Angeles LA USA.

[36] Bishop, Christopher M and Nasrabadi, Nasser M. *Pattern recognition and machine learning*. Vol. 4. Springer (2006).

[37] Andoni, Alexandr. "Nearest neighbor search: the old, the new, and the impossible." Ph.D. Thesis, Massachusetts Institute of Technology. 2009.

[38] Kulesza, Alex, Taskar, Ben et al. "Determinantal point processes for machine learning." *Foundations and Trends® in Machine Learning* Vol. 5 No. 2–3 (2012): pp. 123–286.

[39] Ahmed, Faez and Fuge, Mark. "Ranking ideas for diversity and quality." *Journal of Mechanical Design* Vol. 140 No. 1 (2018): p. 011101.

[40] Beyer, Kevin, Goldstein, Jonathan, Ramakrishnan, Raghu and Shaft, Uri. "When is "nearest neighbor" meaningful?" *Database Theory—ICDT'99: 7th International Conference Jerusalem, Israel, January 10–12, 1999 Proceedings 7*: pp. 217–235. 1999. Springer.

[41] Aggarwal, Charu C, Hinneburg, Alexander and Keim, Daniel A. "On the surprising behavior of distance metrics in high dimensional space." *Database Theory—ICDT 2001: 8th International Conference London, UK, January 4–6, 2001 Proceedings 8*: pp. 420–434. 2001. Springer.

[42] Weinberger, Kilian Q and Saul, Lawrence K. "Distance metric learning for large margin nearest neighbor classification." *Journal of machine learning research* Vol. 10 No. 2 (2009).

[43] Fefferman, Charles, Mitter, Sanjoy and Narayanan, Hariharan. "Testing the manifold hypothesis." *Journal of the American Mathematical Society* Vol. 29 No. 4 (2016): pp. 983–1049.

[44] Roth, Peter M, Hirzer, Martin, Köstinger, Martin, Beleznai, Csaba and Bischof, Horst. "Mahalanobis distance learning for person re-identification." *Person re-identification* (2014): pp. 247–267.

[45] Lee, Kimin, Lee, Kibok, Lee, Honglak and Shin, Jinwoo. "A simple unified framework for detecting out-of-distribution samples and adversarial attacks." *Advances in neural information processing systems* Vol. 31 (2018).

[46] Bronstein, Michael M, Bruna, Joan, Cohen, Taco and Veličković, Petar. "Geometric deep learning: Grids, groups, graphs, geodesics, and gauges." *arXiv preprint arXiv:2104.13478* (2021).

[47] Lee, John M. *Riemannian manifolds: an introduction to curvature*. Vol. 176. Springer Science & Business Media (2006).

[48] Fletcher, P Thomas, Lu, Conglin, Pizer, Stephen M and Joshi, Sarang. "Principal geodesic analysis for the study of nonlinear statistics of shape." *IEEE transactions on medical imaging* Vol. 23 No. 8 (2004): pp. 995–1005.

[49] Nilsson, Jens, Fioretos, Thoas, Höglund, Mattias and Fontes, Magnus. "Approximate geodesic distances reveal biologically relevant structures in microarray data." *Bioinformatics* Vol. 20 No. 6 (2004): pp. 874–880.

[50] Srivastava, Anuj, Klassen, Eric, Joshi, Shantanu H and Jermyn, Ian H. "Shape analysis of elastic curves in eu-

clidean spaces." *IEEE transactions on pattern analysis and machine intelligence* Vol. 33 No. 7 (2010): pp. 1415–1428.

[51] Tosi, Alessandra, Hauberg, Søren, Vellido, Alfredo and Lawrence, Neil D. "Metrics for probabilistic geometries." *arXiv preprint arXiv:1411.7432* (2014).

[52] Chen, Nutan, Klushyn, Alexej, Kurle, Richard, Jiang, Xueyan, Bayer, Justin and Smagt, Patrick. "Metrics for deep generative models." *International Conference on Artificial Intelligence and Statistics*: pp. 1540–1550. 2018. PMLR.

[53] Linderman, George C and Steinerberger, Stefan. "Clustering with t-SNE, provably." *SIAM Journal on Mathematics of Data Science* Vol. 1 No. 2 (2019): pp. 313–332.

[54] Tenenbaum, Joshua B, Silva, Vin de and Langford, John C. "A global geometric framework for nonlinear dimensionality reduction." *science* Vol. 290 No. 5500 (2000): pp. 2319–2323.

[55] Roweis, Sam T and Saul, Lawrence K. "Nonlinear dimensionality reduction by locally linear embedding." *science* Vol. 290 No. 5500 (2000): pp. 2323–2326.

[56] Goodfellow, Ian, Bengio, Yoshua and Courville, Aaron. *Deep learning*. MIT press (2016).

[57] Coifman, Ronald R and Lafon, Stéphane. "Diffusion maps." *Applied and computational harmonic analysis* Vol. 21 No. 1 (2006): pp. 5–30.

[58] Lee, John. *Introduction to topological manifolds*. Vol. 202. Springer Science & Business Media (2010).

[59] Asinomás. "Composition equals identity function." Mathematics Stack Exchange. URL https://math.stackexchange.com/q/1445743.

[60] Dinh, Laurent, Krueger, David and Bengio, Yoshua. "Nice: Non-linear independent components estimation." *arXiv preprint arXiv:1410.8516* (2014).

[61] Dinh, Laurent, Sohl-Dickstein, Jascha and Bengio, Samy. "Density estimation using real nvp." *arXiv preprint arXiv:1605.08803* (2016).

[62] Kingma, Durk P and Dhariwal, Prafulla. "Glow: Generative flow with invertible 1x1 convolutions." *Advances in neural information processing systems* Vol. 31 (2018).

[63] Zhang, Hongyi, Cisse, Moustapha, Dauphin, Yann N and Lopez-Paz, David. "mixup: Beyond empirical risk minimization." *arXiv preprint arXiv:1710.09412* (2017).

[64] Zhu, Jun-Yan, Park, Taesung, Isola, Phillip and Efros, Alexei A. "Unpaired image-to-image translation using cycle-consistent adversarial networks." *Proceedings of the IEEE international conference on computer vision*: pp. 2223–2232. 2017.

[65] Lee, John M and Lee, John M. *Smooth manifolds*. Springer (2012).

[66] Levina, Elizaveta and Bickel, Peter. "Maximum likelihood estimation of intrinsic dimension." *Advances in neural information processing systems* Vol. 17 (2004).

[67] MacKay, David JC and Ghahramani, Zoubin. "Comments on 'Maximum likelihood estimation of intrinsic dimension'by E." *Levina and P. Bickel* (2005).

[68] Chen, Qiuyi, Pope, Phillip and Fuge, Mark. "Learning Airfoil Manifolds with Optimal Transport." *AIAA SCITECH 2022 Forum*: p. 2352. 2022.

Copyright © 2023 by ASME

[69] Feydy, Jean, Séjourné, Thibault, Vialard, François-Xavier, Amari, Shun-ichi, Trouvé, Alain and Peyré, Gabriel. "Interpolating between optimal transport and mmd using sinkhorn divergences." *The 22nd International Conference on Artificial Intelligence and Statistics*: pp. 2681–2690. 2019. PMLR.

[70] Zhang, Richard, Isola, Phillip, Efros, Alexei A, Shechtman, Eli and Wang, Oliver. "The unreasonable effectiveness of deep features as a perceptual metric." *Proceedings of the IEEE conference on computer vision and pattern recognition*: pp. 586–595. 2018.

Proceedings of the ASME 2023
International Design Engineering Technical Conferences and
Computers and Information in Engineering Conference
IDETC-CIE2023
August 20-23, 2023, Boston, Massachusetts

DETC2023-117216

COUNTERFACTUALS FOR DESIGN: A MODEL-AGNOSTIC METHOD FOR DESIGN RECOMMENDATIONS

Lyle Regenwetter[1],*, Yazan Abu Obaideh[2], Faez Ahmed[1]

[1]Massachusetts Institute of Technology, Cambridge, MA
[2]ProgressSoft, Amman, Jordan

ABSTRACT

We introduce Multi-Objective Counterfactuals for Design (MCD), a novel method for counterfactual optimization in design problems. Counterfactuals are hypothetical situations that can lead to a different decision or choice. In this paper, the authors frame the counterfactual search problem as a design recommendation tool that can help identify modifications to a design, leading to better functional performance. MCD improves upon existing counterfactual search methods by supporting multi-objective queries, which are crucial in design problems, and by decoupling the counterfactual search and sampling processes, thus enhancing efficiency and facilitating objective tradeoff visualization. The paper demonstrates MCD's core functionality using a two-dimensional test case, followed by three case studies of bicycle design that showcase MCD's effectiveness in real-world design problems. In the first case study, MCD excels at recommending modifications to query designs that can significantly enhance functional performance, such as weight savings and improvements to the structural safety factor. The second case study demonstrates that MCD can work with a pre-trained language model to suggest design changes based on a subjective text prompt effectively. Lastly, the authors task MCD with increasing a query design's similarity to a target image and text prompt while simultaneously reducing weight and improving structural performance, demonstrating MCD's performance on a complex multimodal query. Overall, MCD has the potential to provide valuable recommendations for practitioners and design automation researchers looking for answers to their "What if" questions by exploring hypothetical design modifications and their impact on multiple design objectives.

Keywords: Counterfactuals, Human-AI Collaboration, Design Automation, Multi-Objective Optimization

---

*Corresponding author: regenwet@mit.edu

1. INTRODUCTION

Modifying existing designs to generate new ones is an essential aspect of various engineering sectors, such as aerospace, automotive, architecture, pharmaceuticals, consumer goods, and many others. Design modification significantly impacts the performance, efficiency, and safety of engineered systems. Effective methods for design modification can lead to more sustainable and environmentally friendly technologies, better transportation systems, and safer infrastructure. Furthermore, improved design modification methods can enable cost savings and improved efficiency, making products more accessible and affordable for society. However, coming up with good design modifications can be challenging, as it requires navigating huge design spaces and making numerous trade-offs between competing objectives. Often there are too many design attributes and potential modifications to consider. Not surprisingly, designers often struggle with the available choices and may often ask themselves, "What if?".

As a designer, the ability to ask "What if?" questions is crucial in the iterative process of design modification. By exploring hypothetical scenarios, designers can identify opportunities to improve design performance and functionality. However, answering "What if?" questions can be challenging as it requires considering an extensive range of potential modifications and their effects on multiple design objectives. Counterfactuals are a powerful reasoning tool that allows designers to ask such questions by exploring hypothetical design modifications and their impact on multiple design objectives.

A counterfactual is a hypothetical situation that depicts what could have happened if a specific event or action did not occur. It requires envisioning an alternate reality where a different choice or decision was made and analyzing the differences in results. Counterfactuals are often employed in reasoning, decision-making, and causal inference. They aid in comprehending the impact of particular events or actions on outcomes and considering the ramifications of various choices.

Counterfactuals are typically employed to understand how

Copyright © 2023 by ASME

an outcome would change given a different set of actions. This style of counterfactual can be applied to design problems to answer questions like: "How would the performance of this design change if I modified this particular attribute?" There are many tools to predict these 'classic' counterfactuals, such as simulations and predictive models. In this work, we instead consider an 'inverse' counterfactual problem, which states: "What events would have needed to occur to result in this other outcome?" In design contexts, this often equates to the question: "What attributes of my design would I need to change to achieve a particular performance target, design classification, or functional requirement?"

This paper proposes an approach to answer such 'inverse' counterfactual hypotheticals using multi-objective optimization. Our proposed approach, Multi-Objective Counterfactuals for Design (MCD), allows users to input a design and a set of desired attributes, then recommends targeted modifications to the design to achieve these attributes. It identifies these modifications by querying a set of attribute predictors in a directed search procedure dictated by an evolutionary algorithm. We demonstrate how predictors ranging from machine learning regressors to text embedding models can support target attributes ranging from functional performance targets to subjective text requirements.

MCD can be viewed as an AI design assistant that allows users to ask challenging objective and subjective questions about an existing design, such as: "What modifications would it take to make this product 10% lighter?", "What would make my design look like this other concept?", or "How would my design need to change to look more sleek and futuristic?" By enabling designers to interact with AI systems simply and intuitively, counterfactuals open the doors to more successful human-AI collaboration by enhancing and accelerating the design process. A block diagram demonstrating MCD's anticipated usage scenario is shown in Figure 1.

A particularly related body of research to our work is counterfactual explanations, originally developed as a tool to interpret black-box machine learning (ML) models. Counterfactual explanations allow practitioners to understand the behavior of otherwise uninterpretable models by asking questions about counterfactual scenarios. A classic motivating example for counterfactual explanations involves a model that is deciding whether to approve a loan, where the applicant may ask: "What would I need to change for this model to approve my application?" Broadly speaking, these counterfactuals answer a very versatile question: "Hypothetically, what would I need to change about the input to my model for it to predict another outcome?" Many of the common challenges that designers face can be framed as such a question. For example, given a model that predicts the functional performance of a design, a designer can ask how to change the design to achieve some desired functional performance. Despite this, counterfactual explanations have not yet been used in design engineering problems, to the best of our knowledge[1].

In this paper, we showcase our MCD method and demonstrate

---

[1] A search for the term "counterfactual explanations" on the entire ASME digital collection, that includes design venues such as the IDETC conference and the Journal of Mechanical Design, returns zero results on March 10, 2023.

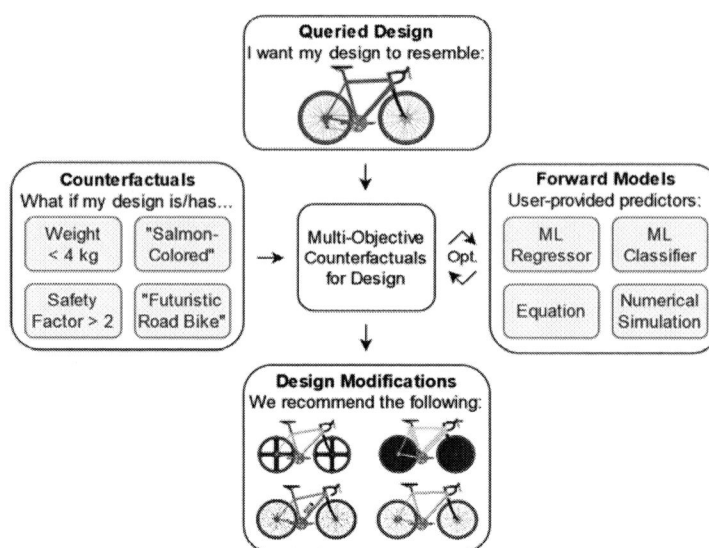

FIGURE 1: MULTI-OBJECTIVE COUNTERFACTUALS FOR DESIGN (MCD) IS A HUMAN-AI COLLABORATIVE DESIGN RECOMMENDATION TOOL. USERS PROVIDE AN INITIAL DESIGN AND A SET OF COUNTERFACTUAL ATTRIBUTES THEY WOULD LIKE TO ACHIEVE. MCD QUERIES A SET OF ATTRIBUTE PREDICTORS TO SEARCH FOR A SET OF DIVERSE MODIFICATIONS TO THE ORIGINAL DESIGN THAT ACHIEVE THE COUNTERFACTUAL ATTRIBUTES.

that counterfactual search is a simple yet powerful AI-driven design tool that real designers can leverage for a variety of tasks. To do so, we make several key contributions, which we summarize as follows:

1. We introduce Multi-Objective Counterfactuals for Design (MCD), a new method to search for counterfactual design modifications to achieve desired outcomes. We formulate MCD as a multi-objective search problem to minimize the magnitude and extent of the modifications, encourage proximity to the data manifold, and satisfy user-provided multimodal requirements.

2. We demonstrate that MCD effectively suggests targeted design modifications to improve the functional performance of query designs, illustrating that counterfactual search could be viewed as an effective design recommendation tool.

3. We present the first text and image-based counterfactual search in design using the Contrastive Language-Image Pre-training (CLIP) method. These cross-modal queries were previously not possible with existing counterfactual methods.

4. We demonstrate that MCD can effectively handle multimodal queries, including a mixed-variable text, image, and parametric query, the first example of multimodal queries to a counterfactual search model, to our knowledge.

2. BACKGROUND

Counterfactuals are a useful tool for investigating causality and forecasting the potential outcomes of different actions.

Copyright © 2023 by ASME

Counterfactuals have been extensively used in various fields, including psychology, philosophy, social sciences, and machine learning as they offer a valuable tool for examining causality and understanding the consequences of actions [1]. In psychology, counterfactual thinking has been studied in relation to emotions, such as regret and disappointment. In philosophy, counterfactuals have been used to explore questions of determinism and free will. In social sciences, counterfactual analysis is widely used to evaluate the impact of policies and interventions. Counterfactual explanations are also gaining traction in the field of machine learning as a means to improve the interpretability and fairness of machine learning models.

In this literature review, we discuss three key areas that relate closely to our work — 1) explainability and counterfactuals in machine learning, 2) multi-objective optimization approaches to counterfactuals, and, 3) a multi-modal, zero-shot machine learning model that enables us to capture user requirements.

2.1 Explainability and Counterfactuals in Machine Learning

Counterfactual explanations are frequently used as a machine learning explainability tool. In machine learning, particularly deep learning, predictions are often mysterious and intractable. To remedy this intractability, a wealth of machine learning 'explainability' tools have been proposed in recent years. One common approach involves determining the sensitivity of the output with respect to the various input parameters (features), a technique known as 'feature importance.' Some popular methods in this category include Local Interpretable Model-Agnostic Explanations (LIME) [2] and Shapley Additive Explanations (SHAP) [3]. In the design automation community, these methods are often used to determine which design parameters have outsized impacts on design performance [4–7] or which parameters are important for relationships between products [8]. Another common approach to explainability involves visualizing a model's decisions in some way. This technique lends itself well to data modalities that are easily appreciated visually, such as images, for which saliency maps are a common explainability method [9–11].

Counterfactuals were first proposed for machine learning (ML) explainability by Wachter *et al.* [12]. Since then, researchers have proposed a wealth of counterfactual explanation approaches, which Verma *et al.* [1] and Guidotti *et al.* [13] review. Among the popular methods are Diverse Counterfactual Explanations (DiCE) [14], Feasible and Actionable Counterfactual Explanations (FACE) [15], and Multi-Objective Counterfactuals (MOC) [16]. Counterfactuals make for a great explainability tool since they allow users to intuitively understand the ML model's internal decision thresholds (i.e. "Where does my model start predicting a different outcome?"). Much like LIME and SHAP, many counterfactuals take a localized approach, with some even fitting local approximations to the data manifold to guide their explanations [17]. Counterfactuals have most commonly been proposed for tabular data, but have also been applied to images [18] and text [19], among other modalities. In general, good counterfactual explanations should typically demonstrate the following properties:

1. **Validity:** First and foremost, a good counterfactual explanation should result in the desired outcome. Depending on the nature of the problem, this desired outcome may be a class, an inequality, a range, an exact equality, or some combination of the above. For example, if we are querying a model that predicts the mass of a design and we specify a range of 2-3 kg, a proposed counterfactual should have a predicted mass in this range.

2. **Sparsity:** Good counterfactuals should be easy to realize, meaning that they should not change many features of the query. Sparsity refers to the number of features that must be modified to realize a counterfactual.

3. **Proximity:** While the number of modifications needed to realize a counterfactual is an important consideration, the extent of these modifications is also important. In simple terms, we would like counterfactuals to be as similar to the query as possible. This is typically quantified as a distance to the original query.

4. **Manifold Proximity:** In the classic usage of counterfactuals as an ML explainer, a predictive model has been trained on a dataset and is being iteratively queried by the counterfactual model. If queries lie too far from the data manifold on which the predictor was trained, predictions (and by extension counterfactuals) will no longer be accurate. In other use cases where the counterfactual is not explaining a statistical model, manifold proximity may not be desirable.

5. **Actionability:** In many problems, certain input parameters may not be changeable, but will nonetheless play a role in the output of the model. For example, the weight of the rider will play a significant role in the structural loading of a bicycle. However, when designing a bicycle, we can't choose to simply make the rider lighter. A good counterfactual explanation should only modify actionable features. Several works, such as [15], have also proposed more nuanced methods to handle actionability.

6. **Causality:** Features in a dataset may be causally linked, implying that changing one feature may necessitate changing another. In general, establishing causality is difficult. However, in design, we may be aware of causal relationships thanks to our fundamental understanding of the physics relating various input variables. For example, selecting a denser material for a given design may necessitate increasing the weight, provided that the geometry remains unchanged. This has clear ramifications for effective counterfactuals, which should ideally capture and respect any known causal relations in the problem.

A strong counterfactual method should undoubtedly generate high-quality counterfactuals. However, good counterfactual methods should also exhibit several properties that may not be reflected in the strength of individual counterfactual examples themselves:

1. **Diverse Sets:** As emphasized in [14], it may be highly desirable to generate diverse sets of counterfactuals. This gives the user a wealth of options, ideally with different actionable requirements to achieve the query objective.

Copyright © 2023 by ASME

2. **Model-Agnosticism:** Ideally, the algorithms used for the generation of counterfactual explanations should treat the model as a black box and interact with the model only through its predict function [1]. These "model-agnostic" algorithms allow for wider applicability and code reuse. Notably, model-agnostic approaches do not rely on gradient information from the predictor but may be less sample-efficient than methods that leverage gradients, when available.

Researchers have also adopted counterfactuals in a recommender-system setting. Tran *et al.* [20] review the use of counterfactual explanations in recommendation systems and propose a method to generate counterfactuals for recommender systems. While related, this work differs slightly from our proposed use case, in which the counterfactual-generating model is the recommender. Regenwetter [21] also briefly investigates leveraging Diverse Counterfactual Explanations [14] for design recommendations, citing challenges due to the limitations of single-objective queries.

2.2 Multi-Objective Counterfactual Explanations

Counterfactual explanations can be viewed as an optimization problem, and can similarly be implemented using an optimization algorithm. Many methods summarize the optimization objective as a weighted sum of the different objectives discussed earlier. However, another approach instead frames the counterfactual search process as a classic multi-objective optimization problem. Dandl *et al.* [16] were the first to formalize this parallel between counterfactual explanations and multi-objective optimization (MOO) in Multi-Objective Counterfactuals (MOC). By handling objectives individually rather than as a single aggregated objective, MOC realizes a key benefit of Multi-Objective Optimization, namely the ability to generate non-dominated sets of counterfactual explanations. Whereas a single-objective approach returns a counterfactual that optimizes for a statically weighted aggregation of objectives, the non-dominated set allows designers to adaptively select counterfactuals based on their specific search priorities, which typically depend on the problem at hand.

Multi-Objective Counterfactuals (MOC) [16] is a primary inspiration for Multi-Objective Counterfactuals in Design (MCD). However, we have expanded on MCD in several key directions. Chiefly, despite its name, MOC does not inherently support multi-objective queries. Furthermore, MOC does not distinguish between hard and soft constraints, despite the fact that this functionality is ingrained in the Non-Dominated Sorting Genetic Algorithm II (NSGA-II) [22] that MOC is built around. MCD addresses these gaps while also decoupling the optimization and sampling steps, and introducing new ways to integrate counterfactuals with a multi-modal, zero-shot machine learning model.

Since the overarching goal of MCD is not to explain predictors, but rather to search for design recommendation counterfactuals, we refer to the problem as 'counterfactual search.' Note that unlike counterfactual explanations, counterfactual search does not require ML predictors and can work with many types of forward models. It also has the additional goals of manifold similarity and meeting multi-objective multi-modal requirements.

2.3 Cross-Modal Design Recommendations

The multitude of data modalities spanned by design data remains a prominent challenge in data-driven design [23–25]. Though a model explained by a counterfactual method may make predictions in one modality, users may instead prefer to query targets in an entirely different modality. We will demonstrate in this paper that MCD can be used in conjunction with rendering pipelines and trained language models to generate counterfactuals for a parametric model using images or even text prompts. In this way, counterfactuals can capture complex and abstract user requirements in a 'zero-shot' fashion, requiring no additional training to understand the context of the prompts. To provide context for this discussion, we will introduce a brief background on relevant subjects in cross-modal learning.

When handling data of modalities like graphs [26], images [27], 3D geometry [28], text [29, 30], and mixed modalities, a common general technique involves mapping datapoints to a vector space. This effectively creates a link from datapoints of the modality to datapoints in the vector space. Two or more modalities can then be linked by creating shared embeddings for the modalities using the same vector space.

Shared text-image embeddings are an example of cross-modal embeddings that have garnered significant attention in recent years [31]. Radford *et al.* [32] propose one of the most widely used models for text-image shared embeddings called Contrastive Language-Image Pretraining (CLIP). CLIP trains a text embedding and image embedding model simultaneously on a dataset of text-image pairs. The models are rewarded for mapping matching pairs to similar embedding vectors and mapping non-matching pairs to dissimilar embedding vectors. In our second and third case studies, we will be leveraging pre-trained CLIP models to query counterfactuals using text prompts. Next, we move on to discuss our methodology.

3. METHODOLOGY

In this section, we discuss the construction of the optimization algorithm behind MCD, emphasizing the constraints, objectives, and operators used. We then present our approach for sampling diverse sets of counterfactuals and discuss how we decouple the optimization from the final sampling step. Finally, we demonstrate the capabilities of MCD on a simple 2D problem.

3.1 Objectives

Optimization algorithms typically seek to find constraint-satisfying solutions that achieve optimal objective scores. We will first discuss how objectives are defined in MCD, then go on to discuss constraints. Broadly, we consider two types of objectives: Objectives related to counterfactual quality and user-specified auxiliary objectives (often used for soft constraints). The former draw on the work of Dandl *et al.* [16], who, among other things, leverage Gower distance [33] and the number of changed features as optimization objectives in MOC.

1. **Gower Distance:** Gower Distance [33] is a metric that indicates the distance between mixed feature data points. Its use as an objective tackles the issue of "proximity" introduced in Sec. 2.1. The Gower distance between d-dimensional counterfactual p and query q is given in terms of their feature

Copyright © 2023 by ASME

values p_i and q_i for $i \in [1...d]$, as:

$$f_{pr}(p,q) = \frac{1}{d}\sum_{i=1}^{d}\delta_G(p_i,q_i) \qquad (1)$$

$\delta_G(p_i,q_i)$ is a function that depends on feature type and is given as:

$$\delta_G(p,q) = \begin{cases} \frac{1}{\hat{R}_i}|p_i - q_i| & \text{if } p_i \text{ is numerical} \\ \mathbb{1}_{p_i \neq q_i} & \text{if } p_i \text{ is categorical} \end{cases} \qquad (2)$$

Here, $\hat{R}_i$ is the range of the feature i observed in the dataset.

2. **Changed Feature Ratio:** This objective calculates the proportion of features that the proposed counterfactual, p, modifies from the query, q. Its use as an objective tackles the issue of "sparsity" introduced in Sec. 2.1.

$$f_{sp}(p,q) = \frac{\|p - q\|_0}{d} = \frac{1}{d}\sum_{i=1}^{d}\mathbb{1}_{p_i \neq q_i} \qquad (3)$$

3. **Average Gower Distance:** To measure the "manifold proximity" discussed in Sec. 2.1, Dandl *et al.* [16] calculate the average Gower distance to the k nearest observed data points $s^i...s^k$ from the dataset S:

$$f_{mp}(p,S) = \frac{1}{k}\sum_{i=1}^{k}\frac{1}{d}\sum_{j=1}^{d}\delta_G(p_j,s_j^i) \qquad (4)$$

4. **Problem-Specific Objectives:** Just as the user may specify non-negotiable requirements for the model outcome (hard constraints), they may also specify objectives $(f_1(p)...f_M(p))$ that they would like to satisfy, and later specify targets for these objectives during sampling. These auxiliary objectives are directly included as optimization objectives in NSGA-II.

3.2 Constraints

In a counterfactual search, a variety of optimization constraints may be present. Constraints are considered non-negotiable and always take precedence over objectives. In practice, many optimization algorithms, including the variant of the NSGA-II algorithm driving MCD, prioritize resolving constraint violations before proceeding to the optimization of objectives. MCD considers several types of constraints:

1. **Variable and Constant Features:** Like many counterfactual models, we implement a mechanism to constrain which features are allowed to be modified by a counterfactual, as specified by the user. This addresses the challenge of "actionability" introduced in Sec. 2.1. We call the set of actionable features A.

2. **Model Output Constraints:** Users querying a counterfactual method may have requirements for the output of their model. In most counterfactual search approaches, these requirements are treated as non-negotiable hard constraints

to satisfy the "validity" property introduced in Sec. 2.1. MCD supports such hard requirements, which are handled as constraints in NSGA-II[2], but does not require them. We consider any output with a constraint as belonging to a set B and require that $L_b \leq f_b(p) \leq U_b \forall b \in B$. Instead, we also allow users to specify soft constraints in the form of additional optimization objectives, paired with targets to be used during sampling.

3. **Domain-Specific Constraint Functions:** There are cases in which certain hard constraints are known a priori. MCD can be configured to respect such hard constraints through user-specified black-box constraint functions. Domain-specific constraints can be used for a variety of different purposes, including encoding causality relations into the optimization as discussed in Sec. 2.1. We specify these constraint functions as $g_1(p)...g_K(p)$ and, for simplicity, assume they are satisfied for $g_k(p) \geq 0$

3.3 Formulation as MOO problem

In summary, we express the multi-objective optimization problem in terms of the variables, sets, and functions defined above as follows:

$$\text{minimize: } f_i(p), \ \forall i \in \{pr, sp, mp, 1, ..., M\} \qquad (5)$$

$$\text{subject to: } f_j(p) - L_j \geq 0, \ U_j - f_j(p) \geq 0, \ \forall j \in B,$$

$$g_k(p) \geq 0, \ \forall k \in \{1, ..., K\},$$

$$p_l = q_l, \ \forall l \notin A$$

3.4 Algorithm

Any gradient- or non-gradient-based multi-objective optimization method could be used in MCD. To demonstrate our results in this paper, we leverage the Non-Dominated Sorting Genetic Algorithm II (NSGA-II) [22] as the backend of MCD. NSGA-II is a multi-objective genetic algorithm that boasts several innovative features, such as non-dominated sorting for elitist selection, crowding distance to encourage diversity, and genetic operators such as tournament selection, simulated binary crossover, and polynomial mutation. We use an implementation of NSGA-II from [34], including the mixed-variable selection, crossover, and mutation functions provided.

The initial population always consists of the query and a set of randomly sampled points from the dataset or the user-specified design space boundaries. In problems with continuous variables, we find that without any precautions to maintain the exact parameter values from the original query, these values tend to get 'lost,' and can never be exactly reconstructed, hurting the sparsity objective of counterfactuals. To allow the algorithm to 'rediscover' the exact parameter values from the query, we introduce a custom operator that randomly reverts individual parameter values back to the query's values with a certain probability.

---

[2]By default, we expect queries in the form of inequalities. Since range and equality constraints (or objectives) can be specified using two inequalities, we find this to be an adequately versatile interface for most types of constraints. In rare cases where users need to specify complex constraints, such as disjoint ranges, they can do so by creating a custom constraint function and passing it in as a black box.

Copyright © 2023 by ASME

3.5 Sampling

Contrary to other counterfactual search approaches, our method decouples the optimization and sampling steps. Conventionally, a user will have to decide on the priorities between various objectives (e.g. proximity, diversity, manifold proximity, etc.) before running the optimization. This is impractical, as these objectives are challenging to select intuitively, and must often be chosen through trial and error. For example, a designer might realize that the generated counterfactuals are much too different from the query to be practically realizable. By avoiding retraining, our method can save significant computational expense and, as we will discuss in Sec. 5, enable users to quickly consider counterfactuals from different regions of the objective landscape. We decouple the search and sampling process as follows:

1. Given a query, a set of constraints, and objectives, the optimizer generates a collection of candidate counterfactuals by running NSGA-II.

2. The sampling algorithm collects a set of objective priority weights and optional targets from the user. By collecting these weights after training, MCD allows rapid counterfactual sampling under different objective weights without the need for retraining, unlike other approaches.

3. Each candidate counterfactual is assigned an aggregate quality score, which is calculated as a sum of individual objective scores, weighted by their priority. For any objectives with specified targets, the Design Target Achievement Index [35] is used to quantify target achievement before factoring into the aggregate score. The aggregate score, S of a counterfactual candidate, p, is given in terms of objective priority weights w_{pr}, w_{sp}, w_{mp} by:

$$S(p) = w_{pr}f_{pr}(p,q) + w_{sp}f_{sp}(p,q) + w_{mp}f_{mp}(p,S)$$
$$+ DTAI(p,t,\alpha,\beta)$$
$$(6)$$

Here, $DTAI(p,t,\alpha,\beta)$ is the Design Target Achievement Index of the candidate given auxiliary objective targets, t, priority weights, α, and decay parameters, β [35].

4. A performance-weighted diversity matrix is calculated using a Gower distance-based similarity kernel to evaluate the similarity between counterfactuals. Matrix entries are calculated as a function of aggregate scores and a diversity parameter, w_d as:

$$D_{i,j} = \delta_G(i,j)\left(S(i)S(j)\right)^{\frac{1}{w_d}}$$
$$(7)$$

5. A diverse set of high-performing counterfactuals is sampled from this matrix using k-greedy diverse sampling [36].

If the user requests only a single counterfactual instead of a diverse set, the candidate with the highest aggregate quality score is returned.

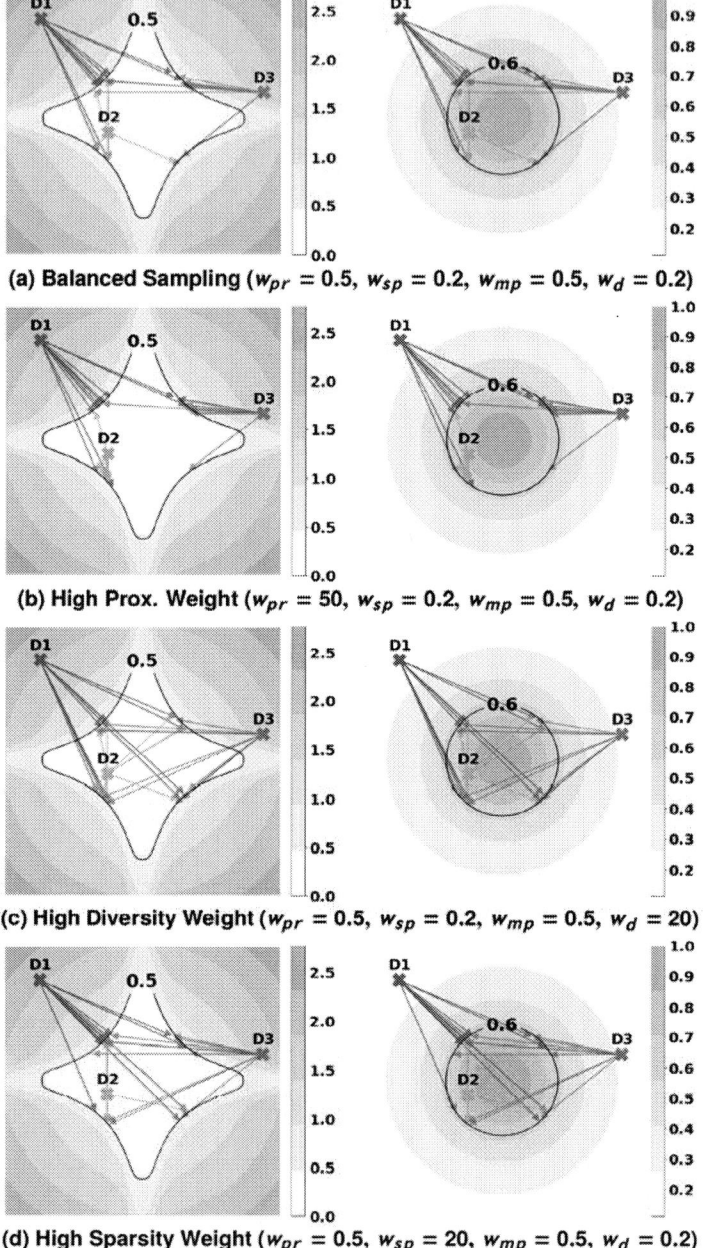

(a) Balanced Sampling ($w_{pr} = 0.5$, $w_{sp} = 0.2$, $w_{mp} = 0.5$, $w_d = 0.2$)

(b) High Prox. Weight ($w_{pr} = 50$, $w_{sp} = 0.2$, $w_{mp} = 0.5$, $w_d = 0.2$)

(c) High Diversity Weight ($w_{pr} = 0.5$, $w_{sp} = 0.2$, $w_{mp} = 0.5$, $w_d = 20$)

(d) High Sparsity Weight ($w_{pr} = 0.5$, $w_{sp} = 20$, $w_{mp} = 0.5$, $w_d = 0.2$)

FIGURE 2: COUNTERFACTUAL SETS RETURNED FOR THREE QUERY DESIGNS UNDER DIFFERENT WEIGHTINGS OF COUNTERFACTUAL QUALITY OBJECTIVES. PERFORMANCE SPACE CONSTRAINTS ARE INDICATED ON THE PLOTS. VALID COUNTERFACTUALS MUST SIMULTANEOUSLY MEET BOTH CONSTRAINTS.

Copyright © 2023 by ASME

3.6 Showcasing Functionality on 2D Examples

Before showcasing the capabilities of MCD on real design datasets, we will first demonstrate its performance on a simple two-dimensional problem for ease of visualization. We select a challenging two-objective problem and sample synthetic data. We then query three different designs, D1-3, and specify the same challenging constraint criterion for each query, which is only satisfiable in four small disjoint regions of the space. Mathematically, we constrain the performance space values Y_1 and Y_2 such that $0.4 \leq Y_1 \leq 0.6$ and $Y_2 \geq 0.6$. In simple terms, any valid counterfactual must lie near the star-shaped contour on the left and strictly within the circle on the right in the contour plots in Fig. 2. We consider four choices of objective weights:

1. First we examine a fairly "balanced" selection of objective weights ($w_{pr} = 0.5$, $w_{sp} = 0.2$, $w_{mp} = 0.5$, $w_d = 0.2$) in Fig. 2a. In this setting, the sampled counterfactual sets achieve a balance of proximity, diversity, and sparsity.

2. Next, we consider a setting where proximity is prioritized over other objectives ($w_{pr} = 50$, $w_{sp} = 0.2$, $w_{mp} = .05$, $w_d = 0.2$) in Fig. 2b. In this setting, most counterfactuals in each set are sampled from the mode nearest the queries, though counterfactuals are still diversified within these modes.

3. We next consider a case where diversity is given precedence over other objectives ($w_{pr} = 0.5$, $w_{sp} = 0.2$, $w_{mp} = 0.5$, $w_d = 20$) in Fig. 2c. In this case, the sampled counterfactual sets are very well distributed across the feasible regions of the space.

4. Finally, we consider the case where sparsity is given the highest priority ($w_{pr} = 0.5$, $w_{sp} = 20$, $w_{mp} = 0.5$, $w_d = 0.2$) in Fig. 2d. Many sampled counterfactuals change only one parameter from the query, when possible.

Each of these subsets is sampled from the same set of counterfactual candidates with no re-optimization necessary. Now, having demonstrated MCD's functionality on a simple 2D problem, we move on to a more complex real-world design problem: Bike frame design.

4. CASE STUDY 1: DESIGN REFINEMENT USING STRUCTURAL PERFORMANCE QUERIES

In our first case study, we consider the counterfactual: "What if my design were 30% lighter?" Specifically, we consider a bicycle frame design problem where we are trying to improve the structural properties and reduce the weight of a query design. We use a regression model trained on the FRAMED dataset consisting of Finite Element (FE) simulation results from 4500 community-designed bike frames [37], including weight, safety factors, and deflections under various loading conditions. The trained regression model is an AutoGluon tabular AutoML regressor [38] intended to predict the structural performance of bicycle frames accurately.

To illustrate MCD's capabilities, we feed it three variants of the same query. The first has a single objective: finding counterfactuals that reduce the predicted mass of a given design.

The second has two competing objectives: Maximize a design's safety factor while minimizing its mass. The third has the same objectives as the second but restricts MCD to only vary a more constrained and actionable set of features. In each example, we query the same design: a steel tube road bike with minor structural inefficiencies. These inefficiencies largely stem from a down tube with insufficient wall thickness, requiring other components to be over-engineered. This bike has a safety factor[3] of 1.24 and a mass of 4.26 kg, so our primary objective is to reduce the mass. Each optimization ran for 100 generations with a population size of 500.

Single objective query: In the first variant, MCD was tasked with finding counterfactuals that reduced the mass of the original design from 4.26 kg to under 3 kg. MCD effectively successfully discovered hundreds of valid counterfactuals and sampled a set of five diverse counterfactuals which had, on average, a mass of 2.3 kg, as tabulated in Table 1. Although MCD succeeded in its explicitly stated objective, a closer look reveals that it did nothing to remedy the wall thickness issue in the down tube, and as a consequence of weight savings in other parts of the sampled frames, the average safety factor across sampled counterfactuals was an abysmal 0.48. This disregard for secondary objectives is quite characteristic of the many existing single-objective counterfactual search algorithms and illustrates why MCD's novel support of multi-objective queries is so essential for design problems. Our next example showcases how to leverage multi-objective queries to avoid these issues.

TABLE 1: GENERATED COUNTERFACTUALS FOR VARIANT 1 (34 COLUMNS OMITTED). LIKE MANY SINGLE-OBJECTIVE COUNTERFACTUAL ENGINES, MCD TENDS TO ACHIEVE SINGLE-OBJECTIVE QUERIES AT THE EXPENSE OF SECONDARY OBJECTIVES. MCD'S UNIQUE SUPPORT OF MULTI-OBJECTIVE QUERIES REMEDIES THIS PROBLEM.

	Material	Stack (mm)	...	Down Tube Thick. (mm)	Safety Factor	Frame Mass (kg)
Query	Steel	565.6	...	0.52	1.24	4.26
CF 1	Steel	570.8	...	0.52	0.52	1.99
CF 2	Steel	565.6	...	0.52	0.27	1.64
CF 3	Steel	565.6	...	0.52	0.76	2.48
CF 4	Steel	565.6	...	0.52	0.64	2.69
CF 5	Aluminum	522.6	...	0.52	0.22	2.70

Bi-objective query: In the second variant, a second objective was introduced: Increase the safety factor to a minimum value of 1.5. Again, MCD successfully discovered numerous counterfactuals, and the diverse 5-bike set sampled this time had an average mass of 2.4 kg and a safety factor of 1.7, as shown in Table 2. This time, MCD realized that the bike could be made significantly more weight-efficient by increasing the down tube wall thickness to relieve structural stress on other components to be lightened. However, it also changed the material of the bike from steel to

---

[3]We use predicted safety factor in FRAMED's in-plane loading scenario [37]

Copyright © 2023 by ASME

aluminum or titanium in four of the five counterfactuals, a modification that would likely carry a significant increase to the cost and may thus be unactionable. In the presence of a cost prediction model, MCD could consider cost as another query objective. However, even without such a model, MCD can be ordered to leave certain design parameters unchanged, as we demonstrate in our final example.

TABLE 2: GENERATED COUNTERFACTUALS FOR VARIANT 2 (34 COLUMNS OMITTED). BY QUERYING MULTIPLE OBJECTIVES SIMULTANEOUSLY, MCD AVOIDED THE SAFETY FACTOR ISSUE THAT OCCURRED IN QUERY 1.

	Material	Stack (mm)	...	Down Tube Thick. (mm)	Safety Factor	Frame Mass (kg)
Query	Steel	565.6	...	0.52	1.24	4.26
CF 1	Aluminum	565.0	...	2.20	1.91	2.81
CF 2	Titanium	561.6	...	2.46	1.82	2.21
CF 3	Aluminum	532.2	...	1.81	1.58	1.75
CF 4	Titanium	563.5	...	3.92	1.60	2.23
CF 5	Steel	565.6	...	2.48	1.65	2.87

Bi-objective query with constraints: In the third variant, MCD was no longer allowed to vary frame material. It proceeded to find tens of valid designs through variations in certain tube diameters, lengths, and other structural configurations. From these valid designs, a 5-bike set was sampled that had an average mass of 2.5 kg and an average safety factor of 1.8, as shown in Table 3.

TABLE 3: GENERATED COUNTERFACTUALS FOR QUERY 3 (34 COLUMNS OMITTED). WHEN RESTRICTED FROM MODIFYING FRAME MATERIAL, MCD IS STILL ABLE TO RECOMMEND DESIGN MODIFICATIONS THAT MEET THE SAFETY FACTOR AND MASS TARGETS.

	Material	Stack (mm)	...	Down Tube Thick. (mm)	Safety Factor	Frame Mass (kg)
Query	Steel	565.6	...	0.52	1.24	4.26
CF 1	Steel	565.6	...	2.44	2.05	2.93
CF 2	Steel	601.7	...	3.38	2.06	2.31
CF 3	Steel	565.6	...	3.22	1.58	2.71
CF 4	Steel	601.7	...	2.12	1.61	1.87
CF 5	Steel	565.6	...	3.35	1.56	2.82

Through these examples, we have attempted to demonstrate that MCD excels at handling multi-objective performance queries and can be used in such a setting to recommend performance-enhancing design modifications. In our next example, we consider a scenario in which more abstract text queries are provided instead of hard performance constraints.

5. CASE STUDY 2: MODIFYING DESIGNS USING CROSS-MODAL TEXT QUERIES

In this case study we examine subjective counterfactuals like: "What if my design looked more 'cyberpunk themed?'"

Classically, counterfactual search requires a query in the same data modality as the predictive model. This can be constraining, since it may be more natural in many cases to place queries in a different data modality, especially if that modality is more intuitive for human users. This is often the case for images or text, which are much more easily understood by humans compared to tabular or parametric data. Accordingly, we demonstrate how we can query MCD in a cross-modal setting using text prompts.

5.1 Methodology: Case Study 2

To enable cross-model queries, we construct an objective evaluation function comprised of several key building blocks:

- To begin, we require a rendered image of a bicycle design. We construct an automated rendering pipeline that works in conjunction with the BikeCAD software to generate an image of a bicycle given a parametric vector.

- We then calculate an embedding for the generated bike image using a pre-trained CLIP model introduced in Sec. 2.3 that maps the generated bike renders to a vector embedding space.

- Next, we compute the embedding vector for a target text prompt using a pre-trained CLIP text embedding model.

- Finally, we calculate cosine similarity between the two 512-dimensional embedding vectors.

In this case study, this entire objective evaluation pipeline serves as the predictor for counterfactual search. By generating counterfactuals that minimize this cosine similarity objective, the optimizer ensures that generated counterfactuals better match the given text prompts.

We select a subset of the BIKED [5] dataset's parameter space to consider during optimization and choose a generic red road bike design as a query design. We choose two text prompts as optimization objectives: "A futuristic black cyberpunk-style road racing bicycle" and "A sturdy compact bright blue mountain bike with thick tires." Because the demands of human designers are often difficult to quantify using traditional parametric methods, the first text prompt was selected to be highly subjective. The second prompt is less subjective, offering details about design features, but stops short of explicit design guidelines. In this context, the user is effectively asking questions like: "How would my red road bike design change if I wanted it to look more like a black cyberpunk-style bike?" We optimize for 400 generations with a population size of 100. Next, we perform a series of sampling operations with different objective weights, as shown in 2.3. By selecting the optimal bikes at a sweep of different objective weights, we can visualize the best bikes under numerous configurations of objective priorities.

Counterfactual quality objective weights in the ith row were chosen as:

$$w_1 = w_2 = w_3 = \frac{0.2}{2^i} \qquad (8)$$

In this way, counterfactuals with better proximity, sparsity, and manifold proximity were prioritized toward the top of the grid, while counterfactuals were given more leeway to deviate from

Copyright © 2023 by ASME

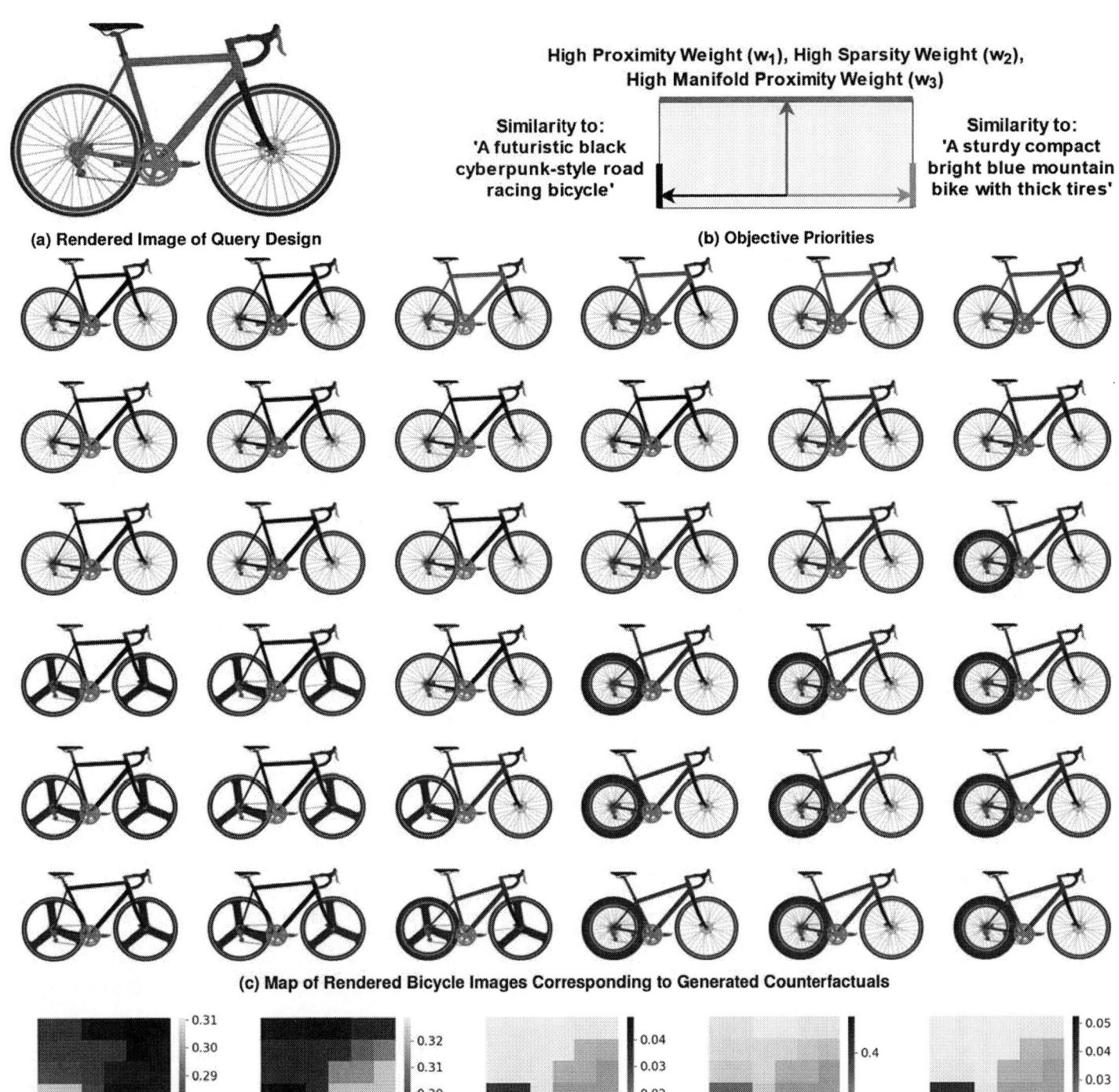

(a) Rendered Image of Query Design

High Proximity Weight (w_1), High Sparsity Weight (w_2), High Manifold Proximity Weight (w_3)

Similarity to: 'A futuristic black cyberpunk-style road racing bicycle'

Similarity to: 'A sturdy compact bright blue mountain bike with thick tires'

(b) Objective Priorities

(c) Map of Rendered Bicycle Images Corresponding to Generated Counterfactuals

(d) Similarity to: 'A futuristic black cyberpunk-style road racing bicycle'

(e) Similarity to: 'A sturdy compact bright blue mountain bike with thick tires'

(f) Counterfactual Proximity (Gower Distance)

(g) Counterfactual Sparsity (Changed Feature Ratio)

(h) Counterfactual Manifold Proximity (Average Gower Distance)

FIGURE 3: VISUALIZATION OF THE OBJECTIVE MANIFOLD FOR CROSS-MODAL COUNTERFACTUAL SELECTION. DESIGNS SAMPLED FROM THE TOP OF THE MANIFOLD PRIORITIZE PROXIMITY, SPARSITY, AND MANIFOLD PROXIMITY. DESIGNS IN THE LEFT AND RIGHT CORNERS PRIORITIZE SIMILARITY TO TWO RESPECTIVE TEXT PROMPTS. HEATMAPS SHOW INDIVIDUAL OBJECTIVE SCORES (LIGHTER IS BETTER).

Copyright © 2023 by ASME

the query design and data manifold toward the bottom. Diversity weight, w_d, was irrelevant, as only one design was sampled for each combination of objective weights.

Similarly, auxiliary objective weights in the jth column were set through the DTAI objective weighting parameter, α, in terms of the number of columns, n (in our case 6), as:

$$\alpha_1 = 1.5^{n-j}, \quad \alpha_2 = 1.5^{j-1} \tag{9}$$

These objectives allowed similarity to the first text prompt to take precedence on the left edge of the grid and similarity to the second text prompt to take precedence on the right.

5.2 Discussion: Case Study 2

As expected, models at the top of the grid are appreciably similar to the red bike; some were essentially indistinguishable. Bikes further down the grid become progressively more visually different, which is corroborated by objective scores, as shown in Figs. 3f-3h.

Bikes in the lower left corner of the grid can be subjectively identified as more similar to "A futuristic black cyberpunk-style road racing bicycle." Among the key modifications are a color change and a shift to tri-spoke wheels, which may be more on-theme for a 'cyberpunk-style' bike. Similarity to the text prompt as evaluated by CLIP agrees, as shown in Fig. 3d.

Likewise, bikes towards the bottom right corner of the grid can be subjectively identified as more similar to "A sturdy compact bright blue mountain bike with thick tires." Bikes in this corner have the slanted down tube which is characteristic of mountain bikes; have the requested color change; and have a thick rear tire. Notably, the model either does not discover a modification to the front tire or does not find that such a modification improves similarity to the prompt. Also, the models maintain the dropped handlebars present on the query, which are characteristic of road bikes. Nevertheless, similarity to the text prompt as evaluated by CLIP was found to be best in this corner, as shown in Fig. 3e.

In this case study we demonstrated that MCD effectively handles multi-objective cross-modal prompts. Next, we move on to consider a challenging multi-modal query case as our final case study.

6. CASE STUDY 3: MODIFYING DESIGNS USING MULTIMODAL TEXT, IMAGE, AND PARAMETRIC QUERIES

In this final case study, we examine hybrid counterfactuals like: "What if my design were lighter, looked more 'cyberpunk themed,' had better structural properties, and looked like this other design?" Having considered multi-objective cross-modal queries in the previous case study, we now present our most challenging case study. This time, we provide a multi-objective multi-modal query consisting of a target text prompt, image, frame safety factor, and frame mass.

6.1 Methodology: Case Study 3

To calculate image and text similarity, we leverage the rendering pipeline and pre-trained CLIP model used in case study 2. To calculate structural performance, we use the AutoGluon [38]

model trained in [37], which was used in case study 1. We again select the same generic red road bike as our query design and select "A futuristic black cyberpunk-style road racing bicycle" as our text prompt. For our target image, we select an image of a Fuji Wendigo 1.1 mountain bike which closely matches the second text description from the previous case study.

Like the last case study, we sample designs in a grid, as shown in Fig. 4 based on a variable objective weighting scheme. We select a spread of DTAI objective weighting parameter (α) values in terms of the ith row and j^{th} column as follows:

$$\alpha_{\text{text}} = 2^{n-j}, \quad \alpha_{\text{image}} = 2^j$$
$$\alpha_{\text{sf}} = 1.5^{n-i-1}, \quad \alpha_{\text{mass}} = 1.5^i \tag{10}$$

This time, we hold the counterfactual quality objective weights constant at:

$$w_1 = w_2 = w_3 = 0.05 \tag{11}$$

6.2 Discussion: Case Study 3

As in the previous case study, MCD modifies several components to better match the text query of the models on the left side, including recoloring them black and replacing a regular spoked wheel with a disk wheel. However, due to the proximity, sparsity, and manifold proximity weights being fixed at moderate values, it does not deviate as far from the dataset as some of the most extreme designs in the previous case study.

The bikes on the right side of the grid are visibly more similar to mountain bikes, displaying the characteristic slanted top tube and, in some cases, adding a front suspension to the design. Interestingly, MCD does not generate any blue bikes, indicating that the color of the reference image is not as strongly emphasized as when a color is explicitly stated in a text prompt.

Structural modifications of the bike frame are challenging to appreciate in renderings because the largest drivers of structural performance are tube wall thickness parameters and material, none of which have a visual signature in the rendering. However, Figs. 4g and 4h indicate that bikes at the top and bottom prioritize safety factor and weight, respectively, as intended.

From Fig. 4i, we can see that the bikes toward the top of the grid fall far outside of the data manifold. Unsurprisingly, we see various design infeasibilities in these bikes, such as colliding components. Unless explicitly prevented using constraints, such infeasibilities are typically more common as counterfactuals fall further from the data manifold.

Though some of the generated counterfactuals suffer from infeasibilities, we have demonstrated that MCD can provide meaningful counterfactuals in high-dimensional (i.e., 4 auxiliary objectives and 3 counterfactual quality objectives) and multimodal objective spaces. Next, we proceed to discuss MCD's limitations.

7. LIMITATIONS

MCD makes several key contributions to counterfactual optimization methods for designers, such as incorporating multiple objectives. However, it also has a few limitations. In model-agnostic configurations, MCD must use a gradient-free optimizer, preventing it from leveraging gradient information, even if some of the predictive models are differentiable. While this gradient-free approach allows MCD to support nondifferentiable predictors

Copyright © 2023 by ASME

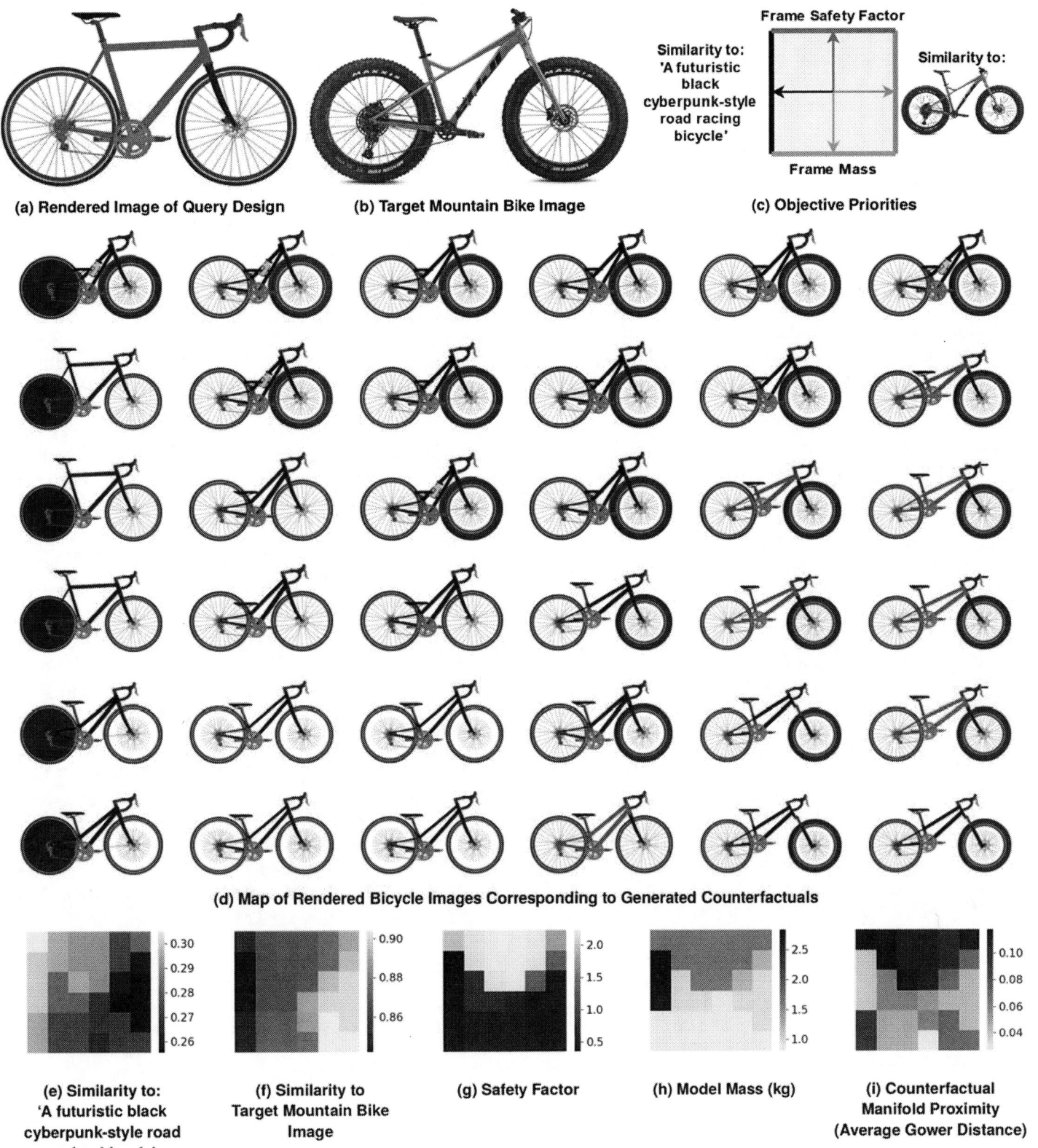

(a) Rendered Image of Query Design

(b) Target Mountain Bike Image

(c) Objective Priorities

Frame Safety Factor

Similarity to: 'A futuristic black cyberpunk-style road racing bicycle'

Similarity to:

Frame Mass

(d) Map of Rendered Bicycle Images Corresponding to Generated Counterfactuals

(e) Similarity to: 'A futuristic black cyberpunk-style road racing bicycle'

(f) Similarity to Target Mountain Bike Image

(g) Safety Factor

(h) Model Mass (kg)

(i) Counterfactual Manifold Proximity (Average Gower Distance)

FIGURE 4: VISUALIZATION OF THE OBJECTIVE MANIFOLD FOR MULTIMODAL COUNTERFACTUAL SELECTION. DESIGNS SAMPLED TO-WARDS THE TOP AND BOTTOM OF THE MANIFOLD PRIORITIZE SAFETY FACTOR AND WEIGHT RESPECTIVELY. DESIGNS SAMPLED TOWARDS THE LEFT AND RIGHT EDGES PRIORITIZE SIMILARITY TO A TARGET TEXT PROMPT AND TARGET IMAGE, RESPECTIVELY. HEATMAPS SHOW INDIVIDUAL OBJECTIVE SCORES (LIGHTER IS BETTER). DESIGNS THAT FALL FAR OUTSIDE OF THE DATA MANIFOLD STRUGGLE WITH COMPONENT OVERLAP AND OTHER INFEASIBILITY ISSUES.

Copyright © 2023 by ASME

and avoid local minima, it potentially makes MCD less sample-efficient than similar gradient-based approaches.

Another key limitation stems from the difficulty of genetic algorithms in handling a large number of objectives. Because MCD adds three counterfactual quality objectives to the objective space, it slightly exacerbates the dimensionality issue of multi-objective genetic algorithms. Future work will explore MCD variants that leverage gradient information and many-objective optimization methods to address these limitations.

Additionally, we would like to acknowledge certain limitations with the text-based queries presented in the last two case studies. Though CLIP embeddings can capture more abstract and subjective ideas, they struggle to capture fine-grained technical details of designs. As such, we recommend that users with highly technical constraints specify them parametrically, instead of through text. However, as machine learning models continue to improve, querying counterfactual models for precise technical details through text and images may improve significantly.

8. CONCLUSION

In this paper, we have introduced Multi-Objective Counterfactuals for Design (MCD), a specialized counterfactual optimization method for design tasks. We first discussed previous counterfactual optimization approaches, many stemming from machine learning explainability literature. We then identified key limitations with existing works, particularly their inability to sample multi-objective queries and the inherent coupling of the optimization and sampling process. Next, we demonstrated using 2D examples how MCD solves these two challenges.

We presented a bicycle frame optimization problem and showed how MCD's support of multi-objective queries allows it to recommend meaningful modifications to a query design which improves structural performance. We then identified that although previous counterfactual search models have not supported cross-modal queries, advancements in multi-modal learning reasonably allow counterfactuals to be queried in different data modalities. Next, we showcased how MCD can be queried with text prompts, and illustrated how MCD's decoupling of optimization and sampling allows it to visualize complex objective manifolds without re-optimization. Finally, we asked MCD to generate counterfactuals given a multimodal text, image, and parameter query. By effectively recommending design modifications to match these queries, MCD demonstrated that it can support complex multimodal queries.

All in all, MCD is a valuable tool for designers looking to optimize their designs and design automation researchers looking to interact intuitively with their models. We are excited to release our code and examples at http://decode.mit.edu/projects/counterfactuals/ and anticipate a variety of interesting use cases across the community.

9. ACKNOWLEDGMENTS

We would like to thank Amin Heyrani Nobari for his contributions to the image rendering pipeline that enabled much of the cross-modal work presented. We would also like to thank Tyler Butler for his feedback and edits.

REFERENCES

[1] Verma, Sahil, Boonsanong, Varich, Hoang, Minh, Hines, Keegan E, Dickerson, John P and Shah, Chirag. "Counterfactual Explanations and Algorithmic Recourses for Machine Learning: A Review." *arXiv preprint arXiv:2010.10596* (2020).

[2] Ribeiro, Marco Tulio, Singh, Sameer and Guestrin, Carlos. ""Why should i trust you?" Explaining the predictions of any classifier." *Proceedings of the 22nd ACM SIGKDD international conference on knowledge discovery and data mining*: pp. 1135–1144. 2016.

[3] Lundberg, Scott M and Lee, Su-In. "A unified approach to interpreting model predictions." *Advances in neural information processing systems* Vol. 30 (2017).

[4] Joung, Junegak and Kim, Harrison M. "Approach for importance–performance analysis of product attributes from online reviews." *Journal of Mechanical Design* Vol. 143 No. 8 (2021).

[5] Regenwetter, Lyle, Curry, Brent and Ahmed, Faez. "BIKED: A dataset for computational bicycle design with machine learning benchmarks." *Journal of Mechanical Design* Vol. 144 No. 3 (2022).

[6] Rodríguez-Pérez, Raquel and Bajorath, Jürgen. "Interpretation of machine learning models using shapley values: application to compound potency and multi-target activity predictions." *Journal of computer-aided molecular design* Vol. 34 (2020): pp. 1013–1026.

[7] Dachowicz, Adam, Mall, Kshitij, Balasubramani, Prajwal, Maheshwari, Apoorv, Raz, Ali K, Panchal, Jitesh H and DeLaurentis, Daniel A. "Mission engineering and design using real-time strategy games: An explainable AI approach." *Journal of Mechanical Design* Vol. 144 No. 2 (2022).

[8] Ahmed, Faez, Cui, Yaxin, Fu, Yan and Chen, Wei. "Product Competition Prediction in Engineering Design Using Graph Neural Networks." *ASME Open Journal of Engineering* Vol. 1 (2022). DOI 10.1115/1.4054299. URL https://doi.org/10.1115/1.4054299. 011020.

[9] Simonyan, Karen, Vedaldi, Andrea and Zisserman, Andrew. "Deep inside convolutional networks: Visualising image classification models and saliency maps." *Proceedings of the International Conference on Learning Representations (ICLR)* (2014).

[10] Zeiler, Matthew D and Fergus, Rob. "Visualizing and understanding convolutional networks." *Computer Vision–ECCV 2014: 13th European Conference, Zurich, Switzerland, September 6-12, 2014, Proceedings, Part I 13*: pp. 818–833. 2014. Springer.

[11] Zhou, Bolei, Sun, Yiyou, Bau, David and Torralba, Antonio. "Interpretable basis decomposition for visual explanation." *Proceedings of the European Conference on Computer Vision (ECCV)*: pp. 119–134. 2018.

[12] Wachter, Sandra, Mittelstadt, Brent and Russell, Chris. "Counterfactual explanations without opening the black box: Automated decisions and the GDPR." *Harv. JL & Tech.* Vol. 31 (2017): p. 841.

[13] Guidotti, Riccardo. "Counterfactual explanations and how

Copyright © 2023 by ASME

to find them: literature review and benchmarking." *Data Mining and Knowledge Discovery* (2022): pp. 1–55.

[14] Mothilal, Ramaravind K, Sharma, Amit and Tan, Chenhao. "Explaining machine learning classifiers through diverse counterfactual explanations." *Proceedings of the 2020 conference on fairness, accountability, and transparency*: pp. 607–617. 2020.

[15] Poyiadzi, Rafael, Sokol, Kacper, Santos-Rodriguez, Raul, De Bie, Tijl and Flach, Peter. "FACE: feasible and actionable counterfactual explanations." *Proceedings of the AAAI/ACM Conference on AI, Ethics, and Society*: pp. 344–350. 2020.

[16] Dandl, Susanne, Molnar, Christoph, Binder, Martin and Bischl, Bernd. "Multi-objective counterfactual explanations." *Parallel Problem Solving from Nature–PPSN XVI: 16th International Conference, PPSN 2020, Leiden, The Netherlands, September 5-9, 2020, Proceedings, Part I*: pp. 448–469. 2020. Springer.

[17] Guidotti, Riccardo, Monreale, Anna, Giannotti, Fosca, Pedreschi, Dino, Ruggieri, Salvatore and Turini, Franco. "Factual and counterfactual explanations for black box decision making." *IEEE Intelligent Systems* Vol. 34 No. 6 : pp. 14–23.

[18] Goyal, Yash, Wu, Ziyan, Ernst, Jan, Batra, Dhruv, Parikh, Devi and Lee, Stefan. "Counterfactual visual explanations." *International Conference on Machine Learning*: pp. 2376–2384. 2019. PMLR.

[19] Hendricks, Lisa Anne, Hu, Ronghang, Darrell, Trevor and Akata, Zeynep. "Generating Counterfactual Explanations with Natural Language." *ICML Workshop on Human Interpretability in Machine Learning*: pp. 95–98. 2018.

[20] Tran, Khanh Hiep, Ghazimatin, Azin and Saha Roy, Rishiraj. "Counterfactual explanations for neural recommenders." *Proceedings of the 44th International ACM SIGIR Conference on Research and Development in Information Retrieval*: pp. 1627–1631. 2021.

[21] Regenwetter, Lyle. "Data-Driven Bicycle Design using Performance-Aware Deep Generative Models." Ph.D. Thesis, Massachusetts Institute of Technology. 2022.

[22] Deb, Kalyanmoy, Pratap, Amrit, Agarwal, Sameer and Meyarivan, TAMT. "A fast and elitist multiobjective genetic algorithm: NSGA-II." *IEEE transactions on evolutionary computation* Vol. 6 No. 2 (2002): pp. 182–197.

[23] Regenwetter, Lyle, Nobari, Amin Heyrani and Ahmed, Faez. "Deep generative models in engineering design: A review." *Journal of Mechanical Design* Vol. 144 No. 7 (2022): p. 071704.

[24] Song, Binyang, Zhou, Rui and Ahmed, Faez. "Multimodal Machine Learning in Engineering Design: A Review and Future Directions." *arXiv preprint arXiv:2302.10909* (2023).

[25] Regenwetter, Lyle, Srivastava, Akash, Gutfreund, Dan and Ahmed, Faez. "Beyond Statistical Similarity: Rethinking Metrics for Deep Generative Models in Engineering Design." *arXiv preprint arXiv:2302.02913* (2023).

[26] Cai, Hongyun, Zheng, Vincent W and Chang, Kevin Chen-Chuan. "A comprehensive survey of graph embedding: Problems, techniques, and applications." *IEEE Transactions on Knowledge and Data Engineering* Vol. 30 No. 9 (2018): pp. 1616–1637.

[27] Faghri, Fartash, Fleet, David J, Kiros, Jamie Ryan and Fidler, Sanja. "Vse++: Improving visual-semantic embeddings with hard negatives." *arXiv preprint arXiv:1707.05612* (2017).

[28] Dai, Guoxian, Xie, Jin and Fang, Yi. "Siamese cnn-bilstm architecture for 3D shape representation learning." *IJCAI*: pp. 670–676. 2018.

[29] Devlin, Jacob, Chang, Ming-Wei, Lee, Kenton and Toutanova, Kristina. "Bert: Pre-training of deep bidirectional transformers for language understanding." *arXiv preprint arXiv:1810.04805* (2018).

[30] Cer, Daniel, Yang, Yinfei, Kong, Sheng-yi, Hua, Nan, Limtiaco, Nicole, John, Rhomni St, Constant, Noah, Guajardo-Cespedes, Mario, Yuan, Steve, Tar, Chris et al. "Universal sentence encoder." *arXiv preprint arXiv:1803.11175* (2018).

[31] Rombach, Robin, Blattmann, Andreas, Lorenz, Dominik, Esser, Patrick and Ommer, Björn. "High-resolution image synthesis with latent diffusion models." *Proceedings of the IEEE/CVF Conference on Computer Vision and Pattern Recognition*: pp. 10684–10695. 2022.

[32] Radford, Alec, Kim, Jong Wook, Hallacy, Chris, Ramesh, Aditya, Goh, Gabriel, Agarwal, Sandhini, Sastry, Girish, Askell, Amanda, Mishkin, Pamela, Clark, Jack et al. "Learning transferable visual models from natural language supervision." *International conference on machine learning*: pp. 8748–8763. 2021. PMLR.

[33] Gower, John C. "A general coefficient of similarity and some of its properties." *Biometrics* (1971): pp. 857–871.

[34] Blank, J. and Deb, K. "pymoo: Multi-Objective Optimization in Python." *IEEE Access* Vol. 8 (2020): pp. 89497–89509.

[35] Regenwetter, Lyle and Ahmed, Faez. "Design target achievement index: A differentiable metric to enhance deep generative models in multi-objective inverse design." *International Design Engineering Technical Conferences and Computers and Information in Engineering Conference*, Vol. 86236: p. V03BT03A046. 2022. American Society of Mechanical Engineers.

[36] Celis, Elisa, Keswani, Vijay, Straszak, Damian, Deshpande, Amit, Kathuria, Tarun and Vishnoi, Nisheeth. "Fair and diverse DPP-based data summarization." *International Conference on Machine Learning*: pp. 716–725. 2018. PMLR.

[37] Regenwetter, Lyle, Weaver, Colin and Ahmed, Faez. "Framed: Data-driven structural performance analysis of community-designed bicycle frames." *arXiv preprint arXiv:2201.10459* (2022).

[38] Erickson, Nick, Mueller, Jonas, Shirkov, Alexander, Zhang, Hang, Larroy, Pedro, Li, Mu and Smola, Alexander. "Autogluon-tabular: Robust and accurate automl for structured data." *arXiv preprint arXiv:2003.06505* (2020).

Copyright © 2023 by ASME

**Proceedings of the ASME 2023
International Design Engineering Technical Conferences and
Computers and Information in Engineering Conference
IDETC-CIE2023
August 20-23, 2023, Boston, Massachusetts**

DETC2023-113308

BAYESIAN MESH OPTIMIZATION FOR GRAPH NEURAL NETWORKS TO ENHANCE ENGINEERING PERFORMANCE PREDICTION

Jangseop Park[1,2], Namwoo Kang[1,2,*]

[1]Korea Advanced Institute of Science and Technology (KAIST), Daejeon, Republic of Korea
[2]Narnia Labs, Daejeon, Republic of Korea

ABSTRACT

In the field of engineering design, surrogate models for 3D computer-aided design (CAD) have been widely used to replace computationally expensive simulations. However, the conventional surrogate modeling process, which relies on the geometric parameters (or design variables) of CAD, has limitations when dealing with complex structural shapes commonly found in industry datasets. These limitations include information loss in low dimensions and difficulty in parametrization. This study proposes a Bayesian graph neural network (GNN) framework for a 3D deep-learning-based surrogate model that predicts engineering performance by directly learning the geometric features of CAD with mesh representation. Our proposed framework derives the optimal size of the mesh elements, creating a high-accuracy surrogate model with Bayesian optimization. It also solves the heterogeneity problem of 3D CAD data in that 2D images have regular pixel structures, whereas 3D CADs have irregular structures. From the experimental results, the mesh quality is highly correlated with the prediction accuracy of the surrogate model, and there exists an optimal mesh size that satisfies the high-performance requirements of the surrogate model. We expect that our proposed framework has the potential to be applied to mesh-based simulations in various engineering fields, reflecting the physics-based information widely used in computer-aided engineering.

Keywords: Metamodeling, CAD/Features Technology, Design Representation, FEA/Meshing/CAE, Neural Networks

1. INTRODUCTION

Over the past few decades, computer simulations have started playing a crucial role in modeling complex physics-based systems for various engineering problems (e.g., optimization design, uncertainty design, reliability analysis, and robust design). How-

ever, simulations are computationally expensive to run and reflect the detailed representations of real-world systems.

Surrogate models (SMs) and metamodels are replacing expensive simulations. SMs are cheaper and more convenient to evaluate than simulations for value estimation. In the conventional surrogate modeling process, the number of parameters (or design variables) is vital for determining the design shapes, sizes of engineering problems, and computational costs [1]. However, relying only on parameters has two limitations in representing complex 3D computer-aided design (CAD) datasets. First, several parameters of parametric design or topology optimization methods for generating 3D CAD datasets suffer from low-dimensional information in representing the 3D inherent geometric shapes. Second, high-dimensional CAD datasets have difficulty in parameterization to determine the data resolution. Direct learning of intrinsic 3D geometric structures is required necessarily and essentially for 3D deep-learning (DL)-based SMs.

With the significant development of 2D computer vision with convolutional neural networks (CNN), 3D objects and models have increased tremendously, not only with the interest in 3D tasks such as retrieval [2–4], classification [2, 3, 5–9], and segmentation [9–11], but also with 3D Shape Retrieval Contest (SHREC) [12] including CAD models. However, two main challenges remain: 3D data representation and heterogeneity problems with 3D data [13].

The first challenge involves selecting a suitable 3D data representation that affects the accuracy of 3D DL-based SMs with criteria such as simplicity, usability, and efficiency [14]. This includes point clouds [5, 6, 8], voxels [2, 7, 8, 10] (similarly octrees [11, 15]), meshes [3, 9], and graphs [4, 16–19]. The second challenge is the heterogeneity of 3D data. 2D image data have regular pixels; in contrast, 3D object data have the heterogeneity problem of irregular structures (e.g., meshes may differ in the number of vertices and faces). PyTorch toolkits [20] have recently emerged to solve heterogeneous problems and implement

---

*Corresponding author: nwkang@kaist.ac.kr
Documentation for `asmeconf.cls`: Version 1.34, May 15, 2023.

Copyright © 2023 by ASME

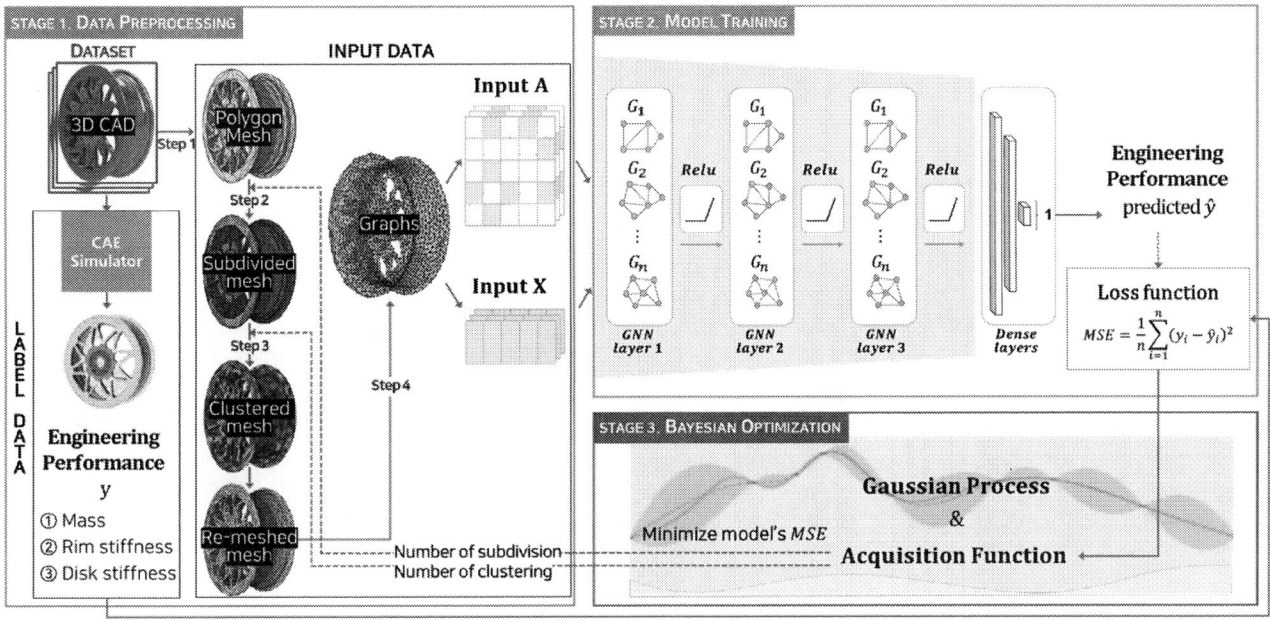

FIGURE 1: PROPOSED BAYESIAN GNN FRAMEWORK

batch operations with graph neural networks (GNN) and graph convolutional networks (GCN).

In this paper, we use the graph representation for three reasons, including addressing the two main challenges mentioned earlier. First, graph representation has advantages in terms of memory efficiency and ease of rendering. Second, graph representation is also helpful in conducting a GNN or GCN, which can address the heterogeneity problem for varying input sizes of 3D CAD datasets. Third, the graph is identical to the mesh representation in terms of data structures with Gauss points (i.e., integration points) of the finite element method (FEM) [16]. Here, we consider graph representation equal to mesh representation to apply the FEM concept, which is a robust numerical method for solving partial differential equations (PDEs) in computer-aided engineering (CAE).

This paper proposes a Bayesian GNN framework for a 3D DL-based surrogate model that predicts engineering performance with high accuracy to replace expensive simulations. The trade-off between the accuracy of the 3D DL-based SM and the efficiency of the mesh resolution can be controlled by determining the optimal size of the mesh elements (i.e., graph) with Bayesian optimization (BO). We expect our proposed framework to be applied to mesh-based simulations in various engineering fields, reflecting the physics-based information widely used in CAE.

The contributions of this study are as follows: (1) A 3D DL-based surrogate model can replace expensive CAE simulations. (2) The optimal mesh element size is derived by satisfying the high accuracy requirements of the 3D DL-based surrogate model with Bayesian optimization. (3) The heterogeneity problem for 3D data (i.e., irregular mesh element sizes as input data) is addressed using the GNN model. (4) The graph representation has expandability because it uses physics-based information of mesh-based simulations.

The remainder of this paper is organized as follows. In Section 2, we present the proposed framework, which consists of three stages: data preprocessing, model training, and Bayesian optimization. Section 3 presents the experimental results, including model performance, comparison with other models, and the relationship between mesh quality and prediction accuracy. Finally, Section 4 concludes this paper.

2. METHODOLOGY

We propose a Bayesian GNN framework that uses a GNN model as a 3D DL-based SM to predict engineering performance with a 3D CAD dataset, as shown in Fig. 1. Our proposed framework aims to derive the optimal size of mesh elements, creating a high-accuracy surrogate model through Bayesian optimization. The description of each stage of the proposed framework is presented as follows:

- *Stage 1.* includes data preprocessing of the label and input data. 3D CAD dataset were analyzed using modal analysis in the label data process to obtain the engineering performance. Four steps were performed to transform a 3D CAD dataset into a graph dataset during the input data process. In step 1, we convert the 3D CAD dataset into a polygon mesh dataset. In steps 2 to 3, we determine mesh element size with two hyperparameters related to subdividing and re-meshing the polygon mesh dataset. Finally, step 4 converts the re-meshed mesh into a graph as the input of the GNN model of stage 2.

- *Stage 2.* is related to the GNN model for predicting the engineering performance. The GNN model was trained to minimize the mean square error (MSE) between the estimated and ground-truth values for the engineering performance. A detailed description of the GNN architecture, including the model hyperparameters, activation functions, dense layers, etc., is presented in Section 2.2.

Copyright © 2023 by ASME

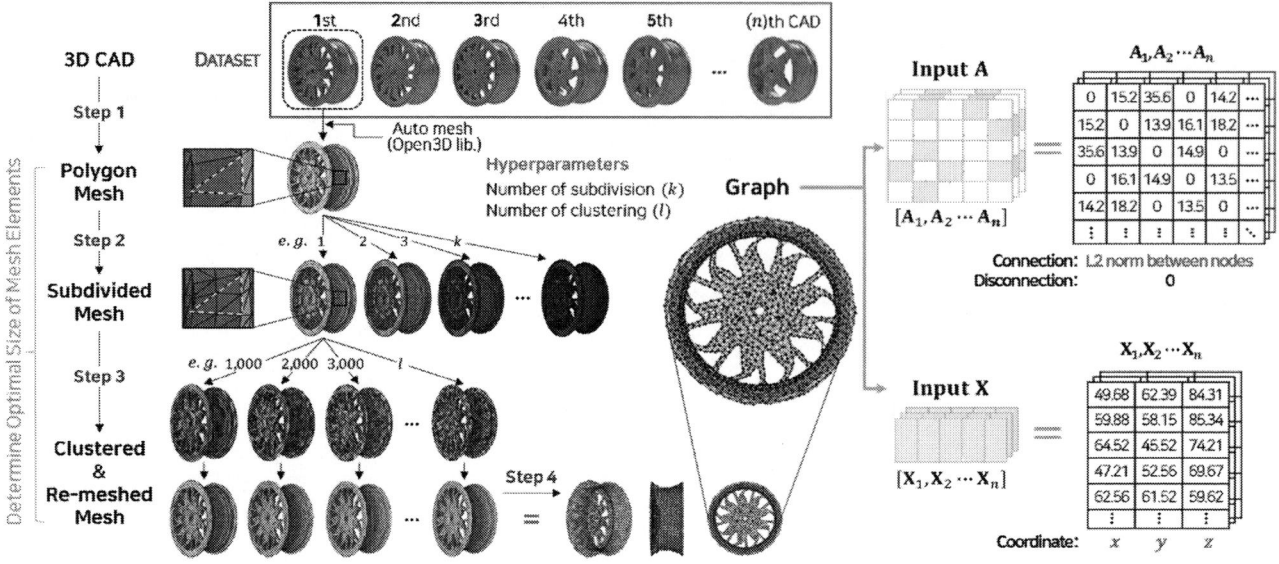

FIGURE 2: PREPROCESSING OF INPUT DATA COMPRISES 4 STEPS

• *Stage 3.* is used to explore the dependent variable space (MSEs of the GNN model) while overlooking the independent variable space (two hyperparameters related to the size of the mesh elements). In other words, we use BO as a sequential strategy for the global optimization of the GNN model, known as a black-box and expensive-to-evaluate function. The acquisition function of BO recommends the two hyperparameters to be used in the next iteration, which conducts stages 1 to 3 sequentially. After several iterations until the predefined criteria are satisfied, we can finally obtain the optimal size of the mesh elements.

2.1 Data Preprocessing (stage 1)

Preprocessing of Label Data. This study utilized 925 3D road wheel CADs from the study of [21] as a 3D CAD dataset. The label data of the 3D CAD dataset consist of the engineering performance. Mode and frequency response analyses were performed on the 3D CAD dataset to obtain the rim and disk stiffness. For CAE automation, the macro function in Altairs Simlab [22] was used to obtain m, f, f_1, and f_2.

$$k_{rim} = (2\pi f)^2 \, m \qquad (1)$$

$$k_{disk} = (2\pi f_2)^2 \left[m - m \left(\frac{f_2^2}{f_1^2} \right) \right] \qquad (2)$$

$$y^p_{scale} = \frac{y^p - y^p_{min}}{y^p_{max} - y^p_{min}} \qquad (3)$$

where m corresponds to the value of the mass for each 3D wheel CAD, f is the natural frequency in the normal mode of modal analysis, and f_1 and f_2 correspond to the resonance frequency and anti-resonance frequency in mode 11 (or the lateral mode) of the frequency response analysis, respectively. p indicates the engineering performance: mass, disk stiffness, or rim stiffness. The rim stiffness and disk stiffness were calculated using Eq. (1)

and Eq. (2). Then, the mass, rim stiffness, and disk stiffness were used as the labels by min-max scaling (i.e., normalization), which adjusts the values to a fixed range (regularly 0 to 1) using Eq. (3).

TABLE 1: POLYGON MESH DATASET IN STEP 1

	min.	max.	avg.
Number of nodes	624.0	1,692.0	1,039.0
Number of edges	1,932.0	5,148.0	3,271.0
Number of faces	1,287.0	4,238.0	2,177.3

2.2 Model Training (stage 2)

Preprocessing of Input Data. The preprocessing of input data for a 3D CAD dataset consists of four steps: (1) polygon mesh, (2) subdivided mesh, (3) clustered and re-meshed mesh, and (4) graph transformation as shown in Fig. 2.

Step 1: The collected 3D CAD dataset is converted into a surface mesh dataset with triangular meshes, which consists of a combination of nodes (or vertices) and faces. The nodes contain a connectivity list that describes how each node is connected. Open3D library with Python API [23] is used to automatically convert the 3D CAD dataset into a polygon mesh dataset with irregular mesh element size as show in Table 1.

Step 2: We utilize the re-meshing algorithm in steps 2–3 with anisotropic discrete Voronoi diagrams (DCVD) [24]. DCVD has two advantages: handling complex models with several million triangles and preserving the features of general objects [25]. Step 2 involves subdividing the meshes to increase the number of meshes when polygon meshes converted from 3D CAD data have fewer nodes. The number of subdivisions k as the first hyperparameter implies that one triangular mesh is divided into 4^k triangular meshes. For example, if k is 1, one triangular mesh is subdivided into four triangular meshes (white dashed lines in Fig. 2).

Copyright © 2023 by ASME

Step 3: The DCVD algorithm minimizes the global energy term by partitioning (or clustering) the input meshes to efficiently distribute the node budget by the number of clustering l as the second hyperparameter. DCVD is a theoretical clustering algorithm for preserving high-quality meshes based on the duality between the Delaunay triangulation (DT) and Voronoi diagrams (VD) which is widely used in geometrics. The clustered mesh can be easily transformed into a re-meshed mesh because the re-meshed mesh (DT) is a straight-line dual of the clustered mesh (VD) [26]. Steps 2 to 3 can be summarized with the re-meshing algorithm by determining two parameters, k and l.

Step 4: The mesh representation is converted into a graph representation as the input form of the GNN model. Re-meshed meshes can be transformed into graphs whose nodes and edges are the same as those of the mesh [16]. The details of the mesh-to-graph conversion are introduced in Section 2.2.

Graph Dataset. We converted the 3D CAD dataset consisting of 925 3D CADs into a re-meshed mesh dataset through steps 1 to 3 in the data preprocessing of the input data. Let $\mathcal{M}$ be the re-meshed dataset as $\mathcal{M} \equiv \{M_1, M_2 \ldots M_{925}\}$. Here, we represent a mesh data M as $M = \{\mathcal{V}, \mathcal{F}\}$ with the node set $\mathcal{V}$ and face (or cell) set $\mathcal{F}$. In step 4 of Section 2.1, one mesh data M is transformed into the graph data G as $G = \{\mathcal{V}, \mathcal{E}\}$ with node set $\mathcal{V}$ and edge set $\mathcal{E}$, which is identical to the mesh in terms of the data structure. Finally, we obtain the graph dataset $\mathcal{G} \equiv \{G_1, G_2 \ldots G_{925}\}$.

One graph G as the input data of a GNN model is represented in two matrix forms with adjacency matrix $\mathbf{A} \in \mathbb{R}^{N \times N}$ and node feature matrix $\mathbf{X} \in \mathbb{R}^{N \times F}$, where N is the number of nodes and F is the number of node features. The adjacency matrix $\mathbf{A}$ includes edge information about the relationships between adjacent nodes. Equally, a node feature matrix $\mathbf{X}$ includes the node coordinates (x, y, z).

One engineering performance label y^p is obtained from one 3D CAD by performing CAE simulations. Given a set of datasets $\mathcal{D}^p = \{(G_i, y_i^p)\}$ $i = 1, \ldots, 925$, for each engineering performance, the goal of a GNN model as a DL-based surrogate model is to learn the relationships $f : \mathcal{G} \rightarrow \mathcal{Y}$, where $y_i^p \in \mathcal{Y}^p$ is the label corresponding to graph $G_i \in \mathcal{G}$.

Even though we adopt the GNN model [18] as the baseline model of our proposed framework, we cover both the GNN and GCN models to compare the model performance without re-meshing in Section 3.2.

Graph Neural Networks. An adjacency matrix $\mathbf{A} \in \mathbb{R}^{N \times N}$ of the GNN model has connective or non-connective values: L2 norm or 0 as shown in Fig. 2. The adjacency matrix $\mathbf{A}$ is symmetrically normalized by $\mathbf{D}^{-\frac{1}{2}} \mathbf{A} \mathbf{D}^{\frac{1}{2}}$, where $\mathbf{D}$ is the diagonal node degree matrix of $\mathbf{A}$.

$$\mathbf{H}^l = \sigma(\mathbf{D}^{-\frac{1}{2}} \mathbf{A} \mathbf{D}^{\frac{1}{2}} \mathbf{H}^{l-1} \mathbf{W}_1^{l-1} + \mathbf{H}^{l-1} \mathbf{W}_2^l) \qquad (4)$$

where $\mathbf{W}_1^{l-1} \in \mathbb{R}^{F \times d}$ and $\mathbf{W}_2^{l-1} \in \mathbb{R}^{F \times d}$ are the trainable weights, and $\mathbf{H}^l$ denotes the (l)th node embedding features computed by the propagation function σ with the node embeddings of the previous ($l-1$)th step. In other words, the combinations with node embedding and propagation are prevalently called "message

passing (MP)." In Eq. (4), $\mathbf{H}^0$ is the node feature matrix at the initial node embedding ($l = 1$) as an initial input. d is the size of the latent vector used to propagate the node embedding of the ($l-1$)th layer to the (l)th layer.

Graph Convolutional Networks. An adjacency matrix $\hat{\mathbf{A}}$ of GCN has connective or non-connective values: 1 or 0. $\hat{\mathbf{A}}$ has self-loops to include the node features of itself, where $\hat{\mathbf{A}} = \mathbf{A} + \mathbf{I}$, $\mathbf{I} \in \mathbb{R}^{N \times N}$ is the identity matrix. The GCN model has a common trainable weight $\mathbf{W}^{l-1} \in \mathbb{R}^{F \times d}$ which makes the convolutional operations possible similar to a CNN model.

$$\mathbf{H}^l = \sigma(\hat{\mathbf{D}}^{-\frac{1}{2}} \hat{\mathbf{A}} \hat{\mathbf{D}}^{\frac{1}{2}} \mathbf{H}^{l-1} \mathbf{W}^{l-1}) \qquad (5)$$

where $\hat{\mathbf{D}} \in \mathbb{R}^{N \times N}$ is the diagonal node degree matrix of $\hat{\mathbf{A}}$ and adjacency matrix $\mathbf{A}$ is symmetrically normalized by $\hat{\mathbf{D}}^{-\frac{1}{2}} \hat{\mathbf{A}} \hat{\mathbf{D}}^{\frac{1}{2}}$. Eq. (5) has an essentially equivalent MP for aggregating node embeddings from adjacent neighborhoods.

Ultimately, the difference between GNN and GCN is the adjacency matrix $\mathbf{A}$ representing the edge connectivity between nodes, and they have the same node feature matrix $\mathbf{X}$ normalized by (6).

$$\mathbf{X}_{scale} = \frac{\mathbf{X} - \mathbf{X}_{min}}{\mathbf{X}_{max} - \mathbf{X}_{min}} \qquad (6)$$

Model Architecture. This subsection describes the proposed GNN model architecture that aims to predict the engineering performance obtained from CAE analysis.

The graph layers serve as graph feature extractions, and dense (or fully connected) layers are used for regression. There are three common numbers for both the graph and dense layers. The graph layers have a fixed latent vector size of 512 dimensions. The dense layers have latent vector sizes of 500, 200, and 25 dimensions, sequentially. A rectified linear unit (ReLU) was used as the activation function, except for the last layer. The proposed GNN model was trained using the Adam optimizer with a learning rate of 0.0002 and a batch size of 1. The epochs were set to 10,000, but the early stopping of callback techniques was used to prevent overfitting. The patience of the early stopping which is used to determine the duration of model training was set to 50 for validation loss. The loss function was set as the mean squared error (MSE) to predict each engineering performance, calculated using Eq. (7). A total of 925 (100%) graphs were split into 740 (80%) training, 92 (10%) validation, and 93 (10%) test sets. The labels of the validation set were not used for training.

$$MSE^p = \frac{1}{n} \sum_{i=1}^{n} (y_i^p - \hat{y}_i^p)^2 \qquad (7)$$

where n is the number of graphs G_i ($G_i \in \mathcal{G}$) and y_i^p is the (i)th ground label for each engineering performance according to the (i)th graph G_i.

2.3 Bayesian Optimization (stage 3)

We describe the BO process to obtain the optimal mesh element size, improving the accuracy of our proposed GNN model. BO has been widely used in machine learning [27] to find hyperparameters to optimize the performance of DL models. In this study, we utilize BO to find the optimal size of mesh elements,

Copyright © 2023 by ASME

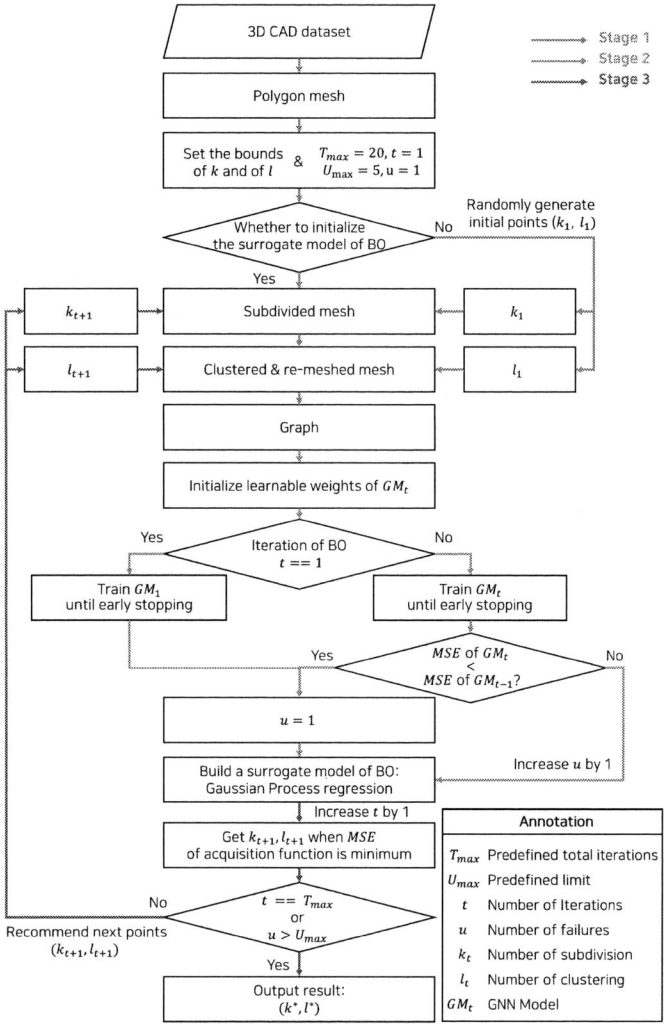

FIGURE 3: FLOWCHART OF THE PROPOSED BAYESIAN GNN FRAMEWORK

l, s.t. $\{k, l \in \mathbb{Z} \,|\, 2 \le k \le 4, \ 3,000 \le l \le 5,000\}$ were set in this study. A flowchart of the entire BO process used to obtain the optimal hyperparameters k and l is illustrated in Fig. 3.

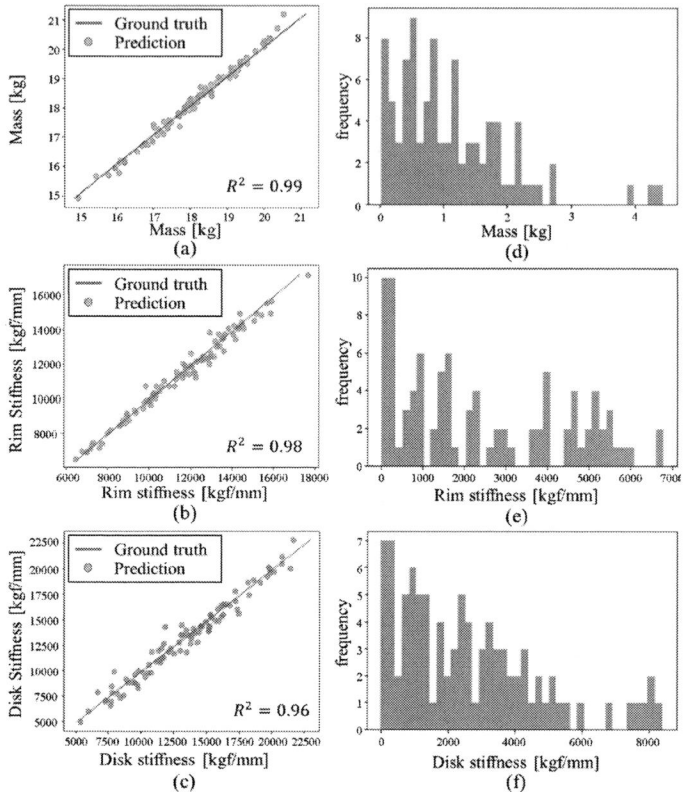

FIGURE 4: (A)-(C) SCATTER PLOTS SHOW THE RELATIONSHIPS BETWEEN BO-EI GNN MODEL PREDICTION AND THE GROUND TRUTH FOR THE TEST SET AND (D)-(F) HISTOGRAM OF ABSOLUTE ERROR (AE) FOR THE TEST SET

that is, the optimal hyperparameters (the numbers of subdivision k and of the clustering l) satisfying the minimum MSE of a GNN model on a bounded hyperparameter set.

3. RESULTS AND DISCUSSION

3.1 Proposed Model Prediction Result

Three things should be considered when performing BO. The first is a prior function that is optimized to approximate a GNN model as a surrogate model of BO. We selected the Gaussian process (GP) as a prior function because GP has high flexibility and tractability to express the assumption for data distribution. The second is an acquisition function that recommends evaluating the next set of hyperparameters. We selected the expected improvement (EI) instead of upper confidence bound (UCB), which is superior in function evaluations and time elapsed as the acquisition function for our proposed framework. Here, we refer to a GNN model with EI as "BO-EI GNN," with UCB as "BO-UCB GNN," and with the Metropolis–Hastings algorithm of Markov Chain Monte Carlo as "MCMC GNN." The comparison of those models is discussed in Section 3.2. Finally, the bounds for k and

$$\left(R^2\right)^P = 1 - \frac{\sum_i^n \left(y_i^P - \hat{y}_i^P\right)^2}{\sum_i^n \left(y_i^P - \bar{y}_i^P\right)^2} \tag{8}$$

$$AE^P = \left|y_i^P - \hat{y}_i^P\right|, i = 1, \ldots, n \tag{9}$$

$$RMSE^P = \sqrt{\frac{\sum_i^n \left(y_i^P - \hat{y}_i^P\right)^2}{n}} \tag{10}$$

$$MAPE^P = 100 \times \frac{1}{n} \sum_{i=1}^n \left|\frac{y_i^P - \hat{y}_i^P}{y_i^P}\right| \tag{11}$$

where n is the number of input data, i is the (i)th input data, y_i^P and $\hat{y}_i^P$ are the (i)th ground truth, and the (i)th predicted engineering performance, respectively.

We evaluate the model prediction accuracy for the train, validation and test set with the R-squared (R^2), the absolute error (AE), root mean square error (RMSE), and mean absolute percentage error (MAPE) calculated using Eq. (8), (9), (10), (11), respectively. The results of the proposed BO-EI GNN model for the each set with the RMSE, MAPE, and R^2 metrics are shown in Table 2. In the case of mass, all metrics are superior to the others

Copyright © 2023 by ASME

TABLE 2: COMPARISON OF 3D CNN, GCN, GNN, AND BO-EI GNN MODELS FOR PREDICTION ACCURACY

*units: Mass (kg), Rim stiffness (kgf/mm), Disk stiffness (kgf/mm)

Methods	Engineering Performance	Training set			Validation set			Test set		
		RMSE	MAPE	R^2	RMSE	MAPE	R^2	RMSE	MAPE	R^2
3D CNN	Mass	0.09	0.41	0.991	0.17	0.69	0.952	0.28	1.22	0.915
	Rim stiffness	240.31	1.74	0.992	269.42	1.73	0.973	488.44	2.77	0.931
	Disk stiffness	1144.87	7.41	0.915	809.63	4.01	0.907	1409.8	6.85	0.728
GCN	Mass	0.45	1.99	0.863	0.46	2.05	0.885	0.62	2.71	0.756
	Rim stiffness	443.71	3.07	0.968	694.58	5.03	0.931	958.08	6.1	0.888
	Disk stiffness	708.2	4.47	0.967	1182.86	8.19	0.924	1404.76	9.37	0.892
GNN	Mass	0.19	0.87	0.982	0.38	1.62	0.924	0.41	1.8	0.894
	Rim stiffness	226.17	1.57	0.993	690.66	4.7	0.931	802.06	5.06	0.923
	Disk stiffness	785.11	4.91	0.962	1163.18	7.62	0.922	1488.39	9.91	0.876
BO-EI GNN (our proposed)	Mass	0.07	0.29	0.997	0.16	0.67	0.986	0.16	0.69	**0.985**
	Rim stiffness	200.82	1.36	0.994	208.93	1.42	0.993	350.13	2.25	**0.978**
	Disk stiffness	494.79	3.13	0.985	593.79	3.49	0.979	758.03	4.94	**0.963**

owing to the variance of the ground-truth distribution. The mass data had the narrowest variance, from approximately 15 to 20. In contrast, the disk stiffness data had the widest variance, from approximately 5,000 to 22,500. Thus, the BO-EI GNN model for mass can be trained with higher accuracy than the disk stiffness model.

TABLE 3: OPTIMAL SIZE OF MESH ELEMENTS FROM OUR PROPOSED BO-EI GNN MODEL AND R^2 FOR THE TEST SET

Optimal size of mesh elements	Mass	Rim stiffness	Disk stiffness
Number of subdivision (k)	3	3	3
Number of clustering (l)	4,557	4,626	3,438
R^2 for the test set	0.985	0.978	0.963

The results of the optimal size of the mesh elements for each engineering performance are shown in Table 3. In the case of mass as an engineering performance, the k and l were derived with values of 3 and 4,557, respectively. Finally, the R^2 value of 0.985 was higher than that of the other engineering performance.

The scatter plots of Fig. 4 (a)-(c) show the predicted and actual values for each engineering performance test conducted on the test set. The solid line indicates the ground truth, and the dots indicate the predicted values. The dots of the mass closest to the solid line had the highest R^2 value. The AE is the error between the predicted and actual values for each engineering performance. The error distribution for the test set is shown in Fig. 4 (d)-(f).

3.2 Comparison in Different Models

Comparison of Re-meshing and Not Re-meshing. This subsection compares our proposed model BO-EI GNN with different models without re-meshing algorithms, including the voxel representation, that is, with the original polygon mesh dataset. Here, we refer to a GNN model without re-meshing as the "GNN

model," the GCN model without re-meshing as the "GCN model," and the 3D CNN without re-meshing as the "3D CNN model."

In this section, the 3D CAD dataset was converted into different 3D data representations. Then, 3D deep learning models suitable for each representation were trained, evaluated, and compared without re-meshing algorithms, except for the BO-EI GNN model. Each representation proceeds with a voxel, point cloud, and graph representation. In this study, the point cloud is used with the PointNet [5] model by converting the classification prediction model into a regression model to predict the engineering performance. However, the point cloud model results were excluded because of the low accuracy of model prediction.

The voxel model was adopted for the 3D CNN model proposed by [8], which predicts the engineering performance mass for 3D printing. After converting 3D CAD data into polygon mesh data, elements are expressed as occupied (value of 1) or non-occupied (value of 0) values by checking the connection information of the adjacent mesh nodes. The 3D CNN model used a voxel size of 64 × 64 × 64, three convolutional layers, and three fully connected layers. The three convolutional layers had 64, 16, and 8 filters, and the same padding was applied. The dimensions of the dense layers were the same (500, 200, and 25) as those in the BO-EI GNN model in this study. Except for the last layer, a rectified linear unit (ReLU) was used as the activation function for the remaining layers. Through a sufficient hyperparameter search, the 3D CNN model used the Adam optimizer, and the learning rate, epochs, patience, and batch size were set as 0.0001, 10,000, 100, and 8, respectively. The 3D CNN model evaluation for each engineering performance is presented in Table 2.

Graph representation is evaluated for three models: GNN, GCN, and BO-EI GNN models. As explained in Section 2.2, the GNN model uses Eq. (4) to update node features through message passing of the graph layers. The GCN model uses Eq. (5) to perform graph convolutional operations. Except for having a different form of the adjacency matrices between the GNN and GCN models, both models used the same values and functions, such as the number of layers, the size of the latent

Copyright © 2023 by ASME

vector, hyperparameters, activation functions, and an optimizer. Both models are trained with a polygon mesh dataset, which is not the optimal size of mesh elements. The results of the experiment show that BO-EI GNN model is excellent in terms of all metrics for each engineering performance, as shown in Table 2.

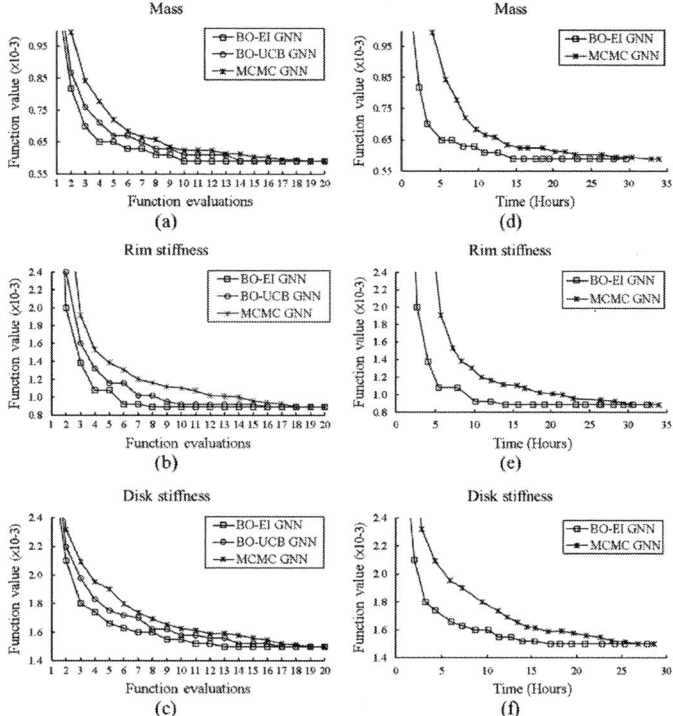

FIGURE 5: (A)-(C) DIFFERENT STRATEGIES OF OPTIMIZATION IN TERMS OF FUNCTION EVALUATIONS AND (D)-(F) COMPARISONS OF BO-EI GNN AND MCMC GNN MODELS IN TERMS OF TIME ELAPSED WITH TRAINING FOR THE VALIDATION TEST

Comparison of BO-EI GNN, BO-UCB GNN, and MCMC GNN Models. We empirically analyzed and compared the optimization strategies in terms of function evaluations and time elapsed in the BO-EI GNN, BO-UCB GNN, and MCMC GNN. We utilize the original graph dataset converted from the polygon mesh dataset as input dataset as mentioned in 1. In Fig. 5, "function value" indicates the values of MSE evaluated by each trained model. For "function evaluations," MCMC GNN model training was performed 10 times for each number of function evaluations, and the mean value was reported. The MCMC GNN model can be evaluated in parallel, in contrast, the other models were run once because they should be trained by sequential iteration steps owing to the Bayesian optimization. The results of the analyses are presented in Fig. 5 (a)-(c) in terms of function evaluations. For all the engineering performance, the BO-EI GNN model is superior to the BO-UCB GNN and significantly outperforms the MCMC GNN, with a minimum MSE of less than half as function evaluations and time costs using one GPU (Geforce RTX 3090) and CPU 64 Cores (four AMD EPYC 7282 16-core processors). In terms of time elapsed, the BO-EI model is compared with the MCMC GNN in Fig. 5 (d)-(f).

3.3 Mesh Quality

High-quality meshes are essential not only in practical applications, such as 3D visualization, numerical simulations, and animation but also in the field of 3D deep learning-based surrogate models. We discuss mesh quality measures with the visualization of the optimal size of the mesh elements and examine the relationship between the mesh quality values using two measures and the values of the model prediction accuracy.

Mesh Quality Measures. In finite element analysis (FEA), there are various measures for evaluating the quality of a mesh, including the minimal angle, maximal angle, aspect ratio, regularity, and feature preservation. This study evaluated the mesh quality using minimal and maximal angle measures. We evaluated the mesh quality for one mesh dataset $\mathcal{M} \equiv \{M_1, M_2 \ldots M_{925}\}$ by applying the minimal and maximal angle measures of $Q(\mathcal{M})_{min}$ and $Q(\mathcal{M})_{max}$, respectively. The equations for mesh quality are as follows:

$$Q\left(\mathcal{M}_{k,l}\right)_{min} = \frac{1}{n}\sum_{i=1}^{n}\frac{1}{s}\sum_{j=1}^{s} min\left[\{\theta \mid \theta \in M_i(\phi_j)\}\right] \quad (12)$$

$$Q\left(\mathcal{M}_{k,l}\right)_{max} = \frac{1}{n}\sum_{i=1}^{n}\frac{1}{s}\sum_{j=1}^{s} max\left[\{\theta \mid \theta \in M_i(\phi_j)\}\right] \quad (13)$$

where θ is the interior angle of the (j)th triangle cell $\phi_j \in M_i$, s is the number of triangular cells ϕ_j, and n is the number of mesh data $M_i \in \mathcal{M}$. One mesh data M_i of mesh dataset $\mathcal{M}_{k,l}$ consists of s cells ϕ_j. One cell ϕ_j includes the values of the three angles $\theta \in M_i(\phi_j)$ because this study uses a cell as a triangle, which means that the sum of the three angles is 180°. Ultimately, the minimal angle measure $Q\left(\mathcal{M}_{k,l}\right)_{min}$ is calculated using Eq. (12) and which is the summation of the smallest angles among the three angles.

For example, in the case of the mass, the re-meshed mesh dataset for the optimal size of mesh elements is expressed as $\mathcal{M}_{3,4557}$ because k and l are 3 and 4,557, respectively, as shown in Table 3. In other words, $\mathcal{M}_{3,4557}$, which is clustered into 4557 and divided three times. Figure 6 shows the visualization with minimal and maximal angle measures for an arbitrary 585th mesh dataset M_{585} of mesh dataset $\mathcal{M}_{3,4557}$. In general, the mesh quality indicates that the closer the triangle cell is to an equilateral triangle, the higher the mesh quality.

In the case of a polygon mesh converted from a 3D CAD in Fig. 6 (a), the colors of the cells are expressed more as blue or red, as shown in Fig. 6 (b) and (d). In contrast, the re-meshed meshes in yellow are distributed more than those in blue (or red part), which indicates closer to 60 °, as shown in Fig. 6 (c) and (d). Thus, we can verify that the optimal size of the mesh elements is higher than that of a polygon mesh.

Relationship between Mesh Quality and Model Prediction Accuracy. We generated 50 re-meshed mesh datasets $\{\mathcal{M}_1, \mathcal{M}_2 \ldots \mathcal{M}_{50}\}$ by sampling randomly with k and l values, which determine the mesh quality. Here, one of the 50 samples contains the optimal size of the mesh elements obtained in Section 3.1. In addition, we derived 50 R^2 scores and the values of the model prediction accuracy by training the GNN model on 50 dataset.

Copyright © 2023 by ASME

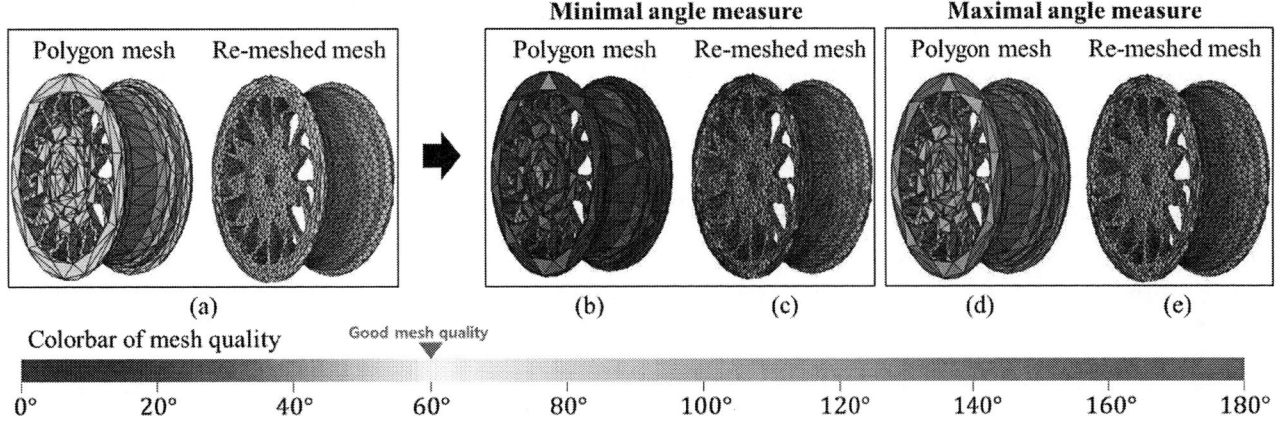

FIGURE 6: VISUALIZATION OF MESH QUALITY FOR MINIMAL AND MAXIMAL ANGLE MEASURES. (A) A POLYGON MESH CONVERTED FROM A 3D CAD, (B)-(C) IN CASE OF MINIMAL ANGLE MEASURE AND (D)-(E) IN CASE OF MAXIMAL ANGLE MEASURE

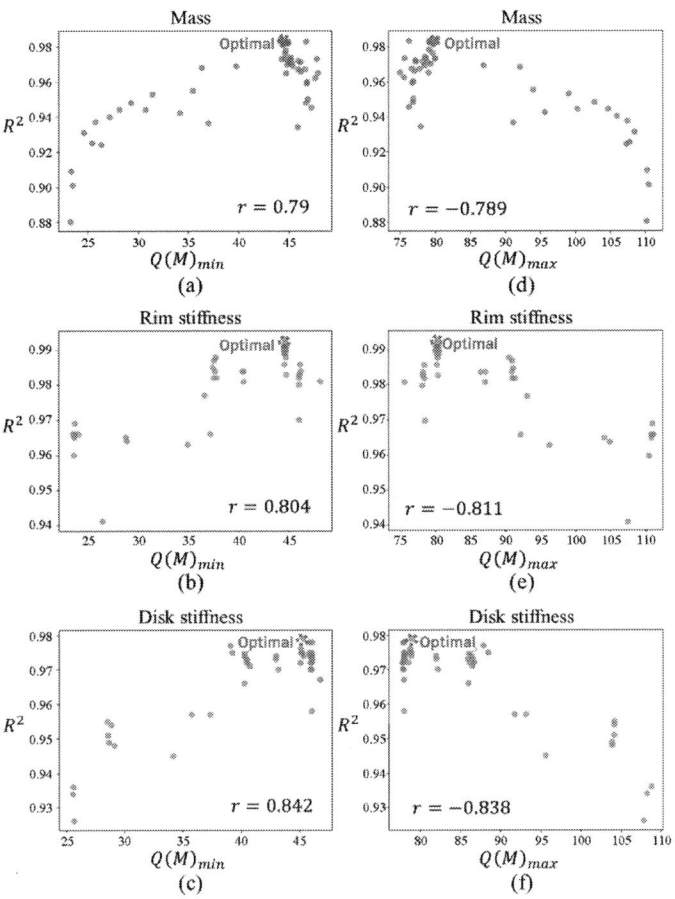

FIGURE 7: PEARSON CORRELATION COEFFICIENTS AND SCATTER PLOTS OF RELATIONSHIPS BETWEEN R^2 AND MESH QUALITY $Q(\mathcal{M})$. (A)-(C) IN CASE OF MINIMAL ANGLE MEASURE AND (D)-(F) IN CASE OF MAXIMAL ANGLE MEASURE.

We used the Pearson correlation coefficient, r, which measures a linear correlation. r is a number between -1 and 1, which measures the strength and direction of the relationship between the two variables. The Pearson correlation coefficient calculated using Eq. (14) was employed for a total of six cases with a combination of three types of the engineering performance and two types of mesh quality measures (minimal and maximal angle), as shown in Fig. 7.

$$r^P = \frac{\sum_{i=1}^{m} [Q(\mathcal{M})_i - \bar{Q}(\mathcal{M})_i][(R^2)_i^P - (\bar{R}^2)_i^P]}{\sqrt{\sum_{i=1}^{m} [Q(\mathcal{M})_i - \bar{Q}(\mathcal{M})_i]^2 \sum_{i=1}^{m} [(R^2)_i^P - (\bar{R}^2)_i^P]^2}} \quad (14)$$

where m is the number of data samples (where m is 50) and i is the number of orders for both the mesh quality of the (i)th re-meshed dataset and the prediction accuracy of the (i)th GNN model out of 50 samples. In the case of the minimal angle measure shown in Fig. 7 (a)-(c), the Pearson correlation coefficient r has values of 0.79, 0.804, and 0.842 for mass, rim stiffness, and disk stiffness, respectively. The minimal angle measure and model prediction accuracy have a high positive correlation because r is above 0.7, which indicates a high correlation. The model prediction accuracy tended to improve as the mesh quality increased. However, the optimal point of the mesh elements with the highest prediction accuracy (red dashed circles in Fig. 7 (a)-(c) was not located in the upper right corner. Figure 7 (d)-(f) shows the results of the maximal angle measure, and each r value shows high negative correlations of -0.789, -0.811, and -0.838 for each engineering performance. Equivalent to the minimal measures, the optimal point that improves the model accuracy obtained through Bayesian optimization is not located in the upper-left corner. Although the model accuracy is highly related to the mesh quality, it is necessary to derive the optimal mesh element size.

4. CONCLUSION

In this paper, we propose a 3D deep learning-based surrogate model that predicts engineering performance by directly learning the intrinsic geometric characteristics of CAD using graph representation. Our proposed BO-EI GNN model derives the optimal size of the mesh elements for creating a high-accuracy surrogate model through Bayesian optimization. We found that the mesh quality is highly correlated with the prediction accuracy of the surrogate model.

Copyright © 2023 by ASME

Our work has the following contributions. First, the proposed Bayesian GNN framework can replace expensive CAE simulations. Our model predicts engineering performance with high accuracy and can provide quick results for a 3D CAD dataset in the same domain. Second, the GNN framework addresses heterogeneity problems, which are prevalent in 3D datasets with irregular input sizes. Using a batch size of one solves this problem, allowing for the extension of our framework to other 3D data representations, including graph representations. Third, we demonstrate the effectiveness of using Bayesian optimization to derive the optimal mesh element size for a 3D CAD dataset.

However, our study had limitations. We used a surface mesh instead of a mesh of CAE simulations. we plan to check whether the proposed framework is equally applicable to CAE meshes of complex shapes. Second, we only used the coordinates of the node feature matrices as input instead of physics-based information. Future research will focus on using physics-based information such as displacement, velocity, and materials to develop dynamic analysis models for 3D CADs. We expect that our proposed Bayesian GNN framework has the potential to be applied to various engineering fields, reflecting the physics-based information widely used in CAE.

ACKNOWLEDGMENTS

This work was supported by the National Research Foundation of Korea grant (2018R1A5A7025409) and the Ministry of Science and ICT of Korea grant (No.2022-0-00969, No.2022-0-00986).

REFERENCES

[1] Alizadeh, Reza, Allen, Janet K and Mistree, Farrokh. "Managing computational complexity using surrogate models: a critical review." *Research in Engineering Design* Vol. 31 (2020): pp. 275–298.

[2] Wu, Zhirong, Song, Shuran, Khosla, Aditya, Yu, Fisher, Zhang, Linguang, Tang, Xiaou and Xiao, Jianxiong. "3d shapenets: A deep representation for volumetric shapes." *Proceedings of the IEEE conference on computer vision and pattern recognition*: pp. 1912–1920. 2015.

[3] Hanocka, Rana, Hertz, Amir, Fish, Noa, Giryes, Raja, Fleishman, Shachar and Cohen-Or, Daniel. "Meshcnn: a network with an edge." *ACM Transactions on Graphics (TOG)* Vol. 38 No. 4 (2019): pp. 1–12.

[4] Agathos, Alexander, Pratikakis, Ioannis, Papadakis, Panagiotis, Perantonis, Stavros, Azariadis, Philip and Sapidis, Nickolas S. "3D articulated object retrieval using a graph-based representation." *The Visual Computer* Vol. 26 (2010): pp. 1301–1319.

[5] Qi, Charles R, Su, Hao, Mo, Kaichun and Guibas, Leonidas J. "Pointnet: Deep learning on point sets for 3d classification and segmentation." *Proceedings of the IEEE conference on computer vision and pattern recognition*: pp. 652–660. 2017.

[6] Qi, Charles Ruizhongtai, Yi, Li, Su, Hao and Guibas, Leonidas J. "Pointnet++: Deep hierarchical feature learning on point sets in a metric space." *Advances in neural information processing systems* Vol. 30 (2017).

[7] Xiang, Yu, Choi, Wongun, Lin, Yuanqing and Savarese, Silvio. "Data-driven 3d voxel patterns for object category recognition." *Proceedings of the IEEE conference on computer vision and pattern recognition*: pp. 1903–1911. 2015.

[8] Maturana, Daniel and Scherer, Sebastian. "Voxnet: A 3d convolutional neural network for real-time object recognition." *2015 IEEE/RSJ international conference on intelligent robots and systems (IROS)*: pp. 922–928. 2015. IEEE.

[9] Feng, Yutong, Feng, Yifan, You, Haoxuan, Zhao, Xibin and Gao, Yue. "Meshnet: Mesh neural network for 3d shape representation." *Proceedings of the AAAI conference on artificial intelligence*, Vol. 33. 01: pp. 8279–8286. 2019.

[10] Tatarchenko, Maxim, Dosovitskiy, Alexey and Brox, Thomas. "Octree generating networks: Efficient convolutional architectures for high-resolution 3d outputs." *Proceedings of the IEEE international conference on computer vision*: pp. 2088–2096. 2017.

[11] Wang, Peng-Shuai, Liu, Yang, Guo, Yu-Xiao, Sun, Chun-Yu and Tong, Xin. "O-cnn: Octree-based convolutional neural networks for 3d shape analysis." *ACM Transactions On Graphics (TOG)* Vol. 36 No. 4 (2017): pp. 1–11.

[12] Frome, Andrea, Jégou, Hervé, Aigrain, Jean and Zisserman, Andrew. "SHREC'11: robust feature detection and description benchmark." *Computer Graphics Forum* Vol. 30 No. 5 (2011): pp. 1777–1787.

[13] Wu, Jia, Chen, Xiu-Yun, Zhang, Hao, Xiong, Li-Dong, Lei, Hang and Deng, Si-Hao. "Hyperparameter optimization for machine learning models based on Bayesian optimization." *Journal of Electronic Science and Technology* Vol. 17 No. 1 (2019): pp. 26–40.

[14] Gezawa, Abubakar Sulaiman, Zhang, Yan, Wang, Qicong and Yunqi, Lei. "A review on deep learning approaches for 3d data representations in retrieval and classifications." *IEEE access* Vol. 8 (2020): pp. 57566–57593.

[15] Riegler, Gernot, Osman Ulusoy, Ali and Geiger, Andreas. "Octnet: Learning deep 3d representations at high resolutions." *Proceedings of the IEEE conference on computer vision and pattern recognition*: pp. 3577–3586. 2017.

[16] Fey, Matthias, Lenssen, Jan Eric, Weichert, Frank and Müller, Heinrich. "Splinecnn: Fast geometric deep learning with continuous b-spline kernels." *Proceedings of the IEEE conference on computer vision and pattern recognition*: pp. 869–877. 2018.

[17] Scarselli, Franco, Gori, Marco, Tsoi, Ah Chung, Hagenbuchner, Markus and Monfardini, Gabriele. "The graph neural network model." *IEEE transactions on neural networks* Vol. 20 No. 1 (2008): pp. 61–80.

[18] Bianchi, Filippo Maria, Grattarola, Daniele and Alippi, Cesare. "Spectral clustering with graph neural networks for graph pooling." *International conference on machine learning*: pp. 874–883. 2020. PMLR.

[19] Kipf, Thomas N and Welling, Max. "Semi-supervised classification with graph convolutional networks." *arXiv preprint arXiv:1609.02907* (2016).

[20] Ravi, Nikhila, Reizenstein, Jeremy, Novotny, David, Gordon, Taylor, Lo, Wan-Yen, Johnson, Justin and Gkioxari,

Copyright © 2023 by ASME

Georgia. "Accelerating 3d deep learning with pytorch3d." *arXiv preprint arXiv:2007.08501* (2020).

[21] Yoo, Soyoung, Lee, Sunghee, Kim, Seongsin, Hwang, Kwang Hyeon, Park, Jong Ho and Kang, Namwoo. "Integrating deep learning into CAD/CAE system: generative design and evaluation of 3D conceptual wheel." *Structural and multidisciplinary optimization* Vol. 64 No. 4 (2021): pp. 2725–2747.

[22] Atair Aerospace, Inc. "SimLab." https://www.atair.com/simlab/ (accessed March 12, 2023).

[23] Zhou, Qian-Yi, Park, Jaesik and Koltun, Vladlen. "Open3D: A Modern Library for 3D Data Processing." *arXiv:1801.09847* (2018).

[24] Valette, Sébastien, Chassery, Jean Marc and Prost, Rémy. "Generic remeshing of 3D triangular meshes with metric-dependent discrete Voronoi diagrams." *IEEE Transactions on Visualization and Computer Graphics* Vol. 14 No. 2 (2008): pp. 369–381.

[25] Khan, Dawar, Plopski, Alexander, Fujimoto, Yuichiro, Kanbara, Masayuki, Jabeen, Gul, Zhang, Yongjie Jessica, Zhang, Xiaopeng and Kato, Hirokazu. "Surface remeshing: A systematic literature review of methods and research directions." *IEEE transactions on visualization and computer graphics* Vol. 28 No. 3 (2020): pp. 1680–1713.

[26] Aurenhammer, Franz, Klein, Rolf and Lee, Der-Tsai. *Voronoi diagrams and Delaunay triangulations.* World Scientific Publishing Company (2013).

[27] Snoek, Jasper, Larochelle, Hugo and Adams, Ryan P. "Practical bayesian optimization of machine learning algorithms." *Advances in neural information processing systems* Vol. 25 (2012).

[28] Deaton, Joshua D and Grandhi, Ramana V. "A survey of structural and multidisciplinary continuum topology optimization: post 2000." *Structural and Multidisciplinary Optimization* Vol. 49 (2014): pp. 1–38.

[29] Gano, Shawn E, Renaud, John E, Martin, Jay D and Simpson, Timothy W. "Update strategies for kriging models used in variable fidelity optimization." *Structural and Multidisciplinary Optimization* Vol. 32 (2006): pp. 287–298.

[30] Shan, Songqing and Wang, G Gary. "Survey of modeling and optimization strategies to solve high-dimensional design problems with computationally-expensive black-box functions." *Structural and multidisciplinary optimization* Vol. 41 (2010): pp. 219–241.

[31] Friedman, Jerome H. "Greedy function approximation: a gradient boosting machine." *Annals of statistics* (2001): pp. 1189–1232.

[32] Hastie, Trevor, Tibshirani, Robert, Friedman, Jerome H and Friedman, Jerome H. *The elements of statistical learning: data mining, inference, and prediction.* Vol. 2. Springer (2009).

[33] Gorissen, Dirk, Couckuyt, Ivo, Demeester, Piet, Dhaene, Tom and Crombecq, Karel. "A surrogate modeling and adaptive sampling toolbox for computer based design."

Journal of machine learning research.-Cambridge, Mass. Vol. 11 (2010): pp. 2051–2055.

[34] Williams, Glen, Meisel, Nicholas A, Simpson, Timothy W and McComb, Christopher. "Design repository effectiveness for 3D convolutional neural networks: Application to additive manufacturing." *Journal of Mechanical Design* Vol. 141 No. 11 (2019).

[35] Yoo, Soyoung and Kang, Namwoo. "Explainable artificial intelligence for manufacturing cost estimation and machining feature visualization." *Expert Systems with Applications* Vol. 183 (2021): p. 115430.

[36] Shin, Seungyeon, Jin, Ah-hyeon, Yoo, Soyoung, Lee, Sunghee, Kim, ChangGon, Heo, Sungpil and Kang, Namwoo. "Wheel impact test by deep learning: prediction of location and magnitude of maximum stress." *Structural and Multidisciplinary Optimization* Vol. 66 No. 1 (2023): p. 24.

[37] Khadilkar, Aditya, Wang, Jun and Rai, Rahul. "Deep learning–based stress prediction for bottom-up SLA 3D printing process." *The International Journal of Advanced Manufacturing Technology* Vol. 102 (2019): pp. 2555–2569.

[38] Cunningham, James D, Simpson, Timothy W and Tucker, Conrad S. "An investigation of surrogate models for efficient performance based decoding of 3D point clouds." *Journal of Mechanical Design* Vol. 141 No. 12 (2019).

[39] Umetani, Nobuyuki and Bickel, Bernd. "Learning three-dimensional flow for interactive aerodynamic design." *ACM Transactions on Graphics (TOG)* Vol. 37 No. 4 (2018): pp. 1–10.

[40] Durasov, Nikita, Lukoyanov, Artem, Donier, Jonathan and Fua, Pascal. "DEBOSH: Deep Bayesian Shape Optimization." *arXiv preprint arXiv:2109.13337* (2021).

[41] Whalen, Eamon and Mueller, Caitlin. "Toward reusable surrogate models: Graph-based transfer learning on trusses." *Journal of Mechanical Design* Vol. 144 No. 2 (2022).

[42] He, Kaiming, Zhang, Xiangyu, Ren, Shaoqing and Sun, Jian. "Deep residual learning for image recognition." *Proceedings of the IEEE conference on computer vision and pattern recognition*: pp. 770–778. 2016.

[43] Simonyan, Karen and Zisserman, Andrew. "Very deep convolutional networks for large-scale image recognition." *arXiv preprint arXiv:1409.1556* (2014).

[44] Chang, Angel X, Funkhouser, Thomas, Guibas, Leonidas, Hanrahan, Pat, Huang, Qixing, Li, Zimo, Savarese, Silvio, Savva, Manolis, Song, Shuran, Su, Hao et al. "Shapenet: An information-rich 3d model repository." *arXiv preprint arXiv:1512.03012* (2015).

[45] Han, Xu, Gao, Han, Pffaf, Tobias, Wang, Jian-Xun and Liu, Li-Ping. "Predicting physics in mesh-reduced space with temporal attention." *arXiv preprint arXiv:2201.09113* (2022).

[46] Williams, Christopher K and Rasmussen, Carl Edward. "Gaussian processes for machine learning, vol. 2." *MA: MIT press Cambridge* (2006).

Copyright © 2023 by ASME

Proceedings of the ASME 2023
International Design Engineering Technical Conferences and
Computers and Information in Engineering Conference
IDETC-CIE2023
August 20-23, 2023, Boston, Massachusetts

DETC2023-114592

ON THE USE OF GEOMETRIC DEEP LEARNING TOWARDS THE EVALUATION OF GRAPH-CENTRIC ENGINEERING SYSTEMS

Anthony Sirico Jr.[*] **Daniel R. Herber**
Department of Systems Engineering
Colorado State University
Fort Collins, CO 80523
Email: {anthony.sirico, daniel.herber}@colostate.edu

ABSTRACT

Many complex engineering systems can be represented in a topological form, such as graphs. This paper utilizes a machine learning technique called Geometric Deep Learning (GDL) to aid designers with challenging, graph-centric design problems. The strategy presented here is to take the graph data and apply GDL to seek the best realizable performing solution effectively and efficiently with lower computational costs. This case study used here is the synthesis of analog electrical circuits that attempt to match a specific frequency response within a particular frequency range. Previous studies utilized an enumeration technique to generate 43,249 unique undirected graphs presenting valid potential circuits. Unfortunately, determining the sizing and performance of many circuits can be too expensive. To reduce computational costs with a quantified trade-off in accuracy, the fraction of the circuit graphs and their performance are used as input data to a classification-focused GDL model. Then, the GDL model can be used to predict the remainder cheaply, thus, aiding decision-makers in the search for the best graph solutions. The results discussed in this paper show that additional graph-based features are useful, favorable total set classification accuracy of 80% in using only 10% of the graphs, and iteratively-built GDL models can further subdivide the graphs into targeted groups with medians significantly closer to the best and containing 88.2 of the top 100 best-performing graphs on average.

Keywords: machine learning, geometric deep learning, graph classification, graph-based design, circuit synthesis

---

[*]Corresponding author, anthony.sirico@colostate.edu

1 INTRODUCTION

Mathematical graphs can be used to represent many systems and decisions because of their ability to capture discrete compositional and relational information. For decades, studies have employed different graph representations to capture their respective problems [1–8], and for well over a century, researchers have utilized graph enumeration to understand engineering design problems to help in decision making [2, 3, 9–11]. Today, many engineering design problems, including the construction of the "system architecture", are increasing in scope and complexity to a point where traditional discrete and continuous presentations are insufficient to represent the system [2].

The system architecture is a conceptual model capturing the structure, behavior, rules, etc., of a product, process, or element [12] and is often the foundation on which it is designed, built, and operated. Many studies have concentrated on the effective representation of system architectures using graph theory [13–15], where the goal is often a set of useful architectures that are feasible with respect to constraints. Furthermore, one or more value metrics (e.g., performance or cost) might be determined for the different architectures so that the designer might sort through the candidates. One method for generating all these options is graph enumeration, where a complete and ordered listing of the potential graphs is produced for some prescribed structure [9, 16, 17]. However, this approach can lead to an enormous amount of potential solutions depending on the problem at hand and its selected representation. Paired with the increasing complexity of modern systems, the result is an exponential increase in computational costs making the decision-making process for the de-

Copyright © 2023 by ASME

signer or system architect increasingly challenging. These challenges drive the need for an approach that facilitates the decision-making for larger and more complex graph-centric design problems.

In this paper, we consider deep learning, specifically *Geometric Deep Learning (GDL)*, as a potential strategy to address these issues. Deep learning is defined as machine learning models composed of multiple processing layers capable of learning data representations with multiple levels of abstraction [18], and "deep" refers to a larger number of hidden layers within the neural network. Now, GDL is an umbrella term encompassing an emerging technique that generalizes neural networks to Euclidean and non-Euclidean domains, such as graphs, manifolds, meshes, or string representations [19], and uses *Graph Neural Networks (GNNs)*. In essence, GDL encompasses approaches that incorporate information on the input variables' structure space and symmetry properties and leverage it to improve the quality of the data captured by the model. GDL has immense potential and is widely used in other scientific communities, including molecular representations [20–22], materials science [23], architecture [24], and the medical field [25, 26].

Perhaps due to the lack of relevant graph-based design datasets, there is limited usage of GDL or other similar graph-based, machine-learning approaches in engineering design, despite the availability of the tools and documented success in other fields [27]. Many studies utilize machine learning for the representation and selection of objects in 3D space, often in computer-aided design (CAD) applications [28, 29]. However, there are some key differences between meshes and other graph representations, including size, structure, locality, and geometric interpretation (or lack thereof). It is the intent of this study to provide another type of graph design example, and data set [30] illustrating how GDL can aid in the engineering design processes.

Here we are particularly motivated by graphic-centric design problems where generating many potential graphs is feasible, but determining the value or performance of each option is too expensive. The proposed approach uses GDL to reduce the computational expense of classifying what might be "good" or "bad" options with a trade-off in classification accuracy.

The remainder of the paper is organized as follows: Section 2 discusses the necessary background information required to have a basic understanding of GDL. In Sec. 3, we go over the proposed approach, including graph classification, the machine learning model, and the metrics used to determine the performance of the models. Section 4 discusses the case study on electric circuit frequency response matching graph design, and Sec. 5 describes several experiments conducted to explore GDL on this problem. Lastly, we conclude in Sec. 6 with the final discussions and future work.

2 BACKGROUND

To better understand the motivation and techniques behind the proposed methodology, this section will highlight additional key aspects of graph theory, system representation through graphs, GDL, and GNNs.

2.1 Graph Theory

A *graph G* is a pair of sets (V, E) where $E \subseteq [V]^2$ (i.e., E is a two-element subset of V). E represents the edges of the graph, while V is the set of its vertices or nodes [31]. A graph has various properties. For example, a graph's order, denoted by $n = |G|$, is the number of vertices in the graph. Each of those vertices has an associated degree value, which is the number of neighbors or edges to a vertex and is denoted by $d_G(v) = d(v)$.

While there are different ways to represent a graph mathematically, here we focus on one known as the *adjacency matrix* $\mathbf{A} = (a_{i,j})_{n \times n}$ and is defined as:

$$a_{i,j} := \begin{cases} 1, & \text{if } (v_i, v_j) \in E \\ 0, & \text{otherwise} \end{cases} \quad (1)$$

For simple, undirected graphs, the matrix is symmetric and the diagonal of $\mathbf{A}$ will be all zeros, which indicates that the graph has no self-loops.

Another important concept is *graph isomorphism*. Say we have two graphs, $G_1 = (V_1, E_1)$ and $G_2 = (V_2, E_2)$. These two graphs are isomorphic, denoted $G1 \cong G_2$, if there is a bijection, φ, from $V_1 \rightarrow V_2$ such that $(v_i, v_j) \in E_1 \leftrightarrow (\varphi(v_i), \varphi(v_j)) \in E_2$ for all $v_i, v_j \in V$ [32]. We also consider a feature or *labeled graph isomorphism*. Here we consider the same two graphs from the previous definition, but they contain a third feature. Specifically, we will consider this feature as a vertex label, denoted X, where $G_1 = (V_1, E_1, X_1)$ and $G_2 = (V_2, E_2, X_2)$. The graphs will be isomorphic as long as the vertex label property is preserved under some valid bijection φ. More generally, a features matrix $\mathbf{X} \in \mathbb{R}^{n \times c}$ can have as many columns as necessary to represent the c features associated with each vertex.

2.1.1 Representing Electric Circuits as Graphs. The concepts from the previous section can be used to represent engineering systems. The particular type of system used in the case study here is electric *RLC* circuits, so they will be used to illustrate their representation as a graph. Looking at the left side of Fig. 1, we can see a basic *RLC* circuit schematic. Then on the right side, we have the same circuit represented pictorially as a graph. Each vertex in the graph is now labeled as the corresponding component where I and O represent the input and output nodes, R is a resistor, L is an inductor, C is a capacitor, and lastly, N represents a voltage node with constant voltage. Also, other graph representations of the same circuit are possible.

For the graph in Fig. 1, an adjacency matrix $\mathbf{A}$ representing

Copyright © 2023 by ASME

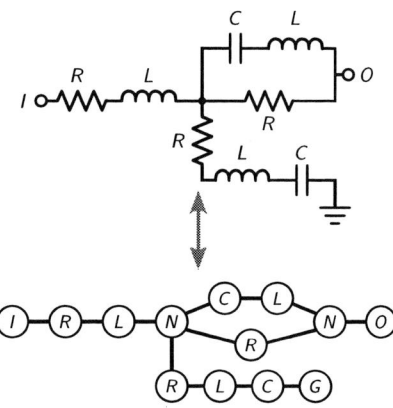

FIGURE 1: Electrical circuit schematic represented as a vertex-labeled graph.

its structure and vertex list $\mathbf{V}$ representing the vertex labels are:

$$\mathbf{A} = \begin{bmatrix} 0 & 1 & 0 & 0 & 0 & 0 & 0 & 0 & 0 & 0 & 0 & 0 & 0 & 0 \\ 1 & 0 & 1 & 0 & 0 & 0 & 0 & 0 & 0 & 0 & 0 & 0 & 0 & 0 \\ 0 & 1 & 0 & 1 & 0 & 0 & 0 & 0 & 0 & 0 & 0 & 0 & 0 & 0 \\ 0 & 0 & 1 & 0 & 1 & 0 & 0 & 0 & 1 & 1 & 0 & 0 & 0 & 0 \\ 0 & 0 & 0 & 1 & 0 & 1 & 0 & 0 & 0 & 0 & 0 & 0 & 0 & 0 \\ 0 & 0 & 0 & 0 & 1 & 0 & 1 & 0 & 0 & 0 & 0 & 0 & 0 & 0 \\ 0 & 0 & 0 & 0 & 0 & 1 & 0 & 1 & 1 & 0 & 0 & 0 & 0 & 0 \\ 0 & 0 & 0 & 0 & 0 & 0 & 1 & 0 & 0 & 0 & 0 & 0 & 0 & 0 \\ 0 & 0 & 0 & 1 & 0 & 0 & 1 & 0 & 0 & 0 & 0 & 0 & 0 & 0 \\ 0 & 0 & 0 & 1 & 0 & 0 & 0 & 0 & 0 & 0 & 1 & 0 & 0 \\ 0 & 0 & 0 & 0 & 0 & 0 & 0 & 0 & 0 & 1 & 0 & 1 & 0 \\ 0 & 0 & 0 & 0 & 0 & 0 & 0 & 0 & 0 & 0 & 1 & 0 & 1 \\ 0 & 0 & 0 & 0 & 0 & 0 & 0 & 0 & 0 & 0 & 0 & 1 & 0 \end{bmatrix} \quad \mathbf{X} = \begin{bmatrix} I \\ R \\ L \\ N \\ C \\ R \\ L \\ N \\ O \\ R \\ L \\ C \\ G \end{bmatrix} \quad (2)$$

where 1 represents a connection between two vertices.

2.2 Graph Enumeration and Optimization

Designers and system architects still often rely on engineering intuition and trial-and-error methods when exploring graph-centric problems. However, these approaches can lead to a lack of novel and satisfactory solutions within time restrictions.

A potential systematic strategy is *graph enumeration*, which are techniques for generating (or sometimes simply counting) nonisomorphic graphs with particular properties [33]. The properties can be quite diverse and are sometimes termed network structure constraints [17]. Some examples include bounding the potential degrees of the vertices in the graph or a maximum cost associated with the graph [10]. It is important to note that the number of valid graphs is usually combinatorial in nature; thus, the computational costs for just generating an enumeration can become quite expensive. However, two different graph enumeration algorithms might produce the desired enumeration, but one might do it more efficiently [10].

In any case, there are several important engineering appli-

cations where the enumeration of graphs, often with thousands to millions of entries, is practical. Even if graph enumeration is not possible, generating a partial listing (either algorithmically or manually) may still be desirable to explore potential solutions. In this work, it is assumed that there is a generated set of graphs, denoted $\mathcal{G}$, that contains the graphs of interest. For the purposes of the case study, this is an enumeration of *RLC* circuit graphs with certain properties based on [1].

While having a list of novel graphs can be useful on its own, in many engineering domains, the graph is only an intermediary representation used to compute one or more value metrics of interest. For example, the graph might correspond to a physics-based system model that is used to determine the comfort and handling of an automotive suspension [34]. For this work, we will consider a single performance metric, denoted $J(G_i)$, that is a function of the graph, and the natural goal is seeking graphs that minimize this "performance" metric:

$$\underset{G_i}{\text{minimize:}} \quad J(G_i) \quad (3)$$

One of the challenges in engineering design with graphs is sometimes the computational cost of $J(G_i)$ is quite expensive, perhaps many orders of magnitude larger than generating the graph G_i itself. The source of this cost can be quite diverse, including high-fidelity simulations, optimization, human-centric evaluation, and physical experiments. The focus of this work is when this cost is prohibitive.

Based on the classification put forth by [35], we define the following three types of graph-centric design problems:

- *Type 0* — All desired graphs can be generated and so can their performance metric $J(G_i)$ within time T
- *Type 1* — All desired graphs can be generated, but only some of the performance metrics $J(G_i)$ can be evaluated within time T; the performance assessment is too expensive
- *Type 2* — All desired graphs cannot be generated within time T

where T is the amount of time allocated to complete the graph design study. This work focuses on methods for Type 1 problems (using data from a large Type 0 study).

2.3 Geometric Deep Learning

Deep learning models have been very successful when training on images [36], text [37], and speech [38], but these contain an underlying Euclidean structure. Graphs are fundamentally different from the more common Euclidean data used in most deep learning applications (language, images, and videos). Euclidean data has an underlying grid-like structure. For example, an image can be translated to an (x, y, z) Cartesian coordinate system, where each pixel is located at an (x, y) coordinate, and z represents the color. But, this approach does not work for all problems because certain operations require many dimensions. This "flat" representation has limitations, such as not being able to represent hierarchies and other direct relationships.

Copyright © 2023 by ASME

This motivated the study of hyperbolic space for graph representation, or non-Euclidean learning [39, 40], which could potentially perform those same operations more flexibly, and GDL methods that can take advantage of the rich structure of such data to learn better representations and make better predictions [19]. For example, many real-world data sets are naturally structured as graphs, where the *relationships* between data points are more critical than individual data points. Euclidean-based methods struggle with such data, as they are designed to operate on flat, unstructured data. On the other hand, GDL methods can directly exploit the structure of the data, leading to better outcomes.

Since its inception, GDL can be applied to all forms of data represented as geometric priors [41]. Geometric priors encode information about the geometry of the data, such as smoothness, the sparsity of the data, or the relationship between the data and other variables, allowing us to work with data of higher dimensionality. Symmetry is essential in GDL and is often described as invariance and equivariance. Invariance is a property of particular mathematical objects that remain unchanged under certain transformations, while equivariance is a property of certain relations whereby they remain in the same relative position to one another under certain transformations [42,43]. For example, consider the graph isomorphism property from Sec. 2.1. Another advantage of geometric deep learning is its flexibility toward a broader range of data types and problems. This flexibility makes it a powerful tool for solving real-world problems that Euclidean methods cannot.

2.4 Graph Neural Networks

Similar to traditional deep learning models that use Convolutional Neural Networks (CNNs), GDL utilizes GNNs, which are used to learn representations of graph-structured data, such as social networks, molecules, and computer programs. GNNs are very similar to CNNs in that they can extract localized features and compose them to construct representations [44]. As previously mentioned, the critical difference between the two is that GNNs can learn on non-euclidean data and handle data that is not evenly structured, like images or text. Lastly, GNNs can learn from data that is not labeled, known as unsupervised learning. This is important because many real-world datasets are not labeled.

Training a GNN for graph classification (i.e., determining if a graph has a characteristic or not) is relatively simple, and there is typically a three-step process that needs to occur: 1) embed each node by performing multiple rounds of message passing, 2) aggregate node embeddings into a unified graph embedding, and then 3) train a final classifier on the graph embedding. We take a closer look at how this occurs in Sec. 3.3, where we discuss each layer of the GNN used here.

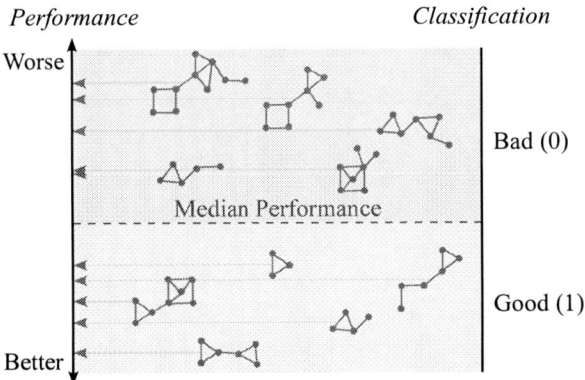

FIGURE 2: Illustration of graph classification based on performance values.

3 METHODOLOGY

The methodology for apply GDL on graph-based design problems will have a graph classification approach. This section will discuss this concept along with the selected model, hyperparameters, and metrics used to evaluate performance of the model.

3.1 Graph Classification

Graph classification refers to classifying graphs based on selected labels. Here, we consider *supervised* graph classification, where we have a collection of graphs $\mathcal{G}$, each with its label $J(G_i)$, used to train a model using additional graph properties, such as the structure, embeddings, and node data. These features help the model discriminate between the graphs, improving the prediction of the labels. An illustration of graph classification is shown in Fig. 2. Here, the median performance value might be used as the determination point to divide our graphs into binary classes representing "good" graphs in the lower half of the observed performance values and "bad" graphs in the upper half.

The rationale for this approach over regression was to better reflect the designer's intention in early-stage conceptual design. Often the goal is not to narrow the potential graphs down to *one* particular graph, but rather a *group* of "good" or promising graphs that would be analyzed further (often at a higher fidelity due to assumptions made in modeling and other areas during graph design). Furthermore, the performance values $J(G_i)$ in the case study have significant variations and many coarse values. Therefore, the intent is to show the "belongingness" of the graphs to a specific group versus assigning specific values to graphs.

There are various methods of graph classification, including Deep Graph Convolutional Neural Networks (DGCNN) [45], hidden layer representation that encodes the graph structure and node features [46], EigenPooling [47], and differentiable pooling [48]. It has been used in many different studies, such as learning molecular fingerprints [49], text categorization [50], encrypted traffic analysis [51], and cancer research [52].

Copyright © 2023 by ASME

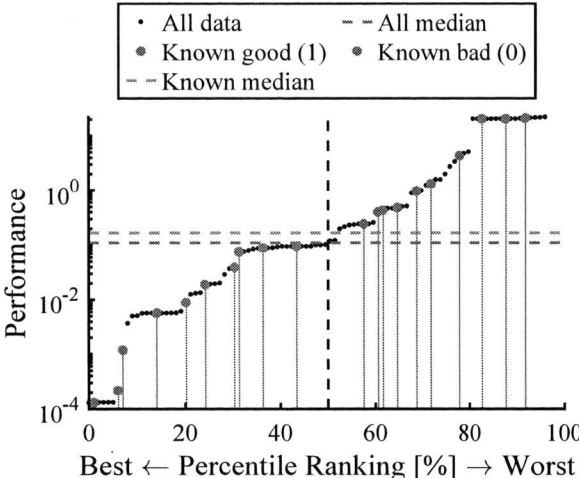

Legend:
- All data
- All median
- Known good (1)
- Known bad (0)
- Known median

Axes: Performance (vertical, 10^{-4} to 10^0); Best ← Percentile Ranking [%] → Worst (horizontal, 0 to 100)

FIGURE 3: Illustration of $\mathcal{G}_{all}$ and $\mathcal{G}_{known}$ for 100 graphs, the performance-based classification of $\mathcal{G}_{known}$, and the potential difference between the medians.

3.2 Datasets

Based on the Type 1 problem classification considered here from Sec. 2.2, we will consider the case when only some of the performance values $J(G_i)$ for $G_i \subset \mathcal{G}$ are known. This will divide the graphs into two sets as follows:

$$\mathcal{G} \equiv \mathcal{G}_{all} = \mathcal{G}_{known} \cup \mathcal{G}_{unknown} \tag{4}$$

where $\mathcal{G}_{known}$ is the set of graphs with known values for $J(G_i)$ and $\mathcal{G}_{unknown}$ represents graphs with unknown $J(G_i)$ values (and this is what the GDL model is for). We will denote the sizes of the sets as $|\mathcal{G}_{all}| = N_{all}$, $|\mathcal{G}_{known}| = N_{known}$, and $|\mathcal{G}_{unknown}| = N_{unknown}$. They also satisfy:

$$N_{all} = N_{known} + N_{unknown} \tag{5}$$

We also note that this is a bit different than other types of datasets used in machine learning as there is a known, finite amount of potential inputs. The goal here is to develop accurate models where $N_{known} \ll N_{unknown}$.

As is typical in machine learning, we will create two subsets from $\mathcal{G}_{known}$:

$$\mathcal{G}_{known} = \mathcal{G}_{training} \cup \mathcal{G}_{validation} \tag{6}$$

where $\mathcal{G}_{training}$ is the training dataset and $\mathcal{G}_{validation}$ is the validation dataset. The model fits using $\mathcal{G}_{training}$, and the fitted model is used to predict the responses for the observations in $\mathcal{G}_{validation}$ [53]. For example, after each iteration, the model will adjust its weights accordingly and test them on the validation set, which can help understand model performance and adjust options as necessary. The unknown set is used when the model is finalized, and you want to now test it on data that the model has not seen.

As shown in Fig. 2, the binary classification of graphs in $\mathcal{G}_{known}$ is based on the median performance value (or potentially some other dividing line using the known data). Since we only know $J(G_i)$ in $\mathcal{G}_{known}$ rather than $\mathcal{G}_{all}$, the median performance value used for the initial labeling might be different than the true value if we had all the performance values for $\mathcal{G}_{all}$. However, as the relative size of $\mathcal{G}_{known}$ increases, this error will become small. The concepts in this section are illustrated in Fig. 3. Here we have 100 graphs in $\mathcal{G}_{all}$ and 20 randomly selected graphs for $\mathcal{G}_{known}$. The true median and $\mathcal{G}_{known}$-based median do differ, and two graphs are between these lines.

3.3 The Model

There are five layers for the model used in this study: three graph convolutional layers [54], a mean pool layer, and a linear layer for the final readout. To accomplish the graph classification goal, the model uses: 1) the convolutional layers to embed each node through message passing, 2) agglomerating the node embeddings to create a graph embedding, then 3) use the graph embedding to train the classifying layer.

3.3.1 Graph Convolutional Layer.
The Graph Convolutional Layer (GCN) layers determine the output features $\mathbf{X}'$ by:

$$\mathbf{X}'_i = \mathbf{W}_1 \mathbf{X}_i + \mathbf{W}_2 \sum_{j \in \mathrm{N}(i)} e_{j,i} \cdot \mathbf{X}_j \tag{7}$$

where $\mathbf{X}$ is the input feature of each node (as discussed in Sec. 2.1), $(\mathbf{W}_1, \mathbf{W}_2)$ represents the weights adjusted after each iteration while training, and $e_{j,i}$ is the edge weight from source node j to target node i [54]. Each GCN layer also uses the ReLU activation function:

$$f(x) = \max(0, x) \tag{8}$$

which returns 0 if it receives any negative input, but for any positive value x, it returns that value. Finally, the outcome of Eq. (7) is passed through the activation function:

$$\mathbf{X}_{i+1} = f(\mathbf{X}_i) \tag{9}$$

which helps us obtain localized node embeddings.

Node embeddings are a mapping of a graph's nodes or vertices into an N-dimensional numerical space. As an example, using the graph from Fig. 1, we can use Node2Vec [55] to compute two node embeddings, visualized in Fig. 4. This process creates numerical features as vectors from the graph structure. Nodes similar to one another will be spaced close together in what are known as communities.

3.3.2 Global Mean Pooling Layer.
The next layer is *Global Mean Pooling*, which takes in the last output from the GCN layers and returns an output $\mathbf{r}$ by averaging the node features $\mathbf{X}$ to include node embeddings for each graph across all nodes N, creating a graph embedding:

$$\mathbf{r}_i = \frac{1}{N_i} \sum_{n=1}^{N_i} \mathbf{X}_n \tag{10}$$

Copyright © 2023 by ASME

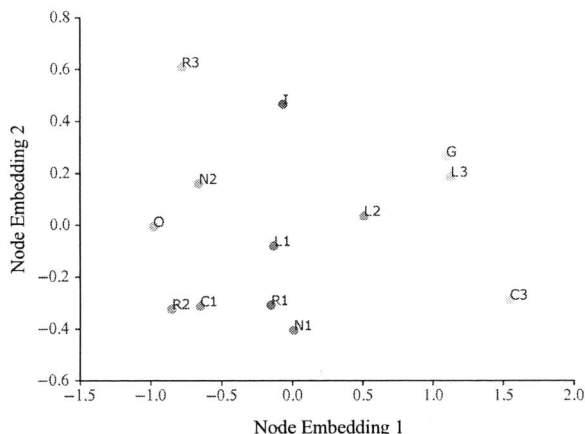

FIGURE 4: Calculated node embeddings for the graph in Fig. 1.

3.3.3 Linear Layer.
The final layer is the classification layer, which takes the mean pooling layer output as its input and applies the following linear transformation:

$$\mathbf{y} = \mathbf{r}\mathbf{A}^T + \mathbf{b} \tag{11}$$

However, before this layer, a *dropout* is applied, which randomly zeros some of the elements of the input tensor by some assigned probability using samples from a Bernoulli distribution [56].

3.4 Model Hyperparameters and Options
For the selected model architecture, there are several items that can be tuned so that the model trains well on the type of data provided. For more information on hyperparameters and tuning, please refer to the following sources [53, 57, 58].

3.4.1 Learning Rate.
The model's learning rate (*LR*) is a positive scalar value that determines the size of the step response to the estimated error as the weights are updated with each epoch and is considered one of the more important hyperparameters [58]. Typically, this value ranges from 0.0 to 1.0. Choosing a value too small can result in a longer training time, and choosing a value too large can result in an unstable training process. For these case studies, we use a *LR* of 0.001.

3.4.2 Number of Epochs.
An epoch refers to the number of training iterations through the entire training dataset [59]. The data is passed through the model during each epoch, updating the weights. Epochs can often range from hundreds to thousands, allowing the model to train until the error in the model is minimized. The number of epochs is explored in the different experiments in the case study.

3.4.3 Optimization Algorithm.
An optimization algorithm is a procedure for finding the input parameters for a function that results in the minimum or maximum of the function [59]. In this model, we chose to use the Adaptive Movement Estimation

TABLE 1: An example confusion matrix with sample values.

		Data	
		Actually Positive (1)	Actually Negative (0)
Model	Predicted Positive (1)	2,050 (T_P)	450 (F_P)
	Predicted Negative (0)	250 (F_N)	2,250 (T_N)

(ADAM) algorithm, a stochastic optimization method that computes individual learning rates for different parameters of the first and second moments of the gradients [60].

3.4.4 Loss Function.
A loss function computes the distance between the current output of the model with the expected out of the model. Then, this metric measures how well the model is performing. There are various loss functions, and the one used in this model is the *cross-entropy* loss function, typically used for classification problems. It measures the difference between two probability distributions for a given variable.

3.4.5 Batch Size.
While training, we can decide on batch size or the number of data points to work through before updating the model weights. This technique is known as mini-batching and is highly advantageous when training a deep-learning model by allowing the model to scale better to large amounts of data. Instead of processing data points one-by-one or all at once, a mini-batch groups a set of data points of intermediary size. This hyperparameter will be explored in the case study.

3.5 The Metrics
This section discusses the metrics used to evaluate the models' performance.

3.5.1 The Confusion Matrix.
A confusion matrix (CM), as seen in Fig. 1, is a two-dimensional matrix where each column contains the samples of the classifier (model) output, and each row contains the sample in the true class (data) [61]. For a binary classifier, the top left box represents the *True Positives* (T_P), the data points that were classified correctly as ones. The top right represents the *False Positives* (F_P), the points that were classified as ones but are, in fact, zeros. The bottom left contains the *False Negatives* (F_N), the points that were classified as zeros but are actually ones. Lastly, the bottom right is the *True Negatives* (T_N), the points that were correctly classified as zeros. The values in this matrix are used for many of the following metrics.

3.5.2 Accuracy.
Accuracy is the number of correct predictions divided by the total number of predictions, see Eq. (12).

$$Accuracy = \frac{T_P + T_N}{N} \tag{12}$$

Using the data from Table 1, we can calculate that our classifier had an accuracy of 0.86.

Copyright © 2023 by ASME

3.5.3 Precision

This metric, also called Positive Predictive Value (PPV), tells us what proportion of positive classifications were actually correct. This metric is particularly useful when your data has a class imbalance, e.g., more zeros than ones. When training on a dataset with more of one class than the other, you risk your model classifying all the data as the most frequent class, thus a "high accuracy". That is where *Precision* comes in because it is a class-specific metric defined as:

$$Precision = \frac{T_P}{T_P + F_P} \qquad (13)$$

Once again, using our example, our classifier has a *Precision* of 0.82, which means when it predicts a data point as one, it is correct 82% of the time. This same metric can also be applied to the negative value, but for this paper, we will use *Precision* since our case studies focus on finding the *best* architecture.

3.5.4 Recall.

This metric is calculated with:

$$Recall = \frac{T_P}{T_P + F_N} \qquad (14)$$

which indicates the proportion of actual positive classifications were identified correctly. Using the data from Table 1, *Recall* is 0.89, which states the model correctly identified 89% of all ones.

3.5.5 F1 Score.

This is a combination of both precision and recall into a single metric by calculating the harmonic mean between the two values as:

$$F1\ Score = 2 \cdot \frac{Precision \cdot Recall}{Precision + Recall} \qquad (15)$$

A model with a high *F1 Score* means that both *Precision* and *Recall* were high. Continuing with the example, the *F1 Score* for this model is 0.85.

3.5.6 Matthews Correlation Coefficient (MCC).

Of the metrics used to evaluate the performance of the classification models, this metric is one of the most important. Once again, using the values from the confusion matrix, the *MCC* produces a high score if the model obtained good results in *all* boxes of the confusion matrix [62, 63]; this means high T_P and T_N values and low F_P and F_N values. MCC is computed with:

$$MCC = \frac{T_P \cdot T_N - F_P \cdot F_N}{\sqrt{(T_P + F_P)(T_P + F_N)(T_N + F_P)(T_N + F_N)}} \qquad (16)$$

MCC ranges from −1 to 1, where 1 indicates the model can make predictions perfectly, −1 indicates that every prediction was incorrect, and 0 indicates that the model is just as good as random chance.

Finally, in our example, the achieved *MCC* for this model is 0.72, which is generally considered a good model.

3.5.7 Total Set Accuracy.

When the datasets are broken down into their respective subsets ($\mathcal{G}_{known}, \mathcal{G}_{unknown}$), we still want to know how well the predictions are when compared to the $\mathcal{G}_{all}$, if it is available. Therefore, we define *Total Set Accuracy* as:

$$Total\ Set\ Accuracy = \frac{T_P^{(u)} + T_N^{(u)} + T_P^{(k)} + T_N^{(k)}}{N_{all}} \qquad (17)$$

where $(T_P^{(u)}, T_N^{(u)})$ are determined on $\mathcal{G}_{unknown}$ using the trained model, and $(T_P^{(k)}, T_N^{(k)})$ are determined using $\mathcal{G}_{known}$, which might not be perfectly correct due to the median difference described in Sec. 3.2. If there is no misclassification based on the different medians, then $T_P^{(k)} + T_N^{(k)} = N_{known}$. This metric captures the outcome of a potential real-world scenario where a designer uses both the known and unknown data to classify all graphs.

4 CASE STUDY: ELECTRIC CIRCUIT FREQUENCY RESPONSE GRAPH-DESIGN PROBLEM

To explore the capabilities of GDL in graph-based engineering design problems, we utilize the results of a frequency response matching study [1]. The dataset and code to replicate these studies are available at Ref. [30].

4.1 Enumeration of Circuit Graphs

As discussed in Sec. 2.2, we will be considering the graph enumeration of electric circuits from Ref. [1]. In particular, the set $\mathcal{G}$ with 43,249 undirected graphs includes all topologies that have up to 6 impedance subcircuits with RC components and a required connection to the ground [1]. Many researchers have utilized the enumeration technique of circuit problems [11, 64, 65] and in other problem areas where the data can be represented as graphs and other enumerable objects [66–68]. As already mentioned, each circuit here is represented by a vertex-labeled graph, meaning every vertex has an associated label representing some circuit concept. The size of each graph varied between 6 and 20 nodes, all of which had guaranteed different topologies (i.e., no labeled graph isomorphisms from Sec. 2.1), which is relatively small for GDL.

4.2 Frequency Response Matching Optimization

The performance value $J(G_i)$ here is the error between the desired frequency response and the one a selected circuit provides, based originally on the study in Ref. [69]. A circuit graph G_i does not have an intrinsic value for J, rather it is a function of the *tunable values* for the RLC coefficients in the given graph and a selected *optimization problem* with an objective and constraints. From Ref. [1], we consider the "set 1" problem defined as:

$$\underset{\mathbf{z}_i = \{\mathbf{R}_i, \mathbf{C}_i\}}{\text{minimize:}} \quad J = \sum_k \left(\log |H_i(j\omega_k, \mathbf{z}_i)| - \log |F(j\omega_k)| \right)^2 \qquad (18a)$$

$$\text{subject to:} \quad 10^{-2} \leq R_j \leq 10^0 \quad \text{for all } R_j \text{ in } G_i \qquad (18b)$$

$$10^{-2} \leq C_j \leq 10^0 \quad \text{for all } C_j \text{ in } G_i \qquad (18c)$$

$$\text{where:} \quad |F(j\omega)| = \sqrt{\frac{2\pi}{10\omega}} \quad 0.2 \leq \frac{\omega}{2\pi} \leq 5 \qquad (18d)$$

Copyright © 2023 by ASME

where $|F(j\omega)|$ is the desired frequency response, $H_i(j\omega_k, \mathbf{z}_i)$ is the frequency response for circuit graph G_i, $\mathbf{z}_i$ is the collection of optimization variables for the resistors R and capacitors C in graph G_i, and ω is sampled at 500 logarithmically-spaced evaluation points.

Solving this nonlinear constrained least squares optimization problem can be expensive; the original study required over 8 hours to optimize each graph to determine $J(G_i)$. Therefore, it is desirable to reduce the computational costs to discern good and bad circuit graphs.

The performance values here have a large range, between 4×10^{-4} to 2×10^2. Classification is done based on a sampled $\mathcal{G}_{known}$ using the median value of the known performance values, as discussed in Sec. 3.2. As we have the performance values for all graphs, this problem can serve an a good example to explore the potential effectiveness of GDL in these kinds of problems, so we can compare to classification using $\mathcal{G}_{all}$.

5 CASE STUDY EXPERIMENTS AND RESULTS

In this section, we describe the experiments conducted on the engineering graph-design problem dataset from Sec. 4. The tools and computing architecture are described in App. A [30].

5.1 Experiment 1: Establish a Baseline

In the first experiment, we start by creating a baseline model for future comparisons. Here we consider a 72% (31,139 graphs) in the training set $\mathcal{G}_{training}$, 18% (7,784) in the validation set $\mathcal{G}_{validation}$, and the remaining 10% (4,324) in $\mathcal{G}_{unknown}$. These are by no means recommended distributions, but rather a scenario with lots of data available for the GDL model that will be used to explore what is possible and what might be done to balance accuracy versus efficiency.

The CM for this experiment is in Table 2 using the 4,324 graphs in $\mathcal{G}_{unknown}$ that we have not used in any way to train the model (and in practice, you would not have this CM). Using Eq. (12), we find a 79.1% accuracy, fairly close to the baseline models' final accuracy value for $\mathcal{G}_{known}$ of 79.5%. Next, using Eq. (13), we calculate that the baseline model has a *Precision* of 77%. Using Eq. (14), we calculate the baseline models' *Recall* to be 79.9%, and lastly, using Eq. (15) and Eq. (16), we calculate the *F1 Score* to be 78.6% and *MCC* to be 58%, respectively.

5.2 Experiment 2: Additional Graph-based Features

Here we consider adding additional graph-based features to $\mathbf{X}$ beyond the current vertex labels of R, C, G, etc. where the additional features should have a greater ability to explain the variance in the training data (known as feature engineering and feature selection). The theory is that these features will improve the model's performance for the same number of training epochs. The two features added here are eigenvector centrality and betweenness centrality.

TABLE 2: Confusion matrix for the *baseline* model predicting $\mathcal{G}_{unknown}$.

		Data	
		Actually Positive (1)	Actually Negative (0)
Model	Predicted Positive (1)	1,667	487
	Predicted Negative (0)	419	1,751

TABLE 3: Confusion matrix for the *3-feature* model predicting $\mathcal{G}_{unknown}$.

		Data	
		Actually Positive (1)	Actually Negative (0)
Model	Predicted Positive (1)	1,901	213
	Predicted Negative (0)	451	1,759

Eigenvector centrality computes a node's centrality based on its neighbors' centrality and is a measure of the influence of a node in a graph where a high eigenvector centrality score implies that a node is connected to many nodes that themselves have high scores. The eigenvector centrality for node v is the i-th normalized element of the vector $\mathbf{v}$ from:

$$\mathbf{A}\mathbf{v} = \lambda\mathbf{v} \tag{19}$$

where $\mathbf{A}$ is the adjacency matrix and λ is the largest eigenvalue [70].

Now, the betweenness centrality of a node is the sum of the fraction of all-pairs shortest paths that pass through that node:

$$c_B(v) = \sum_{s,t \in V} \frac{\sigma(s,t|v)}{\sigma(s,t)} \tag{20}$$

where V is the set of nodes, $\sigma(s,t)$ is the number of shortest paths, and $\sigma(s,t|v)$ is the number of those paths passing through some node v other than (s,t) [71].

We also point out that the computational expense of determining these additional features is quite low compared to calculated $J(G_i)$, so they add a relatively minor cost for each graph. With these two additional features, the new features matrix $\mathbf{X}$ for each individual graph has gone from $\mathbf{X} \in \mathbb{R}^{n \times 1}$ to $\mathbf{X} \in \mathbb{R}^{n \times 3}$. The CM for this experiment is in Table 3. Here we have that by adding the two additional features, the model was able to achieve an accuracy of 85%, *Precision* of 89.9%, *Recall* of 80.8%, *F1 Score* of 85%, and an *MCC* of 70%. Therefore, the GDL model was able to make better predictions on the same $\mathcal{G}_{known}$ dataset as all metrics are the same or better. We will include all three features going forward.

Copyright © 2023 by ASME

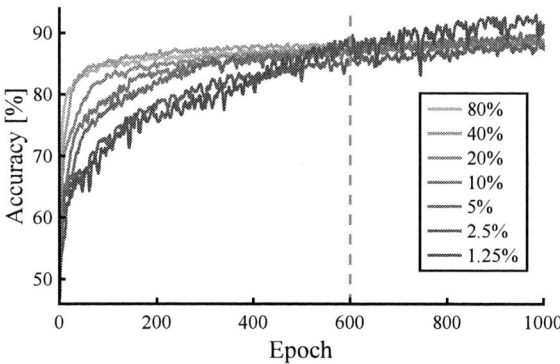

FIGURE 5: Accuracy of seven different models during training using different percentages of the dataset to determine the approximate number of epochs stopping point of 600.

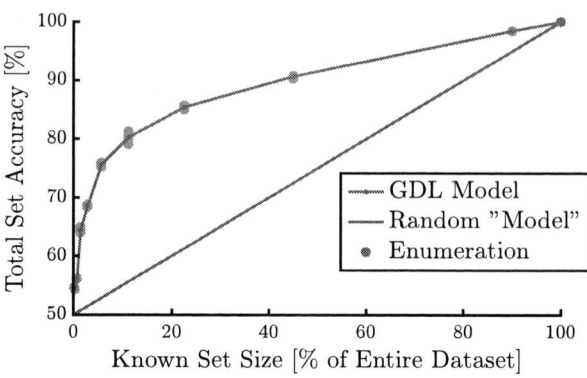

FIGURE 6: Total set accuracy scores averaged over five runs for different known set sizes N_{known}.

5.3 Experiment 3: Number of Epochs & Number of Known Graphs

The previous experiments assumed a large number of epochs, but it is desirable to have insights into how many are actually required to effectively train the model to reduce computation costs. Here we will observe the training behavior up to 1,000 epochs and decide a limit for future experiments.

Additionally, using 80% of the data for training, generally as was done in the previous experiments, will not be desirable, especially when the cost of each $J(G_i)$ is quite high. Therefore, we will also explore in this experiment different percentage values of $\mathcal{G}_{known}$, which represents fewer circuits that are sized using the expensive Eq. (18). The goal is to determine (for this particular dataset) a rule for the relative size of $\mathcal{G}_{known}$ to $\mathcal{G}_{all}$ that maintains the suitable accuracy while reducing the overall computational burden (both through model training time and time to generate $\mathcal{G}_{known}$). We will expect the model's performance to degrade as the amount of training data decreases.

5.3.1 Number of Epochs.
Seven different models were training using different N_{known} spaced between 1.25% and 80%. Observing the results in Fig. 5, there is a clear distinction around 600 epochs where many of the accuracy metrics plateau. For some of the smaller training sets, accuracy continues to increase beyond this point, but it was observed that this was mostly overfitting to the small training dataset. Therefore, a 600 epoch limit is established and will generally result in models that are suitably trained without wasting additional iterations.

5.3.2 Number of Known Graphs.
Using the same setup, we can explore trade-offs in the number of graphs (represented as a percentage here) included in $\mathcal{G}_{known}$. Reducing N_{known} has the consequence that $N_{unknown}$ increases, per Eq. (5), so more graphs will need their classifications predicted. Therefore, this experiment aims to see what minimum amount of data is required

for the models to retain a high level of accuracy.

The results are shown in Table 4 with all metrics computed using the unknown graphs $\mathcal{G}_{unknown}$. As we might expect, the larger N_{known} is, the higher the model's accuracy. Now, simply reducing the amount of data in half from 80% to 40%, we still achieve a high accuracy of $\approx 83\%$ and similar MCC of 0.68. However, a key takeaway is the training time. The 80% model took 4,294 s (or about 72 min), while the 40% model took only 37 minutes; the training time in half by cutting the data in half. We should also consider the impact on computational cost required to construct $\mathcal{G}_{known}$; 40% will nominally take half as long to determine all required $J(G_i)$. Both of these factors are important in understanding what % of N_{all} is desirable when balancing total computational cost and model effectiveness (e.g., accuracy). Looking into the smaller N_{known} values, the results are also as expected: the smaller N_{known}, the lower many of the scores. At the limit of the runs, only 135 graphs (0.3%), the model was only able to achieve an accuracy of $\approx 60\%$ with a much lower MCC of 0.23.

To better understand the trade-offs in using GDL models for the case study, we visualize the results of the metric *Total Set Accuracy* from Eq. (17) in Fig. 6. As there are several stochastic elements to the approach, the mean of five different runs is shown with the different gray dots indicating the values for the individual runs. Overall, the variation is relatively smaller between runs. In addition to this curve, a theoretical random "model" is added where we assume that all known values are predicted correctly, and all others are randomly assigned a good/bad classification. Finally, when the known set size is 100% of the dataset, then we have enumerated all graphs, so the total set *Total Set Accuracy* is naturally 100%. Therefore, the GDL should be compared with respect to these additional standards.

Overall, we see the GDL model greatly outperforms the random model between around 1%–40%, but the gap begins to close as more graphs are known and all approach the 100%/100% enu-

Copyright © 2023 by ASME

TABLE 4: The results for different values of N_{known} with all metrics computed with respect to the unknown graphs $\mathcal{G}_{unknown}$.

N_{known} [# of graphs]	34,599	17,300	8,650	4,325	2,162	1,081	541	270	135
% of N_{all}	80%	40%	20%	10%	5%	2.5%	1.3%	0.6%	0.3%
Training Time [s]	4,294	2,239	1,257	720	464	334	269	243	229
Accuracy [%]	83.80	83.99	82.70	81.73	78.68	76.15	72.28	66.70	60.69
Precision [%]	84.10	88.37	80.29	79.11	77.18	76.51	70.97	70.54	67.70
Recall [%]	83.22	78.52	86.85	86.30	81.37	75.49	75.41	57.39	40.90
F1 Score [%]	83.66	83.15	83.44	82.55	79.22	75.99	73.13	63.29	51.00
MCC	0.68	0.68	0.66	0.64	0.57	0.52	0.45	0.34	0.23

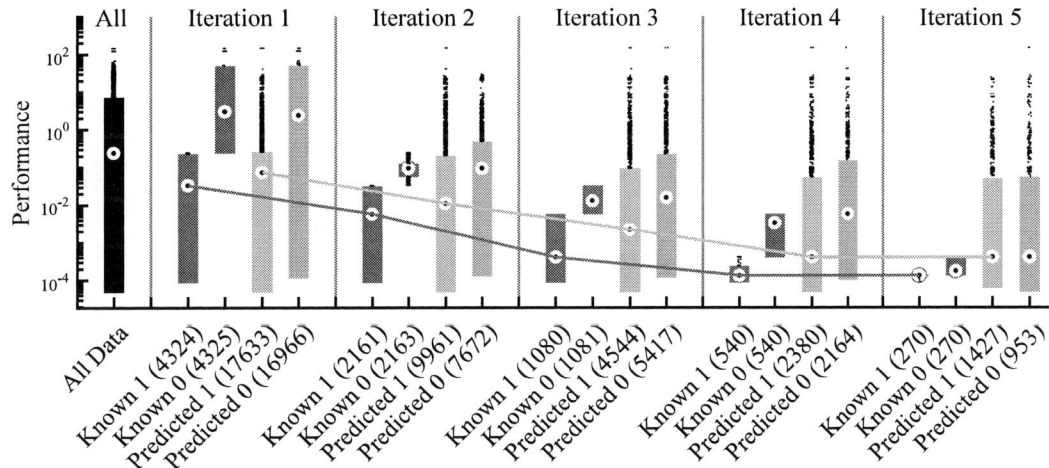

FIGURE 7: Five iterations of the approach described in Sec. 5.4 for retraining a GDL model based on successive performance minimizing-focused subsets of the graphs (with the numbers indicating the number of graphs in the respective set) along with the distribution of all the data points $J(\mathcal{G}_{all})$ for comparison.

meration point as N_{known} increases. Therefore, a general recommendation might be to select N_{known} as 20% of N_{all}, but this value certainly can be problem-specific and depend on the designer's preferences to balance *Total Set Accuracy* vs. computational costs (which is generally proportional to N_{known}).

5.4 Experiment 4: Iterative GDL Classification

Even if with perfect classification, the approach outlined so far would only result in determining the top 50% performing graphs in $\mathcal{G}_{all}$. While this outcome can undoubtedly be useful in practice, it is generally desirable to narrow down the graphs further to the best-performing. However, there is no reason we only need to construct a single classification model.

In this experiment, we seek a smaller, better median performance set of graphs from $\mathcal{G}_{all}$ by iteratively constructing GDL models with these steps:

1. Set $k = 1$ and create an initial $\mathcal{G}_{known}^{k}$.
2. Create a GDL model $m^k(G_i)$ using $\mathcal{G}_{known}^{k}$, which is naturally broken into sets "Known 1" and "Known 0" based on the

median J value of $\mathcal{G}_{known}^{k}$.
3. Predict the classes of the $\mathcal{G}_{unknown}^{k}$ using $m^k(G_i)$, creating "Predicted 1" and "Predicted 0", which are sets of graphs predicted to be good (1) or bad (0), respectively.
4. The goal is to identify good graphs, so we set $\mathcal{G}_{known}^{k+1}$ equal to "Known 1" and $\mathcal{G}_{unknown}^{k+1}$ equal to "Predicted 1" (and the remaining graphs are removed under the assumption that they are bad).
5. Set $k \to k+1$ and repeat Step 2 until $k = n$.

At each iteration, the size of $\mathcal{G}_{known}^{k}$ will be halved in a way that decreases its median performance value, so the good/bad classification threshold of $m^k(G_i)$ should shift as well.

The results of this experiment initialized with 20% of the data known are summarized in Fig. 7. First, the leftmost box plot shows the distribution of $J(\mathcal{G}_{all})$. The next two box plots show the splitting of $\mathcal{G}_{known}$ based on the classification threshold at iteration 1. The final two box plots in iteration 1 are the predicted 1s and 0s based on the GDL model; note that, while

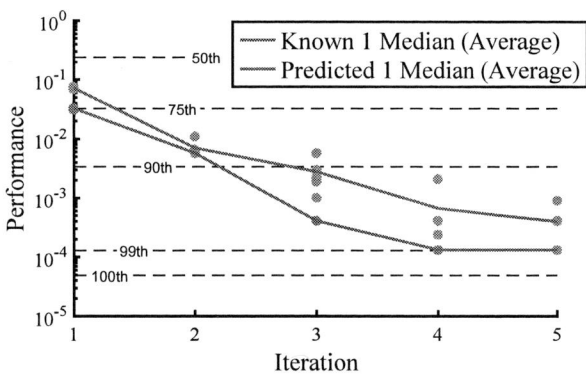

FIGURE 8: Median values of the "Known 1" and "Predicted 1" sets averaged over six runs using the iterative GDL classification approach with different performance percentiles for $J(\mathcal{G}_{all})$ shown.

there are certainly incorrectly identified graphs, the medians are quite separated and consistent with their "Known" counterparts. The subsequent box plots show the results for each iteration, up to $n = 5$. We observe at iteration 5, with $N_{known} = 540$ graphs now, the classification is quite poor as the "Predicted" box plots are similarly distributed. The trend lines for the two medians also convey this point, with substantial decreases until the fifth iteration. This behavior is perhaps expected because we are not adding any new data from the initial $\mathcal{G}^1_{known}$, and the number of graphs for training decreases. Adding more data at each iteration based on "Predicted 1" is future work.

Assessing the outcome at the end of iteration 4 (since the fifth did not classify well), there are a total of 2,920 graphs in the final "good" set with a median of 3×10^{-4}, substantially lower than the initial classification median of 2×10^{-1}. Here we assume a designer would evaluate $J(G_i)$ for all the graphs in "Predicted 1"; thus, there would be a total of 11,029 graphs that were optimized. For this final set, *all* of the top 10, 87 of the top 100, and 727 of the top 1000 graphs still remain at 25% the computational cost of complete enumeration. Furthermore, compared to randomly sampling 11,029 graphs from $\mathcal{G}_{all}$, the iterative GDL model results greatly exceed the expected means of 2.25, 22.51, and 225.13 for the top 10, 100, and 1000 graphs being included in the random set, respectively.

To better understand the statistical significance of this iterative approach, five more randomized runs were completed with different samplings of $\mathcal{G}^1_{known}$ for a total of six runs. Over these runs, the average graphs that would be known or "optimized" was 11282.2 graphs (with a 492.8 standard deviation). For the top 100 graphs, 88.2 (2.5) remained compared to an expected value of 22.96. Finally, for the top 1000 graphs, 751.2 (39.5) remained compared to an expected value of 229.6. Furthermore, the median values of the "Known 1" and "Predicted 1" graph sets

for each iteration are shown in Fig. 8 averaged over the six runs. Here we still see a decrease in the median values indicating the iterative GDL approach is narrowing down the set of potentially "good" graphs to the higher percentiles of performance. In fact, at iteration 4, all medians for "Predicted 1" are better than the 90th percentile, with the mean closer to the 95th. However, as previously discussed in Fig. 7, the benefits from the 5th iteration are minimal.

6 CONCLUSION

In this paper, we have presented a Geometric Deep Learning (GDL) approach for classifying and down-selecting graph-based engineering design problems towards sets of better-performing solutions. Using the electrical circuit graph-design case study where all the graphs were known but not their expensive optimization-based performance values, it was shown the potential capabilities of GDL in helping classify "good" and "bad" graphs based on limited performance data, as well as some insights into the key hyperparameters that need to be tuned. Additionally, the inclusion of the graph-based features eigenvalue centrality and betweenness centrality improved many of the key metrics.

The results showed that GDL can be used as an effective and efficient method for narrowing the graphs sets for this case study as the model was able to achieve a total set classification accuracy of 80% in using only 10% of the graphs. Furthermore, the iterative GDL classification approach identified 9.0 of the top 10, 88.2 of the top 100 graphs, and 751.2 out of the top 1000 at 25% the computational cost of complete enumeration.

Key future work items include iteratively adding new graphs to the dataset as mentioned in Sec. 5.4, and investigating transfer learning on similar problems [1]. There is also the task of investigating this approach on other larger datasets in different engineering fields, such as ones with directed graphs and multiple performance metrics. The promising results shown for this particular large dataset full of small engineering graphs indicate a promising future for iterative classification-based GDL for a wide application in engineering graph-design problems.

REFERENCES

[1] Herber, D. R., 2017. "Advances in combined architecture, plant, and control design". Ph.D. Dissertation, University of Illinois at Urbana-Champaign, Urbana, IL, USA.

[2] Selva, D., Cameron, B., and Crawley, E., 2016. "Patterns in system architecture decisions". *Syst. Eng.,* **19**(6), pp. 477–497. doi: 10.1002/sys.21370

[3] Foster, R. M., 1932. "Geometrical circuits of electrical networks". *Transactions of the American Institute of Electrical Engineers,* **51**(2), pp. 309–317. doi: 10.1109/T-AIEE.1932.5056068

[4] Fan, W., Ma, Y., Li, Q., He, Y., Zhao, E., Tang, J., and Yin, D., 2019. "Graph neural networks for social recommendation". In WWW '19: The World Wide Web Conference, p. 417–426. doi: 10.1145/3308558.3313488

Copyright © 2023 by ASME

[5] Zhou, Y., Liu, S., Siow, J., Du, X., and Liu, Y., 2019. "Devign: effective vulnerability identification by learning comprehensive program semantics via graph neural networks". In Conference on Neural Information Processing Systems.

[6] Cheng, X., Wang, H., Hua, J., Xu, G., and Sui, Y., 2021. "Deepwukong: Statically detecting software vulnerabilities using deep graph neural network". *ACM Trans. Softw. Eng. Methodol., 30*(3), apr.

[7] Yang, W., Ding, H., and Zhang, D., 2018. "New graph representation for planetary gear trains". *ASME J. Mech. Design, 140*(1), p. 012303. doi: 10.1115/1.4038303

[8] Hsu, C.-H., and Lam, K.-T., 1992. "A new graph representation for the automatic kinematic analysis of planetary spurgear trains". *ASME J. Mech. Design, 114*(1), pp. 196–200. doi: 10.1115/1.2916916

[9] Herber, D. R., Guo, T., and Allison, J. T., 2017. "Enumeration of architectures with perfect matchings". *ASME J. Mech. Design, 139*(5), p. 051403. doi: 10.1115/1.4036132

[10] Herber, D. R., 2020. "Enhancements to the perfect matching approach for graph enumeration-based engineering challenges". In ASME International Design Engineering Technical Conferences. doi: 10.1115/DETC2020-22774

[11] Macmahon, P. A., 1994. "The combinations of resistances". *Discret. Appl. Math., 54*(2), pp. 225–228. doi: 10.1016/0166-218X(94)90024-8

[12] Maier, M. W., and Rechtin, E., 2009. *The Art of Systems Architecting*, 3rd ed. CRC Press.

[13] Arney, D. C., and Wilhite, A. W., 2014. "Modeling space system architectures with graph theory". *J. Spacecraft Rockets, 51*(5), pp. 1413–1429. doi: 10.2514/1.A32578

[14] Taft, J., 2018. A mathematical representation of system architectures. Tech. Rep. PNNL-27387, Battelle for the US Department of Energy, Pacific Northwest National Laboratory, Mar.

[15] Potts, M., Pia, S., Johnson, A., and Bullock, S. Hidden structures: using graph theory to explore complex system of systems architectures.

[16] Schmidt, L. C., Shetty, H., and Chase, S. C., 1999. "A graph grammar approach for structure synthesis of mechanisms". *ASME J. Mech. Design, 122*(4), pp. 371–376. doi: 10.1115/1.1315299

[17] Wyatt, D. F., Wynn, D. C., Jarrett, J. P., and Clarkson, P. J., 2012. "Supporting product architecture design using computational design synthesis with network structure constraints". *Res. Eng. Des., 23*, pp. 17–52. doi: 10.1007/s00163-011-0112-y

[18] LeCun, Y., Bengio, Y., and Hinton, G., 2015. "Deep learning". *Nature, 521*, pp. 436–444. doi: 10.1038/nature14539

[19] Bronstein, M. M., Bruna, J., LeCun, Y., Szlam, A., and Vandergheynst, P., 2017. "Geometric deep learning: going beyond euclidean data". *IEEE Signal Process Mag., 34*(4), pp. 18–42. doi: 10.1109/msp.2017.2693418

[20] Atz, K., Grisoni, F., and Schneider, G., 2021. Geometric deep learning on molecular representations. arXiv:2107.12375

[21] Gainza, P., Sverrisson, F., Monti, F., Rodola, E., Boscaini, D., Bronstein, M. M., and Correia, B. E., 2020. "Deciphering interaction fingerprints from protein molecular surfaces using geometric deep learning". *Nature, 17*, pp. 184–192. doi: 10.1038/s41592-019-0666-6

[22] Segler, M. H. S., Kogej, T., Tyrchan, C., and Waller, M. P., 2018. "Generating focused molecule libraries for drug discovery with recurrent neural networks". *ACS Cent. Sci., 4*(1), pp. 120–131. doi: 10.1021/acscentsci.7b00512

[23] Krokos, V., Bordas, S. P. A., and Kerfriden, P., 2022. A graph-based probabilistic geometric deep learning framework with online physics-based corrections to predict the criticality of defects in porous materials. arXiv:2205.06562

[24] Fedorova, S., Tono, A., Nigam, M. S., Zhang, J., Ahmadnia, A.,

Bolognesi, C., and Michels, D., 2021. "Synthetic data generation pipeline for geometric deep learning in architecture". *Int. Arch. Photogramm. Remote Sens. Spat. Inf. Sci., XLIII-B2-2021*, pp. 337–344. doi: 10.5194/isprs-archives-XLIII-B2-2021-337-202

[25] Thiery, A. H., Braeu, F., Tun, T. A., Aung, T., and Girard, M. J. A., 2022. Medical application of geometric deep learning for the diagnosis of glaucoma. arXiv:2204.07004

[26] Sarasua, I., Lee, J., and Wachinger, C., 2021. Geometric deep learning on anatomical meshes for the prediction of alzheimer's disease. arXiv:2104.10047

[27] Regenwetter, L., Nobari, A. H., and Ahmed, F., 2022. "Deep generative models in engineering design: A review". *ASME J. Mech. Design, 144*(7), p. 071704. doi: 10.1115/1.4053859

[28] Ranjan, A., Bolkart, T., Sanyal, S., and Black, M. J., 2018. Generating 3D faces using convolutional mesh autoencoders. arXiv:1807.10267

[29] Cheng, S., Bronstein, M., Zhou, Y., Kotsia, I., Pantic, M., and Zafeiriou, S., 2019. MeshGAN: Non-linear 3D morphable models of faces. arXiv:1903.10384

[30] https://github.com/anthonysirico/GDL-for-Engineering-Design.

[31] Diestel, R., 2017. *Graph Theory*. Springer. doi: 10.1007/978-3-662-53622-3

[32] Godsil, C., and Royle, G., 2001. *Algebraic Graph Theory*. Springer. doi: 10.1007/978-1-4613-0163-9

[33] Borkar, V. S., Ejov, V., and Nguyen, G. T., 2012. *Hamiltonian Cycle Problem and Markov Chains*. Springer. doi: 10.1007/978-1-4614-3232-6

[34] Herber, D. R., and Allison, J. T., 2019. "A problem class with combined architecture, plant, and control design applied to vehicle suspensions". *ASME J. Mech. Design, 141*(10), p. 101401. doi: 10.1115/1.4043312

[35] Guo, T., Herber, D. R., and Allison, J. T., 2018. "Reducing evaluation cost for circuit synthesis using active learning". In ASME International Design Engineering Technical Conferences, no. DETC2018-85654. doi: 10.1115/DETC2018-85654

[36] Krizhevsky, A., Sutskever, I., and Hinton, G. E., 2012. "Imagenet classification with deep convolutional neural networks". *Advances in Neural Information Processing Systems 25*, pp. 1097–1105.

[37] Wang, T., Wu, D. J., Coates, A., and Ng, A. Y., 2012. "End-to-end text recognition with convolutional neural networks". In International Conference on Pattern Recognition, pp. 3304–3308.

[38] Deng, L., Li, J., Huang, J.-T., Yao, K., Yu, D., Seide, F., Seltzer, M., Zweig, G., He, X., Williams, J., Gong, Y., and Acero, A., 2013. "Recent advances in deep learning for speech research at Microsoft". In IEEE International Conference on Acoustics, Speech and Signal Processing, pp. 8604–8608. doi: 10.1109/ICASSP.2013.6639345

[39] Nickel, M., and Kiela, D., 2017. Poincaré embeddings for learning hierarchical representations. arXiv:1705.08039

[40] Chamberlain, B. P., Clough, J., and Deisenroth, M. P., 2017. Neural embeddings of graphs in hyperbolic space. arXiv:1705.10359

[41] Bronstein, M. M., Bruna, J., Cohen, T., and Velickovic, P., 2021. Geometric deep learning: grids, groups, graphs, geodesics, and gauges. arXiv:2104.13478

[42] Cohen, T. S., and Welling, M., 2016. Steerable CNNs. arXiv:1612.08498

[43] Cohen, T. S., Geiger, M., Koehler, J., and Welling, M., 2018. Spherical CNNs. doi: 1801.10130

[44] Lecun, Y., Bottou, L., Bengio, Y., and Haffner, P., 1998. "Gradient-based learning applied to document recognition". *Proc. IEEE, 86*(11), pp. 2278–2324. doi: 10.1109/5.726791

[45] Zhang, M., Cui, Z., Neumann, M., and Chen, Y., 2018. "An end-to-end deep learning architecture for graph classification". In AAAI Conference on Artificial Intelligence, Vol. 32, pp. 4438–4445. doi: 10.1609/aaai.v32i1.11782

[46] Kipf, T. N., and Welling, M., 2016. Semi-supervised classification

Copyright © 2023 by ASME

with graph convolutional networks. arXiv:1609.02907

[47] Ma, Y., Wang, S., Aggarwal, C. C., and Tang, J., 2019. Graph convolutional networks with eigenpooling. arXiv:1904.13107

[48] Ying, R., You, J., Morris, C., Ren, X., Hamilton, W. L., and Leskovec, J., 2018. Hierarchical graph representation learning with differentiable pooling. arXiv:1806.08804

[49] Duvenaud, D., Maclaurin, D., Aguilera-Iparraguirre, J., Gómez-Bombarelli, R., Hirzel, T., Aspuru-Guzik, A., and Adams, R. P., 2015. Convolutional networks on graphs for learning molecular fingerprints. arXiv:1509.09292

[50] Rousseau, F., Kiagias, E., and Vazirgiannis, M., 2015. "Text categorization as a graph classification problem". In Annual Meeting of the Association for Computational Linguistics and the 7th International Joint Conference on Natural Language Processing, pp. 1702–1712. doi: 10.3115/v1/P15-1164

[51] Shen, M., Zhang, J., Zhu, L., Xu, K., and Du, X., 2021. "Accurate decentralized application identification via encrypted traffic analysis using graph neural networks". *IEEE Trans. Inf. Forensics Secur., 16*, pp. 2367–2380. doi: 10.1109/TIFS.2021.3050608

[52] Hashemi, A., and Pilevar, A. H., 2013. "Mass detection in lung CT images by using graph classification". *J. Electr. Electron. Eng., 3*(3).

[53] James, G., Witten, D., Hastie, T., and Tibshirani, R., 2021. *An Introduction to Statistical Learning*. Springer. doi: 10.1007/978-1-0716-1418-1

[54] Morris, C., Ritzert, M., Fey, M., Hamilton, W. L., Lenssen, J. E., Rattan, G., and Grohe, M., 2018. Weisfeiler and Leman go neural: higher-order graph neural networks. arXiv:1810.02244

[55] Grover, A., and Leskovec, J., 2016. node2vec: scalable feature learning for networks. arXiv:1607.00653

[56] Hinton, G. E., Srivastava, N., Krizhevsky, A., Sutskever, I., and Salakhutdinov, R., 2012. Improving neural networks by preventing co-adaptation of feature detectors. arXiv:1207.0580

[57] Bengio, Y., 2012. Practical recommendations for gradient-based training of deep architectures. arXiv:1206.5533

[58] Goodfellow, I., Bengio, Y., and Courville, A., 2016. *Deep Learning*. MIT Press.

[59] Chollet, F., 2018. *Deep Learning with Python*. Manning Publications.

[60] Kingma, D. P., and Ba, J. L., 2015. "Adam: a method for stochastic optimization". In International Conference on Learning Representations.

[61] Simske, S. J., 2013. *Meta-Algorithmics: Patterns for Robust, Low Cost, High Quality Systems*. John Wiley & Sons.

[62] Jurman, G., Riccadonna, S., and Furlanello, C., 2012. "A comparison of MCC and CEN error measures in multi-class prediction". *PLOS One, 7*(8), p. e41882. doi: 10.1371/journal.pone.0041882

[63] Chicco, D., 2017. "Ten quick tips for machine learning in computational biology". *BioData Min., 10*(35). doi: 10.1186/s13040-017-0155-3

[64] Lomnicki, Z. A., 1972. "Two-terminal series-parallel networks". *Adv. Appl. Probab., 4*(1), pp. 109–150. doi: 10.2307/1425808

[65] Isokawa, Y., 2016. "Series-parallel circuits and continued fractions". *Appl. Math. Sci., 10*(27), pp. 1321–1331. doi: 10.12988/ams.2016.63103

[66] Bayrak, A. E., Ren, Y., and Papalambros, P. Y., 2016. "Topology generation for hybrid electric vehicle architecture design". *ASME J. Mech. Design, 138*(8), p. 081401. doi: 10.1115/1.4033656

[67] del Castillo, J. M., 2002. "Enumeration of 1-DOF planetary gear train graphs based on functional constraints". *ASME J. Mech. Design, 124*(4), pp. 723–732. doi: 10.1115/1.1514663

[68] Ma, W., Trusina, A., El-Samad, H., Lim, W. A., and Tang, C., 2009. "Defining network topologies that can achieve biochemical adaptation". *Cell, 138*(4), pp. 760–773. doi: 10.1016/j.cell.2009.06.013

[69] Grimbleby, J. B., 1995. "Automatic analogue network synthesis using genetic algorithms". In First International Conference on Genetic Algorithms in Engineering Systems: Innovations and Applications, pp. 53–58. doi: 10.1049/cp:19951024

[70] Bonacich, P., 1987. "Power and centrality: a family of measures". *Am. J. Sociol., 92*(5), pp. 1170–1182.

[71] Freeman, L. C., 1977. "A set of measures of centrality based on betweenness". *Sociometry, 40*(1), pp. 35–41. doi: 10.2307/3033543

[72] Fey, M., and Lenssen, J. E., 2019. "Fast graph representation learning with PyTorch Geometric". In ICLR Workshop on Representation Learning on Graphs and Manifolds.

[73] Paszke, A., and et al., 2019. "PyTorch: an imperative style, high-performance deep learning library". In Advances in Neural Information Processing Systems, Vol. 32.

[74] Van Rossum, G., and Drake, F. L., 2009. *Python 3 Reference Manual*. CreateSpace, Scotts Valley, CA.

[75] Hagberg, A. A., Schult, D. A., and Swart, P. J., 2008. "Exploring network structure, dynamics, and function using NetworkX". In Python in Science Conference, pp. 11–15.

[76] The pandas development team, 2020. pandas-dev/pandas: Pandas. doi: 10.5281/zenodo.3509134

A TOOLS AND COMPUTING ARCHITECTURE

The primary tool used is PyTorch-Geometric (PyG) [72] because of the extensive applications that PyG can perform on graph data. It is built on top of PyTorch [73], an open-source machine learning framework, which also contains an extensive library of tools for model manipulation and data analysis. Since both tools are used primarily with Python [74], that will be the language of choice for the model. We also include Networkx, a Python package for the creation and manipulation of complex networks [75]. This decision is because PyG requires the data to be of a certain instance, whether a SciPy sparse matrix or a Trimesh instance; it needs to be in a form readable by PyG. Networkx was chosen because of the information that can be added to the graph, which can be easily transformed into a PyG instance. We then use Pandas [76] for its data organization capabilities to import the data and prepare it to be passed on to Networkx and, finally, PyG. For a full list of the tools used and their versions, please see Table 5.

All the tools were utilized on a personal workstation consisting of an Intel Core i9-9900k CPU @ 3.60GHz, 32GB installed memory, and an Nvidia GeForce RTX 2060 Super GPU.

TABLE 5: A list of the primary tools and their versions.

Tool	Version
Python	3.9
Networkx	2.8.7
PyTorch	1.12.1
PyTorch-Geometric	2.1.0
SciPy	1.9.1
Pandas	1.5.0

Copyright © 2023 by ASME

Proceedings of the ASME 2023
International Design Engineering Technical Conferences and
Computers and Information in Engineering Conference
IDETC-CIE2023
August 20-23, 2023, Boston, Massachusetts

DETC2023-115065

AUTOSURF: AUTOMATED EXPERT-GUIDED MESHING WITH GRAPH NEURAL NETWORKS AND CONFORMAL PREDICTIONS

Amin Heyrani Nobari[1,*], Justin Rey[2], Suhas Kodali[2], Matthew Jones[2], Faez Ahmed[1]

[1]Massachusetts Institute of Technology, Cambridge, MA
[2]Lincoln Laboratory, Massachusetts Institute of Technology, Lexington, MA

ABSTRACT

Computational Fluid Dynamics (CFD) is widely used in different engineering fields, but accurate simulations are dependent upon proper meshing of the simulation domain. While highly refined meshes may ensure precision, they come with high computational costs. Similarly, adaptive remeshing techniques require multiple simulations and come at a great computational cost. This means that the meshing process is reliant upon expert knowledge and years of experience. Automating mesh generation can save significant time and effort and lead to a faster and more efficient design process. This paper presents a machine learning-based scheme that utilizes Graph Neural Networks (GNN) and expert guidance to automatically generate CFD meshes for aircraft models. In this work, we introduce a new 3D segmentation algorithm that outperforms two state-of-the-art models, Point-Net++ and PointMLP, for surface classification. We also present a novel approach to project predictions from 3D mesh segmentation models to CAD surfaces using the conformal predictions method, which provides marginal statistical guarantees and robust uncertainty quantification and handling. We demonstrate that the addition of conformal predictions effectively enables the model to avoid under-refinement, hence failure, in CFD meshing even for weak and less accurate models. Finally, we demonstrate the efficacy of our approach through a real-world case study that demonstrates that our automatically generated mesh is comparable in quality to expert-generated meshes, and enables the solver to converge and produce accurate results. The code and data for this project is made publicly available at https://github.com/ahnobari/AutoSurf.

Keywords: CFD, Mesh, Deep Learning, Graph Neural Networks, Conformal Predictions, GNN, Mesh Segmentation, 3D model segementation

---

*Corresponding author: ahnobari@mit.edu

1. INTRODUCTION

Computational Fluid Dynamics (CFD) has revolutionized how engineers analyze fluid flow phenomena, leading to remarkable advances in engineering applications, including aerospace, automotive, and environmental engineering. Accurate and reliable simulations of fluid flow phenomena can lead to safer and more efficient designs, reducing the costs of development and improving overall performance. This can positively impact the broader society by enhancing energy consumption efficiency, reducing environmental pollution, and improving transportation safety. Fast CFD simulations may also lead to the advancement of scientific knowledge in fluid dynamics, leading to further breakthroughs in the field. Faster CFD also enables researchers to perform high-fidelity CFD simulations rather than relying on low-fidelity and less accurate simulations, which is often the case in many computational design approaches for applications such as aerodynamic design [1–3], turbine design [4], and much more. However, accurate simulations of fluid flow require a proper representation of the geometry and the physics of the problem.

The geometry is usually represented as a discrete set of elements, called a mesh, which defines the computational domain and discretizes the equations governing the fluid flow. Therefore, the quality of the mesh has a significant impact on the accuracy and reliability of the CFD simulations. Meshing is the process of creating a finite element mesh. The mesh quality directly affects the accuracy and efficiency of the numerical simulations. While highly refined meshes may seem like a logical approach to ensure precision, this method comes with enormous computational costs. Ideally, the mesh should have enough resolution to capture the details of the fluid flow but, at the same time, should be sufficiently coarse to minimize the computational cost. Therefore, optimal mesh refinements need to be made, typically in a few critical regions of a structure (e.g., an aircraft's nosecone) that impact the fluid flow around it. However, identifying where to apply refinements to generate a good mesh is an intricate task

Copyright © 2023 by ASME

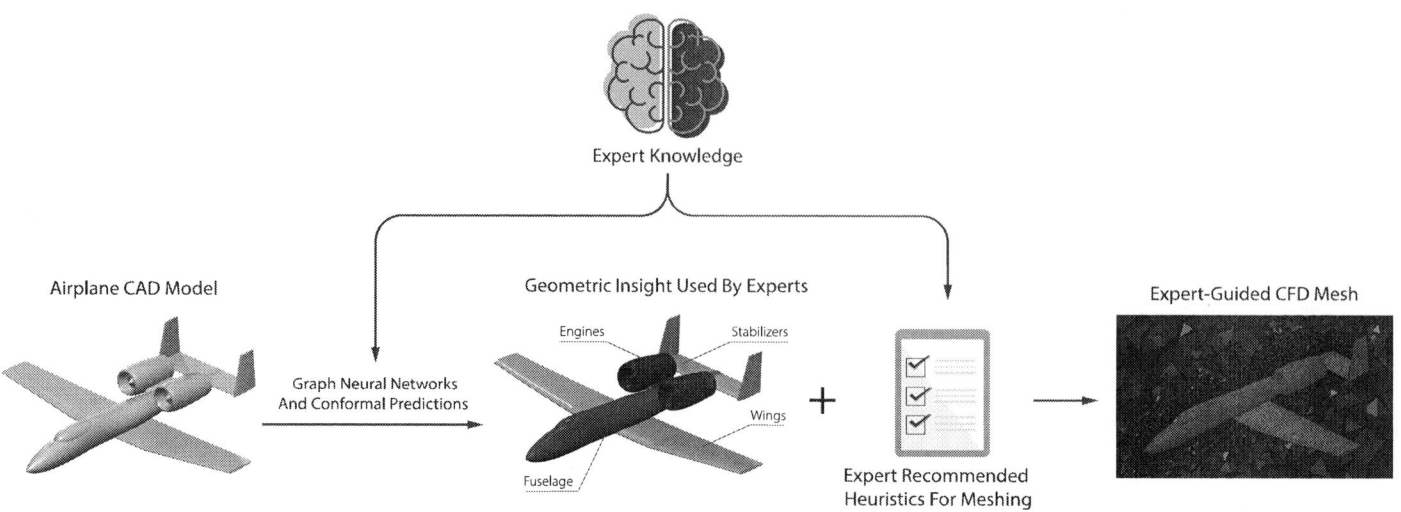

FIGURE 1: EXPERT-GUIDED MESH GENERATION USING GRAPH NEURAL NETWORKS AND CONFORMAL PREDICTIONS

that depends on the specific geometry and flow conditions. This means that significant time and effort are given to the meshing process as it is crucial to the success of a simulation. In case of a poor mesh quality, the entire time-consuming simulation process needs to be repeated. Consequently, generating good computational meshes often requires years of experience from CFD experts. Thus automating the meshing process is of significant value to the overall CFD process and can save significant time and effort and lead to a faster and more efficient design process.

Meshing has been an active area of research for several decades, and a vast number of mesh generation techniques have been proposed. An automated meshing algorithm must generate a mesh that satisfies both accuracy and computational efficiency requirements. In automating mesh generation for CFD, significant research has been conducted on adaptive re-meshing methods [5, 6], where an initial mesh is iteratively refined concurrently with the solver by tracking errors in the domain and refining the mesh based on heuristic rules or optimization [5–11]. However, these iterative adaptive meshing approaches are often slow and incur a high computational cost, making these methods often lead to much longer simulation times which makes them inferior to expert generated meshes that may not be as optimal as adaptive ones (maybe slightly over refined) but will lead to a valid solution in less time.

Despite the progress made in this field, meshing remains a challenging task, especially when dealing with complex geometries. In recent years deep learning models have shown great promise in many different domains of engineering design [12]. Substantial works have been proposed for improving iterative methods such as adaptive re-meshing using deep learning models [13–15]. However, most of these approaches focus on accelerating adaptive re-meshing and ignore the human experts involved in the mesh generation process. Our discussions with CFD experts at the Lincoln Laboratories revealed that, in reality, experts often use specific heuristics or thumb rules to generate meshes instead of adaptive meshing, which can be time-consuming. The rules are specifically handy when the underlying geometries do

not change significantly from one instance to another. Many of these heuristic rules require a deeper understanding of the geometry of the problem. In this paper, we attempt to accelerate this expert-guided meshing process.

We present a machine learning-based scheme for automating the expert-guided mesh generation process for airplane models, with the aim of achieving expert-level mesh quality (Fig.1). Our approach utilizes Graph Neural Networks (GNN) to gain a deeper understanding of geometry and develop heuristic-based algorithms that automatically generate high-quality meshes. We begin with rough surface meshes of airplane models generated by CAD software and use a GNN-based model for mesh segmentation of airplane parts. In the aerospace industry, reliable risk assessment and uncertainty quantification are essential components of design. To address this, we introduce a component to our approach that utilizes conformal prediction methods [16–18] to quantify uncertainties associated with our model and provide a mechanism for risk handling with marginal statistical guarantees. By enabling designers to set an acceptable risk level, our approach ensures that the generated mesh is, at worst over-refined, with under-refinement occurring only at an acceptable level of risk. The last step of our approach involves projecting the model predictions onto the CAD surfaces to identify the aircraft's individual parts. Expert-guided heuristic rules are then applied to determine the necessary mesh fidelity and refinements for each part, which are used to automatically generate CFD meshes for each aircraft model (figure 1). We also release a new annotated dataset of aircraft models, which is created by a new augmenting approach proposed in this paper.

Our research paper provides several significant contributions to the field, including:

1. **Application Impact:** We establish an approach for an expert-guided CFD meshing process for aircraft by combining graph-neural networks based predictions with a rule-based approach.

2. **Method:** We introduce a new 3D segmentation algorithm

Copyright © 2023 by ASME

that outperforms two state-of-the-art models, PointNet++ and PointMLP, for surface classification.

3. **Robust Risk Management:** We present a novel approach to project predictions from 3D mesh segmentation models to CAD surfaces using the conformal predictions method, which provides marginal statistical guarantees and robust uncertainty quantification and handling.

4. **Dataset:** We provide a dataset of realistic airplane models with segmentation labels, which can be used to benchmark future 3D segmentation models.

5. **CFD Case Study:** Through a real-world case study, we demonstrate that our automatically generated mesh is comparable in quality to expert-generated meshes, and enables the solver to converge and produce accurate results.

2. BACKGROUND AND RELATED WORKS

Our approach is based on 3D model segmentation where we take meshes of aircraft models and predict which part of the aircraft each face in the mesh belongs to. We then map these predictions to CAD surfaces, which are then used for determining mesh settings. In this section, we introduce a brief background on 3D segmentation models and conformal predictions, both of which play a major role in our work.

2.1 3D Segmentation and Mesh Segmentation

The popularity of deep learning-based methods has led to the proposal of various learning techniques in the computer vision and computer graphics communities for 3D shape classification and segmentation. Although conventional algorithms exist for this purpose, we will only focus on deep learning-based methods, as they significantly outperform conventional ones.

There are three primary approaches for 3D shape segmentation: 1) Voxelizing 3D shapes and using 3D convolutional neural networks (CNN) to generate 3D semantic segmentation labels, 2) Using point cloud representations and applying point-based approaches to classify each point in the point cloud, 3) Using a surface mesh representation of 3D shapes and classifying either vertices, edges, or faces of the mesh. Our method uses a mesh representation and classifies the faces of the mesh. Mesh is the most physically meaningful segmentation, with faces representing surfaces of the 3D model that would belong to different parts of an object.

In the following sections, we will briefly discuss the state of the art in each of the three approaches to shape classification and highlight some relevant parts of these models that inspired our method.

2.1.1 CNN Based Methods. One straightforward approach for using CNNs in mesh segmentation is to leverage rendered images of 3D objects from different viewpoints and apply existing CNNs that are highly effective in image processing tasks to make predictions on such images. Su et al.[19] were pioneers in this approach, using a multi-view CNN for shape classification, but their method could not be easily adapted for semantic segmentation. Later research developed a more comprehensive

multi-view framework [20] for shape segmentation, where segmentation maps were predicted for each view using CNNs, and the resulting image-level predictions were projected onto 3D models by enforcing label consistency using Conditional Random Fields (CRF) [20].

By converting a three-dimensional shape into a binary voxel format, it is possible to create a grid representation similar to that of a two-dimensional image. This approach enables the same CNN operations used for images to be easily extended to three-dimensional grids, allowing for a seamless transfer of traditional image-based techniques to the realm of shapes. Wu et al. [21] were the first to explore this idea for 3D shapes, and subsequent works have expanded on this approach [22–26]. However, volumetric representations and image-based methods demand significant computational resources and extensive memory usage. Furthermore, binary voxel representations do not readily map to more accurate representations of 3D shapes, such as meshes. Given these limitations, our work departs from image-based and volumetric representations as well as CNN-based methods.

2.1.2 Point cloud Approaches. Point clouds are a common representation method of 3D shapes that can be easily obtained from various other types of 3D shape representations, leading to significant research efforts in utilizing point-based deep learning methods for 3D shape analysis. One of the seminal works in this area is PointNet [27], which utilizes 1×1 convolutions on point cloud coordinates for 3D shape prediction, as well as introducing the Transformation Network (T-Net) to enable model invariance to geometric transformations such as rotations and translations [27]. This concept has proven to enhance learning generalization and can be applied to most representations using point/vertex positions, such as point clouds or meshes. Our approach also employs a variant of T-Net to take advantage of these benefits.

PointNet++ [28], a follow-up to PointNet, partitions points into groups to better capture local structures, resulting in significant performance improvements [28]. Other approaches incorporate information from local neighborhoods to perform dynamic updates by computing the similarity between points based on their euclidean distance in the feature space [29]. Some models, such as RSNet [30], sort points and treat them as sequences, using Recurrent Neural Networks (RNN) to capture local features. PointCNN [31] expands the convolution concept beyond local grids to a χ-convolution that operates on points located within their respective Euclidean neighborhoods. However, PointCNN is not invariant to point permutation, which is addressed in Point-Conv [32], extending the notion of convolution to an operator invariant to order.

One of the current best approach, PointMLP [33], proposed at the International Conference on Learning Representations (2022), removes convolution and instead employs simple multi-layer perceptrons (MLP) as operators. It extends the ideas of PointNet and PointNet++ and outperforms all other models in segmentation and classification tasks, including those using mesh representations [33]. In the results section, we show how our proposed approach outperforms PointMLP.

Most of these approaches capture local features by hierarchically grouping points and learning group-wise features. However,

Copyright © 2023 by ASME

if the mesh structure is available, its edges (i.e., adjacency information) can enable local feature extraction without the need for grouping and complex operations. This is the topic of the next section and the primary reason for our use of mesh representations.

2.1.3 Mesh Based Approaches.
The non-uniform polygonal mesh serves as the foundation for 3D data representation in computer graphics. This approach utilizes a smaller number of large polygons to cover expansive, flat regions, while employing a larger number of polygons to capture intricate details. A mesh accurately captures complex structures and distinguishes them from nearby surfaces by explicitly representing the surface topology. As such, meshes are considered one of the most accurate representations of 3D data. Models that utilize meshes directly are able to capture local features as needed, eliminating the issue of locality present in point clouds. Additionally, mesh representations do not suffer from permutation issues since they capture adjacency rather than a set of unordered points, as seen in point clouds. These advantages make meshes the most efficient and promising representation of 3D data for learning.

Numerous approaches have been explored for mesh-based deep learning models, including the popular MeshCNN approach introduced by Hanocka et al. [34]. This model utilizes mesh edges as the basis for its approach and computes features for each edge in the mesh, which are invariant to transformations. Convolution is then applied to these features in the neighborhood of each edge, determined by the edges sharing faces in the mesh. Despite its success, MeshCNN has limitations due to its reliance on the convolution operator, making it necessary to refine or coarsen meshes to a specific number of edges, thereby limiting its broad applicability.

To overcome this limitation, a more generalizable and versatile approach has been recently explored, namely the graph-based approach. In this class of models, graph neural networks (GNNs) are utilized to analyze 3D meshes, which lend themselves to the graph representation, with vertices serving as nodes. As such, GNNs are well-suited to handle meshes, and recent works have focused on this approach [35–38]. We combine the GNN-based approaches with point cloud-based approaches in our work, to present a new architecture that encompasses the benefits of both classes of models. Further details of our methodology are discussed in the following sections. Next, we shift our focus to a new method, which allows us to make predictions with provable guarantees on the error rates.

2.2 Conformal Predictions
Conformal prediction is a framework in machine learning that provides a measure of confidence in the predictions made by a model without any assumptions on the underlying predictive model. It is based on the idea of constructing prediction regions around each point in the input space, rather than just making a single-point prediction. These prediction regions can be calibrated to provide a desired level of confidence. This is made possible through the principle of exchangeability of the data (i.e. the order of the data does not affect the predictions).

Let's consider a classification task where we have n datapoints $\{(X_i, Y_i)\}_{i=1}^{n}$ with features $X_i \in \mathbb{R}^f$ and a classification label $Y_i \in \mathcal{Y} = \{1, 2, \ldots, C\}$ with C possible discrete classes in the label space. If we train a typical machine learning classifier that outputs a softmax probability distribution for each class we would have a vague notion of the model's certainty as these values are just guesses that the model has made. The reality is that there is no guaranteed accuracy/uncertainty that these distributions are reflective of the true distribution P_{XY}. If the cost of mistakes isn't high, one can just report the single most likely class for any input which comes with no statistical guarantees. However, in our work, each CFD simulation can be time-consuming, and knowing a measure of confidence in the machine learning algorithm can reduce mistakes. When we obtain predictions on mesh faces, we project them onto CAD surfaces. Each CAD surface receives information from many mesh faces associated with it and given the mesh CFD fidelity is determined by these predictions it is important to have some statistical guarantees on the accuracy of the predictions. Conformal predictions enable this by constructing prediction sets $\hat{\mathcal{C}}_\alpha \subseteq \mathcal{Y}$ with some risk level α for each input rather than a single prediction. Most notably, these prediction sets come with a marginal guarantee of obeying:

$$\mathbb{P}\left[Y \in \hat{\mathcal{C}}_\alpha(X)\right] \geq 1 - \alpha \tag{1}$$

These approaches enable such a guarantee by running the model on a test set of data that the model has not seen before and calibrating the uncertainty outputs of the model to a threshold that the model must meet for the above statistical guarantee [17]. This kind of statistical guarantee gives us a much better insight into the uncertainty of the predicted label distributions of the deep learning models which are arbitrary guesses with no . For these kinds of classification problems, many researchers have proposed conformal prediction schemes which can be easily applied to machine learning models [16–18]. In our works we take the approach of Adaptive Prediction Sets (APS) proposed by Romano et al. [17] to construct predictions sets which we then use for a voting algorithm.

Making incorrect predictions for CFD meshing can significantly delay the meshing process. By providing a measure of confidence, conformal prediction can help CFD experts make informed decisions and take appropriate action to prevent failures or minimize risks. This is particularly crucial in simulations that may require very large computational resources which may not be readily accessible at all times. In such cases where failure cannot be tolerated, conformal predictions can be applied with very low-risk acceptance while in cases where failure can be tolerated to some extent, this risk can be relaxed in favor of a less conservative mesh which can speed up the simulation.

3. METHODOLOGY

In this paper, we introduce a new model for mesh segmentation which is designed to work with very small datasets without overfitting while having a complex and deep architecture that takes aspects of GNN-based models as well as point cloud-based models to construct a hybrid model with greater accuracy. The overall architecture of our model can be seen in figure 2. We then propose an approach to take the predictions on the surface mesh and project them to the CAD surfaces, which then can be

Copyright © 2023 by ASME

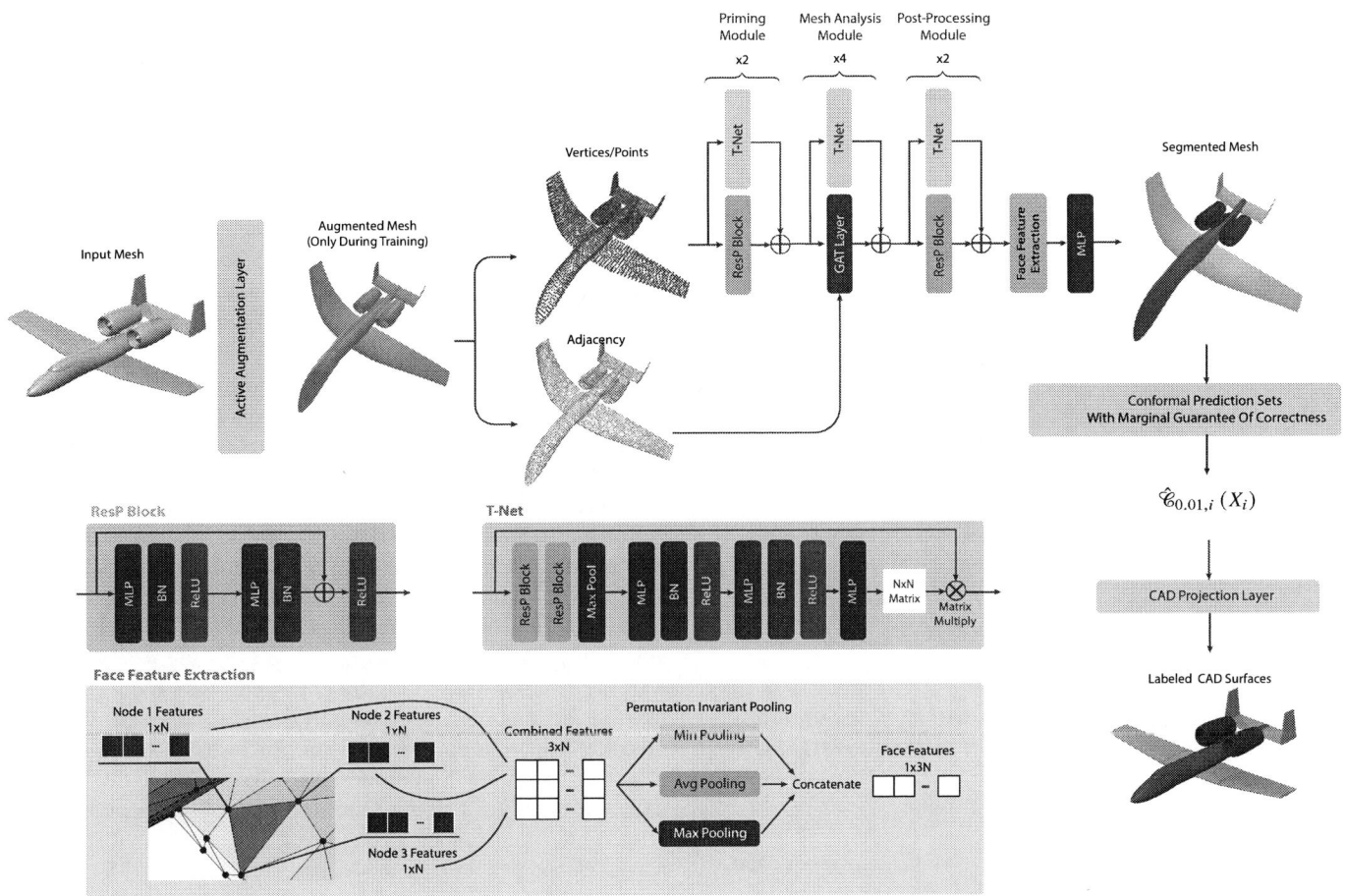

FIGURE 2: THE PROPOSED MESH SEGMENTATION MODEL IS A COMBINATION OF POINT-BASED AND GRAPH NEURAL NETWORK (GNN)-BASED APPROACHES. DURING TRAINING, EACH INPUT FIRST GOES THROUGH AN AUGMENTATION LAYER, THEN THE AUGMENTED MESH IS PROCESSED THROUGH TWO RESP BLOCKS, FOLLOWED BY FOUR GAT LAYERS AND ANOTHER TWO RESP BLOCKS, WITH A T-NET RESIDUAL APPLIED AT EACH LAYER. THE RESULTING FEATURES FOR EACH FACE ARE EXTRACTED AND PASSED THROUGH A CLASSIFIER MLP TO GENERATE SEGMENTATION LABELS FOR THE MESH. ONCE THE MESH IS SEGMENTED, MESH FACES ARE PROJECTED ONTO CAD SURFACES BASED ON CONFORMAL PREDICTION SETS, WHICH PROVIDE MARGINAL STATISTICAL GUARANTEES OF ACCURACY.

used to construct the final CFD mesh. In the sections that follow we will describe the details of our approach.

3.1 Mesh Segmentation Approach

In our model, we focus on developing large models that can be trained on very small datasets. This is particularly important in many engineering design applications as usually the size of datasets in engineering design is very small, which makes the application of large deep learning models difficult [12]. However, it has been demonstrated that proper data augmentation can overcome this challenge and enable the application of large deep-learning models without overfitting [12, 39, 40]. Given this, we introduce an active augmentation layer to our model to enable better generalization of learning and training with few samples. We describe in detail what this layer does and how it is implemented in training in later sections.

After the augmentation, we introduce a novel architecture (figure 2) for mesh face classification using a combination of point-based and graph-based deep learning approaches which combine the effectiveness of point-based models and the versatility of mesh graph-based models to allow for easy and accurate mesh segmentation.

Finally, we use the conformal predictions approach known as Adaptive Prediction Sets (APS) proposed by Romano et al. [17] to construct prediction sets from our segmentation model's output such that a marginal statistical guarantee can be given to the users when it comes to the possible labels each face in the mesh may be associated with. In the following sections, we describe each of the three aspects of our approach in detail.

3.1.1 Scheduled Augmentation Layer. When working with small datasets, large deep learning models tend to overfit the training data and fail to generalize to data the model has not been trained on. This leads to a lack of generalization which makes the models unreliable in real-world applications. There are several ways to overcome this problem, one way is to simply gather more data, but that may be too expensive or even imprac-

Copyright © 2023 by ASME

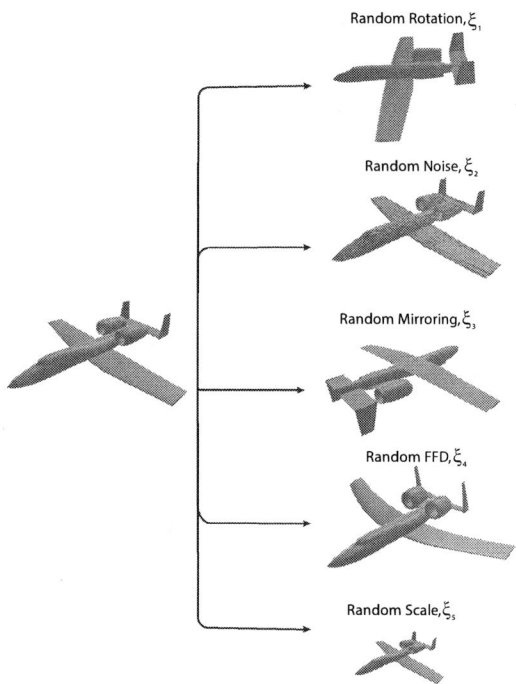

Random Rotation, ξ_1

Random Noise, ξ_2

Random Mirroring, ξ_3

Random FFD, ξ_4

Random Scale, ξ_5

FIGURE 3: VISUAL DEMONSTRATION OF THE FIVE ACTIVE AUGMENTATION APPROACHES WE EMPLOY IN OUR AUGMENTATION LAYER.

tical. Another way to overcome this issue is to take the input data and alter the data slightly so as to prevent the model to overfit or memorize the data. In computer vision and image data, this is typically done by applying random transformation and noise to the data [39]. Here we propose an augmentation layer that applies random changes to the input data to prevent overfitting. In our approach, we introduce a similar method but for 3D data. Our contribution to the augmentation layer is, therefore, two-fold. First, we introduce several augmentation techniques that we find to improve the generalizability of our model. Our second contribution is how we change the extent of our augmentation while the training is going on. In this section, we will first discuss the details of our augmentation pipeline/layer, then we will describe the dynamic augmentation mechanism we use during training.

Augmentation of 3D mesh data: Drawing inspiration from image augmentation methods [39], this paper proposes augmentation mechanisms for 3D mesh data to prevent overfitting in deep learning models. The proposed mechanisms include random rotations in all three directions, adding noise to data by applying random noise to the position of vertices in the mesh, mirroring 3D models with respect to different planes, random application of Free-Form Deformations, and random scaling of the model. One of the common approaches for image augmentation is the random rotation of the images, which is easy to adapt to 3D data by randomly rotating 3D objects in space. Unlike image-based approaches that apply one rotation angle, we apply random rotations in all three directions. Another common augmentation technique is adding noise to data, which we also use in our layer

by applying random noise to the position of vertices in the mesh. Another augmentation mechanism we employ is mirroring 3D models with respect to the XY, XZ, or YZ planes randomly. It also observed that warping images could be an effective form of augmentation in computer vision. We translate that idea to 3D, by using a random application of Free-Form Deformations (FFD) as analogous to the wrapping technique used for images. Finally, we have random scaling of the model, which either enlarges or shrinks a model slightly. These augmentation mechanisms are all demonstrated in Fig.3 to help visualize how they work. However, as we discussed earlier, we intend to control these augmentation mechanisms during training. As such, each augmentation mechanism will be controlled by a specific parameter.

For rotation, we randomly sample a 3D angle, Θ, from a uniform distribution of $\Theta \sim U(0, \xi_1) \in \mathbb{R}^3$ and the control parameter ξ_1 determines the extent of augmentation. Similarly, for the noise, we sample random noise for each vertex (v_i) from a uniform distribution $\epsilon_i \sim U(0, \xi_2) \in \mathbb{R}^3$ and the control parameter ξ_2 determines the extent of augmentation. For mirroring, we apply the mirroring across each of the 3 XY, XZ, and YZ planes randomly with a probability of $0 \leq \xi_3 < 1$, which controls how likely the mirroring is to occur during augmentation. For warping, we apply the FFD by deforming the model randomly in all directions, and we randomly sample the extent to which these deformations occur from a uniform distribution $U(-\xi_4, \xi_4)$, where ξ_4 determines how aggressively the model is deformed. Finally, for scaling, we apply a uniform scale adjustment across all three dimensions, which is sampled from a uniform distribution $SA \sim U(-\xi_5, \xi_5)$, and the model is scaled by $1 - SA$. The ξ_1, ξ_2, ξ_3, ξ_4, and ξ_5 parameters control how aggressively the models are augmented which allows us to control the extent of augmentation during training. More details can be found in the authors' code and data, which will be made available upon acceptance of the paper.

Augmentation Schedule: When training starts, we start with no augmentation and based on the target value of the ξ_1, ξ_2, ξ_3, ξ_4, and ξ_5 parameters the augmentation increases as training goes on. In general, we have an augmentation schedule that increases the extent of augmentation as every epoch goes on. The following equation describes this schedule:

$$\xi_i^T = \xi_i^\infty \times \frac{\left(1 + \cos\left(\frac{T\pi}{\tau} - \pi\right)\right)}{2} \qquad (2)$$

where T is the current epoch of training and ξ_i^∞ is the target value of the augmentation parameter while ξ_i^T is the adjusted augmentation parameter at epoch T and τ is the scaling factor which is set to the maximum number of epochs (the total number of epochs that we plan to train the model).

While training the deep learning model, we increase the level of augmentation gradually until the model no longer overfits the data. To ensure this, we keep track of the validation accuracy at the end of each epoch and save checkpoints of the model's weights when the accuracy is maximized. However, we do not want to make the task too difficult for the model, so we simultaneously decrease the learning rate of the optimizer as the level of augmentation increases. This helps the model adjust smoothly

Copyright © 2023 by ASME

to the more challenging task and prevents it from straying too far from the optimal parameters it has learned. Thus, we strike a balance between augmentation for generalizability and not hindering the model's learning. More implementation details can be found in our code and data, which will be available to the public upon acceptance of this paper.

3.1.2 Mesh Segmentation Model.
As discussed in previous sections, point cloud-based models have shown great success in 3D shape segmentation and classification, even outperforming the latest and best mesh-based and GNN-based models [33]. However, these models neglect the most accurate and flexible representation of 3D data[34], which is the mesh. Instead, they use neighborhood grouping and sampling to extract local relationships. Despite this, even when specialized for specific tasks, the latest point cloud-based model PointMLP [33] is comparable in performance to GNN-based models [35, 36]. Therefore, we cannot overlook the performance advantages of point cloud-based methods despite their limitations. To create a hybrid model that combines the strengths of both, we incorporate the lessons learned from point cloud-based models into GNN approaches, aiming to achieve the benefits of point cloud models and the versatility of GNN models.

Figure 2, shows the overall architecture of our model. There are several stages in our architecture, which we discuss in more detail here.

Priming Module: As it can be seen, the model starts with a priming module which includes an adaptation of the pointMLP's Residual Point (ResP) [33] block with the addition of T-Nets proposed by PointNet [27]. As we mentioned in the background section, the T-Net is a mechanism for transformation invariance, which is important when it comes to 3D data, and this matter was taken care of in PointNet++ [38] and PointMLP [33] by grouping and sampling through their geometric affine modules. Since we intend to use the mesh directly in our model and do not wish to perform geometric grouping and sampling, we introduce the T-Net. Our implementation of T-Net is shown in Figure 2. It takes as input a point set and using a few ResP layers, determines a linear transformation matrix for the point set which is then applied to the point set's features.

In this module, despite only having point-based components, we do not down-sample the vertices in the mesh to a specific number of points, as this would not allow us to apply GNNs to the entire mesh effectively. One of the main limitations of the PointBased models, is that during training, only subsets of all the points will be seen, while in our model, we not only input all the points but make classifications on all of our mesh faces. Another limitation of the point cloud-based model is that predictions will be made at the points, not surfaces. Furthermore, predictions in point cloud-based models are made at the points, while true labels are only meaningful for mesh faces in 3D segmentation. Thus, for point-based models, less accurate labels must be determined for individual points, and combining vertices of the mesh is not feasible, as all the vertices associated with faces may not have been captured during the subsampling of training. Therefore, by inputting all the vertices and using GNNs, we avoid this issue and

properly analyze the entire model.

Mesh Analysis Module: Once the data goes through the point-based priming module, we take the features from this stage as the input features for the mesh analysis module. In the mesh analysis module, we represent our data as graphs with node features from the priming module and adjacency matrix same as the mesh adjacency. In this module, we use a GNN approach called Graph Attention Convolution (GAT) [41]. For the sake of brevity, we will not discuss the details of GAT here and refer the readers to the original body of work for more detail [41]. Simultaneously, we find applying the T-Net from the priming module helps with the model generalization. For the GAT layers, we use 8 attention heads with 64 features each and apply a 10% dropout during training.

Post-processing Module: After the mesh analysis module, the resulting features go through the post-processing module, which is made up of ResP blocks and T-Nets. After that, the features of the mesh vertices are aggregated for each face. That is, the features of the vertices associated with each face are aggregated using an order-invariant face feature extraction function (see figure 2). This aggregation function must be order invariant because if the model sees different permutations of the points associated with a face, the outcome should not change. In our approach, we perform aggregation by min-pooling, average-pooling, and max-pooling and concatenate the three in that order. In this way, the features from all three vertices are captured without loss of information while the order invariance of the aggregation function is maintained. Finally, in the post-processing module, we apply an MLP to each face's features to predict the label for each face.

During training, we use the categorical cross-entropy loss to train the model for segmentation:

$$L_{CLS} = -\frac{1}{N} \sum_{i}^{N} \sum_{j}^{M} y_{ij} \log\left(\hat{y}_{ij}\right) \qquad (3)$$

where L_{CLS} is the classification loss that we minimize during training and N is the number of faces in a batch during training and M is the number of possible labels, and y_{ij} is the true label for the face i and class j while $\hat{y}_{ij}$ is the label predicted by the model. Beyond this loss, we also apply regularization loss to the transformations predicted by every layer of T-Net we have in our model. This is needed as the dimensionality of the features in the model increase, and the predicted transformations become much larger, which leads to overfitting and instabilities in the model. To mitigate this issue, we apply a regularization loss to the transformations predicted by T-Nets:

$$L_{\text{T-reg}} = \sum_{i}^{N} \left\| I - A_i A_i^T \right\|^2 \qquad (4)$$

Where $L_{\text{T-reg}}$ is the regularization loss minimized and N is the number of T-Net layers and A_i is the transformation predicted by the i-th T-Net layer. The total loss that we minimize during training is, therefore:

Copyright © 2023 by ASME

$$L = L_{CLS} + \gamma \times L_{\text{T-reg}} \qquad (5)$$

Where gamma is the weight given to the regularization loss.

3.1.3 Conformal Predictions.

Conformal prediction is a powerful technique for improving the reliability and robustness of machine learning models, making it a useful tool in a wide range of engineering applications. At this point, we have described a standard 3D segmentation model. However, the performance of deep-learning models is dependent on the data, and they simply make point predictions that are best guesses. The predictions for new test cases do not come with any statistical guarantees. This means that if we only base our classification on the model's best guess, we cannot guarantee any accuracy. In general uncertainty quantification with any kind of guarantee based on learning is difficult and most work in this area is restricted to Gaussian Processes. However, we could easily overcome this issue in a model-agnostic way by using conformal predictions. In our work, we use the approach proposed by Romano et al. [17] to construct prediction sets that are marginally guaranteed to include the correct prediction 95% of the time. More formally:

$$\mathbb{P}\left[Y_i \in \hat{\mathscr{C}}_{0.05,i}(X_i)\right] \geq 0.95 \qquad (6)$$

Where $\hat{\mathscr{C}}_{0.05,i}$ is the prediction set for the input sample (mesh face) X_i and Y_i is the true label. We then use these prediction sets to map the predictions on the mesh to the CAD surfaces. One thing to note here is that if the acceptable risk level (i.e., 0.05) were to be lowered, this may lead to the model becoming more conservative, leading to more surfaces having uncertain labels associated with them, which can lead to an over-refined mesh more often, leading to more expensive CFD simulation. As such, great care must be given to defining an appropriate risk level. In this work, we found that for our model a 0.05 risk led to no increase in the number of mistakes by the model but correctly enabled the identification of uncertainties and removed any under-refinements.

3.2 Mapping Mesh Segmentation to CAD Segmentation

In order to generate an effective Computational Fluid Dynamics (CFD) mesh for an airplane model, it is necessary to apply specific mesh settings to different surfaces of the aircraft. As a result, it is not sufficient to classify different parts of the airplane in a single mesh. Instead, each surface of the aircraft must be meshed with specific settings. Thus, the most practical approach to automating the CFD meshing process would involve mapping the classifications from the mesh to the corresponding surfaces in a CAD model. These surfaces can then be meshed in a CFD meshing platform using the appropriate settings for each surface type.

Estimating Mesh to CAD Surface Distance: To map predictions from the mesh to the CAD model, we take the predictions for each face in the mesh associated with the CAD files and calculate the location of the centroid of the face. Then we measure the distance from the centroid of each face to all the CAD surfaces and identify the closest surface in the CAD for each mesh face.

To measure the distance to different CAD surfaces we take the B-Spline surfaces of the CAD file and discretize the surfaces numerically using the control points defining the B-Spline surface and obtain a grid of points in each surface. Then we measure the distance between any given face centroid and all of the points in the discretized surface and take the minimum of all these distances as the distance from a face centroid to any given CAD surface:

$$D_{ij} = \min_k \|x_i - x_{jk}\| \qquad (7)$$

Where D_{ij} is the distance between face i and surface j and x_i is the location of the centroid of face i and x_{jk} is the location of the k-th discretized point of the j-th surface.

Voting Mechanism for CAD Surface Mapping: Once the CAD surface for each face is identified, the faces will vote on the nearest CAD surface based on the conformal predictions made previously, which have a 95% marginal correctness guarantee. The votes for each surface are then tallied, and if a single class obtains the majority vote (i.e., 50%+1 vote), it is selected as the surface type. If this does not occur, we select the surface classification from the top two classes with the most votes that require more refined mesh settings (see section 3.3 for details). This approach ensures that the mesh is not under-refined and can resolve the flow, leading to valid solutions. However, it may result in some surfaces being over-refined, which comes at a higher computational cost for the solver. When the conformal prediction sets have a 95% marginal correctness guarantee, the model can accurately predict a singular majority in most cases. If this is not possible, our model takes a more conservative approach to meshing to avoid any negative outcomes. Therefore, with an accurate enough model, we can practically guarantee valid mesh settings.

3.3 Expert-Guided Automated Meshing

After obtaining the CAD surface classifications through voting, we use expert-guided rules to generate meshes automatically for the CFD simulation under specific flow conditions and simulation types (as shown in Figure 1). These rules would be specific to the CFD simulation type and flow conditions. The primary idea in our work is that a database of rules would be established for each kind of simulation type, and equations will be established for specific mesh settings to be adjusted based on flow conditions. In our work, we focus on a specific setup for CFD simulations and use rules established by experts for mesh generation. Specifically, we develop rules for simulations intended to resolve unsteady flow structure in wake regions for aero-optical work. Furthermore, the mesh generated is meant to be simulated in the FUN3D CFD solver (second-order time accurate simulation), with Detached Eddy Simulation (DES) turbulence model. For this work, we limit the scope of work to one specific flow condition as well. This is done to enable us to demonstrate a proof of concept in this paper, and future work will generalize the rules to more situations. The flow conditions we develop our rules around are flight at sea level, at Mach 0.8, and Angle of Attack (AoA) 0°. We also resize airplane models to have a

Copyright © 2023 by ASME

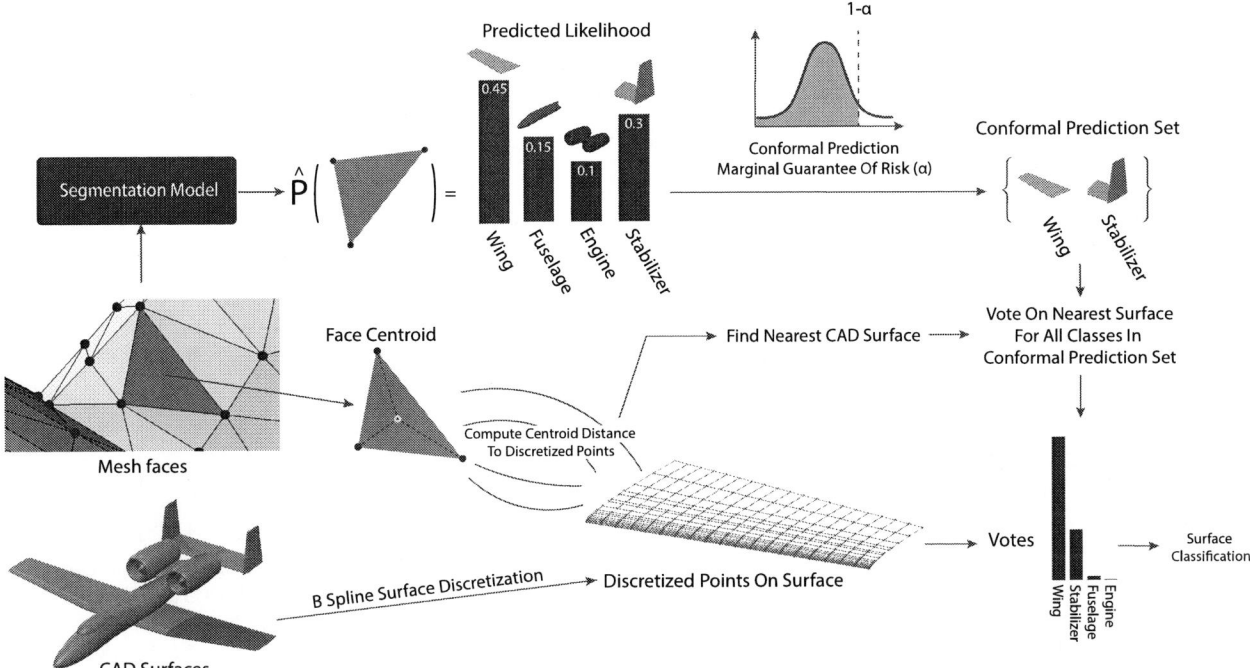

FIGURE 4: OUR APPROACH FOR MAPPING MODEL PREDICTIONS TO CAD SURFACES WITH MARGINAL STATISTICAL GUARANTEES USING CONFORMAL PREDICTIONS.

length of 30m. We then develop rules (based on Pointwise meshing software settings) for different aircraft parts to automate the mesh generation process. We do not go over the specific rules in this section to improve the readability of the paper. These rules are specified in appendix A. Although these rules are established automation of the process is not explicitly discussed further in this paper, as mesh settings databases are established by experts at the Lincoln laboratories, and the sensitive nature of their work limits the amount of information that can be discussed publicly. As such, we only provide the details for the above-mentioned flight conditions and see the process of automation as a black box that involves providing surface predictions to experts who can automate the meshing in their desired meshing software closing the loop of expert-guidance integration (see figure 1).

An important thing to note is that the settings that must be applied to the wing are the most refined, followed by the stabilizers, and finally the fuselage has the least refinements. As such when a majority is not reached in CAD surface classification (see section 3.2) priority is given to the classification that requires the most refined mesh.

4. DATASET

So far, we have discussed a framework for mesh classification and how we can map these predictions to CAD files and use expert guidance to generate CFD meshes. However, there are no publicly available datasets for the purpose of our work. As such to train our model we needed to create a new dataset of airplane models which meet the following criteria. First, the models used for training must be real and accurate aircraft designs and not include examples of toy aircraft and concept designs which are common in datasets such as ShapeNet. This is because, in many

of these cases, the features in the models are not indicative of realistic aircraft designs and therefore, may confuse the model and reduce generalizability to real models, and bias the model towards the features observed in unrealistic samples. The other criterion is that the models in the dataset must have segmentation labels pertaining to different aircraft parts, specifically the wing, stabilizers, fuselage, and engine as these are the parts of the aircraft that are important for mesh settings. The third criterion is that the models must include accurate CAD representations associated with them as well since we need to classify the CAD surfaces to apply the mesh settings properly in meshing software.

To create a model that is practically useful for CFD simulations at Lincoln laboratories, we created a dataset of realistic aircraft using Open Vehicle Sketch Pad (OpenVSP) [42]. OpenVSP, initially developed by NASA, is an open-source tool for creating parametric aircraft geometry that enables the engineering analysis of those models. To create the dataset, we use a few selected aircraft designs from OpenVSP and then apply parametric sampling within OpenVSP and data augmentation techniques outside of OpenVSP. In the following sections, we describe how we went about doing this in a more detailed manner.

4.1 Dataset Backbone

As discussed above, we chose OpenVSP to create a dataset of realistic aircraft models, as their website contains many aerodynamically accurate airplane models within their software, which are made publicly available under the name VSP Hanger[1]. However, not all of these models can be used easily, as we will see. Before we can use any given model, we need to ensure that we can

---
[1]https://hangar.openvsp.org/

Copyright © 2023 by ASME

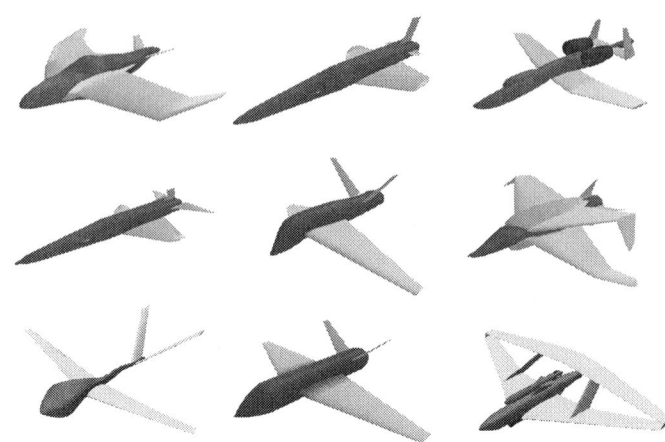

FIGURE 5: THE DATASET IS DERIVED FROM TEN BASE AIRPLANE MODELS, EACH BELONGING TO A DIFFERENT AIRCRAFT TYPE. THE MODELS WERE PROVIDED TO US BY CFD EXPERTS, WHO ENSURED THAT THEY COVER A WIDE RANGE OF AERODYNAMIC DESIGNS.

generate valid geometries for each model and label the different parts of the aircraft. This is a crucial part of the dataset curation, and as a result, we attempt to automate the process as much as possible to enable future dataset expansion.

The initial task is to develop a labeling method for different parts of the aircraft. OpenVSP has a feature that enables users to allocate various parts of the geometry to different sets, which is useful for our goal of segmenting airplane models. We manually assign different parts of the aircraft geometry to various sets, creating lists of sets for each aircraft's fuselage, wing, stabilizers, and engines. We extract separate meshes for each set in the OpenVSP model and use MeshLab's [43] boolean operations to combine them, resulting in a water-tight mesh. As we keep track of which OpenVSP set each face in the mesh belongs to during boolean operations, we can generate labeled meshes using this information and our manual labeling. We will not delve into the intricacies of this process since the dataset is not the primary focus of our work. However, we will make the code we use for data processing and the data itself publicly available and refer interested readers to our code and data for further detail.

The final note on the data is that given the process of manual labeling and verification of geometric validity is manual and labor intensive at this moment; our dataset is limited to ten accurate aircraft models selected by CFD experts at Lincoln laboratories, which are displayed in figure 5. Despite being a small number of models, there is significant variation and diversity in the dataset (figure 5) to capture many different types of aircraft designs. However, this small size presents a major challenge when it comes to applying large deep-learning models, as training a model on eigth aircraft models (assuming two models are set aside for validation) is almost certainly going to lead to overfitting. To overcome this, we propose an augmentation approach which we discuss in the following section.

4.2 Working With Small Datasets In Data-Driven Models

Since the small size of the dataset does not lend itself to deep learning, we employ some data-level augmentation, which is applied at the data-gathering stage rather than later during training. In our dataset, we employ two strategies for augmentation: design alterations, and mesh refinement alteration.

For design alteration, we use OpenVSP's python API. First, we manually analyze the design parameters (geometric parameters) in the OpenVSP graphical interface and determine appropriate design parameters that can be altered without breaking the

FIGURE 6: SHOWN ABOVE ARE NINE RANDOM AIRPLANE MODELS FROM THE AUGMENTED DATASET, WHICH ILLUSTRATES THE SIGNIFICANT DIVERSITY ACHIEVED BY CHANGING DESIGN PARAMETERS. IT IS IMPORTANT TO NOTE THAT THE EFFECTS OF REMESHING ARE NOT VISIBLE IN THESE IMAGES.

models and the range of acceptable values for said parameters. Once the parameters are determined, we use the OpenVSP API to create randomly altered versions of each of the ten models in our dataset by randomly sampling the values of the design parameters from a uniform distribution of acceptable values. We then generate 20 variations of each of our original models extending the dataset to 210 samples.

One important thing to mention, aside from the limited size of the dataset, is the fact that the mesh generated by OpenVSP has specific patterns in different parts of the aircraft, which could be exploited by a machine learning model to overfit these patterns. Consequently, it is important to ensure the model does not develop a bias towards such patterns. Remeshing the OpenVSP-generated meshes will enable two things for us. The first is that by remeshing at different mesh fidelities, we will augment the dataset and increase the size of the dataset, which is crucial to the success of any learning-based method. The second benefit of this would be that the model cannot be biased toward a specific mesh pattern generated by OpenVSP. To achieve both of these goals, we pick an unbiased uniform remeshing algorithm, isotropic explicit remeshing [43, 44], to remesh each model into five different mesh element sizes and fairly uniform elements, which remove the bias created by software-specific patterns in the mesh. The results of the aforementioned augmentations are visualized in figure 6.

5. RESULTS AND DISCUSSION

So far, we have established the details of our approach. In this section, we report the results of applying the method to the dataset and compare the performance of our model to other existing state-of-the-art approaches. But before we continue, it is important to establish the training details of our model and augmentation layer.

Copyright © 2023 by ASME

5.1 Model Configuration

To demonstrate the efficacy of our approach in mesh segmentation, we compare our method to two state-of-the-art methods, PointMLP [33] and PointNet++ [38], that outperform most other models or closely match their performance. These models serve as a benchmark for state-of-the-art. In both of these models, to obtain mesh face classifications we simply add the model's outputs for each vertex on a face and use that as the predictions for the face labels. We also train a naive implementation of a simple graph neural network with six layers of the graph convolution(GCN) [45] layers as the naive baseline. This model serves as a baseline for the simplest possible approach and is trained to demonstrate the notable gap that would exist without better model architectures. For PointNet++ and PointMLP, we use the code provided by the original authors of the paper and only make the final prediction modification without changing anything in their work or model configuration. For details of the naive baseline, we refer you to our code which will be made publicly available upon the acceptance of this paper.

For our model, we use a γ value of 0.1 in our loss (based on Eqn. 5), and for the active augmentation layer, we use target parameters of $\xi_1 = \pi/6$, $\xi_2 = 0.001$, $\xi_3 = 0.2$, $\xi_4 = 0.4$, and $\xi_5 = 0.15$. We find that in our experiments, these values yield the best generalizability for our model configuration. We train our model based on the loss function defined in Eqn. 5 and optimize our model using the Adam optimizer with a learning rate that starts at 10^{-4} and decays 15 times during training at equal intervals and at a rate of 0.65. For further details of the size of layers in our architecture and smaller implementation notes, we refer you to our code which will be made publicly available upon the acceptance of this work. To make the comparisons fair, we train all models for 200 epochs and use the original authors' code for the above benchmark models.

5.2 Mesh Segmentation Results

As mentioned above, we train all models on our dataset. For training, we use 840 models (based on the first eight in figure 5) and leave 210 models generated based on the last two aircraft (The two models at the bottom right of figure 5) for testing and validation. In figure 7, we visually demonstrate the mesh segmentation predictions made by our model. It is impressive that most of the faces on the meshes are predicted correctly. More importantly, this kind of accuracy will only be amplified when it comes to mapping face predictions to CAD surfaces, as it is much more unlikely that the majority of faces associated with CAD surfaces are incorrectly classified, leading to an even more accurate CAD surface classification.

Besides visual verification of the model's performance, it is important to quantify the accuracy of the model and compare it to state-of-the-art. As such, we train our model and competing models ten times each and record the means value of the best validation accuracy (percentage of correctly classified faces based on the model's top guess) for each experiment and model and report the results in Table 1. As evident, our model outperforms all models, including the very recently published and powerful PointMLP, which has established itself as the go-to model for 3D data. Besides that, we do observe that PointNet++ and GCN per-

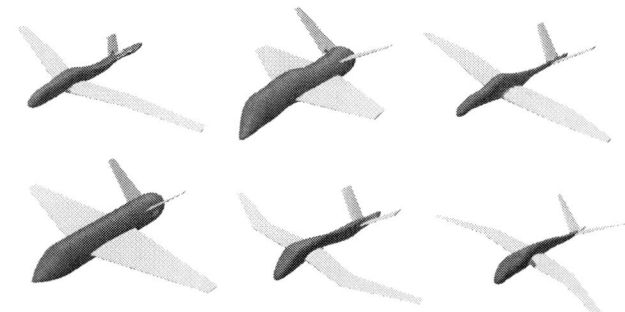

FIGURE 7: WE PRESENT SIX SAMPLE SEGMENTATION PREDICTIONS GENERATED BY OUR MODEL ON PREVIOUSLY UNSEEN VALIDATION DATA. THE HIGH ACCURACY OF OUR MODEL ON THIS DATASET IS APPARENT FROM THE VISUAL RESULTS. NOTABLY, THE MODEL CORRECTLY CLASSIFIES THE MAJORITY OF FACES ASSOCIATED WITH EACH CAD SURFACE, WHICH ENABLES EVEN MORE ACCURATE CAD SURFACE PREDICTIONS.

form poorly, which is expected as PointMLP is an improvement on the concepts introduced in PointNet++, and the naive GCN model, as expected performs much worse than all other models proving that the baseline of what the simplest possible approach leaves a lot to be desired.

TABLE 1: COMPARISON OF THE MEAN ACCURACY OF DIFFERENT MODELS ON THE VALIDATION DATA. THE VALUES AFTER ± SHOW THE STANDARD DEVIATION ACROSS TEN TRAINING RUNS.

Model	Top 1 Accuracy (%)
PointNet++	89.14 ± 0.25
PointMLP	93.92 ± 0.15
Naive GCN	81.23 ± 1.06
Our Model (w/ Augmentation Layer)	$\mathbf{96.65 \pm 0.12}$
Our Model (w/o Augmentation Layer)	92.35 ± 0.97

5.3 Conformal Predictions Supported CAD Surface Classification

For CAD surface classification we take the best-trained model for each method and implement a simple voting algorithm to classify CAD surfaces. This algorithm is very similar to our CAD surface classification (see figure 4) except without conformal predictions such that each face only votes for the top guess of the model. For CAD surface classification we record the number of misclassified surfaces in the validation data for each approach. Furthermore, we calculate the number of misclassifications that would lead to a more refined mesh than necessary (ultimately not failing in simulation) and the number of misclassifications that lead to under-refinement, which can be problematic in simulation. We present our results in table 2. The first observation in these results is the fact that our model which has the most accurate predictions on the mesh as we saw in the prior section, leads to the fewest misclassified surfaces compared to all other models.

Copyright © 2023 by ASME

TABLE 2: COMPARISON OF THE ACCURACY OF DIFFERENT MODELS ON CAD SURFACE PREDICTIONS ON THE VALIDATION DATA.

Model	# Incorrect Surfaces	# Under Refined Surfaces	# Over Refined Surfaces	Accuracy (%)
PointNet++	23	9	14	97.26
PointMLP	13	5	8	98.45
Naive GCN	42	9	33	95.00
Naive GCN + Conformal	116	**0**	116	86.19
Our Model	**7**	2	5	**99.17**
Our Model + Conformal	**7**	**0**	7	**99.17**

The most notable observation is that using the model's predictions directly without the conformal predictions approach leads to under-refined surfaces, which can affect the simulation efficacy and result in inaccurate solutions, ultimately a failure. This demonstrates the importance of applying conformal predictions to quantify uncertainty and apply the most conservative mesh settings when the model is faced with uncertainty in surface classification. More importantly, conformal predictions enable uncertainty quantification with marginal statistical guarantees, which is a very robust approach, and guarantees under refinement in the mesh will be avoided to a measurable extent which can be set by the user depending on the sensitivity of their work. Another significant observation is that adding the more conservative conformal predictions with 95% marginal guarantee has not caused more over-refinement, which is a testament to the accuracy of our model which is out of the box is good enough that even with the conservative conformal predictions remains accurate. On the contrary, we see that when conformal predictions are applied to models with low accuracies, such as the GCN baseline (see table 2), the conformal predictions fail to provide the necessary guarantee easily, leading to much more over-refinement and an overly conservative CFD mesh. Still, it is impressive that conformal prediction still sees no failed CFD mesh with under-refinement even when the model is inaccurate (See GCN+Conformal in table 2). To the best of our knowledge, this work is the first demonstration of model-agnostic conformal predictions in the engineering design community, and we hope that our results will inspire further research in different engineering applications to improve deep learning model robustness using conformal prediction-based approach.

5.4 Automated Meshing: A Case Study

So far, we have demonstrated that our model is very effective at extracting the necessary geometric insights needed by experts to automate the meshing process. We also discussed the specific settings that can be applied for a specific flight condition. To further showcase the potential of our methods, we conduct a case study on one aircraft model. Specifically, we generate the mesh for the aircraft model resembling the A-10 warthog (top left aircraft in figure 5). We use the expert rules for the flight conditions discussed in section 3.3. The results of the rule-based meshing are shown in figure 8. As we discussed, the flow patterns around the wings are the most intricate and require significant refinement compared to the fuselage, and we can see that given the correct classification, the rule-based meshing successfully refined the mesh around the wing while keeping the mesh

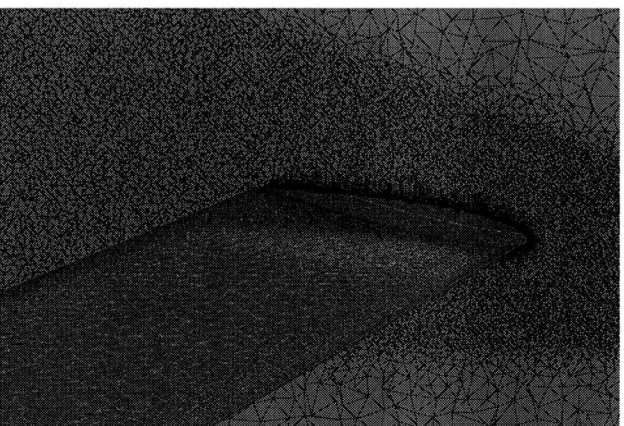

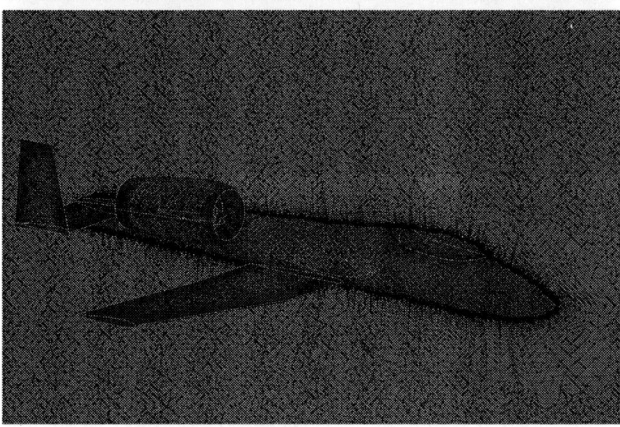

FIGURE 8: THESE IMAGES ILLUSTRATE HOW THE APPLICATION OF DIFFERENT MESHING RULES CAN RESULT IN VARYING LEVELS OF REFINEMENT IN DIFFERENT AREAS OF THE MODEL. SPECIFICALLY, THE TOP TWO IMAGES SHOW HOW OUR MESHING RULES LEAD TO MORE REFINEMENT AROUND THE WINGS, WHILE LESS REFINEMENT IS APPLIED AROUND THE FUSELAGE WHERE FLOW PATTERNS ARE SIMPLER.

refinement lower around the fuselage of the aircraft. In this way, the meshing process for this aircraft can be automated with little input or effort from the users without significant added cost on the solution time. These results show that expert-guided automated meshing for aircraft can be of great value for designers and perhaps provide a viable alternative to very expensive adaptive remeshing approaches or even serve as a great initialization to adaptive remeshing leading to significant time saves in both the meshing process itself and the simulation time.

Copyright © 2023 by ASME

Finally, to verify that this generated mesh is capable of capturing the necessary details in the simulation, we run the simulation for the generated mesh with the flow conditions mentioned in section 3.3. We observe that the solver converges without any issues, and the resulting simulations capture the details necessary for flow analysis. To showcase the success of the meshing when it comes to the solution, we specifically look at areas of the flow where higher detail must be captured. Specifically, we see that the simulation results have successfully captured the details of the flow in the most complex parts of the flow field near the wings. This shows that expert-guided refinements are crucial for successful flow simulations and accurate results. In the benefit of brevity, we refrain from discussing the CFD details, given that the primary focus of our work is on meshing.

6. LIMITATIONS AND FUTURE WORK

While our methods have been successful, there are areas in which we aim to develop and improve in the future. Currently, our approach is limited to rules based solely on 3D model segmentation, meaning that experts are only able to make rules based on surface types. In the future, we plan to collaborate with CFD experts to create more advanced rules for meshing by identifying other relevant features of 3D models, such as engine inlet and outlet, trailing edges of lift-generating surfaces, and more. This will enable expert guidance to become even more insightful, leading to a more effective and highly optimal mesh generation scheme. We believe that our GNN and point-based model will be able to handle this challenge with ease, and we plan to take this step in the near future.

Another area of improvement for our work would be to expand our dataset to encompass a much larger set of models. This will lead to a more powerful and accurate segmentation model, significantly improving the overall approach. As such, we plan to continue adding models to our dataset in the future and make the data publicly available as a living dataset that will receive updates with more models being added over time.

At this stage, our approach relies on third-party meshing software for CFD mesh generation. However, we believe that a more independent CFD mesh generation pipeline would provide our model with more freedom and enable expert guidance to be significantly more detailed. Therefore, significant effort must be devoted to developing such a pipeline to enable methods like ours to become even more versatile in the future.

7. CONCLUSION

We propose an expert-guided CFD meshing method that combines graph neural networks and expert heuristics to generate CFD meshes that not only improve numerical solver convergence but also capture the necessary details of flow patterns in areas with complex flow patterns. To develop our model, we introduce a novel deep-learning approach that combines the strengths of point-based models and mesh-based GNNs. Additionally, we publicly release a new annotated dataset and show how an active augmentation approach enables us to train a large deep-learning model on a small dataset without sacrificing generalization. We demonstrate that our approach outperforms the latest state-of-the-art techniques for 3D model segmentation, establishing our model as the new state-of-the-art.

Furthermore, we introduce a mechanism for mapping predictions from a mesh to CAD surfaces and utilize conformal predictions in CAD surface classification. Our model is capable of quantifying uncertainty and handling the uncertainty with a 95% marginal statistical guarantee of correctness. Based on meshing rules established for different surface types, we utilize expert guidance to generate CFD meshes. We show that our model provides the most accurate predictions on CAD surfaces (over 99% accurate), and our conformal predictions approach leads to correct uncertainty identification and a more conservative approach to meshing with no under-refined surfaces, effectively enabling failure avoidance in the solver.

Finally, we validate the efficacy of our approach by conducting a case-study simulation of one of the aircraft in the dataset. We demonstrate that given accurate surface predictions, the resulting mesh is appropriately refined and leads to successful simulations by the solver, thus establishing the effectiveness of our approach. We make our code and data publicly available for interested readers to examine. The future work includes developing more advanced rules for meshing and expanding the dataset for a more powerful segmentation model. Additionally, efforts will be given to developing an independent CFD mesh generation pipeline for more freedom and detailed expert guidance.

ACKNOWLEDGMENTS

The authors acknowledge the Lincoln Laboratory for providing computational resources that have contributed to the research results reported in this paper. We also thank MathWorks for supporting Amin Heyrani Nobari's studies while working on this project.

REFERENCES

[1] Chen, Wei and Ahmed, Faez. "PaDGAN: Learning to Generate High-Quality Novel Designs." *Journal of Mechanical Design* Vol. 143 No. 3 (2020). DOI 10.1115/1.4048626. URL https://asmedigitalcollection.asme.org/mechanicaldesign/article-pdf/143/3/031703/6632924/md_143_3_031703.pdf, URL https://doi.org/10.1115/1.4048626. 031703.

[2] Chen, Wei and Ahmed, Faez. "MO-PaDGAN: Reparameterizing Engineering Designs for augmented multi-objective optimization." *Applied Soft Computing* Vol. 113 (2021): p. 107909. DOI https://doi.org/10.1016/j.asoc.2021.107909. URL https://www.sciencedirect.com/science/article/pii/S1568494621008310.

[3] Heyrani Nobari, Amin, Chen, Wei and Ahmed, Faez. "PcDGAN: A Continuous Conditional Diverse Generative Adversarial Network For Inverse Design." *Proceedings of the 27th ACM SIGKDD Conference on Knowledge Discovery & Data Mining*: pp. 606–616. 2021.

[4] Langroudi, A Tahadjodi, Afifi, F Zare, Nobari, A Heyrani and Najafi, AF. "Modeling and numerical investigation on multi-objective design improvement of a novel cross-flow lift-based turbine for in-pipe hydro energy harvesting applications." *Energy conversion and management* Vol. 203 (2020): p. 112233.

Copyright © 2023 by ASME

[5] Plewa, Tomasz, Linde, Timur and Weirs, V. Gregory. *Adaptive mesh refinement: Theory and applications: Proceedings of the chicago workshop on adaptive mesh refinement methods, September 3-5, 2003.* Springer (2005).

[6] Fidkowski, Krzysztof J and Darmofal, David L. "Review of output-based error estimation and mesh adaptation in computational fluid dynamics." *AIAA journal* Vol. 49 No. 4 (2011): pp. 673–694.

[7] Becker, Roland and Rannacher, Rolf. "An optimal control approach to a posteriori error estimation in finite element methods." *Acta Numerica* Vol. 10 (2001): p. 1–102. DOI 10.1017/S0962492901000010.

[8] Babuška, I. and Miller, A. "A feedback finite element method with a posteriori error estimation: Part I. The finite element method and some basic properties of the a posteriori error estimator." *Computer Methods in Applied Mechanics and Engineering* Vol. 61 No. 1 (1987): pp. 1–40. DOI https://doi.org/10.1016/0045-7825(87)90114-9. URL https://www.sciencedirect.com/science/article/pii/0045782587901149.

[9] Eriksson, Kenneth, Estep, Don, Hansbo, Peter and Johnson, Claes. "Introduction to Adaptive Methods for Differential Equations." *Acta Numerica* Vol. 4 (1995): p. 105–158. DOI 10.1017/S0962492900002531.

[10] Verfürth, R. "A posteriori error estimation and adaptive mesh-refinement techniques." *Journal of Computational and Applied Mathematics* Vol. 50 No. 1 (1994): pp. 67–83. DOI https://doi.org/10.1016/0377-0427(94)90290-9. URL https://www.sciencedirect.com/science/article/pii/0377042794902909.

[11] Braack, Malte and Ern, Alexandre. "Adaptive Computation of Reactive Flows with Local Mesh Refinement and Model Adaptation." Feistauer, Miloslav, Dolejší, Vít, Knobloch, Petr and Najzar, Karel (eds.). *Numerical Mathematics and Advanced Applications*: pp. 159–168. 2004. Springer Berlin Heidelberg, Berlin, Heidelberg.

[12] Regenwetter, Lyle, Nobari, Amin Heyrani and Ahmed, Faez. "Deep generative models in engineering design: A review." *Journal of Mechanical Design* Vol. 144 No. 7 (2022): p. 071704.

[13] Huang, Keefe, Krügener, Moritz, Brown, Alistair, Menhorn, Friedrich, Bungartz, Hans-Joachim and Hartmann, Dirk. "Machine learning-based optimal mesh generation in computational fluid dynamics." *arXiv preprint arXiv:2102.12923* (2021).

[14] Fidkowski, Krzysztof J. and Chen, Guodong. "Metric-based, goal-oriented mesh adaptation using machine learning." *Journal of Computational Physics* Vol. 426 (2021): p. 109957. DOI https://doi.org/10.1016/j.jcp.2020.109957. URL https://www.sciencedirect.com/science/article/pii/S0021999120307312.

[15] Pfaff, Tobias, Fortunato, Meire, Sanchez-Gonzalez, Alvaro and Battaglia, Peter. "Learning Mesh-Based Simulation with Graph Networks." *International Conference on Learning Representations*. 2021. URL https://openreview.net/forum?id=roNqYL0_XP.

[16] Sadinle, Mauricio, Lei, Jing and Wasserman, Larry. "Least Ambiguous Set-Valued Classifiers With Bounded Error Levels." *Journal of the American Statistical Association* Vol. 114 No. 525 (2019): pp. 223–234. DOI 10.1080/01621459.2017.1395341. URL https://doi.org/10.1080/01621459.2017.1395341, URL https://doi.org/10.1080/01621459.2017.1395341.

[17] Romano, Yaniv, Sesia, Matteo and Candès, Emmanuel J. "Classification with Valid and Adaptive Coverage." *Proceedings of the 34th International Conference on Neural Information Processing Systems*. 2020. Curran Associates Inc., Red Hook, NY, USA.

[18] Angelopoulos, Anastasios, Bates, Stephen, Malik, Jitendra and Jordan, Michael I. "Uncertainty sets for image classifiers using conformal prediction." *arXiv preprint arXiv:2009.14193* (2020).

[19] Su, Hang, Maji, Subhransu, Kalogerakis, Evangelos and Learned-Miller, Erik G. "Multi-view Convolutional Neural Networks for 3D Shape Recognition." *2015 IEEE International Conference on Computer Vision (ICCV)* (2015): pp. 945–953.

[20] Kalogerakis, Evangelos, Averkiou, Melinos, Maji, Subhransu and Chaudhuri, Siddhartha. "3D shape segmentation with projective convolutional networks." *proceedings of the IEEE conference on computer vision and pattern recognition*: pp. 3779–3788. 2017.

[21] Wu, Zhirong, Song, Shuran, Khosla, Aditya, Yu, Fisher, Zhang, Linguang, Tang, Xiaoou and Xiao, Jianxiong. "3D ShapeNets: A deep representation for volumetric shapes." *2015 IEEE Conference on Computer Vision and Pattern Recognition (CVPR)*: pp. 1912–1920. 2015. DOI 10.1109/CVPR.2015.7298801.

[22] Brock, Andrew, Lim, Theodore, Ritchie, James M. and Weston, Nick. "Generative and Discriminative Voxel Modeling with Convolutional Neural Networks." *ArXiv* Vol. abs/1608.04236 (2016).

[23] Tchapmi, Lyne P., Choy, Christopher Bongsoo, Armeni, Iro, Gwak, JunYoung and Savarese, Silvio. "SEGCloud: Semantic Segmentation of 3D Point Clouds." *2017 International Conference on 3D Vision (3DV)* (2017): pp. 537–547.

[24] Hanocka, Rana, Fish, Noa, Wang, Zhenhua, Giryes, Raja, Fleishman, Shachar and Cohen-Or, Daniel. "ALIGNet: Partial-Shape Agnostic Alignment via Unsupervised Learning." *ACM Trans. Graph.* Vol. 38 (2018): pp. 1:1–1:14.

[25] Maturana, Daniel and Scherer, Sebastian. "VoxNet: A 3D Convolutional Neural Network for real-time object recognition." *2015 IEEE/RSJ International Conference on Intelligent Robots and Systems (IROS)*: pp. 922–928. 2015. DOI 10.1109/IROS.2015.7353481.

[26] Sun, Kai, Zhang, Jiangshe, Liu, Junmin, Yu, Ruixuan and Song, Zengjie. "DRCNN: Dynamic Routing Convolutional Neural Network for Multi-View 3D Object Recognition." *Trans. Img. Proc.* Vol. 30 (2021): p. 868–877. DOI 10.1109/TIP.2020.3039378. URL https://doi.org/10.1109/TIP.2020.3039378.

Copyright © 2023 by ASME

[27] Qi, Charles R., Su, Hao, Mo, Kaichun and Guibas, Leonidas J. "PointNet: Deep Learning on Point Sets for 3D Classification and Segmentation." *Proceedings of the IEEE Conference on Computer Vision and Pattern Recognition (CVPR)*. 2017.

[28] Qi, Charles Ruizhongtai, Yi, Li, Su, Hao and Guibas, Leonidas J. "PointNet++: Deep Hierarchical Feature Learning on Point Sets in a Metric Space." Guyon, I., Luxburg, U. Von, Bengio, S., Wallach, H., Fergus, R., Vishwanathan, S. and Garnett, R. (eds.). *Advances in Neural Information Processing Systems*, Vol. 30. 2017. Curran Associates, Inc. URL https://proceedings.neurips.cc/paper/2017/file/d8bf84be3800d12f74d8b05e9b89836f-Paper.pdf.

[29] Wang, Yue, Sun, Yongbin, Liu, Ziwei, Sarma, Sanjay E., Bronstein, Michael M. and Solomon, Justin M. "Dynamic Graph CNN for Learning on Point Clouds." *ACM Transactions on Graphics (TOG)* (2019).

[30] Huang, Qiangui, Wang, Weiyue and Neumann, Ulrich. "Recurrent slice networks for 3d segmentation of point clouds." *Proceedings of the IEEE conference on computer vision and pattern recognition*: pp. 2626–2635. 2018.

[31] Li, Yangyan, Bu, Rui, Sun, Mingchao, Wu, Wei, Di, Xinhan and Chen, Baoquan. "PointCNN: Convolution On X-Transformed Points." Bengio, S., Wallach, H., Larochelle, H., Grauman, K., Cesa-Bianchi, N. and Garnett, R. (eds.). *Advances in Neural Information Processing Systems*, Vol. 31. 2018. Curran Associates, Inc. URL https://proceedings.neurips.cc/paper/2018/file/f5f8590cd58a54e94377e6ae2eded4d9-Paper.pdf.

[32] Wu, Wenxuan, Qi, Zhongang and Fuxin, Li. "PointConv: Deep Convolutional Networks on 3D Point Clouds." *Proceedings of the IEEE/CVF Conference on Computer Vision and Pattern Recognition (CVPR)*. 2019.

[33] Ma, Xu, Qin, Can, You, Haoxuan, Ran, Haoxi and Fu, Yun. "Rethinking Network Design and Local Geometry in Point Cloud: A Simple Residual MLP Framework." *International Conference on Learning Representations*. 2022. URL https://openreview.net/forum?id=3Pbra-_u76D.

[34] Hanocka, Rana, Hertz, Amir, Fish, Noa, Giryes, Raja, Fleishman, Shachar and Cohen-Or, Daniel. "MeshCNN: a network with an edge." *ACM Transactions on Graphics (TOG)* Vol. 38 (2019): pp. 1 – 12.

[35] Lian, Chunfeng, Wang, Li, Wu, Tai-Hsien, Wang, Fan, Yap, Pew-Thian, Ko, Ching-Chang and Shen, Dinggang. "Deep Multi-Scale Mesh Feature Learning for Automated Labeling of Raw Dental Surfaces From 3D Intraoral Scanners." *IEEE Transactions on Medical Imaging* Vol. 39 No. 7 (2020): pp. 2440–2450. DOI 10.1109/TMI.2020.2971730.

[36] Zheng, Youyi, Chen, Beijia, Shen, Yuefan and Shen, Kaidi. "TeethGNN: Semantic 3D Teeth Segmentation with Graph Neural Networks." *IEEE Transactions on Visualization and Computer Graphics* (2022): pp. 1–1DOI 10.1109/TVCG.2022.3153501.

[37] Wen, Tingxi, Zhuang, Jiafu, Du, Yu, Yang, Linjie and Xu, Jianfei. "Dual-Sampling Attention Pooling for Graph Neural Networks on 3D Mesh." *Computer Methods and Programs in Biomedicine* Vol. 208 (2021): p.

106250. DOI https://doi.org/10.1016/j.cmpb.2021.106250. URL https://www.sciencedirect.com/science/article/pii/S0169260721003242.

[38] Qi, Xiaojuan, Liao, Renjie, Jia, Jiaya, Fidler, Sanja and Urtasun, Raquel. "3D Graph Neural Networks for RGBD Semantic Segmentation." *Proceedings of the IEEE International Conference on Computer Vision (ICCV)*. 2017.

[39] Karras, Tero, Aittala, Miika, Hellsten, Janne, Laine, Samuli, Lehtinen, Jaakko and Aila, Timo. "Training Generative Adversarial Networks with Limited Data." (2020). DOI 10.48550/ARXIV.2006.06676. URL https://arxiv.org/abs/2006.06676.

[40] Nobari, Amin Heyrani, Chen, Wei and Ahmed, Faez. "Range-Constrained Generative Adversarial Network: Design Synthesis Under Constraints Using Conditional Generative Adversarial Networks." *Journal of Mechanical Design* Vol. 144 No. 2 (2021). DOI 10.1115/1.4052442. URL https://asmedigitalcollection.asme.org/mechanicaldesign/article-pdf/144/2/021708/6768330/md_144_2_021708.pdf, URL https://doi.org/10.1115/1.4052442. 021708.

[41] Veličković, Petar, Cucurull, Guillem, Casanova, Arantxa, Romero, Adriana, Lio, Pietro and Bengio, Yoshua. "Graph attention networks." *arXiv preprint arXiv:1710.10903* (2017).

[42] McDonald, Robert A. and Gloudemans, James R. *Open Vehicle Sketch Pad: An Open Source Parametric Geometry and Analysis Tool for Conceptual Aircraft Design*: DOI 10.2514/6.2022-0004. URL https://arc.aiaa.org/doi/pdf/10.2514/6.2022-0004, URL https://arc.aiaa.org/doi/abs/10.2514/6.2022-0004.

[43] Cignoni, Paolo, Callieri, Marco, Corsini, Massimiliano, Dellepiane, Matteo, Ganovelli, Fabio and Ranzuglia, Guido. "MeshLab: an Open-Source Mesh Processing Tool." Scarano, Vittorio, Chiara, Rosario De and Erra, Ugo (eds.). *Eurographics Italian Chapter Conference*. 2008. The Eurographics Association. DOI 10.2312/LocalChapterEvents/ItalChap/ItalianChapConf2008/136.

[44] Hoppe, Hugues, DeRose, Tony, Duchamp, Tom, McDonald, John and Stuetzle, Werner. "Mesh optimization." *Proceedings of the 20th annual conference on Computer graphics and interactive techniques*: pp. 19–26. 1993.

[45] Kipf, Thomas N and Welling, Max. "Semi-supervised classification with graph convolutional networks." *arXiv preprint arXiv:1609.02907* (2016).

APPENDIX A. EXPERT-GUIDED RULES FOR DIFFERENT AIRCRAFT SURFACES

In this section, we present the specific rules provided by experts we use for the flight conditions and software mentioned in section 3.3. Below are the specific mesh settings that are applied in PointWise meshing software for different aircraft parts:

1. **Wings:**

   - Set surface mesh dimension = 0.05

Copyright © 2023 by ASME

- Set surface mesh to: quadrilateral cells dominant
- Set volume mesh to: hexahedral cells dominant
- Set initial wall spacing = 4.7e-6
- Set Growth rate = 1.1
- Set Collision Buffer = 2.0
- Grow volume cells from surface cells. Check Y+ value... If > 1, then decrease initial wall spacing proportionally

2. **Stabalizers:**

- Set surface mesh dimension = 0.2
- Set surface mesh to: quadrilateral cells dominant
- Set volume mesh to: hexahedral cells dominant
- Set initial wall spacing = 4.7e-6
- Set Growth rate = 1.1

- Set Collision Buffer = 2.0
- Grow volume cells from surface cells. Check Y+ value... If > 1, then decrease initial wall spacing proportionally

3. **Fuselage:**

- Set surface mesh dimension = 1.0
- Set surface mesh to: quadrilateral cells dominant
- Set volume mesh to: hexahedral cells dominant
- Set initial wall spacing = 4.7e-6
- Set Growth rate = 1.1
- Set Collision Buffer = 2.0
- Grow volume cells from surface cells. Check Y+ value... If > 1, then decrease initial wall spacing proportionally

Copyright © 2023 by ASME

Proceedings of the ASME 2023
International Design Engineering Technical Conferences and
Computers and Information in Engineering Conference
IDETC-CIE2023
August 20-23, 2023, Boston, Massachusetts

DETC2023-116595

DIFFUSING THE OPTIMAL TOPOLOGY: A GENERATIVE OPTIMIZATION APPROACH

Giorgio Giannone[1,2,*], Faez Ahmed[1]

[1]Massachusetts Institute of Technology, Cambridge, MA
[2]Technical University of Denmark, Lyngby, DK

ABSTRACT

Topology Optimization is a widely used method to find optimal designs that meet constraints and maximize system performance. However, traditional iterative optimization methods such as SIMP are computationally expensive and prone to getting stuck in local minima, limiting their effectiveness for complex or large-scale problems. To overcome these challenges, we propose a novel Generative Optimization method that integrates classic optimization (SIMP) as a refining mechanism for the topologies generated by a diffusion model. By introducing a computationally efficient approximation inspired by classic ODE solutions, the need for conditioning on physical fields is eliminated, reducing inference time by half. Our method facilitates the generation of feasible and high-performance topologies while explicitly guiding the process towards regions with superior manufacturability and performance without the need for external auxiliary models or additional labeled data. Notably, our approach overcomes drawbacks encountered by learning-based methods in out-of-distribution constraint configurations, including floating material and high compliance error, as well as the requirement for extensive pre-processing and surrogate models. Experimental results showcase a substantial 20% increase in performance. Overall, the integration of deep generative models, such as Diffusion models and Generative Adversarial Networks, to warm start optimization holds significant promise in advancing the design and optimization of structures in engineering applications, extending its applicability to a broader spectrum of performance-aware engineering design problems.

1. INTRODUCTION

Topology optimization (TO [1, 2]) is an essential engineering tool that finds the optimal material distribution for meeting performance objectives while satisfying constraints.

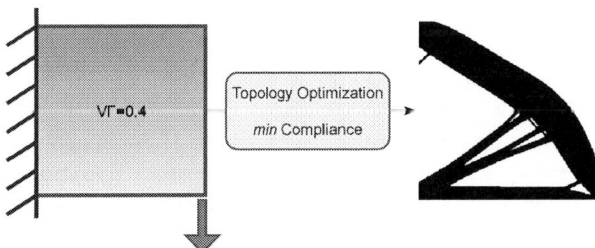

Figure 1: Minimum Compliance Topology Optimization problem. Given a set of constraints (loads, boundary conditions, volume fraction), find the optimal topology minimizing functional performance (structural compliance).

Constraints can take various forms, such as loads, boundary conditions, and volume fractions. Topology optimization can significantly improve the efficiency, safety, reliability, and durability of structures across various fields, including aerospace, automotive, civil, and mechanical engineering. The demand for improvements in topology optimization techniques is ever-growing to enhance design engineering for a new age of engineering discovery. Topology optimization uses Finite Element Analysis (FEA), with gradient-based [1] and gradient-free methods [3] being two primary approaches. The Solid Isotropic Material with Penalization (SIMP) method [4, 5] is one of the most widely used gradient-based optimization algorithms. SIMP and many variants utilize iterative optimization methods over continuous, density-based representation to determine the optimal material distribution. This class of methods iteratively adjusts the material density in a given domain to check whether the design meets the desired constraints.

Iterative optimization-based topology optimization (TO) methods have great potential benefits but face challenges in practical applications, particularly for large-scale problems due to their computational complexity. Iterative algorithms,

---

*Corresponding author: ggiorgio@mit.edu, faez@mit.edu

Copyright © 2023 by ASME

like SIMP, are computationally expensive and prone to getting trapped in local optima, limiting the search for a global optimum. To address these challenges, there has been a recent surge of interest in learning-based methods for topology optimization, which use machine learning (ML) algorithms, such as deep generative models, to speed up the optimization process and generate more diverse structural topologies. By leveraging large datasets of existing designs, ML models can quickly identify patterns and generate new designs that meet specified constraints. Unlike traditional optimization methods, ML methods can handle high-dimensional data, explore a broader range of design solutions, and provide more diverse results that may not have been considered previously. Companies and start-ups[1] are exploring the usage of deep generative topology optimization in their products, intending to add flexibility and creativity for the designers and speed up the design generation phase.

Deep Generative Models for TO. Deep Generative models (DGMs [6]) have shown an impressive ability to model high-dimensional and complex data modalities, such as images or text, with high fidelity and diversity. This capability opens the door for significant improvements in creativity and productivity in various fields in the coming years. Recent advancements in large vision models [7]) and language models [8]) have greatly increased our capacity to process unstructured data, leading to the development of multimodal generation techniques [9]. These models hold great promise for enhancing engineering design [10, 11]. Building on these successes, DGMs have also been applied in the engineering field to problems with constraints, with a focus on improving the design process. However, these applications have primarily focused on statistical metrics like reconstruction quality and often don't fully address the fulfillment of engineering requirements.

Applying Deep Generative Models (DGMs) for topology optimization introduces hurdles in securing high performance (e.g., low compliance) and ensuring manufacturability (e.g., no disconnected components). To help navigate these difficulties, a conditional GAN-based technique, called TopologyGAN [12], has been proposed. This method seeks to align the generated samples more closely with the fundamental engineering problem by conditioning on physics-based information, thereby integrating physical fields like the Von Mises stress, strain energy density, and displacement fields. However, the process of calculating these physical fields demands the execution of a Finite Element Analysis (FEA) routine, which can be time-consuming for each individual load and boundary condition configuration, despite the base model's (a conditional GAN) swift sampling of new topologies.

While TopologyGAN has been popular, it's worth mentioning that its optimization focus is on a pixel-wise, reconstruction-based loss. This could potentially overlook certain engineering requirements such as high performance and feasibility. Furthermore, even with the incorporation

[1]https://www.ntop.com/

of physical fields, some designs generated by these models may include floating materials, which can pose a challenge for manufacturability. Additionally, the diversity and generalizability of these designs when applied out-of-distribution might present limitations.

To address some of these challenges, researchers proposed a conditional diffusion model, named TopoDiff [13]. This model conditions on fields in a manner akin to [12] and incorporates an extra guidance mechanism to steer the generative process towards regions of high manufacturability and superior performance (specifically low compliance in this case). Although it demonstrated notable advancements in compliance error reduction and manufacturability, it was characterized by slow sampling, continued reliance on computationally demanding physical fields, and the introduction of surrogate models for performance and manufacturability considerations. During the inference phase, the model must determine the strain and force fields for each constraint configuration using Finite Element Analysis (FEA). This information is then used to condition the model, empowering it to generate optimized topologies that satisfy the given requirements. This conditioning step is essential in ensuring the model's performance and the generation of feasible and optimized outputs. However, it is also time-consuming and presents challenges when attempting to scale to more complex structures, higher dimensionality, and 3D domains.

To summarize, GAN-based methods such as those proposed in [12, 14] can generate a large number of topologies efficiently but may produce un-manufacturable topologies that violate soft constraints such as volume fraction errors and lead to higher compliance structures. On the other hand, diffusion-based approaches like TopoDiff [13] generate samples that satisfy constraints more accurately but are computationally expensive due to iterative sampling, reliance on physical fields, and surrogate models with auxiliary labeled data. Existing deep learning models also seem to overlook the wide array of existing optimization methods, and often aim to directly generate optimal designs.

To overcome the limitations of existing approaches, we propose a novel method that addresses the issues of slow sampling, reliance on physical fields, and the need for additional surrogate models. Our approach involves reducing the number of steps required for sampling, approximating physical fields using a computationally inexpensive kernel based on classic ODE solutions, and integrating optimization methods like SIMP for refining the generated topology. By explicitly guiding the generated topology to regions with high manufacturability and low compliance with only a few optimization steps, we can efficiently sample good topologies without needing external auxiliary models, FEM solvers for pre-processing, or additional labeled data.

Contribution. Our contributions are the following:

- We introduce a generative optimization technique that integrates a conditional diffusion model with traditional

Copyright © 2023 by ASME

topology optimization algorithms. This method employs the fast conditional diffusion model to predict an initial topology, which is then fine-tuned using established topology optimization strategies within a few steps (5-10 iterations). This methodology enhances manufacturability and boosts performance by 23.81% and 25.64%, respectively, for in- and out-of-distribution constraints.

- We introduce new conditioning techniques, inspired by computationally efficient approximation inspired by classic ODE solutions with Green's function, that eliminates the need to compute force and energy strain fields, which can be a major bottleneck in the optimization process. As a result, we have reduced the inference time for generation by 53.93% compared to the baselines.

- We decrease the computational cost of generative topology optimization while maintaining high performance. To achieve this, we explore more efficient sampling methods for TopoDiff reducing the number of steps required for generation by an order of magnitude while ensuring minimal loss in performance across all models.

2. RELATED WORK

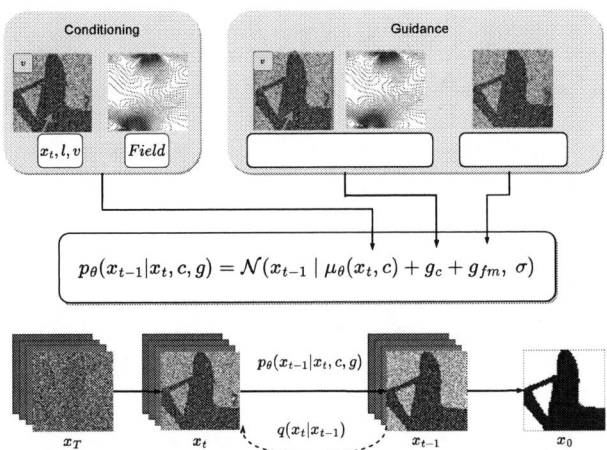

Figure 2: TopoDiff-GUIDED: TopoDiff is a conditional diffusion model guided by a classifier and a regression score. The conditioning mechanism c involves loads l, volume fraction v, and stress and energy fields f. i.e. $c = h(l, v, f)$. The guidance mechanism involves a classifier for the presence or absence of floating material and a regressor to minimize compliance error between generated and optimized topologies.

Topology Optimization. Engineering design is the process of creating solutions to technical problems under engineering requirements [15]. The goal is to create designs that are highly performant given the required constraints. Topology Optimization (TO [1]) is a branch of engineering design and is a critical component of the design process in many industries, including aerospace, automotive, manufacturing, and software development. From the inception of the homogenization method for TO, a number of different approaches have been proposed, including density-based [4, 5], level-set [16], derivative-based [17], evolutionary [18], and others [19]. The density-based methods are widely used and use a nodal-based representation where the level-set leverages shapes derivative to obtain the optimal topology. Topology Optimization has evolved as a computationally intensive discipline, with the availability of efficient open-source code [20, 21]. See [21] for more on this topic and [2] for a comprehensive review of the Topology Optimization field.

Deep Learning for Topology Optimization. Following the success of Deep Learning (DL) in vision, a surging interest arose recently for transferring these methods to the engineering field. In particular, DL methods have been employed for direct-design [22], accelating the optimization process [23], optimizing the shape [24], super-resolution [25, 26], and 3D topologies [27]. Among these methods, Deep Generative Models (DGMs) are especially appealing to improve design diversity in engineering design[28, 29]. Additionally, DGMs have been used for Topology Optimization problems conditioning on constraints (loads, boundary conditions, volume fraction for the structural case), directly generating topologies [30, 31] training dataset of optimized topologies, leveraging superresolution methods to improve fidelity [32], using filtering and iterative design approaches [33] to improve quality and diversity. Methods for 3D topologies have also been proposed [14]. Recently, GAN-based approaches conditioning on constraints and physical information have had success in modeling the TO problem [12]. For a comprehensive review and critique of the field, see [34].

Conditional Diffusion Models. Recently, TopoDiff [13] introduced the idea of using conditional diffusion models for effectively generating topologies that fulfill the constraints and have high manufacturability and high performance. TopoDiff relies on physics information and surrogate models to guide the sampling of novel topologies with good performance. However, it is slow in inference. Improving sampling speed for diffusion models is an active research topic [35]. Recently, distillation has been used to reduce sampling steps [36] significantly. Methods to condition DDPM have been proposed, conditioning at inference time [37], learning a class-conditional score [38], explicitly conditioning on class information [39], set-based features [40], and physical properties [41]. Text-to-image diffusion models [42] have been recently proposed for a guided generation. These approaches have demonstrated that diffusion models can significantly reduce inference time.

3. BACKGROUND

Here we briefly introduce the Topology Optimization problem [1], diffusion models [43], a class of deep generative models, conditioning and guidance mechanisms for diffusion

Copyright © 2023 by ASME

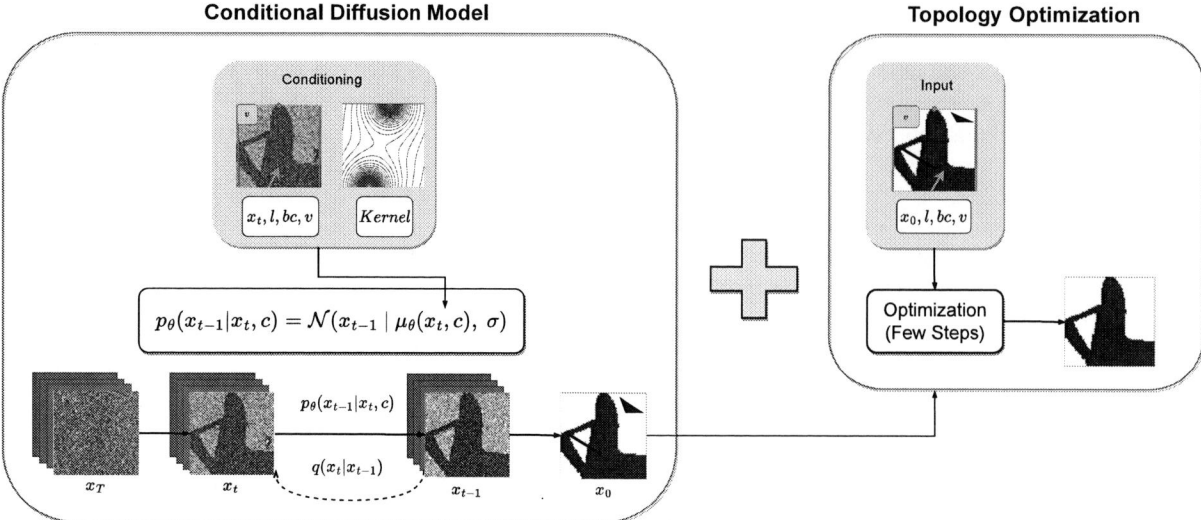

Figure 3: TopoDiff-FF + SIMP. The TopoDiff-FF pipeline is a conditional diffusion model where we condition on a cheap kernel relaxation instead of using expensive FEA to obtain the stress and energy field like in TopoDiff. After the generation step, we can improve the generated topology using few steps of SIMP (5/10 iterations) to remove floating material and explicitly optimize performance (minimize structural compliance).

models, and recent deep generative models for topology optimization [12, 13].

The Topology Optimization Problem. Topology optimization (TO) is a powerful computational design approach used to determine the optimal configuration of a given structure given a set of constraints. The goal of structural topology optimization is to identify the most efficient use of material while ensuring that the structure satisfies specific performance requirements. One common approach to topology optimization is the SIMP (Solid Isotropic Material with Penalization) method [4]. The SIMP method involves modeling the material properties using a density field, where the density represents the proportion of the material present in a given region. The optimization process involves adjusting the density field iteratively, subject to various constraints such as stress and boundary conditions. The objective of a generic minimum compliance problem for a mechanical system is to find the material density distribution $\mathbf{x} \in \mathbb{R}^n$ that minimizes the structure's deformation under the prescribed boundary and loads condition [2, 21]. Given a set of design variables $\mathbf{x} = \{\mathbf{x}_i\}_{i=0}^n$, where n is the domain dimensionality (in our case $n = 64*64$), the minimum compliance problems (Fig. 1) can be written as:

$$\begin{aligned} \min_{\mathbf{x}} \quad & c(\mathbf{x}) = F^T U(\mathbf{x}) \\ \text{s.t.} \quad & v(\mathbf{x}) = v^T \mathbf{x} < \bar{v} \qquad (1) \\ & 0 \le \mathbf{x} \le 1 \end{aligned}$$

where the goal is to find the design variables that minimize compliance $c(\mathbf{x})$ given the constraints. F is the tensor of applied loads and $U(\mathbf{x})$ is the node displacement, solution of the equilibrium equation $K(\mathbf{x})U(\mathbf{x}) = F$ where $K(\mathbf{x})$ is the stiffness matrix and is a function of the considered material. $v(\mathbf{x})$

is the required volume fraction. The problem is a relaxation of the topology optimization task, where the design variables are continuous between 0 and 1. Thresholding mechanisms are employed to assign the presence or absence of material in each node in the design domain. One significant advantage of topology optimization is its ability to create optimized structures that meet specific performance requirements. However, a major drawback of topology optimization is that it can be computationally intensive and may require significant computational resources. Furthermore, topology optimization typically initiates its search process from scratch, disregarding data on previously known optimal structures. Additionally, some approaches to topology optimization may be limited in their ability to generate highly complex geometries and get stuck in local minima.

Diffusion Models. Below, we describe the preliminaries of diffusion models, which are a type of deep generative models. Let $\mathbf{x}$ denote the observed data which is either continuous $\mathbf{x} \in \mathbb{R}^D$ or discrete $\mathbf{x} \in \{0, ..., 255\}^D$. Let $\mathbf{z}_1, ..., \mathbf{z}_T$ denote T latent variables in $\mathbb{R}^D$. We now introduce, the *forward or diffusion process* q, the *reverse or generative process* p_θ, and the objective L. The forward or diffusion process q is defined as [44]:

$$q(\mathbf{z}_{1:T}|\mathbf{x}) = q(\mathbf{z}_1|\mathbf{x}) \prod_{t=2}^{T} q(\mathbf{z}_t|\mathbf{z}_{t-1}), \qquad (2)$$

$$q(\mathbf{z}_t|\mathbf{z}_{t-1}) = \mathcal{N}(\mathbf{z}_t|\sqrt{1-\beta_t}\,\mathbf{z}_{t-1}, \beta_t I) \qquad (3)$$

The beta schedule $\beta_1, \beta_2, ..., \beta_T$ is chosen such that the final latent image $\mathbf{z}_T$ is nearly Gaussian noise. The generative or

Copyright © 2023 by ASME

	Load	BC	Kernel Load	Kernel BC	Force Field	Energy Field	VF	Performance
SIMP [1]	✓	✓	✗	✗	✗	✗	✓	✓
TopologyGAN [12]	✓	✓	✗	✗	✓	✓	✓	✗
TopoDiff [13]	✓	✓	✗	✗	✓	✓	✓	✗
TopoDiff-GUIDED [13]	✓	✓	✗	✗	✓	✓	✓	✓
TopoDiff-FF (ours)	✓	✓	✓	✓	✗	✗	✓	✗
TopoDiff-FF + SIMP (ours)	✓	✓	✓	✓	✗	✗	✓	✓

Table 1: Conditioning and guiding variables for different optimization methods and model configurations. All models condition directly or indirectly on loads, boundary conditions, and volume fraction. TopologyGAN, TopoDiff, and TopoDiff-GUIDED have additional conditioning on stress and energy fields, where TopoDiff-FF conditions are on the kernels. TopoDiff-GUIDED, TopoDiff-FF + SIMP, and SIMP itself are guided by a measure of performance.

inverse process p_θ is defined as:

$$p_\theta(\mathbf{x}, \mathbf{z}_{1:T}) = p_\theta(\mathbf{x}|\mathbf{z}_1)p(\mathbf{z}_T) \prod_{t=2}^{T} p_\theta(\mathbf{z}_{t-1}|\mathbf{z}_t), \quad (4)$$

$$p_\theta(\mathbf{z}_{t-1}|\mathbf{z}_t) = \mathcal{N}(\mathbf{z}_{t-1}|\mu_\theta(\mathbf{z}_t, t), \sigma_t^2 I), \quad (5)$$

where $p(\mathbf{z}_T) = \mathcal{N}(\mathbf{z}_T|0, I)$, and σ_t^2 often is fixed (e.g. to $\sigma_t^2 = \beta_t$). The neural network $\mu_\theta(\mathbf{z}_t, t)$ is shared among all time steps and is conditioned on t. The model is trained with a re-weighted version of the ELBO that relates to denoising score matching [45]. The negative ELBO L can be written as

$$\mathbb{E}_q \left[-\log \frac{p_\theta(\mathbf{x}, \mathbf{z}_{1:T})}{q(\mathbf{z}_{1:T}|\mathbf{x})} \right] = L_0 + \sum_{t=2}^{T} L_{t-1} + L_T, \quad (6)$$

where $L_0 = \mathbb{E}_{q(\mathbf{z}_1|\mathbf{x})} [-\log p(\mathbf{x}|\mathbf{z}_1)]$ is the likelihood term (parameterized by a discretized Gaussian distribution) and, if $\beta_1, ...\beta_T$ are fixed, $L_T = \mathbb{KL}[q(\mathbf{z}_T|\mathbf{x}), p(\mathbf{z}_T)]$ is a constant. The terms L_{t-1} for $t = 2, ..., T$ can be written as:

$$L_{t-1} = \mathbb{E}_{q(\mathbf{z}_t|\mathbf{x})} \left[\mathbb{KL}[q(\mathbf{z}_{t-1}|\mathbf{z}_t, \mathbf{x}) \mid p(\mathbf{z}_{t-1}|\mathbf{z}_t)] \right]$$

$$= \mathbb{E}_{q(\mathbf{z}_t|\mathbf{x})} \left[\frac{1}{2\sigma_t^2} \|\mu_\theta(\mathbf{z}_t, t) - \tilde{\mu}(\mathbf{z}_t, \mathbf{x})\|_2^2 \right] + C_t, \quad (7)$$

where $C_t = \frac{D}{2} \left[\frac{\tilde{\beta}_t}{\sigma_t^2} - 1 + \log \frac{\sigma_t^2}{\tilde{\beta}_t} \right]$. By further applying the reparameterization trick [46], the terms $L_{1:T-1}$ can be rewritten as a prediction of the noise ϵ added to $\mathbf{x}$ in $q(\mathbf{z}_t|\mathbf{x})$. Parameterizing μ_θ using the noise prediction ϵ_θ, we can write

$$L_{t-1,\epsilon}(\mathbf{x}) = \mathbb{E}_{q(\epsilon)} \left[w_t \|\epsilon_\theta(\mathbf{z}_t(\mathbf{x}, \epsilon)) - \epsilon\|_2^2 \right], \quad (8)$$

where $w_t = \frac{\beta_t^2}{2\sigma_t^2 \alpha_t (1-\bar{\alpha}_t)}$, which corresponds to the ELBO objective [47].

Conditioning. Similarly, introducing a conditioning variable $\mathbf{c}$, conditional modeling can be written as

$$p_\theta(\mathbf{x}, \mathbf{z}_{1:T}|\mathbf{c}) = p_\theta(\mathbf{x}|\mathbf{z}_1, \mathbf{c})p_\theta(\mathbf{z}_T|\mathbf{c}) \prod_{t=2}^{T} p_\theta(\mathbf{z}_{t-1}|\mathbf{z}_t, \mathbf{c}), \quad (9)$$

and we can similarly write the per-layer loss with noise prediction as:

$$L_{t-1,\epsilon}^c(\mathbf{x}) = \mathbb{E}_{q(\epsilon)} \left[w_t \|\epsilon_\theta(\mathbf{z}_t(\mathbf{x}, \epsilon), \mathbf{c}) - \epsilon\|_2^2 \right]. \quad (10)$$

We will now discuss different ways to condition the model.

Guidance. Guidance is used to improve sample fidelity vs. mode coverage in conditional diffusion models post-training [48, 49]. Classifier guidance [48] is based on Bayesian inversion and the fact that inverse problems can be tackled relatively easily in diffusion models. In particular, the goal of the guidance is to shift the model mean μ_θ closer to the target guidance $\mathbf{c}$. The general idea in classifier guidance is to leverage Bayes inversion

$$p(\mathbf{z}|\mathbf{g}) \propto p(\mathbf{g}|\mathbf{z})p(\mathbf{z}), \quad (11)$$

and notice that training the conditional model $p(\mathbf{z}|\mathbf{g})$ is equivalent to training an unconditional model $p(\mathbf{z})$ and a classifier $p(\mathbf{g}|\mathbf{z})$. Then the training signal will be a composition of unconditional and conditional scores for the model. Taking the gradients, we can see that classifier guidance is shifting the unconditional score for each class [48]. We can leverage similar ideas for feasible vs. unfeasible designs. With similar reasoning, regression guidance introduced in [13], aims to guide the sampling process towards configurations with low regression error as the target. In particular, the regression target for generating topologies is low compliance error between predicted (with a regression model) and real (obtained running a FEM solver) compliance. Guidance is useful for reducing high-compliance configurations generated by the model.

TopoDiff. Conditional diffusion models have been adapted for constrained engineering problems with performance requirements. TopoDiff [13] proposes to condition on loads, volume fraction, and fields similarly to [12] to learn a constrained generative model (Fig. 2). In particular, the generative model can be written as:

$$p_\theta(\mathbf{x}_{t-1}|\mathbf{x}_t, \mathbf{c}, \mathbf{g}) = \mathcal{N}(\mathbf{x}_{t-1} \mid \mu_\theta(\mathbf{x}_t, \mathbf{c}) + \mathbf{g}_c + \mathbf{g}_{fm}, \sigma), \quad (12)$$

where $\mathbf{c}$ is a conditioning term and is a function of the loads l, volume fraction v, and fields f, i.e $\mathbf{c} = h(l, v, f)$. The fields considered are the Von Mises stress $\sigma_{vm} = (\sigma_{11}^2 - \sigma_{11}\sigma_{22} + \sigma_{22}^2 + 3\sigma_{12}^2)^{1/2}$ and the strain energy density field $W = (\sigma_{11}\epsilon_{11} + \sigma_{22}\epsilon_{22} + 2\sigma_{12}\epsilon_{12})/2$. Here σ_{ij} and ϵ_{ij} are the stress and energy components over the domain. $\mathbf{g}$ is a guidance term, containing information to guide the sampling process toward regions with low floating material (using a classifier and $\mathbf{g}_{fm}$) and regions with low compliance error, where the generated topologies are close to optimized one

Copyright © 2023 by ASME

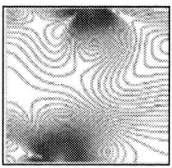 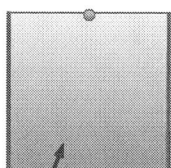 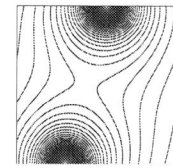

Figure 4: Field (left) vs Kernel (right) conditioning. Where computing the fields requires solving an expensive iterative FEA problem for each new configuration, computing the kernel relaxation is a single-step, computationally inexpensive approximation that does not rely on domain knowledge and scales to any resolution or domain structure.

Class	Metrics	Goal	Challenging?
Hard-constraint	Loads Disrespect	Feasibility	No
Hard-constraint	Floating Material	Manufacturability	Yes
Soft-constraint	Volume Fraction	Min Cost	No
Functional Performance	Compliance Error	Max Performance	Yes
Modeling Requirements	Sampling Time	Fast Inference	Yes

Table 2: Design and Modelling requirements for a constrained generative model for topology optimization. Our goal is to improve the requirements that are challenging to fulfill. In this work, we focus on improving Floating Material, reducing Compliance Error, and reducing Sampling Time.

(using a regression model and $\mathbf{g}_c$). Where conditioning $\mathbf{c}$ is always present and applied during training, the guidance mechanism $\mathbf{g}$ is optional and applied only at inference time.

Limitations. TopoDiff is an effective method for generating topologies that satisfy constraints and have low compliance errors. However, the generative model is computationally expensive as hundreds of layers need to be sampled for each sample. Furthermore, the model requires preprocessing of configurations using a FEM solver, which is also computationally expensive and time-consuming. This approach relies heavily on fine-grained knowledge of the domain input, limiting its applicability to more challenging topology problems. Additionally, the model requires training of two surrogate models (a classification and a regression model), which can be helpful for out-of-distribution configurations. However, additional optimal and suboptimal topologies are needed to train the regression model, assuming access to the desired performance metric on the train set. Similarly, the classifier requires additional labeled data to be gathered. This work attempts to address these limitations of TopoDiff.

4. METHOD

Our primary goals encompass three key areas:

1. Enhancing inference efficiency and reducing the sampling time of diffusion-based topology generation, while ensuring the fulfillment of design requirements with minimal compromise on performance.

2. Reducing dependency on force and strain fields as conditioning information, thereby diminishing the computational load during inference and eliminating the need

for ad-hoc conditioning mechanisms for each problem and domain.

3. Integrating learning-based and optimization-based methods, with the aim of refining the topology generated using a conditional diffusion model, thus enhancing the final solution in terms of manufacturability and performance.

Our overarching objective is to advance generative models with constraints and eradicate the need for expensive FEM solutions as conditioning mechanisms. We call our approach TopoDiff-FF, an acronym for Topology-Diffusion-Fields-Free, and when coupled with optimization-based refinement, we refer to our method as TopoDiff-FF + SIMP.

Conditioning on Hard and Soft Constraints. All models are subject to conditioning based on loads, boundary conditions, and volume fractions. In addition, TopoDiff and TopoDiff-GUIDED undergo conditioning based on force field and energy strain, while TopoDiff-FF is conditioned based on kernel relaxation, which is discussed in the following paragraph (see Fig. 4).

Green's Functions. To improve the efficiency of diffusion-based topology generation and minimize reliance on force and strain fields, we aim to relax boundary conditions and loads by leveraging kernels as approximations for the way such constraints act on the domain. One possible choice of kernel structure is inspired by Green's method [50], which defines integral functions that are solutions to the time-invariant Poisson's Equation [51], a generalization of Laplace's Equation for point sources excitations. Poisson's Equation can be written as $\nabla_x^2 f(x) = h$, where h is a forcing term and f is a generic function defined over the domain $\mathfrak{X}$. This equation governs many phenomena in nature, and a special case is a forcing part $h = 0$, which yields the Laplace's Equation formulation commonly employed in heat transfer problems. Green's method is a mathematical construction to solve partial differential equations without prior knowledge of the domain. The solutions obtained with this method are known as Green's functions [52]. While solutions obtained with this method can be complex in general, for a large class of physical problems involving constraints and forces that can be approximated with points, a simple functional form can be derived by leveraging the idea of source and sink. Consider a laminar domain (e.g., a beam or a plate) constrained in a feasible way. If a point source is applied to this domain (e.g., a downward force on the edge of a beam or on the center of a plate) in x_f, such force can be described using the Dirac delta function, $\delta(x - x_f)$. The delta function is highly discontinuous but has powerful integration properties. In particular $\int f(x)\delta(x - x_f)dx = f(x_f)$ over the domain $\mathfrak{X}$. The solution of the time-invariant Poisson's Equation with point concentrated forces can be written as a Green's function solution, where the solution depends only on the distance from the force application point. In particular:

Copyright © 2023 by ASME

$$\mathcal{G}(x, x') = -\frac{1}{4\pi} \frac{1}{|x - x'|}, \qquad (13)$$

where $r = |x - x'| = \sqrt{|x_i - x_i'|^2 + |x_j - x_j'|^2}$. We propose to approximate the forces and loads applied to our topologies using a kernel relaxation built using Green's functions. While this formulation may not provide a correct solution for generic loads and boundary conditions, it allows us to provide computationally inexpensive conditioning information that respects the original physical and engineering constraints. By leveraging these ideas, we aim to increase the amount of information provided to condition the model, ultimately improving generative models with constraints.

Kernel Relaxation. We can use these kernels to construct a kernel relaxation method to condition generative models (Fig. 4). The idea is to use the kernels as approximations of the way boundary conditions and loads act on the domain. Specifically, we can use the kernels to represent the effects of the boundary conditions and loads as smooth functions across the domain. This approach avoids the need for computationally expensive and time-consuming finite element Analysis (FEA) to provide conditioning information. To apply the kernel relaxation method, we first determine the locations of the sources and sinks in the domain. Then, we compute the corresponding kernels for the loads and boundary conditions, respectively. Finally, we use the resulting kernels as conditioning information for generative models. In particular, we consider loads as sources and boundary conditions as sinks. For a load or source p, and $r = |x - x^P| = \sqrt{|x_i - x_i^P|^2 + |x_j - x_j^P|^2}$ we have:

$$K_l(x, x^P; \alpha, \beta) \propto (1 - e^{-\alpha/r^\beta}) \, \bar{p}, \qquad (14)$$

where $\bar{p}$ is the module of a generic force in 2D. Notice how, for $r \to 0$, $K_l(x, x^P) \to p$, and $r \to \infty$, $K_l(x, x^P) \to 0$. For a boundary condition or sink:

$$K_{bc}(x, x^{bc}; \alpha, \beta) \propto e^{-\alpha/r^\beta}. \qquad (15)$$

We notice how closer to the boundary the kernel is null, and farther from the boundary the kernel tends to 1. Note that the choice of α and β parameters in the kernels affect the smoothness and range of the kernel functions. By adjusting these parameters, we can control the trade-off between accuracy and computational cost. Furthermore, one limitation of these kernels is that they are isotropic, meaning that they do not depend on the direction in which they are applied. Overall, the kernel relaxation method offers a computationally inexpensive way to condition generative models on boundary conditions and loads, making them more applicable in practical engineering and design contexts.

5. EXPERIMENTS

Setup. We train all the models for 200k steps on 30000 optimized topologies on a 64x64 domain. We set the hyperparameters, conditioning structure, and training routine as

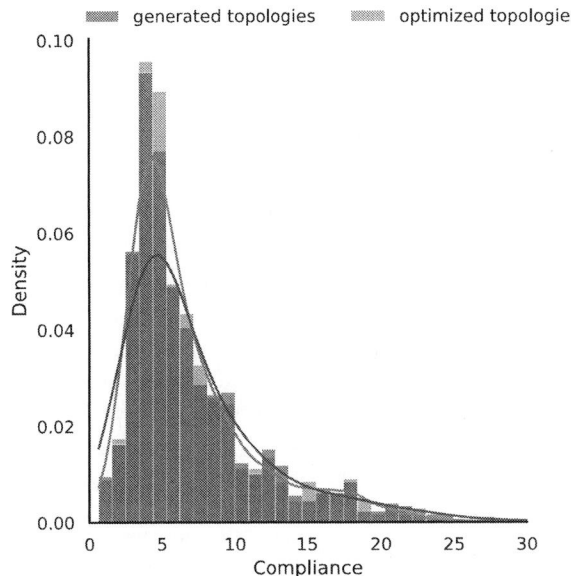

Figure 5: Histogram empirical distribution generated and optimized compliances for task-2 (unknown constraints). We see that the generated topologies match well the distribution of compliance for unseen constraint configurations.

proposed in [13]. For all the models we condition on volume fraction and loads. For TopoDiff, we condition on additional stress and energy fields. For TopoDiff-FF we condition on boundary conditions and kernels. TopoDiff-GUIDED leverages a compliance regressor and floating material classifier guidance (CLS+REG), as proposed in [13]. We use a reduced number of sampling steps for all the experiments as we found that it does not significantly impact compliance error. See Table 2 for an overview of conditioning and guidance variables used for different models and methods considered in this work.

Dataset. We use the dataset of optimized topologies gathered using SIMP proposed in [13]. For each topology, we have

Cut-off	10	25	50	100
$F(x)$	0.9871	0.9963	0.9985	0.9990

Table 3: Cumulative Density Function for compliance cut-off on the train set. We can see that more than 99.90% of the training set has compliance lower than 100.

information about the loading condition, boundary condition, volume fraction, and optimal compliance. Additionally, for each configuration, a pre-processing step computes the force and strain energy fields (see Fig. 4).

Evaluation. We employ both engineering and generative metrics to evaluate the effectiveness of our model. Our metrics encompass physical, engineering, and modeling requirements to determine the model's overall performance. We

Copyright © 2023 by ASME

measure the Volume Fraction Error (VFE), which calculates the deviation between the generated topology volume and the prescribed volume. We also use the presence of Floating Material (FM) to assess the manufacturability of the model. For performance, we use the structural compliance metric [2], which measures the overall displacement under specified constraints. Minimizing compliance is one of the primary objectives of topology optimization. We calculate the average and median compliance (Avg C, Mdn C) for all tasks. We also use the Compliance Error (CE) to evaluate the engineering performance, which determines the deviation between the compliance calculated by the SIMP optimization method and the compliance generated by the diffusion model. Additionally, we analyze the inference time, which includes preprocessing and sampling time, to determine the model's speed in creating fast concept designs. To assess the performance of the generative models, we design three test tasks: task-1, task-2, and task-3. We divide the test configurations into various groups based on their constraints and expected optimal performance. Table 4 presents these tasks. We notice that all the models have null load disrespect (LD), showing that fulfilling this hard constraint is easy for deep generative models.

- Task-1: the constraints in the task-1 test set are identical to those in the training set. To remove outliers, we filter out generated configurations with extremely high compliance when measuring performance on this task.

- Task-2: the constraints in the task-2 test set differ from those in the training set. To remove outliers, we filter out generated configurations with extremely high compliance when measuring performance on this task.

- Task-3: the constraints in the task-3 test set are the same as those in task-2. We consider all the generated configurations with low and high compliance when measuring performance on this task.

The purpose of these tasks is to evaluate the generalization capability of the machine learning models in- and out-of-distribution. By testing the models on different test sets with varying levels of difficulty, we can assess how well the models can perform on new, unseen data. More importantly, we want to understand how important the role of the force field and energy strain is with unknown constraints.

task	constraints	performance	avg C	mdn C
task-1	in-distro	in-distro	7.73	5.54
task-2	out-distro	in-distro	6.84	5.29
task-3	out-distro	out-distro	6.85	5.31

Table 4: Tasks in order of challenge for the model. We expect the models to perform well on task-1, acceptably on task-2, and fail on task-3. We also report the average and median compliance over the different sets as a measure of optimality for the generated topologies.

5.1 Performance with In-Distribution Constraints

Table 5 presents an evaluation of different model variants on in-distribution constraints for Task 1, with various measures of compliance, compliance error, volume fraction error, and floating material. Lower is better for all the metrics. The models evaluated include TopoDiff, TopoDiff-GUIDED, TopoDiff-FF, and TopoDiff-FF + SIMP (10), with varying types of constraints and guidance. TopoDiff and TopoDiff-GUIDED condition of stress and energy fields, with TopoDiff-GUIDED leveraging surrogate models for guidance (classifier and regressor); TopoDiff-FF conditions on kernel relaxation. The table shows that TopoDiff and TopoDiff-GUIDED have lower values for most measures of compliance, compliance error, volume fraction error, and floating material compared to TopoDiff-FF. The TopoDiff-FF model has higher compliance error as expected but comparable VFE and FM, indicating that a lightweight kernel relaxation is a viable conditioning option. Additionally, TopoDiff-FF + SIMP (10) has the lowest values for compliance error, volume fraction error, and floating material. The table also reports the time taken for sampling, pre-processing, and inference for each model. TopoDiff-FF, even reducing the number of sampling steps for TopoDiff to 100, has significantly lower times for pre-processing and overall inference, because of the absence of FEA pre-processing. Overall, the results suggest that the TopoDiff and TopoDiff-GUIDED models perform better than TopoDiff-FF on in-distribution constraints, with TopoDiff-FF (and TopoDiff-FF+SIMP) being much faster at sampling and a viable data-driven alternative when FEA is impractical or computationally unfeasible. The results for TopoDiff-FF+SIMP suggest that using a combination of deep generative models and a simplified iterative method (SIMP) can lead to better results in terms of compliance and design requirements, even if the gap with TopoDiff is not so relevant when dealing with the in-distribution scenario.

5.2 Performance with Out-of-Distribution Constraints

Table 6 reports results on task-2 with out-of-distribution constraints. The considered models, conditioning, and guidance mechanisms are the same as explained in Subsec. 5.1. The results show that in out-of-distribution scenarios, TopoDiff-GUIDED model performs better than the TopoDiff, achieving a lower CE and VFE. TopoDiff-FF performs well in terms of VFE and FM, but poorly in terms of compliance, with high CE. However, the TopoDiff-FF + SIMP (10) model significantly improves the performance on Task-2, achieving the lowest CE, VFE, and FM values among all models. As for task-1, the inference time for TopoDiff-FF and TopoDiff-FF+ SIMP is much better than for the full model. Overall, the table shows that the use of guidance and optimization techniques, such as SIMP, can significantly improve the performance of topology optimization models, especially on out-of-distribution constraints. See Fig 9 for a visual overview with confidence intervals over five runs.

Copyright © 2023 by ASME

	Task	Steps	Constraints	Guidance	Avg C ↓	Mdn C ↓	% CE ↓	% VFE ↓	% FM ↓	Sampling ↓	Processing ↓	Inference ↓
TopoDiff	task-1	100	FIELD	COND	8.01	5.60	5.46	1.47	5.79	2.23	3.31	5.54
TopoDiff-GUIDED	task-1	100	FIELD	CLS+REG	8.15	5.63	5.93	1.49	5.82	2.46	3.31	5.87
TopoDiff-FF	task-1	100	KERNEL	COND	9.76	6.31	24.90	2.05	8.15	2.23	0.12	**2.35**
TopoDiff-FF + SIMP (10)	task-1	100	KERNEL	COND	8.05	5.77	**4.16**	**1.16**	**5.61**	2.23	0.32	2.55

Table 5: Evaluation of different model variants on in-distribution constraints. We remove samples with compliance >100. C: compliance. CE: compliance error. VFE: volume fraction error. FM: floating material. We use 100 sampling steps for all models. SIMP (10) means that we run SIMP for 10 iterations. Results averaged over 5 runs. For a general overview with confidence intervals see the left side of Fig. 9.

	Task	Steps	Constraints	Guidance	Avg C ↓	Mdn C ↓	% CE ↓	% VFE ↓	% FM ↓	Sampling ↓	Processing ↓	Inference ↓
TopoDiff	task-2	100	FIELD	COND	7.80	5.49	12.02	1.49	6.65	2.23	3.31	5.54
TopoDiff-GUIDED	task-2	100	FIELD	CLS+REG	7.80	5.57	10.55	1.47	7.39	2.46	3.31	5.87
TopoDiff-FF	task-2	100	KERNEL	COND	11.65	6.94	58.36	1.97	7.86	2.23	0.12	**2.35**
TopoDiff-FF + SIMP (10)	task-2	100	KERNEL	COND	7.65	5.70	**7.84**	**1.29**	6.53	2.23	0.32	2.55

Table 6: Evaluation of different model variants on out-of-distribution constraints. We remove samples with compliance >100. C: compliance. CE: compliance error. VFE: volume fraction error. FM: floating material. We use 100 sampling steps for all models. SIMP (10) means that we run SIMP for 10 iterations. Results averaged over 5 runs. For a general overview with confidence intervals see the right side of Fig. 9.

5.3 Performance with Out-of-Distribution Constraints and Outliers

Table 7 presents results on task-3 with out-of-distribution and without filtering out outliers, i.e configurations with high-compliance (low performance). High-compliance configurations (>50/100) are out-of-distribution performance for the training set compliance distribution. In particular in Table 3 where we show that more than 99.8% of the configurations have compliance lower than 50. This means that when we obtain generated samples with such high compliance, the model has partially failed to fulfill the design constraints. The considered models, conditioning, and guidance mechanisms are the same as explained in Subsec. 5.1. The results in this table are similar to task-2, where TopoDiff-GUIDED performs better than TopoDiff in terms of CE and TopoDiff-FF achieves high compliance. However, using TopoDiff-FF+SIMP we not only improve all the metrics, but we completely get rid of such outlier configurations, being easy for the optimization process to fix the generated high-compliance configurations in a few iterations. Contrarily, for all other methods, there are always outliers in terms of topologies with high compliance that cannot easily be fixed, making a strong case for unifying deep generative models and optimization for topology optimization in challenging scenarios.

5.4 Compliance Analysis

Here we discuss and study how compliance influences the generated topology quality. Compliance magnitude seems to be the most important factor for the generation quality and manufacturability of proposed designs. In Fig. 5 we show qualitative information, providing empirical compliance distribution for generated vs optimized topologies. In Fig. 7 and Fig. 6 The growth of compliance error with compliance indicates that the model struggles to generate high-compliance topologies that fulfill the design requirements. This is further supported by the fact that very few configurations have high compliance, as indicated by the dataset. These findings highlight the difficulty of modeling compliant structures with high accuracy and precision and suggest that additional techniques or modifications to the model may be necessary to address this challenge. Moreover, the limited ability of the model to generate high-compliance topologies could have important implications for real-world applications, where compliant structures are often critical components of engineering designs. Therefore, improving the model's ability to generate high-compliance topologies is an important research direction for advancing the state-of-the-art in this field.

High-Compliance Configurations. High-compliance configurations are often associated with mechanically unstable structures, meaning that they can collapse or deform easily under load. As a result, practical engineering designs often prioritize achieving low-compliance solutions that are stable, efficient, and lightweight. However, it is still important to study high-compliance configurations as they can provide valuable insights into the behavior and limitations of structural systems, as well as inspire new design concepts and approaches. We show some examples of high-compliance and high-compliance errors in Fig. 10. For low-compliance and low-compliance errors, see Fig. 12.

5.5 Inference Time

Where in [13] has been shown that TopoDiff-based models largely outperform GAN-based models on design performances (CE, VFE, FM) for in- and out-distribution scenarios, inference time for GAN-based model is still much better than for Diffusion-based models. Here we present our approach to improving the efficiency of inference for TopoDiff-FF, measured in terms of both pre-processing and sampling time. Specifically, we focus on two key areas of optimization: I) Reducing the number of steps required for sampling to improve overall sampling time. II) Leveraging the kernel relaxation

Copyright © 2023 by ASME

	Task	Steps	Constraints	Guidance	Avg C ↓	Mdn C ↓	% CE ↓	% VFE ↓	% FM ↓	Sampling ↓	Processing ↓	Inference ↓
TopoDiff	task-3	100	FIELD	COND	8.91	5.52	31.04	1.49	<u>6.61</u>	2.23	3.31	5.54
TopoDiff-GUIDED	task-3	100	FIELD	CLS+REG	8.23	5.58	<u>17.69</u>	<u>1.47</u>	7.36	2.46	3.31	5.87
TopoDiff-FF	task-3	100	KERNEL	COND	19.01	7.22	144.23	2.00	8.70	2.23	0.12	**2.35**
TopoDiff-FF + SIMP (10)	task-3	100	KERNEL	COND	7.65	5.70	**7.84**	**1.29**	6.53	2.23	0.32	<u>2.55</u>

Table 7: Evaluation of different model variants on out-of-distribution constraints. We do not filter out samples with high compliance. C: compliance. CE: compliance error. VFE: volume fraction error. FM: floating material. We use 100 sampling steps for all models. SIMP (10) means that we run SIMP for 10 iterations. Results averaged over 5 runs.

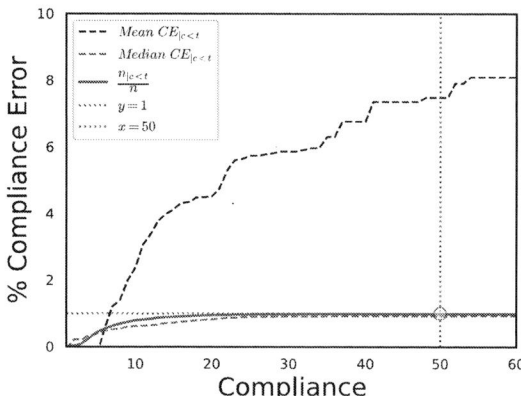

Figure 6: Mean and Median Compliance error for different compliance thresholds. We see that more than 99.8 % of the data has compliance lower than 50 and how to mean compliance error increases with compliance, where the median compliance error plateau.

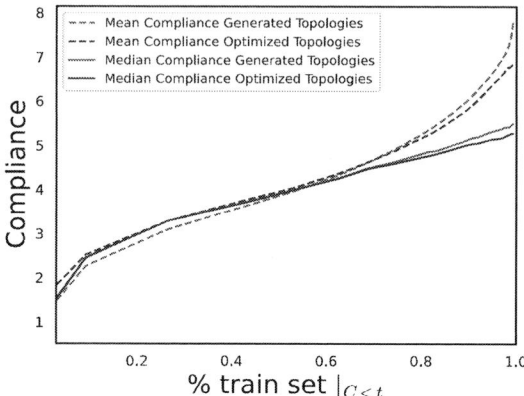

Figure 7: Mean and Median Compliance vs % train set for task-2 (unknown) constraints. The generated topologies using TopoDiff perform well for low compliance values. Most of the loss in terms of compliance happens for high compliance values.

technique to reduce pre-processing time by avoiding the computation of force and energy fields. To demonstrate the effectiveness of our approach, we compare the pre-processing and sampling times of TopologyGAN, TopoDiff, and TopoDiff-FF in Fig. 8. Our results show that TopoDiff-FF is able to generate high-quality samples that satisfy prescribed constraints and exhibit strong performance, all while requiring significantly less time for both pre-processing and sampling compared to TopoDiff and slightly less time than TopologyGAN. This improved efficiency is a key goal of TopoDiff-FF, as it simplifies the conditioning process and enables faster inference.

5.6 Measuring the Relative Gap in Performance

To better summarize the results on in-distribution and out-of-distribution constraints, we propose to evaluate the models in terms of the relative design gap between the models, using a global metric that accounts for all the requirements presented in Table 2. We also consider relative ranking (Avg Rank) between the considered models. The gap and the rank are global metrics that account for compliance, compliance error, volume fraction error, floating material, processing, and inference time. For the average gap (Avg Gap), we use average compliance and compliance error. For the median gap (Mdn Gap), the corresponding median values are used. The design gap is a proxy to quantify the distance from an optimal design, where all the requirements are perfectly satisfied,

we have compliance error null, and inference time is negligible. Table 2 presents a comparison of different conditioning configurations for generative designs in terms of their performance on task-1 and task-2. The table also shows whether the configuration includes kernels and field constraints. Overall, the models perform similarly in terms of the average and median gap on task-1, with TopoDiff-FF+SIMP (10) performing the best overall. On task-2, TopoDiff-FF loses in terms of the average gap because of the large CE on out-of-distribution. Interestingly, when considering average ranking, the gap is reduced because of the fast processing and inference time for TopoDiff-FF. As noted with the previous experiments, TopoDiff-FF with SIMP outperforms all other models in terms of both design quality and relative performance on both tasks, which suggests that the use of SIMP in conjunction with deep generative models is a promising approach for generating high-quality designs for topology optimization.

5.7 Merging Deep Generative Models and Optimization

Table 9 shows the results of experiments conducted with three algorithms, SIMP, TopoDiff-FF, and TopoDiff-FF + SIMP, and tested on a particular data split. The table presents the values of two metrics, average C and Compliance Error, for each algorithm, with varying numbers of iterations. It is clear that increasing the number of iterations resulted in improved performance for all algorithms. Moreover, the combination of TopoDiff-FF and SIMP performed better than

Copyright © 2023 by ASME

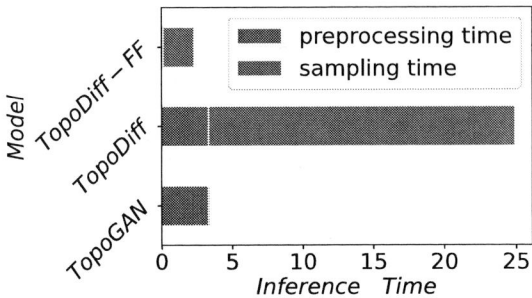

Figure 8: Sampling time. In this figure, we compare the total inference time (pre-processing and sampling time) for TopoDiff, TopologyGAN, and TopoDiff-FF. We can see that we completely remove the pre-processing time component using the kernel relaxation. And reducing the number of sampling steps reduces, even more, the sampling time, making TopoDiff-FF relatively fast at the price of generating configurations with higher compliance than TopoDiff and TopoDiff-GUIDED.

	Task	Kernels	Fields	Mdn Gap ↓	Avg Gap ↓	avg Rank ↓
TopoDiff	task-1	✗	✓	3.12	4.38	2.2
TopoDiff-GUIDED	task-1	✗	✓	3.16	4.52	3.2
TopoDiff-FF	task-1	✓	✗	4.37	7.86	2.8
TopoDiff-FF + SIMP	task-1	✓	✗	**2.69**	**3.57**	**1.6**
TopoDiff	task-2	✗	✓	3.35	5.58	2.2
TopoDiff-GUIDED	task-2	✗	✓	3.55	5.54	3.2
TopoDiff-FF	task-2	✓	✗	5.86	13.69	2.8
TopoDiff-FF + SIMP	task-2	✓	✗	**3.08**	**4.27**	**1.6**

Table 8: Global overview of different conditioning configurations showing how far the generative designs are from an ideal optimal design. On in-distribution configurations, our approach is extremely effective. On out-of-distribution configuration, there is still a gap in performance. Adding a few steps of SIMP (10 steps) as refinement is effective in improving the design quality.

the other algorithms, achieving the lowest values of average C and CE in both experiments. These results suggest that using the combination of TopoDiff-FF and SIMP is a way to impose the performance constraints in the model without the need for surrogate models or guidance. Also, we are sure that some physical properties are respected, reducing the FM, VFE, and CE using a fast and relatively cheap refinement (5/10 iterations).

5.8 Manufacturability

We define a topology as manufacturable if it satisfies all constraints, has no floating material, and compliance is below 100. In Figure 11, we depict how manufacturability, measured as the percentage of configurations that meet these criteria, varies with performance. We evaluate out-of-distribution constraints and compare models trained on raw loads and boundary conditions with those trained on kernel relaxation. Our results show that manufacturability improves with performance, i.e., lower average compliance, of generated topologies. Notably, we observed a higher fraction of manufacturable topologies when conditioning on the kernel

	Task	Iter	Avg C ↓	% CE ↓
SIMP	task-2	+5	9.87	35.13
TopoDiff-FF	task-2	-	19.01	58.36
TopoDiff-FF + SIMP	task-2	+5	**8.67**	**20.34**
SIMP	task-2	+10	8.57	17.61
TopoDiff-FF	task-2	-	19.01	58.36
TopoDiff-FF + SIMP	task-2	+10	**7.65**	**7.84**

Table 9: Comparison of SIMP, TopoDiff-FF and TopoDiff-FF + SIMP. We see that by merging together the topology generated using a conditional diffusion model and SIMP for refinement we obtain the best results in terms of average compliance and compliance error with just 5 or 10 SIMP iterations.

relaxation, supporting our hypothesis that the kernel is an effective approximation for loads and boundaries.

Kernel Ablation. The table presents the results of a kernel ablation study for the TopoDiff-FF model. We train the models using short runs. Four different inverse distance kernels with different exponents are evaluated, and their performance is measured in terms of compliance error (CE), volume fraction error (VFE), floating material (FM), load disrespect (LD), and the percentage of compliant samples with a compliance value smaller than 100 ($conf|c < 100$). Lower values for CE, VFE, FM, and LD indicate better performance, while higher values for $conf|c < 100$ indicate better compliance with the given constraints. The results show that the $1/r^2$ kernel performs well overall, with the lowest CE and VFE values and the highest $conf|c < 100$ percentage when learning the power using β. The $1/r^\beta$ kernel also performs well, with the lowest FM value, while the $1/r$ kernel performs the worst in all metrics.

	CE ↓	VFE ↓	FM ↓	LD ↓	$conf|_{c<t}$ ↑
$K(1/r)$	189.76	15.54	8.60	1.4	48.80
$K(1/r^2)$	47.62	2.38	9.23	0.0	94.32
$K(1/r^4)$	55.01	2.35	8.15	0.0	95.90
$K(1/r^\beta)$	36.20	2.22	8.63	0.0	95.66

Table 10: Kernel Ablation for TopoDiff-FF. All metrics in **%**. We train models for a shorter time on a subset of the training data.

6. DISCUSSION

It is essential to acknowledge the limitations of purely data-driven approaches, such as TopoDiff-FF, which struggle to generalize to out-of-distribution scenarios since these models lack a way to generalize to new situations not encountered during training. In contrast, TopoDiff incorporates additional conditioning information obtained from numerical analysis and the underlying physics of the problem, which enables the model to generalize to regions of the domain not included in the training data. The kernel proposed for TopoDiff-FF is versatile and can be easily applied to any domain, resolution,

Copyright © 2023 by ASME

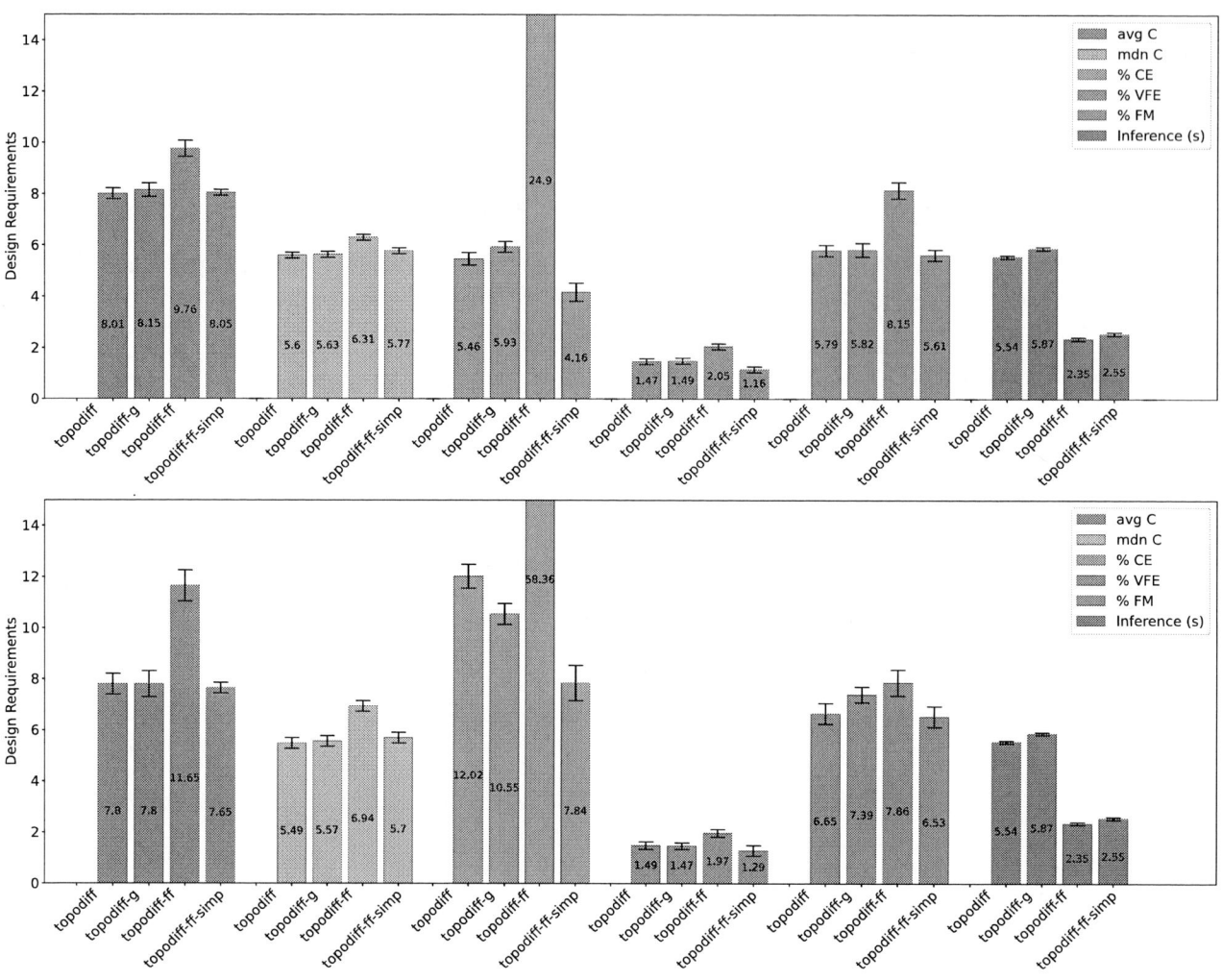

Figure 9: Overview design requirements for the considered models for task-1 (in-distribution constraints, top) and task-2 (out-of-distribution constraints, bottom). From left to right: average compliance in blue. Median compliance in orange. Compliance error in green. Volume fraction error in red. Floating material in purple. Inference time in brown. Lower is better for all the metrics.

and material, making it a valuable tool for engineering design. Contrary, FEM conditioning requires a thorough understanding of the problem, and it does not scale well with resolution and problem complexity. Therefore, TopoDiff-FF represents a powerful alternative for engineering design, particularly in situations where FEM conditioning is not feasible or effective. Furthermore, our results demonstrate that combining the topology generated by TopoDiff-FF with a few iterations of SIMP can produce better results than either data-driven or physics-informed methods, indicating a way to increase expressivity for such models in engineering design.

Limitations. Deep Generative models for topology optimization are a promising direction to improve efficiency, scalability, and variety in engineering design but they still present limitations [34]. In general, three main objectives can be identified for data-driven approaches: direct design, speedup, and upsampling. The authors in [34] are critical of end-to-end approaches for direct design, stating that most of the meth-

ods proposed in the literature produce poor designs and are expensive and limited in the variety of problems and mesh resolutions they can handle. This is indeed a good critique because most methods focus on a specific domain and resolution. The authors argue that the iterative nature of topology optimization is not what makes it impractical but rather the computational load of the costly components within each iteration. They suggest that the focus should shift towards alleviating this computational load. The authors also suggest that the main contribution of the learning methods should be in speeding up the computations in the intermediate steps of the iterative optimization process or post-processing optimized results for manufacturability. Where most of these critiques are accurate, there is a new wave of generative models that are tackling such challenges and solving or alleviating most of these issues, from sampling efficiency to reliance on a specific domain; to optimization-based refinement and direct design with performance awareness [10].

Copyright © 2023 by ASME

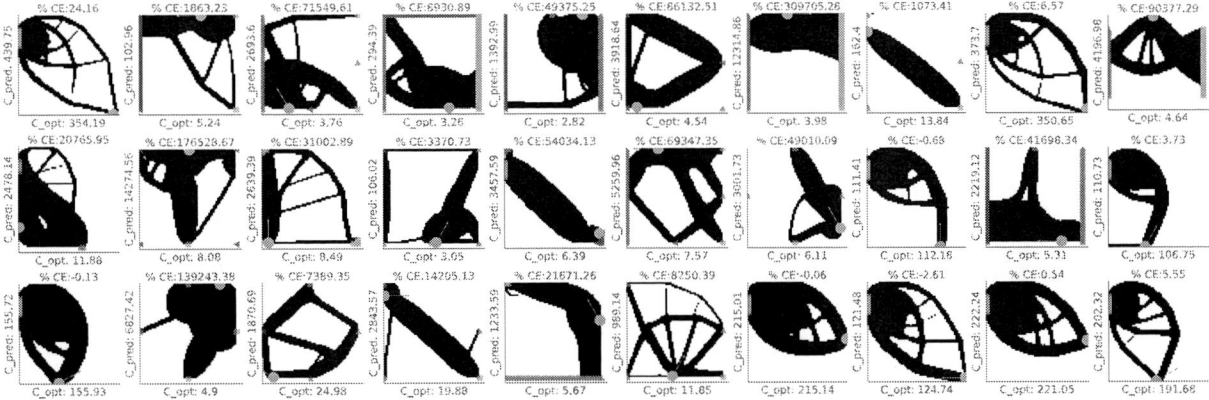

Figure 10: Examples of generated topologies with bad performance.

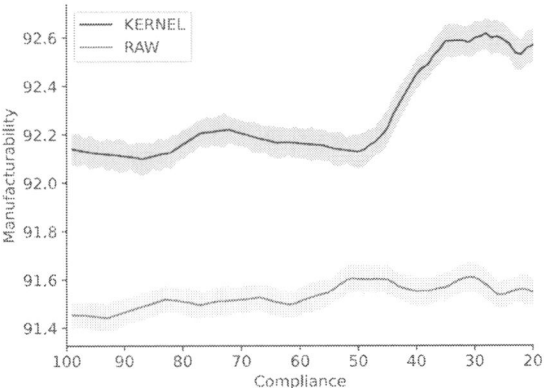

Figure 11: Manufacturability as measured in terms of absence of floating material and load disrespect, and compliance lower than 100 for out-of-distribution constraints.

7. CONCLUSION

In summary, our method of utilizing kernel approximation as conditioning for generative models and refining with optimization techniques has shown promise in enhancing precision, constraint satisfaction, manufacturability, and performance. However, as with any promising methodology, there are still areas for improvement and future exploration. One potential avenue for enhancement could be to devise strategies that preserve diversity while maintaining high levels of precision and constraint satisfaction. This is a crucial aspect for the generation of a wide range of effective and feasible design solutions. Moreover, an important aspect that warrants further investigation is the scalability of our approach. Expanding the applicability of our method to larger and more complex problems remains a significant challenge. Successfully addressing this could further extend the utility of our method and contribute to solving intricate design problems on a larger scale. In light of these results, we remain optimistic about the future of generative optimization in topology optimization. The integration of deep generative models with traditional optimization techniques, as shown in our work, signals a new and promising direction for the field, potentially broadening the scope of its applicability and effectiveness in

performance-aware engineering design problems

REFERENCES

[1] Bendsøe, Martin Philip and Kikuchi, Noboru. "Generating optimal topologies in structural design using a homogenization method." *Computer methods in applied mechanics and engineering* Vol. 71 No. 2 (1988): pp. 197–224.

[2] Sigmund, Ole and Maute, Kurt. "Topology optimization approaches: A comparative review." *Structural and Multidisciplinary Optimization* Vol. 48 No. 6 (2013): pp. 1031–1055.

[3] Ahmed, Faez, Bhattacharya, Bishakh and Deb, Kalyanmoy. "Constructive solid geometry based topology optimization using evolutionary algorithm." *Proceedings of seventh international conference on bio-inspired computing: theories and applications (BIC-TA 2012)*: pp. 227–238. 2013. Springer.

[4] Bendsøe, Martin P. "Optimal shape design as a material distribution problem." *Structural optimization* Vol. 1 (1989): pp. 193–202.

[5] Rozvany, George IN, Zhou, Ming and Birker, Torben. "Generalized shape optimization without homogenization." *Structural optimization* Vol. 4 (1992): pp. 250–252.

[6] Bond-Taylor, Sam, Leach, Adam, Long, Yang and Willcocks, Chris G. "Deep generative modelling: A comparative review of vaes, gans, normalizing flows, energy-based and autoregressive models." *IEEE transactions on pattern analysis and machine intelligence* (2021).

[7] Rombach, Robin, Blattmann, Andreas, Lorenz, Dominik, Esser, Patrick and Ommer, Björn. "High-Resolution Image Synthesis with Latent Diffusion Models." *arXiv preprint arXiv:2112.10752* (2021).

[8] Brown, Tom B, Mann, Benjamin, Ryder, Nick, Subbiah, Melanie, Kaplan, Jared, Dhariwal, Prafulla, Neelakantan, Arvind, Shyam, Pranav, Sastry, Girish, Askell, Amanda et al. "Language models are few-shot learners." *arXiv preprint arXiv:2005.14165* (2020).

[9] Ramesh, Aditya, Pavlov, Mikhail, Goh, Gabriel, Gray, Scott, Voss, Chelsea, Radford, Alec, Chen, Mark and Sutskever, Ilya. "Zero-shot text-to-image generation." *International Conference on Machine Learning*: pp. 8821–8831. 2021. PMLR.

[10] Regenwetter, Lyle, Nobari, Amin Heyrani and Ahmed, Faez. "Deep generative models in engineering design: A review." *Journal of Mechanical Design* Vol. 144 No. 7 (2022): p. 071704.

Copyright © 2023 by ASME

[11] Song, Binyang, Zhou, Rui and Ahmed, Faez. "Multi-modal Machine Learning in Engineering Design: A Review and Future Directions." *arXiv preprint arXiv:2302.10909* (2023).

[12] Nie, Zhenguo, Lin, Tong, Jiang, Haoliang and Kara, Levent Burak. "Topologygan: Topology optimization using generative adversarial networks based on physical fields over the initial domain." *Journal of Mechanical Design* Vol. 143 No. 3 (2021).

[13] Mazé, F. and Ahmed, F. "Diffusion Models Beat GANs on Topology Optimization." *Proceedings of the AAAI Conference on Artificial Intelligence (AAAI).* 2023. Washington, DC.

[14] Behzadi, Mohammad Mahdi and Ilieş, Horea T. "Real-Time Topology Optimization in 3D via Deep Transfer Learning." *Computer-Aided Design* Vol. 135 (2021): p. 103014.

[15] Shigley, Joseph Edward, Mitchell, Larry D and Saunders, H. "Mechanical engineering design." (1985).

[16] Allaire, Grégoire, Jouve, François and Toader, Anca-Maria. "A level-set method for shape optimization." *Comptes Rendus Mathematique* Vol. 334 No. 12 (2002): pp. 1125–1130.

[17] Sokolowski, Jan and Zochowski, Antoni. "On the topological derivative in shape optimization." *SIAM journal on control and optimization* Vol. 37 No. 4 (1999): pp. 1251–1272.

[18] Xie, Y Mike, Steven, Grant P, Xie, YM and Steven, GP. *Basic evolutionary structural optimization.* Springer (1997).

[19] Bourdin, Blaise and Chambolle, Antonin. "Design-dependent loads in topology optimization." *ESAIM: Control, Optimisation and Calculus of Variations* Vol. 9 (2003): pp. 19–48.

[20] Hunter, William et al. "Topy-topology optimization with python." (2017).

[21] Liu, Kai and Tovar, Andrés. "An efficient 3D topology optimization code written in Matlab." *Structural and Multidisciplinary Optimization* Vol. 50 (2014): pp. 1175–1196.

[22] Abueidda, Diab W., Koric, Seid and Sobh, Nahil A. "Topology optimization of 2D structures with nonlinearities using deep learning." *Computers & Structures* Vol. 237 (2020): p. 106283.

[23] Banga, Saurabh, Gehani, Harsh, Bhilare, Sanket, Patel, Sagar and Kara, Levent. "3D Topology Optimization using Convolutional Neural Networks." *Preprint* (2018).

[24] Hertlein, Nathan, Buskohl, Philip R., Gillman, Andrew, Vemaganti, Kumar and Anand, Sam. "Generative adversarial network for early-stage design flexibility in topology optimization for additive manufacturing." *Journal of Manufacturing Systems* Vol. 59 (2021): pp. 675–685.

[25] Elingaard, Martin Ohrt, Aage, Niels, Bærentzen, Jakob Andreas and Sigmund, Ole. "De-homogenization using convolutional neural networks." *Computer Methods in Applied Mechanics and Engineering* Vol. 388 (2022): p. 114197.

[26] Napier, Nicholas, Sriraman, Sai-Aksharah, Tran, Huy T. and James, Kai A. "An Artificial Neural Network Approach for Generating High-Resolution Designs From Low-Resolution Input in Topology Optimization." *Journal of Mechanical Design* Vol. 142 No. 1 (2020).

[27] Kench, Steve and Cooper, Samuel J. "Generating three-dimensional structures from a two-dimensional slice with generative adversarial network-based dimensionality expansion." *Nature Machine Intelligence* Vol. 3 No. 4 (2021): pp. 299–305.

[28] Jiang, Jiaqi, Chen, Mingkun and Fan, Jonathan A. "Deep neural networks for the evaluation and design of photonic devices." *Nature Reviews Materials* Vol. 6 No. 8 (2021): pp. 679–700.

[29] Rawat, Sharad and Shen, M. H. Herman. "A novel topology design approach using an integrated deep learning network architecture." *Preprint* (2018).

[30] Rawat, Sharad and Shen, MH Herman. "Application of adversarial networks for 3d structural topology optimization." Technical report no. SAE Technical Paper. 2019.

[31] Sharpe, Conner and Seepersad, Carolyn Conner. "Topology design with conditional generative adversarial networks." *International Design Engineering Technical Conferences and Computers and Information in Engineering Conference*, Vol. 59186: p. V02AT03A062. 2019. American Society of Mechanical Engineers.

[32] Yu, Yonggyun, Hur, Taeil, Jung, Jaeho and Jang, In Gwun. "Deep learning for determining a near-optimal topological design without any iteration." *Structural and Multidisciplinary Optimization* Vol. 59 No. 3 (2019): pp. 787–799.

[33] Berthelot, David, Schumm, Thomas and Metz, Luke. "Began: Boundary equilibrium generative adversarial networks." *arXiv preprint arXiv:1703.10717* (2017).

[34] Woldseth, Rebekka V, Aage, Niels, Bærentzen, J Andreas and Sigmund, Ole. "On the use of artificial neural networks in topology optimisation." *Structural and Multidisciplinary Optimization* Vol. 65 No. 10 (2022): p. 294.

[35] Kong, Zhifeng and Ping, Wei. "On fast sampling of diffusion probabilistic models." *arXiv preprint arXiv:2106.00132* (2021).

[36] Meng, Chenlin, Gao, Ruiqi, Kingma, Diederik P, Ermon, Stefano, Ho, Jonathan and Salimans, Tim. "On distillation of guided diffusion models." *arXiv preprint arXiv:2210.03142* (2022).

[37] Choi, Jooyoung, Kim, Sungwon, Jeong, Yonghyun, Gwon, Youngjune and Yoon, Sungroh. "Ilvr: Conditioning method for denoising diffusion probabilistic models." *arXiv preprint arXiv:2108.02938* (2021).

[38] Song, Yang, Sohl-Dickstein, Jascha, Kingma, Diederik P, Kumar, Abhishek, Ermon, Stefano and Poole, Ben. "Score-based generative modeling through stochastic differential equations." *arXiv preprint arXiv:2011.13456* (2020).

[39] Nichol, Alexander Quinn and Dhariwal, Prafulla. "Improved denoising diffusion probabilistic models." *International Conference on Machine Learning*: pp. 8162–8171. 2021. PMLR.

[40] Giannone, Giorgio, Nielsen, Didrik and Winther, Ole. "Few-shot diffusion models." *arXiv preprint arXiv:2205.15463* (2022).

[41] Xie, Tian, Fu, Xiang, Ganea, Octavian-Eugen, Barzilay, Regina and Jaakkola, Tommi. "Crystal Diffusion Variational Autoencoder for Periodic Material Generation." *arXiv preprint arXiv:2110.06197* (2021).

[42] Ramesh, Aditya, Dhariwal, Prafulla, Nichol, Alex, Chu, Casey and Chen, Mark. "Hierarchical text-conditional image generation with clip latents." *arXiv preprint arXiv:2204.06125* (2022).

[43] Sohl-Dickstein, Jascha, Weiss, Eric, Maheswaranathan, Niru and Ganguli, Surya. "Deep Unsupervised Learning using Nonequilibrium Thermodynamics." Bach, Francis and Blei, David (eds.). *Proceedings of the 32nd International Conference on Machine Learning*, Vol. 37: pp. 2256–2265. 2015. PMLR, Lille, France.

[44] Ho, Jonathan, Jain, Ajay and Abbeel, Pieter. "Denoising Diffusion Probabilistic Models." *Advances in Neural Information Processing Systems 33.* 2020.

[45] Song, Yang and Ermon, Stefano. "Generative Modeling by Estimating Gradients of the Data Distribution." *Advances in Neural Information Processing Systems 32.* 2019.

Copyright © 2023 by ASME

[46] Kingma, Diederik P and Welling, Max. "Auto-encoding variational bayes." *arXiv preprint arXiv:1312.6114* (2013).

[47] Jordan, Michael I, Ghahramani, Zoubin, Jaakkola, Tommi S and Saul, Lawrence K. "An introduction to variational methods for graphical models." *Machine learning* Vol. 37 No. 2 (1999): pp. 183–233.

[48] Dhariwal, Prafulla and Nichol, Alexander. "Diffusion models beat gans on image synthesis." *Advances in Neural Information Processing Systems* Vol. 34 (2021).

[49] Ho, Jonathan and Salimans, Tim. "Classifier-Free Diffusion Guidance." *NeurIPS 2021 Workshop on Deep Generative Models and Downstream Applications*. 2021.

[50] Garabedian, PR. "Partial differential equations with more than two independent variables in the complex domain." *Journal of Mathematics and Mechanics* (1960): pp. 241–271.

[51] Hale, Jack K and Lunel, Sjoerd M Verduyn. *Introduction to functional differential equations*. Vol. 99. Springer Science & Business Media (2013).

[52] Keldysh, Mstislav Vsevolodovich. "On the characteristic values and characteristic functions of certain classes of non-self-adjoint equations." *Dokl. Akad. Nauk SSSR*, Vol. 77. 1: pp. 11–14. 1951.

APPENDIX A. EXAMPLES OF GENERATED TOPOLOGIES

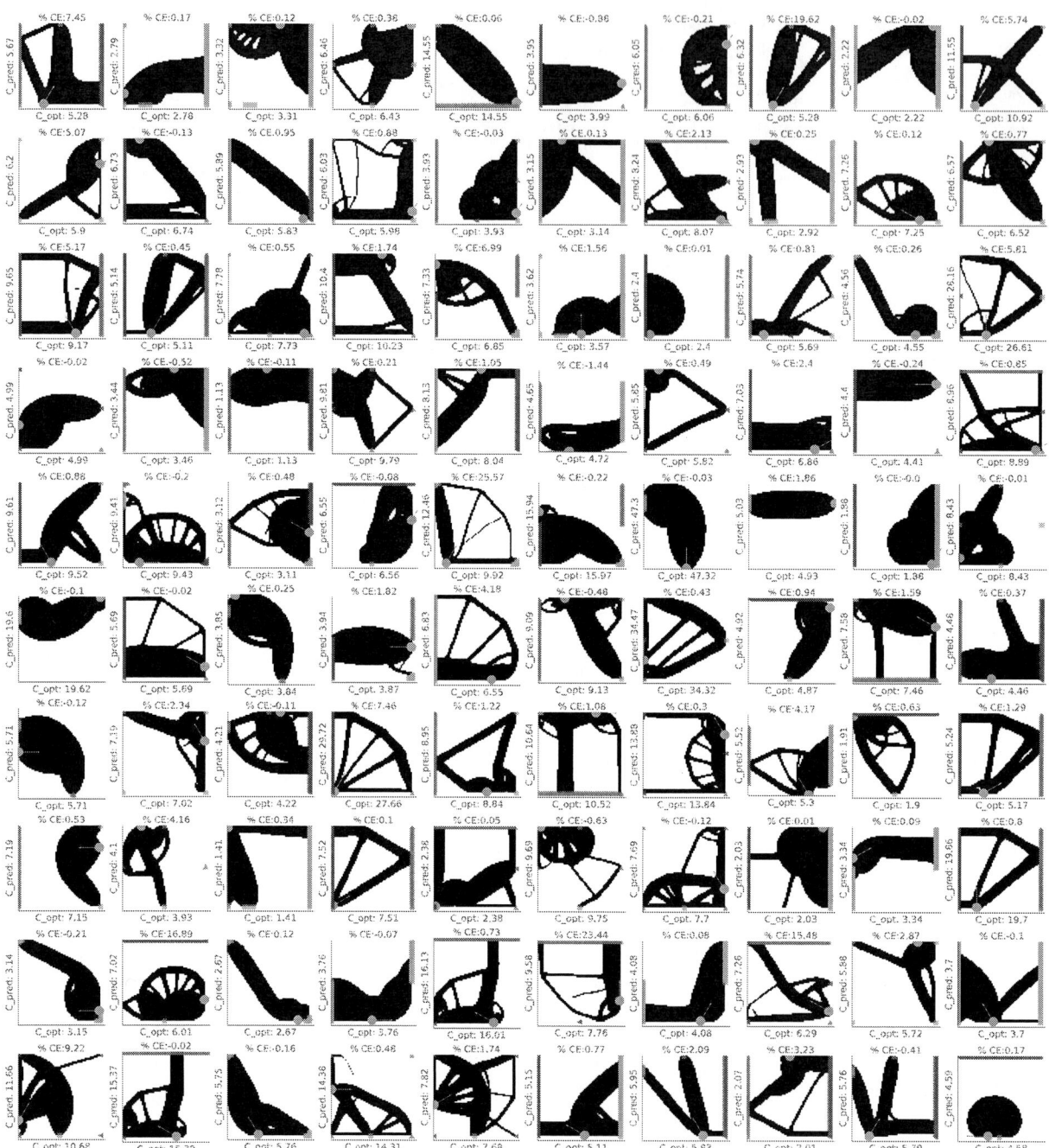

Figure 12: Examples of generated topologies with good performance.

Proceedings of the ASME 2023
International Design Engineering Technical Conferences and
Computers and Information in Engineering Conference
IDETC-CIE2023
August 20-23, 2023, Boston, Massachusetts

DETC2023-116746

DOCUMENT UNDERSTANDING-BASED DESIGN SUPPORT: LANGUAGE MODEL BASED DESIGN KNOWLEDGE EXTRACTION

Yunjian Qiu
IMPACT Laboratory
Dept. of Aerospace & Mechanical Engineering
University of Southern California
Los Angeles, CA, 90089
yunjianq@usc.edu

Yan Jin*
IMPACT Laboratory
Dept. of Aerospace & Mechanical Engineering
University of Southern California
Los Angeles, CA, 90089
yjin@usc.edu
(*corresponding author)

ABSTRACT

Design knowledge in the vast amount of design reports and documents can be a great resource for designers in their practice. However, capturing such domain-specific information embedded in long-length unstructured texts is always time-consuming and sometimes difficult. Therefore, it is highly desirable for a computer system to automatically extract the main knowledge points and their corresponding inner structures from given documents. In this study of document understanding for design support (DocUDS), a design-perspective knowledge extraction approach is proposed that uses phrase-level domain-specific labeled datasets to finetune a Bidirectional Encoder Representation from Transformers (BERT) model so that it can extract design knowledge from documents. The BERT model finetuning attempts to blend in the domain-specific knowledge of well-recognized domain concepts and is based on the datasets generated from design reports. The model is utilized to map the captured sentences to the main design entities <requirement>, <function>, and <solution>. In addition, this approach uncovers inner relationships among the sentences and constructs overall structures of documents to enhance understanding. The definitions of design perspectives, inter-perspective relations, and intra-perspective relations are introduced, which together capture the main design knowledge points and their relations and constitute an understanding of the design domain knowledge of a text. The case study results have demonstrated the proposed approach's effectiveness in understanding and extracting relevant design knowledge points.

Keywords: Design-perspective knowledge extraction, document understanding, phrase-level dataset, language model, text classification, sentence-design entity mapping

1 INTRODUCTION

As the amount of textual material grows daily, there is a great need to reduce unstructured text data to shorter, more focused information while maintaining significant meanings. For example, text summarization using automatic text processing techniques has been developed recently in domains such as medical biology, healthcare, and engineering [1]. This summary-generating process can efficiently capture the main information of the text and significantly reduce the problem of overwhelming data [2]. Researchers have used sentence features like term frequency to summarize information underlying the text [3]. The contextual embedding method has also been applied to capture texts' main content [4]. The recent progress of language model research has made it possible to finetune a language model with domain-specific datasets to carry out domain-specific tasks, such as generating a summary for a given design report [5].

In addition to general text summarization, researchers in the engineering design domain have focused on design knowledge retrieval and extraction to distill the important information from long-length texts. Different knowledge retrieval methods have been proposed to quickly retrieve critical information from engineering documents. Traditional keyword-based methods [6] and ontology-based retrieval methods [7][8] have been used to capture the concepts and their relationships. However, this literal matching and "flat" search tend to lose the semantic information within the text. As the development of natural language processing (NLP) techniques accelerates, researchers have started exploring semantic-level knowledge extraction in engineering design using NLP methods. Most existing methods focus on word-level applications by extracting important entities and their associations [9][10][11][12]. Knowledge reuse and

Copyright © 2023 by ASME

exploration remain in the attribute-level application [13]. As a result, knowledge extraction may lead to insufficient information capture and a discontinuity in knowledge representation.

Through the engineering design process, applying domain knowledge is critical to successful design activities such as design ideation, analysis, and evaluation. In many cases, designers record their ideas and solutions in various design documents. It will be of great value if the information underlying these document texts can be extracted and reused for similar design situations. Further, given the overwhelming number of documents, using automatic processes to capture engineering-*focused views* or *perspectives* from the documents can increase efficiency and allow designers to have much broader access. As the evolution of machine learning (ML) techniques and NLP methods provides the possibility to capture domain knowledge from texts, how to extract design knowledge effectively by making the design support system "understand" and learn domain-specific knowledge has become a challenge for the design research community.

The issue is how to construct a design knowledge extraction framework that can possess domain perspectives with specific design knowledge rather than only general language skills. Since most language models are pretrained based on large general datasets, extracting meaningful domain-perspective knowledge using language models alone can be difficult. Although there has been research on the roles of key information extraction and knowledge capturing, little work has been done aiming at building an automatic document processing tool for augmenting engineering designers with a clearly defined document understanding framework. The fundamental issue is the missing link between the NLP language models and the knowledge engineering tradition in the engineering design domain.

In this research, a document understanding-based design support (DocUDS) framework is proposed that combines the language model with the ontological knowledge found in knowledge-based engineering (KBE), focusing on the engineering design process and knowledge. Several research questions arise: *What relevance and meaningful information can be captured from engineering documents? Can we transfer domain knowledge learned from phrase-level datasets to sentence-level data? How can we extract the hierarchical structure and uncover the inner sentence relationships?*

This study focuses on design knowledge extraction and inner structure elicitation from engineering documents. It investigates the applicability and robustness of finetuned BERT model using a phrase-level dataset to sentence-level text. In the rest of this paper, the related work is reviewed in Section 2, and a systematic approach to dealing with design knowledge extraction is described in Section 3. Section 4 presents a comparative study of contextual embeddings generated by finetuned BERT model together with a comprehensive design-perspective generation scheme using several case studies. The insights obtained are discussed in Section 5, followed by the conclusions and future work in Section 6.

2 RELATED WORK
2.1 Applications of NLP in Engineering Design
NLP techniques have been applied in engineering design to support designers in the design process. Knowledge underlying past design documents, such as design reports or related papers, can be captured using NLP-based tools and reused by designers. By capturing information within unstructured texts, NLP-based methods can be applied to topic discovery, ontology extraction, keyword recommendation, and text generation [16]. Specifically, as one of the popular applications in engineering design, NLP techniques are applied for word-level knowledge discovery in the design process. Lin et al. [17] created a passage retrieval method called "OntoPassages," using a domain ontology to map user needs to domain concepts. Hou et al. [13] developed an automatic way to identify and structure product affordance from a user's review using the rule-based NLP method for discovering customer needs. Liu et al. [18] constructed a data-driven concept network using ML methods to extract design concepts and meaningful combinations of concepts from web documents and literature. Han et al. [19] proposeusing the BERT model to elicit attribute-level user needs from online review text. Zhu et al. [20] proposed a biologically inspired design concept generation method using GPT-3.

Although word-based domain knowledge extraction can be applied to building knowledge graphs or networks to extract the information in documents, high dimensional sentence-level information capturing can contain more useful knowledge underlying the unstructured text. Scholars in engineering design also explored sentence-level knowledge applications containing requirement classifications. Brisco et al. [21] collected text data from students' global design projects. They classified the sentences in unstructured data into different design concept categories like requirements, technologies, and technical functionalities. Ye and Lu [22] presented an NLP-based scheme to translate engineering design knowledge to semantic rules by neural machine translation techniques. Siddharth et al. [23] created a patent retrieval method using sentence embedding generated by the BERT model and graph embedding. However, exploring sentence relationships inside one document needs further investigation.

2.2 Engineering Design Processes
In engineering design, the design process can be depicted as a systematic process with standard workflow moving from the conceptual phase to the product phase [24]. Underlying the process, design knowledge and design ideation for the specific problem can be implicit and uncovered. It is helpful and efficient for engineering designers to address similar design problems by accessing previous design documents for useful information. Knowledge-based engineering (KBE) has potential solutions to effectively capture and reuse the domain knowledge from design experts [25]. This approach can be applied to realize knowledge elicitation and encoded into tools that can support the transformation of knowledge into computer code [26]. The design process is the key to reducing ambiguity, complexity, and uncertainty of design by defining and standardizing distinct phases and activities. While allowing some freedom to exist within each activity or phase, information, knowledge, and insights applied in the design process can be considered systematic methods and reused for similar design problems. Mainly, several design process theories aim at characterizing design into related sections: the design problem, the design

Copyright © 2023 by ASME

process, and the design outputs [27]. Pahl and Beitz [28] proposed a systematic design process that contains various phases such as task clarification, conceptual design, embodiment design, and detail design. After clarifying the task and elaborating on respective specifications, problem identification, concept variant generation, and design optimization will be carried out. Ullman [29] presented a mechanical design process model to explain how the design process can be divided into project definition and planning, conceptual design, and product design. Suh [30] introduced an axiomatic design theory for mechanical systems, which provided a theoretical framework for the design process from customer needs, function requirements, and design parameters to process variables.

The different definitions of the engineering design process reveal the major conceptual entities that the designers manipulate: customer needs (or requirements), functional requirements, and design parameters (or solutions). Decisions made during the design of these design entities significantly influence inevitable factors such as cost, performance, and reliability. Therefore, in this research, customer needs clarification, functional requirements analysis, and concept variant generation are considered major steps in engineering design. The information within these steps will be captured and reused for model training and knowledge structuring.

2.3 Design Knowledge Extraction

With the acceleration of artificial intelligence (AI) and ML, implementing AI and ML algorithms with KBE becomes possible. Besides largely increasing the working efficiency and reducing the costs of the knowledge-capturing process [31], the interactions between knowledge systems and designers can be more flexible. As one subfield of ML, NLP plays a critical role in several different tasks like ontology extraction [32], document structuring [33], keyword recommendation[34], etc. Researchers often focus on information retrieval to capture and reuse the domain knowledge. If captured from the relevant documents, such knowledge can be applied for design support, including design ideation. Traditional keyword-based retrieval models [6] [7] and ontology-based models [8][34][35] can be used for literal matching and ontological concepts and their relationship capturing. In engineering design, it was not until the significant improvement of methodology and computational models in the NLP area that these tasks became realizable for extracting design knowledge and processing unstructured data [12]. As NLP techniques become maturing, researchers in the engineering design area tend to retrieve design entities and their relations from a large number of documents to conclude engineering facts and reuse them in the future. Martinez-Rodriguez et al. [10] proposed a knowledge graph construction approach to capture name entities and their binary relations from unstructured text. Sarica & Luo. [11] utilized a network graph displaying the design-related entities and their relations as a summary of the design description to help engineers understand the new design context. Siddharth et al. [16] produced a large engineering knowledge graph comprising engineering rules as <entity, relation, entity> triples. Akay & Kim [38] introduced an automated method using the BERT model to generate individual functional requirement trees and compile the functional requirements from several documents. However, document

analysis and knowledge elicitation based on sentence relations remained unfinished.

3 A DOCUMENT UNDERSTANDING APPROACH TO DESIGN KNOWLEDGE EXTRACTION

For designers reading a document, it can take a long time to grasp the key points about the design, its process, and the inner structure of the document. The design information and the reasoning about customer needs, functional requirements, and design solution concepts can be scattered throughout the document, and their relations can be implicitly embedded. Therefore, capturing key information and understanding it by mapping it to designers' understanding can assist designers in focusing on the main points of documents.

3.1 "Understanding" in DocUDS

In his landmark book "Comprehension: A Paradigm of Cognition," Kintch [58] proposed a construction and integration model (C&I model) for human text comprehension and suggested it be a way to treat human cognition. In this model, as illustrated in **Figure 1**, a reader reads words and sentences (e.g., *the little boy chopped wood*), generates propositions and networks (e.g., [little, boy]–[chop, boy, wood]), and then forms final text representations. This mental process has two phases. In the first phase, approximate and possibly inaccurate representations are constructed (hence the construction phase) via context-insensitive construction rules, and then these mental representations are integrated (hence the integration phase) via a process of spreading activation through the use of the context information and the reader's background knowledge. Only the comprehended and active propositions will remain in the final text representation.

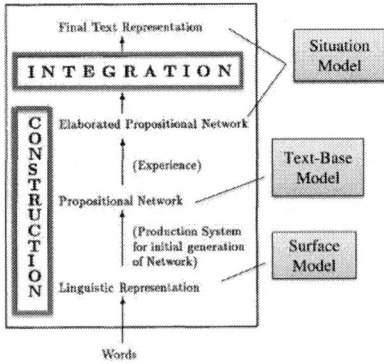

Figure 1: An illustration of the C&I model [59]

According to the C&I model, the formation of the *surface model* and *text-base model* requires little mental effort, but the creation of the *situation model* involves integrating the reader's background knowledge and contextual information with the content of the text-base, which may take effort or even problem-solving skills. The true understanding of human text reading happens when the situation model is built in one's mind.

Recent progress in language models (LM), especially the generative large language models (LLM), has led to the rising interest in applying LMs to text processing problems. While LLMs have exceptional formal language skills, it has been

Copyright © 2023 by ASME

pointed out that their functional language skills and "understanding" are very limited [60]. From a perspective of providing engineering support, two issues must be addressed for successfully applying LMs; one relates to the generality of LMs' training datasets and the specificity of engineering domains, and the other is LMs' lack of "understanding" of the text from a domain's perspective. The first issue can be addressed properly with the finetuning techniques [5]. The second issue requires devising engineering knowledge to make an LM hold a certain domain and its perspectives for understanding.

It is understood that the C&I model captures the essence of human text comprehension at the proposition and mental representation level. However, the concept of forming the situation model has led us to consider building the "domain understanding" into the pretrained BERT model by embedding the domain ontological knowledge during the finetuning process. A classic view of design objects and processes for information extraction is FBS (Function–Behavior–Structure) ontology [15], which can be used to represent any class of process in engineering design. Following this ontology, in this research, an RFS (Requirement–Function–Solution) perspective framework is applied to capture domain knowledge:

- Requirement (R) in a document is defined as the problem statement, customer needs, or background of a design problem.
- Function (F) is defined as the functional property of a proposed product in the conceptual design process.
- Solution (S) is defined as generated concepts and solutions based on customer needs and proposed functions.

Given an LM, e.g., BERT, and the RFS ontology, an engineering document understanding system can be built by using the BERT model to build the *surface model* shown in Figure 1. A non-propositional *text-based model* can also be constructed by the BERT model with little effort. In order to build the *situation-level model* (i.e., the understanding of the document), the BERT model will be finetuned by a "domain-ontology embedded in the finetuning dataset."

After being finetuned, the BERT model can be applied to process and "understand" target documents: (1) generate sentence embeddings of the text, (2) classify sentences from unstructured texts into design-related entities, namely, <requirement>, <function>, and <solution>, (3) uncover the inner relationships among the design entities and the entire document's structure based on the high-dimensional embedding space. **Figure 2** illustrates the proposed DocUDS framework.

As shown in **Figure 2**, DocUDS comprises two main modules: the BERT model finetuning module and an application module. The three major inputs are the pretrained BERT model, a domain knowledge-based ontology, and target unstructured document datasets. The outputs of DocUDS are structured domain-specific documents for the designers to look up or search from. In the rest of this section, the related details about data collection and preprocessing are illustrated in Sections 3.2 and 3.3. The BERT model finetuning is described in Section 3.4. Section 3.5 introduces the sentence classification process and generation of contextual embeddings of sequence, and Section 3.6 presents the definitions of design perspectives together with intra-perspective and inner-perspective relationships.

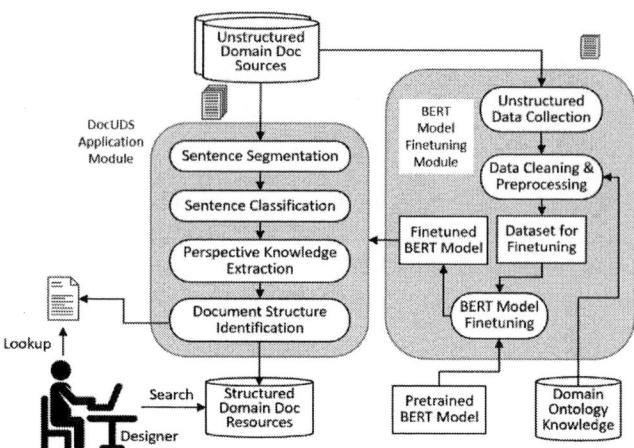

Figure 2. DocUDS: A Document Understanding Based Design Support Framework

3.2 Data Collection and Labeling

The first step to finetune the BERT model for text classification is to collect relevant documents and produce a labeled dataset. In this study, 43 design project reports written by different student groups from a senior-level engineering design class of the Aerospace and Mechanical Engineering Department at the University of Southern California are collected and considered data resources. Each report is generated by a team of 4-5 students that records their entire product design process. In the report, they needed to apply design knowledge they had learned from the classes to solve their selected design problem and document the entire design process for product design, from conceptual design to embodiment design and detail design. Following the conceptual design process, customers' requirements, functional diagrams, and morphology charts for concept generations are created and listed by the design teams for their design problems.

These design-related contents are chosen as domain-specific knowledge and gathered as a raw dataset. To instill the domain knowledge into the pretrained BERT model through finetuning, the RFS ontology mentioned above is used to label the raw dataset. Specifically, the content of customers' requirements, functional statements and diagrams, and variant generations are captured from students' reports and labeled as <requirement>, <function>, and <solution> correspondingly. Respectively, labels <requirement>, <function> and <solution> are encoded as {0, 1, 2}. To enhance the efficiency of manual labeling, sequences from paragraphs are first segmented using the NLTK tool [39]. Then the sequence related to different attributes is annotated with corresponding labels using an open-source annotation tool Doccano [40].

In addition, to meet the BERT model's requirements, for each sentence, tokens [CLS] and [SEP] are inserted at the start and the end of the sentence, respectively. Based on these labeling rules, there are 1247 sequences labeled in the raw dataset. Among the dataset, 452 sequences are labeled as <requirement>, 393 sequences are labeled as <function>, and 402 sequences are labeled as <solution>. Therefore, the raw dataset can be considered as a balanced dataset.

Copyright © 2023 by ASME

3.3 Data Cleaning and Preprocessing

Data cleaning and data preprocessing need to be conducted after raw dataset collection. All text is converted to lowercase. Also, duplicate spaces and mathematical symbols in the text are erased to reduce the noise. Stop words are deleted from the original text to remove low-level information and focus on the most important text. Furthermore, since duplicates can happen during the labeling process, the same sequences are removed from the dataset.

Although the dataset can be easily labeled and generated based on the sections and titles from students' reports, to guarantee the quality of the content and the correctness of information for labeled dataset creation, a consistency check experiment of this dataset is conducted to verify labeling results. Two research assistants with engineering design backgrounds were invited to label a 10% randomly generated dataset sample. Following the consistency check procedure using Cohen's Kappa statistic [41], the research assistants were required to label the sample individually and independently without knowing the original labeling results. Then Kappa statistic was applied to measure the agreement among research assistants. Based on Equation 1 below, the Kappa statistic can show the possibility of an agreement among raters.

$$\kappa = \frac{p_o - p_e}{1 - p_e} = 1 - \frac{1 - p_o}{1 - p_e} \qquad (1)$$

Respectively, the results between research assistants and original labelers are 0.917 and 0.923. According to the interpretations of Kappa statistics from [42], the results showed almost perfect agreement among different raters, which implies that the quality and consistency of the dataset can be guaranteed.

3.4 BERT Model Finetuning

BERT [43] is a language model based on a transformer developed by the Google AI platform. The finetuning process of the BERT model can be unstable, depending on the chosen hyperparameters and optimization methods [43][45]. In this research, the hyper-parameters are empirically set, as shown below, for finetuning the BERT model for the best performance.

(1) *Preprocessing of long sentences.* The maximum sequence length of the BERT model was set to 256 and then padded with zeros.

(2) *Selection of layers.* There are many different pretrained BERT models provided by the Google AI platform. In this study, an uncased BERT-base model, consisting of an embedding layer, 12 encoding layers, and a pooling layer with 768 hidden sizes, is selected as the pretrained model. Two additional fully connected layers for flattening the output and adding dropouts and a SoftMax activation function layer were added for the multi-class text classification task to predict the labels out of {0, 1, 2}.

(3) *Selection of an optimizer.* Adam optimizer was used with a 2e-5 learning rate in the finetuning process. This optimizer is an optimization algorithm for stochastic gradient descent during deep learning model training and can be applied to deal with sparse gradients on the noisy problem (Kingma & Ba, 2014).

(4) *Hyperparameter setting.* The authors of the BERT model recommend using 2-4 epochs to train the BERT model [43]. In this study, four epochs are chosen to finetune the BERT model. Moreover, the learning rate is 2e-5, and the batch size is set to 16.

3.5 Sequence Classification and Perspective Knowledge Extraction

Some researchers in the biomedical domain tried to consider phrases captured from health-related reviews as sequences and finetune the BERT model with phrases to complete text classification [47]. In this study, the finetuning dataset is at the phrase level, meaning that the data points are phrases instead of sentences. The data are mainly collected from bullet points and highlighted sections from students' reports. Over half of the data points contain 2 to 3 words, while very few include 10 or more.

It is worth mentioning that the phrases of each label can have different lengths, and two sequences of the same length can be given different labels. For instance, the two-word sequence "accept materials" is labeled as <function>, while the eight-word "confirm sample mass less than max sample mass" is also labeled as <function>, but the sequence with two words "aesthetically pleasing" is labeled as <requirement>. The eight-word sequence "external harness attached to the deployed rope by crane" is labeled as <solution>. Therefore, the phrase-level datasets will be fed into the BERT model as whole sequences and used to complete the multi-class text classification downstream task without the length restriction.

The application of language models involves using the model to create a high-dimensional space and then casting, or embedding, the sequence of the sentences in such a high-dimensional space for further processing. There are several methods to generate such high-dimensional contextual embedding from the BERT model. One is average embedding, which averages all the word embedding in one sentence as a fixed-length vector and then considers it as sentence embedding for the whole sentence [49] [50]. However, this method may result in a loss of information during the averaging process. Another method is to apply Sentence-BERT or S-BERT [51], which utilizes the Siamese network to directly generate sentence embedding and achieve significant performance in semantic similarity tasks. From a finetuning point of view, designing domain-specific training datasets for S-BERT is complicated and time-consuming. The third method, i.e., considering [CLS] embedding as the contextual embedding for an entire sequence, is also utilized by researchers in different domains [47][52]. Since the [CLS] token is a special token prepended to an input sequence for the BERT model, this token can be used as a summary token for the input sequence for classification tasks [45]. Therefore, in this study, after considering the data collection and high-dimensional contextual embedding generation, the third method [CLS] token is applied as sequence embedding for further clustering and sequence mapping tasks.

Based on the findings in [5], sentence embedding with higher dimensions is more suitable for acquiring complete information. Therefore, 768D sequence embeddings are created from the BERT model in further experiments. 768D is the maximum dimensions that the BERT model can generate,

3.6 Document Knowledge Structure Identification

Extracting key information and mapping that information with main entities in the design process are important steps for design-perspective knowledge capture and understanding. However, making computers do "reading" and "understanding" human-understandable texts requires more effort. Specifically,

Copyright © 2023 by ASME

capturing the meaning of design perspectives by defining proper design perspectives is important. In addition to key information extractions, the inner structural relationships among the design entities ought to be captured, and a complete knowledge structure underlying the document needs to be identified.

Therefore, two types of design perspective relations are introduced to represent the design knowledge structure underlying documents. *Intra-perspective relations* capture the inner connections between the same design entities, and *inter-perspective relations* are among different design entities. Some example relationships between the same design entities and among different design entities are illustrated in **Figure 3.**

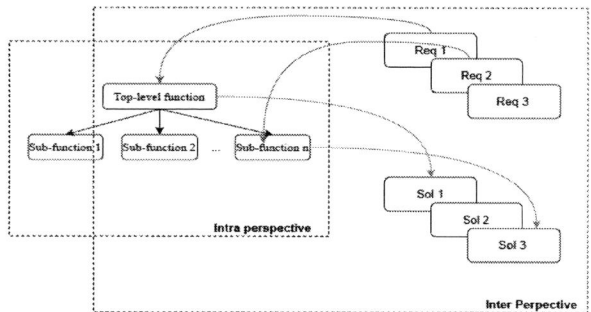

Figure 3. Example of intra-perspective of <function> and inter-perspective <requirement, function> and <function, solution>.

Intra-perspective Relations: Intra-perspective relations are identified by generating clusters among the same properties and capturing the inner structure of the same entities. For instance, <function> can be generated due to different targets and resources such as <signal>, <material>, and <energy> (Ogot, 2004). Therefore, clustering methods can be applied to create a relation tree to extract the inner structures of design perspectives. Since it can be difficult to predefine the number of clustering, the hierarchical clustering [54] method is applied. There are two types of hierarchical clustering: agglomerative (bottom-up) [55] and divisive (top-down) [56]. Agglomerative clustering is utilized to cluster the same properties to guarantee computational efficiency. Euclidean distance between each input sequence is calculated using contextual embedding generated from the BERT model. Respectively, the entities with similar focus or target are clustered first, and the most general attributes are included in the last step. One simple example based on function structure is shown in **Figure 4.**

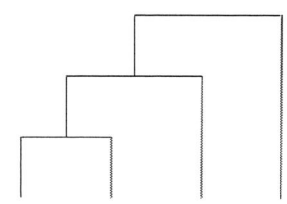

Figure 4. An example bottom-up hierarchy clustering for <function> design entity

Inter-perspective Relations: Unlike intra-perspective relations, which focus on the hierarchy inside the same properties, inter-perspective relations represent the logical and semantic mapping between different design perspectives. For instance, mapping a function to its corresponding solutions or mapping a requirement to its corresponding functions can signify a "better understanding" of the design-perspective knowledge. The structural information underlying documents can be uncovered and featured using semantic similarity and score threshold settings to select the best mapping pair between different design entities. One example of mapping among <requirement>, < function>, and <solution> is illustrated in **Figure 5.**

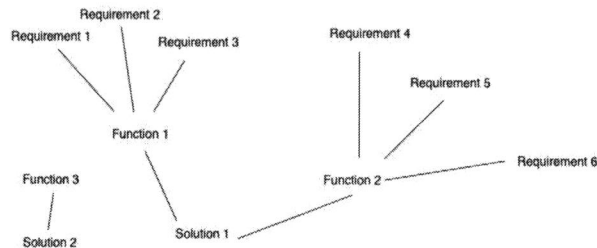

Figure 5. An example of mapping between different design entities

4 CASE STUDIES

The proposed document understanding approach to design knowledge extraction described above has been tested through several case studies on different design documents. The results have led to useful findings and insights. In the following subsections, the finetuning results of the BERT model for the text classification task are illustrated, and contextual embeddings of sequences are generated from the finetuned BERT model and then utilized for design-perspective knowledge extraction. After that, the results of five case studies on knowledge extraction from different documents are presented.

4.1 BERT Model Finetuning and Sentence Embedding Generation

The original raw training dataset is collected from 43 student design project reports from a USC senior-level engineering design class. The dataset contains 1247 sequences of different lengths. Generally, the proportions of the training, validation and testing datasets are 8:1:1, which is applied to this study. At the end of the training process over 4 epochs, the training accuracy reaches 0.952, and the validation accuracy is around 0.908.

As aforementioned, the dataset in this study is collected at the phrase level from project reports completed by student groups of an engineering design class at USC. The reports' consistency and relatively clear organization have resulted in a relatively simple and straightforward process of gathering clean and domain-specific data. However, it can be tedious and time-consuming for humans to collect unstained datasets for different domains in the real world. To address this issue, one must investigate how effective a finetuned BERT model can be when applied to documents of slightly different domains. In the following case studies, the finetuned BERT model is applied for processing sentence-level sequences for various engineering design documents that differ from the finetuning dataset resource. This way, the studies can help investigate the *robustness* of the finetuned BERT model under distinct circumstances.

Copyright © 2023 by ASME

4.2 Design-Perspective Knowledge Extraction

For representing intra-perspective relations, *hierarchical clustering* is applied to construct the relationships among the same design entities. Meanwhile, *semantic mapping* is utilized to capture the best-matching entities under different design concept categories.

The data points used for case studies are extracted from two different resources to verify the BERT model's robustness and transferability. The first data resource is the executive summary of students' design project reports, which summarizes the completed design process from problem abstraction to final concept generation. Sentences in this resource are different from the training dataset but are written under the same project. Further, the second data resource is gathered from case studies conducted by IDEO's HCD Connect platform, which precisely recorded the completed design process for each human-centered design case. To investigate the transferability of finetuned BERT model, all the data points are segmented from documents at the sentence level using the NLTK tool. Each summary may contain different focuses and formats compared to the training dataset. The results of the case study are shown in the following subsection.

To evaluate the performance of finetuned BERT model in distinct types of documents and verify its robustness, five case studies are conducted and compared from four angles, as shown in **Table 1**. Specifically, "Source Type" shows the data source of the original document, and "Design Concept Category" means the labeling results of sentences in the documents.

Some documents may not include <function> specification in classification results due to biased content. "Problem Domain" represents the design type described in the document, such as mechanical design. Meanwhile, "Content Focus" displays a simple content summarization of the focus of each document.

Table 1. Comparison of Five Case Studies

Case #	Source Types	Design Concept Category	Problem Domain	Content Focus
Case #1	Student Report	Req, Fun, Sol	Mechanical Design	Design Process
Case #2	Student Report	Req, Sol	Mechanical Design	Design Process
Case #3	Professional Record	Req, Fun, Sol	Software Design	Ideation Process
Case #4	Professional Record	Req, Sol	Mechanical Design	Performance of Design Team
Case #5	Professional Record	Req, Fun, Sol	App Design	Ideation Process

4.3 Cases #1 & #3: Different Source Types

Case #1 is a summary extracted from students' reports in an engineering design class, which described the entire design process, including background and conclusions at the sentence level. And Case #3 is captured from one of the case studies by IDEO.org about designing a virtual budget toolkit in professional record format. These two cases both have balanced data with <requirement>, <function>, and <solution> specifications and include a completed design process. However, Case #1 focuses on mechanical design, while Case #3 concentrates on software app design. Therefore, the following results show comparisons between two case studies under *different problem domains*.

The original document of Case #1 depicted the background of the design problem and the conceptual design and solutions. Specifically, there are 19 sentences included. After being classified by the finetuned BERT model, 7 sentences are classified as <function>, 10 sentences are classified as <requirement>, and 2 are classified as <solution>. Most of the background descriptions of the design are classified as requirements, while illustrations of the designed product are considered solutions. More detailed knowledge elicitation and mapping are described below.

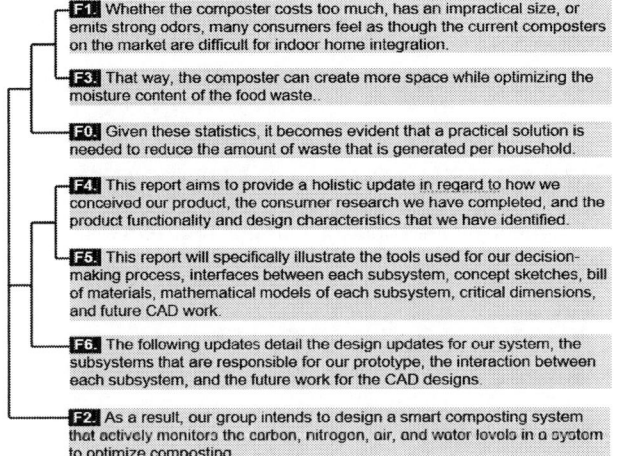

Figure 6. Clustering results of classified functions in case #1

Intra-perspective relations of <**function**>: Seven sentences are classified as <function> in this document, and their dendrogram shows in **Figure 6**. Specifically, F1 and F3 described specific functions the product needs/should have, like "indoor home integration," "create more space," and "optimize the moisture content of the food waste." Those short sequences are precisely captured by the finetuned BERT model and encoded in sequence embeddings. Meanwhile, F4, F5, and F6 present the subsequent design plan of the designers, which are also classified as <function>, but these sentences cannot be categorized into any design entities in the real world, so they are clustered into different groups from F1 and F3.

Inter perspective relations: <**requirement–function**> and <**function–solution**>: **Figure 7** shows mapping results between requirements and their best matching functions. It depicted that only F1 and F3, presenting real functions of the product, are matched with different requirements. For example, R4 and R5 described the need for a rational air ratio inside the composter, which matched with F3, "optimizing the moisture content of the food waste." In addition, R6 and R7 illustrate the current limitations of the composter about "simply take too long to operate" which can be mapped with the function outlined in F1, "indoor home integration."

Figure 8 shows the matching results between functions and corresponding solutions. From the table, it is clear that F1 and F3 are both mapped with S0, which talks about the "practice of composting at home," while F4, F5, and F6, which are not about real functions of products, got relatively low semantic similarity score with S1. This result exactly aligned with the clustering results of functions in intra-perspective relations.

Copyright © 2023 by ASME

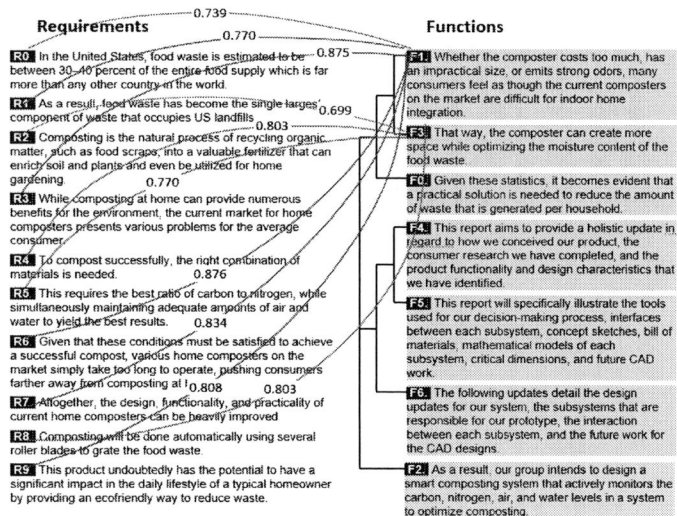

Figure 7. Mapping relationships between requirements and functions in case #1

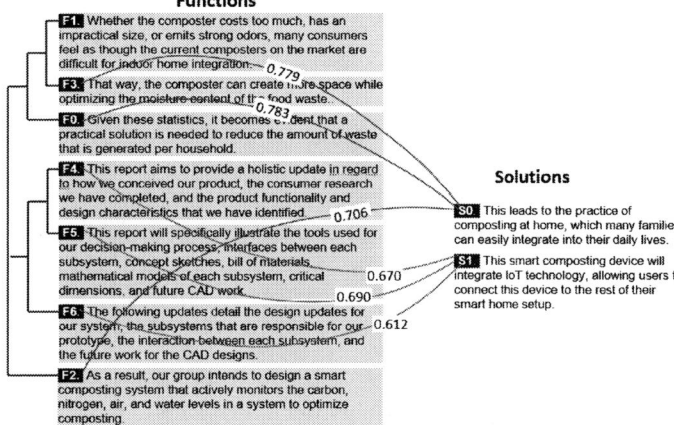

Figure 8. Mapping relationships between functions and solutions in case #1

Case #3 document describes a completed human-centered software design process, including background, ideation process, and conclusions. Three sentences are classified as <requirement>, five sentences are classified as <function>, and seven sentences are classified as <solution>. It is clear that the structure of this document is different from students' reports since more solutions are captured. Sentences labeled as <requirement> are related to background and design principles. For example, R2 introduced an applied design principle called "Orient design around life aspirations." Meanwhile, sentence categories such as <solution> illustrate the final design of this toolkit, such as "developed a financial coaching toolkit that provides a fresh take on an otherwise intimidating ritual" in S3.

Intra-perspective relations of <function>: As **Figure 9** shows, sentences classified as <function> are mainly grouped in one cluster, while F0 and F4 have the closest relationship and F3 is the most irrelevant. F0 and F4 described the way that the design team applied to customers in order to define functions such as "ask more difficult financial questions" or "help build long-term financial strategies" using collected feedback. Meanwhile, F1 and F2 are about the collected feedback, which

is precisely grouped with F0. And F3 presented a specific function of the toolkit, which is about "images represented Mission District accurately."

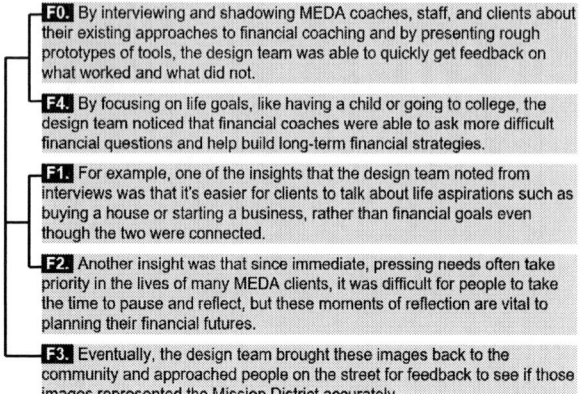

Figure 9. Clustering results of classified functions in case #3

Inter-perspective relations <Requirement–Function>: As **Figure 10** displays, requirements are matched with F0, F3, and F4, which are the main functions shown in intra-perspective relations. For instance, R1 introduced an applied design principle, "Orient design around life aspirations," in the design process, which is matched with F4, "ask more difficult financial questions" and "help build long-term financial strategies" implemented in the product under the mentioned design principle. In addition, R0 about "a visual language" is parallel with F3, "images represented the Mission District accurately" of the designed toolkit.

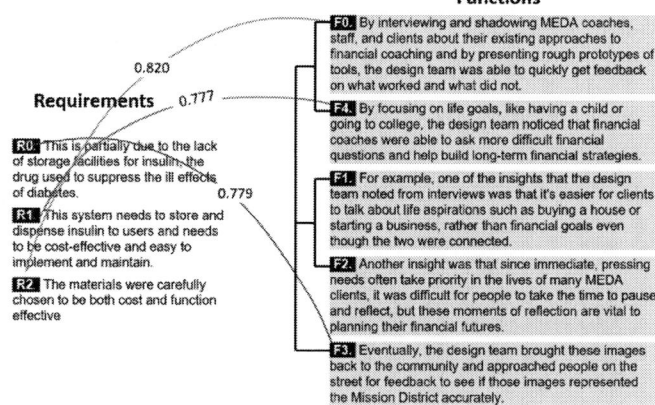

Figure 10. Mapping relationships between requirements and functions in case #3

Inter-perspective relations <Function – Solution>: **Figure 11** shows the mapping results between functions and solutions. F0 and F4 about "help build financial strategies" are both mapped to S2, illustrating the specific design about "the design team decided to build and run prototypes of different tools that linked the DISC model to life situations and goals." In addition, F3 about "images represented the Mission District accurately" is matched with S1 reviewing the determined design solution "by taking photos of murals, colors, decorations on the streets of the Mission."

According to the above results, finetuned BERT model shows a strong robustness ability in dealing with a different

Copyright © 2023 by ASME

source type. Encountering a completely distinct structure and writing style, the model can still utilize the domain knowledge embedded in high dimensional embedding and complete knowledge elicitation tasks.

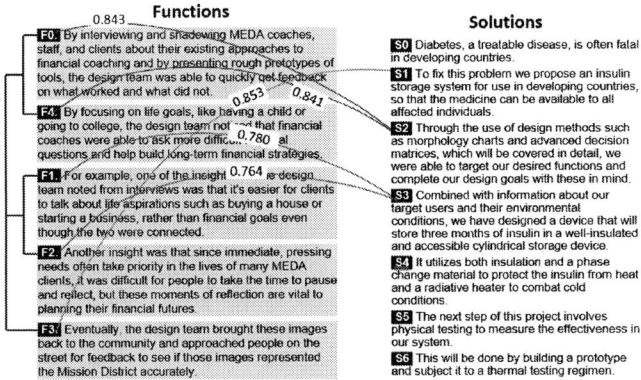

Figure 11. Mapping relationships between functions and solutions in case #3

4.4 Case #2: Varying Concept Categories

Case #2 is a summary extracted from students' reports in engineering design class containing eleven sentences that describes the entire design process, including background and conclusions at the sentence level. Four sentences are classified as <requirement>, and seven sentences are classified as <solution>. Different from Case #1, it can be seen that no function specifications are present in this document. Therefore, intra-perspective would be conducted using <requirement>, and inter-perspective would be conducted using <Requirement-Solution>. In addition, most of the sentences are classified as <requirement>, illustrating physical and functional requirements such as "cost-effective" in R1 and "function effective" in R2. Most of the solutions depicted the designed product in detail, such as "propose an insulin storage system" in S1 and "designed a device that will store three months of insulin" in S3. For the sequences that cannot be classified into any entities, finetuned BERT model categorized them as <solution> like S5 and S6.

Intra-perspective relations of <requirement>: Intra-perspectives are conducted using <requirement> since requirements can be classified under different aspects, such as physical requirements and functional requirements. As **Figure 12** shows, four requirements in this document are clustered in one group where R1 and R2, describing requirements of "to store and dispense insulin," "cost-effective," and "easy to implement and maintain" are the closest, while R3 depicting the future plan of the team is mostly irrelevant.

Inter-perspective relations of <Requirement–Solution>: The inter-perspective relations of <Requirement – Solution> mapping results are shown in **Figure 13**. Most of the requirements are mapped with Solution 1, which provides most of the details of the product. The requirements requiring "cost-effective" and "easy to implement" are mapped with S1 about the "insulin storage system available to all affected individuals." R3, which describes a potential advantage of the product, is matched with S2, illustrating a detailed product design process.

Although the training dataset used for finetuning is balanced, the results indicate that the finetuned BERT model can still handle imbalanced new data points. By uncovering the inner

structure inside design entities, the model expresses its robustness in dealing with imbalanced data.

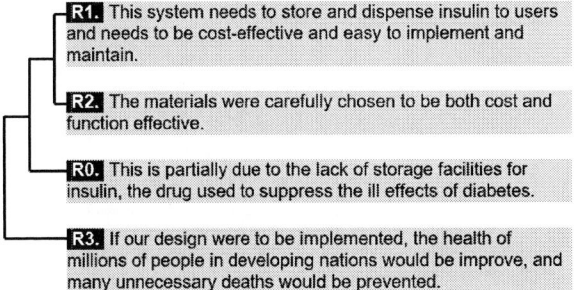

Figure 12. Clustering results of classified requirements: case #2

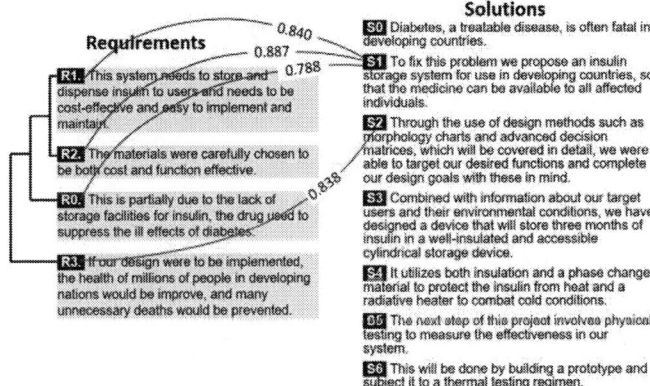

Figure 13. Mapping relationships between requirements and solutions in case #2

4.5 Case #5: Different Problem Domains

Case #5, with the same source types as Case #3, describes a different design process of a selfie application containing interviewing process, ideation method, and final solution. Twenty sentences are included with one < function> describing "receiving money" and "plan purchases," eight <Requirement> and eleven <Solution>. Since only one function specification is displayed, only requirement-related design perspective relations are generated.

Intra-perspective relations of < requirement>: The intra-perspective relations of <requirement> are shown in **Figure 14**. R2 and R5 are both relevant to customer needs on "financial challenges," which are closest. R0 is about the background and method to investigate the design problem where the content of R2 is an observation from that of R0. R3, R4, and R6 describe different findings and unsolved needs during the design process, such as "they resisted the temptation to spend," "without a social element, the students weren't particularly engaged," or "Other students could then like and comment on the photos."

Inter-perspective relations <Requirement – Solution>: Mapping results between < requirement> and < solution> are shown in **Figure 15**. Most of the requirements which presented needs about "social elements," "like and commented on the photos" are mapped with S9, which illustrates the resolved method for customer needs "classroom version of this exercise would have involved looking at photos and talking about etiquette, the app-enabled the experience to be interactive, personal, and fun." Moreover, specific challenges "to capture the fun and the currency" in R1, it is solved by "sharing and getting

Copyright © 2023 by ASME

feedback and affirmation" in S1. Compared to Case #2, the similarity score is much higher, which shows the confidence of the BERT model for mapping.

It can be seen that, even with the different problem domains, the common design knowledge embedding in sentence embedding can still be applied to different design branches. The commonality of knowledge that humans may not realize can be captured and elicited by the language model.

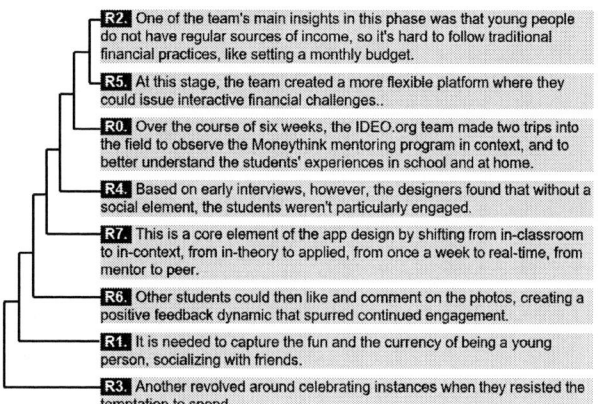

Figure 14. Clustering results of classified requirements: case #5

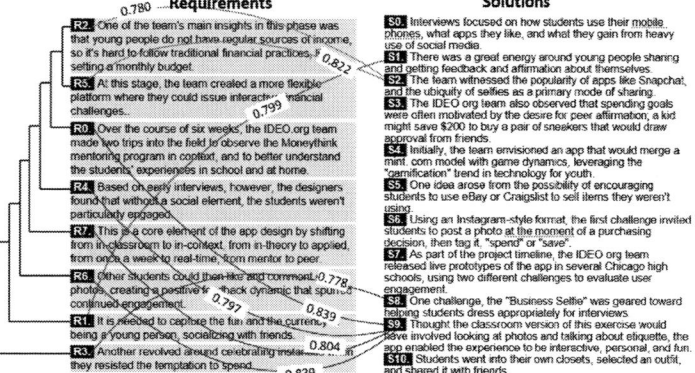

Figure 15. Mapping relationships between requirements and solutions in case #5

4.6 Case #4: Different Content Focuses

Case #4, extracted from IDEO.org's case studies, is an example of the mechanical design of an outstanding design team for an LED phototherapy device. The content of this document is largely different from other design-related documents shown above since it is generated from the perspective of the design team other than customers. Rather than describing a detailed design process, this document focuses more on the applied design methods and achievement of the design team. 13 sentences are included depicting the way that the team followed to solve a design problem about improving an LED phototherapy device. 11 sentences are categorized as <requirement>, and 2 sentences are categorized as <solution>. No function specifications are present in this document, so only requirement-related design perspective relation will be conducted.

Intra-perspective relations of <requirement>: The intra-perspective relations of <requirement> are shown in **Figure 16**. This document is written from the perspective of a design team. Therefore, many sentences cannot be considered as any of the

three classes. By using finetuned BERT model, other than sentences like R2 about customer requirements, "fewer components than previous phototherapy models, making it lighter, cheaper, and easier to maintain," are classified as a requirement, sentences like R9 introducing the background of the design story are also categorized in the same class. However, aside from the classification result of sentences, the clustering of < requirement> is matched with the original documents. For instance, R1 and R5 describe the way that the design team started the design such as "In the case of Brilliance, D-Rev did not invent a totally new technology" and "team incorporated feedback from everyone who touches Brilliance" are parallel with each other.

Inter-perspective relations <Requirement – Solution>: Mapping results between < requirement> and < solution> are shown in **Figure 17**. It can be seen that the similarity score between < requirement> and < solution> is much lower than the results of other case studies. Due to the specialty of the content of this document which is about a thinking process of a design team, it is hard for the model to extract specific solutions for the designed product. It is evident that the finetuned model is not confident with the mapping results and provides a low score matching. However, regardless of the classification results, the mappings between sentences are observable. R6, presenting the needs for a low-cost device like "country with the highest need for a low-cost device," are matched with S1 about a potential expectation of the product "nobody had yet designed a version that was affordable in low-income markets and could be delivered efficiently."

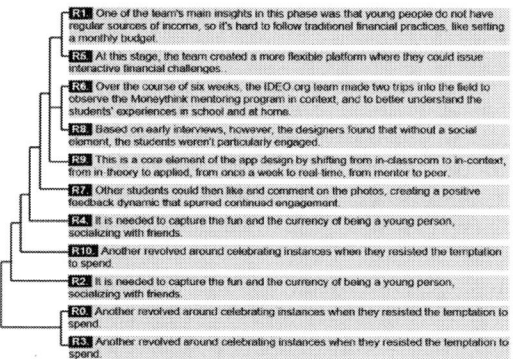

Figure 16. Clustering results of classified requirements: case #4

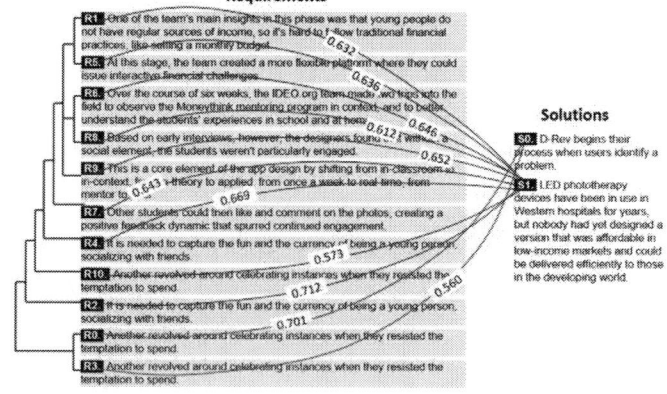

Figure 17. Mapping relationships between requirements and solutions in case #4.

Copyright © 2023 by ASME

When the content focus is completely different from the training dataset, finetuned BERT model shows its limitation. Since the labels defined by a human are limited, the BERT model is not able to recognize content outside of the defined domain. However, the low similarity score among design entities implies low confidence in dealing with the input document from the language model. From this signal, designers can quickly seize the hint that this document is largely distinct before they read the entire text and may need more human efforts to uncover the information.

5 DISCUSSION

Effectiveness of the domain knowledge capturing. The BERT model can be used to categorize unseen sequences and generate contextual embedding with good performance. With a small-size dataset, it can still be finetuned to generate relatively high classification accuracy and meaningful sequence embeddings. Using [CLS] token, which has better performance on sequence classification compared to averaging embedding methods, sequence embeddings created by the finetuned BERT model are applied for design-perspective conductions and knowledge elicitation with good performance.

Document understanding mechanism. Applying domain ontology knowledge for finetuning dataset labeling is an effective way to devise domain knowledge in a language model, such as BERT, so that it can "understand" the text from the domain's perspective. Such an "understanding" follows Kintch's C&I model in two ways. First, the language model's powerful formal language skills are applied to generate high-dimensional embeddings corresponding to the surface and text-base models. Secondly, the domain ontology instilled finetuning dataset provides background knowledge and contextual information to bring the process closer to "understanding." Further modeling work is needed to expand the ontological framework and test a variety of finetuning datasets for domain knowledge blending into the language model.

Robustness of language model. Although the domain-specific dataset is collected from bullet points or highlighted sections of students' reports at the phrase level, after being finetuned, the BERT model can complete text classification tasks and create meaningful high-dimensional embeddings in the sentence-level dataset. Instead of transferring from phrase level to sentence level, the finetuned BERT model is verified to be able to transfer learning and understanding between different types of documents, as shown by the results of case studies. The model performs well when encountering text about the design process from the perspective of customers on both students' project reports and other application design processes from IDEO.org. Even when processing untypical design documents where predefined classes cannot be appropriately applied, the BERT model can still uncover the inner relationships among design entities.

Design perspective knowledge extraction. The document-level knowledge extraction and elicitation on the basis of design perspective are realized by high-dimensional embeddings using hierarchical clustering and semantic similarity mapping. The results of sequence classifications have shown that the finetuned BERT model can focus on some specific range of text and classify it as short-length design entities like <function>. In addition, sometimes, the classification results of language models can be different from humans. For example, sentences describing functional requirements can also be considered as <function>, or sentences describing details in solutions can be considered as <function>. Both perspectives make sense from a human's point of view. But the sentence can only be classified as exactly one category by the language model. Besides, hierarchical clustering without setting a number of clusters in advance can successfully group the same design entities from an intra-perspective and provide clear relationships within the entity. And semantic similarity matching among different design entities can not only reflect the mapping information between two entities from the inter-perspective but also align with the clustering results from the intra-perspective.

6 CONCLUSIONS AND FUTURE WORK

In this paper, a document understanding-based design support framework is proposed. A descriptive language model, the BERT model, finetuned by design-related datasets, is utilized to classify sequences in documents and generate high-dimensional contextual embeddings. Meanwhile, design perspectives, together with intra-perspective and inter-perspective relations, are introduced to elicit design knowledge underlying the unstructured text and uncover the hierarchical structure of documents. Through the model-building process and the case study results, the conclusions can be drawn as follows:

- By finetuning at the phrase-level with a domain-specific dataset, the BERT model can effectively capture knowledge from texts and classify unseen sequences in different formats and content accurately.
- The BERT model finetuned with a certain predefined dataset can also classify sequences in different formats and content as well as generate high-dimensional embeddings containing design knowledge, demonstrating the robustness of the language model-based approach.
- Hierarchical clustering and semantic similarity mapping are effective for uncovering perspective structures underlying the documents.

The framework presented in this paper is limited in several ways. First, the scale is still small. To process large-scale documents and extract a wide range of domain knowledge, a much more efficient method is needed to prepare finetuning datasets. To develop such a method, the required "level of cleanness" of the dataset needs to be investigated. Second, given any ontological concepts for finetuning, there will be some sequences that may not belong to any of the categories, and a method is needed to filter unrelated sequences. Thirdly, the current finetuning dataset is cleanly labeled at the phrase level. It remains unclear whether the sentence-level labeling will hinder or enhance the performance of the framework. Our ongoing work addresses these issues.

REFERENCES

[1] Fleuren, W. W., & Alkema, W. (2015). Application of text mining in the biomedical domain. Methods, 74, 97-106

[2] Ferreira, R., de Souza Cabral, L., Lins, R. D., e Silva, G. P., Freitas, F., Cavalcanti, G. D., Lima, R., Simske, S. J., & Favaro, L. (2013). Assessing sentence scoring techniques

Copyright © 2023 by ASME

for extractive text summarization. Expert Systems with Applications, 40(14), 5755-5764.

[3] Mishra, R., Bian, J., Fiszman, M., Weir, C. R., Jonnalagadda, S., Mostafa, J., & Del Fiol, G. (2014). Text summarization in the biomedical domain: a systematic review of recent research. Journal of Biomedical Informatics, 52, 457-467.

[4] Camacho-Collados, J., & Pilehvar, M. T. (2018). From word to sense embeddings: A survey on vector representations of meaning. Journal of Artificial Intelligence Research, 63, 743-788.

[5] Qiu, Y. and Jin, Y. "Engineering Document Summarization: A Bidirectional Language Model-Based Approach" in J. of Computer and Information Science for Engineering, Vol.22, April 2022. 061010-1 – 16.

[6] Beigbeder, M., & Mercier, A. (2005). An information retrieval model using the fuzzy proximity degree of term occurences. Paper presented at the Proceedings of the 2005 ACM Symposium on Applied Computing, 1018-1022.

[7] Castells, P., Fernandez, M., & Vallet, D. (2006). An adaptation of the vector-space model for ontology-based information retrieval. IEEE Transactions on Knowledge and Data Engineering, 19(2), 261-272.

[8] Zhang, X., Hou, X., Chen, X., & Zhuang, T. (2013). Ontology-based semantic retrieval for engineering domain knowledge. *Neurocomputing, 116*, 382-391.

[9] Shi, F., Chen, L., Han, J., & Childs, P. (2017). A data-driven text mining and semantic network analysis for design information retrieval. Journal of Mechanical Design, 139(11)

[10] Martinez-Rodriguez, J. L., López-Arévalo, I., & Rios-Alvarado, A. B. (2018). Openie-based approach for knowledge graph construction from text. Expert Systems with Applications, 113, 339-355.

[11] Sarica, S., Luo, J., & Wood, K. L. (2020). TechNet: Technology semantic network based on patent data. Expert Systems with Applications, 142, 112995.

[12] Siddharth, L., Blessing, L., Wood, K. L., & Luo, J. (2022). Engineering Knowledge Graph from Patent Database. Journal of Computing and Information Science in Engineering, 22(2)

[13] Hou, Tianjun & Yannou, Bernard & Leroy, Yann & Poirson, Emilie. (2018). Mining changes of user expectations over time from online reviews.

[14] Kintsch, W. (2018). Revisiting the construction—integration model of text comprehension and its Implications for Instruction. Theoretical models and processes of literacy (pp. 178-203). Routledge.

[15] Gero, J. S., & Kannengiesser, U. (2014). The function-behaviour-structure ontology of design. An Anthology of Theories and Models of Design: Philosophy, Approaches and Empirical Explorations, , 263-283.

[16] Siddharth, L., Blessing, L., & Luo, J. (2022). Natural language processing in-and-for design research. Design Science, 8

[17] Lin, H., Chi, N., & Hsieh, S. (2012). A concept-based information retrieval approach for engineering domain-specific technical documents. Advanced Engineering Informatics, 26(2), 349-360.

[18] Liu, Q., Wang, K., Li, Y., & Liu, Y. (2020). Data-driven concept network for inspiring designers' idea generation. Journal of Computing and Information Science in Engineering, 20(3)

[19] Han, Y., & Moghaddam, M. (2021). Eliciting Attribute-Level User Needs From Online Reviews With Deep Language Models and Information Extraction. Journal of Mechanical Design, 143(6)

[20] Zhu, Q., Zhang, X., and Luo, J. (January 17, 2023). "Biologically Inspired Design Concept Generation Using Generative Pre-Trained Transformers." ASME. J. Mech. Des. April 2023; 145(4): 041409.

[21] Brisco, R., Whitfield, R. I., & Grierson, H. (2020). A novel systematic method to evaluate computer-supported collaborative design technologies. Research in Engineering Design, 31(1), 53-81.

[22] Ye, X., & Lu, Y. (2020). Automatic extraction of engineering rules from unstructured text: a natural language processing approach. Journal of Computing and Information Science in Engineering, 20(3)

[23] Siddharth, L., Li, G., and Luo, J., "Enhancing patent retrieval using text and knowledge graph embeddings: a technical note," Journal of Engineering Design, vol. 33, no. 8–9, pp. 670–683, Nov. 2022.

[24] Hubka, V., & Eder, W. E. (2012). Theory of technical systems: a total concept theory for engineering design. Springer Science & Business Media.

[25] Quintana-Amate, S., Bermell-Garcia, P., & Tiwari, A. (2015). Transforming expertise into Knowledge-Based Engineering tools: A survey of knowledge sourcing in the context of engineering design. Knowledge-Based Systems, 84, 89-97.

[26] Verhagen, W. J. C., Bermell-Garcia, P., van Dijk, R. E. C., & Curran, R. (2012). A critical review of Knowledge-Based Engineering: An identification of research challenges. Advanced Engineering Informatics, 26(1), 5-15. https://doi-org.libproxy1.usc.edu/10.1016/j.aei.2011.06.004

[27] Howard, T. J., Culley, S. J., & Dekoninck, E. (2008). Describing the creative design process by the integration of engineering design and cognitive psychology literature. Design Studies, 29(2), 160-180.

[28] Beitz, W., Pahl, G., & Grote, K. (1996). Engineering design: a systematic approach. MRS Bulletin, 71

[29] Ullman, D. G. (1992). The mechanical design process. McGraw-Hill New York.

[30] Suh, N. P. (1995). Axiomatic design of mechanical systems.

[31] Rocca, G. L. (2012). Knowledge based engineering: Between AI and CAD. Review of a language based technology to support engineering design. Advanced Engineering Informatics, 26(2), 159-179. https://doi-org.libproxy1.usc.edu/10.1016/j.aei.2012.02.002

[32] Bouhana, A., Zidi, A., Fekih, A., Chabchoub, H., & Abed, M. (2015). An ontology-based CBR approach for personalized itinerary search systems for sustainable urban freight transport. Expert Systems with Applications, 42(7), 3724-3741.

Available from Https://Github.Com/Doccano/Doccano,

[33] Morkos, B., Mathieson, J., & Summers, J. D. (2014). Comparative analysis of requirements change prediction

models: manual, linguistic, and neural network. Research in Engineering Design, 25(2), 139-156.

[34] Zhang, Z., Liu, L., Wei, W., Tao, F., Li, T., & Liu, A. (2017). A systematic function recommendation process for data-driven product and service design. *Journal of Mechanical Design, 139*(11)

[35] Sanya, I. O., & Shehab, E. M. (2015). A framework for developing engineering design ontologies within the aerospace industry. International Journal of Production Research, 53(8), 2383-2409.

[36] Zhang, C., Zhou, G., Lu, Q., & Chang, F. (2017). Graph-based knowledge reuse for supporting knowledge-driven decision-making in new product development. *International Journal of Production Research, 55*(23), 7187-7203.

[37] Sarica, S., & Luo, J. (2021). Design knowledge representation with technology semantic network. Proceedings of the Design Society, 1, 1043-1052.

[38] Akay, H., & Kim, S. (2021). Extracting functional requirements from design documentation using machine learning. Procedia CIRP, 100, 31-36.

[39] Loper, E., & Bird, S. (2002). Nltk: The natural language toolkit. arXiv Preprint Cs/0205028,

[40] Nakayama, H., Kubo, T., Kamura, J., Taniguchi, Y., & Liang, X. (2018). doccano: Text annotation tool for human. Software Available from Https://Github.Com/Doccano/ Doccano

[41] Cohen, J. (1960). A coefficient of agreement for nominal scales. Educational and Psychological Measurement, 20(1), 37-46.

[42] McHugh, M. L. (2012). Interrater reliability: the kappa statistic. Biochemia Medica, 22(3), 276-282.

[43] Devlin, J., Chang, M., Lee, K., & Toutanova, K. (2018). Bert: Pre-training of deep bidirectional transformers for language understanding. arXiv Preprint arXiv:1810.04805

[44] Vaswani, A., Shazeer, N., Parmar, N., Uszkoreit, J., Jones, L., Gomez, A. N., Kaiser, Ł., & Polosukhin, I. (2017). Attention is all you need. Advances in Neural Information Processing Systems, 30

[45] Zhang, T., Wu, F., Katiyar, A., Weinberger, K. Q., & Artzi, Y. (2020). Revisiting few-sample BERT finetuning. *arXiv Preprint arXiv:2006.05987*

[46] Kingma, D. P., & Ba, J. (2014). Adam: A method for stochastic optimization. arXiv Preprint arXiv:1412.6980

[47] Kalyan, K. S., & Sangeetha, S. (2021). BertMCN: Mapping colloquial phrases to standard medical concepts using BERT and highway network. Artificial Intelligence in Medicine, 112, 102008.https://doi-org.libproxy1.usc.edu/ 10.1016/j.artmed.2021.102008

[48] Qiu, Y., & Jin, Y. (2022). Engineering Document Summarization: A Bidirectional Language Model-Based Approach. Journal of Computing and Information Science in Engineering, 22(6), 061004. [50] Ullman, D. G. (1992). The mechanical design process. McGraw-Hill New York.

[49] Miller, D. (2019). Leveraging BERT for extractive text summarization on lectures. arXiv Preprint arXiv:1906.04165 [50] Alghanmi, I., Espinosa-Anke, L., & Schockaert, S. (2020). Combining BERT with static word embeddings for categorizing social media.

[51] Reimers, N., & Gurevych, I. (2019). Sentence-bert: Sentence embeddings using siamese bert-networks. arXiv Preprint arXiv:1908.10084

[52] Cheng, X. (2021). Dual-view distilled bert for sentence embedding. Paper presented at the Proceedings of the 44th International ACM SIGIR Conference on Research and Development in Information Retrieval, 2151-2155.

[53] Ogot, M. (2004). EMS models: adaptation of engineering design black-box models for use in TRIZ. Paper presented at the Proceedings of the ETRIA TRIZ Future Conference, No

[54] Abbas, O. A. (2008). Comparisons between data clustering algorithms. International Arab Journal of Information Technology (IAJIT), 5(3)

[55] Beeferman, D., & Berger, A. (2000). Agglomerative clustering of a search engine query log. Paper presented at the Proceedings of the Sixth ACM SIGKDD International Conference on Knowledge Discovery and Data Mining, 407-416.

[56] Savaresi, S. M., Boley, D. L., Bittanti, S., & Gazzaniga, G. (2002). Cluster selection in divisive clustering algorithms. Paper presented at the Proceedings of the 2002 SIAM International Conference on Data Mining, 299-314.

[57] Van der Maaten, L., & Hinton, G. (2008). Visualizing data using t-SNE. Journal of Machine Learning Research, 9(11)

[58] Kintch, W. (1998) *Comprehension: A Paradigm of Cognition*, Cambridge University Press, Cambridge UK.

[59] Wharton, C., & Kintsch, W. (1991). An overview of construction-integration model: a theory of comprehension as a foundation for a new cognitive architecture. *SIGART Bull., 2*, 169-173.

[60] Mahowald, K., AA Ivanova, I. A. Blank, N. Kanwisher, J. B. Tenenbaum, E. Fedorenko (2023), Dissociating language and thought in large language models: a cognitive perspective, arXiv:2301.06627v1 [cs.CL] 16 Jan 2023

Copyright © 2023 by ASME

Proceedings of the ASME 2023
International Design Engineering Technical Conferences and
Computers and Information in Engineering Conference
IDETC-CIE2023
August 20-23, 2023, Boston, Massachusetts

DETC2023-116937

TOPOLOGY OPTIMIZATION USING NEURAL NETWORKS WITH CONDITIONING FIELD INITIALIZATION FOR IMPROVED EFFICIENCY

Hongrui Chen   Aditya Joglekar   Levent Burak Kara[*]
Department of Mechanical Engineering
Carnegie Mellon University
Pittsburgh, PA, 15213, USA

ABSTRACT

We propose conditioning field initialization for neural network based topology optimization. In this work, we focus on (1) improving upon existing neural network based topology optimization, (2) demonstrating that by using a prior initial field on the unoptimized domain, the efficiency of neural network based topology optimization can be further improved. Our approach consists of a topology neural network that is trained on a case by case basis to represent the geometry for a single topology optimization problem. It takes in domain coordinates as input to represent the density at each coordinate where the topology is represented by a continuous density field. The displacement is solved through a finite element solver. We employ the strain energy field calculated on the initial design domain as an additional conditioning field input to the neural network throughout the optimization. The addition of the strain energy field input improves the convergence speed compared to standalone neural network based topology optimization.

1 INTRODUCTION

There has been a recent increase in machine learning driven topology optimization approaches, particularly using neural networks for performing topology optimization. Both data-driven and online training based approaches have been explored. Data-driven approaches require large training database generation and a long training time. They perform instant optimal topology generation during inference time. Online training approaches use the neural network to represent the density field of a single to a small subset of designs for better parameterization. The online training approaches require similar or more time compared to conventional topology optimization approaches like SIMP (Solid Isotropic Material with Penalisation) [1, 2]. We find that the results of the online training approaches, particularly the convergence speed, can be improved through insights derived from the mechanical aspects of the problem.

Machine learning driven topology optimization approaches offer the advantage of being easily able to accommodate additional insights in the form of pre-computed fields. The usage of these fields has been explored in data-driven approaches such as TopologyGAN [3], which use physical fields such as von Mises stress and strain energy density for achieving better results. However, there has been no work incorporating these physical fields in the online training topology optimization setting. In this work, we further improve upon TOuNN (Topology Optimization using Neural Networks), an online training approach proposed by Chandrasekhar and Suresh [4], by adding a strain energy field in addition to the domain coordinates as a conditioning input to the neural network. We show that this improves the convergence speed and can give a better compliance. With the additional strain energy field as a conditioning input, the neural network not only learns a mapping function between the domain coordinates to the density field output but also between the strain energy field to the density field output. Ideally, if the conditioning field is the same as the converged topology, then the neural network only needs to learn a constant function which is the identity function. However, the converged topology is not known at the

---

[*]Address all correspondences to lkara@cmu.edu

Copyright © 2023 by ASME

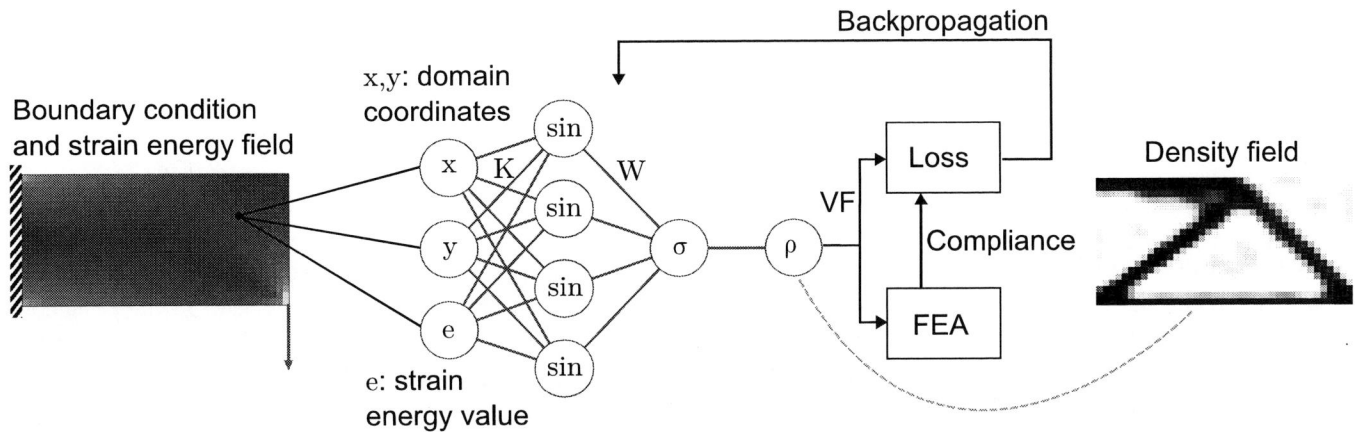

FIGURE 1: The strain energy field is calculated at the beginning of the optimization based on the boundary condition. The strain energy conditioning field is fixed throughout the training. Domain coordinates and the strain energy value at each coordinate point is used as the input to the neural network. The neural network outputs density ρ at each coordinate point. By sampling coordinate point across the design domain, we obtain the density field. From the density field, we calculate the current volume fraction and the compliance from a FEA solver. The compliance and volume fraction is then formulated as a loss function which is used in back propagation of the training process until convergence.

beginning of the optimization. Thus, the strain energy field is used as a good alternative since it can be computed through a single function call of Finite Element Analysis (FEA) prior to the online training of the neural network. We verify the performance increase obtained with this additional conditioning input across parametric experiments with varying boundary conditions and volume fractions.

The code for running the experiments in this paper can be found at: https://github.com/HongRayChen/Hybrid-TopOpt

2 RELATED WORK

Conventional topology optimization: Bendsøe and Kikuchi [5] introduced the homogenization approach for topology optimization. The SIMP method [1, 2] considers the relative material density in each element of the Finite Element (FE) mesh as design variables, allowing for a simpler interpretation and optimised designs with more clearly defined features. Other common approaches to topology optimization include the level-set method [6, 7] and evolutionary algorithms [8].

All these methods use an iterative process to create a complex mapping from problem characteristics (supports, loads and objective function) to an optimised structure, where each iteration has an expensive FEA calculation involved. A more accurate and detailed solution can be obtained with greater number of elements in the FE mesh, however this increases the computational cost. Therefore, current developments within the field

are strongly motivated by the desire to either limit the number of iterations needed to obtain an optimised structure or the computational cost of completing an iteration [9]. Recent advances in deep learning, particularly for image analysis tasks, have showed potential for removing the expensive FEA iterations required until the convergence of the topology in the conventional topology optimization approaches. Hence, various topology optimization approaches that utilize neural networks have been proposed. Woldseth et al. [9] provide an extensive overview on this topic.

Data-driven topology optimization: We refer to data-driven topology optimization methods as those that aim to learn a neural network model from a database of topology optimization results for instant prediction of the optimal topology. Many methods rely on Convolutional Neural Networks (CNN) for their capabilities to learn from a large set of image data. Banga et al. [10] used a 3D encoder-decoder CNN to generate 3D topology results and show that interpolating the final output using the 3D CNN from the initial iterations obtained from the 'TopOpt' [11] solver, offers a 40% reduction in time over the conventional approach of using the solver alone. Yu et al. [12] use a conditional generative adversarial network (cGAN) in addition to CNN based encoder-decoder network. However, the results indicate there sometimes there may be disconnections present in the predicted topology which may drastically affect the compliance values. Nakamura and Suzuki [13] improve on the results with their direct design network and with a larger dataset, however, disconnections are still observed in some solutions. Behzadi and Ilieş [14] used deep

Copyright © 2023 by ASME

transfer learning with CNN. Zheng et al. [15] used U-net CNN for 3D topology synthesis. Nie at al. [3] used various physical fields computed on the original, unoptimized material domain, as inputs to the generator of a cGAN and achieved a 3 times reduction in mean square error as compared to a baseline cGAN. Mazé and Ahmed [16] show that diffusion models can outperform GANs for this task. They use regressor and classifier guidance to ensure that the generated structures are manufacturable and mechanical compliance has been minimized.

All these data-driven approaches aim to reduce optimal topology prediction time but face difficulties in generalization. Though over the years there have been improvements on the generalization capability, suitable training dataset generation is not trivial, especially for the 3D domain, and satisfactory and reliable results have not been achieved yet for direct use in real-world problems.

Online training topology optimization: We refer to online training topology optimization methods as those which do not use any prior data, rather train a neural network in an self-supervised manner for learning the optimal density distribution/topology. Chandrasekhar and Suresh [4] explored a online approach where the density field is parameterized using a neural network. Fourier projection based neural network for length scale control [17] and application for multi-material topology optimization [18] has also been explored . Deng and To [19] propose topology optimization with Deep Representation Learning, with a similar concept of re-parametrization, and demonstrate the effectiveness of proposed method on minimum compliance and stress-constrained problems. Deng and To [20] also propose a neural network based method for level-set topology optimization, where the implicit function of level-set is described by a fully connected deep neural network. Zehnder et al. [21] effectively leverage neural representations in the context of mesh-free topology optimization and use multilayer perceptrons to parameterize both density and displacement fields. It enables self-supervised learning of continuous solution spaces for topology optimization problems. Mai et al. [22] develop a similar approach for optimum design of truss structures. Hoyer et al. [23] use CNNs for density parametrization and directly enforce the constraints in each iteration, reducing the loss function to compliance only. They observe that the CNN solutions are qualitatively different from the baselines and often involve simpler and more effective structures. Zhang at al. [24] adopt a similar strategy and show solutions for different optimization problems including stress-constrained problems and compliant mechanism design.

Generalization is not an issue with all these online training topology optimization methods. However, the computational time and cost is similar to traditional topology optimization approaches. An advantage offered is that the density representation is independent of the FE mesh and because of the analytical density-field representation, sharper structural boundaries can be obtained [4]. We show that by adding an initial condition field as an extra input, we can improve the convergence speed and get better results.

3 PROPOSED METHOD

In our proposed method, the density distribution of the geometry is directly represented by the topology neural network. The strain energy field and the compliance used for backpropagation is calculated from an FE solver. The program is implemented in Python and backpropagation of the loss function into each module is handled by the machine learning package TensorFlow [25].

3.1 Neural network

The topology network $T(\mathbf{X})$ (Figure 1), learns a density field in a different manner as compared to typical topology optimization which represents the density field as a finite element mesh. The topology neural network takes in domain coordinates x, y, as well as the strain energy value e at coordinate x, y. The strain energy value gets concatenated with the domain coordinates to form the input to the topology network, $\mathbf{X} = [x, y, e]$. The domain coordinates are normalized between -0.5 to 0.5 for the longest edge. It outputs the density value ρ at each coordinate point. The domain coordinates represent the center of each element in the design domain. During topology optimization, a batch of domain coordinates that correspond to the mesh grid and the corresponding strain energy field is fed into the topology network. The output is then sent to the Finite Element Analysis (FEA) solver. The solver outputs the compliance which is combined with the volume fraction violation as a loss. The loss is then backpropagated to learn the weights of the topology network.

For the topology network design, we employed a simple architecture that resembles the function expression of $f(x) = \mathbf{w}sin(\mathbf{k}x + \mathbf{b})$. Similar neural network architectures have been used to control the length scale of geometry in topology optimization [17]. The conditioned domain coordinates are multiplied with a kernel $\mathbf{K}$. The kernel $\mathbf{K}$ regulates the frequency of the sine function. We add a constant value of 1 to break the sine function's rotation symmetry around the origin. We use a Sigmoid function to guarantee the output is between 0 and 1. The topology network can be formulated as follows:

$$T(\mathbf{X}) = \sigma(\mathbf{W}\sin(\mathbf{K}\mathbf{X} + 1)) \tag{1}$$

where:

$\mathbf{X}$: Domain coordinate input, $\mathbf{X} = (x, y, e)$

σ: Sigmoid activation function

$\mathbf{K}$: Trainable frequency kernels, initialized in $[-25, 25]$

$\mathbf{W}$: Trainable weights, initialized to 0

Copyright © 2023 by ASME

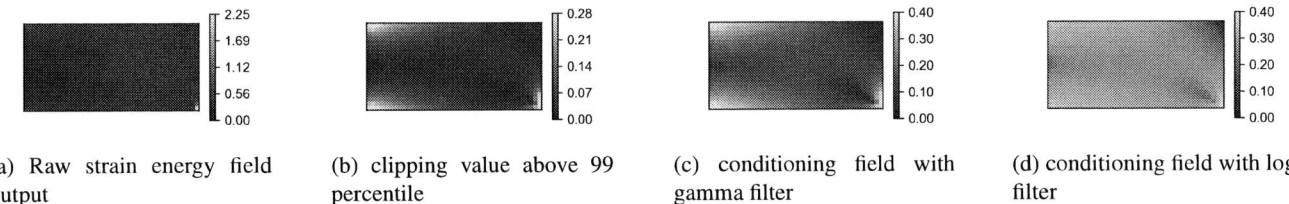

(a) Raw strain energy field output

(b) clipping value above 99 percentile

(c) conditioning field with gamma filter

(d) conditioning field with log filter

FIGURE 2: Two method is evaluated in terms of processing of the strain energy conditioning field. We used a gamma filter in (a-c) and a log filter in (d)

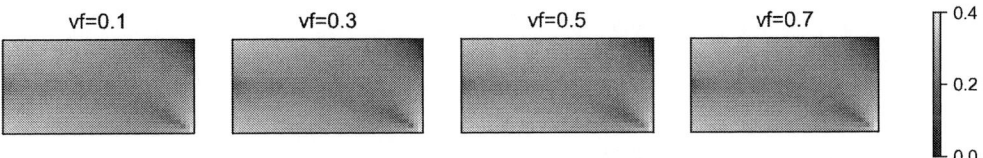

FIGURE 3: For the gamma filtering of the conditioning field, we adjust the gamma based on the volume fraction target of the optimization

We can upsample the 3D coordinate input or only sample specific regions of the density field to manipulate the resolution of the discretized visualization. Due to the strain energy conditioning field computed from the finite element mesh grid, interpolation needs to be used to calculate the intermediate values when upsampling the domain coordinates.

3.2 Strain energy conditioning field

The strain energy conditioning field is used to augment the domain coordinate input. We calculate the conditioning field from the initial homogeneous density domain. In topology optimization, for a 2D problem with n elements of four nodes each, the strain energy field $\mathbf{E}$ can be calculated as follows:

$$\mathbf{E} = \sum (\mathbf{U}_e \times \mathbf{S}_e) \circ \mathbf{U}_e \qquad (2)$$

where:

$\mathbf{U}_e$: the displacement matrix, $n \times 8$
$\mathbf{S}_e$: the element stiffness matrix, 8×8

The summation is along the axis containing the values for each element.

In most topology optimization implementations, the compliance is then calculated by summation of the above strain energy for all elements.

The strain energy field can vary greatly in range depending on the problem domain size, boundary condition, and geometry constraints. Therefore, normalization needs to be done to regulate the value range of the strain energy field. Otherwise, the range of the strain energy field will deviate from the normalized

range of the domain coordinates. Furthermore, a simple normalization will not suffice as the high max value of the strain energy field reduces the amplitude of other relevant features and patterns (Figure 2 (a)). We explore gamma and logarithmic filtering to normalize the strain energy field. For the gamma filtering, we clip the strain energy field by using the 99th percentile, P_{99}. After clipping, more details of the field $\mathbf{E}_c$ can be seen (Figure 2 (b)). We also further adjust the feature of the strain energy field by using gamma correction. The gamma value is set to be the complement of the target volume fraction V^* for the optimization ($\gamma = 1 - V^*$). The effect of the gamma correction based on the volume fraction is illustrated in Figure 3. As the volume fraction increases, the edge feature in the strain energy field is more and more pronounced. Finally, after the gamma correction step, the strain energy field is normalized between 0 and 0.4 to obtain the processed field $\mathbf{E}_p$. The processing step on the gamma filtering of strain energy field can be summarized in the following equation:

$$\mathbf{E}_c = min(\mathbf{E}, P_{99}) \qquad (3)$$

$$\mathbf{E}_\gamma = 0.4 \left\{ \frac{\mathbf{E}_c - min(\mathbf{E}_c)}{max(\mathbf{E}_c) - min(\mathbf{E}_c)} \right\}^\gamma \qquad (4)$$

For the logarithmic filtering, we do not clip the value, instead, the log filter is directly applied to the strain energy field

Copyright © 2023 by ASME

and then normalized between 0 and 0.4. We determine this range empirically to give the best results.

$$\mathbf{E}_{\log} = 0.4 \frac{\log \mathbf{E} - min(\log \mathbf{E})}{max(\log \mathbf{E}) - min(\log \mathbf{E})} \quad (5)$$

3.3 Online topology optimization with neural network

During optimization, the topology network outputs the density value at the center for each element. These density values are then sent to the finite element solver to calculate compliance based on the SIMP interpolation.

The finite element solver is treated as a black box within the neural network. It takes in the density of each element and outputs the compliance and the sensitivity for each element with respect to the compliance. Variables that are being optimized are the weights $\mathbf{W}$ and kernels $\mathbf{K}$ of the neural network. Adam [26] is used to train the neural network. The constrained optimization problem needs to be transferred into unconstrained minimization problem for neural network. We adopt the loss function formulated by Chandrasekhar and Suresh [4] of compliance minimization and volume fraction constraint. The combined loss function is

$$L = \frac{c}{c_0} + \alpha(\frac{\bar{\rho}}{V^*} - 1)^2 \quad (6)$$

In the optimization, the target volume fraction V^* is an equality constraint and $\bar{\rho}$ is the volume fraction of the current design. When α increased to infinity, the equality constraint is satisfied. We assign a maximum value of 100 for α with initial value of 1 and gradually increase α every iteration. c is the current compliance and c_0 is the initial compliance calculated on the design domain with the uniform volume fraction V^*.

4 RESULTS AND DISCUSSIONS

The possible combinations of boundary conditions, problem size, and configurations is enormous. It is impossible for us to cover all. To demonstrate the effectiveness of our proposed approach, we explore both a beam problem and a parametric study in 2D. In the beam problem, we showcase the convergence of the network's output and the convergence history. In the parametric study, problems across different boundary conditions and volume fractions are explored. We report the compliance value where subscript FENN represents Finite Element (FE) compliance solver with Neural Network (NN) as topology representation, and FENNCF as neural topology optimization with strain energy Conditioning Field (CF). For these two experiments, the problem size is 40×20 pixels. Comparison with SIMP is not the main focus of this paper and a more in-depth comparison can be found in TOuNN [4]. All experiments are run on a PC with i7-12700K as processor, 32 GB of RAM, and Nvidia RTX3080 GPU.

4.1 Beam example

Our first experiment is the beam example. The left side of the domain is fixed and a downward point load is on the center-right side. The boundary condition illustration and the strain energy conditioning field are shown in Figure 4. The target volume fraction is 0.3. We run the online topology optimization for a total of 1000 epochs.

The convergence history plot is illustrated in Figure 4 (c). We observe that by epoch 50, with the strain energy conditioning field, the network's compliance takes over the lead and maintains lower compliance all the way to the end of the training epochs. We also show the density field snapshot and the corresponding compliance during training in Figure 4 (b). Analyzing the geometry of the neural network with the conditioning field, we observe that there is a subtle difference compared to without the conditioning field. The neural network with conditioning field shares greater similarities to the strain energy field where the top and bottom edges are shorter. We can also observe that most of the geometry convergence happens between 0 and 400 epochs. Between 400 to 1000 epochs, the geometry remained relatively unchanged. The only change being a darker tone of red, showing the density values get pushed closer towards 1. In both of the examples, the final volume fraction is within 1% error of the given target volume fraction. Therefore, we do not include the volume fraction convergence plot.

4.2 Parametric study

We set up a parametric study to analyze the effectiveness of the gamma and log filter of the conditioning field. The boundary condition setup is illustrated in Figure 5 (a). The bottom right loading point is varied across the region highlighted in green which accounts for 50 load conditions. We also vary the target volume fraction between 0.2 to 0.5 with an increment of 0.1. In total, this sums up to 200 total combinations. In the previous beam example, we observe that geometries do not change significantly after 400 epochs, therefore we limit the total epochs for the parametric study to 400 epochs.

The parametric study result is summarized in Figure 6. In Figure 6 (a), we sort with respect to the compliance of topology optimization without conditioning field and show the compliance from both methods. We observe that the overall conditioning field converged at lower compliance. The improvement of the conditioning field is more significant when the compliance is higher. The higher compliance occurs when the volume fraction is low. To visualize the convergence speed increase, Fig-

Copyright © 2023 by ASME

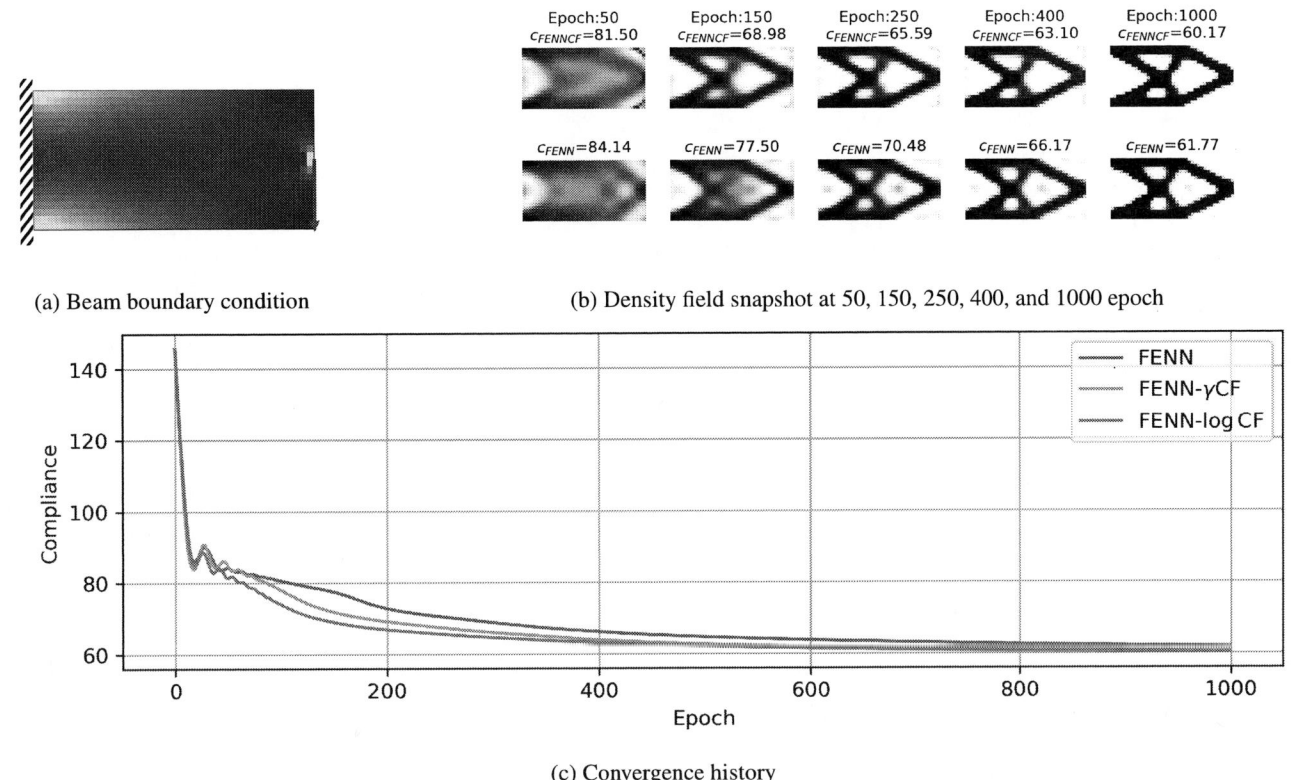

(a) Beam boundary condition

(b) Density field snapshot at 50, 150, 250, 400, and 1000 epoch

(c) Convergence history

FIGURE 4: Comparing the convergence history for a beam example for with and without strain energy conditioning field. The result presented is using the gamma filtering. For FENN-logCF took 22.5s while FENN took 22.1s.

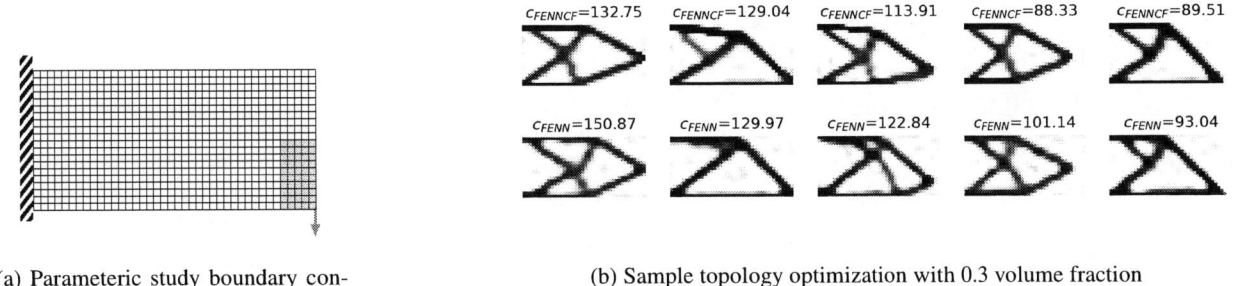

(a) Parameteric study boundary conditions

(b) Sample topology optimization with 0.3 volume fraction

FIGURE 5: Boundary conditions and some sample topology optimization results with 0.3 volume fraction within the parametric study examples

ure 6 shows the percentage improvement with the conditioning field. The percentage improvement is calculated by identifying the epoch at which the conditioning field reaches a lower compliance compared to the final compliance of the optimization without the conditioning field. The average performance increase with gamma filter is 37.6% and with log filter is 44.7%. With both filters, the performance increase is more pronounced with

lower volume fraction examples. The log filter has a better overall performance increase across all solutions compared to gamma filter.

We compare our result against the result of "88-lines" by Andreassen et. al. [27] with a filtering radius of 1.5 to accommodate the problem size. We observe that when the compliance is low, FENN performed slightly better than SIMP. This is also con-

Copyright © 2023 by ASME

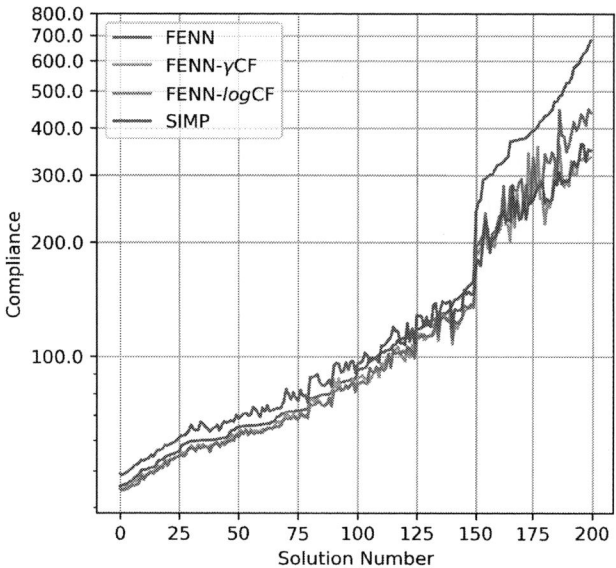

(a) Compliance comparison of with and without strain energy field

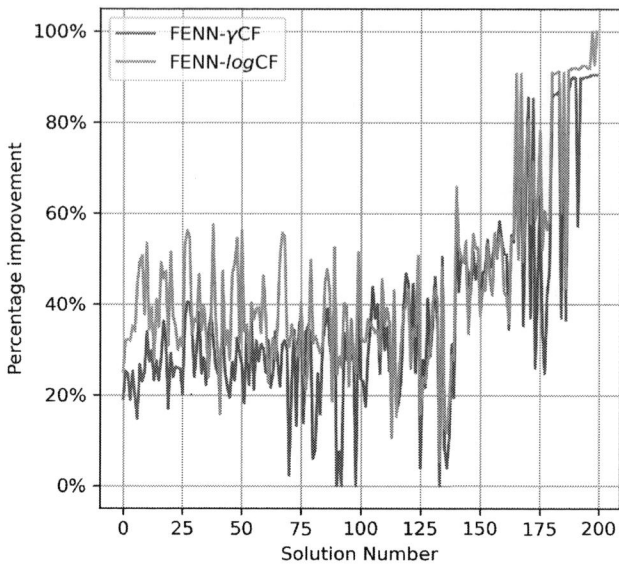

(b) Percentage convergence speed improvement

FIGURE 6: Comparing the final compliance and the speed of convergence for parametric study examples for with gamma and log filter and without the conditioning field. We also run the same problem configuration with "88-lines" by Andreassen et al. [27] denoted by the legend "SIMP" in the figure.

sistent with the result reported by Chandrasekhar and Suresh [4]. For problems with relatively higher compliance, we observe that FENN with conditioning field can in some cases converge to a lower compliance than "88-lines". We note that in general, the Matlab code [27] takes around 0.2 to 1.5s to run whereas FENN and FENN with either conditioning field takes around 10s. However, a definite time comparison is difficult to establish as "88-lines" runs on Matlab whereas FENN runs on Python. In "88-lines" the optimizer is optimality criteria whereas FENN rely on Adam with a learning rate of 0.002.

We also observe that within the 200 examples with gamma filter, there are four cases where the conditioning field does not improve convergence speed. When plotting out example results in Figure 5, the examples with the load on the right bottom edge have lower performance increase with the conditioning field. On the other hand the examples with the load close to the center have a greater performance increase and a bigger gap in compliance. Our hypothesis is that the conditioning field approach performs best when the topology is complex. The complexity in geometry can occur based on the volume fraction constraint or the configuration of the boundary conditions. As the volume fraction decrease, thinner members are required which increase complexity of the structure. Whereas the geometries in Figure 5 (b) showed that for the same volume fraction, the length scale of the part is also dependent on the boundary condition.

4.3 Additional examples

In Figure 7, we demonstrate the improvements resulting from the conditioning field on 4 complex boundary conditions in 2D. Cases 2, 3 and 4 in Figure 7 have obstacle regions (passive elements). Furthermore, in Figure 8, we analyze the impact of increasing the problem resolution (i.e. the FE mesh size) for the boundary conditions of case 1 in Figure 7, and observe similar improvements. We also show the improvements seen for a 3D problem in Figure 9.

5 LIMITATIONS AND FUTURE WORK

We exploit the ability of neural networks as a universal function approximator to learn the additional mapping from the strain energy conditioning field to the density field output. Currently, the improvement with the conditioning field is not stable across all possible boundary condition configurations. More tuning and testing is required. Another aspect is that the current conditioning field remains fixed during optimization. This is due to the neural network's inability to encode temporal features. The strain energy field changes throughout the optimization, without the ability to capture the temporal feature of the changing strain energy field. As such, the neural network has difficulty providing stable optimization results.

This work also demonstrates promising results using a conditioning field for online neural topology optimization. The

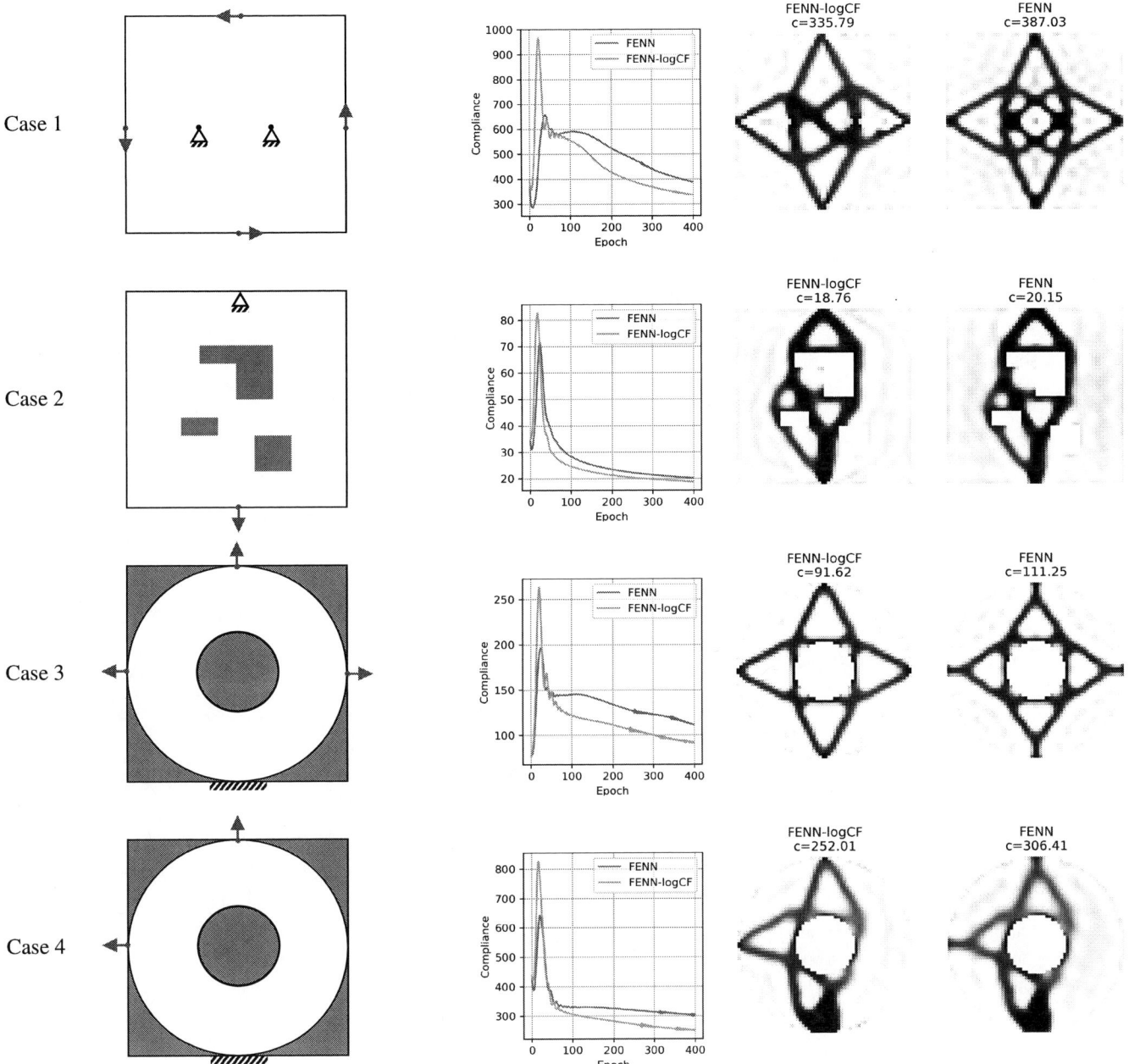

FIGURE 7: Four additional test cases across varying boundary conditions and passive elements, all using 0.2 target volume fraction. Each example is 60×60 in resolution and takes around 30 seconds to run with no significant difference between with and without conditioning field. Log filtered conditioning field demonstrates good convergence speed increase.

strain energy field may not be the best conditioning field out there and future work may focus on trying out different combinations of conditioning fields similar to TopologyGAN [3]. This conditioning field approach may demonstrate great synergy with the existing data-driven approach. Using the output of data-driven topology optimization as the conditioning field, online optimization can exploit a conditioning field that is much closer to the final solution. This reduces the complexity of the map-

Case 1 with 120×120 resolution

Case 1 with 180×180 resolution

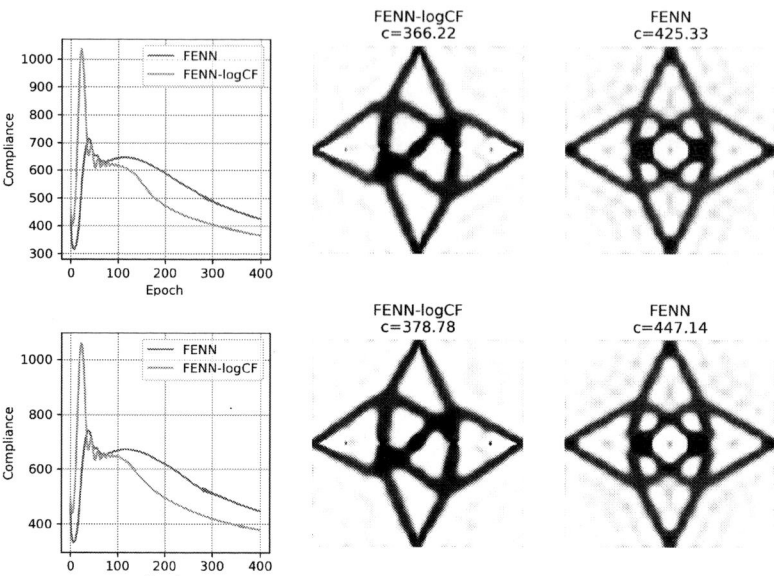

FIGURE 8: We run the same boundary condition for Case 1 with two and three times the resolution. The runtime for 120×120 is 3 min and for 180×180 is 20 min

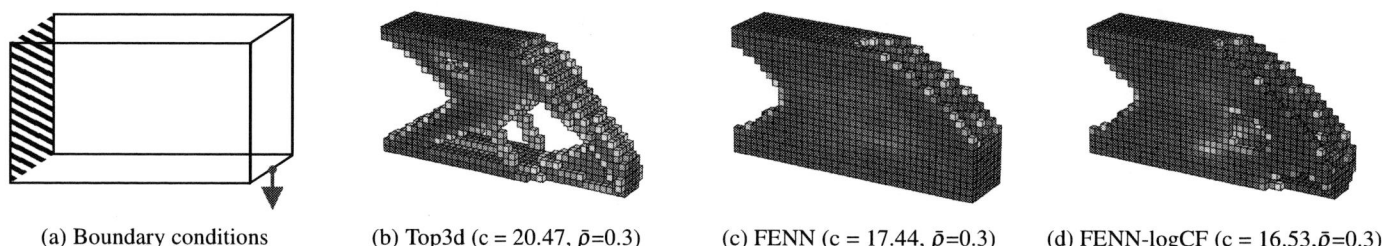

(a) Boundary conditions   (b) Top3d (c = 20.47, $\bar{\rho}$=0.3)   (c) FENN (c = 17.44, $\bar{\rho}$=0.3)   (d) FENN-logCF (c = 16.53, $\bar{\rho}$=0.3)

FIGURE 9: Comparing the results for a 3D cantilever beam example. All examples are run for 200 epochs. b) Top3d [28](standard 3d topology optimization code using SIMP). c) Using a neural network for density parametrization. d) Using a neural network for density parametrization and additional initial strain energy input with log filtering. We observe that FENN and FENN-logCF choose to create a shell around both side which gives an illusion that the volume fraction is higher. However, the volume fraction is also very close to the target volume fraction of 0.3 (both converged to 0.3003 specifically).

ping function for which the neural network needs to learn. Since most data-driven approaches lack the guarantee of compliance minimization, online optimization can serve as the final post-processing step to connect disconnected edges and truly minimize the compliance.

In this work, we also compare our result against SIMP using "88-lines" [27]. However, it may be not possible to determine which one is definitively better or worse. As each program is tuned for different platforms and the possible combinations of problem configuration is endless. Covering all possible problem configurations to reach a conclusion may not be possible. There

are exciting possibilities with neural network-based topology optimization, for example, since the design density field is represented by a continuous function, one can infinitely upsample the result to obtain very crisp boundaries [4]. We can also use the same neural network architecture with physics-informed neural networks to conduct mesh-free topology optimization without a FE solver [29] to name a few.

6 CONCLUSIONS

We have proposed a novel approach for improving neural

Copyright © 2023 by ASME

network based topology optimization using a conditioning field. Our method involves using a topology neural network that is trained on a case-by-case basis to represent the geometry for a single topology optimization problem. By incorporating the strain energy field calculated on the initial design domain as an additional conditioning field input to the neural network, we have demonstrated faster convergence speed can be achieved. Our results suggest that the efficacy of neural network based topology optimization can be further improved using a prior initial field on the unoptimized domain. We believe that our proposed conditioning field initialization approach could have broad applications in the field of topology optimization, particularly for problems that involve complex geometries.

REFERENCES

[1] Bendsøe, M. P., 1989. "Optimal shape design as a material distribution problem". *Structural optimization, 1*, pp. 193–202.

[2] Zhou, M., and Rozvany, G., 1991. "The coc algorithm, part ii: Topological, geometrical and generalized shape optimization". *Computer methods in applied mechanics and engineering, 89*(1-3), pp. 309–336.

[3] Nie, Z., Lin, T., Jiang, H., and Kara, L. B., 2021. "Topologygan: Topology optimization using generative adversarial networks based on physical fields over the initial domain". *Journal of Mechanical Design, 143*(3).

[4] Chandrasekhar, A., and Suresh, K., 2021. "Tounn: Topology optimization using neural networks". *Structural and Multidisciplinary Optimization, 63*.

[5] Bens0e, M., and Kikuchi, N., 1988. "Generating optimal topologies in structural design using a homogenization method, comp". *Meths. Appl. Mechs. Engng, 71*, pp. 197–224.

[6] Allaire, G., Jouve, F., and Toader, A.-M., 2002. "A level-set method for shape optimization". *Comptes Rendus Mathematique, 334*(12), pp. 1125–1130.

[7] WANGM, Y., and WANG, X., 2003. "Guo d ma level set method for structural topology optimizations". *Computer Methods in Applied Mechanics and Engineering, 192*(1/2), pp. 227–246.

[8] Xie, Y. M., Steven, G. P., Xie, Y., and Steven, G., 1997. *Basic evolutionary structural optimization*. Springer.

[9] Woldseth, R. V., Aage, N., Bærentzen, J. A., and Sigmund, O., 2022. "On the use of artificial neural networks in topology optimisation". *Structural and Multidisciplinary Optimization, 65*(10), p. 294.

[10] Banga, S., Gehani, H., Bhilare, S., Patel, S., and Kara, L., 2018. "3d topology optimization using convolutional neural networks". *arXiv preprint arXiv:1808.07440*.

[11] Aage, N., Andreassen, E., and Lazarov, B. S., 2015. "Topology optimization using petsc: An easy-to-use,

fully parallel, open source topology optimization framework". *Structural and Multidisciplinary Optimization, 51*, pp. 565–572.

[12] Yu, Y., Hur, T., Jung, J., and Jang, I. G., 2019. "Deep learning for determining a near-optimal topological design without any iteration". *Structural and Multidisciplinary Optimization, 59*(3), pp. 787–799.

[13] Nakamura, K., and Suzuki, Y., 2020. "Deep learning-based topological optimization for representing a user-specified design area". *arXiv preprint arXiv:2004.05461*.

[14] Behzadi, M. M., and Ilieş, H. T., 2021. "Real-time topology optimization in 3d via deep transfer learning". *Computer-Aided Design, 135*, p. 103014.

[15] Zheng, S., He, Z., and Liu, H., 2021. "Generating three-dimensional structural topologies via a u-net convolutional neural network". *Thin-Walled Structures, 159*, p. 107263.

[16] Mazé, F., and Ahmed, F. "Diffusion models beat gans on topology optimization".

[17] Chandrasekhar, A., and Suresh, K., 2021. "Length scale control in topology optimization using fourier enhanced neural networks". *CoRR, abs/2109.01861*.

[18] Chandrasekhar, A., and Suresh, K., 2021. "Multi-material topology optimization using neural networks". *CAD Computer Aided Design, 136*.

[19] Deng, H., and To, A. C., 2020. "Topology optimization based on deep representation learning (drl) for compliance and stress-constrained design". *Computational Mechanics, 66*(2), pp. 449–469.

[20] Deng, H., and To, A. C., 2021. "A parametric level set method for topology optimization based on deep neural network". *Journal of Mechanical Design, 143*(9).

[21] Zehnder, J., Li, Y., Coros, S., and Thomaszewski, B., 2021. "Ntopo: Mesh-free topology optimization using implicit neural representations". *Advances in Neural Information Processing Systems, 34*, pp. 10368–10381.

[22] Mai, H. T., Mai, D. D., Kang, J., Lee, J., and Lee, J., 2023. "Physics-informed neural energy-force network: a unified solver-free numerical simulation for structural optimization". *Engineering with Computers*, pp. 1–24.

[23] Hoyer, S., Sohl-Dickstein, J., and Greydanus, S., 2019. "Neural reparameterization improves structural optimization". *arXiv preprint arXiv:1909.04240*.

[24] Zhang, Z., Li, Y., Zhou, W., Chen, X., Yao, W., and Zhao, Y., 2021. "Tonr: An exploration for a novel way combining neural network with topology optimization". *Computer Methods in Applied Mechanics and Engineering, 386*, p. 114083.

[25] Abadi, M., Agarwal, A., Barham, P., Brevdo, E., Chen, Z., Citro, C., Corrado, G. S., Davis, A., Dean, J., Devin, M., Ghemawat, S., Goodfellow, I., Harp, A., Irving, G., Isard, M., Jia, Y., Jozefowicz, R., Kaiser, L., Kudlur, M., Levenberg, J., Mane, D., Monga, R., Moore, S., Murray, D.,

Copyright © 2023 by ASME

Olah, C., Schuster, M., Shlens, J., Steiner, B., Sutskever, I., Talwar, K., Tucker, P., Vanhoucke, V., Vasudevan, V., Viegas, F., Vinyals, O., Warden, P., Wattenberg, M., Wicke, M., Yu, Y., and Zheng, X., 2016. "Tensorflow: Large-scale machine learning on heterogeneous distributed systems".

[26] Kingma, D. P., and Ba, J., 2014. "Adam: A method for stochastic optimization". *arXiv preprint arXiv:1412.6980.*

[27] Andreassen, E., Clausen, A., Schevenels, M., Lazarov, B. S., and Sigmund, O., 2011. "Efficient topology optimization in matlab using 88 lines of code". *Structural and Multidisciplinary Optimization,* **43**, pp. 1–16.

[28] Liu, K., and Tovar, A., 2014. "An efficient 3d topology optimization code written in matlab". *Structural and Multidisciplinary Optimization,* **50**, pp. 1175–1196.

[29] Joglekar, A., Chen, H., and Kara, L. B., 2023. Dmf-tonn: Direct mesh-free topology optimization using neural networks.

Proceedings of the ASME 2023
International Design Engineering Technical Conferences and
Computers and Information in Engineering Conference
IDETC-CIE2023
August 20-23, 2023, Boston, Massachusetts

DETC2023-111609

DATED: GUIDELINES FOR CREATING SYNTHETIC DATASETS FOR ENGINEERING DESIGN APPLICATIONS

Cyril Picard[1,*], Jürg Schiffmann[2], Faez Ahmed[1]

[1] Massachusetts Institute of Technology, Cambridge, MA
[2] École polytechnique fédérale de Lausanne (EPFL), Lausanne, Switzerland

ABSTRACT

Exploiting the recent advancements in artificial intelligence, showcased by ChatGPT and DALL-E, in real-world applications necessitates vast, domain-specific, and publicly accessible datasets. Unfortunately, the scarcity of such datasets poses a significant challenge for researchers aiming to apply these breakthroughs in engineering design. Synthetic datasets emerge as a viable alternative. However, practitioners are often uncertain about generating high-quality datasets that accurately represent real-world data and are suitable for the intended downstream applications. This study aims to fill this knowledge gap by proposing comprehensive guidelines for generating, annotating, and validating synthetic datasets. The trade-offs and methods associated with each of these aspects are elaborated upon. Further, the practical implications of these guidelines are illustrated through the creation of a turbo-compressors dataset. The study underscores the importance of thoughtful sampling methods to ensure the appropriate size, diversity, utility, and realism of a dataset. It also highlights that design diversity does not equate to performance diversity or realism. By employing test sets that represent uniform, real, or task-specific samples, the influence of sample size and sampling strategy is scrutinized. Overall, this paper offers valuable insights for researchers intending to create and publish synthetic datasets for engineering design, thereby paving the way for more effective applications of AI advancements in the field. The code and data for the dataset and methods are made publicly accessible at https://github.com/cyrilpic/radcomp.

1. INTRODUCTION

Recently popular *artificial intelligence* (AI) tools like ChatGPT and DALL-E have given a taste of "intelligence" to the general public due to their significant impact on a variety of fields, including art, design, and entertainment. In the field of natural language processing, these models can be used to generate realistic text, which has the potential to be used in a wide range of

applications, including chatbots, automated content creation, and even generating news articles. In art, these models allow artists to generate new images, videos, and music. This has left many engineers wondering how it will impact their field.

The leap of machine learning in general and of large language models (LLMs) in particular was made possible by a combination of new model architectures and a significant increase in model size (about 175 billion parameters for GPT-3 [1]) associated with the availability of very large datasets. Indeed, LLMs have demonstrated that machine learning becomes capable of solving new tasks with upward scaling of computation (FLOPs), model size (number of parameters), and data size [2]. An example of such *emergent behavior* is the ability of generative models to draw correct and legible text on images [3]. Beyond these examples, though, data size is often cited as a bottleneck [4].

In engineering design, generative AI and LLMs have already been applied to a variety of tasks, e.g., aircraft shape design [5], linkage mechanism generation [6], and concept sketch analysis [7]. In 2019, researchers urged the community to release more datasets, including multi-modal ones [8]. Yet a 2022 review still highlights that datasets in the field remain scarce and small [9]. While very large datasets are being published, e.g., 100 million sized LINKS dataset [6], they remain rare and not always easily accessible. This impacts the reproducibility and comparability of published *machine learning* (ML) methods for engineering design. It also impairs the field's ability to benefit from the advances in AI and experience its own *emergent behavior* moments.

Data can generally be categorized as either synthetic or real-world data. The latter is generated by real-world events, e.g., bank transactions, images posted on public web pages, texts written on Wikipedia, patients undergoing X-ray imaging, or autonomous cars driving on the streets. The related challenges include the logistics of collecting that data, processing it into a shared format, assessing its quality and biases, and labeling it [4]. The milk

---

*Corresponding author: cyrilp@mit.edu

Copyright © 2023 by ASME

frother dataset [7] comprising of sketches made by students,[1] or the BIKED dataset comprising of user-uploaded bike designs [10] are examples of real-world datasets for engineering design. In contrast, synthetic data is typically purpose-driven and generated *artificially*, such as by exhaustively enumerating all design options (e.g., periodic cellular structures [11]), by sampling from a design space (e.g., [6]), or by augmenting existing designs or real datasets through perturbations (e.g., ship hulls [12]).

Real datasets in engineering design are unlikely to fulfill the need for large and available datasets for ML research. They tend to be small (e.g., about 1000 sketches, or 4500 bike designs). Further, these designs tend to be clustered in specific areas of the design space, limiting the capacity for design exploration. Indeed, existing real-world designs are often deliberately of good quality. From an optimization lens, they can be considered optimal and thus near the edge of the design space [13], or they are the result of iterative refinement and share attributes with their parents. Finally, while larger datasets do exist, they are closely guarded by the industry, i.e., not openly available at scale.

Conversely, synthetic data could play a central role in addressing the shortage of large datasets through its capacity to be generated at scale. To quantify the performance of synthetic designs, one could leverage a plethora of tools developed over the past decades in different fields that enable performance quantification for a growing range of engineering designs and systems. However, there are two questions that arise: how should one collect or sample a dataset, and how should one quantify if that dataset is useful or not? It is important to note that not all datasets are equally useful for design and ML research. For instance, the ship hull dataset contains 300 shapes resulting from data augmentation methods applied to a single existing hull design. It is therefore limited in both size and scope to a narrow domain of single-hull variants. Conversely, LINKS contains 100 million one-degree-of-freedom linkage mechanisms and more than a billion coupler curves, but as the authors noted, most of the resulting coupler curves are arcs and circles, which are of limited practical interest.

Creating a good dataset is thus not a trivial task and can be subject to conflicting objectives, yet comprehensive methods to support researchers in this process are not discussed in the literature.

To address some of these shortcomings, in this paper we:

1. Provide guidelines for researchers and practitioners on factors to consider while collecting synthetic data for machine learning applications,

2. Illustrate common issues with a case study of turbocompressor dataset generation, and

3. Release the code and data used in the case study for researchers to build upon.

The guidelines cover three key areas of research: (1) design representation and data generation, (2) modeling and simulation, and (3) validation and verification. In the first area, researchers

[1]https://sites.psu.edu/creativitymetrics/2018/07/18/milkfrother/

Step 1. Design Representation Selection

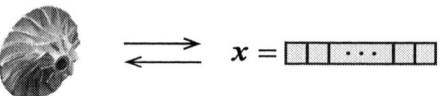

Step 2. Data Generation via:

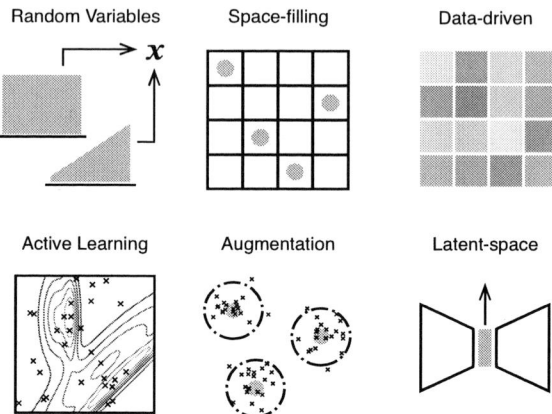

FIGURE 1: OVERVIEW OF THE IMPORTANT STEPS FOR DATA SELECTION AND THE GENERAL DATA GENERATION APPROACHES.

must carefully select the source data and properly characterize its properties. They must select the parameters they want to sample, and the sampling method, and understand the different objectives they need to consider while sampling. In the second area, researchers must choose appropriate modeling and simulation techniques to generate and characterize synthetic data that realistically represent the source data. In the third area, researchers must validate and verify the synthetic dataset to ensure that it accurately reflects the source data or proves useful for certain data-driven ML applications. Finally, while not the scope of this paper, researchers must also properly document and share the synthetic dataset to enable future reuse and collaboration (refer to for example [14]).

2. DATA REPRESENTATION AND SELECTION

The first step in creating a dataset is to define what the data will be. This requires answering (i) how the data is represented, and (ii) how the individual data points are generated. The key topics discussed in this section are visually summarized in Fig. 1.

2.1 Representation and Feature Selection

This study focuses on tabular data, which are common in engineering. In tabular design data, each row represents a design variant, and each column is a feature or a label. For example, the UIUC airfoil dataset[2] contains $N = 1600$ real-world airfoils that form a table with about N rows, 384 features—the x,y-coordinates of 192 surface points—and one label—the associated lift-to-drag ratio. Formally, a design vector or the vector of features denoted as x is the numerical representation of a design within $\mathbb{R}^d$, Eq. (1), where d is called the dimensionality of the space.[3] The collec-

[2]https://m-selig.ae.illinois.edu/ads/coord_database.html
[3]Note: some features can be integers or encoded categories.

Copyright © 2023 by ASME

tion of all design vectors forms the design space $\mathcal{D}$, Eq. (2), from which new design samples will be drawn in the next step. Labels refer to information that can be derived given a design vector, such as text descriptions, graphical representations, ranks, performance measures, or constraints. The process of gathering labels, i.e., data annotation, is discussed in Section 3.

$$\text{design}_i \longleftrightarrow \boldsymbol{x}_i = [x_1, x_2, \ldots, x_d] \in \mathbb{R}^d \qquad (1)$$

$$\mathcal{D} = \{\boldsymbol{x}_1, \boldsymbol{x}_2, \ldots\} \subset \mathbb{R}^d \qquad (2)$$

As in the case of the airfoil dataset, features are often related to parameters in a parametric design framework but they need not be strictly related to geometrical aspects of design. Features can represent colors, materials, or operating conditions. In the context of dataset creation, it is obviously central to carefully choose the link between a design (an airfoil) and its design vector (the x-y coordinates of control points). While that decision is highly context-dependent, the selection of features should in general be:

1. **Compatible** with any existing data and with the relevant models and simulations;

2. **Complete** ensuring that all necessary design characteristics are captured;

3. **Compact** to reduce feature collinearities and improve processing efficiency;

4. **General** enough to enable reuse in different applications and amortize the cost of creating a dataset.

The second and third points should not be seen as directly contradictory. If features correspond to a parameterized design, for example, it may be possible to choose a more compact parameterization without affecting completeness. In such cases, the more compact version should be preferred. Otherwise, if uncertain about a feature, it is better to include it and not use it rather than having to add a feature later. Methods exist to quantify the importance of features in a model, which can aid in the selection process once the dataset exists. Some examples are provided in Section 4. Finally, it is important to create a dataset that has broader applications, which may include multi-modal learning or using a representation that can generate images or meshes for a design.

The design space associated with a representation may be bounded—i.e., each feature is contained within a range and the resulting space can be defined as in Eq. (3).

$$\mathcal{D} = \{\boldsymbol{x} \in \mathbb{R}^d \,|\, x_i^{(L)} \leq x_i \leq x_i^{(U)} \,\forall\, i = 1, 2, \ldots, d\} \qquad (3)$$

where $\boldsymbol{x}^{(L)}$ and $\boldsymbol{x}^{(U)}$ are the lower and upper bound vectors respectively. In the simplest case, all bounds are independent, but it is also quite common for the bounds to depend upon others. Defining dependent bounds is particularly interesting if many variable combinations yield "nonsense" designs, since this may otherwise make the generation of meaningful designs harder. This is quite common in mechanical engineering where bounds are given on ratios (e.g., the length-to-diameter ratio of holes should be in a given range). Note, however, that this distorts the space, and should be kept in mind when sampling (e.g., very long holes in absolute terms are harder to get at random since it depends on the diameter). In addition, in some cases, upper and/or lower bounds cannot be defined and the design space is then called unbounded. Again, this results in special challenges that need to be considered in the data generation part.

2.2 Data Generation

Once the representation and the design space are defined, designs within it can be generated. This section discusses the objectives of the process, the question of the dataset size, as well as the many methods that can generate the actual data.

Objectives. Before discussing the hows, it is important to know what are the objectives of the datasets. For example, in metamaterial dataset design, the objective may be to get metamaterials that cover a preset property range, or the objective could be to differentiate between manufacturable and unmanufacturable metamaterials. In general terms, the goal of data generation is to collect a set of *relevant* points, where *relevant* is context-dependent. It is often important to ensure that the dataset has sufficient diversity. Data-driven methods usually have degraded performance for out-of-distribution (OOD) data. So, having good coverage of the space is important to improve the generalizability of machine learning models trained on those datasets. Likewise, a machine learning model's accuracy is often highest in regions with a high density of points.

Dataset Size. How many data points are needed? The answer first depends on the dimensionality d. If one divides, for example, all features into three values, one already needs 3^d points. For thirteen features ($d = 13$), that is more than 1.5 million points. This phenomenon is referred to as the curse of dimensionality. It becomes very clear that achieving a high density of points across the whole design space is impossible. There is therefore a trade-off between space coverage (for design exploration) and high local density in regions of interest. Still, in general, more data is preferred for improving the performance of machine learning methods [15]. If needed, it is always possible to downsample for specific tasks. The bottleneck for dataset size, however, is usually the required budget for data annotation (some advice will be provided in Section 3).

Sampling. In the context of synthetic dataset generation from parameters in given ranges, sampling refers to the process of randomly selecting values for each parameter from their respective ranges. This approach is commonly used to create synthetic datasets for machine learning or other types of modeling where real-world data may be limited or not available. As a reminder, the parameters can represent various characteristics or features of the data, such as numerical values, categorical variables, or distributions. For example, if we are generating a synthetic dataset for a regression model for bicycle structural performance, the parameters might include continuous variables such as the length of the top tube or bottom tube, each with a specified range of values or categorical options.

In this work, sampling methods have been grouped into six categories: (1) random sampling, (2) space-filling sampling, (3)

Copyright © 2023 by ASME

data-driven sampling, (4) active-learning sampling, (5) data augmentation, and (6) latent-space sampling, summarized graphically in Fig. 1.

Below, we discuss these six different sampling methods, highlighting their advantages and disadvantages.

Random Sampling. In this approach, each sample is drawn from a random distribution. The uniform distribution is the most commonly used in this setting. When the design space is bounded, drawing uniform sampling is fast and if enough points are considered, it will generally yield a good overall coverage. Not surprisingly, many datasets are collected that way. It is also possible to go beyond uniform distributions to achieve more tailored behaviors. For example, many distributions are unbounded, e.g., normal distributions, and can thus also be applied to unbounded spaces. For features that cover several orders of magnitudes, uniform sampling will generate most samples on the largest scale. If for example, a feature corresponds to a pressure that can span from kPa to MPa, most samples will be in the order of MPa. Power distributions or uniform distributions in the logarithmic space can be used to ensure that more samples are spread more evenly within each scale. For categorical features, values can be selected randomly from a list of given categories. While random sampling is fast to run, it may not always ensure that the design or feature space has good coverage. It may also ignore relationships between features, and a significant number of samples may get wasted, as they may not lead to any feasible design.

Space-filling Sampling. Space-filling sampling approaches aim to provide good coverage over the design space. Classical approaches (e.g., factorial design) put a grid with fixed partitions onto the space and allocate a point to each node. The resulting number of points can be extremely large due to the curse of dimensionality (k^d) detailed previously. More recent approaches focus on covering the space best with a fixed but choosable number of points. Examples include latin hypercube sampling (LHS) [16] or Sobol sequences [17]. They work by partitioning the space into smaller regions but follow point placement rules. In particular, Sobol sequences can be advantageous since they guarantee certain coverage properties while being computationally efficient [18]. Also, the code to generate such sequences is readily available in major ML frameworks. However, the number of samples is typically restricted to powers of two, and adding points sequentially while maintaining the space-filling properties can be challenging,

Data-driven Sampling. In data-driven sampling, the sample is selected based on the characteristics of the data, such as its distribution, variability, or patterns. Sometimes similar in intent to space-filling approaches, data-driven sampling methods do however not manipulate space, but data. As such, some data must already exist or have been generated by another method. One common approach to data-driven sampling is cluster sampling, where the data is first partitioned into clusters based on similarity, and then a sample is selected from each cluster. This approach can be useful for identifying patterns or trends within the data, as well as for reducing the computational burden of analyzing large datasets. Also popular, determinantal point processs (DPPs), which given a similarity matrix, model the likelihood

of selecting a diverse subset [19]. The definition of similarity controls the selection preference. Multiple preferences can be blended together with a parameter w using Eq. (4). An approach successfully applied in [20] to have samples diverse in the design and in the performance space.

$$L = (1 - w)L_1 + wL_2 \qquad (4)$$

Data-driven sampling can be advantageous over other sampling methods because it allows for a more nuanced and flexible approach to selecting the sample. However, it can also lead to bias or inaccurate conclusions if the data is not representative of the larger design space.

Active-learning Sampling. In an active learning context, a model of the data is built iteratively. New sample points are selected by optimizing a utility (or acquisition) function over this model. To incorporate performance information, new samples are usually annotated as they are sampled. Bayesian optimization is a popular active-learning flavor, but other approaches are possible. For example in t-METASET, a DPP-based active-learning sampling framework is presented that considers design and performance diversity in general while being able to tailor the distribution of points according to task-specific needs [21]. The availability of labels and the diversity of possible utility functions make it a very powerful tool. For example, when large regions of the design space yield invalid/infeasible designs—i.e., designs that violate some constraints—active learning can discover the feasibility boundaries and allocate samples more effectively [22]. The added information comes obviously at a price. Recent advances have tried to alleviate those issues by enabling batches of samples to be selected [23] and running the active-learning model on GPU [24].

Data Augmentation. When a small dataset, for example of existing designs, is already available, data augmentation techniques can be applied to generate more similar samples. In data augmentation-based sampling, the original data is used as a starting point, and various transformations or augmentations are applied to generate new samples. For example, in the context of image data, augmentations can include flipping the image, cropping a part of the image to create a copy, rotating the image, changing the brightness or contrast, or adding noise to the image. In engineering design, augmentation is often achieved by applying a small perturbation to the designs, e.g., [12]. In practice, this can be done by sampling from a normal distribution whose mean is set to the actual design and the variance to a fraction, e.g., 1%, of the length of the design space. Other approaches include using the mutation and crossover operators from genetic algorithms [25]. By generating new samples through data augmentation, the size of the training dataset can be increased without the need for additional data or annotation collection. However, augmentations should often be carefully collected using domain knowledge.

Latent-space Sampling. With the development of generative AI, researchers have started using variational autoencoders (VAEs), generative adversarial networks (GANs) or other deep generative models to generate new designs by sampling in the latent space [26]. As the latent space is a reduced dimensional space, sampling there reduces issues such as collinearity/correlations between features. In addition, deep generative models can be trained to favor certain properties such as diversity and novelty [9]. They can also be conditioned to generate designs with certain performances [26]. The obvious challenge to this approach is that a deep generative model with the desired properties needs to exist or be trained first.

3. MODELING AND SIMULATION

While covering all aspects of modeling and simulation is out of scope for this work, there are several aspects and trade-offs that need to be discussed in the context of dataset creation.

As previously stated, labels (also referred to as ground truth, tags, metadata, performance metrics, or constraints) can be very different and can include numerical performance measures (e.g., efficiency, power output, drag coefficient), images (e.g., rendering of the design or flow fields), points clouds (e.g., 3D representations), or rankings (e.g., design A is preferred over design B). While obtaining some may require complex finite-element methods, others may be straightforward to calculate (e.g., calculating the cross-sectional area of an airfoil).

When deciding which labels to include and how to obtain them, researchers and practitioners should consider (i) the level of detail of the data representations, (ii) the needed accuracy for the application, and (iii) the available solvers. Here again to promote reuse, as many labels as possible should be considered to have a multitask-ready dataset. Researchers should also consider purposefully including different label types to create rich datasets.

When collecting dataset labels using computational simulations or analytical methods, it is important to carefully select an appropriate model, tune the model parameters, validate the accuracy of the model, optimize computational efficiency, and format the labels in a consistent and usable way. This process requires careful consideration and validation to ensure that the generated data is reliable and accurate.

If the desired simulations are too expensive for a large dataset, it may be possible to compromise on the accuracy by running a high-fidelity solver for fewer iterations, or by using a reduced-order model instead [27, 28]. If high accuracy is nonetheless desired, a multi-fidelity approach might offer a better trade-off. In multi-fidelity modeling, models of different fidelity are available and an active learning strategy searches for the optimal computational budget allocation strategy [29]. Alternatively, weak labeling approaches can be considered [4]. Weak labeling makes use of proxy labels for the originally desired annotation. For example, a performance ranking procedure can be used in place of an expensive evaluation of the exact performance (e.g., [30]). Similarly, weak labeling works best if some samples can be annotated with the original method, and active learning approaches should then be applied to decide which sample to annotate with which method.

Despite the associated computational cost, we want to emphasize that the value of a dataset comes from its labels. As a counter-example, ShapeNet is a large-scale 3D shape database [31], which contains many engineering-relevant designs (amongst others: 4,043 airplane models). However, its use for

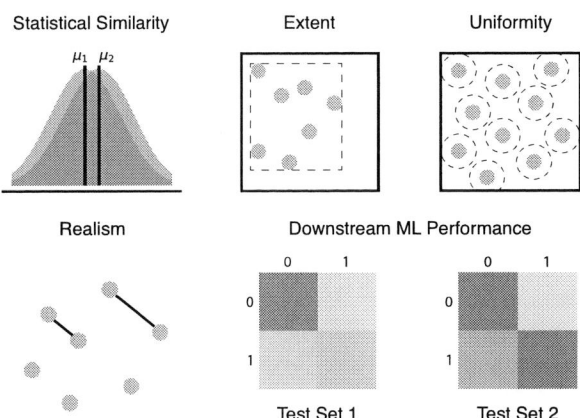

FIGURE 2: OVERVIEW OF THE APPROACHES THAT CAN BE USED TO VALIDATE THE CHARACTERISTICS OF A DATASET, AND VERIFY THE DATASET'S SUITABILITY.

engineering design is limited since these shapes have no associated performance label. Researchers that want to use for example those airplane models, first need to evaluate their own labels.

Finally, for a high-quality dataset, the results generated in this step should be carefully post-processed to properly label errors and handle unlikely values.

4. VALIDATION AND VERIFICATION

With the data generated and annotated, there remains an important, and sometimes overlooked, step: validation and verification. The collected data should be accurate, complete, and representative of the real-world system being modeled or observed. It is also important to ensure that the data is relevant and appropriate for the intended use case or application, while also staying free from biases that could impact the results or conclusions of downstream analyses. Finally, ensuring that the data is consistent and reliable across different samples, experiments, or observations is important. This can involve comparing the data to known or expected results, or evaluating the data for consistency over time or across different data sources. The goal of validation and verification is to validate that the dataset has the desired characteristics and to verify that those characteristics transfer into good performance on downstream tasks. An illustrative overview of the validation and verification approaches discussed in this section is provided in Fig. 2.

4.1 Characterization

This section is all about available methods that can be used to obtain a qualitative and quantitative understanding of a large dataset. Before diving into the methods themselves, it is important to keep the two following points in mind. First, most methods have underlying assumptions—e.g., linear relationships or normality. Therefore, it is advised to always combine several methods. Second, many characterization methods rely on the existence of a similarity and/or distance measure between two designs. If a context-specific method exists, that method should be preferred. For distance measures, common choices include

Copyright © 2023 by ASME

the Euclidean distance (or L^2 distance), the Manhattan distance (or L^1 distance), or the Hausdorff distance (which measures the distance between sets of points and would, for example, be appropriate to measure the distance between airfoil profiles). Similarity measures are often built by "inverting" a distance measure, for example, with a radial basis function kernel where the similarity $S(i, j)$ between sample i and j can be expressed as a function of their distance $d(i, j)$, $S(i, j) = \exp\left(-0.5d(i, j)^2\right)$. Another common, similarity measure is cosine similarity, which is the cosine of the angle between two samples. Lastly, it is worth mentioning that in most measures, multiple features are being summed in some way. So, large differences in scale can lead to biases towards certain features. For this reason, it may be interesting to consider calculating these similarity/distance measures in the normalized or standardized design space, or in a learned embedding (e.g., [20]). For a broader discussion on metrics, readers can refer to the discussion in [32], which can be transposed to the context of dataset creation.

Statistical Methods. Typical in ML, datasets are characterized using a collection of statistical approaches. For individual or pairs of features/labels, this includes looking at histograms to assess their distribution, quantifying unbalance for categorical data, or calculating descriptive statistics (e.g., mean, variance, median, interquartile range,...). The objectives are to (i) validate that the dataset has produced the desired feature distributions, (ii) investigate the resulting label distributions, and (iii) identify extreme values and potentially unrealistic values. The latter is particularly important when using numerical models which may not converge correctly, and involves making sure quantities such as weights or pressures are positive, or that efficiencies are between zero and one.

Further, relations within features or between features and labels can be evaluated, for example, by calculating correlations (linear model: Pearson's r or nonparametric: Spearman's ρ) or performing an analysis of variance (ANOVA) or a principle component analysis (PCA). Strongly correlated features can be indicative of collinearities (redundant features) that would cause issues when training ML models. It is however important to note that some degree of correlation may also reflect biases in the sampling. For example, in a dataset of holes, the diameters and the lengths may be correlated because deep holes are more frequently associated with large diameters without it implying collinearity. In addition, ANOVA or PCA can provide a first insight into the relations and importance of features with respect to labels.

Clustering methods can also support finding patterns that could be indicative of biases in the data. K-means, for example, is such an algorithm and it will try to identify n groups of equal variance within the data. Clustering can be used in combination with dimensionality reduction techniques, such as PCA or t-distributed stochastic neighbor embedding (t-SNE), to visualize in a low-dimensional space high-dimensional data.

Diversity. Diversity can be considered both in the design and the performance space. It refers to how well a dataset covers the extent of that space and how uniformly the data points are spaced out. These two components can either be quantified separately or together in a single metric. The extent also called spread, is often measured in terms of bounding volume. This typically means finding the smallest box or sphere that fully encloses the data points and calculating its volume. Depending on the distribution of points, bounding boxes and spheres tend to overestimate the extent. An alternative is to calculate the volume of the convex hull of the data points. The convex hull will usually be much tighter around the points since it is similar to wrapping a plastic film around them. In all cases, larger volumes mean more spread. Uniformity is typically measured by looking at the distances between neighboring points. In a uniform dataset, these distances will be similar across all points. Whereas if there are differences in distances, points with large distances to nearest neighbors would indicate a low-density area and small distances would indicate a high-density area. Clustering can be used in such cases to identify these different areas. There are also a few metrics that combine both uniformity and extent. Some common metrics are the Shannon entropy index, or the DPP diversity score [19]. The latter calculates a diversity score based on the eigenvalue decomposition of a pairwise similarity matrix.

Realism. In this work, realism refers to the representativeness of the dataset with respect to real-world data. It can be considered in a statistical similarity sense, where realism would mean that the biases of real-world data are also present in the dataset. The Kullback-Leibler divergence is an example of a statistical similarity measure. Further, realism can be understood as a proximity measure. Datasets whose points are close to real data would have a high realism. The difference with respect to the statistical similarity is that the frequency of data points close to real data is less important. Typical set-to-set distance measures include the Hausdorff or the Chamfer distance. Both work by calculating the distance to the closest point from each set to the other, and then either taking the maximum or the average.

4.2 Measuring Dataset Usefulness

The final step of the creation of a dataset is to use it with some data-driven model and assess whether its characteristics enable the desired downstream performance.

Representativeness of Test Sets. As data-driven models are optimized towards having good performance on the samples provided for training, their performance should be assessed on data never seen before by the model. This set of data is called a test set. It plays a central role whether comparing the performance of different models or in this context, verifying the suitability of the created dataset.

Most commonly, test sets are created by randomly splitting the data into train and test sets—many ML methods also use a validation set, derived from the train set, to monitor the training progress. The resulting test set will most likely have a similar distribution to the training data. While certainly always good to have, it may not be a good predictor of the performance of the model if the conditions are different when the model is deployed. For example, if the goal is to create a surrogate model to be used in an optimization routine, the inputs will rapidly be biased towards designs with high performance. As such, one may want

Copyright © 2023 by ASME

to have a test set to specifically test the accuracy of a model on high-performing designs.

Test sets are key to the assessment of ML models. They should be sufficiently large to reduce random effects and be representative of easier to more complex ML tasks. Further, we recommend defining several test sets with different characteristics and providing them along with the data. That way, they can be indicative of the strength and limitations of various ML methods, but also of the dataset itself. As such, in addition to the standard test set similar to the training data, researchers should consider test sets that (i) verify the generalizability in the design and performance space, (ii) assess the realism, and (iii) are representative of the deployed context. For example, the following test sets could cover those objectives:

- Similar to the training data (the standard approach);

- Diverse in the design space;

- Diverse in the performance space;

- Similar to real designs;

- Composed of designs outside or at the edge of the design space;

- Task-specific (e.g., high-performing designs only).

In practice, bespoke data-driven sampling methods can be created to obtain most of these desired goals, while others may require domain knowledge.

Verification. With the train set and the test sets fully defined, the final task is to actually evaluate different models on various tasks and see how state-of-the-art methods perform on the newly created dataset. Here as well, multiple models considering multiple tasks should be trained to verify the dataset's suitability. In addition to the common supervised classification and regression tasks, researchers should also apply their dataset to unsupervised autoencoding tasks, or self-supervised or supervised generative tasks.

Once a model is trained, feature importance analysis methods, such as SHAP [33] or permutation approaches [34], can provide the weight of each feature in the predicted outcome. Thus, it becomes possible to loop back to the question of data representation and feature selection and reevaluate if features are missing or should be removed. On this final note, readers are reminded that while the process has been described in a linear fashion, the task of creating a synthetic dataset is often an iterative process.

5. CASE STUDY: CENTRIFUGAL COMPRESSOR DESIGN

To illustrate some of the trade-offs and concepts presented in the previous sections, we consider the design of centrifugal compressors, see Fig. 3.

Centrifugal compressors play an important role in increasing the performance of internal combustion engines. Due to their efficiency and high power density, they are increasingly used for heat pump and refrigeration applications [35], or waste-heat recovery [36]. Their design undergoes a long process that

FIGURE 3: PICTURE OF A SMALL-SCALE CENTRIFUGAL COMPRESSOR SUPPORTED ON GAS BEARINGS (BY THE LABORATORY FOR APPLIED MECHANICAL DESIGN AT EPFL).

starts by defining key dimensions, such as tip diameter and operating speed, and ends with complex three-dimensional flow analyses. Previous work has shown that data-driven methods can greatly support and speed up the first stages of their design process [28, 37]. These contributions use a mean-line analysis model validated using experimental data [38] to evaluate the feasibility and the performance of each design in less than a second. The model has been converted to Python and is publicly released for the first time along with the generated data.[4]

This case study offers the advantage of being complex enough—it has 21 features and three labels—while still being computationally affordable. Further, we have collected the designs of 14 real-world compressor geometries from the literature [35, 38–42]. They form the dataset of *real* designs and their performance is evaluated with the same model as synthetic samples.

5.1 Methods and Experimental Conditions

The features correspond to the geometrical parameters relevant in the predesign stage, and to the operating conditions at which each compressor will be evaluated. They are listed along with their range in Table 1 in Appendix. The ranges are set to cover the set of existing designs and follow general recommendations for centrifugal compressors [43].

For the purpose of this work, a total of 22 million samples have been generated: 19 million by random sampling, two million by data augmentation of the 14 real-world compressors, and, in a second step, another million by random sampling of a restricted part of the design space (discussed in the next section). The random number generator routines of *numpy*[5] were used for sampling. All features were sampled uniformly within the specified ranges, except for the reduced inlet pressure $P_{r,1}$ which was sampled from a power distribution with $\alpha = 5$. The data-augmented

---
[4]https://github.com/cyrilpic/radcomp
[5]Numpy 1.23.3 and Python 3.9.13

Copyright © 2023 by ASME

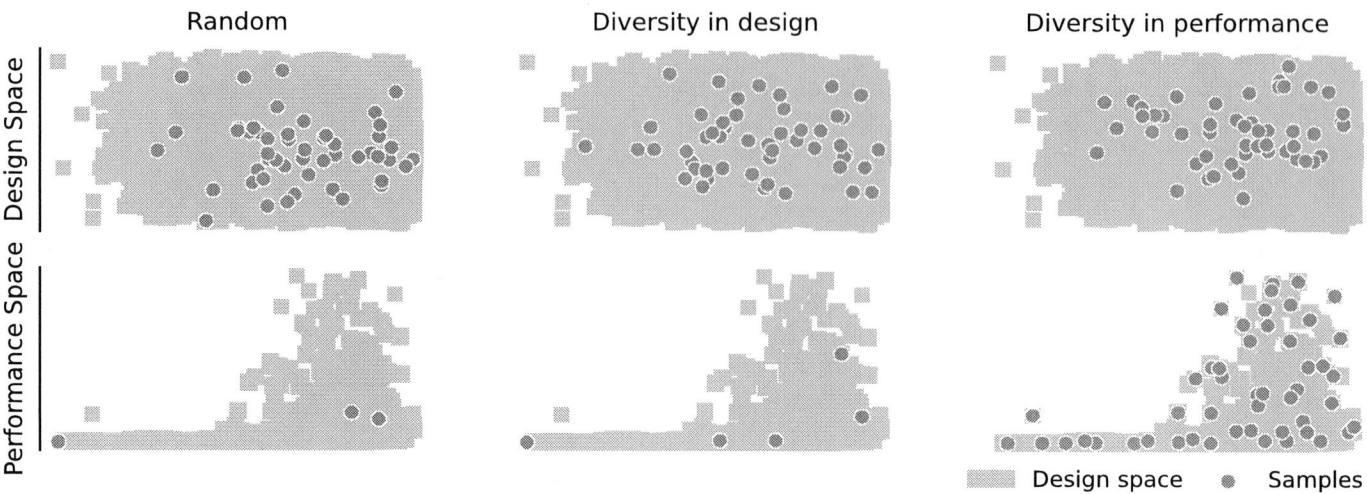

FIGURE 4: COMPARISONS OF THE DISTRIBUTION OF 50 DATA POINTS IN THE PCA-EMBEDDED DESIGN SPACE (TOP) AND IN THE PERFORMANCE SPACE (BOTTOM) SELECTED AT RANDOM (LEFT), DIVERSELY SAMPLED IN THE DESIGN SPACE (MIDDLE), AND DIVERSELY SAMPLED IN THE PERFORMANCE SPACE.

designs were obtained by applying a normal noise with σ of 1% of the design space to the 14 real-world compressors.

The samples were then annotated using the aforementioned mean-line analysis model. Given geometrical parameters and operating conditions (inlet pressure and temperature, mass flow, and rotational speed), the model calculates the input and output velocity triangles at the impeller, and the velocity and flow angle at the inducer and the vaneless diffuser. The mean-line flow is corrected by accounting for losses using established correlations. The performance of the compressor is then given by the total-to-total isentropic efficiency and the pressure ratio. In addition, the model outputs a flag indicating whether the compressor is in *working* condition. *Non-working* conditions include condensation, choke, or surge. When *non-working*, no meaningful efficiency or pressure ratio can be calculated, and their values are set to 0 and 1, respectively. The model evaluations were performed on MIT SuperCloud [44].

To verify the performance and suitability of the dataset, the classification into *working/non-working* is selected as the target ML task. Consequently, the following three test sets are defined:

1. Uniform: random samples from within each class from the random dataset (balanced 150,000 samples);

2. Real: random samples from within each class from the augmented dataset (balanced 40,000 samples);

3. Specialized: samples obtained by evaluating real designs on a dense grid of operating conditions and by retaining only points that are near the edge of the *working* boundary (balanced 40,000 samples).

The samples included in a test set are removed from the train sets. In the next sections, the combination of both random sample sets is referred to as the random dataset (19.4 million samples), while the augmented set refers to the samples generated through data augmentation (1.37 million samples).

As ML modeling is not the focus of this work, we choose the state-of-the-art AutoGluon's tabular predictor [45]. AutoGluon trains an ensemble of different ML models and tunes their hyperparameters to create an aggregated predictor. For all cases, we have used the "good quality" preset and used one GPU for training. No time budget was set and the default accuracy metric was set as the tuning criterion.

5.2 Discussion

In this section, we will first present some of the characteristics of the obtained dataset and discuss how diverse sampling in different spaces impacts the selected samples. Second, we train a set of classifiers on train sets of different sizes and of different compositions and report the resulting performance on different test sets. Finally, we give an example of feature importance analysis.

How to sample diverse points?. Analyzing the 19 million random samples, one characteristic strikes out immediately: only about 8% have been labeled as *working*. While the sampling bounds have been set to maximize the number of *working* compressors, it remains that random combinations of compressor geometry and operating conditions result mostly in *non-working* data points. Consequently, this dataset is highly unbalanced, and by extension, training a classifier on that is difficult. Further, it also means that 92% of the samples have an efficiency and a pressure ratio set at 0 and 1, respectively. Feeding the dataset to a regressor to predict either one will equally not result in a high-performing model. This is a common situation that many researchers have probably encountered, and we discuss some approaches to this problem.

While there are ways to tackle the imbalance at the model level—by adding different weights to each sample for example—there will always be the issue of information unbalance. So instead, the effect of different sampling approaches is considered here. Figure 4 shows the distribution in the design and performance space of points sampled randomly, considering diversity

Copyright © 2023 by ASME

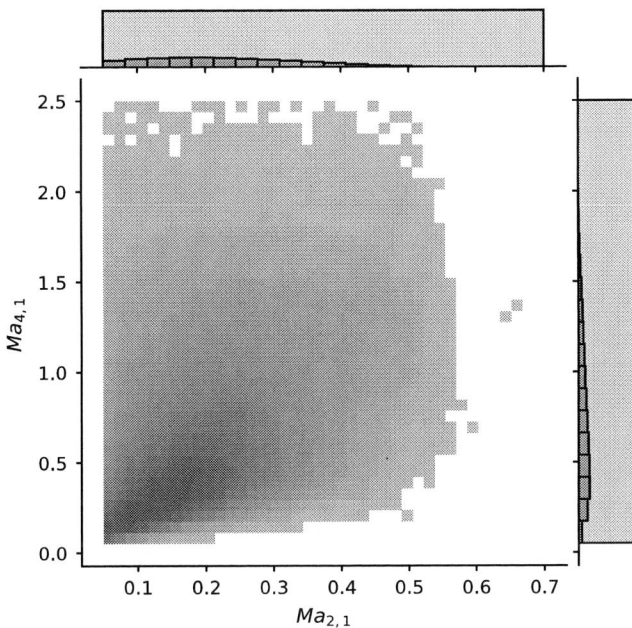

FIGURE 5: DENSITY PLOT OF WORKING DATA POINTS IN THE $(Ma_{2,1}, Ma_{4,1})$ SPACE, WITH MARGINAL DISTRIBUTIONS DISPLAYING THEIR SHARE WITH RESPECT TO NON-WORKING POINTS, HIGHLIGHTING THE AREA WHERE MOST WORKING SAMPLES CAN BE FOUND.

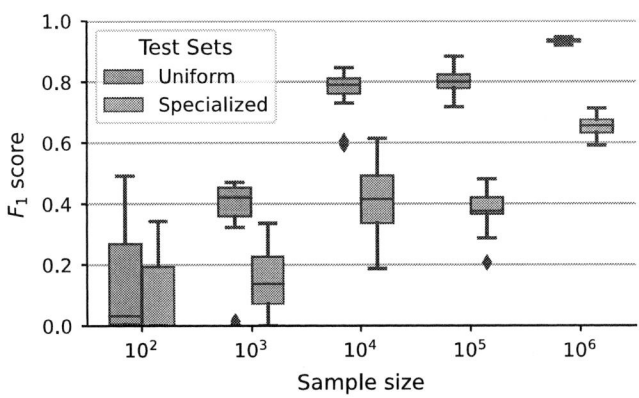

FIGURE 6: EVOLUTION OF THE CLASSIFICATION PERFORMANCE (F_1 SCORE) ON THE UNIFORM AND SPECIALIZED TEST SETS OF AUTOGLUON MODELS TRAINED WITH VARYING SAMPLE SIZES. FOR EACH, TEN INDEPENDENT MODELS, ARE TRAINED WITH SAMPLES DRAWN AT RANDOM FROM THE RANDOM DATASET.

in design and considering diversity in performance. It highlights that having good diversity in design space is not correlated with good diversity in performance space. As a consequence, using a space-filling sampling alone would not avoid the imbalance. Figure 4 also demonstrates that the design diverse sampling has fewer clusters and might be beneficial for covering the design space more effectively with fewer samples. An overall better diversity could be achieved by using a data-driven method that blends in the two objectives, see for example the work in METASET [20].

Alternatively, it is sometimes also possible to directly use domain knowledge to infer where more *working* samples can be found. In the case of a compressor, the inlet and impeller tip Mach numbers, $Ma_{2,1}$ and $Ma_{4,1}$ proportional to the mass flow and the rotational speed respectively, are important drivers. Indeed, Fig. 5 shows the density of *working* samples in the random dataset with respect to these two variables. While *working* samples can be found almost everywhere, there is a high-density area in the lower-left corner. Using that information, a second batch of random samples was drawn over a reduced range with $Ma_{2,1} \in [0.15, 0.25]$ and $Ma_{4,1} \in [0.35, 0.7]$. In that batch, 44% of the samples are *working*. Figure 5 also illustrates the kind of characteristics that can be learned by looking at feature distributions.

Overall, it is important to consider coverage over both design and performance space, as a lack of it can lead to degraded ML model performance. Visualization methods can help better understand the characteristics of these spaces and inform changes to sampling methods.

How to choose the sample size?. The common intuition in dataset collection is that more data is always better for machine learning performance. In this experiment, we test this intuition. We use the trained classifiers to investigate the relationship between classification performance and the train set size. The models are tested with the uniform and the specialized test sets for sample sizes from 10^2 to 10^6 drawn from the random dataset. The F_1 score (5) is used as the performance metric. The results are shown in Fig. 6. Sample sizes smaller than 10^4 lead to high variability and low performance in general. Starting with 10^4, most trained classifiers have an F_1 classification score greater than 0.6. With the largest sample size, very good performance with marginal variations is measured with the uniform test set. It is interesting to note that the performance of models on the uniform test set is always better than the specialized test set. The specialized test set evaluates precisely the edge of the *working* domain, and the performance on it does not reach very good performance levels even with a million samples.

$$F_1 = \frac{2 \text{ true positives}}{2 \text{ true positives} + \text{false positives} + \text{false negatives}} \quad (5)$$

The experiment highlights the importance of having several test sets and ensuring they fit the application. Indeed, if in practice, such a classifier would work with data similar to the uniform test set, then one could conclude that 10^6 is a sufficiently large dataset and one could even settle with a smaller dataset if memory-constrained. However, the selected uniform training data sampling method does not seem a good fit in a scenario where the data are similar to the specialized test set.

In general, it is a good practice to see how the performance of a model evolves as the dataset is reduced.

Copyright © 2023 by ASME

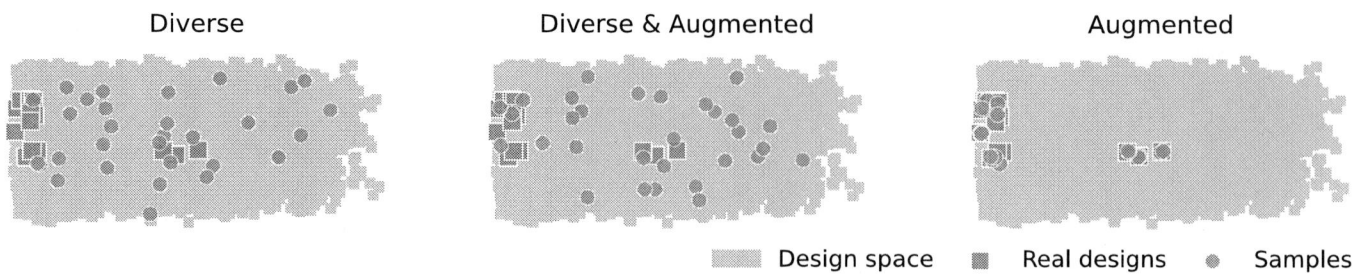

Diverse	Diverse & Augmented	Augmented

Design space ▪ Real designs ● Samples

FIGURE 7: COMPARISONS WITHIN THE PCA-EMBEDDED GEOMETRICAL PARAMETER SPACE OF THE DISTRIBUTION OF REAL DESIGNS AGAINST 30 DATA POINTS SAMPLED CONSIDERING DIVERSITY (LEFT) DIVERSITY AND AUGMENTATION (MIDDLE), AND AUGMENTATION ONLY (RIGHT).

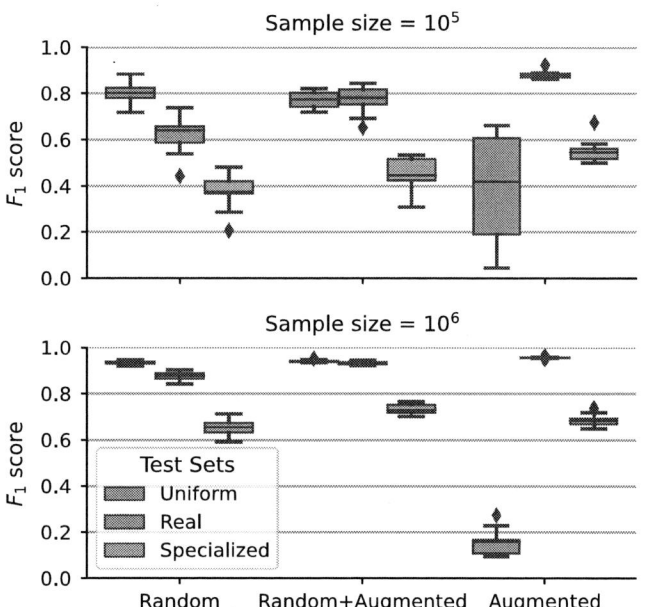

FIGURE 8: EFFECT OF DIFFERENT SAMPLING APPROACHES WITH DIFFERENT BLENDS OF AUGMENTED REAL DESIGNS ON THE CLASSIFICATION PERFORMANCE (F_1 SCORE) OF AUTO-GLUON MODELS EVALUATED ON THE UNIFORM, REAL, AND SPECIALIZED TEST SETS FOR SAMPLE SIZES OF 10^5 (TOP) AND 10^6 (BOTTOM). FOR EACH SAMPLING APPROACH, TEN INDEPENDENT MODELS ARE TRAINED WITH SAMPLES DRAWN AT RANDOM FROM THE RANDOM AND/OR AUGMENTED DATASETS RESPECTIVELY.

How relevant is realism in a dataset?. Taking another perspective on the representativeness of a dataset, we look at the question of realism. Figure 7 shows the relative distribution of three different sub-sample sets compared to the design space and more importantly to real designs. First, Fig. 7 confirms that real designs are indeed clustered, and most of them, are located near the edge of the design space. Consequently, while the diverse sub-set has samples near real designs, their density near real design clusters is low. In contrast, the augmented-only sub-set has a high density near real design, but it has no samples in most areas of the design space. In between, the sub-set combining diverse samples with samples from the augmented dataset display a good balance between design space and real-design cluster coverage. In many real-world applications of machine learning, the model may be tested on new real-world designs. It is quite likely that features of new real-world designs look more similar to other existing real-world designs and not necessarily to synthetic samples randomly generated from a large design space. This may create an issue, as machine learning models trained on the latter may perform poorly on real-world designs, although, they may get very good validation performance. Hence, we propose considering realism as an important factor in both the sampling and testing of datasets.

To confirm the interest in including augmented data, we verify how this translates to the performance of classifiers. For this, we consider train sets composed of (i) random samples, (ii) 70% random samples and 30% augmented samples, and (iii) only augmented samples. Figure 8 shows the F_1 score of the trained classifier for each dataset combination and for all three test sets. It highlights that the combined dataset enables very good performance on both the uniform and the real test sets, with only minor differences compared to their respective reference case. The combined dataset also slightly improves the classification performance on the specialized test set, which is expected as this test set also derives from real designs. Interestingly, Fig. 8 also shows that the augmented-only dataset can lead to over-fitting. Indeed, the model trained on 10^6 samples has seen the whole dataset and has a lower F_1 score on the uniform test set than the models trained with smaller subsets of that dataset.

In summary, we highlight the importance of considering "realism" as an objective for sampling methods to create machine learning datasets. Adding realism as an objective of dataset sampling methods in machine learning means prioritizing data that is representative of the real-world scenarios that the model is intended to operate in. This can be achieved by collecting data from real-world sources. However, if there is little data available from real-world sources, then it is also important to cover the design space. The objective is to create a dataset that reflects the variability and complexity of the real-world designs as well as all possible variations of a design, helping to ensure that the model is well-suited to real-world applications and can perform well on a diverse range of test cases.

Copyright © 2023 by ASME

6. CONCLUSION

This paper has discussed the many facets and challenges of synthetic tabular dataset creation in engineering design. We have provided guidelines for data representation and selection, modeling and simulation, and characterization and verification, all of which are critical steps in ensuring that the generated dataset is accurate, reliable, and representative of the real-world system being modeled or observed. For each step, we have provided methods and examples to support researchers, using a case study of the centrifugal compressor design dataset. We have highlighted that achieving diversity in design, diversity in performance and *realism* are sometimes competitive objectives, that should and can be factored in when selecting data generation methods. Further, we have illustrated the critical role of well-crafted test sets to ensure a relevant performance assessment of machine learning models.

Overall, the guidelines and considerations presented in this paper provide a valuable resource for researchers and practitioners involved in synthetic tabular dataset creation in engineering design. By following these guidelines, it is possible to create high-quality datasets that accurately reflect the real-world system being modeled or observed, and that are well-suited for use in machine learning and other data-driven applications.

Future work in this area includes the development of standardized procedures for creating and evaluating synthetic datasets, the exploration of new modeling and simulation techniques, and the investigation of ways to reduce bias in synthetic datasets.

ACKNOWLEDGMENTS

The Swiss National Science Foundation is acknowledged for its financial support (grant P500PT_206937). The authors also acknowledge the MIT SuperCloud and Lincoln Laboratory Supercomputing Center for providing high-power computing resources that have contributed to the research results reported within this paper.

REFERENCES

[1] Brown, Tom, Mann, Benjamin, Ryder, Nick, Subbiah, Melanie, Kaplan, Jared D, Dhariwal, Prafulla, Neelakantan, Arvind, Shyam, Pranav, Sastry, Girish, Askell, Amanda, Agarwal, Sandhini, Herbert-Voss, Ariel, Krueger, Gretchen, Henighan, Tom, Child, Rewon, Ramesh, Aditya, Ziegler, Daniel, Wu, Jeffrey, Winter, Clemens, Hesse, Chris, Chen, Mark, Sigler, Eric, Litwin, Mateusz, Gray, Scott, Chess, Benjamin, Clark, Jack, Berner, Christopher, McCandlish, Sam, Radford, Alec, Sutskever, Ilya and Amodei, Dario. "Language Models Are Few-Shot Learners." *Advances in Neural Information Processing Systems*, Vol. 33: pp. 1877–1901. 2020. Curran Associates, Inc.

[2] Wei, Jason, Tay, Yi, Bommasani, Rishi, Raffel, Colin, Zoph, Barret, Borgeaud, Sebastian, Yogatama, Dani, Bosma, Maarten, Zhou, Denny, Metzler, Donald, Chi, Ed H., Hashimoto, Tatsunori, Vinyals, Oriol, Liang, Percy, Dean, Jeff and Fedus, William. "Emergent Abilities of Large Language Models." *Transactions on Machine Learning Research* (2022).

[3] Yu, Jiahui, Xu, Yuanzhong, Koh, Jing Yu, Luong, Thang, Baid, Gunjan, Wang, Zirui, Vasudevan, Vijay, Ku, Alexander, Yang, Yinfei, Ayan, Burcu Karagol, Hutchinson, Ben, Han, Wei, Parekh, Zarana, Li, Xin, Zhang, Han, Baldridge, Jason and Wu, Yonghui. "Scaling Autoregressive Models for Content-Rich Text-to-Image Generation." (2022). DOI 10.48550/arXiv.2206.10789. Accessed 2023-03-07, URL arXiv:2206.10789.

[4] Shani, Chen, Zarecki, Jonathan and Shahaf, Dafna. "The Lean Data Scientist: Recent Advances Toward Overcoming the Data Bottleneck." *Communications of the ACM* Vol. 66 No. 2 (2023): pp. 92–102. DOI 10.1145/3551635.

[5] Shu, Dule, Cunningham, James, Stump, Gary, Miller, Simon W., Yukish, Michael A., Simpson, Timothy W. and Tucker, Conrad S. "3D Design Using Generative Adversarial Networks and Physics-Based Validation." *Journal of Mechanical Design* Vol. 142 No. 7 (2019). DOI 10.1115/1.4045419.

[6] Heyrani Nobari, Amin, Srivastava, Akash, Gutfreund, Dan and Ahmed, Faez. "LINKS: A Dataset of a Hundred Million Planar Linkage Mechanisms for Data-Driven Kinematic Design." *ASME 2022 International Design Engineering Technical Conferences and Computers and Information in Engineering Conference*. 2022. American Society of Mechanical Engineers Digital Collection. DOI 10.1115/DETC2022-89798.

[7] Song, Binyang, Miller, Scarlett and Ahmed, Faez. "Attention-Enhanced Multimodal Learning for Conceptual Design Evaluations." *Journal of Mechanical Design* Vol. 145 No. 4 (2023). DOI 10.1115/1.4056669.

[8] "Special Issue: Machine Learning for Engineering Design." *Journal of Mechanical Design* Vol. 141 No. 11 (2019). DOI 10.1115/1.4044690.

[9] Regenwetter, Lyle, Heyrani Nobari, Amin and Ahmed, Faez. "Deep Generative Models in Engineering Design: A Review." *Journal of Mechanical Design* Vol. 144 No. 7 (2022). DOI 10.1115/1.4053859.

[10] Regenwetter, Lyle, Curry, Brent and Ahmed, Faez. "BIKED: A Dataset for Computational Bicycle Design With Machine Learning Benchmarks." *Journal of Mechanical Design* Vol. 144 No. 3 (2021). DOI 10.1115/1.4052585.

[11] Lumpe, Thomas S. and Stankovic, Tino. "Exploring the Property Space of Periodic Cellular Structures Based on Crystal Networks." *Proceedings of the National Academy of Sciences* Vol. 118 No. 7 (2021): p. e2003504118. DOI 10.1073/pnas.2003504118.

[12] Wang, Yuyang, Joseph, Joe, Aniruddhan Unni, T. P., Yamakawa, Soji, Barati Farimani, Amir and Shimada, Kenji. "Three-Dimensional Ship Hull Encoding and Optimization via Deep Neural Networks." *Journal of Mechanical Design* Vol. 144 No. 10 (2022). DOI 10.1115/1.4054494.

[13] Papalambros, Panos Y. and Wilde, Douglass J. *Principles of Optimal Design: Modeling and Computation*, 3rd ed. Cambridge University Press (2017). DOI 10.1017/9781316451038.

[14] Gebru, Timnit, Morgenstern, Jamie, Vecchione, Briana, Vaughan, Jennifer Wortman, Wallach, Hanna, Daumé III,

Hal and Crawford, Kate. "Datasheets for Datasets." (2021). DOI 10.48550/arXiv.1803.09010.

[15] Sorscher, Ben, Geirhos, Robert, Shekhar, Shashank, Ganguli, Surya and Morcos, Ari S. "Beyond Neural Scaling Laws: Beating Power Law Scaling via Data Pruning." (2022). DOI 10.48550/arXiv.2206.14486. URL 2206.14486v5.

[16] Mckay, Michael D., Beckman, Richard J. and Conover, William J. "A Comparison of Three Methods for Selecting Values of Input Variables in the Analysis of Output From a Computer Code." *Technometrics* Vol. 42 No. 1 (2000): pp. 55–61. DOI 10.1080/00401706.2000.10485979.

[17] Joe, Stephen and Kuo, Frances Y. "Constructing Sobol Sequences with Better Two-Dimensional Projections." *SIAM Journal on Scientific Computing* Vol. 30 No. 5 (2008): pp. 2635–2654. DOI 10.1137/070709359.

[18] Kucherenko, Sergei, Albrecht, Daniel and Saltelli, Andrea. "Exploring Multi-Dimensional Spaces: A Comparison of Latin Hypercube and Quasi Monte Carlo Sampling Techniques." (2015). DOI 10.48550/arXiv.1505.02350. Accessed 2023-03-03, URL arXiv:1505.02350.

[19] Kulesza, Alex and Taskar, Ben. "Determinantal Point Processes for Machine Learning." *Foundations and Trends® in Machine Learning* Vol. 5 No. 2–3 (2012): pp. 123–286. DOI 10.1561/2200000044.

[20] Chan, Yu-Chin, Ahmed, Faez, Wang, Liwei and Chen, Wei. "METASET: Exploring Shape and Property Spaces for Data-Driven Metamaterials Design." *Journal of Mechanical Design* Vol. 143 No. 3 (2020). DOI 10.1115/1.4048629.

[21] Lee, Doksoo, Chan, Yu-Chin, Chen, Wei (Wayne), Wang, Liwei, van Beek, Anton and Chen, Wei. "T-METASET: Task-Aware Acquisition of Metamaterial Datasets Through Diversity-Based Active Learning." *Journal of Mechanical Design* Vol. 145 No. 3 (2022). DOI 10.1115/1.4055925.

[22] Bryan, Brent, Nichol, Robert C., Genovese, Christopher R, Schneider, Jeff, Miller, Christopher J. and Wasserman, Larry. "Active Learning For Identifying Function Threshold Boundaries." *Advances in Neural Information Processing Systems*, Vol. 18. 2005. MIT Press.

[23] Wilson, James, Hutter, Frank and Deisenroth, Marc. "Maximizing Acquisition Functions for Bayesian Optimization." *Advances in Neural Information Processing Systems*, Vol. 31. 2018. Curran Associates, Inc.

[24] Balandat, Maximilian, Karrer, Brian, Jiang, Daniel, Daulton, Samuel, Letham, Ben, Wilson, Andrew G and Bakshy, Eytan. "BoTorch: A Framework for Efficient Monte-Carlo Bayesian Optimization." *Advances in Neural Information Processing Systems*, Vol. 33: pp. 21524–21538. 2020. Curran Associates, Inc.

[25] Hadka, David and Reed, Patrick. "Borg: An Auto-Adaptive Many-Objective Evolutionary Computing Framework." *Evolutionary Computation* Vol. 21 No. 2 (2013): pp. 231–259. DOI 10.1162/EVCO_a_00075.

[26] Heyrani Nobari, Amin, Chen, Wei and Ahmed, Faez. "PcDGAN: A Continuous Conditional Diverse Generative Adversarial Network For Inverse Design." *Proceedings of the 27th ACM SIGKDD Conference on Knowledge Discovery & Data Mining*: p. 606–616. 2021. Association for Computing Machinery, New York, NY, USA. DOI 10.1145/3447548.3467414.

[27] Benner, Peter, Gugercin, Serkan and Willcox, Karen. "A Survey of Projection-Based Model Reduction Methods for Parametric Dynamical Systems." *SIAM Review* Vol. 57 No. 4 (2015): pp. 483–531. DOI 10.1137/130932715.

[28] Massoudi, Soheyl, Picard, Cyril and Schiffmann, Jürg. "Robust Design Using Multiobjective Optimisation and Artificial Neural Networks with Application to a Heat Pump Radial Compressor." *Design Science* Vol. 8 (2022/ed). DOI 10.1017/dsj.2021.25.

[29] Sarkar, Soumalya, Mondal, Sudeepta, Joly, Michael, Lynch, Matthew E., Bopardikar, Shaunak D., Acharya, Ranadip and Perdikaris, Paris. "Multifidelity and Multiscale Bayesian Framework for High-Dimensional Engineering Design and Calibration." *Journal of Mechanical Design* Vol. 141 No. 12 (2019). DOI 10.1115/1.4044598.

[30] Chaudhary, P., D'Aronco, S., Leitão, J. P., Schindler, K. and Wegner, J. D. "Water Level Prediction from Social Media Images with a Multi-Task Ranking Approach." *ISPRS Journal of Photogrammetry and Remote Sensing* Vol. 167 (2020): pp. 252–262. DOI 10.1016/j.isprsjprs.2020.07.003.

[31] Chang, Angel X., Funkhouser, Thomas, Guibas, Leonidas, Hanrahan, Pat, Huang, Qixing, Li, Zimo, Savarese, Silvio, Savva, Manolis, Song, Shuran, Su, Hao, Xiao, Jianxiong, Yi, Li and Yu, Fisher. "ShapeNet: An Information-Rich 3D Model Repository." (2015). DOI 10.48550/arXiv.1512.03012. URL arXiv:1512.03012.

[32] Regenwetter, Lyle, Srivastava, Akash, Gutfreund, Dan and Ahmed, Faez. "Beyond Statistical Similarity: Rethinking Metrics for Deep Generative Models in Engineering Design." (2023). DOI 10.48550/arXiv.2302.02913. URL arXiv:2302.02913.

[33] Lundberg, Scott M and Lee, Su-In. "A Unified Approach to Interpreting Model Predictions." *Advances in Neural Information Processing Systems*, Vol. 30. 2017. Curran Associates, Inc.

[34] Breiman, Leo. "Random Forests." *Machine Learning* Vol. 45 No. 1 (2001): pp. 5–32. DOI 10.1023/A:1010933404324.

[35] Javed, Adeel, Arpagaus, Cordin, Bertsch, Stefan and Schiffmann, Jürg. "Small-Scale Turbocompressors for Wide-Range Operation with Large Tip-Clearances for a Two-Stage Heat Pump Concept." *International Journal of Refrigeration* Vol. 69 (2016): pp. 285–302. DOI 10.1016/j.ijrefrig.2016.06.015.

[36] Demierre, Jonathan, Rubino, Antonio and Schiffmann, Jürg. "Modeling and Experimental Investigation of an Oil-Free Microcompressor-Turbine Unit for an Organic Rankine Cycle Driven Heat Pump." *Journal of Engineering for Gas Turbines and Power* Vol. 137 No. 3 (2014): pp. 032602–032602. DOI 10.1115/1.4028391.

[37] Mounier, Violette, Picard, Cyril and Schiffmann, Jurg. "Data-Driven Predesign Tool for Small-Scale Centrifugal

Copyright © 2023 by ASME

Compressor in Refrigeration." *Journal of Engineering for Gas Turbines and Power* Vol. 140 No. 12 (2018): pp. 121011–121018. DOI 10.1115/1.4040845.

[38] Schiffmann, Jürg and Favrat, Daniel. "Design, Experimental Investigation and Multi-Objective Optimization of a Small-Scale Radial Compressor for Heat Pump Applications." *Energy* Vol. 35 No. 1 (2010): pp. 436–450. DOI 10.1016/j.energy.2009.10.010.

[39] Javed, Adeel, Olivero, Mattia, Pecnik, Rene and van Buijtenen, Jos P. "Performance Analysis of a Microturbine Centrifugal Compressor From a Manufacturing Perspective." *ASME 2011 Turbo Expo: Turbine Technical Conference and Exposition*: pp. 2155–2166. 2011. American Society of Mechanical Engineers Digital Collection. DOI 10.1115/GT2011-46374.

[40] De Bellis, Vincenzo, Bozza, Fabio, Bevilacqua, Marco, Bonamassa, Guido and Schernus, Christof. "Validation of a 1D Compressor Model for Performance Prediction." *SAE International Journal of Engines* Vol. 6 No. 3 (2013): pp. 1786–1800. DOI 10.4271/2013-24-0120.

[41] Meroni, Andrea, Zühlsdorf, Benjamin, Elmegaard, Brian and Haglind, Fredrik. "Design of Centrifugal Compressors for Heat Pump Systems." *Applied Energy* Vol. 232 (2018): pp. 139–156. DOI 10.1016/j.apenergy.2018.09.210.

[42] Olmedo, Luis Eric. "High-Speed Turbocompressors for Natural Fluid Heat-Pumps: From Integrated Design to Digital-Twin." Ph.D. Thesis, EPFL, Lausanne. 2023. DOI 10.5075/epfl-thesis-8970.

[43] Baines, Nicholas C. *Fundamentals of Turbocharging.* Concepts NREC (2005).

[44] Reuther, Albert, Kepner, Jeremy, Byun, Chansup, Samsi, Siddharth, Arcand, William, Bestor, David, Bergeron, Bill, Gadepally, Vijay, Houle, Michael, Hubbell, Matthew, Jones, Michael, Klein, Anna, Milechin, Lauren, Mullen, Julia, Prout, Andrew, Rosa, Antonio, Yee, Charles and Michaleas, Peter. "Interactive Supercomputing on 40,000 Cores for Machine Learning and Data Analysis." *2018 IEEE High Performance Extreme Computing Conference (HPEC)*: pp. 1–6. 2018. DOI 10.1109/HPEC.2018.8547629.

[45] Erickson, Nick, Mueller, Jonas, Shirkov, Alexander, Zhang, Hang, Larroy, Pedro, Li, Mu and Smola, Alexander.

"AutoGluon-Tabular: Robust and Accurate AutoML for Structured Data." (2020). DOI 10.48550/arXiv.2003.06505.

APPENDIX A. PARAMETERS OF THE COMPRESSOR MODEL

Table 1 lists the geometrical and operating condition parameters needed to run the mean-line analysis model for centrifugal compressors. In addition to falling in the provided range, the inlet pressure and the temperature are also adjusted depending on the selected working fluid.

TABLE 1: DESCRIPTIONS OF THE PARAMETERS OF THE COMPRESSOR MODEL INCLUDING ACCEPTABLE RANGE AND UNITS

	Independent geometrical parameters		
r_4	Impeller tip radius	$[5, 250]$	mm
β_2	Mean-line impeller inlet blade angle	$[-60, 0]$	°
β_4	Mean-line impeller outlet blade angle	$[-70, -35]$	°
e_b	Mean blade thickness	$[0.1, 3]$	mm
Z_b	Number of blades	$[\![5, 20]\!]$	-
	Dependent geometrical parameters		
r_1	Inducer inlet radius	$[r_{2s}, r_4]$	mm
r_{2h}	Impeller hub radius	$[0.1r_4, 0.5r_4]$	mm
r_{2s}	Impeller shroud radius	$[1.2r_{2h}, 0.8r_4]$	mm
r_5	Diffuser outlet radius	$[r_4, 4r_4]$	mm
b_4	Impeller outlet blade height	$[0.015r_4, 0.3r_4]$	mm
b_5	Diffuser channel width	$[0.5b_4, 1.5b_4]$	mm
β_{2s}	Impeller inlet shroud blade angle	$[\beta_2 - 20, \beta_2]$	°
e_{tp}	Tip clearance	$[0.01b_4, 0.15b_4]$	mm
e_{bk}	Impeller backface clearance	$[0.001r_4, 0.15r_4]$	mm
l_{ind}	Inducer length	$[r_4, 4r_4]$	mm
Z_s	Number of splitter blades	0 if $Z_b > 11$ else Z_b	-
	Fixed geometrical parameters		
Ra	Inducer/Impeller surface roughness	1.2×10^{-5}	m
c_b	Blockage coefficients	1	-
	Independent operating condition parameters		
	Working fluid: air, ammonia, isobutane, pentane, propane, R1234yf, R134a, R245fa		
$Ma_{2,1}$	Inlet Mach number	$[5 \times 10^{-2}, 0.7]$	-
$Ma_{4,1}$	Impeller tip Mach number	$[5 \times 10^{-2}, 2.5]$	-
	Dependent operating condition parameters		
T_1	Inducer inlet temperature	$[170, 400]$	K
$P_{r,1}$	Reduced inducer inlet pressure	$[1, 100]$	-

Copyright © 2023 by ASME

Proceedings of the ASME 2023
International Design Engineering Technical Conferences and
Computers and Information in Engineering Conference
IDETC-CIE2023
August 20-23, 2023, Boston, Massachusetts

DETC2023-114464

DEEP GENERATIVE MODEL-BASED SYNTHESIS OF FOUR-BAR LINKAGE MECHANISMS CONSIDERING BOTH KINEMATIC AND DYNAMIC CONDITIONS

Sumin Lee[1,†], Jihoon Kim[1,†], Namwoo Kang[1,2,*]

[1]Korea Advanced Institute of Science and Technology, Daejeon, Republic of Korea
[2]Narnia Labs, Daejeon, Republic of Korea

ABSTRACT

Mechanisms are essential components designed to perform specific tasks in various mechanical systems. However, designing a mechanism that satisfies both kinematic and dynamic requirements is a challenging task. The kinematic requirements may include the workspace of a mechanism, where the maximum scalar displacement of the mechanism can be derived from this region. On the other hand, the dynamic requirements of a mechanism may include its torque transmission, which refers to the ability of the mechanism to transfer power and torque effectively. In this paper, we propose a deep generative model that can generate multiple crank-rocker four-bar linkage mechanism samples that satisfy both the kinematic and dynamic requirements aforementioned. The proposed model is based on a conditional generative adversarial network (cGAN) with some modifications for mechanism synthesis, which is trained to learn the relationship between the requirements of a mechanism with respect to linkage lengths, and generates multiple mechanism samples that satisfy given requirements. The results demonstrate that our proposed method can successfully generate multiple mechanisms that satisfy specific requirements. Our approach has several advantages over traditional design methods. It enables designers to explore a larger design space and efficiently generate multiple diverse and feasible designs. Also, the proposed method considers both the kinematic and dynamic requirements, which can lead to more efficient and effective mechanisms for real-world use, making it a promising tool for linkage mechanism design.

Keywords: Generative Design, Mechanism Synthesis

1. INTRODUCTION

A mechanism design is essential in many mechanical systems to perform tasks. For example, many different forms of a mechanism are used in locking pliers, suspensions, robots, and more. In particular, a mechanism with linkages is widely used in industries due to its ability to transfer a simple circular motion (e.g., from a motor) into multi-joint movements that draw a unique path or a workspace. Therefore, a linkage mechanism can be very useful if its design is carefully considered and tested for one's requirements. Nevertheless, it is very challenging to design a multi-linkage mechanism that satisfies specific requirements, since its behavior changes significantly with minor changes in its linkage lengths. Although it is possible to analytically calculate the lengths of a linkage mechanism that passes a few given points, e.g., Burmester Theory [1], real-world problems and requirements are often more complicated.

There has been significant effort in solving this issue in the field. The numerical methods have been studied to find the lengths of linkages that follow a specific target path with feed-forward optimization processes [2–5]. These methods are effective in providing a good solution if there is a linkage mechanism that can definitely satisfy these conditions. However, their iterative processes can be time-consuming, especially in high resolution, which is more problematic when these processes must be repeated for different given conditions. Further, these methods only can provide a single solution to the problem where there could be many other linkage mechanisms with similar conditions.

The data-driven methods are also an active research area [6]. Especially, [7] created a big dataset of planar linkage mechanisms for data-driven kinematic design. Their data includes the paths drawn by the mechanisms, which provides a unique opportunity to simply find mechanisms from the dataset that satisfy required kinematic conditions. Nevertheless, they mentioned that the generation of the dataset is very computationally expensive, and there still may not be a mechanism that perfectly satisfies the target path. The deep learning-based [8] and also image-based [9] generative methods have been studied to synthesize paths of multiple linkage mechanisms. These methods are use-

---

†Joint first authors
*Corresponding author: nwkang@kaist.ac.kr
Documentation for `asmeconf.cls`: Version 1.34, May 14, 2023.

Copyright © 2023 by ASME

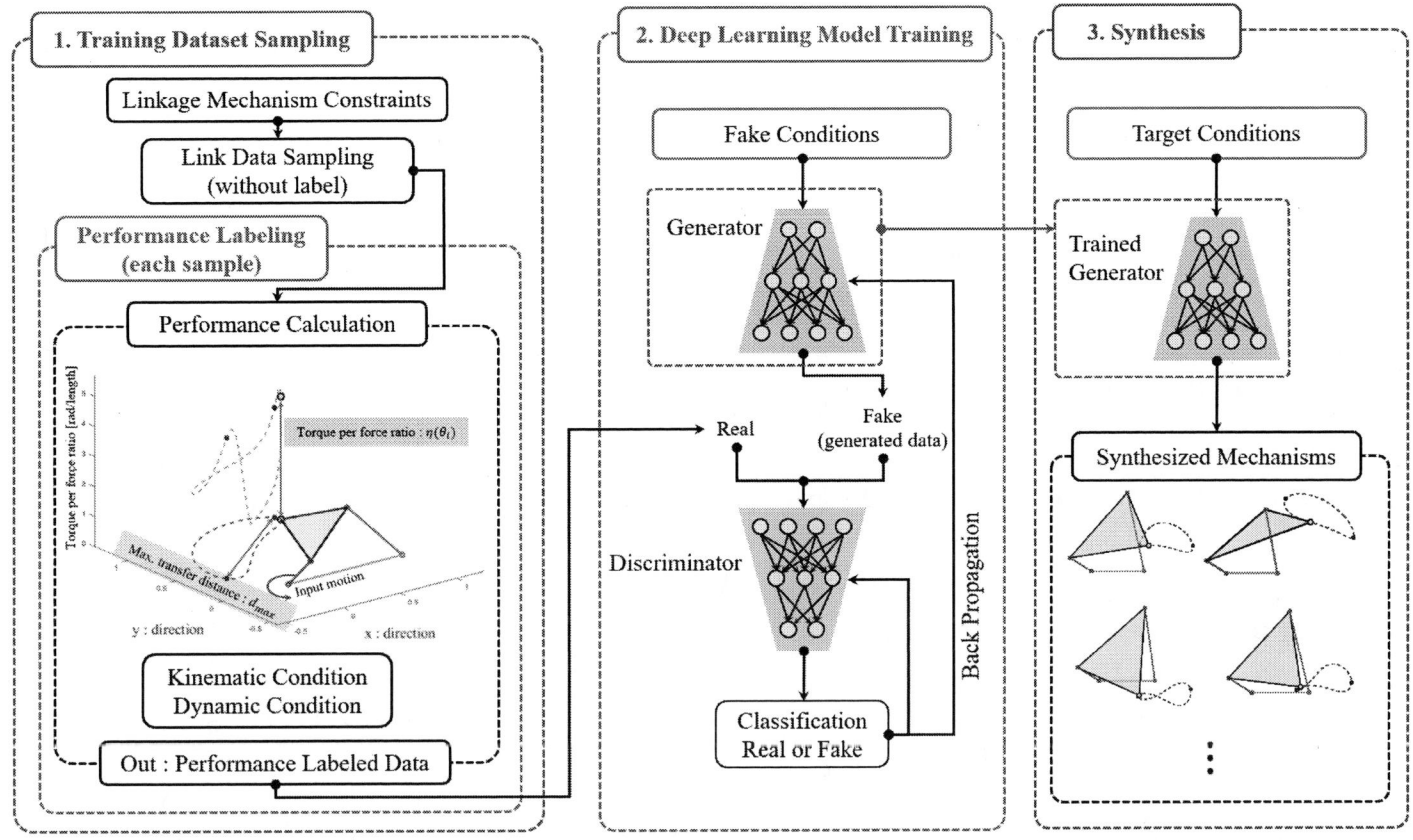

FIGURE 1: DEEP GENERATIVE MODEL-BASED MECHANISM SYNTHESIS CONSIDERING KINEMATIC AND DYNAMIC CONDITIONS

ful in that they are able to generate samples more quickly than traditional optimization methods discussed above. Also, they are able to generate a wide range of solutions, whereas most of the optimization-based methods can only generate a single sample. However, the quality of the generated samples may be inferior compared to the optimization-based methods as it relies on the limited amount of the dataset it was trained on.

Most of the studies in four-bar linkage mechanism synthesis focus on generating a mechanism that can follow a specific target path [5, 7, 8, 10]. However, it may be more useful to have multiple mechanisms that satisfy various requirements that are required to consider for real-world use cases than only considering their paths. For example, mechanism synthesis models that only consider kinematic conditions, such as paths, may create mechanisms that require very high torque to perform tasks. Thus, it is important to design a mechanism that considers kinematic and dynamic conditions simultaneously. All the synthesis methods mentioned above do not consider any dynamic conditions in linkage mechanisms (e.g., torque transmission, maximum payload, etc.) that are essential to consider for a mechanism to perform meaningful tasks in the real world. Therefore, we propose a deep generative model that synthesizes multiple distinct linkage mechanism samples that satisfy the kinematic and dynamic requirements given as the conditions. Designers then will be able to choose a mechanism that suits their needs from multiple samples that satisfy the requirements.

Figure 1 shows the overall flow for the training of the proposed generative model and the linkage mechanism synthesis. This paper is structured as follows: 1) the mechanism samples are generated for the training of the deep generative model (Section 2). 2) the model is trained with this dataset (Section 3.1). 3) the synthesis performance of the model is evaluated (Section 3.2). The paper is summarised in Section 4.

2. GENERATION OF LINKAGE MECHANISMS

2.1 Crank-rocker Four-bar Linkage Mechanism

It is crucial that the generated samples satisfy the constraints of a four-bar linkage mechanism. While there are many types of four-bar linkage mechanisms (e.g., drag-link, crank-rocker, double-rocker, etc.), this paper only considers a crank-rocker mechanism as its crank can fully revolve and continuously draw a path as shown in Fig. 2. This provides the data needed to calculate the kinematic and dynamic conditions for the inverse design of mechanisms.

According to Grashof's Law, in order for a crank-rocker system to rotate continuously, the sum of the shortest and longest link lengths is less than or equal to the sum of the other two link lengths. Applying Reuleaux's Law, it must satisfy all the conditions in Eq. (1). Since l_1 is fixed to 1.0, the other linkage lengths can be considered as ratios with respect to l_1.

Copyright © 2023 by ASME

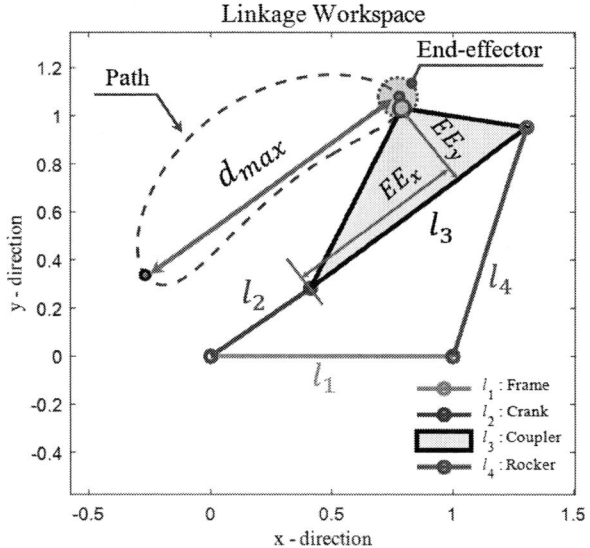

FIGURE 2: A LABELED EXAMPLE OF A CRANK-ROCKER FOUR-BAR LINKAGE MECHANISM

$$l_1 - l_2 \le l_3 + l_4$$
$$l_1 - l_2 \ge l_3 - l_4$$
$$l_1 + l_2 \ge l_3 - l_4 \quad (1)$$
$$l_1 + l_2 \le l_3 + l_4$$

In theory, any fully-rotating crank-rocker mechanisms can be produced if they satisfy the conditions in Eq. (1). However, it is not viable to create a mechanism too long that would be infeasible or even impossible to produce in the real world, as there are infinite number of link sets that can satisfy Eq. (1). Therefore, the range of the link lengths for the synthesis was arbitrarily selected as shown in Table 1.

TABLE 1: THE RANGES OF LINKAGE LENGTHS SET TO GENERATE MECHANISMS, IN METERS

	l_2	l_3	l_4	EE_x	EE_y
max.	0.95	2.00	3.00	2.50	1.50
min.	0.05	0.05	0.05	-0.50	-1.50

2.2 Conditions

Due to the difference in linkage lengths, mechanisms naturally have different mechanical properties that may be used to assess their performance. This study aims to create numerous and distinct linkage mechanisms that satisfy the given performance conditions. In other words, linkage mechanisms produced given similar conditions must be performance-compliant, yet still different to each other. Thus, a means to evaluate the performance and uniqueness of the link sets is required. This section explains the kinematic and dynamic conditions of the four-bar linkage mechanisms we considered in this paper.

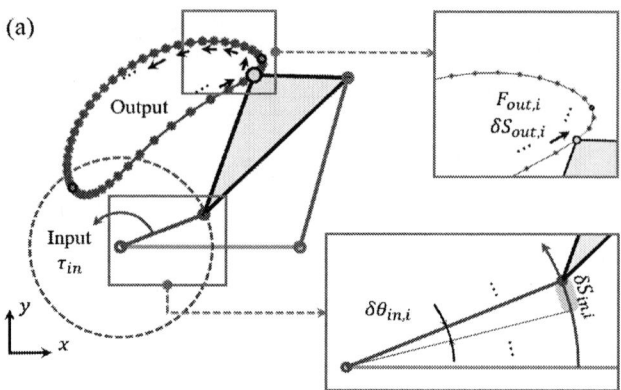

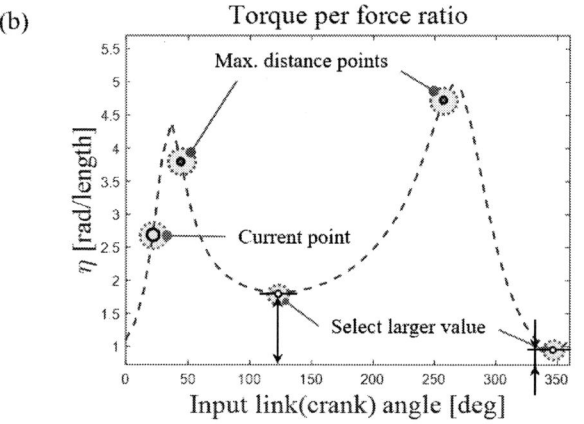

FIGURE 3: (A) THE PARAMETERS NEEDED FOR THE CALCULATION OF η_{min} AND (B) THE PLOT OF η ALONG THE PATH OF A MECHANISM

Kinematic Condition. As shown in Fig. 2, the maximum distance available in the path a mechanism draws (d_{max}) can be derived by numerically calculating the Euclidean distances between the points along the path. Figure 5 shows that there are many different samples that satisfy the same kinematic condition, despite their different linkage lengths and paths.

Dynamic Condition. In the real world, there are many cases where the dynamic aspects of a mechanism must also be considered, e.g., delivering a particular load from one position to another in a desired path or workspace. In order to determine the maximum payload of a mechanism for a task, we calculate the torque per force ratio between the input link (crank) and the end-effector along a path.

The forces acting in a workspace is determined by the transmission of input torque, and their relationships can be known using the work conservation law. Assuming the energy is conserved within the mechanism, the relationship between the input and the output work is defined as $|W_{in}| = |W_{out}|$, where W_{in} is the input work at the crank and W_{out} is the output work at the end-effector. As $W = F \cdot S$ where F is the force and S is the displacement, this equation can be converted to Eq. (2).

$$|F_{in} \cdot \delta S_{in}| = |F_{out} \cdot \delta S_{out}| \quad (2)$$

Copyright © 2023 by ASME

As $F \cdot S = \tau\theta$ where τ is torque and θ is angular displacement, it can further be derived as Eq. (3).

$$|\tau_{in} \cdot \delta\theta_{in}| = |F_{out} \cdot \delta S_{out}| \qquad (3)$$

where $\delta\theta_{in}$ is the incremental angular displacement at the crank and δS_{out} is the incremental displacement at the end-effector. Since we defined η as the torque per force ratio, η can be expressed as Eq. (4).

$$\eta = \left| \frac{\delta S_{out}}{\delta\theta_{in}} \right| \qquad (4)$$

Equation (4) can now be used to calculate the η at the end-effector where a payload will be attached to. Figure 3 (a) shows that η of a mechanism is determined by the incremental angular displacement of the crank ($\delta\theta$) and the incremental displacement of the end-effector (δS_{out}). As this study solely considers a fully-revolving crank-rocker mechanism, η along its full path can be calculated as shown in Fig. 3 (b). If we divide the path by two at the maximum distance points (red points in Fig. 3 (b)), two different η_{min} can be determined at each path (blue in Fig. 3 (b)). Although a crank-rocker can fully revolve, it can also move only in one of these two paths. In other words, one of the two paths that gives the higher minimum η (η_{min}) can be selected to maximize the payload.

By using this measure, we can determine the minimum torque requirement of a motor to lift desirable payload, and vice versa.

2.3 Data Generation for Model Training

The dataset of the linkage mechanisms used for the training of a generative model and synthesis was produced using Latin Hypercube Sampling (LHS) [11] within the linkage length ranges defined in Table 1. The goal of using LHS is to ensure that the samples are representative of the distribution across all ranges of linkage lengths, while minimizing correlation between them. Any generated samples that do not satisfy the crank-rocker conditions given in Eq. (1) were removed. Due to the high computational costs required to generate mechanisms and to calculate the conditions, only 1,000 valid samples were generated in total.

We denote this dataset X_{train}. The generation of X_{train} was done in MATLAB. Figure 4 shows the distribution of d_{max} and η_{min} calculated from X_{train}. As shown, d_{max} and η_{min} have a non-linear relationship that was not apparent before the generation. A few of the generated mechanisms with similar conditions are visualized in Fig. 5 to show their diversity in linkage lengths.

3. SYNTHESIS WITH DEEP GENERATIVE MODEL

A conditional Generative Adversarial Network (cGAN) [12, 13], a popular deep learning-based generative model, was used to synthesize mechanism samples for given conditions. cGAN has a strength compared to other deep generative models in that it can generate multiple samples that satisfy given conditions. This section explains the model training with X_{train}, the mechanism synthesis, and evaluates the trained model.

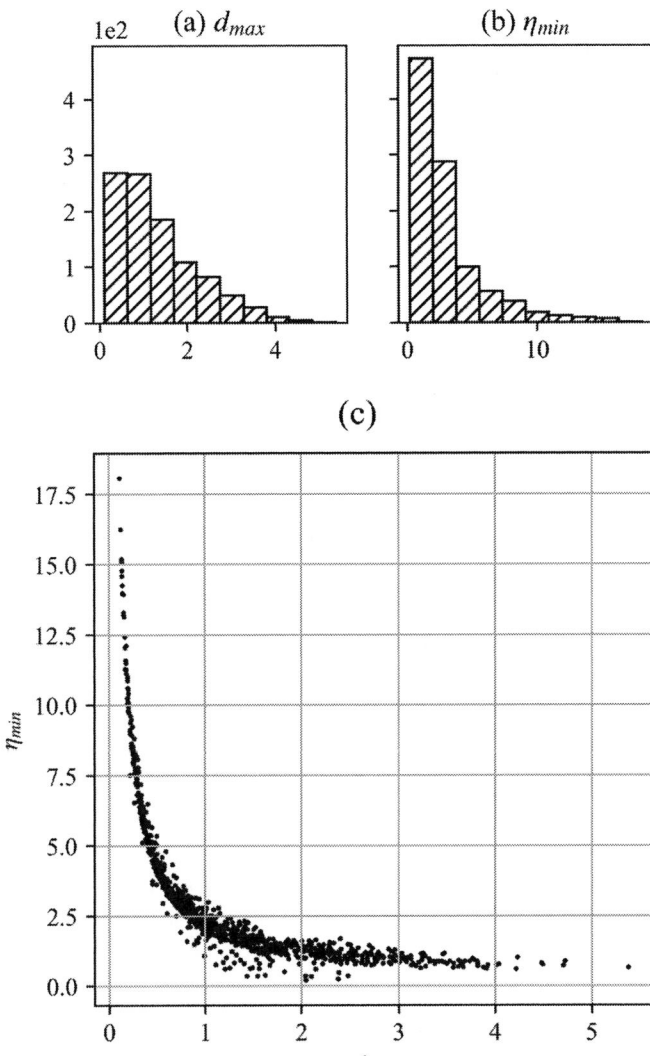

FIGURE 4: (A), (B) HISTOGRAMS AND (C) A SCATTER PLOT OF d_{max} AND η_{min} OF X_{train}

3.1 Conditional Generative Adversarial Network

Training. The training procedure of the cGAN model is shown in Fig. 6. The generator (G) generates synthetic design samples for given random conditions, where the conditions are d_{max} and η_{min}. The random noise input is to ensure that random samples are generated. The discriminator (D) then classifies unlabeled real and synthesized (fake) samples on whether they are real or fake. This process is looped, and G and D are adversarially trained, which enables the generator to synthesize realistic samples. The samples consist of link lengths (l_2, l_3, l_4, EE_x and EE_y) and the conditions. The difference between real and fake samples is that the real conditions are the calculated ground truths for the given link lengths, whereas link lengths are estimates that depend on randomly generated conditions for fake samples.

Structure. G is a multi-layer perceptron (MLP) with five fully-connected layers with 20 neurons each. Note that the number of layers and neurons are hyperparameters, and the best ones

Copyright © 2023 by ASME

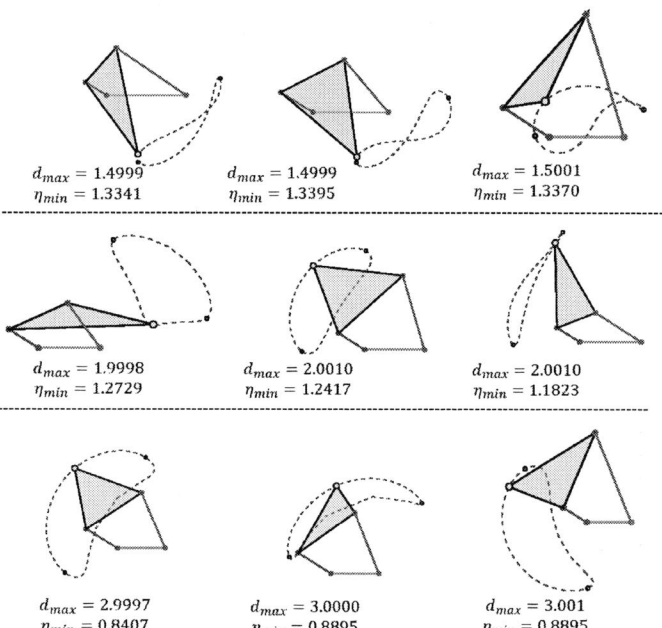

$d_{max} = 1.4999$
$\eta_{min} = 1.3341$

$d_{max} = 1.4999$
$\eta_{min} = 1.3395$

$d_{max} = 1.5001$
$\eta_{min} = 1.3370$

$d_{max} = 1.9998$
$\eta_{min} = 1.2729$

$d_{max} = 2.0010$
$\eta_{min} = 1.2417$

$d_{max} = 2.0010$
$\eta_{min} = 1.1823$

$d_{max} = 2.9997$
$\eta_{min} = 0.8407$

$d_{max} = 3.0000$
$\eta_{min} = 0.8895$

$d_{max} = 3.001$
$\eta_{min} = 0.8895$

FIGURE 5: VISUALIZATION OF FEW MECHANISMS WITH SIMILAR CONDITIONS GENERATED FOR THE TRAINING

that perform reasonable inverse design were found by trial-and-error. A rectifier linear unit (ReLU) [14] activation function was used between the layers. D is identical to G, except that it has an additional sigmoid activation function at the end to enforce the output between 0 and 1 to calculate classification losses. A binary cross-entropy (BCE) loss was used for D ($\mathcal{L}_D$). The batch size was set to 100.

Modifications on cGAN. The baseline cGAN was first used to generate mechanism samples. However, the model could not generate good samples due to mode collapse. To ensure that G generates adequate and distinct samples, additional methods were used on the baseline cGAN model.

A condition predictor (P) was used to predict the estimated actual d_{max} and η_{min} (d_r and η_r) for given samples. This was used to minimize the gap between the d_{max} and η_{min} given as target conditions (d_t and η_t) to G and d_r and η_r of the samples that G synthesized. The gap was measured with mean squared error (MSE) loss ($\mathcal{L}_P$) of a batch as shown in Eq. (5), which was used as a part of loss functions for G. P is also an MLP identical to G, except that its inputs are the linkage mechanism samples, and the outputs are the estimated conditions. It is worth noting that P can be simply replaced with the script that was used to calculate the conditions of X_{train} in Section 2.2. However, this will make the training of cGAN too slow as the script takes a long time to calculate. Therefore, P was trained using the existing data (X_{train}) and the corresponding conditions that have been already calculated. P was trained prior to the training of cGAN and was fixed with no updates afterwards.

$$\mathcal{L}_P = \frac{1}{N} \sum_{i=1}^{N} \left[(d_{r,i} - d_{t,i})^2 + (\eta_{r,i} - \eta_{t,i})^2 \right] \quad (5)$$

Similarity loss ($\mathcal{L}_s$) was used to ensure that G generates distinct samples. The variance of the synthesized samples was calculated, and the inverse of it was used as a loss function as shown in Eq. (6) to maximize the variance between samples, thus encouraging the generation of more distinct mechanism sets for similar given conditions. L is a set of linkages where their number is the batch size set while training the model and N is the order, e.g., if the batch size is 100, L_2 is a set of 100 generated l_2 values. Thus, $\mathcal{L}_s$ is a mean of variances of synthesized linkage lengths (cranks, coupler, rockers, and end-effectors).

$$\mathcal{L}_s = \frac{1}{N} \sum_{i=1}^{N} \text{Var}(L_N) \quad (6)$$

Therefore, the loss function used to train G is as shown in Eq. (7), where w_P and w_s are the weights for $\mathcal{L}_P$ and $\mathcal{L}_s$, respectively, and were adjusted accordingly to stabilize the training.

$$\mathcal{L}_{GPs} = \mathcal{L}_G + w_P \mathcal{L}_P + w_s \mathcal{L}_s \quad (7)$$

After the training, the samples can be generated given the values of d_t and η_t, and the respective d_r and η_r can be calculated.

Sampling of Fake Conditions. It is important to sample realistic fake conditions for G, as D could simply discriminate fake samples based on their unrealistic fake conditions. A random sampling of fake conditions was initially conducted to sample fake conditions for the training of the generative model. However, as shown in Fig. 4 ©, the real conditions have Pareto-like distribution and make random sampling unreasonable. This necessitated a more sophisticated sampling strategy. Synthetic minority over-sampling technique (SMOTE) [15] is widely used to mitigate imbalanced classification problems by oversampling based on real samples. SMOTE works by creating synthetic examples of the minority class by interpolating between existing examples. The same algorithm was used here to create more realistic fake conditions based on real conditions instead, considering all the real condition samples as the minority class. As shown in Fig. 7, SMOTE was able to generate the fake condition samples that are much closer to the real condition samples.

3.2 Evaluation

Figure 8 shows the evaluation procedure of the trained model. The procedure is as follows: **(a)** The number of i fake conditions are sampled.

$$[\mathcal{D}_t, \mathcal{H}_t] = \left[\{d_{t,1}, d_{t,2}, ..., d_{t,i}\}, \{\eta_{t,1}, \eta_{t,2}, ..., \eta_{t,i}\} \right] \quad (8)$$

Equation (8) states a set of fake (target) conditions $[\mathcal{D}_t, \mathcal{H}_t]$. **(b)** The linkage mechanism length samples are synthesized with G with $[\mathcal{D}_t, \mathcal{H}_t]$. We denote this dataset X_{syn}. © d_r and η_r values of X_{syn} are calculated.

$$[\mathcal{D}_r, \mathcal{H}_r] = \left[\{d_{r,1}, d_{r,2}, ..., d_{r,i}\}, \{\eta_{r,1}, \eta_{r,2}, ..., \eta_{r,i}\} \right] \quad (9)$$

Equation (9) states a set of actual conditions $[\mathcal{D}_r, \mathcal{H}_r]$ calculated from X_{syn}. **(d)** Finally, the model is evaluated by calculating errors between $[\mathcal{D}_t, \mathcal{H}_t]$ and $[\mathcal{D}_r, \mathcal{H}_r]$.

The trained model was tested with two cases. Firstly, multiple fake conditions with 1,000 samples ($i = 1,000$) were sampled

Copyright © 2023 by ASME

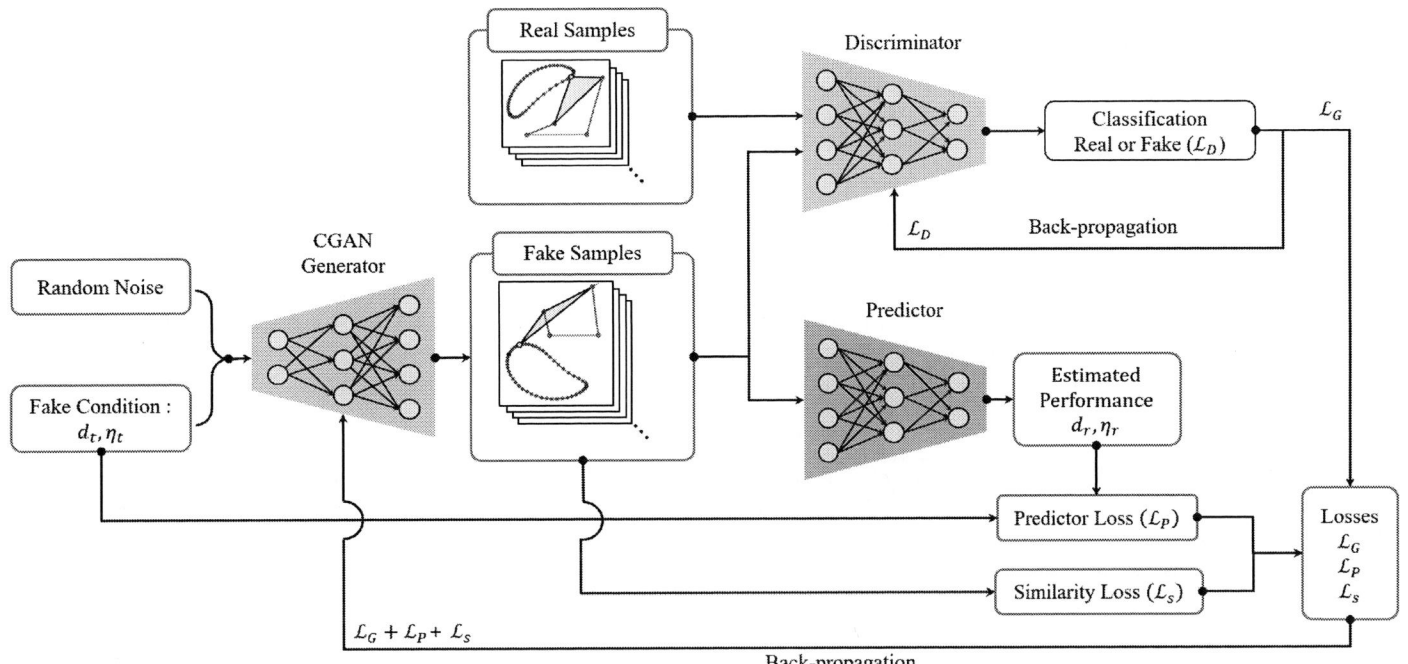

FIGURE 6: THE TRAINING PROCEDURE (LOOP FROM LEFT TO RIGHT) OF THE GENERATIVE METHOD

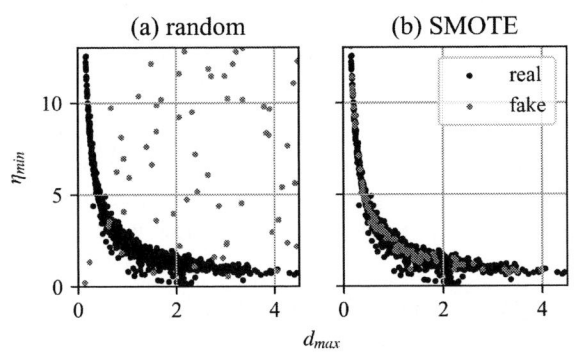

FIGURE 7: A COMPARISON OF THE SAMPLING STRATEGIES FOR FAKE CONDITIONS: (A) RANDOM AND (B) SMOTE

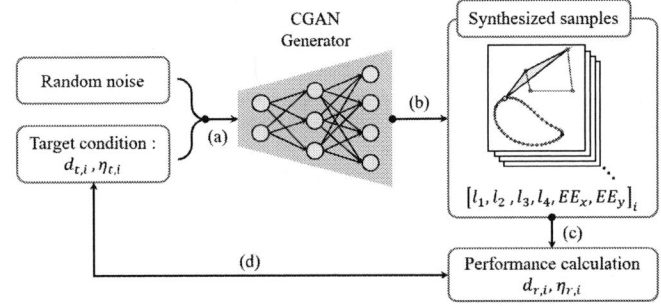

FIGURE 8: THE EVALUATION PROCEDURE OF G

from X_{train} with SMOTE to test the model performance in generating mechanism samples that satisfy the given target conditions. Secondly, 100 samples ($i = 100$) with a single (same) condition were used to test the ability of the model to produce diverse mechanism samples.

TABLE 2: THE PERFORMANCE OF THE TRAINED MODEL FOR MULTIPLE CONDITIONS

	$[\mathcal{D}_t, \mathcal{D}_r]$	$[\mathcal{H}_t, \mathcal{H}_r]$
RMSE	0.147	0.894
MAE	0.110	0.398
R^2	0.958	0.850

Multiple Conditions. Root mean squared error (RMSE), mean absolute error (MAE), and R^2 scores were used to evaluate the performance of the trained cGAN model in Table 2. As shown, G was able to generate synthetic linkage mechanisms that satisfy both of the target conditions with marginal errors.

Figure 9 shows the scatter plots between the target and the actual conditions. G was mostly able to produce the samples that satisfy the target conditions. However, it struggled to match η_t when $\eta_t > 6$. We suspect this is due to the lack of data in this region; as shown in Fig. 4 (b), X_{train} naturally has much more data where $\eta_{min} < 6$, even though X_{train} was generated with random linkage lengths. This means that G would have been trained to synthesize better where d_{max} and η_{min} are mostly concentrated, as the fake conditions were generated based on the distribution of the conditions of X_{train}. Moreover, d_{max} gets extremely low as η_{min} increases, and vice versa, which may not be useful

Copyright © 2023 by ASME

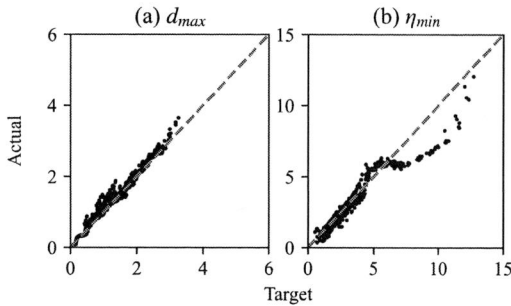

FIGURE 9: THE SCATTER PLOTS OF (A) d_t VS d_r AND (B) η_t VS η_r. THE DASHED LINE IN RED IS $y = x$.

for real-world use. Therefore, one may want a mechanism that reasonably satisfies both of the requirements, in which proposed model provides. Nevertheless, a more sophisticated sampling technique for fake conditions or a variable weighting of samples while training G might mitigate this issue.

TABLE 3: THE PERFORMANCE OF THE TRAINED MODEL FOR THE SINGLE CONDITION

	$[\mathscr{D}_t, \mathscr{D}_r]$	$[\mathscr{H}_t, \mathscr{H}_r]$
RMSE	0.265	0.295
MAE	0.174	0.262

TABLE 4: THE STANDARD DEVIATION OF THE SYNTHESIZED SAMPLE LENGTHS FOR THE SINGLE CONDITION

l_2	l_3	l_4	EE_x	EE_y
0.041	0.036	0.138	0.644	0.168

Single Condition. The condition set was fixed to $d_t = 1.0, \eta_t = 2.0$ to evaluate the model performance in generating diverse samples. In other words, $[\mathscr{D}_t, \mathscr{H}_t]$ now have the same values regardless of i. Table 3 shows the error metrics of the model given the same condition. Note that the R^2 metric is not used as the conditions are the same in this case. The standard deviation of the synthesized sample lengths is shown in Table 4. Figure 10 shows some of the synthesized samples with diversity. From these examples and the metrics, it can be seen that the model is able to synthesize diverse mechanism samples while satisfying the target conditions with marginal errors.

4. CONCLUSION

In this paper, we proposed a modified cGAN that generates multiple mechanism samples that satisfy the kinematic and dynamic requirements. To sample the data for cGAN training, we considered the crank-rocker four-bar linkage mechanism constraints, and d_{max} and η_{min} were considered as the performance metrics for a mechanism. We trained the cGAN with the limited amount of training data due to the computational constraints, and evaluated the model by generating 1,000 samples with random

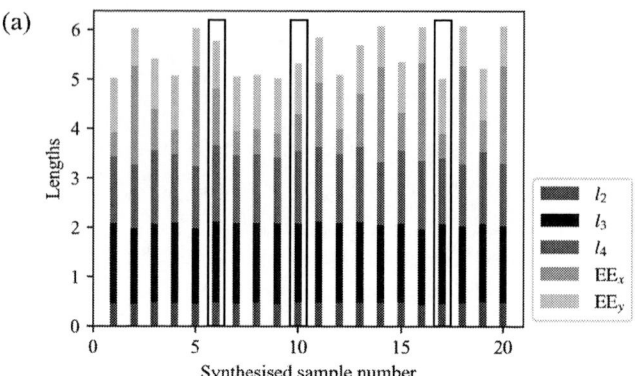

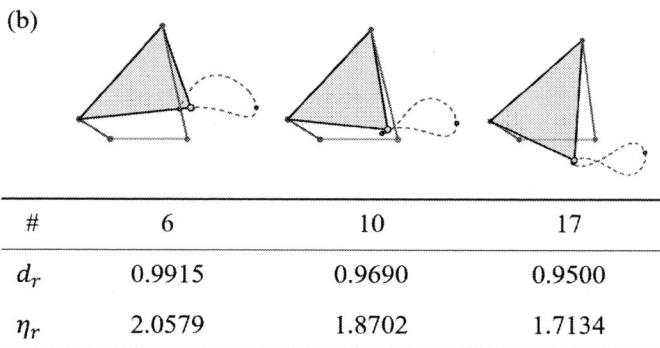

#	6	10	17
d_r	0.9915	0.9690	0.9500
η_r	2.0579	1.8702	1.7134

FIGURE 10: (A) A STACKED BAR PLOT OF THE SYNTHESIZED LINKAGE LENGTHS FOR THE SINGLE CONDITION AND (B) THE VISUALIZATION OF THE SAMPLES WITH THE RESPECTIVE d_r AND η_r

condition sets and 100 samples with a single condition set. It was shown that the proposed generative model can synthesize four-bar linkage mechanism samples that satisfy the given condition set with marginal errors.

Limitations. The mechanism synthesis method proposed in this study has several limitations. First, this study assumes a rigid linkage mechanism, and its mass and inertia were ignored, as this study aims to promote a straightforward synthesis of more diverse mechanism design with cGAN. Thus, this study may be considered as a solution for a quasi-static problem rather than dynamic. Nevertheless, this study may be improved by considering mass and inertia to help the synthesis of mechanism design considering more realistic dynamic conditions.

Future Work. We performed preliminary research on generating linkage mechanisms using cGAN. Although the model performed well in synthesizing a simple four-bar crank-rocker linkage mechanism with one degree-of-freedom (1-DOF) that was solely considered in this paper, we anticipate that this will be more challenging for mechanisms with two or more DOF or with less constraints than a crank-rocker. Thus, we plan to conduct research on mechanisms with higher degrees of freedom or more complex mechanisms where multiple linkages are intertwined, and test if a deep generative model-based mechanism synthesis method, similar to the one proposed in this paper, will still be effective in the rapid and effective generation of diverse

Copyright © 2023 by ASME

mechanisms.

Also, other cGAN variants, e.g., PcDGAN [16] or Mo-PadGAN [17], may show better performance than the baseline cGAN for this inverse design. Further empirical studies with improved cGAN variants may be conducted to produce more compliant and diverse mechanisms.

ACKNOWLEDGMENTS

This work was supported by the National Research Foundation of Korea grant (2018R1A5A7025409) and the Ministry of Science and ICT of Korea grant (No.2022-0-00969, No.2022-0-00986).

REFERENCES

[1] Burmester, Ludwig. *Lehrbuch der Kinematik*. Leipzig (1888).

[2] Kang, Seok Won, Kim, Suh In and Kim, Yoon Young. "Topology optimization of planar linkage systems involving general joint types." *Mechanism and Machine Theory* Vol. 104 (2016): pp. 130–160.

[3] Yim, Neung Hwan, Kang, Seok Won and Kim, Yoon Young. "Topology optimization of planar gear-linkage mechanisms." *Journal of Mechanical Design* Vol. 141 No. 3 (2019): p. 032301.

[4] Yu, Jeonghan, Han, Sang Min and Kim, Yoon Young. "Simultaneous shape and topology optimization of planar linkage mechanisms based on the spring-connected rigid block model." *Journal of Mechanical Design* Vol. 142 No. 1 (2020).

[5] Han, Sang Min and Kim, Yoon Young. "Topology optimization of linkage mechanisms simultaneously considering both kinematic and compliance characteristics." *Journal of Mechanical Design* Vol. 143 No. 6 (2021).

[6] Yim, Neung Hwan, Lee, Jongjun, Kim, Jungho and Kim, Yoon Young. "Big data approach for the simultaneous determination of the topology and end-effector location of a planar linkage mechanism." *Mechanism and Machine Theory* Vol. 163 (2021): p. 104375.

[7] Heyrani Nobari, Amin, Srivastava, Akash, Gutfreund, Dan and Ahmed, Faez. "LINKS: A Dataset of a Hundred Million Planar Linkage Mechanisms for Data-Driven Kinematic Design." *International Design Engineering Technical Conferences and Computers and Information in Engineering Conference*, Vol. 86229: p. V03AT03A013. 2022. American Society of Mechanical Engineers.

[8] Deshpande, Shrinath and Purwar, Anurag. "Computational creativity via assisted variational synthesis of mechanisms using deep generative models." *Journal of Mechanical Design* Vol. 141 No. 12 (2019).

[9] Deshpande, Shrinath and Purwar, Anurag. "An Image-Based Approach to Variational Path Synthesis of Linkages." *Journal of Computing and Information Science in Engineering* Vol. 21 No. 2 (2021).

[10] Deshpande, Shrinath and Purwar, Anurag. "A machine learning approach to kinematic synthesis of defect-free planar four-bar linkages." *Journal of Computing and Information Science in Engineering* Vol. 19 No. 2 (2019).

[11] McKay, MD, Beckman, RJ and Conover, WJ. "Comparison of Three Methods for Selecting Values of Input Variables in the Analysis of Output from a Computer Code." *Technometrics* Vol. 21 No. 2 (1979): pp. 239–245.

[12] Goodfellow, Ian, Pouget-Abadie, Jean, Mirza, Mehdi, Xu, Bing, Warde-Farley, David, Ozair, Sherjil, Courville, Aaron and Bengio, Yoshua. "Generative adversarial networks." *Communications of the ACM* Vol. 63 No. 11 (2020): pp. 139–144.

[13] Mirza, Mehdi and Osindero, Simon. "Conditional generative adversarial nets." *arXiv preprint arXiv:1411.1784* (2014).

[14] Glorot, Xavier, Bordes, Antoine and Bengio, Yoshua. "Deep sparse rectifier neural networks." *Proceedings of the fourteenth international conference on artificial intelligence and statistics*: pp. 315–323. 2011. JMLR Workshop and Conference Proceedings.

[15] Chawla, Nitesh V, Bowyer, Kevin W, Hall, Lawrence O and Kegelmeyer, W Philip. "SMOTE: synthetic minority over-sampling technique." *Journal of artificial intelligence research* Vol. 16 (2002): pp. 321–357.

[16] Heyrani Nobari, Amin, Chen, Wei and Ahmed, Faez. "Pcdgan: A continuous conditional diverse generative adversarial network for inverse design." *Proceedings of the 27th ACM SIGKDD conference on knowledge discovery & data mining*: pp. 606–616. 2021.

[17] Chen, Wei and Ahmed, Faez. "Mo-padgan: Reparameterizing engineering designs for augmented multi-objective optimization." *Applied Soft Computing* Vol. 113 (2021): p. 107909.

Proceedings of the ASME 2023
International Design Engineering Technical Conferences and
Computers and Information in Engineering Conference
IDETC-CIE2023
August 20-23, 2023, Boston, Massachusetts

DETC2023-114660

GRADES: AN AI-DRIVEN GRAPHIC DESIGN SUPPORT SYSTEM FOR DESIGN STYLE ANALYSIS

Jinyu Song[1,2], Weitao You[1,2,*], Shuhui Shi[1], Ziwei Tu[3], Juntao Ji[1], Kaixin Han[1], Lingyun Sun[1]

[1]College of Computer Science and Technology, Zhejiang University, Hangzhou, China
[2]Alibaba-Zhejiang University Joint Research Institute of Frontier Technologies, Zhejiang University, Hangzhou, China
[3]College of Arts, Sichuan University, Chengdu, China

ABSTRACT

The style of graphic design is an important design factor that influences the memorability of designs. Graphic designers routinely analyze the latest designs, capture the style trends, and create designs that match the style trends to appeal to a larger audience. Nonetheless, the lack of quantitative style analysis techniques can lead to an inefficient analysis process and introduce subjectivity and bias. To expedite designers' understanding of design style trends and make the analysis more objective, we propose GradeS, an AI-driven graphic design support system that facilitates multifaceted quantitative analysis of graphic design style. The system was designed and developed in collaboration with designers and comprises four primary interfaces: GradeS:S, GradeS:Q, GradeS:C, and GradeS:T, each serving specific needs identified through interviews with designers. We leveraged the Vision Transformer to model the one-to-many relationship between designs and styles and implemented all interfaces based on the quantitative style representation learned by the model. To train the model, we built a graphic design dataset with carefully designed coarse-grained style labels. We have released the dataset to the community to promote research in data-driven design. To demonstrate the effectiveness of our study, we evaluated both the model and the system. Our model exhibits superior performance in style classification compared to CLIP. Through a user study involving six designers, our system's effectiveness in supporting designers in analyzing style quantitatively, capturing style trends comprehensively and quickly, and further stimulating creative thinking was demonstrated.

Keywords: Graphic Design Style, Design Intelligence, Data-Driven Design, Dataset

---
*Corresponding author: weitao_you@zju.edu.cn

1. INTRODUCTION

Graphic design is ubiquitous in every corner of modern life. By organically integrating texts, symbols, images, shapes and other visual elements (e.g. 3D model) together, graphic design ensures aesthetic quality while containing the desired messages and serving a variety of purposes whether commercial, social, cultural, educational or personal [1]. For example, presentation slides convey ideas through graphic design, serving educational purposes. Advertisements promotes the goods or services through graphic design, serving commercial purpose. To convey ideas or messages more effectively, thus serving purposes better, especially in this age of information explosion, graphic designers need to carefully design the overall visual perception of their work to attract the user's attention through the first impression and prompt the user to understand the content that the graphic design is intended to convey. This overall visual perception is also known as image style [2]. Image style is a high-level descriptor that describes how an image is viewed and perceived, determining the first impression [3] of the image of the user and further affects the memorability [4], which is closely related to whether the information can be effectively conveyed. Therefore, designers usually analyze designs' style implicitly to get the latest cognition of style, such as the latest style trends.

Many studies provide assistance for graphic design from design factors such as font [5], color [6, 7], and layout [8]. However, design aids related to style are rarely studied. Due to the lack of quantitative style analysis approaches and style-orientated aids, designers have to analyze style on their own, which is inefficient and subjective. In this paper, we propose GradeS(**Gra**phic **De**sign **S**tyle), an AI-driven graphic design support system that enables quantitative design style analysis. GradeS was designed in collaboration with designers. We first identified four design goals for our system through interviews with designers. Then, we implemented the basis of our system — an AI model that provides quantitative style representations, making quantitative

Copyright © 2023 by ASME

style analysis possible. We adopted *Vision Transformers* [9] to model the one-to-many relationship between design images and design styles. To train the model, we built a 13K graphic design dataset and labeled it with coarse-grained style labels through crowdsourcing. Each image was labeled at least three times to reduce subjectivity in style recognition. After training, we used its output (i.e., the probability of each style) and the feature at the *CLS* token to represent the style of the design image. Based on the model, we provided four quantitative style analytical interfaces, including *GradeS:S*, *GradeS:Q*, *GradeS:C*, and *GradeS:T*. Each interface was designed to achieve a specific system design goal. All interfaces are complementary to each other and work together to support designers in the style analysis process. Details will be discussed in Section 5.

To demonstrate the effectiveness of the proposed system, we evaluated both the model and the system. To evaluate the model performance, we compared our model with OpenAI CLIP [10]. In addition, we did a ablation study to evaluate the implementation details. To evaluate the system, we conducted a user study with six designers of varying experience (3 to 10 years). We designed four tasks for them to accomplish using our system to evaluate usability and learnability through the system usability scale (SUS). We further asked them to answer a survey to evaluate the overall effectiveness of supporting style analysis. The results indicate that our system is designer-friendly and achieves the system design goals by providing effective style-based search, quantitative style representation and comparison, and objective and reliable style trend information. The contributions of this paper are as follows.

- We proposed an AI-driven graphic design support system, GradeS, which provides designers with quantitative style analysis approaches. Our system is an exploration of how to provide style analysis aids for designers.
- We trained a Vision Transformer (ViT) [9] to model the one-to-many relationship between designs and design styles. Our model achieved the best evaluation performance, 87.60%, which was significantly higher than CLIP. The style representation learned by the model can be widely used in the related application of design style.
- We built a 13K graphic design dataset with coarse-grained style labels through crowdsourcing. Each design image was labeled at least three times to prevent subjectivity. We have released the dataset to facilitate research in the fields of data-driven design and design style. [1]

2. RELATED WORK

The task of graphic design is an ill-defined problem that requires designers to explore the design factor space and make rational choices among the infinite number of decisions that each design factor may face to complete the design process. To alleviate this burden, many studies have been conducted in the field of graphic design. These studies have contributed to helping designers design more efficiently and scientifically by understanding design factors, promoting creative thinking, providing guiding principles, presenting design support systems, etc. Our research

[1]To get the dataset, please email: songjinyu@zju.edu.cn.

aligns with these goals, but focusing specifically on graphic design style. We propose an AI-driven design support system that offers designers a quantitative approach to analyze styles, capture style trends, and ultimately create better designs. In the following sections, we organize our literature review into three subsections: graphic design, image style, and dataset.

2.1 Graphic Design

Graphic design comprises factors like color, layout, and design elements(e.g., font, image, etc.). [6] modeled the relationship between color palettes and mood, [7] investigated color compatibility theories using a quantitative lasso regression model, and [11] presented an automatic pattern coloring algorithm.[12] studied automatic recoloring of graphic designs, while [13] and [14] looked into color contrast and emotional effects in specific contexts. For design elements, [5] presented font selection interfaces considering high-level similarity, [15] improved search suggestions for images and colors through a compatibility model, and [16] focused on generating compound icons from text. Regarding layout, [8] introduced an interactive system for layout suggestions. [17] proposed generating layouts based on images, keywords, and categories. [18] implemented a layout generative model that supports continuous layout style changes. [19] developed a composition-aware method for visual-textual poster layouts.

The literature reviewed above addresses individual design factors, but some studies focus on overall features of graphic design, such as visual importance [20, 21] and style [22–25]. [20] proposed a visual importance dataset and trained neural network for predicting visually important regions, while UMSI [21] further predicted visual importance on various images. In terms of style, [22] explored a web design style search tool using hand-crafted features, [23] measured style similarity between infographics, and recent studies like [24] and [25] utilized deep features to represent style. [24] conducted a model-based analysis to figure out what determines the perceived personalities of graphic design. [25] tackled the problem of learning the client's implicit style preferences through deep style features learned from a few positive and negative samples. This paper extends previous work by collecting a dataset with coarse-grained style labels and proposing an AI-driven system to provide designers with graphic design support around style through quantitative representations of design style.

2.2 Image Style

Style is an important descriptor for both natural and artificial images, describing how images are viewed and perceived [26]. The meaning of visual style varies across image subdomains, such as macro style photos, simple-style posters, and Cubism-style paintings. Many studies have used computational methods to represent style. Low-level features was used to represent the style of web pages [22] and infographics [23]. [26] leveraged deep feature to represent the style of photos and paintings. [27] studied painting style classification using deep pre-trained features and discussed the impact of network structure and data augmentation.[28] examined painting style classification with deep correlation features, while [29] proposed a deep multi-patch

Copyright © 2023 by ASME

aggregation network to address ambiguity in style classification. Advances in computer vision have improved style understanding, with [30] learning image style representation using triplet loss, and [31] improving painting artistic style classification accuracy using EnAET [32]. CLIP [10] connected textual and visual concepts. With CLIP, DALLE·2 [33] demonstrated potential in generating specific artistic styles, indicating that CLIP has learned the artistic style concepts both textually and visually.

In our study, we used a learning-based approach to explore style representations, focusing on the design style of graphic design rather than photography or artistic styles in photos and paintings. Our work shares a common subject with [24], but our objectives differ significantly. We aimed to provide designers with quantitative style analytical interfaces, while [24] investigated how various design factors impact design styles.

2.3 Dataset

Several studies utilizing computational models to investigate graphic design have introduced datasets tailored to specific applications. [20, 21] presented datasets centered on visual importance, while [17, 19, 34] proposed datasets focusing on graphic design layouts for advertising posters, magazines, and slides, respectively. [7] introduced a color palettes dataset. Though these datasets contribute to the development of computational graphic design, they are not related to style.

Research in image style has presented a number of datasets. [35] contributes a dataset serving aesthetic computing. As their dataset primarily contains photographs, the style labels consist of photographic concepts like HDR, macro, and motion blur. [26] presented two datasets, *80K Flickr photographs* and *85K paintings*, for style recognition in photographs and paintings. For *80K Flickr photographs* their 20 photography styles were identified from optical techniques, atmosphere, mood, etc. For *85K paintings*, they labeled paintings with 25 historical concepts. Other datasets focus on paintings [36], portraits [37], contemporary art images [38], and photos with content themes [39]. These datasets primarily address style-related tasks for photographs or paintings, and their style definitions may not be suitable for graphic design images. [24] collected a graphic design dataset with style labels. Unfortunately, the dataset is not public. To fill the gap and train our graphic design style classification model, we built a 13K graphic design dataset with coarse-grained style labels in our work.

3. FORMATIVE INTERVIEW

To gather design style analysis requirements in the current design process, we interviewed 4 graphic design professionals (3 females, 1 male) with visual communication backgrounds and over 4 years of experience each. Due to COVID-19, interviews were conducted via Tencent Meeting, individually with each designer to ensure independent viewpoints. Each interview lasted approximately 30 minutes, covering the importance of style, design style analysis significance and challenges, and potential AI solutions. Based on the interview content, we summarized four system design goals:

Goal 1 provides style-based search to improve the efficiency of seeking examples of certain styles.

Goal 2 provides the ability of quantitative style analysis of a single graphic design to understand design style from a new perspective.

Goal 3 provides quantitative comparison of designs for designers to gain more specific, in-depth style understanding.

Goal 4 provides designers with objective statistical information regarding style trends to help designers capture style trends more easily and comprehensively.

4. METHODS

To achieve our system design goals, we needed a computational model capable of mapping each design to a quantitative style representation. We trained an end-to-end multi-label graphic design style classification model based on Vision Transformers [9]. Since existing datasets lacked graphic design images or suitable style labels, we created a 13K graphic design dataset with design style labels, identified in collaboration with designers. To facilitate data cleaning and annotation, we developed a data annotation platform, where we completed the data cleaning and style annotation task through crowdsourcing. Due to style recognition's subjectivity, each graphic design image was labeled by at least three annotators. We will discuss the methods in two aspects: dataset and model.

4.1 Dataset

4.1.1 Style Label Definition. One of our system design goals is to capture and quantify style trends. Such a goal poses a challenge for the definition of style. Design style can be considered a combination of design factors, but the design factor space is large and complex. If we consider each new combination of design factors as a new style, the number of styles would be too enormous to be effectively recognized by the computational method. To simplify the problem, we proposed an empirical hypothesis that *the design style is hierarchical* with high-level (coarse-grained) styles containing many lower-level (fine-grained) styles. For example, retro as a coarse-grained style includes fine-grained styles like Chinese retro and American retro. We believe that designers usually make creative attempts under a coarse-grained style, and the designs created through these creative attempts may fall within the scope of the existing fine-grained style or may produce a new fine-grained style. The emerging fine-grained styles will not escape the scope of the coarse-grained style. Figure 1 graphically depicts this hypothesis, with white circles denoting fine-grained styles, large colored circles representing coarse-grained styles, and a black arrows showing style evolution (or creative attempts). Based on the hypothesis, we leveraged coarse-grained styles to simplify the problem of capturing and quantifying style trends. Using coarse-grained styles, our dataset aimed to lead the model to learn high-level style representations for each design, improving generalization and effectively quantifying unseen design styles.

In collaboration with designers, we proposed coarse-grained styles that were high-level enough without conceptual overlapping. First, we collected more than fifty fine-grained design styles (e.g., flat design) and their representative design images. Then, we conducted a workshop with designers to propose coarse-grained style labels through manual clustering. We provided

Copyright © 2023 by ASME

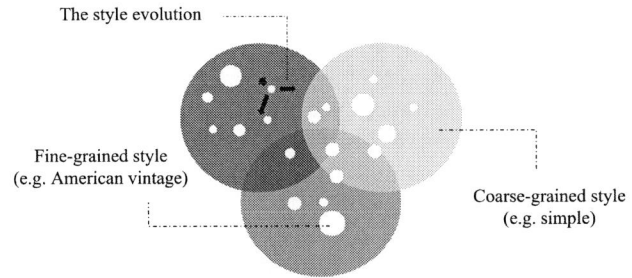

FIGURE 1: A SIMPLE ABSTRACTION OF OUR HYPOTHESIS.

FIGURE 2: EXAMPLES OF EACH STYLE OF OUR DATASET. IMAGES IN SIMPLE, FUTURISTIC, RETRO, CARTOONISH, COMPLEX, REALISTIC, AND TYPOGRAPHICAL STYLE ARE PRESENTED FROM LEFT TO RIGHT.

some candidate styles we came up with early to prevent idea fixation. In the end, we derived 7 coarse-grained style labels, including **simple**, **futuristic**, **retro**, **cartoonish**, **complex**, **realistic**, and **typographical** style. Each label is a representative adjective of a set of keywords for that category, describing the overall visual feeling of the design. Figure 2 shows the examples of each style. To verify the reasonableness of these style labels, we conducted a validation experiment with 21 students, including 2 design students and 19 non-design students. We primarily invited non-design students to verify that the proposed labels could be understood by non-majors. During the experiment, we asked each participant to label 200 graphic design images with one or multiple style labels we proposed. The experiment was conducted on our data annotation platform. Afterward, we collected feedback through questionnaires. The results show that $80.95\%(17/21)$ participants thought the style labels were reasonable.

4.1.2 Data Collection. We collected 50K (50,434) publicly accessible images from Behance for graphic design research, maintaining proper copyright information. The original data is quite dirty, which is mainly caused by two reasons. First, style-based search is not supported. Second, designers prefer to use mockups to share their works. We cleaned it by labeling design images as valid, invalid, or partially valid on our data annotation platform. We ended up with 13K (13,245) relatively clean design images, accepting some noise as inevitable.

To annotate these images with style labels, we assigned one or more labels to each design, considering that an image could exhibit multiple styles simultaneously. We recruited 16 participants to annotate images on our platform, with each image labeled by at least three participants for accuracy and objectivity. An "other" option was available for images not fitting the proposed labels. The result showed that only $2.45\%(324/13245)$ images were labeled "other". The annotated dataset is now suitable for training and other tasks related to graphic design styles and data-driven design. The distribution of our dataset is displayed in Figure 3,

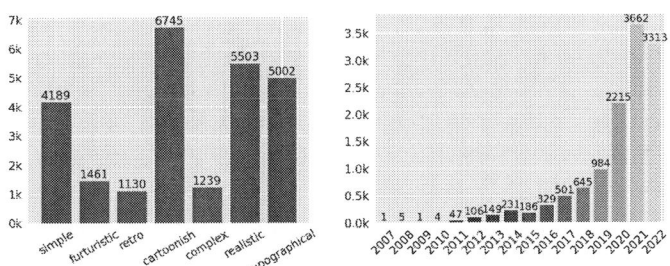

FIGURE 3: THE STYLE DISTRIBUTION (LEFT) AND CREATION TIME DISTRIBUTION (RIGHT) OF OUR DATASET.

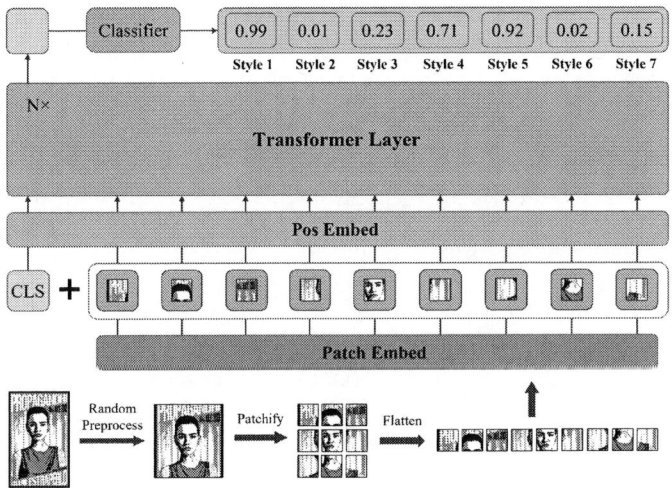

FIGURE 4: THE ARCHITECTURE OF OUR MODEL.

with the sum of individual style counts exceeding the total number of images due to multiple labels per image. There are 866 images in the dataset without a creation date.

4.2 Model

4.2.1 Model Architecture. Since style describes the overall look and feel and how images are viewed and perceived, we believe that a model that explicitly captures global features would be more conducive to the representation and classification of style. Thus, we leverage Vision Transformer (ViT) [9] to learn the coarse-grained style representation in a supervised manner. The model outputs a vector representing the probability of an input image belonging to each coarse-grained style, as depicted in Figure 4. Specifically, given an original design image I of arbitrary resolution, the following steps are executed.

Preprocess. Original image I is preprocessed into $\hat{I} \in \mathbb{R}^{C \times H \times W}$, with $H = W = 224$ and $C = 3$. The data augmentation methods we used in preprocess step during training are illustrated in Section 4.2.2. In the evaluation stage, we simply resize I into 224×224.

Patchify. The image $\hat{I}$ is segmented into small patches of size $P \times P$ that are encoded into feature vectors $F_0 \in \mathbb{R}^{L \times D}$ where $L = \lfloor \frac{H}{P} \rfloor \times \lfloor \frac{W}{P} \rfloor$ and D is the feature dimension of our model. Segmentation and encoding are implemented using a convolutional layer and transposition operation, with kernel size and stride size set to P, and feature dimension set to D. We use $P = 16$ and $D = 768$.

Copyright © 2023 by ASME

Positional Embedding. A CLS token is prepended to F_0, and positional information is added via Pos Embed module to form the transformer-ready vector F_{inp}. We utilize learned position encoding instead of sinusoidal ones since the learned ones better adapt to the input sequence of different lengths and structures.

Transformer. We use $N = 12$ transformer layers to extract the feature of image. The feature dimension of transformer layer is D. Each transformer layer consists of a feed forward layer and a multi-head attention layer (num_head=12). The feature vector of the CLS token location (i.e. the first location) F_s is used to represent the style feature of the design image.

Classifier. The style feature F_s is input into the classifier, yielding a style representation $S \in \mathbb{R}^7$. The classifier is just a single linear layer, with a sigmoid non-linearity to make each value of S in $[0, 1]$.

4.2.2 Implementation Detail. We initialized our model with ImageNet1K pretrained parameters and trained it on our dataset, allocating 81% for training, 9% for validation, and another 10% for testing. Style labels were normalized to obtain style probabilities as training targets.

Due to the naturally imbalanced style distribution, we employed multi-label focal loss as the training objective function, which is a custom variant of the original focal loss [40] supporting multi-label classification, to prevent overfitting the style distribution. The focal loss is calculated according to Eq.1, where $\hat{y}$ stands for the predicted possibility computed by the sigmoid, y stands for the target possibility, and γ stands for the focusing parameter. We omitted the hyperparameter α from the original focal loss as it has minimal impact on performance without careful tuning.

$$\mathscr{L}_{FL}(\hat{y}) = \begin{cases} -(1 - \hat{y})^{\gamma} log(\hat{y}), & if \ y >= 0.5 \\ -\hat{y}^{\gamma} log(1 - \hat{y}), & if \ y < 0.5 \end{cases} \quad (1)$$

To enhance the generalization performance of the model, we utilized various data augmentations during training. Since the style of the graphic design is relevant to the color, color-related data augmentations (e.g., color jitter, contrast adjustment, solarize, etc.) are ignored. In practice, we only use data augmentations related to geometric transformation, including random resize and crop, random flip both horizontally and vertically, and random perspective transformation. All the data augmentations we used were implemented by Torchvision. We swept hyperparameters (e.g., base learning rate, scheduler type, drop path rate, optimizer, etc.) to find the best setting.

5. GRADES

In this section, we will introduce GradeS in detail. GradeS is an AI-driven graphic design support system that provides designers with quantitative analytical interfaces for style. GradeS:S, GradeS:Q, GradeS:C, and GradeS:T were designed for different analysis scenarios and needs. As a whole, GradeS enables designers to efficiently retrieve graphic design images based on style, objectively analyze graphic design styles, easily capture the latest style trends, and get insights of style from a quantitative perspective, which promotes design creativity.

5.1 GradeS:S (Goal 1)

GradeS:S (S stands for search) provides designers with a style-based search for graphic design. The current content-based search is not suitable for looking for examples of certain styles. We implemented the style-based search by measuring style similarity based on the style probability representation computed by our model. The interface is shown in Figure 5a. The top of the interface has some interactive buttons, the bottom right shows the search results in a list, and the bottom left shows the selected search result in a larger resolution and provides some basic image manipulations. There are three style-based search modes supported by our interface: quick mode, quantitative mode, and image mode. In quick mode, designers can search for graphic designs of specific styles by clicking the style button. Figure 5a shows an example of searching for a simple style. We allowed designers to select multiple styles to get more meaningful search results. The quantitative mode was designed for a more accurate search. In quantitative mode, there is a slider (hereinafter called the probability slider) under each style button that indicates the degree of style. By dragging the slider, designers can set specific probabilities for each style. Figure 6a shows an example of searching images that are 100% simple and 82.16% typographical. To further improve the ability of style-based search to play a role in more scenarios, we added image mode. In image mode, designers can search for graphic design images that match the design style of the uploaded graphic design image. Figure 6b shows an example of image mode.

5.2 GradeS:Q (Goal 2)

GradeS:Q (Q stands for quantification) provides designers with the ability to quantitatively analyze the style of a single graphic design. Designers usually analyze the style of graphic design in a qualitative and subjective manner. We hope to provide a quantitative perspective on graphic design as a supplement. As shown in Figure 5b, GradeS:Q contains three sub-functions. First, the style probability of each graphic design image is visualized. Second, the style importance map is generated to visualize which part of the design led to the style. We use the method described in [41] to generate the style importance map. The default style importance map is calculated according to the style that has the highest classification probability, which is the cartoonish style in the figure. Designers can view the style importance map of the selected style by clicking the style button. It's also possible to generate the style importance map according to multiple styles. At last, designers can view other graphic design images that are similar to the current image through the "Find Similar" button in the GradeS:S interface. As we said before, all interfaces are complementary and work together to provide support. GradeS:Q can be visited by clicking the image in GradeS:S. In addition to analyzing the graphic design images in the search results, GradeS:Q also supports analyzing uploaded graphic design images.

5.3 GradeS:C (Goal 3)

GradeS:C (C stands for comparison) provides designers with the ability to quantitatively compare the style of graphic designs to gain a more specific, in-depth style understanding. Comparing the difference in style between designs is usually hard and subjective,

Copyright © 2023 by ASME

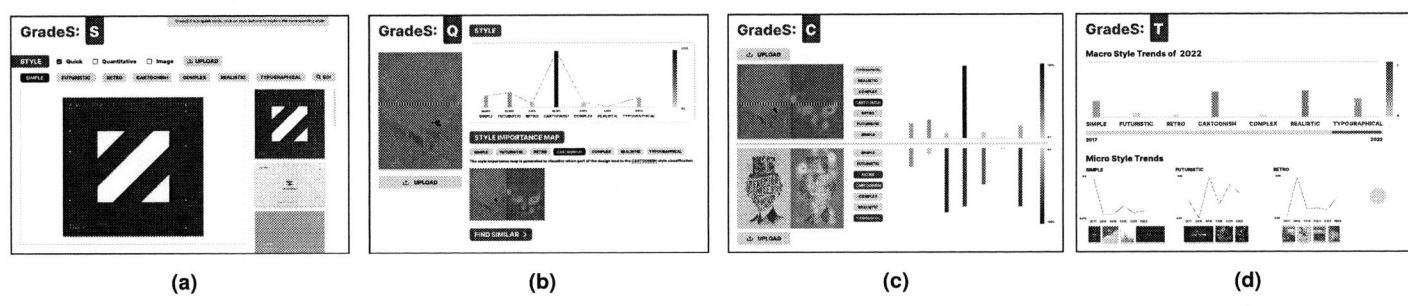

FIGURE 5: THE USER INTERFACES OF GRADES. (A) GRADES:S IN QUICK MODE. (B) GRADES:Q. (C) GRADES:C. (D) GRADES:T.

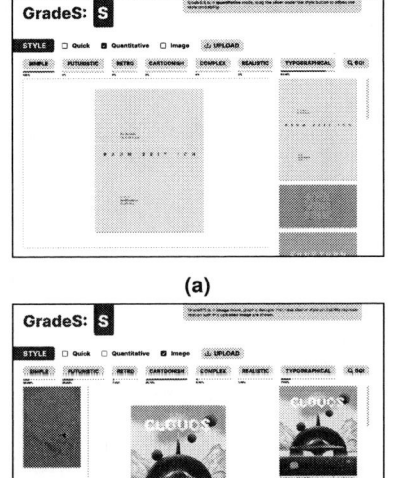

(a)

(b)

FIGURE 6: THE VARIANTS OF GRADES:S. (A) QUANTITATIVE MODE. (B) IMAGE MODE.

which results in a vague conclusion. We believe comparison is a good way to gain insights. Therefore, GradeS:C was designed to provide a quantitative comparison of style. As shown in Figure 5c, we compare the style of graphic design through the style importance map and style probability. Since the style importance map can be generated according to any style, designers can select the same style for both graphic design images to generate style importance maps for comparison.

5.4 GradeS:T (Goal 4)

GradeS:T (T stands for trend) provides quantitative information about style trends to help designers capture style trends more easily and comprehensively. Trend is a descriptor that indicates the popularity or the frequency of a style over a given period of time. To quantitatively analyze style trends, we first defined the quantitative style trend. We called it the style trend indicator. The style trend indicator I of the style s in a given time period t is calculated by dividing the number of graphic design images C of the style s in the time period t by the total number of graphic design images A in the time period t according to Eq.2. The user interface of GradeS:T is shown in Figure 5d. GradeS:T contains two parts, macro style trend and micro style trend. The macro

Model name	Accuracy	Recall	Precision
CLIP-RN50	54.70%	65.36%	31.84%
CLIP-RN101	61.04%	51.72%	33.02%
CLIP-ViT-B/32	61.28%	43.93%	33.68%
CLIP-ViT-B/16	61.80%	40.67%	32.06%
Ours	**87.60%**	**74.21%**	**78.43%**

TABLE 1: THE MULTI-LABEL STYLE CLASSIFICATION PERFORMANCE OF OUR MODEL AND CLIP.

style trend shows the style trend indicators for each style in a given year, indicating the distribution of style in the year. The micro style trend focuses on the trend of each style, using a line chart to show the transformation of the style. Under the chart, some randomly selected examples created in the specific year are shown. By default, these examples are all from the current year. Designers can change examples by clicking the year text button under the chart.

$$I_t^s = \frac{C_t^s}{A_t} \quad (2)$$

6. EVALUATION

To demonstrate the effectiveness of our study, we evaluate both the model and the system.

6.1 Model Evaluation

We used the pre-made test set from Section 4.2.2 as the evaluation dataset with a seed of 42 for reproducibility. To evaluate our model's performance, we compared our model with CLIP, a large-scale pretrained model that was proved to have good style representation ability [33] and plays an vital role in the most advanced AIGC systems [33, 42, 43]. Since CLIP was pre-trained on 400 million image text pairs, we evaluated it in a zero-shot manner (i.e., without finetuning) like other research [33, 42, 44]. Due to the fact that CLIP can only output similarity between images and text, using CLIP for multi-label classification tasks is not straightforward. To address this, we simulate multi-label classification by using 7 binary classification tasks for different styles. For each task, we iput two texts including "style of $\mathbb{S}$" (the first) and "not style of $\mathbb{S}$" (the second) and one design image into CLIP to calculate the cosine similarity. The $\mathbb{S}$ is a placeholder for 7 coarse-grained styles mentioned in Section 4.1. We perform a softmax operation on the cosine similarities between the image and texts to get the probability that the image belongs to each text, where the probability of the first text represents the probability that the image belongs to $\mathbb{S}$. We set the probability threshold to 0.5, which means if the cosine similarity between

Copyright © 2023 by ASME

the image and the first text is greater than the cosine similiarity between the second text, CLIP considers the image to belong to that style. By concatenating the probabilities of image styles obtained from the 7 binary classification tasks described above, we create a pseudo-multi-label classification result. In this way, we utilized CLIP for multi-label style classification tasks. As for metrics, accuracy, precision, and recall were used to quantitatively evaluate the performance of multi-label style classification. Accuracy denotes the percentage of images with correctly identified styles, encompassing both the accurate recognition of images belonging to particular styles and the accurate exclusion of images not pertaining to those styles. Recall indicates how many images belonging to particular styles are correctly identified as those styles. Precision, on the other hand, represents the proportion of correctly identified images for specific styles out of all the images identified as those styles. The performance of four CLIP variants and our model is displayed in Table 1. With relatively higher accuracy, recall, and precision, our model's performance in style recognition is demonstrated.

We also did an ablation study where we compared different model architectures, data augmentation strategies and loss functions. We reported the quantitative results in Table 2, where *RRC*, *RHF*, *RVF*, *RP* and *RA* stands for *RandomResizedCrop*, *RandomHorizontalFlip*, *RandomVerticalFlip*, *RandomPerspective* and *RandomAffine*. The scale parameter of *RandomResizedCrop* is $(0.9, 1.0)$. The parameters of *RandomPerspective* keep default. The parameters of *RandomAffine* are consistent with [27]. As for loss function, both Focal and BCE with pos_weights are used to mitigate the impact of data imbalance. The results demonstrate the effectiveness of implementation details. For example, comparing No.1 with No.4 and No.5, the ViT is more suitable for style classification task than CNN-based. Comparing No.1 and No.3, RP provides better data augmentation in style classification task than RA, result in better performance.

6.2 User Study

To demonstrate the effectiveness of our system, we conducted a user study with six designers.

6.2.1 Experiment Design.
We conducted a user study with six designers who had three to ten years of design experience, using the interface discussed in Section 5. We designed four tasks to cover each analytical interface, helping designers familiarize themselves with our system quickly. After completing all tasks, the System Usability Scale (SUS) was used to evaluate the system's usability. To further verify the system's effectiveness in supporting style analysis, we asked each designer to complete a survey with questions listed in Appendix A. The first eight questions focused on specific interface effectiveness, while the last two targeted the entire system. We used 7-point Likert scales for all questions (1: Strongly Disagree; 7: Strongly Agree).

6.2.2 Results.
SUS: The SUS scores given by six participants (C1-C6) are presented in Figure 7a (mean=77.65, std=4.488). According to [45], SUS scores of no less than 50.9, 71.4, 85.5, and 90.9 are considered 'ok', 'good', 'excellent', and 'best imaginable', respectively. One participant (C3) rated our system as "Excellent". Other participants (C1, C2, C4, C5, C6)

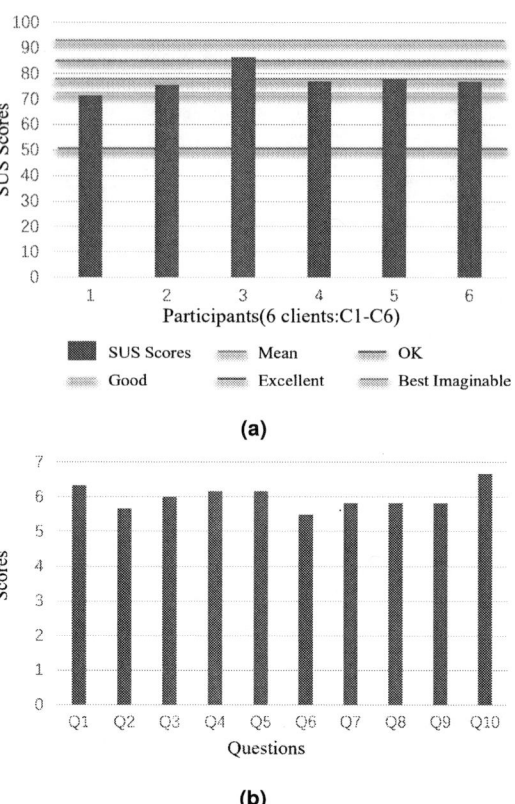

FIGURE 7: THE USER STUDY RESULTS. (A) THE SCORES OF SUS. (B) THE SCORES OF SURVEY.

rated our system as "good". The average SUS score of our user interface falls between "good" and "excellent", indicating that the overall usability and learnability of GradeS are satisfactory. **Survey:** The average scores of questionnaire questions are shown in Figure 7b (mean=6.0, std=0.325). For each analytic interface, GradeS:S, GradeS:Q, GradeS:C, GradeS:T got 6.04, 5.83, 5.67, and 5.83, respectively. The highest-scoring interface is GradeS:S, suggesting that we provide a relatively good example of style-based search. The lowest-scoring interface is GradeS:C, showing the imperfection of this interface. We believe this could be improved by providing more quantitative metrics and adding explanatory text. Q10 received the highest score, indicating that our system effectively supports the current design process. The score of Q6 (5.5) is the lowest of all questions. This may be because the style importance map is still too abstract for designers. The overall average score of 6.0 indicates that our system accomplishes the system design goals.

7. DISCUSSION

GradeS is an exploration of how to provide designers with style-oriented support. In the following, we discuss the limitations and future work.

7.1 Dataset
Style Label. The coarse-grained style labels defined in Section 4.1.1 were carefully crafted through collaboration with designers, but there is room for improvement. One issue is the inconsistency in the conceptual scope of each style. In future work, we aim to

Copyright © 2023 by ASME

No	Model Architecture	Data Augmentation	Loss	Epoch	Accuracy	Recall	Precision
1	vit-base-patch16-224	RRC + RHF + RVF + RP	Focal	35	**87.60%**	**74.21%**	78.43%
2	vit-base-patch16-224	RRC + RHF + RVF + RP	BCE with pos_weights	35	85.10%	73.47%	73.01%
3	vit-base-patch16-224	RRC + RHF + RVF + RA	Focal	35	84.94%	54.70%	85.91%
4	resnet152d	RRC + RHF + RVF + RP	Focal	35	86.02%	57.54%	**87.89%**
5	resnet50d	RRC + RHF + RVF + RP	Focal	35	84.78%	53.34%	86.68%

TABLE 2: THE ABLATION STUDY RESULTS.

address this problem by concentrating on refining the definition of style labels.

Data Annotation. Although we have worked hard in data collection and cleaning, we still got a not perfectly clean dataset, which means a small number of images are not graphic designs and a few designs were not labeled correctly. To address the problems, in future work, we plan to reduce the noise of style labels by increasing the requirements of annotators, increasing the number of annotations for each graphic design, and setting checkpoints in the system to prevent cognitive bias.

Data Collection. As we mentioned in Section 4.1.2, we only collected data from Behance, which means platform bias inevitably exists. To build a larger, more reliable, and diversified dataset in the future, we plan to collect data from multiple design websites and utilize an AI model for semi-automated data cleaning.

7.2 Model

As mentioned in Section 4.2, we adopted Vision Transformer to do the multi-label style classification task. As a research aimed at helping designers in style analysis, our method is simple and sufficient. However, there are indeed very many improvements proposed in studies of computer vision that can be used to enhance our model, such as multi-patch aggregation [29] and EnAET [32] and MAE [46]. We leave improving the model for future work.

7.3 System

The user study indicates the effectiveness of our system. Nevertheless, there are still certain aspects of our system that require refinement in order to more effectively cater to the needs of designers. For example, the style-based search provided by GradeS:S is a good supplement to content-based search but not enough for some practical design scenarios. During the user study, designers reported that it was more common to search for a certain style of design for a certain product or service, indicating that a style-content hybrid search is more practical. In addition, our system currently has no mechanism to handle cases of misclassification. In our future work, we plan to improve our system in terms of enhancing the stability of the current system, improving each existing interface to better suit the needs of designers, and providing more useful, analytical interfaces to improve the efficiency of designers.

8. CONCLUSION

In this paper, we propose an AI-driven graphic design support system, GradeS, for automatically analyzing design style. We conducted interviews with four graphic designers to identify the problems and needs of style analysis and identified four

system design goals to guide the system's implementation. Our system provides four quantitative style analytical interfaces, including GradeS:S for efficient style-based search, GradeS:Q and GradeS:C for objectively understanding style through quantification and quantitative comparison, and GradeS:T for providing objective style trends statistical information. We adopted vision transformers to learn style representation. We contributed 13K graphic design datasets with coarse-grained style labels to train the model and encourage more research on graphic design style and data-driven design. To demonstrate the effectiveness of our system, we conducted a model evaluation and a user study to evaluate the model's performance and the effectiveness of our system in supporting style analysis. The results show that our system accomplishes the system design goals by providing effective style-based search, quantitative style representation and comparison, and objective and reliable style trend information. Our research is an exploration of using computational methods to provide style-related support for designers. Future work will focus on improving the performance of the model, releasing a better and larger graphic design dataset, and providing more useful analytical interfaces for designers.

ACKNOWLEDGMENTS

This work is supported by the National Key R&D Program of China (2022YFB3303301), and the National Natural Science Foundation of China (No. 62107035 and No. 62006208).

REFERENCES

[1] Landa, Robin. *Graphic design solutions*. Cengage Learning (2018).

[2] Karayev, Sergey, Trentacoste, Matthew, Han, Helen, Agarwala, Aseem, Darrell, Trevor, Hertzmann, Aaron and Winnemoeller, Holger. "Recognizing image style." *arXiv preprint arXiv:1311.3715* (2013).

[3] Reinecke, Katharina, Yeh, Tom, Miratrix, Luke, Mardiko, Rahmatri, Zhao, Yuechen, Liu, Jenny and Gajos, Krzysztof Z. "Predicting users' first impressions of website aesthetics with a quantification of perceived visual complexity and colorfulness." *Proceedings of the SIGCHI conference on human factors in computing systems*: pp. 2049–2058. 2013.

[4] Isola, Phillip, Xiao, Jianxiong, Torralba, Antonio and Oliva, Aude. "What makes an image memorable?" *CVPR 2011*: pp. 145–152. 2011. IEEE.

[5] O'Donovan, Peter, Lībeks, Jānis, Agarwala, Aseem and Hertzmann, Aaron. "Exploratory font selection using

Copyright © 2023 by ASME

crowdsourced attributes." *ACM Transactions on Graphics (TOG)* Vol. 33 No. 4 (2014): pp. 1–9.

[6] Csurka, Gabriela, Skaff, Sandra, Marchesotti, Luca and Saunders, Craig. "Learning moods and emotions from color combinations." *Proceedings of the Seventh Indian Conference on Computer Vision, Graphics and Image Processing*: pp. 298–305. 2010.

[7] O'Donovan, Peter, Agarwala, Aseem and Hertzmann, Aaron. "Color compatibility from large datasets." *ACM SIGGRAPH 2011 papers* (2011): pp. 1–12.

[8] O'Donovan, Peter, Agarwala, Aseem and Hertzmann, Aaron. "Designscape: Design with interactive layout suggestions." *Proceedings of the 33rd annual ACM conference on human factors in computing systems*: pp. 1221–1224. 2015.

[9] Dosovitskiy, Alexey, Beyer, Lucas, Kolesnikov, Alexander, Weissenborn, Dirk, Zhai, Xiaohua, Unterthiner, Thomas, Dehghani, Mostafa, Minderer, Matthias, Heigold, Georg, Gelly, Sylvain et al. "An image is worth 16x16 words: Transformers for image recognition at scale." *arXiv preprint arXiv:2010.11929* (2020).

[10] Radford, Alec, Kim, Jong Wook, Hallacy, Chris, Ramesh, Aditya, Goh, Gabriel, Agarwal, Sandhini, Sastry, Girish, Askell, Amanda, Mishkin, Pamela, Clark, Jack, Krueger, Gretchen and Sutskever, Ilya. "Learning Transferable Visual Models From Natural Language Supervision." Meila, Marina and Zhang, Tong (eds.). *Proceedings of the 38th International Conference on Machine Learning*, Vol. 139: pp. 8748–8763. 2021. PMLR. URL https://proceedings.mlr.press/v139/radford21a.html.

[11] Lin, Sharon, Ritchie, Daniel, Fisher, Matthew and Hanrahan, Pat. "Probabilistic color-by-numbers: Suggesting pattern colorizations using factor graphs." *ACM Transactions on Graphics (TOG)* Vol. 32 No. 4 (2013): pp. 1–12.

[12] Zhao, Nanxuan, Zheng, Quanlong, Liao, Jing, Cao, Ying, Pfister, Hanspeter and Lau, Rynson WH. "Selective region-based photo color adjustment for graphic designs." *ACM Transactions on Graphics (TOG)* Vol. 40 No. 2 (2021): pp. 1–16.

[13] You, Fang, Li, Yaru, Hansen, Preben, Li, Liping, Fu, Mengting, Yang, Yifan, Jin, Xin and Wang, Jianmin. "Interface Color Design of Intelligent Vehicle Central Console." *International Conference on Applied Human Factors and Ergonomics*: pp. 784–792. 2021. Springer.

[14] Chen, Qinyue, Yan, Yuchun and Suk, Hyeon-Jeong. "Bubble coloring to visualize the speech emotion." *Extended Abstracts of the 2021 CHI Conference on Human Factors in Computing Systems*: pp. 1–6. 2021.

[15] Kovacs, Balazs, O'Donovan, Peter, Bala, Kavita and Hertzmann, Aaron. "Context-aware asset search for graphic design." *IEEE transactions on visualization and computer graphics* Vol. 25 No. 7 (2018): pp. 2419–2429.

[16] Zhao, Nanxuan, Kim, Nam Wook, Herman, Laura Mariah, Pfister, Hanspeter, Lau, Rynson WH, Echevarria, Jose and Bylinskii, Zoya. "Iconate: Automatic compound icon generation and ideation." *Proceedings of the 2020 CHI Confer-*

ence on Human Factors in Computing Systems: pp. 1–13. 2020.

[17] Zheng, Xinru, Qiao, Xiaotian, Cao, Ying and Lau, Rynson WH. "Content-aware generative modeling of graphic design layouts." *ACM Transactions on Graphics (TOG)* Vol. 38 No. 4 (2019): pp. 1–15.

[18] Ueno, Michihiko and Satoh, Shin'ichi. "Continuous and Gradual Style Changes of Graphic Designs with Generative Model." *26th International Conference on Intelligent User Interfaces*: pp. 280–289. 2021.

[19] Zhou, Min, Xu, Chenchen, Ma, Ye, Ge, Tiezheng, Jiang, Yuning and Xu, Weiwei. "Composition-aware Graphic Layout GAN for Visual-textual Presentation Designs." *arXiv preprint arXiv:2205.00303* (2022).

[20] Bylinskii, Zoya, Kim, Nam Wook, O'Donovan, Peter, Alsheikh, Sami, Madan, Spandan, Pfister, Hanspeter, Durand, Fredo, Russell, Bryan and Hertzmann, Aaron. "Learning visual importance for graphic designs and data visualizations." *Proceedings of the 30th Annual ACM symposium on user interface software and technology*: pp. 57–69. 2017.

[21] Fosco, Camilo, Casser, Vincent, Bedi, Amish Kumar, O'Donovan, Peter, Hertzmann, Aaron and Bylinskii, Zoya. "Predicting visual importance across graphic design types." *Proceedings of the 33rd Annual ACM Symposium on User Interface Software and Technology*: pp. 249–260. 2020.

[22] Ritchie, Daniel, Kejriwal, Ankita Arvind and Klemmer, Scott R. "d.tour: Style-based exploration of design example galleries." *Proceedings of the 24th annual ACM symposium on User interface software and technology*: pp. 165–174. 2011.

[23] Saleh, Babak, Dontcheva, Mira, Hertzmann, Aaron and Liu, Zhicheng. "Learning style similarity for searching infographics." *arXiv preprint arXiv:1505.01214* (2015).

[24] Zhao, Nanxuan, Cao, Ying and Lau, Rynson WH. "What characterizes personalities of graphic designs?" *ACM Transactions on Graphics (TOG)* Vol. 37 No. 4 (2018): pp. 1–15.

[25] Lin, David Chuan-En and Martelaro, Nikolas. "Learning Personal Style from Few Examples." *Designing Interactive Systems Conference 2021*: pp. 1566–1578. 2021.

[26] Karayev, Sergey, Trentacoste, Matthew, Han, Helen, Agarwala, Aseem, Darrell, Trevor, Hertzmann, Aaron and Winnemoeller, Holger. "Recognizing Image Style." *Proceedings of the British Machine Vision Conference*. 2014. BMVA Press. DOI http://dx.doi.org/10.5244/C.28.122.

[27] Lecoutre, Adrian, Negrevergne, Benjamin and Yger, Florian. "Recognizing art style automatically in painting with deep learning." *Asian conference on machine learning*: pp. 327–342. 2017. PMLR.

[28] Chu, Wei-Ta and Wu, Yi-Ling. "Deep correlation features for image style classification." *Proceedings of the 24th ACM international conference on Multimedia*: pp. 402–406. 2016.

[29] Lu, Xin, Lin, Zhe, Shen, Xiaohui, Mech, Radomir and Wang, James Z. "Deep multi-patch aggregation network for image style, aesthetics, and quality estimation." *Pro-*

Copyright © 2023 by ASME

ceedings of the IEEE international conference on computer vision: pp. 990–998. 2015.

[30] Gairola, Siddhartha, Shah, Rajvi and Narayanan, PJ. "Unsupervised image style embeddings for retrieval and recognition tasks." *Proceedings of the IEEE/CVF Winter Conference on Applications of Computer Vision*: pp. 3281–3289. 2020.

[31] Joshi, Akshay, Agrawal, Ankit and Nair, Sushmita. "Art Style Classification with Self-Trained Ensemble of AutoEncoding Transformations." *arXiv preprint arXiv:2012.03377* (2020).

[32] Wang, Xiao, Kihara, Daisuke, Luo, Jiebo and Qi, Guo-Jun. "EnAET: A self-trained framework for semi-supervised and supervised learning with ensemble transformations." *IEEE Transactions on Image Processing* Vol. 30 (2020): pp. 1639–1647.

[33] Ramesh, Aditya, Dhariwal, Prafulla, Nichol, Alex, Chu, Casey and Chen, Mark. "Hierarchical text-conditional image generation with clip latents." *arXiv preprint arXiv:2204.06125* (2022).

[34] Xie, Yuxi, Huang, Danqing, Wang, Jinpeng and Lin, Chin-Yew. "CanvasEmb: Learning Layout Representation with Large-scale Pre-training for Graphic Design." *Proceedings of the 29th ACM International Conference on Multimedia*: pp. 4100–4108. 2021.

[35] Murray, Naila, Marchesotti, Luca and Perronnin, Florent. "AVA: A large-scale database for aesthetic visual analysis." *2012 IEEE conference on computer vision and pattern recognition*: pp. 2408–2415. 2012. IEEE.

[36] Khan, Fahad Shahbaz, Beigpour, Shida, Van de Weijer, Joost and Felsberg, Michael. "Painting-91: a large scale database for computational painting categorization." *Machine vision and applications* Vol. 25 No. 6 (2014): pp. 1385–1397.

[37] Shaik, Sadat, Bucher, Bernadette, Agrafiotis, Nephele, Phillips, Stephen, Daniilidis, Kostas and Schmenner, William. "Learning portrait style representations." *arXiv preprint arXiv:2012.04153* (2020).

[38] Wilber, Michael J, Fang, Chen, Jin, Hailin, Hertzmann, Aaron, Collomosse, John and Belongie, Serge. "Bam! the behance artistic media dataset for recognition beyond photography." *Proceedings of the IEEE international conference on computer vision*: pp. 1202–1211. 2017.

[39] Singhal, Trisha, Liu, Junhua, Blessing, Lucienne and Lim, Kwan Hui. "Photozilla: A Large-Scale Photography Dataset and Visual Embedding for 20 Photography Styles." *arXiv preprint arXiv:2106.11359* (2021).

[40] Lin, Tsung-Yi, Goyal, Priya, Girshick, Ross, He, Kaiming and Dollár, Piotr. "Focal loss for dense object detection." *Proceedings of the IEEE international conference on computer vision*: pp. 2980–2988. 2017.

[41] Chefer, Hila, Gur, Shir and Wolf, Lior. "Transformer interpretability beyond attention visualization." *Proceedings of the IEEE/CVF Conference on Computer Vision and Pattern Recognition*: pp. 782–791. 2021.

[42] Ramesh, Aditya, Pavlov, Mikhail, Goh, Gabriel, Gray, Scott, Voss, Chelsea, Radford, Alec, Chen, Mark and Sutskever, Ilya. "Zero-shot text-to-image generation." *International Conference on Machine Learning*: pp. 8821–8831. 2021. PMLR.

[43] Rombach, Robin, Blattmann, Andreas, Lorenz, Dominik, Esser, Patrick and Ommer, Björn. "High-resolution image synthesis with latent diffusion models." *Proceedings of the IEEE/CVF Conference on Computer Vision and Pattern Recognition*: pp. 10684–10695. 2022.

[44] Crowson, Katherine, Biderman, Stella, Kornis, Daniel, Stander, Dashiell, Hallahan, Eric, Castricato, Louis and Raff, Edward. "Vqgan-clip: Open domain image generation and editing with natural language guidance." *Computer Vision–ECCV 2022: 17th European Conference, Tel Aviv, Israel, October 23–27, 2022, Proceedings, Part XXXVII*: pp. 88–105. 2022. Springer.

[45] Bangor, Aaron, Kortum, Philip T and Miller, James T. "An empirical evaluation of the system usability scale." *Intl. Journal of Human–Computer Interaction* Vol. 24 No. 6 (2008): pp. 574–594.

[46] He, Kaiming, Chen, Xinlei, Xie, Saining, Li, Yanghao, Dollár, Piotr and Girshick, Ross. "Masked autoencoders are scalable vision learners." *Proceedings of the IEEE/CVF Conference on Computer Vision and Pattern Recognition*: pp. 16000–16009. 2022.

APPENDIX A. THE SURVEY QUESTIONS

The survey questions we used to evaluate our system are as follows:

Q1 Does GradeS:S help you quickly find graphic design examples of a particular style?

Q2 Does the quantitative mode in GradeS:S help you search for graphic design images more accurately?

Q3 Does a quantitative style search inspire you to try mixing styles in your creations?

Q4 Do you think that style-based search is as useful as content-based search in design creation?

Q5 Does the quantitative information on style in the GradeS:Q interface play a supporting role in the analysis of the style of the work?

Q6 Does the style importance map help you understand style from a new perspective?

Q7 Does GradeS:C help you understand style better?

Q8 Does the style trend information provided in GradeS:T meet your needs?

Q9 Does GradeS as a whole improve your efficiency of style analysis and design creation by providing style-based search and quantitative style information?

Q10 Would you like to use GradeS in your design process?

Copyright © 2023 by ASME

**Proceedings of the ASME 2023
International Design Engineering Technical Conferences and
Computers and Information in Engineering Conference
IDETC-CIE2023
August 20-23, 2023, Boston, Massachusetts**

DETC2023-114806

ON-THE-FLY DUAL REDUCTION METHOD ON TRANSIENT FLUID TOPOLOGY OPTIMIZATION

Tianye Wang, Xiaoping Qian*

Department of Mechanical Engineering, University of Wisconsin - Madion, Madison, WI

ABSTRACT

Transient topology optimization (TO) requires huge computational resources in terms of both computing time and storage. In this study, we have applied the proper orthogonal decomposition (POD) based on-the-fly dual reduction method on TO problems for time-dependent incompressible fluid channel optimization. The fluid motion was governed by time-dependent Navier-Stokes equation. Total energy dissipation over a given period was considered as the cost function, and total fluid volume was constrained during optimization process. POD based reduced order modeling (ROM) was applied to both the primal equations and the adjoint equations for reduction. We examined the reduction efficacy though examples including U-bend channel and nozzle. Tolerance and snapshot set size were studied as hyper-parameters to improve the reduction performance. The best cases can reach 99% of skip ratios, 50% reduction in computational time, and 96% saving of memory storage.

Keywords: topology optimization, reduced-order modeling, time-dependent, fluid, on-the-fly

1. INTRODUCTION

Topology optimization (TO) becomes an important design method with the additive manufacturing technology emerging and becoming mature in the last several decades [1]. The density based TO method has beed studied throughtly in a large range of topics such as structure optimization, heat conduction, and steady state fluid problems. And the optimization results have been successfully converted to industrial products thorugh additive manufacturing in recent years. However, when coming to the transient fluid optimization problems, the long computational time and huge data storage requirements, which are caused by nonlinearity and time-dependent features of the fluid problem itself, make the TO method impractical for real 3D applications. Therefore, reduction methods are desired for transient fluid TO problems.

In the fluid related optimization field, TO method was first introduced and investigated in steady state problems. Borrvall and Petersson studied the steady Stokes problem which ignores the nonlinear convection term, proposed the optimization formulation, and solved the problems numerically by finite element method [2]. Guest and Prévost also studied the Darcy-Stokes flow problem and investigated channel, nozzle, and blunt body drag optimizations [3]. Gersborg-Hansen et al. first took the convection term into simulations and optimizations [4]. Yoon applied TO to steady state fluid structure interaction (FSI) problems [5]. Further, TO method was also proposed and developed for transient fluid problems. Kreissl et al. studied the transient fluid optimization problem and proposed the adjoint senstivity analysis method [6]. In the same year, Deng et al. also investigated a similar transient fluid TO problem for channel optimization [7]. Transient fluid structure interaction optimization was explored by Andreasen and Sigmund later [8]. Abdelwahed and Hassine optimized 2D and 3D pipes with the similar approach [9]. Besides the finitie element method (FEM), the lattice Boltzmann method (LBM) was also chosen for transient fluid TO problem because of its explicit scheme and fast computational speed. Yaji et al. derived adjoint methods for LBM equations, and obtained optimizations for 2D pure fluid and thermal fluid coupled problems [10]. Later, they applied their LBM based method on 3D geometries [11, 12]. Even though plenty of work on transient fluid TO has been done in the past decades, few of them studied 3D cases, especially for the FEM based method, because of the computational cost required by the implicit solver.

To make it computationally feasible, reduction methods are indispensable for the transient fluid TO method. Galerkin proper orthogonal decomposition (POD) method came up for fluid dynamics by Kunisch and Volkwein in 2002 [13]. Meyer and Matthies used Karhunen-Loeve expansion for model reduction in nonlinear dynamics [14]. Rathinam and Petzold investigated the basic properties of the POD method when it is applied to model reduction of finite dimensional nonlinear systems [15]. The traditional POD method has two stage: the offline stage and

*Corresponding author: qian@engr.wisc.edu

Copyright © 2023 by ASME

the online stage. One needs to prepare full order model (FOM) solutions and train the model in the offline stage, and uses the reduced order model (ROM) to speed up the solution process in the online stage [16]. The on-the-fly reduction method, on the other hand, integrates the data preparation and training process into the prediction process, merging the previous 2 stages into 1. Gogu proposed the on-the-fly ROM construction for steady state linear elasticity TO problem [17]. Zhao and Wang applied it to the dynamic response TO problem which solved transient problem [18]. Qian stuided an elastodynamics TO problem and proposed the on-the-fly dual reduction method to both the primal and the adjoint equations to achieve dual reduction for both computational time and memeory storage [19].

From the above introduction of the related previous work, the TO method for fluid problem has been developed from linearized Stokes problem [2, 3] to nonlinear Navier-Stokes problem [4, 5], from steady state problem to time-dependent problem [6, 7]. In this study, we studied the full version of the fluid governing equation, i.e. the time-dependent nonlinear Navier-Stokes equations with a similar TO approach that was developed by the pioneers. Different from using the LBM to reduce the computational cost [11, 12], we used the FEM and a POD-Galerkin projection method to reduce the spatial computational cost which was proposed in [13]. Inspired by the "on-the-fly" idea [17] and the "dual-reduction" idea [19] which were first proposed for elasticity and elastodynamic problems, we implemented the similar reduction scheme to the transient fluid TO problem which has seldomly studied previously. Meanwhile, the total energy dissipation over a given time period was minimized as the cost function in optimization. Adjoint method was derived and implemented to obtain the sensitivities. Volume constraint was applied to the fluid part. the Method of Moving Asymptote (MMA) [20] was employed to solve the optimization numerically. The on-the-fly Galerkin-POD reduction method was applied to both the primal and adjoint equations. Several numerical examples including U-bend pipe (from [7]), nozzle (from [6]), and blunt body flow (from [5]) were investigated for testing the reduction performance. The optimized designs were compared with the benchmark results from the references. Both the computational time and memory storage were observed to be reduced at a level above 95%. When coherent structures come into the problem, however, the reduction performance drops dramatically.

This paper is organized in the following structure. In section 2, the physical model and TO formulation will be introduced. In section 3, detailed information about the employed on-the-fly dual-reduction method will be discussed. In section 4, 4 studied 2D examples will be shown with optimized structures and reduction statistics. And in section 5, the conclusion will be drawn.

2. TRANSIENT FLUID MODEL AND TOPOLOGY OPTIMIZATION FORMULATION

2.1 Physical Model of the Transient Fluid Problem

The studied geometry of the transient fluid optimization problem is shown in Fig. 1. Incompressible fluid comes into the geometry through the inlet boundary Γ_{in}, and goes out from the outlet boundary Γ_{out}. The remained surfaces, Γ_{rm}, are applied

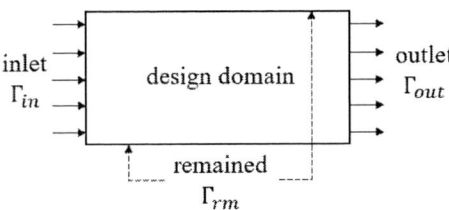

FIGURE 1: BOUNDARIES AND DESIGN DOMAIN CONFIGURATION

with no-slip boundary conditions. Inside the design domain, a density field γ is used to represent the design structures, where $\gamma = 0$ symbolizes solid and $\gamma = 1$ symbolizes fluid channel.

The transient Navier-Stokes equations for incompressible fluid with appropriate boundary conditions are shown in Eq. (1)

$$\frac{\partial u}{\partial t} + u \cdot \nabla u + \nabla p - \nabla \cdot \left(\frac{1}{Re} \nabla u \right) + \alpha (\gamma) u = 0 \quad \text{in } \Omega \quad (1a)$$

$$\nabla \cdot u = 0 \quad \text{in } \Omega \quad (1b)$$

$$u = u_{in} \quad \text{on } \Gamma_{in} \quad (1c)$$

$$n \cdot \nabla u = 0 \quad \text{in } \Gamma_{out} \quad (1d)$$

$$u = 0 \quad \text{on } \Gamma_{rm} \quad (1e)$$

$$n \cdot \nabla p = 0 \quad \text{on } \Gamma_{in} \quad (1f)$$

$$p = 0 \quad \text{on } \Gamma_{out} \quad (1g)$$

$$n \cdot \nabla p = 0 \quad \text{on } \Gamma_{rm} \quad (1h)$$

where u and p are velocity and pressure of fluid field, Re is the Reynolds number, u_{in} is the inlet speed, and α is the Brickmann friction interpolation for TO formulation, which is defined by Rational Approximation of Material Properties (RAMP) method [21] Eq. (2):

$$\alpha (\gamma) = \alpha_{min} + (\alpha_{max} - \alpha_{min}) \frac{q (1 - \gamma)}{q + \gamma} \quad (2)$$

In Eq. (2), α_{min} is set as 0 and α_{max} is calculated by $\alpha_{max} = \frac{1}{Da \cdot Re}$. q is the penalty factor. With this interpolation, in fluid region ($\gamma = 1$), the α value becomes to 0 and Eq. (1a) returns to the normal Navier-Stokes equation. In the solid region ($\gamma = 0$) α will have a large value and therefore Eq. (1a) will be simplified as $u = 0$.

The time-dependent governing equations Eq. (1) are first discretized in temporal domain. Here, we used explicit scheme for convection term and implicit scheme for viscous term. By that, Eq. (1a) is discretized as Eq. (3):

$$\frac{u^{(n)} - u^{(n-1)}}{\Delta t} + u^{(n-1)} \cdot \nabla u^{(n)} + \nabla p^{(n)}$$

$$- \nabla \cdot \left(\frac{1}{Re} \nabla u^{(n)} \right) + \alpha (\gamma) u^{(n)} = 0 \quad (3)$$

2.2 Topology Optimization Formulation

In the optimization problem, the quantity that represents the energy loss from mechanical form to heat is the total energy

Copyright © 2023 by ASME

dissipation given by Eq. (4):

$$c = \int_0^T \int_\Omega \left[\varepsilon\left(\boldsymbol{u}\right) : \varepsilon\left(\boldsymbol{u}\right) + \alpha\left(\gamma\right) \boldsymbol{u} \cdot \boldsymbol{u} \right] dV dt \qquad (4)$$

where T is the end time of the temporal domain, and ε is the strain tensor which is calculated by Eq. (5):

$$\varepsilon\left(\boldsymbol{u}\right) = \frac{1}{2}\left(\nabla\boldsymbol{u} + \boldsymbol{u}^T\right) \qquad (5)$$

Besides, we also have an inequality constraint on fluid channel volume, which is given by Eq. (6)

$$g = \int_\Omega \gamma dV \leq V_0 \qquad (6)$$

where V_0 is the upper threshold.

Combining cost function, governing equations, and volume constraint together, we get the full optimization expressed in Eq. (7)

$$\min_\gamma \quad c = \int_0^T \int_\Omega \left[\varepsilon\left(\boldsymbol{u}\right) : \varepsilon\left(\boldsymbol{u}\right) + \alpha\left(\gamma\right) \boldsymbol{u} \cdot \boldsymbol{u} \right] dV dt \qquad (7a)$$

$$s.t. \ \text{Eq.}(1a) - (1h) \qquad (7b)$$

$$g = \int_\Omega \gamma dV \leq V_0 \qquad (7c)$$

$$0 \leq \gamma \leq 1 \qquad (7d)$$

To solve the optimization problem Eq. (7), the augmented Lagrange method is used to derive the adjoint equations and the sensitivities, with the weak form of the governing equations. The augmented Lagrange function $\mathscr{L}$ is obtained in Eq. (8)

$$\mathscr{L} = \sum_{n=1}^{N_t} \int_\Omega \varepsilon\left(\boldsymbol{u}^{(n)}\right) : \varepsilon\left(\boldsymbol{u}^{(n)}\right) dV\Delta t + \sum_{n=1}^{N_t} \int_\Omega \boldsymbol{u}^{(n)} \cdot \boldsymbol{u}^{(n)} dV\Delta t$$

$$+ \sum_{n=1}^{N_t} \frac{1}{\Delta t} \int_\Omega \boldsymbol{u}^{(n)} \cdot \boldsymbol{v}^{(n)} dV - \sum_{n=1}^{N_t} \frac{1}{\Delta t} \int_\Omega \boldsymbol{u}^{(n-1)} \cdot \boldsymbol{v}^{(n)} dV$$

$$+ \sum_{n=1}^{N_t} \int_\Omega \boldsymbol{u}^{(n-1)} \cdot \nabla\boldsymbol{u}^{(n)} \cdot \boldsymbol{v}^{(n)} dV + \sum_{n=1}^{N_t} \int_\Omega \boldsymbol{v}^{(n)} \cdot \nabla p^{(n)} dV$$

$$+ \sum_{n=1}^{N_t} \int_\Omega \frac{1}{Re} \nabla\boldsymbol{u}^{(n)} \cdot \nabla\boldsymbol{v}^{(n)} dV + \sum_{n=1}^{N_t} \int_\Omega \alpha\left(\gamma\right) \boldsymbol{u}^{(n)} \cdot \boldsymbol{v}^{(n)} dV$$

$$- \sum_{n=1}^{N_t} \int_\Omega q^{(n)} \left(\nabla \cdot \boldsymbol{u}^{(n)}\right) dV \qquad (8)$$

where $\boldsymbol{u}^{(n)}$ and $q^{(n)}$ ($n = 1..N_t$) are the adjoint variables.

The variation of the augmented lagrange function Eq. (8) is derived as Eq. (9).

$$\delta\mathscr{L} = \sum_{n=1}^{N_t} \Delta t \int_\Omega 2\varepsilon\left(\delta\boldsymbol{u}^{(n)}\right) : \varepsilon\left(\boldsymbol{u}^{(n)}\right) dV$$

$$+ \sum_{n=1}^{N_t} \Delta t \int_\Omega 2\delta\boldsymbol{u}^{(n)} \cdot \boldsymbol{u}^{(n)} dV + \sum_{n=1}^{N_t} \frac{1}{\Delta t} \int_\Omega \delta\boldsymbol{u}^{(n)} \cdot \boldsymbol{v}^{(n)} dV$$

$$- \sum_{n=1}^{N_t-1} \frac{1}{\Delta t} \int_\Omega \delta\boldsymbol{u}^{(n)} \cdot \boldsymbol{v}^{(n+1)} dV + \sum_{n=1}^{N_t} \int_\Omega \boldsymbol{v}^{(n)} \cdot \nabla\delta p^{(n)} dV$$

$$+ \sum_{n=1}^{N_t-1} \int_\Omega \delta\boldsymbol{u}^{(n)} \cdot \nabla\boldsymbol{u}^{(n+1)} \cdot \boldsymbol{v}^{(n+1)} dV$$

$$+ \sum_{n=1}^{N_t} \int_\Omega \boldsymbol{u}^{(n-1)} \cdot \nabla\delta\boldsymbol{u}^{(n)} \cdot \boldsymbol{v}^{(n)} dV$$

$$+ \sum_{n=1}^{N_t} \int_\Omega \frac{1}{Re} \nabla\delta\boldsymbol{u}^{(n)} \cdot \nabla\boldsymbol{v}^{(n)} dV + \sum_{n=1}^{N_t} \int_\Omega \frac{d\alpha}{d\gamma} \boldsymbol{u}^{(n)} \cdot \boldsymbol{v}^{(n)} dV$$

$$+ \sum_{n=1}^{N_t} \int_\Omega \alpha\left(\gamma\right) \delta\boldsymbol{u}^{(n)} \cdot \boldsymbol{v}^{(n)} dV - \sum_{n=1}^{N_t} \int_\Omega q^{(n)} \left(\nabla \cdot \delta\boldsymbol{u}^{(n)}\right) dV$$

$$(9)$$

where $\delta\boldsymbol{u}^{(n)}$ and $\delta p^{(n)}$ are the perturbation functions of primal variables $\boldsymbol{u}^{(n)}$ and $p^{(n)}$.

Then, collecting all the terms containing $\delta\boldsymbol{u}^{(N_t)}$ or $\delta p^{(N_t)}$ will give us the adjoint equation at time step N_t as Eq. (10)

$$\Delta t \int_\Omega 2\varepsilon\left(\hat{\boldsymbol{v}}\right) : \varepsilon\left(\boldsymbol{u}^{(N_t)}\right) dV + \Delta t \int_\Omega 2\hat{\boldsymbol{v}} \cdot \boldsymbol{u}^{(N_t)} dV$$

$$+ \frac{1}{\Delta t} \int_\Omega \hat{\boldsymbol{v}} \cdot \boldsymbol{v}^{(N_t)} dV + \int_\Omega \boldsymbol{v}^{(N_t)} \cdot \nabla\hat{q} dV$$

$$+ \int_\Omega \boldsymbol{u}^{(N_t-1)} \cdot \nabla\hat{\boldsymbol{v}} \cdot \boldsymbol{v}^{(N_t)} dV + \frac{1}{Re} \int_\Omega \nabla\hat{\boldsymbol{v}} \cdot \nabla\boldsymbol{v}^{(N_t)} dV$$

$$+ \int_\Omega \alpha\left(\gamma\right) \hat{\boldsymbol{v}} \cdot \boldsymbol{v}^{(N_t)} dV - \int_\Omega q^{(N_t)} \left(\nabla \cdot \hat{\boldsymbol{v}}\right) dV = 0 \qquad (10)$$

where $\hat{\boldsymbol{v}}$ and $\hat{q}$ are the test functions required by weak form.

Similarly, collecting all the terms containing $\delta\boldsymbol{u}^{(n)}$ or $\delta p^{(n)}$ ($n = 1..N_t - 1$) will give us the adjoint equation at time step $n = 1..N_t - 1$ as Eq. (11)

$$\Delta t \int_\Omega 2\varepsilon\left(\hat{\boldsymbol{v}}\right) : \varepsilon\left(\boldsymbol{u}^{(n)}\right) dV + \Delta t \int_\Omega 2\hat{\boldsymbol{v}} \cdot \boldsymbol{u}^{(n)} dV$$

$$+ \frac{1}{\Delta t} \int_\Omega \hat{\boldsymbol{v}} \cdot \boldsymbol{v}^{(n)} dV - \frac{1}{\Delta t} \int_\Omega \hat{\boldsymbol{v}} \cdot \boldsymbol{v}^{(n+1)} dV$$

$$+ \int_\Omega \boldsymbol{v}^{(n)} \cdot \nabla\hat{q} dV + \int_\Omega \hat{\boldsymbol{v}} \cdot \nabla\boldsymbol{u}^{(n+1)} \cdot \boldsymbol{v}^{(n+1)} dV$$

$$+ \int_\Omega \boldsymbol{u}^{(n-1)} \cdot \nabla\hat{\boldsymbol{v}} \cdot \boldsymbol{v}^{(n)} dV + \frac{1}{Re} \int_\Omega \nabla\hat{\boldsymbol{v}} \cdot \nabla\boldsymbol{v}^{(n)} dV$$

$$+ \int_\Omega \alpha\left(\gamma\right) \hat{\boldsymbol{v}} \cdot \boldsymbol{v}^{(n)} dV - \int_\Omega q^{(n)} \left(\nabla \cdot \hat{\boldsymbol{v}}\right) dV = 0 \qquad (11)$$

Eliminating Eq. (10) and Eq. (11) from Eq. (9), we get the sensitivity of the augmented Lagrange function as shown in Eq. (12)

$$\delta\mathscr{L} = \sum_{n=1}^{N_t} \int_\Omega \frac{d\alpha}{d\gamma} \boldsymbol{u}^{(n)} \cdot \boldsymbol{v}^{(n)} dV \qquad (12)$$

Copyright © 2023 by ASME

In practice, the Helmholtz-type PDE filter [22] and the Heaviside projection [23] are utilized to reduce the checkerboard issue which is quite common in TO problems. The original design density field γ is filtered to $\tilde{\gamma}$ by solving a PDE as Eq. (13)

$$-R^2 \nabla^2 \tilde{\gamma} + \tilde{\gamma} = \gamma \tag{13}$$

And the intermediate density field $\tilde{\gamma}$ is projected to the physical design field $\bar{\gamma}$ by Eq. (14)

$$\bar{\gamma} = \begin{cases} \eta \left[e^{-\beta \left(1 - \frac{\tilde{\gamma}}{\eta}\right)} - \left(1 - \frac{\tilde{\gamma}}{\eta}\right) e^{-\beta} \right] & , \tilde{\gamma} \in [0, \eta] \\ (1-\eta) \left[1 - e^{-\beta \frac{\tilde{\gamma}-\eta}{1-\eta}} + \frac{\tilde{\gamma}-\eta}{1-\eta} e^{-\beta} \right] + \eta & , \tilde{\gamma} \in (\eta, 1] \end{cases} \tag{14}$$

where η is the Heavisde threshold and β is the Heaviside steepness. the Method of Moving Asymptote [20] is used to update the design density field.

3. ON-THE-FLY DUAL REDUCTION METHOD
3.1 Galerkin-POD reduction method

For a given data set, the Galerkin-POD reduction method can be applied as follows. The snapshot set $U = \{U_1, .., U_{N_b}\}$ is consisted of a series of known solutions of the parameterized PDE. In this study, the parameterized PDE is the Navier-Stokes equations Eq. (1) with the design density field γ as the parameter. And the snapshot set U comes from the FOM solution of Eq. (1). The goal of reduction is to find a n_m dimensional subspace $\Phi = span\{\phi_1, .., \phi_{n_m}\}$ such that the snapshot set can be represented by this subspace with minimum least-square residual as shown in Eq. (15)

$$L = \sum_{i=1}^{N_b} \left\| U_i - \sum_{j=1}^{n_m} \langle U_i, \phi_j \rangle \phi_j \right\|^2 \tag{15}$$

where $\langle ., . \rangle$ is the inner product defined in the linear space.

The solution of this least-square minimization problem can be obtained by singular value decomposition (SVD) of snapshot matrix U, which can then be converted to such an eigenvalue problem shown in Eq. (16)

$$\left(U^T \cdot U \right) \psi_i = \lambda_i \psi_i \tag{16}$$

where λ_i and ψ_i are the corresponding eigenvalues and eigenvectors. After obtaining λ_i and ψ_i, the basis vector ϕ_i can then be evaluated as Eq. (17)

$$\phi_i = \frac{1}{\sqrt{\lambda_i}} \mathbb{S} \psi_i, \quad i = 1..n_m \tag{17}$$

With the POD basis Φ, the FOM solving process can then be reduced to the ROM space. At each time step, the PDE system will first be assembled to a linear system in the FOM space as shown in Eq. (18):

$$A_F \cdot U_F = b_F \tag{18}$$

where A_F and b_F are the FOM matrix and the FOM right-hand-side vector. The ROM matrix A_R and vector b_R can be computed by projecting A_F and b_F to the ROM space as Eq. (19)

$$A_R = \Phi^T \cdot A_F \cdot \Phi \tag{19a}$$
$$b_R = \Phi^T \cdot b_F \tag{19b}$$

Then, the solution in the ROM space, U_R, is solved from the reduced equation Eq. (20)

$$A_R \cdot U_R = b_R \tag{20}$$

The FOM solution can then be obtained by projecting the ROM solution back to the FOM space as Eq. (21)

$$U_F = \Phi \cdot U_R \tag{21}$$

The residual described in Eq. (22) is usually used as a criterion to assess the deviation between the FOM solution and the ROM solution. If the residual is large, the snapshot matrix U should be reconstructed.

$$e = \cdot \frac{\|A_F \cdot U_F - b_F\|^2}{\|b_F\|^2} \tag{22}$$

3.2 On-the-fly reduction scheme

In the on-the-fly scheme, the snapshot matrix U is updated during the optimization process. We split the whole snapshot matrix into 2 parts, the latest solution set U_f and the time-related set U_t. The first set U_f is updated by the residual criterion during the optimization process. At some time steps, if the residual evaluated by Eq. (22) does not meet the tolerance requirement, FOM solution at the current time step will be computed again through the FOM linear system Eq. (18). Then, the snapshot subset U_f will drop the oldest solution and append the latest FOM solution into it. With this updating scheme, U_f will always keep the latest solution to improve the reduction performance.

The second snapshot subset, U_t, is updated by time step criterion. It is usually the case in the time-dependent problems that some time steps have influence to the whole temporal domain. Hence, only taking the latest FOM solution like subset U_f is not enough. To improve the reduction further, solutions at some pre-determined time steps will be taken into the snapshot subset U_t no matter whether the residual criterion is met or not. During the optimization process, when it goes to the pre-determined step, the new solution will replace the ole solution at the same time step in U_t. In this study, the pre-determined time steps are chosen uniformly in the temporal domain.

The whole snapshot matrix is the union of the above-mentioned 2 subsets, as shown in Eq. (23). And the "on-the-fly" feature requires updating the snapshot matrix from time to time according to the criteria discussed before.

$$U = U_f \cup U_t \tag{23}$$

3.3 Flow chart for the whole optimization solving

The POD based on-the-fly reduction method discussed in section 3.1 and 3.2 are applied to the primal equations Eq. (1) and the adjoint equations Eq. (10) and (11) respectively. That is the reason why this scheme for TO problem is called "dual" reduction. A flow chart for the whole TO process with the on-the-fly dual reduction method is shown in Fig. 2.

Through the whole reduction process shown in Fig. 2, the main difference between the current on-the-fly reduction method and the traditional POD based reduction method is that this

Copyright © 2023 by ASME

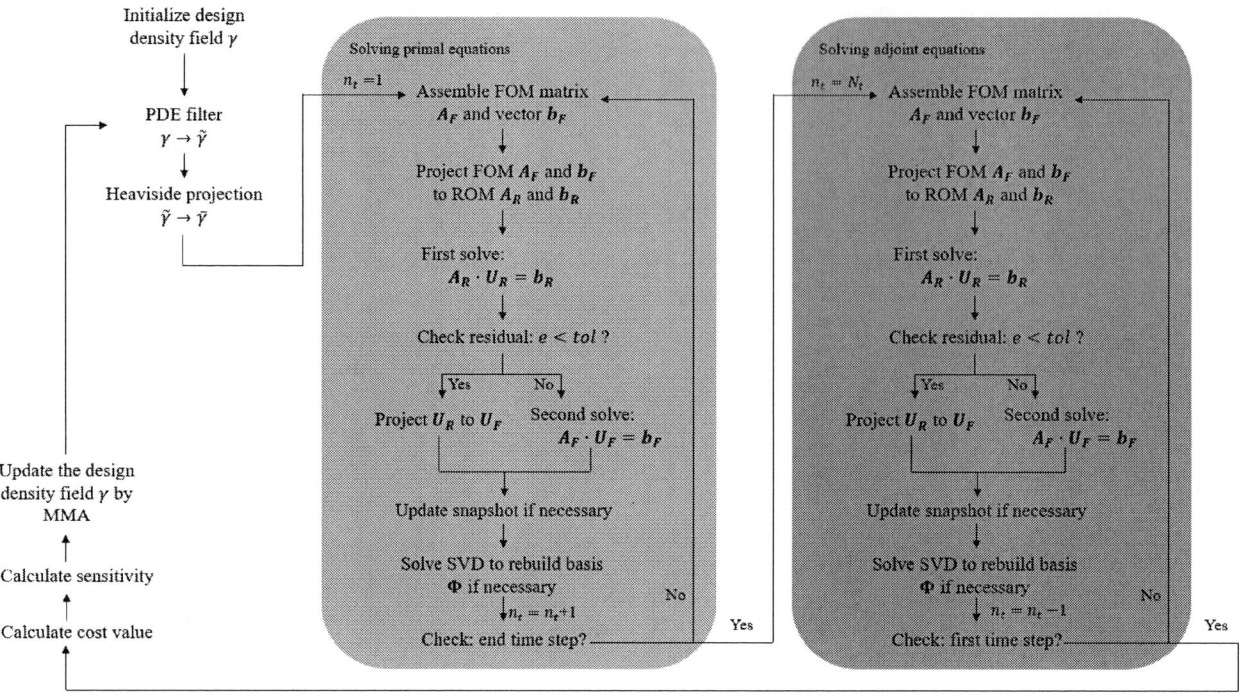

FIGURE 2: FLOW CHART FOR THE WHOLE OPTIMIZATION PROCESS

method does not depend on a prescribed POD basis. The on-the-fly scheme will update the snapshot sets from time to time so that the POD basis is not fixed. Given that the topology and geometry of the design usually change dramatically during the optimization process, finding such a fixed POD basis suitable for the whole optimization is a hard task. Therefore, the current method in which the basis is updated on-the-fly is more appropriate, compared with the fixed basis ROM method. It should also not be ignored that the arificial neural network (ANN) based has got a fast development in the recent decade. The surrogate-based ANN has been applied to plenty of areas that need raw performance prediction. However, the main limit of the surrogate model is the training time it needs for a good convergence. Since the on-the-fly scheme needs to change the basis frequently, the re-traning process of the surrogate model will need a lot of additional time and is therefore adverse to time reduction.

4. NUMERICAL RESULTS

We implemented the transient TO problem mentioned in section 2 with an open source FEM library, FEniCS. A mixed finite element space is used for the velocity-pressure coupled problem. The 1st-order and 2nd-order Lagrange polynomial basises are chosen for pressure and velocity respectively so that the LBB stable condition can be met. The direct LU-decomposition solver is used to solve the assembled linear system. The eigenvalue problems required by the on-the-fly dual reduction scheme is fulfilled by the open source Python package slepc4py.

Some hyper-parameters such as Brickmann friction interpolation penalty factor q, Heaviside steepness β, and ROM residual tolerance tol are updated with a continuum scheme during the optimization. q starts with the value 0.01 at the initialization, and

is multiplied by 10 after each 50 TO iterations. β is initialized with 1 and becomes to 1.5 times after each 20 iterations. tol is reduced half after each 20 TO iterations until it is below than 0.001. Snapshot size of U_f and U_t is 10.

The total iteration number of the optimization process is set 200 in all the cases. The stop criterion of the TO process is that the max absolute change of the design density field γ between two recent steps is lower than 0.01. This stop criterion is commonly used by the TO field, and from testing and experience, 200 is a large enough iteration number for a good convergence of the design field in 2D fluid TO problem.

In this section, we will introduce several numerical examples. In section 4.1, the U-bend configuration is discussed. A similar nozzle example is shown in section 4.2, and the blunt body flow optimization results are displayed in section 4.3.

4.1 U-bend channel Configuration

The design domain and boundary condition settings for the U-bend configuration are shown in Fig. 3. In this case, inlet boundary is on the top left region and the outlet boundary is on the bottom right region. A parabolic distribution is used for the inlet velocity, as shown in Eq. (24).

$$u_{in} = 4 \frac{\left(H_{in,max} - y\right)\left(y - H_{in,min}\right)}{\left(H_{in,max} - H_{in,min}\right)^2} \qquad (24)$$

where y is the vertical coordinate, $H_{in,max}$ and $H_{in,min}$ are the upper and lower position of the inlet. Reynolds number is set as $Re = 100$ in this case for a laminar flow. Time step size is $\Delta t = 0.0001$ and total step number is $N_t = 1000$ in this case.

The fully FOM solving process was also studied for a comparison with the on-the-fly dual reduction process. The FOM and

Copyright © 2023 by ASME

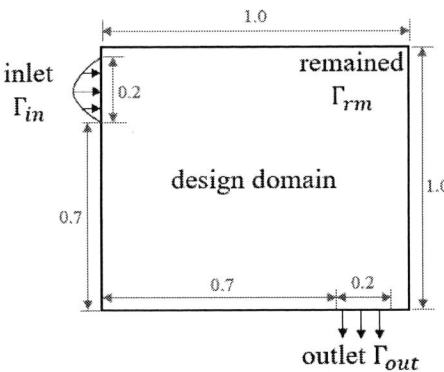

FIGURE 3: CONFIGURATION OF U-BEND CHANNEL

TABLE 1: REDUCTION PERFORMANCE OF U-BEND CASES

Performance	FOM	ROM 1	ROM 2	ROM 3
Final tol	–	10^{-1}	10^{-2}	10^{-3}
skip ratio (primal)	–	99.85%	99.26%	98.83%
skip ratio (adjoint)	–	98.46%	96.22%	94.82%
assemble time [s]	19427	20699	20719	20745
solving time [s]	112557	3702	4876	5661
residual time [s]	–	44780	44306	43928
basis SVD time [s]	–	173	316	401
avg FOM time [s]	0.563	0.405	0.418	0.420
avg ROM time [s]	–	0.015	0.015	0.015
total time [s]	137140	70344	71208	71723
memory usage [MB]	2152	21	46	63

ROM optimized structures are shown in Fig. 4. Comparing the FOM optimized design in Fig. 4 with the optimized design in reference [7] Fig. 15(c) (benchmark design), we can see that both the two designs have the similar "U-shape" bending. The difference of the curvature can be explained by the different Reynolds number used in the studies. In Fig. 4, it can be observed that the optimized structures from the ROM solving process are quite similar to the FOM results. This shows the liability of the on-the-fly dual-reduction method.

To test the performance of the reduction, we investigated several different final residual tolerances. The FOM solution skip ratio, time reduction, and memory saving are shown in Table. 1.

From Table. 1, for each ROM cases, the skip ratios of both primal and adjoint equations reach above 95%. Hence, the time needed for solving the linear system is reduced dramatically from about 110,000s to about 4500s. However, since the assembling process and residual calculation process still need to deal with the FOM matrices, the time for these 2 parts are not reduced. It can also be observed from Table. 1 that the assemble time and residual time are the 2 main source in the ROM process. Compared to them, the time spent on basis building and SVD is negligible. The average time required by a single FOM/ROM solve is also compared in Table. 1, which shows a huge reduction from FOM to ROM. Using the on-the-fly reduction method, nearly 50% of total execution time can be reduced, and 95% of memory storage can be saved.

TABLE 2: REDUCTION PERFORMANCE OF SNAPSHOT SIZE TEST

Number of U_f/U_t	Total time [s]	Skip ratio (primal/adjoint)
U_f:10, U_t:10	128996	98.43% / 92.38%
U_f:20, U_t:20	135218	98.39% / 89.67%
U_f:30, U_t:30	155156	98.07% / 86.86%

TABLE 3: REDUCTION PERFORMANCE OF DOF NUMBER TEST

DOF	avg time(FOM/ROM)[s]	memory(FOM/ROM)[MB]
29203	377 / 355	552 / 25
65403	985 / 676	1619 / 148.93
88903	1439 / 903.14	2352 / 88

Besides, we also tested the reduction scheme with different sizes of the snapshot set U_f and U_t. The total computational time and skip ratios are listed in Table. 2. From Table. 2, increasing the size of the snapshot set can have little improvement in skip ratios, and the computational time becomes even worse due to the additional computational requirement according to the larger basis. This indicates that the snapshot size $U_f = 10$, $U_t = 10$ is already a good choice and will be kept in the remained work of this study.

After that, the scaling test in which different numbers of degree of freedom (DOF) were studied. Three different coarse and fine meshes were investigated and the results are listed in Table. 3. From Table. 3, the average time per TO iteration can be futher reduced when the DOF number was increased, since the POD basis size is fixed at 20. This indicates that the current reduction method is especially advantage to the 3D TO problems, because number of DOF in 3D geometry is much larger than 2D and therefore can have huge time reduction.

4.2 Nozzle Configuration

The nozzle configuration shown in Fig. 5 was also studied in a similar way in this study. Inlet velocity is set by Eq. (24). Reynolds number is $Re = 100$. Time step gap is set as $\Delta t = 10^{-4}$, and total time steps are $N_t = 1000$. The residual tolerance is initially set as $tol = 10^{-1}$, and is reduced to 10^{-1}, 10^{-2}, and 10^{-3} gradually at the end.

The optimized structures by FOM and 3 ROM cases with different final tolerance are shown in Fig. 6. Comparing the FOM optimized design in Fig. 6 with the optimized design in reference [6] Fig. 14 which was treated as a benchmark result, we can see that the design structures are quite similar, which can validate our TO process. The first row is the FOM results at different TO iterations, while the second row is the ROM results with final tolerance $tol = 10^{-1}$. At this level of tolerance, the final design is quite far away from the FOM design, which indicates that the large residuals remain in the primal and adjoint equations may affect the final design. On the other hand, the results of ROM with final $tol = 10^{-3}$, which are displayed in the third row, can converge to the FOM design at the end, although it has bad designs in the early stages such as 10 and 50 iterations. This is because the tolerance is reduced half after each 20 iterations.

Copyright © 2023 by ASME

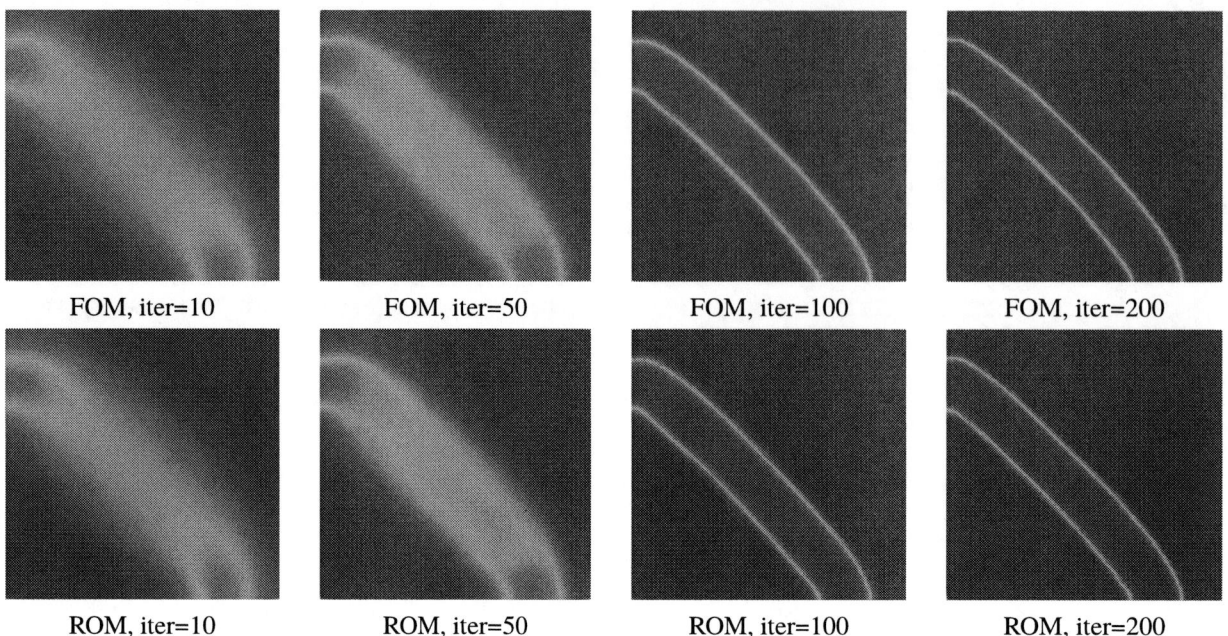

FIGURE 4: FOM AND ROM OPTIMIZED RESULTS OF THE U-BEND CASE

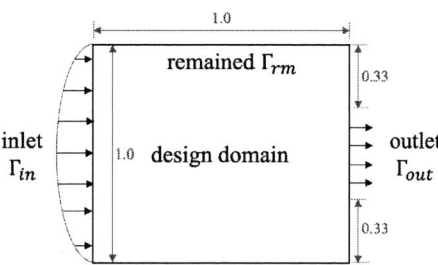

FIGURE 5: CONFIGURATION OF NOZZLE

Therefore, choosing an appropriate tolerance is quite important in the optimization process.

The reduction performance of this nozzle case is listed in Table. 4. Like the U-bend case, the skip ratios of primal and adjoint equations reach 95%. The memory saving is propotional to the skip ratio, therefore, the memory reduction by the ROM method is also above 95%. The average solving time of single ROM solution is nearly 1/30 of the average single FOM solution. Taken the assembling time and residual calculation time into consideration, the actural reduction time is about 50%. Some technique such as pre-factorization can be used to futher reduce the assembling time in the ROM process.

4.3 Blunt Body Configuration

We also studied a blunt body flow optimization problem using the on-the-fly dual reduction method to reduce the computational time. The geometry and boundaries are shown in Fig. 7. In this case, the design domain is not the entire computational domain anymore. Instead, the design domain is restricted in a rectangular region in the upstream of the channel. The initial design is a uniform grey density field so that it is a blunt body

TABLE 4: REDUCTION PERFORMANCE OF NOZZLE CASES

Performance	FOM	ROM 1	ROM 2	ROM 3
Final tol	–	10^{-1}	10^{-2}	10^{-3}
skip ratio (primal)	–	99.56%	98.32%	97.49%
skip ratio (adjoint)	–	98.45%	91.51%	87.64%
assemble time [s]	38358	38217	38405	39859
solving time [s]	219561	9742	18903	24035
residual time [s]	–	108480	112670	106851
basis SVD time [s]	–	489	1401	1921
avg FOM time [s]	0.549	0.567	0.565	0.561
avg ROM time [s]	–	0.019	0.019	0.018
total time [s]	279019	165637	191104	198850
memory usage [MB]	2149	22	110	161

sitting in the channel. Vortexes (coherent structures) will shed from the blunt body boundary.

In this case, Reynolds number is $Re = 500$. Time step size is $\Delta t = 0.1$. Total time step number is $N_t = 200$. A parabolic inlet velocity boundary condition as Eq. (24) is applied to the inlet Γ_{in}. Final tolerance $tol = 10^{-2}$ and $tol = 10^{-3}$ are studied. FOM and ROM optimized structures are displayed in Fig. 8.

Comparing the FOM optimized design in Fig. 8 with the optimized design in reference [5] Fig. 22(c) (benchmark design), we can observe that the designs in the both studies converge to an airfoil shape. From Fig. 8, both the FOM and ROM results converge to an airfoil shape structures at the end of the optimization. In the early TO stages (before iter=50), both the FOM and ROM structures are symmetric along the vertical direction. The FOM solution loses its symmetry a little bit in the late TO stage (iter>100). A possible reason for this is that the Heaviside steepness has been raised to high value in the late stage, making the

Copyright © 2023 by ASME

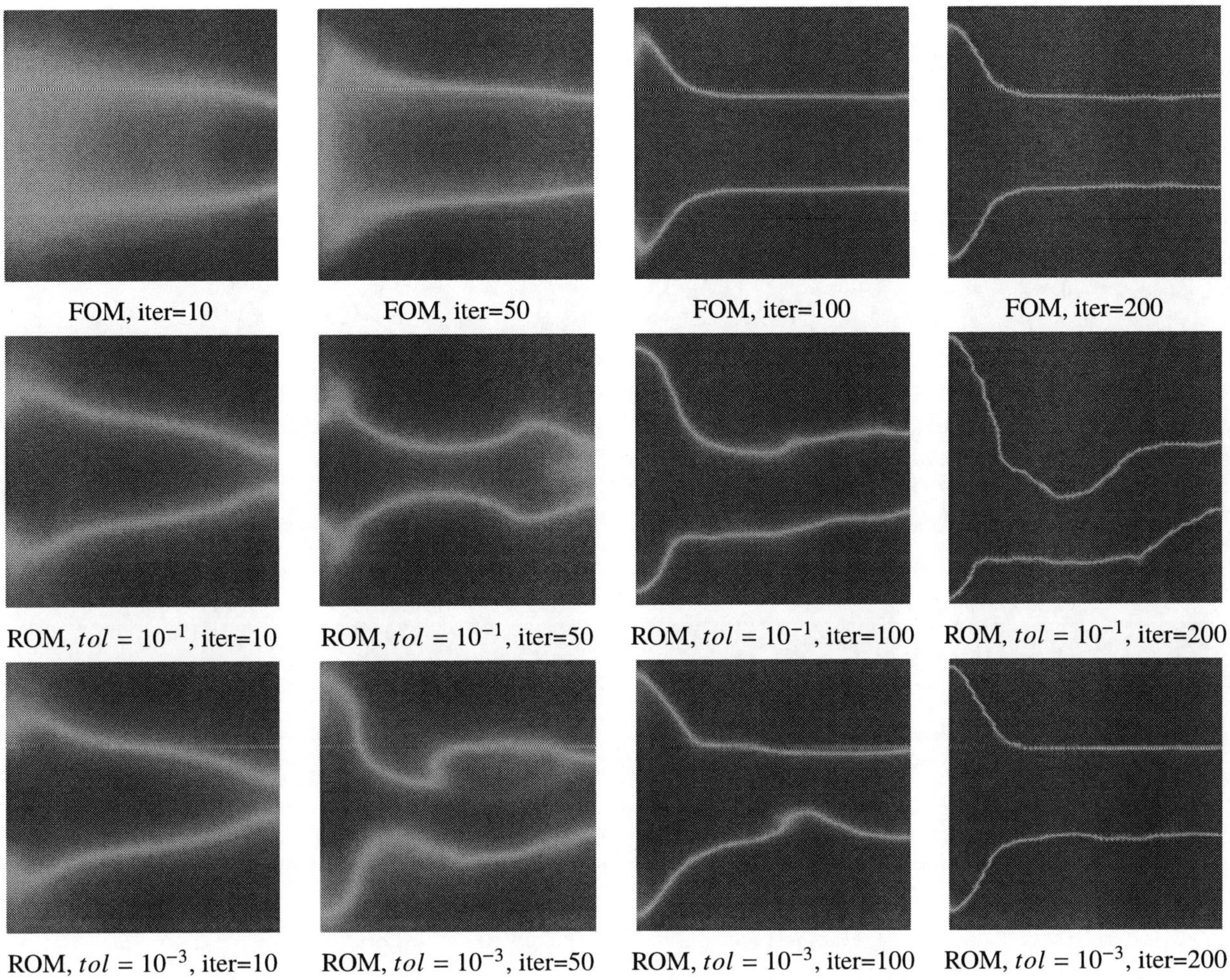

| FOM, iter=10 | FOM, iter=50 | FOM, iter=100 | FOM, iter=200 |

| ROM, $tol = 10^{-1}$, iter=10 | ROM, $tol = 10^{-1}$, iter=50 | ROM, $tol = 10^{-1}$, iter=100 | ROM, $tol = 10^{-1}$, iter=200 |

| ROM, $tol = 10^{-3}$, iter=10 | ROM, $tol = 10^{-3}$, iter=50 | ROM, $tol = 10^{-3}$, iter=100 | ROM, $tol = 10^{-3}$, iter=200 |

FIGURE 6: FOM AND ROM OPTIMIZED RESULTS OF THE NOZZLE CASE

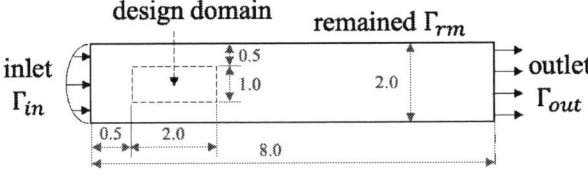

FIGURE 7: CONFIGURATION OF BLUNT BODY FLOW

TABLE 5: REDUCTION PERFORMANCE OF BLUNT BODY CASES

Performance	FOM	ROM 1	ROM 2
Final tol	–	10^{-2}	10^{-3}
skip ratio (primal)	–	69.12%	80.73%
skip ratio (adjoint)	–	8.88%	18.02%
assemble time [s]	4934	5150	5231
solving time [s]	39956	26941	22728
residual time [s]	–	14538	14076
basis SVD time [s]	–	1763	1506
avg FOM time [s]	0.499	0.529	0.532
avg ROM time [s]	–	0.014	0.014
total time [s]	47130	61732	54397
memory usage [MB]	430	261	217

problem more nonlinear. The ROM solution keeps the symmetry during the whole optimization. Despite the symmetry, the ROM structures look very closed to the FOM structures.

The reduction performance of this blunt body case is shown in Table. 5. In this case, the skip ratios of primal and adjoint equations drop dramatically. Especially, the reduction method becomes ineffective in adjoint equation, only having a skip ratio lower than 20%. As a consequence, the ROM cases used even more computational time than the FOM case, because ROM also needs to calculate residual and build POD basis. From the statistical results, the average single ROM solving time is still lower than the average single FOM solving, but since ROM solutions

Copyright © 2023 by ASME

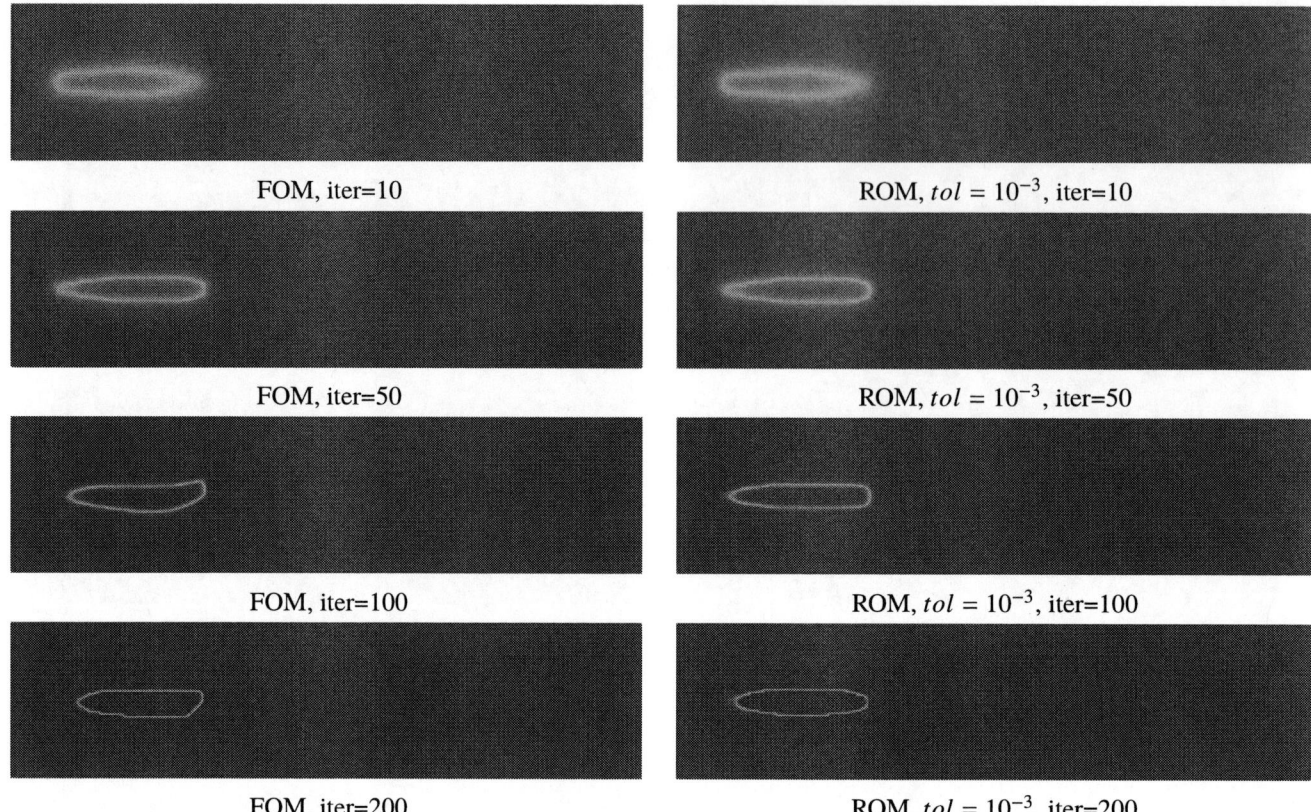

FIGURE 8: FOM AND ROM OPTIMIZED RESULTS OF THE NOZZLE CASE

are adopted in little steps, the total computational time does not show the advantage of ROM. Besides, the saved memory is about 50% which is not as good as the previous 2 cases. The reason for the failure is that in the blunt body cases, the vortexes sheding from the body and traveling downstream make the fluid patterns much more complex than the previous 2 channel cases. For visualization, the streamline fields of the FOM optimized design are plotted at different time steps in Fig. 9. As an example, the vortex in the red circle at $t = 69.5s$ starts to shed from the solid structure. The vortex moves to the donstream and as a result the center position of the vortex at time $t = 70.0s$ and $t = 70.5s$ goes to rigth. At $t = 71.0s$ the current vortex decays and new vortex starts to form in the upstream. Those coherent structures emerge periodically, making the flow complex. Some new methods that can improve the reduction performance should be proposed and applied in the future work.

5. CONCLUSION

In this study, we have applied the on-the-fly dual reduction method to the transient incompressible fluid TO problems. The proposed method was implemented with FEM and investigated through 3 examples, U-bend channel, nozzle, and blunt body flow. In the first 2 cases, the ROM method can provide similar final designs with FOM results, with an appropriate final tolerance. The total reduction time can reach above 50%, and the total reduction in memory can reach about 95%. This suggests that the on-the-fly dual reduction method is useful for transient fluid TO.

However, in the blunt body flow where coherent structures emerge in the flow field, the current scheme does not provide substantial reduction. Future work would be focusing on extending this method on the designs under complex flow.

ACKNOWLEDGMENTS

The authors want to acknowledge the support of NSF grant #2210031.

REFERENCES

[1] Bendsoe, Martin Philip and Sigmund, Ole. *Topology optimization: theory, methods, and applications*. Springer Science & Business Media (2003).

[2] Borrvall, Thomas and Petersson, Joakim. "Topology optimization of fluids in Stokes flow." *International journal for numerical methods in fluids* Vol. 41 No. 1 (2003): pp. 77–107.

[3] Guest, James K and Prévost, Jean H. "Topology optimization of creeping fluid flows using a Darcy–Stokes finite element." *International Journal for Numerical Methods in Engineering* Vol. 66 No. 3 (2006): pp. 461–484.

[4] Gersborg-Hansen, Allan, Sigmund, Ole and Haber, Robert B. "Topology optimization of channel flow problems." *Structural and multidisciplinary optimization* Vol. 30 (2005): pp. 181–192.

[5] Yoon, Gil Ho. "Topology optimization for stationary fluid–structure interaction problems using a new monolithic for-

Copyright © 2023 by ASME

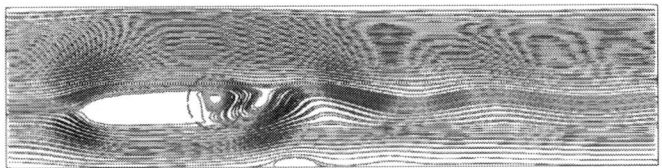
$t = 69.5\ s$

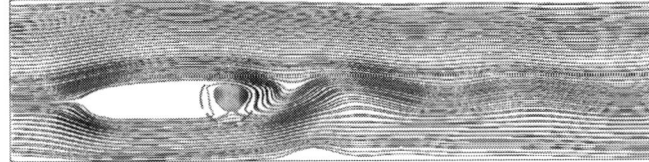
$t = 70.0\ s$

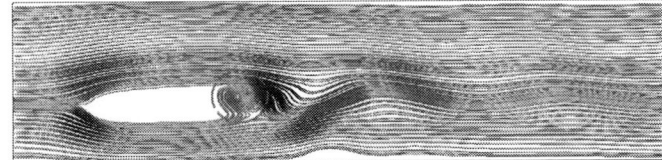
$t = 70.5\ s$

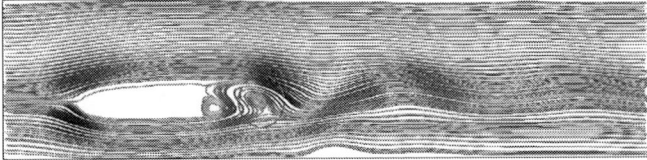
$t = 71.0\ s$

FIGURE 9: STREAMLINE FIELDS OF THE FOM OPTIMIZED DESIGN

mulation." *International journal for numerical methods in engineering* Vol. 82 No. 5 (2010): pp. 591–616.

[6] Kreissl, Sebastian, Pingen, Georg and Maute, Kurt. "Topology optimization for unsteady flow." *International Journal for Numerical Methods in Engineering* Vol. 87 No. 13 (2011): pp. 1229–1253.

[7] Deng, Yongbo, Liu, Zhenyu, Zhang, Ping, Liu, Yongshun and Wu, Yihui. "Topology optimization of unsteady incompressible Navier–Stokes flows." *Journal of Computational Physics* Vol. 230 No. 17 (2011): pp. 6688–6708.

[8] Andreasen, Casper Schousboe and Sigmund, Ole. "Topology optimization of fluid–structure-interaction problems in poroelasticity." *Computer Methods in Applied Mechanics and Engineering* Vol. 258 (2013): pp. 55–62.

[9] Abdelwahed, Mohamed and Hassine, Maatoug. "Topology optimization of time dependent viscous incompressible flows." *Abstract and Applied Analysis*, Vol. 2014. 2014. Hindawi.

[10] Yaji, Kentaro, Yamada, Takayuki, Yoshino, Masato, Matsumoto, Toshiro, Izui, Kazuhiro and Nishiwaki, Shinji. "Topology optimization in thermal-fluid flow using the lattice Boltzmann method." *Journal of Computational Physics* Vol. 307 (2016): pp. 355–377.

[11] Chen, Cong, Yaji, Kentaro, Yamada, Takayuki, Izui, Kazuhiro and Nishiwaki, Shinji. "Local-in-time adjoint-based topology optimization of unsteady fluid flows using the lattice Boltzmann method." *Mechanical Engineering Journal* Vol. 4 No. 3 (2017): pp. 17–00120.

[12] Yaji, Kentaro, Ogino, Masao, Chen, Cong and Fujita, Kikuo. "Large-scale topology optimization incorporating local-in-time adjoint-based method for unsteady thermal-fluid problem." *Structural and Multidisciplinary Optimization* Vol. 58 (2018): pp. 817–822.

[13] Kunisch, Karl and Volkwein, Stefan. "Galerkin proper orthogonal decomposition methods for a general equation in fluid dynamics." *SIAM Journal on Numerical analysis* Vol. 40 No. 2 (2002): pp. 492–515.

[14] Meyer, Marcus and Matthies, Hermann G. "Efficient model reduction in non-linear dynamics using the Karhunen-Loeve expansion and dual-weighted-residual methods." *Computational Mechanics* Vol. 31 No. 1 (2003): pp. 179–191.

[15] Rathinam, Muruhan and Petzold, Linda R. "A new look at proper orthogonal decomposition." *SIAM Journal on Numerical Analysis* Vol. 41 No. 5 (2003): pp. 1893–1925.

[16] Guo, Mengwu and Hesthaven, Jan S. "Data-driven reduced order modeling for time-dependent problems." *Computer methods in applied mechanics and engineering* Vol. 345 (2019): pp. 75–99.

[17] Gogu, Christian. "Improving the efficiency of large scale topology optimization through on-the-fly reduced order model construction." *International Journal for Numerical Methods in Engineering* Vol. 101 No. 4 (2015): pp. 281–304.

[18] Zhao, Junpeng and Wang, Chunjie. "Dynamic response topology optimization in the time domain using model reduction method." *Structural and Multidisciplinary Optimization* Vol. 53 (2016): pp. 101–114.

[19] Qian, Xiaoping. "On-the-fly dual reduction for time-dependent topology optimization." *Journal of Computational Physics* Vol. 452 (2022): p. 110917.

[20] Svanberg, Krister. "The method of moving asymptotes—a new method for structural optimization." *International journal for numerical methods in engineering* Vol. 24 No. 2 (1987): pp. 359–373.

[21] Stolpe, Mathias and Svanberg, Krister. "An alternative interpolation scheme for minimum compliance topology optimization." *Structural and Multidisciplinary Optimization* Vol. 22 No. 2 (2001): pp. 116–124.

[22] Lazarov, Boyan Stefanov and Sigmund, Ole. "Filters in topology optimization based on Helmholtz-type differential equations." *International Journal for Numerical Methods in Engineering* Vol. 86 No. 6 (2011): pp. 765–781.

[23] Guest, James K, Prévost, Jean H and Belytschko, Ted. "Achieving minimum length scale in topology optimization using nodal design variables and projection functions." *International journal for numerical methods in engineering* Vol. 61 No. 2 (2004): pp. 238–254.

Copyright © 2023 by ASME

Proceedings of the ASME 2023
International Design Engineering Technical Conferences and
Computers and Information in Engineering Conference
IDETC-CIE2023
August 20-23, 2023, Boston, Massachusetts

DETC2023-114966

INVESTIGATING THE EFFECT OF A BRAND FACTOR IN PRODUCT DESIGN BASED ON A DATA-DRIVEN APPROACH USING ONLINE REVIEWS

Seyoung Park, Harrison Kim[*]
Department of Industrial and Enterprise Systems Engineering
University of Illinois Urbana-Champaign
Urbana, Illinois 61801
Email: seyoung7@illinois.edu, hmkim@illinois.edu

ABSTRACT

Recently, online user-generated data has emerged as a valuable source for consumer product research. However, most studies have neglected the brand effect, although it is a significant factor in conventional market research. This paper demonstrates the importance of brands in data-driven design using online reviews. Specifically, the study utilizes game theory and suggests a game setting representing market competition. Elements of the game are determined based on online data analysis. The proposed approach consists of three stages. The first stage divides online customers into different segments and analyzes them to extract the feature importance of each brand in each segment. The importance is based on the term frequency of each feature, and it becomes the customer's partial utility for each feature. The second stage defines the specification of product candidates and calculates their costs. This study refers to real market datasets (Bill of Materials) available online. At this point, the game is all set. The final stage finds the Nash Equilibrium of the designed game and compares the optimal strategy for a product portfolio with and without brand consideration. The suggested approach was tested on smartphone reviews from Amazon. The result shows that the lack of brand consideration leads a company to choose a non-optimal product strategy, illustrating the significance of the brand factor.

---

[*]Corresponding Author

1 INTRODUCTION

A brand is a significant factor in today's marketplace. The company's success is greatly dependent on the power of its brand [1], which is measured by various indices such as brand awareness, brand image, brand preference, and brand relevance [2]. Na et al. [3] explained a brand power model as a combination of brand awareness and brand image. Aaker [2] pointed out that the most common basis of market competition is to win the brand preference battle [2]. A renowned American business magazine, Forbes, publishes the brand values of global companies every year [4]. Fig. 1 shows the list of the most valuable brands in 2020. The American Customer Satisfaction Index (ACSI), an influential survey published by the University of Michigan, reports satisfaction benchmarks for brands in various industry sectors [5].

In industry, companies consider their brand indices when developing new products. They usually hire a market research firm [6] and conduct surveys to obtain data, which quantifies the market player's brand power in various aspects. The company recognizes its market position based on the result and devises the strategy for its products. For example, in the automobile market, Toyota has strengths in the affordable price and design that appeals to young customers. On the other hand, Mercedes has strengths in that the brand has a touch of class, and its design appeals to older people [7]. Therefore, these two brands will target different customer segments and set up different strategies to maximize the value of their products for the target customers.

Copyright © 2023 by ASME

	Rank	Brand	Brand Value	1-Yr Value Change	Brand Revenue	Company Advertising	Industry
	#1	Apple	$241.2 B	17%	$260.2 B	--	Technology
	#2	Google	$207.5 B	24%	$145.6 B	$6.8 B	Technology
	#3	Microsoft	$162.9 B	30%	$125.8 B	$1.6 B	Technology
	#4	Amazon	$135.4 B	40%	$260.5 B	$11 B	Technology
	#5	Facebook	$70.3 B	-21%	$49.7 B	$1.6 B	Technology

FIGURE 1. Forbes 2020 Most Valuable Brands in the World [4]

Recently, many studies have been utilizing online user-generated data in their research. In data-driven design [8], these studies analyze online data to understand customers' preferences and draw design implications. The resultant implications include feature importance [9, 10], usage [11], spec guidance [12], and ideas for new features [13, 14]. However, the previous studies did not consider brand influence while it is a significant factor in the industry. They analyzed the whole product base assuming that there is no difference between brands, i.e., different brands have similar strengths and weaknesses. It is a significant gap between the industry and research.

This paper aims to demonstrate the brand effects in data-driven design by illustrating changes in product portfolio strategies. The rest of the paper is organized as follows. Section 2 reviews relevant studies and section 3 explains the details of the proposed methodology. In section 4, the methodology is tested on real-world datasets. Section 5 compares the resultant product portfolio with and without the brand factor and discusses the influence of the brand. Finally, section 6 summarizes the contribution of this research and discusses future works.

2 LITERATURE REVIEW
2.1 Data-Driven Design
Data-driven design means that the design is based on the use of data science algorithms supporting specific phases of the product development process [15]. Product designers can harness their organization's competitive edge by uncovering patterns and novel insights from huge and highly contextualized data [8]. Among various types of data sources in this area, online user-generated data has become a popular resource for consumer product design. Many studies utilize online data because of its strength in time and cost-efficiency compared to conventional methods such as surveys and interviews. There exist various approaches to extracting design implications from online data.

Chaklader & Parkinson [12] proposed a methodology to extract proper size specifications for headset products from online reviews. They selected reviews with positive sentiments for the product size and calculated the average rating of these reviews. The authors suggested proper specs by comparing this value with the average rating of total reviews.

Some studies focused on the different importance of product features. Suryadi & Kim [9] analyzed online reviews to understand the influence of product features on product sales ranking. First, they identified product features in data using Word2Vec [16] and clustering. Then, the authors quantified each reviewer's sentiment for the identified features. Each review has a set of {feature: sentiment score} pairs. The scores became input data to linear regression, and the output data is sales ranking. In the regression result, the coefficients indicate the effect of each product feature on sales ranking. Wang & Chen [17] studied the effect of product features on customers' purchase behavior. The authors generated choice sets using online user data and constructed MNL (multinomial logit) models [18]. The coefficients in the result show the influence of product attributes on the customer's purchase decision. Joung & Kim [10] suggested a methodology to identify the importance of product features based on review ratings. They extracted feature keywords from online reviews using LDA (Latent Dirichlet Allocation) [19] and analyzed each customer's sentiment for each feature. Then, the authors built a neural network model, where the input data is the sentiment scores, and the output data is the customer's rating for the product. By interpreting the trained model with SHAP (SHapley Additive exPlanations) [20], the authors obtained influence scores of product features on the review ratings.

Another approach is to discover new product features from user-generated data. Tuarob & Tucker [13] suggested a methodology to extract ideas for new features from social media data. They extracted ground-truth features from product spec documents and user-discussed features from Twitter data. Then, the authors identified latent features and detected lead users on Twitter based on these features. The suggested methodology discovered new smartphone features by analyzing the lead users' Twitter mentions. Goldberg & Abrahams [14] presented a method that sources product innovation ideas from online reviews. They adopted and revised the attribute mapping framework. It differentiates product attributes based on customer sentiments (positive/ negative/ neutral) and attribute types (basic/ discriminators/ energizers). The reviews mentioning product features were analyzed and assigned to one of the categories in the framework. The result suggested candidates for new product features and their priorities.

As shown above, the previous studies proposed various methods for extracting design implications from online user-generated data. However, they have a limitation since they disregarded a brand factor assuming the same characteristics for all products. This study incorporates the brand factor when analyzing the user data and shows the difference in product strategies made by brands.

Copyright © 2023 by ASME

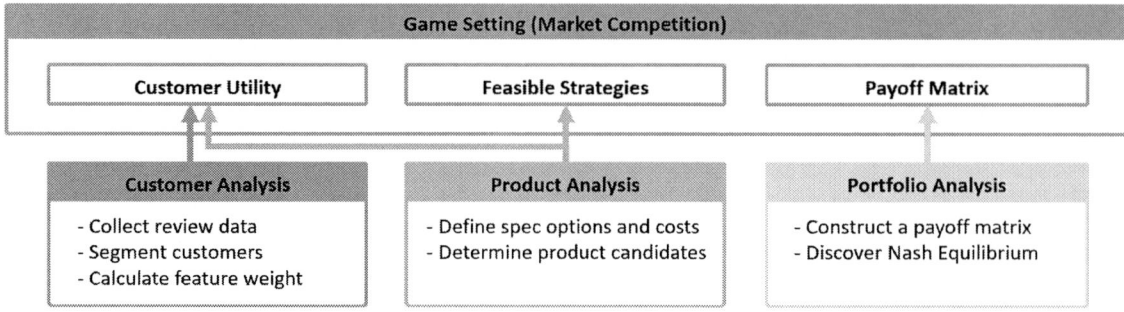

FIGURE 2. Overview of The Proposed Methodology

2.2 Brands in User-Generated Data

It can be questioned whether the brand effect is reflected in user-generated data. Regarding this question, some studies showed the existence of brand effects on user data by analyzing online sources. Jin et al. [21] analyzed Amazon reviews and extracted customer sentiments for mobile phone attributes. They compared products of different brands and analyzed whether one product is more favorable than a competitive one at the feature level. The result shows that each brand has different strengths and weaknesses in terms of product features. Nuortimo & Harkonen [22] suggested a method that extracts a brand index from text data on social media. The authors analyzed the sentiment in user opinions mentioning the target brand and computed the percentual share of negative opinions. They showed that different brands have different percentage values, and a lower percentage means a higher brand index. The resultant brand index was validated by the consistency with the Forbes brand index [4]. Alzate et al. [23] proposed a method that extracts brand image and brand positioning from online reviews. They analyzed the text data using the lexicon-based Linguistic Inquiry and Word Count (LIWC) program and clustered brands in the cosmetic industry. The resultant brand positioning map illustrates differences in four brand clusters.

Although the above studies showed that the brand factor is reflected in user-generated data, the influence of the brand in industrial applications has been rarely discussed. This study investigates how the brand factor affects product design or a company's strategy. Specifically, this paper compares the company's product portfolio with and without brand consideration and demonstrates the importance of brands in data-driven design.

3 METHODOLOGY

This study analyzes the influence of the brand in data-driven design, especially using online user-generated data. The game theory approach was adopted to compare the company's product portfolio with and without brand consideration. Fig. 2 shows an overview of the proposed methodology. This paper first intro-

TABLE 1. Game Settings

Players	Product	Cost	Utility (U_{ij}^b)				
			S_1	S_2	S_3	S_4	S_5
Brand 1	P_1^1	C_1^1					
	P_2^1	C_2^1					
	P_3^1	C_3^1					
	P_4^1	C_4^1					
Brand 2	P_1^2	C_1^2					
	P_2^2	C_2^2					
	P_3^2	C_3^2					
	P_4^2	C_4^2					

* P_i^b=Product candidate i for brand b, S_j=Customer segment j
* The utility values are calculated by Eq 1.

duces the game setting for market competition and then explains three stages corresponding to elements of the game setting. In the customer analysis stage, online user data is collected. The method partitions the collected customer base into segments and analyzes feature weights that each segment has for product features. The obtained data is used in calculating customer utility for product candidates. In the product analysis stage, available spec options and their costs are defined. Based on this data, the method determines product candidates with different spec configurations and costs. The result is used in customer utility computation and feasible strategy decisions. In the final stage, the method constructs a payoff matrix and determines the optimal product portfolio strategy by discovering Nash Equilibrium.

Regarding the market competition, the game setting in Ref. [24] is modified and used in this study. Table 1 shows the framework of the game setting. There are two players, and each player has four product candidates for their portfolios. For each product, the cost is required to calculate a customer's utility per unit price. It is assumed that customers in different segments have different utilities for the same product. The customer utility is calculated based on Eq. 1, where b, i, j, and k represent the brand, product, customer segment, and product feature respectively. w_{jk} is the weight (importance) that segment j has for product feature k. x_{ik}^b represents the spec value of feature k of

Copyright © 2023 by ASME

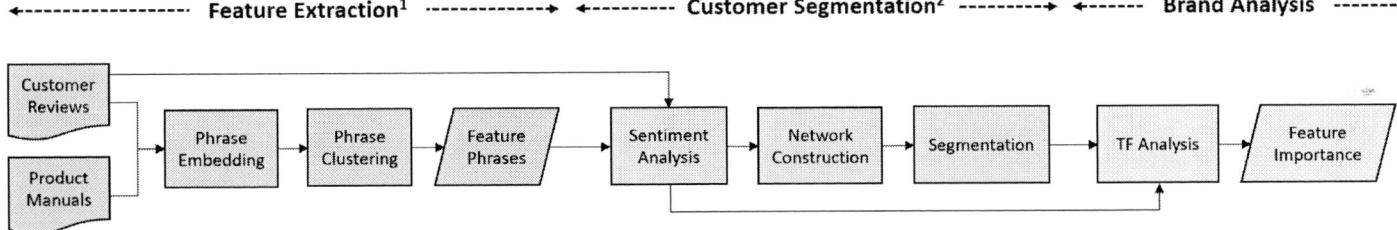

1. Park & Kim (a) (2022) [25], 2. Park & Kim (b). (2022) [26]

FIGURE 3. Flowchart of Customer Analysis

product i in brand b.

$$U_{ij}^b = \sum_{k=1}^{K} w_{jk}^b \times x_{ik}^b \qquad (1)$$

Filling out the utility column in Table 1 requires three types of datasets: i) segmented customers; ii) feature weights by brand in each segment; (iii) product spec configurations. Among them, i) and ii) belong to customer analysis, and iii) is determined in product analysis.

3.1 Customer Analysis

Fig. 3 shows the process of customer analysis, which consists of three stages. In the first stage, product features of customer interests are extracted from online user-generated data. Next, customers' sentiments for these features are analyzed, and customers with similar interests are grouped. In the final stage, feature weights for each brand are calculated based on the previously obtained sentiment analysis. The first and second stages are based on the authors' previous works. It is considered necessary to summarize the method while the details are available in Ref. [25, 26].

The first stage uses phrase embedding and clustering [25]. The words in the online review data are embedded into vectors by Word2Vec [16]. Next, the method extracts phrases from the review data and embeds the phrases into a vector space using word vectors and product manuals. Finally, the phrase vectors are grouped into clusters. The phrases in the feature-related clusters represent sub-features mentioned by customers.

The second stage adopts customer segmentation based on network analysis [26]. A customer's attribute is defined as the reviewer's sentiment for product features extracted from the previous stage. For example, let us assume a reviewer says *"Great phone for the **price**. This phone is easy to use and feels like an expensive smartphone despite the cheap track phone **price**. The **camera** is very nice and the interaction between commands is smooth."* The reviewer is satisfied with the price, so the attribute

value for the price+ (P+) is 1. Also, the review expresses positive sentiment for the camera feature. Therefore, camera+ (C+) becomes 1. Since no other features are mentioned in this review, the rest of the attributes have a value of 0, as shown below.

S+	S-	A+	A-	...	C+	C-	...	P+	P-
0	0	0	0	...	1	0	...	1	0

The method constructs a customer network by connecting reviewers with certain levels of commonality in interests. Then, network clustering is conducted to divide a customer base into several groups with similar characteristics. The optimal number of segments is automatically determined by modularity clustering [27, 28].

The final stage is brand analysis, in which feature weights for each brand are analyzed. The goal of this research is to demonstrate the influence of brands rather than to develop a new method for analyzing feature weights for different brands. Therefore, this study uses the term frequency (TF), a simple method for importance evaluation [29]. Specifically, the number of positive reviews for each feature is counted and normalized, as shown in Eq. 2. TF_{jl}^b represents the number of reviews expressing positive sentiments for a specific feature l of products in brand b among customers in segment j. In the same manner, w_{jl}^b is the feature weight that customers in segment j have for a specific feature l of products in brand b. The TF for a certain feature is zero in some segments. Therefore an offset of 1 is applied to all TF. The final value is the normalized TF for each feature with offset.

$$w_{jl}^b = \frac{(TF_{jl}^b + 1)}{\sum_{k=1}^{K}(TF_{jk}^b + 1)} \qquad (2)$$

3.2 Product Analysis

The remaining part of the game setting is product data. Table 1 requires spec configurations for product candidates (P_i^b)

Copyright © 2023 by ASME

and their costs (C_i^b). In the industry, available spec options for each feature are determined by the internal sourcing department of a company. The cost of each component is dependent on the estimated sales volume. Therefore, the data for product candidates may be different by brand (company). In this study, available spec options are defined based on the released product in the market. Then, this paper suggests three scenarios for product candidates.

Case 1: Two brands have the same product candidates.
Case 2: Two brands have the same product candidates with different spec configurations from Case 1.
Case 3: Two brands have different product candidates.

Once spec configurations are determined, the utility of each product candidate in each customer segment (U_{ij}^b) can be calculated based on Eq. 1.

3.3 Portfolio Analysis

This study aims to demonstrate the changes in the companies' optimal strategies when they consider a brand factor. A game theory approach is a proper way to mimic the competition of different brands. Therefore, the proposed method determines the best strategy for each brand by finding Nash Equilibrium (NE) in the payoff matrix. The process consists of three steps: (i) define feasible strategies; (ii) construct a payoff matrix; (iii) find NE. First, feasible strategies should be defined. For example, Brand 1 wants to release two new products this year. In this case, the feasible strategies are { (P_1^1, P_2^1), (P_1^1, P_3^1), (P_1^1, P_4^1), (P_2^1, P_3^1), (P_2^1, P_4^1), (P_3^1, P_4^1) }.

Next, the payoff matrix needs to be constructed. It shows the benefit of each brand when two brands choose certain strategies. This study adopts the function used in Ref [24] to calculate the payoff for both brands. In Eq. 3, $f_b(Z_x^1, Z_y^2)$ represents the expected shared surplus of brand b when two brands choose strategies x and y respectively. $\frac{U_{ij}^b}{C_i^b}$ is segment j's utility for product i of brand b per unit cost. The next term computes the market share that product i of barnd b has in segment j, where I_{com} is the number of competing products under (Z_x^1, Z_y^2) and μ is a scaling parameter. Q_j is the market size of segment j.

$$f_b(Z_x^1, Z_y^2) = \sum_{j=1}^{J} \sum_{i=1}^{I'} \left(\frac{U_{ij}^b}{C_i^b} \times \frac{e^{\mu U_{ij}^b}}{\sum_{c=1}^{I_{com}} e^{\mu U_{cj}}} \times Q_j \right) \quad (3)$$

Finally, NE in the payoff matrix is discovered. Eq. 4 shows the conditions for the NE. A is the payoff matrix for brand 1, and B is that for brand 2. x and y are unit vectors indicating strategy choice. The first equation means that the selected x gives the best payoff for brand 1 among all the possible strategies $\tilde{x}$. The second equation means the same condition for brand 2. If

---

Algorithm 1 Pure Nash Equilibrium

1: **for** $y = 1, 2, \ldots, N$ **do**
2: Find x that maximizes $f_1(Z_x^1, Z_y^2)$
3: Find k that maximizes $f_2(Z_x^1, Z_k^2)$
4: **if** $k = y$ **then**
5:   Print (x, y)
6: **end if**
7: **end for**

---

the size of the payoff matrix is small, NE can be found manually. However, finding NE becomes complex when the matrix size gets large. This study implements a simple algorithm to identify NE efficiently. The strategy discussed here is the product portfolio, so only Pure Nash Equilibrium (PNE) is applicable. The pseudo-code for finding PNE is shown in Algorithm 1. Z_a^b indicates brand b chooses strategy a, and the algorithm prints out the optimal strategy (x, y), which leads to the PNE.

$$x^T A y \geq \tilde{x}^T A y, \quad \forall \tilde{x} \in S_1$$
$$x^T B y \geq x^T B \tilde{y}, \quad \forall \tilde{y} \in S_2 \quad (4)$$

4 CASE STUDY

The proposed methodology was tested on real-world data. A smartphone was chosen as the target product because the market fits the game setting presented in Section 3. Specifically, the US smartphone market is dominated by two major brands - Apple and Samsung [30]. Also, the Bill of Materials (BoM) can be easily obtained online [31].

4.1 Customer Analysis

In this section, the customer analysis in Fig 3 was tested on smartphone review data, and the result is presented.

4.1.1 Data Collection and Preprocessing
The review data was collected from Amazon.com. Among the top 100 items in the cellphone category, 85 products were selected (feature phones and duplicated items were filtered out). The data contains 44,691 reviews written from July 10, 2017 to March 24, 2022. Only the reviews marked as 'Verified Purchase' and written in the United States were considered. The collected reviews went through preprocessing. Symbols, numbers, and punctuation marks except a period were removed. Upper cases were converted to lower cases, and then all the words in the review data were lemmatized.

4.1.2 Feature Extraction
Product features of customer interests were extracted from the collected review data

Copyright © 2023 by ASME

TABLE 2. Cue Phrases for Product Features

Screen	screen display, screen size, inch display, screen resolution, screen brightness, screen sensitivity, screen ratio, oled screen, etc.
Processor	fast processor, slow processor, snapdragon processor, exynos processor, process speed, processing speed
Memory	gb memory, storage capacity, internal memory, more memory, extra memory, expandable memory, gb ram, extra storage, etc.
Camera	front camera, selfie camera, rear camera, main camera, mp camera, camera lens, camera quality, camera, camera shutter, etc.
Battery	battery capacity, mah battery, battery charge, battery life, battery percentage, battery saver, battery health, removable battery, etc.
Unlock	fingerprint reader, fingerprint sensor, fingerprint scanner, finger reader, iris scanner, face recognition, facial recognition, etc.
Price	price range, price difference, price tag, decent price, affordable price, awesome price, perfect price, cheap price, half price, etc.

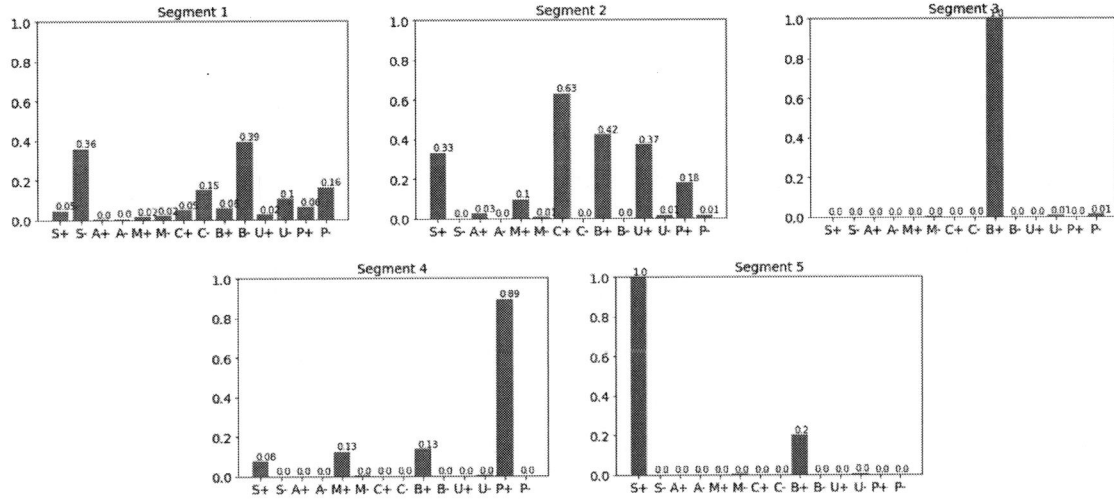

FIGURE 4. Properties of Customer Segments

based on the methodology in Ref [25]. In this case study, seven feature categories were detected: screen, application processor (AP), memory, camera, battery, unlock, and price. Table 2 shows the corresponding cue phrases for each feature category.

4.1.3 Customer Segmentation Next, customers' interests and sentiments for product features were analyzed based on these cue phrases. Reviews not mentioning any product features were removed. The number of filtered customers was 13,961. Then, people with similar interests were connected by the networking rule presented in Ref [26]. Modularity clustering was applied to this network and divided the customer base into different segments. Fig. 4 shows the characteristics of each segment. The x-axis indicates the feature and sentiment. For example, S+ means the positive sentiment for the Screen feature, and B- means the negative sentiment for the Battery feature. The y-axis represents the percentage of customers expressing each sentiment for features. People in different segments have different properties, i.e., interests and sentiments for product features. Specifically, customers in segment 1 have complaints about overall features. On the other hand, customers in segment 2 are satisfied with most features. In segment 3, people are interested in

the battery feature only. In segment 4, most people care about the price. Customers in segment 5 have a high interest in the screen.

4.1.4 Brand Analysis Since segments have different properties, it can be inferred that they have different partial utilities, i.e., the utility for each feature category. Table 3 shows the partial utility calculated by Eq. 2. Since the US smartphone market is dominated by Apple and Samsung, this study analyzed the partial utility of these two brands. The third item in the brand column is 'Total', which means the partial utility when the brand factor is disregarded. The result shows that partial utilities vary by brand. Moreover, utilities with and without the brand factor are not the same.

4.2 Product Analysis

In the game setting of Table 1, each brand has four product candidates. Therefore, four sets of spec configurations need to be determined. As a preliminary work for this, available spec values for each product feature were defined, as shown in Table 4. For the simplicity of the simulation, this study considers one sub-feature for each feature category. The screen size, AP speed, memory ROM, number of rear cameras, battery capacity, unlock

Copyright © 2023 by ASME

TABLE 3. Partial Utility by Segment

Brand	Segment	Screen	AP	Memory	Camera	Battery	Unlock	Price
Apple	1	0.205	0.004	0.043	0.137	0.258	0.096	0.258
	2	0.179	0.003	0.038	0.276	0.229	0.199	0.077
	3	0.001	0.001	0.001	0.001	0.997	0.001	0.001
	4	0.058	0.001	0.093	0.001	0.122	0.001	0.725
	5	0.825	0.001	0.001	0.001	0.172	0.001	0.001
Samsung	1	0.195	0.017	0.078	0.212	0.174	0.084	0.238
	2	0.153	0.019	0.055	0.331	0.175	0.173	0.094
	3	0.002	0.002	0.002	0.002	0.988	0.002	0.002
	4	0.070	0.001	0.105	0.001	0.074	0.001	0.747
	5	0.840	0.002	0.002	0.002	0.147	0.002	0.002
Total	1	0.181	0.011	0.060	0.189	0.221	0.096	0.242
	2	0.161	0.013	0.048	0.307	0.204	0.180	0.087
	3	0.000	0.000	0.000	0.000	0.998	0.000	0.000
	4	0.061	0.000	0.102	0.000	0.110	0.000	0.725
	5	0.830	0.001	0.001	0.001	0.167	0.001	0.001

* The utility values are summed up to 1 for each row.

TABLE 4. Spec Options for Product Features

Feature	Spec values				Cost ($)			
	Option 1	Option 2	Option 3	Option 4	Option 1	Option 2	Option 3	Option 4
Screen size	5.6	6.0	6.4	6.9	61.5	71.5	81.5	91.5
AP speed	2.8	2.9	3.0	3.1	47.0	52.0	57.0	62.0
Memory ROM	64	128	256	512	56.5	61.5	66.5	71.5
Camera count	1	2	3	4	40.0	50.0	60.0	70.0
Battery capacity	3600	4000	4500	5000	4.5	7.2	9.9	12.6
Unlock type	0	1	2	3	10.0	13.0	16.0	19.0
Price	0	1	2	3	-	-	-	-

TABLE 5. Game Settings with Data (Case 1)

Players		Product (P_i^b)							Cost	Utility (U_{ij}^b)				
		S	A	M	C	B	U	P		S_1	S_2	S_3	S_4	S_5
Apple	P_1^1	6.4	3.1	512	4	3600	1	2	299.5	0.484	0.479	0.002	0.616	0.514
	P_2^1	5.6	2.8	256	2	4500	3	2	250.9	0.465	0.439	0.639	0.605	0.111
	P_3^1	6.0	2.8	64	1	5000	3	4	243.6	0.385	0.418	0.997	0.140	0.428
	P_4^1	6.9	2.8	64	1	4500	1	3	254.9	0.455	0.351	0.639	0.375	0.935
Samsung	P_1^2	6.4	3.1	512	4	3600	1	2	299.5	0.589	0.562	0.008	0.651	0.530
	P_2^2	5.6	2.8	256	2	4500	3	2	250.9	0.432	0.424	0.637	0.595	0.100
	P_3^2	6.0	2.8	64	1	5000	3	4	243.6	0.291	0.339	0.990	0.097	0.410
	P_4^2	6.9	2.8	64	1	4500	1	3	254.9	0.385	0.296	0.635	0.364	0.935

type, and price level. The costs of spec options were estimated based on the BoM of Samsung smartphones [31, 32].

This study configured the specifications of product candidates (P_i^b) in Table 1 by randomly selecting spec options from Table 4. The three scenarios mentioned in Section 3.2 were tested based on the data obtained in the previous sections. Specifically, Table 5 shows the game setting for Case 1. The utility by segment (U_{ij}^b) was filled out based on the partial utility from Section 4.1 and the specs of product candidates. Since each feature category has a different scale, the spec values were normalized by the min_max_scaling. In other words, the spec values range from 0 to 1. For example, the screen size of 6.9" becomes 1, and

the size of 5.6" becomes 0. These scaled values were the input data for x_{ik}^b in Eq. 1. Because higher prices decrease customer utility, the price data was converted to (1-P) and plugged into the utility function. Table 5 shows the calculated customer utility by segment. The result demonstrates that customer utility varies by brand for the same product.

4.3 Product Portfolio Strategy

In this study, the feasible strategies were defined as 'up to 3 products out of 4 candidates'. Therefore, each brand has 14 strategies as follows. Z_a^b indicates the ath strategy for brand b.

Copyright © 2023 by ASME

$$
\begin{array}{ll}
Z_1^1 = \{P_1^1\} & Z_1^2 = \{P_1^2\} \\
Z_2^1 = \{P_2^1\} & Z_2^2 = \{P_2^2\} \\
Z_3^1 = \{P_3^1\} & Z_3^2 = \{P_3^2\} \\
Z_4^1 = \{P_4^1\} & Z_4^2 = \{P_4^2\} \\
Z_5^1 = \{P_1^1, P_2^1\} & Z_5^2 = \{P_1^2, P_2^2\} \\
Z_6^1 = \{P_1^1, P_3^1\} & Z_6^2 = \{P_1^2, P_3^2\} \\
Z_7^1 = \{P_1^1, P_4^1\} & Z_7^2 = \{P_1^2, P_4^2\} \\
Z_8^1 = \{P_2^1, P_3^1\} & Z_8^2 = \{P_2^2, P_3^2\} \\
Z_9^1 = \{P_2^1, P_4^1\} & Z_9^2 = \{P_2^2, P_4^2\} \\
Z_{10}^1 = \{P_3^1, P_4^1\} & Z_{10}^2 = \{P_3^2, P_4^2\} \\
Z_{11}^1 = \{P_1^1, P_2^1, P_3^1\} & Z_{11}^2 = \{P_1^2, P_2^2, P_3^2\} \\
Z_{12}^1 = \{P_1^1, P_2^1, P_4^1\} & Z_{12}^2 = \{P_1^2, P_2^2, P_4^2\} \\
Z_{13}^1 = \{P_1^1, P_3^1, P_4^1\} & Z_{13}^2 = \{P_1^2, P_3^2, P_4^2\} \\
Z_{14}^1 = \{P_2^1, P_3^1, P_4^1\} & Z_{14}^2 = \{P_2^2, P_3^2, P_4^2\}
\end{array}
$$

The payoff for each feasible strategy was calculated by Eq. 3, and the resultant payoff matrix is shown in Fig. 5. Algorithm 1 for finding PNE was implemented in Python and applied to this payoff matrix. The PNE is highlighted in Fig. 5.

5 RESULTS & DISCUSSION

The goal of this study is to demonstrate the influence of the brand factor in data-driven design using online user data. For this, the study compares the company's product portfolio with and without brand consideration. The baseline model is the game setting with the partial utility of 'Total' brands in Table 3. In this setting, two brands (Apple and Samsung) have the same importance for each product feature. The comparative model is a new game setting with partial utility reflecting brand effects. Specifically, the partial utility of 'Apple' and 'Samsung' in Table 3 is used. In this setting, the two brands have different weights for product features. The payoff matrices for the baseline and comparative models are constructed, and then the PNE of the two models are compared. As mentioned in Section 4.2, this study tests three cases.

5.1 Case 1

In the first case, two brands have identical product candidates ($P_i^1 = P_i^2$). The payoff matrices with and without a brand factor are shown in Fig. 5 where PNE is highlighted in yellow. When the brand factor is considered, PNE is (Z_{14}^1, Z_{12}^2). Therefore, the product portfolio for Apple is $\{P_2^1, P_3^1, P_4^1\}$ and the product strategy for Samsung is $\{P_1^2, P_2^2, P_4^2\}$. On the other hand, when a brand factor is disregarded, PNE is (Z_{14}^1, Z_{14}^2). Samsung's portfolio is changed to $\{P_2^2, P_3^2, P_4^2\}$. This study assumes that the payoff with the brand factor is the true one. Therefore, the lack of consideration for the brand effect leads to Samsung choosing a strategy that is not optimal. In specific, Samsung chooses P_{14}^2 with a payoff of 0.092 instead of P_{12}^2 with a payoff of 0.093.

5.2 Case 2

In the second case, two brands again have the same product candidates ($P_i^1 = P_i^2$) but with spec configurations different

w/ Brand	1	2	3	4	5	6	7	8	9	10	11	12	13	14
1	(0.069, 0.083)	(0.074, 0.095)	(0.078, 0.093)	(0.073, 0.098)	(0.047, 0.121)	(0.049, 0.125)	(0.047, 0.124)	(0.051, 0.125)	(0.049, 0.130)	(0.051, 0.126)	(0.037, 0.140)	(0.036, 0.141)	(0.037, 0.142)	(0.038, 0.143)
2	(0.097, 0.085)	(0.095, 0.090)	(0.098, 0.087)	(0.098, 0.096)	(0.064, 0.119)	(0.065, 0.122)	(0.065, 0.123)	(0.065, 0.121)	(0.065, 0.127)	(0.066, 0.123)	(0.048, 0.137)	(0.048, 0.140)	(0.034, 0.103)	(0.049, 0.140)
3	(0.104, 0.089)	(0.100, 0.092)	(0.097, 0.082)	(0.099, 0.094)	(0.070, 0.121)	(0.067, 0.120)	(0.069, 0.123)	(0.066, 0.119)	(0.067, 0.127)	(0.066, 0.120)	(0.051, 0.137)	(0.052, 0.140)	(0.051, 0.138)	(0.050, 0.139)
4	(0.104, 0.085)	(0.105, 0.092)	(0.104, 0.085)	(0.100, 0.091)	(0.071, 0.120)	(0.069, 0.120)	(0.068, 0.121)	(0.071, 0.120)	(0.069, 0.125)	(0.068, 0.120)	(0.053, 0.137)	(0.052, 0.138)	(0.052, 0.138)	(0.052, 0.138)
5	(0.114, 0.057)	(0.114, 0.062)	(0.117, 0.062)	(0.115, 0.066)	(0.085, 0.090)	(0.086, 0.093)	(0.086, 0.098)	(0.087, 0.093)	(0.086, 0.097)	(0.088, 0.095)	(0.069, 0.111)	(0.068, 0.113)	(0.069, 0.113)	(0.070, 0.114)
6	(0.123, 0.058)	(0.121, 0.062)	(0.121, 0.059)	(0.121, 0.064)	(0.092, 0.091)	(0.091, 0.092)	(0.091, 0.098)	(0.091, 0.091)	(0.091, 0.096)	(0.091, 0.093)	(0.073, 0.111)	(0.073, 0.112)	(0.072, 0.112)	(0.072, 0.113)
7	(0.119, 0.057)	(0.120, 0.063)	(0.121, 0.061)	(0.118, 0.063)	(0.090, 0.090)	(0.090, 0.092)	(0.088, 0.092)	(0.092, 0.093)	(0.090, 0.096)	(0.090, 0.093)	(0.073, 0.111)	(0.071, 0.112)	(0.072, 0.112)	(0.073, 0.113)
8	(0.133, 0.059)	(0.132, 0.061)	(0.131, 0.056)	(0.132, 0.064)	(0.100, 0.090)	(0.099, 0.090)	(0.101, 0.098)	(0.099, 0.089)	(0.099, 0.095)	(0.099, 0.091)	(0.080, 0.109)	(0.080, 0.112)	(0.080, 0.111)	(0.080, 0.111)
9	(0.135, 0.057)	(0.135, 0.061)	(0.136, 0.058)	(0.134, 0.062)	(0.102, 0.090)	(0.102, 0.090)	(0.101, 0.091)	(0.102, 0.090)	(0.101, 0.094)	(0.102, 0.091)	(0.082, 0.109)	(0.081, 0.111)	(0.081, 0.111)	(0.082, 0.111)
10	(0.137, 0.059)	(0.137, 0.062)	(0.135, 0.055)	(0.134, 0.061)	(0.105, 0.091)	(0.103, 0.090)	(0.103, 0.092)	(0.103, 0.089)	(0.102, 0.094)	(0.101, 0.089)	(0.083, 0.109)	(0.083, 0.111)	(0.082, 0.110)	(0.082, 0.110)
11	(0.140, 0.044)	(0.139, 0.046)	(0.140, 0.044)	(0.139, 0.049)	(0.112, 0.072)	(0.112, 0.073)	(0.112, 0.075)	(0.112, 0.073)	(0.111, 0.077)	(0.112, 0.075)	(0.093, 0.092)	(0.093, 0.094)	(0.093, 0.093)	(0.093, 0.094)
12	(0.139, 0.043)	(0.140, 0.046)	(0.141, 0.046)	(0.139, 0.048)	(0.112, 0.072)	(0.112, 0.073)	(0.111, 0.074)	(0.113, 0.074)	(0.111, 0.076)	(0.112, 0.075)	(0.093, 0.092)	(0.093, 0.093)	(0.093, 0.093)	(0.094, 0.094)
13	(0.143, 0.044)	(0.143, 0.047)	(0.143, 0.044)	(0.142, 0.047)	(0.115, 0.073)	(0.114, 0.073)	(0.114, 0.074)	(0.115, 0.073)	(0.114, 0.076)	(0.114, 0.073)	(0.096, 0.092)	(0.095, 0.093)	(0.095, 0.092)	(0.095, 0.093)
14	(0.152, 0.044)	(0.152, 0.046)	(0.152, 0.042)	(0.151, 0.047)	(0.123, 0.072)	(0.122, 0.072)	(0.122, 0.074)	(0.122, 0.071)	(0.122, 0.075)	(0.121, 0.072)	(0.102, 0.091)	(0.102, 0.093)	(0.102, 0.092)	(0.102, 0.092)

w/o Brand	1	2	3	4	5	6	7	8	9	10	11	12	13	14
1	(0.076, 0.076)	(0.079, 0.097)	(0.083, 0.100)	(0.079, 0.100)	(0.052, 0.118)	(0.053, 0.124)	(0.052, 0.120)	(0.055, 0.130)	(0.053, 0.132)	(0.055, 0.132)	(0.040, 0.140)	(0.039, 0.140)	(0.040, 0.142)	(0.041, 0.148)
2	(0.097, 0.079)	(0.093, 0.093)	(0.095, 0.094)	(0.096, 0.099)	(0.063, 0.117)	(0.064, 0.121)	(0.065, 0.121)	(0.063, 0.127)	(0.063, 0.131)	(0.064, 0.129)	(0.047, 0.139)	(0.048, 0.139)	(0.034, 0.104)	(0.048, 0.146)
3	(0.100, 0.083)	(0.094, 0.095)	(0.090, 0.090)	(0.092, 0.097)	(0.066, 0.119)	(0.063, 0.121)	(0.066, 0.121)	(0.062, 0.126)	(0.063, 0.131)	(0.061, 0.127)	(0.048, 0.138)	(0.049, 0.140)	(0.048, 0.140)	(0.046, 0.145)
4	(0.100, 0.079)	(0.099, 0.096)	(0.097, 0.092)	(0.095, 0.095)	(0.068, 0.118)	(0.066, 0.121)	(0.065, 0.119)	(0.066, 0.127)	(0.065, 0.129)	(0.064, 0.127)	(0.050, 0.139)	(0.050, 0.138)	(0.049, 0.139)	(0.049, 0.145)
5	(0.118, 0.052)	(0.117, 0.063)	(0.119, 0.066)	(0.118, 0.068)	(0.088, 0.088)	(0.089, 0.092)	(0.088, 0.091)	(0.089, 0.097)	(0.088, 0.099)	(0.089, 0.100)	(0.071, 0.112)	(0.070, 0.112)	(0.071, 0.114)	(0.071, 0.119)
6	(0.124, 0.053)	(0.121, 0.064)	(0.121, 0.063)	(0.121, 0.066)	(0.092, 0.089)	(0.091, 0.091)	(0.092, 0.091)	(0.090, 0.096)	(0.090, 0.099)	(0.090, 0.097)	(0.073, 0.111)	(0.073, 0.112)	(0.073, 0.113)	(0.072, 0.117)
7	(0.120, 0.052)	(0.121, 0.065)	(0.121, 0.066)	(0.119, 0.065)	(0.091, 0.088)	(0.091, 0.092)	(0.089, 0.089)	(0.091, 0.098)	(0.090, 0.098)	(0.090, 0.098)	(0.073, 0.112)	(0.072, 0.111)	(0.072, 0.113)	(0.073, 0.118)
8	(0.130, 0.055)	(0.127, 0.063)	(0.126, 0.062)	(0.127, 0.066)	(0.097, 0.089)	(0.096, 0.090)	(0.098, 0.091)	(0.095, 0.095)	(0.095, 0.098)	(0.095, 0.097)	(0.077, 0.111)	(0.078, 0.112)	(0.077, 0.112)	(0.076, 0.117)
9	(0.132, 0.053)	(0.131, 0.063)	(0.131, 0.063)	(0.129, 0.065)	(0.099, 0.088)	(0.099, 0.090)	(0.098, 0.090)	(0.098, 0.095)	(0.097, 0.097)	(0.097, 0.096)	(0.079, 0.111)	(0.079, 0.111)	(0.078, 0.112)	(0.078, 0.116)
10	(0.132, 0.055)	(0.129, 0.064)	(0.127, 0.061)	(0.127, 0.064)	(0.100, 0.089)	(0.097, 0.090)	(0.098, 0.090)	(0.097, 0.095)	(0.096, 0.097)	(0.095, 0.095)	(0.079, 0.111)	(0.079, 0.111)	(0.077, 0.111)	(0.077, 0.116)
11	(0.140, 0.040)	(0.139, 0.047)	(0.138, 0.048)	(0.139, 0.050)	(0.112, 0.071)	(0.111, 0.073)	(0.112, 0.073)	(0.111, 0.077)	(0.111, 0.079)	(0.111, 0.079)	(0.093, 0.093)	(0.093, 0.093)	(0.093, 0.094)	(0.092, 0.098)
12	(0.140, 0.039)	(0.139, 0.048)	(0.140, 0.049)	(0.138, 0.050)	(0.112, 0.070)	(0.112, 0.073)	(0.111, 0.072)	(0.112, 0.078)	(0.111, 0.079)	(0.111, 0.079)	(0.093, 0.093)	(0.093, 0.093)	(0.093, 0.094)	(0.093, 0.098)
13	(0.142, 0.040)	(0.141, 0.048)	(0.140, 0.048)	(0.139, 0.049)	(0.114, 0.071)	(0.113, 0.073)	(0.113, 0.072)	(0.112, 0.077)	(0.112, 0.078)	(0.111, 0.077)	(0.094, 0.093)	(0.094, 0.093)	(0.093, 0.093)	(0.093, 0.097)
14	(0.148, 0.041)	(0.146, 0.048)	(0.145, 0.046)	(0.145, 0.049)	(0.119, 0.071)	(0.117, 0.072)	(0.118, 0.073)	(0.117, 0.076)	(0.116, 0.078)	(0.116, 0.077)	(0.098, 0.092)	(0.098, 0.093)	(0.097, 0.093)	(0.097, 0.097)

FIGURE 5. Payoff Matrix - Case 1

TABLE 6. Game Settings with Data (Case 2)

Brand		S	A	M	C	B	U	P
Apple	P_1^1	6.4	3.1	256	3	3600	1	4
	P_2^1	6.9	2.8	128	1	4500	2	1
	P_3^1	5.6	2.9	256	2	4500	3	3
	P_4^1	6.0	2.8	64	1	5000	3	1
Samsung	P_1^2	6.4	3.1	256	3	3600	1	4
	P_2^2	6.9	2.8	128	1	4500	2	1
	P_3^2	5.6	2.9	256	2	4500	3	3
	P_4^2	6.0	2.8	64	1	5000	3	1

TABLE 7. Game Settings with Data (Case 3)

Brand		S	A	M	C	B	U	P
Apple	P_1^1	6.4	2.9	256	3	4000	1	4
	P_2^1	5.6	3.0	128	1	4500	3	2
	P_3^1	6.0	2.8	256	2	4500	3	3
	P_4^1	6.9	3.1	64	1	5000	3	1
Samsung	P_1^2	6.4	2.9	256	3	3600	1	4
	P_2^2	5.6	3.0	128	1	4500	3	2
	P_3^2	6.0	2.8	256	2	4500	3	3
	P_4^2	6.9	3.1	64	1	5000	3	1

from Case 1. The details of the game setting are shown in Table 6. Considering the brand factor, PNE is (Z_{14}^1, Z_{13}^2). Therefore, the product portfolio for Apple is $\{P_1^1, P_3^1, P_4^1\}$ and the product strategy for Samsung is $\{P_2^2, P_3^2, P_4^2\}$. When the brand factor is not considered, PNE is (Z_{13}^1, Z_{13}^2). In this case, Apple's product portfolio is changed to $\{P_1^1, P_3^1, P_4^1\}$. Therefore, disregarding the brand effect leads to Apple selecting a non-optimal strategy. Apple chooses P_{13}^1 with a payoff of 0.096 rather than P_{14}^1 with a payoff of 0.098.

5.3 Case 3

In Case 3, two brands have different product candidates ($P_i^1 \neq P_i^2$ for some i). The game setting is described in Table 7. With the brand factor, PNE is (Z_{12}^1, Z_{12}^2). Therefore, the product portfolio for Apple is $\{P_1^1, P_2^1, P_4^1\}$ and the product strategy for Samsung is $\{P_1^2, P_2^2, P_4^2\}$. Without the brand factor, PNE is (Z_{12}^1, Z_{14}^2), and Samsung's portfolio is changed to $\{P_2^2, P_3^2, P_4^2\}$. Disregarding brands, Samsung results in choosing P_{14}^2 with a payoff of 0.093 instead of P_{12}^2 with payoff of 0.095.

6 CONCLUSION & FUTURE WORKS

This study focused on the neglected brand effect in data-driven design based on online user-generated data. Online data has been a popular resource for customer analysis due to its strength in time and cost-efficiency compared to conventional data collection methods such as surveys and interviews. However, previous studies utilizing online data disregarded brand effects while it is a significant factor in the industry. In the field, companies research various brand indexes to identify their strengths and weaknesses and devise proper strategies for market competition. Therefore, the brand factor needs to be taken into account in relevant research.

This paper proposed a game theory-based approach to investigate the influence of the brand in product strategy based on user-generated online data. The approach consists of three stages: (i) customer analysis, (ii) product analysis, and (iii) product portfolio analysis. In the first stage, the customer base was divided into segments based on the online review data. Then, the method analyzed each segment's partial utility for product

features. In the second stage, the method defined spec options for each feature category and determined spec configurations for product candidates. Finally, the game setting in Table 1 that represents market competition was filled in based on the results from (i) and (ii). The feasible strategies and corresponding payoffs were established, and Pure Nash Equilibrium (PNE) was discovered. This study compared the resultant PNE with and without brand consideration. As discussed in Section 5, disregarding brand effects resulted in a company choosing a non-optimal strategy for its product portfolio. In all three cases presented in this paper, the brand factor altered PNE, the optimal strategy for companies. These results demonstrate the importance of the brand factor in data-driven design using online data. The proposed method can be applied to other product domains, such as laptops and headphones. The review data for these products are available online, and the market is dominated by a few brands. The result helps companies devise optimal strategies for their product line-up by reflecting market competition in the real world.

The limitation of this paper is that it considers the competition of two brands only. Finding NE becomes more complex when more than two players exist in the game setting. Also, this study did not discuss the case where PNE does not exist. In future works, more diverse cases will be tested, and the cases with no PNE will be discussed. Also, the partial utility obtained in Section 3.1 can be applied to design applications other than the product portfolio. For example, companies can adopt this result for designing a new product that gives maximum customer utility. The previous studies utilizing online data provided various design applications, and the feature weight by brand obtained in this study can be applied to those applications. The new application can further demonstrate the importance of the brand factor.

NOMENCLATURE

U_{ij}^b Utility that segment j has for product i of brand b

w_{jk}^b Weight that segment j has for feature k in brand b

x_{ik}^b Spec value for feature k of product i of brand b

TF_{jk}^b Term Frequency of feature k among segment j in brand b

Z_x^1 Strategy x of brand 1

Z_y^2 Strategy y of brand 2

Copyright © 2023 by ASME

$f_b(Z_x^1, Z_y^2)$ Payoff for brand b with strategy (x, y)
C_i^b Cost of product i of brand b
I_{com} Number of competing products under (Z_x^1, Z_y^2)
μ Scaling parameter
Q_j Market size of segment j

REFERENCES

[1] Oh, T. T., Keller, K. L., Neslin, S. A., Reibstein, D. J., and Lehmann, D. R., 2020. "The past, present, and future of brand research". *Marketing Letters, 31*(2), pp. 151–162.

[2] Aaker, D. A., 2012. "Win the brand relevance battle and then build competitor barriers". *California Management Review, 54*(2), pp. 43–57.

[3] Na, W. B., Marshall, R., and Keller, K. L., 1999. "Measuring brand power: validating a model for optimizing brand equity". *Journal of product & brand management*.

[4] Forbes, 2020. "The world's most valuable brands". *https://www.forbes.com/powerful-brands/list/3/*.

[5] Businesswire, 2022. "Hp sets its sights on apple in the pc industry, customers remain dissatisfied with household appliances, acsi data show". *https://www.businesswire.com/news/home/20210921005181/en*.

[6] Help, S. T., 2022. "Top 10 market research companies". *https://www.softwaretestinghelp.com/market-research-companies/*.

[7] Rekettye, G., and Liu, J., 2001. "Segmenting the hungarian automobile market brand using perceptual and value mapping". *Journal of Targeting, Measurement and Analysis for Marketing, 9*(3), pp. 241–253.

[8] Kim, H. H. M., Liu, Y., Wang, C. C., and Wang, Y., 2017. "Data-driven design (d3)". *Journal of Mechanical Design, 139*(11).

[9] Suryadi, D., and Kim, H., 2018. "A systematic methodology based on word embedding for identifying the relation between online customer reviews and sales rank". *Journal of Mechanical Design, 140*(12), Dec.

[10] Joung, J., and Kim, H. M., 2021. "Automated keyword filtering in latent dirichlet allocation for identifying product attributes from online reviews". *Journal of Mechanical Design, 143*(8).

[11] Yang, B., Liu, Y., Liang, Y., and Tang, M., 2019. "Exploiting user experience from online customer reviews for product design". *International Journal of Information Management, 46*, Jun, pp. 173–186.

[12] Chaklader, R., and Parkinson, M. B., 2017. "Data-driven sizing specification utilizing consumer text reviews". *Journal of Mechanical Design, 139*(11).

[13] Tuarob, S., and Tucker, C. S., 2015. "Automated discovery of lead users and latent product features by mining large scale social media networks". *Journal of Mechanical Design, 137*(7).

[14] Goldberg, D. M., and Abrahams, A. S., 2022. "Sourcing product innovation intelligence from online reviews". *Decision Support Systems, 157*, p. 113751.

[15] Bertoni, A., 2020. "Data-driven design in concept development: systematic review and missed opportunities". In Proceedings of the Design Society: DESIGN Conference, Vol. 1, Cambridge University Press, pp. 101–110.

[16] Mikolov, T., Sutskever, I., Chen, K., Corrado, G. S., and Dean, J., 2013. "Distributed representations of words and phrases and their compositionality". *Advances in neural information processing systems, 26*.

[17] Wang, M., and Chen, W., 2015. "A data-driven network analysis approach to predicting customer choice sets for choice modeling in engineering design". *Journal of Mechanical Design, 137*(7), p. 071410.

[18] Chen, W., Hoyle, C., and Wassenaar, H. J., 2012. *Decision-Based Design*. Springer.

[19] Blei, D. M., Ng, A. Y., and Jordan, M. I., 2003. "Latent dirichlet allocation". *Journal of machine Learning research*, Jan.

[20] Lundberg, S. M., and Lee, S., 2017. "A unified approach to interpreting model predictions". *Advances in Neural Information Processing Systems 30*, pp. 4765–4774.

[21] Jin, J., Liu, Y., Ji, P., and Liu, H., 2016. "Understanding big consumer opinion data for market-driven product design". *International Journal of Production Research, 54*(10), pp. 3019–3041.

[22] Nuortimo, K., and Harkonen, J., 2019. "Establishing an automated brand index based on opinion mining: analysis of printed and social media". *Journal of Marketing Analytics, 7*(3), pp. 141–151.

[23] Alzate, M., Arce-Urriza, M., and Cebollada, J., 2022. "Mining the text of online consumer reviews to analyze brand image and brand positioning". *Journal of Retailing and Consumer Services, 67*, p. 102989.

[24] Sadeghi, A., and Zandieh, M., 2011. "A game theory-based model for product portfolio management in a competitive market". *Expert Systems with Applications, 38*(7), pp. 7919–7923.

[25] Park, S., and Kim, H. M., 2022. "Phrase embedding and clustering for sub-feature extraction from online data". *Journal of Mechanical Design, 144*(5).

[26] Park, S., and Kim, H. M., 2022. "Finding social networks among online reviewers for customer segmentation". *Journal of Mechanical Design, 144*(12), p. 121703.

[27] Blondel, V. D., Guillaume, J.-L., Lambiotte, R., and Lefebvre, E., 2008. "Fast unfolding of communities in large networks". *Journal of statistical mechanics: theory and experiment, 2008*(10), p. P10008.

[28] Dinh, T. N., Li, X., and Thai, M. T., 2015. "Network clustering via maximizing modularity: Approximation algorithms and theoretical limits". In 2015 IEEE International Conference on Data Mining.

[29] Kim, J., Park, S., and Kim, H. M., 2022. "Optimal modular remanufactured product configuration and harvesting planning for end-of-life products". *Journal of Mechanical Design, 144*(4).

[30] Counterpoint, T., 2022. "Us smartphone market share: By quarter". *https://www.counterpointresearch.com/us-market-smartphone-share/*.

[31] GSMArena, 2020. "Samsung galaxy note20 ultra 5g bom". *https://www.gsmarena.com/samsung_galaxy_note20_ultra_5g_bom_is_550_calculate_analysts-news-45209.php*.

[32] Businesswire, 2018. "Galaxy s9+ materials cost $43 more than previous versions, ihs markit teardown shows". *https://www.businesswire.com/news/home/20180321005364/en*.

Copyright © 2023 by ASME

Proceedings of the ASME 2023
International Design Engineering Technical Conferences and
Computers and Information in Engineering Conference
IDETC-CIE2023
August 20-23, 2023, Boston, Massachusetts

DETC2023-115029

A VISUAL REPRESENTATION OF ENGINEERING CATALOGS USING VARIATIONAL AUTOENCODERS

Saketh Sridhara[1], Krishnan Suresh[1,*]

[1]Department of Mechanical Engineering
University of Wisconsin-Madison, Madison, WI, United States

ABSTRACT

Catalogs have been used for over a century for designing engineering systems. While catalogs are excellent repositories of engineering information, they are difficult to navigate, specifically to spot clusters, gaps, substitutes, and outliers. Inspired by Ashby charts for material selection, we propose here, a visual representation of engineering catalogs using neural networks. In particular, we employ variational autoencoders (VAEs) to project catalog data onto a lower-dimensional latent space. The latent space can then be visualized to explore the underlying structure of the catalog. Specifically, creators can use this visual representation to identify gaps and outliers in their data, while end users can benefit from this representation to compare catalogs from competitors, and to find substitutes. Contours can be superimposed on the charts to enable selection based on user-defined attributes; these contours are generalization of design indices associated with Ashby charts. Various examples of catalogs across engineering disciplines, ranging from materials and bearings to motors and batteries are illustrated using the proposed method. Using these examples, we (1) study the impact of the latent space dimension on the representational error, (2) illustrate how designers can easily choose alternate configurations based on their design requirements, and (3) gaps in catalog offerings can be clearly identified, providing a stimulus for new product development.

Keywords: Catalogs, representation, visualization, neural networks, latent space

1. INTRODUCTION

Representations form the backbone of engineering advances. For example, geometric representations such as constructive solid geometry and boundary representation [1] have led to breakthroughs in computer-aided design (CAD), computer-aided analysis (CAE), and computer-aided manufacturing (CAM). Yet, we

rely on a century-old representation of engineering data, namely, *catalogs* for designing almost any engineering system; see Figure 1. As engineering systems become more complex, this approach has significant limitations that pose challenges for catalog creators and end users.

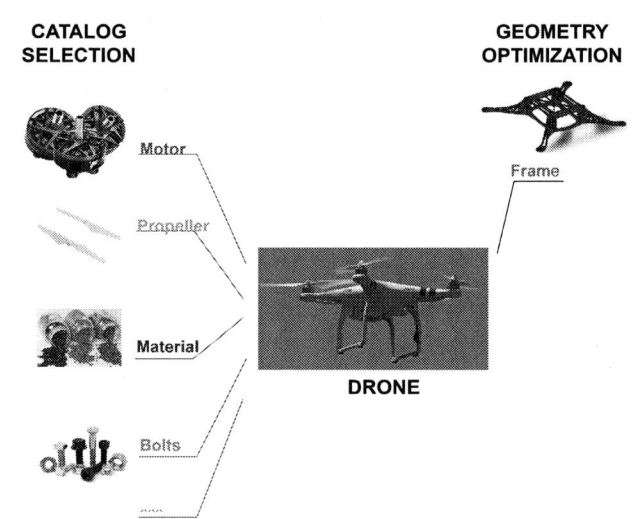

FIGURE 1: ENGINEERING DESIGN INVOLVES THE SIMULTANEOUS TASK OF SELECTING COMPONENTS FROM DESIGN CATALOGS AND DESIGNING CUSTOM GEOMETRIES.[2, 3].

From the creators' perspective, it is difficult to identify *outliers* and *gaps* within these catalogs:

- An outlier is an entry that deviates markedly from other entries in the catalog [4]. Identification of outliers is crucial for two reasons. First, if the data is erroneous, it must be removed or corrected. Second, if they truly reflect exceptional characteristics, additional validation may be required.

- Gaps in the catalog are regions for innovation and provide a

---

*Corresponding author: ksuresh@wisc.edu

Copyright © 2023 by ASME

stimulus for new product development.

Unfortunately, both outliers and gaps are difficult to identify from tabular data. From the catalog user's perspective:

- Tabular data is ill-suited for finding substitutes [5]. Identification of clusters and nearby designs [6] can be immensely valuable as it enables reuse.

- Similarly, it is not easy to determine from a catalog if two different types of, say, ball bearings have similar attributes.

- Finally, it is difficult to compare catalogs from competitors since the tabular schemes can be vastly different.

1.1 Contributions

To address the aforementioned challenges, we propose here a visual representation of engineering catalogs that would enable: (1) catalog creators to rapidly screen for outliers, identify gaps and new form factors for product development, and (2) end-users to gain structural insight into the data, make well-informed design decisions [7] and choose from alternate design configurations.

2. PROPOSED METHOD

The proposed framework draws inspiration from the pioneering work of Dr. Ashby on the visualization of *material catalogs* [8, 9].

2.1 Ashby Charts for Materials

Materials dictate the performance of all engineered structures and devices. There are well over 50,000 engineering materials [8], and this number is constantly growing. A typical subset of a material catalog is shown in Table 1, with material properties (attributes) such as Young's modulus, cost, mass density, and yield strength.

TABLE 1: A CURATED SUBSET OF MATERIALS AND THEIR PROPERTIES.

Mat Name	Class	E $[10^9 Pa]$	C $[\$/kg]$	ρ $[kg/m^3]$	Y $[10^7 Pa]$
A286 Iron	Steel	201	5.18	7920	62.0
AISI 304	Steel	190	2.40	8000	51.7
Gray C Iron	Steel	66.2	0.65	7200	15.2
3003-H16	Al	69.0	2.18	2730	18.0
5052-O	Al	70.0	2.23	2680	19.5
7050-T7651	Al	72.0	2.33	2830	55.0
Acrylic	Plastic	3.00	2.80	1200	7.3
ABS	Plastic	2.00	2.91	1020	3.0
PE HD	Plastic	1.07	2.21	952	2.2

Observe that the attributes are poorly correlated, i.e., a higher Young's modulus does not imply a higher cost, and the values can vary by several orders of magnitude. More importantly, selecting a suitable material for a particular application can be challenging since material selection is tightly coupled with other design decisions. To address these challenges, Ashby's charts were proposed in [8, 9]. These charts represent two properties at a time on a two-dimensional plot. For example, Ashby's chart for Young's modulus versus density is illustrated in Figure 2. Observe the natural clustering of materials that the Ashby chart reveals. Furthermore, under certain conditions, design indices can be associated with these charts to help select suitable materials for a particular application [10]. Ashby charts effectively address the limitations of catalogs described earlier for up to three attributes, and have been extended to hybrid materials and composites [11].

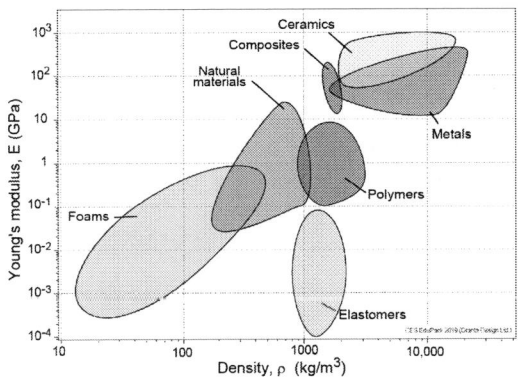

FIGURE 2: ASHBY'S CHARTS FOR YOUNG'S MODULUS VS DENSITY; CHART CREATED USING CES EDUPACK 2019, ANSYS GRANTA © 2020 GRANTA DESIGN.

2.2 Variational Autoencoders

Drawing inspiration from such Ashby charts, the proposed framework relies on variational autoencoders (VAEs) to create charts of arbitrary catalogs, with any number of attributes through nonlinear *dimensional reduction*. Dimensional reduction techniques aim to decrease the number of dimensions in high-dimensional data and convert it to a lower-dimensional space, often 2D or 3D, in order to improve human understanding. Data visualization provides an intuitive understanding of the underlying structure, in order to gain insight and to develop hypotheses. It has been shown that dimensional reduction can help highlight biases and pervasive noise in the data [12]. Several researchers have leveraged dimensionality reduction methods for visualization in engineering design: examples include multidimensional scaling [13, 14], self-organizing maps [15, 16], design manifolds through kernel PCA [17, 18] to name a few. The visual representation in [14] does not guarantee an accurate approximation of the design space and is susceptible to crowding. Self-organizing maps are limited in their ability to capture complex data distributions and the projection to a 2D planar grid allows only a restricted number of different distances to be represented [19]. The performance of Kernel PCA methods is highly dependent on the choice of the kernel function [20], thereby requiring multiple trials and experiments.

Copyright © 2023 by ASME

VAEs are special neural network configurations with wide applicability in dimensional reduction, data compression, semi-supervised learning, and data interpolation; for a detailed review please see [21]. VAEs have been employed to synthesize new samples similar to the input as well as anomaly detection; popular examples include the generation of synthetic human faces [22] and outlier detection [23, 24]. In science and engineering, VAEs have been used to generate new designs by exposing them to large databases containing images of micro-structures [25], photonic crystals [26], heat-conduction materials [27], discovering partial differential equations [28], and synthesizing new chemical compounds [29].

In this work, we do not use VAEs to synthesize new data, instead *we leverage the VAE's ability to map uncorrelated data onto an abstract, visualizable latent space*. To understand the construction of the VAE, let us consider a material catalog, such as the one in Table 1. A simplified VAE architecture for this data is illustrated in Figure 3. In this instance, it consists of:

1. A four-dimensional *input* module corresponding to the four properties in Table 1, namely the Young's modulus (E), cost (C), mass density (ρ) and yield strength (Y). The input set is denoted by ζ.

2. An *encoder* $\mathscr{E}$ consisting of a fully-connected network of, say, 250 neurons, associated with activation functions and weights [30].

3. A two-dimensional *latent space*, denoted by z_0, z_1 that lies at the heart of the VAE.

4. A *decoder* $\mathscr{D}$, that is similar to the encoder, consisting of a fully-connected network.

5. A four-dimensional *output* denoted by $\hat{\zeta}$ corresponding to the same four properties.

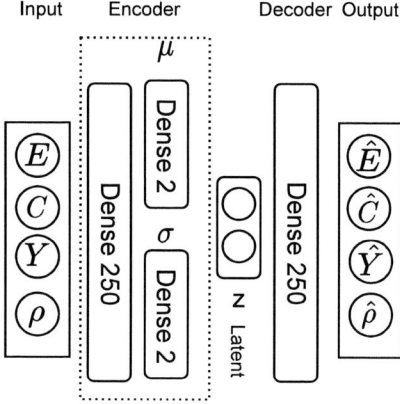

FIGURE 3: ARCHITECTURE OF A VARIATIONAL AUTOENCODER.

The VAE's primary task is to match the output to the input as closely as possible via an intermediate low-dimensional latent space. This is done through an optimization process (also referred to as training), using the weights associated with the encoder and decoder as optimization parameters. In other words, the

VAE minimizes the *reconstruction loss* $||\zeta - \hat{\zeta}||$. Additionally, a *regularization loss* is included to ensure that latent space is spatially compact [31]. Thus, VAEs are a generalized form of principal component analysis (PCA) for dimensional reduction, where the nonlinear nature of the neural networks allows for far greater flexibility [21, 32] and higher accuracy [33] when compared to classical nonlinear dimension reduction methods such as Isomaps [34].

2.3 Training procedure

The training procedure is described in Algorithm 1, with the input data being the material database. Encoder $\mathscr{E}$ is a neural network that takes in the set of material data ζ and encodes the four-dimensional data to the two-dimensional latent space denoted by z_0, z_1 through a probabilistic latent distribution governed by μ, σ. The VAE loss (or error) is then used to drive the training per Equation 1 until a sufficiently high representational accuracy is attained. The first term represents the reconstruction loss $||\zeta - \hat{\zeta}||$, i.e., the difference between the input and reconstructed data. The second term represents the KL divergence loss that is imposed to ensure that latent space is compact and resembles a standard Gaussian distribution $z \sim \mathcal{N}(\mu = 0, \sigma = 1)$ [21]; the two losses are combined with a weight factor β. After the training, the encoder is discarded, and the decoder $\mathscr{D}$ is retained for data retrieval. The decoder takes the latent space coordinates as input and returns the predicted material properties.

$$L = ||\zeta - \hat{\zeta}|| + \beta \, KL(z||\mathcal{N}) \qquad (1)$$

Algorithm 1 ENCODE MATERIALS

1: **procedure** MATENCODE(ζ)   ▷ Input: Training data
2:     epoch = 0   ▷ iteration counter
3:     **repeat**   ▷ VAE training
4:        $\mathscr{E}(\zeta) \rightarrow \{\mu, \sigma\}$ ▷ Forward prop. encoder
5:        $\{\mu, \sigma\} \rightarrow z$ ▷ Reparameterization [21]
6:        $\{\mu, \sigma\} \rightarrow KL(z||\mathcal{N})$ ▷ KL loss
7:        $\mathscr{D}(z) \rightarrow \hat{\zeta}$ ▷ Forward prop. decoder
8:        $\{\zeta, \hat{\zeta}, KL\} \rightarrow L$ ▷ VAE Loss
9:        $w + \Delta w(\nabla L) \rightarrow w$ ▷ Update VAE wts: Adam[35]
10:        epoch ++
11:     **until** error is acceptable   ▷ Iterate
12:     **return** $\mathscr{D}$ ▷ Trained decoder
13: **end procedure**

---

3. VISUAL REPRESENTATION OF MATERIALS

In this work, we used PyTorch [36] to model the VAE, and the gradient-based Adam optimizer [35] to minimize Equation 1 with recommended values of learning rate 0.002 for $20,000$ epochs, and $\beta = 5 * 10^{-5}$. A threshold for total loss may also be specified for termination of training.

Copyright © 2023 by ASME

3.1 Material latent space

The material data consisting of 92 materials was sourced from SolidWorks material library [37]. The VAE training took approximately 22 seconds on a 6-core Intel i7 CPU with 32GB RAM running Ubuntu. The resulting two-dimensional latent space is illustrated in Figure 4. Observe that each material can now be represented as a pair z_0, z_1. For example, annealed AISI 1020 is represented by the pair $(-0.6, 2.1)$ while Acrylic is represented by $(0.6, -0.2)$. The VAE clusters similar materials together in the latent space, as can be observed in Figure 4. Conceptually, the latent space in Figure 4 is similar to the popular Ashby charts seen in Figure 2.

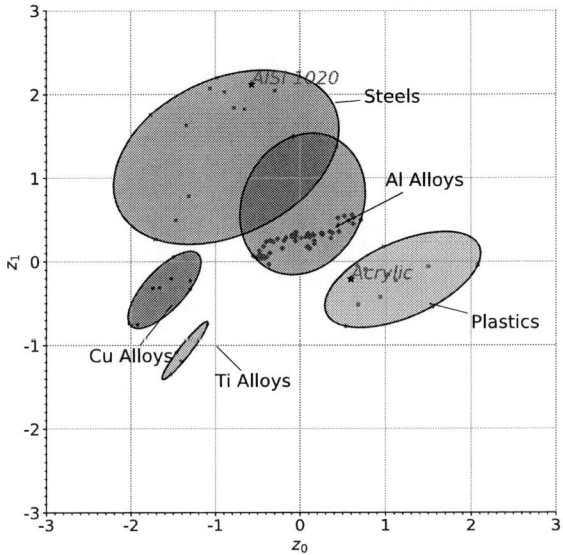

FIGURE 4: MATERIALS REPRESENTED ON A 2D LATENT SPACE

3.2 Convergence

The convergence of the total loss, reconstruction loss and regularization loss are illustrated in Figure 5.

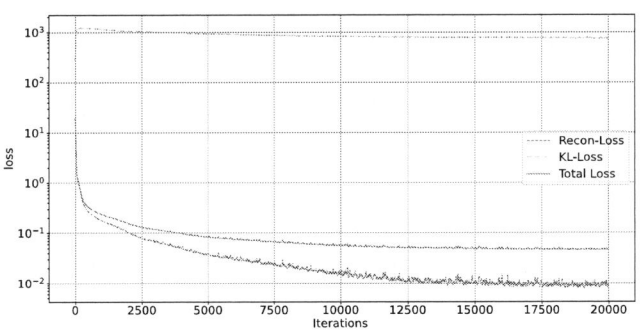

FIGURE 5: CONVERGENCE OF THE LOSS DURING TRAINING

3.3 Property contours

One can overlay the latent space in Figure 4 with specific material properties (or their combination) to gain further insight. This is illustrated in Figure 6 where the contour plots of cost are

superimposed on the latent space. This allows designers to identify the class of materials that meet one or more requirements, as seen in Figure 7, where materials whose cost per kg lies between $2 and $5 are identified.

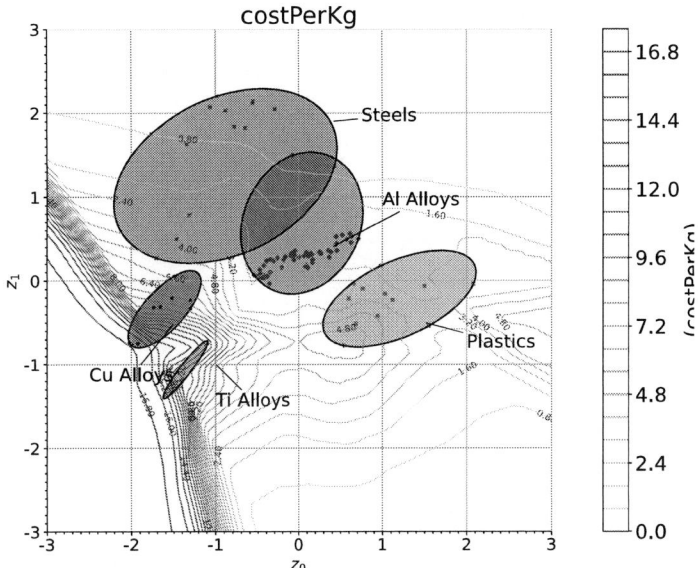

FIGURE 6: COST PER KG CONTOUR PLOT SUPERIMPOSED ON THE LATENT SPACE OF MATERIALS

3.4 Outlier Detection

VAEs have been increasingly used in recent years for outlier detection, a crucial task in anomaly detection, fraud detection, and other applications where detecting rare or unexpected events is important. We exploit the same for engineering data. For example, in the material database that we sourced from [37], we identify a specific outlier, namely, Aluminum 4032-T6, as indicated in Figure 8; it is far away from other Aluminum members, and is closer to steels [1]. Upon further inspection, we identified that this is due to an unusually low cost associated with Aluminum 4032-T6. Catalog creators can leverage this framework to validate their data.

3.5 Representational Accuracy

As with any dimensional reduction method, it is expected that the values obtained through the decoder will differ slightly from the input data [13]. For example, one can compute the properties of Acrylic predicted by the decoder, and compare it against the true values. Table 2 summarizes the errors for each attribute for some of the materials. The following observations are worth noting:

1. Despite the lack of correlation between material properties, and two orders of magnitude difference in values, the VAE captures the entire database of 92 materials reasonably well, using a simple two-dimensional latent space.

---

[1]Prof. Morlier's team at ISAE-SUPAERO (https://ica.cnrs.fr/author/jmorlier/) brought this to our attention.

Copyright © 2023 by ASME

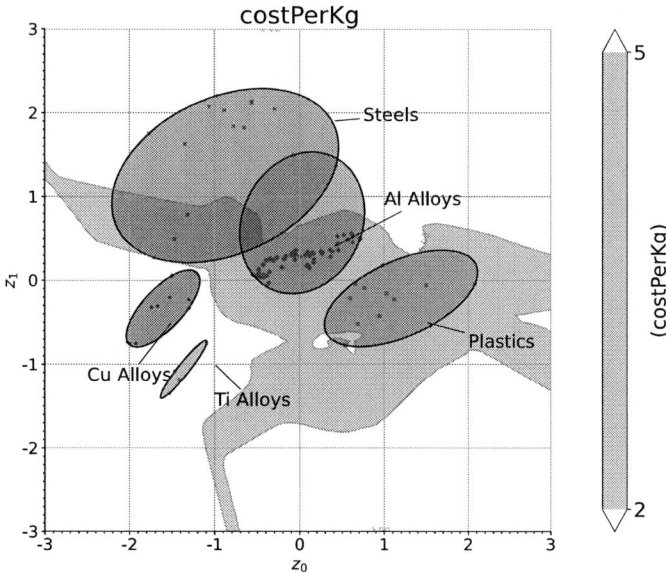

FIGURE 7: FEASIBLE SPACE IS MARKED IN GREEN, CHOICES LYING IN THIS SPACE MEET DESIGN REQUIREMENT

2. The error can be further reduced by either increasing the dimension of the latent space or by tuning the size and optimization parameters of the VAE. One such case is explored in Section 4.

TABLE 2: PERCENTAGE ERROR BETWEEN ACTUAL AND DECODED DATA.

Material	ΔE %	ΔC %	$\Delta \rho$ %	ΔY %
A286 Iron	1.0	0.6	2.3	1.6
ABS	1.7	2.2	0.0	0.7
AISI 304	3.3	0.7	0.3	2.8
Gray Cast Fe	1.3	2.7	2.1	2.9
3003-H16	0.3	0.3	0.3	2.3
5052-O	0.8	2.4	1.2	1.3
7050-T7651	0.2	2.3	0.4	4.0
Acrylic	1.6	0.5	0.1	1.4
PE HD	3.2	0.3	2.0	3.9
Max error	**6.1**	**5.2**	**3.2**	**9.0**

4. APPLICATIONS AND CASE-STUDIES

The proposed method generalizes to a variety of catalogs. We use several examples below to illustrate potential applications of the proposed framework.

4.1 Cross sections: Clustering on a line

Engineering structures are often built using beams with predefined cross-sections, such as rectangles and I-shapes. Consider a typical catalog of cross sections [38], such as the one in Table 3.

The visual representation of this catalog is shown in Figure 9. Observe that the disks are clustered onto a line in the 2D space.

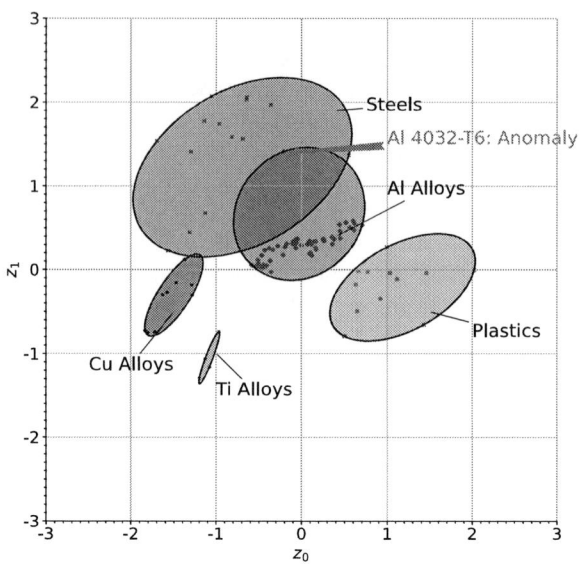

FIGURE 8: ANOMALY DETECTION IN THE MATERIAL CATALOG WITH VAE. THE AL 4032-T6 ENTRY IS AN OUTLIER, FURTHER AWAY FROM OTHER AL ALLOYS

TABLE 3: A CURATED SUBSET OF CROSS SECTIONAL AREAS AND THEIR PROPERTIES.

Section ID	Section Type	Area A $[10^{-4}m^2]$	I_x $[m^4]$	I_y $[m^4]$
1	Rectangle	4.00	1.33E-08	1.33E-08
2	Rectangle	4.44	1.83E-08	1.48E-08
3	Rectangle	4.89	2.43E-08	1.63E-08
101	Disk	12.6	1.26E-07	1.26E-07
102	Disk	12.8	1.31E-07	1.31E-07
103	Disk	13.1	1.36E-07	1.36E-07
201	Hollow Rect	1.44	7.87E-09	7.87E-09
202	Hollow Rect	1.60	1.24E-08	9.17E-09
203	Hollow Rect	1.76	1.82E-08	1.05E-08
309	Annulus	2.39	4.32E-08	4.32E-08
310	Annulus	2.64	4.72E-08	4.72E-08
311	Annulus	2.88	5.11E-08	5.11E-08
409	I-Beam	2.50	1.21E-08	6.77E-09
410	I-Beam	2.70	1.96E-08	6.81E-09
411	I-Beam	2.90	2.93E-08	6.85E-09

This implies that all attributes, for disc shapes, are functions of a single variable (in this case, radius r). Such intrinsic properties are often hard to detect from tabular data.

4.2 Bearings: Higher dimensional latent spaces

Next we consider a catalog of bearings from Timken [39]. Bearings are widely used in engineering applications, ranging from machinery and vehicles to aerospace and medical equipment. There are different types of bearings, including ball bearings, roller bearings, thrust bearings etc., each suitable for a certain rated speed and load - a subset of the database is shown in Figure 10. We choose six attributes for each catalog entry,

Copyright © 2023 by ASME

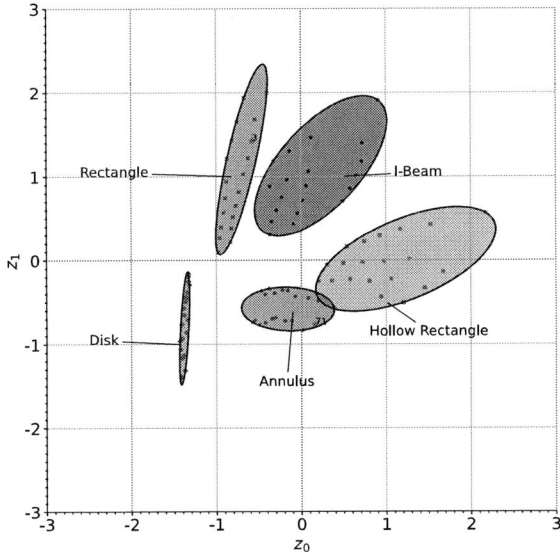

FIGURE 9: LATENT SPACE OF CROSS SECTIONS.

namely bore diameter (mm), outer diameter (mm), bearing width (mm), rated RPM, rated dynamic load (kN), and weight (kg).

The visual representation of this catalog with 100 different bearings is seen in Figure 11, overlaid with contours indicating the RPM rating of the bearings.

Next, we changed the latent dimension to three; the resulting latent space is illustrated in Figure 12. As expected, the features of the 2D representation extend to 3D as well in terms of clusters, outliers, and gaps.

For the same training parameters, we compare the representational accuracy of 2D and 3D latent spaces. We report the maximum and average reconstruction errors in Table 4. Observe that the 3D latent space has a lower average reconstruction error across every attribute.

TABLE 4: MAX AND AVG ERROR PERCENTAGES IN 2D AND 3D LATENT SPACES FOR BEARINGS.

Parameter	2D Latent		3D Latent	
	Max(%) Error	Avg(%) Error	Max(%) Error	Avg(%) Error
Bore Dia	2.8	0.78	3.6	0.75
Outer Dia	2.4	0.63	2.4	0.44
Width	3.4	0.55	1.2	0.34
Load Rating	9.1	2	4.3	1.17
RPM	6	1.23	4.2	0.69
Weight	9.6	2.37	6.5	1.49

4.3 Motors: Selecting alternate configurations

We next consider a catalog of motors from GE [40]. GE produces a variety of motors for industrial and commercial applications, with different configurations to suit various needs of torque and speeds. We consider 4 attributes: horsepower, RPM,

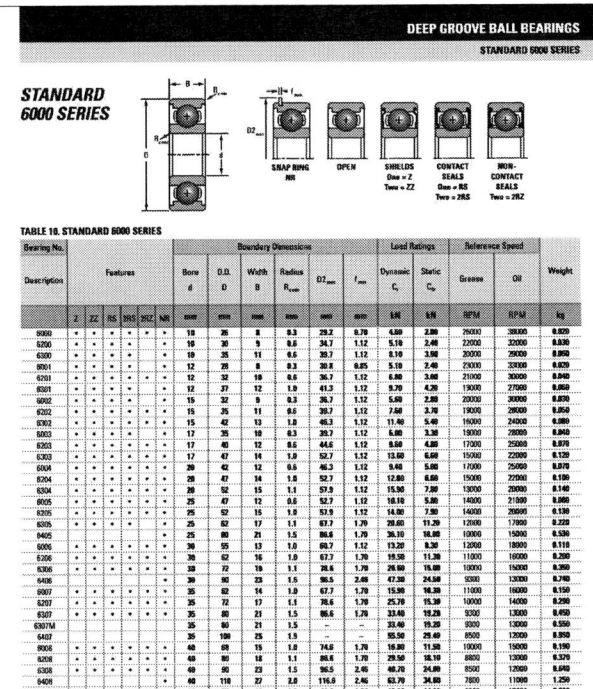

FIGURE 10: DEEP GROOVE BALL BEARINGS: TIMKEN [39]

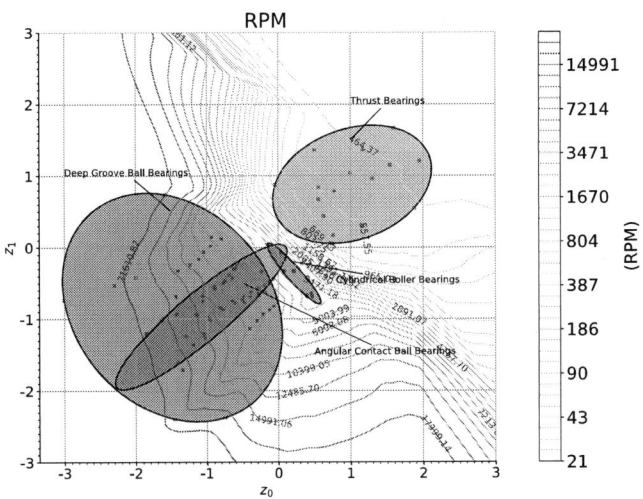

FIGURE 11: RPM CONTOUR PLOT SUPERIMPOSED ON THE LATENT SPACE OF BEARINGS

operating voltage, and cost for each entry across 4 types of motors: horizontal shaft, vertical shaft, keyless shaft and DC motors; see Figure 13, for a total of 168 motors.

The visual representation of this catalog is shown in Figure 14. The overlaps indicate multiple configurations are available with similar performance characteristics. For example, based on horsepower requirements, the design teams can choose from either a horizontal or a vertical shaft motor for their product, or switch to a DC power source from AC. The visual latent space provides insights into possible substitutes or alternate design choices for the end user.

Copyright © 2023 by ASME

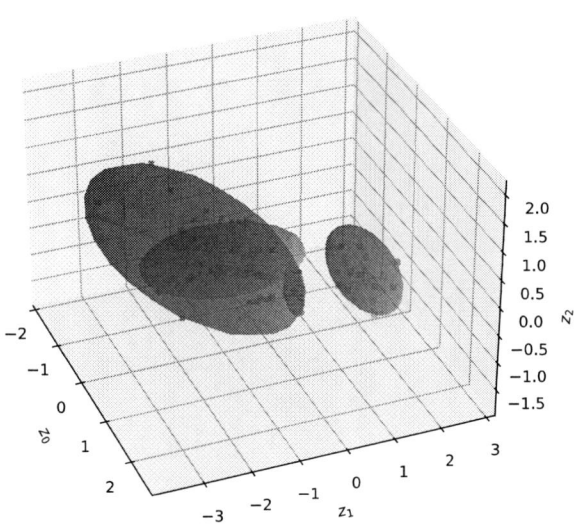

FIGURE 12: VISUALIZING THE BEARING CATALOG IN A 3D LATENT SPACE

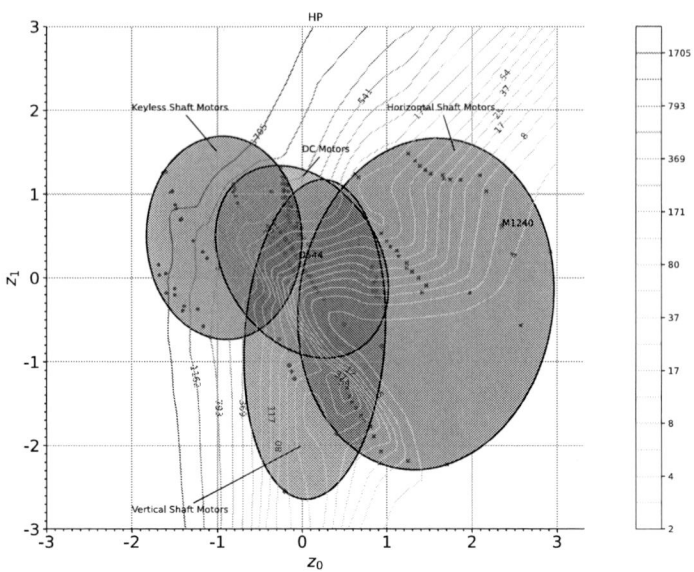

FIGURE 14: HORSEPOWER CONTOUR PLOT SUPERIMPOSED ON THE LATENT SPACE OF MOTORS

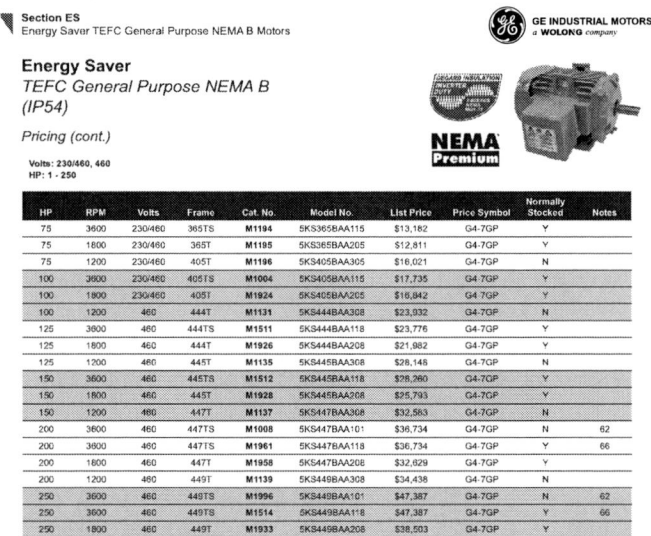

FIGURE 13: HORIZONTAL SHAFT MOTORS CATALOG, SOURCE: GE INDUSTRIAL MOTORS [40]

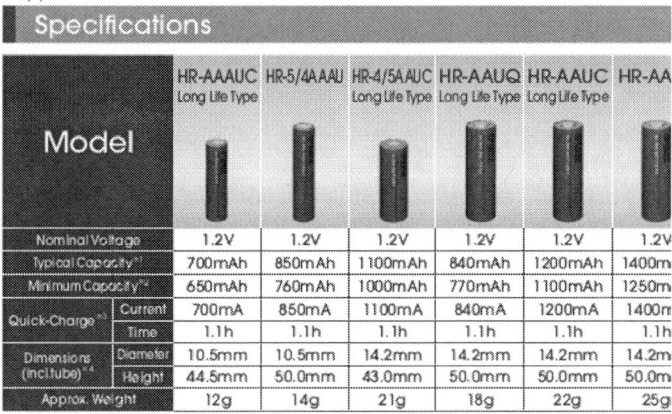

FIGURE 15: NIMH BATTERIES CATALOG, SOURCE: FDK BATTERIES [41]

4.4 Batteries: Identifying new form factors

Lastly, we consider a catalog of 44 batteries from FDK [41]. There are several types of batteries available today, each with its own unique set of properties and characteristics. Some of the most common types of batteries include Alkaline, Lithium-ion, Nickel-metal hydride etc., each with different energy densities, operating voltages, and capacity. A sample of the catalog is shown in Figure 15. For visualization, we consider six classes of batteries and four attributes: voltage (V), current (mA), capacity (mAh) and weight (g).

The visual representation of batteries is shown in Figure 16, overlaid with contours of battery capacity (mAh). Upon generating contours of capacity in the latent space, the catalog creators can identify gaps to place newer offerings, and contrast their existing gaps with the competitors' catalogs to develop new products that meet the needs of customers.

Copyright © 2023 by ASME

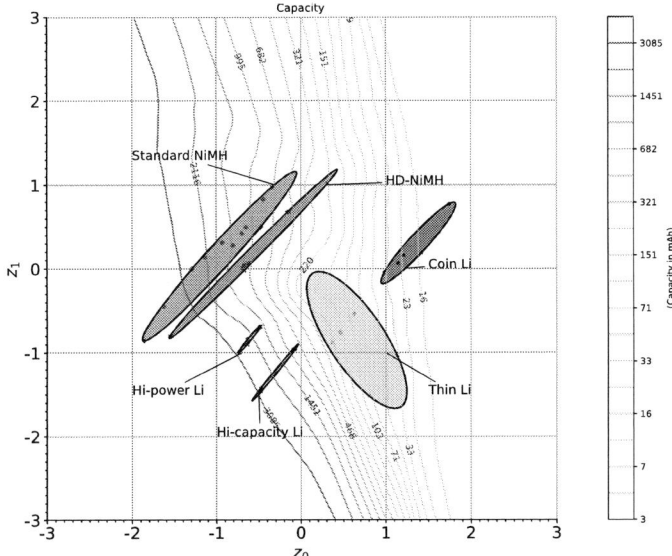

FIGURE 16: CAPACITY CONTOURS SUPERIMPOSED ON THE LATENT SPACE OF BATTERIES

5. CONCLUSION

The proposed visual representation of catalogs addresses challenges faced by creators in identifying gaps and outliers in their data. End users can benefit from this representation in identifying clusters for substitutes, reuse, and configuration changes. While the method has shown success in representing a variety of catalogs, a few challenges remain. The first is to improve the representational accuracy and reduce reconstruction error by incorporating recent advancements in VAEs [42, 43]. Secondly, the framework needs to be validated on larger catalogs consisting of thousands of entries such as the material database [44, 45]. Third, the abstract latent space variables, which are complex analytical functions of the attributes, currently lack interpretability [46]; this can be a challenge to practicing engineers. Finally, the method could benefit from better tuning of the networks and optimal choice of the VAE architecture [47, 48].

The latent space offers other advantages that are being explored. The latent space of the VAE is differentiable - this would allow seamless integration of catalog data into gradient-based design optimization problems [49, 50], leading to optimally designed systems. Further, the generative capability of the latent space can be leveraged for innovation [29, 51, 52], and will be explored in future work. Finally, another interesting research direction is incorporating data uncertainties within the latent space.

ACKNOWLEDGMENTS

The authors would like to thank the support of the National Science Foundation through grant CMMI 1561899, and the U. S. Office of Naval Research under PANTHER award number N00014-21-1-2916 through Dr. Timothy Bentley.

CONFLICT OF INTEREST

The authors declare that they have no conflict of interest.

DATA AVAILABILITY STATEMENT

The data that support the findings of this study are available from the corresponding author upon reasonable request.

REFERENCES

[1] Requicha, Aristides G. "Representations for rigid solids: Theory, methods, and systems." *ACM Computing Surveys (CSUR)* Vol. 12 No. 4 (1980): pp. 437–464.

[2] Shigley, Joseph E, Mischke, Charles R and Brown Jr, Thomas Hunter. *Standard handbook of machine design.* McGraw-Hill Education (2004).

[3] Martinez Leon, AS, Rukavitsyn, AN and Jatsun, SF. "UAV Airframe Topology Optimization." *International Conference on Industrial Engineering*: pp. 338–346. 2021. Springer.

[4] NIST/SEMATECH. "Exploratory Data Analysis (EDA)." (2013). URL https://www.itl.nist.gov/div898/handbook/eda/section3/eda35h.htm. [Online; accessed 4-March-2023].

[5] Ashby, Michael F and CEBON, David. "Materials selection in mechanical design." *Le Journal de Physique IV* Vol. 3 No. C7 (1993): pp. C7–1.

[6] Pham, DT and Afify, AA. "Clustering techniques and their applications in engineering." *Proceedings of the Institution of Mechanical Engineers, Part C: Journal of Mechanical Engineering Science* Vol. 221 No. 11 (2007): pp. 1445–1459.

[7] Keim, Daniel A. "Information visualization and visual data mining." *IEEE transactions on Visualization and Computer Graphics* Vol. 8 No. 1 (2002): pp. 1–8.

[8] Ashby, Michael F and Jones, David RH. *Engineering materials 1: an introduction to properties, applications and design.* Vol. 1. Elsevier (2011).

[9] Ashby, Michael F and Johnson, Kara. *Materials and design: the art and science of material selection in product design.* Butterworth-Heinemann (2013).

[10] Ananthasuresh, GK and Ashby, MF. "Concurrent design and material selection for trusses." *Workshop: Optimal Design, Laboratoire de Mécanique des Solides, Ecole Polytechnique Palaiseau France* (2003).

[11] Ashby, Mike. "Designing architectured materials." *Scripta Materialia* Vol. 68 No. 1 (2013): pp. 4–7.

[12] Rudin, Cynthia, Chen, Chaofan, Chen, Zhi, Huang, Haiyang, Semenova, Lesia and Zhong, Chudi. "Interpretable machine learning: Fundamental principles and 10 grand challenges." *Statistic Surveys* Vol. 16 (2022): pp. 1–85.

[13] Kruskal, Joseph B and Wish, Myron. *Multidimensional scaling.* 11, Sage (1978).

[14] Knerr, Nathan and Selva, Daniel. "Cityplot: visualization of high-dimensional design spaces with multiple criteria." *Journal of Mechanical Design* Vol. 138 No. 9 (2016): p. 091403.

[15] Kohonen, Teuvo. "The self-organizing map." *Proceedings of the IEEE* Vol. 78 No. 9 (1990): pp. 1464–1480.

Copyright © 2023 by ASME

[16] Thole, Sidhant Pravinkumar and Ramu, Palaniappan. "Design space exploration and optimization using self-organizing maps." *Structural and Multidisciplinary Optimization* Vol. 62 No. 3 (2020): pp. 1071–1088.

[17] Mika, Sebastian, Schölkopf, Bernhard, Smola, Alex, Müller, Klaus-Robert, Scholz, Matthias and Rätsch, Gunnar. "Kernel PCA and de-noising in feature spaces." *Advances in neural information processing systems* Vol. 11 (1998).

[18] Chen, Wei, Fuge, Mark and Chazan, Jonah. "Design manifolds capture the intrinsic complexity and dimension of design spaces." *Journal of Mechanical Design* Vol. 139 No. 5 (2017).

[19] Flexer, Arthur. "Limitations of self-organizing maps for vector quantization and multidimensional scaling." *Advances in neural information processing systems* Vol. 9 (1996).

[20] Schölkopf, Bernhard, Smola, Alexander and Müller, Klaus-Robert. "Kernel principal component analysis." *Artificial Neural Networks—ICANN'97: 7th International Conference Lausanne, Switzerland, October 8–10, 1997 Proceeedings*: pp. 583–588. 2005. Springer.

[21] Kingma, Diederik P and Welling, Max. "An introduction to variational autoencoders." *arXiv preprint arXiv:1906.02691* (2019).

[22] MacDorman, Karl F, Green, Robert D, Ho, Chin-Chang and Koch, Clinton T. "Too real for comfort? Uncanny responses to computer generated faces." *Computers in human behavior* Vol. 25 No. 3 (2009): pp. 695–710.

[23] Chen, Jinghui, Sathe, Saket, Aggarwal, Charu and Turaga, Deepak. "Outlier detection with autoencoder ensembles." *Proceedings of the 2017 SIAM international conference on data mining*: pp. 90–98. 2017. SIAM.

[24] Xu, Haowen, Chen, Wenxiao, Zhao, Nengwen, Li, Zeyan, Bu, Jiahao, Li, Zhihan, Liu, Ying, Zhao, Youjian, Pei, Dan, Feng, Yang et al. "Unsupervised anomaly detection via variational auto-encoder for seasonal kpis in web applications." *Proceedings of the 2018 world wide web conference*: pp. 187–196. 2018.

[25] Wang, Liwei, Chan, Yu-Chin, Ahmed, Faez, Liu, Zhao, Zhu, Ping and Chen, Wei. "Deep generative modeling for mechanistic-based learning and design of metamaterial systems." *Computer Methods in Applied Mechanics and Engineering* Vol. 372 (2020): p. 113377.

[26] Li, Xiang, Ning, Shaowu, Liu, Zhanli, Yan, Ziming, Luo, Chengcheng and Zhuang, Zhuo. "Designing phononic crystal with anticipated band gap through a deep learning based data-driven method." *Computer Methods in Applied Mechanics and Engineering* Vol. 361 (2020): p. 112737.

[27] Guo, Tinghao, Lohan, Danny J, Cang, Ruijin, Ren, Max Yi and Allison, James T. "An indirect design representation for topology optimization using variational autoencoder and style transfer." *2018 AIAA/ASCE/AHS/ASC Structures, Structural Dynamics, and Materials Conference*: p. 0804. 2018.

[28] Champion, Kathleen, Lusch, Bethany, Kutz, J Nathan and Brunton, Steven L. "Data-driven discovery of coordinates and governing equations." *Proceedings of the National Academy of Sciences* Vol. 116 No. 45 (2019): pp. 22445–22451.

[29] Noh, Juhwan, Kim, Jaehoon, Stein, Helge S, Sanchez-Lengeling, Benjamin, Gregoire, John M, Aspuru-Guzik, Alan and Jung, Yousung. "Inverse design of solid-state materials via a continuous representation." *Matter* Vol. 1 No. 5 (2019): pp. 1370–1384.

[30] Schmidhuber, Jürgen. "Deep learning in neural networks: An overview." *Neural networks* Vol. 61 (2015): pp. 85–117.

[31] Doersch, Carl. "Tutorial on Variational Autoencoders." *https://arxiv.org/abs/1606.05908* (2016)DOI 10.48550/ARXIV.1606.05908.

[32] Goodfellow, Ian, Bengio, Yoshua and Courville, Aaron. *Deep Learning*. MIT Press (2016).

[33] Fournier, Quentin and Aloise, Daniel. "Empirical comparison between autoencoders and traditional dimensionality reduction methods." *2019 IEEE Second International Conference on Artificial Intelligence and Knowledge Engineering (AIKE)*: pp. 211–214. 2019. IEEE.

[34] Tenenbaum, Joshua B, Silva, Vin de and Langford, John C. "A global geometric framework for nonlinear dimensionality reduction." *science* Vol. 290 No. 5500 (2000): pp. 2319–2323.

[35] Kingma, Diederik P. and Ba, Jimmy Lei. "Adam: A method for stochastic optimization." *3rd International Conference on Learning Representations, ICLR 2015 - Conference Track Proceedings*. 2015. International Conference on Learning Representations, ICLR. URL 1412.6980.

[36] Paszke, Adam, Gross, Sam, Massa, Francisco, Lerer, Adam, Bradbury, James, Chanan, Gregory, Killeen, Trevor, Lin, Zeming, Gimelshein, Natalia, Antiga, Luca, Desmaison, Alban, Kopf, Andreas, Yang, Edward, DeVito, Zachary, Raison, Martin, Tejani, Alykhan, Chilamkurthy, Sasank, Steiner, Benoit, Fang, Lu, Bai, Junjie and Chintala, Soumith. "PyTorch: An Imperative Style, High-Performance Deep Learning Library." *Advances in Neural Information Processing Systems 32*. Curran Associates, Inc. (2019): pp. 8024–8035.

[37] Dassault Systèmes SE. "Simulation Materials." Accessed March 8, 2023, URL https://help.solidworks.com/2021/english/SolidWorks/cworks/r_Simulation_Materials.htm.

[38] Gao, Huanhuan, Breitkopf, Piotr, Coelho, Rajan Filomeno and Xiao, Manyu. "Categorical structural optimization using discrete manifold learning approach and custom-built evolutionary operators." *Structural and Multidisciplinary Optimization* Vol. 58 No. 1 (2018): pp. 215–228.

[39] "Timken Catalog." Accessed March 8, 2023, URL https://catalog.timken.com/.

[40] "GE Motors Wolong - Products." Accessed March 8, 2023, URL https://www.gemotorswolong.com/products.

[41] FDK Corporation. "Lithium Ion Battery Data Sheet." Accessed March 8, 2023, URL https://www.fdk.com/battery/lithium_e/data_sheet/FDK_battery_en.pdf.

[42] Higgins, Irina, Matthey, Loic, Pal, Arka, Burgess, Christopher, Glorot, Xavier, Botvinick, Matthew, Mohamed,

Copyright © 2023 by ASME

Shakir and Lerchner, Alexander. "beta-vae: Learning basic visual concepts with a constrained variational framework." (2016).

[43] Razavi, Ali, Van den Oord, Aaron and Vinyals, Oriol. "Generating diverse high-fidelity images with vq-vae-2." *Advances in neural information processing systems* Vol. 32 (2019).

[44] Jain, Anubhav, Ong, Shyue Ping, Hautier, Geoffroy, Chen, Wei, Richards, William Davidson, Dacek, Stephen, Cholia, Shreyas, Gunter, Dan, Skinner, David, Ceder, Gerbrand et al. "Commentary: The Materials Project: A materials genome approach to accelerating materials innovation." *APL materials* Vol. 1 No. 1 (2013): p. 011002.

[45] The Materials Project. "The Materials Project." (2021). Accessed March 10, 2023, URL https://materialsproject.org/.

[46] Spinner, Thilo, Körner, Jonas, Görtler, Jochen and Deussen, Oliver. "Towards an interpretable latent space: an intuitive comparison of autoencoders with variational autoencoders." *IEEE VIS 2018*. 2018.

[47] Leung, Frank Hung-Fat, Lam, Hak-Keung, Ling, Sai-Ho and Tam, Peter Kwong-Shun. "Tuning of the structure and parameters of a neural network using an improved genetic algorithm." *IEEE Transactions on Neural networks* Vol. 14 No. 1 (2003): pp. 79–88.

[48] Hou, Xianxu, Shen, Linlin, Sun, Ke and Qiu, Guoping. "Deep feature consistent variational autoencoder." *2017 IEEE Winter Conference on Applications of Computer Vision (WACV)*: pp. 1133–1141. 2017. IEEE.

[49] Chandrasekhar, Aaditya, Sridhara, Saketh and Suresh, Krishnan. "Integrating material selection with design optimization via neural networks." *Engineering with Computers* Vol. 38 No. 5 (2022): pp. 4715–4730.

[50] Chandrasekhar, Aaditya, Sridhara, Saketh and Suresh, Krishnan. "Auto: a framework for automatic differentiation in topology optimization." *Structural and Multidisciplinary Optimization* Vol. 64 No. 6 (2021): pp. 4355–4365.

[51] Bostanabad, Ramin, Chan, Yu-Chin, Wang, Liwei, Zhu, Ping and Chen, Wei. "Globally approximate gaussian processes for big data with application to data-driven metamaterials design." *Journal of Mechanical Design* Vol. 141 No. 11 (2019).

[52] Wang, Liwei, Tao, Siyu, Zhu, Ping and Chen, Wei. "Data-driven topology optimization with multiclass microstructures using latent variable Gaussian process." *Journal of Mechanical Design* Vol. 143 No. 3 (2021).

Proceedings of the ASME 2023
International Design Engineering Technical Conferences and
Computers and Information in Engineering Conference
IDETC-CIE2023
August 20-23, 2023, Boston, Massachusetts

DETC2023-115103

INTERPRETABLE NEURAL NETWORK ANALYSES FOR UNDERSTANDING COMPLEX PHYSICAL INTERACTIONS IN ENGINEERING DESIGN

Tuba Dolar[1], Doksoo Lee[1], Wei Chen[1]

[1]Northwestern University
Evanston, IL

ABSTRACT

In engineering design, global sensitivity analysis (GSA) is used for analyzing the effects of the system inputs on the model response. Common GSA methods use analytical or surrogate models to study the relationships between model inputs and outputs. However, the accuracy of such system models depends on the complete conformance to all model assumptions. Even so, they are not flexible and fail to capture nonlinear behaviors in complex systems. Besides these GSA approaches, interpretable machine learning would also identify the relationships between system variables, eliminating the disadvantages of common GSA implementations. Apart from studying the independent variables individually, the evaluation of groups of them is likewise valuable. One example motivation in engineering design for performing GSA with groups of input variables would be managing the design space complexity in programmable material systems (PMS) development. In this article, we employ a flexible, interpretable artificial neural network model to uncover individual as well as grouped global sensitivity indices for understanding complex physical interactions in engineering design. The employed model allows the investigation of the feature importance of the main effects and pairwise interaction effects in GSA according to functional analysis of variance (ANOVA) decomposition. To draw a higher-level understanding, we further use a subset decomposition method to analyze the significance of the groups of input variables. Using PMS as an example, we demonstrate the use of our approach for understanding the impact of material, architecture, and stimulus variables as well as their interactions for a programmable photonic metasurface system. This information lays the foundation for deriving design guidelines for PMS development.

Keywords: interpretable machine learning, artificial neural networks, global sensitivity analysis, grouped global sensitivity indices

1. INTRODUCTION

Global sensitivity analysis (GSA) is widely used in engineering design to better understand complex models. GSA identifies the contribution of system inputs to the uncertainty of the output. It supports analyzing the impacts of design inputs on the model output. Model verification, model simplification, and establishment of research priorities for identifying the critical model components are examples that benefit from sensitivity analysis [1].

Numerous statistical methods have been developed for studying the relationship between the model inputs and outputs. Sensitivity analysis is a task related to uncertainty quantification and has been extensively studied in statistics. The first historical approaches to sensitivity analysis focused on revealing the impact of small input perturbations on system response, which is referred to as local sensitivity analysis (LSA) [2]. Following that, methods that consider the variation of the entire model parameters were developed in a statistical framework under GSA approaches [2]. Among them, regression coefficients are used for sensitivity analysis purposes where a simple linear model is first fit, then its regression coefficients are regarded as sensitivity indexes [3]. Variance-based methods decompose the output variance into terms caused by combinations of input variables and/or input variables groups. Here, sensitivity is assessed with the amount of output variance explained by an input [4]. Surrogate models can also be built as relatively simple models to approximate the relationship between dependent and

Copyright © 2023 by ASME

independent variables [5]. The accuracy of GSA with surrogate models depends on the sampling strategy used for constructing the model and the difference between the original and surrogate models. There are also other methods for conducting sensitivity analysis including design of experiments, graphical methods, Fourier amplitude sensitivity test (FAST), mutual information index, mathematical approximation strategies such as polynomial chaos expansion (PCE), and so on [6].

Despite the availability of diverse techniques, statistical methods use analytical or surrogate models to study the relationships between model inputs and outputs. However, using such models has its challenges created by the model assumptions and limitations. When working with a linear regression model, for example, we assume that the residuals are normally distributed, they have equal variance, samples are independent, and the true relationship is linear. Such conditions have to be satisfied for the model to be valid, restricting the flexibility to accurately identify complex interactions. Most real-world problems consist of numerous subsystems interacting with each other, leading to an increase in model size and complexity. In such cases, system models are difficult to define and even if they are built, they fail to capture nonlinear behaviors. Lacking accurate models, statistical sensitivity analysis is not applicable to high-dimensional complex problems. On contrary, machine learning models emerge as valuable tools for obtaining accurate models of complex systems where the physics behind the model behavior is not identified mathematically. They allow researchers to form black-box models with the availability of sufficient data [7].

Even though black-box machine learning models do not serve the objectives of GSA, interpretable machine learning, a growing research area under machine learning, focuses on transforming black-box models into glass-box models which can provide insights from a sensitivity point of view [8]. "Interpretability is the degree to which a human can understand the cause of a decision" [9]. The more interpretable a model is, the easier for researchers to understand the its inner workings, which perfectly aligns with the purposes of sensitivity analysis. Similar to sensitivity analysis goals, various interpretable machine learning techniques are available for local and global analysis. Local interpretability explains why an individual prediction is made whereas global interpretability describes the entire model behavior [8].

Considering its benefits in miscellaneous frameworks, interpretable machine learning is broadly adopted in many disciplines. Medical image analysis [10], policy-making [11], diversified portfolio construction in finance [12], cellular imaging in computational biology [13], and discovering brain functions in neuroscience [14] are a few examples that reveal the potential of interpretable methods on a multidisciplinary scale. Design engineers widely use data-driven design when it is impossible or expensive to work with system models. However, they fail to utilize a machine learning model's potential for further analyzing the system behavior. In the engineering design context, interpretable machine learning would allow researchers to identify the relationships between model variables,

eliminating the necessity of expensive statistical sensitivity analysis implementations. To give an example, an interpretable machine learning model can reveal the contribution of input variables while making predictions, which can be depicted as global sensitivity indices of these variables. Different from the conventional statistical analysis-based GSA, interpretable machine learning does not require an analytical model, can handle complex design spaces with combinations of categorical and continuous variables, and manage large sample sizes and number of variables.

Apart from uncovering the importance of design variables, another need while analyzing high-dimensional engineering design problems is the evaluation of the group effects. It is valuable to divide the independent variables into groups and study the relationships between these, for example in cases where the model is complex with a large number of variables or where there are explicitly meaningful variable groups. To better illustrate, robust design aims at designing systems that meet the performance requirement regardless of the many sources of variation. In robust design, the minimization of the performance variations caused by variations in noise factors and the minimization of the performance variations caused by control factors are the two broad problems [15]. In this framework, it is valuable to evaluate groups of uncontrollable parameters and design variables. Studying the importance and effect of these two groups helps with mitigating the effects of the uncontrollable parameters.

One example motivation in engineering design for performing GSA with subsets of input variables is managing the design space complexity in PMS development. PMS are manufactured with smart materials which are responsive to an external stimulus such as magnetic field, temperature, or humidity. This property brings the opportunity of programming PMS to change their shapes and dimensions for performing sophisticated functions, controlled drug release for example [16]. For such complex structures with high-dimensional representations of spatially diverse material composition, topological architecture, and external stimulus; interpretable machine learning promotes the investigation of the role of material, architecture, and stimulus while performing mission-specific tasks (Fig. 1). This information then can be used as an input for managing the design complexity and deriving design guidelines for PMS development by providing answers to; 1) is it possible to obtain a set of functional states with just one material or is a spatially-varying combination of multiple materials necessary, 2) is it possible to obtain the target performance with just designing the architecture variables or is the concurrent design of architecture and stimuli required, 3) which architectural structures provides a highly diversified set of functional states when programmed with the stimulus?

Copyright © 2023 by ASME

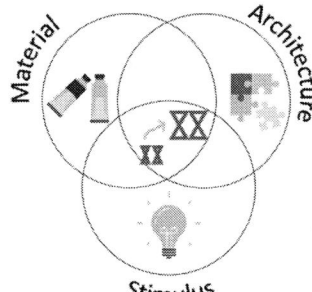

FIGURE 1: INTERACTIONS BETWEEN MATERIAL-ARCHITECTURE-STIMULUS ARE THE KEY FACTORS IN THE TRANSFORMATIONS BETWEEN VARIOUS FUNCTIONAL STATES IN PMS.

In this article, we use interpretable neural networks for examining complex physical interactions in engineering design problems. We use an interpretable neural network architecture based on generalized additive models with structured interactions (GAMI-Net) [17]. The model is inherently interpretable and eliminates the necessity of first training a machine learning model, following that using additional post hoc interpretability tools. The employed model allows the investigation of the feature importance of the main effects and pairwise interaction effects between the input variables. To draw a higher-level understanding, we further use a subset decomposition method [18] to analyze the significance of the groups of input variables. The combination of these techniques allows us to:

1) improve an interpretable neural network analysis, which works with individual input variables, to further interpret grouped effects,

2) reveal the complex physical interactions in engineering design problems,

3) represent an application and the benefits of data-driven global sensitivity analysis in the engineering design context.

In Section 2, we start by introducing the interpretable neural network model we use. This model is then combined with the subset decomposition approach as a contribution of this study to serve the need for the global sensitivity analysis of grouped effects. Then, Section 3 demonstrates our approach in a synthetic example problem and a photonic metasurface design problem. Following that, we break down the complex physical interactions and compile the design rules for programmable photonic metasurface design in Section 4. Finally, Section 5 concludes the article and discusses future work (Fig. 2).

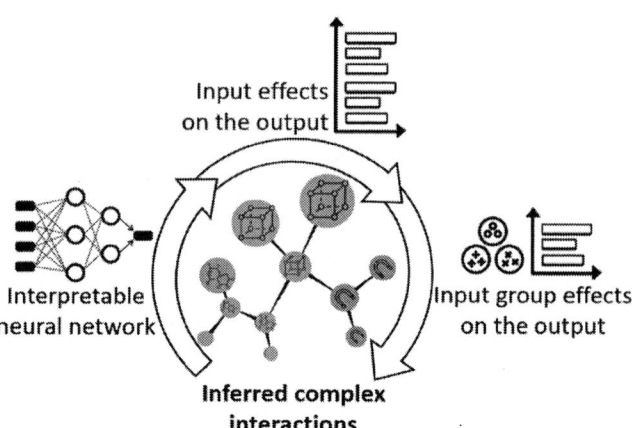

FIGURE 2: DATA-DRIVEN INTERPRETABLE NEURAL NETWORK ANALYSES FOR REVEALING COMPLEX INTERACTIONS. FIRST, AN INTERPRETABLE NEURAL NETWORK MODEL IS DEVELOPED FOR GSA OF THE MODEL INPUTS. FOLLOWING THAT, INPUT GROUPS ARE DEFINED AND THEIR EFFECTS ARE ANALYZED.

2. METHODS

2.1 Interpretable Neural Networks

Data-driven design methods for interpretable analysis are advantageous in engineering where an analytical model does not exist. Engineering design problems often involve the simultaneous design of multiple design entities including; material properties such as modulus of elasticity, structural features including dimensions and shapes, and even the design of stimulus, location and magnitude of external factors like load or magnetic field in design of programmable material systems. Designing material, architecture, and stimulus synchronously raises the complexity level of the system of interest for which a physical model becomes unattainable. In such cases, data-driven approaches allow researchers to build comparably accurate models with the guidance of data.

Yet, obtaining a machine learning model comes with its downsides as it is a black-box, meaning it does not reveal any information about the model's internal workings. For example, deep neural networks have complex model structures with a varying number of layers, all sorts of connections, and numerous neurons with millions of weight and bias parameters. They provide exceptional predictive accuracy with the availability of quality data but it is difficult to understand the role of each neuron and which input contributes to the output at which level. Data-driven design builds such black-box models where design engineers cannot infer the model's reasoning.

Understanding model behavior is exceptionally valuable in engineering design as it allows the transferability of the model to be used for the development of other similar systems. Design engineers need to understand how a model arrives at a specific result so that this knowledge can be readdressed for future decisions. This advantage is a natural consequence of model transparency offered by interpretable machine learning.

Copyright © 2023 by ASME

Engineers can question an interpretable data-driven model to see whether it is consistent with the prior physics-based knowledge about the system or not.

Interpretable machine learning resolves the disadvantages of black-box models and allows us to understand how a model works. Interpretability research involves two types of interpretability: post hoc and intrinsic which is determined by whether interpretability is achieved by using additional methods after a predictive model is obtained or whether it is achieved during the training phase [8]. Intrinsically interpretable models have simple structures which make it easy to evaluate model behavior. For example, sparse linear models and decision trees are easy to interpret without any additional effort [8]. Post hoc interpretability however includes adopting an external method after the model training is completed. This additional interpretability analysis can aim for understanding a local decision or drawing global inferences. Post hoc interpretability tools are model agnostic and can be adapted to any model [8]. Some examples of post hoc methods for global interpretability include partial dependence plots (PDP) [19], accumulated local effects (ALE) [20] plots, and permutation feature importance (PFI) [21]. PDP and ALE are visual tools that show how one or two features affect the model prediction. PFI evaluates an input's importance by varying it and observing the change in the model's prediction error; the more important the input is, the higher the model error becomes. Besides, post hoc interpretability of local decisions can be achieved with individual conditional expectation (ICE) [22] plots, local interpretable model-agnostic explanations (LIME) [23], and Shapley additive explanations (SHAP) [24]. ICE plots are equivalent to PDP with the only difference being constructed for individual data points. An ICE plot consists of separate lines per instance that show how the instance's prediction changes with respect to a feature. LIME involves training a simple local surrogate model around the area of interest to explain the individual prediction. SHAP is adopted from game theory and explains the contribution of each input on the model prediction for individual data points.

In this article, we use an inherently interpretable deep neural network model for GSA instead of first training a model and then using post hoc interpretability tools. A tradeoff between predictive performance and interpretability is reported in the literature [25-28]. That is, inherently interpretable models limit the model complexity, and that hurts the predictive accuracy. In such cases where predictive performance is more critical than interpretation, post hoc methods are favored. However, the model we use achieves competitive predictive accuracy while allowing to:

1) utilize data-driven design when an analytical or surrogate model is unavailable or expensive,
2) manage large datasets and high-dimensional problems,
3) avoid any model limitations as it is flexible to capture nonlinear system behavior,
4) work with mixed input spaces consisting of continuous and categorical variables,
5) perform GSA and LSA inherently, and also further post hoc interpretability analysis.

One approach for developing inherently interpretable models is using functional analysis of variance (ANOVA) decomposition [29].

Let x_1 $\forall i = \{1, 2, \dots, M\}$ be an independent random variable with probability density functions $p_i(x_i)$, first-order effects $\varphi_i(x_i)$, and second-order effects $\varphi_{i_1 i_2}(x_{i_1}, x_{i_2})$:

$$
\begin{aligned}
f(x_1, &\dots, x_M) \\
&= f_0 + \sum_{i=1}^{M} \varphi_i(x_i) + \sum_{i_1=1}^{M} \sum_{i_2=i_1+1}^{M} \varphi_{i_1 i_2}(x_{i_1}, x_{i_2}) + \cdots \\
&\quad + \varphi_{1\dots M}(x_1, \dots x_M) \quad (1)
\end{aligned}
$$

where the model is decomposed into a constant mean, first order-effects, second-order effects, and so on [29]. Preferably, a few low-order terms will be sufficient to approximate f.

$$
f_0 = \int f(\boldsymbol{x}) \prod_{i=1}^{M} [p_i(x_i) dx_i] \quad (2)
$$

$$
\varphi_i(x_i) = \int f(\boldsymbol{x}) \prod_{j \neq i} [p_j(x_j) dx_j] - f_0 \quad (3)
$$

$$
\begin{aligned}
\varphi_{i_1 i_2}(x_{i_1}, x_{i_2}) = &\int f(\boldsymbol{x}) \prod_{j \neq i_1, i_2} [p_j(x_j) dx_j] \\
&- \varphi_{i_1}(x_{i_1}) - \varphi_{i_2}(x_{i_2}) - f_0 \quad (4)
\end{aligned}
$$

The interpretable neural network model, GAMI-Net [17], is a generalized additive model based on functional ANOVA where the dependent variable is the sum of a combination of variables. The method captures the main effects of the independent variables as well as the effects of pairwise interactions:

$$
\begin{aligned}
g(\mathbb{E}(y|\boldsymbol{x})) = \mu &+ \sum_{j \in S_1} h_j(x_j) \\
&+ \sum_{(j,k) \in S_2} f_{jk}(x_j, x_k) \quad (5)
\end{aligned}
$$

$$
\int h_j(x_j) dF(x_j) = 0, \forall j \in S_1 \quad (6)
$$

$$
\int f_{jk}(x_j, x_k) dF(x_j, x_k) = 0, \forall (j,k) \in S_2 \quad (7)
$$

$$
\int h_j(x_j) f_{jk}(x_j, x_k) dF(\boldsymbol{x}) = 0 \quad (8)
$$

Copyright © 2023 by ASME

where $F(x_j)$, $F(x_j, x_k)$, and $F(\boldsymbol{x})$ are the respective cumulative distributions. Decomposition terms average to zero (Equation (6), Equation (7)) and main and pairwise interaction effects are orthogonal (Equation (8)).

The GAMI-Net architecture contains two submodules for first capturing the main effects, $h_j(x_j)$ in Equation (5), then interaction effects, $f_{jk}(x_j, x_k)$ in Equation (5). For the main effects, a fully connected subnetwork is modeled from scratch between each individual input variable and the output. Each main effect is normalized to have zero means, then they are ranked according to their contributions and added to the model in descending order. Following that, fully-connected subnetworks fitted between two input variables for which the interaction is investigated and the residuals which were obtained after modeling the main effects in the first submodule. Similar to the main effects, each interaction effect is normalized to have zero means, then they are ranked and added to the model with respect to their contributions (Fig. 3). GAMI-Net also uses one-hot encoding to support the analysis of a combination of continuous and categorical variables. To improve interpretability and limit model complexity, GAMI-Net prunes insignificant main and interaction effects, includes an interaction effect only if at least one main effect parents are significant and kept in the model, and makes sure that main effects and their child interactions are not correlated. Finally, all network parameters are fine-tuned during the model-tuning stage.

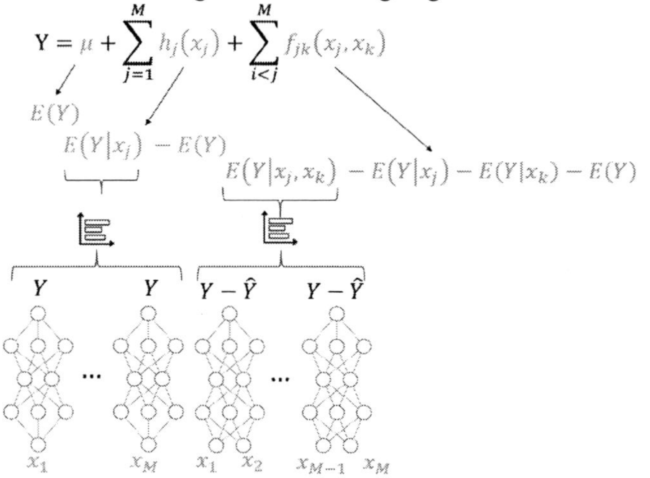

$$Y = \mu + \sum_{j=1}^{M} h_j(x_j) + \sum_{i<j}^{M} f_{jk}(x_j, x_k)$$

FIGURE 3: GAMI-NET MODEL STRUCTURE. FIRST, SUBNETWORKS ARE FITTED FOR THE MAIN EFFECTS AND THEY ARE ADDED TO THE MODEL WITH RESPECT TO THEIR CONTRIBUTIONS. THEN, SUBNETWORKS ARE FITTED FOR THE PAIRWISE INTERACTION EFFECTS AND THEY ARE ADDED TO THE MODEL ACCORDING TO THEIR CONTRIBUTION RANKING. IN THE FINAL STAGE, ALL THE NETWORK PARAMETERS ARE FINE-TUNED.

The importance of main and interaction effects are then assessed based on the amount of output variation they explain:

$$D(h_j) = \frac{1}{n-1} \sum h_j^2(x_j) \tag{9}$$

$$D(f_{jk}) = \frac{1}{n-1} \sum f_{jk}^2(x_j, x_k) \tag{10}$$

where n is the sample size. Then the contribution of each feature is further calculated as in Equation (11), Equation (12) where $IR(j)$ corresponds to the importance ratio of input variable j and $IR(j, k)$ refers to the importance ratio of the interaction between the input variables j and k, respectively:

$$IR(j) = \frac{D(h_j)}{\sum_{j \in S_1} D(h_j) + \sum_{(j,k) \in S_2} D(f_{jk})} \tag{11}$$

$$IR(j, k) = \frac{D(f_{jk})}{\sum_{j \in S_1} D(h_j) + \sum_{(j,k) \in S_2} D(f_{jk})} \tag{12}$$

2.2 Variable Set Decomposition

Engineering design problems often involve simultaneously designing multiple systems to achieve a target function. Thus, problems contain numerous input variables, which increases the model size and makes it difficult to analyze the effect of a variety of individual inputs.

In many cases, it is desirable to divide the model inputs into meaningful groups and perform analysis for these instead of separate, individual variables. Forming groups for design and noise variables in robust design and forming groups of material, architecture, and stimulus variables in programmable metamaterial design are examples where this approach is valuable.

Statistical methods are already available for the sensitivity analysis of a group of variables in the literature. Sobol indices [30], a very well-known variance-based sensitivity analysis method, can handle either individual inputs or sets of inputs as the output variance can be decomposed with respect to input groups. Morris method with grouping [31] also allows analyzing groups of variables by varying the variables within a group simultaneously along a trajectory, then observing the change in the system response. This method fails to distinguish low and high-order interactions. Derivative-based global sensitivity measures (DSGM) [32] calculates the average of local derivatives of the variables from the same group. This approach involves working with gradients, which is not applicable to problems with categorical variables.

In this article, we use subset decomposition [18] which is a variance-based sensitivity analysis method similar to the functional ANOVA decomposition for individual variables. Variance decomposition is used in global sensitivity analysis to partition the total variance into input variables. Assuming that the groups are statistically independent, we use the same logic for the decomposition of the input variables to partition the total variance into groups of them:

Copyright © 2023 by ASME

$$f(x)$$
$$= f_0 + \sum_r \hat{\varphi}_{U_i}(x_{U_i})$$
$$+ \sum_{i_1=1}^{r} \sum_{i_2=i_1+1}^{r} \hat{\varphi}_{U_{i_1} U_{i_2}}\left(x_{U_{i_1} U_{i_2}}, x_{U_{i_1} U_{i_2}}\right)$$
$$+ \hat{\varphi}_{U_1 \dots U_T}(x_{U_1}, \dots, x_{U_T}) \tag{13}$$

where "$\wedge$" refers to the decomposition items belonging to the same group. Total variation now can be decomposed into the sum of variances caused by the groups:

$$V = \sum \hat{V}_{U_i} + \sum_{i_1 < i_2} \hat{V}_{U_{i_1} U_{i_2}} + \dots + \hat{V}_{U_1 \dots U_T} \tag{14}$$

The importance of each subset then represents how much of the total variability each group accounts for:

$$\hat{S}_{U_{i_1} \dots U_{i_s}} = \frac{\hat{V}_{U_{i_1} \dots U_{i_s}}}{V} \tag{15}$$

Obtained group importance values correspond to global sensitivity indices which can be used for model interpretability with respect to input groups.

3. RESULTS

3.1 Synthetic Example

In the first example, we analyze an analytical function with 10 input variables (Equation (16)). The defined synthetic function involves a variety of interactions of varying shapes and strengths between the variables. It is a high-dimensional problem with 10 independent variables, involving terms with intuitive subsets of them. The problem of interest is a complex function with different mathematical operations including trigonometric, logarithmic, and exponential calculations. We hope to analyze the effect of different mathematical operation terms on the function output. This first example is intentionally designed to be intuitive for validating the method.

$$f(x) = \tanh(x_1 x_2 + x_3 x_4)\sqrt{|x_5|} + 0.3e^{x_5 + x_6}$$
$$+ \log((x_6 x_7 x_8)^2 + 1) + 2x_5 x_{10}$$
$$+ \frac{1}{|3x_9| + |3x_{10}|} \tag{16}$$

The importance of the main effects are not easy to predict before the analysis as it highly depends on the distribution of the variables as well as the respective mathematical operation type. Still, pairwise interactions between variables within the same term, such as x_5 and x_6 from the exponential term and x_6, x_7, and x_8 from the logarithmic term, are expected.

3.1.1 Dataset Construction

The function contains ten continuous independent variables where $x_i \sim U(-1, 1), \forall i$ and one continuous dependent variable. Identical uniform distributions are selected for all input variables to see the effects of each one across the design space. The Latin hypercube sampling method is used for generating 10,000 samples.

3.1.2 Global Sensitivity Analysis of Individual Variables

We build an interpretable artificial neural network model for modeling Equation (16) using GAMI-Net [33]. The prepared dataset is split into training and test sets with a 0.2 test set ratio. In the network architecture, we consider the main effects and the most meaningful first 20 pairwise interaction effects (Fig. 4(a)). Each subnetwork has 5 ReLU hidden layers with 40 nodes per layer. The batch size is set to 512 while the maximum number of epochs for the main effects, interaction effects, and model tuning is 1000. The learning rates of the Adam optimizer for the main effects, interaction effects, and model tuning is all 0.0001. Model accuracy is evaluated with mean squared error. Trained interpretable neural network model achieved 0.0001 mean squared error and 0.7346 R-squared on the test set.

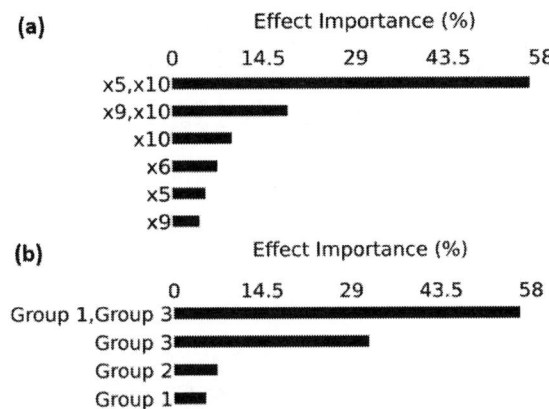

FIGURE 4: IMPORTANCE OF THE MAIN EFFECTS AND PAIRWISE INTERACTION EFFECTS OF THE MATH FUNCTION FOR: (A) INPUTS, (B) INPUT SUBSETS.

The interaction between the variables x_5 and x_{10} has the highest impact on the output, accounting for 56.2% of the total variation. Following that comes another interaction effect between x_9 and x_{10} but with only 18.3% importance. All main effect contributions are less than 10% with x_{10}, x_6, x_5, and x_9 standing out. Significant interactions between x_5, x_{10} and x_9, x_{10} are mathematically valid with the terms $2x_5 x_{10}$ and $1/(|3x_9| + |3x_{10}|)$ in Equation (16). Considering the sampling distributions for the respective variables, multiplication and multiplicative inverse of absolute value sums mathematical operations are expected to have high impacts on the function output.

Copyright © 2023 by ASME

We additionally used Sobol sensitivity analysis for the same problem [34]. Saltelli sampling method is used for generating 131,000 samples from the same distributions defined previously. Most meaningful effects are x_5, x_{10} interaction with 0.3368 and x_9, x_{10} interaction with 0.2099. The following effects are x_{10}, x_9, x_6, x_5, x_1 and x_2 interaction, x_3 and x_4 interaction, and finally x_5 and x_6 interaction all having negligible sensitivity indices compared to the two dominant interactions x_5, x_{10} and x_9, x_{10}. GSA with Sobol sensitivity analysis results match the data-driven sensitivity analysis results in means of the ranking of the important effects. There are slight differences between the two methods regarding the unimportant terms in means of ranking and the strength of the effect. This is an expected result as GAMI-Net prunes insignificant effects and pairwise effects if none of the parent main effects are significant for limiting model complexity. Overall, the Sobol sensitivity analysis results validate the results of our data-driven GSA method.

3.1.3 Global Sensitivity Analysis of Variable Groups

We further analyze the importance of grouped effects instead of individual consideration of each variable as in Fig. 4(a). We sampled every variable from independent distributions, and no correlations between the inputs exist. Thus, we can form groups of the input space using the subset decomposition method.

Equation (16) contains 5 terms making up the function. Intuitively, we are interested in analyzing 3 groups of variables such that $Group_1 = \{x_1, x_2, x_3, x_4, x_5\}$, $Group_2 = \{x_6, x_7, x_8\}$, $Group_3 = \{x_9, x_{10}\}$ (Fig. 4(b)).

$$
\begin{aligned}
f(\boldsymbol{x}) = \ & \tanh(x_1 x_2 + x_3 x_4)\sqrt{|x_5|} + 0.3 e^{x_5 + x_6} \\
& + \log((x_6 x_7 x_8)^2 + 1) + 2x_5 x_{10} \\
& + \frac{1}{|3x_9| + |3x_{10}|}
\end{aligned}
\tag{16}
$$

Similar to the results shown in Fig. 4(a), interaction effects between the groups dominate the group main effects. Variables from $Group_1$ and $Group_2$, similarly from $Group_1$ and $Group_3$ appear together within the same terms in Equation (16). Thus, interactions are expected between $Group_1, Group_2$ and $Group_1, Group_3$. The interaction between $Group_1$ and $Group_3$ accounts for 56.2% importance, while individual effects of $Group_1$ and $Group_3$ are 5.1% and 31.8% respectively. This is an interesting outcome as the interaction effect significantly exceeds the separate, individual effects of the parent subsets. This result sets an example for the benefit of analyzing the interplay between variable groups. It is crucial in this problem to analyze how these two subsets affect each other and create a joint effect that is more critical than the parent variables. It is worth studying the shape of the interaction, the relationship between the subsets, as well as the strength of it.

3.2 Programmable Photonic Metasurface Design

3.2.1 Background

Photonic metasurfaces are artificially engineered structures that can support sophisticated light-matter interaction through subwavelength inclusions [35-36]. Advancements in design and fabrication of photonic metasurfaces enabled remarkable functionalities, ranging from nano to microscales, such as perfect absorption, super-resolution imaging, sensing, waveguiding, and invisibility cloak.

Programmable photonic metasurfaces are a special type of photonic metasurfaces that can transform between different functional states as a response to external stimulus [37-38]. To enable programmable photonic metasurfaces, a diverse array of physical mechanisms has been reported in the photonic communities; examples include mechanical [39-40], thermal [41-42], electric [43-45], chemical [46-47], and light [48-50].

Light-based programmable photonic metasurfaces, which is covered in this article, are open for multiple design entities; namely material, architecture, and stimulus. Stimulus, an input electromagnetic loading at a high level, deserves separate attention as it allows sufficient design freedom in terms of amplitude, phase, and polarization. In some prior work, the whole two-dimensional incident field was viewed as the stimulus to be designed. Ideally; material, architecture, and stimulus should be modeled and designed concurrently to ensure transparency and avoid suboptimality. Nevertheless, the common practice has been specifying the material (e.g., dielectric; metallic) and stimulus (e.g., a single frequency or a frequency band; polarization type) a priori and then only designing the architecture. Instead of managing the large design space with such simple approaches, in this study we analyze the complex interactions between material, architecture, and stimulus and reveal which ones deserve the design efforts.

3.2.2 Design Problem

In this article, we develop an interpretable neural network model for a photonic metasurface system (Fig. 5(a)). We are particularly interested in analyzing architecture and stimuli with respect to transmission. It is a key system response that quantifies the energy transport from input to output in a two-port system. We aim to reveal the relationships between architecture-transmission, stimulus-transmission, and their interaction on transmission. Under the stimulus, we consider excitation frequency (Fig. 5(b)) and polarization (Fig. 5(c)). As for the architecture, we consider four different types of geometric families, each of which is allowed for parametric variation specified by two continuous variables (Fig. 5(d)). The entire design space contains both categorical and continuous variables.

Copyright © 2023 by ASME

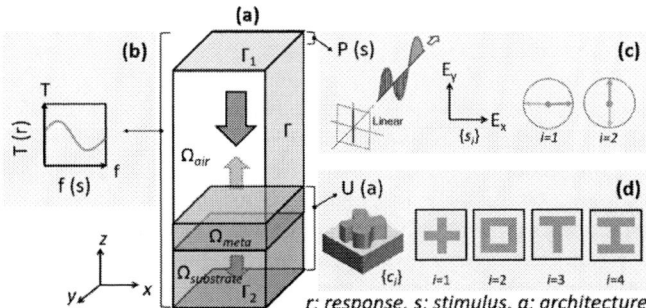

FIGURE 5: ILLUSTRATION OF THE PROGRAMMABLE PHOTONIC METASURFACE: (A) A SCHEMATIC OF THE WAVE ANALYSIS, (B) A TRANSMISSION SPECTRUM AS SYSTEM RESPONSE AND FREQUENCY AS STIMULUS, (C) POLARIZATION AS STIMULUS, (D) UNIT CELL TYPES WITH PARAMETRIC VARIATION AS ARCHITECTURE.

Frequency (Fig. 5(b)) is a continuous input variable in our model which belongs to the stimulus group. Importantly, excitation frequency is inversely proportional to wavelength λ. The ratio between λ and a characteristic length scale of the system (e.g., periodicity Λ in a metasurface), primarily governs the light-matter interaction. Depending on the order of the ratio λ/Λ, the associated behavior of light-matter interaction tends to vary significantly.

The second input variable in the stimulus group is polarization (Fig. 5(c)). Polarization is a property of transverse waves and it characterizes the orientation of the field oscillations. An electromagnetic wave, an instance of transverse waves, contains electric field E and magnetic field H, both of which have orthogonal directions to the wave propagation direction k. Polarization is conventionally described by stating the electric field direction. In this article, we consider two types of polarizations; x-directional and y-directional linear polarization which are treated as categorical variables in the model.

Under the architecture subset, we examine unit cell designs (Fig. 5(d)). Photonic metasurfaces contain periodic subwavelength features as the major building blocks in the architecture design. The cross-sectional geometry of the building blocks can take free form without any restrictions. In this article, we are particularly interested in four canonical classes of unit cells reported in photonics communities. Two continuous unit cell design parameters create parametric variation within each unit cell class. Due to the complicated light-matter interaction, predicting the influence of unit cell design, which is defined by the unit cell type and additional parameters, on the response is challenging.

The target response we focus on is transmission (Fig. 5(b)). The power transport between the ports is described by an S-parameter matrix, whose individual components correspond to pairwise power transports. Given an n-port system with port k as the input port, the power transport S_{ij} from port i to port j of the electromagnetic system is computed with [51]:

$$S_{ij} = \begin{cases} \dfrac{\int_{A_i} \left((E_c - E_i).E_i^\dagger \right) dA_i}{\int_{A_i} \left(E_i.E_i^\dagger \right) dA_i} & i = k \\[2em] \dfrac{\int_{A_i} \left(E_c.E_i^\dagger \right) dA_i}{\int_{A_i} \left(E_j.E_i^\dagger \right) dA_i} & otherwise \end{cases} \quad (17)$$

where E_c is the computed electric field that includes both excitation and scattered field, A_i is the face of port i, and $\dagger$ is the conjugate operator. This simulation can be viewed as a two-port network with the excitation port at the top face (Γ_1 in Fig. 5(a)) and the listener port at the bottom face ((Γ_2 in Fig. 5(a))). From the S-parameter matrix, transmission is formulated as:

$$T(\omega) = |S_{21}(\omega)|^2 \quad (18)$$

where ω is the excitation angular frequency.

3.2.3 Dataset Construction

The incident wave, stimulus in the design problem, can be viewed as an electromagnetic loading condition and is illuminated from the top face, propagating along the -z direction. It is a plane wave specified by two input conditions: excitation frequency $f \in I_F = [30, 60]$ THz and two polarization types.

As for the architecture in the design problem, we consider four different families. The cross-section is extruded along the z-direction with the height $H = 1000$ nm. The periodicity Λ of the analysis domain is set as $\Lambda = 2800$ nm. Assumed to be lossless, the refractive index n of the metasurface and the SiO2 substrate is set as $n = 5$ and $n = 1.45$, respectively. All the lateral faces Γ are subject to periodic boundary conditions. This setting effectively mimics the periodical tessellation of identical, infinitely many building blocks on the xy-plane.

The full-wave analysis is conducted by the RF module of COMSOL Multiphysics® [51]. The simulations for the data generation process involve solving the Maxwell's equation, with Equation (17) applied to the excitation and listener ports of the system as the boundary condition. The analysis domains illustrated in Fig. 5(a) are discretized into about 10,000 tetrahedral elements of 65,837 degrees of freedom. Each analysis simulation involves 31 different excitation frequencies and takes about two to three minutes on a computer with 3.60 GHz Intel dual core processors and 16GB of RAM.

The input space is constructed as: $U \times G \times P \times f \rightarrow T$ where $U = \{c_j | j = 1,2,3,4\}$ is the set of unit cell types, $g = \{g_j | j = 1,2\} \in G \subset \mathbb{R}^2$ is the continuous vector that specifies parametric variation within G with respect to a given unit cell type c_j, $P = \{s_i | i = 1,2\}$ is the set of polarization types, $f \in I_F = [30, 60]$ THz is the excitation frequency, $T \in [0,1]$ is the transmission.

To generate a dataset D, 20 space-filling designs are sampled from G using Latin hypercube sampling method. The

Copyright © 2023 by ASME

frequency band in concern I_F is discretized with a spacing of $\Delta f = 1$ THz. As a result, D includes $|D| = |U| \times |G| \times |P| \times (|I_F|/\Delta f) = 4{,}960$ observations.

3.2.4 Global Sensitivity Analysis of Individual Variables

We use GAMI-Net [33] to build an interpretable artificial neural network model for this problem. The test to train dataset ratio is 0.2. In the network architecture, we consider the main effects and the most meaningful first 20 pairwise interaction effects (Fig. 6(a)). Each subnetwork has 5 ReLU hidden layers with 40 nodes per layer. The batch size is set to 512 while the maximum number of epochs for the main effects, interaction effects, and model tuning is 1000. The learning rates of the Adam optimizer for the main effects, interaction effects, and model tuning is all 0.0001. Model accuracy is evaluated with mean squared error. The trained interpretable neural network model achieved 0.0022 mean squared error and 0.9793 R-squared on the test set.

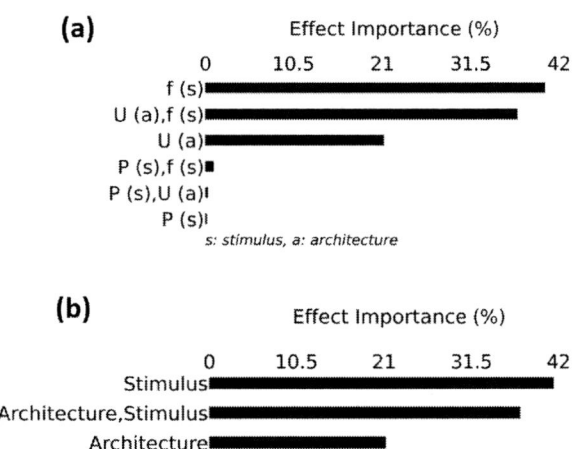

FIGURE 6: IMPORTANCE OF THE MAIN EFFECTS AND PAIRWISE INTERACTION EFFECTS OF THE PHOTONIC METASURFACE FOR: (A) INPUTS, (B) INPUT SUBSETS.

Excitation frequency steps up with 40.3% of the total effect on transmission. The following main effect is unit cell type with 21.2% importance, with even higher importance (37.1%) of its interaction with frequency. Polarization on its own and its interactions with other variables all show negligible significance.

3.2.5 Global Sensitivity Analysis of Variable Groups

We obtain groups of the input variables to further analyze the importance of grouped main and interaction effects instead of individual consideration of each input variable as in Fig. 6(a). Since we sampled every variable from independent distributions, no correlations between the inputs exist. Thus, we can form groups in the variable space using subset decomposition.

Photonic metasurface design problem naturally contains variables related to architecture and stimulus. Here, we are interested in inferring the importance of these groups as well as

any interactions between them such that $Architecture = \{U, G\}, Stimuli = \{P, f\}$ (Fig. 6(b)).

Stimulus is the most important variable group for explaining the metasurface response with 41.4% importance. The other defined group, architecture, has a value of 21.2%. Surprisingly, designing architecture on its own or similarly designing stimulus in isolation would be an inadequate approach for this case as the interaction between stimulus and architecture has a substantial effect on the response with 37.4%. Thus, it is imperative to further analyze the architecture-stimulus relation and their combined influence on the system response. This result is expected and it stems from the complex interactions in PMS problems.

4. DISCUSSION

Figure 6(b) indicates the value of further investigating stimulus, architecture, and stimulus-architecture interaction [37]. For this, we use the PDPs of polarization (Fig. 7(a)) and frequency (Fig. 7(b)) for stimulus, the PDP of unit cell type (Fig. 7(c)) for architecture, and the PDPs of polarization-unit cell type (Fig. 7(d)) and frequency-unit cell type (Fig. 7(e)) for stimulus-architecture interactions.

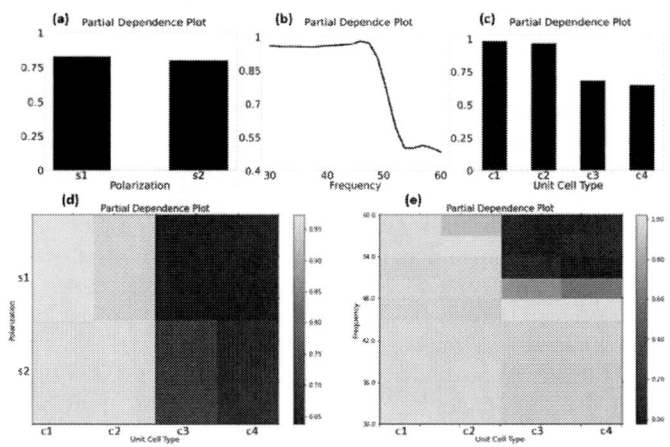

FIGURE 7: PROGRAMMABLE PHOTONIC METASURFACE PDPS FOR THE SIGNIFICANT SUBSETS AND SUBSET INTERACTIONS FOR: (A) POLARIZATION (B) FREQUENCY, (C) UNIT CELL TYPE, (D) POLARIZATION-UNIT CELL TYPE (E) UNIT CELL TYPE-FREQUENCY.

Two polarization types covered in this article, s_1 and s_2, have similar effects on the system response (Fig. 7(a)). Hence, polarization type should not be a concern for this problem.

When the frequency is below 47 THz, transmission reaches the maximum with a value of 1, whereas it displays a steep decrease for [47,52] THz, and finally settles to 0.5 when the frequency is above 52 THz (Fig. 7(b)).

Among the four different unit cell types, c_1 and c_2 display identical effects as they maximize the transmission. A similar situation is observed for c_3 and c_4 resulting in a transmission of

Copyright © 2023 by ASME

0.7. Therefore, c_1 - c_2 and c_3 - c_4 can be interchangeably used without significant changes in the system response (Fig. 7(c)).

As for the interactions, the polarization-unit cell type does not offer any unexpected insights in addition to the individual polarization PDP and unit cell type PDP. Again, polarization type does not have any meaningful contribution while unit cell types c_1 - c_2 and c_3 - c_4 have similar effects (Fig. 7(d)).

When it comes to frequency-unit cell type interaction, insightful observations appear (Fig. 7(e)). When the excitation frequency is set below 52 THz, the transmission response of the photonic metasurface becomes maximum with a value of 1. This system output is valid regardless of the unit cell type, that is no matter what architecture design is preferred, the maximum system response is attained. When the frequency is set above 52 THz, the system response becomes either maximum or minimum depending on the unit cell type. To elaborate, c_1 and c_2 maximize the transmission while c_3 and c_4 minimize it. For frequencies higher than 52 THz, architecture plays a critical role. This result validates the necessity of simultaneous consideration of stimulus and architecture since the interaction between them brings about a unique effect on the model response.

TABLE 1: DESIGN GUIDELINES FOR DEVELOPING A PROGRAMMABLE PHOTONIC METASURFACE THAT ACHIEVES TWO FUNCTIONAL STATES.

Maximum response: $T = 1$		Minimum response: $T = 0$	
Stimulus	Architecture	Stimulus	Architecture
$f > 52$ THz	c_1, c_2	$f < 52$ THz	c_3, c_4
$f < 52$ THz	c_1, c_2, c_3, c_4		

Table 1 summarizes the conclusions we derive for achieving two functional states. To set the programmable system to maximum transmission, two alternative configurations are available. For the other functional state when the transmission is minimum, just one configuration is convenient.

$f < 52$ THz allows achieving both states, thus could be preferable when using two different unit cells is inexpensive. Similarly, c_1 or c_2 also can result in both states and can be a better option when using two different excitation frequencies is inexpensive.

Our method helped us to identify one frequency and two unit cells or two frequencies and one unit cell for designing a PMS with two functional states. We reduced a complex problem with a large input space with many parameters to just a few items.

Our framework is promising for constructing transferable design rules as follows:

1) In case where both stimulus and architecture are open for design, it is better to conduct concurrent design in the joint design space of architecture and stimulus. Separate, independent modeling for each would be invalid as the associated interaction is as significant as the main effects.
2) If we are interested in designing a system that exhibits distinct transmission according to polarization, the geometric classes that are sensitive to polarization shift can be taken.

3) The global sensitivity analysis can be harnessed for pre-specifying a set of unit cell types when generating a high-quality, diverse dataset for data-driven metamaterial design.
4) The framework reveals the impact of parametric variation within a particular class.
5) The analysis offers the impact of each group of variables. This enables subset screening in concurrent design, as an extension to variable screening, which is useful for defining a compact design space that captures major variations.
6) Optionally, the architecture space can be extended through multiclass blending. In this case, our trained model can serve as an initial surrogate for active learning.

5. CONCLUSION

In this article, we study the effects of material, architecture, and stimulus as well as interactions between them for a programmable photonic metasurface system. The employed method performed the functional ANOVA decomposition of a machine learning model for partitioning the output variation into terms associated with the inputs. We used our GSA analysis for understanding the complex physical interactions in engineering design. With the help of adopting interpretable machine learning over classical statistical methods, we were able to work with a mixed variable set containing both continuous and categorical variables, work with a flexible machine learning model, and analyze a high-dimensional problem with a large sample size. Apart from presenting these advantages of the utilized method in the engineering design context; we improved the interpretable neural network analysis to further interpret the grouped effects we revealed the complex interactions in engineering design problems, and we present an interpretable machine learning application for global sensitivity analysis in engineering design.

Considering the results of this article, we identify some future efforts with potential benefits. First, the model we used in this article only covers the interpretability of the main effects and pairwise interaction effects. In both problems we presented in this article, these results were able to capture almost 100% of the total effects on the system response. However, adjusting the interpretable neural network design for also considering higher-order interactions can generate a more accurate model with the cost of increased model complexity and computation time. Similarly, the model we use is suitable for single-target problems where only one dependent variable exists. Considering that many engineering problems involve multiple outputs, such as tensors of elastics constants in mechanics, adjusting the interpretable neural network structure for multi-output problems is a promising effort. A simple approach of building separate models for each output would be inaccurate in cases where the output variables are correlated. Also, working with one model for multi-outputs would be computationally efficient as it allows training just one model instead of separate ones for each output. Finally, we plan on analyzing more complex PMS problems where there are more input variables with interactions between all three groups of material, architecture, and stimulus. In this way, we hope to obtain more accurate design guidelines and ensure the transferability of the rules we obtained in this article.

Copyright © 2023 by ASME

ACKNOWLEDGEMENTS

This work is supported by the NSF BRITE fellow program (award number 2227641) and the NSF CSSI program (award number 1835782).

REFERENCES

[1] Saltelli, Andrea, Marco Ratto, Terry Andres, Francesca Campolongo, Jessica Cariboni, Debora Gatelli, Michaela Saisana, and Stefano Tarantola. Global sensitivity analysis: the primer. John Wiley & Sons (2008).

[2] Iooss, Bertrand, and Paul Lemaître. "A review on global sensitivity analysis methods." Uncertainty management in simulation-optimization of complex systems: algorithms and applications (2015): 101-122.

[3] Hadi, Ali S., and Samprit Chatterjee. Sensitivity analysis in linear regression. John Wiley & Sons (2009).

[4] Saltelli, Andrea, Paola Annoni, Ivano Azzini, Francesca Campolongo, Marco Ratto, and Stefano Tarantola. "Variance based sensitivity analysis of model output. Design and estimator for the total sensitivity index." Computer physics communications 181, no. 2 (2010): 259-270.

[5] Cheng, Kai, Zhenzhou Lu, Chunyan Ling, and Suting Zhou. "Surrogate-assisted global sensitivity analysis: an overview." Structural and Multidisciplinary Optimization 61 (2020): 1187-1213.

[6] Christopher Frey, H., and Sumeet R. Patil. "Identification and review of sensitivity analysis methods." Risk analysis 22, no. 3 (2002): 553-578.

[7] Panchal, Jitesh H., Mark Fuge, Ying Liu, Samy Missoum, and Conrad Tucker. "Machine learning for engineering design." Journal of Mechanical Design 141, no. 11 (2019).

[8] Molnar, Christoph. Interpretable machine learning. Lulu. com (2020).

[9] Miller, Tim. "Explanation in artificial intelligence: Insights from the social sciences." Artificial intelligence 267 (2019): 1-38.

[10] Litjens, Geert, Thijs Kooi, Babak Ehteshami Bejnordi, Arnaud Arindra Adiyoso Setio, Francesco Ciompi, Mohsen Ghafoorian, Jeroen Awm Van Der Laak, Bram Van Ginneken, and Clara I. Sánchez. "A survey on deep learning in medical image analysis." Medical image analysis 42 (2017): 60-88.

[11] Brennan, Tim, and William L. Oliver. "Emergence of machine learning techniques in criminology: implications of complexity in our data and in research questions." Criminology & Pub. Pol'y 12 (2013): 551.

[12] Jaeger, Markus, Stephan Krügel, Dimitri Marinelli, Jochen Papenbrock, and Peter Schwendner. "Interpretable machine learning for diversified portfolio construction." The Journal of Financial Data Science 3, no. 3 (2021): 31-51.

[13] Angermueller, Christof, Tanel Pärnamaa, Leopold Parts, and Oliver Stegle. "Deep learning for computational biology." Molecular systems biology 12, no. 7 (2016): 878.

[14] Vu, Mai-Anh T., Tülay Adalı, Demba Ba, György Buzsáki, David Carlson, Katherine Heller, Conor Liston et al. "A shared vision for machine learning in neuroscience." Journal of Neuroscience 38, no. 7 (2018): 1601-1607.

[15] Chen, Wei, Janet K. Allen, Kwok-Leung Tsui, and Farrokh Mistree. "A procedure for robust design: minimizing variations caused by noise factors and control factors." (1996): 478-485.

[16] Fan, X., Chung, J. Y., Lim, Y. X., Li, Z., & Loh, X. J. (2016). Review of adaptive programmable materials and their bioapplications. ACS applied materials & interfaces, 8(49), 33351-33370.

[17] Yang, Zebin, Aijun Zhang, and Agus Sudjianto. "GAMI-Net: An explainable neural network based on generalized additive models with structured interactions." Pattern Recognition 120 (2021): 108192.

[18] Chen, Wei, Ruichen Jin, and Agus Sudjianto. "Analytical variance-based global sensitivity analysis in simulation-based design under uncertainty." Journal of mechanical design 127, no. 5 (2005): 875-886.

[19] Friedman, Jerome H. "Greedy function approximation: a gradient boosting machine." Annals of statistics (2001): 1189-1232.

[20] Apley, Daniel W., and Jingyu Zhu. "Visualizing the effects of predictor variables in black box supervised learning models." Journal of the Royal Statistical Society Series B: Statistical Methodology 82, no. 4 (2020): 1059-1086.

[21] Breiman, Leo. "Bagging predictors." Machine learning 24 (1996): 123-140.

[22] Goldstein, Alex, Adam Kapelner, Justin Bleich, and Emil Pitkin. "Peeking inside the black box: Visualizing statistical learning with plots of individual conditional expectation." journal of Computational and Graphical Statistics 24, no. 1 (2015): 44-65.

[23] Ribeiro, Marco Tulio, Sameer Singh, and Carlos Guestrin. "" Why should i trust you?" Explaining the predictions of any classifier." In Proceedings of the 22nd ACM SIGKDD international conference on knowledge discovery and data mining, pp. 1135-1144. 2016.

[24] Lundberg, Scott M., and Su-In Lee. "A unified approach to interpreting model predictions." Advances in neural information processing systems 30 (2017).

[25] Marchese Robinson, Richard L., Anna Palczewska, Jan Palczewski, and Nathan Kidley. "Comparison of the predictive performance and interpretability of random forest and linear models on benchmark data sets." Journal of chemical information and modeling 57, no. 8 (2017): 1773-1792.

[26] Baryannis, George, Samir Dani, and Grigoris Antoniou. "Predicting supply chain risks using machine learning: The trade-off between performance and interpretability." Future Generation Computer Systems 101 (2019): 993-1004.

[27] Dunnington, Dewey W., Benjamin F. Trueman, William J. Raseman, Lindsay E. Anderson, and Graham A. Gagnon. "Comparing the Predictive performance, interpretability, and accessibility of machine learning and physically based models for water treatment." ACS ES&T Engineering 1, no. 3 (2020): 348-356.

Copyright © 2023 by ASME

[28] Johansson, Ulf, Cecilia Sönströd, Ulf Norinder, and Henrik Boström. "Trade-off between accuracy and interpretability for predictive in silico modeling." Future medicinal chemistry 3, no. 6 (2011): 647-663.

[29] Hooker, Giles. "Generalized functional anova diagnostics for high-dimensional functions of dependent variables." Journal of Computational and Graphical Statistics 16, no. 3 (2007): 709-732.

[30] Sobol, Ilya M. "Global sensitivity indices for nonlinear mathematical models and their Monte Carlo estimates." Mathematics and computers in simulation 55, no. 1-3 (2001): 271-280.

[31] Campolongo, Francesca, Jessica Cariboni, and Andrea Saltelli. "An effective screening design for sensitivity analysis of large models." Environmental modelling & software 22, no. 10 (2007): 1509-1518.

[32] Sobol, I. M., and S. Kucherenko. "Derivative based global sensitivity measures." Procedia-Social and Behavioral Sciences 2, no. 6 (2010): 7745-7746.

[33] Sudjianto, Agus, Zhang, Aijun, Yang, Zebin, Su, Yu, Zeng, Ningzhou, & Nair, Vijay. (2022). PiML: A Python toolbox for interpretable machine learning model development and validation.

[34] Hansen, David J., Jennifer T. McGuire, and Binayak P. Mohanty. "Enhanced biogeochemical cycling and subsequent reduction of hydraulic conductivity associated with soil-layer interfaces in the vadose zone." Journal of environmental quality 40, no. 6 (2011): 1941-1954.

[35] Bukhari, Syed S., J. Vardaxoglou, and William Whittow. "A metasurfaces review: Definitions and applications." Applied Sciences 9, no. 13 (2019): 2727.

[36] Chen, Hou-Tong, Antoinette J. Taylor, and Nanfang Yu. "A review of metasurfaces: physics and applications." Reports on progress in physics 79, no. 7 (2016): 076401.

[37] Yang, Huanhuan, Xiangyu Cao, Fan Yang, Jun Gao, Shenheng Xu, Maokun Li, Xibi Chen, Yi Zhao, Yuejun Zheng, and Sijia Li. "A programmable metasurface with dynamic polarization, scattering and focusing control." Scientific reports 6, no. 1 (2016): 1-11.

[38] Nemati, Arash, Qian Wang, Minghui Hong, and Jinghua Teng. "Tunable and reconfigurable metasurfaces and metadevices." Opto-Electronic Advances 1, no. 5 (2018): 180009.

[39] Specht, Marius, Matthew Berwind, and Chris Eberl. "Adaptive Wettability of a Programmable Metasurface." Advanced Engineering Materials 23, no. 2 (2021): 2001037.

[40] Liu, Shuo, Lei Zhang, Guo Dong Bai, and Tie Jun Cui. "Flexible controls of broadband electromagnetic wavefronts with a mechanically programmable metamaterial." Scientific Reports 9, no. 1 (2019): 1-9.

[41] Lor, Chhunheng, Ratanak Phon, Minjae Lee, and Sungjoon Lim. "Multi-functional thermal-mechanical anisotropic metasurface with shape memory alloy actuators." Materials & Design 216 (2022): 110569.

[42] Yin, Haoyu, Qingxuan Liang, Yubing Duan, Jun Fan, and Zhaohui Li. "3D printing of a thermally programmable conformal metasurface." Advanced Materials Technologies 7, no. 7 (2022): 2101479.

[43] Shirmanesh, Ghazaleh Kafaie, Ruzan Sokhoyan, Pin Chieh Wu, and Harry A. Atwater. "Electro-optically tunable multifunctional metasurfaces." ACS nano 14, no. 6 (2020): 6912-6920.

[44] Wan, Xiang, Mei Qing Qi, Tian Yi Chen, and Tie Jun Cui. "Field-programmable beam reconfiguring based on digitally-controlled coding metasurface." Scientific reports 6, no. 1 (2016): 20663.

[45] Fu, Xiaojian, Lei Shi, Jun Yang, Yuan Fu, Chenxi Liu, Jun Wei Wu, Fei Yang, Lei Bao, and Tie Jun Cui. "Flexible terahertz beam manipulations based on liquid-crystal-integrated programmable metasurfaces." ACS Applied Materials & Interfaces 14, no. 19 (2022): 22287-22294.

[46] Dong, Shi, Kai Zhang, Zhiping Yu, and Jonathan A. Fan. "Electrochemically programmable plasmonic antennas." Acs Nano 10, no. 7 (2016): 6716-6724.

[47] Li, Jianxiong, Yiqin Chen, Yueqiang Hu, Huigao Duan, and Na Liu. "Magnesium-based metasurfaces for dual-function switching between dynamic holography and dynamic color display." ACS nano 14, no. 7 (2020): 7892-7898.

[48] Kao, Tsung-Sheng, E. T. F. Rogers, Jun-Yu Ou, and N. I. Zheludev. ""Digitally" addressable focusing of light into a subwavelength hot spot." Nano letters 12, no. 6 (2012): 2728-2731.

[49] Buijs, Robin D., Tom AW Wolterink, Giampiero Gerini, Ewold Verhagen, and A. Femius Koenderink. "Programming Metasurface Near-Fields for Nano-Optical Sensing." Advanced Optical Materials 9, no. 15 (2021): 2100435.

[50] Lee, Doksoo, Shizhou Jiang, Oluwaseyi Balogun, and Wei Chen. "Dynamic control of plasmonic localization by inverse optimization of spatial phase modulation." ACS Photonics 9, no. 2 (2021): 351-359.

[51] COMSOL Multiphysics® v. 6.1. www.comsol.com. COMSOL AB, Stockholm, Sweden.

Copyright © 2023 by ASME

Proceedings of the ASME 2023
International Design Engineering Technical Conferences and
Computers and Information in Engineering Conference
IDETC-CIE2023
August 20-23, 2023, Boston, Massachusetts

DETC2023-115206

INVESTIGATE CUSTOMER PREFERENCES USING ONLINE VIDEO REVIEWS - PRELIMINARY RESULTS

Kangcheng Lin, Harrison Kim
Enterprise Systems Optimization Laboratory
Department of Industrial and Enterprise Systems Engineering
University of Illinois at Urbana-Champaign
Urbana, Illinois 61801
Email: {klin14, hmkim}@illinois.edu

ABSTRACT

The wealth of online reviews has grown exponentially, attracting the attention of many researchers who recognize their potential as a valuable source of customer feedback. Leveraging online reviews enables product designers to gain a deeper understanding of customer preferences and make informed decisions to improve product design. Traditionally, the major source of online reviews comes from e-commerce websites, such as Amazon and eBay. However, as social media platforms gain popularity, video reviews from these platforms are becoming an increasingly important source of customer feedback. Video reviews have several advantages over traditional textual reviews. They are more comprehensive, less likely to be fake, have greater coverage of the customer base, and contain more interaction in the form of comments. The paper presents a four-stage methodology to analyze video reviews from social media platforms as an alternative source of customer feedback. This involves collecting and preprocessing video reviews, extracting product features using latent Dirichlet allocation (LDA) models, analyzing sentiment using the valence aware dictionary and sentiment reasoner (VADER) package, and computing feature importance using SHAP values. To the best knowledge of the authors, very few literatures have investigated the feasibility of using video reviews as a substitute for traditional online reviews. Therefore, this paper contributes to the growing literature by demonstrating the viability of video reviews as an alternative source of online re-

views to understand customer preferences and their implications for engineering design.

1 INTRODUCTION

Customer preferences have always been an essential part of decision-making process in engineering design. Traditionally, surveys and questionnaires were the major avenues for customer feedback. Nonetheless, these methods were slow, expensive, geometrically restricted and potentially biased. Therefore, in the recent decades, online reviews have been intensively studied to understand customer preferences [1–6]. One of the most common sources of online reviews is from e-commerce websites, such as Amazon and eBay. As new generation of Internet users appears, however, e-commerce websites are no longer the only go-to place for customers to express their opinions about products. Alternatively, social media platforms, such as Youtube, Facebook, and Tiktok, are gaining an increasing amount of attention, because not only do they offer a large audience, but also they allow customers to communicate their thoughts in a variety of content formats, such as video reviews, unboxing videos, and product comparisons.

Video reviews, as compared to textual reviews from e-commerce websites, have several advantages. Firstly, they are usually more well-structured and comprehensive than textual reviews. Textual reviews can sometimes be too succinct: cus-

Copyright © 2023 by ASME

tomers may point out a single product feature that either impresses or frustrates them, without describing a thorough experience about the product; for example, *"very nice looking phone. Only returning because I wanted a bigger screen"*. Secondly, video reviews have a better chance of avoiding fake reviews, as the actual product is commonly shown in the video. Thirdly, video reviews can provide greater coverage of customer base. For example, if we are collecting reviews of cellphones from Amazon, then we will miss out the feedback from customers purchasing brand-new iPhones, because brand-new iPhones are not sold via Amazon. In contrast, social media platforms are readily available to anyone. Lastly, video reivews contain much more interaction, in form of comments, than traditional textual reviews, allowing a deeper analysis with the additional layers of discussion.

This paper intends to investigate video reviews from social media platforms as an alternative source of online customer feedback. Specifically, our methodology comprises four stages: (i) collect and preprocess video reviews from social media platforms, (ii) extract product features of customer interests from video reviews using the latent dirichelet allocation (LDA) models, (iii) analyze their corresponding sentiment using the valence aware dictionary and sentiment reasoner (VADER) package, and (iv) compute feature importance using SHAP values.

The paper is organized as follows. Section 2 will discuss relevant researches associated to the main topics in this paper. Section 3 will elaborate in details the proposed methodology and experiment design. Section 4 will present the data and results for the case study using the proposed methodogy and experiment. Section 5 will provide a summary and discussion of future work before concluding the paper.

2 LITERATURE REVIEW

In this section, we will present three main topics related to the paper, namely *Customer Preference Elicitation*, *Topic Modeling*, *Sentiment Analysis*, and *Importance Analysis*.

2.1 Customer Preference Elicitation

In recent decades, numerous studies have explored ways to enhance product design by leveraging online reviews. Traditional approaches such as questionnaires and surveys can be costly to scale up and may suffer from temporal and geographical biases. Alternatively, methodologies that use online reviews are more economical and easier to implement, potentially yielding less biased results. Chen et al. [7] proposed analytical discrete choice models to comprehend different customer tastes and predict their purchase decisions. Similarly, Tuarob et al. [8] developed a rule-based method to extract features using pre-defined rules and seed features. Additionally, text mining has been used in several studies to identify significant customer reviews from a

product designer's perspective, such as those presented in [9–11].

While various approaches have been proposed in these studies, the overall systems typically consist of three primary components: (i) topic modeling to extract relevant product keywords and features from natural language reviews, (ii) sentiment analysis to quantify customer preferences regarding extracted product features, and (iii) importance analysis, which is used to identify the product features that a company should focus on based on their quantified importance and performance. The following subsections provide further details on previous literatures concerning these main components.

2.2 Topic Modeling

Topic modeling is a widely-used analytical tool for data analysis. Among the various algorithms utilized for text analysis across multiple domains, some of the most prominent ones are latent semantic analysis, non-negative matrix factorization, probabilistic latent semantic analysis, and latent dirichlet allocation (LDA) [12].

Latent semantic analysis is an algebraic technique that relies on singular value decomposition. It was first proposed by Landauer and Dumais in the 1990s [13] and has since been used in diverse areas such as information retrieval, natural language processing, and modeling of human language knowledge [14–16].

Non-negative matrix factorization and probabilistic latent semantic analysis are both dimension-reduction methods. The former was originally proposed for environmental data [17] but has since been adapted in numerous other research domains such as cancer identification using molecular gene expression datasets [18, 19]. The latter is a technique that is based on the bag-of-words approach and uses a probabilistic framework to detect the semantic co-occurrence of terms within a corpus [20].

LDA [21] is a generative statistical model that captures the statistical structures and distribution of documents across various topics. Each document consists of different words, and each topic can be associated with some of these words. Similar topics are assumed to have a cluster of similar words. LDA has been applied in numerous research areas such as e-commerce [6, 22–28].

Despite different methods of topic modeling developed based on the context of datasets, LDA is often preferred due to its flexibility and adaptability compared to other techniques.

2.3 Sentiment Analysis

Sentiment analysis is a text mining technique that quantifies the opinions, sentiments, and subjectivity of a given text. As a continuously evolving research field, numerous algorithms have been developed [29–33]. There are typically two types of sentiment analysis methods, namely unsupervised (e.g., lexicon-based) [34–36] and supervised [37–40] learning methods. While the latter methods are usually more accurate in their respective

Copyright © 2023 by ASME

domains, the former methods require less time and have lower memory complexity [41].

Recently, more studies have been utilizing sentiment analysis to extract customer preferences from online reviews. For instance, Zhang et al. [42] proposed an opinion mining algorithm to discover relationships between product attributes. Another study by Jiang et al. [43] developed a method to infer the importance of future product features using a fuzzy time series model. Additionally, a framework based on the social perception score of a brand and the polarity of reviews was proposed by Bag et al. [44] to build a prediction model of customer purchase intention. Lastly, to investigate the relation between online customer reviews and sales rank, Suryadi et al. [3] developed a systematic methodology based on word embedding and sentiment analysis using a dependency tree.

2.4 Importance Analysis

Importance analysis is a common technique used to strategically improve customer satisfaction. As another ongoing research field in text mining, various approaches have been developed for importance analysis. One approach to measure the importance of product features is by analyzing the frequency of product-related words. High-frequency product features and high-frequency low-sentiment-score product features are considered to have high importance, according to studies such as [43, 45]. Another classification of importance is self-stated (relevance) and implicit (determinance) importance, as proposed by Mikuli et al. [46]. Self-stated importance is measured directly by asking customers, while implicit importance requires different handling. Decker et al. [47] proposed a negative binomial regression method based on sentiment scores and user ratings to estimate the implicit importance values of the pros and cons in a review. More recently, Bi et al. [48] presented a neural network for evaluating implicit importance in natural language text. Moreover, Joung et al. [49] developed a method to use information fusion-based Shapley additive explanation to understand product attribute importance by applying it to a deep neural network.

3 METHODOLOGY

In this seciton, we will present our methodology, which inccludes four stages: *data collection and preprocessing, identification of features of interests, feature sentiment,* and *feature importance.* The overall framework is demonstrated in Fig. 1.

3.1 Data Collection and Preprocessing

Video reviews of products can be collected from popular social media platforms such as Youtube, Facebook, and Tiktok. The data collected contain video titles, numbers of views, video release dates, video lengths, comments, and videos. Videos contain two major components: visual and audio. In this study, we only extract the audio component and convert it to textual form. Then, similar to the previous literature [1, 3, 5, 23, 50–52], we will preprocess the data by stripping off punctuations, removing emojis, and converting all letters to lower-case [53]. After that, we will extract all the nouns and noun phrases from the data.

However, some of the extracted nouns and noun phrases may not be relevant to our analysis: they may be unrelated to the product, such as *youtube channel* and *subscription*; they may be too general and contain no information about any specific product attributes, such as *samsung* and *iphone*. Therefore, the collection of nouns and noun phrases from video reveiws will be compared against relevant product manuals [3, 50, 51], which can be collected from the official websites of manufacturers or online e-commerce websites. Words that do not appear in the product manuals are removed, whereas the remaining ones are considered product keywords.

3.2 Identification of Features of Interests

In this stage, product features of customer interests are obtained from the list of product keywords in the previous stage. For this purpose, LDA, a probablistic topic model that summarizes enormous textual data by finding hidden topics, is selected. LDA makes use of a generative statistical model that creates a list of common topics that categorize all product reviews [54]. In LDA models, each product review can be summarized by a set of topics in probability, and each topic in the set can be summarized by a set of (product) keywords in probability. The number of topics is determined based on the measure of topic coherence, and the output of LDA is a topic-keyword matrix. Each topic is labeled based on the keywords and typical reviews associated wtih the topic. These labels represent product features of customer interests [4, 6, 51].

Once product features are identified, its keywords can be extended by adding synonyms. This study employs word embedding [55] for synonym extraction. To begin with, feature-relevant keywords are selected from the top-30 nouns in each topic. Subsequently, the top-20 words that are most similar to the selected feature-relevant keywords are extracted from word vectors. The union of these word sets generates the extended set feature keywords.

3.3 Feature Sentiment

This study utilizes an unsupervised method for sentiment analysis, which involves defining target words and collecting sentiment indices associated with these words. The feature keywords obtained in Section 3.2 serve as the target words for this study. The unsupervised method for sentiment analysis is preferred in this study due to its efficiency and speed, as it does not require labeling. In particular, VADER [33], an unsupervised

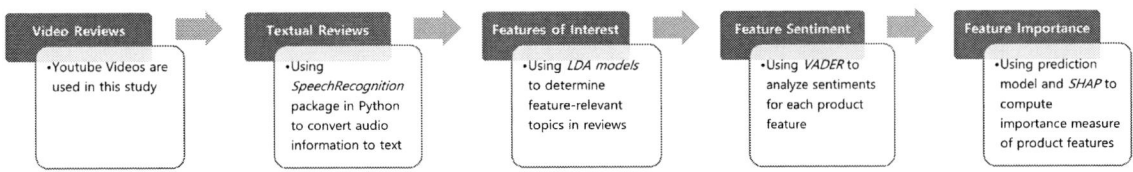

FIGURE 1: A visual demonstration of our methodology.

machine learning model based on lexicons and rules, is used to estimate customer sentiments for product features. This model does not require manual labeling for training data, making it easily applicable to other products and fields.

VADER sentiment analysis computes the polarity and intensity of a given sentence. To account for the possibility of a reviewer repeating the same feature, this study calculates the average sentiment score of relevant sentences. For example, if a reviewer comments on camera multiple times in the video, then the sentiment score for camera is computed as the average of the sentiment scores obtained for the relevant sentences. Sentiment scores are calculated for all product features mentioned in the review, and if a feature is not mentioned, its sentiment remains empty.

3.4 Feature Importance

To assess the importance of product features, this stage employs the SHAP-based method recommended by [49]. SHAP [56] is a method that evaluates the impact of all input variables on the output and can interpret black box neural network (NN) models. The method for importance estimation is presented in Fig. 2. The overall sentiments based on video titles and videos are used as ratings. As such, each review contains a vector of sentiment scores for product features and a rating score. The sentiment scores are obtained as explained in Section 3.3 and ranged from -1 to 1, where -1 indicates very negative, 0 is neutral, and 1 is very positive. To avoid empty input data for the prediction model, the sentiment for feature K is set to zero when a reviewer does not mention it. Reviews with zero sentiment scores for all features are discarded. Ratings, ranged from -1 to 1, are discretized by using a threshold: everything above threshold is categorized as positive, whereas everything below threshold is categorized as negative. Then, a model is trained to predict ratings based on customer sentiments for product features. This study tests four different models: support vector machine (SVM), light gradient boosting machine (LGBM), extreme gradient boosting (XGBoost), and neural network. The model with the best performance (F1 score) is selected. Finally, the selected trained model is interpreted using SHAP, one of the most popular interpretable machine learning methods, and the resulting SHAP values are analyzed. This study calculates the average of the absolute SHAP values for each product feature, which provides an indication of the importance of each feature.

4 CASE STUDY

This section presents a case study using video reviews of cellphones from Youtube.

4.1 Data Collection

Two types of data, online video reviews and product manual documents, were collected for this study. The cellphone video review data were gathered from Youtube. To ensure consistent search results, a fixed pattern of search keywords is used: *Brand + Model + 'Reviews'*. For example, *Apple iPhone 13 Reviews* were the search keywords to obtain video reviews about iPhone 13 from Youtube. In total, 1,240 video reviews with an average 9.6 minutes of video lengths were collected for four brands: Apple, Samsung, Lenovo, and Motorola. Detailed information can be found in Table 1. Then, with the use of *SpeechRecognition* package from Python, the audio components of video reviews were extracted and translated to texts. Due to the limited processing capacity of the package, videos longer than 20 minutes were not included in the study.

For this study, manual documents for seven cellphone products were used. These documents are provided by manufacturers and can be accessed online. While the number of manual documents used was smaller than the number of smartphones included in the study, it was assumed that the features and functions of the cellphones were similar. Therefore, analyzing the cellphone manuals for these seven models was considered sufficient to cover the representative features of cellphones. Words in the review data that were not present in the manual documents were not included in the subsequent analysis.

4.2 Results

4.2.1 Features of Customer Interests Table 2 presents the LDA-generated topics. The first column contains topic labels representing product features mentioned by reviewers. The second column lists feature-relevant keywords, with words not appearing in the manual documents removed from the review data. The third column records the total number of videos

Copyright © 2023 by ASME

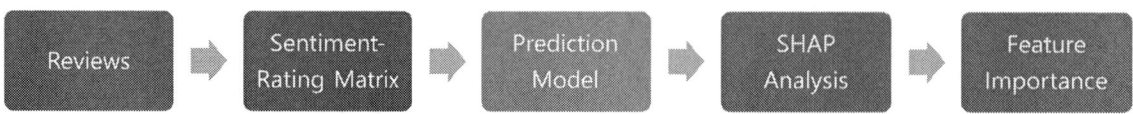

FIGURE 2: Flowchart of feature importance estimation.

TABLE 1: Video data by brands

Brand	# Videos	Video Length (Minutes)	# Views
Apple	407	4,405	336,774,837
Samsung	305	2,873	45,744,534
Lenovo	247	2,274	60,042,971
Motorola	281	2,348	30,763,023

TABLE 2: Smartphone features of customer interest

Topic label	Keywords	# Videos	Ratio
Screen	screen, screen_protector, display, resolution, screen_size, oled_display, etc.	1,110	90%
Camera	camera, picture, photo, video, selfie, hdr, 4k_video, front_camera, rear_camera, etc.	1,239	99%
Battery	battery, battery_life, power, charger, wireless_charging, fast_charging, etc.	1,085	88%
Security	fingerprint, sensor, face_recognition, privacy, scanner, biometrics, data_transfer, etc.	913	74%
Technical Specifications	performance, power, game, ram, storage, memory, temperature, heat, multitasking, etc.	1,145	92%
Network	sim, card, service, carrier, message, signal, network, connectivity, 5g_connectivity, etc.	1,039	84%

that have mentioned the corresponding product features. The last column computes the percentage of videos (out of total 1,240 videos) that have mentioned the corresponding product features.

4.2.2 Feature Sentiment
Fig. 3 (a) demonstrates the sentiment scores of overall product features of all brands. As seen in the figure, *Camera*, *Screen*, and *Technical Specifications* have the highest sentiment scores, indicating on average customers are excited about these features when reviewing their cellphones. On the other hand, Fig. 3 (b) displays a side-by-side comparison of the sentiment analysis of each brand. Noticeably, Apple stands out in all product features, followed by Samsung, while Lenovo and Motorola have received more reserved opinions in many of their features, as compared to the former two brands.

4.2.3 Feature Importance
In addition to analyzing the sentiments of product features, this study also examined their importance. One of the major hurdles during the computation of importance score was the extreme imbalance present in the data. Out of total 1,240 video reviews, there were 1,068 positive data points but only 172 negative data points. To rectify the data imbalance, the technique of undersampling positive data points was used to ensure approximately 1-to-1 ratio between positve and negative data points. For each epoch of undersampling, a set of

importance scores was computed as explained in section 3.4. The process was repeated 500 times. The final results were shown in Table 3 and Fig. 4.

As seen from the results, *Technical Specifications* is the most important feature, followed by *Camera*. This indicates that in the reviews with positve ratings, *Technical Specifications* and *Camera* are strongly emphasized. These two features can be easily quantified: the former can be represented by CPU power and storage size, whereas the latter can be represented by pixels and zoom ranges. The quantification allows these features to stand out effortlessly, demonstrating the advantages and improvement of the particular models in a simple manner, and thus easily appealing to the customers. On the other hand, *Screen* has the least importance score. One of the possible explanations is that *Screen*, while remaining to be a crucial component of cellphones, has become so standardized nowadays that many customers are taking it for granted.

5 CONCLUSION
This paper has investigated the viability of video reviews as an alternative source of online reviews to understand customer preferences and their implications for engineering design, by (i) collecting and preprocessing video reviews from social media platforms, (ii) extracting product features of customer interests from video reviews using LDA models, (iii) analyzing their

Copyright © 2023 by ASME

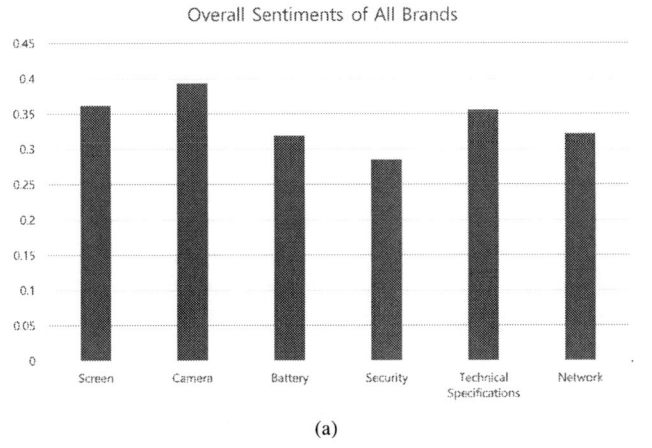

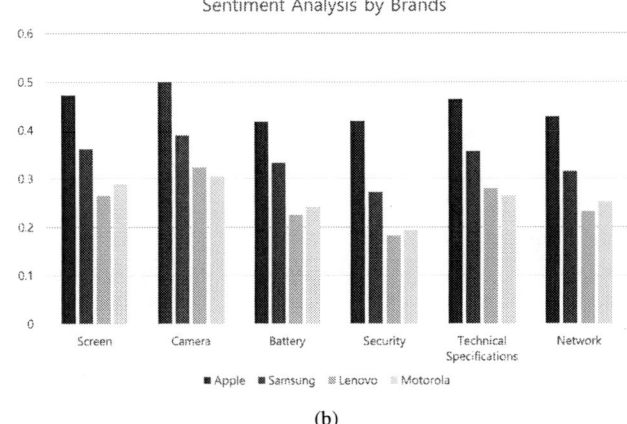

FIGURE 3: (a). Overall product feature sentiment of all brands, and (b). sentiment analysis of each brand.

TABLE 3: Feature Importance (Model: XGBoost, Epochs: 500)

Product Features	Average Feature Importance	Standard Deviation
Screen	0.663	0.183
Camera	1.065	0.164
Battery	0.861	0.263
Security	1.005	0.253
Technical Specifications	1.471	0.248
Network	0.847	0.200

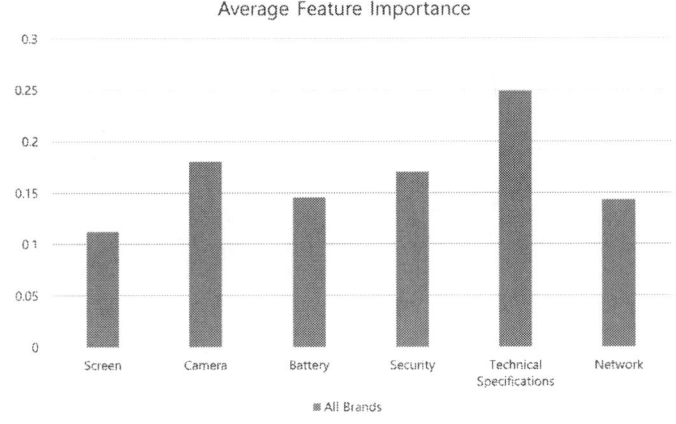

FIGURE 4: Normalized average importance scores of cellphone features.

corresponding sentiment using VADER package, and (iv) compute feature importance using SHAP values. In addition, this paper has demonstrated the feasibility of its methodology framework by presenting a case study of cellphone video reviews from Youtube.

5.1 Limitations

However, since the study of video reviews is still in its burgeoning state, it has some limitations:

Missing visual component - Video reviews commonly contain two components: audio and visual. So far, this study has only worked with the audio component of video reviews. The visual component is equally important, and should not be left behind without investigation. Recently, an impressive performance has been observed in the field of object detection in video [57–60]. Application of the techniques and models from these fields may allow us to initiate a preliminary analysis on the visual component of video reviews.

Translation issue - Some video reviews are made by non-native English speakers, who may have such particular English accents that common audio-to-text packages cannot accurately translate. This is analogous to the issue in textual reviews, in which the contents are written in poor English.

Sponsored reviews - Some Youtubers (people who upload Youtube videos), usually called *influencers* in their communities, are paid to make videos advertising products. This frequently happens to Youtubers who have a considerable number of subscribers. Due to the natures of these videos, they usually focus on the pros of the products, thus generating biased or even ex-

Copyright © 2023 by ASME

aggerated reviews. Sponsored reviews exist in both texts and videos. They are hard to detect and eradicate due to the lack of information.

5.2 Future Work

At last, this paper proposes some directions for future work:

Investigate the interaction between video and its corresponding comments - Comments are the unique aspect of video reviews; text reviews from e-commerce websites seldom have this feature. Comments can either be an independent piece of information, in which commenters are sharing their own product experiences; or, they can be a hierarchical information, in which commenters are reacting/responding to the corresponding video. This unique interaction between video and its comments can provide a greater depth of analysis. For example, commenters may disapprove the video if its content is fake, exaggerating, or misleading.

Choice set construction using comparison videos - Traditionally, choice sets using online reviews are constructed with product specifications [61, 62]; similar products are grouped together and considered substitutes to one another if they share similar product specifications. The process of making choice sets can be improved using comparison videos, which are a special kind of video reviews. For example, a video review *"Galaxy S22 vs iPhone 13 - which to choose"* compares two competing models from different brands. Comparison videos usually compare products of similar specifications that are in the same level of competition. This allows researchers to easier and reliably obtain choice set information.

Identify lead users/experts - Lead users play a vital role in next generation product development, as they help designers discover relevant product feature preferences months or even years before they are desired by the general customer base [63]. Some influencers, as mentioned earlier in this section, may share many similar functions as lead users. For example, a famous photographer may share his/her opinions about different product features and why he/she prefers one over another. These opinions from the influencers may reshape, or *influence*, the preferences of many customers. Therefore, the ability to pinpoint the tastes and preferences of these lead users/experts may allow product designers to quickly capture the general picture of the market.

REFERENCES

[1] Tuarob, S., and Tucker, C. S., 2014. "Discovering next generation product innovations by identifying lead user preferences expressed through large scale social media data". *Volume 1B: 34th Computers and Information in Engineering Conference.*

[2] Zhang, H., Sekhari, A., Ouzrout, Y., and Bouras, A., 2016. "Jointly identifying opinion mining elements and fuzzy

measurement of opinion intensity to analyze product features". *Engineering Applications of Artificial Intelligence,* **47**, p. 122–139.

[3] Suryadi, D., and Kim, H. M., 2018. "A systematic methodology based on word embedding for identifying the relation between online customer reviews and sales rank". *ASME. J. Mech. Des.*

[4] Jeong, B., Yoon, J., and Lee, J.-M., 2019. "Social media mining for product planning: A product opportunity mining approach based on topic modeling and sentiment analysis". *International Journal of Information Management,* **48**, p. 280–290.

[5] Suryadi, D., and Kim, H. M., 2019. "A data-driven methodology to construct customer choice sets using online data and customer reviews". *ASME. J. Mech. Des.,* 11.

[6] Zhou, F., Ayoub, J., Xu, Q., and Yang, X. J., 2019. "A machine learning approach to customer needs analysis for product ecosystems". *Journal of Mechanical Design,* **142**, 08, p. 1.

[7] Chen, W., Hoyle, C., and Wassenaar, H. J., 2013. *Decision-based design: integrating consumer preferences into engineering design.* Springer.

[8] Tuarob, S., and Tucker, C. S., 2015. "Quantifying product favorability and extracting notable product features using large scale social media data". *Journal of Computing and Information Science in Engineering,* **15**(3).

[9] Wang, W., Li, Z., Tian, Z., Wang, J., and Cheng, M., 2018. "Extracting and summarizing affective features and responses from online product descriptions and reviews: A kansei text mining approach". *Engineering Applications of Artificial Intelligence,* **73**, pp. 149–162.

[10] Singh, A., and Tucker, C. S., 2017. "A machine learning approach to product review disambiguation based on function, form and behavior classification". *Decision Support Systems,* **97**, pp. 81–91.

[11] Liu, Y., Jin, J., Ji, P., Harding, J. A., and Fung, R. Y., 2013. "Identifying helpful online reviews: A product designer's perspective". *Computer-Aided Design,* **45**(2), pp. 180–194. Solid and Physical Modeling 2012.

[12] Kherwa, P., and Bansal, P., 2019. "Topic modeling: A comprehensive review". *EAI Endorsed Transactions on Scalable Information Systems,* **7**(24), 7.

[13] Deerwester, S., Dumais, S. T., Furnas, G. W., Landauer, T. K., and Harshman, R., 1990. "Indexing by latent semantic analysis". *Journal of the American society for information science,* **41**(6), pp. 391–407.

[14] Berry, M. W., and Martin, D. I., 2005. "Principal component analysis for information retrieval". In *Handbook of parallel computing and statistics.* Chapman and Hall/CRC, pp. 415–430.

[15] Buckley, C., Allan, J., and Salton, G., 1994. "Automatic routing and ad-hoc retrieval using smart: Trec 2". *NIST*

Copyright © 2023 by ASME

SPECIAL PUBLICATION SP, pp. 45–45.

[16] Kherwa, P., and Bansal, P., 2017. "Latent semantic analysis: an approach to understand semantic of text". In 2017 International Conference on Current Trends in Computer, Electrical, Electronics and Communication (CTCEEC), IEEE, pp. 870–874.

[17] Paatero, P., and Tapper, U., 1994. "Positive matrix factorization: A non-negative factor model with optimal utilization of error estimates of data values". *Environmetrics, 5*(2), pp. 111–126.

[18] Lee, D. D., and Seung, H. S., 1999. "Learning the parts of objects by non-negative matrix factorization". *Nature, 401*(6755), pp. 788–791.

[19] Wainwright, M. J., and Jordan, M. I., 2005. "11 a variational principle for graphical models".

[20] Hofmann, T., 2013. "Probabilistic latent semantic analysis". *arXiv preprint arXiv:1301.6705.*

[21] Blei, D. M., Ng, A. Y., and Jordan, M. I., 2003. "Latent dirichlet allocation". *Journal of machine Learning research, 3*(Jan), pp. 993–1022.

[22] Joung, J., and Kim, H. M., 2021. "Approach for importance–performance analysis of product attributes from online reviews". *Journal of Mechanical Design, 143*(8).

[23] Joung, J., and Kim, H. M., 2021. "Automated Keyword Filtering in Latent Dirichlet Allocation for Identifying Product Attributes From Online Reviews". *Journal of Mechanical Design, 143*(8), 02. 084501.

[24] Bongini, P., Osborne, F., Pedrazzoli, A., and Rossolini, M., 2022. "A topic modelling analysis of white papers in security token offerings: Which topic matters for funding?". *Technological Forecasting and Social Change, 184*, p. 122005.

[25] Yang, Z., Wu, Q., Venkatachalam, K., Li, Y., Xu, B., and Trojovský, P., 2022. "Topic identification and sentiment trends in weibo and wechat content related to intellectual property in china". *Technological Forecasting and Social Change, 184*, p. 121980.

[26] Wu, Z., Duan, C., Cui, Y., and Qin, R., 2023. "Consumers' attitudes toward low-carbon consumption based on a computational model: Evidence from china". *Technological Forecasting and Social Change, 186*, p. 122119.

[27] Ma, T., Zhou, X., Liu, J., Lou, Z., Hua, Z., and Wang, R., 2021. "Combining topic modeling and sao semantic analysis to identify technological opportunities of emerging technologies". *Technological Forecasting and Social Change, 173*, p. 121159.

[28] Zhang, H., Daim, T., and Zhang, Y. P., 2021. "Integrating patent analysis into technology roadmapping: A latent dirichlet allocation based technology assessment and roadmapping in the field of blockchain". *Technological Forecasting and Social Change, 167*, p. 120729.

[29] Yan-Yan, Z., Bing, Q., and Ting, L., 2010. "Integrating intra-and inter-document evidences for improving sentence sentiment classification". *Acta Automatica Sinica, 36*(10), pp. 1417–1425.

[30] Hu, Y., and Li, W., 2011. "Document sentiment classification by exploring description model of topical terms". *Computer Speech & Language, 25*(2), pp. 386–403.

[31] Kang, H., Yoo, S. J., and Han, D., 2012. "Senti-lexicon and improved naïve bayes algorithms for sentiment analysis of restaurant reviews". *Expert Systems with Applications, 39*(5), pp. 6000–6010.

[32] Ptaszynski, M., Dokoshi, H., Oyama, S., Rzepka, R., Kurihara, M., Araki, K., and Momouchi, Y., 2013. "Affect analysis in context of characters in narratives". *Expert Systems with Applications, 40*(1), pp. 168–176.

[33] Hutto, C., and Gilbert, E., 2014. "Vader: A parsimonious rule-based model for sentiment analysis of social media text". In Proceedings of the international AAAI conference on web and social media, Vol. 8, pp. 216–225.

[34] Zhang, L., Ghosh, R., Dekhil, M., Hsu, M., and Liu, B., 2011. "Combining lexicon-based and learning-based methods for twitter sentiment analysis". *HP Laboratories, Technical Report HPL-2011, 89*, pp. 1–8.

[35] Zagibalov, T., 2010. "Unsupervised and knowledge-poor approaches to sentiment analysis". PhD thesis, University of Sussex.

[36] Augustyniak, Ł., Szymański, P., Kajdanowicz, T., and Tuligłowicz, W., 2015. "Comprehensive study on lexicon-based ensemble classification sentiment analysis". *Entropy, 18*(1), p. 4.

[37] Gonçalves, P., Araújo, M., Benevenuto, F., and Cha, M., 2013. "Comparing and combining sentiment analysis methods". In Proceedings of the first ACM conference on Online social networks, pp. 27–38.

[38] Shi, H.-X., and Li, X.-J., 2011. "A sentiment analysis model for hotel reviews based on supervised learning". In 2011 International Conference on Machine Learning and Cybernetics, Vol. 3, IEEE, pp. 950–954.

[39] Anjaria, M., and Guddeti, R. M. R., 2014. "A novel sentiment analysis of social networks using supervised learning". *Social Network Analysis and Mining, 4*(1), pp. 1–15.

[40] Vilares, D., Alonso, M. A., and Gómez-Rodríguez, C., 2017. "Supervised sentiment analysis in multilingual environments". *Information Processing & Management, 53*(3), pp. 595–607.

[41] Mukhtar, N., Khan, M. A., and Chiragh, N., 2018. "Lexicon-based approach outperforms supervised machine learning approach for urdu sentiment analysis in multiple domains". *Telematics and Informatics, 35*(8), pp. 2173–2183.

[42] Zhang, H., Sekhari, A., Ouzrout, Y., and Bouras, A., 2016. "Jointly identifying opinion mining elements and fuzzy

Copyright © 2023 by ASME

measurement of opinion intensity to analyze product features". *Engineering Applications of Artificial Intelligence, 47*, pp. 122–139.

[43] Jiang, H., Kwong, C. K., and Yung, K. L., 2017. "Predicting future importance of product features based on online customer reviews". *Journal of Mechanical Design, 139*(11).

[44] Bag, S., Tiwari, M. K., and Chan, F. T., 2019. "Predicting the consumer's purchase intention of durable goods: An attribute-level analysis". *Journal of Business Research, 94*, pp. 408–419.

[45] Rai, R., 2012. "Identifying key product attributes and their importance levels from online customer reviews". In International Design Engineering Technical Conferences and Computers and Information in Engineering Conference, Vol. 45028, American Society of Mechanical Engineers, pp. 533–540.

[46] Mikulić, J., and Prebežac, D., 2012. "Accounting for dynamics in attribute-importance and for competitor performance to enhance reliability of bpnn-based importance–performance analysis". *Expert Systems with Applications, 39*(5), pp. 5144–5153.

[47] Decker, R., and Trusov, M., 2010. "Estimating aggregate consumer preferences from online product reviews". *International Journal of Research in Marketing, 27*(4), pp. 293–307.

[48] Bi, J.-W., Liu, Y., Fan, Z.-P., and Zhang, J., 2019. "Wisdom of crowds: Conducting importance-performance analysis (ipa) through online reviews". *Tourism Management, 70*, pp. 460–478.

[49] Joung, J., and Kim, H. M., 2021. "Explainable neural network-based approach to kano categorisation of product features from online reviews". *International Journal of Production Research*, pp. 1–21.

[50] Suryadi, D., and Kim, H., 2019. "Automatic identification of product usage contexts from online customer reviews". *Proceedings of the Design Society: International Conference on Engineering Design, 1*(1), p. 2507–2516.

[51] Joung, J., and Kim, H. M., 2020. "Importance-performance analysis of product attributes using explainable deep neural network from online reviews". *Volume 11A: 46th Design Automation Conference (DAC)*.

[52] Joung, J., and Kim, H. M., 2021. "Approach for Importance–Performance Analysis of Product Attributes From Online Reviews". *Journal of Mechanical Design, 143*(8), 02. 081705.

[53] Denny, M. J., and Spirling, A., 2018. "Text preprocessing for unsupervised learning: Why it matters, when it misleads, and what to do about it". *Political Analysis, 26*, pp. 168–189.

[54] Blei, D., Ng, A., and Jordan, M., 2003. "Latent dirichlet allocation". *Journal of Machine Learning Research, 3*, 05,

pp. 993–1022.

[55] Mikolov, T., Sutskever, I., Chen, K., Corrado, G. S., and Dean, J., 2013. "Distributed representations of words and phrases and their compositionality". *Advances in neural information processing systems, 26*.

[56] Lundberg, S. M., and Lee, S.-I., 2017. "A unified approach to interpreting model predictions". *Advances in neural information processing systems, 30*.

[57] Kang, K., Ouyang, W., Li, H., and Wang, X., 2016. "Object detection from video tubelets with convolutional neural networks". Proceedings of the IEEE Conference on Computer Vision and Pattern Recognition (CVPR).

[58] Shang, X., Ren, T., Guo, J., Zhang, H., and Chua, T.-S., 2017. "Video visual relation detection". In Proceedings of the 25th ACM International Conference on Multimedia, MM '17, Association for Computing Machinery, p. 1300–1308.

[59] Bertasius, G., Torresani, L., and Shi, J., 2018. "Object detection in video with spatiotemporal sampling networks". Proceedings of the European Conference on Computer Vision (ECCV).

[60] Du, X., Wang, X., Gozum, G., and Li, Y., 2022. "Unknown-aware object detection: Learning what you don't know from videos in the wild". Proceedings of the IEEE/CVF Conference on Computer Vision and Pattern Recognition (CVPR).

[61] Lin, K., and Kim, H. M., 2021. "Investigate the Influence of Online Ratings and Reviews in Purchase Behavior Using Customer Choice Sets". Vol. Volume 3A: 47th Design Automation Conference (DAC) of *International Design Engineering Technical Conferences and Computers and Information in Engineering Conference*. V03AT03A017.

[62] Suryadi, D., and Kim, H. M., 2019. "A Data-Driven Methodology to Construct Customer Choice Sets Using Online Data and Customer Reviews". *Journal of Mechanical Design, 141*(11), 09. 111103.

[63] Tuarob, S., and Tucker, C. S., 2015. "Automated Discovery of Lead Users and Latent Product Features by Mining Large Scale Social Media Networks". *Journal of Mechanical Design, 137*(7), 07. 071402.

Copyright © 2023 by ASME

Proceedings of the ASME 2023
International Design Engineering Technical Conferences and
Computers and Information in Engineering Conference
IDETC-CIE2023
August 20-23, 2023, Boston, Massachusetts

DETC2023-116429

HEAT SINK DESIGN OPTIMIZATION VIA GAN-CNN COMBINED DEEP-LEARNING

Nathan Flynn, Xiaoping Qian*
n8flynn@outlook.com

Department of Mechanical Engineering
University of Wisconsin-Madison, Madison, WI

ABSTRACT

This work proposes a combined deep learning based approach to improve thermal component heat sinks involving turbulent fluid flow. A Generative Adversarial Network (GAN) is trained to learn and recreate the new ellipse based heat sinks. Simulation data for new designs is efficiently generated using OpenFOAM 7 (Open Source Computational Fluid Dynamics software) along with high throughput computing. To improve the speed of design evaluation, a Convolutional Neural Network (CNN) is trained to predict the entire temperature field for a given design. The trained CNN is able to predict the entire temperature field for the design with a mean average error of 1.140 degrees kelvin in 0.04 seconds (22,500 times faster than the simulation). A combined model is formed using the trained CNN and GAN networks to create and simulate new designs. The combined model optimizes the latent representation of 64 random designs on a Graphical Processing Unit (GPU) in ten minutes. The optimized designs perform fourteen degrees kelvin better on average than the non-optimized designs. The highest preforming design outperforms any design in the training data by 1.83 degrees kelvin.

1. INTRODUCTION

Improving the performance of heat sinks via design optimization is important to be able to more effectively pull heat away from demanding electrical components such as processors. Typical approaches to this problem involve using an optimization scheme such as shape [1] or topology optimization [2] to improve the performance of a given design. Shape and topology optimization involves optimizing a design using an objective, such as minimizing pressure drop in a fluid channel. Shape optimization optimizes the shape of an initial design, while topology optimization creates large topological changes from the initial design. To evaluate the objective function the design must be simulated whenever the design is updated, which can be computationally expensive over many iterations.

Another approach to optimizing fluid related problems involves the use of deep-learning methods such as regression based surrogate models which allow the prediction of an output based on an input such as the channel length in a fluid problem [3]. Surrogate models use machine learning models to learn how to relate the input of a simulation to an output to replace the simulation. The surrogate models are then paired with a genetic algorithm or optimization method to improve the performance of the design by changing some portion of the geometry, such as optimizing the length of the baffles in a micro-channel [4]. The drawback of this approach is that full simulations are required to be able to predict only a few of the desired outputs (temperature of the outlet, pressure drop, etc) but not other desired outputs (entire solution field for temperature or pressure) and the geometries have to be relatively simple.

Recent advances in deep learning have created models such as the Generative Adversarial Networks (GAN) which are able to create new designs by learning the distribution of a given set of input designs [5]. With GAN's ability to generate new data and designs, they have been recently paired with topology optimization-based approaches to produce new optimal structures. In another paper, a GAN is used in combination with transfer learning to create topologically-optimized designs to seen and un-seen boundary conditions while testing the model [6]. In another paper, previously generated topological designs are used alongside new designs generated by a GAN to generate aesthetically optimal and high performance designs [7]. Lastly, a Variational Auto-Encoder GAN (VEGAN) is used alongside a database of optimized airfoil designs to be able to create new designs that have equivalent or slightly better performance than the training data [8]. The issues with these TO-based models is that rely on high performance data, either from a data base or data created through topology optimization.

Recently, a GAN and Convolutional Neural Network (CNN) combined based approach is able to produce optimal microstructure based designs by using inverse design based method to achieve a desired compliance [9]. The method developed by

---
*Corresponding author: Email: qian@engr.wisc.edu

Copyright © 2023 by ASME

Tan et al. [9] is successfully used to generate training micro structure based images from scratch, generate new realistic data with a GAN and then predict the performance of a given design using a Le-Net CNN architecture. This method is applied to a relatively simple problem and inverse design produces desired structures.

In this paper, a similar combined deep learning approach is applied to a conjugate heat-transfer design optimization problem. The problem involves optimizing the design of a heat sink using both steady state and turbulent air flow. The GAN is used to create new designs from a set of randomly generated fin structures with elliptical cross sections using a Python library called "ellipse packing" [9]. The CNN evaluates the performance of the design by predicting the entire temperature field for a given design. The two models are separately trained and then combined into a single model. The combined model is used along side an optimizer to directly improve the performance of a batch of 64 randomly generated heat sink designs.

2. DESIGN PROBLEM

The design problem is a turbulent ($Re = 4520$), steady state, conjugate heat-transfer problem with conduction and convection. The fluid domain is a 74 x 66 mm rectangle and the design domain is a 64 x 64 mm square that is centered within the fluid domain. A graphical representation of the problem, initial and boundary conditions is shown in Fig. 1.

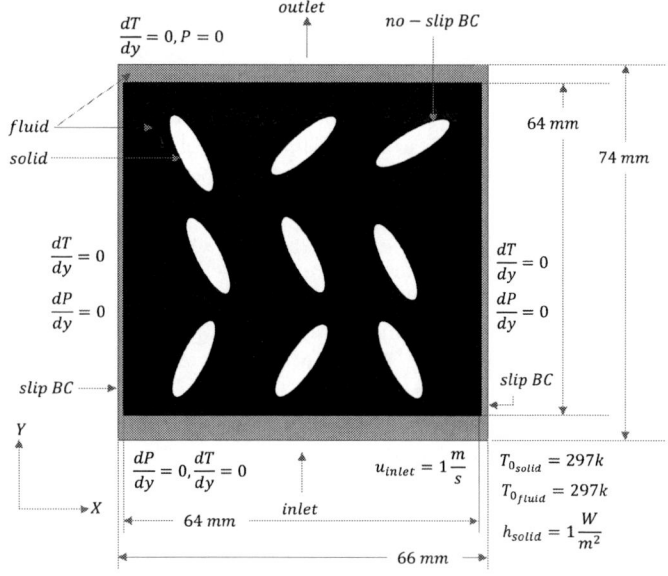

FIGURE 1: DESIGN PROBLEM PARAMETERS

Fluid flows from the inlet at the bottom to the outlet at the top. The white ellipses within the design domain represent the solid fins of the heat sink. The goal of this project is to generate optimal structures that dissipate a $1 \; W/m^2$ load on the heat sink. The fluid is air and the solid is 6061 Aluminum Alloy (Al). The material properties for the solid and fluid are shown in Table 1. Performance of a given design is measured by calculating the average solid temperature. The best design has the lowest average solid temperature.

TABLE 1: DESIGN PROBLEM MATERIAL PROPERTIES

Parameters	Solid	Fluid
Material	6061 Al	Air
Molecular Weight (g/mol)	63.5	28.966
$\rho \; (kg/m^3)$	2719	1.225
$c_p \; (J/kg/K)$	871	1006.43
$\mu \; (kg/m/s)$	-	$1.7884e - 05$
Pr	-	0.71

3. PROPOSED METHOD

The proposed method to solve this problem involves using a combined deep-learning optimization framework composed of a design generator and a design evaluator. The design generator (g) creates new designs (X) from a given input latent space vector (a compressed representation of an image) z shown in Eq. (1).

$$X = g(z) \tag{1}$$

The design evaluator, f, uses the design (X) as the input and outputs the temperature field (Y) for that design shown in Eq. (2).

$$Y = f(X) \tag{2}$$

The combined model consists of both the design generator (g) and the design evaluator (f) used in combination shown in Eq. (3).

$$Y = f(g(z)) \tag{3}$$

The loss function for the combined model is the average temperature shown in Eq. (4).

$$Loss = \frac{\sum (Y \cdot X)}{\sum X} \tag{4}$$

The Wasserstein Generative Adversarial Network with Gradient Penalty (WGAN-GP) [10] creates the generator model in Eq. (1) and the U-NET CNN model [11] creates the evaluator model in Eq. (2). Training of both models is done separately. The WGAN-GP is trained on images of randomly generated fin designs with elliptical cross sections. After training, the WGAN-GP is used to create new designs by passing a random latent space vector (z) into the GAN. Simulation is preformed on the new designs using OpenFOAM 7 (an open-source computational fluid dynamics package) to produce the temperature field for a given design. The U-NET maps the designs to their corresponding temperature field. Once training for both models is complete, a combined model is formed using the generator of the GAN and the U-NET CNN. Optimization is performed on the input to the generator (z) using Average Stochastic Gradient Decent (ASGD) within a Python library called PyTorch to minimize Eq. (4)

3.1 Design generation

To generate heat sink designs, an open source Python library called "ellipse packing" from GitHub creates random elliptical structures [12]. The number of mesh points in the x and y direction and the scale of the major and minor axes is controlled with the library. Examples shown in Fig. 3. The placement of the

Copyright © 2023 by ASME

ellipses is random and is created by producing an evenly space grid of points. Next, Delaunay Triangulation is used to form a mesh of triangles. Finally, formation of Steiner In-ellipses is accomplished by inscribing the ellipses within the triangles shown in Fig. 2.

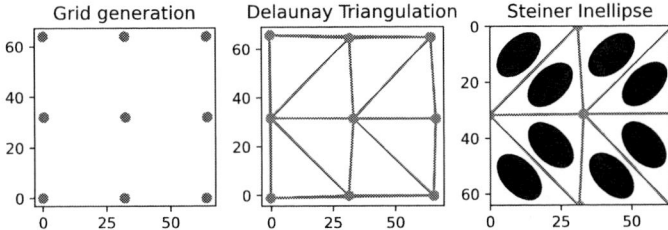

FIGURE 2: DESIGN GENERATION METHOD

Each image is gray-scale and has a resolution of 64 x 64 pixels. To generate 100,000 unique designs, randomization is done using the NumPy random function between the following settings; number of grid points in the x and y direction between 2 and 10, the scale of the major axis between 0.5 and 2.75, and the minor axis between 0.5 and 1.25.

FIGURE 3: EXAMPLE ELLIPSE FIN DESIGNS

3.2 Design simulation

Temperature field data for the design evaluator is generated using simulations from OpenFOAM 7 to produce the temperature field for each design. The kEpsilon model is used to simulate turbulent fluid flow. Conversion of the designs into a mesh friendly format for OpenFOAM is accomplished by filtering the gray-scale images to a solid (1) and fluid (0) representation. The gray-scale images have pixel value ranges from 0 to 255, where 255 and 0 represent a solid and fluid pixel respectively. Any pixel value

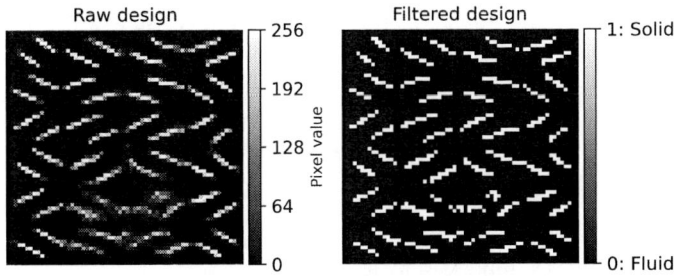

FIGURE 4: IMAGE FILTERING

greater than or equal to 90 is changed to 1 and any pixel value less than 90 is changed to 0. An example is shown in Fig. 4. The solid/fluid threshold of 90 is selected by trial and error. The purpose of this threshold is to ensure that no fluid pixels ended up inside of the solid domain. If this happens, the simulation fails to run. Image processing is applied on-top of filtering to fill in potential voids using the SciPy Python library. The binary fill

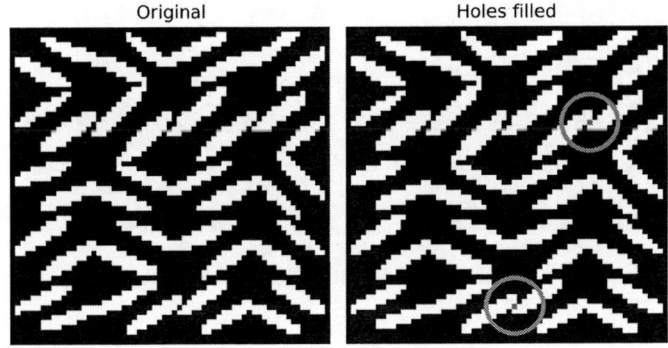

FIGURE 5: FILLING VOID EXAMPLE

holes function within SciPy invades the complement of the image from the boundary of the image using binary dilation. Once the invasion step is complete, only the holes remain in the image because there is no connection from them to the boundary. A final step is done to fill in the holes to create a void free design [13] shown in Fig. 5 where the red pixels denote filled in holes. The design (64 x 64) is then added to the center of the fluid domain (74 x 66) and then is meshed in OpenFOAM 7. To improve the

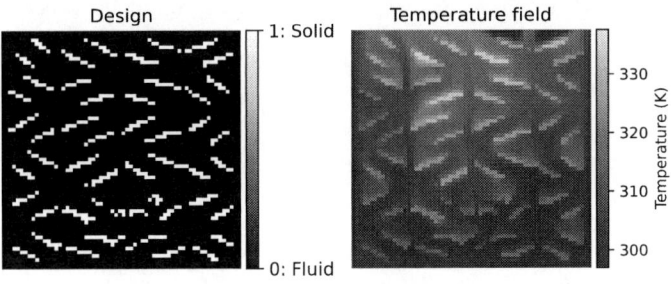

FIGURE 6: DESIGN AND TEMPERATURE FIELD

accuracy of the simulation, the mesh is refined to increase the number of cells from 4884 (74 x 66) to 19,536 (148 x 132). The output temperature field of the simulation is then mapped back

Copyright © 2023 by ASME

into the 74 x 66 domain and the fluid region is trimmed off to leave the 64 x 64 temperature field. An example of the input and output of the simulation is shown in Fig. 6.

3.3 Design performance metric and benchmark

The average solid temperature is used to record performance, EQ. (5). In this equation, Y represents the temperature field prediction, and X_{solid} is the filtered version of X only containing solid pixels, example is shown in Fig. 4.

$$T_{avg_{solid}} = \frac{\sum (Y \cdot X_{solid})}{\sum X_{solid}}, \quad X_{solid} = \begin{cases} 1 & X \geq \frac{90}{255} \\ 0 & X < \frac{90}{255} \end{cases} \quad (5)$$

A simple parallel fin design is constructed and simulated as a benchmark to compare to the results from the combined model. The parallel fin design and performance is shown in Fig. 7.

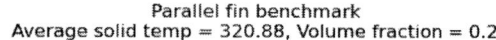

Parallel fin benchmark
Average solid temp = 320.88, Volume fraction = 0.2

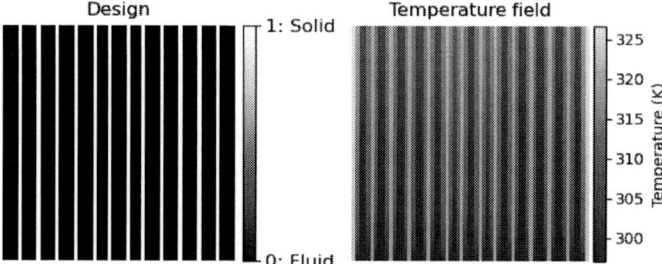

FIGURE 7: PARALLEL-FIN BENCHMARK

3.4 Scaling simulation runs

Each simulation takes about 2 minutes to run on a single core (Intel 6700K at 4.00 GHz). Assuming 2 minutes per run, conducting 500,000 simulations using a single core would take around 2 years. The Center for High Throughput Computing (CHTC) at the University of Wisconsin-Madison is used to generate simulation data. The Center has over 20,000 cores available and is managed using HTCondor software. Using this resource for data generation made it possible to complete this project in a reasonable amount of time [14]. An OpenFOAM 7 Docker image is used to run the simulation at the CHTC [15]. To balance out the workload on all of the servers nodes, the simulations need to be broken up into smaller amounts of work. For a batch of 500,000 simulations, creation of 5,000 jobs is required, with each job containing 100 simulations worth of work. Each job runs on a single node of the server, allowing results for 500,000 simulations to be generated in less than a day.

3.5 Design generator

A GAN is used to create new designs from a set of existing designs. A traditional GAN is made up of two components, a generator and a discriminator. The discriminator is trained using existing images and the generator produces images that look similar to the existing images. The discriminator scores the generated images on the probability that the image came from the existing image data set [5]. Training of the two models is accomplished by playing a min-max game. If the generator creates an image that fools the discriminator, the discriminator learns from that image and improves the discriminator model. Similarly, the generator model improves when the discriminator rejects images created by the generator due to the generated images being too different from the training set of images. GANs take a random uniform distribution vector between 0 and 1 called the latent space z. GANs learn how to decompress this latent space into the full design space (latent space vector 100D to 64x64 image). Designs can be changed by moving within the latent space and certain areas of this space are attributed to certain features in the images. In this paper the latent space is a 100D vector similar to the work done by Lipton et al. [16].

The Wasserstein Generative Adversarial Network with Gradient Penalty (WGAN-GP) [10] is used to generate new designs (implementation by [17] in Pytorch). The WGAN-GP is the successor to the Deep Convolutional GAN (DCGAN) which aims at eliminating common problems with the network; model collapse and vanishing gradients. Model collapse is a common issue within GAN's and causes the model to produce the same group of images. The WGAN-GP reduces the chance of model collapse by using a different loss formulation. Wasserstein loss uses Earth Mover Distance (the minimum cost to transport the mass of one distribution into another to make them the same). The gradient penalty (GP) is added to the loss formulation to add more stability than the original WGAN alone [10].

Use of the conventional loss function for the generator and the critic is unreliable to gauge model convergence due to both networks playing a min-max game during training causing the loss function for both the generator and the discriminator to fluctuate. The other way to gauge model performance is through visual inspection of the images. The challenge with visual inspection is that it is a qualitative approach. A new set of quantitative metrics emerged to remedy this issue called Inception Score (IS) [18] and Fréchet Inception Distance (FID) [19]. The Inception Score is used to evaluate the quality of the images being produced. Fréchet Inception Distance is used to measure the diversity of the images. A Python library called pytorch-gan-metrics is used to calculate both the FID and IS scores [20].

To train the GAN, simulation is preformed on 100,000 designs and evaluated using Eq. 5. From the pool of designs, the GAN training set of designs is made up of 50K randomly selected designs. Examples of training data is shown in Fig. 3. The original implementation from [17] is used along side a learning rate value of $1e-4$ for both the critic and generator. The decreased learning rate along with the other original settings is found to produce the best diversity in the designs.

Once the WGAN-GP is trained the generator from the model is separated and the last layer of the model is changed from the hyperbolic tangent function (tanh) to the sigmoid function. The tanh function is used during training to restrict the output of the GAN between -1 and 1. The sigmoid function replaces the tanh function to restrict the output between 0 and 1. A comparison is done for the same image using the Sigmoid function and scaling the tanh functions output between 0 and 1 show in Fig. 9. The Sigmoid function produces similar images to the scaled tanh function and reduces the number of computational steps to scale

Copyright © 2023 by ASME

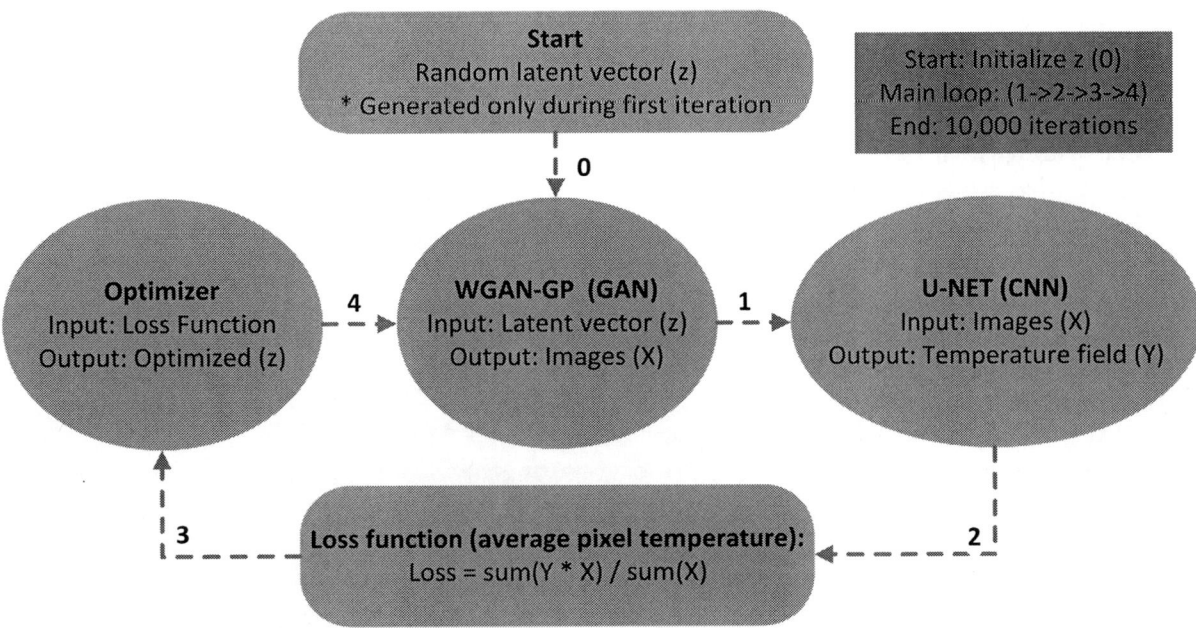

FIGURE 8: COMBINED MODEL FLOWCHART

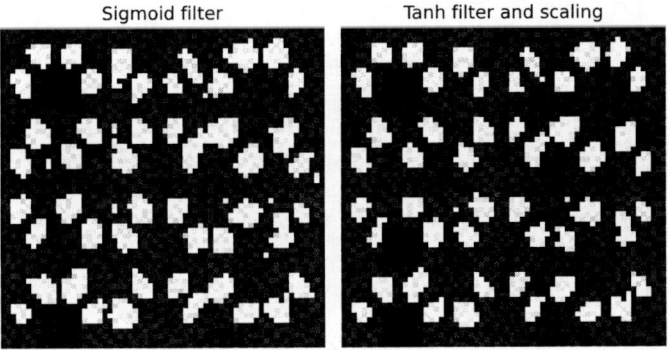

FIGURE 9: IMAGE FILTERING COMPARISON

the image between 0 and 1. A random latent space vector is generated between 0 and 1 using the PyTorch random function and then is fed into the generator to create new designs. In total, the generator created 750,000 new designs to train the evaluator with the designs corresponding simulation data.

3.6 Design evaluator

To evaluate all of the WGAN-GP's generated designs a CNN is used to predict the temperature field of the design. The U-NET CNN architecture is selected for its ability to execute pixel by pixel based regression. It was originally designed for fast biomedical pixel based segmentation tasks [11]. Since its inception, the U-NET architecture has been used for a variety of engineering based tasks. Recently, the U-NET is used to predict a temperature field for various heat source intensities, sizes, and layouts within a 200 x 200 grid problem [21]. Later, the same model is used along side an optimization framework to create the best heat-source layout for various electronics [22]. In this work, the U-NET CNN takes the 64 x 64 input design in the form of an image and outputs a 64

x 64 temperature field prediction. The implementation from [21] is used to create the model within PyTorch.

Mean absolute error (MAE) is used to keep track of the U-NET's performance during training, shown in Eq. (6) where N represents the mesh size. In this case is 64 from the 64 x 64 design. MAE represents the average prediction error per pixel in terms of Kelvin (K). In Chen et al. [21] they were able to achieve a MAE between 0.02 and 0.5 K depending on the complexity of problem.

$$MAE = \frac{1}{N^2} \sum_{i=1}^{N} \sum_{j=1}^{N} \left| \hat{Y}_{ij} - Y_{ij} \right| \qquad (6)$$

Before training the network, the temperature field data is normalized to aid in training. The temperature field data has values that range from 298K to 1100K. Using equation Eq. (7) with a $X_m = 100$ and $X_{std} = 297\ K$, the data is normalized between 0 and 1. A similar approach is done by Chen eta. [21] to train the U-NET on a heat source layout problem.

$$X_0 = \frac{X - X_m}{X_{std}} \qquad (7)$$

To train the network, simulation of the 750K GAN generated designs is done using HTC to produce temperature field data. Of the 750K sets of data, only 550K sets is used to train the U-NET CNN. With 50K of the data being used to test the model after training. Of the 500K sets of data, 80% (400K) is used to directly train the model and the remaining 20% (100K) is used to evaluate the MAE during training (validation data). A surplus of 200K designs is used as a margin of safety. Occasionally, the GAN produces designs with fluid inclusions in solid regions, causing the simulation to fail without image processing to fill in the voids.

Copyright © 2023 by ASME

3.7 Combined Model GAN + CNN

Once training of both WGAN-GP and U-NET is completed, the generator from the WGAN-GP and the U-NET model is combined to form the model shown in Fig. 8. The combined model functions as follows: a random latent space variable is generated with 64 designs and then the latent space variable z is passed into generator of the GAN. The output of the GAN (a set of images) is passed directly into the U-NET as an input as shown in Eq. (3). To optimize the latent space variable (z) the loss function below in Eq. (4) is used. This loss function approximates the average temperature of each pixel. In this formulation, the design X is not filtered to decrease computational time. The ASGD optimizer from PyTorch is used with a learning rate of $9e^{-3}$ to optimize the latent representation of the designs. The loss is back propagated into z and z is optimized to minimize the loss function. Stochastic gradient clipping is employed on the latent space variable after each optimization step to improve convergence [23]. Post processing is preformed on the designs using the binary fill holes function to fill in any voids in the designs.

4. RESULTS

4.1 GAN

Training the GAN with 50K images took about 8 hours on an RTX 4080. To monitor convergence of the GAN, Inception Score (IS) and Frechet Inception Distance (FID) values are computed every epoch as shown in Fig. 10. Inception Score measures the diversity of the images, while Fréchet Inception Distance measures the quality of the images. The convergence of training the GAN is monitored using both metrics (IS and FID). Training of the GAN is stopped when both metrics plateau. The GAN is trained for a total of 200 epochs with an image batch size of 64. A version of the model parameters is saved for every iteration and the final model version is selected based on the model having the smallest IS and highest FID score possible. This happens to be the model at iteration 155, which is denoted by the blue star in Fig. 10.

Once training the WGAN-GP is completed, the generator

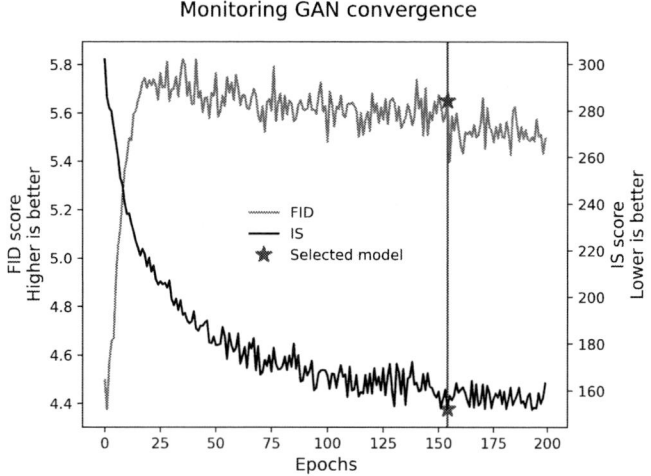

FIGURE 10: GAN TRAINING EVALUATION

model is separated from the Network and generates 750K new designs by feeding random noise into the generator via the latent space z, examples shown in Fig. 11. The images appear to be

FIGURE 11: GAN GENERATED DESIGNS

nosier that the original training data because GANs approximate the original training data. Image filtering on the images still produces meaningful and high performance designs.

The volume fractions of the designs produced by the GAN and the training data is compared to training data in Fig. 12, computed using EQ (8). A large portion of the

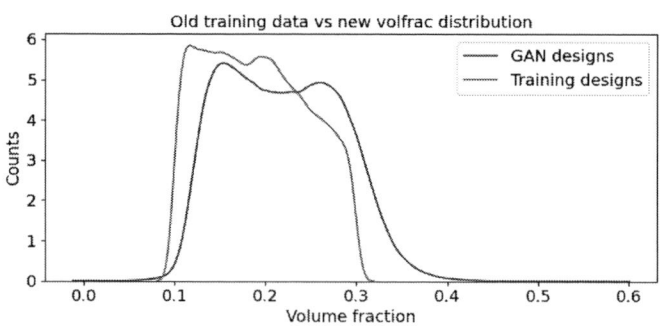

FIGURE 12: VOLUME FRACTION DISTRIBUTION

designs in the training data have a volume fraction between 0.1 and 0.3. While the GAN's designs seem to have a larger emphasis on volume fractions falling in the range of 0.15 to 0.35. This is most likely due to the GAN's images being more grainy than the original images, causing the volume fraction to increase. The GAN produced a small amount of designs outside of the training data volume fraction range (0 to 0.1 and 0.3 to 0.6).

$$Vol_{frac} = \frac{\sum X_{solid}}{64*64}, \quad X_{solid} = \begin{cases} 1 & X \geq \frac{90}{255} \\ 0 & X < \frac{90}{255} \end{cases} \quad (8)$$

Copyright © 2023 by ASME

The average solid temperatures of the training data and the output of the GAN is shown in Table 2. The GAN is able to create

TABLE 2: GAN DESIGN PERFORMANCE

Average solid temperature	Average [K]	Best [K]	Worst [K]
Training data	333.62	312.56	499.87
GAN	335.27	312.23	1117.75

designs that preformed two degrees worse on average compared to training data and produced a design that is slightly better than the best design in the training data. The volume fraction of the design versus its temperature is shown in Fig. 13 (the data is truncated below 500 K to zoom in on the majority of the data). The scatter plot shows that volume fraction isn't a good indicator of design performance and that placement, spacing, and orientation of the ellipses is important.

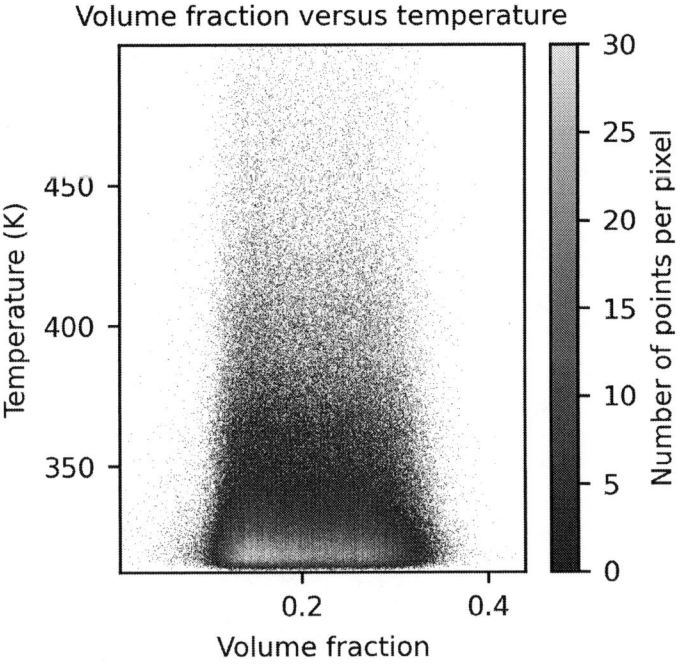

FIGURE 13: DESIGN VOLUME FRACTION VS TEMPERATURE

Velocity plots for the worst (avg temp of 1053.69 K), middle (avg temp of 325.79 K), and top 3 designs (avg temp of 312.39 K) from each category shown in Fig. 14. The top row is the best designs, the middle row is the mediocre designs, and the bottom row represents the worst designs. The red coloring indicates high fluid velocity and blue represents low fluid velocity. The worst designs comprise of very few fins and did not take advantage of the fluid flow, as most of the fluid went past the fins. The middle preforming designs made better use of the fluid flow and had smaller fins staggered evenly. The best preforming designs had small evenly spaced fins with even flow throughout the heat sink.

4.2 U-NET

Training the CNN took around 8 hours on an RTX 4080. Mean average error (MAE) is employed on the validation data to

Velocity for worst, middle, and best designs

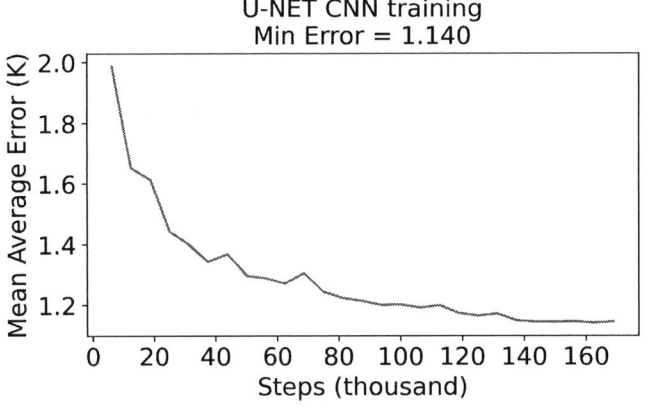

FIGURE 14: VELOCITY FOR THE GAN'S BEST, MEDIOCRE, AND WORST DESIGNS

keep track of how the model is doing. The loss for the training model is shown in Fig. 15. The model is able to achieve an

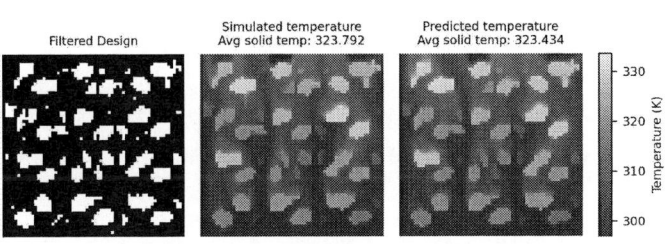

FIGURE 15: CNN TRAINING RESULTS

MAE of 1.140 K, about double the results for the complicated heat source layout case from [21]. Once training the model is complete, it can produce accurate predictions shown in Fig. 16. The left most figure is the design itself, the middle figure is the true temperature field, and the right most image is the predicted

FIGURE 16: EXAMPLE OUTPUT OF U-NET CNN

Copyright © 2023 by ASME

temperature field. Averaged over the 10,000 iterations, the CNN took about 0.004 seconds to make a single temperature field prediction. On average, the simulation took about 90 seconds to run. The CNN ran about 22,500X faster than the simulation, making it an excellent tool to run optimizations (optimization requires many solutions) or explore the performance of many designs quickly.

4.3 Optimization results

Using the combined architecture shown in Fig. 8, the WGAN-GP generates 64 designs by passing a randomly generated latent variable z from the PyTorch rand function between 0 and 1 (latent space has a shape of 64x100x1x1) into the GAN. The GAN then decompresses the latent variable z into full scale designs (from 64x100x1x1 to 64x1x64x64). Optimization is preformed on the latent space representations of the designs using ASGD within PyTorch. Optimization takes 10 minutes on the GPU over the course of 10,000 iterations. Simulation is preformed on the original and optimized designs to get the before and after temperatures shown in Table 3.

TABLE 3: DESIGN OPTIMIZATION RESULTS

Average solid temperature	Average [K]	Best [K]	Worst [K]
Before optimization	336.05	315.35	441.88
After optimization	321.96	310.73	453.11

On average, the optimized designs improved by fourteen degrees kelvin. Optimization is also able to produce a design better than the training data with an average temperature of 310.73K. The best design from optimization out-preformed the best design in the 550K designs generated by the GAN to train the CNN or the GANs training data by 1.83 degrees K. The worst design ended up getting worse after optimization going from 441.88K to 453.11 K. A visual representation for the optimization improvement can be seen in Fig. 17 where optimization improves the performance of the majority of the designs.

Extraction is preformed on highest preforming design from the 64 optimized designs. The before and after design is shown

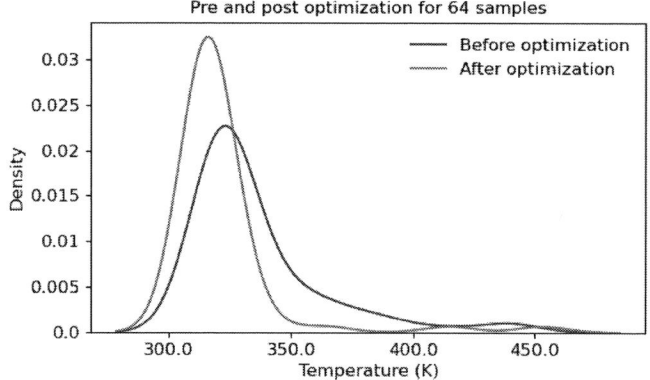

FIGURE 17: TEMPERATURE DISTRIBUTION BEFORE AND AFTER OPTIMIZATION

in Fig. 18. The left side of the figure is the design before optimization and the right side is after optimization. The design in the top changes to minimize the hot-zone in the original design. The hot zone size after optimization is decreased and the overall scale of the temperatures is decreased as well.

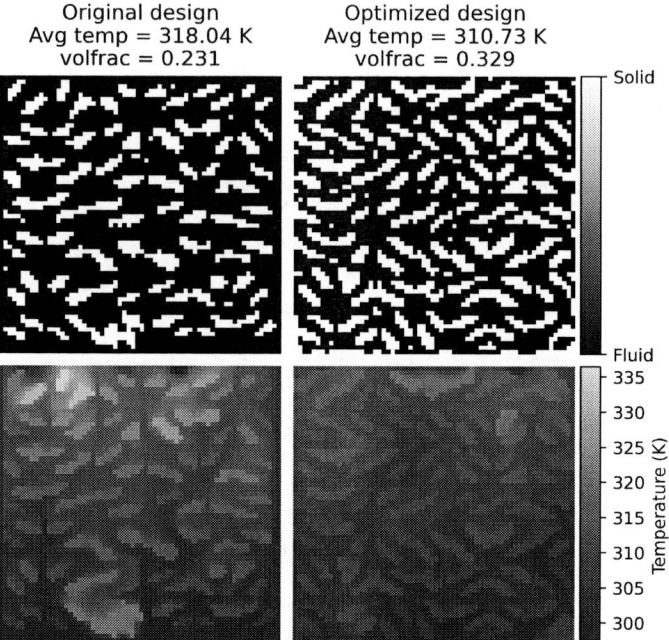

FIGURE 18: BEST DESIGN OPTIMIZATION BEFORE AND AFTER

5. CONCLUSION

In this work, we propose a combined GAN and CNN approach based on a deep learning model that is capable of producing improved designs from poor-performance training data on a conjugate heat transfer optimization problem. The WGAN-GP model learns from a wide distribution of images from the training data with acceptable quality without experiencing model collapse. The U-NET CNN model is able to predict the corresponding temperature field with an MAE of 1.140K with a 22,500x speed up compared to the simulation model. Combining the generator from the GAN and CNN form the combined model and optimization is preformed on the latent space representation of the designs with an average improvement of fourteen degrees over 64 non-optimized designs. The combined model is able to produce designs that out-preformed the training data. The best design had an average temperature of 310.73 K, 1.83 K better than any of the designs in the training data.

In future work, the combined model could be applied to a higher resolution images 512 x 512, or even 3D designs by changing the layers in the GAN and CNN. Finally, this combined model could be applied to a different set of fluid or engineering related problems.

ACKNOWLEDGMENTS

The authors want to acknowledge the support from NSF grants #1941206 and #2219931.

Copyright © 2023 by ASME

REPRODUCIBILITY

Code can be found at https://github.com/FlynnDesigns/Research

REFERENCES

[1] Zhang, Ruochun and Qian, Xiaoping. "Parameter-free Shape Optimization of Heat Sinks." *2020 19th IEEE Intersociety Conference on Thermal and Thermomechanical Phenomena in Electronic Systems (ITherm)*: pp. 756–765. 2020. DOI 10.1109/ITherm45881.2020.9190501.

[2] Sun, Sicheng, Liebersbach, Piotr and Qian, Xiaoping. "3D topology optimization of heat sinks for liquid cooling." *Applied Thermal Engineering* Vol. 178 (2020): p. 115540. DOI https://doi.org/10.1016/j.applthermaleng.2020.115540. URL https://www.sciencedirect.com/science/article/pii/S1359431120330222.

[3] Jiang, Ping, Zhou, Qi and Shao, Xinyu. *Surrogate model-based engineering design and optimization.* Springer (2020).

[4] Shi, Haoning, Ma, Ting, Chu, Wenxiao and Wang, Qiuwang. "Optimization of inlet part of a microchannel ceramic heat exchanger using surrogate model coupled with genetic algorithm." *Energy Conversion and Management* Vol. 149 (2017): pp. 988–996. DOI https://doi.org/10.1016/j.enconman.2017.04.035. URL https://www.sciencedirect.com/science/article/pii/S019689041730345X.

[5] Goodfellow, Ian J., Pouget-Abadie, Jean, Mirza, Mehdi, Xu, Bing, Warde-Farley, David, Ozair, Sherjil, Courville, Aaron and Bengio, Yoshua. "Generative Adversarial Networks." (2014). DOI 10.48550/ARXIV.1406.2661. URL https://arxiv.org/abs/1406.2661.

[6] Behzadi, Mohammad Mahdi and Ilieş, Horea T. "GANTL: Toward Practical and Real-Time Topology Optimization With Conditional Generative Adversarial Networks and Transfer Learning." *Journal of Mechanical Design* Vol. 144 No. 2 (2021). DOI 10.1115/1.4052757. URL https://asmedigitalcollection.asme.org/mechanicaldesign/article-pdf/144/2/021711/6806350/md_144_2_021711.pdf, URL https://doi.org/10.1115/1.4052757. 021711.

[7] Oh, Sangeun, Jung, Yongsu, Kim, Seongsin, Lee, Ikjin and Kang, Namwoo. "Deep Generative Design: Integration of Topology Optimization and Generative Models." *Journal of Mechanical Design* Vol. 141 No. 11 (2019). DOI 10.1115/1.4044229. URL https://asmedigitalcollection.asme.org/mechanicaldesign/article-pdf/141/11/111405/6578473/md_141_11_111405.pdf, URL https://doi.org/10.1115/1.4044229. 111405.

[8] Wang, Yuyang, Shimada, Kenji and Farimani, Amir Barati. "Airfoil GAN: Encoding and Synthesizing Airfoils forAerodynamic-aware Shape Optimization." (2021). DOI 10.48550/ARXIV.2101.04757. URL https://arxiv.org/abs/2101.04757.

[9] Tan, Ren Kai, Zhang, Nevin L. and Ye, Wenjing. "A deep learning-based method for the design of microstructural materials." *STRUCTURAL AND MULTIDISCIPLINARY OPTIMIZATION* Vol. 61 No. 4 (2020): pp. 1417–1438. DOI 10.1007/s00158-019-02424-2.

[10] Gulrajani, Ishaan, Ahmed, Faruk, Arjovsky, Martin, Dumoulin, Vincent and Courville, Aaron. "Improved Training of Wasserstein GANs." (2017). DOI 10.48550/ARXIV.1704.00028. URL https://arxiv.org/abs/1704.00028.

[11] Ronneberger, Olaf, Fischer, Philipp and Brox, Thomas. "U-Net: Convolutional Networks for Biomedical Image Segmentation." (2015). DOI 10.48550/ARXIV.1505.04597. URL https://arxiv.org/abs/1505.04597.

[12] https://github.com/nicoguaro/ellipse_packing.

[13] https://docs.scipy.org/doc/scipy/reference/generated/scipy.ndimage.binary_fill_holes.html.

[14] "Center for High Throughput Computing (CHTC)." DOI https://doi.org/10.21231/GNT1-HW21. URL https://chtc.cs.wisc.edu.

[15] https://hub.docker.com/r/natsumizu/myopenfoam7.

[16] Lipton, Zachary C. and Tripathi, Subarna. "Precise Recovery of Latent Vectors from Generative Adversarial Networks." DOI 10.48550/ARXIV.1702.04782. URL https://arxiv.org/abs/1702.04782.

[17] https://github.com/aladdinpersson/Machine-Learning-Collection.

[18] Barratt, Shane and Sharma, Rishi. "A Note on the Inception Score." (2018). DOI 10.48550/ARXIV.1801.01973. URL https://arxiv.org/abs/1801.01973.

[19] Heusel, Martin, Ramsauer, Hubert, Unterthiner, Thomas, Nessler, Bernhard and Hochreiter, Sepp. "GANs Trained by a Two Time-Scale Update Rule Converge to a Local Nash Equilibrium." (2017)DOI 10.48550/ARXIV.1706.08500. URL https://arxiv.org/abs/1706.08500.

[20] https://pypi.org/project/pytorch-gan-metrics/.

[21] Chen, Xianqi, Zhao, Xiaoyu, Gong, Zhiqiang, Zhang, Jun, Zhou, Weien, Chen, Xiaoqian and Yao, Wen. "A Deep Neural Network Surrogate Modeling Benchmark for Temperature Field Prediction of Heat Source Layout." (2021). DOI 10.48550/ARXIV.2103.11177. URL https://arxiv.org/abs/2103.11177.

[22] Sun, Jialiang, Zheng, Xiaohu, Yao, Wen, Zhang, Xiaoya, Zhou, Weien and Chen, Xiaoqian. "Heat Source Layout Optimization Using Automatic Deep Learning Surrogate and Multimodal Neighborhood Search Algorithm." (2022). DOI 10.48550/ARXIV.2205.07812. URL https://arxiv.org/abs/2205.07812.

[23] Mai, Vien V. and Johansson, Mikael. "Stability and Convergence of Stochastic Gradient Clipping: Beyond Lipschitz Continuity and Smoothness." (2021). URL 2102.06489.

Copyright © 2023 by ASME

Proceedings of the ASME 2023
International Design Engineering Technical Conferences and
Computers and Information in Engineering Conference
IDETC-CIE2023
August 20-23, 2023, Boston, Massachusetts

DETC2023-116574

ON THE CONNECTEDNESS OF THE TOPOLOGY OPTIMIZATION PREDICTORS

Mohammad Mahdi Behzadi[1], Horea T. Ilies[1,*],

[1]University of Connecticut, Storrs, CT

ABSTRACT

Deep learning-based topology optimization predictors have been shown to be effective in generating optimal designs. However, these predictors are prone to topological errors, particularly for high-resolution domains. Although various methods have been developed to enhance the accuracy of predicted structures, such as using large training datasets, complex networks, and physics-based loss functions, they do not include topological metrics in the deep learning models. Similar issues arise in other applications, such as blood vessels, neurons, or road segmentation from images, and several modifications to typical loss functions have been proposed to improve the topological validity of the predictions. In this study, we evaluate and compare four distinct topological loss functions to explore their influence on the performance of deep learning-based topology optimization predictors. Our findings offer insights into the advantages and limitations of these modified loss functions and provide a basis for future research and development aimed at improving the accuracy and efficiency of deep learning predictors in topology optimization.

Keywords: Topology Optimization, Topological loss functions, Deep learning

1. INTRODUCTION

Various deep learning algorithms have been used in topology optimization (TO) to efficiently predict optimal designs. These approaches have successfully generated exemplary designs through gradient-based TO and have shown some level of generalizability in exploring the design space. However, a major drawback of these methods is that the predicted structures often suffer from topological errors, resulting in weaker structural performance and decreased manufacturability compared to the ground truths obtained from gradient-based optimization methods [1]. To address this challenge, different approaches have been proposed, such as using large training datasets [2], em-

*Corresponding author: horea.ilies@uconn.edu

ploying complex networks [3], and incorporating a physical loss function into the training process [4]. Although these methods show promise in improving the quality of predicted structures, they are computationally inefficient and do not integrate topological metrics into the deep learning model. The premise of this work is that by integrating topological information into the model, topologically correct structures can be generated without requiring large amounts of training data or complex networks.

Of course, topological errors are not confined to topology optimization (TO) predictors. Segmentation algorithms, such as those used for road and blood vessel segmentation, are also susceptible to this type of errors. Various methods have been developed to address this issue in segmentation models, including the use of topological loss functions [5–7] and of complex networks [8, 9]. For further information, please refer to [10]. Despite extensive efforts to integrate topological features into deep learning models in the segmentation literature, this issue has not received much attention in the TO community. Our recent work [11] demonstrates that the incorporation of topological loss terms improves the connectivity of the predicted optimal structures. Specifically, in [11] we have shown that using persistence homology-based loss functions can generate structures with twice better connectivity compared to the baseline predictions.

The existing loss functions are essentially global metrics that assess the overall fit for a domain, typically represented as a uniform grid in 2D or 3D. Despite incorporating topology-aware terms, these loss functions generally fail to account for localized geometric changes, rendering them ineffective in penalizing the unwanted "artifacts" in the predictions produced by the generative models. Developing metrics that capture local geometric changes, which in turn induce desired topological properties, remains a challenge.

This paper aims to take a step in addressing this limitation by investigating the effectiveness of various topological loss functions in enhancing the accuracy of deep learning-based topology optimization predictors and by assessing the corresponding advantages and limitations. Specifically, a convolutional neu-

Copyright © 2023 by ASME

ral network similar to the one used in [12] is trained with four different topological loss functions: clDice [5], the TopoLoss [6], the Wasserstein loss [13], and the Bottleneck loss function [11]. Moreover, we also explore the limitations of each loss function, such as the associated computational complexity, and evaluate their effectiveness in generating accurate and well-connected structures. Overall, this study contributes to the ongoing efforts to improve the accuracy and efficiency of deep learning-based topology optimization predictors, and its findings can be used to guide future research and development in this area.

2. BACKGROUND

Gradient-based topology optimization algorithms are computationally expensive, which limits their ability to explore the design space efficiently. In recent years, deep learning-based topology optimization predictors have emerged as a potential alternative because, once the models are trained, they can efficiently generate optimal designs [14–17]. However, one major challenge of all these methods is that the predicted structures often suffer from topological errors, which in turn leads to weaker structural performance and decreased manufacturability compared to the optimal solutions predicted by gradient-based optimization methods.

Various approaches have been proposed to improve the topological quality of the structures predicted by deep learning models. For example, the authors in [2] showed that by using 296,000 training cases for their CNN model, they could obtain the structures with better pixel-wise accuracy compared to their benchmark method [18] where the authors used 80,000 cases for training. As expected, using a larger training dataset leads to better accuracy. However, generating this type of data is a very expensive process because every solution requires often hundreds of Finite Element simulations, and the cost increases exponentially with the resolution and the dimension of the space. Moreover, these models are trained to predict material occupancy for each cell of the grid (pixel in 2D or voxel in 3D), and do not include any topological information in the development of the model. Some other studies used more complex models, e.g., Generative Adversarial Networks, to obtain better structures [19, 20]. Although these models have shown a better performance and generalizability compared to CNNs, these models are often difficult to train, need a large amount of data, and often produce structures with disconnected members. At the same time, the severity of these errors increases rapidly as the boundary conditions move away from those included in the training distribution. Physical constraints have been added to the loss function in [4], and these constraints are based on a compliance error defined through the pixel-wise loss function in the training process of the deep learning model. Despite some improvements in the quality of the predicted structures, the proposed method estimates the compliance at every step through a finite element analysis each time the loss function is calculate, which is time-consuming. Although these approaches can improve the quality of the predicted structures, they do not incorporate topological information into the training process.

The inclusion of topological information into the training process did not receive significant attention in the TO community. Nevertheless, numerous studies have focused on the topological connectedness of segmentation algorithms, which encounter similar challenges. Topological errors in segmentation can lead to inaccurate identification of regions, thereby affecting the accuracy of subsequent analysis. Consequently, several methods have been developed to address this problem, and the vast majority of the existing methods included topology-aware terms into their loss functions. For example, in one of the early attempts to use topological loss functions, [6] proposed a modified Wasserstein distance based on persistent homology [21, 22]. This loss function enforced the segmented image to exhibit the same Betti error as the ground truth. Results from this study demonstrated that the loss function effectively reduced Betti errors on various 2D datasets containing natural and biomedical images. Another work by [23] introduced a topological loss function that utilizes persistent homology barcodes. They showed that their loss function improved the Betti error by a factor of 10 in the multi-class segmentation of 2D short axis and 3D whole heart cardiac magnetic resonance (CMR) images compared to a baseline U-Net model. In a different approach, [5] proposed a novel connectivity-aware similarity measure called center-lineDice (clDice) for image segmentation problems. Their loss function involved intersecting the segmentation masks with morphological skeletons. Results presented in their study, encompassing five public 2D and 3D datasets, demonstrated that clDice effectively improved topological errors in the segmented images. Moreover, [13] employed the Wasserstein distance in the loss function to minimize topological discrepancies in 3D reconstruction. They revealed that incorporating the topological loss function alongside the pixel-wise loss function improved the quality of the reconstructed images. Numerous other studies have also integrated topological information into segmentation problems, such as those conducted in [24–30]. Interested readers can explore these studies for more detailed information.

It's important to consider that while many of the existing topological loss functions have proven to be versatile in their ability to handle both 2D and 3D images, they often come with a significant computational burden, especially when dealing with 3D images. Specifically, approaches based on persistence homology may be too computationally expensive for high-resolution images, which can limit their usefulness in certain scenarios. Therefore, it is crucial to continue exploring alternative approaches or implementations for integrating topological features into deep learning methods' training process. This includes investigating topology optimization predictors, with a specific emphasis on enhancing their scalability and computational efficiency when working with 3D image datasets.

3. CONTRIBUTIONS AND OUTLINE

While the importance of topological connectedness in topology optimization problems cannot be overstated, only one study [11] has examined the utilization of topological loss functions during the training phase. However, this previous work did not comprehensively evaluate the significance and limitations of different topological loss functions. To address this gap, the present paper investigates the impact of various existing topological loss functions, including those proposed by [5, 6, 11, 13], on the performance of deep learning-based topology optimization pre-

Copyright © 2023 by ASME

dictors. Our investigation includes a detailed analysis of the advantages and limitations of each loss function, taking into account factors such as computational complexity. Additionally, we assess the effectiveness of each loss function in generating accurate and connected structures. This research study contributes to ongoing efforts aimed at enhancing the accuracy and efficiency of deep learning-based topology optimization predictors by exploring the application of diverse topological loss functions. The findings of this study can provide valuable guidance for future research and development endeavors in this field.

4. METHOD

4.1 Topological loss functions

A CNN model similar to the one employed in [12] has been trained using $2,250$ structures at a resolution of 120×240, as described in [11], and all model hyperparameters are similar to those used in [12]. The model underwent an initial training of 100 epochs, followed by fine-tuning for another 150 epochs using four different topological loss functions: TopoLoss, clDice, Wasserstein distance, and Bottleneck distance. During the fine-tuning process of all but the clDice loss function, the topological losses were incorporated as an additional term with an appropriate coefficient into the binary cross-entropy loss. The clDice loss function uses the Dice score instead of binary cross-entropy as its primary loss function. In the following section, we provide a detailed explanation of each topological loss function.

4.1.1 TopoLoss [6]. Persistent homology is a fundamental concept in the field of topological data analysis, which aims to capture the essential topological properties of an object. It achieves this by examining how the homology groups of a space change as a parameter, such as scale or distance, varies. Let's consider a continuous image domain denoted as $\Omega \subset R^2$, along with a predicted image function $f : \Omega \rightarrow R$ and a segmentation image $X \subset R$ obtained by applying a threshold to the predicted image. The d-dimensional topological structure of X represents a class of d-manifolds that can be transformed into each other within X. Simply put, 0-dimensional structures correspond to connected components, while 1-dimensional structures correspond to handles. To capture the complete range of topological information within f, persistent homology involves applying different thresholds to the predicted image f. As the threshold decreases, the topology of f undergoes changes, leading to the "birth" of new topological structures and the elimination of existing ones. This information is often visualized using a persistence diagram, denoted as $Dgm(f)$, which provides a graphical representation of the birth and death of topological features across different parameter values [21, 22].

TopoLoss [6] describes a modified version of the Wasserstein distance that is specifically designed for comparing persistence diagrams. Given the persistence diagrams of an image prediction $(Dgm(f))$ and the ground truth $(Dgm(g))$, TopoLoss aims to find the optimal one-to-one mapping between the points in $Dgm(f)$ and $Dgm(g)$, and then calculate the total squared distance between the corresponding point sets. The matching algorithm proceeds as follows. The points from $Dgm(g)$ are placed at the upper-left corner $p_{ul} = (0, 1)$, with $birth(p_{ul}) = 1$ and $death(p_{ul}) = $

0. Then, TopoLoss identifies the points in $Dgm(f)$ that are closest to p_{ul} and maps them to the corresponding points in $Dgm(g)$. The algorithm computes the squared distances from all points in $Dgm(f)$ to p_{ul}, and then sorts these distances. In summary, TopoLoss can be formulated as:

$$L_{topo}(f, g) = \sum_{p \in dgm(f)} [birth(p) - birth(\gamma^*(p))]^2$$
$$+ [death(p) - death(\gamma^*(p))]^2 \quad (1)$$

Where γ^* is the optimal matching between two different point sets. For further information about TopoLoss, we direct the reader to the original paper [6].

4.1.2 Center-LineDice (clDice) [5]. The clDice loss function utilizes morphological thinning, achieved through min- and max-pooling as a substitute for morphological erosion and dilation, on both the predicted and ground truth images. Given the ground truth (G) and the predicted image (P) as a binary image and their corresponding skeletons S_G and S_P, the clDice loss function can be obtained from the following formula:

$$clDice(G, P) = 2 \times \frac{T_{prec}(S_P, G) \times T_{sens}(S_G, P)}{T_{prec}(S_P, G) + T_{sens}(S_G, P)} \quad (2)$$

Where T_{prec} and T_{sens} are the Topology Precision and Topology Sensitivity, respectively, that can be obtained from:

$$T_{prec}(S_P, G) = \frac{|S_P \cap G|}{|S_P|} \quad (3)$$

$$T_{sens}(S_G, P) = \frac{|S_G \cap P|}{|S_G|} \quad (4)$$

The crucial step to calculate the clDice loss is to find the skeleton of the image. Importantly, the process of obtaining the skeleton should be differentiable in order to be used in the loss function. The authors in [5] proposed an iterative process based on min- and max-pooling that are used as a proxy for morphological erosion and dilation to calculate the skeleton of the image. The proposed skeletonization algorithms allow the use of clDice as a fully differentiable, real-valued, optimizable measure. Finally, the loss function used in the training process is described by:

$$L_{clDice} = (1 - \alpha)(1 - Dice) + \alpha(1 - clDice) \quad (5)$$

Where $\alpha \in [0, 0.5]$.

4.1.3 Wasserstein distance [13]. The q-Wasserstein distance measures the similarity between two persistence diagrams, which is defined as the minimum value achieved by a perfect matching between the points of the two diagrams. If $Dgm(f)$ and $Dgm(g)$ are the persistence diagrams of the predicted and the ground truth images, respectively, the q-Wasserstein distance can be obtained from:

$$W_q(Dgm(f), Dgm(g)) = \left[\inf_{\eta} \sum_{p \in Dgm(f)} ||p - \eta(p)||_\infty^q \right]^{\frac{1}{q}} \quad (6)$$

Where η is a perfect matching between the diagrams. The Wasserstein distance can be obtained by solving the optimal transport algorithms [31]. For the purpose of this study, the value of q is assumed to be 1.

Copyright © 2023 by ASME

4.1.4 Bottleneck distance [11]. The Bottleneck distance is the fourth metric that can be employed to assess the similarity between two persistence diagrams. More precisely, it determines the minimum distance required to achieve a perfect matching between the points in the two diagrams. Alternatively, the Bottleneck distance can be computed as the maximum weight in the solution to the minimum weight perfect matching problem. Similar to the Wasserstein distance, the Bottleneck distance can be defined as:

$$W_\infty(Dgm(f), Dgm(g)) = \inf_{\eta} \sup_{p \in Dgm(f)} ||p - \eta(p)||_\infty \quad (7)$$

4.2 Evaluation Metrics

We utilize multiple evaluation metrics to assess the performance of each loss function. These metrics include the pixel-wise accuracy, which is defined as the proportion of correctly classified pixels, as well as the Betti number errors, which directly measure the topological difference (e.g., the number of handles) between the predicted and ground truth images. The 0th Betti number is calculated by counting the number of connected components in the image, while the 1st Betti number is obtained by counting the number of holes in the structure. The Betti number error is calculated as follows:

$$i^{th} Betti\ error = \frac{B^i(p) - B^i(g)}{B^i(g)} \quad (8)$$

where B^i is the ith Betti number, p is short for predicted structure and g stands for ground truth.

Two other metrics that we used to measure the performance of the loss functions are the Compliance Error (CE), which compares the performance of the structures in terms of structural compliance, and the Volume Fraction Error (VFE), which compares the amount of material used in the prediction an the ground truth. The compliance error can be calculated from the following:

$$CE = \frac{|C(p) - C(g)|}{C(g)} \quad (9)$$

where $C(p)$ and $C(g)$ are the compliance of the predicted and the ground truth structure, respectively. Similarly, VFE is obtained from:

$$VF\ error = \frac{|VF(p) - VF(g)|}{VF(g)} \quad (10)$$

where $VF(p)$ and $VF(g)$ are the volume fraction of the prediction and the ground truth, respectively.

5. RESULTS AND DISCUSSION

5.1 Overall Results

We started by conducting an initial training session for our convolutional neural network (CNN). This session involved a maximum of 100 epochs, utilizing the binary cross-entropy loss function and early stopping to identify the best-performing network. Afterward, we proceeded to fine-tune the network by incorporating various topological loss functions. During the fine-tuning process, the model underwent training for a maximum of 150 epochs, with early stopping used to select the model that

achieved the lowest topological loss value. All model parameters were considered trainable throughout the fine-tuning stage. The topological loss function was integrated into the binary cross-entropy loss, except for clDice, as described above. Each term in the loss function was assigned an appropriate weight, determined through grid search to optimize performance. To assess the effectiveness of these different approaches, we present a comparison of the original CNN prediction with structures generated using the fine-tuned network employing various topological loss functions (refer to Figure 1). Additionally, we collected evaluation data using performance metrics outlined in Section 4.2 and summarized our findings in Table 1. It's worth noting that both the models trained with and without topological loss terms were capable of making predictions in real-time, with minimal disparity in the prediction time.

The performance analysis indicates that the utilization of different topological loss functions has the potential to improve the connectivity of predicted structures. Specifically, when employing the TopoLoss function, both the 0th and 1st Betti errors showed notable enhancements, decreasing from 95.78% to 77.1% and from 50.1% to 27.2%, respectively. This improvement in Betti errors also resulted in lower compliance errors, indicating better structural performance, along with slightly better accuracy. Comparing the predicted structures using TopoLoss (shown in Figure 1) with the original predictions clearly demonstrates the significant improvement achieved through the use of TopoLoss. Similarly, the clDice function also improved the 0th and 1st Betti errors, reducing them from 95.78% to 46.82% and from 50.1% to 41.25%, respectively. From figure 1, it can be seen that in some cases clDice filled the holes in the structure, which contributed to higher 1st Betti error, compliance error, volume fraction error, and lower accuracy compared to other topological loss functions.

Both the Wasserstein and Bottleneck distances demonstrated significant improvements in Betti errors, structural performance, and accuracy. Although the Wasserstein distance did not substantially improve the 0th Betti error, it outperformed other loss functions in terms of 1st Betti error, compliance error, and volume fraction error. The better compliance error and volume fraction error, despite having nearly the highest 0th Betti error, can be attributed to the Wasserstein distance's ability to connect floating material in the original prediction to the main body through the correct path, albeit at the cost of introducing additional floating members in the structure. A comparison of the results shown in the last row of Figure 1 for the Wasserstein distance with the original prediction and ground truth highlights this trade-off. The Bottleneck distance outperformed the Wasserstein distance in terms of 0th Betti error, but it exhibited higher 1st Betti error, compliance error, and volume fraction error. The high volume fraction error in the Bottleneck distance could be attributed to the addition of extra material that connect the unwanted part to the main body, as well as the failure to remove unwanted material from the original prediction. For instance, comparing the structures shown in the fourth row of Figure 1 for the Bottleneck distance with those for the Wasserstein distance and TopoLoss reveals that the latter two losses attempted to remove the extra material from the original prediction, while the Bottleneck loss connected that part to the main body. This difference explains the

Copyright © 2023 by ASME

TABLE 1: THE QUALITATIVE COMPARISON BETWEEN THE PREDICTED STRUCTURES WITH DIFFERENT TOPOLOGICAL LOSS FUNCTIONS AND THE ORIGINAL MODEL WITHOUT ANY TOPOLOGICAL LOSS FUNCTION. THE DOMAINS HAVE A RESOLUTION OF 120 × 240 AND THE SEEN BOUNDARY CONDITIONS REPORTED IN [11].

Metrics	Baseline	TopoLoss	clDice	Wasserstein distance	Bottleneck distance
0th Betti error	95.78 %	77.1 %	46.82 %	94.2 %	86.5 %
1st Betti error	50.1 %	27.2 %	41.25 %	25.5 %	26.9 %
Compliance Error	13.96 %	11.30 %	14.50 %	10.5 %	11.2 %
Volume Fraction Error	0.41 %	0.45 %	1.67 %	0.28 %	1.01 %
Accuracy	94.34 &	94.74 %	94.25 %	94.65 %	94.36 %

TABLE 2: THE COMPARISON BETWEEN THE TRAINING TIME REQUIRED BY EACH LOSS FUNCTION. (EPOCH/HOUR)

Time	BCE	TopoLoss	clDice	Wasserstein distance	Bottleneck distance
Training time	40	1.8	30	0.41	0.41

higher volume fraction error observed in the Bottleneck distance.

Upon examining the data presented in table 1, it is evident that achieving a superior 0th Betti error does not necessarily lead to improved structural performance. Nevertheless, the findings strongly indicate a significant correlation between the 1th Betti error and compliance error since structures with lower 1th Betti error exhibit correspondingly lower compliance errors. The 1th Betti error compares the main body of the predicted structure (which contributes to compliance error) with the ground truth, disregarding completely disconnected members. Consequently, a lower 1th Betti error suggests that the predicted structure's main body aligns more closely with the ground truth, resulting in decreased compliance error. Hence, the 1th Betti error is a more critical factor in determining the best loss function. In this regard, the Wasserstein distance outperforms other loss functions, although all loss functions can improve the quality of predicted structures by TO predictors.

5.2 Computational Cost

Table 2 provides a breakdown of the training time for each loss function on a UConn HPC equipped with an NVIDIA GeForce RTX 2080Ti GPU. The results reveal that the topological losses employing persistence homology are computationally expensive. Specifically, the training for TopoLoss is almost 20 times slower than binary cross-entropy and 15 times slower than clDice. However, TopoLoss is more than 4 times faster than the Wasserstein and Bottleneck distance, as the latter requires solving the optimal matching problem to find the optimal matching between the points in the diagrams of the ground truth and the predicted structures in addition to obtaining the persistence diagram, which is a time-consuming process. The computational cost of clDice comes from the iterative skeletonization process, which uses min-max pooling operations that can be applied to higher-resolution images. However, the process may require more iterations for higher-resolution domains, slightly increasing its training time. On the other hand, the persistence-based loss functions face challenges when working with high-resolution domains, as the cost of calculating the persistence diagram increases significantly as the image size grows.

5.3 Limitations

The application of persistence homology-based loss functions to enhance the accuracy of deep learning models is constrained by various factors. One significant limitation is their high computational cost, rendering them impractical for handling high-resolution images. As image size grows, the computation cost associated with calculating the persistence diagram also escalates considerably, thereby restricting the use of these loss functions in numerous real-world scenarios. Additionally, fine-tuning their contribution in the training process is necessary for different networks and datasets, leading to time-consuming procedures demanding substantial time and computational resources.

One limitation of all topological loss functions is their lack of generalizability, which means they cannot be directly employed with diverse networks and tasks without undergoing training. Thus, given their computational demands, their application may be unfeasible in many contexts. In addition, this study has also revealed that the utilization of topological loss functions may introduce noise into predicted structures, requiring subsequent post-processing. Figure 2 provides an illustration of the noise generated by TopoLoss and the Bottleneck loss function integrated with the relatively simple CNN model described above. We note here that the results presented in section 5.1 have undergone noise removal procedures, and that the extent of noise created by the loss functions varies depending on their weight during the training process. At the same time, the network's architecture itself may impact the presence of noise, as evidenced by the fact that the GAN used in [11] did not exhibit noise in the generated structures. Nevertheless, additional work is required to better understand the source of noise, although this work is outside the scope of this paper.

6. CONCLUSION

This study investigates the effectiveness of various topological loss functions in improving the quality of predicted structures generated by topology optimization (TO) predictors. The performance of the Wasserstein distance, Bottleneck distance, clDice, and TopoLoss was evaluated based on their impact on the 0th and 1st Betti errors, compliance error, volume fraction error, and accuracy. It should be noted that the CNN model used in this research was deliberately selected to be simple, and that the training was conducted with a relatively small dataset in order to manage the computational costs. Nonetheless, these loss functions can be applied effectively to more complex models like GANs,

Copyright © 2023 by ASME

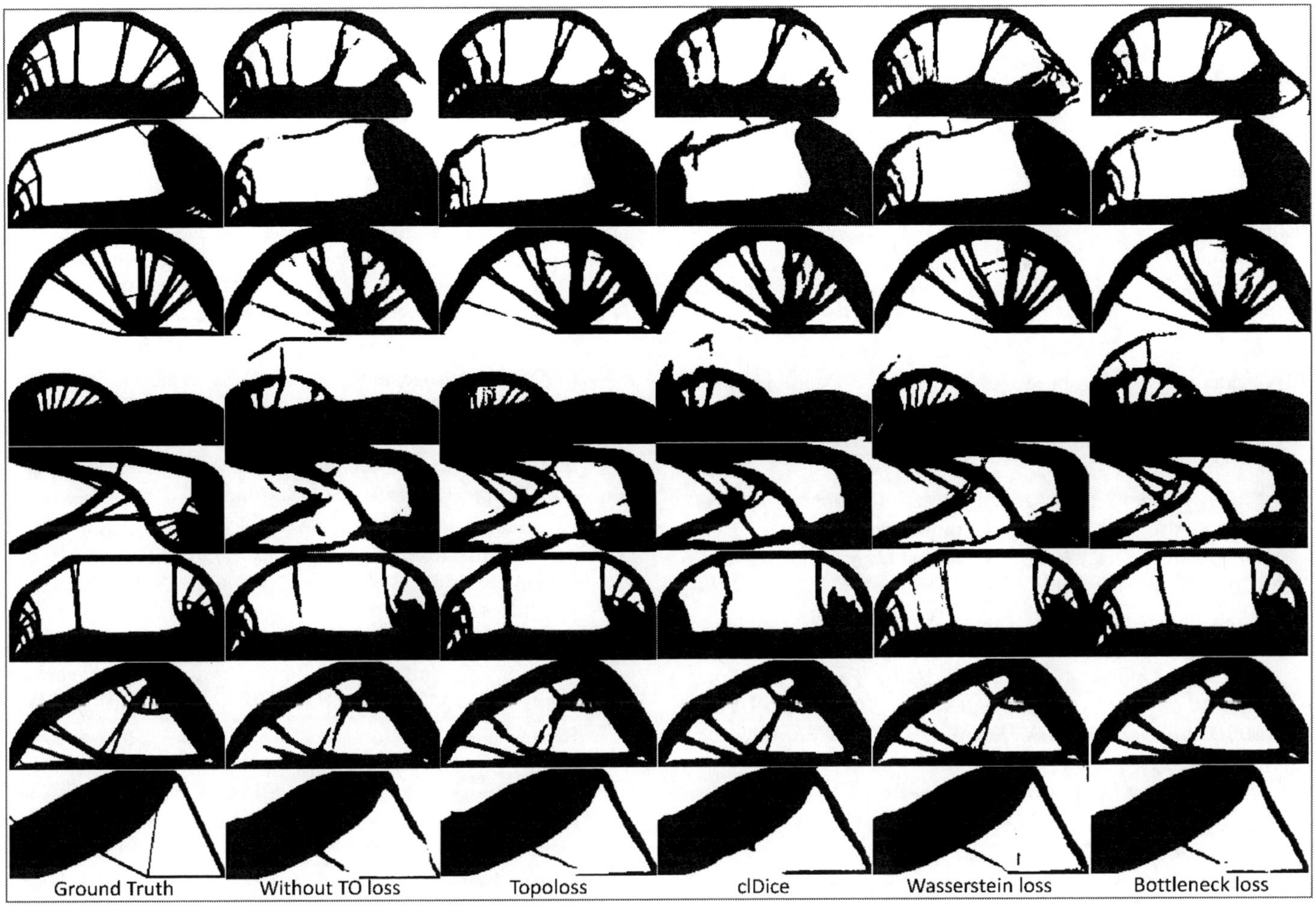

FIGURE 1: COMPARISON BETWEEN THE GROUND TRUTH (SIMP OPTIMIZED) 2D STRUCTURES, THE PREDICTED STRUCTURES BY CNN WITHOUT TOPOLOGICAL LOSS, AND THE CNN PREDICTIONS WITH DIFFERENT TOPOLOGICAL LOSSES.

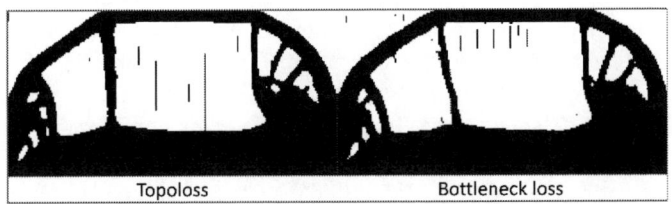

Topoloss   Bottleneck loss

FIGURE 2: EXAMPLE OF NOISE GENERATED BY TOPOLOSS AND BOTTLENECK LOSS FUNCTIONS USING THE CNN. OBSERVE THAT THIS TYPE OF NOISE FOR THE BOTTLENECK DISTANCE WAS NOT SEEN PREVIOUSLY IN [11].

and volume fraction error. However, the Bottleneck distance outperformed the Wasserstein distance in terms of the 0th Betti error, albeit with higher values for the 1st Betti error, compliance error, and volume fraction error.

A key advantage of topological loss functions is their ability to be integrated into the training process of any deep learning model, such as CNNs and GANs, as demonstrated in [11], enabling improvements in the topological connectivity of images across various applications. However, it should be noted that these loss functions do not guarantee the connectivity of the resulting predictions. One significant drawback of persistence homology-based loss functions is the computational cost associated with persistent homology, which increases significantly with larger image sizes. Consequently, such loss functions are currently impractical for many real-world applications that require high-resolution 2D or 3D images. Considering that topological loss functions represent global measures of connectivity, an alternative approach for future research could involve leveraging more local constructs like connectivity graphs.

as demonstrated in [11], with larger training datasets to achieve improved prediction accuracy.

The results indicate that employing topological loss functions can significantly enhance the connectivity of predicted structures, resulting in better Betti errors, compliance errors, volume fraction errors, and accuracy. TopoLoss and the Wasserstein distance were identified as the most effective, with the latter being particularly effective in improving the 1st Betti error, compliance error,

Copyright © 2023 by ASME

ACKNOWLEDGEMENTS

The authors acknowledge the support from the National Science Foundation grant CMMI-2232612 and the Office of Naval Research/NIUVT. The authors would like to thank the anonymous reviewers for their constructive feedback. The responsibility for any errors and omissions lies solely with the authors.

REFERENCES

[1] Woldseth, Rebekka V, Aage, Niels, Bærentzen, J Andreas and Sigmund, Ole. "On the use of artificial neural networks in topology optimisation." *Structural and Multidisciplinary Optimization* Vol. 65 No. 10 (2022): p. 294.

[2] Nakamura, Keigo and Suzuki, Yoshiro. "Deep learning-based topological optimization for representing a user-specified design area." *arXiv preprint arXiv:2004.05461* (2020).

[3] Mazé, François and Ahmed, Faez. "Diffusion Models Beat GANs on Topology Optimization." (2022). DOI 10.48550/ARXIV.2208.09591. URL https://arxiv.org/abs/2208.09591.

[4] Luo, Jiaxiang, Li, Yu, Zhou, Weien, Gong, Zhiqiang, Zhang, Zeyu and Yao, Wen. "An Improved Data-Driven Topology Optimization Method Using Feature Pyramid Networks with Physical Constraints." *CMES-COMPUTER MODELING IN ENGINEERING & SCIENCES* Vol. 128 No. 3 (2021): pp. 823–848.

[5] Shit, Suprosanna, Paetzold, Johannes C, Sekuboyina, Anjany, Ezhov, Ivan, Unger, Alexander, Zhylka, Andrey, Pluim, Josien PW, Bauer, Ulrich and Menze, Bjoern H. "clDice-a novel topology-preserving loss function for tubular structure segmentation." *Proceedings of the IEEE/CVF Conference on Computer Vision and Pattern Recognition*: pp. 16560–16569. 2021.

[6] Hu, Xiaoling, Li, Fuxin, Samaras, Dimitris and Chen, Chao. "Topology-preserving deep image segmentation." *Advances in neural information processing systems* Vol. 32 (2019).

[7] Ngoc, Minh On Vu, Chen, Yizi, Boutry, Nicolas, Fabrizio, Jonathan and Mallet, Clément. "BuyTheDips: PathLoss for improved topology-preserving deep learning-based image segmentation." *arXiv preprint arXiv:2207.11446* (2022).

[8] Zhang, Han and Lui, Lok Ming. "Topology-preserving segmentation network: A deep learning segmentation framework for connected component." *arXiv preprint arXiv:2202.13331* (2022).

[9] He, Yufan, Carass, Aaron, Yun, Yeyi, Zhao, Can, Jedynak, Bruno M, Solomon, Sharon D, Saidha, Shiv, Calabresi, Peter A and Prince, Jerry L. "Towards topological correct segmentation of macular OCT from cascaded FCNs." *Fetal, Infant and Ophthalmic Medical Image Analysis: International Workshop, FIFI 2017, and 4th International Workshop, OMIA 2017, Held in Conjunction with MICCAI 2017, Québec City, QC, Canada, September 14, Proceedings 4*: pp. 202–209. 2017. Springer.

[10] Zia, Ali, Khamis, Abdelwahed, Nichols, James, Hayder, Zeeshan, Rolland, Vivien and Petersson, Lars. "Topological Deep Learning: A Review of an Emerging Paradigm." *arXiv preprint arXiv:2302.03836* (2023).

[11] Behzadi, Mohammad Mahdi and Ilieş, Horea T. "Gantl: Toward practical and real-time topology optimization with conditional generative adversarial networks and transfer learning." *Journal of Mechanical Design* Vol. 144 No. 2 (2022).

[12] Behzadi, Mohammad Mahdi and Ilieş, Horea T. "Real-time topology optimization in 3D via deep transfer learning." *Computer-Aided Design* Vol. 135 (2021): p. 103014.

[13] Waibel, Dominik JE, Atwell, Scott, Meier, Matthias, Marr, Carsten and Rieck, Bastian. "Capturing shape information with multi-scale topological loss terms for 3D reconstruction." *Medical Image Computing and Computer Assisted Intervention–MICCAI 2022: 25th International Conference, Singapore, September 18–22, 2022, Proceedings, Part IV*: pp. 150–159. 2022. Springer.

[14] Abueidda, Diab W, Koric, Seid and Sobh, Nahil A. "Topology Optimization of 2D Structures with Nonlinearities using Deep Learning." *Computers & Structures* Vol. 237 (2020): p. 106283.

[15] Sosnovik, Ivan and Oseledets, Ivan. "Neural Networks for Topology Optimization." *Russian Journal of Numerical Analysis and Mathematical Modelling* Vol. 34 No. 4 (2019): pp. 215–223.

[16] Wang, Dalei, Xiang, Cheng, Pan, Yue, Chen, Airong, Zhou, Xiaoyi and Zhang, Yiquan. "A Deep Convolutional Neural Network for Topology Optimization with Perceptible Generalization Ability." *Engineering Optimization* Vol. 54 No. 6 (2022): pp. 973–988.

[17] Banga, Saurabh, Gehani, Harsh, Bhilare, Sanket, Patel, Sagar and Kara, Levent. "3D Topology Optimization using Convolutional Neural Networks." *arXiv preprint arXiv:1808.07440* (2018).

[18] Yu, Yonggyun, Hur, Taeil, Jung, Jaeho and Jang, In Gwun. "Deep Learning for Determining a Near-Optimal Topological Design without any Iteration." *Structural and Multidisciplinary Optimization* Vol. 59 No. 3 (2019): pp. 787–799.

[19] Nie, Zhenguo, Lin, Tong, Jiang, Haoliang and Kara, Levent Burak. "TopologyGAN: Topology Optimization using Generative Adversarial Networks based on Physical Fields over the Initial Domain." *Journal of Mechanical Design* Vol. 143 No. 3 (2021).

[20] Li, Baotong, Huang, Congjia, Li, Xin, Zheng, Shuai and Hong, Jun. "Non-iterative Structural Topology Optimization using Deep Learning." *Computer-Aided Design* Vol. 115 (2019): pp. 172–180.

[21] Edelsbrunner, Herbert, Harer, John et al. "Persistent Homology-A survey." *Contemporary Mathematics* Vol. 453 (2008): pp. 257–282.

[22] Cohen-Steiner, David, Edelsbrunner, Herbert and Harer, John. "Stability of Persistence Diagrams." *Proceedings of the twenty-first Annual Symposium on Computational Geometry*: pp. 263–271. 2005.

[23] Byrne, Nick, Clough, James R, Valverde, Israel, Montana, Giovanni and King, Andrew P. "A Persistent Homology-

Based Topological Loss for CNN-Based Multiclass Segmentation of CMR." *IEEE Transactions on Medical Imaging* Vol. 42 No. 1 (2022): pp. 3–14.

[24] Stucki, Nico, Paetzold, Johannes C, Shit, Suprosanna, Menze, Bjoern and Bauer, Ulrich. "Topologically faithful image segmentation via induced matching of persistence barcodes." *arXiv preprint arXiv:2211.15272* (2022).

[25] de Dumast, Priscille, Kebiri, Hamza, Dunet, Vincent, Koob, Mériam and Cuadra, Meritxell Bach. "Multi-dimensional topological loss for cortical plate segmentation in fetal brain MRI." *arXiv preprint arXiv:2208.07566* (2022).

[26] Panconi, Luca, Makarova, Maria, Lambert, Eleanor R, May, Robin C and Owen, Dylan M. "Topology-based fluorescence image analysis for automated cell identification and segmentation." *Journal of Biophotonics* (2022).

[27] Songdechakraiwut, Tananun, Shen, Li and Chung, Moo. "Topological learning and its application to multimodal brain network integration." *Medical Image Computing and Computer Assisted Intervention–MICCAI 2021: 24th International Conference, Strasbourg, France, September 27–October 1, 2021, Proceedings, Part II 24*: pp. 166–176. 2021. Springer.

[28] Hüseyin, Furkan. "Shape-Preserving Loss in Deep Learning for Cell Segmentation." Ph.D. Thesis, Bilkent Universitesi (Turkey). 2020.

[29] Hu, Xiaoling, Wang, Yusu, Fuxin, Li, Samaras, Dimitris and Chen, Chao. "Topology-aware segmentation using discrete Morse theory." *arXiv preprint arXiv:2103.09992* (2021).

[30] Gupta, Saumya, Hu, Xiaoling, Kaan, James, Jin, Michael, Mpoy, Mutshipay, Chung, Katherine, Singh, Gagandeep, Saltz, Mary, Kurc, Tahsin, Saltz, Joel et al. "Learning Topological Interactions for Multi-Class Medical Image Segmentation." *Computer Vision–ECCV 2022: 17th European Conference, Tel Aviv, Israel, October 23–27, 2022, Proceedings, Part XXIX*: pp. 701–718. 2022. Springer.

[31] Flamary, Rémi, Courty, Nicolas, Gramfort, Alexandre, Alaya, Mokhtar Z, Boisbunon, Aurélie, Chambon, Stanislas, Chapel, Laetitia, Corenflos, Adrien, Fatras, Kilian, Fournier, Nemo et al. "Pot: Python optimal transport." *The Journal of Machine Learning Research* Vol. 22 No. 1 (2021): pp. 3571–3578.

Proceedings of the ASME 2023
International Design Engineering Technical Conferences and
Computers and Information in Engineering Conference
IDETC-CIE2023
August 20-23, 2023, Boston, Massachusetts

DETC2023-116678

WHEN IS IT ACTUALLY WORTH LEARNING INVERSE DESIGN?

Milad Habibi[1], Jun Wang[2], Mark Fuge[1],*

[1]Center for Risk and Reliability, Department of Mechanical Engineering, University of Maryland, College Park, MD
[2]Department of Mechanical Engineering, Santa Clara University, Santa Clara, CA

ABSTRACT

Design optimization, and particularly adjoint-based multiphysics shape and topology optimization, is time-consuming and often requires expensive iterations to converge to desired designs. In response, researchers have developed Machine Learning (ML) approaches—often referred to as Inverse Design methods—to either replace or accelerate tools like Topology optimization (TO). However, these methods have their own hidden, non-trivial costs including that of data generation, training, and refinement of ML-produced designs. This begs the question: when is it actually worth learning Inverse Design, compared to just optimizing designs without ML assistance?

This paper quantitatively addresses this question by comparing the costs and benefits of three different Inverse Design ML model families on a Topology Optimization (TO) task, compared to just running the optimizer by itself. We explore the relationship between the size of training data and the predictive power of each ML model, as well as the computational and training costs of the models and the extent to which they accelerate or hinder TO convergence. The results demonstrate that simpler models, such as K-Nearest Neighbors and Random Forests, are more effective for TO warmstarting with limited training data, while more complex models, such as Deconvolutional Neural Networks, are preferable with more data. We also emphasize the need to balance the benefits of using larger training sets with the costs of data generation when selecting the appropriate ID model. Finally, the paper addresses some challenges that arise when using ML predictions to warmstart optimization, and provides some suggestions for budget and resource management.

1. INTRODUCTION

Shape and Topology Optimization, or design optimization more generally, can be time-consuming and require expensive iterations to reach optimal results. For example, in Topology Optimization, solving forward problems requires an iterative algorithm to minimize objective functions typically via an additional

adjoint-based backward pass to perform gradient-based optimization at each iteration. While these can be efficient for large-scale problems, existing gradient-based methods do still require multiple passes through an often computationally expensive simulator, and can get trapped in local optima, owing to their gradient-based nature. To overcome these challenges, researchers have studied methods that combine or supplement TO with inverse design and Machine learning methods. These methods can have significant time and resource savings compared to optimizing a design for each input condition, when compared with solving the forward model. For example, Chen *et al.* [15] showed that ML-based Inverse Design methods could predict an optimal airfoil shape to within 96% of the optimal efficiency compared to gradient-based methods, and more importantly, further warm-start Shape Optimization on the ML-predicted solution produced, on average, better solutions than gradient-based methods could achieve by themselves.

The use of ML-based design methods, however, come with an important, often unaddressed cost: one has to either find or generate data with which to train such models. For example, Woldseth *et al.* [34] investigated several previous studies in terms of the computational effort associated with generating training samples, running the learning algorithm, and applying the proposed procedure to obtain the optimal solution. They showed, for instance, Nakamura *et al.* [26] sampled 300,000 optimized structures for their training and validating data. They concluded that each Inverse Design model should be applied to at least 333,001 new similarly-sized problems to be "worth it" which they viewed as a high computational cost relative to just running a TO procedure [34].

This concern about the data needed for ML-based methods leads to an important, though often ignored question: when is it actually worth learning Inverse Design? That is, under what conditions would it actually make sense to go through the trouble of collecting the data for and then training any kind of ML method, as opposed to just expending the high computational cost needed to run a high-quality TO method for that same problem? This paper attempts to quantitatively address that question, by

*Corresponding author: fuge@umd.edu

Copyright © 2023 by ASME

specifically contributing the following:

1. We study how changing the amount of training data provided to several classes of Inverse Design models impacts both the Instantaneous and Cumulative Optimality Gaps of the generated designs on a specific 2D Heat Diffusion SIMP Topology Optimization problem. We show that, as expected, increasing training data size produces lower optimality gaps, and that the extent of this improvement is model and problem dependent.

2. We introduce several cost measures that quantify and differentiate different ID models and provide quantitative comparisons of those costs on an example ID problem. This allows us to broadly analyze the Return-on-Data-Investment (RODI) for a given model, taking into account costs of data generation, training, and warmstart optimization of the ML predicted solution.

3. We use these investment measures to quantify a "breakeven point" where Inverse Design produces positive return, compared to just running TO in isolation. We demonstrate how this breakeven point changes depending on the type of model and amount of training data used.

2. BACKGROUND AND RELATED WORK

2.1 Inverse Design

An inverse design problem belongs to the broader category of inverse problems. Inverse problems use observed measurements to infer model parameters [31]. The measurements can be obtained either by observing physical systems or by simulating them. The formulation for inverse design can be described as follow:

$$y = F(x) + \epsilon \qquad (1)$$

where $y \in Y$ measurement data, $x \in X$ parameter of interests, F is the forward model which maps the parameter of interest to the measurement data, and ϵ is the noise of observed data. One of the main essential needs for using inverse problems is that the underlying governing equation on many applications are unknown or costly [4]. Therefore developing a surrogate model to map between observed data and parameters of interests are one of the main strategies for solving inverse problems.

Researchers have studied Inverse design in the field of mechanical engineering for a variety of problems including material design [3, 28], hydrodynamic design [32, 36], and aerodynamic shape design [15, 22].

Inverse design is also used extensively for warm starting optimization process [19, 20], as opposed to just using the ML-provided solution as-is. In warm start optimization, the ML-predicted solution is used as an initial point to reduce the running time of iterative methods. A common hypothesis is that if the initial ML-predicted solution is close to the true optimum, this warm start will significantly reduce the running time of traditional optimization [14, 17]. In comparison, this paper attempts to quantify such gains as a function of the amount of training data.

2.2 Topology optimization

Topology optimization (TO) determines how to place material in a design domain to optimize one or more objectives while satisfying constraints [29]. One of the most widely used formulations of TO problems are pseudo-density-based approaches such as the Solid Isotropic Material with Penalization method (SIMP) [5]. Both Topology and Shape Optimization are widely used for finding local and global optima for a variety of design tasks [5, 6]. Gradient-based topology optimization has been widely studied to place material within a design domain. However, computing the solutions to such optimization tasks can be computationally prohibitive, motivating the use of fast, Machine Learning based approximations of the optimal solution [13, 21, 24, 33]. Recently, promising work in applying ML to Optimization problems has focused on Deep Generative Models [16, 27], and specifically, the incorporation of physics or other engineering constraints into those models [23]. Beyond direct Inverse Design prediction as a supervised learning problem, some research has alternatively formulated this as a reinforcement learning problem [10].

The above works demonstrate a variety of methods for integrating deep learning into design optimization. However, all are either supervised or unsupervised algorithms that rely on the use of training data to learn the ML model. In contrast to this paper, none rigorously or quantitatively address the important question of when such models are worth the return on time or cost needed to generate the training data or how the performance improves or degrades as we change that cost.

2.3 Return-on-Data-Investment

While the above-mentioned papers have shown that using ML-based methods for Inverse Design can be possible and performant, they come with a non-trivial cost: the need to generate training data to feed into the model. To understand under what conditions generating this data is "worth it," we need to define a way to measure the impact on a model's performance as we generate additional data. We refer to this throughout the paper as the Return-on-Data-Investment (RODI), after common Return on Investment (ROI) measures used in economics.

Finding the ROI parameter plays a key role in decision-making in engineering and business [11]. As we use more complex ML models to improve ID results, the data investment costs often increase nonlinearly. There is, however, a point at which the quality and performance of a given ML model saturates, implying that there is no benefit to adding more complexity or data to the models [18]. For instance, Deshpande et al. [18], analyzed Random Forest and Bidirectional Encoder Representations from Transformers, based on the accuracy and ROI for two publicly available data sets and they recommend selecting ML classification not solely based on performance, but also given how much data one has available.

Calculating a Return on data investment is one of the most important parameters that can help us address when Inverse Design is actually worth it. In the TO community, this motivation has not received much attention and there are few studies on it. For example, Woldseth et al. [34], introduced the computational costs in general terms for ML-based TO including the actual solu-

Copyright © 2023 by ASME

tion time, collecting data, and quality needed for the performance. They compared some methods in the reviewed literature via the perceived generalization ability to when speed-up is achieved, and defined a qualitative breakeven threshold. To make the analysis of that breakeven point more precise, one of this paper's main contributions is to study the exact relationship between three driving costs on a concrete common example: (1) TO iterations, which also drives Data Generation Cost; (2) ML model choice, which drives Model Training Cost; and (3) a model's ID performance quality, which drives its Relative Performance Improvement (defined below) with respect to TO. We compare these across a range of models and on common problems such that we can compare them fairly. The next section further details our test problem and exact cost measures.

3. METHODOLOGY

To address the contributions mentioned in the introduction, our methodology is divided into the following sections: (1) Defining our Heat condition topology optimization problem and how we generate the data set, (2) how we train and optimize our models, and (3) how we measure and evaluate the performance and cost results.

3.1 Heat conduction topology optimization problem

While research literature uses a variety of TO test problems, we chose to use a fast-to-evaluate and simple to replicate 2D Heat Conduction test problem that might serve as a useful baseline (§5 addresses other or more complex problems). Topology optimization has been successfully applied to determine the material distribution in a variety of heat conduction applications [25], and the purpose of these studies was to find the optimal material distribution for a heat transfer problem inside a region of interest while satisfying design constraints. For instance, Yang *et al.* investigated the optimality of structures optimization of heat conduction structures for minimum thermal compliance while satisfying minimum-maximum temperature conditions [35].

In this study, we consider a 2D heat sink problem subjected to pure conduction. This problem can be described as finding the material distribution that minimizes the integral of the temperature when the amount of highly conducting material is limited. This Test case was derived from a demo example described in the dofin-adjoint solver [1] and which we had used in a prior study of Inverse Design methods [8]. The goal of this problem is to minimize the thermal compliance subjected to the Poisson equation with mixed Dirichlet–Neumann conditions [1], which formulate as follows:

$$\int_\Omega fT + \alpha \int_\Omega \nabla a \cdot \nabla a \qquad (2)$$

where f is the heat source term (here is a constant 10^{-2}), T is the temperature, Ω is the region of interest (unit square), α is a regularization term, and a is the mass distribution function (a(x) = 1 for material, a(x) = 0 for no material). It is subjected to a control constraint over the domain:

$$\int_\Omega a \leq V \qquad (3)$$

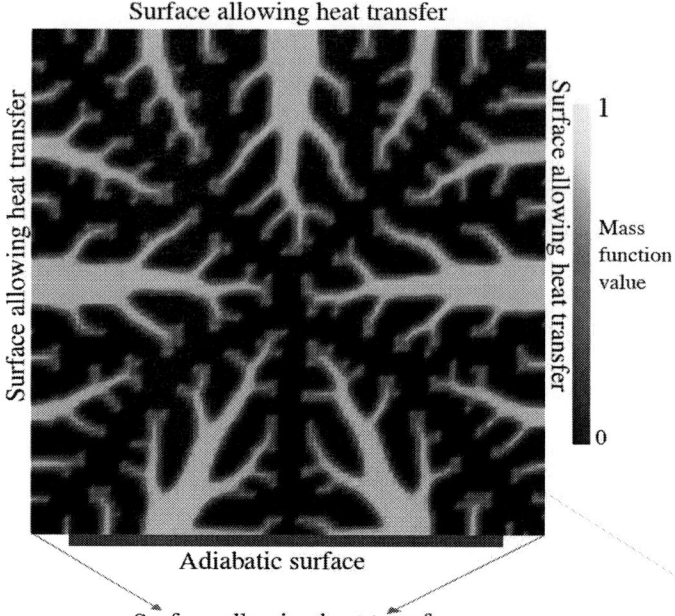

FIGURE 1: THE PHYSICAL LAYOUT OF THE TOPOLOGY OPTIMIZATION PROBLEM WE USED IN THIS PAPER. THE COLOR BAR REPRESENTS THE VALUE OF THE MASS FUNCTION AT ANY POINT.

$$a \in [0, 1] \qquad (4)$$

Where V is the volume bound over the region of the interest domain Ω.

3.1.1 Dataset and Preprocessing. In this topology optimization problem, we aim to minimize the compliance of the domain upon a given limit on the volume of conducting material and length of the adiabatic region as design parameters similar to our previous study [8]. The adiabatic region is shown with a blue color in Fig 1.

To generate the data set needed for this study, we used two input parameters (design parameters): the volume limits on the material distributions and the length of the adiabatic region. The physical layout of this topology optimization problem is shown in Fig 1. We chose adiabatic length which is shown with a blue region at the bottom of the geometry between 0 and 1 and bounded the upper volume limit between 0.3 and 0.6, since our interior solver (IPOPT) cannot produce converged results outside of these ranges. Each of these parameters was divided into 20 equal segments, which resulted in 21 values for each. Hence, using 21*21 design parameters, 421 different optimized designs were generated.

In order to show the details without increasing the computational time, a mesh 100 x 100 was used as the design domain. In addition, every combination of design parameters was run for 100 iterations until IPOPT satisfied the tolerance of 1.0e-100. For every combination design space, five values are collected for each point, including x-coordinate, y-coordinate, volume bound, adiabatic region length, and mass function value.

Copyright © 2023 by ASME

3.2 Machine Learning Models Compared in this Paper

Understanding under what conditions data collection and model training is actually worthwhile for Inverse Design clearly depends on the quality and sample efficiency of the underlying Machine Learning model one might use. While there are a large number of possible models, we selected a representative subset of supervised learning models spanning a range of complexities to compare in this paper. These range across three main model families of non-linear Supervised Learning approaches from fairly simple prototype-based methods, such as K-Nearest Neighbors (KNN), to Ensembles such as Random Forests, and to Adaptive Basis Function methods, such as Deconvolutional Neural Networks (DeCNN). This section briefly reviews each of these models and provides citations to further reading for interested readers.

K-Nearest Neighbors: K-nearest neighbors (KNN) is a non-parametric and supervised learning algorithm used for classification and regression. We used a regression model of KNN, whose predictions for a given test point are weighted averages of the k-closest training data points. This algorithm uses different types of distances to compute that proximity, such as Euclidean distance, Manhattan distance, and Minkowski distance, and can modify the number of neighbors (k) over which it takes a weighted average [30]. We chose to use a KNN model for comparison, since it is straightforward to implement, fast to evaluate for a small number of samples, has a limited range of hyper-parameters requiring tuning, and has fast and deterministic training procedures. In this sense, while it is not as capable of generalization compared to other models, it represents a model with comparatively "low-cost."

Random Forests: A Random Forest (RF) is an ensemble learning method for classification and regression which builds a number of decision trees using independent bootstrapping samples of the dataset [9]. We used the RF regression model which predicts a given test point based on averaging over multiple decision trees employed over training data sets. In this algorithm, trees are run in parallel, meaning there is no interaction between them. We used an RF model for comparison since it is one of the most widely used ensemble learning models, while also possessing a limited number of hyperparameters that require tuning, thus limiting required computational costs.

Deconvolutional Neural Network: Deconvolutional neural networks (DeCNNs) are also called transposed convolution neural networks, which use a special case of convolution to perform weight-sharing within a feedforward network. At the time of writing, Deconvolutional Neural Network architectures were among the most commonly used by contemporary papers attempting to do Inverse Design of Topologically or Shape Optimized problems, and thus represent a natural comparison case [37]. Their popularity is due to the comparatively large hypothesis class of functions that DeCNNs can learn, owing to their large model capacity. This has led to generally lower test Mean Squared Error on Inverse Design tasks compared to simpler models, although DeCNN's possess their own shortcomings. For example, they have many possible hyper-parameters that need to be optimized, such as training epochs, learning rates, regularization strengths, and neural network architectures, making them harder to train and optimize with stable variance given a small number of training samples. In addition, the non-convexity of the training loss surface often necessitates additional diagnostic checks or multiple restarts during training, further increasing the cost and complexity of training such models. A common criticism of such DeCNN models in Inverse Design is that, while their performance may appear to be relatively strong, their needed training time and complexity, along with their dependence on a larger training sample size may not make them "worth it" [34]. This paper helps illuminate under what conditions that conjecture might be true.

3.2.1 Model Training, Hyperparameter Optimization, and Data Postprocessing.
To conduct our subsequent experiments, we had to perform several steps for creating a cross-validation set, optimizing each model's hyperparameters, downsampling the training datasets, and pre-processing the input data to make it compatible with each model.

Cross Validation and Hyperparameter Optimization: To perform cross-validation and hyperparameter optimization, we randomly selected two unique values of adiabatic length and volume limits from the dataset and excluded them from training data to act as a test set. After this, we randomly selected half of the excluded data as validation data and the other half as test data. To find the point-wise mean squared error for hyperparameter optimization, we tested each models' predictive abilities on validation data points corresponding to the topologies defined in the excluded validation dataset.

We used the KNN and RF implementations from the scikit-learn library [12]. In the KNN model, hyperparameter optimization involves both the weighting and number of neighbors. In the RF model, we optimize the number of estimators as well as the minimum number of samples in newly created leaves. We implemented the DeCNN model using Tensorflow [2] and optimized the learning rate based on the maximum batch sizes possible for each training size.

All models were trained on the different training datasets described above, and each model's hyperparameters were optimized on the common validation data. Therefore, hyperparameters are chosen for each model such that it produced the best performance, as measured by point-wise mean square error (PMSE) on a validation dataset. We exclude the test data from both the training and validation datasets, such that the test errors are not influenced by our hyperparameter selection approaches.

Training Dataset Downsampling: To study how each model performs when trained on varying amounts of data, we gradually reduce the size of the training dataset. After excluding the validation and test dataset—which we will keep common across all models—the largest training data set includes 361 designs. Every subsequent training set is obtained by randomly removing data from the next largest set. For example, the training dataset with a size of 200 is obtained by removing 50 data from the training dataset with a size of 250. In this way, the 200 size dataset is a strict subset of the 250 size dataset. This ensures that any difference in performance between models of different sizes is due only to additional training points. Below, we evaluate the models on training data sets ranging in size from 2 to 361 designs.

Data Postprocessing: During our model evaluation process, the thermal compliance (the objective value) and trajectory are normalized and averaged. The specific postprocessing procedure is:

- The objective value of each model/data size is normalized with respect to the optimal value obtained in the control trajectory (*i.e.*, a uniform initialization with no ML warm-starting).

- The number of iterations of each trajectory extended to the maximum number of iterations if the run that converged prior to the maximum. Taking the optimal value that reached the trajectory and extrapolating it to the remaining iterations, makes it easier to visually compare different trajectories.

- The average objective values and the 95% Empirical CI on the average percentile are calculated for each trajectory and shown as a solid line and shaded color, respectively.

3.2.2 Return in Data Investment Measures. To define an actual breakeven point where actually ML-based Inverse Design is worth it, this section defines the overall computational costs of generating training samples, computational effort associated with running the learning model, and the desired performance quality. These cost measures allow us to talk about trade offs among them. Our first step is to introduce the costs associated with iterative solvers and training models. Then we describe the relative percentage performance improvement, which is essentially a measure of solution quality. Understanding return on data investment requires describing trade-offs among all the above-mentioned parameters.

Data Generation Cost: The computational cost of ML-based inverse design is not only related to the machine learning model, but also time spent on generating sample training data. Creating training samples is usually expensive and depends on the type of solvers used, the quality needed (such as the resolution or fidelity) and convergence criteria, as well as the computational resources used to prepare the data. In this study, we fix the quality needed for training data samples such as resolution and convergence criteria as described in Sec. 3.1.1 which quantifies the number of iterations needed for the solver to converge the desired criteria. To establish a common basis among different examples, this cost is converted into the time needed for simulating each problem on an Intel(R) E5-1620 CPU. Herein, we convert the generation data cost to time spent on the mentioned CPU for this generation. Later sections will discuss how hardware changes may affect our results.

Model training cost: The model training cost plays a key role in choosing among ML models. We used various ML models, each of which has different computational costs for training. We measure this cost via training each model on a single CPU E5-1620 Intel(R) similar to data generation costs. Therefore, for various amounts of training data and ML-models we report the training cost of each model with time values measured on a single CPU. Later sections will discuss how hardware changes may affect our results.

Relative performance improvement: The other important factor in calculating our return data investment is how well the

trained model produces high quality (*i.e.*, high performing) solutions. We will use the relative percentage improvement (RPI) in each problem's objective function to measure this, formulated as follows:

$$RPI = \frac{C_{CONTROL}^{Init} - C^{ML}}{C_{CONTROL}^{Init} - C_{CONTROL}^{OPT}} \times 100\% \quad (5)$$

where $C_{CONTROL}^{Init}$ is initial compliance (first iteration) calculated based on a uniform initialization of the SIMP method, $C_{CONTROL}^{OPT}$ is optimal compliance (last iteration) calculated based on the final converged SIMP method, C^{ML} is the compliance of the design computed based on ML models prediction. Therefore, the return on data investment can be interpreted as a tradeoff between the RPI and the data generation and model training costs.

4. EXPERIMENTAL RESULTS AND DISCUSSION

Given this methodology, below we first demonstrate the impact of different training sizes for different ID models on the warm start initialization performance. The shaded parts of these plots represent 95% confidence intervals for corresponding plotted functions. Following that, we plot the return on data investment to determine when different ID methods can find designs with lower Instantaneous and Cumulative Optimality Gaps as a function of additional training data samples. As we detail further below, this curve, coupled with the cost associated with generating additional training data, helps us determine the return on data investment.

4.1 How well do ID models warmstart TO?

KNN: using hyperparameters optimized for each size of training data, we now compare how well a simple KNN model can predict the optimal geometry, as well as that prediction's usefulness as a warm start for further topology optimization. As a baseline, we compare it with uniform initialization commonly used in TO (We label this baseline "Control").

On average, we found that a KNN model with all sizes of training data from 15 to 361 samples outperformed the control condition. Furthermore, when the size of training data exceeds two samples, the KNN models produced predictions with thermal compliance values that were significantly lower than the control model. As well as improving prediction compliance, we found that increasing the number of training sizes accelerates convergence to optimal compliance value. Figure 2 plots a subset of the training data sizes for clarity's sake, and plots containing the full set of results are located in the Supplemental Material. As Fig. 2 shows, in all warm start trajectories the optimizer increases the thermal compliance in early iterations of warm starting (see around iteration 7), and then it accelerates the convergence to optimal compliance values compared to the control condition. These early "spikes" in the objective function value (roughly between iterations two to ten) are related to the IPOPT solver, and we detail this behavior in the later discussion section.

RF: we use hyperparameters optimized for each size of training data to predict the optimal geometries for test data in order to compare the effects of changing training size on RF predictions.

Copyright © 2023 by ASME

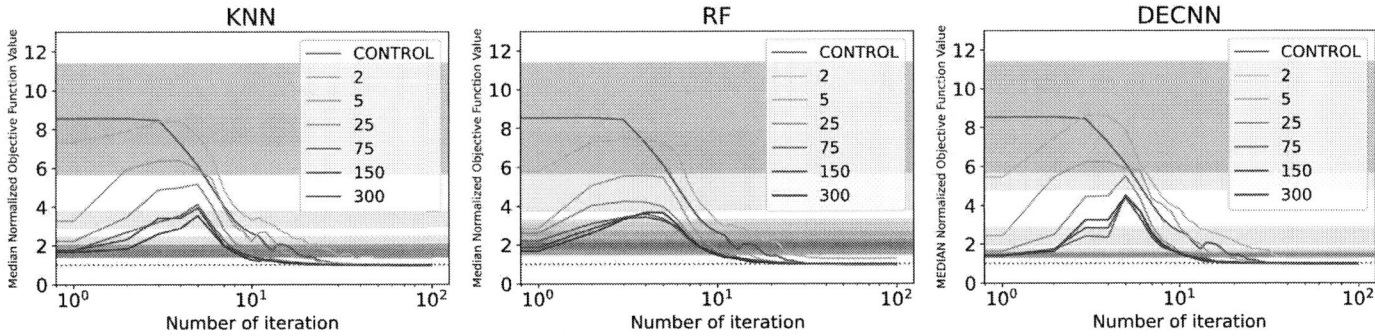

FIGURE 2: 2D HEAT CONDUCTION OPTIMIZATION TRAJECTORIES WARMSTARTED WITH ALL THREE ID MODELS, TRAINED ON THE NUMBER OF DATA NOTED IN THE LEGEND. THE 'CONTROL' TRAJECTORY IS INITIALIZED WITH A CONSTANT DISTRIBUTION SET TO THE VOLUME FRACTION. TRAJECTORIES OF FURTHER DATA SIZES CAN BE FOUND IN THE SUPPLEMENTAL MATERIAL.

Then, we compare the performance of each training size prediction as optimal geometry and a warm start for further topology optimization.

On average, Fig. 2 shows that the RF model with training set size of more than seventy five converges faster to the optimal design compared to the control condition. Fig. 2 shows that the same increasing trend in compliance occurs in the early stages of RF warm start trajectories as it does for KNN trajectories. In addition, the performance of RF models improves as training sizes are increased.

DeCNN: we use the hyperparameter optimized for each size of training data to train the deconvolutional neural network model. We found that on average the DeCNN model predicts geometries with lower initial optimality gap compared to the RF and KNN models. As above, we compare the performance of different training sizes at predicting the optimal geometry and as a warm start for further topology optimization. Figure 2 provides a subset of training sizes, with the full version provided in the Appendix. Figure 2 shows that the DeCNN model with training sizes beyond twenty five design samples outperforms the control condition. Also, the DeCNN model can predict optimal designs with significantly lower thermal compliance with respect to the control condition. In the early stages of the warm start trajectory, compliance increases similarly to the KNN and RF models.

Overall, our results in Fig. 2 show that, with a limited amount of training data ($\approx 5+$), the KNN model is the most effective in terms of convergence speed. However, when the amount of training data increases ($\approx 25+$), the DeCNN model outperforms the other models in terms of both the initial design prediction as well as convergence acceleration, although simpler models like KNN and RF also improve with increased training data. Our findings suggest (perhaps somewhat expectedly) that when warmstarting TO with limited training sizes, simpler models can be more effective, whereas with more data more complex models like DeCNN produce better warmstarts.

4.2 How well do ID models predict the optimal design without warmstarting?

What if we do not have access to a TO solver to warmstart further optimization and wish instead to just use directly the geometry or design output by the ID model?

Figure 3 relates the size of training data to the Normalized Initial Optimality Gap, which measures how close in performance to the Control (TO) solution the initial ID model's prediction gets before further optimization. Figure 3 shows the initial optimally gap reduces, on average, when we increase the amount of training data.

As expected, we observe that increasing the amount of training data decreases the initial optimally gap, that is, more data improves the ML models' prediction accuracy. In this measure, the DeCNN model outperforms the other models, on average. This suggests that if we only care about an ID model's prediction without further optimization, the DeCNN model is the best choice among the models studied in this paper. Notably, none of the models ever achieved complete parity in performance with the Control (TO-only) solution.

4.3 How much data do ID models actually need to accelerate optimization?

The whole point of attempting Inverse Design is to reduce needed optimization effort on subsequent design cases. As such, one of the most important costs we need to track is how increasing the data budget reduces the computational budget we need to spend on further optimization of new or unseen problems. If investing in an ID method significantly reduces subsequent optimization costs, then perhaps the return on data and model investment is worth it; otherwise, it may not be.

To show the effect of data set investment versus further optimization cost, we measured the relative performance improvement of three machine learning models versus the time cost of warm start optimization—that is, how much further time did we need to spend to optimize the ID provided solution to achieve results similar to TO without the ML-based warmstart? Figure 4 varies the amount of training data given to each model.

Expectedly, Fig. 4 depicts that as each ID model is given a larger training size, each outputs designs that have better Relative Performance Improvement (Eqn. 5), *i.e.*, are closer to the optimal thermal compliance found by the TO-only Control solution. Among the models, the DeCNN outperformed the other two models in terms of initial relative performance. The DeCNN model with 300 training data produces the best initial relative performance improvement, whereas the KNN model with 5 training

Copyright © 2023 by ASME

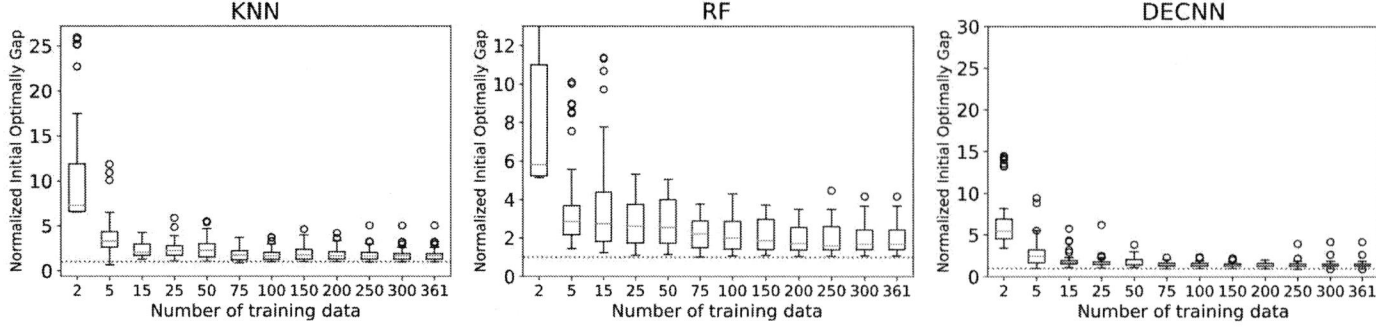

FIGURE 3: THE NORMALIZED INITIAL OPTIMALITY GAP OF THE KNN, RF, AND DECNN MODELS AS A FUNCTION OF TRAINING DATA SIZE. THE ORANGE LINES REPRESENT IS THE MEDIAN OF EACH BOX-PLOT AND THE DASHED GREEN LINE REPRESENTS THE OPTIMAL VALUE ACHIEVED UNDER UNIFORM INITIALIZATION (THE CONTROL CONDITION).

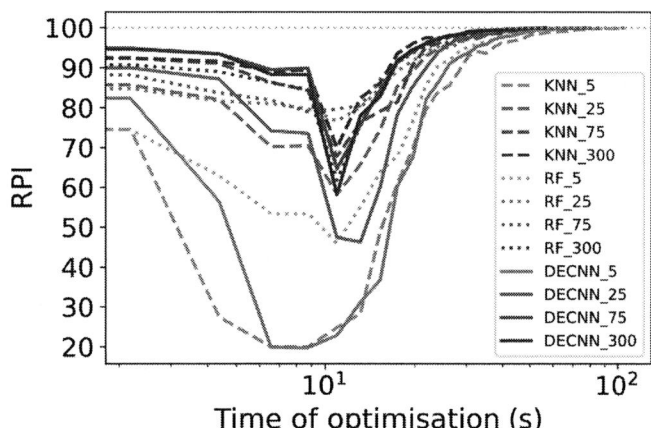

FIGURE 4: RPI TRAJECTORIES USING WARMSTART OPTIMIZA-TION, WHERE THE BLUE/GREEN/RED CURVES CORRESPOND TO THE KNN/RF/DECNN MODELS, RESPECTIVELY, AT THE NOTED TRAINING DATA SIZE.

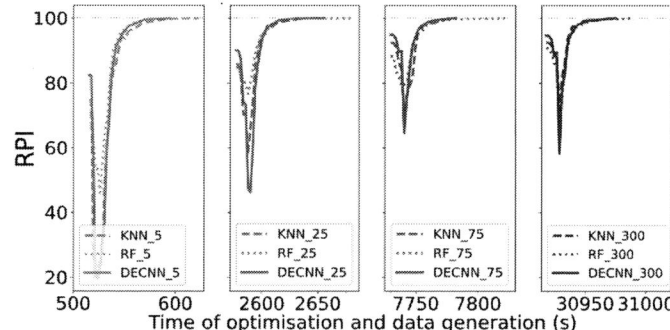

FIGURE 5: COMPARISON BETWEEN OF THE TIME SPENT ON OP-TIMIZATION AND DATA GENERATION FOR WARMSTARTED OPTI-MIZATION TRAJECTORIES WITH VARIOUS MODELS AND TRAIN-ING DATA SIZES.

data shows the lowest initial relative performance improvement. Furthermore, when used to warm start further optimization, the DeCNN trained with 300 training data requires the fewest itera-tions to converge to the optima, compared to the control solution. This implies that, when given at least some small amount of training data, ID methods can meaningfully accelerate TO con-vergence.

4.4 What about the cost of data generation?

While our earlier results would seem to imply that a DeCNN with 300 training samples is best, this ignores both the data gen-eration cost and training costs. To capture data generation cost, Fig. 5 compares the time spent on both warmstart optimization and data generation. Factoring in data generation, the total cost increases as the amount of training data grows—note in particular the different x-axis scales. As expected, the models that predict designs with the best initial RPI also have the highest training data costs.

Figure 5 suggests that the amount of data used to train ID models has a significant positive impact on their performance.

Past literature also uses large dataset sizes to train ID models, such as in [26] where the authors create 300,000 optimized struc-tures for their training and validation data. As expected, using larger training sets leads to better ID-predicted designs and im-proved relative performance improvement. But at what cost? Existing papers often ignore the non-trivial sunk cost of gener-ating this data. Specifically, the models that produced the best initial relative performance improvement also had the highest data generation costs, implying a trade off between the compute cost we wish to "invest" into data generation versus the amount the trained ID model would "save" on downstream optimization. For example, while Fig. 4 showed that a DeCNN model trained with a large number of samples accelerated optimization the fastest, when we include the time needed to generate training data in Fig. 5 the hidden time cost of this data generation becomes plain to see.

4.5 What about the cost of training time?

From Fig. 5 alone, it would again appear that subject to similar training data amounts, a DeCNN model outperforms oth-ers; however this ignores the model training cost/time. Figure 6 provides a comparison of various inverse design optimization trajectories that were warm-started while adding in the amount of training time required for each model. (Note: in Fig. 6 we

Copyright © 2023 by ASME

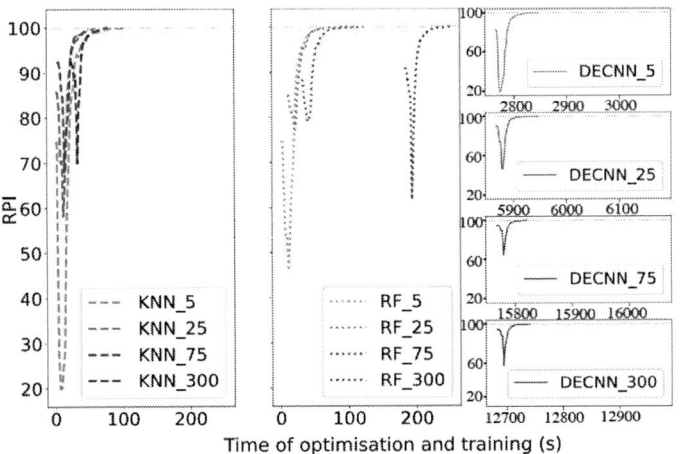

FIGURE 6: COMPARISON BETWEEN OF THE TIME SPENT ON OPTIMIZATION AND MODEL TRAINING FOR WARM STARTED OPTIMIZATION TRAJECTORIES WITH VARIOUS MODELS AND TRAINING DATA SIZES.

have removed the data generation costs of Fig. 5 so as to only demonstrate the differences in training time.) The results illustrate that the DeCNN has the most significant training cost when compared to other models and that cost depends strongly on the training data amount.

Several studies have emphasized the importance of training cost in model selection. For instance, Bengio *et al.* found that the cost of training deep neural networks is a significant barrier to the adoption of deep learning in the industry [7]. Figure 6 demonstrates similar effects in ID models, where we observe that the DeCNN model outperforms other models in terms of initial relative performance, but it has far higher training cost compared to inexpensive models such as KNN and RF (for simplicity, the figure ignores the data generation costs show in Fig. 5). In fairness, the time costs here were all normalized to CPU time for consistency across the paper, and in practice one could take greater care to leverage distributed training across GPU resources as we describe later, but even under those cases the training costs would remain non-trivial. We suspect this result is painfully obvious to any readers who have actually tried training a DeCNN model on these types of problems, but hopefully it is illustrative to readers who have not.

4.6 What if I had different cost tradeoffs between data generation, optimization, and training?

Our above comparisons are limited to our chosen example problem and, more specifically, the fact that the optimization, data generation, and training time costs are all treated equally as normalized to common CPU time for our particular computing platform. How would things change if the relative costs of optimization, data-generation, and training differed? For example, if we already had an existing dataset, or if we could parallelize the training costs? To consider cases where these relative costs change, Fig. 7 compares different relative weights between model training, data generation, and warmstart optimization (*i.e.*, running further TO on the ID model provided solution).

In that figure, we simplify each model's optimization trajectory by creating a line connecting the initial design predicted by the ID method with the corresponding design after completing warmstart optimization—as such, these end points will generally have an RPI of 100% unless the warmstart traps the TO solver in a local optima. The total time cost on the x-axis is calculated based on a weighted average of the three costs described above: optimization cost (the cost of warmstart iterations), data generation cost, and model training cost. The simplex on the left-hand side of Figure 7 allows us to define different weights for each time cost using Barycentric coordinates, and the right-hand side compares different models under the corresponding cost weight.

Figure 7.A describes the situation we have used thus far throughout the paper where the model training, data generation, and optimization costs have equal weight. As we saw above, in this case the KNN and RF models with lower training data size outperform the more complex DeCNN model, largely due to their lower model training cost and the relative inexpense of running additional TO optimization iteratons for this simple 2D thermal compliance problem.

Figure 7.B describes the situation where data generation is effectively free, while the model training and optimization costs have equal weight. In practice, this could occur when an existing dataset is available, but a researcher or company would need to bear the burden of model training and further optimization. In such cases, a KNN with a larger training size outperforms other models due largely to its lower training cost.

Figure 7.C describes the situation where we ignore the model training cost (*i.e.*, training is free), and optimization and data generation have equal weight. In practice, this could occur when training is hardware parallelized or where the training cost is borne by a third party. In such cases, complex models such as the DeCNN with smaller amounts of training data produce better return compared to simple models such as a KNN.

Figure 7.D describes the situation where we heavily discount the costs of both data generation and model training, compared to warmstart optimization. In practice, this could occur in cases where we use high quality pre-trained models that are fine-tuned on small set of domain-specific training samples, but where the optimization task itself is extremely costly—such as a 3D multiphysics optimization problem with multiple constraints. This is the most common setting in "Transfer Learning" approaches to Inverse Design. In such cases, Fig. 7.D suggests that even simple models trained on small-to-moderate amounts of training data may produce greater return than investing in larger training datasets or complex models.

Note the large difference between point D in Figure 7 compared to the nearby edge of the simplex where both data generation and model training costs are considered "free"—the earlier Fig. 4 showed this exact case where complex DeCNN models with large amounts of training data appeared to dominate all other models.

4.7 When do ID methods break even compared to just using TO?

One of this paper's main objectives is to determine the break-even point where a machine learning (ML) based inverse design method generates positive returns on investment, compared to

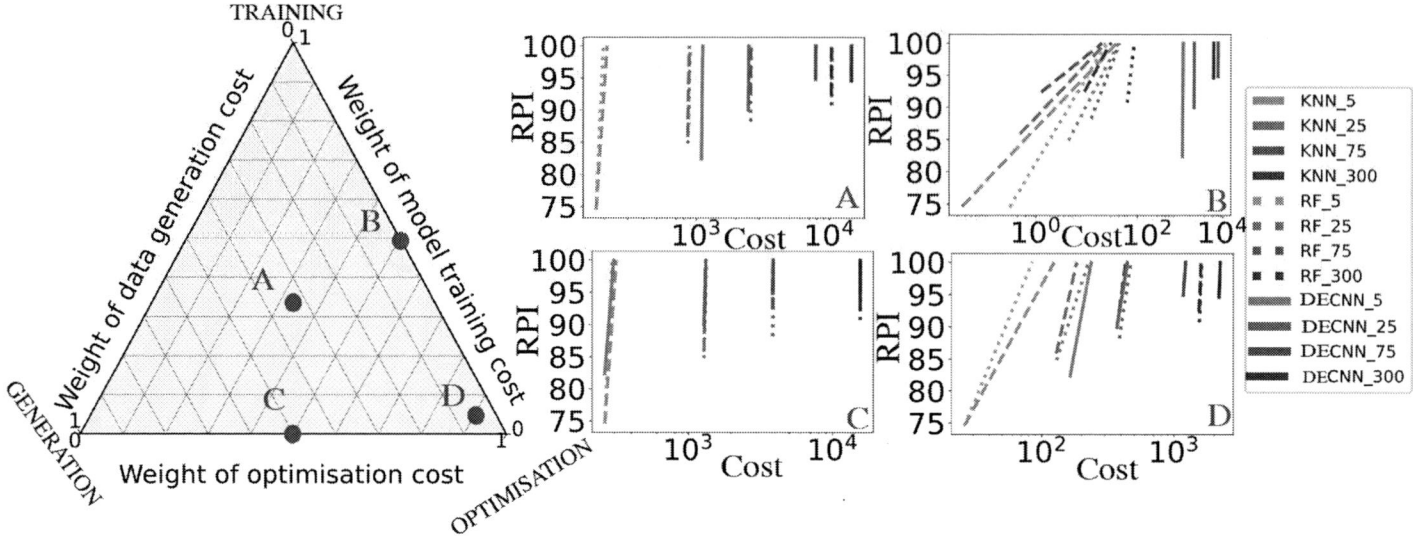

FIGURE 7: LEFT: SIMPLEX DIAGRAM OF RELATIVE WEIGHTS BETWEEN DATA GENERATION, OPTIMIZATION, AND MODEL TRAINING COSTS. RIGHT: COST VERSUS RPI FOR DIFFERENT MODELS AND TRAINING DATA SIZES. A: EQUAL WEIGHT. B: DATA GENERATION HAS NO COST AND MODEL TRAINING AND OPTIMIZATION HAVE THE SAME WEIGHT. C: MODEL TRAINING HAS NO COST AND DATA GENERATION AND OPTIMIZATION HAVE THE SAME WEIGHT. D: OPTIMIZATION HAS 20 TIMES THE WEIGHT OF DATA GENERATION AND TRAINING.

only using topology optimization. Up until now our focus has been solely on comparing the cost of generating a single predicted solution. This is an unfair comparison. In practice, no sane person would invest in data generation and model training unless they were able to *amortize* that fixed cost over many future design tasks to reduce their variable costs.

To analyze this trade-off between an upfront fixed investment and a reduction in variable costs, we first calculated the median number of warmstart optimization iterations each model required—this represents the variable cost of using a model to predict and then optimize a new design. To compute the fixed costs, we added each model's data generation cost and training cost. For comparison, we calculated the median number of iterations and corresponding costs for the control method (TO-only), which represents the variable costs of the TO solver, and we note that the control condition has no fixed cost, since there is no data or training.

Figure 8 plots both the fixed and variable costs of the control condition and the various models under different amounts of training data cost. Our results demonstrate that after a certain breakeven threshold of the number of new designs, some ML-based inverse design models outperform the control in terms of cost. For example, in the top left diagram, if you "invest" in generating 5 training data points, then an RF model would be less costly than the Control condition if you needed to compute 47 or more new designs. In the top right figure, after investing in 25 training data points, a KNN model breaks even with the Control condition after 168 designs. Furthermore, the bottom left diagram demonstrates that investing in 75 training points leads the DeCNN and KNN models to break even with Control after 1074 and 294 new designs, respectively. Lastly, the bottom right diagram shows that DeCNN and KNN models require you to use the model for 2492 and 1768 new designs, respectively,

to recoup the 300 training data investment. Interestingly, even though complex models trained on large datasets have faster TO convergence—seen as a lower slope and thus variable cost in Fig. 8—they have larger breakeven points, owing to the fixed costs required for data generation and training.

5. DISCUSSION AND LIMITATIONS

5.1 When do ML models provide suboptimal warmstart conditions?

One challenge we observed is that in a small subset of cases an ML model's warmstart initialization was suboptimal compared to uniform initialization, leading the IPOPT solver to get stuck in a local optima. This resulted in a small handful of cases that highly skewed the mean optimization trajectories for each model, which is why we chose to report each model's median value for the optimization trajectories since it is less sensitive to outliers. Nevertheless, it is important to discuss here the fact that in a small percentage of cases ID methods might produce warm start initializations that are worst than uniform initialization.

5.2 Effect of hardware parallelism

Using hardware parallelism, GPUs can significantly improve both the training and hyper-parameter optimization time of machine learning models. While studying parallelization was not a rigorous focus of this paper, anecdotally, we observed that using GPUs could reduce the DeCNN training costs by about one order of magnitude. This case can be considered similar to Fig. 7.C in which we assume that model training is hardware parallelized, and as a result, the DeCNN model with smaller amounts of training data performs better compared to simple models such as the KNN. This suggests that parallelization plays a crucial role in making complex models more viable for inverse design and improves the return on investment in such models.

Copyright © 2023 by ASME

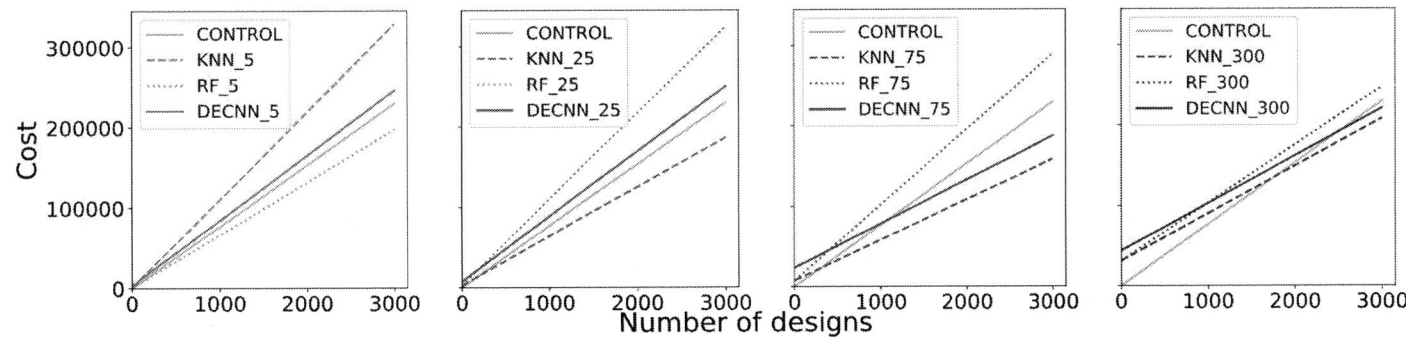

FIGURE 8: THE COMPARISON BETWEEN THE CONTROL AND KNN/RF/DECNN COSTS, RELATIVE TO THE NUMBER OF NEW DESIGNS TO BE OPTIMIZED, FOR DIFFERENT TRAINING DATA SIZES—5, 25, 75, & 300 DATA POINTS.

Similarly, in Fig. 7.D, where both data generation and model training costs are heavily discounted compared to warm start optimization, it is also possible that parallelism in terms of data generation (for example, on a compute cluster) could have a significant impact on the results.

5.3 Why does the thermal compliance increase after warm-starting?

In our previous paper, we presented a visual demonstration of designs that exhibit an early increase in compliance when using warm-starting [8]. To gain insight into the cause of this behavior, we conducted a comparison of the IPOPT and scipy solvers using the sequential least squares programming (SLSQP) method which the results are located in the Supplemental Material. We observed the scipy solver exhibits a gradual, monotonic decrease, while the trajectory of the warm start optimization with IPOPT displays an initial increase in compliance. We hypothesize that IPOPT behaves this way due to its solution method, which numerically approximates the hessian matrix during the initial solver iterations, and thus may take early sub-optimal steps. We hypothesize that this explains initial "bump" in the convergence trajectory.

5.4 Limitations in the test problem

This paper focused on a 2D heat diffusion problem with a limited volume fraction range of 0.3 to 0.6 and an adiabatic region width that can vary from 0.0 to 1.0. While this choice helped us obtain useful insights into this problem, it is natural to ask whether our results would generalize to three dimensional or more complex problems. Future work could explore the extent to which each model's cost-benefit behavior changes under increased complexity. Additionally, to assess the transferability of our claims to other problems, it would be beneficial to compare our results across multiple Inverse Design problems.

Relatedly, this paper did not consider other optimizers beyond the IPOPT solver, and less-efficient TO solvers will shift the costs-benefit trade-off substantially by increasing the costs of both the optimizer and the data-generation reducing the comparative effect of training costs. There are also additional hidden costs such as hyperparameter optimization of ML models, which can add significantly to training-time costs. We did not consider

those rigorously here for sake of compactness and brevity, but future work can investigate the impact of these hidden costs on the performance of the model and explore alternative optimization strategies to address them.

6. CONCLUSION

So, when is it actually worth learning Inverse Design? Our answer ranges from "Almost Never" to "Worth Experimenting With" to "Almost Always" depending on the design situation.

Almost Never If the only performance metric we care about is the ability of an ID model to accurately and reliably predict close to the optimal design without further warmstarting, then §4.2 and Fig. 7 imply that only a complex model (such as the DeCNN or better) trained on large quantities of training data (75+ in our example), can achieve an RPI even close to 95% when compared to a TO-only control condition. Such models incur large training and data generation costs, and in cases where we only intend to evaluate a few new design cases, this cost will almost never be worth the large fixed cost. Thus, such methods may only apply to problems where effective Transfer Learning can take place or where data and training are effectively "free" as in Fig. 4 and 7.D. In this sense, our results mirror the somewhat pessimistic view taken by the conclusions in [34].

Worth Experimenting With However, if we relax the above conditions only slightly in one of several directions, then the picture changes significantly enough to warrant initial experiments in new problem domains. For example, if achieving up to an 80% RPI is acceptable for an initial design case, such as an approximate trade-space analysis or rough downselection of alternatives, then our results suggest that simple ID methods are efficient for this, breaking even with TO at anywhere from a small handful to a few hundred cases (on our admittedly simple example). Likewise, if your domain possesses existing datasets or specialized hardware that reduces the relative cost of data generation and training relative to optimization cost, then experimenting with possible gains from ID methods may be worthwhile.

Almost Always We saw several conditions under which training and using ID methods was almost always a good idea. Most importantly, if we allow the ID method to work in concert with existing TO solvers as a warm start mechanism, there is comparatively little disadvantage to using ID methods, except in all but the most expensive of cases, and much to be gained. Figure 2's log-scaled x-axis shows that even simple models trained on a few data points can meaningfully accelerate TO convergence, even if their initial RPI is not great. In cases where we need to evaluate many possible new designs, Fig. 8 demonstrates clearly how ID methods reduce variable costs compared to only using TO, breaking even with them within a few hundred cases. While the specific break-even numbers we report are idiosyncratic to our example (and may be low compared to more complex problems), we suspect that similar behavior will occur in other problems, and that there will always exist a threshold wherein ID methods produce positive return on data investment.

Looking forward, we expect that ongoing and future work in improving both the sample efficiency and generalization ability ID methods will continue to shift these break even points lower over time. While there may be niche cases where investing in ID methods does not make sense—such as in large one-off optimization problems wherein generating data is neither practical nor warranted—we suspect that, given design's iterative nature, further ID advances will create a supportive interplay with existing optimizers that will be greater than the sum of their parts.

ACKNOWLEDGEMENTS

We acknowledge the support from the National Science Foundation through award #1943699 as well as ARPA-E award DE-AR0001216. We also thank Jamie Guest, Katie Kirsch, and Amit Bhatia for helpful conversations on the chosen TO test problem and some of the cost measures.

REFERENCES

[1] Topology optimisation of heat conduction problems governed by the poisson equation. http://www.dolfin-adjoint.org/en/latest/documentation/poisson-topology/poisson-topology.html. Accessed: 2022-02-10.

[2] Martín Abadi, Ashish Agarwal, Paul Barham, Eugene Brevdo, Zhifeng Chen, Craig Citro, Greg S. Corrado, Andy Davis, Jeffrey Dean, Matthieu Devin, Sanjay Ghemawat, Ian Goodfellow, Andrew Harp, Geoffrey Irving, Michael Isard, Yangqing Jia, Rafal Jozefowicz, Lukasz Kaiser, Manjunath Kudlur, Josh Levenberg, Dandelion Mané, Rajat Monga, Sherry Moore, Derek Murray, Chris Olah, Mike Schuster, Jonathon Shlens, Benoit Steiner, Ilya Sutskever, Kunal Talwar, Paul Tucker, Vincent Vanhoucke, Vijay Vasudevan, Fernanda Viégas, Oriol Vinyals, Pete Warden, Martin Wattenberg, Martin Wicke, Yuan Yu, and Xiaoqiang Zheng. TensorFlow: Large-scale machine learning on heterogeneous systems, 2015. Software available from tensorflow.org.

[3] Ovo Adagha, Richard M Levy, and Sheelagh Carpendale. Towards a product design assessment of visual analytics in decision support applications: a systematic review. *Journal of Intelligent Manufacturing*, 28(7):1623–1633, 2017.

[4] Simon Arridge, Peter Maass, Ozan Öktem, and Carola-Bibiane Schönlieb. Solving inverse problems using data-driven models. *Acta Numerica*, 28:1–174, 2019.

[5] Martin P Bendsøe. Optimal shape design as a material distribution problem. *Structural optimization*, 1(4):193–202, 1989.

[6] Martin Philip Bendsøe and Noboru Kikuchi. Generating optimal topologies in structural design using a homogenization method. *Computer methods in applied mechanics and engineering*, 71(2):197–224, 1988.

[7] Yoshua Bengio, Andrea Lodi, and Antoine Prouvost. Machine learning for combinatorial optimization: a methodological tour d'horizon. *European Journal of Operational Research*, 290(2):405–421, 2021.

[8] Shai Bernard, Jun Wang, and Mark Fuge. Mean squared error may lead you astray when optimizing your inverse design methods. In *International Design Engineering Technical Conferences and Computers and Information in Engineering Conference*, volume 86229, page V03AT03A004. American Society of Mechanical Engineers, 2022.

[9] Leo Breiman. Random forests. *Machine learning*, 45(1):5–32, 2001.

[10] Nathan K Brown, Anthony P Garland, Georges M Fadel, and Gang Li. Deep reinforcement learning for engineering design through topology optimization of elementally discretized design domains. *Materials & Design*, 218:110672, 2022.

[11] Wojtek Buczynski, Fabio Cuzzolin, and Barbara Sahakian. A review of machine learning experiments in equity investment decision-making: Why most published research findings do not live up to their promise in real life. *International Journal of Data Science and Analytics*, 11(3):221–242, 2021.

[12] Lars Buitinck, Gilles Louppe, Mathieu Blondel, Fabian Pedregosa, Andreas Mueller, Olivier Grisel, Vlad Niculae, Peter Prettenhofer, Alexandre Gramfort, Jaques Grobler, Robert Layton, Jake VanderPlas, Arnaud Joly, Brian Holt, and Gaël Varoquaux. API design for machine learning software: experiences from the scikit-learn project. In *ECML PKDD Workshop: Languages for Data Mining and Machine Learning*, pages 108–122, 2013.

[13] Aaditya Chandrasekhar and Krishnan Suresh. Tounn: topology optimization using neural networks. *Structural and Multidisciplinary Optimization*, 63(3):1135–1149, 2021.

[14] Qiuyi Chen, Jun Wang, Phillip Pope, Wei Chen, and Mark Fuge. Inverse design of two-dimensional airfoils using conditional generative models and surrogate log-likelihoods. *Journal of Mechanical Design*, 144(2):021712, 2022.

[15] Qiuyi Chen, Jun Wang, Phillip Pope, Mark Fuge, et al. Inverse design of two-dimensional airfoils using conditional generative models and surrogate log-likelihoods. *Journal of Mechanical Design*, 144(2), 2022.

[16] Wei Chen, Kevin Chiu, and Mark D Fuge. Airfoil design parameterization and optimization using bézier generative adversarial networks. *AIAA journal*, 58(11):4723–4735, 2020.

Copyright © 2023 by ASME

[17] Bo-Yu Chu, Chia-Hua Ho, Cheng-Hao Tsai, Chieh-Yen Lin, and Chih-Jen Lin. Warm start for parameter selection of linear classifiers. In *Proceedings of the 21th ACM SIGKDD international conference on knowledge discovery and data mining*, pages 149–158, 2015.

[18] Gouri Deshpande, Guenther Ruhe, and Chad Saunders. How much data analytics is enough? the roi of machine learning classification and its application to requirements dependency classification. *arXiv preprint arXiv:2109.14097*, 2021.

[19] Ravi Hegde. Sample-efficient deep learning for accelerating photonic inverse design. *OSA Continuum*, 4(3):1019–1033, 2021.

[20] Martin Klaučo, Martin Kalúz, and Michal Kvasnica. Machine learning-based warm starting of active set methods in embedded model predictive control. *Engineering Applications of Artificial Intelligence*, 77:1–8, 2019.

[21] Hunter T Kollmann, Diab W Abueidda, Seid Koric, Erman Guleryuz, and Nahil A Sobh. Deep learning for topology optimization of 2d metamaterials. *Materials & Design*, 196:109098, 2020.

[22] Ruiwu Lei, Junqiang Bai, Hui Wang, Boxiao Zhou, and Meihong Zhang. Deep learning based multistage method for inverse design of supercritical airfoil. *Aerospace Science and Technology*, 119:107101, 2021.

[23] Mo-How Herman Shen Liang Chen. A New Topology Optimization Approach by Physics-Informed Deep Learning Process. *Advances in Science, Technology and Engineering Systems Journal*, 6(4):233–240, 2021.

[24] Qiyin Lin, Jun Hong, Zheng Liu, Baotong Li, and Jihong Wang. Investigation into the topology optimization for conductive heat transfer based on deep learning approach. *International Communications in Heat and Mass Transfer*, 97:103–109, 2018.

[25] Danny J Lohan, Ercan M Dede, and James T Allison. A study on practical objectives and constraints for heat conduction topology optimization. *Structural and Multidisciplinary Optimization*, 61(2):475–489, 2020.

[26] Keigo Nakamura and Yoshiro Suzuki. Deep learning-based topological optimization for representing a user-specified design area. *arXiv preprint arXiv:2004.05461*, 2020.

[27] Sangeun Oh, Yongsu Jung, Seongsin Kim, Ikjin Lee, and Namwoo Kang. Deep generative design: Integration of topology optimization and generative models. *Journal of Mechanical Design*, 141(11), 2019.

[28] John D Perkins, Tula R Paudel, Andriy Zakutayev, Paul F Ndione, Philip A Parilla, DL Young, Stephan Lany, David S Ginley, Alex Zunger, Nicola H Perry, et al. Inverse design approach to hole doping in ternary oxides: Enhancing p-type conductivity in cobalt oxide spinels. *Physical Review B*, 84(20):205207, 2011.

[29] Ole Sigmund and Kurt Maute. Topology optimization approaches. *Structural and Multidisciplinary Optimization*, 48(6):1031–1055, 2013.

[30] Asmita Singh, Malka N Halgamuge, and Rajasekaran Lakshmiganthan. Impact of different data types on classifier performance of random forest, naive bayes, and k-nearest neighbors algorithms. *International Journal of Advanced Computer Science and Applications*, 8(12), 2017.

[31] Albert Tarantola. *Inverse problem theory and methods for model parameter estimation*. SIAM, 2005.

[32] Vicenc Torra. Trends in information fusion in data mining. *Information fusion in data mining*, pages 1–6, 2003.

[33] Dalei Wang, Cheng Xiang, Yue Pan, Airong Chen, Xiaoyi Zhou, and Yiquan Zhang. A deep convolutional neural network for topology optimization with perceptible generalization ability. *Engineering Optimization*, 54(6):973–988, 2022.

[34] Rebekka V Woldseth, Niels Aage, J Andreas Bærentzen, and Ole Sigmund. On the use of artificial neural networks in topology optimisation. *Structural and Multidisciplinary Optimization*, 65(10):1–36, 2022.

[35] Suna Yan, Fengwen Wang, and Ole Sigmund. On the non-optimality of tree structures for heat conduction. *International Journal of Heat and Mass Transfer*, 122:660–680, 2018.

[36] Junlian Yin and Dezhong Wang. Review on applications of 3d inverse design method for pump. *Chinese Journal of Mechanical Engineering*, 27(3):520–527, 2014.

[37] Matthew D Zeiler, Dilip Krishnan, Graham W Taylor, and Rob Fergus. Deconvolutional networks. In *2010 IEEE Computer Society Conference on computer vision and pattern recognition*, pages 2528–2535. IEEE, 2010.

Copyright © 2023 by ASME

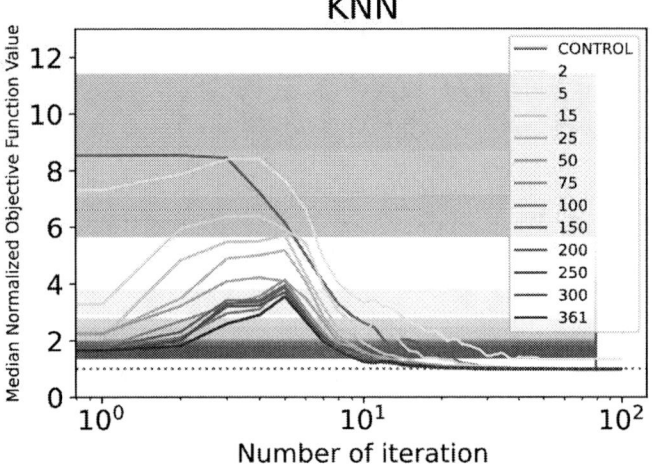

FIGURE 9: THE EVOLUTION OF 2D HEAT CONDUCTION DESIGNS OVER THE COURSE OF THE OPTIMIZATION PROCESS. HERE, TRAJECTORY 2, 5, 15,....,300, AND 361 BELONGS TO THE TRAJECTORY INITIALIZED WITH THE PREDICTION OF THE KNN MODEL WITH THE CORRESPONDING SIZE OF TRAINING DATA.

SUPPLEMENTAL MATERIAL

Figure 9, 10 and 11 show the trajectories 2, 5, 15,....,300, and 361 belong to the trajectory initialized with the prediction of the KNN and RF model with the corresponding size of training data, respectively.

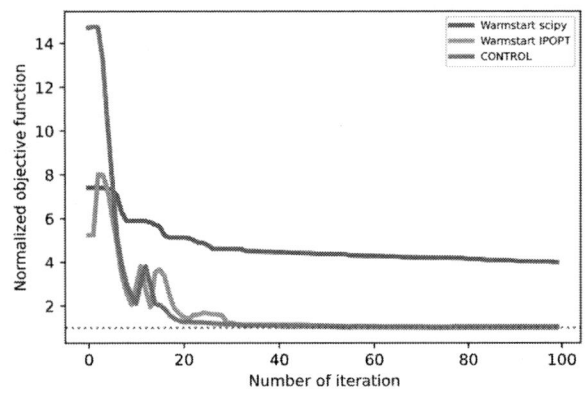

FIGURE 12: COMPARISON OF DIFFERENT WARMSTART TO SOLVERS COMPARED TO THE CONTROL CONDITION.

We used a KNN model prediction to warm start the optimiz-

ers for a design with a volume limit of 0.4 and an adiabatic length of 0.35. The comparison of warm start optimization with scipy and warm start optimization with IPOPT is shown in Fig 12.

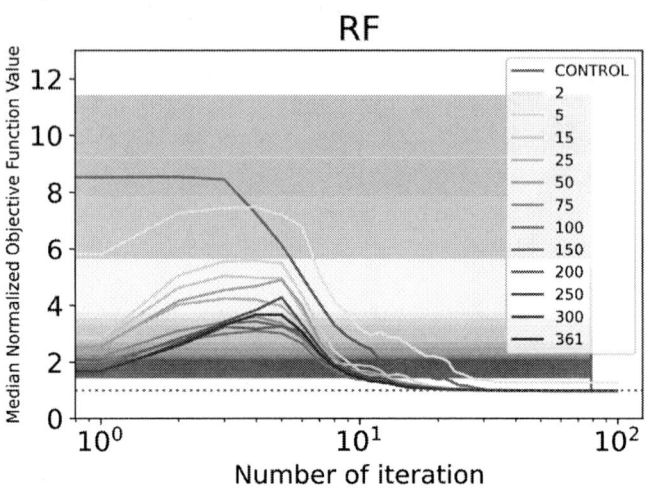

FIGURE 10: THE EVOLUTION OF 2D HEAT CONDUCTION DESIGNS OVER THE COURSE OF THE OPTIMIZATION PROCESS. HERE, TRAJECTORY 2, 5, 15,....,300, AND 361 BELONGS TO THE TRAJECTORY INITIALIZED WITH THE PREDICTION OF THE RF MODEL WITH THE CORRESPONDING SIZE OF TRAINING DATA.

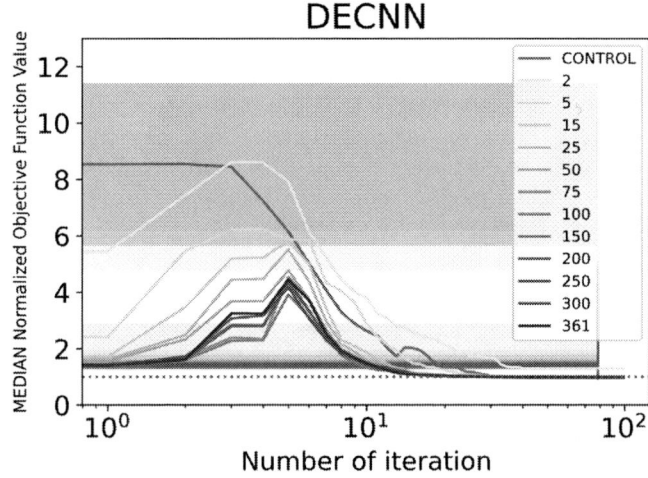

FIGURE 11: THE EVOLUTION OF 2D HEAT CONDUCTION DESIGNS OVER THE COURSE OF THE OPTIMIZATION PROCESS. HERE, TRAJECTORY 2, 5, 15,....,300, AND 361 BELONGS TO THE TRAJECTORY INITIALIZED WITH THE PREDICTION OF THE DECNN MODEL WITH THE CORRESPONDING SIZE OF TRAINING DATA.

Copyright © 2023 by ASME

Proceedings of the ASME 2023
International Design Engineering Technical Conferences and
Computers and Information in Engineering Conference
IDETC-CIE2023
August 20-23, 2023, Boston, Massachusetts

DETC2023-116703

DESIGN OF SELF-ORGANIZING SYSTEMS USING MULTI-AGENT REINFORCEMENT LEARNING AND THE COMPROMISE DECISION SUPPORT PROBLEM CONSTRUCT

Mingfei Jiang
BS Student
Electric Engineering
Beijing Institute of Technology, Beijing, China

Zhenjun Ming
Assistant Professor
Beijing Institute of Technology, Beijing, China

Chuanhao Li
MS Student
Mechanical Engineering
Beijing Institute of Technology, Beijing, China

Farrokh Mistree
L.A. Comp Chair and Professor
The Systems Realization Laboratory @ OU
University of Oklahoma, Norman, OK, USA

Janet K. Allen[1]
John and Mary Moore Chair and Professor
The Systems Realization Laboratory @ OU
University of Oklahoma, Norman, OK, USA

ABSTRACT

How can multi-robot self-organizing systems be designed so that they show the desired behaviors and are able to perform tasks specified by designers?

Multi-robot self-organizing systems, e.g., swarm robots, have great potential for adapting when performing complex tasks in a changing environment. However, such systems are difficult to design due to the stochasticity of the system performance and the non-linearity between the local actions/interaction and the desired global behavior. In order to address this, in this paper we propose a framework for designing self-organizing systems using Multi-Agent Reinforcement Learning (MARL) and the compromise Decision-Support Problem (cDSP) construct.

The design framework consists of two stages – preliminary design and design improvement. In the preliminary design stage, MARL is used to help designers train the robots so that they show stable group behavior for performing the task. In the design improvement stage, the cDSP construct is used to enable designers to explore the design space and identify satisfactory solutions considering several performance indicators based on the trained system established in the previous stage. Between the two stages, surrogate models are used to map the relationship

between local parameters and global performance indicators utilizing the data generated in preliminary design. The surrogate models represent the goal functions in the cDSP. A multi-robot box-pushing problem is used as an example to test the efficacy of the proposed framework. The framework is general and can be extended to design other multi-robot self-organizing systems. Our focus in this paper is in describing the framework.

Keywords: Self-organizing System, Compromise Decision Support Problem, Box-Pushing Problem

1. FRAME OF REFERENCE

Group behavior is widespread with phenomena like ant colonies, fish swarms, bird colonies, and so on. Industry has been inspired by this and developed techniques such as reinforcement learning to make it possible for multi-robot systems to perform complex group tasks. It is now evident that engineered self-organizing systems, consisting of a large number of simple individuals, tend to have better flexibility as well as lower manufacturing costs. However, a group of robots is only one example of a self-organizing system, and we will consider the characteristics of the general self-organizing systems.

---

[1] Corresponding author: janet.allen@ou.edu

Copyright © 2023 by ASME

First, we recognize that a self-organizing system does not require the intervention of external forces to establish order, increase order, and can even change from one state of organization to another on its own. In other words, this self-organizing system is no longer a traditional expert system that requires external rules or knowledge, but one that is achieved by designing the end goal and thus establishing internal actions. From the design point of view, a designer usually does not impose some external forces on the system to force it to develop in a certain direction, but designs some characteristics of the self-organizing system itself to guide it toward some good emerging behaviors.

Next, we consider multi-robot systems to be distributed, which means that each robot acts independently in actually performing its task. It represents not only decentralized control but also a concept that is locally knowable and globally unknown, which means that the individual robot can only perceive local information about its surroundings and has the ability to interact only with its surrounding local robots and the static environment. The global behavior of the system is hard to predict [1]. Designing such systems is a challenging task because of the difficulty in defining the micro-macro link [2]. The global behavior of the system is a result of the emerging process of local behavior of each robot and interactions among them. In other words, when multiple simple robots form a system, they may have emergent properties similar to natural organisms with capabilities and behaviors that individual robots do not have until the system is formed. We suggest that we lack a holistic framework and local-to-global linkage for designing such self-organizing systems.

In this paper, we propose a framework to design a self-organizing system, explore the solution space, and test the efficacy of the framework by solving a box-pushing problem as an example. The problem involves multiple small independent robots forming a self-organizing system and pushing the box to the destination, in the middle of which the robots encounter many obstacles or crevices, and they have to learn to move the box around or through them. We first model this problem and design the system, using reinforcement learning, to allow multiple robots to better perform the task. Then, we draw inferences by some global performance, from which those key local parameters of the multi-robot system are identified as design variables. We use these local parameters as hyperparameters of the model and specify the performance of the model as the output. For better design, we need to obtain the relationships between local and global behavior, and we construct surrogate models with these parameters, and the results of these surrogate models are used to effectively represent the corresponding relationships of the model. Ultimately, we expect to achieve satisfactory performance of the whole system through the design of local parameters, and we completely explore the solution space using the compromise Decision Support Problem construct [3]. We take the synergistic performance exhibited by the multi-robot system in the box-pushing problem as the goal of multi-objective design mathematical model. With cDSP construct, the deviation function is used to actually design the corresponding parameters required for the envisioned performance.

The rest of this paper is organized as follows. In Section 2, we critically review relevant papers on system design and identify the research gap. In Section 3, we briefly introduce our framework and demonstrate the way of designing a system. In Section 4, we show the results generated by the framework and give some examples. Finally, we conclude this paper with contributions and future work in Section 5.

2. CRITICAL REVIEW OF RELATED WORK

Self-organizing systems evolve over time. In this section we review the literature on designing the multi-robot self-organizing systems and identify the gaps that we address in this paper.

2.1 Self-organizing Systems

There are many kinds of self-organizing systems. In this paper we focus mainly on multiple robots for industry. We clarify the definition of self-organizing systems, and then go into deeper discussion and comparison. Gershenson et al. conclude that self-organization is a way of observing systems, not an absolute class of systems [4]. Banzhaf concludes that elements of self-organizing systems are able to manipulate or organize other elements of the same system in a way that stabilizes either structure or function of the whole against external fluctuations [5]. We define a self-organizing system as a system composed of many simple individuals that can produce complex global behaviors through coordinated movements among individuals and can adapt to a more complex external environment to accomplish tasks rationally.

The most important thing we notice is that in 1952 Turing published a paper regarding the mathematical theory of pattern formation in biology, and found that global order in a system can arise from local interactions [6]. Howard concludes that self-organization leads to emergent properties, meaning that the whole system has characteristics that differ qualitatively from those of the component parts without the interactions [7]. Hence, we suggest that fostering emergent behavior in self-organizing systems is an effective way to get things done. Since local actions may generate some global behavior, so it is worthwhile to do some research on local actions.

Self-organizing systems exist in nature, including non-living as well as the living world, they exist in human-made systems, but also in the world of abstract ideas [5]. We believe that the self-organizing system, in each case, contains similar mechanisms, but with different manifestations. So, it is inspiring to use nature's organizing mechanism to achieve the goal.

2.2 Machine Learning in Self-organizing Systems

With the rapid development of machine intelligence, related methods are used by researchers to explore self-organizing systems. The limitations of traditional algorithms in the field of swarm robots are also identified by researchers. Although the machine intelligence methods used include means like genetic algorithms [8], most of them are still studied with the reinforcement learning approach. In particular, the performance of multiple reinforcement learning methods, using the box-pushing problem as an example, are summarized and compared in [9].

Researchers have described different directions on the task goals addressed by multi-robot self-organizing systems. Long et

Copyright © 2023 by ASME

al. use deep multi-agent reinforcement learning to train a group of robots for obstacle avoidance tasks in complex environments [10]. Kakish et al. used multi-agent reinforcement learning to design a control strategy for a "leader" agent that moves a group of "follower" agents neatly from one location to another (the target area).[11]. Bae et al. used multi-agent reinforcement learning to implement a population robot path planning problem [12]. In this paper, we focus on a box-pushing problem.

In our paper, we use a reinforcement learning method - Multi-agent Deep Deterministic Policy Gradient (MADDPG) [13] to train the robots. Lowe et al. show the strength of this approach compared to existing methods in cooperative as well as competitive scenarios, where robot populations are able to discover various physical and informational coordination strategies [13]. We take advantage of the overall training and divisional execution of MADDPG to handle multi-robot problems.

2.3 Design of Self-organizing Systems

There are many stages for implementing a multi-robot self-organizing system, and we mainly focus on the design stage. In fact, the design of self-organizing systems is a research gap by itself.

In this paper is a box-pushing problem, for which a related design has been developed is used by way of example [8]. A deep multi-agent reinforcement learning algorithm was devised as a mechanism to train self-organizing system robots for the acquisition of the task field and social rule knowledge, and the scalability property of this learning approach was investigated with respect to the changing team sizes and environmental noises [14].

We find that most of the systems described in the literature are designed to accomplish the task of pushing the box in a new way to accomplish the goal, not that the multi-robot self-organizing system itself was designed to have some emergent behavior. Since the push-box problem is a problem with high task difficulty and deep spatial complexity, we recognize previous work and propose a more general design framework for self-organizing systems. Therefore, in our framework, we temporarily discard the ideas in literature of exploring different action spaces through enumeration, and instead fundamentally design the desirable overall performance by identifying proper local parameters mathematically, and, in this way, we propose to reduce the time and effort spent in the process of "trial and error".

2.4 Gaps and Contributions

Based on the discussion of existing research, we suggest that the gaps that need to be bridged to design multi-robot self-organizing systems (the box-pushing problem in particular) are as follows:

Gap 1: Lack of in-depth multi-agent control theory.

For multi-agent control, researchers have mostly focused on studying formation, collaboration, and adversarial tasks. Researchers have mostly utilized expert actions, potential field interaction relations, and reinforcement learning under an *a priori* rule systems to accomplish preconceived task goals. Some of these approaches require more elaborate designs resulting in less robustness, some are unable to accomplish more complex task missions, and some have poor stability to guarantee task completion rates. Therefore, there is currently no universally available multi-agent control method that can be efficient and robust.

Gap 2: Lack of theories of emergent behavior.

There are many publications in the field of robotics on the various actions of individual robots, but the problem of how they interact optimally in the field of swarm robots, in particular, remains to be solved. While the performance of individual robots has been fully exploited, the actions of multiple robots are prone to conflict and do not necessarily achieve a "1+1>2" effect. Emergent behaviors are difficult to predict directly, and the corresponding action requirements cannot be given directly for non-expert actions. When expert actions, most designers are concerned with either the whole or local actions, or to set goals for intelligent algorithms to find their way. Hence there is a gap as to how to reach, interpret, and predict expert actions.

Gap 3: Lack of multi-objective optimization for multi-robot self-organizing systems

In research publications on multi-robot self-organizing systems, most of them are devoted to understanding how to go about improving the overall performance of the system at the algorithm level, or by designing some special rule actions to reach the efficient task goals. Few people consider the various performances of such systems as goals to reach the optimum through some intra-system parameter adjustments. In fact, from a designer's point of view, it is more practical to achieve better (satisficing) results with the existing system than with a new one. Therefore, this gap may be the key to the lack of overall effectiveness in terms of stability, etc.

Gap 4: Lack of general framework for designing multi-robot self-organizing system.

In the literature on multi-robot self-organizing system design, most of the design frameworks are focused on mimicking nature (e.g., ants' collective construction behavior) or elaborating robot's actions for some specific team tasks (e.g., box-pushing). There is a lack of a framework that is independent of a nature phenomenon and extends to cross-task applications, based on which designers can achieve the desired global system performance by specifying local robot action or interaction related parameters given the task requirements.

To fill the aforementioned gaps, we propose a framework for the design of self-organizing systems. Our contributions in this paper are summarized as follows:

- We develop an understanding of emergent behavior using a surrogate model, that constitutes a link from local parameters to overall performance. When emergent behavior is difficult to explain and the training process is somewhat unknown, we can use surrogate models to fit the design space well. This allows us to obtain the connection between the individual parameters and the overall emergent behavior, which not only reveals the actual connection of the corresponding effectiveness, but also helps the design process afterwards.

- We use the cDSP construct to design of self-organizing systems to satisfice multiple goals. This allows us to design local parameters to achieve the overall performance desired by a designer. With an agent model that establishes the relationship between local parameters and performance, the

Copyright © 2023 by ASME

cDSP structure is used to explore the solution points that may lead to a more efficient and rational operation of the system. We argue that this approach can be effective in exploring satisfactory task solutions and improving the system according to a designer's preference, especially in exploring points of unknown performance and achieving a reasonable configuration of system performance.

- We propose a holistic framework for designing self-organizing systems, which facilitates the design of a multi-robot self-organizing system with desirable performance. By initially designing the environment, designing the reinforcement learning algorithm, and then using an agent model to relate individual parameters to the overall performance, a satisfactory design can be determined using the cDSP construct. This framework can be used to some extent to guide the study of similar systems that are not directly predictable and can be designed with local parameters to perform more difficult tasks and form more efficient overall emergent behavior.

3 FRAMEWORK AND CONSTRUCTS

In this section, we introduce the framework for the design of self-organizing systems, and how we can use reinforcement learning to design local actions. Then we describe how we go about correlating overall performance with local actions and designing global actions using cDSP.

3.1 Framework for Design of Self-organizing Systems

We show the overall framework for designing a self-organizing system, which is the most important part of our paper and will sort out the whole design idea, as shown in Figure 1. In this framework we illustrate the process of moving from individual robot elements to machine learning methods to overall system performance, with a design space exploration process consisting of surrogate models and compromise Decision Support Problem construct in between.

We first need to make clear that the entire intention lies in designing emergent behavior, so it is necessary to start from the basic parameters of the system, and gradually make the system improve with multiple evaluations.

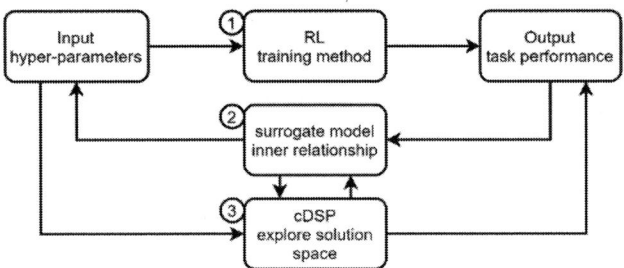

FIGURE 1. SCHEMATIC DIAGRAM OF THE FRAMEWORK

Pre-processing: Identification. The whole design process starts from local parameter identification and needs to go through an important hyperparameter extraction to clarify the important parameters that can fundamentally change the emergent behavior of the system. (These hyperparameters are not the traditional training parameters in machine learning, but functional

parameters that are locally relevant to the system.). It is also necessary to clarify the performance metrics of the system output as a good judge of emergent behavior. We believe that this part must be defined in advance, depending on the specific task.

Step1: Training. A system is trained by appropriate machine learning methods so that it can achieve some corresponding tasks and try multiple sets of input parameters and get the corresponding outputs according to the preconceived design spatial range, and in this regard the task completion is evaluated in a relevant way.

Step2: Connection. The surrogate model is used to establish the connection between the local parameters and the task performance. The corresponding relational model is built based on the results of the training required to explore the space, so that we can understand the relationship between the basic parameters and the emergent behavior.

Step 3: Exploration. Using the obtained connections from the local to the global, it is possible to construct an exploration solution space using cDSP and find a satisfactory solution, resulting in a self-organizing system with good emergent properties.

3.2 Step1 Figure 1: Training with Multi-agent Reinforcement Learning

The robot-learning approach we use in this paper is reinforcement learning, which we believe is closer to reality and more robust compared to other intelligent algorithms such as GA. Although it is still a black box at the theoretical level, we use surrogate models and the cDSP construct for the overall tuning to achieve better results. It is important to note that the reinforcement learning approach is closer to the system's configuration and no longer relies on expert knowledge or rules and can be developed completely from the basic parameter design to different emergent behaviors.

In this paper, we use a reinforcement learning method called Multi-Agent Deep Deterministic Policy Gradient (MADDPG). We use MADDPG mainly because it is a multi-agent decentralized actor, centralized Critic approach. As indicated in Figure 2, each agent performs its own action independently, and then rewards and trains on the overall result when the whole is completed. In this way, each agent can learn

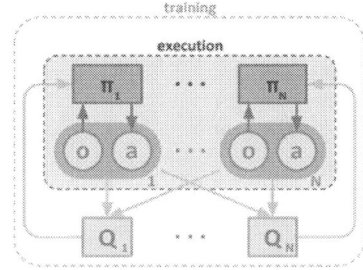

FIGURE 2. OVERVIEW OF OUR MULTI-AGENT DECENTRALIZED ACTOR, CENTRALIZED CRITIC APPROACH [13]

its behavior and still achieve the best possible overall performance. Each agent learns through its actions and has a global performance reward, and the overall optimum can be

Copyright © 2023 by ASME

achieved through various iterations. Further, the inputs of the system are the basic parameters in multi-agent reinforcement learning. The principles of MADDPG are not repeated here, for details see [13].

We note that an agent can get different reward values through multiple training sessions. The agent achieves the goal like a baby toddler. So it means that we use multi-agent reinforcement learning as a training tool to allow multi-robot self-organizing systems to achieve the best possible solution for the case of input parameters. From a designer's point of view the whole process of reinforcement learning can be divided into initial settings called hyperparameters, and rewards, states, and actions.

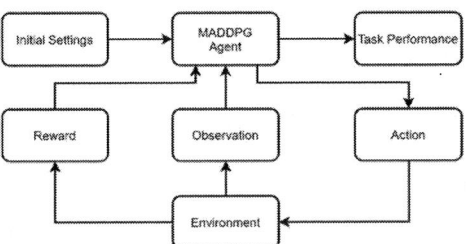

FIGURE 3. GENERAL FLOW CHART OF REINFORCEMENT LEARNING UNDER DESIGNERS' VIEW

In Figure 3, we show the overall process of reinforcement learning, however, for the design process, it first needs to go through pre-processing in the framework to get the corresponding exact inputs and outputs, i.e., the parameter inputs to the system and the task evaluation criteria. When we are faced with a design task, we first need to specify the design goals and design the appropriate reward functions based on the corresponding goals.

$$Reward = \sum E_{lement} \cdot W_{eight}$$

The reward function is a scoring function given after the reinforcement learning session for each action. This is the most important from a designer's point of view, who will directly determine whether a system is able to learn the action that completes the task. It is defined as the cumulative sum of the performance outputs multiplied by the weights. Although it contains some output elements, the importance of each factor is an important input. This directly affects the direction and outcome of the training and influences the performance of the group.

In addition to the reward function, the parameters in the machine learning system that make a significant difference to the results are the action space, observation range, and initial matrix of each agent. The agent's action space may be continuous or discrete, and, in the absence of expert knowledge of the setting, we assume that the agent will only move in the most basic way - back and forth and not in a fixed combination of actions. It is also important to note that since the multi-robot self-organizing system is a distributed system, when training the Critic with parameter updates, the input values of one agent are the actions and observations of all other agents, i.e., there are locally all state values. Moreover, the initial matrix can be considered to have some behaviors that each agent is initially better at. Some settings they are born with may be important in the group after some training. We note that new emergent behaviors may appear in different task designs.

Many researchers have mentioned the instability of reinforcement learning during training, which we think is acceptable. Although the results may be different for each training, the expected results will converge after several training sessions.

3.3 Step2 Figure 1: Connecting Local and Global Behaviors Using Surrogate Models

From a holistic point of view, it is important to determine the relationship between the underlying fundamental parameters and the emergent behavior of the upper layers. In the design problem, we consider the input parameters to be some hyperparameters that have a great impact on the system, and the output emergent behavior as each different performance indicator. Since the training process is the result of a black-box-like reinforcement learning, we do not obtain the corresponding connections directly by logical inference. So we consider a surrogate model to fit the actual logic function by fitting.

In this paper, we apply Box-Behnken designs (BBD) response surface analysis [15] as a surrogate model. BBD are a class of rotatable or nearly rotatable second-order designs based on three-level incomplete factorial designs. For three factors their graphical representation is shown in Figure 4.

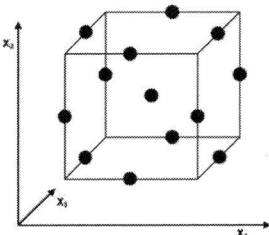

FIGURE 4. THE CUBE FOR BBD [15]

Another reason why we adopt this method is that it fits better with the idea of design space, using the boundary values of the design space as well as intermediate values to determine the accuracy of the overall design. We the training method of machine learning, which consumes a long time and the data points are not so easy to collect, so here we need less data to get good results, and the method also takes into account the overall point fluctuations and repeats the experiment at important points. Therefore, with the data points generated in the design space, we can fit the response surface to obtain the relationship between the input and output. In this way, we can predict the impact of local changes on the overall performance and thus further improve the design.

3.4 Step3 Figure 1: Design of the Global Behavior Using the cDSP Construct

In the design process, once we have the relationship between the parameter inputs and the performance outputs, we should use certain design methods to tune the input parameters to make them optimal. We hope that this emergent behavior will lead to more comprehensive performance, allowing individuals in the system to work together to achieve advantages that are not possible for individuals. Of course, it is impossible to achieve

Copyright © 2023 by ASME

entirely optimal system performance, there will be some loss, and we may have to make some trade-offs. As shown in Figure 5, in the design of a multi-robot self-organizing system, we use the cDSP to help a designer explore the solution space and make a compromise decision among multiple objectives. Compromise Decision Support Problems (cDSP) are used to model engineering decisions involving multiple trade-offs [3].

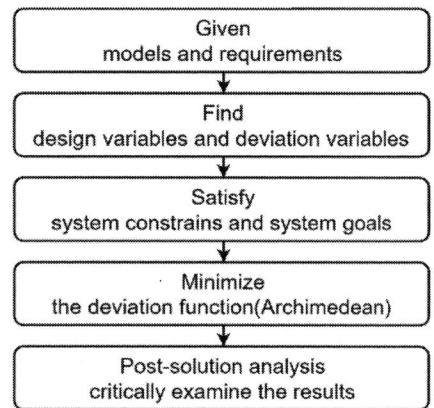

FIGURE 5. SCHEMATIC OF THE CDSP CONSTRUCT

The mathematical form of the cDSP is shown in Table 1 and the steps for utilizing cDSP in design of multi-robot self-organizing systems is shown in Figure 5. First, the model to be designed is carefully studied, the parameters and boundary conditions are determined, and the requirements for the design are identified. After that, specific system design variables are determined, as well as the deviation variables from the system objective. Then, these variables are required to satisfy the system constraints and system objectives, and a system objective function is constructed for the specific objective values. Next, we use the deviation variables to construct a deviation function come as close as possible to achieving all the objectives simultaneously. After that, the results can be critically examined and verified for correctness.

Based on the mathematical representation of the cDSP construct, we go ahead and fill in the required parts of the framework to obtain reasonable design results. Then, the solution space can be further explored using a tailored computational environment called DSIDES with the incorporated Adaptive Linear Programming (ALP) algorithm [16] to find satisfactory design solutions. We can then vary the performance weights according to the design expectations of the target within a predefined solution space, and thus obtain the corresponding solutions.

4 EXAMPLE AND RESULT ANALYSIS

In this section, we test the efficacy of the framework proposed in Section 3 using an example of the box-pushing problem. We analyze the results and discuss the solutions.

TABLE 1. MATHEMATICAL FORM OF CDSP CONSTRUCT.
[13]

Given
n	Number of system design variables
p	Number of inequality constraints
q	Number of equation constraints
p+q	Number of system design constraints
m	Number of system design goals
$g_i(X)$	System design constraint function
$f_k(d_i)$	Target deviation variable minimization function

Find

X_i i = 1, ···, n   Variables

d_i^+, d_i^- i = 1, ···, m   Deviation variables

Satisfy

   Restraints

     $g_i(X) = 0$ i = 1, ···, p

     $g_i(X) \geq 0$ i = p + 1, ···, p + q

   Goals

     $A_i(X) + d_i^- - d_i^+ = G_i$ i = 1, ···, m

   Boundaries

     $X_i^{min} \leq X_i \leq X_i^{max}$ i = 1, ···, n

     $d_i^+, d_i^- \geq 0; \; d_i^+ + d_i^- = 0$

Minimize

   $Z = \sum_{i=1}^{m} w_i(d_i^-, d_i^+)$ $\sum w_i = 1; \; w_i \geq 0$

4.1 The Box-Pushing Problem

The problem of pushing a box is a focal problem in multi-robot systems, that involves multiple robots hitting a box through certain synergistic actions, thus making the box perform a series of actions. In this section, we focus on two synergistic difficulties, one is to push the box through a set obstacle slit, and the other is to be able to push the box to the specified target endpoint. These are problems that have not been explored by previous researchers and we want to start our research from here. We do not use expert actions in this paper, but let the robots can learn the actions of pushing the box by themselves.

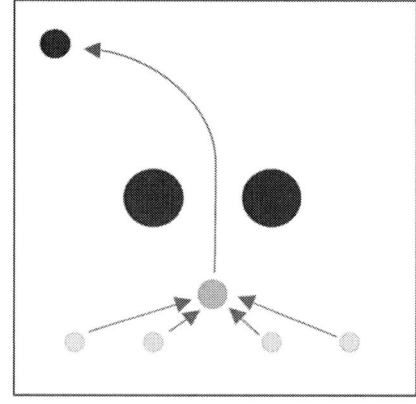

FIGURE 6. THE MPE ENVIRONMENT FOR BOX-PUSHING

From an overall perspective, we seek to design the parameters of the robots to achieve the desired global behavior. To begin with, we construct a reinforcement learning

Copyright © 2023 by ASME

environment for multi-robot using the Multiple Particle Environment.

As shown in Figure 6, the entire environment consists of a blue endpoint, two black barriers, a red box, and some green robots. The positions of the endpoints, obstacles, and boxes are fixed in this environment, but we increase the number of robots. It is also necessary to declare some important parameter indicators first. Since it is a virtual environment, there is no specific unit of length, but only an overall relative relationship. The overall environment has two-unit lengths for the sides, 0.15 for the radius of the barriers, and 0.075 for the radius of the boxes. Between the barriers, the distance from the centers of the circles is 0.6.

4.2 Machine Learning Datasets and Performance Surrogate Models

4.2.1 Parameter determination (Pre-processing)

From the design point of view, we should first perform the preprocessing part of the framework, that is, the selection of the parameters of the design and the evaluation indicators of the performance. Our ultimate goal is to design the emergent behavior of the self-organizing system, and then we need to first characterize which parameters may change this type of behavior more substantially. Through our analysis, we believe that the overall design focuses on the number, size, and speed of the robots. Only these local parameters related to the robot can fundamentally change the performance of the system, resulting in different emergent behaviors. In this experiment, we design the parameters of each robot to achieve the overall performance of the push box.

TABLE 2. DEFINITIONS OF PARAMETERS

Name	Parameter
Number	A
Size	B
Velocity	C
Collision (in robots)	D_1
Collision(box-robot)	D_2
Robot distance	D_3
Box distance	D_4
Time	D_5

4.2.2 Machine learning datasets (Step1 Figure 1)

After designing the overall environment, the next step is to collect data by training the robot in the environment. In Figure 1, we plan the first step as training the corresponding system to achieve the task goal, and in order to do this, we need to design the goal and reward function for training the robot. As described in the machine learning section of Section 3, our training approach here is MADDPG, where the action space is simply moving in a certain direction, that is, directional information, and the observation environment is the global information of initial input and the location information of teammates. Therefore, we focus here on designing a reward function as a way to make a small robot trained to accomplish the task goal. The goal is to push the box to the end. The reward function is related to the goal and is defined as follows.

Distance Reward: The reward for pushing the box is represented as D_{ab} and is the distance between robot and box at the end of each time step. The reward for pushing the box closer to the goal position is represented as D_{change}, and is:.

$$D_{change} = 10 \times (D_{old} - D_{new})$$

D_{old} and D_{new} is the distance between the robot and the box from different time steps.

Rotation Reward: For the robot to learn to push at the back of the box towards the end, we deliberately used the learning about the direction of the force.

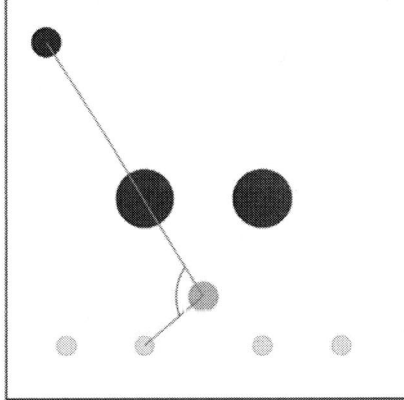

FIGURE 7. THE ROTATION REWARD FOR BOX-PUSHING

$Cos \, \theta$ is the angle formed by the endpoint, the box, and the robot as shown in Figure 7. If it is relatively in front of the box, it is negative and if in the back it is positive. So, the reward should be negative.

Collision Reward: The reward for pushing the box is represented as C_{ab}, which is defined by the collision between robot and box.

$$C_{ab} = \begin{cases} 5 & \text{if collision occurs} \\ 0 & \text{if no collision occurs} \end{cases}$$

The reward for the obstruction of box travel is represented as C_{aa}, which is defined by the collision between robots. Such behavior is something we don't want, so it's a negative value.

$$C_{aa} = \begin{cases} -5 & \text{if collision occurs} \\ 0 & \text{if no collision occurs} \end{cases}$$

The reward for pushing the box correctly is represented as $C_{D>0}$, which is defined by whether the collision helps solve the tasks.

$$C_{D>0} = \begin{cases} 5 & \text{if the distance becomes shorter} \\ 0 & \text{if the distance becomes longer} \end{cases}$$

Goal Reward: The reward for finishing the task is represented as a Flag which is defined by whether the robot achieves the goal of reaching the destination.

$$Flag = \begin{cases} 1000 & \text{if reaching the goal} \\ 0 & \text{if not reaching the goal} \end{cases}$$

Copyright © 2023 by ASME

Bound Reward: The reward for hitting the wall is represented as Bound, which is defined by whether robots go beyond the wall.

$$Bound = \begin{cases} -10 & if \ reaching \ the \ bound \\ 0 & if \ not \ reaching \ the \ bound \end{cases}$$

The total reward is the sum of all these rewards, as shown in the equation below:

$$R_{eward} = D_{change} - D_{ab} - \cos\theta + C_{ab} + C_{aa} + C_{D>0} + Flag$$
$$+ Bound$$

After we complete the overall training design, we carry out the corresponding training. We can directly train the system using the reward as described in the formula above. The initial conditions of the training process need to be adjusted according to the data points required in the design method, which will be described in the next section. In this process, we use a 3060 GPU for training, and each data point is trained for 10,000 rounds in a step size of 60, and several iterations of the experiment are performed.

4.2.3 Surrogate models (Step2 Figure 1)

Before the overall learning of the robots, we specify the overall performance with five aspects, collision between robots, the collision between robots and boxes, sum of the distance traveled by robots, distance traveled by boxes, and task completion time. Again, according to our design flow in Figure 1, we should explore the connection between hyperparameters and the performance of emergent behavior. Therefore, we use a Box-Behnken approach with a three-factor-five-response design, and collect a total of 17 data points. Further, we make full use of these 17 sets of orthogonal data to complete the response surface analysis, and the results are as follows.

For a better representation, we denote the three input parameters as **A number, B size, and C velocity.** Here we show the results generated by all surrogate models.

Collision D_1: $D_1 = -23.574 + 0.034 \times A + 336.696 \times B + 7.803 \times C - 34.780 \times AB - 0.239 \times AC - 24.263 \times BC + 0.586 \times A^2 - 551.775 \times B^2 - 0.731 \times C^2$

As indicated by the color of the data points in Figure 8, we observe that the larger the number, the more collisions between robots. As for the speed, to a certain extent the faster the speed is better for collision avoidance. The effect of size is less significant.

Box Collision D_2: $D_2 = 76.313 - 7.785 \times A + 168.294 \times B - 12.802 \times C + 6.717 \times AB + 0.348 \times AC - 46.576 \times BC + 0.723 \times A^2 + 657.923 \times B^2 + 1.329 \times C^2$

By looking at the response surface data in Figure 9, we can clearly see that as the speed becomes faster, the collision between the box and robot decreases, which we believe is due to an increase in the kinetic energy given, allowing it to reach the end with fewer contacts. We observe that the larger the robot is the more collisions there are, which we think is because the size increases the chance of collision.

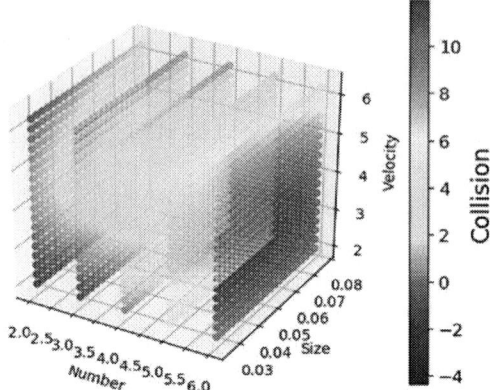

FIGURE 8. THE RESPONSE SURFACE OF THE COLLISION BETWEEN ROBOTS

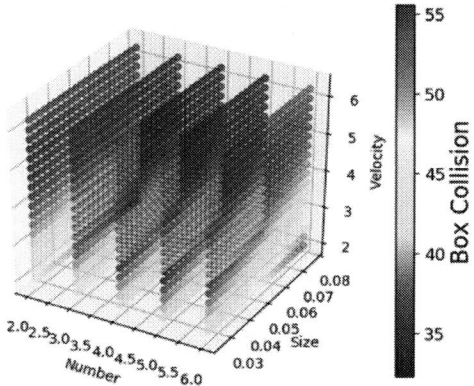

FIGURE 9. THE RESPONSE SURFACE FOR THE COLLISION BETWEEN THE ROBOT AND THE BOX

Robot Distance D_3: $D_3 = -0.630 + 1.885 \times A - 17.441 \times B + 0.623 \times C$

The data points in Figure 10 indicate that the more robots there are, the longer the paths they walkthrough. We note that the faster and smaller the size, the longer the walk will be, which we believe that here too the robot's control over the box becomes weaker, making it requires more paths to correct the direction.

Box Distance D_4: $D_4 = 1.441 - 0.000865 \times A - 1.067 \times B + 0.0829 \times C$

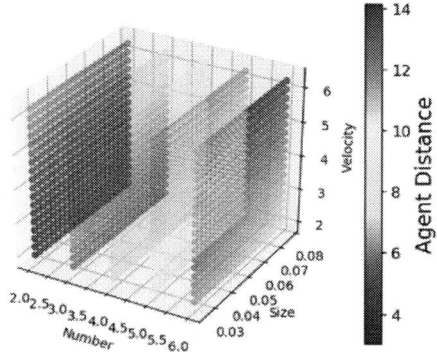

FIGURE 10. THE RESPONSE SURFACE FOR THE ROBOT DISTANCE

Copyright © 2023 by ASME

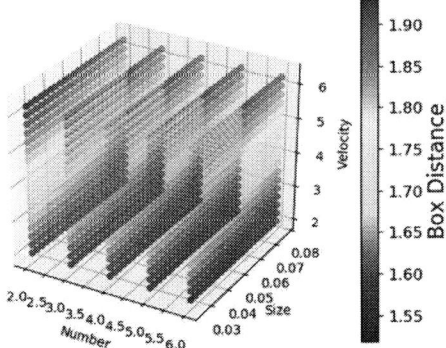

FIGURE 11. THE RESPONSE SURFACE FOR THE BOX DISTANCE

By looking at the response surface data in Figure 11, we observe that the greater the speed, the longer the distance traveled by the box, and we believe that it is the speed that makes the robot's control of the box's path weaker.

Time D_5: $D_5 = 1.152 - 0.0126 \times A - 0.881 \times B - 0.115 \times C$

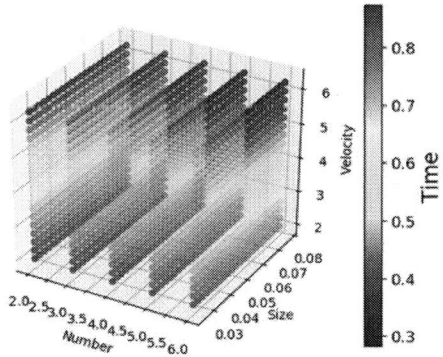

FIGURE 12. THE RESPONSE SURFACE FOR THE TIME

The response surface of time is shown in Figure 12. We observe that the change of number of robots can hardly affect the task completion time. We suggest that speed has a positive impact on the completion time: the greater the speed, the shorter the completion time.

These response surface results are theoretically significant and consistent with our actual perceptions. Therefore, we suggest that this model represents the connection between the local robot parameters and the overall task performance well.

4.3 Exploration of The Design Space (Step3 Figure 1)

Following the cDSP construct, we identify the initial settings shown in Table 2.

TABLE 2. CONSTRAINTS OF PARAMETERS

Name	Goal	Limit	Importance
Number	In range	[2, 6]	/
Size	In range	[0.025, 0.075]	/
Velocity	In range	[2, 6]	/
Collision a-a	Minimize	[0, 11]	w_1
Collision a-b	Maximize	[31, 52]	0
Agent distance	Minimize	[3.673, 13.536]	0
Box distance	Minimize	[1.527, 1.952]	w_4
Time	Minimize	[0.341, 0.918]	w_5

We defined a range of design variables and assigned initial relative weights to the objectives based on specific experimental data. Although there are five output metrics here, after observing the data, we summarized some important metrics among them. In fact for the length of the path taken by the robot is clearly related to the number of robots, fewer robots will also have less loss, which is not very helpful for improving control or efficiency. Further, we believe that the actual number of times the robot pushes the box is related to the robot's decision method and not so close to the actual control efficiency. The main factors we consider here are the number of collisions between robots, the distance traveled by the box, and the time. We believe that in the real world, efficiency is more important, which means that there should be less loss of collisions, less distance traveled by the boxes, and faster time. In summary, we use these three metrics as our design goals. After specifying the objectives, we formulate a cDSP using the relevant surrogate models, as shown in Table 3. We use the cDSPs to explore the solution space with their target values as follows.

$$D1_{Target} = 3.0$$

$$D4_{Target} = 1.53$$

$$D5_{Target} = 0.35$$

Although after the analysis, we know which response is relatively important, it is difficult to judge the good or bad design directly from the output of the system, and the relative proportions of the whole exploration space needs to be judged again. In effect the importance between each indicator also needs to be judged again in the specific task. Therefore, in our design, we roughly estimated the response positions of data points with different weights and studied the values of several corresponding points from which to judge good or bad. Also, it is a very important part of the design to consider not only the weight of deviation, but also to analyze more the results of the response surface to find the specific design goal. This will directly affect the results and guide a designer to explore the direction of the solution space. Based on the above analysis, we take the response of collision, box distance and time as an example to make some trade-offs for the decision model. The solution space is very large and there are many solutions available to meet the requirements.

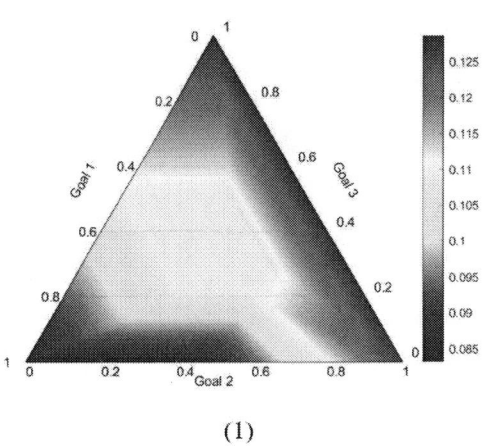

(1)

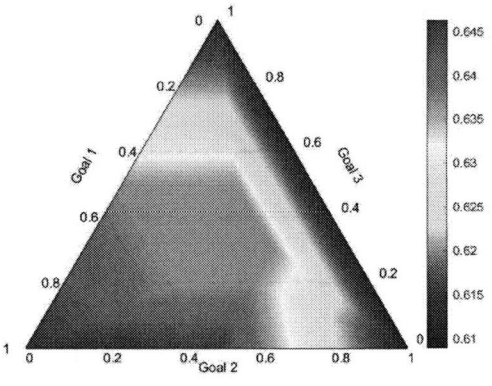

(2)

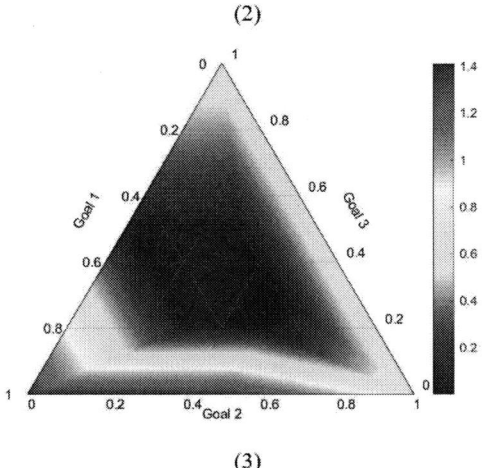

(3)

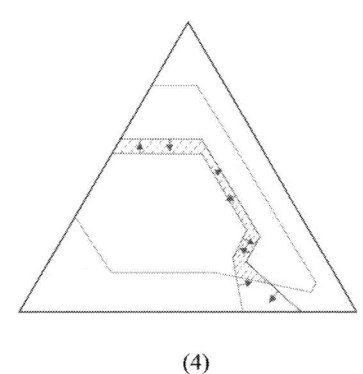

(4)

FIGURE 13. THE VALUES OF GOALS UNDER
DIFFERENT WEIGHTS AND THE DECISION DIAGRAM (BOX
DISTANCE IS GOAL 1, TIME IS GOAL 2, AND COLLISION
BETWEEN ROBOTS IS GOAL 3)
(1) VALUE OF BOX DISTANCE WITH DIFFERENT
WEIGHTS
(2) VALUE OF TIME WITH DIFFERENT WEIGHTS
(3) VALUE OF COLLISION WITH DIFFERENT WEIGHTS
(4) THE FEASIBLE DOMAIN OF THE SOLUTION

TABLE 4. CDSP CONSTRUCT FOR IDENTIFYING THE
SOLUTION SPACE

Given	
3	Number of system design variables
6	Number of inequality constraints
0	Number of equation constraints
6	Number of system design constraints
5	Number of system design goals
$g_i(A,B,C)$	System design constraint function
$D_k(d_i)$	Target deviation variable minimization function
Find	
A B C	Variables
d_i^+, d_i^- $i = 1, \cdots, m$	Deviation variables
Satisfy	

Restraints (in the previous table)

$$A - 6 \leq 0$$
$$A - 2 \geq 0$$
$$B - 0.075 \leq 0$$
$$B - 0.025 \geq 0$$
$$C - 6 \leq 0$$
$$C - 2 \geq 0$$

Goals (in the previous surrogate model)

$$\frac{D1(A,B,C)}{D1_{Target}} + d_1^- - d_1^+ = 1 \qquad Goal\ (1)$$
$$\frac{D4(A,B,C)}{D4_{Target}} + d_4^- - d_4^+ = 1 \qquad Goal\ (2)$$
$$\frac{D5(A,B,C)}{D5_{Target}} + d_5^- - d_5^+ = 1 \qquad Goal\ (3)$$

Boundaries

$$A, B, C^{min} \leq A, B, C \leq A, B, C^{max} \qquad i = 1, \cdots, n$$
$$d_i^+, d_i^- \geq 0; \ d_i^+ + d_i^- = 0$$

Minimize

$$Z = \sum_{i=1}^{m} w_i(d_i^-, d_i^+) \qquad \sum w_i = 1; \ w_i \geq 0$$

In Figure 13, we show the results of three different decision goals with different weights in ternary plots, starting from the left with inter-robot collisions, distance traveled by the box and time. These plots represent the solutions of the three different objectives with the corresponding different weights, whose values are obtained from the cDSP in Table 4, and the color of the dots indicates the relative magnitude of this value. To make a decision, we superimpose the plots shown in Figure 13. In the superimposed Figure (4), a'' '' 'ons identified for the overall performance goals and o (4) iirements are combined to identify a single region that satisfies all the requirements. In analyzing Figure 13, in (3) ion area that satisfies the minimum box distance, tl ꜱonding time is exactly the longest, and the path in the shortest time area is just the longest, and the collision in the area where they are all compromised is the least. Therefore, a designer is faced with the dilemma of choosing the solution that satisfies the objectives among the time and box distance regions. To make a decision, we first identify some solution points from the superposition diagram and analyze how well the objective is satisfied. We identify five solution

Copyright © 2023 by ASME

points from the ternary space, and the results associated with each solution point are summarized in Table 5.

It is particularly important to note that in making this range of decisions, the choice actually needs to be made based on the specific task requirements as well as the design preferences of a designer. Here we have chosen a narrow region, so the space where the design goals are located is not large. We mainly consider a balance among several aspects, as well as special consideration to the higher requirements for control in real systems, where no errors can occur, so we need to go after efficiency on the basis of stability. In this regard, the corresponding design area is drawn.

TABLE 5. SOLUTION POINT SELECTED

Weight Of Goals [1, 2, 3}	Solution (Size$\times 10^{-2}$)	Deviation
[0.4,0,0.6]	[2,5.555483,2.000001]	0.0435388
[0.2,0.2,0.6]	[2,5.555483,2.000001]	0.0217696
[0.2,0.55,0.25]	[2,5.555482,2.000002]	0.0217695
[0.3,0.55,0.15]	[2,5.555482,2.000002]	0.0326540
[0.3,0.6,0.1]	[2,5.555482,2.000002]	0.0326539

We obtain the feasible solution as *"number = 2 size = 0.0555 velocity = 2.0"*. This result is also similar to the one obtained in the reinforcement learning simulation environment, both in that the lower the number of robots, the lower the overall consumption and collisions. In our goal, which is control on the one hand and effectiveness on the other, the optimal number of robots is likely to be 2 if both robots can perform the task well. In the context of the size and speed tradeoff, we believe it is the improvement in control that results in much fewer collisions, despite the fact that some time is lost at low speed. The specific advantages of the results will be shown in the next section. This section is mainly to show the method of self-organizing system solution.

4.4 Discussion – Efficacy of the Solution

From the results, we suggest that the overall design is reasonable and satisfactory. More factors can be added at a later stage to increase the fidelity of the model and improve the resulting performance. In this way, we can find the desired target results by designing the system from global to local and explore the solution space using the cDSP construct. Unlike general optimal design, the cDSP helps designers to find relatively good and satisfactory solutions by trading off between different performance metrics, rather than the globally prescribed optimal solution. Here, we are not pursuing the best in terms of time and control performance, but to maximize the control performance of the robot while maintaining a certain time consideration, so that the robot can reduce collisions and distances. In Tables 6 and 7, we show the results with the surrogate model and the simulation, where the surrogate model gives the regular level of performance in the solution region and the simulation is the real level with volatility during the experiment.

TABLE 6. DATA FROM THE SURROGATE MODEL

Num	Size	Velocity	Collision	Box distance	Time
2	0.025	4	**0.349**	1.744	0.645
2	0.05	2	**0.115**	**1.552**	0.853
2	0.05	6	1.170	1.883	**0.393**
2	0.075	4	6.095	1.691	0.601
4	0.025	2	**0.089**	1.577	0.850
4	0.025	6	1.659	1.908	**0.390**
4	0.05	4	5.277	1.716	0.598
4	0.075	2	4.783	**1.523**	0.806
4	0.075	6	1.500	1.855	**0.346**
6	0.025	4	11.935	1.741	0.594
6	0.05	2	10.135	**1.548**	0.802
6	0.05	6	7.366	1.880	**0.342**
6	0.075	4	10.725	1.687	0.550
2	0.0556	2.000	**1.006**	**1.546**	0.848

It is observed that in the surrogate model, we basically get the expected desired solution, which is a better solution in both collision and box distance, and only time may be slightly high. In the real simulation, we observe that the difference is not too big, but basically fluctuates up and down around the solution of the surrogate model, which also reflects that the model is more reasonable and reliable; Table 7. The bold part are the four values that perform best in the overall data. We also note that the results we obtained are the best in both control aspects, and they are not bad in terms of time.

The system is slightly inferior regarding the time, because the results from the solutions we get are slower, as we can know according to our surrogate model, so that overall it will be slower to finish. The reason for this composition is, on the one hand, that in Figure 13 when we choose the solution region, the time considered is the slower region, and we hope that the application in industry can be more controlled and consider other factors a little more. On the other hand, is that even if the speed is slower, efficiency is actually very high because here only two robots are used to complete the task. We can also observe in Figure 14 that different speeds show different control when the system has the same number and size of robots, and obviously the slower the speed the better the control, as in the case of the blue line in the figure which is the best path we expected. The result we got is just like the blue line, and it can be clearly observed that it is better than other parameters.

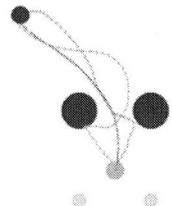

FIGURE 14. THE LINES OF BOX UNDER DIFFERENT VELOCITY (BLUE V=2 ORANGE V=4 GREY V=6)

TABLE 7. DATA FROM THE SIMULATION

Num	Size	Velocity	Collision	Box distance	Time
2	0.025	4	**0.16**	1.730	0.512
2	0.05	2	0.4	**1.558**	0.918
2	0.05	6	**0**	1.926	0.492
2	0.075	4	7.194	1.753	0.494
4	0.025	2	**0**	**1.527**	0.876
4	0.025	6	3.048	1.875	**0.403**
4	0.05	4	5.281	1.703	0.599
4	0.075	2	3.405	1.564	0.849
4	0.075	6	1.6	1.760	**0.342**
6	0.025	4	10.841	1.811	0.542
6	0.05	2	11.310	**1.537**	0.823
6	0.05	6	7.091	1.952	**0.376**
6	0.075	4	10.919	1.652	**0.474**
2	0.0556	2.000	**0**	**1.540**	0.931

5 CLOSING REMARKS

Multi-robot self-organizing systems are a promising area for research and application in industry. In this paper, we propose a design framework that combines multi-agent reinforcement learning and surrogate models with cDSP constructs to answer the following question:

How can multi-robot self-organizing systems be designed so that they show the desired behaviors and are able to perform tasks specified by designers?

We suggest that based on an analysis of the results in Section 4, that our framework is promising in designing a multi-robot self-organizing box-pushing system. Specifically, first we use a multi-agent reinforcement learning algorithm to train the system to perform in a way that the robots can well accomplish the given task. Then we use surrogate models to approximate the mapping relationship between the local parameters to the global performance indicators. Next, with the surrogate models representing the goal functions, we formulate the cDSP and explore the solution space using the associated ALP algorithm to identify satisfactory solutions. Finally, designers make a decision by making trade-offs among several conflicting performance indicators.

In future, we plan to combine more design content with local action spaces that are of interest (e.g., the robots may have two wheels and the speeds of which are different so as to change directions when moving). In this regard, we can explore the solution space in more realistic environments, and the overall performance of the robots will be better than simple collision. Not only can the system be applied to more shapes of boxes, but it can also be applied to more environments and can be directly designed to push boxes in a way that reduces training time and increases system stability.

In conclusion, we propose a promising framework for designing multi-robot self-organizing systems. The framework is general and can also apply to other problems in addition to multi-robot box-pushing. The benefits to different stakeholders are summarized as follows:

1) The benefit to designers. The proposed framework is expected to help designers of multi-robot self-organizing systems to identify designs that are capable of delivering wanted global behaviors and performances in the early design stage.
2) The benefit to researchers. Researchers can more deeply utilize or draw on the framework for more local-to-whole self-organizing system design to better exploit multi-agent emergent behavior.

We are happy to share our data with others who may wish to use it. To obtain the data, please contact the second author.

ACKNOWLEDGEMENTS

Zhenjun Ming acknowledges support from the National Natural Science Foundation of China (51805033), Beijing Municipal Science & Technology Foundation (3222020) and Beijing Institute of Technology Research Fund Program for Young Scholars (3030011182037). Janet K. Allen acknowledges funding from the John and Mary Moore Chair. Farrokh Mistree acknowledge the funding from L.A. Comp Chair at the University of Oklahoma.

REFERENCES

[1] J. Werfel, K. Peterse and R. Nagpal, "Designing collective behavior in a termite-inspired robot construction team," Science, 343 (2014) 754-758.

[2] A. Reina, Engineering swarm systems: A design pattern for the best-of-n decision problem, PhD Dissertation, Ecole Polytechnique do Bruxelles (2016).

[3] B. Bras and F. Mistree, Robust design using compromise decision support problems, Engineering Optimization, 21 (1993) 213-239.

[4] C. Gershenson and F. Heylighen, When can we call a system self-organizing? European Conference on Artificial Life, Springer, 2003, pp. 606-614.

[5] W. Banzhaf, Self-organizing systems, Encyclopedia of complexity and systems science, 14 (2009) 589.

[6] A.M. Turing, The chemical basis of morphogenesis, Bulletin of Mathematical Biology, 52 (1990) 153-197.

[7] J. Howard, B. Fuller, Self-organization in biology, Research Perspectives, (2010) 28-29.

[8] Humann, N. Khani, and Y. Jin, Adaptability tradeoffs in the design of self-organizing systems, International Design Engineering Technical Conferences and Computers and Information in Engineering Conference, American Society of Mechanical Engineers, Paper Number DETC2016-60053 (2016).

[9] M. Rahimi, S. Gibb, Y. Shen, and H.M. La, A comparison of various approaches to reinforcement learning algorithms for multi-robot box pushing, International Conference on Engineering Research and Applications, Springer, 2018, pp. 16-30.

[10] P. Long, T. Fan, X. Liao, W. Liu, H. Zhang, and J. Pan, Towards optimally decentralized multi-robot collision avoidance via deep reinforcement learning, 2018 IEEE International Conference on Robotics and Automation (ICRA), IEEE, 2018, pp. 6252-6259.

Copyright © 2023 by ASME

[11] Z. Kakish, K. Elamvazhuthi, and S. Berman, Using reinforcement learning to herd a robotic swarm to a target distribution, International Symposium Distributed Autonomous Robotic Systems, Springer, 2021, pp. 401-414.

[12] H. Bae, G. Kim, J. Kim, D. Qian, and S. Lee, Multi-robot path planning method using reinforcement learning, Applied Sciences, 9 (2019) 3057.

[13] R. Lowe, Y.I. Wu, A. Tamar, J. Harb, O. Pieter Abbeel, and I. Mordatch, Multi-agent actor-critic for mixed cooperative-competitive environments, Advances in Neural Information Processing Systems, 30 (2017).

[14] H. Ji, and Y. Jin, Designing self-organizing systems with deep multi-agent reinforcement learning, International Design Engineering Technical Conferences and Computers and Information in Engineering Conference, American Society of Mechanical Engineers, Paper Number DETC2019-98286 (2019).

[15] S.C. Ferreira, R. Bruns, H. Ferreira, G. Matos, J. David, G. Brandão, E.P. da Silva, L. Portugal, P. Dos Reis, and A. Souza, Box-Behnken design: an alternative for the optimization of analytical methods, Analytica chimica acta, 597 (2007) 179-186.

[16] F. Mistree, O.F. Hughes, and B. Bras, Compromise Decision Support Problem and the adaptive linear programming algorithm, in Structural Optimization: Status and Promise, 247-286 (M.P. Kamat, editor) (1993) AIAA, Washington, D.C.

Proceedings of the ASME 2023
International Design Engineering Technical Conferences and
Computers and Information in Engineering Conference
IDETC-CIE2023
August 20-23, 2023, Boston, Massachusetts

DETC2023-116751

USING MACHINE LEARNING TO PREDICT THE ADOPTION OF BUILDING ELECTRIFICATION TECHNOLOGIES IN US HOUSEHOLDS

Andrew Majowicz
Graduate Research Assistant
School of Systems and Enterprises
Stevens Institute of Technology
Hoboken, NJ 07030
Email: amajowic@stevens.edu

Philip Odonkor
Assistant Professor
School of Systems and Enterprises
Stevens Institute of Technology
Hoboken, NJ 07030
Email: podonkor@stevens.edu

ABSTRACT

This paper explores the use of machine learning to predict the adoption of building electrification technologies within US households. This is important due to the increasing prevalence of building electrification as a pathway to addressing climate change, which inadvertently poses a threat to the energy resilience of households during power outages. A non-intrusive, data-driven means of predicting the level of technology adoption can help guide mitigation and adaptation strategies aimed at minimizing the risks vulnerable households may face when power outages are compounded by extreme weather events. This study develops machine learning models based on the energy consumption dynamics of US households to predict the presence of critical electric appliances, including furnace, water heater, induction stove, cooling system, and solar panels. The models are trained using a large dataset of building end-use load consumption for buildings located in New Jersey. The results show that the models are reasonably accurate in predicting the presence of appliances in homes, although there is still significant potential for improvement in model accuracy.

Keywords : load profile, building electrification, machine learning.

1 INTRODUCTION

Buildings are significant contributors to climate change, responsible for nearly 40% of global greenhouse gas emissions [1].

Yet, despite recent investments towards advancing their energy efficiency [2], data shows that building emissions have surged sharply from their pandemic lows to record highs [3]. At the same time, the frequency and intensity of extreme weather events [4] have raised concern and ignited a renewed sense of urgency in addressing the impact of buildings on the environment. The adoption of building electrification, particularly in US homes, has emerged as an attractive pathway for tackling climate change [5]. It is a decarbonization strategy that replaces fuel-burning home appliances with electric alternatives such as heat pumps, solar photovoltaics, electric vehicle charging, and battery technologies. Indeed, building electrification is among the five transformative pathways prioritized by the US government in achieving a net-zero economy by 2050 [6]. This strategic focus not only aims to enhance environmental sustainability, but also aims to improve the overall wellbeing and equity of all Americans.

However, as we move towards a more electrified future, the increasing demand for electricity is exerting strain on the power grid, elevating the risk of power outages [7]. Furthermore, extreme weather events are increasing the occurrence and severity of power outages across the country. The confluence of these trends presents a grim outlook for electrified buildings. Without electricity, fully electrified households may experience difficulty in maintaining a comfortable indoor climate and supporting essential activities such as cooking. This reality was painfully evident during the winter blizzard that caused major power outages in Buffalo, NY, in December 2022 [8]. Amidst sub-zero tempera-

Copyright © 2023 by ASME

tures, households that relied exclusively on electricity for warmth were left vulnerable, resulting in a number of tragic fatalities. As power outages become more prevalent, and homes become more electrified, their ability to maintain some degree of operational efficacy during outages becomes critical. Consequently, there is a need for tools that can not only inform investments in grid resilience and restoration, but also support emergency response systems to ensure that vulnerable households can be easily identified both prior to and during climate-related crises.

The objective of this study is to employ data-driven methodologies to predict the ability of US homes to withstand power outages. In this context, resilience refers to a household's capacity to sustain critical functions like heating, cooling, and cooking during power disruptions. The paper aims to enhance our understanding of how machine learning algorithms, combined with data on the energy-use patterns of US households, can be used to accurately predict the types of appliances present in a household.

2 LITERATURE REVIEW

As climate change intensifies, the likelihood of heatwaves, hurricanes, and wildfires increase [9]. These climate hazards can significantly disrupt electricity access, resulting in energy insecurity and related hardships for individuals, households, and communities [10]. The literature on these impacts highlights their far-reaching and often devastating consequences. A brief sample reveals that climate-induced power outages disproportionately impact low-income and marginalized communities, resulting in adverse health outcomes, financial burdens, and diminished quality of life [11]. A prime example showcasing these impacts was Hurricane Katrina, a destructive storm that struck the Gulf Coast of the United States in 2005. The storm exposed vulnerabilities in the US energy infrastructure [12], resulting in extensive power outages across regions such as Louisiana and Mississippi, depriving millions of people of electricity. The loss of electricity disrupted daily routines, hampering the ability to prepare and store food, regulate indoor temperature, and maintain social cohesion.

Similarly, the 2017 Hurricane Maria, which struck Puerto Rico, caused catastrophic power disruptions that lasted several months [13]. The hurricane damaged the island's electrical infrastructure, leaving the entire population without access to electricity. The lack of power had a significant impact on daily life, disrupting critical services such as healthcare, water supply, and transportation. In recent years, extreme weather events such as wildfires, droughts, and heat waves have also caused significant energy insecurity in parts of the United States. For example, the wildfires in California in 2017 and 2018 caused widespread power outages as utility companies shut off electricity to prevent sparking fires [14]. Another example is the Texas winter storm in 2021 [11]. The winter storm caused extremely low temperatures and snowfall across Texas, which caused significant stress on the

state's energy grid. As a result, millions of people were left without power and heat for several days, leading to a widespread energy crisis [15]. The preceding events underscore the urgent need for tools and methodologies to effectively diagnose and understand the dynamics of household vulnerability during climate-induced outages.

Despite power outages presenting significant challenges to building electrification, both in terms of public perception and technical obstacles, there is a growing emphasis on enhancing the reliability, resilience, and affordability of building electrification technologies. In 2022, the Building Electrification Initiative was launched by the US Department of Energy (DoE) with the goal of expediting the integration of building electrification technologies [16]. The program offers technical assistance, research, and outreach to accelerate US adoption of building electrification technologies. Non-profit organizations such as the Rocky Mountain Institute [17] and the National Resource Defense Council (NRDC) [18] have also been active in demonstrating the feasibility and benefits of electrifying buildings. Complementary to these technical efforts are the initiatives led by NRDC advocating for policies that support the building electrification. Their work focuses on a wide array of topics including building codes and incentives to promote the adoption of electrification technologies.

Research efforts in the building electrification domain are currently focused on a variety of topics including the development of advanced heat pump systems [19], the use of building automation systems to optimize energy efficiency [20], the integration of renewable energy sources into building systems [21], and the exploration of new materials and building designs to improve energy performance. Additionally, there is growing interest in the use of machine learning (ML) to predict power outages and improve building energy performance [22]. For example, machine learning algorithms can predict which areas are most vulnerable to specific climate hazards [23], thereby facilitating effective climate mitigation and adaptation strategies. Moreover, ML has proved useful in improving the energy performance of buildings [24], leveraging its proficiency in analyzing historical energy consumption patterns to pinpoint operational inefficiencies and recommend interventions.

Although machine learning holds great promise for advancing building electrification efforts, there remain significant challenges to be addressed. One major obstacle is the lack of comprehensive datasets on the adoption of electrification technologies, particularly in underserved communities. Previous work [25] by the author successfully mapped technology adoption across the state of NY; however, this mapping relied on reported data, which may not accurately reflect the true adoption rates due to voluntary reporting. Hence, in the event of climate-induced power disruptions, it is presently unfeasible to accurately identify all vulnerable households. This insight is critically important to efficiently strategize aid initiatives, prioritize relief efforts, and construct proactive measures to mitigate and adapt to future challenges.

Copyright © 2023 by ASME

To overcome this challenge, this study analyzes the dynamics of electricity consumption within US households with the aim of discerning the types of domestic appliances installed, along with the primary energy sources that drive them, whether it be electricity or fossil fuels. For US homes, particularly those in colder climates, there are three key appliances whose electrification may pose resilience risks, namely the Heating, Ventilation, and Air Conditioning (HVAC) system, the water heater, and the kitchen stove. To this end, the models developed in this work seek to answer the following questions:

1. Does the home use electricity for heat?
2. Does the home use electricity for cooking?
3. Does the home use electricity for heating water?
4. Does the home have a cooling system?
5. Does the home have solar panels?

The first three questions try to understand vulnerability to power outages by determining how many of the critical household appliances are electrified. The fourth question tries to understand resiliency during heat waves, and finally, the last question tries to detect energy self-sufficiency. Electrified homes with generation capabilities may fair better during power outages than electrified homes without the ability to generate their own electricity. Collectively, the answers to these questions provide an idea of how vulnerable a household may be to a power outage during an extreme weather event. The outcome of this work is the creation of a scalable and non-intrusive method for understanding technology adoption in US households, and its impact on household energy resiliency.

The remainder of the paper details how this was achieved and is structured as follows; Section 2 provides an overview of the methodology, focusing on the methods and assumptions used. Section 3 presents the results, including key performance metrics observed. Furthermore, it provides a discussion of the results and addresses its limitations. Finally, Section 4 closes the paper with concluding remarks along with a brief discussion of future direction.

3 METHODOLOGY

3.1 Aggregating Residential End-Use Load Profiles

End-use load profiles serve as a valuable tool in identifying the timing and quantifying the magnitude of electricity consumption in buildings, while also shedding light on the behaviors and usage patterns of the building occupants [26]. However, due to privacy concerns, there exists a paucity of comprehensive and readily accessible datasets on residential end-use load profiles.

Until recently, sourcing authentic residential load profiles had posed numerous obstacles. However, in 2019, a team of experts from the National Renewable Energy Laboratory (NREL), Lawrence Berkeley National Laboratory (LBNL), and Argonne National Laboratory addressed this challenge by creating veri-

fied, physics-based building models that accurately characterize energy consumption across US building stock. Known as ResStock [27], this comprehensive dataset covers a wide range of regions across the country and enables detailed examination of energy consumption in buildings. Moreover, the models have the versatility to be applied to buildings in climates and regions with traditionally limited data coverage. The supplementary ResStock metadata provides an extensive breakdown of electricity usage among various installed household appliances. This crucially provides us with the ground truth of technology adoption at a granularity that is often unavailable with such datasets. Consequently, with the aid of the end-use load profile, not only can we scrutinize the energy consumption patterns of each individual household within the dataset, but we can also categorize these patterns according to the distinct household appliances that contribute to them.

The practice of categorizing household appliances based on end-use load profiles is not a novel concept. In fact, an entire field of research known as Non-Intrusive Load Monitoring (NILM) [28] has been established with the specific aim of achieving this objective. NILM seeks to disaggregate the total energy consumption of a household into the energy consumption of individual appliances, without requiring physical metering of each appliance. However, the proposed work distinguishes itself from NILM in that, unlike NILM, there is no prior knowledge of the appliances installed in a household. Additionally, our approach strives to predict the primary fuel source of the installed devices, whereas in NILM, electricity is assumed to be the de facto fuel source.

3.2 Machine Learning Pipeline

In this work, a machine learning pipeline is implemented as a comprehensive framework to manage the flow of information into, and predictions from, the machine learning models. The pipeline is carefully designed to follow a structured process that encompasses several key stages, including data pre-processing, feature engineering, model selection and training, and model evaluation. In the subsequent subsections, we delve into the specifics of how each stage of the data pipeline was constructed to address the posed research questions.

3.2.1 Data Acquisition and Pre-processing The initial phase of the pipeline development process involved acquiring the pertinent end-use load profiles from the ResStock dataset. For project feasibility, we restricted our geographic focus solely to the state of New Jersey (NJ). This decision was made for two reasons: firstly, the NJ climate is characterized by moderately cold winters and warm, humid summers, which allows for a diverse range of installed home appliances and fuel types. Secondly, as our home institution is located in NJ, our familiarity and understanding of the area provided a strong starting point. Overall, we sampled end-use load profiles from more than

Copyright © 2023 by ASME

14,700 households in NJ. Table 1 provides an outline of the various types of residential properties included in this analysis. The load profiles utilized in this study represent consumption patterns during the year 2018 and were recorded at 15-minute intervals. The focus of this paper is on the year 2018 as it allows for a clear assessment of energy use dynamics that existed before the pandemic.

TABLE 1: Inventory of Residential Building Types Used In This Study.

Building Type	Quantity
Single-Family Detached	7908
Single-Family Attached	1422
Mobile Home	138
Multifamily with 2-4 units	2293
Multifamily with 5+ units, 1-3 stories	2032
Multifamily with 5+ units, 4-7 stories	610
Multifamily with 5+ units, 8+ stories	380

Given the varying characteristics of residential buildings analyzed in this study, it was necessary to standardize all input features employed in this work. This required transforming all feature values to achieve a mean of 0 and a standard deviation of 1. This transformation was done using the following equation:

$$z = \frac{x - \mu}{\sigma} \qquad (1)$$

where x represents the original value, μ is the feature mean, σ represents the standard deviation of the feature across the dataset, and output z represents the standardized value of the feature after scaling. According to existing literature, the absence of feature standardization can significantly impair the performance of developed machine learning models, making it an important step in our pipeline process.

3.2.2 Feature Engineering

The subsequent phase in the pipeline involved feature engineering, where we engineered features that enabled us to effectively address the five questions that this work aims to answer. As noted earlier, the ResStock dataset comes with a metadata file that provides a comprehensive breakdown of electricity consumption on a per-appliance basis for every household in the dataset. This enabled us to analyze the data fields associated with the furnace, cooling system, induction cooker, photovoltaic system, and water heater, thereby enabling us to determine the presence or absence of these appliances. Additionally, the metadata file provided supplementary information on the primary fuel source utilized by each appliance. With this in mind, we augmented the dataset by appending five new binary fields that indicated the presence (or absence) of each of the aforementioned electrical appliances. In addition, we also aggregated energy consumption from a 15-minute interval to a monthly interval. This decision was guided by previous experience, where we found that data granularity beyond a certain point yields diminishing returns, especially when drawing insights over an annual time frame. This approach enabled us to considerably reduce the dimensionality of the data. Specifically, for each household, we were able to distill the data from a little over 35,000 data points to 12.

It is worth highlighting another decision that was made in this study, which pertains to the use of an annual time span. Although not strictly a feature engineering task, this strategic decision was taken to facilitate the learning process. An annual timeframe allowed electricity consumption trends across all four seasons, namely summer, fall, winter, and spring, to be observed. We are of the opinion that the points of transition between seasons provide valuable insights about the types of appliances installed in most households. For instance, let us consider a home that employs electricity for cooling during the summer and natural gas for heating during the winter. In such a case, there would be a marked shift in electricity consumption following the summer-winter transition.

3.2.3 Model Selection and Training

The next phase involved model selection and training. This step required us to identify a suitable machine learning (ML) approach and algorithms to employ for the prediction task. Since we generated output labels via feature engineering, supervised learning proved to be the most logical choice. Moreover, the research questions lent themselves well to binary classification, making techniques like logistic regression, decision trees, support vector machines (SVMs), neural networks, and random forests compelling options. For this study, we focused on the use of neural networks.

As per the standard practice, we adopted an 80/20 train-test split to create training and validation datasets. The training dataset was used to train the prediction models, and the validation dataset was used to evaluate model performance. Two unique approaches were considered for model training. Initially, we developed a single machine learning model capable of answering all five questions using a single output. For this, we encoded the question responses as a hot-encoded binary string, and the final node of the ML model was a multi-classification layer with 32 unique outputs, representing all the possible binary combinations of the five questions. Despite its relative simplicity, this approach did not produce promising outcomes. Hence, we shifted to a second approach, where we trained independent models to address each unique question. The results presented later in this paper correspond to our findings using this approach.

The models implemented in this study were developed using the Python programming environment and relied on the Tensor-

Copyright © 2023 by ASME

Flow software library to support machine learning. The sequential class available in the Keras library was used to construct the models, and while each model had a unique architecture, they primarily utilized rectified linear units (ReLU) for activation. To ensure output values remained within the range of 0 to 1, output layers were constructed with sigmoid functions. The Adam optimizer was employed to optimize all models. The imbalanced-learn library was used to address class imbalance challenges encountered during the modeling process, which will be further elaborated upon in the results section. Training time for all models were in the sub 5-minute range. Table 2 summarizes the model configurations employed in this study.

TABLE 2: Model Details for Each Classification Model.

Classification Model	Hidden Layers	Batch Size	Learning Rate	Epochs	Activation Neurons	Total Neurons
Cooling	4	32	0.0001	100	128	1153
Cooking	6	32	0.0001	100	128	1665
Solar	4	2	0.001	1000	13	782
Water Heating	2	32	0.0001	100	128	385
Heating	5	8	0.0001	800	13	1038

3.2.4 Model Evaluation

The performance evaluation of the trained models was conducted using the validation dataset. We followed the standard practice of comparing the predicted outcomes of the model with the actual outcomes to determine model accuracy, precision, and recall. Additionally, we visualized the performance of each model using graphs and confusion matrices. Accuracy was computed as the ratio of correct classifications made by the model to the total number of classifications performed. Precision, on the other hand, measured how many of the positive identifications made by the model were actually correct. Finally, recall measured the percentage of true positives that were correctly identified by the model. These evaluation metrics allowed us to assess the performance of our models and compare their effectiveness in addressing the five questions posed in this work.

4 RESULTS

4.1 Predicting Electric Heating

The initial model aimed to forecast the use of electricity for heating in households. In general, electric heating necessitates substantial electricity consumption, rendering it relatively easy to identify. However, tackling this problem proved challenging due to significant class imbalances. This refers to a situation in machine learning where the distribution of classes in the training set is uneven, often resulting in training bias. Specifically, in our dataset, 96% of households relied on fossil fuels for heating, resulting in the first few models we developed being heavily biased towards the majority class, with poor performance in recognizing households that used electricity for heating. To overcome this problem, we adopted a strategy of under-sampling the majority class. The primary goal was to create a class balance, which would enable the model to learn from a more representative set of samples. However, it should be noted that random under-sampling can potentially lead to loss of information from the majority class, which may result in bias towards the minority class in extreme cases. Future work will explore alternative approaches, such as incorporating class weights, to preserve as much original data as possible while addressing the data imbalance issue. Following the removal of samples from the majority class and training of the model over 800 epochs, the resulting model performance is illustrated in Figure 1.

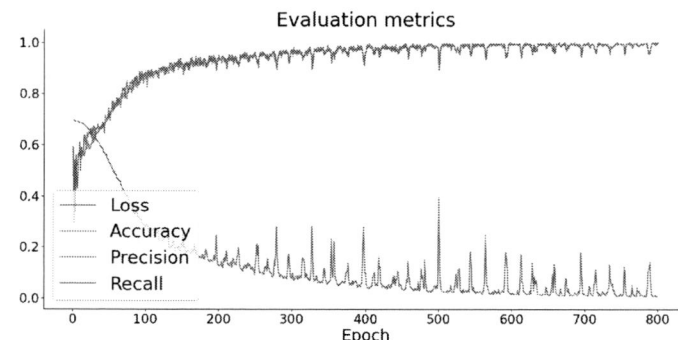

FIGURE 1: *Model performance for predicting electric heating.*

In general, the model achieved a 63% accuracy, 60% precision, and 79% recall. We deem these results as acceptable considering the highly imbalanced dataset we had to work with. To counter the problem of imbalanced classes, one possible solution is to extend the scope of study to consider regions with a strong electrification of heat, such as California. Furthermore, refining the model architecture and parameters through experimentation may lead to improved performance. The comprehensive breakdown of the classifications is illustrated in the confusion matrix shown in Figure 2.

4.2 Predicting Electric Cooking

The second model aimed to forecast the prevalence of induction stoves in New Jersey households. As with the previous prediction task, the dataset was also found to be severely imbalanced, with gas stoves dominating the dataset (88% prevalence). To counteract this imbalance, we utilized the Synthetic Minority Over-Sampling Technique (SMOTE), which generates synthetic instances of the minority class to balance the class distribution. Specifically, SMOTE creates synthetic samples by interpolating between existing minority class samples. It should be noted that we did not employ this technique in the previous prediction task

Copyright © 2023 by ASME

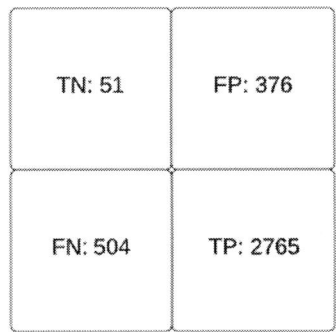

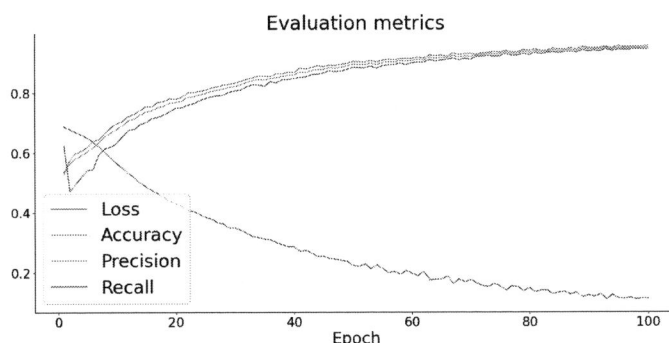

FIGURE 2: *Confusion Matrix for predicting electric heating. TN - True Negative, FP - False Positive, FN - False Negative, TP - True Positive.*

because of insufficient minority class samples. One potential risk of using SMOTE is the possibility of overfitting the data. After generating synthetic samples to increase the count of the minority class, we trained a supervised learning model over 100 epochs. The resulting model performance is illustrated in Figure 3. We tested the model using a test size of 3696 households.

FIGURE 3: *Model performance for predicting induction stove.*

Overall, the model delivered promising results, with an accuracy rate of 76%, a precision score of 88%, and a recall rate of 85%. For a more comprehensive understanding of the classifications, refer to the confusion matrix illustrated in Figure 4. While the findings are encouraging, we recognize the limitations of our dataset may be limiting model accuracy. Therefore, we intend to expand our dataset scope to include a more balanced representation of classes to improve the model's overall performance. Furthermore, the model currently learns from an aggregated monthly interval of electricity consumption. This may lack enough granularity for this specific task. Therefore, future research will explore the impact of using weekly and bi-weekly electricity consumption intervals on model performance.

FIGURE 4: *Confusion Matrix for predicting presence of induction stove. TN - True Negative, FP - False Positive, FN - False Negative, TP - True Positive.*

4.3 Predicting Electric Water Heater

The third model developed in this study aimed to predict the utilization of electric water heaters in New Jersey households. This task was also plagued with significant class imbalances, where 99% of households had electric water heating. To address this issue, we utilized class weighting to mitigate the class imbalance problem. Class weighting involves assigning higher weights to the underrepresented class during the model training process, thereby instructing the model to prioritize learning from the minority class. Specifically, we increased the weight of the minority class during the model fitting process. Although this approach is not inherently superior to our previous methods (e.g. SMOTE), it provides an alternative solution worth exploring. After implementing class weighting, we trained a supervised learning model for 100 epochs, and illustrated the model's performance in Figure 5.

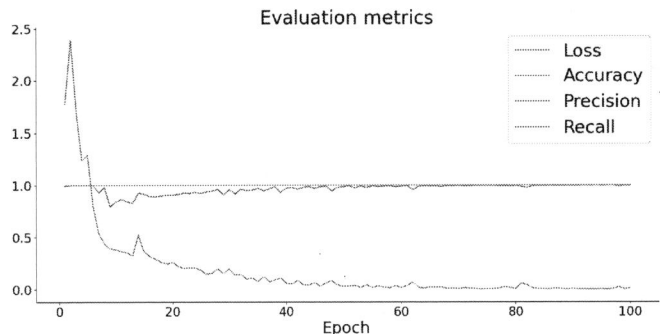

FIGURE 5: *Model performance for predicting electric water heater.*

The model that was trained attained an impressive 97% accuracy, 100% precision, and 97% recall. A detailed classification breakdown is provided in the confusion matrix shown in Figure

Copyright © 2023 by ASME

6. This is another case where data from other geographical regions is needed to better balance the classes and potentially improve the model.

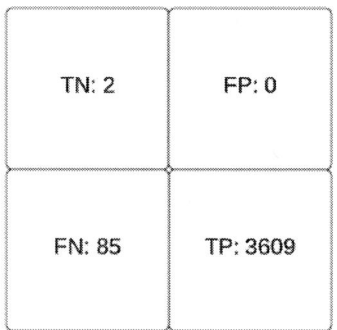

FIGURE 6: *Confusion Matrix for predicting presence of an electric water heater. TN - True Negative, FP - False Positive, FN - False Negative, TP - True Positive.*

4.4 Predicting Presence of Cooling System

The fourth model developed in this research aimed to predict the prevalence of forced-air cooling systems in households. Again, this task was characterized by a significant class imbalance, where 89% of households had cooling systems. To address this issue, we employed the Synthetic Minority Over-sampling Technique (SMOTE). This was motivated by the fact that it surpassed other methods in its ability to preserve valuable information from the original dataset. In this particular case, the data exhibited an imbalance, with the minority class containing sufficient diversity to enable the random selection of instances for the creation of synthetic samples, thereby balancing the overall dataset. This technique involves identifying the nearest neighbors of the selected points and adjusting the class distribution to enhance performance on the minority class. It is essential to note that we applied SMOTE only to the training dataset, while the testing dataset retained only the original values (test sample size: 2956 households). The resulting model performance over 100 epochs is illustrated in Figure 7.

The trained model demonstrated an accuracy of 74%, a precision of 89%, and a recall of 81%. A detailed classification distribution is depicted in the confusion matrix presented in Figure 8. However, the disparity in the class distribution again underscores the need for incorporating data from other geographical regions to achieve greater balance among the classes.

4.5 Predicting Presence of Solar Panels

The final model developed aimed to forecast the presence of installed solar photo-voltaic (PV) systems across New Jersey's building stock. Of the 14,783 households studied, fewer than 2%

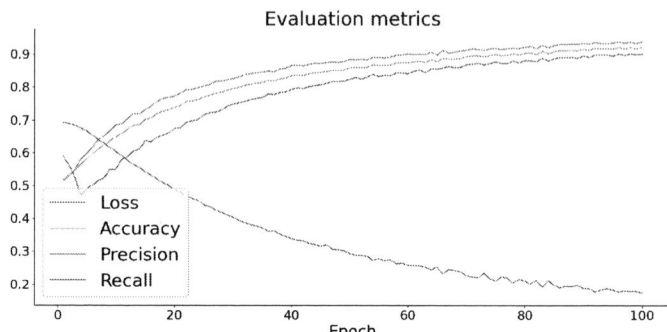

FIGURE 7: *Model performance for predicting electric cooling.*

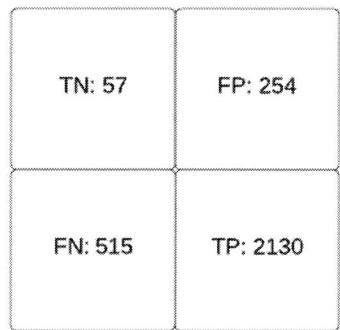

FIGURE 8: *Confusion Matrix for predicting presence of electric cooling. TN - True Negative, FP - False Positive, FN - False Negative, TP - True Positive.*

had installed solar panels. Similar to the heating classification problem, the majority under-sampling technique was employed (over 800 epochs)because the limited amount of data hindered the effectiveness of SMOTE in generating appropriate synthetic samples for the model's training. The resulting model performance is depicted in Figure 9.

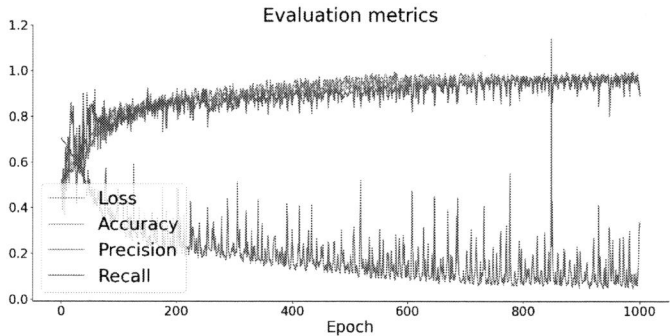

FIGURE 9: *Model performance for predicting presence of solar panels.*

Copyright © 2023 by ASME

In summary, our trained model achieved an accuracy of 63%, a precision of 60%, and a recall of 78%. A detailed account of the model's classification performance is presented in the confusion matrix, as shown in Figure 10. It is worth noting that the learning process was quite noisy (see Figure 9). This observation hints at a sub-optimal model architecture and/or model parameter choices.

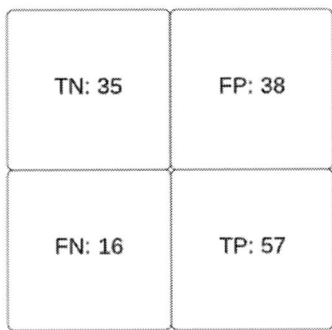

FIGURE 10: *Confusion Matrix for predicting presence of solar panels. TN - True Negative, FP - False Positive, FN - False Negative, TP - True Positive.*

4.6 Discussion

Table 3 summarizes the model performance across all five prediction tasks. Notably, the models could reasonably forecast technology adoption in most cases, which is indeed a promising start. However, there is still considerable room for improvement. As the predictions from each model are independent and equally weighted, we can estimate the accuracy of the combined prediction by multiplying their individual accuracies. The resulting combined model accuracy is approximately 22%. Although this figure may seem low, it is important to recognize that there is currently no benchmark for such predictions. Thus, we lack guidance on how to evaluate the results' quality. Moreover, this estimate assumes that the residuals from the individual models are uncorrelated, a supposition that has not been tested yet. Nonetheless, we strongly believe there is a lot of room for improvement of the individual models. As previously mentioned, the skewed distribution of classes in the dataset was undoubtedly a significant factor contributing to the relatively low accuracy scores observed across the prediction tasks. Notably, one potential drawback of utilizing under-sampling methods to address this problem is the potential loss of informative data points that could otherwise facilitate more accurate class labeling. Moving forward, we aim to overcome this limitation by incorporating end-use load profiles obtained from various geographic regions to improve class balancing in the dataset.

TABLE 3: Summary of model performance for each prediction task

Classification	Model Accuracy
Does the home use electricity for heat?	63%
Does the home use electricity for cooking?	76%
Does the home use electricity for heating water?	97%
Does the home have a cooling system?	74%
Does the home have solar panels?	63%

4.7 Limitations

Although the preliminary findings of this research are promising, it is crucial to acknowledge the study's limitations. Firstly, this investigation was confined to fewer than 15,000 households in New Jersey. As such, the dataset severely lacks diversity, a finding that was contrary to our original belif given NJ's moderate climate. Despite our efforts to avoid both under and overfitting, it is plausible that valuable data was overlooked during the under-sampling process. Another shortcoming of this research is the absence of exhaustive empirical experiments to explore different model architectures and parameters. Additionally, it would be insightful to compare the outcomes across multiple prediction techniques such as decision trees and logistic regression models. These limitations are currently being addressed in future work as we strive to enhance the current benchmarks established in this study.

5 CONCLUSION

This paper presents a novel data-driven approach to diagnose the technology adoption in US households. By analyzing the electricity end-use load profiles of households in New Jersey, we developed and validated five distinct models to forecast the presence of various essential household appliances. Our findings underscore the potential of machine learning as a useful tool to facilitate decision-making processes related to building electrification technologies and to support efforts to identify vulnerable households during climate-related disasters. The approach we adopted in this research harnesses the extensive ResStock dataset to create predictive models that would have otherwise been challenging to develop using conventional, siloed datasets. Future studies should expand the scope of our work and assess the generalizability of our findings to other regions and building types.

REFERENCES

[1] World Green Building Council, accessed 2023. Embodied carbon. https://worldgbc.org/advancing-net-zero/embodied-carbon/.

[2] International Energy Agency, 2019. Energy end-use and efficiency. https://www.iea.org/reports/world-energy-investment-2019/energy-end-use-and-efficiency. [Online; accessed March 1, 2023].

Copyright © 2023 by ASME

[3] UNEP, 2021. Co2 emissions from buildings and construction hit new high, leaving sector behind in meeting climate goals.

[4] Wang, D., Chen, Y., Jarin, M., and Xie, X., 2022. "Increasingly frequent extreme weather events urge the development of point-of-use water treatment systems". *Npj Clean Water, 5*(1), p. 36.

[5] Deason, J., and Borgeson, M., 2019. "Electrification of buildings: potential, challenges, and outlook". *Current Sustainable/Renewable Energy Reports, 6*, pp. 131–139.

[6] of the Federal Chief Sustainability Officer, O., 2021. Net-Zero Emissions Buildings by 2045, including a 50% reduction by 2032. On the WWW, Retrieved Jan 2023.

[7] House, T. W., 2022. A White House call for real-time, standardized, and transparent power outage data. . On the WWW, Retrieved Jan 2023. URL https://bit.ly/3lgZIbB.

[8] Corey Kilgannon, Lola Fadulu, H. M., and Nir, S. M., 2022. "How the Buffalo Blizzard Became So Deadly". *The New York Times*, December.

[9] Trenberth, K. E., 2018. "Climate change caused by human activities is happening and it already has major consequences". *Journal of energy & natural resources law, 36*(4), pp. 463–481.

[10] Graff, M., Konisky, D. M., Carley, S., and Memmott, T., 2022. "Climate change and energy insecurity: a growing need for policy intervention". *Environmental Justice, 15*(2), pp. 76–82.

[11] Flores, N. M., McBrien, H., Do, V., Kiang, M. V., Schlegelmilch, J., and Casey, J. A., 2023. "The 2021 texas power crisis: distribution, duration, and disparities". *Journal of Exposure Science & Environmental Epidemiology, 33*(1), pp. 21–31.

[12] Chow, E., and Elkind, J., 2005. "Hurricane katrina and us energy security". *Survival, 47*(4), pp. 145–160.

[13] Kwasinski, A., Andrade, F., Castro-Sitiriche, M. J., and O'Neill-Carrillo, E., 2019. "Hurricane maria effects on puerto rico electric power infrastructure". *IEEE Power and Energy Technology Systems Journal, 6*(1), pp. 85–94.

[14] Zanocco, C., Flora, J., Rajagopal, R., and Boudet, H., 2021. "When the lights go out: Californians' experience with wildfire-related public safety power shutoffs increases intention to adopt solar and storage". *Energy Research & Social Science, 79*, p. 102183.

[15] Lee, C.-C., Maron, M., and Mostafavi, A., 2022. "Community-scale big data reveals disparate impacts of the texas winter storm of 2021 and its managed power outage". *Humanities and Social Sciences Communications, 9*(1), pp. 1–12.

[16] Department of Energy, 2022. Biden-harris administration announces steps to electrify and cut emissions from federal buildings. https://www.energy.gov/articles/biden-harris-administration-announces-steps-electrify-and-cut-emissions-federal-buildings. [Online; accessed March 1, 2023].

[17] Rocky Mountain Institute, 2023. Rmi. https://rmi.org/. [Online; accessed March 11, 2023].

[18] Natural Resources Defense Council, 2023. Nrdc. https://www.nrdc.org/. [Online; accessed March 12, 2023].

[19] Valancius, R., Singh, R. M., Jurelionis, A., and Vaiciunas, J., 2019. "A review of heat pump systems and applications in cold climates: Evidence from lithuania". *Energies, 12*(22), p. 4331.

[20] O'Grady, T., Chong, H.-Y., and Morrison, G. M., 2021. "A systematic review and meta-analysis of building automation systems". *Building and Environment, 195*, p. 107770.

[21] Biyik, E., Araz, M., Hepbasli, A., Shahrestani, M., Yao, R., Shao, L., Essah, E., Oliveira, A. C., Del Cano, T., Rico, E., et al., 2017. "A key review of building integrated photovoltaic (bipv) systems". *Engineering science and technology, an international journal, 20*(3), pp. 833–858.

[22] Eskandarpour, R., and Khodaei, A., 2016. "Machine learning based power grid outage prediction in response to extreme events". *IEEE Transactions on Power Systems, 32*(4), pp. 3315–3316.

[23] Yousefi, S., Pourghasemi, H. R., Emami, S. N., Pouyan, S., Eskandari, S., and Tiefenbacher, J. P., 2020. "A machine learning framework for multi-hazards modeling and mapping in a mountainous area". *Scientific Reports, 10*(1), p. 12144.

[24] Odonkor, P., and Lewis, K., 2019. "Automated design of energy efficient control strategies for building clusters using reinforcement learning". *Journal of Mechanical Design, 141*(2).

[25] Odonkor, p., and Preziuso, D., 2022. "Towards an equitable grid interactive efficient building landscape: Analyzing technology adoption in the state of new york", 2022 aceee summer study on energy efficiency in buildings". *2022 Summer Study on Energy Efficiency in Buildings*.

[26] Preziuso, D., Kaminski, G., and Odonkor, P., 2021. "Understanding the energy behavior of building occupants through the chronology of their energy interactions". In International Design Engineering Technical Conferences and Computers and Information in Engineering Conference, Vol. 85383, American Society of Mechanical Engineers, p. V03AT03A014.

[27] National Renewable Energy Laboratory, 2023. Resstock. https://www.https://resstock.nrel.gov/. [Online; accessed Jan 03, 2023].

[28] Angelis, G.-F., Timplalexis, C., Krinidis, S., Ioannidis, D., and Tzovaras, D., 2022. "Nilm applications: Literature review of learning approaches, recent developments and challenges". *Energy and Buildings*, p. 111951.

Copyright © 2023 by ASME

Proceedings of the ASME 2023
International Design Engineering Technical Conferences and
Computers and Information in Engineering Conference
IDETC-CIE2023
August 20-23, 2023, Boston, Massachusetts

DETC2023-117003

SHIP-D: SHIP HULL DATASET FOR DESIGN OPTIMIZATION USING MACHINE LEARNING

Noah J. Bagazinski *
Department of Mechanical Engineering
Massachusetts Institute of Technology
Cambridge, Massachusetts, 02139
Email: noahbagz@mit.edu

Faez Ahmed
Department of Mechanical Engineering
Massachusetts Institute of Technology
Cambridge, Massachusetts, 02139
Email: faez@mit.edu

ABSTRACT

Machine learning has recently made significant strides in reducing design cycle time for complex products. Ship design, which currently involves years-long cycles and small batch production, could greatly benefit from these advancements. By developing a machine learning tool for ship design that learns from the design of many different types of ships, trade-offs in ship design could be identified and optimized. However, the lack of publicly available ship design datasets currently limits the potential for leveraging machine learning in generalized ship design. To address this gap, this paper presents a large dataset of 30,000 ship hulls, each with design and functional performance information, including parameterization, mesh, point-cloud, and image representations, as well as 32 hydrodynamic drag measures under different operating conditions. The dataset is structured to allow human input and is also designed for computational methods. Additionally, the paper introduces a set of 12 ship hulls from publicly available CAD repositories to showcase the proposed parameterization's ability to accurately reconstruct existing hulls. A surrogate model was developed to predict the 32 wave drag coefficients, which was then implemented in a genetic algorithm case study to reduce the total drag of a hull by 60% while maintaining the shape of the hull's cross section and the length of the parallel midbody. Our work provides a comprehensive dataset and application examples for other researchers to use in advancing data-driven ship design.

INTRODUCTION

Recent advancements in machine learning for engineering design have shown the ability to create novel designs [1], and high-performing systems level designs with significantly reduced cycle time [2]. The design of ships can greatly benefit from these advancements in machine learning methods as they have years long design cycles and are produced as one-off designs or in small batches. A well-designed machine learning tool for ship design could learn design trade-offs for ships through the continual design of many different types of ships. This can streamline the ship design process, which currently requires large teams of naval architects to balance all the trade-offs in a single ship's design. The current lack of a publicly available dataset for designing ships impinges this possibility. In order to create a machine learning tool capable of generalized ship design, a dataset of ships is needed that represents the vast array of current existing ships. Literature review was unable to find engineering datasets for machine learning for ship design that encompassed the full spectrum of ship shapes needed to generalize the ship design process. The lack of available public datasets is likely due to prior computational limitations, which have now been solved with time as computation has become cheaper and faster.

This paper presents groundwork for the creation of a dataset of diverse ship hulls to implement machine learning methods for ship hull design. Hull design was chosen as a starting point for the creation of a dataset as this is the traditional starting point in ship design [3]. The hull shape affects several key aspects of a ship's performance, including the buoyancy, upright stability,

---

*Address all correspondence to this author.

Copyright © 2023 by ASME

hydrodynamics, and general arrangements of the ship. In addition, the shape of the hull has a direct impact on over 70% of the cost of a ship [4]. A ship's hull has a significant impact on many aspects of an overall ship system, making it a great candidate to apply machine learning methods to its design to balance overall design trade-offs with a data driven approach.

The following sections detail the literature review of previous work, the methodology for generating a dataset of ships, measures of the dataset, optimization of a ship hull using a trained surrogate model from the dataset, and a discussion on the impact of the work. The dataset of hulls is largely dependent on a parameterization that can represent the broad spectrum of geometric features seen across many traditional hull forms and allows for human and computer inputs to exist together in the same data frame. This parameterization allowed for the creation of a dataset of ship hulls that includes the .stl mesh, images, and hydrodynamic resistance measures of these ship hulls. The key contributions of this paper are:

1. Creation of a novel ship hull parameterization to represent a broad spectrum of hull geometries.
2. Compilation of a set of twelve ship hulls from publicly available CAD repositories to showcase the proposed parameterization's ability accurately reconstruct existing ship hulls, which can be used as a benchmark for future studies in generalized ship hull design representation.
3. A publicly available dataset of ship hulls to implement data driven approaches in ship design. Each hull in the dataset has the parametric representation, meshes, images, and hydrodynamics drag measurements.
4. A case study demonstrating surrogate based optimization using a residual neural network and genetic algorithms for ship hulls.

PREVIOUS WORK

This section reviews previous work that informed the work presented in the remainder of the paper. The first subsection provides a background in generating datasets for engineering problems, providing inference to the size, scope, and contents of good engineering datasets. The second subsection investigates prior work in ship hull design representation, showing that a single design representation for the diversity of ships needs to be created. The third subsection reviews methods for machine learning for hull design, showcasing that current work in the field focuses on hydrodynamic optimization. To enable the current practices of the field with this dataset, the final subsection overviews different methods of predicting the hydrodynamic resistance, or total drag, of ships, showcasing that linear potential flow solvers for predicting wave drag balance accuracy and computational efficiency for use in dataset generation.

Dataset Generation for Engineering Design

The creation of a dataset is paramount for data driven design and surrogate modeling of a design's performance. The two critical components of a dataset for engineering design are a design representation and performance metrics for each sample in the dataset. Publicly available engineering datasets found in a literature search include bicycles [5, 6], linkage systems [7], meta-materials [8, 9], and ships [10]. The subsection on design representation will continue to explore this dataset of ships as well. Other work has shown that multi-modal information can lead to improved accuracy in training a machine learning model [11]. In order to provide the best possible results for future work for machine learning for ship design, multi-modal information on the representation, shape, and performance will be included in this dataset. Sample size among these public datasets range from several hundred [10] to over a hundred million [7]. In order to create the most impact for the design of ship hulls with machine learning techniques, a dataset that can encompass most traditional ship hull designs will need to be construed. The goal of the number of samples in this paper's dataset is to provide broad coverage of the entire feasible domain of ship hulls. An analysis on this is provided in the Results and Discussion Sections.

Design Representation

A dataset of ship hulls will need a design representation that is comprehensive enough to cover the broad spectrum of traditional ship hull forms. A literature review has shown multiple representations for complex designs, including graphs [2, 7], images [9, 12], parameterized vectors [1, 8, 10, 13–21], and free form deformation techniques [22–26]. The most common representation found for ship hulls was vectored parameterization. Many of these human defined parameterizations allow human users to create their own designs with few (<10) inputs. This created a lack of diversity among the geometric features found in the design space of these parameterized representations. Meanwhile, free-form deformation techniques allow the greatest diversity in geometry, but they require an initial design to seed the deformation process, limiting shape diversity. In order to allow for reasonable human input in ship hull design, a parameterization with a broad definition of geometric features will need to be developed to represent a large design space that encompasses most traditional hulls. Dimensionality analysis performed on different parent hulls by Wang et al. and Kahn et al found that 32 and 27 learned parameters can reasonably reconstruct complex surface features on hulls, respectively. As these two sources only analyzed the dimensionality of single hull forms, it is likely that the hull parameterization's dimensionality will require more than 32 dimensions to cover the desired diversity, although it is likely within a similar order of magnitude. The Methods Section will show that the dimensionality of the developed parameterization for a diverse spread of hull forms is 45 parameters.

Copyright © 2023 by ASME

Machine Learning for Engineering Design

The goal of releasing the dataset produced in this paper is to provide it for machine learning researchers train their models for the improvement in data driven hull design. Outside the realm of ship hulls, machine learning methods applied for design purposes include classification [27], reinforcement learning [2, 27], data augmentation [1, 12], and surrogate modeling for optimization [1, 12, 27]. Within applications for ship hull design, machine learning practices have primarily focused on surrogate modeling of the hull's hydrodynamics [10, 15, 16, 22–24, 28]. Additional work in machine learning for ship hulls noted in the previous subsection is dimensionality analysis of geometric features on ship hulls. In order to continue the prominent current direction of research in machine learning for hull design, hydrodynamic measures of hulls will be captured in the dataset to enable surrogate modeling of a hull's hydrodynamics. The next subsection details the different methods of measuring total drag to weigh computational effort versus simulation accuracy.

Hydrodynamic Resistance Prediction

Several methods for measuring the hydrodynamic resistance of ship hulls were considered. This section details these methods.

Traditionally, before the advent of computational tools, the resistance of a ship hull was measured in a towing tank with a scaled model of the hull. Guidance by the International Towing Tank Conference (ITTC) gives the process of scaling the total drag of a model hull into total drag estimate of a full sized ship [29]. Measuring the drag of with a scaled physical model test is the most accurate method of predicting the drag of a full sized hull as the hull is measured in real water as opposed to a simulation's model for water and fluid dynamics. However, this method is too time and cost intensive to produce a dataset for machine learning, since each individual hull would need to be constructed for testing. Towing tank tests, however, are used to benchmark computational predictions of total drag [30–34]. Included in this paper are the parameterized reconstruction of two of these benchmark hulls: The Wigley Hull [31], and the DTMB 5415 Hull [35]. Computational models of ship drag, on the other hand, provide cost effective and accurate measures of drag.

Computational models vary in computational effort and accuracy. Computational fluid dynamics (CFD) solvers are commonly used to create accurate simulations of drag on a ship at the expense of high computation time. CFD solvers have been used in several prior works for design optimization [18, 26] and machine learning for ship design [22]. Other predictions of drag rely on empirical predictions of drag such as Savitsky's method [36] and Hollenbach's method [37, 38]. These computational models are not good for a dataset generation as these models are limited to specific types of ships These models take geometric measures as inputs, but not the 3D model of the hull itself, limiting the scope of applicability for dataset generation. Many papers used empirical regression models in ship design optimization to arrive at principle dimensions of a hull [13, 20, 21, 39–41] and in machine learning applications to improve performance prediction [28].

Another simulation method, linear wave solvers, provide accurate results to drag measurement with reduced computational effort relative to CFD. These solvers use potential flow to simulate the waves produced by a ship in steady forward motion to estimate drag from propagating waves. Different linear wave solvers including Michell's Integral [42, 43], Rankine Panel Methods [44], Neumann-Kelvin Theory (also called Dawson's Method) [45], and Neumann-Michell Theory [30, 32, 34]. These potential flow solvers input the 3D geometry of a hull and provide accurate measures of drag at typical operating speeds of a hull. As seen in the literature review, the balance of computational speed and accurate results make potential flow solvers great candidates for both optimization and machine learning methods for ship hull design [10, 14–16, 19, 23–25] Among the available solvers, the Michell Integral was chosen as the simulation for hulls in this dataset.

METHODS

This section details methods used to define the hull parameterization for the dataset, validate the parameterization's ability to construct a diversity of ship forms, and train the surrogate model for drag prediction.

Hull Parameterization

This section details the hull parameterization and methods for generating aspects of the data set, such as the meshes and images of each hull.

Parameterization Terms The proposed parameterization encompasses broad features seen in traditional ship hull geometries. As mentioned in the prior work section, prior parameterizations used for ship hull analysis characterized ship hulls with traditional measures of ships hulls, such as block coefficient, midship coefficient, and waterplane coefficient. While these traditional characteristics can allow for the rapid generation of some geometric aspects of a hull, they cannot fully represent a diversity of hull forms, nor do they contain enough information to generate a final hull form. The proposed parameterization characterizes geometric features found on traditional hull forms using measures of angles and length ratios, which are applied to a set of algebraic equations to define points on the hull's surface.

The proposed hull parameterization is made up of 45 terms. These terms were construed through analyzing and characterizing the shape and curvature of many different publicly available hull geometries. Some of these hulls were chosen to be a part of the set of target hulls seen in Figure 5. The following breakdown

Copyright © 2023 by ASME

of the parameters also follows the process of first generalizing the shape of the hull, breaking down the ship into sections, and defining specific parameters that encompass the geometric features seen in each section. The first seven terms define the main principal dimensions of the hull. These terms include the length overall, the beam at the main deck, the beam at the stern, and the depth of the hull. The next four terms define the cross section of the parallel midbody of the hull. The cross section terms are the deadrise angle, the chine radius, the keel radius, and the beam of the chine. Twenty terms define the geometry of the bow and stern taper of the hull. These terms characterize the shape of the bow and stern rake, the keelrise, the transition from the taper to the parallel midbody, the drift angle from the bow across the hull's depth, and the cross section of the transom. The final fourteen terms define the geometry of bulbs at the bow and stern, which were inspired by parameterizations of bulbous bows found in literature [17, 18]. These parameters characterize the dimensions, vertical asymmetry, and fillet to transition the bulb into the hull. Overall, these 45 terms are characterized using a human understanding of ship hull geometry and are labeled to allow for human input, in addition to having a vectored structure for computer generated input as well. These 45 terms are intended to characterize a diversity of curvature and shapes seen across large ships to small recreational boat hulls, so that the design of most hulls can be all characterized in the same design representation. A study on the accuracy of reconstructing existing hulls is described later in the Methods Section.

These parameters populate a set of algebraic equations that define the surface of the hull. By characterizing the shape of the hull with a set of equations, the hull can be characterized and measured at any fidelity, which allows for a large range of computational opportunities to characterize and measure the hulls. The section on meshing later in the Methods section details the construction of the surface of a hull.

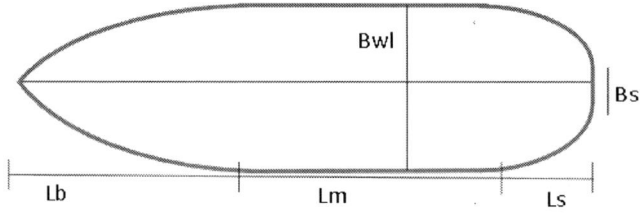

FIGURE 1. Seven terms define the principal dimensions of the hull, including the length, beam, depth, draft, and tapers at the ends of the hull.

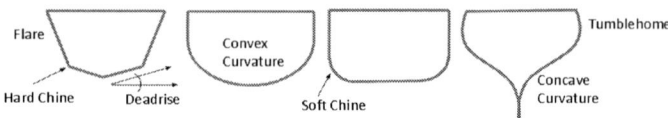

FIGURE 2. Four terms define the cross section of the hull in the parallel midbody. These terms can create cross sections seen on traditional hulls ranging from chines, bilges, flare, tumblehome, and S-chines.

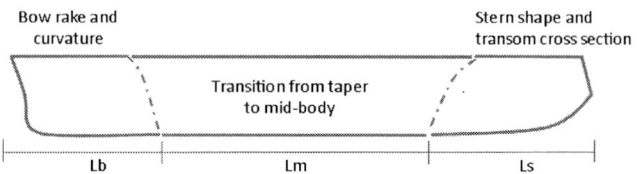

FIGURE 3. Twenty terms define the tapered regions at the bow and stern of the hull. These terms define features such as drift angle, keelrise, transom cross section, rake, and the transition from the taper to the parallel midbody.

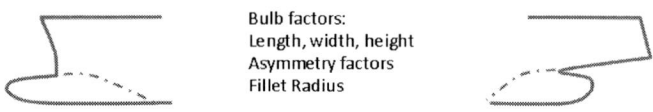

FIGURE 4. Fourteen terms define the bow and stern bulb geometries, including terms that define the size, vertical asymmetry, and the fillet transition of the bulb into the hull.

Constraint Definitions While the parameterization can define a large design space of hull geometries, constraints on the parameterization are needed to ensure that a feasible hull will be produced by a specific set of parameters. To satisfy a "feasible" hull shape, the hull's surface only needs to satisfy two criteria:

1. The hull is watertight, meaning that there are no holes in the surface of the hull.
2. The hull surface is not self-intersecting.

As the hull surface is defined by a set of equations with constants dictated by the parameter values, conditions to determine whether a hull's surface satisfies the two main feasibility criteria can be solved algebraically. The advantage to algebraically solving these conditions is significantly reduced computational effort to algebraically check hull feasibility compared to feasibility checks with mesh generation. After searching through the design space of the hull parameterization and examining the equations that define the hull surface, a set of 49 constraints were defined

Copyright © 2023 by ASME

to determine if a hull surface produced from a specific parameterization satisfies the two feasibility criteria. Mesh generation and feasibility checks are computed in $O(Nlog(N))$, where N is the number of vertices on the mesh. For comparison, on an Intel Core i9-10980XE processor, the construction and check of a hull mesh with approximately 80000 vertices is 1.77 seconds, while the algebraic constraints check feasibility in 0.000199 seconds. This is a 10^4 increase in speed for checking hull feasibility with the algebraic constraints.

Surface Generation and Meshing In conjunction with the parameterization terms, a set of equations was developed to generate the surface of the hull. Terms expressing the cross section define a set of lines that are tangent to circular curves to create the keel and chine. With this, terms for the bow shape, drift angle, and taper endpoints for a given waterline height give four boundary conditions to define a cubic polynomial to define the (X, Y) points along the bow curvature for a given Z position. This is similarly true for the stern taper, although there is extra consideration for the transom cross section. Additionally, terms for the bulbs define ellipsoid surfaces that can be controlled with the parameterization terms. The bubs merge to the hull via fourth order polynomials to fillet the bulb to the remainder of the hull.

With the set of equations, any point cloud with custom spacing of X, Y, and Z can define the hull's surface. The meshes of the hulls provided in the data set were constructed from these point clouds of the hulls with even spacing between the X and Z coordinates. Further provided with the dataset are five images of each hull mesh:

1. Front View
2. Profile View
3. Plan View
4. Three-Quarter Starboard Bow View
5. Three-Quarter Port Stern View

Dataset Generation

The following section details the generation of the hull parameterizations in the dataset. The hulls generated in the dataset were randomly generated and made up of three distinct subsets of hulls. Each term in the parameterization was sampled uniformly from its range of possible values. Many of the terms in the parameterization are relational and have limits between 0 and 1. Other terms rely on user defined inputs to ensure that the generated designs are similar to realistic ship hulls. For example, the term related to the beam-length ratio of the hull was limited to be between 0.0833 and 0.333 to ensure that the beam-length ratio of the dataset hulls encompasses that of typical ship hulls. This is similarly true for the term related to the depth-length ratio, which was limited to be between 0.05 and 0.25. Additionally,

the term related to the deadrise angle of the cross section was limited to be between $0°$ and $45°$. After generating a random parameterization, the forty nine algebraic constraints were checked for each given random parameterization. If the randomly generated parameterization led to a feasible hull shape, then it was added to the dataset. This process was repeated until each subset contained ten thousand hulls, for a total of thirty thousand hulls in the dataset. The purpose of randomly generating the parameterized hulls was to create a dataset that fully encompasses the possible design space of ship hulls that meet the feasibility criteria so that a machine learning model can learn the relative performance of a hull's geometric features in isolation and in combination. The training of a machine learning model to predict the drag coefficient of hulls is detailed in a later section of the paper.

The dataset is comprised of three subsets of ten thousand hulls. The first subset of the dataset contains hull forms generated from the full possible range of each term in the parameterization. These hulls contain all the possible combinations of all the geometric features defined by the hull parameterization. The second subset of hulls is constrained so that they do not contain bulbs. By allowing the full range of all the terms except those that define bulbs, geometric features that are typically seen in smaller hull forms are more prominent in this subset. Smaller hull forms include hulls that are typically less than fifty meters in length, such as tugboats, fishing trawlers, ferries, yachts, and recreational watercraft. Such geometric features include deadrise, concave cross sections, and chines. Meanwhile, the third subset contains hulls that are biased towards features seen in larger hulls. These hulls have a keel radius that is strictly positive and have a zero degree deadrise angle. By eliminating these features and allowing for the presence or absence of bulbs, this subset contains hulls with geometric features that are more prominent in larger hull forms. Larger hull forms include those seen on warships, cargo ships, cruise ships, and research vessels. The slight biases introduced within the latter two subsets ensures that there exist samples with features akin to realistic hulls. This will give surrogate models trained on the whole dataset more information to accurately predict the performance of realistic hulls. This will ensure surrogate models trained on this dataset can yield reasonable predictions for any hull and accurate predictions for realistic hulls.

Chamfer Distance Comparison of Hulls

The proposed parameterization is intended to represent a large diversity of hull forms. However, how does one validate if a proposed parameterization is sufficiently expressive? One solution is to collect a diverse set of real-world hull forms and check if the proposed parameterization is capable of recreating those hulls. This strategy is used to validate the usefulness of the proposed parameterization. The proposed method gathers a

Copyright © 2023 by ASME

small set of realistic hulls and generates their surfaces as point clouds. Then, custom parameterized hulls are also constructed to match the point cloud of the target hulls with point clouds generated by the hull parameterization. The set of twelve target hulls listed in Table 1.

The first ten hulls were gathered from GrabCAD, an online repository of 3D CAD models. The final two, the Wigley Hull and the DTMB 5415 Hull, are hulls commonly used as benchmarks in hydrodynamics computation and tow tank testing [30–33, 35, 43]. These hulls represent a large diversity of ship hull shapes and scales, including recreational watercraft, commercial ships, and warships. Parameterized reconstruction of these twelve hulls is detailed in the Results and Discussion Sections.

The metric used to evaluate the match between two point clouds is the bidirectional mean of squared Chamfer distances. For two point clouds, A and B, the Chamfer distance finds the distances from each point in A to its nearest neighbor in B. The distance metric used is the squared Euclidean distance between two points. Bidirectional Chamfer distance is calculated for all points in A to B and for all points in B to A. The sum of all the square distances is then averaged to form the bidirectional mean of squared Chamfer distances. The formula for this evaluation metric is shown below:

$$CD = \frac{1}{N_A + N_B}\left(\sum_{n=1}^{N_A} ||A_n - B_{n*}||^2 + \sum_{n=1}^{N_B} ||B_n - A_{n*}||^2\right) \quad (1)$$

where B_{n*} is the nearest neighbor of the point A_n in B and A_{n*} is the nearest neighbor to the point B_n in A. N_A and N_B are the total number of points in A and B, respectively. While the parameterization is designed to be manipulated by a human designer, it is difficult to manipulate the parameterization by hand to match another hull. In order to reconstruct the set of target hulls as parameterized hulls, a genetic algorithm was written to minimize the bidirectional mean of the square of Chamfer distances. The reconstructed target hulls generated from the optimization are shown in the Results Section.

Hull Resistance Calculation

As noted in the Background Section, the method to simulate the total drag of each hull in the dataset will be a linear wave solver to simulate the wave drag and with the ITTC regression line to predict viscous drag on the hull. The following subsections detail the method of calculating the wave drag, validating the simulation, and training a surrogate model to predict the wave drag coefficient of a hull from its parameterization.

Michell Integral The Michell Integral was chosen to simulate wave drag over other linear wave methods for its relative computational efficiency for the accuracy it provides. The Michell integral is a linear estimate of wave drag of a slender ship in forward motion. The model performs a Fourier series analysis on the waves that propagate from the hull and thus, this model is not time dependent, leading to its computational efficiency. The Michell Integral is defined by the following equation [42, 43]:

$$R_w = \frac{A\rho g^2}{\pi U^2} \int_1^\infty (I^2 + J^2) * \frac{\lambda^2}{\sqrt{\lambda^2 - 1}} d\lambda \quad (2)$$

where ρ is the density of water, g is gravity, U is the ship speed, and $A, I, and J$ are integrated terms relating to the surface normal across the hull and the direction of wave propagation. Further insight into these terms is in Michell's paper form 1898 [42].

Using the Michell Integral, thirty two wave drag coefficients were calculated for each hull across four different draft and speed operating conditions. The four drafts were 25%, 33%, 50%, and 67% of the hull's total depth. The eight speed conditions were normalized to Froude numbers between 0.15 through 0.45 in steps of 0.05. These Froude numbers correspond to typical operating conditions of traditional displacement hulls [29, 33]. The Froude number is the relative scaling between inertial and gravitational forces described in the equation below:

$$F_n = \frac{U}{\sqrt{gL}} \quad (3)$$

Where U is the hull speed, g is gravity and L is a length scale. The length used in simulating the 32 speed-draft conditions of the hulls was the length of the waterline at the tested draft mark. This way, thirty two unique conditions were measured. As the wave drag is a function of the hull geometry and the interference a propagating wave makes with the hull, a full spectrum of speed and draft marks were calculated for the dataset. For the purposes of applying machine learning to this dataset, including a full spectrum of speed-drag conditions in the dataset allows a machine learning model to predict the drag at multiple operating conditions as opposed to only one operating condition. Providing all this information allows the model to learn the effects of drag due to changing submerged geometry with draft and speed. In addition to scaling the relative speed and draft conditions for the hulls, the wave drag is also scaled using the following equation:

$$C_w = \frac{R_w}{\frac{1}{2}\rho U^2 LOA^2} \quad (4)$$

Typical drag coefficients of hulls are scaled by the wetted surface area of the hull. Within the dataset, however, the wetted surface

Copyright © 2023 by ASME

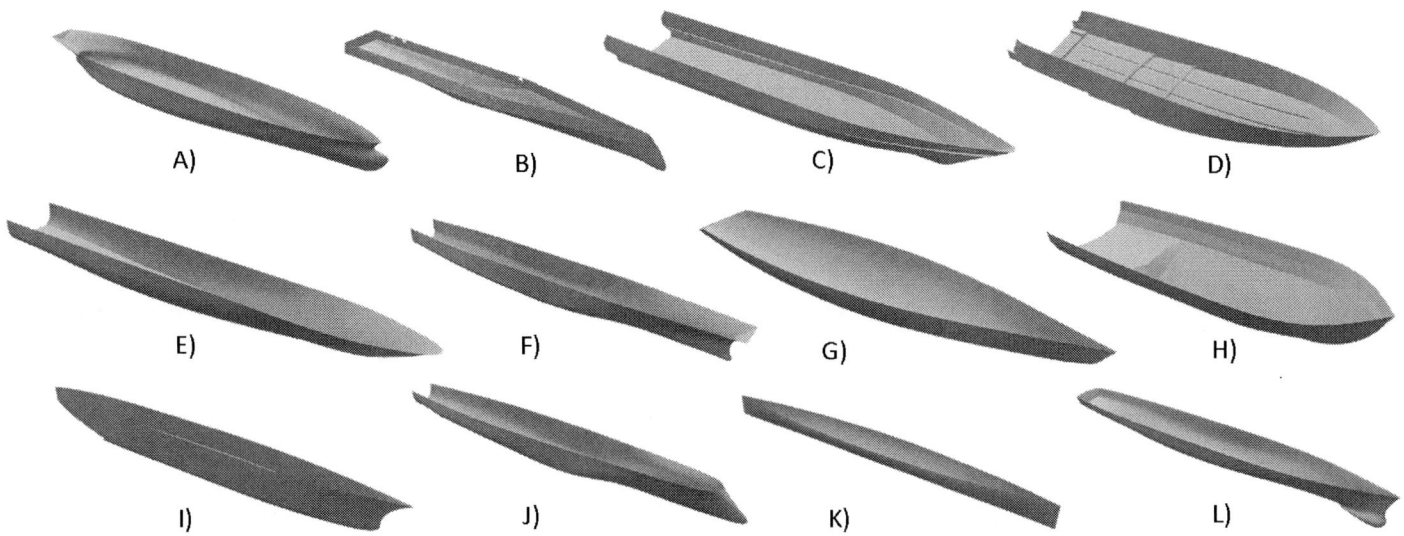

FIGURE 5. The set of twelve target hulls used for validating the capability of the hull parameterization to reconstruct a diverse set of realistic hull forms.

area of the hulls can vary greatly. Instead, the Length-Overall (LOA) is used instead as this is the first term in the parameterization. For the purposes of machine learning with the dataset, the wave drag coefficient can be characterized by the remaining 44 terms in the parameterization and the hull's relative speed and draft. With the thirty two wave drag coefficients, any speed-draft condition within the range of the dataset conditions can be interpolated. The calculation of the thirty two wave drag coefficients for the thirty thousand hulls in the dataset was performed in parallel on an Intel Core i9-10980XE processor. The average computation time for an individual hull was 72 seconds for the thirty two wave drag coefficients.

Total Resistance   For displacement hulls, the principle characteristics of drag are defined as the sum of residual drag and skin friction drag. Skin friction is viscous drag due to boundary layer effects of water across the surface of the hull. In traditional naval architecture, the skin friction drag is approximated from a series of regression tests performed by the ITTC [29]. The skin friction coefficient regression is:

$$C_f = \frac{0.075}{(Log_{10}(Re) - 2)^2} \quad (5)$$

where Re is the Reynolds number of the hull, scaled with its forward velocity, and length. As the skin friction coefficient scales with the wetted surface area of the hull, the total skin friction

drag scales with:

$$R_f = \frac{1}{2} C_f \rho U^2 A_{ws} \quad (6)$$

where R_f is the skin friction resistance and A_{ws} is the wetted surface are of the hull. The other component of ship drag, residual drag, is the sum of viscous pressure drag and wave drag. As ships' hulls are considered slender, the contributions of viscous pressure drag are negligible relative to the scale of wave drag. With this consideration, the total Resistance, R_t is the sum of wave drag and skin friction drag:

$$R_t = R_w + R_f \quad (7)$$

The following subsection details the validation of this assumption with the DTMB 5415 hull form.

Wave Drag Validation   Two validation checks were conducted to ensure that the numerical Michell Integral simulation used in the dataset of wave drag coefficients is reasonable. The first test checked the accuracy of the wave drag numerical prediction to the analytical evaluation of wave drag using the Michell Integral. This check was performed using the Wigley hull, a hull with parabolic curvature. Figure 6 plots the numerical wave drag versus several hull speeds calculated by integrating over 301 discrete points along the length of the hull for 51 waterlines along the displaced volume. Also included in this graph is the analytic

Copyright © 2023 by ASME

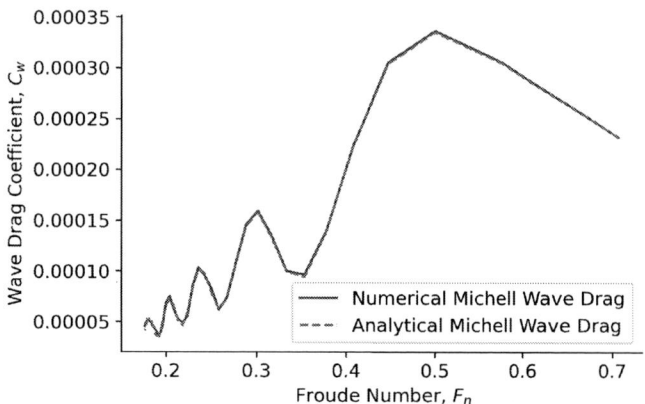

FIGURE 6. The numerical calculation of the Michell integral measured using 301 discrete lengthwise points over 51 discrete waterlines is well resolved to the analytic solution to the Michell integral of the Wigley hull.

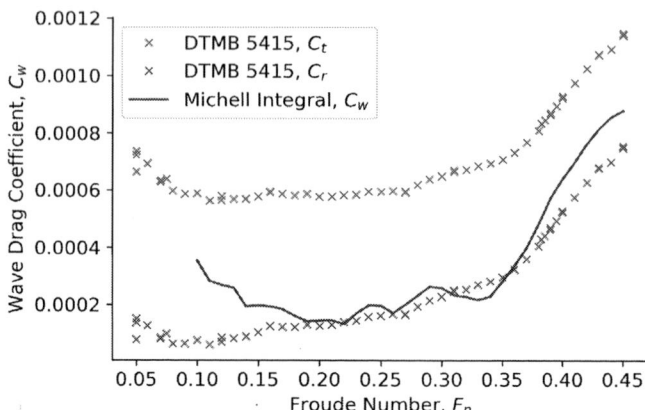

FIGURE 7. The numerical calculation of wave drag coefficient using the Michell integral is well resolved to towing tank measures of the residual resistance coefficient of the DTMB 5414 hull at speeds between $F_n = 0.15$ AND $F_n = 0.45$.

solution to the Wigley hull at the same speed conditions [31]. The graph shows that the numerical solution is well resolved to the analytic solution to the Michell Integral. For this reason, the wave drag evaluation for the hull dataset was also computed over a grid of 301 waterline points and 51 waterlines. The second validation performed was to compare the wave drag calculated by the Michell Integral at several speeds to towing tank testing results of a real hull form. Figure 7 shows that the residual resistance coefficient of the DTMB 5415 hull [35] is well resolved to the numerical calculation of the wave drag coefficients using the Michell Integral.

Surrogate Model for Resistance Prediction   A major benefit of creating a dataset is that a surrogate model can be trained to predict the drag of a hull with increased computational speed for a small loss in accuracy. As opposed to constructing a 3D hull from the parameterization and simulating the wave drag, a surrogate model predicts the wave drag coefficients directly from the parameterization vector. The increased speed in drag prediction enables design optimization of a hull on a time scale many orders of magnitude faster than directly simulating the hull. Since the dataset contains hull parameterizations that fall within the full range of feasible hull forms, any feasible hull can have its wave drag coefficients predicted by the surrogate model from its parameterization. The surrogate model used in a regression to predict wave drag coefficients from a hull's parameterization was a residual neural network, chosen for both its speed and the ability to fully differentiate the model. This means that the derivative of the wave drag coefficient can be taken against any of the terms in the parameterization.

In the training of the surrogate model, two considerations

relating to the distribution of the data were implemented in the training. In the Results Section, Figure 11 shows that the distribution of wave drag coefficients spans several orders of magnitude. Instead of predicting the wave drag coefficient, the surrogate model will predict the $Log_{10}(C_w)$ for the thirty two speed and draft conditions to normalize the final prediction layer. The second consideration implemented in training was to up-sample the instances of hulls that had a wave drag coefficient less than one standard deviation below the mean of samples in the dataset by a factor of four. This up-sampling is intended to increase the prediction accuracy of wave drag prediction for low drag hulls during hull optimization using the trained surrogate model.

After experimenting with different neural network structures, a residual neural network with four hidden layers and 256 nodes in each layer was found to have the greatest prediction accuracy, with an R^2 value equal to 0.969. This network structure is seen in Figure 8. Immediately prior to the final prediction of the wave drag coefficients, the values in the first hidden layer are summed with the values in the final hidden layer. This assists in boosting the gradients across the network during training to improve the accuracy of the network as a regression model. This residual network (ResNet) predicts the common logarithm of the thirty two wave drag estimates for a given hull in 0.15 seconds, a 480x speed up in for the prediction and is fully differentiable. The ResNet surrogate model is used in a later section of the paper as a tool in hull optimization to minimize drag.

RESULTS

This section details the results of the dataset generation and evaluation. This first subsection provides results for optimizing

Copyright © 2023 by ASME

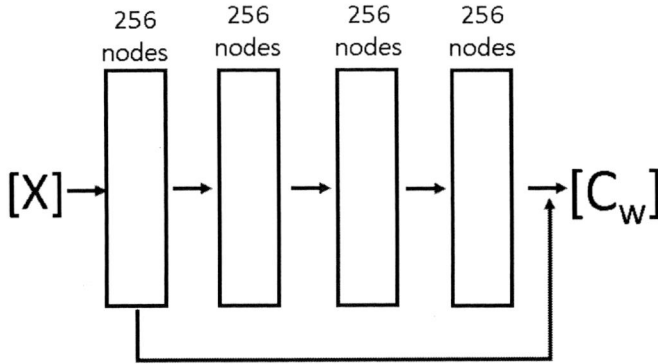

FIGURE 8. The residual neural network trained to predict 32 wave drag coefficients contains 4 hidden layers with 256 nodes in each layer. The input to the ResNet is the hull parameterization and the output is the 32 wave drag coefficients.

Label	Target Hull Name	RMS of CD
A)	Container Ship [1]	0.343%
B)	*USS Zumwalt* [2]	0.252%
C)	Fast Ferry [3]	0.277%
D)	Recreational Fishing Boat [4]	0.505%
E)	*USS Freedom* [5]	0.402%
F)	*USS Nimitz* [6]	0.342%
G)	Sailing Yacht [7]	0.397%
H)	Tug Boat [8]	0.559%
I)	*USS Indianapolis* [9]	0.308%
J)	X-Bow Ship [10]	0.390%
K)	Wigley Hull [31]	0.0802%
L)	DTMB 5415 Hull [11]	0.401%

TABLE 1. List of twelve hulls used for reconstruction validation and their corresponding normalized root-mean-square of Chamfer distances. These values are all less than 0.51% of the hull length, measuring accurate reconstruction. Original hulls are shown in Figure 5. Reconstructed hulls are shown in Figure 9

the parameterization terms to match a set of target hulls. The second subsection provides some statistics and analysis on the dataset of thirty thousand hulls. Finally, the third subsection provides the results of the optimization of the hull parameterization to minimize drag on a hull.

Hull Matching Validation

This subsection provides the results of the parameterized hull matching to the set of twelve target hulls. Figure 9 shows twelve parameterized hulls that were constructed by minimizing the bidirectional mean of the squared Chamfer distance between each parameterized hull and its corresponding target hull from Figure 5. Table1 lists each of these reconstructed hulls and the normalized square root of the bidirectional mean squared Chamfer distance (RMS of CD). The results of each of these reconstructed hulls vary greatly in scale, so the RMS of CD is listed as a percentage of the LOA of each reconstructed hull.

Dataset of Parameterized Hulls

After generating thirty thousand hulls and computing the thirty two wave drag coefficients, some analysis was performed

[1] https://grabcad.com/library/general-cargo-ship-1
[2] https://grabcad.com/library/ddg-1000
[3] https://grabcad.com/library/fast-passenger-monohull-ferry-mh35-1
[4] https://grabcad.com/library/10-meters-fishing-boat-1
[5] https://grabcad.com/library/littoral-combat-ship-1
[6] https://grabcad.com/library/hull-of-carrier-of-nimitz-class-1
[7] https://grabcad.com/library/36-meter-sailing-yacht-1
[8] https://grabcad.com/library/renko-dangar-marine-steel-boat-project-1
[9] https://grabcad.com/library/uss-indianapolis-ca-35-1
[10] https://grabcad.com/library/x-bow-hull-1
[11] http://www.simman2008.dk/5415/5415_geometry.htm

on the dataset to better understand the parameterization distribution across the feasible region in the design space and extreme minimum values associated with measurements of drag. One measure of the distribution of the spread of hull samples is the average Euclidean distance each hull parameterization vector is to its nearest neighbor. Figure 10 shows this measure with increasing sample size in the dataset. Further analysis of the spread of samples in the dataset is provided in the Discussion Section.

Another measure of the dataset of hulls is to measure the spread of the wave drag coefficients of the samples. Figure 11 shows the distribution of the wave drag coefficients across the three subsets in the dataset. It is important to note that the scale of the Y-axis of this chart is on a logarithmic scale as the distribution of wave drag coefficients spans multiple orders of magnitude. Please see the Discussion Section for an analysis of this result.

Minimum values of different measures of drag in the dataset were also collected. One measure of the dataset of hulls was to find the hulls with the minimum total drag when scaled to different volumetric displacements. Table 2 showcases that two different hulls standout among the dataset as having the lowest total drag. These two hulls can be seen in Figure 13. Hull 1-3715 has the lowest total drag for displacements ranging from 5000 cubic meters to 100000 cubic meters, while Hull 1-1340 has the lowest drag for displacements of 500 cubic meters and 1000 cubic meters. As scale decreases, viscous forces increase suggesting

Copyright © 2023 by ASME

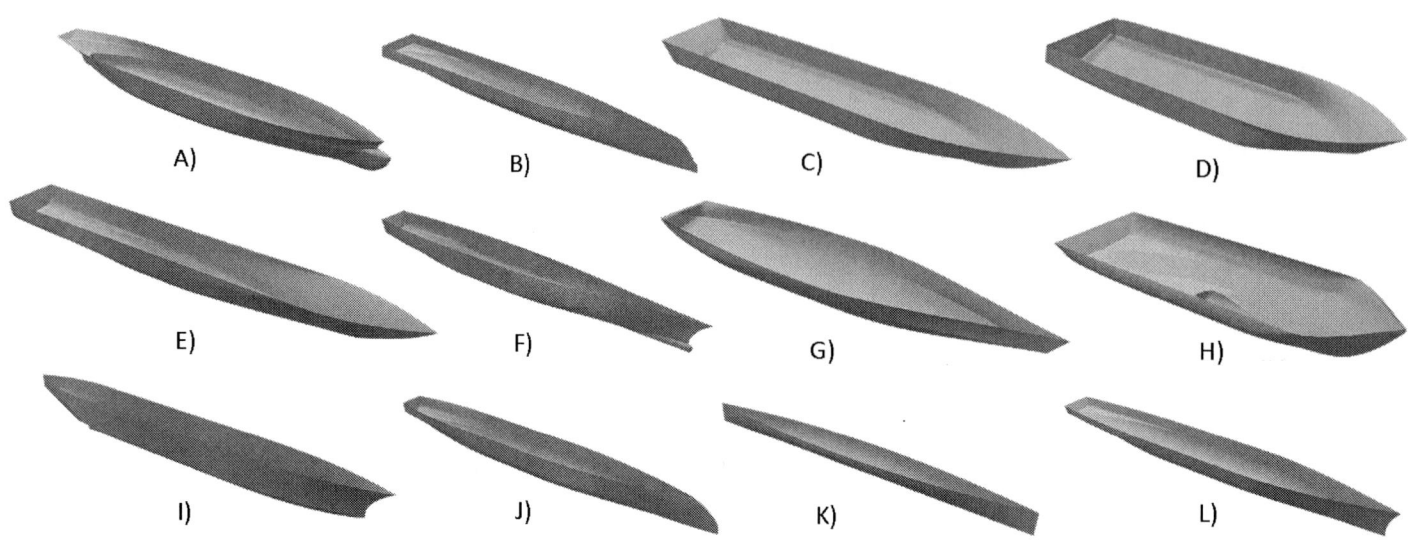

FIGURE 9. The set of twelve parameterized hulls visualize accurate reconstruction of the twelve target hulls.

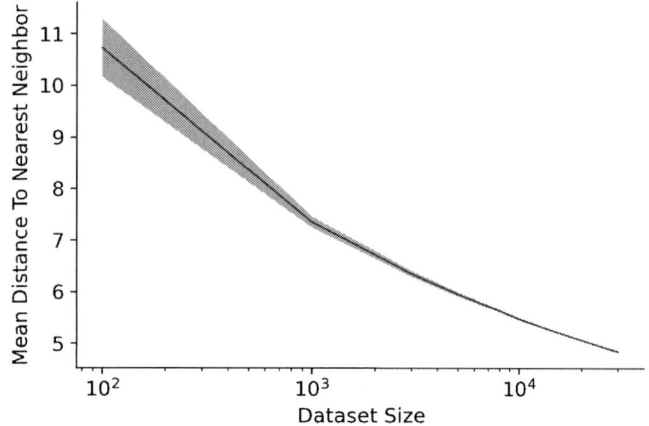

FIGURE 10. Mean Euclidean distance to nearest neighbor with increasing sample size. The shaded region shows 2 standard deviations from the mean collapses with increasing sample size.

DISP. VOLUME	R_T	HULL	LOA
$500m^3$	$67.04kN$	Set 1, #1340	$109.38m$
$1000m^3$	$96.97kN$	Set 1, #1340	$137.81m$
$5000m^3$	$284.0kN$	Set 1, #3715	$225.09m$
$10000m^3$	$338.2kN$	Set 1, #3715	$283.59m$
$100000m^3$	$1046kN$	Set 1, #3715	$611.0m$

TABLE 2. Tabulated data of the dataset hulls with minimum total drag when scaled to different volumetric displacements. Skin friction and wave drag scale differently, leading to different minimal drag hulls at different scales.

why different hulls at different scaled displacements have minimum drag. In addition to looking at total drag, hulls from the dataset with the lowest coefficients were also found. The first case was the hull with the lowest drag at a speed-draft condition of ($F_N = 0.3, T/Dd = 0.5$). This hull had a wave drag coefficient equal to $1.53 * 10^{-5}$. The second measure of minimum wave drag was to aggregate the wave drag coefficients across all tested speeds at a $T/Dd = 0.5$ condition. Both hulls are shown in Figure 12. Analysis of the dataset as it pertains to drag is provided in the Discussion Section.

Hull Form Optimization Via ResNet Surrogate for Wave Drag Coefficient

Optimization of the hull parameters saw significant reductions in total drag for three test cases. The first optimization of the parameters was constrained so that the hull had a volumetric displacement of 100000 cubic meters and a speed of 25 knots. The optimized hull had a total drag of 785kN, which is a 25% Reduction in total drag compared to the hull in the dataset with the lowest drag. Results were computed with NSGA2 to minimize both R_t and the interpolated C_w for the speed/draft condition. The hulls in the final population were then constructed and total drag was measured with the Michell Integral for the calculation of wave drag. The hull shown in Figure 14 shows this optimized hull, which had a total length of 685.24 meters. It is important to note that this hull is not the hull with the minimum

Copyright © 2023 by ASME

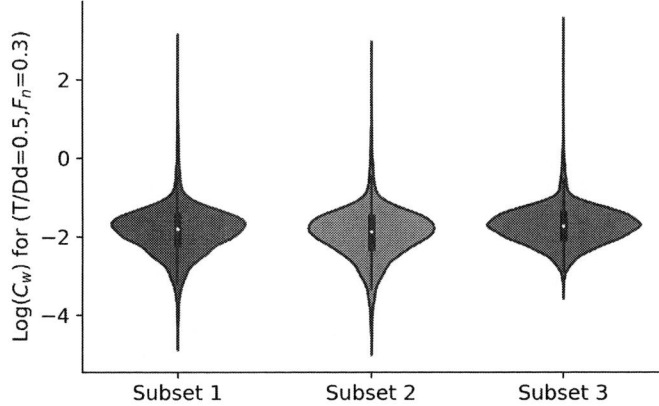

FIGURE 11. Violin plots of wave drag coefficients of dataset hulls for a draft of 1/2 the depth of the hull and a Froude number = 0.3 (($F_N = 0.3, T/Dd = 0.5$)). Results are separated into the dataset subsets

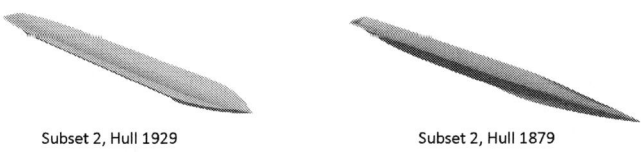

Subset 2, Hull 1929   Subset 2, Hull 1879

FIGURE 12. Hulls from the dataset with the minimum wave drag coefficients. The left hull has the lowest wave drag coefficient at the ($F_N = 0.3, T/Dd = 0.5$) condtions. The right hull has the lowest aggregated wave drag coefficients across all speed conditions at $T/Dd = 0.5$.

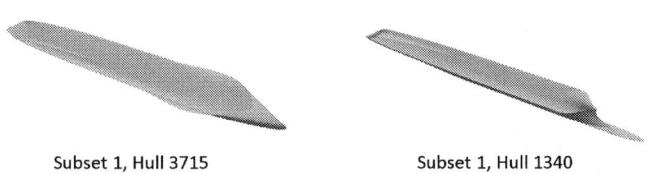

Subset 1, Hull 3715   Subset 1, Hull 1340

FIGURE 13. Hulls from the dataset with the lowest total drag for $U = 25 knots$ when scaled to different volumetric displacements. The left hull has the lowest total drag when scaled to displacement volumes of 5000, 10000, and 100000 cubic meters. The right hull has the lowest total drag when scaled to displacement volumes of 500 and 1000 cubic meters.

drag predicted by the ResNet, but it did belong to the final population of optimized hulls.

The second and third optimizations were modeled after problems found in the literature. The goal was to optimize a hull while maintaining the aspect of the geometry of an initial hull form. This initial hull the container ship hull from Figure 9

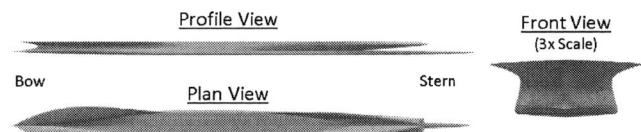

FIGURE 14. Optimized hull that displaces 100000 cubic meters has 25% reduction in drag compared to the hull in the dataset with the minimum drag in the same conditions

was re-proportioned to have an LOA of 200 meters and a bow and stern taper over the forward-most and aft-most 30% of the hull. The total drag on this hull with an operating speed of 25 knots and a draft of 12.5 meters is $2.694 * 10^7 N$. With the same optimization process described above, two optimizations were performed. The first optimization only manipulated the parameters related to the bulb geometries. This optimization reduced the total drag on the hull by 54.3% in the same speed and draft condition. A second optimization manipulated the parameters associated with the bow, stern, and bulbs while maintaining the body of the midship. This optimization reduced the total drag of the hull by 60.7% for the same speed and draft condition. Figure 15 shows the original container ship hull, the same hull with optimized bulbs, and the optimized hull with the same parallel midbody as the initial hull. Further analysis of the hull optimization with the ResNet is provided in the Discussion Section.

DISCUSSION

This section analyzes the findings from the Results Section. The first subsection provides insight into the reconstruction of target hulls as parameterized hulls. The second subsection analyzes the dataset of parameterized hulls and the distribution of information provided in the set. The third subsection provides an analysis on the training of the ResNet to predict the wave drag coefficients of a hull at multiple operating conditions. In conjunction with the third subsection, the fourth subsection analyzes the ResNet's ability as a regression model to be used with an optimization algorithm to minimize drag.

Parameterized Reconstruction of Target Hulls

Based on the visual and computational results of this study, the parameterization scheme proposed in this paper can accurately reconstruct a large variety of classical hull forms. The creation of this parameterization was specifically aimed at closing two specific gaps found in the literature:

1. A comprehensive design representation of ship hulls that encompasses a large diversity of geometric features seen on hulls

Copyright © 2023 by ASME

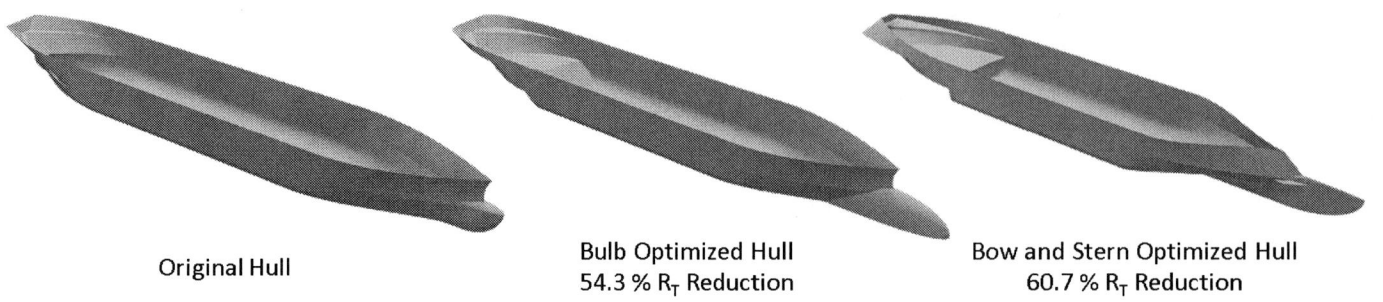

Original Hull

Bulb Optimized Hull
54.3 % R_T Reduction

Bow and Stern Optimized Hull
60.7 % R_T Reduction

FIGURE 15. Optimization of the modified container ship created constrained to only optimizing the bulbs on the hull lead to a 54.3% reduction in total drag. Optimization parameters relating to the shape of the bow taper, stern taper, and bulbs lead to a 60.7% reduction in total drag.

2. A design representation for ship hulls that allows for human manipulation and computational design methods

The parametric scheme was designed to allow human input in terms of concrete geometric features seen in ship hulls. Additionally, this scheme allows for computer generated inputs so that a ship hull can be computationally designed in the same data frame as the human defined design. Due to this, future machine learning models leveraging this parametric scheme for ship hull design can learn from both human and computer generated inputs. These results will serve as a model for future benchmark studies for comprehensive parametric reconstruction of ship hulls.

The parametric scheme is certainly not a panacea to parametric ship hull design, as there are a few features seen in the target hulls that this parameterization is able to reasonably reconstruct. Most notably, the target hull for a fishing boat (image D in Figures 5 and 9) has a stepped hull and longitudinal strakes for assisting in the hull's hydrodynamics. The surface of the reconstructed parameterized hull of the fishing boat is smooth as the parameterization cannot create these features. Another example, the target hull for the *USS Nimitz*(image F in Figures 5 and 9) has a flared, non-pointed, bow to support structure for the aircraft runway on the ship's deck. This geometric feature is impossible with the proposed parameterization. The optimized reconstructed parameterized hull of the *USS Nimitz* minimized the Chamfer distance over a majority of the hull surface but had a pointed bow form as defined with the parameterization. A third notable feature is that the parameterization had difficulty reconstructing the bulb in the DTMB 5415 (image L in Figures 5 and 9) target hull has the bulb protruded below the baseline of the hull. This positioning of the bulb is not possible with the parameterization. It is important to note that the reconstructed parameterized DTMB hull does not contain a bulb at all.

Overall, this parameterization can comprehensively design diverse hulls for human and machine learning design processes. The 45 parameters defined in this parameterization provide suffi-

cient information to reconstruct a large array of hull geometries, yet does not over-complicate the design space in a way that inhibits its usefulness in a computational model due to the curse of dimensionality. This parameterization is shown to design hull geometries seen across the full spectrum of traditional hull designs and scales, from small recreational watercraft to large naval hulls. One note to add is that the geometry of a hull is defined by 44 parameters and the final parameter is the hull's length overall. This way, in theory, the geometry of the recreational fishing boat (Hull D) could exist as any length by only changing the length overall parameter. Additionally, the parameterization allows for future machine learning work performed on the dataset to focus on tools and methods for performance prediction and design generation without the need to create models for representing the design of a hull given only its mesh. Future work in the hull representation will look at reconstructing minute features that are specific to certain types of ships.

Data Set Generation

The dataset of thirty thousand ship hulls was generated to cover the full range of feasible samples for the hull parameterization. In order to ensure that the total feasible design space of ship hulls is encompassed in the dataset, the Euclidean distance to the nearest neighbor of each parameterized hull was measured with increasing dataset size. Figure 10 shows that as the number of samples approaches thirty thousand the slope of this curve and the standard deviation significantly decays, suggesting that the entire design space of hulls is well sampled with thirty thousand samples. More samples will certainly yield better results; however the current sampling achieved a point of diminishing returns, where any further reduction in the mean distance will require a significant increase in the number of samples.

While the dataset covers the design space of feasible hulls well, this produced a large range of wave drag coefficients in the performance space, spanning several orders of magnitude.

Copyright © 2023 by ASME

Figure 11 shows the distribution of wave drag coefficients across the three subsets of hulls in the dataset, showing that they all have similar means. It is important to note that the distribution of wave drag coefficients for subset 3 is narrower and skewed higher than that of subsets 1 and 2. From a design perspective, this is a reasonable outcome. The hulls in subset 3 are intended to represent large hulls, especially for shipping. The design of larger hulls is dependent on both the hydrodynamics the ability to carry large amounts of cargo. Due to this required balance, it makes sense that a set of designs skewed to represent cargo ships would have higher drag coefficients as there are additional design criteria for these hull forms. In addition, subset 2 was skewed to represent hull forms seen on smaller ships. So, it also makes sense that these hulls might have lower drag coefficients as the design of smaller craft is dominated by speed, whether this is for competition, leisure, or other purposes.

The design space coverage and large distribution of wave drag coefficients led to the successful training of the ResNet to learn how the geometric features of a hull defined by the parameterization can accurately predict the drag of a hull. The next subsection discusses the implications of the surrogate model training for design evaluation and optimization.

Hull Form Optimization for Drag via ResNet Using the ResNet to quickly predict drag within an optimization of a hull led to the generation of hulls with significantly reduced drag. Minimizing total drag with few constraints produces a hull with a 25% reduction in drag compared to the hull in the dataset with the minimal drag with a displaced volume of 100000 cubic meters. Unfortunately, the optimized hull is too long to be a real ship. For reference, one of the target hulls provided in the paper, the *USS Nimitz*, has a displacement of approximately 1000000 cubic meters, but is only 332 meters in length. Due to constraints on the global infrastructure of ports, very few ships exceed 350 meters in length. This optimized hull is more than double the length of the *USS Nimitz*, yet displaces approximately the same volume, suggesting that there are harder constraints on the design of real ships than only drag. This same argument was detailed in the analysis of the distributions of data across the three subsets in the previous section. An additional consideration in the design of a ship with the optimized hull is the increased structural need of supporting bending moments on this hull, further limiting its mission capabilities. Neglecting that the optimization produced a practically infeasible hull, the optimization with the ResNet had two important findings:

1. The ResNet was able to learn how geometric features of hulls affect drag
2. Hull optimization with the ResNet was able to find geometric features in combination and was able to reduce drag more than any of the hulls existing in the training data.

Additionally, constrained optimization of an initial hull form showed that wave drag coefficient predictions with the ResNet were able to consider the individual geometric features of the hull parameterization and optimize parameters to produce ship hulls with significantly reduced drag compared to the initial hull form. This is true for optimizing local features such as bulbs, or global features such as the entire bow and stern of the hull. It is important to note that the optimization of a bulbous bow is a delicate balancing act. The bulb creates destructive interference in a hull's wave, reducing the wave drag; however, this is at the cost of increasing skin friction. The optimization with the ResNet balanced this trade-off. A reduction in 5-10% of total drag is no small feat for a human hull designer to accomplish. Computational tools reducing the drag of a hull upwards of 60% will certainly produce greatly improved outcomes in the cost of shipping and fossil fuel emissions. The current limit in this potential is in creating ship hulls that are practically feasible for real-world use.

CONCLUSION

This paper describes the creation of a ship hull dataset for computational and data-driven design. The dataset was generated using a novel parameterization method that comprehensively covers the vast design space of ship hulls, including traditional geometries. This parameterization method allows for both human and computer-generated designs to exist within the same data frame. It accurately reconstructs 12 distinct ship hulls with a diversity of geometric features, with a normalized root-mean-square of Chamfer distance of less than 0.51% of the target hull's length. The resulting dataset contains 30,000 ship hulls covering the full design space of feasible hull geometries, with some bias towards realistic hull features. This dataset is over 42 times larger than any other publicly available ship hull dataset and characterizes more geometric features. A surrogate model based on ResNet architecture is trained on this dataset that accurately predicts the wave drag coefficients for 32 speed/draft conditions, with an R-squared value of 0.969. A case study of surrogate-based optimization to minimize total drag of a hull in constrained conditions demonstrates the model's ability to predict the influence of individual geometric features and the influence of high-quality geometric features in combination, resulting in reductions of up to 60% in total hull drag.

Future work will involve characterizing the performance of ship hulls using additional performance metrics, including geometric, hydrostatic, and hydrodynamic measures. In addition, a future study into surrogate modeling with multi fidelity simulation will explore surrogate prediction accuracy and computational effort to produce datasets for early stage data driven design. Another consideration for future work is that many hulls in the current dataset have high drag. This indicates that the random sampling of parameters may not be an effective approach

Copyright © 2023 by ASME

for generating high-performing hulls. Future work will aim to generate a larger dataset that considers hull performance in addition to geometric feasibility, enabling the training of surrogate models with greater accuracy in the regions of the total design space containing high-performing hulls similar to existing real hull forms.

ACKNOWLEDGEMENTS

We thank the National Defense Science and Engineering Graduate Fellowship program for supporting Noah Bagazinski's studies while working on this project. Additionally, we thank MIT Supercloud [46] for providing the computational resources needed for the work performed in this paper. The dataset, code, and documentation for this work is at `https://github.com/noahbagz/ShipD`.

REFERENCES

[1] Chen, W., and Ahmed, F., 2021. "Padgan: Learning to generate high-quality novel designs". *Journal of Mechanical Design, 143*(3).

[2] Mirhoseini, A., Goldie, A., Yazgan, M., Jiang, J. W., Songhori, E., Wang, S., Lee, Y.-J., Johnson, E., Pathak, O., Nazi, A., et al., 2021. "A graph placement methodology for fast chip design". *Nature, 594*(7862), pp. 207–212.

[3] Evans, J. H., 1959. "Basic design concepts". *Journal of the American Society for Naval Engineers, 71*(4), pp. 671–678.

[4] Lin, C.-K., and Shaw, H.-J., 2017. "Feature-based estimation of preliminary costs in shipbuilding". *Ocean Engineering, 144*, pp. 305–319.

[5] Regenwetter, L., Curry, B., and Ahmed, F., 2022. "Biked: A dataset for computational bicycle design with machine learning benchmarks". *Journal of Mechanical Design, 144*(3).

[6] Regenwetter, L., Weaver, C., and Ahmed, F., 2023. "Framed: An automl approach for structural performance prediction of bicycle frames". *Computer-Aided Design, 156*, p. 103446.

[7] Heyrani Nobari, A., Srivastava, A., Gutfreund, D., and Ahmed, F., 2022. "Links: A dataset of a hundred million planar linkage mechanisms for data-driven kinematic design". In International Design Engineering Technical Conferences and Computers and Information in Engineering Conference, Vol. 86229, American Society of Mechanical Engineers, p. V03AT03A013.

[8] Chan, Y.-C., Ahmed, F., Wang, L., and Chen, W., 2021. "Metaset: Exploring shape and property spaces for data-driven metamaterials design". *Journal of Mechanical Design, 143*(3).

[9] Lee, D., Chan, Y.-C., Chen, W., Wang, L., van Beek, A., and Chen, W., 2023. "t-metaset: Task-aware acquisition of metamaterial datasets through diversity-based active learning". *Journal of Mechanical Design, 145*(3), p. 031704.

[10] Read, D., 2009. *A drag estimate for concept-stage ship design optimization.* The University of Maine.

[11] Song, B., Miller, S., and Ahmed, F., 2023. "Attention-enhanced multimodal learning for conceptual design evaluations". *Journal of Mechanical Design*, pp. 1–38.

[12] Mazé, F., and Ahmed, F., 2022. "Topodiff: A performance and constraint-guided diffusion model for topology optimization". *arXiv preprint arXiv:2208.09591*.

[13] Brown, A., and Salcedo, J., 2003. "Multiple-objective optimization in naval ship design". *Naval Engineers Journal, 115*(4), pp. 49–62.

[14] Feng, Y., el Moctar, O., and Schellin, T., 2022. "Parametric hull form optimization of containerships for minimum resistance in calm water and in waves". *Journal of Marine Science and Applications*, January.

[15] Khan, S., Kaklis, P., Serani, A., Diez, M., and Kostas, K., 2022. "Shape-supervised dimension reduction: Extracting geometry and physics associated features with geometric moments". *Computer-Aided Design, 150*, p. 103327.

[16] Khan, S., Kaklis, P., Serani, A., and Diez, M., 2022. "Geometric moment-dependent global sensitivity analysis without simulation data: application to ship hull form optimisation". *Computer-Aided Design, 151*, p. 103339.

[17] Zhang, Y., Kim, D.-J., and Bahatmaka, A., 2018. "Parametric method using grasshopper for bulbous bow generation". In 2018 International Conference on Computing, Electronics & Communications Engineering (iCCECE), IEEE, pp. 307–310.

[18] Chrismianto, D., and Kim, D.-J., 2014. "Parametric bulbous bow design using the cubic bezier curve and curve-plane intersection method for the minimization of ship resistance in cfd". *Journal of Marine Science and Technology, 19*, pp. 479–492.

[19] Lu, Y., Chang, X., and Hu, A.-k., 2016. "A hydrodynamic optimization design methodology for a ship bulbous bow under multiple operating conditions". *Engineering Applications of Computational Fluid Mechanics, 10*(1), pp. 330–345.

[20] Knight, J. T., Zahradka, F. T., Singer, D. J., and Collette, M. D., 2014. "Multiobjective Particle Swarm Optimization of a Planing Craft with Uncertainty". *Journal of Ship Production and Design, 30*(04), 11, pp. 194–200.

[21] Knight, J. T., Singer, D. J., and Collette, M. D., 2015. "Testing of a spreading mechanism to promote diversity in multi-objective particle swarm optimization". *Optimization and Engineering, 16*, June, pp. 279–302.

[22] Wang, Y., Joseph, J., Aniruddhan Unni, T., Yamakawa, S., Barati Farimani, A., and Shimada, K., 2022. "Three-dimensional ship hull encoding and optimization via deep

Copyright © 2023 by ASME

neural networks". *Journal of Mechanical Design,* **144**(10), p. 101701.

[23] Ao, Y., Li, Y., Gong, J., and Li, S., 2021. "An artificial intelligence-aided design (aiad) of ship hull structures". *Journal of Ocean Engineering and Science.*

[24] Ao, Y., Li, Y., Gong, J., and Li, S., 2022. "Artificial intelligence design for ship structures: A variant multiple-input neural network-based ship resistance prediction". *Journal of Mechanical Design,* **144**(9), p. 091707.

[25] Peri, D., Rossetti, M., and Campana, E. F., 2001. "Design optimization of ship hulls via cfd techniques". *Journal of ship research,* **45**(02), pp. 140–149.

[26] Demo, N., Tezzele, M., Mola, A., and Rozza, G., 2021. "Hull shape design optimization with parameter space and model reductions, and self-learning mesh morphing". *Journal of Marine Science and Engineering,* **9**(2), p. 185.

[27] Li, J., Du, X., and Martins, J. R., 2022. "Machine learning in aerodynamic shape optimization". *Progress in Aerospace Sciences,* **134**, p. 100849.

[28] Marlantes, K., and Maki, K., 2021. "Modeling vertical planing boat motions using a neural-corrector method".

[29] Zubaly, R., 1996. *Applied Naval Architecture.* Cornell Maritime Press.

[30] Noblesse, F., Huang, F., and Yang, C., 2013. "The neumann–michell theory of ship waves". *Journal of Engineering Mathematics,* **79**(1), pp. 51–71.

[31] Noblesse, F., and McCarthy, J., 1983. Proceedings of the dtnsrdc (david w. taylor naval ship research and development center) workshop on ship wave-resistance computations (2nd). Tech. rep., David W Taylor Naval Ship Research and Development Center, November.

[32] Huang, F., Yang, C., and Noblesse, F., 2013. "Numerical implementation and validation of the neumann–michell theory of ship waves". *European Journal of Mechanics-B/Fluids,* **42**, pp. 47–68.

[33] Newman, J. N., 2018. *Marine hydrodynamics.* The MIT press.

[34] Yang, C., Huang, F., and Noblesse, F., 2013. "Practical evaluation of the drag of a ship for design and optimization". *Journal of Hydrodynamics,* **25**(5), pp. 645–654.

[35] Olivieri, A., Pistani, F., Avanzini, A., Stern, F., and Penna, R., 2001. Towing tank experiments of resistance, sinkage and trim, boundary layer, wake, and free surface flow around a naval combatant insean 2340 model. Tech. rep., Iowa Univ Iowa City Coll of Engineering.

[36] Savitsky, D., 1964. "Hydrodynamic Design of Planing Hulls". *Marine Technology and SNAME News,* **1**(04), 10, pp. 71–95.

[37] Hollenbach, K. U., 1998. "Estimating resistance and propulsion for single-screw and twin-screw ships-ship technology research 45 (1998)". *Schiffstechnik,* **45**(2), p. 72.

[38] Hollenbach, U., and Friesch, J., 2007. "Efficient hull forms–what can be gained". In Proceedings of the 1st International Conference on Ship Efficiency, Hamburg, Germany, pp. 8–9.

[39] Hart, C. G., and Vlahopoulos, N., 2010. "An integrated multidisciplinary particle swarm optimization approach to conceptual ship design". *Structural and Multidisciplinary Optimization,* **41**, pp. 481–494.

[40] Diez, M., and Peri, D., 2010. "Robust optimization for ship conceptual design". *Ocean Engineering,* **37**(11-12), pp. 966–977.

[41] Daniels, A., and Parsons, M., 2008. "A hybrid agent — genetic algorithm approach to general arrangements". *Ship Technology Research,* **55**(2), pp. 78–86.

[42] Michell, J. H., 1898. "Xi. the wave-resistance of a ship". *The London, Edinburgh, and Dublin Philosophical Magazine and Journal of Science,* **45**(272), pp. 106–123.

[43] Tuck, E. O., 1989. "The wave resistance formula of jh michell (1898) and its significance to recent research in ship hydrodynamics". *The ANZIAM Journal,* **30**(4), pp. 365–377.

[44] Mantzaris, D. A., 1998. "A rankine panel method as a tool for the hydrodynamic design of complex marine vehicles". PhD thesis, Massachusetts Institute of Technology.

[45] Dawson, C., 1977. "A practical computer method for solving ship-wave problems". In Proceedings of Second International Conference on Numerical Ship Hydrodynamics, pp. 30–38.

[46] Reuther, A., Kepner, J., Byun, C., Samsi, S., Arcand, W., Bestor, D., Bergeron, B., Gadepally, V., Houle, M., Hubbell, M., Jones, M., Klein, A., Milechin, L., Mullen, J., Prout, A., Rosa, A., Yee, C., and Michaleas, P., 2018. "Interactive supercomputing on 40,000 cores for machine learning and data analysis". In 2018 IEEE High Performance extreme Computing Conference (HPEC), pp. 1–6.

Copyright © 2023 by ASME

**Proceedings of the ASME 2023
International Design Engineering Technical Conferences and
Computers and Information in Engineering Conference
IDETC-CIE2023
August 20-23, 2023, Boston, Massachusetts**

DETC2023-117400

SURROGATE MODELING OF CAR DRAG COEFFICIENT WITH DEPTH AND NORMAL RENDERINGS

Binyang Song[1],*, Chenyang Yuan[2], Frank Permenter[2], Nikos Arechiga[3], Faez Ahmed[1]

[1] Department of Mechanical Engineering, Massachusetts Institute of Technology, Cambridge, MA
[2] Toyota Research Institute, Cambridge, MA
[3] Toyota Research Institute, Los Altos, CA

ABSTRACT

Generative AI models have made significant progress in automating the creation of 3D shapes, which has the potential to transform car design. In engineering design and optimization, evaluating engineering metrics is crucial. To make generative models performance-aware and enable them to create high-performing designs, surrogate modeling of these metrics is necessary. However, the currently used representations of three-dimensional (3D) shapes either require extensive computational resources to learn or suffer from significant information loss, which impairs their effectiveness in surrogate modeling. To address this issue, we propose a new two-dimensional (2D) representation of 3D shapes. We develop a surrogate drag model based on this representation to verify its effectiveness in predicting 3D car drag. We construct a diverse dataset of 9,070 high-quality 3D car meshes labeled by drag coefficients computed from computational fluid dynamics (CFD) simulations to train our model. Our experiments demonstrate that our model can accurately and efficiently evaluate drag coefficients with an R^2 value above 0.84 for various car categories. Moreover, the proposed representation method can be generalized to many other product categories beyond cars. Our model is implemented using deep neural networks, making it compatible with recent AI image generation tools (such as Stable Diffusion) and a significant step towards the automatic generation of drag-optimized car designs. We have made the dataset and code publicly available at https://decode.mit.edu/projects/dragprediction/.

.

Keywords: design representation, drag coefficient, car design, surrogate modeling

---
*Corresponding author: binyangs@mit.edu

1. INTRODUCTION

Engineers often need to work with three-dimensional (3D) representations of an object for design, evaluation, and optimization. At the same time, computer vision researchers have developed powerful deep-learning techniques for various 3D tasks [1–10], including automatic generation of novel 3D objects. Applying these techniques to design tasks requires evaluating performance metrics at scale. Traditionally, performance evaluation relies on physical simulation, which is time-consuming and computationally expensive. Data-driven surrogate models provide more scalable alternatives. This paper develops a surrogate model for evaluating the aerodynamic drag of 3D vehicles, aiming toward the eventual performance-guided generation of vehicle designs.

A key challenge in developing a surrogate model is representing shapes in a computationally efficient way that also captures the structure needed to accurately estimate relevant performance metrics. In machine learning, commonly used 3D shape representation methods include voxels, point clouds, and meshes, each affording different advantages and disadvantages. For example, 3D convolutional neural networks (CNNs) are commonly applied to learn structured voxel data [11, 12], while graph neural networks (GNNs) [4, 5] and CNNs generalized to irregular spaces [13–15] can learn unstructured 3D meshes. In recent years, diffusion models have successfully been leveraged for learning point clouds for 3D shape generation [6–10]. These direct 3D representations are computationally limited to low-resolution shapes, which in turn limits their applications to practical engineering problems.

In addition to direct 3D representations, abstract representations in terms of two-dimensional (2D) renderings have also been explored. Since technologies for recognizing and generating 2D data is older and more mature than that for learning 3D data, several studies employ 2D renderings or point coordinate matrices to represent 3D shapes [16–18]. Parametric representations are another option to simplify the representation of 3D shapes [19–21].

Copyright © 2023 by ASME

These simplified 2D and parametric representations, however, suffer from varying degrees of information loss, and cannot provide sufficient information to reconstruct the corresponding 3D shapes. Accordingly, we propose a new image-based representation of 3D shapes that augments traditional 2D renderings with surface normal and depth information.

We use our representation to train a surrogate model for vehicle drag coefficient prediction, which is a key performance metric that affects not only fuel efficiency but also vehicle aesthetics. As we show, this enables fast and accurate estimation of 3D drag from 2D input. Our contributions are summarized as follows.

1. We construct and share a large and diverse set of high-quality car 3D meshes labeled with drag coefficients computed by a fluid dynamics simulation.

2. We propose a 2D image representation of 3D shapes that annotates 2D renderings with depth and surface normal information using pixel values.

3. We develop a high-performing surrogate model using the proposed representation for car drag coefficient prediction. Leveraging our 2D image representation, we base this model on powerful pre-trained neural networks for image processing tasks.

In total, these contributions are a step towards the automatic, performance-aware generation of vehicle body designs. The surrogate model for car drag coefficient prediction also offers an efficient alternative to expensive 3D fluid dynamic simulations. We also hope that our dataset will facilitate the development of various deep-learning techniques for car body design, evaluation, and optimization.

The remainder of this paper is organized as follows. Section 2 provides a detailed review of the relevant literature. In Section 3, we describe our dataset, our novel representation of 3D shapes, and our surrogate model for drag coefficient prediction. Section 4 reports and discusses the effectiveness of the proposed representation and the performance of the surrogate model, and also summarizes the limitations of our approach.

2. LITERATURE REVIEW

The two research areas most relevant to our contributions are 3D object representation and data-driven prediction of drag coefficients.

2.1 3D Shape Representation and Learning

In machine learning, 3D shapes are commonly represented as voxels, point clouds, or meshes. Different representations are often matched with different learning algorithms since different algorithms are better suited to exploit the advantages of each representation. For example, similar to CNNs that employ 2D kernels to learn visual features from images, 3D CNNs utilize 3D kernels to capture geometric features from structured 3D spatial data in Euclidean spaces. They are a popular option to learn voxels [11, 12] and occupancy grids for 3D shape recognition [1] and generation [2]. Since point clouds and meshes are unstructured, prior studies have explored transforming them into regular voxel grids [11, 22] or other canonicalized formats [23]. However, the sparsity of most 3D data representations makes the computation of the naïve 3D convolutional learning challenging. Researchers have proposed a few approaches to mitigate this issue. For example, multiple-resolution 3D CNNs can learn multi-scale features from multi-level voxels [24], while OctNet [2] represents its volumetric output as an octree with improved resolutions in the later levels. The voting [25] or probing [26] schemes in neural networks have been developed to assign varying amounts of computational effort to different regions of sparse data inputs.

In contrast, a more diverse set of deep learning models have been developed to learn unstructured 3D representations in non-Euclidean spaces (e.g., meshes, manifolds, and point clouds). Inspired by conventional CNNs, a group of researchers developed a variety of CNN variants to learn irregular representations, including localized spectral CNNs [13], anisotropic CNNs [14], spline-based CNNs [27], geodesic CNNs [15], and others. Beyond that, GNNs have been applied to learn both point clouds and meshes for 3D shape recognition [3] and generation [4, 5]. More recently, diffusion models are becoming an area of active research interest. They have been applied to generate 3D shapes represented by point clouds or similar representations [6–10]. Due to computational cost, the 3D point clouds or meshes generated by these models still present low resolutions, impairing their applications in engineering domains. Prior studies have also explored simple multi-layer perceptrons (MLPs) for mesh texture editing [28, 29].

2D representations have also been explored to represent 3D shapes. A few studies look into representing 3D shapes using 2D images or renderings, which can be processed by standard image learning algorithms [16, 17]. Despite the improved computational efficiency, such methods often suffer from information loss. Alternatively, Achlioptas et al. [18] proposed a representation that uses the point coordinates of a point cloud as a matrix and trains a generative model with the 2D matrix representation. This approach, however, can only work with point clouds that have a fixed number of points. Another set of studies maps 3D shapes to 2D parameter domains, then trains GANs to generate samples in the 2D domains, and finally converts them to 3D meshes [30–33]. Additionally, implicit representations have also been explored for machine learning tasks. Implicit representations take a latent embedding of a shape and point coordinates as input and assign a value to each point which indicates if this point is inside or outside the shape [34, 35]. These representations are often used for 3D shape generation [36, 37]. A group of other studies exploits parametric representations which seek to convey the control points or other prominent features of 3D shapes for machine learning tasks [19–21].

In summary, the recognition, evaluation, and generation of 3D shapes using machine learning rely on effective and accurate 3D geometric feature learning. Existing representations of 3D shapes are still greatly limited by their high computational costs, while alternative 2D, implicit, and parametric representations suffer from information loss and may not capture sufficient geometric features for downstream tasks. In this paper, we show that a new representation of 3D shapes using stacked depth and normal renderings is a promising approach, which helps signifi-

Copyright © 2023 by ASME

cantly in the downstream task of predicting the drag coefficients of 3D cars.

2.2 Data-Driven Drag Coefficient Evaluation

Performance evaluation of 3D shapes is critical in engineering design and optimization. Among them, drag coefficient prediction is critical for car body design. It is traditionally conducted through simulations by solving the nonlinear Navier-Stokes equations for many iterations, which are time-consuming and computationally expensive. The solution methods are too slow to run in conjunction with a generative design or optimization process, which needs to evaluate a large number of candidate designs. To mitigate this issue, researchers have explored combining differentiable partial differential equations (PDEs) solvers with deep learning models to accelerate the simulation results without sacrificing the simulation accuracy significantly [38]. These differentiable PDE solvers often simulate the problem at a coarse resolution and the neural networks are employed to infer the results at higher resolutions. Such an approach speeds up the simulation process but is still too slow to be implemented during the deep generative process. As an alternative to differentiable PDE solvers, data-driven surrogate modeling is a desirable alternative to the simulation approaches in deep learning, and previous work has explored surrogate models for drag coefficient evaluation.

Parametric representation is commonly used in surrogate modeling of vehicle drag. For instance, Gunpinar et al. [19] represent a car using the coordinates of a set of control points from the 2D car silhouette and trained computational models to predict its drag coefficient in 2D settings. Their model first reduces the dimension of the representation using principal component analysis and then employs regression models or neural networks to learn the low-dimensional representation for drag coefficient prediction. Likewise, Rosset et al. [39] predicted the pressure field along the car silhouette to optimize 2D car designs. Umetani and Bickel [20] employed a parameterization method to represent simplified cars as vectors that indicate the position of control points and projection heights of the surface points. Then, they learned the representation using regression models, neural networks, or the Gaussian process for drag coefficient prediction. These studies reported that the regression or the Gaussian process models achieved higher explanatory power than the neural network models. Badias et al. [21] used locally linear embeddings to parameterize 3D cars and employed dimensionality reduction and interpolation to predict the drag coefficient of a new car. Limited by their parametric representations, these studies attempt to predict the drag coefficients of simplified cars, such as 2D car silhouettes or 3D cars with mirrors, wheels, and other details removed. This simplification may hinder the applications of such models in practical design contexts.

Another set of surrogate models learns 2D or 3D car representations to predict drag coefficients. For example, MeshSDF [40] learns 3D point clouds obtained from an implicit representation using an irregular CNN (i.e., spline-based CNNs [27]), which applies to drag coefficient prediction. Similarly, Baque et al. [41] exploited a geodesic CNN [15] to obtain a latent representation of 3D car meshes for drag coefficient prediction. Another model learns 2D slices of 3D point clouds using regular CNNs [42], while DEBOSH [43] learns meshes using GNNs for the same purpose. Additionally, another class of models obtains the latent representations of 2D [44] or 3D shapes [45] through reconstruction using generative models like variational autoencoders (VAEs) to predict the pressure fields and drag coefficients. Other surrogate models focus on drag prediction of general 3D shapes beyond cars [46–48]. Due to high computational costs, such models can only work with low-resolution 3D representations or simplified 2D representations. This paper focuses on surrogate modeling using the proposed representation of 3D shapes to circumvent the issues of the reviewed approaches.

3. DATA AND METHOD

In this section, we detail our main contributions in this paper: A high-quality dataset of 3D car meshes and their drag coefficients computed through computational fluid dynamics (CFD) simulations, a 2D representation generated from 3D car meshes tailored to capturing features important for predicting drag coefficients, and a series of surrogate models trained to predict drag coefficients from the 2D representation as a regression task [1].

3.1 Car Data and CFD Simulation

First, we detail our 3D car dataset and CFD simulations for obtaining drag coefficients from 3D mesh data.

3.1.1 Car Data. The 3D car meshes used in this paper are initially from the ShapeNet V1 dataset [49], which contains 7,497 3D car meshes with varying surface qualities. A substantial percentage of the original car meshes from ShapeNet are not watertight, with unsealed areas or holes on the surfaces. We need high-surface-quality car meshes in order to achieve reliable CFD simulation results when computing car drag coefficients. Therefore, we manually checked the surface quality of each car mesh from ShapeNet and selected a subset of 2,474 high-quality car meshes. Since most of the selected meshes are still imperfect, we further repaired them using the repair module in Autodesk Netfabb Premium. It should be noted that this dataset covers a variety of car configurations, such as pick-up trucks, sedans, sport utility vehicles, wagons, and combat vehicles. The diversity helps our learned surrogate models generalize across all cars.

In addition, we employed two different approaches to augment the original dataset. First, we resized the width of each car using a random coefficient between 0.83 (i.e., $1/1.2$) to 1.2. The resizing augmentation created another 2,474 cars with slightly different widths and drag coefficients from the original cars, resulting in a dataset of 4,958 different cars in total. Second, since the car meshes are not perfectly bilaterally symmetric but their drag coefficients are invariant to bilateral flipping, we employed a flipping augmentation to create another 4,948 cars, which have exactly the same drag coefficients as the cars without this augmentation. After the augmentations, we obtain a dataset of 9,896 cars. To avoid data leakage, we only treat the 2,474 unique cars from the original dataset as independent samples when splitting the dataset to train the surrogate model. For every car in any of

---

[1]The dataset and the surrogate models introduced in this paper can be found: Github link

Copyright © 2023 by ASME

the training, validation, or test sets, all of its resized and flipped versions belong to the same set.

3.1.2 CFD Simulation. The drag coefficient of each car is computed by a CFD simulation using OpenFOAM. During mesh preparation, all cars are normalized to have the same length of 3.5 meters to ensure the defined computational domain is suitable for all cars. The computational domain for simulation is then created, serving as a virtual wind tunnel to simulate the airflow around a car, as shown in 1-A. The height, width, and length of the virtual tunnel are 8 meters, 14 meters, and 54 meters, respectively. In order to simulate flow dynamics around the car body more accurately, the computational domain is refined to a smaller mesh size, which becomes coarse away from the car surface, as shown in 1-B. This meshing strategy is applied to all car configurations (e.g., sedans, sports utility vehicles, combat cars, and pick-up trucks) in our dataset.

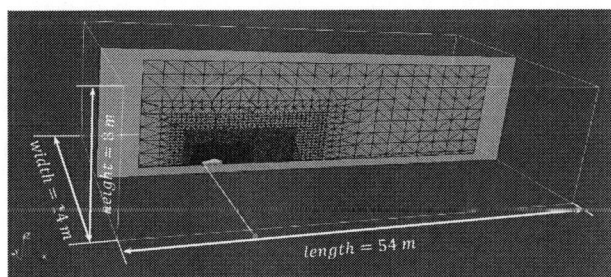

(A) Computational Domain

(B) Refinement regions

FIGURE 1: THE COMPUTATIONAL DOMAIN FOR THE CFD SIMULATION

On this basis, the inlet velocity and turbulence parameters are set as the inlet conditions, while outlet pressure is specified as the outlet condition. The car surface and road are set to be stationary walls. The sides and top of the computational domain are specified as symmetry boundaries. The steady-state "SimpleFoam" solver and the fluid flow "PotentialFoam" solvers are selected for the simulation. The primary boundary conditions and solver settings are listed in Tables 1 and 2. During the simulation, 300 iterations were conducted for each car, which can achieve the required accuracy for concept-level studies [50]. Since the drag coefficient outputs may fluctuate during the simulation process, we use the average value from the last 50 iterations as the final output from the simulation.

3.2 2D Representation of 3D Shapes

In prior work, voxels, point clouds, and meshes are commonly used to represent 3D shapes. They each require different

TABLE 1: BOUNDARY CONDITIONS

Tunnel inlet	Velocity inlet, velocity = 40km/h
Tunnel outlet	Pressure outlet
Tunnel sides	Symmetry
Tunnel top	Symmetry
Tunnel road	No slip wall, with prism layer
Car body	No slip wall

TABLE 2: SOLVER SETTINGS

Gradient scheme	Linear
Divergence scheme (momentum)	Linear upwind
Divergence scheme (turbulence)	Upwind
Laplacian scheme	Linear
Interpolation scheme	Linear
Pressure solver	GMAG
Velocity solver	Smooth solver
No of Non-orthogonal corrections	2

deep neural networks to learn and rely on intensive computational resources to capture fine-grained, high-resolution 3D features. For car body design, we only focus on the surface of the car and ignore any interior architecture. In this paper, we aim to propose a more information-efficient method to represent 3D shapes like car bodies, which supports learning 3D information more effectively and affordably for drag coefficient prediction.

Since machine learning methods for 2D data learning are more explored than those for 3D data learning, 2D renderings have become an option to represent 3D shapes in many studies. However, the commonly used perspective 2D renderings 2-A are generated through perspective projection 2-D, which causes geometric distortion and information loss for machine learning. Accordingly, we propose a new 2D representation of 3D shapes that consists of two types of renderings, namely the normal rendering 2-B and the depth rendering 2-C, generated through orthographic projection 2-E. The points facing the cameras are first projected to the image space through a projection defined by Eq. 1. Herein, P_{camera} and P_{world} represent point coordinates (i.e., x, y) in the rendering and real-world space, respectively. $Scale_{\text{x}}$ and $Scale_{\text{y}}$ denote the scaling factors that are determined by the position and angle of the camera and the size of the rendering. Specifically, the pixel values of the normal rendering encode the unit normal vector at each point of the mesh, with the x (Norm_x), y (Norm_y), and z (Norm_z) coordinates mapped to the red (Color_R), green (Color_G), and blue (Color_B) color channels, respectively, as shown by Eq. 2. The pixel values of the depth rendering encode the depth of each point, i.e., the distance (Dist) between the camera and the point, as formulated by Eq. 3.

$$P_{\text{camera}} = P_{\text{world}} \times \begin{bmatrix} Scale_{\text{x}} & 0 \\ 0 & Scale_{\text{y}} \end{bmatrix}, \tag{1}$$

$$\text{Color}_R = \text{Norm}_x, \ \text{Color}_G = \text{Norm}_y, \ \text{Color}_B = \text{Norm}_z, \tag{2}$$

Copyright © 2023 by ASME

$$Color_R = Color_G = Color_B = Dist. \qquad (3)$$

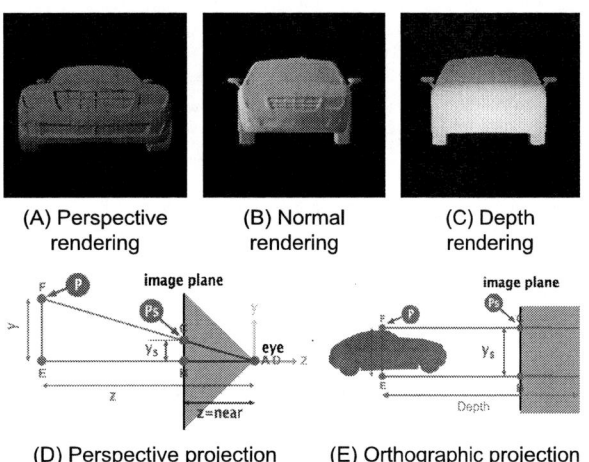

(A) Perspective rendering (B) Normal rendering (C) Depth rendering

(D) Perspective projection (E) Orthographic projection

FIGURE 2: THE PROPOSED 2D REPRESENTATION OF 3D SHAPES

According to the definition, the depth and normal renderings capture the point-wise positional and surface information of 3D shapes respectively. In order to capture the geometric features of a car comprehensively, we generate the normal and depth renderings from six orthographic views: front, rear, top, bottom, left, and right. Then, the six single-view renderings are integrated into a single image. With the combined information from all six single-view renderings, the integrated 2D representation conveys 3D geometric information and be potentially converted back to corresponding 3D shapes. Figure 3 describes the process using the depth rendering of a car. Building on the render module of the kaolin python package developed by NVIDIA[2], we develop a differentiable render for 3D to 2D rendering and a separate module for six view integration to produce the 2D representation for each car. The integrated normal and depth renderings are used as the 2D representation of 3D shapes in this paper. We verify the effectiveness of our proposed representation by developing surrogate models to predict car drag coefficients from our 2D representation.

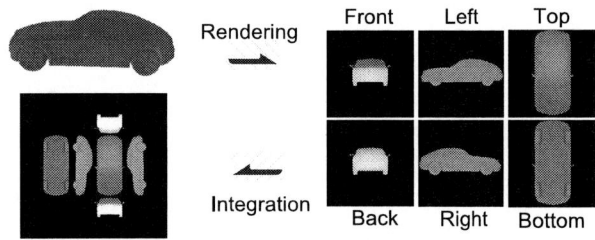

FIGURE 3: THE CONVERSION PROCESS FROM A 3D MESH TO 2D RENDERINGS, AND TO AN INTEGRATED REPRESENTATION

[2]https://github.com/NVIDIAGameWorks/kaolin.

3.3 Surrogate Model

Our proposed 2D representation enables us to represent 3D shapes using 2D pixel data. We next develop and compare three surrogate models that take the 2D representation of a car as input and predict its drag coefficient. In this paper, the 2D representations of all cars in our dataset are images with 3 color channels and a dimension of 384×384, which is the input to all of our surrogate models.

We explore both the CNN-based and transformer-based computer vision models to learn features from the 2D representation of cars. In a set of pilot experiments, we first compare a few different pre-trained CNN-based models, including InceptionV3 [51], ResNet [52], and ResNeXt [53]. In general, they perform similarly after careful hyper-parameter tuning, and ResNeXt is selected in our study because it performs slightly better than the others. The proposed representation integrates six single-view renderings, which exhibit correspondence and convey complementary information for drag coefficient prediction. This characteristic of the representation motivates us to involve attention mechanisms in the surrogate model. Furthermore, since transformer-based image models can capture the interactions between different image regions through the embedded self-attention mechanism, we also compare the CNN-based models against one transformer-based model, the vision transformer (ViT) [54].

The first model (Figure 4-A) employs the pre-trained ResNeXt "$101_32 \times 8d$" module to embed the image input. The output from the ResNeXt embedding module exhibits a dimension of $12 \times 12 \times 2,048$, which is flattened. Following that, a linear layer with 128 neurons is attached before the output layer. We name this model "ResNeXt" in this paper.

The second model (Figure 4-B) applies a self-attention mechanism to enhance the learning of the interactions between different image regions. Specifically, it reshapes the output from the ResNeXt embedding module to $144 \times 2,048$, which is seen as a set of 144 latent features with a dimension of 2,048. A self-attention mechanism with a latent dimension of 128 is applied to capture the interactions between the image regions. Then, the output from the self-attention mechanism is flattened and projected to a lower dimension (128) through a linear layer as the final embedding to predict the car drag coefficient. This model is referred to as "attn-ResNeXt" hereafter.

The third model (Figure 4-C) utilizes a pre-trained ViT module to embed the image input. We compare two different-sized ViT models, including the "vit-large-patch32-384" model and the "vit-base-patch16-224" model, and achieve slightly better performance from the former. Accordingly, "vit-large-patch32-384" was selected for building the third surrogate model. The pooled output from the transformer embedding module is used as the final embedding to predict the car drag coefficient. We call this model "ViT" in this paper.

Since the surrogate models introduced above can learn from only one of the normal/depth renderings at a time, we further explore if fusing the features of the normal and depth renderings can improve the prediction performance. After fine-tuning the hyperparameters of all three surrogate models, we select the best among the three for this exploration, which is the attn-ResNeXt

Copyright © 2023 by ASME

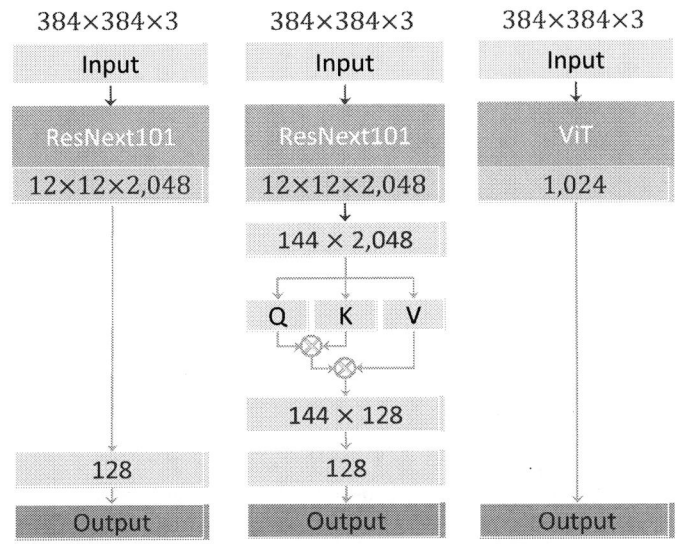

FIGURE 4: THE ARCHITECTURES OF THE THREE SURROGATE MODELS USING DIFFERENT EMBEDDING MODULES OR DIFFERENT ATTENTION MECHANISMS

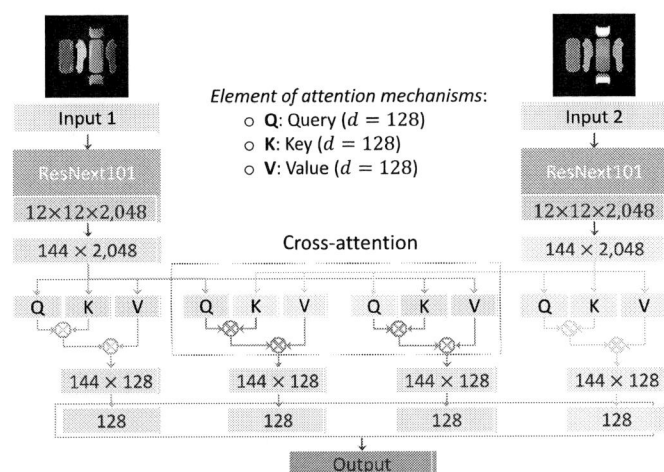

FIGURE 5: THE SURROGATE MODEL FUSING FEATURES OF BOTH THE NORMAL AND DEPTH RENDERINGS USING A SYMMETRIC CROSS-ATTENTION MECHANISM

model in this study. Specifically, we fuse two attn-ResNeXt models respectively pre-trained on the normal and depth renderings using a symmetric cross-attention mechanism, as shown in Figure 5. The cross-attention mechanism is expected to capture the interactions between the regions respectively from the normal and depth renderings. Then, the outputs from the self-attention and cross-attention mechanisms are flattened and projected to a lower dimension (128) through linear layers, which are then concatenated as the final embedding to predict the car drag coefficient. During training, the fused model is initialized with the pre-trained weights from both the normal rendering model and the depth rendering model to transfer the knowledge learned from the single types of renderings to the fused model. This approach has been proven beneficial for avoiding modality failure [55, 56]. We refer to this model as "fused" hereafter.

The hyperparameters of these surrogate models are determined through a set of pilot experiments. In the experiments, all the trainable parameters are unfrozen. The pre-trained ResNeXt and ViT image embedding modules are fine-tuned on our data. We split the entire dataset into the training, validation, and test sets following a ratio of 0.7:0.15:0.15. All models are trained on the same training-validation-test split for easy comparisons. We employ different learning rates ranging from 2×10^{-5} to 8×10^{-5} to train different models with different image inputs. We also apply a decay of 0.96 to schedule the learning rate during the training process. We end the training process if the validation loss does not decrease for 20 consecutive epochs.

4. RESULTS AND DISCUSSION

This section describes our CFD simulation results and compares the performances of different surrogate models based on the proposed 2D representation. To evaluate the models, we report the coefficient of determination (R^2 value) and the mean squared prediction error (MSE). To illustrate sensitivity to initialization, we train each model five times and report the average values of these metrics. We also compare our best surrogate model against two baseline models from prior studies.

4.1 CFD Simulation Results

As described in the last section, our dataset originates from 4,948 car meshes obtained from ShapeNet. Drag coefficients were successfully simulated for 4,535 of these meshes using OpenFOAM. To increase the size of the dataset, we flip each car left to right (which leaves the drag coefficient unchanged), giving a total of $4,535 \times 2 = 9,070$ training examples.

The computed drag coefficients range from 0.175 to 0.907. Figure 6 shows their distribution and three sample vehicle images from different drag coefficient regimes. The data is concentrated on the interval [0.28, 0.65].

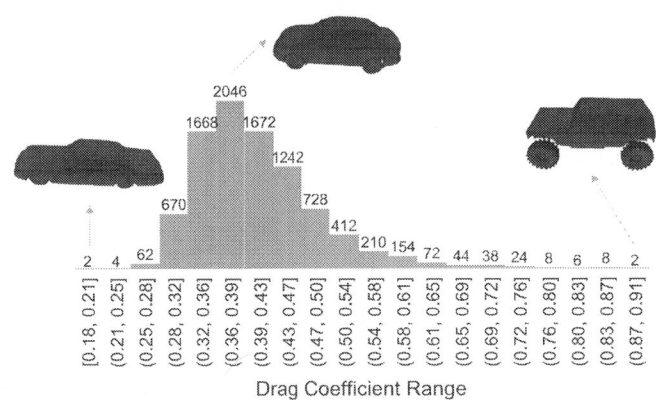

FIGURE 6: THE DISTRIBUTION OF THE DRAG COEFFICIENTS AND THREE EXAMPLE CARS FROM THE LOWEST, BIGGEST, AND HIGHEST DRAG COEFFICIENT CATEGORIES, RESPECTIVELY

Copyright © 2023 by ASME

4.2 Performance of Different Surrogate Models

We first compare the drag coefficient prediction of six different surrogate models. Each model employs one of the three architectures depicted in Figure 4 and is trained on either depth or surface normal renderings. Figure 7 illustrates the performance of each model. Among the three architectures, the attn-ResNeXt model achieves the highest R^2 values and the lowest MSE values. The comparison between ResNeXt and attn-ResNeXt suggests that the self-attention mechanism improves the fusion of information from different image regions. Both ResNeXt and attn-ResNeXt outperform the ViT model. A possible reason is that ResNeXt contains far fewer trainable parameters than the ViT model (about 86 million vs about 2 billion) and overfits our relatively small dataset to a lesser degree.

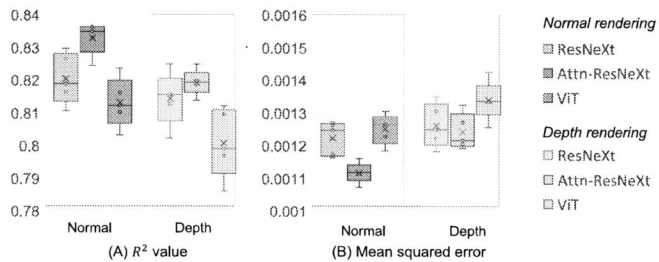

FIGURE 7: THE PERFORMANCE COMPARISON AMONG THE THREE SURROGATE MODELS USING DIFFERENT RENDERING INPUTS

We next illustrate that combining the normal and depth information enhances the performance of the surrogate model. We fuse these features using a symmetric cross-attention mechanism as depicted in Figure 5. Moreover, we train this fused model using the *transfer learning* paradigm; that is, we initialize the training of the fused model using the weights of the attn-ResNeXt models respectively pre-trained on the normal and depth renderings. Figure 8 illustrates the superior performance of the fused model, and significantly reduced sensitivity to initialization of the training procedure, as indicated by the variance of the R^2 values and MSE values.

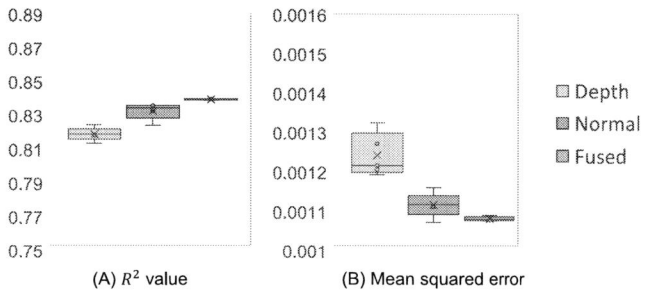

FIGURE 8: THE PERFORMANCE COMPARISON AMONG THE TWO ATTN-RESNEXT MODELS RESPECTIVELY USING THE NORMAL AND DEPTH RENDERINGS AND THE FUSED MODEL USING BOTH RENDERINGS

Figure 9 illustrates how the accuracy of the model depends on the ground-truth drag coefficient. As indicated, the prediction

TABLE 3: THE VARIATION OF THE PREDICTION ERROR WITH THE SIMULATED DRAG COEFFICIENT

Drag Coefficient Range	Average Prediction Error
[0.18,0.3]	0.032
(0.3, 0.4]	0.021
(0.4, 0.5]	0.023
(0.5, 0.6]	0.029
(0.6, 0.7]	0.021
(0.7, 0.8]	0.092
(0.8,0.91]	0.218

exhibits increasing deviations in the lowest and highest drag coefficient ranges. One major reason is that we have much fewer car samples with very low or high drag coefficients in our dataset (Figure 6). Accordingly, the model exhibits higher average prediction errors in the lowest and highest drag coefficient ranges, as listed in Table 3.

Evaluation of the surrogate models is also significantly faster than drag coefficient computation via CFD simulation. Indeed, it takes in total 20 seconds to evaluate the drag coefficients for 1,362 cars using an NVIDIA RTX A5000 GPU. In comparison, the CFD simulation of a single car takes about 6 minutes on average using a Lambda computer with 12 Intel Xeon(R) E5-1650 CPUs. Finally, the surrogate models are also auto-differentiable and hence more easily incorporated into optimization routines.

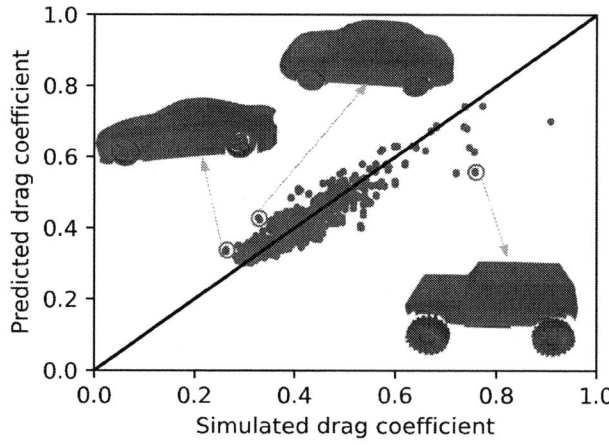

FIGURE 9: THE COMPARISON BETWEEN PREDICTED AND GROUND-TRUTH VALUES

4.3 Effectiveness of the Proposed Representation

In this subsection, we verify the effectiveness of the proposed representation by comparing its informativeness with single-view renderings and the perspective renderings. Beyond that, we also compare the performance of our surrogate model with the baseline models from two prior studies. The best surrogate model identified in the last subsection, attn-ResNeXt, is used for the following experiments.

Copyright © 2023 by ASME

Compared to the single-view renderings, the integrated rendering is more informative for car drag coefficient evaluation. In this set of experiments, the attn-ResNeXt model takes the single-view normal renderings and the integrated normal renderings as input, respectively. Figure 10 depicts their R^2 and MSE values. The model taking the integrated renderings as input exhibits the highest R^2 value and lowest MSE value compared to all other models taking the single views as input. It is intuitive that the integrated renderings contain the geometric information of a car more comprehensively than any single-view rendering.

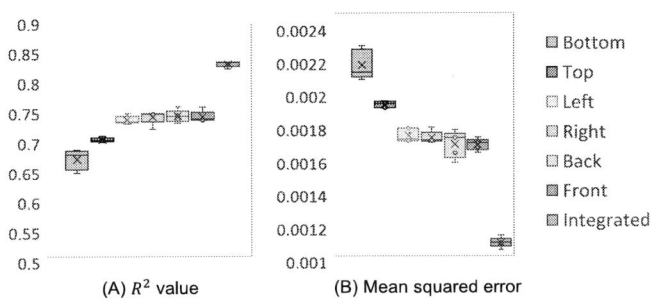

(A) R^2 value (B) Mean squared error

FIGURE 10: THE PERFORMANCE OF THE SURROGATE MODELS USING THE SINGLE-VIEW NORMAL RENDERINGS AND THE INTEGRATED NORMAL RENDERINGS, RESPECTIVELY

Among all single-view renderings, the front, back, left, and right views provide similar amounts of information for drag coefficient evaluation, leading to similar R^2 and MSE values. The bottom view is least informative for this task. In car body design, the streamlined design and the frontal area of a car affect the car's drag coefficient significantly. Every single view contains part of the information. For example, the front and back views reflect the frontal area and the front or rear part of the streamlined design, while the left and right views show the entire streamlined design from two directions. The top view describes the top half of the streamlined design, which is often more informative than the bottom half depicted by the bottom view. The amount of relevant information conveyed by each view greatly determines the explanatory power of the corresponding model.

The proposed representation is also more informative than the perspective renderings as input for car drag coefficient evaluation. In this set of experiments, the attn-ResNeXt model takes the 2D perspective renderings and the proposed normal and depth renderings as input, respectively. Figure 11 depicts their R^2 and MSE values. The models using the normal renderings and depth renderings achieve significantly higher R^2 values and lower MSE values than that using the 2D perspective renderings. That is, the normal and depth information conveyed by the proposed representation enables the model to capture more informative features for drag coefficient prediction.

Additionally, the normal renderings are more informative than the depth renderings for this task when used separately. Two possible reasons can explain this. First, the normal renderings reflect the surface features directly, while the depth renderings provide the positional information from which the surface features can be inferred in a less straightforward way. Since the aerodynamic performance of a car is determined by its surface

features, this difference probably makes the normal renderings more informative for drag coefficient prediction. Second, the three color channels of the normal renderings store different information regarding the normal vectors along the x, y, and z coordinates, respectively. In comparison, the three color channels of the depth rendering store the same information regarding the distance between the camera and a certain point. The richness of the color channels may also allow the surrogate models to capture more information from the normal renderings.

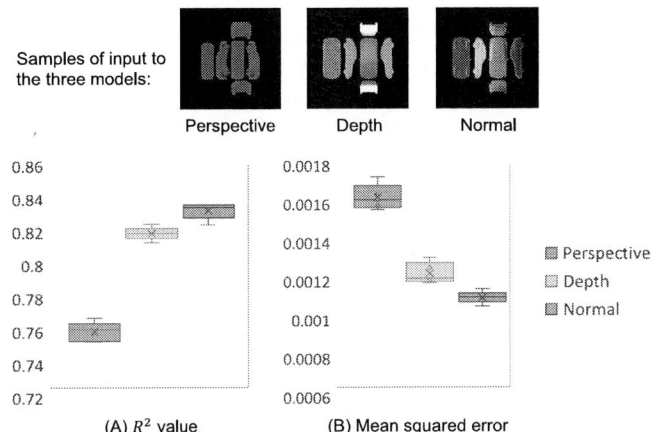

(A) R^2 value (B) Mean squared error

FIGURE 11: THE PERFORMANCES OF THE SURROGATE MODELS USING THE PROPOSED REPRESENTATION AND THE COMMONLY USED PERSPECTIVE RENDERINGS, RESPECTIVELY

Then, we compare our surrogate model with the baseline models from two prior studies. The first study [19] ran 2D CFD simulations with car silhouettes. The second study [20] ran 3D CFD simulations with simplified car designs with certain detailed features (e.g., wheels and mirrors) removed. Moreover, each car in their dataset was only simulated for 10 seconds, which might not return converged and reliable simulation results. That is, their simulations are rough compared to ours. Both baseline models employed parametric car representations to predict the simulated drag coefficients. As shown in Table 4, this study has advantages over the two baseline studies from three perspectives. First, unlike the other two studies targeting at simplified car designs, this study aims to predict the drag coefficients of full-featured car designs. The authenticity of the cars in our dataset makes the associated surrogate model more applicable to the practical car design process. Second, since our dataset covers different types of cars from a wider range of drag coefficients, our surrogate model trained on it is more likely to be generalized to different car categories. Third, our model achieves a lower MSE compared to the first model and a comparable average prediction error with the second model. Since our dataset covers a wider drag coefficient range, the MSE and average error of our model could be lower when it is tested within their drag coefficient ranges, as shown in Figure 9.

The above comparisons verify the effectiveness of the proposed representation. The proposed representation integrating six single-view renderings contains more comprehensive geometric information than any single-view rendering. Moreover,

Copyright © 2023 by ASME

TABLE 4: THE COMPARISON BETWEEN THE BEST MODEL FROM THIS STUDY AND TWO PRIOR STUDIES

	Ours	Study 1 [19]	Study 2 [20]
Input to CFD Simulation	3D meshes	2D silhouettes	3D meshes
Input to surrogate models	2D normal and depth renderings	Parametric representation	Parametric representation
Authenticity	Original	Simplified	Simplified
Drag Coefficient Range	0.17-0.85	0.21-0.51	0.2-0.6
Mean Squared Error	8.2×10^{-4}	1.84×10^{-3}	
Average Error	0.024		0.013-0.021

the proposed representation is more informative than the 2D perspective renderings for two reasons. First, the proposed normal and depth renderings convey the geometric information regarding the surface normal and positional features of each point of a 3D shape. Second, the orthographic projection used to generate the proposed representation avoids geometric distortion compared to the perspective projection. These advantages of the proposed representation allow us to reconstruct 3D shapes from them without any learning process, while it is challenging to accurately reconstruct 3D shapes from the 2D perspective renderings without a learning process. Moreover, the proposed representation method is generalizable to broader 3D shape categories whose major geometric information can be captured from the six orthographic views, such as airplanes, ships, bottles, chairs, and so forth.

The proposed 2D representation has the potential to promote 3D shape generation, evaluation, and optimization using deep learning models. The 3D representations of 3D shapes are either sparse or redundant in many cases. For example, only surface information is needed to represent a car body design. When it is represented as voxels, all the voxels inside the surface are redundant. The redundancy and sparsity of 3D representations make it highly computationally expensive to learn 3D shapes. With limited computational power, deep learning models struggle to handle high-resolution 3D shapes represented by voxels, meshes, or point clouds. Accordingly, these models do not allow for the generation, evaluation, and optimization of 3D shapes with plenty of geometric details, hindering their applications to real-world problems. Moreover, as AI technologies are more explored to handle 2D data as of now, the proposed 2D representation enables us to handle 3D shapes with more powerful 2D AI technologies. It is much easier and less expensive to increase the resolution of the 2D representation of 3D shapes than doing that with the 3D representations directly. Therefore, the proposed representation is promising to enable 3D shape generation, evaluation, and optimization at a higher resolution with less computational power needed.

4.4 Limitations and Future Work

While the proposed 2D representation, dataset, and surrogate model are promising, they have limitations and leave room for further improvement. First, the proposed 2D representation is insufficient to model more complex geometric structures, such as lattice cubes and flowers. Moreover, although the proposed representation is informative for machine learning, it is less intuitive for human perception compared to 3D representations, such as meshes and point clouds. Second, the dataset introduced in this

paper is far smaller than the training sets typically used for deep learning models. In particular, the number of samples with high drag coefficients is low. The small dataset leads to significant over-fitting during the training process. Our hope is to expand this dataset with help from the community. We aim to improve and verify its reliability by training and testing the developed model with a sizable dataset. Third, while we show that the integrated renderings are more informative than the single-view renderings, alternative integration techniques may be more effective. We will explore such alternatives in future work. Fourth, the approach proposed in this paper for drag coefficient prediction is a purely data-driven approach, which does not leverage any physics knowledge regarding CFD simulations. The performance of the surrogate model depends on the quality and quantity of the data, and is unlikely to perform well on inputs far from the training set. We will attempt to incorporate physics into our surrogate model in future work. Lastly, the surrogate model developed in this paper can only make predictions using the proposed 2D representations of cars and does not apply to common car images. A promising future direction is to associate the proposed 2D representations with real images so that the surrogate model can make predictions using easily accessible car images.

CONCLUSION

Drag coefficient evaluation is an indispensable element of the aerodynamic design of cars, which has a critical influence on car fuel efficiency. In this paper, we develop a surrogate model that enables accurate, fast, and differentiable drag coefficient evaluation. This surrogate model is built on a new two-dimensional (2D) representation of three-dimensional (3D) shapes. This representation embeds depth and surface normal information into 2D renderings and combines information from six orthographic views. The results of this study suggest that our proposed representation is more effective and informative than simple 2D perspective renderings for drag coefficient prediction. To train our model, we also assemble a diverse dataset of high-quality 3D car meshes labeled by their drag coefficients, as computed by computational fluid dynamics (CFD) simulations. This dataset, upon public release, can drive the development of other data-driven design approaches. In total, our contributions facilitate the data-driven design of 3D aerodynamic cars and can be readily combined with generative AI techniques to automate design creation.

ACKNOWLEDGMENTS

This research was supported in part by the Toyota Research Institute. Additionally, we thank Mr. Hanqi Su for helping us

Copyright © 2023 by ASME

select high-quality car meshes from ShapeNet.

REFERENCES

[1] Garcia-Garcia, A., Gomez-Donoso, F., Garcia-Rodriguez, J., Orts-Escolano, S., Cazorla, M. and Azorin-Lopez, J. "PointNet: A 3D Convolutional Neural Network for real-time object class recognition." (2016). DOI 10.1109/IJCNN.2016.7727386.

[2] Tatarchenko, Maxim, Dosovitskiy, Alexey and Brox, Thomas. "Octree Generating Networks: Efficient Convolutional Architectures for High-Resolution 3D Outputs." *Proceedings of the IEEE International Conference on Computer Vision (ICCV),*: pp. 2088–2096. 2017. URL https://github.com/lmb-freiburg/ogn.

[3] Li, Jiaxin, Chen, Ben M. and Lee, Gim Hee. "SO-Net: Self-Organizing Network for Point Cloud Analysis." *Proceedings of the IEEE Computer Society Conference on Computer Vision and Pattern Recognition* (2018): pp. 9397–9406DOI 10.1109/CVPR.2018.00979.

[4] Li, Xingang, Xie, Charles and Sha, Zhenghui. "A Predictive and Generative Design Approach for Three-Dimensional Mesh Shapes Using Target-Embedding Variational Autoencoder." *Journal of Mechanical Design* Vol. 144 No. 11 (2022). DOI 10.1115/1.4054906. URL https://asmedigitalcollection.asme.org/mechanicaldesign/article/144/11/114501/1141958/A-Predictive-and-Generative-Design-Approach-for.

[5] Wang, Nanyang, Zhang, Yinda, Li, Zhuwen, Fu, Yanwei, Liu, Wei and Jiang, Yu Gang. "Pixel2Mesh: Generating 3D Mesh Models from Single RGB Images." *Lecture Notes in Computer Science (including subseries Lecture Notes in Artificial Intelligence and Lecture Notes in Bioinformatics)* Vol. 11215 LNCS (2018): pp. 55–71. DOI 10.48550/arxiv.1804.01654. URL https://arxiv.org/abs/1804.01654v2.

[6] Luo, Shitong and Hu, Wei. "Diffusion Probabilistic Models for 3D Point Cloud Generation." *Proceedings of the IEEE Computer Society Conference on Computer Vision and Pattern Recognition* (2021): pp. 2836–2844DOI 10.48550/arxiv.2103.01458. URL https://arxiv.org/abs/2103.01458v2.

[7] Zhou, Linqi, Du, Yilun and Wu, Jiajun. "3D Shape Generation and Completion through Point-Voxel Diffusion." *Proceedings of the IEEE International Conference on Computer Vision* (2021): pp. 5806–5815DOI 10.48550/arxiv.2104.03670. URL https://arxiv.org/abs/2104.03670v3.

[8] Zeng, Xiaohui, Vahdat, Arash, Williams, Francis, Gojcic, Zan, Litany, Or, Fidler, Sanja and Kreis, Karsten. "LION: Latent Point Diffusion Models for 3D Shape Generation." (2022)DOI 10.48550/arxiv.2210.06978. URL https://arxiv.org/abs/2210.06978v1.

[9] Nichol, Alex, Jun, Heewoo, Dhariwal, Prafulla, Mishkin, Pamela and Chen, Mark. "Point-E: A System for Generating 3D Point Clouds from Complex Prompts." (2022)DOI 10.48550/arxiv.2212.08751. URL https://arxiv.org/abs/2212.08751v1.

[10] Nichol, Alex, Dhariwal, Prafulla, Ramesh, Aditya, Shyam, Pranav, Mishkin, Pamela, McGrew, Bob, Sutskever, Ilya and Chen, Mark. "GLIDE: Towards Photorealistic Image Generation and Editing with Text-Guided Diffusion Models." (2021)DOI 10.48550/arxiv.2112.10741. URL https://arxiv.org/abs/2112.10741v3.

[11] Prokhorov, Danil. "A convolutional learning system for object classification in 3-D lidar data." *IEEE Transactions on Neural Networks* Vol. 21 No. 5 (2010): pp. 858–863. DOI 10.1109/TNN.2010.2044802.

[12] Maturana, Daniel and Scherer, Sebastian. "VoxNet: A 3D Convolutional Neural Network for real-time object recognition." *2015 IEEE/RSJ International Conference on Intelligent Robots and Systems (IROS)*: pp. 922–928. 2015. IEEE. DOI 10.1109/IROS.2015.7353481.

[13] Qi, Charles R., Su, Hao, Niessner, Matthias, Dai, Angela, Yan, Mengyuan and Guibas, Leonidas J. "Volumetric and Multi-View CNNs for Object Classification on 3D Data." *Proceedings of the IEEE Computer Society Conference on Computer Vision and Pattern Recognition* Vol. 2016-Decem (2016): pp. 5648–5656. URL https://arxiv.org/abs/1604.03265v2.

[14] Wang, Chu, Pelillo, Marcello and Siddiqi, Kaleem. "Dominant Set Clustering and Pooling for Multi-View 3D Object Recognition." *British Machine Vision Conference 2017, BMVC 2017* (2019)URL https://arxiv.org/abs/1906.01592v1.

[15] Masci, Jonathan, Boscaini, Davide, Bronstein, Michael M. and Vandergheynst, Pierre. "Geodesic convolutional neural networks on Riemannian manifolds." *Proceedings of the IEEE International Conference on Computer Vision* Vol. 2015-Febru (2015): pp. 832–840. URL https://arxiv.org/abs/1501.06297v3.

[16] Ghadai, Sambit, Lee, Xian Yeow, Balu, Aditya, Sarkar, Soumik and Krishnamurthy, Adarsh. "Multi-resolution 3D CNN for learning multi-scale spatial features in CAD models." *Computer Aided Geometric Design* Vol. 91 (2021): p. 102038. DOI 10.1016/J.CAGD.2021.102038.

[17] Su, Hang, Maji, Subhransu, Kalogerakis, Evangelos and Learned-Miller, Erik. "Multi-view convolutional neural networks for 3D shape recognition." *Proceedings of the IEEE International Conference on Computer Vision,* Vol. 2015 Inter: pp. 945–953. 2015. DOI 10.1109/ICCV.2015.114. URL https://www.cv-foundation.org/openaccess/content_iccv_2015/html/Su_Multi-View_Convolutional_Neural_ICCV_2015_paper.html.

[18] Achlioptas, Panos, Diamanti, Olga, Mitliagkas, Ioannis and Guibas, Leonidas. "Learning Representations and Generative Models for 3D Point Clouds." *35th International Conference on Machine Learning, ICML 2018* Vol. 1 (2017): pp. 67–85. DOI 10.48550/arxiv.1707.02392. URL https://arxiv.org/abs/1707.02392v3.

[19] Gunpinar, Erkan, Coskun, Umut Can, Ozsipahi, Mustafa and Gunpinar, Serkan. "A Generative Design and Drag Coefficient Prediction System for Sedan Car Side Silhouettes based on Computational Fluid Dynamics." *Com-*

Copyright © 2023 by ASME

put. *Aided Des.* Vol. 111 (2019): pp. 65–79. DOI 10.1016/J.CAD.2019.02.003.

[20] Umetani, Nobuyuki and Bickel, Bernd. "Learning three-dimensional flow for interactive aerodynamic design." *ACM Transactions on Graphics (TOG)* Vol. 37 No. 4 (2018): p. 10. DOI 10.1145/3197517.3201325. URL https://dl.acm.org/doi/10.1145/3197517.3201325.

[21] Badías, Alberto, Curtit, Sarah, González, David, Alfaro, Icíar, Chinesta, Francisco and Cueto, Elías. "An augmented reality platform for interactive aerodynamic design and analysis." *International Journal for Numerical Methods in Engineering* Vol. 120 No. 1 (2019): pp. 125–138. DOI 10.1002/NME.6127. URL https://hal.science/hal-02457443https://hal.science/hal-02457443/document.

[22] Wu, Zhirong, Song, Shuran, Khosla, Aditya, Yu, Fisher, Zhang, Linguang, Tang, Xiaoou and Xiao, Jianxiong. "3D ShapeNets: A Deep Representation for Volumetric Shapes." (2015). URL http://3dshapenets.cs.princeton.edu.

[23] Wang, Chu, Samari, Babak and Siddiqi, Kaleem. "Local Spectral Graph Convolution for Point Set Feature Learning." *Lecture Notes in Computer Science (including subseries Lecture Notes in Artificial Intelligence and Lecture Notes in Bioinformatics)* Vol. 11208 LNCS (2018): pp. 56–71. URL https://arxiv.org/abs/1803.05827v1.

[24] Boscaini, Davide, Masci, Jonathan, Rodolà, Emanuele and Bronstein, Michael. "Learning shape correspondence with anisotropic convolutional neural networks." *Advances in Neural Information Processing Systems* Vol. 29 (2016).

[25] Wang, Dominic Zeng and Posner, Ingmar. "Voting for voting in online point cloud object detection." *Robotics: Science and Systems* Vol. 11 (2015). DOI 10.15607/RSS.2015.XI.035.

[26] Li, Yangyan, Pirk, Soeren, Su, Hao, Qi, Charles R. and Guibas, Leonidas J. "FPNN: Field Probing Neural Networks for 3D Data." *Advances in Neural Information Processing Systems* (2016): pp. 307–315URL https://arxiv.org/abs/1605.06240v3.

[27] Fey, Matthias, Lenssen, Jan Eric, Weichert, Frank and Müller, Heinrich. "SplineCNN: Fast Geometric Deep Learning with Continuous B-Spline Kernels." *Proceedings of the IEEE Computer Society Conference on Computer Vision and Pattern Recognition* (2017): pp. 869–877URL https://arxiv.org/abs/1711.08920v2.

[28] Michel, Oscar, Bar-On, Roi, Liu, Richard, Benaim, Sagie and Hanocka, Rana. "Text2Mesh: Text-Driven Neural Stylization for Meshes." (2021)DOI 10.48550/arxiv.2112.03221. URL https://arxiv.org/abs/2112.03221v1.

[29] Jetchev, Nikolay. "ClipMatrix: Text-controlled Creation of 3D Textured Meshes." (2021)DOI 10.48550/arxiv.2109.12922. URL https://arxiv.org/abs/2109.12922v1.

[30] Maron, Haggai, Galun, Meirav, Aigerman, Noam, Trope, Miri, Dym, Nadav, Yumer, Ersin, Kim, Vladimir G. and Lipman, Yaron. "Convolutional neural networks on surfaces via seamless toric covers." *ACM Transactions on Graphics (TOG)* Vol. 36 No. 4 (2017). DOI 10.1145/3072959.3073616. URL https://dl.acm.org/doi/10.1145/3072959.3073616.

[31] Ben-Hamu, Heli, Maron, Haggai, Kezurer, Itay, Avineri, Gal and Lipman, Yaron. "Multi-chart Generative Surface Modeling." *SIGGRAPH Asia 2018 Technical Papers, SIGGRAPH Asia 2018* (2018)DOI 10.1145/3272127.3275052. URL http://arxiv.org/abs/1806.02143http://dx.doi.org/10.1145/3272127.3275052.

[32] Saquil, Yassir, Xu, Qun Ce, Yang, Yong Liang and Hall, Peter. "Rank3DGAN: Semantic mesh generation using relative attributes." *AAAI 2020 - 34th AAAI Conference on Artificial Intelligence* (2020): pp. 5586–5594DOI 10.1609/AAAI.V34I04.6011.

[33] Alhaija, Hassan Abu, Dirik, Alara, Knörig, André, Fidler, Sanja and Shugrina, Maria. "XDGAN: Multi-Modal 3D Shape Generation in 2D Space." (2022)DOI 10.48550/arxiv.2210.03007. URL https://arxiv.org/abs/2210.03007v1.

[34] Chen, Zhiqin and Zhang, Hao. "Learning Implicit Fields for Generative Shape Modeling." *Proceedings of the IEEE Computer Society Conference on Computer Vision and Pattern Recognition* Vol. 2019-June (2018): pp. 5932–5941. DOI 10.48550/arxiv.1812.02822. URL https://arxiv.org/abs/1812.02822v5.

[35] Park, Jeong Joon, Florence, Peter, Straub, Julian, Newcombe, Richard and Lovegrove, Steven. "DeepSDF: Learning Continuous Signed Distance Functions for Shape Representation." *Proceedings of the IEEE Computer Society Conference on Computer Vision and Pattern Recognition* Vol. 2019-June (2019): pp. 165–174. DOI 10.48550/arxiv.1901.05103. URL https://arxiv.org/abs/1901.05103v1.

[36] Alwala, Kalyan Vasudev, Gupta, Abhinav and Tulsiani, Shubham. "Pre-train, Self-train, Distill: A simple recipe for Supersizing 3D Reconstruction." (2022): pp. 3763–3772DOI 10.48550/arxiv.2204.03642. URL https://arxiv.org/abs/2204.03642v1.

[37] Liu, Zhengzhe, Dai, Peng, Li, Ruihui, Qi, Xiaojuan and Fu, Chi-Wing. "ISS: Image as Stepping Stone for Text-Guided 3D Shape Generation." (2022)DOI 10.48550/arxiv.2209.04145. URL https://arxiv.org/abs/2209.04145v4.

[38] de Avila Belbute-Peres, Filipe, Economon, Thomas D. and Kolter, J. Zico. "Combining Differentiable PDE Solvers and Graph Neural Networks for Fluid Flow Prediction." *Proceedings of the 37th International Conference on Machine Learning.* 2020. JMLR.org.

[39] Rosset, Nicolas, Cordonnier, Guillaume, Duvigneau, Regis, Bousseau, Adrien and Rosset Guillaume Cordonnier Regis Duvigneau Adrien Bousseau, Nicolas. "Interactive design of 2D car profiles with aerodynamic feedback." *Computer Graphics Forum* Vol. 42 No. 2 (2023): pp. 1–11. URL https://inria.hal.science/hal-03975369https://inria.hal.science/hal-03975369/document.

[40] Remelli, Edoardo, Lukoianov, Artem, Richter, Stephan R., Guillard, Benoît, Bagautdinov, Timur, Baque, Pierre and

Copyright © 2023 by ASME

Fua, Pascal. "MeshSDF: Differentiable Iso-Surface Extraction." *Advances in Neural Information Processing Systems* Vol. 2020-December (2020). URL https://arxiv.org/abs/2006.03997v2.

[41] Baque, Pierre, Remelli, Edoardo, Fleuret, Francois and Fua, Pascal. "Geodesic Convolutional Shape Optimization." *35th International Conference on Machine Learning, ICML 2018* Vol. 2 (2018): pp. 797–809. DOI 10.48550/arxiv.1802.04016. URL https://arxiv.org/abs/1802.04016v1.

[42] Jacob, Sam Jacob, Mrosek, Markus, Othmer, Carsten and Köstler, Harald. "Deep Learning for Real-Time Aerodynamic Evaluations of Arbitrary Vehicle Shapes." *SAE International Journal of Passenger Vehicle Systems* Vol. 15 No. 2 (2021): pp. 77–90. DOI 10.4271/15-15-02-0006. URL http://arxiv.org/abs/2108.05798http://dx.doi.org/10.4271/15-15-02-0006.

[43] Durasov, Nikita, Lukoyanov, Artem, Donier, Jonathan and Fua, Pascal. "DEBOSH: Deep Bayesian Shape Optimization." (2021)URL https://arxiv.org/abs/2109.13337v1.

[44] Thuerey, Nils, Weissenow, Konstantin, Prantl, Lukas and Hu, Xiangyu. "Deep Learning Methods for Reynolds-Averaged Navier-Stokes Simulations of Airfoil Flows." *AIAA Journal* Vol. 58 No. 1 (2018): pp. 25–36. DOI 10.2514/1.j058291. URL http://arxiv.org/abs/1810.08217http://dx.doi.org/10.2514/1.j058291.

[45] Saha, Sneha, Rios, Thiago, Minku, Leandro L., Stein, Bas Vas, Wollstadt, Patricia, Yao, Xin, Back, Thomas, Sendhoff, Bernhard and Menzel, Stefan. "Exploiting Generative Models for Performance Predictions of 3D Car Designs." *2021 IEEE Symposium Series on Computational Intelligence, SSCI 2021 - Proceedings* (2021)DOI 10.1109/SSCI50451.2021.9660034.

[46] Xin, Dajun, Zeng, Junsheng and Xue, Kun. "Surrogate drag model of non-spherical fragments based on artificial neural networks." *Powder Technology* Vol. 404 (2022): p. 117412. DOI 10.1016/J.POWTEC.2022.117412.

[47] TAO, Jun, SUN, Gang, GUO, Liqiang and WANG, Xinyu. "Application of a PCA-DBN-based surrogate model to robust aerodynamic design optimization." *Chinese Journal of Aeronautics* Vol. 33 No. 6 (2020): pp. 1573–1588. DOI 10.1016/J.CJA.2020.01.015.

[48] Sun, Gang and Wang, Shuyue. "A review of the artificial neural network surrogate modeling in aerodynamic design." DOI 10.1177/0954410019864485.

[49] Chang, Angel X., Funkhouser, Thomas, Guibas, Leonidas, Hanrahan, Pat, Huang, Qixing, Li, Zimo, Savarese, Silvio, Savva, Manolis, Song, Shuran, Su, Hao, Xiao, Jianxiong, Yi, Li and Yu, Fisher. "ShapeNet: An Information-Rich 3D Model Repository." (2015)DOI 10.1145/3005274.3005291.

[50] Biswas, Kundan, Gadekar, Ganesh and Chalipat, Sujit. "Development and Prediction of Vehicle Drag Coefficient Using OpenFoam CFD Tool." *SAE Technical Papers* Vol. 2019-Janua No. January (2019). DOI 10.4271/2019-26-0235. URL https://www.sae.org/publications/technical-papers/content/2019-26-0235/.

[51] Szegedy, Christian, Vanhoucke, Vincent, Ioffe, Sergey, Shlens, Jon and Wojna, Zbigniew. "Rethinking the Inception Architecture for Computer Vision." *Proceedings of the IEEE Computer Society Conference on Computer Vision and Pattern Recognition*, Vol. 2016-Decem: pp. 2818–2826. 2016. DOI 10.1109/CVPR.2016.308.

[52] He, Kaiming, Zhang, Xiangyu, Ren, Shaoqing and Sun, Jian. "Deep residual learning for image recognition." *Proceedings of the IEEE Computer Society Conference on Computer Vision and Pattern Recognition* Vol. 2016-Decem (2016): pp. 770–778. DOI 10.1109/CVPR.2016.90.

[53] Xie, Saining, Girshick, Ross, Dollár, Piotr, Tu, Zhuowen and He, Kaiming. "Aggregated Residual Transformations for Deep Neural Networks." *Proceedings - 30th IEEE Conference on Computer Vision and Pattern Recognition, CVPR 2017* Vol. 2017-Janua (2016): pp. 5987–5995. DOI 10.48550/arxiv.1611.05431. URL https://arxiv.org/abs/1611.05431v2.

[54] Dosovitskiy, Alexey, Beyer, Lucas, Kolesnikov, Alexander, Weissenborn, Dirk, Zhai, Xiaohua, Unterthiner, Thomas, Dehghani, Mostafa, Minderer, Matthias, Heigold, Georg, Gelly, Sylvain, Uszkoreit, Jakob and Houlsby, Neil. "An Image is Worth 16x16 Words: Transformers for Image Recognition at Scale." (2020)DOI 10.48550/arxiv.2010.11929. URL https://arxiv.org/abs/2010.11929v2.

[55] Du, Chenzhuang, Li, Tingle, Liu, Yichen, Wen, Zixin, Hua, Tianyu, Wang, Yue and Zhao, Hang. "Improving Multi-Modal Learning with Uni-Modal Teachers." (2021)DOI 10.48550/ARXIV.2106.11059. URL https://arxiv.org/abs/2106.11059.

[56] Song, Binyang, Associate, Postdoctoral, Miller, Scarlett and Ahmed, Faez. "ATTENTION-ENHANCED MULTIMODAL LEARNING FOR CONCEPTUAL DESIGN EVALUATIONS." *Journal of Mechanical Design* (2023): pp. 1–38DOI 10.1115/1.4056669. URL https://asmedigitalcollection.asme.org/mechanicaldesign/article/doi/10.1115/1.4056669/1156042/ATTENTION-ENHANCED-MULTIMODAL-LEARNING-1

[57] Rios, Thiago, Sendhoff, Bernhard, Menzel, Stefan, Back, Thomas and Van Stein, Bas. "On the Efficiency of a Point Cloud Autoencoder as a Geometric Representation for Shape Optimization." *2019 IEEE Symposium Series on Computational Intelligence, SSCI 2019* (2019): pp. 791–798DOI 10.1109/SSCI44817.2019.9003161.

[58] Klokov, Roman and Lempitsky, Victor. "Escape from Cells: Deep Kd-Networks for the Recognition of 3D Point Cloud Models." *Proceedings of the IEEE International Conference on Computer Vision* Vol. 2017-Octob (2017): pp. 863–872. URL https://arxiv.org/abs/1704.01222v2.

[59] Qi, Charles R., Su, Hao, Mo, Kaichun and Guibas, Leonidas J. "PointNet: Deep Learning on Point Sets for 3D Classification and Segmentation." *Proceedings - 30th IEEE Conference on Computer Vision and Pattern Recognition, CVPR 2017* Vol. 2017-Janua (2016): pp. 77–85. URL https://arxiv.org/abs/1612.00593v2.

[60] Kanezaki, Asako, Matsushita, Yasuyuki and Nishida, Yoshifumi. "RotationNet: Joint Object Categorization

and Pose Estimation Using Multiviews from Unsupervised Viewpoints." *Proceedings of the IEEE Computer Society* *Conference on Computer Vision and Pattern Recognition* (2018): pp. 5010–5019DOI 10.1109/CVPR.2018.00526.

Proceedings of the ASME 2023
International Design Engineering Technical Conferences and
Computers and Information in Engineering Conference
IDETC-CIE2023
August 20-23, 2023, Boston, Massachusetts

DETC2023-115259

MODELING THE DYNAMICS OF CUSTOMER DEMAND TO DETERMINE THE OPTIMAL TIME TO RELEASE PRODUCT UPDATES: A COGNITIVE APPROACH

Ian Walter
Elmore Family School of Electrical
and Computer Engineering
Purdue University
West Lafayette, IN 47907 USA

Philip E. Paré
Elmore Family School of Electrical
and Computer Engineering
Purdue University
West Lafayette, IN 47907 USA

Jitesh H. Panchal
School of Mechanical Engineering
Purdue University
West Lafayette, IN 47907 USA

ABSTRACT

The nature of modern products is changing, and design processes need to change accordingly. As products become increasingly software-enabled and with the rise of software-as-a-service models, customer-facing updates can be released more frequently than was previously possible. The development pipelines for software-enabled products are shifting towards agile processes, which emphasize regular product updates, choosing the timing of these updates is important, and requires an understanding of how demand will be influenced by each release. However, the existing discrete-choice models used to predict demand do not capture the dynamic aspects of consumer decision making. To address this gap, we propose a dynamic demand model and demonstrate how it can be used to determine the optimal time to release product updates. The demand model is based on decision field theory (DFT), which enables the modeling of the dynamic and time-varying behavior of human decision makers. The primary contribution in this paper is a product strategy for update release schedules which accounts for the dynamic and time-varying behavior of potential customers. We demonstrate the method using simulations of multiple scenarios.

Keywords: Product design strategy, customer preferences, demand modeling, decision field theory.

Nomenclature

$\eta^i(t_0, t_1, \tau_i^k)$ The expected sales of Product i between times t_0 and t_1, given the kth update of Product i occurs at τ_i^k

$\mathbf{a}_{y,j}(t, z(t))$ Affinity vector for Individual y in Customer Segment j at time-step t for alternative set $z(t)$

α A free variable that affects how long on average it takes for a decision to be made

$\mathbf{C}$ Contrast matrix

$\mathbf{M}_j^{z(t)}$ Subjective evaluation matrix for individuals in Customer Segment j for alternative set $z(t)$

$\mathbf{S}_j^{z(t)}$ Feedback matrix for individuals in Customer Segment j for alternative set $z(t)$

$\phi_{z(t)}^2$ The variance of the change in affinity between time-steps

$\mathcal{T}_{max}^{z_0}$ The first time-step t where $z(t) > z_0$

$\mathcal{T}_{min}^{z_0}$ The first time-step t where $z(t) = z_0$

τ_i^k The time-step Product i is updated for the kth time

$\mathbf{v}_{y,j}(t, z(t))$ Valence vector for Individual y in Customer Segment j at time-step t for alternative set $z(t)$

$\varepsilon(t)$ Noise vector at time-step t

$\mathbf{w}_y^{z(t)}(t)$ Attention proportion vector of individual y at time-step t for alternative set $z(t)$

h The size of the time-step interval

$z(t)$ Index of all product versions at time t

1 Introduction

The nature of modern products is changing, and design processes need to change accordingly. As products shift toward having increasingly greater functionality controlled by general-purpose centralized processing units, companies now have the

Copyright © 2023 by ASME

option to continuously upgrade products after the initial release through software updates. By improving the product through regular updates recurring income can be generated through subscriptions, new customers may be attracted by improved features, and product lifespans can be extended. Consider automobiles for example: today, automobiles are transitioning to become software-defined vehicles [1]. Modern automobiles can often have 100 or more sensors and electronic components. Releasing new yearly versions of a vehicle was necessary to add new functionality, as each component had low compute power and did only exactly what it was intended to do. In software-defined vehicles, many individual sensors and electronic devices are replaced by a few centralized computers with significant computational power [2]. This change in structure means that so long as the data required for a functionality can be collected by existing hardware in the vehicle, new functionality can be added through software updates at any point during the car's lifespan [2].

As cars and other products become software-enabled devices that can be continuously updated even after purchase, the corresponding design and development processes must also be adapted to explicitly account for such updates [1]. Instead of releasing a new product once a year, companies can now develop a common hardware base and then release frequent and regular software updates to add, improve, or change existing functionality [1]. Continuing with the automotive example: Tesla has taken this approach and, instead of releasing regular major hardware revisions, they release software updates roughly once a month for each car [3]. As computers are implemented to control more and more functionality of products, the importance and impact of software is becoming as critical to product design as hardware has always been.

In modern software development, *agile processes* are adopted to release regular and frequent software updates that add or change functionality [4, 5]. Agile development processes emphasize continuous delivery of software updates, and are able to adapt to changing requirements at any point in the development cycle [4]. Products such as cars and airplanes have not traditionally been developed in this manner. The traditional design processes are top-down where a fixed set of requirements is established, the product is designed for those requirements, and active development on that product ends once production begins [6]. While these top-down design processes may still be applicable to the hardware platforms of modern products, with the ability to decouple the software from the hardware and provide over-the-air updates, design and development of the software can continue throughout the entire lifecycle of a product [1].

The lack of ability to support agile development is not the only gap in current design processes. The models used by designers to predict demand based on potential product features fall short in modeling real-world customer behavior. To design both the hardware and software of new products, product designers must be able to accurately predict customer de-

mand. The discrete-choice models (DCM) currently used for demand prediction treat customers as beings with static preferences whose objective is to maximize their utility functions [7]. While the term 'preference' is often used to describe how positively/negatively a decision maker feels about each alternative, in this paper that term refers exclusively to the underlying attribute utility functions. Thus instead of 'preference,' the term 'affinity' is used to describe the quantification of how a decision maker feels about each alternative.

According to the cognitive science of decision making [8], customers gather information and shift their attention between product features throughout their decision-making processes, causing their affinity towards each product to change dynamically based on their subjective feelings [9]. Their decisions are determined by their affinities towards alternatives, either due to a deadline forcing a purchase of the highest-affinity alternative at that time or due to the affinity towards one product exceeding some threshold. Due to the dynamic nature of the decision-making process, the product with the highest affinity at any given time may not be the product that maximizes the underlying preference structure of that decision maker. For example, consider the effect of advertising campaigns which often accompany new product releases. Through advertisements, it is possible to make a decision maker focus on the strong points of one product over other features. While advertisements do not change the utility of each feature to the customer, they are more likely to consider the advertised features of the alternatives they are considering. This shift in attention is not captured in techniques such as the discrete choice analysis [10] because it is assumed that decision makers are essentially rational beings who maximize their (static) utility functions. The dynamics of decision making processes are also not captured in non-rational models of decision making such as cumulative prospect theory because they also assume a static preference structure. This paper is motivated by this research gap.

To address the time-varying nature of customer decision making, one could assume that humans are always aiming to maximize some utility and adapt DCA to time-varying utility functions. However, this approach would not address the deviations from rationality such as strong stochastic transitivity, independence of irrelevant alternatives, and regularity [11]. Instead of taking a utility-maximization perspective, we take an alternate approach developed in cognitive science for modeling how people make decisions over time because of its ability to capture both the dynamics of the decision making process and its ability to model deviations from rational behavior.

The decision making model chosen is based on decision field theory (DFT) and its subsequent extension multialternative decision field theory (MADFT) [8, 12, 13]. DFT is a psychological model that captures many of the time-varying aspects of human decision-making processes. It was created to explain the real-world behavior of decision makers, which often violated the as-

Copyright © 2023 by ASME

sumptions of rationality [12, 13]. Unlike DCM, DFT is a cognitive model of decision-making that does not rely on the quantification of a utility function. Instead, it directly quantifies a person's affinity towards alternatives and how that affinity changes over time as they deliberate about the decision. The quantitative model provided by DFT explains choice response time [14], approach avoidance conflict [15], preference reversals [16], the similarity effect [17], the attraction effect [18, 19], and the compromise effect [20, 21], all of which are often seen in the real-world.

There are gaps in MADFT which make it unsuitable for use as a demand prediction model as it is currently constituted. MADFT assumes that the alternatives are fixed from the beginning of the deliberation process, preventing products from being revised or updated. Additionally, MADFT is typically used to determine what the final decision will be for decision makers within a fixed time-frame, without putting much emphasis on exactly when the decisions would be made. This temporal component is critical to predict demand, as knowing when to expect surges and lapses in demand is necessary for product development. There are other phenomena that affect the choices of consumers such as the context of product use [22], the influence of social networks [23], and the set of products being considered [24]. The model proposed in this paper will serve as a foundation and enable us to analyze these effects in the future.

1.1 Contributions

The primary contribution of this paper is a computational model built on DFT which is used to come up with a product strategy for when different features should be released. We extend the DFT model in two ways. First, we allow for the introduction of new product versions or revisions at arbitrary times. Second, we enable modeling the decision making process for any number of individuals by partitioning a large population into sub-populations (called customer segments) based on similarities in preferences for product attributes. Using this model, we propose a strategy for choosing when to release product updates.

To choose the product release time, we construct an optimization problem for maximizing predicted sales within a specified time window. We then find the optimal release time for some example scenarios via simulations and an exhaustive search. We also propose analytical results for the daily demand, cumulative demand, and expected purchase time as well as highlighting their limitations and potential use-cases.

The paper is organized as follows. In Section 2, the MADFT model of decision making is summarized; further details are available in [11]. The proposed model built on MADFT is then detailed in Section 3 alongside the analytical results. The optimization problem of maximizing sales within a finite time window by selecting the time at which a new version is released is formally defined in Section 4. Applications of this optimization

problem as a product strategy are shown via simulations in Section 5. Finally, closing thoughts are presented in Section 6.

2 CUSTOMER DECISION MAKING MODEL BASED ON DECISION FIELD THEORY

Decision field theory is a psychological model that explains how the attractiveness of different alternatives change throughout the deliberation process [10, 12]. It provides a mathematical framework for analyzing decision-making processes and has been used in cognitive psychology [25] and transportation choice research [26]. DFT is unique among decision-making models because (i) it models the cognitive process of decision making, and (ii) it can explain several empirically observed deviations from rationality, such as the similarity effect, attraction effect, and compromise effect [11]. The extension of DFT known as multi-alternative decision field theory allows for the modeling of decision-making processes with multiple alternatives [13]. There have been numerous extensions of DFT/MADFT proposed which further extend the model's capabilities and range of application [26–30]. By the nature of DFT being a cognitive model of decision making, models built on it are able to capture how individuals adapt to changing circumstances and incorporate new information into their decision-making processes.

DFT provides a model for the deliberation process a decision maker goes through, which captures the temporal dynamics of decisions. By utilizing the fundamental ideas of DFT, it is possible to predict not only which decision alternative will be chosen, but also when the decision will be made and how time pressure affects that decision. Capturing these dynamic behaviors is important to product designers as then can then predict which product a consumer will purchase and when those purchases will likely be made. If there is a customer segment of significant size with similar utilities for product attributes, the demand profile for that segment can thus be predicted. Although individuals may be grouped into customer segments, DFT is applied as a disaggregate demand model in this paper, similar to DCM.

As per the definition of MADFT, there is a set of product alternatives being considered by the decision maker. For each product, a decision maker has an affinity relative to the other products, which is time dependent. The following equations model how these affinities change over a time-step:

$$\mathbf{a}(t+h) = \mathbf{S}\mathbf{a}(t) + \mathbf{v}(t+h) \tag{1a}$$

$$\mathbf{v}(t+h) = \mathbf{C}\mathbf{M}\mathbf{w}(t+h) + \mathbf{C}\varepsilon(t+h). \tag{1b}$$

After each time-step of size h, the decision maker compares some aspect of the alternatives and may receive new information, both of which will influence the relative attractiveness of

Copyright © 2023 by ASME

each alternative. In the column vector of affinities $\mathbf{a}(t)$, each element is the decision maker's affinity towards a specific product. These affinity values at the next time-step depend on the previous affinity values scaled by the feedback matrix $\mathbf{S}$. The $\mathbf{S}$ matrix is called the feedback matrix because it captures both the comparison and memory of the affinity values at the previous time-step. The diagonal elements of $\mathbf{S}$ determine the memory of the decision maker, or how much the previous affinity values influence current and future values. The off-diagonal elements of $\mathbf{S}$ determine the competitive influence of the alternatives on each other, and the strength of these values come from the concept of lateral inhibition. In essence, lateral inhibition is the idea that an alternative will be compared more harshly to similar alternatives than distinct alternatives [31].

The scaled affinity $\mathbf{S}\mathbf{a}(t)$ is combined with the 'input' term $\mathbf{v}(t+h)$. The $\mathbf{v}(t)$ vector, which is based entirely on new information and defined in (1b), is referred to as the *valence* and is the result of the decision maker's instantaneous comparison of the alternatives based on their subjective preferences. Each row of the $\mathbf{M}$ matrix contains the decision maker's personal, subjective, evaluation of the corresponding alternative's features. At each time-step, the decision maker dedicates their attention to a subset of features for comparison. The jth element of the stochastic column vector $\mathbf{w}(t)$, $\mathbf{w}(t)[j]$, is the proportion of time spent comparing the feature j between the alternatives. The values of the elements change at each time-step according to a Markov process, with the constraints that for each entry j, $0 \leq \mathbf{w}(t)[j] \leq 1$, and the sum of all elements in $\mathbf{w}(t)$ is less than or equal to one. The sum may be less than 1 as some portion of time may be spent not comparing the alternatives. The selected features are then compared through the contrast matrix $\mathbf{C}$, which is defined for n alternatives as

$$\mathbf{C} = \mathbf{I} - \frac{1}{n-1}(\mathbf{1} - \mathbf{I})$$

where $\mathbf{I}$ is the $n \times n$ identity matrix and $\mathbf{1}$ is the $n \times n$ matrix filled with ones. By utilizing a contrast matrix of this structure, the sum of all affinities will never change. The resulting effect is that in order for affinity towards one alternative to increase, affinity for at least one other alternative must decrease. Finally, there is additional randomness in the decision-making process which can stem from the acquisition of new information or alternative features not modeled. This randomness is approximated by the zero-mean random noise term $\varepsilon(t)$. This random noise is also multiplied by the contrast matrix $\mathbf{C}$, which is required to ensure the valence sums to 0.

The model in (1) can be used to predict a decision outcome under multiple situations. For a decision where there is no hard deadline, such as deciding on the purchase of a non-essential product, the individual makes a purchase once their affinity for

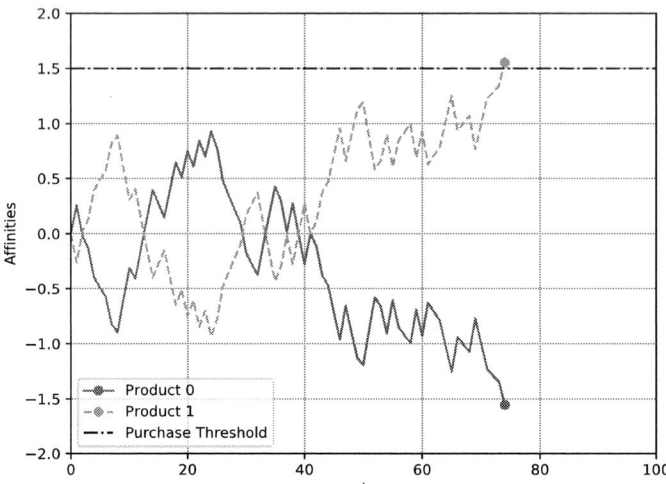

FIGURE 1: An example of how the elements of the affinity vector $\mathbf{a}(t)$ change over time until a decision is made. The decision in this example is made at $t = 74$ when the affinity for Product 1 exceeds the purchase threshold of 1.5.

one alternative exceeds some threshold value. If there is a specific time at which a decision must be made, then the decision maker would choose whichever alternative they have the highest affinity for at that time. Otherwise, if there is a deadline but they can also make a decision ahead of time (e.g., accepting/rejecting school offers) then it would be a combination of the first two scenarios. If at some point before the deadline the affinity for one alternative exceeds a threshold, then the decision is made; otherwise, the decision is made at the final time-step based on the highest affinity. An example decision-making process can be seen in Figure 1. In the figure, a decision maker is considering two competing products, with no fixed deadline. The affinity for each product starts at 0 and thus the affinity for Product 0, $\mathbf{a}^0(t)$, is equal to the negative affinity for Product 1, $\mathbf{a}^1(t)$. After comparing the two products over 74 time-steps, the affinity for Product 1 passes the affinity threshold for making a purchase of 1.5, and thus the decision is made to purchase Product 1 at $t = 74$. If this decision instead had a deadline of $t = 20$, for this individual neither affinity exceeded the purchase threshold before this time and the decision maker would have purchased Product 0 as $\mathbf{a}^0(20) > \mathbf{a}^1(20)$.

3 EXTENSION OF DFT

We extend the computational decision-making model from DFT in two ways. First, we allow for products to be revised at arbitrary times, with the revised product replacing the previous version. Second, we enable predicting demand profiles from a population composed of discrete individuals with different subjec-

Copyright © 2023 by ASME

tive evaluations of the same products (i.e., different underlying preferences and/or perceptions of probability). Finally, we show analytical approximations of the model which provide formulas relating to the probability of individuals purchasing a specified product and the time of its occurrence.

3.1 Preference Dynamics Extension of DFT

The population of potential customers for any product consists of diverse individuals, each with their own process for evaluating a product. Individuals with similar preferences will subjectively evaluate the features of competing products similarly. The subjective evaluations capture the importance of attributes to an individual as well as their subjective probabilities of the states of nature (i.e., the probability they feel different scenarios will occur, such as a product being broken vs functioning fine). For predicting demand from relatively large populations, individuals with similar subjective evaluations can be grouped together into a single customer segment. The assumption is that individuals in the same customer segment not only place similar importance on attributes but also have similar subjective probabilities over the states of nature. There will be minor variations between individuals, but these can be accounted for by the noise term $\varepsilon(t)$. The more diverse the set of individuals being grouped together, the larger the variance of $\varepsilon(t)$ is. The subjective evaluation structure of a decision maker is captured by the $\mathbf{M}$ matrix and therefore an $\mathbf{M}$ matrix must be defined for each customer segment. For a Customer Segment j, the common $\mathbf{M}$ matrix used is denoted by $\mathbf{M}_j$.

Additionally, each product considered can be updated or revised at any time-step throughout the decision making process. The time-step at which Product i is revised for the kth time is represented by τ_i^k. The initial version of a product is referred to as the 0th revision, and thus, if a product is under consideration at the start of the deliberation period, (i.e., the time-frame over which the alternatives are being considered), $\tau_i^0 = 0$. In this work it is assumed that there are at most two versions of each product (the original version and one revision). This assumption is made to simplify notation for the purpose of explaining the concept, but the model itself has no limit to the number of revisions. To track the current state of all n products, we define the variable $z(t)$ as the binary index of the number of revisions per product at time t. Each bit i of $z(t)$ is 1 if the corresponding Product i has changed once. For instance, with two products, prior to any product changes $z(t) = 00$, after Product 0 changes the first time at τ_0^1, $z(\tau_0^1) = 01$, and then if Product 1 changes at time τ_1^1 we get $z(\tau_1^1) = 11$. Note that $z(t)$ is non-decreasing and if $\alpha < \beta$, then $\tau_{max}^\alpha \le \tau_{min}^\beta$. Additionally, it is necessary to be able to indicate when the set of products entered or exited a particular set of revisions. To denote the time at which the set of products entered the state z_0 the notation $\mathscr{T}_{min}^{z_0}$ is used, and the time t_1 at which at least one product is revised causing $z(t_1) > z_0$

is denoted by $\mathscr{T}_{max}^{z_0}$. Formally these terms are defined as

$$\mathscr{T}_{min}^{z_0} = \inf\{t \in [0,T] \mid z(t) \ge z_0\}, \tag{2}$$

$$\mathscr{T}_{max}^{z_0} = \inf\{t \in [\mathscr{T}_{min}^{z_0}, T] \mid z(t) > z_0\}. \tag{3}$$

If $\mathscr{T}_{min}^{z_0} = \mathscr{T}_{max}^{z_0}$, then the set of revisions specified by z_0 occurred for exactly 0 time-steps. If the set of revisions z_0 occurs for 0 time-steps, it is either skipped (e.g., $z: 01 \to 11$ skips the state 10) or is never reached (e.g., if no products are revised then $\mathscr{T}_{min}^{z_0} = \mathscr{T}_{max}^{z_0} = T$ for $z_0 = 01, 10, 11$).

When a product is revised, the features of that product are updated and thus, the subjective value the decision maker assigns each product's features may also change. Hence, when any product is revised, the $\mathbf{M}_j$ matrix for all customer segments j will update. As $\mathbf{M}_j$ may change with each product revision, denote $\mathbf{M}_j^{z(t)}$ as the $\mathbf{M}$ matrix for Customer Segment j under the product versions corresponding to $z(t)$. The probability each feature is considered by each customer segment may also change based on what features are available in the considered products. If a new feature is introduced, causing customers to evaluate a new feature that was previously nonexistent, the probability of each feature being considered will change. As such, define $\mathbf{w}_j^{z(t)}(t)$ as the random attribute consideration vector for Customer Segment j under product versions $z(t)$.

The last component of (1) that may change as the product versions change is the feedback matrix $\mathbf{S}$. The matrix $\mathbf{S}_j$ (the feedback matrix for Customer Segment j) is constructed by

$$\mathbf{S}_j = \mathbf{I} - h\Gamma_j \tag{4}$$

where $\mathbf{I}$ is the identity matrix and Γ_j is the similarity and memory-loss matrix. The off-diagonal elements of Γ_j are decreasing functions of the distance between the corresponding alternatives based on the subjective attribute evaluations, and after being subtracted from $\mathbf{I}$ result in competitive interactions. The diagonal elements of Γ_j determine how much the affinity values from the previous time-step should be forgotten. When the attributes of products are updated at each revision, the distance between products will change accordingly and thus, the resulting $\mathbf{S}_j$ matrix will update. Denote the feedback matrix that changes based on the product versions as $\mathbf{S}_j^{z(t)}$. Additionally, to prevent the changes in results that occur when the time-step size h is changed, for the proposed extension h is fixed at one following the authors of MADFT [13].

Using the newly-defined notation, the affinity dynamics for Individual y who is a member of Customer Segment j are defined by

$$\mathbf{a}_{y,j}(t, z(t)) = \mathbf{S}_j^{z(t)} \mathbf{a}_{y,j}(t-1, z(t-1)) + \mathbf{v}_{y,j}(t, z(t)), \tag{5a}$$

$$\mathbf{v}_{y,j}(t, z(t)) = \mathbf{C}\mathbf{M}_j^{z(t)} \mathbf{w}_y^{z(t)}(t) + \varepsilon(t). \tag{5b}$$

Copyright © 2023 by ASME

To use the extended affinity model in (5) for demand prediction, the same process described for the original DFT model applies. If there are N_j individuals grouped under Customer Segment j, then N_j independent numerical simulations are carried out using the same parameters. Conclusions about the demand profile can then be drawn via Monte Carlo simulations and observing when each individual from each customer segment make purchases of each competing product.

3.2 Analysis of DFT Extension

Some analytical results relating to the expected decisions for the 'optional stopping time' decision-making scenario, which is the scenario most applicable to customers who may or may not purchase any product in a category, are derived in this section. The affinity dynamics process proposed is a discrete-time process with infinite states: the affinity values cannot be guaranteed to fall on any regularly-spaced grid interval. Drawing analytical conclusions about when such processes will reach or exceed some threshold value, also known as the first hitting time, is difficult. However, drawing conclusions about the first hitting time for discrete-time random walk processes with finite state spaces is a well-studied problem. Thus the analytical results in this section come from approximating the affinity dynamics process as a discrete-time finite state space random walk process. In order to approximate the affinity dynamics process, an 'intermediate' conversion to a continuous time diffusion process is used. This intermediate step is taken because the affinity dynamics process converges in distribution to the diffusion process known as the Ornstein-Uhlenbeck (OU) process as the time-step interval h goes to 0, and this diffusion process can be approximated as a random walk [32]. Thus, for this analysis, h is not fixed at one.

Before the affinity dynamics process (5) can be converted into a diffusion process, the dynamics are split into different time windows based on the states (i.e., versions) of the products. Notice that for some arbitrary time t_0, at any time $t \in [\mathcal{T}_{min}^{z(t_0)}, \mathcal{T}_{max}^{z(t_0)})$ the variable $z(t)$ is constant at some value notated by z_0. For a fixed z_0, the only values that will change are the stochastic consideration vector $\mathbf{w}_j^{z_0}(t)$ and the error term $\varepsilon(t)$, as is the case in the original DFT model. Within this time period, each customer segment can be analyzed in isolation as every individual's affinity is assumed to evolve independently. Thus, for a single Individual y in Customer Segment j, their affinity dynamics evolve according to the stochastic process

$$\mathbf{a}_{y,j}(t+h, z_0) = \mathbf{S}_j^{z_0} \mathbf{a}_{y,j}(t, z_0) + \mathbf{v}_{y,j}(t+h, z_0), \quad t \in [\mathcal{T}_{min}^{z_0}, \mathcal{T}_{max}^{z_0}). \tag{6}$$

For the special case with two alternatives, the terms $\mathbf{S}_j^{z_0}(t, z_0)$, $\mathbf{a}_{y,j}(t, z_0)$, and $\mathbf{v}_{y,j}(t, z_0)$ can be written as

$$\mathbf{S}_j^{z_0} = \begin{bmatrix} 1 - h\gamma_1 & -h\gamma_2 \\ -h\gamma_2 & 1 - h\gamma_1 \end{bmatrix}, \quad \mathbf{a}_{y,j} = \begin{bmatrix} \mathbf{a}_{y,j}^0 \\ -\mathbf{a}_{y,j}^0 \end{bmatrix}, \quad \mathbf{v}_{y,j} = \begin{bmatrix} \mathbf{v}_{y,j}^0 \\ -\mathbf{v}_{y,j}^0 \end{bmatrix}$$

where $\mathbf{a}_{y,j}^0$ is the affinity towards the first alternative and $\mathbf{v}_{y,j}^0$ is the valence on the first alternative. Using these definitions, the affinity dynamics can be written as the one-dimensional stochastic equation

$$\mathbf{a}_{y,j}^0(t+h, z_0) = [1 - h(\gamma_1 - \gamma_2)]\mathbf{a}_{y,j}^0(t, z_0) + \mathbf{v}_{y,j}^0(t+h, z_0). \tag{7}$$

This affinity state process converges in distribution to the OU diffusion process as $h \to 0$ (see [32]). The parameters which quantify this OU process are the drift rate $\mu_1(x)$ and the diffusion rate ϕ^2, where ϕ is the standard deviation of the OU process. To determine these parameters, the variance and the expected value of the change in affinity at each time-step are required. The one-dimensional affinity dynamics in (7) are thus rearranged into the following stochastic difference equation for the change in affinity between time-steps:

$$\Delta \mathbf{a}_{y,j}^0(t+h, z_0) = [\mathbf{a}_{y,j}^0(t+h, z_0) - \mathbf{a}_{y,j}^0(t, z_0)] \tag{8}$$
$$= \mathbf{v}_{y,j}^0(t+h, z_0) - h \cdot (\gamma_1 - \gamma_2)\mathbf{a}_{y,j}^0(t, z_0).$$

Finally, the expected value and the variance of the change in affinity at each time step are found to be

$$\mathbb{E}\left[\Delta \mathbf{a}_{y,j}^0(t+h, z_0) \mid \mathbf{a}_{y,j}^0(t, z_0) = x\right] = \tag{9a}$$
$$h\left(\mathbb{E}[\mathbf{v}_{y,j}^0(t+h, z_0) \mid \mathbf{a}_{y,j}^0(t, z_0) = x] - (\gamma_{11} - \gamma_{12})x\right) = h\mu_1^{z_0}(x)$$

$$\mathrm{Var}(\Delta \mathbf{a}_{y,j}^0(t+h, z_0) \mid \mathbf{a}_{y,j}^0(t, z_0) = x) = \tag{9b}$$
$$\mathbb{E}\left[(\mathbf{v}_{y,j}^0(t+h, z_0))^2\right] - \left(h\mu_1^{z_0}(x)\right)^2 = h\phi_{z_0}^2$$

where, again, $\mu_1(x)$ is the drift rate and ϕ^2 is the diffusion rate of the OU process.

The increments of the random walk must be specified before the transition probabilities can be determined, and are set to be $\lambda = 0$, $\lambda = \alpha\phi_{z_0}\sqrt{h}$, or $\lambda = -\alpha\phi_{z_0}\sqrt{h}$. The parameter $\alpha \geq 1$ is a free parameter which affects the temporal behavior of the random walk. Given that the step sizes have intervals with magnitude $\alpha\phi_{z_0}\sqrt{h}$, the total number of states in the random walk is

$$l = 2\left\lceil \frac{\theta}{\alpha\phi_{z_0}\sqrt{h}} \right\rceil + 1 \tag{10}$$

where θ is the purchase threshold of the affinity process and $\lceil \cdot \rceil$ is the ceiling function. The transition probabilities for a positive, negative, and neutral step are $p_{z_0}(x)$, $q_{z_0}(x)$, and $r_{z_0}(x)$, respectively, and are found by equating the normalized moments of the random walk and OU process:

Copyright © 2023 by ASME

$$\frac{\mathbb{E}\left[\Delta \mathbf{a}^0_{y,j}(t+h,z_0) \mid \mathbf{a}^0_{y,j}(t)=x\right]}{h} = \frac{\lambda}{h} \cdot \left[p_{z_0}(x) - q_{z_0}(x)\right] = \mu^{z_0}_1(x) \tag{11}$$

$$\frac{\mathrm{Var}\left(\Delta \mathbf{a}^0_{y,j}(t+h,z_0) \mid \mathbf{a}^0_{y,j}(t)=x\right)}{h} = \tag{12}$$

$$\frac{\lambda^2}{h} \cdot \left[p_{z_0}(x) - q_{z_0}(x)\right] - h(\mu^{z_0}_1(x))^2 = \phi^2_{z_0}.$$

Assuming h is sufficiently small, the second-order term $h(\mu^z_1(x))^2$ in (12) can be ignored. The resulting transition probabilities are

$$p_{z_0}(x) = \frac{1 + \frac{\alpha\sqrt{h}\mu^z_1(x)}{\phi_{z_0}}}{2\alpha^2} \tag{13a}$$

$$q_{z_0}(x) = \frac{1 - \frac{\alpha\sqrt{h}\mu^z_1(x)}{\phi_{z_0}}}{2\alpha^2} \tag{13b}$$

$$r_{z_0}(x) = 1 - \alpha^{-2} \tag{13c}$$

with the constraints $0 \le p_{z_0}(x), q_{z_0}(x), r_{z_0}(x) \le 1$ and $p_{z_0}(x) + q_{z_0}(x) + r_{z_0}(x) = 1$ satisfied. The $l \times l$ transition matrix $\mathbf{T}_{z_0}$ is constructed as

$$\mathbf{T}_{z_0}(i,j) = \begin{cases} p_{z_0}(i\lambda) & \text{if } j = i+1 \\ q_{z_0}(i\lambda) & \text{if } j = i-1 \\ r_{z_0}(i\lambda) & \text{if } j = i. \end{cases} \tag{14}$$

The row-stochastic transition matrix $\mathbf{T}_{z_0}$ is partitioned into a sub-matrix and vectors. Let $\mathbf{Q}_{z_0}$ be the inner $(l-2) \times (l-2)$ transition matrix, $\mathbf{Q}_{z_0} = \mathbf{T}_{z_0}[2:l-1, 2:l-1]$, representing the transition probabilities between the transient states; $\mathbf{k}_{i,z_0}$ be the $(l-2) \times 1$ column vector containing the transition probabilities from all transient states to the 'purchased' state corresponding to Product i: $\mathbf{k}_0 = \mathbf{T}_{z_0}[2:l-1, l]$, and $\mathbf{k}_1 = \mathbf{T}_{z_0}[2:l-1, 1]$. Additionally, let $\mathbf{p}_t$ be the $(l-2) \times 1$ affinity probability distribution column vector at time t. These partitions are used in the following propositions to provide equations for the demand probabilities and expectations.

Proposition 1. *The probability that an Individual y, a member of Customer Segment j, will make a purchase of Product i at time t, defined as $\mathbb{P}(Z^i_{y,j}(t)=1)$, is*

$$\mathbb{P}(Z^i_{y,j}(t)=1) = \mathbf{p}'_0 \left(\prod_{z=00}^{z(t)-01} \mathbf{Q}_z^{\frac{\mathcal{T}^z_{max}-\mathcal{T}^z_{min}}{h}} \right) \mathbf{Q}_{z(t)}^{\frac{t-\mathcal{T}^{z(t)}_{min}}{h}} \mathbf{k}_{i,z(t)}. \tag{15}$$

The probability of an Individual y making a purchase of Product i at time t can be summed over all individuals in all customer segments to get the expected number of purchases at time t, or, in other words, the expected demand. Letting c be the number of customer segments and N_j be the number of individuals in Customer Segment j, the expected demand for Product i is

$$\eta^i(t,t) = \sum_{j=0}^{c} \sum_{y=1}^{N_j} \mathbb{P}(Z^i_{y,j}(t)=1). \tag{16}$$

Due to the time-steps of the random walk being very small relative to the original time-step size of 1, the cumulative demand is more applicable when comparing this predicted demand against the demand simulated by the original affinity process. The cumulative demand distribution for a single individual is provided by the following proposition.

Proposition 2. *Let $X^i_{y,j}(T)$ be an indicator variable of whether Individual y in Customer Segment j has purchased Product i at or before time T. The probability of a purchase being made at or before a finite time T is given by*

$$\mathbb{P}(X^i_{y,j}(T)=1) =$$

$$\mathbf{p}'_0 \sum_{t=0}^{T} \left[\left(\prod_{z=00}^{z(t)-01} \mathbf{Q}_z^{\frac{\mathcal{T}^z_{max}-\mathcal{T}^z_{min}}{h}} \right) \mathbf{Q}_{z(t)}^{\frac{t-\mathcal{T}^{z(t)}_{min}}{h}} \mathbf{k}_{i,z(t)} \right]. \tag{17}$$

The last product change occurs at time $\mathcal{T}^* = \lim_{t\to\infty} \mathcal{T}^{z(t)}_{min}$ and thus

$$\lim_{T\to\infty} \mathbb{P}(X^i_{y,j}(T)=1) =$$

$$\mathbf{p}'_0 \sum_{t=0}^{\mathcal{T}^*} \left[\left(\prod_{z=00}^{z(t)-01} \mathbf{Q}_z^{\frac{\mathcal{T}^z_{max}-\mathcal{T}^z_{min}}{h}} \right) \mathbf{Q}_{z(t)}^{\frac{t-\mathcal{T}^{z(t)}_{min}}{h}} \mathbf{k}_{i,z(t)} \right]$$

$$+ \mathbf{p}'_{\mathcal{T}^*} \left(\mathbf{I} - \mathbf{Q}_{z(\mathcal{T}^*)} \right)^{-1} \mathbf{k}_{i,z(\mathcal{T}^*)}. \tag{18}$$

The probability of a purchase being made, $\mathbb{P}(X^i_{y,j}(T)=1)$, is equivalent to the expected number of purchases made by the assumption that each individual may only ever make one purchase. Hence the expected number of purchases by time T from the entire population can be computed as

$$\eta^i(0,T) = \sum_{j=0}^{c} \sum_{y=1}^{N_j} \mathbb{P}(X^i_{y,j}(T)=1). \tag{19}$$

Proposition 3. *Let u^i_y be the time at which Individual y chooses Product i. If the individual never chooses Product i then $u^i_y = \infty$, otherwise u^i_y is the time such that $Z^i_{y,j}(u^i_y) = 1$. Considering only individuals who will choose Product i within the time*

Copyright © 2023 by ASME

window $[0,T]$, the expected decision time u_y^i is given by

$$\mathbb{E}[u_y^i \in [0,T] \mid X_{y,j}^i(T) = 1] =$$

$$\mathbf{p}_0' \sum_{t=1}^{T} \left[t \cdot \left(\prod_{z=00}^{z(t)-01} \mathbf{Q}_z^{\frac{\mathscr{T}_{max}^z - \mathscr{T}_{min}^z}{h}} \right) \mathbf{Q}_{z(t)}^{\frac{t - \mathscr{T}_{min}^{z(t)}}{h}} \mathbf{k}_{i,z(t)} \right]. \quad (20)$$

This proposition provides an equation that, after averaging over all individuals and customer segments, directly translates to the market share of each product.

3.2.1 Limitations of the Random Walk Approximation

The method followed to derive the random walk approximation used in Propositions 1-3 was inspired by an approach used in [11]. Due to how the random walk approximation is constructed, there are multiple limitations of these analytical results. Recall that, prior to defining the affinity dynamics in (5), the assumption is made that the time-step interval h is fixed to the value of 1. In deriving the covariance matrix of $\mathbf{a}$ from the dynamics in (5) it was found that the covariance of $\mathbf{a}$ depends on h^2. However, in [11], it was assumed that the covariance depends linearly on h. The only non-zero value of h which enables the derived and assumed covariance matrices to be equivalent is $h = h^2 = 1$. However, in order to reach the random walk approximation, h is allowed to become arbitrarily small which results in behavior that differs from the original process. Additionally, the way the resulting random walk's affinity value step sizes are determined almost always result in a difference between the original threshold value and the value of the equivalent state for the random walk.

Another issue is presented by the α parameter. In the context of predicting demand, *when* individuals make their purchasing decisions matters, not just *what* decision they make. The free parameter α controls the probability that the random walk remains in the same state between time-steps, effectively controlling the speed at which the affinity changes. However, there is no effective way to make sure that the random walk's behavior matches the original process with respect to time, other than by simulating both models and calculating the error.

The choices of both h and α change the magnitude of the random walk step-size λ. Different step sizes cause the affinity value of the absorbing state, $\lambda \cdot (l-1)/2$, to change. The difference between the absorbing state's affinity value and the original threshold value θ will hence change for every pair of h and α values selected. The limitations discussed here are detailed further in Section 5.1.

3.2.2 Applications

While there are a number of limitations when it comes to applying the analytical results, they are helpful in providing rapid approximations. Once the h and α values are calibrated for each customer segment, the demand from multiple customer segments of various sizes can be predicted easily. Additionally, if the cumulative demand and the demand distribution do not need to be exact but instead the general behavior of the demand profiles is required, most pairs of h and α values result in reasonable approximations. These approximations would be useful for quickly checking which product the customer segments would prefer, or how the demand profile changes when a product is updated at various times.

4 PRODUCT RELEASE STRATEGY

For any corporate entity, the purpose of releasing a new product or revision is to generate sales which correspond to generating revenue. For simplicity, we assume that the best time to update a product is the time that will result in the most sales in a specified time window. Given this goal, a strategy for product updates which optimizes the number of product sales within a fixed time frame is proposed. The optimal time to release a product is determined by solving the optimization problem constructed in this section.

The model proposed in Section 3 is general, but the optimization problem constructed here is restricted for demonstration. In the construction of this optimization problem, we are not considering the cost of producing or releasing the updated design. Additionally, the optimizing firm has perfect knowledge of their competitor's product and sales. The decision variable, or what the corporation has control over, is τ_0^1, the time at which the revision or update of Product 0 is released. We assume that at most one revision of Product 0 can be released in the time frame considered, and that it is ready and finalized at the beginning of the considered release window. This assumption means that the features of the updated product will be the same no matter when the product is released. The revision replaces the original product, and we assume that the profits do not depend on the time a product is sold or whether the sale is before or after the update is released.

Following the DFT, the predicted number of sales between times t_0 and t_1 can be found by summing the predicted sales at each time-step in this window. An individual makes a purchase once their affinity for a product exceeds a threshold denoted by θ. The probability that Individual y makes a purchase of Product i at time t is equivalent to the probability that Individual y's affinity for Product i exceeds the purchase threshold θ, and they have made no purchase at a previous time-step. The affinity dynamics at each time-step t, defined in (5), will change based on the product versions available. For values $t < \tau_0^1$ the system is in state $z(t) = 00$, while for $t \geq \tau_0^1$ the state is $z(t) = 01$. Using an indicator variable to indicate whether a purchase is made or not,

Copyright © 2023 by ASME

for time-steps $t < \tau_0^1$, we have

$$
Z_{y,j}^i(t,00) = \begin{cases} 1 & \begin{aligned} & \mathbf{a}_{y,j}^i(t,00) \geq \theta, \\ & \mathbf{a}_{y,j}^g(k,00) < \theta \quad \forall g \in [0,n-1],\, k < t \end{aligned} \\ 0 & \text{otherwise,} \end{cases}
$$

(21)

and, for time-steps $t \geq \tau_0^1$, we have

$$
Z_{y,j}^i(t,01) = \begin{cases} 1 & \begin{aligned} & \mathbf{a}_{y,j}^i(t,01) \geq \theta, \\ & \mathbf{a}_{y,j}^g(k,01) < \theta \quad \forall g \in [0,n-1],\, \tau_0^1 \leq k < t, \\ & \mathbf{a}_{y,j}^g(k,00) < \theta \quad \forall g \in [0,n-1],\, k < \tau_0^1 \end{aligned} \\ 0 & \text{otherwise.} \end{cases}
$$

(22)

The probability of Individual y making a purchase of Product i within a specified time window $t \in (t_0, t_1]$ is then found by summing up the purchase probabilities at each time. The expected value of $Z_{y,j}^i(t)$ is the expected number of purchases of Product i from Individual y at time t. The number of purchases are limited to either 0 or 1, and thus the expected number of purchases is equivalent to the probability of a purchase. Thus for Individual y the probability a purchase is made within the time window $t \in [t_0, t_1]$, equivalent to the expected number of purchases made, is

$$
X_{y,j}^i(t_0, t_1, \tau_0^1) = \sum_{t=t_0}^{\tau_0^1} Z_{y,j}^i(t,00) + \sum_{t=\tau_0^1}^{t_1} Z_{y,j}^i(t,01). \quad (23)
$$

To find the total number of expected sales by all individuals in Customer Segment j, we can scale the expected number of purchases for one individual in said segment the size of that segment, due to the assumption that each individual's purchasing decision is independent from all others. Lastly, to get the total number of expected sales, the expected number of sales from each segment j are summed together:

$$
\eta^i(t_0, t_1, \tau_0^1) = \sum_{\forall j} N_j X_{y,j}^i(t_0, t_1, \tau_0^1) = \sum_{\forall j} \sum_{y=1}^{N_j} X_{y,j}^i(t_0, t_1, \tau_0^1), \quad (24)
$$

where N_j is the number of individuals in Customer Segment j. The optimization problem of maximizing sales by time T is for-

mally defined as

$$
\begin{aligned}
\max_{\tau_0^1} \quad & \eta^i(0, T, \tau_0^1) \\
\text{s.t.} \quad & \mathbf{a}_{y,j}(t,z(t)) = \mathbf{S}_j^{z(t)} \mathbf{a}_{y,j}(t-1, z(t-1)) \\
& \qquad + \mathbf{CM}_j^{z(t)} \mathbf{w}_y^{z(t)}(t) + \varepsilon(t).
\end{aligned}
$$

(25)

With the above optimization problem, the optimal time to release the product update can be obtained. If the affinity dynamics approximation is used to find the optimal release time, η^i in (25) must be replaced with (19).

5 SIMULATIONS

Consider an individual starting a vehicle search, comparing two competing alternatives. Initially the individual may not lean towards either alternative, but over time they will lean towards one alternative or the other, with the measure of this called the individual's *affinity* toward each alternative. At some point during this decision-making process, a major update occurs in one alternative which improves some aspect, such as safety, at the cost of convenience. A more general version of this scenario is simulated and analyzed in this section, where generic products are used with arbitrary attributes. In the following simulations, there are four attributes assumed to be the most important features to the customer segments modeled. The quantitative representations of each attribute for both products and their updated versions are listed in Table 1a. Each customer segment subjectively evaluates these features using the equations listed in Table 1b. The equation used to quantify the difference between products (i.e., the off-diagonal elements of Γ) is $0.025 e^{-\|\mathbf{M}_0 - \mathbf{M}_1\|_2}$, where $\mathbf{M}_0$ is the first row of the $\mathbf{M}$ matrix corresponding to Product 0's evaluation and $\mathbf{M}_1$ is Product 1's evaluation, and $\|\cdot\|_2$ is the two-norm function. The likelihood each feature is considered at each time-step is uniform over the features currently available at that time-step. Explicitly, at time-steps $t < \tau_0^1$ when $z(t) = 00$ or 10 the stationary distribution of $\mathbf{w}_y^{z(t)}(t)$ is $[1/3,\ 1/3,\ 1/3,\ 0]^\mathrm{T}$ as neither product has the fourth feature. However, at time-steps $t \geq \tau_0^1$ the stationary distribution of $\mathbf{w}_y^{z(t)}(t)$ is $[1/4,\ 1/4,\ 1/4,\ 1/4]^\mathrm{T}$. Each entry of $\varepsilon(t)$ is distributed according to the normal distribution $\mathcal{N}(0, 0.01)$.

5.1 Comparison between simulations and analytical results

To compare the analytical predictions from (17) with the simulations of (5), an example scenario is constructed with customers from Segments A and B, detailed in Table 1, comparing the original version of Product 0 (P0) and Product 1 (P1). The cumulative demand predicted using (17) is compared to the cumulative demand from simulating the affinity dynamics defined

Copyright © 2023 by ASME

(a)

Product	Att 0	Att 1	Att 2	Att 3
P0 Ver. 0	0.5	0.3	0.8	0
P0 Ver. 1	0.6	0.4	0.7	0.3
P1 Ver. 0	0.6	0.6	0.4	0
P1 Ver. 1	0.7	0.8	0.2	0

(a) Quantified values of the different attributes/features of the original and updated versions of Product 0 and 1.

(b)

Cust. Segment	Att 0	Att 1	Att 2	Att 3
Segment A	x_0^2	x_1^2	$\sqrt{x_2}$	$\sqrt{x_3}$
Segment B	x_0	$\sqrt{x_1}$	x_2^2	x_3^2
Segment C	$0.1 \cdot x_0$	$0.1 \cdot x_1$	$0.3 \cdot x_2$	$-0.2 \cdot x_3$
Segment D	$0.1 \cdot x_0$	$0.1 \cdot x_1$	$0.8 \cdot x_2$	$-0.7 \cdot x_3$

(b) Equations used by the customer segments to subjectively evaluate the product features.

TABLE 1: These tables contain the information required to create the M matrices used in simulations. The values in Table 1a could, for instance, be the ratings for each attribute given by a magazine read by each individual. The equations in Table 1b describe how individuals of each customer segment subjectively evaluate these reported (or otherwise collected) values. The specific values in Table 1a and the equations in Table 1b were chosen arbitrarily for demonstration purposes.

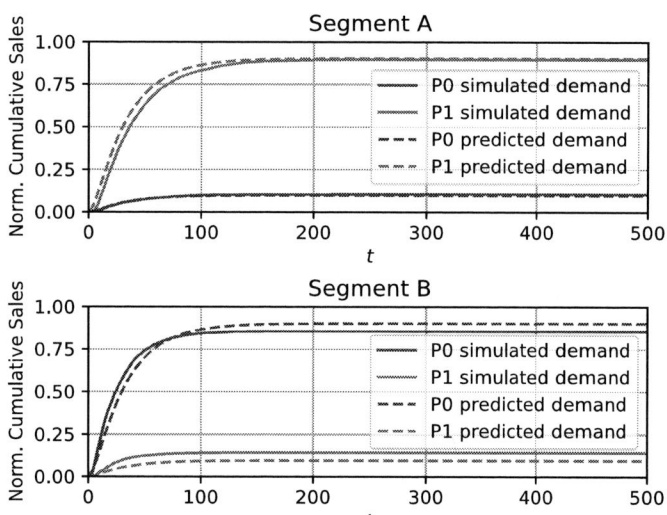

FIGURE 2: Cumulative sales vs time, normalized relative to the total possible sales, from simulations of the affinity dynamics (5) and the predicted cumulative sales from the analytical approximation (17), for two different customer segments and two products. The parameters used in the analytical approximation were $h = 0.43$, $\alpha = 4.95$.

in (5). The normalized cumulative demands for one pair of (h, α) values are shown in Figure 2. The predicted demand and the simulated demand do not match perfectly for both customer segments and both products. However, the demand profile of each customer segment can be approximated regardless. For instance, it is true that from Customer Segment B the demand for Product 0 will start to plateau around time-step 100, and somewhere between $75 - 100\%$ of individuals from Segment B will purchase Product 0.

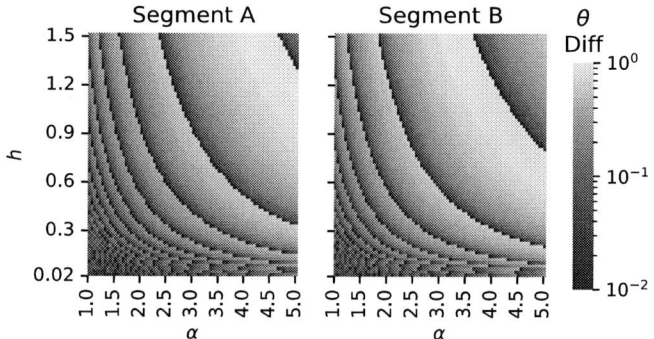

FIGURE 3: Heat maps of the difference between the value of the 'purchased' state in the random walk approximation and the 'real' purchase threshold of 1.5.

To determine the conditions under which the predicted and simulated demand profiles differ, it is necessary to recall how the predicted demand is constructed. Summarizing what is explained in detail in Section 3.2, the affinity dynamics are approximated by defining a continuous-time diffusion process with the same behavior and then approximating that diffusion process with a discrete-time finite-state random walk. The analytical results are then derived using this random walk. This random walk has steps of size $\pm \alpha \phi_{z(t)} \sqrt{h}$ which do not necessarily divide evenly into the threshold value used. Recall that $\phi_{z(t)}^2$ is the variance of the change in affinity between time-steps, α is a free variable that affects how long on average it takes for a decision to be made, and h should be small (according to the assumption used to reach the analytical results).

Heat maps showing the difference between the original purchase threshold of $\theta = 1.5$ and the value of the corresponding 'purchased' state in the random walk are shown in Figure 3. Cus-

348

Copyright © 2023 by ASME

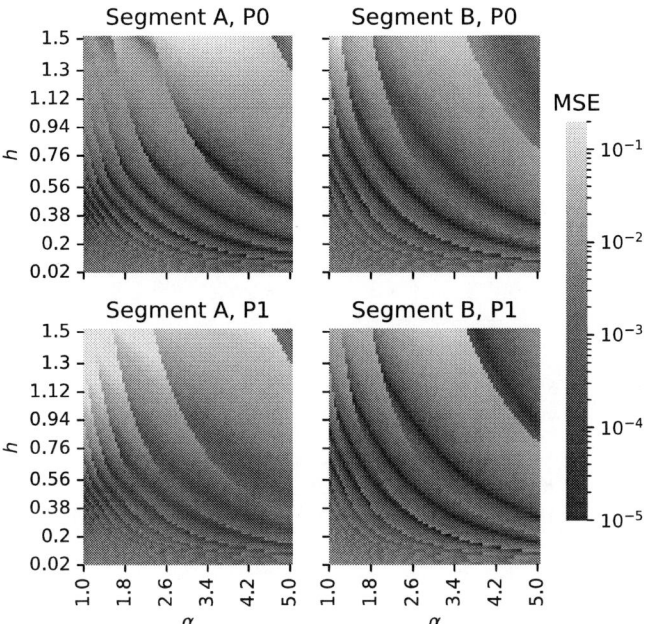

FIGURE 4: Heat maps of the mean squared error between the cumulative predicted sales and cumulative simulated sales for each customer segment and product, over the time interval $t \in [0, 500]$, for varying pairs of h, α.

tomer Segments A and B have different threshold gaps for the same pairs of (h, α) because the variance of the original process is different for each customer segment. The covariance of the affinity dynamics depends on the valence term in (5b) whose random terms are $\varepsilon(t)$ and $\mathbf{w}_y^{z(t)}(t)$. The covariance matrix of the valence thus depends directly on the matrix product $\mathbf{M}_j^{z(t)} \mathbf{w}_y^{z(t)}(t)$, and the $\mathbf{M}_j^{z(t)}$ matrix is different for each customer segment and changes for every distinct set of competing products. Thus, since $\phi_{z(t)}$ depends on the covariance of the original process and that covariance changes for every customer segment, the random walk step sizes and step probabilities will be different as well. Essentially, one pair of α^* and h^* values which results in accurate predictions for one customer segment may result in a relatively poor prediction for the demand from another customer segment. An example of this behavior can be seen in Figure 2, where the values $(h^*, \alpha^*) = (0.43, 4.95)$ result in accurate predictions for Customer Segment A with Product 0, but poor predictions for Customer Segment B with both products. Even if the time-step interval h is kept the same for all customer segments, a different α value could be specified for each. Selecting different α values would result in more accurate overall predictions, though it adds additional work for each customer segment being modeled.

Heat maps showing the mean-square error (MSE) between

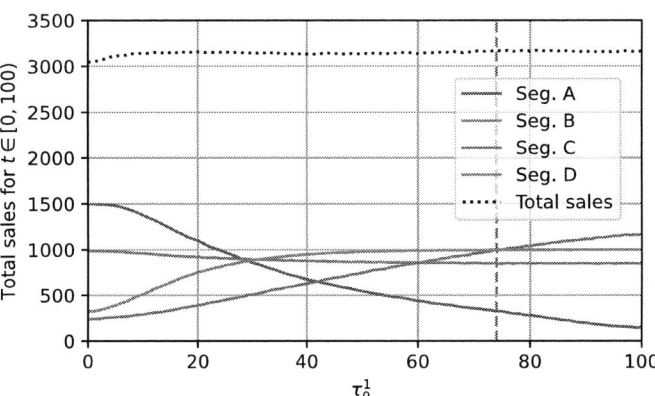

FIGURE 5: This plot shows the total sales of Product 0 in the time frame $t \in [0, 100]$ for all times Revision 1 could be introduced, $\tau_0^1 \in [0, 100]$, averaged over 50 iterations. The population sizes for customer segments (A, B, C, D) are $(1500, 1000, 2000, 1000)$. The variance of the noise term is $\sigma^2 = 0.01$. The optimal time to release the Product 0 revision is $\tau_0^{1*} = 74$, indicated by the vertical dashed line, which resulted in an average of 3174 sales.

the predicted cumulative demand and the simulation results for a range of (h, α) pairs can be seen in Figure 4. For smaller h values, the MSE tends to increase overall. This trend occurs due to approximating the discrete-time process with a continuous-time process, and is in large part due to the delay before any purchases are made in the discrete-time process which is not captured by the continuous-time approximation. An example of the delay can be seen in the simulated demand for Product 1 from Customer Segment A in Figure 2.

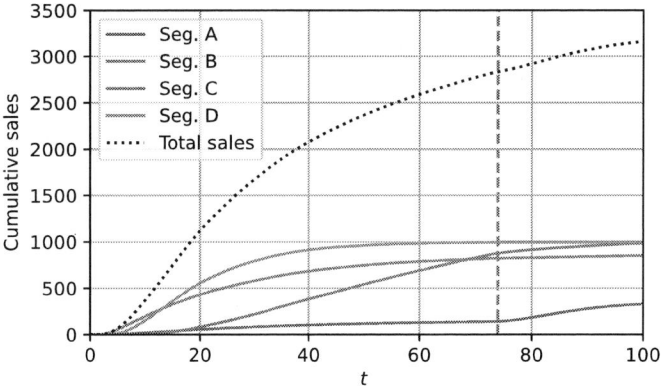

FIGURE 6: Plot showing the cumulative sales over time of Product 0 when it is revised at the optimal time $\tau_0^{1*} = 74$ found from Fig 5.

Copyright © 2023 by ASME

5.2 Optimal release time

The optimal time to release this update, τ_i^{1*}, is determined by solving the optimization problem in (25) by exhaustive search. The first scenario simulated contains four customer demographics considering making a purchase of either Product 0 or 1. The company in charge of Product 0 is concerned about sales within the next $T = 100$ time-steps. Product 0 has an update ready at the start of the simulation and can thus be released at any point, and the company wants to release this revision at the time that maximize sales by $T = 100$. It is assumed that Product 1 has no updates that can be released within the considered time-frame. The plot in Figure 5 shows the sales that occur by time $T = 100$ for all possible update release times τ_0^1. For this scenario, the optimal release time τ_0^{1*} is 74, and results in 3174 sales on average.

The simulated cumulative demand given $\tau_0^1 = \tau_0^{1*} = 74$ can be seen in Figure 6. By τ_0^{1*}, sales from Customer Segments B and D have nearly plateaued. As can be seen in Figure 5, delaying or hastening the release would not cause the sales from these segments to change significantly. Customers from Segment C preferred the original version of Product 0, evident by both the drop in sale rate after the update in Figure 6 as well as the fact that sales from this segment increase the longer the update is delayed, from Figure 5. However, the drop in sales from Segment C is made up for by the increase from Segment A.

5.3 Competing updates

Consider the case where Product 1 also has an update which could be released at any point within this time-window. Only the first two customer segments, Segments A and B, with equal sizes of 1000 are considered. If the features of the competitor's revision are known ahead of time, the optimal time to release the revision of Product 0 can be determined for each of the different revision times of Product 1. If the time that Product 1's update will be released, τ_1^1, is fixed ahead of time and known, the approach becomes identical to that in Section 5.2. The total sales of Product 0 over all τ_0^1 is shown for 11 different values of τ_1^1 in Figure 7. The scenario is identical from the company in charge of Product 1's perspective. If the features of Product 0's update are known ahead of time, the optimal time τ_1^1 can be determined for each τ_0^1. The total sales of Product 1 over all τ_1^1 is shown for 11 different values of τ_0^1 in Figure 8. The total sales of both products for all pairs of (τ_0^1, τ_1^1) values can be seen in Figure 9.

If the competitor's release time is unknown ahead of time, the two companies must play a competitive game. For Company 0, the optimal τ_0^1 is equal to τ_1^1. This behavior is seen in Figure 7 where the '×' markers indicate $\tau_0^1 = \tau_1^1$, as well as in Figure 9. Hence, if possible, the optimal strategy is to delay releasing the update until the competition releases their product update. However, if the release date must be selected in advance and τ_1^1 is unknown, the company may attempt to predict the behavior of Company 1. From the results in Figure 8, we see the

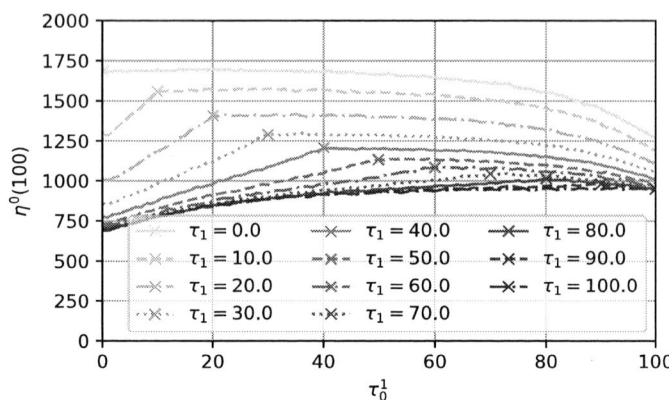

FIGURE 7: Total sales of Product 0 by time $T = 100$ over all times Revision 1 of Product 0 could be introduced, $\tau_0^1 \in [0, T]$, for 11 different Product 1 revision times. If τ_1^1 is known ahead of time, the optimal strategy for Company 0 in this scenario would be to release Product 0's revision immediately after the revision for Product 1 is released.

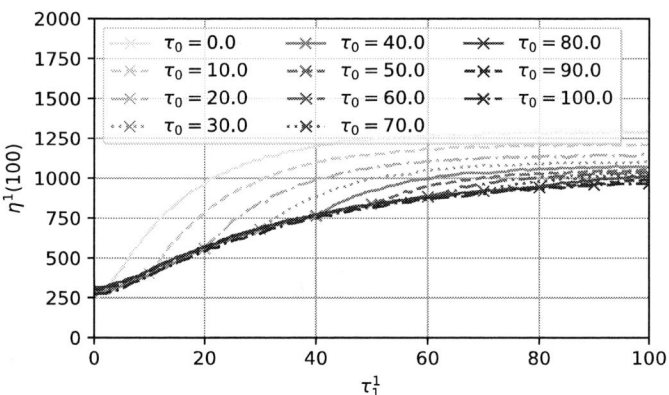

FIGURE 8: Total sales of Product 1 by time $T = 100$ over all possible $\tau_1^1 \in [0, T]$, for 11 different Product 0 revision times. The different values of τ_1^1 are the release times that Company 1 can choose, while the different lines show the sales based on the value of τ_0^1. Regardless of whether or not τ_0^1 is known ahead of time, the optimal strategy for Company 1 in this scenario is to set $\tau_1^1 = 100$.

optimal τ_1^1 is $\tau_1^{1*} = 100$ for the 11 different values of τ_0^1. The optimal time $\tau_1^{1*} = 100$ can be confirmed for the other values of τ_0^1 through Figure 9. From this prediction, Company 0 would select $\tau_0^1 = \tau_1^1 = 100$.

Copyright © 2023 by ASME

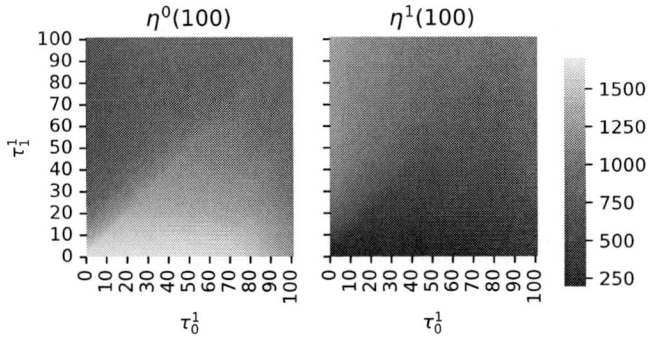

FIGURE 9: Total sales of Product 0 (left) and Product 1 (right) by time $T = 100$ over all times each product could be revised.

6 CONCLUSION

In this paper, we proposed a model for predicting demand built on the cognitive decision-making model of decision field theory. Analytical equations for demand distribution and expected purchase time are provided, using a random walk approximation of the model. Potential use cases and limitations of these analytical equations are provided via simulations. A product revision release strategy is proposed, where the optimal release time is defined as the time that results in the maximum number of total sales within a specified time-frame. There are many factors that influence what the optimal time to release a product will be, such as the costs of producing the update, the costs of releasing the update, and the price differences before and after the update. The implementation of the proposed model in this paper only captures the effect of market saturation when determining the optimal release time. If one customer segment prefers the original version and another prefers the updated version, then the optimal release time may be between the extremes. Otherwise, the optimal release time for the simplified scenario will be either immediately (e.g., the updated version is strongly preferred) or never (e.g., the updated version is worse than the original).

We highlight the potential benefits for demand modeling of using a cognitive-based model such as DFT, which captures many deviations from rationality commonly seen in real-world experiments, over the contemporary discrete-choice models, which assume decision makers are rational. The state-of-the-art works using DFT-based models focus on the final decision of an individual given a fixed decision time. In this paper we instead focus on the optional stopping time problem where individuals have control over if and when they make a decision, as this scenario is more relevant when predicting demand. While the current model and analytical approximation can be useful in various scenarios as constructed in this work, there is room for improvement in future work for both the capabilities of the affinity model and the analytical methods used.

REFERENCES

[1] Sonatus, 2022, "the journey to software-defined vehicles" White Paper, May.

[2] Becker, J., 2022, "Operating system for software-defined vehicles," *ATZelectronics Worldwide,* **17**(5), pp. 40–45.

[3] Digital, and Analytical, 2022, Tesla software updates and release notes Accessed: 2022-12-05.

[4] Beck, K., Beedle, M., van Bennekum, A., Cockburn, A., Cunningham, W., Fowler, M., Grenning, J., Highsmith, J., Hunt, A., Jeffries, R., Kern, J., Marick, B., Martin, R. C., Mellor, S., Schwaber, K., Sutherland, J., and Thomas, D., 2001, "Manifesto for agile software development".

[5] Abrahamsson, P., Salo, O., Ronkainen, J., and Warsta, J., 2017, Agile software development methods: Review and analysis.

[6] Pessôa, M. V. P., and Trabasso, L. G., 2017, *The Lean Product Design and Development Journey* Springer, ch. The Product Development System.

[7] Wassenaar, H. J., Chen, W., Cheng, J., and Sudjianto, A., 2004, "Enhancing discrete choice demand modeling for decision-based design," *Journal of Mechanical Design,* **127**(4), 08, pp. 514–523.

[8] Busemeyer, J. R., and Townsend, J. T., 1992, "Fundamental derivations from decision field theory," *Mathematical Social Sciences,* **23**(3), pp. 255–282.

[9] Carbon, C.-C., 2019, "Psychology of design," *Design Science,* **5**, p. e26.

[10] Tversky, A., and Kahneman, D., 1974, "Judgment under uncertainty: Heuristics and biases," *Science,* **185**(4157), pp. 1124–1131.

[11] Busemeyer, J. R., and Diederich, A., 2002, "Survey of decision field theory," *Mathematical Social Sciences,* **43**(3), pp. 345–370.

[12] Busemeyer, J. R., and Townsend, J. T., 1993, "Decision field theory: A dynamic-cognitive approach to decision making in an uncertain environment," *Psychological Review,* **100**(3), p. 432.

[13] Roe, R. M., Busemeyer, J. R., and Townsend, J. T., 2001, "Multialternative decision field theory: A dynamic connectionst model of decision making.," *Psychological Review,* **108**(2), p. 370.

[14] Haaijer, R., Kamakura, W., and Wedel, M., 2000, "Response latencies in the analysis of conjoint choice experiments," *Journal of Marketing Research,* **37**(3), pp. 376–382.

[15] Corr, P. J., 2013, "Approach and avoidance behaviour: Multiple systems and their interactions," *Emotion Review,* **5**(3), pp. 285–290.

[16] Slovic, P., and Lichtenstein, S., 1983, "Preference reversals: A broader perspective," *The American Economic Review,* **73**(4), pp. 596–605.

[17] Burton, S., and Zinkhan, G. M., 1987, "Changes in con-

Copyright © 2023 by ASME

sumer choice: Further investigation of similarity and attraction effects," *Psychology & Marketing,* **4**(3), pp. 255–266.

[18] Huber, J., Payne, J. W., and Puto, C., 1982, "Adding asymmetrically dominated alternatives: Violations of regularity and the similarity hypothesis," *Journal of Consumer Research,* **9**(1), 06, pp. 90–98.

[19] Huber, J., Payne, J. W., and Puto, C. P., 2014, "Let's be honest about the attraction effect," *Journal of Marketing Research,* **51**(4), pp. 520–525.

[20] Simonson, I., 1989, "Choice based on reasons: The case of attraction and compromise effects," *Journal of Consumer Research,* **16**(2), 09, pp. 158–174.

[21] Chernev, A., 2004, "Extremeness aversion and attribute-balance effects in choice," *Journal of Consumer Research,* **31**(2), 09, pp. 249–263.

[22] He, L., Chen, W., Hoyle, C., and Yannou, B., 2012, "Choice modeling for usage context-based design," *Journal of Mechanical Design,* **134**(3), 02.

[23] He, L., Wang, M., Chen, W., and Conzelmann, G., 2014, "Incorporating social impact on new product adoption in choice modeling: A case study in green vehicles," *Transportation Research Part D: Transport and Environment,* **32**, pp. 421–434.

[24] Wang, M., Chen, W., Huang, Y., Contractor, N. S., and Fu, Y., 2016, "Modeling customer preferences using multidimensional network analysis in engineering design," *Design Science,* **2**, p. e11.

[25] Qin, H., Guan, H., and Wu, Y.-J., 2013, "Analysis of park-and-ride decision behavior based on decision field theory," *Transportation Research Part F: Traffic Psychology and Behaviour,* **18**, pp. 199–212.

[26] Hancock, T. O., Hess, S., Marley, A., and Choudhury, C. F., 2021, "An accumulation of preference: Two alternative dynamic models for understanding transport choices," *Transportation Research Part B: Methodological,* **149**, pp. 250–282.

[27] Lee, S., Son, Y.-J., and Jin, J., 2008, "Decision field theory extensions for behavior modeling in dynamic environment using bayesian belief network," *Information Sciences,* **178**(10), pp. 2297–2314.

[28] Hotaling, J. M., Busemeyer, J. R., and Li, J., 2010, "Theoretical developments in decision field theory: Comment on tsetsos, usher, and chater (2010)," *Psychological Review,* **117**, pp. 1294–1298.

[29] Hancock, T. O., Hess, S., and Choudhury, C. F., 2018, "Decision field theory: Improvements to current methodology and comparisons with standard choice modelling techniques," *Transportation Research Part B: Methodological,* **107**, pp. 18–40.

[30] Hotaling, J. M., 2020, "Decision field theory-planning: A cognitive model of planning on the fly in multistage decision making," *Decision,* **7**, pp. 20–42.

[31] Walley, R. E., and Weiden, T. D., 1973, "Lateral inhibition and cognitive masking: A neuropsychological theory of attention," *Psychological Review,* **80**, pp. 284–302.

[32] Bhattacharya, R. N., and Waymire, E. C., 1990, *Stochastic Processes with Applications* SIAM.

Appendix A: Proposition Proofs

For the convenience of the reader, we re-state the notation definitions used in the following proofs. Recall that $\mathbf{T}_{z(t)}$ is the row-stochastic transition matrix for the random walk approximation of (5) given the product states $z(t)$. This $\mathbf{T}_{z(t)}$ is partitioned into the following sub-matrix and column vectors. Let $\mathbf{Q}_{z(t)} \in \mathbb{R}^{(l-2) \times (l-2)}$ be the inner transition matrix defined by $\mathbf{Q}_{z(t)} = \mathbf{T}_{z(t)}[2 : (l-1), 2 : (l-1)]$, containing the transition probabilities between the transient states; $\mathbf{k}_{i,z(t)} \in \mathbb{R}^{l-2}$ be the column vector containing the transition probabilities from all transient states to the 'purchased' state corresponding to Product i: $\mathbf{k}_0 = \mathbf{T}_{z(t)}[2 : (l-1), l]$, and $\mathbf{k}_1 = \mathbf{T}_{z(t)}[2 : (l-1), 1]$. Additionally, let $\mathbf{p}_t \in \mathbb{R}^{l-2}$ be the affinity probability distribution column vector at time t. Recall that by definition the possible values of t, and by extension $\mathscr{T}_{max}^{z(t)}$ and $\mathscr{T}_{min}^{z(t)}$, are separated by time-step intervals of h starting from 0; thus, t/h, $\mathscr{T}_{max}^{z(t)}/h$, and $\mathscr{T}_{min}^{z(t)}/h$ are always integers.

Proof for Proposition 1

Proof. The probability that an Individual y will make a purchase at time $t > 0$ is given by

$$\mathbb{P}(Z_{y,j}^i(t) = 1) = \mathbf{p}_{t-h}' \mathbf{k}_{i,z(t)} \quad t \geq h. \tag{26}$$

First, consider the case where $t \leq \tau_{max}^0$, such that no product revisions have been released prior to time t. The preference distribution vector $\mathbf{p}_t$ can then be written as

$$\mathbf{p}_t' = \mathbf{p}_0' \mathbf{Q}_{z(0)}^{\frac{t}{h}}. \tag{27}$$

Let us emphasize that the above equation holds even when $z(t) > 0$, so long as $t = \tau_{max}^0 = \mathscr{T}_{min}^{z(t)}$. If $t = \tau_{max}^0 = \mathscr{T}_{min}^{z(t)}$, clearly the following are all equivalent:

$$\mathbf{p}_{\mathscr{T}_{min}^{z(t)}}' = \mathbf{p}_{\tau_{min}^{z(0)}}' \mathbf{Q}_{z(0)}^{\frac{\tau_{max}^{z(0)}}{h}} = \mathbf{p}_{\tau_{min}^{z(t-h)}}' \mathbf{Q}_{z(t-h)}^{\frac{\tau_{max}^{z(t-h)} - \tau_{min}^{z(t-h)}}{h}}. \tag{28}$$

Copyright © 2023 by ASME

The following generalization of (27) can be constructed by recursively writing out $\mathbf{p}'_{\mathcal{T}_{min}^{z(t)}}$ in (28) in terms of $\mathbf{p}'_{\tau_{min}^{z(t)-1}}$:

$$\mathbf{p}'_t = \mathbf{p}'_{\mathcal{T}_{min}^{z(t)}} \mathbf{Q}_{z(t)}^{\frac{t-\mathcal{T}_{min}^{z(t)}}{h}}$$

$$= \left(\mathbf{p}'_0 \prod_{z=00}^{z(t)-01} \mathbf{Q}_z^{\frac{\mathcal{T}_{max}^z - \mathcal{T}_{min}^z}{h}} \right) \mathbf{Q}_z^{\frac{t-\mathcal{T}_{min}^{z(t)}}{h}}. \tag{29}$$

When release combinations between 00 and $z(t)$ do not occur, $\mathcal{T}_{min}^z = \mathcal{T}_{max}^z$ and thus $\mathbf{Q}_z^{(\mathcal{T}_{max}^z - \mathcal{T}_{min}^z)/h} = \mathbf{Q}_z^0 = \mathbf{I}$. Plugging (29) into (26) we get

$$\mathbb{P}(Z_{y,j}^i(t) = 1) = \left(\mathbf{p}'_0 \prod_{z=00}^{z(t)-01} \mathbf{Q}_z^{\frac{\mathcal{T}_{max}^z - \mathcal{T}_{min}^z}{h}} \right) \mathbf{Q}_{z(t)}^{\frac{t-\mathcal{T}_{min}^{z(t)}}{h}} \mathbf{k}_{i,z(t)}. \tag{30}$$

Proof for Proposition 2

Proof. The probability that an Individual y will make a purchase at time t is given by

$$\mathbb{P}(Z_{y,j}^i(t) = 1) = \left(\mathbf{p}'_0 \prod_{z=00}^{z(t)-01} \mathbf{Q}_z^{\frac{\mathcal{T}_{max}^z - \mathcal{T}_{min}^z}{h}} \right) \mathbf{Q}_{z(t)}^{\frac{t-\mathcal{T}_{min}^{z(t)}}{h}} \mathbf{k}_{i,z(t)} \tag{31}$$

according to (30). Then, the probability that an Individual y will make a purchase by a time $T < \infty$ is given by

$$\mathbb{P}(X_{y,j}^i(T) = 1) = \sum_{t=0}^{T} \mathbb{P}(Z_{y,j}^i(t) = 1) \tag{32}$$

$$= \sum_{t=0}^{T} \left(\mathbf{p}'_0 \prod_{z=00}^{z(t)-01} \mathbf{Q}_z^{\frac{\mathcal{T}_{max}^z - \mathcal{T}_{min}^z}{h}} \right) \mathbf{Q}_{z(t)}^{\frac{t-\mathcal{T}_{min}^{z(t)}}{h}} \mathbf{k}_{i,z(t)} \tag{33}$$

$$= \mathbf{p}'_0 \sum_{t=0}^{T} \left(\prod_{z=00}^{z(t)-01} \mathbf{Q}_z^{\frac{\mathcal{T}_{max}^z - \mathcal{T}_{min}^z}{h}} \right) \mathbf{Q}_{z(t)}^{\frac{t-\mathcal{T}_{min}^{z(t)}}{h}} \mathbf{k}_{i,z(t)}. \tag{34}$$

As $T \to \infty$, let the final product update occur at time $\mathcal{T}^* = \lim_{t \to \infty} \mathcal{T}_{min}^{z(t)}$. The equation can be split into a finite and infinite sum:

$$\lim_{T \to \infty} \mathbb{P}(X_{y,j}^i(T) = 1) = \lim_{T \to \infty} \sum_{t=0}^{T} \mathbb{P}(Z_{y,j}^i(t) = 1) \tag{35}$$

$$= \sum_{t=0}^{\mathcal{T}^*} \mathbb{P}(Z_{y,j}^i(t) = 1) + \lim_{T \to \infty} \sum_{t=\mathcal{T}^*}^{T} \mathbb{P}(Z_{y,j}^i(t) = 1). \tag{36}$$

The finite sum is equivalent to (34), and the infinite sum is equivalent to

$$\lim_{T \to \infty} \sum_{t=\mathcal{T}^*}^{T} \mathbb{P}(Z_{y,j}^i(t) = 1) = \mathbf{p}'_{\mathcal{T}^*} \lim_{T \to \infty} \left(\sum_{t=\mathcal{T}^*}^{T} \mathbf{Q}_{z(\mathcal{T}^*)}^{\frac{t-\mathcal{T}^*}{h}} \right) \mathbf{k}_{i,z(\mathcal{T}^*)} \tag{37}$$

$$= \mathbf{p}'_{\mathcal{T}^*} \left(\mathbf{I} - \mathbf{Q}_{z(\mathcal{T}^*)} \right)^{-1} \mathbf{k}_{i,z(\mathcal{T}^*)} \tag{38}$$

where $\mathbf{I}$ is the identity matrix. The result in (38) requires the following facts: $\mathbf{T}_{z(t)}$ is a row-stochastic transition matrix; row-stochastic transition matrices have one eigenvalue equal to one and the rest of smaller magnitude; and a sub-matrix has a spectral radius less than the original matrix. Hence, all eigenvalues of $\mathbf{Q}_{z(\mathcal{T}^*)}$ are less than or equal to one in magnitude, and thus $(\mathbf{I} - \mathbf{Q}_{z(\mathcal{T}^*)})$ is invertible. The total probability, substituting (34) and (38) into (36), is

$$\lim_{T \to \infty} \mathbb{P}(X_{y,j}^i(T) = 1) =$$

$$\mathbf{p}'_0 \sum_{t=0}^{\mathcal{T}^*} \left[\left(\prod_{z=00}^{z(t)-01} \mathbf{Q}_z^{\frac{\mathcal{T}_{max}^z - \mathcal{T}_{min}^z}{h}} \right) \mathbf{Q}_{z(t)}^{\frac{t-\mathcal{T}_{min}^{z(t)}}{h}} \mathbf{k}_{i,z(t)} \right.$$

$$\left. + \mathbf{p}'_{\mathcal{T}^*} \left(\mathbf{I} - \mathbf{Q}_{z(\mathcal{T}^*)} \right)^{-1} \mathbf{k}_{i,z(\mathcal{T}^*)}. \tag{39}$$

Proof for Proposition 3

Proof. The expected time that Individual y in Customer Segment j purchases Product i, given that they make a purchase of Product i between times 0 and T, is

$$\mathbb{E}[u_y^i \in [0, T] \mid X_{y,j}^i(T) = 1] = \sum_{t=0}^{T} t \cdot \mathbb{P}(Z_{y,j}^i(t) = 1). \tag{40}$$

Substituting the definition of $\mathbb{P}(Z_{y,j}^i(t) = 1)$ from (15) into (40), we have

$$\mathbb{E}[u_y^i \in [0, T] \mid X_{y,j}^i(T) = 1] =$$

$$\mathbf{p}'_0 \sum_{t=1}^{T} \left[t \cdot \left(\prod_{z=00}^{z(t)-01} \mathbf{Q}_z^{\frac{\mathcal{T}_{max}^z - \mathcal{T}_{min}^z}{h}} \right) \mathbf{Q}_{z(t)}^{\frac{t-\mathcal{T}_{min}^{z(t)}}{h}} \mathbf{k}_{i,z(t)} \right]. \tag{41}$$

Proceedings of the ASME 2023
International Design Engineering Technical Conferences and
Computers and Information in Engineering Conference
IDETC-CIE2023
August 20-23, 2023, Boston, Massachusetts

DETC2023-117145

A FRAMEWORK TO SUPPORT MULTILEVEL ROBUST CO-DESIGN OF MANUFACTURING SUPPLY NETWORKS

Mathew Baby
Doctoral Student
The Systems Realization Laboratory @ FIT
Florida Institute of Technology, Melbourne, FL, USA

Akshay Guptan
B.S Student
The Systems Realization Laboratory @ FIT
Florida Institute of Technology, Melbourne, FL, USA

Jacob Broussard
B.S Student
The Systems Realization Laboratory @ FIT
Florida Institute of Technology, Melbourne, FL, USA

Janet K. Allen
John and Mary Moore Chair and Professor
The Systems Realization Laboratory @ OU
University of Oklahoma, Norman, OK, USA

Farrokh Mistree
L.A. Comp Chair and Professor
The Systems Realization Laboratory @ OU
University of Oklahoma, Norman, OK, USA

Anand Balu Nellippallil[1]
Assistant Professor
The Systems Realization Laboratory @ FIT
Florida Institute of Technology, Melbourne, FL, USA

ABSTRACT

Decision-making in the design of Manufacturing Supply Networks (MSNs) is complex owing to the need to account for decisions made by groups at multiple levels and the interactions that include potential conflicts and uncertainties. Decisions are made based on information available from simulations that employ computational models. These models are abstractions of reality and are therefore sources of uncertainty. This necessitates that the focus be placed on design space exploration to identify a robust satisficing solution set. Hence, the need to support the efficient exploration of multilevel design/solution spaces, simultaneously.

Many of the frameworks that facilitate multilevel design employ optimization-based iterative approaches that are computationally expensive and ill-suited for design exploration. Frameworks that support robust design exploration are limited by their capability to support the efficient exploration of multi-level design/solution spaces, simultaneously. Hence, the requirement to facilitate the 'multilevel robust co-design and solution space exploration' of MSNs.

In this paper, we present a framework that allows designers to: i) model robust decision-making and their interactions across multiple levels, and ii) visualize and systematically carry out co-

design exploration of multilevel design/solution spaces. In the framework, we combine a Preemptive formulation of coupled Decision Support Problem with robust design constructs and interpretable-Self-Organizing Maps (iSOM) based visualization to facilitate multilevel robust co-design. We use a steel MSN problem with decisions being made at two levels to test the framework. Using the problem, we demonstrate the utility of the framework in supporting designers in modeling multilevel decision-making under uncertainty and their interactions, and efficient co-design exploration of the robust design/solution spaces in MSNs. The proposed framework is based on information and decision flows, making it generic and capable of supporting designers in the robust co-design of multilevel systems.

Keywords: Multilevel systems, Robust Co-design, Manufacturing Supply Networks

GLOSSARY

Manufacturing Supply Network (MSN): A network of independent, interconnected enterprises that work collectively to physically realize the products and deliver them to the customers to satisfy their needs.

Group: We define a group as the collection of all the enterprises that perform the same role in the MSN. Example: The collection

---

[1] Corresponding author, Email: anellippallil@fit.edu

Copyright © 2023 by ASME

of all suppliers that deliver raw materials and components for the manufacturers together constitutes the 'Supplier Group.'

Level: We define a '*level*' as a group or set of groups, characterized by concurrent and independent decision-making.

Co-design: We define co-design as a design approach that supports independent decision-makers, distributed across multiple levels, to work collectively in ensuring the performance of multilevel systems, by supporting multilevel, independent decision-making and facilitating management of conflicts.

Robust Satisficing Solutions: Solutions that are relatively insensitive to uncertainties, while still meeting the requirements of the designer.

Robust Co-design: We define robust co-design as a design approach that supports independent decision-makers, distributed across multiple levels, to work collectively in ensuring the performance of multilevel systems, by supporting multilevel decision-making under uncertainty and conflict management.

Service Level (SL): In this paper, the service level is defined as a measure of the ability to meet delivery expectations in terms of lead times. Mathematically, SL is defined as the ratio of the expected lead time to the actual lead time (actual lead time is computed as the sum of the time for transporting raw materials/products from their source to their destination and the order processing time).

1. FRAME OF REFERENCE

The manufacturing sector is one of the key contributors to the economy of the United States of America (U.S.), contributing a total of $2.3 trillion to the U.S. GDP amounting to 12.0 % of the total U.S. GDP [1]. As per 2021 statistics, manufacturing is a sector that contributes to nearly 10% of total U.S. employment by providing jobs for 14.7 million people in the U.S. [1]. Hence, it is vital to ensure the manufacturing sector's performance to ensure economic stability and growth. The enterprises in the manufacturing sector interact and work closely with various other independent enterprises to realize the products and deliver them to the customers. These independent, interconnected enterprises together form a network, which in this paper is referred to as the Manufacturing Supply Network (MSN). The enterprises in the MSN play various roles like a) suppliers and b) manufacturers. The suppliers are involved in the supply of raw materials, components, sub-assemblies, and other consumable resources required to realize the product based on the manufacturer's demand. The manufacturer processes the raw materials and utilizes the components sourced from the suppliers (in certain cases, from customers in the form of returns, end-of-life products, or scrap) to realize the product according to demand from the customers. The products produced by the manufacturer are then delivered to the customers. Therefore, customers are also an integral part of the MSNs as they are the drivers of demand in the MSN. We refer to the collection of all the enterprises that perform the same role in the MSN as 'groups.' Hence the MSN is composed of the manufacturer, supplier, and customer groups. In this paper, we consider decisions to be made by the 'managers' of the different groups, either concurrently or hierarchically, depending on the relations among the groups in the MSN. Hierarchical decision-making results in the creation of different levels within the MSN, where a '*level*' is defined as a group or set of groups where decisions are made in a concurrent and independent manner. Hence, in this paper, the decisions made at the groups in a level are not interrelated with each other, and the groups are considered to be at the same level based on their direct relationship with the group/groups at other levels. We consider a case where the manufacturer group is the lead decision-maker in the MSN with the supplier and customer group decisions being made concurrently and independently, after the manufacturer decisions. Consequently, the manufacturer group will be located at the upper level of the decision hierarchy (Level 1) followed by the supplier and customer groups at the lower level (Level 2) (see Figure 1). Hence, the MSN is a '*multilevel*' system where decisions are made at different levels by the supplier, manufacturer, and customer groups. The groups/levels interact with each other and are connected by information and material flows as discussed above and shown in Figure 1. Given the interactions of the manufacturer, supplier, and customer groups across different levels, the performance of each group is very likely to be directly influenced by other interacting groups. Hence, the focus needs to be on ensuring the performance of the Manufacturing supply network (MSNs) as a whole.

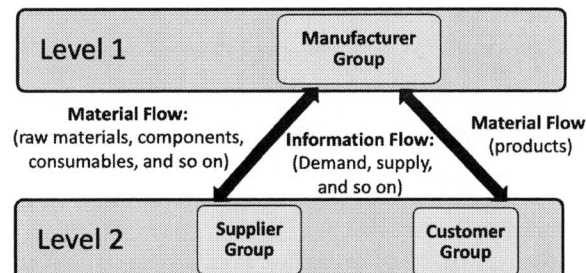

FIGURE 1: A potential configuration of the MSN with the manufacturer, supplier, and customer groups located across two levels and their interactions in terms of material and information flows

The performance of MSN is determined by the numerous decisions made by different groups located across multiple levels of MSN. Decisions by one group are made independently of other groups and are directed toward achieving group-specific goals and ensuring their performance. Given the interactions among the groups/levels in the MSN (see Figure 1), the decisions by a group will impact the decisions and performances of other groups, thereby impacting the performance of the MSN. As a result, the performance of MSNs cannot be ensured by ensuring the performance of individual groups in isolation and there is a need to account for the interactions across different levels to ensure MSN performance. The independent decisions made by different groups of the interrelated levels in the MSN can potentially result in conflicts, where the decision made at a level may not align with the decisions at the interacting level. These conflicts adversely impact the performance of MSN. Hence, there also exists the need to manage conflicts that arise from

Copyright © 2023 by ASME

interactions of group decisions across multiple levels, to ensure MSN performance. The design of MSNs, therefore, requires the facilitation of 'multilevel co-design,' that supports multilevel, independent decision-making while accounting for their interactions and management of conflicts. In this paper, our focus is specifically on the simulation-supported design of MSNs, where designers make use of computational models that are approximations of reality.

From a systems design perspective, we consider design as a goal-oriented, decision-based process that is supported by simulations. We, therefore, follow the Decision-Based Design (DBD) paradigm that has been advocated by Mistree and co-authors [2] and Hazelrigg [3]. Both advocate designing to be a decision-making process wherein designers make a series of decisions, some sequentially while others concurrently. There are several instantiations of DBD. We anchor our work in the Decision Support Problem Technique [4, 5]. Unlike DBD advocated by Hazelrigg, the Decision Support Problem Technique is anchored in the notion of bounded rationality proposed by Herbert A. Simon [6]. Given that the models employed in simulations are incomplete, inaccurate, of different fidelity, and are approximations of reality, designers seek 'satisficing solutions' for the design problem at hand by exploring the solution space. A satisficing solution [7] is one that 'satisfy' and 'suffice' the designers' requirements for the conflicting goals present. The compromise Decision Support Problem (cDSP) [8] is a well-established construct in the literature that is used to explore satisficing solutions for multiple conflicting goals. The use of coupled DSP construct to model multi-level decisions (with multiple conflicting goals at each level) and their interactions have been discussed in the literature [9]. The coupling can be vertical (if the decisions are made hierarchically) or horizontal (if the decisions are made concurrently). In coupled DSPs, the 'deviation function', is modeled using a Preemptive or Archimedean formulation, based on the coupling between the multilevel decision. For the DSPs, the solution space generated is explored to identify satisficing solutions. These satisficing solutions are subject to various uncertainties that could be of different forms and arising from different sources [10] the natural uncertainty inherent in a system in the form of random noises, the uncertainty associated with model parameters, the inherent uncertainty in the models, and the propagation of these uncertainties as information flows in a system. The two approaches typically employed in dealing with these uncertainties are: (i) mitigating uncertainty and (ii) managing uncertainty. The focus when employing uncertainty mitigation approaches is on reducing/mitigating the uncertainty and developing 'perfect' models, by collecting more data and performing expensive computations to quantify the uncertainties. Given that models are never truly perfect, and the computationally expensive nature makes the mitigation approaches less preferable. Therefore, we look at managing the uncertainties by designing the system to be relatively insensitive to the source of uncertainties without reducing or eliminating them, which is termed robust design. Three types of robust designs - Type I, Type II, and Type III are discussed in the

literature to deal with different uncertainties; see [10]. In this paper, we look at Type II robust design, where the focus is on managing the uncertainties associated with design variables. We, therefore, seek 'robust satisficing solutions' that satisfy and suffice the designers' requirements for the conflicting goals present while being relatively insensitive to noise and uncertainties in the design variables. The use of robust design indices, namely, the Design Capability Index (DCI) [11] and Error Margin Index (EMI) [12] in conjunction with the DSP construct have been proposed to help designers identify robust satisficing solutions. DCI is employed for Type I and II robust designs, whereas EMI is employed for Type III robust designs. Given the uncertainty involved in multilevel decisions, the need to account for the interactions among the multilevel decisions, and potential design conflicts that need to be managed, the design of MSNs to ensure system performance is a non-trivial task. Hence, the design of multilevel MSNs requires the facilitation of the 'multilevel robust co-design,' which supports multilevel, independent decision-making under uncertainty while accounting for their interactions, and management of conflicts.

Different approaches have been proposed in the literature to support the co-design of multilevel systems. From the Multi-disciplinary Optimization (MDO) [13] domain, approaches like Bi-Level integrated system synthesis (BLISS), Analytical Target Cascading (ATC), and Collaborative optimization (CO) have been proposed to support the co-design of multilevel systems. Sobieski and co-authors [14-16] present BLISS, an approach where the design of multilevel engineering systems is carried out by decomposing the system-level optimization into many subsystem optimizations that seek to minimize their contribution to the system-level objective under local constraint. In BLISS, the coordination between the subsystem is controlled by the behavior of the derivatives of the local design variables with respect to shared design variables. Kim and co-authors propose the ATC [17] approach, that embodies a hierarchical multilevel optimization formulation where the objective at each level is to minimize the discrepancy between the target (the optimal values calculated at the previous level) and the response at the level. Kroo and co-authors [18] present CO, where multilevel systems are modeled using a bi-level optimization formulation consisting of system-level and subspace optimizations, with the sub-space objectives related to the system objective being satisfied while also satisfying constraints locally. MDO approaches are computationally expensive [19] and result in high costs and time due to repeated iterations arising from the passing of single-point solutions between the levels. Hence, these approaches are not suitable to support the exploration of the design/solution spaces, especially in the early stages of design when the information is incomplete and inaccurate and models are not of equal fidelity. Additionally, All-In-One (AIO) optimization formulations [20, 21], is also proposed to design MSNs, where the design of the multiple levels of the MSN is carried out simultaneously and in an integrated manner. In the AIO approach authors fail to account for decisions made by groups (at different levels) relatively independently of each other. All the cited approaches are based

Copyright © 2023 by ASME

on optimization formulations where the fundamental assumption is that the models used are complete, all the required information is available and the objective function is perfect. Given that the models employed are incomplete and inaccurate, the information available is incomplete, and the objective functions are imperfect, our focus is on 'satisficing' rather than 'optimizing' and we seek a ranged set of satisficing solutions.

Different approaches have been proposed in the literature that supports multilevel robust co-design by the identification of satisficing solution sets. Choi and co-authors propose the Inductive Design Space Exploration (IDEM) [22] approach that is based on the propagation of a ranged set of robust solutions between the multiple levels in a sequential manner to support robust co-design of multilevel systems. There are some limitations to the IDEM as discussed in [23], such as restrictions on the number of design variables that can be considered, errors in discretization, increased computational expense for improved accuracy, and limited flexibility in design. Nellippallil and co-authors [24] present an inverse robust design method (Goal-oriented Inverse Design- GoID) for hierarchical process chains that support the multilevel co-design of material, product, and associated manufacturing processes. In the GoID approach, the focus is on exploring the design/solution spaces at the individual levels separately to identify satisficing solutions and propagating these solutions as targets in an inverse manner along the hierarchical process chain. Hence, this approach does not support the exploration of multiple levels simultaneously. Sharma and co-authors [25] propose the use of coupled DSPs to support the co-design of multi-level engineered systems that are coupled both concurrently and/or hierarchically. In the above work, the design/solution spaces at the different levels are visualized using ternary plots, and solution space exploration is carried out individually at the different levels for each multilevel design scenario to identify satisficing solutions. The above approach is limited by its inability to support the co-design exploration of the design/solution space across multiple levels.

In this paper, our focus is on supporting decision-makers in the simulation-based design of multilevel MSNs operating under uncertainty by i) modeling independent, multilevel decision-making under uncertainty and their interactions in the MSN, and ii) facilitating the co-design exploration of the design/solution spaces across multiple levels. Given the multiple conflicting goals at each level, the independent nature of multi-level decision-making, and their interactions, the multilevel decision-making in MSNs is modeled as a coupled cDSP that uses a Preemptive formulation to facilitate the modeling of independent decision-making at different levels The DCI metric is employed in coupled cDSP formulation to identify robust satisficing solutions across multiple levels. The multilevel design/solution spaces are visualized in an integrated manner using interpretable Self Organizing Maps (iSOM) [26], to aid the co-design exploration of the multilevel design/solution spaces. From a decision-based design perspective, we hypothesize that the multilevel robust co-design of MSNs can be realized using a decision support framework that supports: i) Robust decision-making at the individual levels while considering their interactions in terms of information and decision flows, and ii) effective, co-design exploration of the robust design/solution space and decision-making across multiple levels. In this paper, we present a framework that enables designers to i) model robust decision-making across the multiple levels and their interactions, ii) effectively visualize the robust design/solution space across the multiple levels, and iii) systematically explore the design/solution spaces across the multiple levels simultaneously to support co-design.

The outline of this paper is as follows. In Section 2, a description of the problem is presented. The framework to support the robust co-design of multilevel MSNs is presented in Section 3. In Section 4, we showcase the efficacy of the framework in supporting co-design and managing the inherent uncertainties, using a steel MSN test problem. In the test problem, we focus on the interactions between supplier and manufacturing-level decisions. We end the paper with our key findings and closing remarks in Section 5. We present the models used in the coupled manufacturer-supplier level cDSP formulations in Appendix A.

2. PROBLEM DESCRIPTION

The manufacturing supply networks (MSNs) are composed of multiple independent but interconnected groups across multiple levels, that work collectively in realizing various products that satisfy customer needs, as described previously, and depicted in Figure 1. Considering an MSN with two levels: i) Level 1: composed of the Manufacturer group (j) and ii) Level 2: composed of the Supplier group (i) and Customer group (k), its design involves decision-making across the different groups at the two levels. The decision-making by each group takes place independently and is directed toward achieving the goals specific to the group. But the decisions at the 3 groups are interrelated by the flow of information and decisions within and between the groups/levels as described below and depicted in Figure 2. In Figure 2, the flow of information/decisions within a group is indicated by arrows at Level 1 and the arrows connecting Levels 1 and 2 indicate the flow of information/decisions between groups/levels.

The manufacturer group (Level 1), based on the estimates of the customer demand for products (D_k^e) and expected product delivery lead times (LTk_k^e) from the customers, makes decisions related to production, raw-material (m) sourcing, and product distribution to customers. The production decision involves the estimation of the production quantity (P). The raw-material sourcing decisions include the estimation of the quantity of different raw materials to be procured (Q_{ij}^m, Q_{ij}^s, Qk_{kj}^s), the selection transportation mode for raw materials procurement from customers (Y_{jk}^y), and the estimation of the prices at which the raw materials should be purchased ($s_m^{i,e}$). The product distribution decisions include estimating the sale price for products ($Price$), determining the product supply quantities (Q_j^k), and selecting the mode of transportation to deliver the products to the customers (Y_{jk}^y) (assuming that the raw material is transported using the same mode as products). At the supplier

Copyright © 2023 by ASME

group (Level 2), the sales decisions made are related to the choice of mode of transportation ($y_{ij}^{m,y}$) and sale prices (s_m^i) for the various raw materials. The above decisions are made based on the expected delivery lead times for raw materials from suppliers (LT_{ij}^{me}), expected prices for raw materials ($s_m^{i,e}$), and order quantities for different raw materials from the manufacturer level (Q_{ij}^m, Q_{ij}^s). Hence, the supplier and manufacturer groups at Levels 2 and 1 respectively, are interrelated by the flow of information and decisions (see Figure 2). At the customer group, sales decisions are made with regard to the prices at which products at end of life/scrap are returned to the manufacturer (sk_s^k). Based on the price that the product is sold to the customers by the manufacturer (*Price*) and the quantity available for sale from the manufacturers (Q_j^k), the customers make purchase decisions in terms of quantity of product to be purchased (qk_{jk}^a). Hence, the manufacturer and customer groups at Level 1 and 2 respectively, are interrelated by the flow of information and decisions (see Figure 2). Given the relations between the independent decisions made across different levels in the MSN, decisions at one level can potentially have adverse impacts on the decision of another level. Hence there is a need to facilitate the '*multilevel co-design*' by accounting for the relations between the independent multilevel decisions. A summary of the variables for each group is provided in Table 1.

The decisions made at each level are subject to various uncertainties like production uncertainties (due to machine breakdowns, labor shortages, and so on) and supply uncertainties (due to material shortages at suppliers, transportation delays, and so on), and so on. These uncertainties can impact the achievement of the goals at the different levels and the performance of MSN as a whole. Hence, there is also the need to manage these uncertainties across multiple interrelated levels by identifying '*robust solutions*' that are relatively insensitive to uncertainties.

TABLE 1: List of variables, their notations and description for the different groups

Supplier Group
1. i (supplier index) ϵ I, set of all suppliers
2. $m \epsilon$ M, set of all raw materials
3. $y_{ij}^{m,y}$, selection of transportation mode for raw materials 'm' from supplier 'i' to manufacturer 'j'
4. s_m^i, sale price for raw materials 'm' at supplier 'i'
Manufacturer Group
1. j (manufacturer index) ϵ J, set of all manufacturers
2. P, production quantity
3. Q_{ij}^m, quantity of raw material 'm' to be procured from supplier 'i' by manufacturer 'j'
4. Q_{ij}^s, quantity of scrap 's' to be procured from supplier 'i' by manufacturer 'j'
5. Qk_{kj}^s, quantity of scrap 's' to be procured from customer 'k' by manufacturer 'j'
6. Y_{jk}^y, selection of transportation mode for: i) raw materials procurement from customer 'k', and   ii) product delivery to the customer 'k'
7. $s_m^{i,e}$, estimated price for raw materials 'm' sourced from supplier 'i'
8. $Price$, estimated sale price of the product
9. Q_j^k, product supply quantities to customer 'k' from manufacturer 'j'
10. LT_{ij}^{me}, expected delivery lead times for raw material 'm' from supplier 'i' by manufacturer 'j'
Customer Group
1. k (customer index) ϵ K, set of all customers
2. D_k^e, estimated demand for products from customer 'k'
3. LTk_k^e, expected product delivery lead times from the customers 'k'
4. sk_s^k, estimated sale prices of scrap 's' by customers 'k'
5. qk_{jk}^a, the quantity of product purchased by customer 'k' from manufacturer 'j'

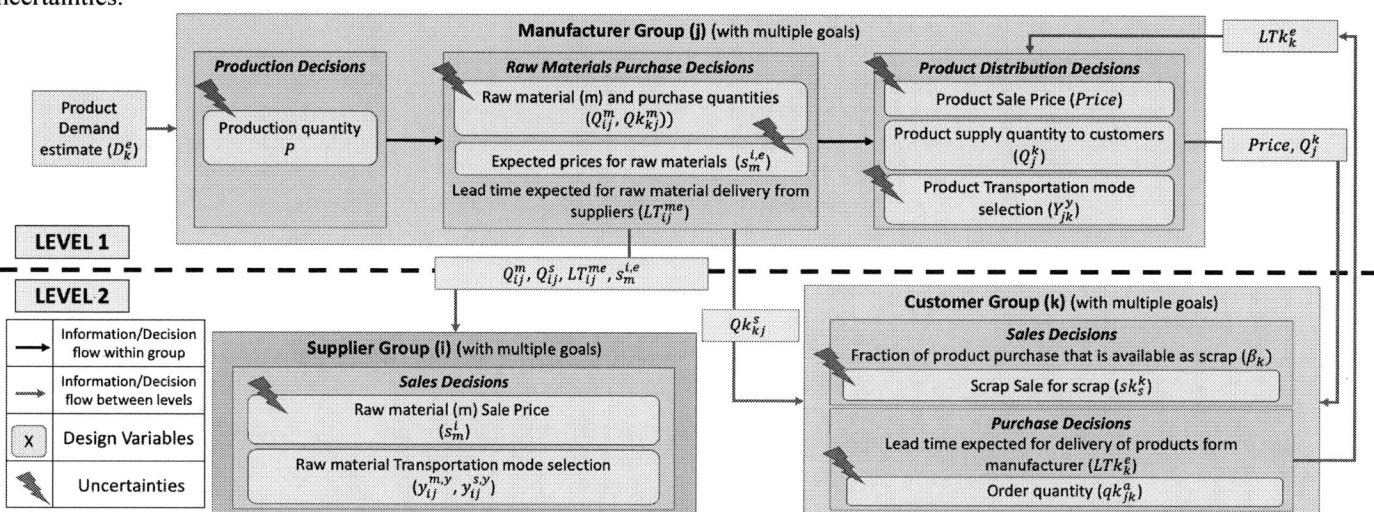

FIGURE 2: Information and Decision flows connecting the Manufacturer, Supplier, and Customer Levels in a Manufacturing Supply Network (MSN)

Copyright © 2023 by ASME

During the simulation-based design of MSNs operating under uncertainty, the focus is on exploring the design/solution spaces to identify a set of '*robust satisficing solutions*' across multiple interrelated levels. This necessitates an approach that aids designers in effectively performing the co-design exploration of the design/solution spaces in multilevel MSNs. Therefore, the need is for a systematic approach that supports the '*multilevel robust co-design*' and facilitates the co-design exploration of design/solution spaces across the multiple levels of MSNs.

3. A FRAMEWORK TO SUPPORT MULTILEVEL ROBUST CO-DESIGN OF MANUFACTURING SUPPLY NETWORKS

The framework to support the multilevel robust co-design of MSNs is presented in this section. The various constructs and tools used in the framework are discussed first. This is followed by a discussion on decision support using the framework.

3.1 Constructs and tools used in the framework

In the proposed framework we make use of three major constructs/tools namely the coupled DSP construct, the DCI robust design construct, and the interpretable Self-Organizing Map (iSOM) visualization tool. They are discussed in detail as follows.

i.   The coupled DSP construct.

The coupled DSP [25] is a DSP construct that is used to model the relations/interactions between decisions (either a selection or compromise) some of which are made hierarchically and others concurrently. The relations between the multilevel decisions in the coupled DSP construct are modeled as either a vertical or horizontal coupling [9]. Vertical coupling is employed for decisions that are made hierarchically, whereas horizontal coupling is used for decisions that are made concurrently. In this paper, we use a coupled cDSP to model the decision-making of the interacting levels in multilevel systems, given the multiple conflicting goals at each level that require compromises.

In coupled DSPs, the 'deviation function', is modeled using a Preemptive or Archimedean formulation (see [8] for more details), based on the coupling between the multilevel decision. A Preemptive formulation is used when the coupling is vertical, and this permits the designers to assign different levels of priority for the goals at different levels of decision-making in the hierarchy.

A higher priority level signifies the need for the goals at that specific level to be achieved first, before looking at the achievement of the goals at any of the lower priority levels (if any). All the goals at a specific priority level can be formulated using an Archimedean formulation by assigning different weights to these goals, with a higher weight value indicating a higher preference, at the specific priority level. An Archimedean formulation is also employed when the coupling is horizontal with all the goals at the same priority level, but with differences in their relative importance. In this framework, we make use of the Preemptive formulation to model the interactions between the multi-level decisions. The Archimedean formulation is used

to account for the differences in the relative importance of various goals at a given level.

ii.   Design Capability Index (DCI) robust design construct

The DCI [11] is a construct that aids designers account for uncertainty in design variables. For a larger is better case where the designer aims to keep the mean response away from a lower requirement limit (see Figure 3), the DCI value is computed as per Equation 1. A value of DCI ≥ 1 indicates that the solutions identified will result in the system being robust against design variable uncertainties. The higher the value of DCI, the higher the measure of safety against failure due to uncertainties in design variables.

$$DCI = \frac{\mu_y - LRL}{\Delta Y} \qquad (1)$$

where,

ΔY - response variation for small variations in design variables

μ_y– Mean responses

LRL – Lower requirement limit

ΔY is computed as per Equation 2.

$$\Delta Y = \sum_{i=1}^{n} |\frac{\partial f}{\partial x_i}| . \Delta x_i \qquad (2)$$

where,

i = 1, 2, 3, ..., n (index of design variables)

Δx_i – variation (uncertainty) in design variable x_i

$\frac{\partial f}{\partial x_i}$ – variation of the response, f, with respect to the design variable x_i

The equations for computing DCI value for the smaller is better and nominal is better cases are shown in Figure 3. We make use of the DCI construct in conjunction with the coupled cDSP construct to identify robust satisficing solutions across multiple levels.

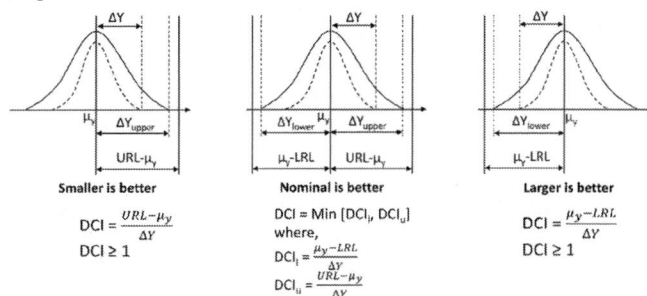

FIGURE 3: Mathematical constructs of DCI [24]

iii.   interpretable Self-Organizing Maps (iSOM)

iSOM [26] is used as a tool to visualize high-dimensional data in lower dimensions, specifically in 2D. iSOM is an unsupervised learning algorithm (artificial neural network) and is a modified form of the conventional SOM (cSOM) [27], which results in the component planes being inherently interpretable and avoiding self-intersections. iSOM helps to visualize the relationship between inputs (design variables) and outputs (responses) by means of the 2D component plots generated for both inputs and outputs. The above ability helps designers carry

Copyright © 2023 by ASME

out forward (from inputs to outputs) and inverse (from outputs to inputs) design space exploration. The utility of the tool in visualizing i) high dimensional design/solution spaces and ii) the relations between inputs and outputs in multilevel systems is demonstrated in the work by Sushil and co-authors [28]. More details on iSOM and its application for decision support in design can be found in [29] [30]. In the proposed framework, we make use of iSOM to support the co-designing exploration of the design/solution spaces across multiple levels for various multi-level design scenarios.

3.2 Decision support using the framework

The use of the decision support framework is described in this section. The framework is executed in three steps as depicted in Figure 4. Each of these steps is described in detail below. In the framework, we consider the interactions between two levels only, for demonstrating the idea.

a. *Step 1*: In the first step, the decision-making at two levels (Level 1 and 2) and their interactions are modeled as a coupled DSP. The formulation of the coupled DSP requires the designer to first establish the individual DSPs of the two interacting levels. This involves identifying the problem-specific information at the 2 levels using the keywords of the cDSP construct - Given, Find, Satisfy, and Minimize as shown in Step 1, Figure 4. The problem-specific information

at each level includes the following: i) design variables - their bounds and variability (uncertainty estimate), ii) goals and goal targets (DCI goal formulations used for a goal are impacted by design variable uncertainty), and iii) level-specific constraints.

Next, the designer establishes the information flow connecting the two DSPs of the individual levels, by identifying any shared design variables between the two levels and sharing the same between the levels (see dashed arrow connecting Levels 1 and 2 in Step 1, Figure 4). At the lower level (Level 2), a copy of the shared design variables in the formulation is used as the level-specific design variable. In the framework, the designer seeks to propagate a ranged set of shared design variables rather than single-point values, to allow increased design flexibility during design. To facilitate the same, '*shared design variable relaxation constraints*' are added into the DSP of the lower of the two interacting levels (Level 2) by the designer, which allows the copy of the shared design variable at the lower level (Level 2) to take a value between a predefined upper and lower percentage of the original shared design variable value (as per designer preference) identified at the upper level (Level 1). The shared design variable relaxation constraint is mathematically represented as in Equation 3.

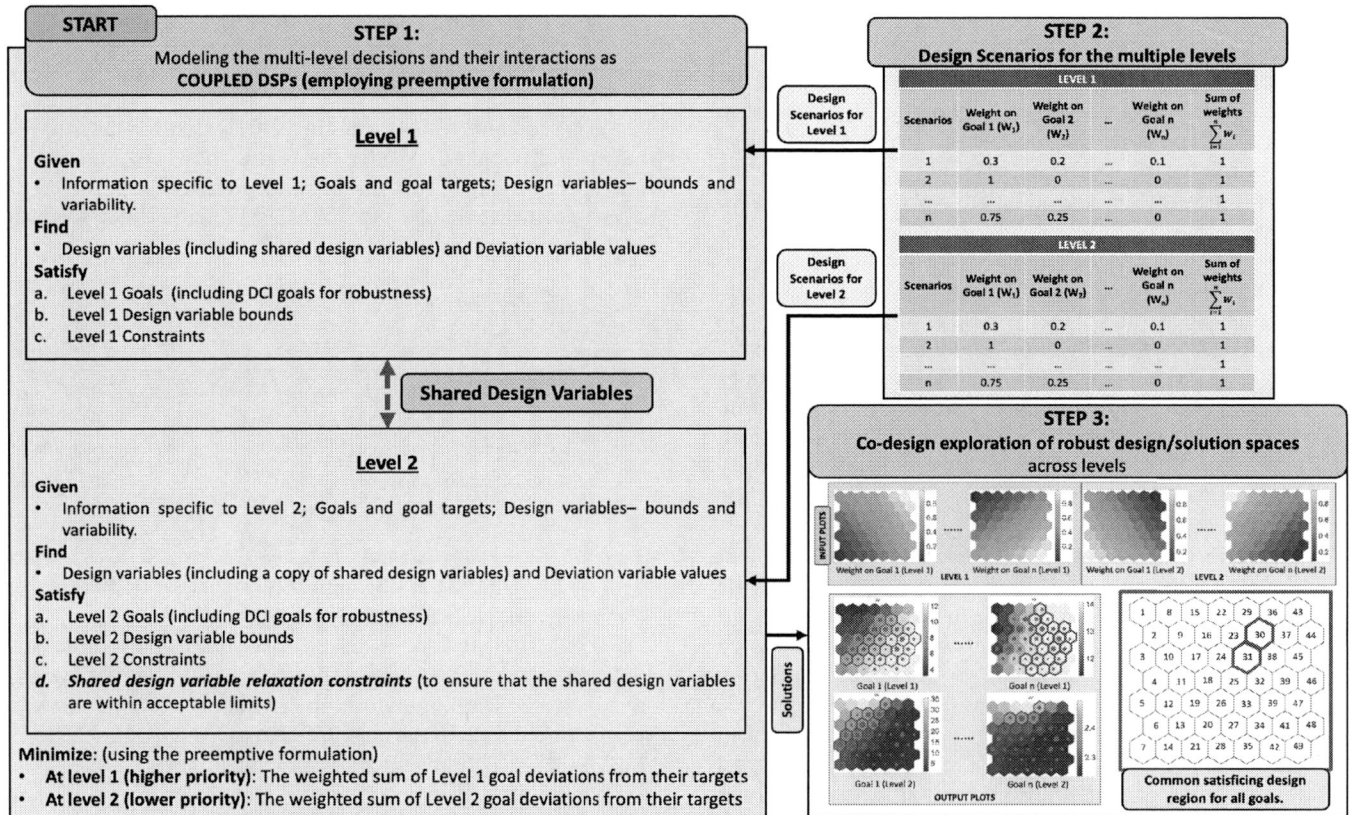

FIGURE 4: Framework for Robust Multilevel Co-Design Exploration (FRoMCoDE)

Copyright © 2023 by ASME

$$J_z * X_{z, \, shared} \leq X_{z, \, shared}^{copy} \leq K_z * X_{z, \, shared} \qquad (3)$$

where,

z – index of the shared design variable, z = 1, 2, …, n

$X_{z, \, shared}$ – shared design variable at level 1

$X_{z, \, shared}^{copy}$ – copy of the shared design variable at level 2

J_z – lower relaxation bound multiplier ($0 < J_z < 1$)

K_z – upper relaxation bound multiplier ($K_z > 1$)

The deviation function of the coupled DSP is then modeled by the designer using the Preemptive or Archimedean formulations, based on the nature of decision-making in the multilevel system, which is either hierarchical or concurrent. In the framework represented in Figure 4, decision-making is considered to take place hierarchically. Hence, the designer employs the Preemptive formulation, where the achievement of the goals at the two levels is given two separate levels of priority. Level 1 decisions are given higher priority as these are made first in the hierarchy, followed by the decisions at Level 2, that is given lower priority. The difference in preferences of the multiple goals at a level is modeled by the designer using the Archimedean formulation, where different weights (values between 0 and 1, and summing up to 1) are assigned to the different goals to indicate varying degrees of preferences amongst the goals at a level.

b. *Step 2*: The coupled DSP formulated in Step 1 is executed for different multilevel (two in the framework depicted in Figure 4) design scenarios. The multilevel design scenarios are created by the designer by combining the design scenarios for the individual levels (created using uniform sampling).

Different design scenarios are created at each level by assigning different weights (values between 0 and 1 and summing up to 1) to the different goals. The weights assigned are indicative of the varying degrees of preferences amongst the goals at a level. The designer creates design scenarios for the two levels of the system (see, Step 2 in Figure 4). The design scenarios at each level are combined in all possible combinations to generate the multilevel design scenario. If the design scenarios for both levels are composed of n design distinct design scenarios, then the multilevel (two) design scenarios will be composed of n^2 distinct design scenarios in total.

c. *Step 3*: The design solutions generated after running the coupled DSP for the different multilevel design scenarios as discussed in Step 2, the design and solution spaces are visualized using iSOM. Using iSOM, component plots for both the inputs (weights for the different goals in the design scenarios) and outputs (goals) are generated. Using the output component plots, the designer explores the solution space to identify satisficing solution regions (identified by the iSOM grid points) for the individual goals across both levels (see output plots in Step 3, Figure 4), by setting satisficing limits for each goal. Only the grid points with design scenarios mapped against them (indicated by the dots on the iSOM grid points) are considered. A larger size of the dot at the center of the hexagonal iSOM grid points indicates a larger number of input points being mapped to that specific iSOM grid point.

The designer seeks a common satisficing region for all the goals. If no common satisficing solutions are identified for all the goals, the designer makes necessary relaxations to the satisficing limits of different goals till a common satisficing region for all the goals is realized (see the plot labeled 'common satisficing design region for all goals' in Step 3, Figure 4). Based on the common region identified, the designer can then identify the common satisficing design scenarios that are mapped to the common region and the corresponding design variable values. The input and output component plots can also be used by the designer to identify the effect of varying the weights on the different goals, on the goals at both levels.

4. TEST PROBLEM: STEEL MANUFACTURING SUPPLY NETWORK

The test problem is a steel manufacturing supply network (MSN). The steel MSN is composed of the steel manufacturer group (j) that produces steel slabs at Level 1, the supplier group composed of the suppliers (i) of raw materials (m) for steel production and the customer group composed of the customers (k) for the steel slabs produced by the manufacturer, at Level 2 (see Figure 5). The steel manufacturer purchases raw materials from suppliers to produce steel slabs, which are then sold directly to the customers. The steel production quantity is determined based on an estimate of expected customer demand.

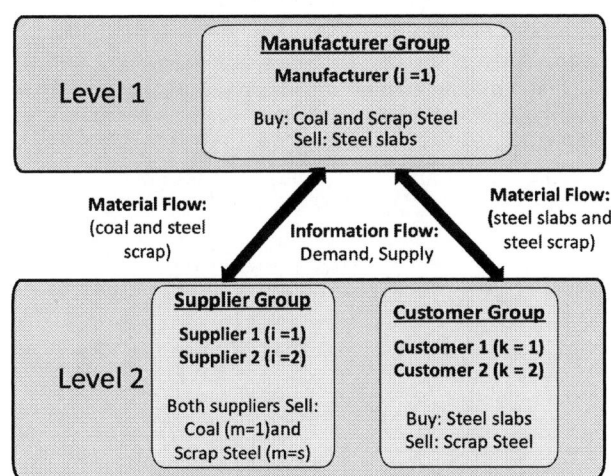

FIGURE 5: Material and information flow between levels in the Steel Manufacturing Supply Network

In this problem, we assume that the customer demand estimates are known. In the steel MSN test example, we consider the supplier group at Level 2 to be composed of 2 suppliers ($i = 1, 2$) capable of supplying both coal and steel scrap. At Level 1, we consider a single manufacturer ($j = 1$) operating a single manufacturing facility that manufactures steel slabs by employing the integrated Blast Furnace (BF) – Basic Oxygen Furnace (BOF) steel production technology [31, 32]. We consider the customer group at Level 2 to be constituted by 2 customers ($k = 1, 2$), who purchase steel slabs from the manufacturer and sell the steel scrap produced after their use, back to the manufacturer. The material flow between the levels is facilitated by employing logistics services, the cost for which is borne by one of the interacting levels. In this test problem, the decision-makers have a choice between 2 modes of transportation to choose from: a) Road (faster but relatively expensive), and b) Rail (less expensive but slower).

In the steel MSN, the decisions made at Levels 1 and 2 by the manufacturer, supplier and customer groups take place independently with each group having its own goals and constraints. Hence, the steel MSN considered is characterized by multilevel independent decisions made across Levels 1 and 2. The independent multilevel decisions are connected by means of information and material flows (see Figure 5). Hence, the decisions made at one level will impact the decisions at another level and therefore, it becomes vital to consider the *interactions across multiple levels*. The interactions of the independent multilevel decision can also result in *conflicts*, where the decisions made at a level may not align with the decisions at the interacting level. The conflicts can adversely impact the performance of MSN and hence there is a need to manage the conflicts to ensure steel MSN performance. This requires the facilitation of '*multilevel co-design*' of the multiple levels in the steel MSN. These multilevel decisions are impacted by uncertainties in the design variables that arise due to various reasons like production delays, quality control issues, damage to products in transit, and so on. Hence, there is a need to *manage uncertainties* in decisions at multiple levels. Hence, there exists a need to support the '*multi-level robust co-design*' of the steel MSN, that will help manage the uncertainties and conflict across the multilevel decisions.

In this paper, we demonstrate the utility of the framework in facilitating the multilevel robust co-design of MSNs by considering the interactions between the manufacturer and the supplier groups across Levels 1 and 2. We also consider the decision-making in MSN to be hierarchical in nature with the manufacturer group at Level 1 being the lead decision-maker.

4.1 Decisions made across multiple levels in the steel MSN and their interactions

The description of the decision made by the manufacturer and supplier groups at Level 1 and 2 respectively and their interactions are discussed below.

4.1.1 Decisions made at Level 1

The manufacturer group at Level 1 is the most important group as it is involved in the manufacture and supply of the products required by the customers, by using raw materials sourced from the supplier and customers. Hence, interactions at this level occur with both the supplier and customer groups at Level 2. These interactions are represented by the information and decision flows as depicted in Figure 6. The arrows within the manufacturer group depict the flow of information/decisions within Level 1. The arrows connecting Level 1 and 2 depict information flow from external sources, including other interacting groups. The manufacturer makes various decisions that include production, materials sourcing, and product distribution decisions (see manufacturer group at Level 1 in Figure 6). These decisions at the manufacturer level are aimed at fulfilling 3 level-specific goals: i) G_1 - maximization of service level (SL), ii) G_2 - maximization of profits, and iii) G_3 - minimization of greenhouse gas (GHG) emissions. These goals are conflicting in nature. When the focus is on maximizing the SL, it results in the reduction of profits and an increase in GHG emissions. The values indicated in the dashed boxes at the manufacturer group in Figure 6 depict the design variables at Level 1. There is variability in all the continuous design variables as described in Section 2 and it is assumed to be 2 percent of the range of values of the variable. The basic assumptions for decisions made at Level 1 are listed below.

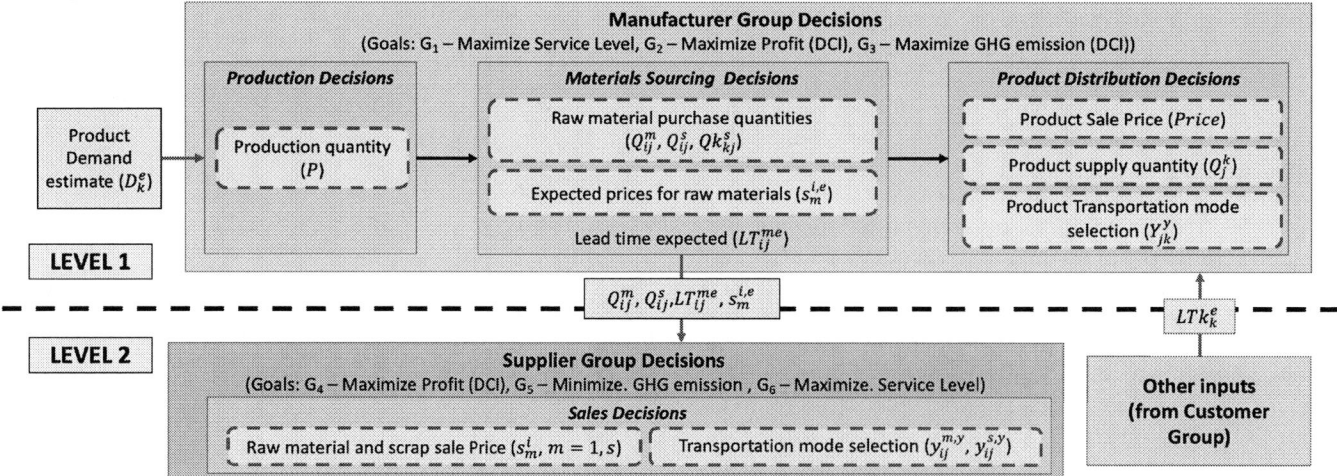

FIGURE 6: Information and Decision flows connecting the manufacturer and supplier groups at Level 1 and 2 respectively in the Steel Manufacturing Supply Network

Copyright © 2023 by ASME

a) The manufacturer makes products based on an estimate of customer demand. Hence, they employ the Made-to-Stock (MTS) approach to manufacturing.
b) The cost of transportation of products to customers is borne by the manufacturer.
c) The cost of transportation of the steel scrap from the customers is borne by the manufacturer.
d) The mode of transport used to deliver products is used to transport scrap purchased from customers.
e) There is sufficient capacity at the suppliers to meet the manufacturer's demand.
f) Scrap availability from customers is 10% of the sale quantity to customers.
g) All modes of transportation have sufficient capacity to supply the required quantity of products together.

Next, we describe the decisions made by the supplier group at Level 2.

4.1.2 Decisions made at Level 2 (supplier group)

The supplier group at Level 2 is involved in supplying the required raw materials (coal and steel scrap) to the manufacturer as per demand. Hence, here the supplier group interacts only with the manufacturer group at Level 1. These interactions are represented by the information and decision flows as depicted in Figure 6. The arrow flowing into the supplier group depicts information flow from Level 1. The decisions at Level 2 are aimed at fulfilling 3 level-specific goals: i) G_4 - maximization of profit, ii) G_5 - maximization of service level (SL), and iii) G_6 - minimization of greenhouse gas (GHG) emissions. These goals are conflicting in nature. When the focus is on maximizing the SL, it results in the reduction of profits and an increase in GHG emissions. The values indicated in the dashed boxes at the supplier group in Figure 6 depict design variables at Level 2.

The basic assumptions for decisions made at Level 2 are listed below.
a) There is sufficient capacity at the suppliers to meet the manufacturer's demand.
b) The cost of transportation of raw materials to the manufacturer is borne by the suppliers.
c) All the suppliers can supply both coal and steel scrap as required.
d) All modes of transportation have sufficient capacity to supply the required quantity of raw materials together.

The manufacturer and supplier groups at Levels 1 and 2 in the steel MSN are related by the shared design variables (prices for the raw materials) and propagated parameters (raw material purchase quantities and expected lead times) (see the arrow connecting the manufacturer and supplier groups in Figure 6).

4.2 Steel MSN: Decision support using the framework

We demonstrate the utility of the framework in supporting multilevel robust co-design of MSNs by applying it to the steel MSN test problem described in Sections 4 and 4.1. The designer starts with Step 1 (see Figure 4), where the decisions at Levels 1 and 2 and their interactions are modeled as a coupled DSP.

4.2.1 Step 1

Given the hierarchical nature of the decision-making in the steel MSN with the manufacturer group (Level 1) making the decisions first followed by the supplier group (Level 2), and the conflicting nature of the goals at each level, the decisions made at Levels 1 and 2 and their interaction are modeled as a vertically coupled cDSP (see Step 1 of the FRoMCoDE framework in Figure 4) in conjunction with the robust design construct DCI to account for uncertainties in the design variables. The hierarchical nature of decision-making also results in the deviation function being formulated using a Preemptive formulation with the Level 1 goals taking higher priority than the supplier-level goals. The word formulation of the coupled cDSP follows.

Given
a) Information specific to Level 1 (see Appendix A3 for details)
   - Manufacturer group (j) Information, and
   - Customer group (k) Information
b) Information specific to Level 2 (see Appendix A3 for details)
   - Manufacturer group (j) Information, and
   - Supplier group (k) Information
c) Design variables, their bounds, and variability
   - At Level 1: There are 16 continuous variables and 4 binary variables.
     Assuming a variability of +/- 2% of the range of values of the continuous design variables.
   - At Level 2: There are 4 continuous variables and 8 binary variables.
     Assuming a variability of +/- 2% of the range of values of the continuous design variables.

   The details of the design variables and their bounds at Levels 1 and 2 are provided in Appendix A4.
d) End requirements at Level 1
   i. Maximize Service Level (SL)
   ii. Maximize Profit (in $)
   iii. Minimize GHG emissions (in kgs of CO_2)

   Corresponding requirements on the cDSP goals (Gi)
   i. Goal G_1: Maximize Service Level (SL)
   ii. Goal G_2: Maximize DCI for Profit
   iii. Goal G_3: Maximize DCI for GHG emission

   End requirements at Level 2
   i. Maximize Profit (in $)
   ii. Minimize GHG emissions (in kgs of CO_2)
   iii. Maximize Service Level (SL)

   Corresponding requirements on the cDSP goals (Gi)
   i. Goal G_4: Maximize DCI for Profit
   ii. Goal G_5: Minimize GHG emission
   iii. Goal G_6: Maximize Service Level (SL)

   The equations for the end requirements at Levels 1 and 2 are provided in Appendix A1 and A2 respectively.
e) At Level 1: Lower Requirement Limit (LRL) for profit goal ($400,000), and Upper Requirement Limit (URL) for GHG emission goal (3,250 tons of CO_2)

   At Level 2: Lower Requirement Limit (LRL) for the profit goal ($575,000).

Copyright © 2023 by ASME

Find

At Level 1:

Values of

a) <u>Design variable values</u>: (16 continuous and 4 binary variables)
  i. <u>Continuous</u>: Production quantity in tons (P), Coal and scrap purchase quantities from suppliers in tons (Q_{ij}^m), Scrap purchase quantities from customers in tons (Qk_{kj}^s), Product supply quantities to customers in tons (Q_j^k), Steel selling price (*Price*) in \$, **Estimated selling price of material 'm' at supplier 'i' in \$ (s_m^{ie})**, and estimated selling price of material 'm' at customer 'k' in \$ (sk_m^{ke})
  ii. <u>Binary</u>: Transportation mode selection (Y_{jk}^y)
b) <u>Deviation variable values</u>: (d_i^+, d_i^-) for all $i = 1, 2, 3$

At Level 2:

Values of

a) <u>Design variable values</u>: (4 continuous and 8 binary variables)
  i. <u>Continuous</u>: **Selling price of material 'm' at supplier 'i' (s_m^i)** for all $m = 1$, s and $i = 1, 2$
  ii. <u>Binary</u>: Transportation mode selection (Y_{ij}^{my}) for all $m = 1$, s, $i = 1, 2$, and $y = 1, 2$
b) <u>Deviation variable values</u>: (d_i^+, d_i^-) for all $i = 4, 5, 6$

Shared Design variables: s_m^{ie} and s_m^i are the shared design variables. s_m^i is a copy of s_m^{ie} at the supplier group (Level 2)

Satisfy

At Level 1

a) <u>Constraints</u>: (10 linear constraints and 4 non-linear constraints)
  i. Total production less than production capacity: $P \leq Capacity$
  ii. Total production greater than total demand/demand forecast: $P \geq \sum_{k=1}^2 D_k^e$
  iii. and iv. Scrap purchased from customers less than the scrap available at customers: $Qk_{kj}^s \leq B_k Q_j^k$, for $k = 1, 2$
  v. Minimum amount of coal ($m = 1$) to be purchased: $\sum_{i=1}^2 Q_{ij}^m \geq A_m * P$
  vi. Minimum amount of scrap ($m=s$) to be purchased: $\sum_{i=1}^2 Q_{ij}^m + \sum_{k=1}^2 Qk_{kj}^s \geq A_m * P$
  vii. and viii Product supply quantity equal to demand/demand forecast: $Q_j^k = D_k^e$, for $k = 1, 2$
  ix. and x. Only 1 mode of transportation can be selected for product/scrap shipment (to both customers): $\sum_{y=1}^2 Y_{jk}^y = 1$, for $k = 1, 2$
  xi. Minimum DCI for Profit, $G1 \geq 1$
  xii. Minimum DCI for GHG emission, $G2 \geq 1$
  xiii. Minimum value of Profit = \$400,000

xiv. Maximum value of GHG emissions = 3,250 tons of CO_2

b) <u>Design variable bounds</u>: For all 20 design variables (as specified in the given section)

At Level 2

a) <u>Constraints</u>: (12 linear constraints and 2 non-linear constraints)
  i. to iv. Only 1 mode of transportation can be selected for coal/steel scrap shipment: $\sum_{y=1}^2 Y_{ij}^{my} = 1$, for i = 1, 2 and all $m=1$
  v. to xii. Maximum and minimum value for all shared design variables **(Shared design variable relaxation constraints: Assuming +/- 10 % relaxation)**
  - $0.9 s_1^{1e} \leq s_1^1 \leq 1.1 s_1^{1e}$
  - $0.9 s_1^{2e} \leq s_1^2 \leq 1.1 s_1^{2e}$
  - $0.9 s_s^{1e} \leq s_s^1 \leq 1.1 s_s^{1e}$
  - $0.9 s_s^{2e} \leq s_s^2 \leq 1.1 s_s^{2e}$
  xiii. Minimum DCI for Profit, $G1 \geq 1$
  xiv. Minimum value of Profit = \$575,000

b) <u>Design variable bounds</u>: For all 12 design variables (as specified in the given section)

Minimize Deviation Function for Preemptive formulation at 2 levels.

Priority 1: Level 1

The deviation function given below needs to be minimized.

$Z_1 = \sum_{i=1}^3 W_i (d_i^+ + d_i^-)$ where, $\Sigma W_i = 1$ and i = 1,2,3

Priority 2: Level 2

The deviation function given below needs to be minimized.

$Z_2 = \sum_{i=4}^6 W_i (d_i^+ + d_i^-)$ where, $\Sigma W_i = 1$ and i = 4,5,6

4.2.2 Step 2

The multi-level design scenarios for executing the coupled cDSP is created using a uniform sampling across the manufacturer and supplier levels (see Step 2 of FRoMCoDE framework in Figure 4). A total of 132 design scenarios are created and the coupled cDSP formulation established in Step 1 (see Section 4.2.1.) is then executed for these 132 design scenarios. Some sample design scenarios are listed in Table 2.

TABLE 2: Sample Multi-level Design scenarios

Scenario number	Level 1 Weights			ΣW_i	Level 2 Weights			ΣW_i
	W_1	W_2	W_3		W_4	W_5	W_6	
1	1	0	0	1	1	0	0	1
2	1	0	0	1	0	1	0	1
-	-	-	-	-	-	-	-	-
55	0	0.5	0.5	1	0	0.75	0.25	1
56	0.5	0	0.5	1	1	0	0	1
-	-	-	-	-	-	-	-	-
131	0.33	0.34	0.33	1	0	0.25	0.75	1
132	0.33	0.34	0.33	1	0	0.75	0.25	1

Copyright © 2023 by ASME

4.2.3 Step 3

This step involves the visualization and co-design exploration of the design/solution space at the manufacturer and supplier levels (see Step 3 of the FRoMCoDE framework in Figure 4). The solutions of the coupled cDSP, corresponding to the 132 multi-level design scenarios (see, Table 2) are visualized using iSOM. Using iSOM, 6 input component plots corresponding to the weights in the design scenarios, W1 to W6 (see Figure 7a.), and 6 output component plots corresponding to the 6 goals across Levels 1 and 2 (see Figure 7b) are generated.

The exploration of the design/solution space starts with the designer identifying the satisficing limits for each of the goals to identify satisficing solution regions for each goal that are indicated by the hexagons highlighted with a red border as shown in Figure 7b. With the robust DCI goals at Levels 1 and 2, the designer focuses on regions with higher DCI goal values to ensure a greater degree of safety against uncertainties in the design variables. The designer seeks to maximize the SL and hence picks regions on the output plot where the values are high. Regions of low values of GHG emission are preferred at Level 2 as the focus is on reducing GHG emissions. Hence, initially, the satisficing limits for the goals at Levels 1 and 2 are set as follows.

At Level 1
  i. Service Level (SL) at Level 1, $G_1 \geq 2.5$
  ii. DCI Profit at Level 1, $G_2 \geq 12$
  iii. DCI GHG emissions at Level 1, $G_3 \geq 14$
At Level 2
  i. DCI Profit at Level 2, $G_4 \geq 20$
  ii. GHG emissions at Level 2, $G_5 \leq 20000$
  iii. SL at Level 2, $G_6 \geq 2.4$

With the satisficing limits set to the above values, no common regions are identified for all 6 goals across Levels 1 and 2. Hence, the designer looks at relaxing the satisficing limits for the goals. The satisficing limits for the goals after relaxation are as follows (<u>relaxed limits in bold</u>):
At Level 1
  i. Service Level (SL) at Level 1, **$G_1 \geq 2.2$**
  ii. DCI Profit at Level 1, **$G_2 \geq 9$**
  iii. DCI GHG emissions at Level 1, $G_3 \geq 14$
At Level 2
  i. DCI Profit at Level 2, **$G_4 \geq 10$**
  ii. GHG emissions at Level 2, $G_5 \leq 20000$
  iii. SL at Level 2, **$G_6 \geq 2.3$**

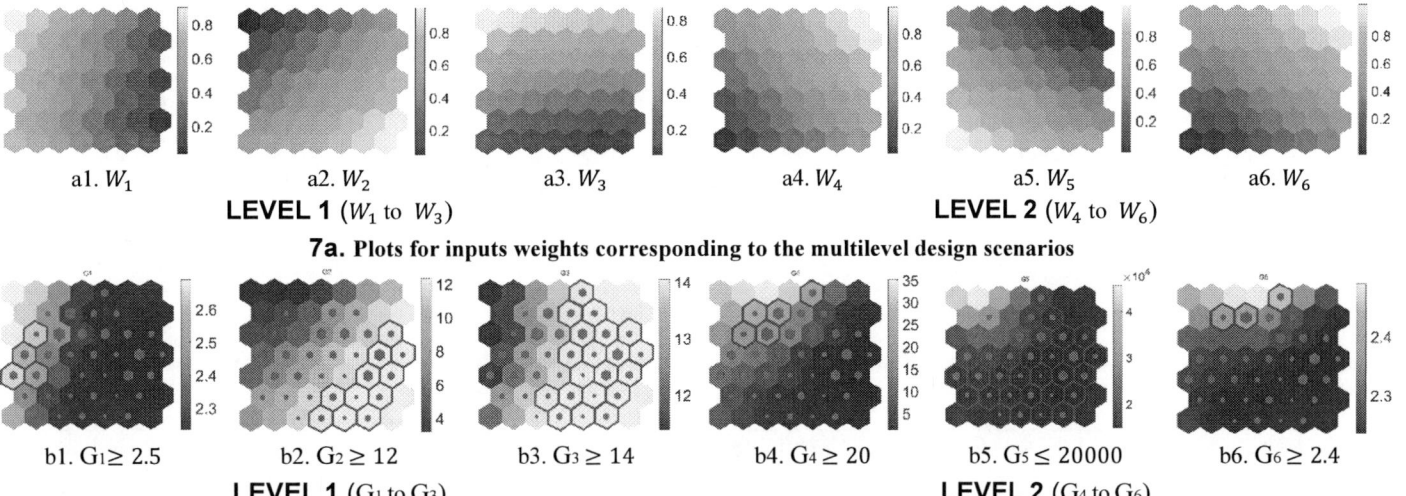

a1. W_1 a2. W_2 a3. W_3 a4. W_4 a5. W_5 a6. W_6

LEVEL 1 (W_1 to W_3)   **LEVEL 2** (W_4 to W_6)

7a. Plots for inputs weights corresponding to the multilevel design scenarios

b1. $G_1 \geq 2.5$ b2. $G_2 \geq 12$ b3. $G_3 \geq 14$ b4. $G_4 \geq 20$ b5. $G_5 \leq 20000$ b6. $G_6 \geq 2.4$

LEVEL 1 (G_1 to G_3)   **LEVEL 2** (G_4 to G_6)

7b. Satisficing solutions plots for goals before the relaxation of satisficing limits (highlighted using red hexagons)

FIGURE 7: The iSOM plots for input (design scenario weights, W_i) and output (Goals, G_i) for the steel MSN problem before the relaxation of satisficing limits.

NOTE: For DCI Goals G_2, G_3, and G_4, the yellow regions indicate regions of high robustness, and the blue regions indicate regions of comparatively lower robustness

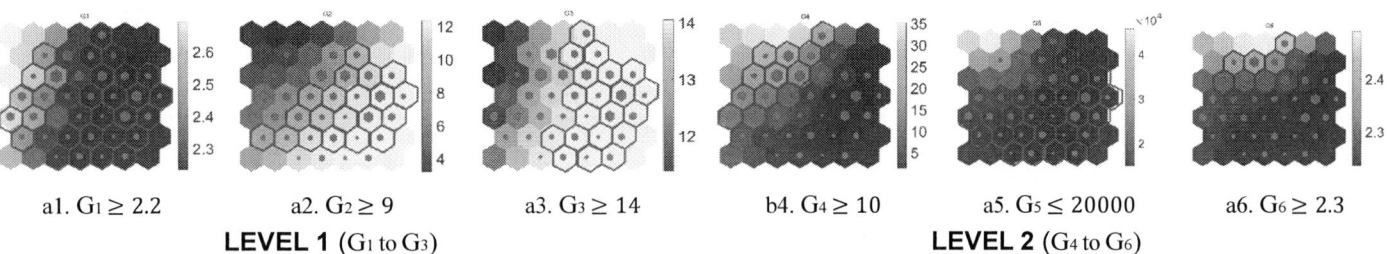

a1. $G_1 \geq 2.2$ a2. $G_2 \geq 9$ a3. $G_3 \geq 14$ b4. $G_4 \geq 10$ a5. $G_5 \leq 20000$ a6. $G_6 \geq 2.3$

LEVEL 1 (G_1 to G_3)   **LEVEL 2** (G_4 to G_6)

FIGURE 8: The satisficing solutions iSOM plots for goals after relaxation of satisficing limits (highlighted using red hexagons)

Copyright © 2023 by ASME

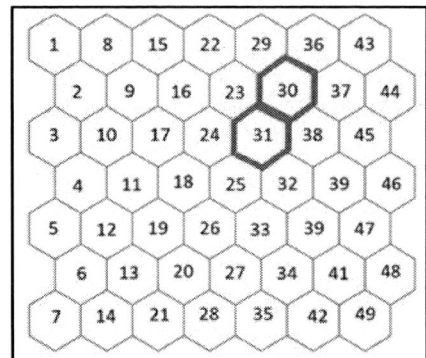

FIGURE 9: Common satisficing solution region for all goals after the relaxation of satisficing limits (Indicated by the hexagons highlighted in red)

NOTE:
- iSOM grid point locations are indicated by the number in the hexagons.
- Common satisficing regions are identified by the iSOM grid numbers of the common grid points (30 and 31, in this example)

With the relaxed satisficing limits, the updated satisfying regions for all the goals are identified (see Figure 8) With the

updated satisficing regions, two points (30 and 31) on the iSOM grid (see Figure. 9) are identified to be common satisficing region for all the goals. Grid points 30 and 31 have two and seven design scenarios mapped against them respectively. Hence a total of 9 common robust satisficing design solutions are identified for Levels 1 and 2 of the steel MSN design problem. These 9 design scenarios and the corresponding goal values at Levels 1 and 2 are listed in Table 3. The design variables values corresponding to the 9 common robust satisficing design solutions are listed in Tables 4 and 5 respectively. When analyzing the goal values for the manufacturer in Table 3, it is evident that the relaxed satisficing limits for all the goals (Goals 1, 2, and 3) at Level 1 were satisfied, due to the higher priority assigned in the Preemptive formulation. From the goal values achieved for Level 2 (lower priority) in Table 3, it is observed that the values for Goals 5 and 6 meet the relaxed satisficing limits. However, Goal 4 (Profit DCI maximization goal) failed to satisfy the relaxed satisficing limit (DCI $G_4 \geq 10$). This is expected due to the high lower requirement limit (LRL) set at the supplier group which results in low values for DCI. But the solutions identified are still robust solutions since their DCI values are greater than 1.

TABLE 3. Goal values at the manufacturer and supplier groups for the common robust satisficing solutions

Scenario	W_1	W_2	W_3	W_4	W_5	W_6	G_1	G_2	G_3	G_4	G_5 (Kgs of CO_2)	G_6
105	0	0.25	0.75	0	0	1	2.23	12.35	14.05	1.37	14793.16	2.23
107	0	0.25	0.75	0.25	0	0.75	2.23	12.35	14.05	1.37	14793.16	2.23
108	0	0.25	0.75	0	0.25	0.75	2.23	12.35	14.05	1.61	14787.02	2.23
123	0.33	0.34	0.33	1	0	0	2.23	12.36	14.05	1.09	14777.69	2.23
125	0.33	0.34	0.33	0	0	1	2.23	12.35	14.05	1.37	14793.16	2.23
127	0.33	0.34	0.33	0.5	0	0.5	2.23	12.35	14.05	1.19	14782.91	2.23
129	0.33	0.34	0.33	0.25	0	0.75	2.23	12.35	14.05	1.37	14793.16	2.23
130	0.33	0.34	0.33	0.75	0	0.25	2.23	12.36	14.05	1.09	14777.69	2.23
131	0.33	0.34	0.33	0	0.25	0.75	2.23	12.35	14.05	1.36	14793.37	2.23

TABLE 4. Design variable values at the manufacturer group corresponding to robust satisficing solutions identified

Scenario	P (tons)	Q_{11}^1 (tons)	Q_{21}^1 (tons)	Q_{11}^s (tons)	Q_{21}^s (tons)	Qk_{11}^s (tons)	Qk_{21}^s (tons)	Q_1^1 (tons)	Q_1^2 (tons)	Price ($/ton)	s_1^{1e} ($/ton)	s_1^{2e} ($/ton)	s_s^{1e} ($/ton)	s_s^{2e} ($/ton)	sk_s^{1e} ($/ton)	sk_s^{2e} ($/ton)	Y_{11}^1	Y_{11}^2	Y_{12}^1
105	2040	1.6	1610.5	0.6	612.0	0.1	0.1	1020.0	1020.0	1199.9	310.0	310.0	210.0	210.0	30.0	40.0	1	0	1
107	2040	1.6	1610.5	0.6	612.0	0.1	0.1	1020.0	1020.0	1199.9	310.0	310.0	210.0	210.0	30.0	40.0	1	0	1
108	2040	0.1	1612.6	0.0	612.0	0.0	0.0	1020.0	1020.0	1200.0	310.0	310.0	210.0	210.0	30.0	40.0	1	0	1
123	2040	0.4	1610.7	0.1	612.0	0.0	0.0	1020.0	1020.0	1200.0	310.0	310.0	210.0	210.0	30.0	40.0	1	0	1
125	2040	1.6	1610.5	0.6	612.0	0.1	0.1	1020.0	1020.0	1199.9	310.0	310.0	210.0	210.0	30.0	40.0	1	0	1
127	2040	0.8	1610.6	0.3	612.0	0.0	0.0	1020.0	1020.0	1200.0	310.0	310.0	210.0	210.0	30.0	40.0	1	0	1
129	2040	1.6	1610.5	0.6	612.0	0.1	0.1	1020.0	1020.0	1199.9	310.0	310.0	210.0	210.0	30.0	40.0	1	0	1
130	2040	0.4	1610.7	0.1	612.0	0.0	0.0	1020.0	1020.0	1200.0	310.0	310.0	210.0	210.0	30.0	40.0	1	0	1
131	2040	1.6	1610.5	0.6	611.9	0.1	0.2	1020.0	1020.0	1199.9	310.0	310.0	210.0	210.0	30.0	40.0	1	0	1

TABLE 5. Design variable values at the supplier group corresponding to robust satisficing solutions identified

Scenario	s_1^{1e} ($/ton)	s_1^{2e} ($/ton)	s_s^{1e} ($/ton)	s_s^{2e} ($/ton)	Y_{11}^1	Y_{11}^2	Y_{12}^1	Y_{12}^2	Y_{11}^1	Y_{11}^2	Y_{12}^1	Y_{12}^2
105	319.9	330.0	210.0	220.0	1	0	1	0	1	0	1	0
107	319.9	330.0	210.0	220.0	1	0	1	0	1	0	1	0
108	300.0	330.0	209.9	220.0	1	0	1	0	1	0	1	0
123	320.0	330.0	210.0	220.0	1	0	1	0	1	0	1	0
125	319.9	330.0	210.0	220.0	1	0	1	0	1	0	1	0
127	320.0	330.0	210.0	220.0	1	0	1	0	1	0	1	0
129	319.9	330.0	210.0	220.0	1	0	1	0	1	0	1	0
130	320.0	330.0	210.0	220.0	1	0	1	0	1	0	1	0
131	319.9	330.0	210.0	220.0	1	0	1	0	1	0	1	0

Copyright © 2023 by ASME

The design variable values corresponding to the common robust satisficing design solutions identified are observed to be similar to each other (see Tables 4 and 5). For the manufacturer group at Level 1, Goals 2 and 3 are robust DCI goals with larger satisficing limits. To meet these satisficing limits, the designers look at design scenarios with higher weightage on Goals 2 and 3. This results in the designers choosing design solutions that have similar design variable values. For the supplier group at Level 2, the decisions made at Level 1 by the manufacturer group limit or restrict the design/solutions space at Level 2. In addition, the high lower requirement limit set for Goal 4 (DCI Profit) will further restrict the design/solution space at Level 2. This will lead to the choice of design solutions that will result in similar design variable values.

Therefore, by exercising the framework presented, designers are able to model, independent decision-making at Levels 1 and 2, while accounting for their interactions (in terms of shared design variables and propagated parameters) and uncertainties in the design variables (as a coupled robust cDSP). Shared design variable relaxation constraints were used to facilitate the sharing of a ranged set of values from Level 1 to Level 2, for the shared design variable (s_m^{ie}) in the coupled cDSP. By exercising the coupled cDSP model for various multilevel design scenarios, the designer is able to generate the solution space for all goals for the 2-level steel MSN problem. Using iSOM, the designer is able to visualize the solution space of all the goals (including robust solution space for DCI goals). By setting satisficing limits for each goal and exploring the solutions spaces visualized using the iSOM output plots, the designer is able to identify 'satisficing/robust satisficing solutions' for all the goals across Levels 1 and 2 (see Figure 7b). When no common satisficing solutions are identified for all goals at Levels 1 and 2, a conflict occurs that needs to be managed to realize multilevel co-design. The designer manages this conflict by relaxing the satisficing limits for the goals (as deemed appropriate) and exploring the solution spaces till a common satisficing/robust satisficing solution region is identified for all the goals across the two interacting levels in the steel MSN. Hence, using the framework, the multilevel robust co-design of the steel MSN is realized.

5. CLOSING REMARKS

Computational models employed in the simulation-based design of MSNs are abstractions of reality and are sources of uncertainties. The design of multilevel MSNs operating under uncertainty requires the consideration of the uncertainty involved in decision-making at different levels and the interactions between the independent multilevel decisions. Failure to account for the interrelations between the independent multilevel decisions will result in conflicts, where the decision made at one level is not in alignment with the decisions at a related level. Such conflicts can adversely impact the performance of MSN. Any lapses in dealing with the uncertainties involved in the decision-making at the individual levels will also have an impact on MSN performance. Hence it is vital to consider the interrelation between multilevel decisions and manage the uncertainties during the design of MSNs to ensure the performance of the MSN. Specifically, support for the simulation-based design of MSNs necessitates a focus on design exploration. This requires the facilitation of visualization and systematic co-design exploration of the multilevel design/solution spaces.

In this paper, we present 'FRoMCoDE', where the Preemptive formulation of the coupled DCP construct is combined with DCI robust design construct and iSOM based visualization to support the 'multilevel robust co-design.' Using the framework, the designer is able to i) model the robust multilevel decision-making and their interrelations and ii) visualize and carry out co-design exploration of the multilevel design/solution spaces. The framework enhances a human designer's ability to systematically model robust multi-level decision-making and their interactions in the MSN, using coupled cDSP construct in conjunction with the DCI construct. Using the Preemptive formulation helps designer's model hierarchical relationships in MSNs. The Archimedean formulation permits the designers to model concurrent decisions. Using the framework, the designer is able to visualize the multi-level design/solution space in an integrated manner (using iSOM) and systematically explore it to identify common robust satisficing solutions for all the interacting levels in the MSN and thereby manage conflicts. The framework is tested for the above functionalities using the steel MSN problem. Using the framework, robust co-design of the steel MSN with interactions between the manufacturer and supplier groups at different levels (Levels 1 and 2) is demonstrated. The conflicts arising from the independent, hierarchically related decision-making between the manufacturer and supplier groups at different levels are managed by systematically exploring the solution space (visualized using iSOM plots) to identify common robust satisficing design solutions for the goals at these levels. The generic nature of the framework is made evident by the generic nature of the constructs and tools employed. Using the framework, we facilitate the robust co-design of multilevel systems, characterized by relations that are by nature hierarchical, concurrent, or a combination of both.

Some avenues for improvement of the framework are identified and discussed below. In the framework presented, a uniform sampling approach is employed to create multi-level design scenarios to cover the multi-level design/solution spaces. However, uniform sampling will require a considerably large number of design scenarios to effectively cover the entire design/solution space. This limitation could be addressed by using Optimal Latin Hypercube Sampling (OLHS) designs that are more efficient in capturing large design/solution spaces effectively. Secondly, even though design variable uncertainties are considered, their propagation along the interacting levels in the MSN is not considered. In the future, this could be addressed by formulating the design problem using a Type IV robust design formulation, which allows for the management of uncertainties associated with noise factors, design variables, models, and uncertainty propagation between interacting levels [33, 34].

Copyright © 2023 by ASME

ACKNOWLEDGEMENTS

Mathew Baby and Anand Balu Nellippallil thank the Department of Mechanical and Civil Engineering, Florida Institute of Technology, for the support. The authors thank Dr. Palaniappan Ramu, and Ms. Rashmi Rama Sushil, ADOPT Laboratory, IIT Madras for their help with iSOM visualization. The authors also thank Mr. Gehendra Sharma, Center for Advance Vehicular Systems, Mississippi State University for the assistance with the formulation of coupled DSPs. Janet K. Allen and Farrokh Mistree gratefully acknowledge the John and Mary Moore Chair and the L.A. Comp Chair at the University of Oklahoma.

REFERENCES

1. Thomas, D., 2022., "Annual Report on U.S. Manufacturing Industry Statistics: 2022," *NIST Advanced Manufacturing Series.*
2. Mistree, F., Smith, W. F., Bras, B. A., Allen, J. K., and Muster, D., 1990, "Decision-Based Design. A Contemporary Paradigm for Ship Design," *Transactions of the Society of Naval Architects and Marine Engineers*, vol. 98, pp. 565–597.
3. Hazelrigg, G. A., 1998, "A Framework for Decision-Based Engineering Design," *Journal of Mechanical Design*, vol. 120, no. 4, pp. 653–658.
4. Muster, D., and Mistree, F., 1988, "The Decision Support Problem Technique in Engineering Design," *International Journal of Applied Engineering Education*, vol. 4, no. 1, pp. 23-33.
5. Mistree, F., and Allen, J. K., 1997, "Position Paper: Optimization in Decision-Based Design," *Optimization in Industry*, Palm Coast, FL.
6. Simon H. A., 1947, "Administrative Behavior", *Mcmillan, New York.*
7. Simon, H. A., 1956, "Rational Choice and the Structure of the Environment," *Psychological Review*, vol. 63, no. 2, pp. 129-138.
8. Mistree, F., Hughes, O. F., and Bras, B., 1993, "Compromise Decision Support Problem And The Adaptive Linear Programming Algorithm," *Structural Optimization: Status Promise*, pp. 247-286.
9. Sharma, G., Allen, J. K., and Mistree, F., 2022, "Designing Concurrently and Hierarchically Coupled Engineered Systems," *Engineering Optimization*, pp. 1-21.
10. Nellippallil, A.B., Allen, J.K., Gautham, B.P., Singh, A.K., Mistree, F., 2020, "Robust Concept Exploration of Materials, Products, and Associated Manufacturing Processes," *Architecting Robust Co-Design of Materials, Products, and Manufacturing Processes*, Springer International Publishing, Cham, pp. 263–296.
11. Chen, W., Simpson, T., Allen, J., and Mistree, F., 1999, "Satisfying Ranged Sets of Design Requirements using Design Capability Indices as Metrics," *Engineering Optimization*, vol.31, pp. 615-639.
12. Choi, H.-J., Austin, R., Allen, J. K., McDowell, D. L., Mistree, F., and Benson, D. J., 2005, "An Approach for Robust Design of Reactive Power Metal Mixtures Based on Non-deterministic Micro-scale Shock Simulation," *Journal of Computer-Aided Materials Design*, vol.12, no.1, pp. 57-85.
13. Martins, J. R. R. A., and Lambe, A. B., 2013, "Multidisciplinary Design Optimization: A Survey of Architectures," *AIAA Journal*, vol. 51, no. 9, pp. 2049-2075.
14. Sobieszczanski-Sobieski, J., and Kodiyalam, S., 2001, "BLISS/S: A New Method for Two-Level Structural Optimization," *Structural and Multidisciplinary Optimization*, vol. 21, no.1, pp. 1-13.
15. Sobieszczanski-Sobieski, J., Agte, J., and Robert Sandusky, J., 1998, "Bi-level Integrated System Synthesis (BLISS)," *7th AIAA/USAF/NASA/ISSMO Symposium on Multidisciplinary Analysis and Optimization*, St. Louis, M.O., U.S.A, Paper No: AIAA-98-4916, pp. 1543-1557.
16. Sobieszczanski-Sobieski, J., Altus, T., Phillips, M., and Sandusky, R., 2002, "Bi-Level Integrated System Synthesis (BLISS) for Concurrent and Distributed Processing," *9th AIAA/ISSMO Symposium on Multidisciplinary Analysis and Optimization*, Atlanta, Georgia, U.S.A, Paper No: AIAA 2002-5409, pp. 1-11.
17. Kim, H. M., Rideout, D. G., Papalambros, P. Y., and Stein, J. L., 2003, "Analytical Target Cascading in Automotive Vehicle Design," *Journal of Mechanical Design*, 125(3), pp. 481-489.
18. Kroo, I., Altus, S., Braun, R., Gage, P., and Sobieski, I., "Multidisciplinary Optimization Methods for Aircraft Preliminary Design," *5th Symposium on Multidisciplinary Analysis and Optimization*, Panama City Beach, F.L., U.S.A., AIAA-94-4325-CP, pp. 697-707
19. Allen, J. K., Seepersad, C., Choi, H., and Mistree, F., 2006, "Robust Design for Multiscale and Multidisciplinary Applications," *Journal of Mechanical Design*, vol. 128, no. 4, pp. 832-843.
20. Pourmehdi, M., Paydar, M. M., and Asadi-Gangraj, E., 2020, "Scenario-Based Design of a Steel Sustainable Closed-Loop Supply Chain Network Considering Production Technology," *Journal of Cleaner Production*, vol. 277, pp. 123298.
21. Singh, S., Mohanty, A., Rai, R., Mahanty, B., and Tiwari, M. K., 2022, "An Optimization Framework for Operational-Level Resource Composition in an Inclusive Manufacturing System," *Journal of Computing and Information Science in Engineering*, vol. 22, no. 5.
22. Choi, H.-J., McDowell, D. L., Allen, J. K., and Mistree, F., 2008, "An Inductive Design Exploration Method for Hierarchical Systems Design under Uncertainty," *Engineering Optimization*, vol. 40, no. 4, pp. 287-307.
23. Nellippallil, A. B., Mohan, P., Allen, J. K., and Mistree, F., 2020, "An Inverse, Decision-Based Design Method for Robust Concept Exploration," *Journal of Mechanical Design*, vol. 142, no. 8, pp. 081703.
24. Nellippallil, A. B., Rangaraj, V., Gautham, B., Singh, A. K., Allen, J. K., and Mistree, F., 2018, "An inverse, decision-

based design method for integrated design exploration of materials, products, and manufacturing processes," *Journal of Mechanical Design*, vol. 140, no. 11., pp. 111403

25. Sharma, G., Allen, J. K., and Mistree, F., 2021, "A Method for Robust Design in a Coupled Decision Environment," *Design Science*, vol. 7, pp. e23.

26. Thole, S. P., and Ramu, P., 2020, "Design Space Exploration and Optimization Using Self-Organizing Maps," *Structural and Multidisciplinary Optimization*, vol. 62, no. 3, pp. 1071-1088.

27. Richardson, T., Kannan, H., Bloebaum, C., and Winer, E., 2014, "Incorporating Value-Driven Design into the Visualization of Design Spaces Using Contextual Self-Organizing Maps: A Case Study of Satellite Design," *15th AIAA/ISSMO Multidisciplinary Analysis and Optimization Conference*, Atlanta, Georgia, U.S.A.

28. Sushil, R. R., Baby, M., Sharma, G., Balu Nellippallil, A., and Ramu, P., 2022, "Data Driven Integrated Design Space Exploration Using iSOM," *ASME 2022 International Design Engineering Technical Conferences and Computers and Information in Engineering Conference,* Paper. No. DETC2022-89895.

29. Nagar, D., Pannerselvam, K., and Ramu, P., 2022, "A Novel Data-Driven Vsualization of n-Dimensional Feasible Region using interpretable Self-Organizing Maps (iSOM)," *Neural Networks*, vol. 155, pp. 398-412.

30. Nagar, D., Ramu, P., and Deb, K., 2023, "Visualization and Analysis of Pareto-Optimal Fronts using interpretable Self-Organizing Map (iSOM)," *Swarm and Evolutionary Computation*, vol. 76, pp. 101202.

31. Madhavan, N., Brooks, G., Rhamdhani, M. A., and Bordignon, A., 2022, "Contribution of CO_2 Emissions from Basic Oxygen Steelmaking Process," *Metals*, vol. 12, no. 5, pp. 797.

32. Ruth, M., 2004, "Steel Production and Energy," *Encyclopedia of Energy*, C. J. Cleveland, ed., Elsevier, New York, pp. 695-706.

33. Allen, J. K., Panchal, J., Mistree, F., Singh, A. K., and Gautham, B. P., 2015, "Uncertainty Management in the Integrated Realization of Materials and Components," *Proceedings of the 3rd World Congress on Integrated Computational Materials Engineering (ICME 2015),* Springer International Publishing, Cham, pp. 339-346.

34. McDowell, D. L., Panchal, J. H., Choi, H.-J., Seepersad, C. C., Allen, J. K., and Mistree, F., 2010, "Chapter 6 - Robust Design of Materials—Design Under Uncertainty," *Integrated Design of Multiscale, Multifunctional Materials and Products*, Butterworth-Heinemann, Boston, pp. 113-145.

APPENDIX A - Models for the end requirements, level specific information and design variables at Levels 1 and 2 of the coupled cDSP formulation (for the steel MSN problem)

What follows is an amplification of what is presented in Section 4.2. The mathematical models for the end requirements

of the manufacturer and supplier groups at Levels 1 and 2 of the steel MSN problem are listed below in AI and A2 respectively. In A3, the information specific to Levels 1 and 2 of the coupled cDSP formulation is provided. In A4, we list the design variables and their bounds for Levels 1 and 2 of the coupled cDSP formulation.

A1. Models for the end requirements of the Manufacturer Group (Level 1) of the steel MSN

i. Maximize Service Level (SL)
$$G_1 = \sum_{k=1}^{2}\{ LTk_k^e / (\sum_{y=1}^{2} \frac{D_{jk}^y}{Speed^y} * Y_{jk}^y + LT_o^j)\}$$

ii. Maximize Profit (in $)
$$G_2 = (Price * \{\sum_{k=1}^{2} Q_j^k\}) - (P * Cost^n + \sum_{m=1,s}\sum_{i=1}^{2} s_m^i Q_{ij}^m + \sum_{k=1}^{2} sk_s^k Qk_{kj}^s + \sum_{i=1}^{2}\sum_{y=1}^{2} Q_j^k D_{jk}^y T_{jk}^y Y_{jk}^y + \sum_{k=1}^{2}\sum_{y=1}^{2} Qk_{kj}^s D_{jk}^y T_{jk}^y Y_{jk}^y)$$

iii. Minimize GHG emissions (in kgs of CO_2)
$$G_3 = (P * E + \sum_{i=1}^{2}\sum_{y=1}^{2} Q_j^k D_{jk}^y E_{jk}^y Y_{jk}^y + \sum_{k=1}^{2}\sum_{y=1}^{2} Qk_{kj}^s D_{jk}^y E_{jk}^y Y_{jk}^y)$$

A2. Models for the end requirements of the Supplier Group (Level 2)

i. Maximize Profit (in $)
$$G_4 = \sum_{m=1,s}(\sum_{i=1}^{2} s_m^i q_{ij}^m - (C_m^i q_{ij}^m + \sum_{y=1}^{2} d_{ij}^{my} q_{ij}^m t_{ij}^{my} Y_{ij}^{my}))$$

ii. Minimize GHG emissions (in kgs of CO_2)
$$G_5 = \sum_{m=1,s}\sum_{i=1}^{2}\sum_{y=1}^{2} d_{ij}^{my} q_{ij}^m e_{ij}^{my} Y_{ij}^{my}$$

iii. Maximize Service Level (SL)
$$G_6 = \{\sum_{i=1}^{2}(\frac{1}{m} * \sum_{m=1,s}(\frac{LT_{ij}^{me}}{\sum_{y=1}^{2}(\frac{D_{ij}^{my}}{Speed^y} * Y_{ij}^{my} + LT_o^i)})\}$$

A3. Level Specific information for the coupled cDSP
At Level 1

- Manufacturer group (j) Information: Set of manufacturers ($j = 1$), Production capacity (Capacity) in tons, Production cost ($Cost$) in $ per ton, the raw material (m) requirement in tons per ton of steel produced $\{A_m\}$ (coal, $m=1$ and steel scrap, $m=s$), transportation information – (modes $\{y = 1,2\}$, speed in km/hr. $\{Speed^y\}$, distance to customers in km $\{D_{jk}^y\}$, transportation costs in $ per ton per km $\{T_{jk}^y\}$, Greenhouse gas $\{GHG\}$ emission in kgs of CO_2 per ton transported per km $\{E_{jk}^y\}$), and demand estimate at customer k (D_k^e).

- Customer group (k) Information: Set of customers ($k = 1, 2$), Steel scrap availability – fraction of tons of steel purchased (B_k), steel scrap prices in $ per ton (sk_s^k), and expected lead time in hrs. (LTk_k^e).

Copyright © 2023 by ASME

At Level 2

- **Manufacturer group Information:** Actual order quantity of material m (Q_{ij}^m), expected lead time for material m in hrs (LT_{ij}^{me}) from supplier i
- **Supplier group Information:** Set of suppliers ($i = 1, 2$), Materials supplied {m, $= 1$ (coal), $= s$ (scrap)}, forecasted demand estimate for material 'm' at manufacturer in tons (d_m^e), Material cost of material 'm' in \$ per ton, (C_m^i), transportation information – (modes {$y = 1,2$}, speed in km/hr {$Speed^y$}, distance to customers in km {d_{ij}^{my}}, transportation costs in \$ per ton per km {t_{ij}^{my}}, Greenhouse gas {GHG} emission in kgs of CO_2 per ton transported per km {e_{ij}^{my}})

A4. Design variables and their bounds at Levels 1 and 2 for the coupled cDSP formulation

At Level 1

- $2000 \leq$ Production quantity in tons (P) ≤ 4000
- Coal and steel scrap purchase quantities from suppliers in tons (Q_{ij}^m), where $i = 1,2; j = 1;$ and $m = 1, s$
  - $0 \leq Q_{11}^1, Q_{21}^1 \leq 3200$
  - $0 \leq Q_{11}^s, Q_{21}^s \leq 1600$
- Scrap purchase quantities from customers in tons (Qk_{kj}^s), where $k = 1,2; j = 1$
  - $0 \leq Qk_{11}^s, Qk_{21}^s \leq 500$
- Product supply quantities to customers in tons (Q_j^k), where $k = 1,2; j = 1$
  - $1000 \leq Q_1^1, Q_1^2 \leq 1200$
- $1000 \leq$ Steel selling price in \$ per ton ($Price$) ≤ 1200
- The estimated selling price of material 'm' at supplier 'i' (s_m^{ie}) in \$ per ton, for all $m = 1, s$ and $i = 1,2$
  - $310 \leq s_1^{1e}, s_1^{2e} \leq 330$
  - $210 \leq s_s^{1e}, s_s^{2e} \leq 230$
- The estimated selling price of material 'm' at customer 'k' (sk_m^{ke}) in \$ per ton, for all $m = s$ and $k = 1,2$
  - $30 \leq sk_s^{1e} \leq 50$
  - $40 \leq sk_s^{2e} \leq 60$
- Transportation mode selection (Y_{jk}^y) for transporting the products to customers and steel scrap from customers, where $k = 1,2; j = 1;$ and $y = 0,1$
  - $Y_{11}^1, Y_{11}^2, Y_{12}^1, Y_{12}^2 = 0, 1$

At Level 2

- The selling price of material 'm' at supplier 'i' (s_m^i) in \$ per ton, for all $m = 1, s$ and $i = 1,2$
  - $300 \leq s_1^1 \leq 320$
  - $310 \leq s_1^2 \leq 330$
  - $190 \leq s_s^1 \leq 210$
  - $200 \leq s_s^2 \leq 220$
- Transportation mode selection (Y_{ij}^{my}) for transportation of materials to the manufacturer, for all $m = 1, s; i = 1,2,$ and $y = 1,2$

- $Y_{11}^{11}, Y_{11}^{12}, Y_{11}^{s1}, Y_{11}^{s2} = 0, 1$
- $Y_{21}^{11}, Y_{21}^{12}, Y_{21}^{s1}, Y_{21}^{s2} = 0, 1$

Copyright © 2023 by ASME

Proceedings of the ASME 2023
International Design Engineering Technical Conferences and
Computers and Information in Engineering Conference
IDETC-CIE2023
August 20-23, 2023, Boston, Massachusetts

DETC2023-111360

IMPROVING NONUNIFORM UTILIZATION OF LI-ION POUCH CELLS USING TAPERED ELECTRODES THROUGH CALENDERING

Changik Cho[1,*], Seth Kelley[1], Joe Tylka[2], Miao He[2], Naresh N Nandola[2], Christopher D. Rahn[1,*]

[1]The Pennsylvania State University, University Park, PA
[2]Siemens Technology, Princeton, NJ, USA

ABSTRACT

Large format lithium-ion pouch cells can have fewer tabs, less packaging material and interconnects, and lower cost but nonuniform material utilization reduces the achievable power and energy densities. The impedance varies across the cell with active material farthest from the tabs having the highest impedance. As a result, these underutilized parts of the cell do not become fully charged or discharged, especially at high current rates, and the energy storage capability of the cell is not fully realized. The current collector thickness can be thickened to increase material utilization, but cost and energy density suffer. In this paper, we propose and simulate linearly tapered electrode coatings to decrease thickness (and increase impedance) in low impedance areas, improving the impedance uniformity and material utilization across the cell. The taper is introduced in the calendering process, by tilting the rollers to change thickness and porosity across the width of the cell. A multiple particle (MP) model is developed and validated in COMSOL to optimize the slope of the taper based on Brute-force algorithm. Higher impedance near the tab balances the higher impedance in the current collector away from the tab. The analysis results show that a 10% linearly tapered positive electrode can improve uniformity by 8 times with 3C charging and increase capacity by 13% with 5C charging.

Keywords: Li-ion pouch cell, Nonuniform utilization, Linearly tapered electrode, Calendering

NOMENCLATURE

Roman letters

A	Cell surface area
A_{cc}	Current collector cross-sectional area
a_s	Specific interfacial area
a_{slope}	Taper slope
c	Li concentration
D	Diffusion coefficient
F	Faraday's constant
I	Current
i_0	Initial current density
j	Current density
k	Linear coefficient between δ and ϵ
n	Number of nodes
R	Universal gas constant
R_{cc}	Current collector resistance
R_{ct}	Charge transfer resistance
R_f	Contact resistance
R_s	Particle radius
U	Equilibrium potential
V	Terminal voltage

Greek letters

α	Transfer coefficient
Δ	Difference of value at the top and at the bottom of electrode
δ	Thickness
ϵ	Porosity
ϵ_s	Electrode volume fraction
η	Overpotential
ρ	Current collector resistivity
ϕ	Equivalent potential

Superscripts and subscripts

a	anode
c	cathode
i	i^{th}
s	solid phase
s, e	solid/electrolyte interface
+	positive electrode
−	negative electrode

1. INTRODUCTION

In recent years, efforts to alleviate the effects of climate change and global warming have driven the development of

*Corresponding author: cmc7589@psu.edu,cdr10@psu.edu

Copyright © 2023 by ASME

lithium-ion (Li-ion) battery technology. Researchers have successfully uncovered the nonuniform utilization effect on Li-ion cells with thick electrodes and large format pouch cells [1–6]. The impedance in Li-ion pouch cells varies across the cell having the highest impedance at the farthest from the tabs. As a result, these underutilized parts of the cell do not become fully charged or discharged. Nonuniform current distribution is measured experimentally with in-situ measurement method and fast charging impairs uniform current distribution having only 60% of capacity [1]. Effects of various tab locations are also investigated to reduce nonuniform utilization of Li-ion cells [2].

Large format cells can have fewer tabs, less packaging material and interconnects, and lower cost. Nonuniform material utilization, however, reduces the achievable power and energy densities. The distance in large pouch cells from the terminal is longer and it varies local resistance bigger. Thus, the local current tends to be higher around the tab while the furthest electrode from the tab gets lower. The discrepancy of local currents drives local temperature difference [3–6], and this causes local aging faster.

Nowadays, a high C rate charging and discharging profile is preferable in electric Vertical Take-Off and Landing (eVTOL) aircraft reducing operating costs while achieving high vehicle utilization rates to maximize revenues. The mission of these aircraft is 20 minutes with 150 km range requiring 1C to 5C discharging [7]. Large format cells can be a good choice for it satisfies high capacity and energy density with eVTOL mission, but nonuniform utilization in pouch cell is still a huge difficulty. Fast charging in large format cells can aggravate the energy storage capability of the cell and aging becomes progressively faster since active material is not fully realized and certain local active material is repeatedly used [8]. The worst scenario is the local internal short with increased local ohmic resistance by fast charging [8]. The current collector thickness can be increased to improve material utilization, but cost and energy density suffer.

In this paper, we focus on solving non-uniform utilization for large format Li-ion pouch cell through a novel configuration of positive electrodes. There are not many papers alleviating nonuniform utilization. Battery separators with variable porosity are introduced to reduce nonuniform utilization [9]. Higher porosity far from the tab and lower porosity near the tab in the separator enables the lateral balance of Li transportation through the separator. The porosity in electrodes along the thickness (lateral) direction is controlled to improve the battery performance for especially thick electrodes, not tab (lateral) direction [10]. The goal of this research is to develop linearly varying tapered electrodes for augmenting the uniform utilization of Li-ion batteries. Contributions of this paper are as followed:

1. This paper suggests linearly varying tapered positive electrodes in the vertical direction manipulating thickness and porosity increasing uniform utilization.

2. We develop a multiple particle (MP) model to optimize how much slope of linearly tapered electrodes needs to be designed, which can represent the thickness and porosity effect along the vertical direction.

Firstly, the concept of linearly varying tapered electrodes is introduced to increase the uniform utilization of Li pouch cells and correlate with battery manufacturing processes. An equivalent circuit model based on MP model is presented to show how we model and represent the thickness and porosity effect. MP model is validated with average voltage response with COMSOL model to reduce computing time for optimization of how the slope of tapered electrode is chosen. We optimize a tapered slope of a positive electrode with an MP model based on electro-chemical and apply results to COMSOL model, a true model. Optimized results of Li concentration on each location at COMSOL are discussed.

2. LINEARLY VARYING TAPERED ELECTRODE

The concept for the tapered electrode is presented and correlated with one of the battery manufacturing processes, calendering process. Manufacturing processes for Li-ion battery production comprise mixing, coating, drying, solvent recovery, calendering, slitting, and formation and aging [11]. The calendering process is the process using two rollers to settle down the active material, conductive additive (Carbon black), and binder (PVDF) onto the current collector in both positive and negative electrodes [12–22]. It is well known that it changes electrochemical parameters including porosity, thickness, tortuosity, electronic conductivity, and ionic conductivity reducing resistance affecting battery performances.

Especially, porosity and thickness have a significant impact on the battery performance and are often used as optimized parameters to design higher energy density cells [23–28]. After the calendering process, porosity and thickness decrease and improve battery performance decreasing impedance [29]. Thickness and porosity values can be normally controlled up and down to 22% during calendering process.

We suggest linearly tapered electrodes easily fabricated by calendering process, which cancels out the non-uniform current distribution. With higher impedance near the tab, the resistance and the impedance balance out around the whole electrode along the vertical direction. A schematic of the conventional and tapered electrode is illustrated in Fig. 1. Electrodes are thicker near the tab and they get thinner as it gets further from the tab. It is noted that the pore structure (porosity) in the electrode is also bigger near the tab and the size of pores is decreasing when the location is far from the terminal. By tilting the rollers or controlling roller gaps, the calendering process enables this to change thickness and porosity across the width of the cell [30, 31].

3. ELECTRO-CHEMICAL MODEL

In this section, we develop a MP model to capture and represent the vertical resistance effect from the current collector for the purpose of computational simplicity. Three electro-chemical models are widely used: equivalent circuit model (ECM), pseudo two dimensional (P2D), and single particle (SP) model. ECMs are widely used in battery management system due to its simplicity of resistance and capacitance [32–36]. Electro-chemical impedance spectroscopy (EIS) is commonly used to find the parameters of Li batteries experimentally such as

Copyright © 2023 by ASME

TABLE 1: GOVERNING EQUATION OF ELECTRO-CHEMICAL MODEL

Li+ Conservation

$$\frac{\partial c_s}{\partial t} = \frac{D_s}{r^2} \frac{\partial}{\partial r}\left(r^2 \frac{\partial c_s}{\partial r}\right) \ for \ r \ \in (0, R_s) \tag{1}$$

$$\text{B.C.s: } \frac{\partial c_s}{\partial r}\bigg|_{r=0} = 0, \ D_s \frac{\partial c_s}{\partial r}\bigg|_{r=R_s} = -\frac{j}{a_s F}$$

$$\frac{C_{s,e}(s)}{I(s)} = \frac{C_{s,e}(s)}{J(s)} \frac{J(s)}{I(s)} = \frac{1}{a_s F A \delta} \frac{R_s}{D_s} \frac{tanh(\beta)}{tanh(\beta) - \beta} \tag{2}$$

$$\text{where } \beta = R_s \sqrt{s/D_s}, \ a_s = 3\epsilon_s/R_s, \ a_s = 3(1-\epsilon)/R_s$$

Butler-Volmer Equation

$$j = a_s i_0 \left\{ exp\left[\frac{\alpha_a F}{RT}\eta\right] - exp\left[-\frac{\alpha_c F}{RT}\eta\right] \right\} \tag{3}$$

Linearlization of the Butler-Volmer Equation

$$\eta = \frac{R_{ct}}{a_s} j, \quad \text{where } R_{ct} = \frac{RT}{i_0 F(\alpha_a + \alpha_c)} \tag{4}$$

Overpotential

$$\eta = \phi_s - U \tag{5}$$

Cell Voltage

$$V(t) = \phi_s(L, t) - \phi_s(0, t) - \frac{R_f}{A} I(t) \tag{6}$$

Transcendental Transfer Function

$$\frac{V(s)}{I(s)} = Z(s) = K + \frac{\partial U_+}{\partial c_{s+}} \frac{1}{A\delta_+} \frac{R_{s+}}{a_{s+} F D_{s+}\delta_+} \frac{tanh(\beta_+)}{tanh(\beta_+) - \beta_+}$$

$$+ \frac{\partial U_-}{\partial c_{s-}} \frac{1}{A\delta_-} \frac{R_{s-}}{a_{s-} F D_{s-}\delta_-} \frac{tanh(\beta_-)}{tanh(\beta_-) - \beta_-} \tag{7}$$

Padé Approximation Model

$$Z(s) = K + \frac{\alpha_1 s^2 + 60\alpha_1\alpha_2 s + 495\alpha_1\alpha_2^2}{s^3 + 189\alpha_2 s^2 + 3465\alpha_2^2 s} + \frac{\beta_1 s^2 + 60\beta_1\beta_2 s + 495\beta_1\beta_2^2}{s^3 + 189\beta_2 s^2 + 3465\beta_2^2 s} \tag{8}$$

where

$$K = -\frac{R_{ct+}}{a_{s+}} \frac{1}{A\delta_+} - \frac{R_{ct-}}{a_{s-}} \frac{1}{A\delta_-} - \frac{R_f}{A}$$

$$\alpha_1 = 21 \frac{\partial U_+}{\partial c_{s,e+}} \frac{1}{AF a_{s+}\delta_+ R_{s+}} \qquad \alpha_2 = \frac{D_{s+}}{[R_{s+}]^2}$$

$$\beta_1 = 21 \frac{\partial U_-}{\partial c_{s,e-}} \frac{1}{AF a_{s-}\delta_- R_{s-}} \qquad \beta_2 = \frac{D_{s-}}{[R_{s-}]^2}$$

Copyright © 2023 by ASME

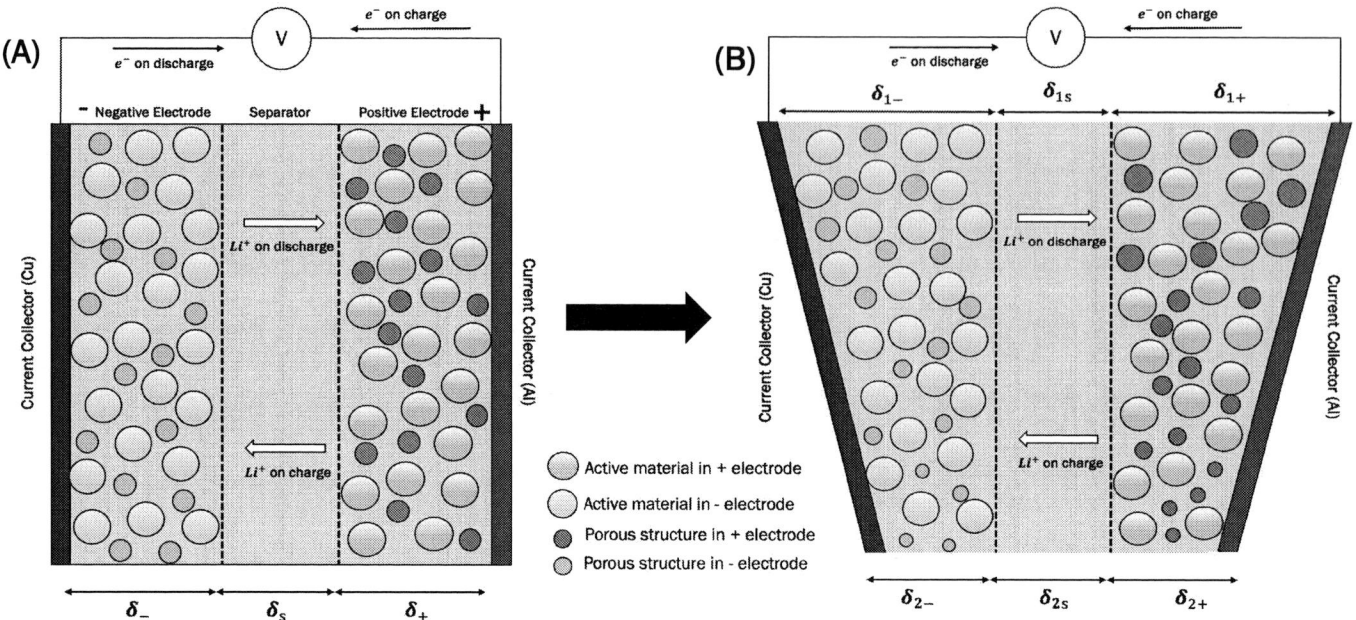

FIGURE 1: SCHEMATIC OF (A) CONVENTIONAL (UNIFORM) AND (B) TAPERED SHAPED LI-ION POUCH CELL.

resistance and capacitance [33]. However, it is hard to quantify and apply characteristics of electro-chemical parameters directly into the model. P2D model is considered one of the most accurate model [37–44]. COMSOL is widely used to solve P2D model with the finite element method (FEM). It requires high computational power and time, so it's not suitable for optimizing how the slope of the tapered electrode is designed. Thus, we develop MP model based on SP model and convert it into ECM to optimize the slope of tapered electrodes and employ optimized value in P2D model on COMOSL.

3.1 Multiple Particle Model

ECM based MP model is developed based on SP model. The benefit of this model is that it has calendering related parameters, thickness (δ) and porosity (ϵ). Figure 1(A) shows a Li-ion cell composed of the positive electrode, separator, and negative electrode. During charging, Li-ion is moving from the positive electrode to the negative electrode through a separator as electrons are transferred via tab (terminal) and the process is reversed when discharging. The beauty of the SP model is the simplicity since only one partial differential equation needs to be solved neglecting electrolyte diffusion, which is more important in fast charging or high power applications such as electric aircraft. Electrolyte model can be introduced into SP model if it's needed [45, 46]. Table 1 shows governing equations of SP model based on electro-chemical phenomena in Li-ion battery cells. Conservation of Li determined by Fick's law is utilized with two boundary conditions and taking the Laplace transform of Eq.(1) yields Eq.(2), transcendental Li-ion concentration transfer function [47]. Butler-Volmer equation is represented as Eq.(3) and it is linearized to simplify the relationship between current density and overpotential. By substituting Eqs.(2), (4), and (5) into Eq.(6), transcendental transfer function of Li-ion battery cell impedance

is obtained. Due to the complexity of the transcendental transfer function, the first principle (5th order) equivalent circuit equation is obtained based on Pade' approximation [48, 49].

Equation (8) is the first principle equivalent circuit model including electro-physical parameters such as thickness and porosity. Simplicity and electro-chemical parameter updates such as diffusivity, which can capture aging effect, lead to aging model [50–55].

MP model is a parallel equivalent circuit model of combining first principle equivalent circuit model as each node with additional resistance, which can explain the spatial effect of electrode since it comes from SP model. A schematic of equivalent circuit model assuming multiple particle model along the vertical axis is illustrated in Fig 2 and each parameter is represented in Table 2. Impedance of each (i) node is represented as

$$Z_i(s) = K_i + \frac{\alpha_{1,i}s^2 + 60\alpha_{1,i}\alpha_{2,i}s + 495\alpha_{1,i}\alpha_{2,i}{}^2}{s^3 + 189\alpha_{2,i}s^2 + 3465\alpha_{2,i}{}^2 s}$$
$$+ \frac{\beta_{1,i}s^2 + 60\beta_{1,i}\beta_{2,i}s + 495\beta_{1,i}\beta_{2,i}{}^2}{s^3 + 189\beta_{2,i}s^2 + 3465\beta_{2,i}{}^2 s} \tag{9}$$

where $K_i = -\frac{R_{ct+,i}}{a_{s+,i}}\frac{1}{A\delta_{+,i}} - \frac{R_{ct-,i}}{a_{s-,i}}\frac{1}{A\delta_{-,i}} - \frac{R_{f,i}}{A_i}$. Local resistance of each node is calculated with the resistivity of the current collector,

$$R_{cc,i} = \frac{\rho L_i}{A_{cc}}. \tag{10}$$

where, L_i is the length from the tab.

Since the final transfer function has "$a_s\delta$", which corresponds "$3\delta(1-\epsilon)/R_s$", we only consider thickness and porosity change and assume that the nominal porosity and the nominal thickness are decreasing linearly while calendering. When porosity (ϵ) decreases, the solid interfacial area (a_s) increases. Thus, there is a tradeoff between porosity and thickness. Depending

Copyright © 2023 by ASME

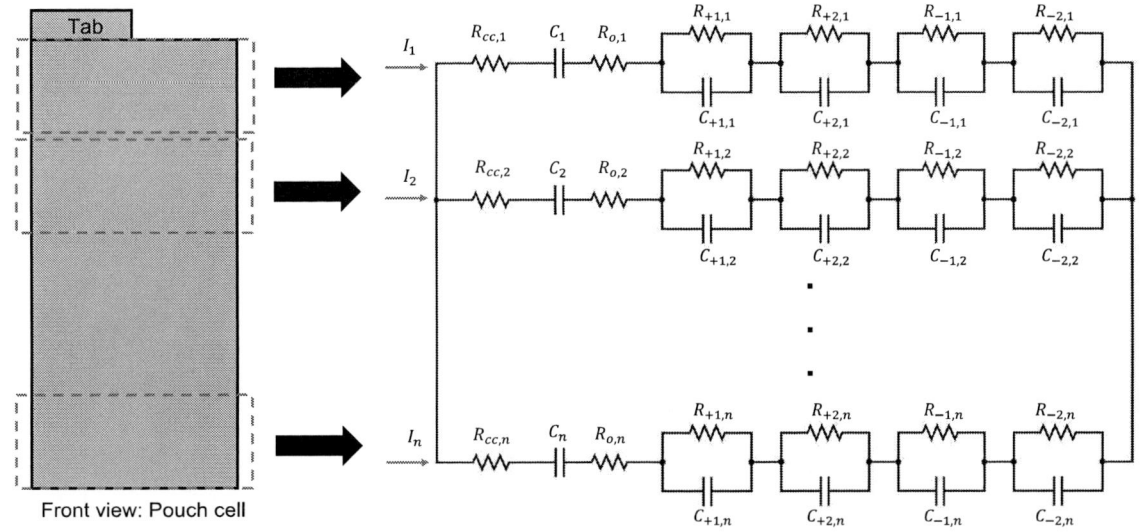

FIGURE 2: EQUIVALENT CIRCUIT MODEL OF MULTIPLE PARTICLE MODEL.

TABLE 2: PARAMETERS FOR LI-ION CELL MODEL

Resistor	Resistance	Capacitor	Capacitance
$R_{o,i}$	K_i	C_i	$\frac{7}{\alpha_{1,i}+\beta_{1,i}}$
$R_{+1,i}$	$\frac{0.0051\,\alpha_{1,i}}{\alpha_{2,i}}$	$C_{+1,i}$	$\frac{9.6246}{\alpha_{1,i}}$
$R_{+2,i}$	$\frac{0.0045\,\alpha_{1,i}}{\alpha_{2,i}}$	$C_{+2,i}$	$\frac{1.3277}{\alpha_{1,i}}$
$R_{-1,i}$	$\frac{0.0051\,\beta_{1,i}}{\beta_{2,i}}$	$C_{-1,i}$	$\frac{9.6246}{\beta_{1,i}}$
$R_{-2,i}$	$\frac{0.0045\,\beta_{1,i}}{\beta_{2,i}}$	$C_{-2,i}$	$\frac{1.3277}{\beta_{1,i}}$

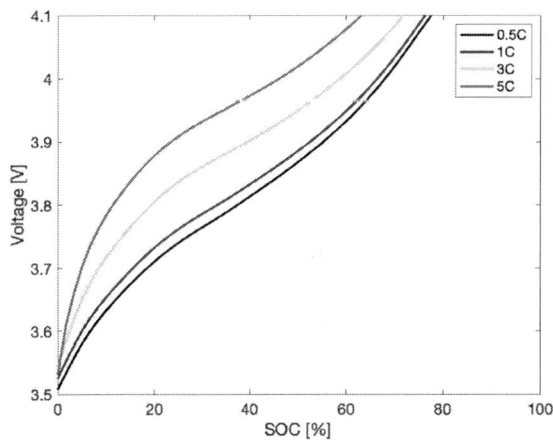

FIGURE 3: CELL PERFORMANCE AT VARIOUS CHARGE C RATES IN COMSOL

on the strong impact of each parameter, nonuniform resistance is canceled out.

1.5Ah Li-ion pouch cell with single-side coated electrodes ($50cmx20cm$) is used for simulation. Values of electro-chemical parameters used in MP model are in Table 3. Due to the change of local resistance of each node from the current collector in the conventional electrode, current inputs from the total charging input are calculated with Kirchhoff's circuit laws. Finally, the electro-chemical battery model is described as a simple parallel equivalent circuit model.

4. OPTIMIZATION

To optimize the slope of the tapered electrode, average and local Li-ion concentration is used as

$$
\begin{aligned}
\underset{c_{s,i}}{\text{minimize}} \quad & \sum_{i=1}^{n}(c_{s,avg} - c_{s,i})^2 \\
\text{subject to} \quad & \frac{\Delta\epsilon}{\epsilon} = k\frac{\Delta\delta}{\delta} \\
& -0.15 \leq a_{slope} \leq 0.15 \\
& 0.2 \leq \epsilon \leq 0.6
\end{aligned} \tag{11}
$$

where $c_{s,avg} = \frac{\sum_{i=1}^{n} c_{s,i}}{n}$ and Δ is the difference between the value of each parameter at the top and at the bottom.

Each constraint is chosen based on assumptions that calendering process can only change 22% of thickness and porosity and it has a linear relationship between thickness and porosity. The slope of the tapered positive electrode is sweeping from -15% to 15% in MATLAB with 60 samples using Brute-force algorithm, known as generate and test. When the slope is changing, the impedance, summation of resistance and capacitance, such as $R_{+1,i}, R_{+2,i}, C_{+1,i}, C_{+2,i}, C_i, R_{o,i}$ varies in each local node and the average current input divides into the local current input. With calculated local current, local Li-ion concentration c_i is obtained to find optimized slope. Using MP model to find the optimal slope of tapered electrodes takes only 2 minutes with Brute-force algorithm with 60 iterations. On the other hand, P2D model with COMSOL needs about 5 hours since it takes 5 minutes/iteration. This paper, we take the calculated optimal slope from MATLAB

Copyright © 2023 by ASME

TABLE 3: PARAMETERS FOR LI-ION CELL MODEL

Parameter	Negative	Separator	Positive
Thickness, δ (cm)	60×10^{-4}	30×10^{-4}	60×10^{-4}
Particle radius, R_s (cm)	2×10^{-4}	–	5×10^{-4}
Polymer phase volume fraction, ϵ_s	0.54	0.6	0.6
Porosity, ϵ	0.46	0.4	0.4
Electrode plate area, A (cm^2)		1000	
Current collector contact resistance (Ωcm^2)		43	
Current collector thickness (cm)	10×10^{-4}		10×10^{-4}
Maximum solid phase concentration $c_{s,max}$ $(molcm^{-3})$	31.5×10^{-3}		22.8×10^{-3}
Stoichiometry at 0% SOC, $x_{0\%}$	0		0.995
Stoichiometry at 100% SOC, $x_{100\%}$	0.98		0.175
Exchange current density, i_0 (Acm^{-2})	2.95×10^{-5}		2.8×10^{-4}
Charge transfer coefficients, α_a, α_c	0.5		0.5
Solid phase Li diffusion coefficients, D_s $(cm^2 s^{-1})$	1.45×10^{-13}		5×10^{-13}

using MP model and apply it into COMSOL to compare the baseline and optimized results.

5. RESULTS

Figure 3 shows cell performance at various charging inputs calculated by COMSOL. With higher current input, the voltage gets the cutoff voltage earlier, available capacity decreases down to 65% in 5C from 78% in 0.5C charging. Figure 4 shows the lumped voltage response, positive and negative concentration with COMSOL and MP model in 1C charging input. Voltage response from 10 min to 50 min fits well. The discrepancy both under 10 min and over 50 min is the linearization of OCV curve at 50% not capturing the nonlinear OCV curve. Li-ion concentration response in both positive and negative electrodes at 1C charging matches very well, thus, MP model is validated with COMSOL based on P2D model.

Figure 5 shows nominal current at (A) various charge inputs and (B) local current at 5C charge in COMSOL. $n = 1$ and $n = 8$ mean the location close to the tab and the furthest location from the tab, respectively. Higher current charging aggravates local current distribution and lower capacity causing local aging. Figure 5(B) shows the local current is higher near the tab and it decreases as the location gets further with the lowest current at the furthest location ($n = 8$).

Li-ion maximum nominal difference at various rates with conventional and linearly tapered electrodes are compared in Fig. 6. 4.1 voltage is chosen as the cutoff voltage and SOC (capacity) with each charging input is obtained by dividing the total capacity from OCV curve. Linearly tapered electrode gives better utilization at higher C rate, but it does not delay the time of reaching cutoff lumped voltage. Li concentration difference reduces with a tapered electrode from 12% to 4% in 5C charging.

Figure 7 shows the contour of positive concentration in pouch cell at t=6 min during 5C discharging in COMSOL. With uniform electrodes, Li distribution is from 45.3% (far from the terminal)

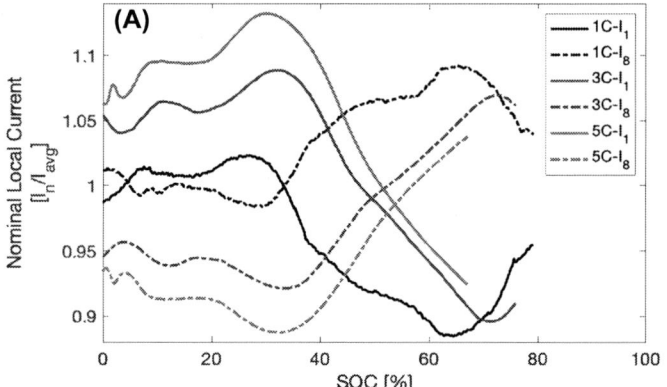

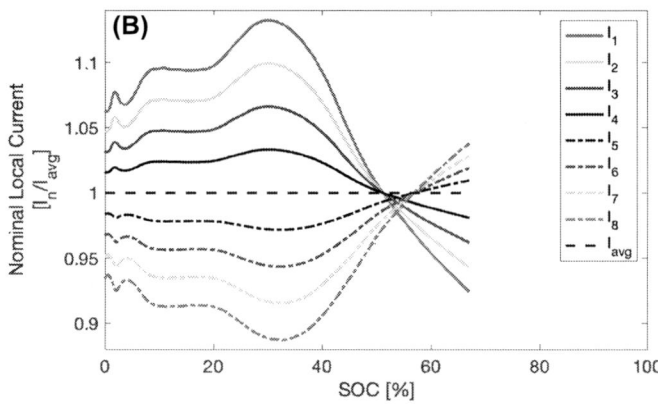

FIGURE 4: (A) AVERAGE VOLTAGE RESPONSE (B) LI-ION POSITIVE AND (C) NEGATIVE CONCENTRATION AT 1C CHARGE.

Copyright © 2023 by ASME

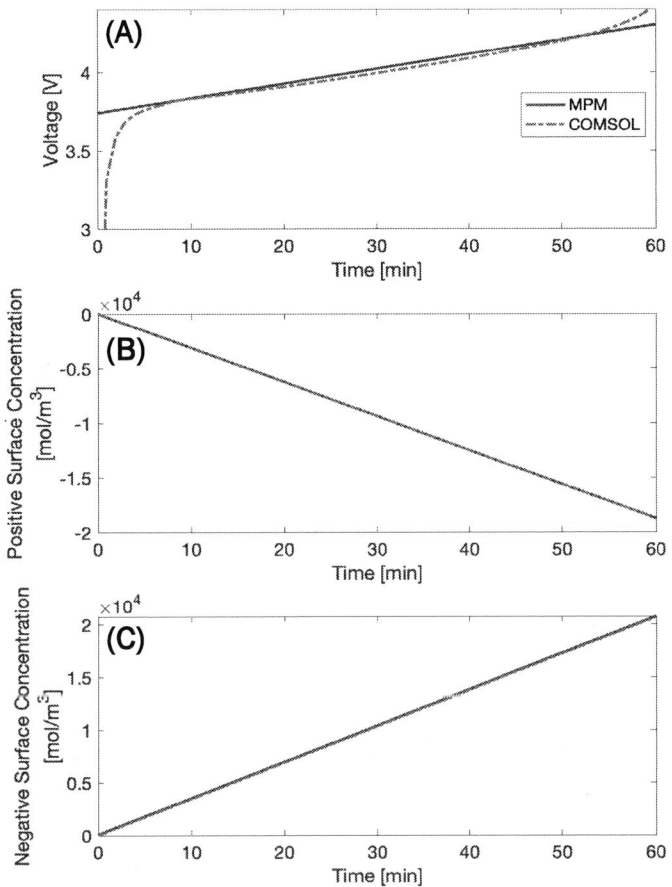

FIGURE 5: (A) NOMINAL LOCAL CURRENT AT VARIOUS CHARGE C RATES (B) WITH 8 LOCATIONS AT 5C CHARGE

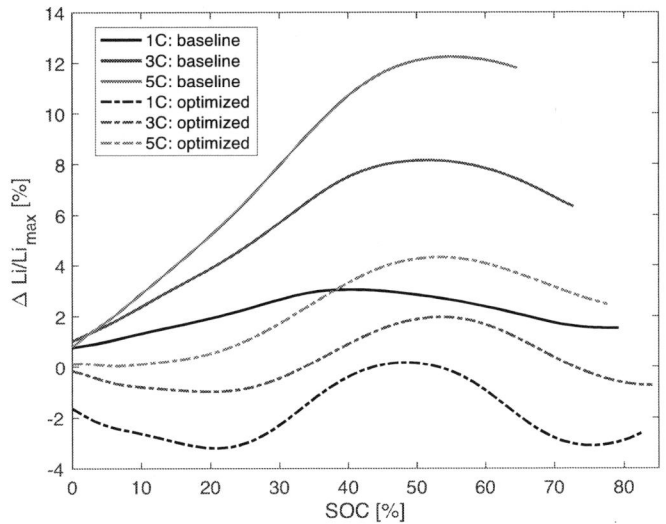

FIGURE 6: LI NOMINAL DIFFERENCE AT VARIOUS C RATES WITH CONVENTIONAL AND 10% LINEARLY TAPERED ELECTRODES.

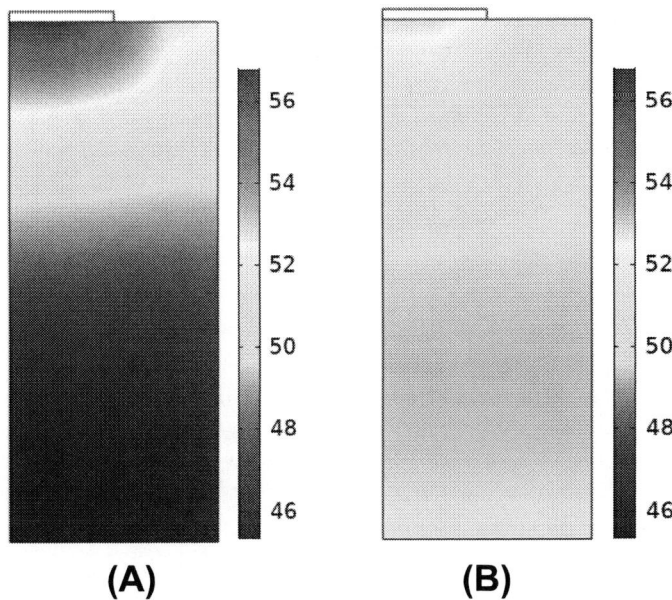

FIGURE 7: NORMALIZED LI-ION CONCENTRATION DISTRIBUTION (A) CONVENTIONAL (B) LINEARLY TAPERED ELECTRODE WITH 5C AT 6MIN FROM COMSOL.

TABLE 4: NOMINAL CAPACITY WITH VARIOUS CHARGING INPUT

	1C	3C	5C
Uniform	77.6 %	72.8 %	64.6 %
Tapered	82.5 %	84.1 %	77.6 %

to 56.8% (at the terminal) while tapered electrodes have 51% Li average concentration. Table 4 compares maximum capacity between uniform and tapered electrodes. Maximum nominal capacity is calculated by integrating local Li-ion concentrations when the lumped voltage reaches the cutoff voltage. A pouch cell with an optimally tapered positive electrode can increase capacity by 13% with high C rates.

6. CONCLUSION

Tapered electrodes in Li-ion batteries to improve energy density and nonuniform utilization are first investigated in this paper. A multiple particle (MP) model is developed and validated with P2D model in COMSOL using FEM to optimize the slope of the taper based on the tab configuration. Higher impedance near the tab balances the higher impedance in the current collector away from the tab. The analysis results show that a small taper (10%) can improve uniformity by eight times in 3C charging and increase capacity by 13% with 5C charging. Optimally designed tapered electrodes in Li-ion batteries can improve battery performance, cycle life, and safety by alleviating nonuniform temperature and current distribution.

ACKNOWLEDGMENTS

This research is funded by Siemens Technology.

Copyright © 2023 by ASME

REFERENCES

[1] Zhang, Guangsheng, Shaffer, Christian E, Wang, Chao-Yang and Rahn, Christopher D. "In-situ measurement of current distribution in a Li-Ion cell." *Journal of The Electrochemical Society* Vol. 160 No. 4 (2013): p. A610.

[2] Zhang, Guangsheng, Shaffer, Christian E, Wang, Chao-Yang and Rahn, Christopher D. "Effects of non-uniform current distribution on energy density of Li-ion cells." *Journal of The Electrochemical Society* Vol. 160 No. 11 (2013): p. A2299.

[3] Kwon, Ki Hyun, Shin, Chee Burm, Kang, Tae Hyuk and Kim, Chi-Su. "A two-dimensional modeling of a lithium-polymer battery." *Journal of Power Sources* Vol. 163 No. 1 (2006): pp. 151–157.

[4] Song, Jung-Hoon, You, Seung-Jae and Jeon, Dong Hyup. "Numerical modeling and experimental validation of pouch-type lithium-ion battery." *Journal of Applied Electrochemistry* Vol. 44 No. 9 (2014): pp. 1013–1023.

[5] Kim, Ui Seong, Shin, Chee Burm and Kim, Chi-Su. "Effect of electrode configuration on the thermal behavior of a lithium-polymer battery." *Journal of Power Sources* Vol. 180 No. 2 (2008): pp. 909–916.

[6] Kim, Ui Seong, Shin, Chee Burm and Kim, Chi-Su. "Modeling for the scale-up of a lithium-ion polymer battery." *Journal of Power Sources* Vol. 189 No. 1 (2009): pp. 841–846.

[7] Yang, Xiao-Guang, Liu, Teng, Ge, Shanhai, Rountree, Eric and Wang, Chao-Yang. "Challenges and key requirements of batteries for electric vertical takeoff and landing aircraft." *Joule* Vol. 5 No. 7 (2021): pp. 1644–1659.

[8] Tomaszewska, Anna, Chu, Zhengyu, Feng, Xuning, O'Kane, Simon, Liu, Xinhua, Chen, Jingyi, Ji, Chenzhen, Endler, Elizabeth, Li, Ruihe, Liu, Lishuo et al. "Lithium-ion battery fast charging: A review." *ETransportation* Vol. 1 (2019): p. 100011.

[9] Baker, Daniel R and Verbrugge, Mark W. "Temperature and current distribution in thin-film batteries." *Journal of the Electrochemical Society* Vol. 146 No. 7 (1999): p. 2413.

[10] Zhang, Xiao, Hui, Zeyu, King, Steven T, Wu, Jingyi, Ju, Zhengyu, Takeuchi, Kenneth J, Marschilok, Amy C, West, Alan C, Takeuchi, Esther S, Wang, Lei et al. "Gradient Architecture Design in Scalable Porous Battery Electrodes." *Nano Letters* Vol. 22 No. 6 (2022): pp. 2521–2528.

[11] Liu, Yangtao, Zhang, Ruihan, Wang, Jun and Wang, Yan. "Current and future lithium-ion battery manufacturing." *Iscience* Vol. 24 No. 4 (2021): p. 102332.

[12] Meyer, Chris, Bockholt, Henrike, Haselrieder, Wolfgang and Kwade, Arno. "Characterization of the calendering process for compaction of electrodes for lithium-ion batteries." *Journal of Materials Processing Technology* Vol. 249 (2017): pp. 172–178.

[13] Bockholt, Henrike, Indrikova, Maira, Netz, Andreas, Golks, Frederik and Kwade, Arno. "The interaction of consecutive process steps in the manufacturing of lithium-ion battery electrodes with regard to structural and electrochemical properties." *Journal of Power Sources* Vol. 325 (2016): pp. 140–151.

[14] Billot, Nicolas, Günther, Till, Schreiner, David, Stahl, Ralf, Kranner, Jakob, Beyer, Moritz and Reinhart, Gunther. "Investigation of the adhesion strength along the electrode manufacturing process for improved lithium-ion anodes." *Energy Technology* Vol. 8 No. 2 (2020): p. 1801136.

[15] Mei, Wenxin, Chen, Haodong, Sun, Jinhua and Wang, Qingsong. "The effect of electrode design parameters on battery performance and optimization of electrode thickness based on the electrochemical–thermal coupling model." *Sustainable energy & fuels* Vol. 3 No. 1 (2019): pp. 148–165.

[16] Park, Keemin, Myeong, Seungcheol, Shin, Donghyeok, Cho, Chae-Woong, Kim, Soo Chan and Song, Taeseup. "Improved swelling behavior of Li ion batteries by microstructural engineering of anode." *Journal of industrial and engineering chemistry* Vol. 71 (2019): pp. 270–276.

[17] Lenze, Georg, Bockholt, Henrike, Schilcher, Christiane, Froböse, Linus, Jansen, Dietmar, Krewer, Ulrike and Kwade, Arno. "Impacts of variations in manufacturing parameters on performance of lithium-ion-batteries." *Journal of The Electrochemical Society* Vol. 165 No. 2 (2018): p. A314.

[18] Dreger, Henning, Haselrieder, Wolfgang and Kwade, Arno. "Influence of dispersing by extrusion and calendering on the performance of lithium-ion battery electrodes." *Journal of Energy Storage* Vol. 21 (2019): pp. 231–240.

[19] Zheng, Honghe, Tan, Li, Liu, Gao, Song, Xiangyun and Battaglia, Vincent S. "Calendering effects on the physical and electrochemical properties of Li [Ni1/3Mn1/3Co1/3] O2 cathode." *Journal of Power Sources* Vol. 208 (2012): pp. 52–57.

[20] Meyer, Chris, Weyhe, Matthias, Haselrieder, Wolfgang and Kwade, Arno. "Heated calendering of cathodes for lithium-ion batteries with varied carbon black and binder contents." *Energy Technology* Vol. 8 No. 2 (2020): p. 1900175.

[21] Schreiner, David, Oguntke, Maximilian, Günther, Till and Reinhart, Gunther. "Modelling of the Calendering Process of NMC-622 Cathodes in Battery Production Analyzing Machine/Material–Process–Structure Correlations." *Energy Technology* Vol. 7 No. 11 (2019): p. 1900840.

[22] Günther, Till, Schreiner, David, Metkar, Ajinkya, Meyer, Chris, Kwade, Arno and Reinhart, Gunther. "Classification of calendering-induced electrode defects and their influence on subsequent processes of lithium-ion battery production." *Energy Technology* Vol. 8 No. 2 (2020): p. 1900026.

[23] Xu, Meng, Reichman, Benjamin and Wang, Xia. "Modeling the effect of electrode thickness on the performance of lithium-ion batteries with experimental validation." *Energy* Vol. 186 (2019): p. 115864.

[24] Taleghani, Sara Taslimi, Marcos, Bernard, Zaghib, Karim and Lantagne, Gaétan. "A study on the effect of porosity and particles size distribution on Li-ion battery performance." *Journal of The Electrochemical Society* Vol. 164 No. 11 (2017): p. E3179.

[25] Heubner, C, Nickol, A, Seeba, J, Reuber, S, Junker, N, Wolter, M, Schneider, M and Michaelis, A. "Understanding thickness and porosity effects on the electrochemical performance of LiNi0. 6Co0. 2Mn0. 2O2-based cathodes for

Copyright © 2023 by ASME

high energy Li-ion batteries." *Journal of Power Sources* Vol. 419 (2019): pp. 119–126.

[26] Priyono, S, Hardiyani, S, Syarif, N, Subhan, A and Suhandi, A. "Electrochemical performanceof LiMn2O4 with varying thickness of cathode sheet." *Journal of Physics: Conference Series*, Vol. 1191. 1: p. 012022. 2019. IOP Publishing.

[27] Ramadesigan, Venkatasailanathan, Methekar, Ravi N, Latinwo, Folarin, Braatz, Richard D and Subramanian, Venkat R. "Optimal porosity distribution for minimized ohmic drop across a porous electrode." *Journal of The Electrochemical Society* Vol. 157 No. 12 (2010): p. A1328.

[28] Dai, Yiling and Srinivasan, Venkat. "On graded electrode porosity as a design tool for improving the energy density of batteries." *Journal of The Electrochemical Society* Vol. 163 No. 3 (2015): p. A406.

[29] Lenze, Georg, Röder, Fridolin, Bockholt, Henrike, Haselrieder, Wolfgang, Kwade, Arno and Krewer, Ulrike. "Simulation-supported analysis of calendering impacts on the performance of lithium-ion-batteries." *Journal of The Electrochemical Society* Vol. 164 No. 6 (2017): p. A1223.

[30] Diener, Alexander, Ivanov, Stoyan, Haselrieder, Wolfgang and Kwade, Arno. "Evaluation of Deformation Behavior and Fast Elastic Recovery of Lithium-Ion Battery Cathodes via Direct Roll-Gap Detection During Calendering." *Energy Technology* Vol. 10 No. 4 (2022): p. 2101033.

[31] Mayr, Andreas, Schreiner, David, Stumper, Benedikt and Daub, Rüdiger. "In-line Sensor-based Process Control of the Calendering Process for Lithium-Ion Batteries." *Procedia CIRP* Vol. 107 (2022): pp. 295–301.

[32] Ren, Hongbin, Zhao, Yuzhuang, Chen, Sizhong and Yang, Lin. "A comparative study of lumped equivalent circuit models of a lithium battery for state of charge prediction." *International Journal of Energy Research* Vol. 43 No. 13 (2019): pp. 7306–7315.

[33] Oldenburger, Marc, Beduerftig, Benjamin, Gruhle, Andreas, Grimsmann, Florian, Richter, Ernst, Findeisen, Rolf and Hintennach, Andreas. "Investigation of the low frequency Warburg impedance of Li-ion cells by frequency domain measurements." *Journal of Energy Storage* Vol. 21 (2019): pp. 272–280.

[34] Hu, Xiaosong, Li, Shengbo and Peng, Huei. "A comparative study of equivalent circuit models for Li-ion batteries." *Journal of Power Sources* Vol. 198 (2012): pp. 359–367.

[35] Chen, Min and Rincon-Mora, Gabriel A. "Accurate electrical battery model capable of predicting runtime and IV performance." *IEEE transactions on energy conversion* Vol. 21 No. 2 (2006): pp. 504–511.

[36] Lin, Xinfan, Perez, Hector E, Mohan, Shankar, Siegel, Jason B, Stefanopoulou, Anna G, Ding, Yi and Castanier, Matthew P. "A lumped-parameter electro-thermal model for cylindrical batteries." *Journal of Power Sources* Vol. 257 (2014): pp. 1–11.

[37] Ai, Weilong, Kraft, Ludwig, Sturm, Johannes, Jossen, Andreas and Wu, Billy. "Electrochemical thermal-mechanical modelling of stress inhomogeneity in lithium-ion pouch cells." *Journal of The Electrochemical Society* Vol. 167 No. 1 (2019): p. 013512.

[38] Chen, Chang-Hui, Planella, Ferran Brosa, O'regan, Kieran, Gastol, Dominika, Widanage, W Dhammika and Kendrick, Emma. "Development of experimental techniques for parameterization of multi-scale lithium-ion battery models." *Journal of The Electrochemical Society* Vol. 167 No. 8 (2020): p. 080534.

[39] Mohtat, Peyman, Lee, Suhak, Sulzer, Valentin, Siegel, Jason B and Stefanopoulou, Anna G. "Differential expansion and voltage model for li-ion batteries at practical charging rates." *Journal of The Electrochemical Society* Vol. 167 No. 11 (2020): p. 110561.

[40] Kim, Gi-Heon, Smith, Kandler, Lee, Kyu-Jin, Santhanagopalan, Shriram and Pesaran, Ahmad. "Multi-domain modeling of lithium-ion batteries encompassing multi-physics in varied length scales." *Journal of the electrochemical society* Vol. 158 No. 8 (2011): p. A955.

[41] Prada, Eric, Di Domenico, D, Creff, Y, Bernard, J, Sauvant-Moynot, Valérie and Huet, François. "A simplified electrochemical and thermal aging model of LiFePO4-graphite Li-ion batteries: power and capacity fade simulations." *Journal of The Electrochemical Society* Vol. 160 No. 4 (2013): p. A616.

[42] Han, Sangwoo, Tang, Yifan and Rahimian, Saeed Khaleghi. "A numerically efficient method of solving the full-order pseudo-2-dimensional (P2D) Li-ion cell model." *Journal of Power Sources* Vol. 490 (2021): p. 229571.

[43] Smith, Kandler A, Rahn, Christopher D and Wang, Chao-Yang. "Control oriented 1D electrochemical model of lithium ion battery." *Energy Conversion and management* Vol. 48 No. 9 (2007): pp. 2565–2578.

[44] Prada, Eric, Di Domenico, D, Creff, Yann, Bernard, J, Sauvant-Moynot, Valérie and Huet, François. "Simplified electrochemical and thermal model of LiFePO4-graphite Li-ion batteries for fast charge applications." *Journal of The Electrochemical Society* Vol. 159 No. 9 (2012): p. A1508.

[45] Ramadass, P, Haran, Bala, Gomadam, Parthasarathy M, White, Ralph and Popov, Branko N. "Development of first principles capacity fade model for Li-ion cells." *Journal of the Electrochemical Society* Vol. 151 No. 2 (2004): p. A196.

[46] Marquis, Scott G, Sulzer, Valentin, Timms, Robert, Please, Colin P and Chapman, S Jon. "An asymptotic derivation of a single particle model with electrolyte." *Journal of The Electrochemical Society* Vol. 166 No. 15 (2019): p. A3693.

[47] Jacobsen, Torben and West, Keld. "Diffusion impedance in planar, cylindrical and spherical symmetry." *Electrochimica acta* Vol. 40 No. 2 (1995): pp. 255–262.

[48] Prasad, Githin K and Rahn, Christopher D. "Development of a first principles equivalent circuit model for a lithium ion battery." *Dynamic Systems and Control Conference*, Vol. 45301: pp. 369–375. 2012. American Society of Mechanical Engineers.

[49] Prasad, Githin K and Rahn, Christopher D. "Model based identification of aging parameters in lithium ion batteries." *Journal of power sources* Vol. 232 (2013): pp. 79–85.

[50] Tanim, Tanvir R, Rahn, Christopher D and Wang, Chao-Yang. "State of charge estimation of a lithium ion cell based

Copyright © 2023 by ASME

on a temperature dependent and electrolyte enhanced single particle model." *Energy* Vol. 80 (2015): pp. 731–739.

[51] Yang, Xiao-Guang, Leng, Yongjun, Zhang, Guangsheng, Ge, Shanhai and Wang, Chao-Yang. "Modeling of lithium plating induced aging of lithium-ion batteries: Transition from linear to nonlinear aging." *Journal of Power Sources* Vol. 360 (2017): pp. 28–40.

[52] Tanim, Tanvir R and Rahn, Christopher D. "Aging formula for lithium ion batteries with solid electrolyte interphase layer growth." *Journal of Power Sources* Vol. 294 (2015): pp. 239–247.

[53] Xu, Shanshan, Chen, Kuan-Hung, Dasgupta, Neil P, Siegel, Jason B and Stefanopoulou, Anna G. "Evolution of dead lithium growth in lithium metal batteries: experimentally validated model of the apparent capacity loss." *Journal of The Electrochemical Society* Vol. 166 No. 14 (2019): p. A3456.

[54] Tanim, Tanvir R, Rahn, Christopher D and Wang, Chao-Yang. "A temperature dependent, single particle, lithium ion cell model including electrolyte diffusion." *Journal of Dynamic Systems, Measurement, and Control* Vol. 137 No. 1 (2015).

[55] Dey, Satadru and Ayalew, Beshah. "Real-time estimation of lithium-ion concentration in both electrodes of a lithium-ion battery cell utilizing electrochemical–thermal coupling." *Journal of Dynamic Systems, Measurement, and Control* Vol. 139 No. 3 (2017).

Proceedings of the ASME 2023
International Design Engineering Technical Conferences and
Computers and Information in Engineering Conference
IDETC-CIE2023
August 20-23, 2023, Boston, Massachusetts

DETC2023-114607

SYSTEM LEVEL TECHNO-ECONOMIC AND ENVIRONMENTAL DESIGN OPTIMIZATION FOR OCEAN WAVE ENERGY

Rebecca McCabe[1]*, Madison Dietrich[1], Alan Liu[2], Maha Haji[1]
[1]Sibley School of Mechanical and Aerospace Engineering
[2]Department of Economics
Cornell University
Ithaca, New York 14853
Email: {rgm222, mjd429, asl259, maha}@cornell.edu

ABSTRACT

Ocean waves have the potential to provide clean energy to large swaths of society. Despite their promise, wave energy converters (WECs) are prohibitively expensive, partly due to the lack of design convergence in the WEC industry. A system-level optimization facilitates cost reduction, systematic evaluation of various architectures, and ultimately design convergence. This paper outlines an optimization framework for wave energy at the industry level, starting with the selection of a suite of metrics that capture the value proposition of wave energy more fully than traditional metrics. This includes metrics that encompass environmental sustainability and economic value in addition to economic cost. Two processes for metric weighting are proposed: a heuristic process for emerging off-grid markets and a capacity expansion model for the grid market. As a demonstration of the weighting process, results from three existing studies are synthesized into weights for the levelized cost of energy, capacity factor, and standard deviation of capacity factor. Finally, the optimization design variables, parameters, objectives, and constraints are formulated, and methods to handle the many categorical or integer design variables and parameters are discussed. The process articulated here can also be applied to other emerging energy technologies, ultimately advancing decarbonization in the energy sector.

---

*Corresponding Author

1 INTRODUCTION

Mitigating the climate crisis requires full decarbonization of the energy sector within the next few decades, with the Biden administration aiming for a 100% clean grid by 2035. Mature technologies like wind and solar dominate most plans, but immature renewables like ocean wave energy converters (WECs) still hold promise. WECs can potentially decrease energy storage requirements and improve energy security, both of which are key bottlenecks to a 100% clean grid [1]. Other appealing aspects of wave energy include temporal characteristics like predictability, a complementary seasonal profile, and more consistent availability; spatial characteristics like proximity to demand and the growth of co-located offshore markets; and other characteristics like unique applications and resource diversification [1]. Indeed, wave energy could fulfill up to 34% of US energy demand [2]. However, WECs have high costs, a long design and testing cycle, and a high-risk investment environment. Consequently, WEC deployment has been limited, and a dominant design has yet to emerge, resulting in a major bottleneck to wide-spread investment in wave energy.

Design optimization can address this bottleneck and help determine the most high-performing, economically-viable, and environmentally-friendly WEC to ensure effort aligns with intent. Renewable energy is inherently a techno-economic-environmental system, as success requires not only generating energy but also achieving market competitiveness and reducing carbon emissions on a climate-relevant scale and timeline. Most WEC design optimization to date is either purely techni-

Copyright © 2023 by ASME

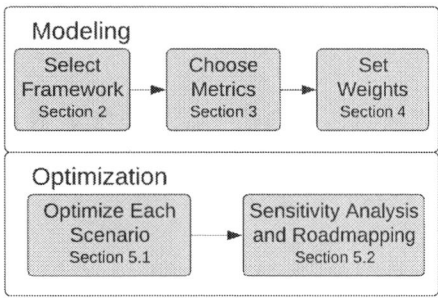

FIGURE 1. Flowchart depicting the framework process and the paper section organization.

cal or techno-economic with a focus on cost. However, value is equally important as cost in determining economic competitiveness, and environmental impact is the true bottom line for climate-motivated technologies. A failure to consider value and environmental effects during the design process can result in a misguided end product that will not achieve market success or climate impact. Therefore, this paper proposes a framework for analyzing and optimizing coupled techno-economic and environmental considerations in WEC design.

First, in Section 2 the paper covers the modeling used. Reasoning for the metrics selected and a process to develop weightings for metrics are discussed in Sections 3 and 4. Section 5 describes the details of the optimization scheme, including considerations for sensitivity analysis. Finally, Section 6 concludes and offers suggestions for future work. Figure 1 lays out this process.

2 MODEL FRAMEWORK

The proposed framework to be described in this section combines aspects of multidisciplinary design optimization, control co-design, techno-economic analysis, capacity expansion modeling, and life cycle analysis.

2.1 Overall Optimization

WEC power production, cost, value, and environmental impact are governed by coupled technical domains including hydrodynamics, controls, powertrain, structures, and economics. Multidisciplinary Design Optimization (MDO) and Control Co-Design (CCD) are emerging techniques that depart from the standard sequential design process by considering subsystem interactions early on. MDO is an optimization framework for interdisciplinary design problems that has been used successfully in the automotive, energy, and aerospace sectors [3]. CCD is the use of control principles to inform device design, and is appropriate because control has a large effect on WEC power production,

structural loads, and power variation in time [4]. Only a few studies [5, 6, 7] have applied these methods to wave energy due to novelty and computational costs. MDO and CCD are important aspects of the optimization strategy proposed here.

The tradeoff between techno-economic viability and environmental sustainability will be assessed using multi-objective optimization. Different stakeholders may place different value on the two, with developers and investors expected to lean economic, national labs and academia expected to balance equally, and government regulators and conservationists expected to lean environmental. A multi-objective approach allows both objectives to be optimized simultaneously without an a-priori prioritization between the two. Understanding this design tradeoff could important for groups like the US Department of Energy when determining policy incentives and R&D funding allocation.

2.2 Techno-Economic Integration

A technology is techno-economically viable if its value exceeds its cost. Estimating the cost of emerging technologies is routinely done with Techno-Economic Analysis. Much WEC R&D effort has previously focused on costs, so it is not described in detail here. Economic value is the neglected portion to be investigated here.

For grid-connected devices, economic value is readily assessed with energy sector optimization tools. These include capacity expansion models (CEMs) to determine the long-term resource mix which minimizes investment cost and economic dispatch (ED) solvers to determine the near-term power allocations which minimizes operating cost. Both are typically formulated as linear programs but can contain millions of decision variables and constraints, depending on the spatial and temporal scope and resolution. This size and resulting computational cost makes it unappealing to simply embed the CEM (or ED) into the design optimization, as shown in the top of Figure 2.

One alternative (not shown in the figure) would be to merge the linear CEM (or ED) and the nonlinear design optimization into a single optimization. This approach has some precedent: [8] linearizes the design problem for a geothermal plant and incorporates the linearization into a CEM, and [9] incorporates a linear aircraft operations problem into a nonlinear aircraft design problem and solves the whole system with a nonlinear solver, creating a surrogate model to manage computational cost. For the present application, linearizing would remove important features of the WEC design space, solving the CEM nonlinearly is infeasible due to its size, and in either case fully integrating the two problems would be difficult because CEMs are typically pre-built with their own solvers. Nonetheless, the surrogate modeling aspect of that approach is still an option, and is presented in the bottom of Figure 2. In the proposed framework of this paper, the CEM (or ED) is run ahead of time to generate the surrogate, removing its computational cost from the design optimization.

Copyright © 2023 by ASME

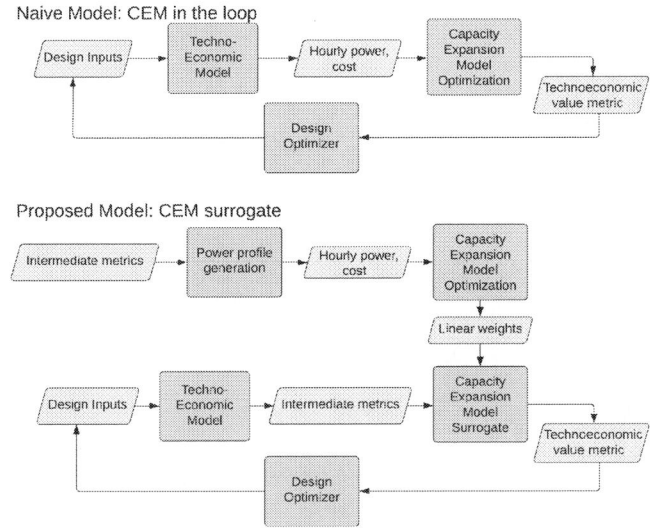

FIGURE 2. Two possible modeling structures for the inclusion of a Capacity Expansion Model (CEM) in design optimization.

Determining economic value for non-grid offshore applications, often known as Powering the Blue Economy (PBE) markets, is more difficult than for the grid because of the immaturity of these markets. One process that could yield systematic CEM-style predictions for PBE markets is System Dynamics. System Dynamics is a method of applying engineering dynamics and feedback systems theory to complex social and economic systems, and it is well-suited to emerging sectors since it does not rely on historic data like some economic models do [10]. However, no relevant literature could be found applying system dynamics to marine energy or the PBE sector, and creating a system dynamics model is outside the scope of this work. Therefore, PBE value is addressed heuristically in Section 4.1.

2.3 Environmental Model

Climate change mitigation, and the corresponding drive for carbon-free energy, remain the primary motivations for renewable energy development. Accordingly, it is necessary to consider environmental impacts in the design process to realize the vision for global decarbonization. Additionally, negative environmental impacts could interfere with the marine permitting process. Notable efforts to catalog WEC environmental effects include the Triton Initiative [11], the Ecological Risk Assessment tool [12], and DTOcean+ [13]. These quantify the pressures on the surrounding marine ecosystem due to WECs, and DTOcean+ additionally incorporates maintenance vessel emissions. However, these tools have not been integrated into design optimization and neglect other life cycle environmental aspects such as raw materials and manufacturing. Life Cycle Assessment (LCA) incorporates these aspects and has been performed on several WECs, but LCA requires detailed design information and is not ideal for actively informing early design. One way to make LCA feasible for early-stage projects is to use the pre-determined eco-costs weighting system introduced in [14]. Finally, perhaps the most important environmental impact of WECs is the emissions reduction associated with the avoided fossil fuel energy production. This can be found by analyzing the sector-wide emissions via the CEM discussed above. Overall, the present work combines marine ecosystem effects, eco-cost weightings, and system-wide avoided emissions to obtain a single comprehensive WEC sustainability objective for optimization.

3 CHOICE OF METRICS

A major gap in the WEC design process is that typical metrics fail to fully address techno-economic potential and environmental impacts across markets. The metric typically used to assess WEC techno-economics is the levelized cost of energy (LCOE), or a proxy thereof [15]. However, LCOE fails to capture some of wave energy's key value propositions like power consistency and geographic proximity to energy demand [1]. Additionally, LCOE mainly applies to utility-scale projects and provides an incomplete picture of emerging Powering the Blue Economy (PBE) markets like offshore aquaculture, desalination, and ocean observation, which have different constraints and priorities [16].

The authors of [17] call for the development and widespread use of meaningful early-stage metrics representing the values that drive investment decisions. Careful choice of metrics is important because prior studies indicate that optimal WEC geometries are highly sensitive to the optimization objective [18]. One techno-economic metric is a cost-conscious variant of the technology readiness level (TRL) called technology performance level (TPL) [19]. However, TPL is a discrete value intended more to classify the development stage than as an optimization metric. A second alternative metric, and the one chosen in this paper, is a value-conscious variant of LCOE called net value of energy (NVOE) [20]. This work proposes a methodology to capture traditional and non-traditional value sources like temporal and locational complementarity, power consistency and predictability, robustness to uncertainty, and global deployability using NVOE and other metrics to be introduced in this section. We also consider the requirements and relative importance of metrics in both utility and the PBE markets, either via heuristic weightings or, for grid-connected devices, using capacity expansion model (CEM) outputs. Using CEM sensitivities to derive WEC metric weighting factors appears to be a novel use case.

The ultimate goal for WECs is to produce clean, cheap, and reliable energy. Achieving this requires analysis of the economic, environmental, and technical systems that govern the WEC design and implementation process. A flowchart for the calculation

Copyright © 2023 by ASME

of the metrics explored in this paper is presented in Figure 3.

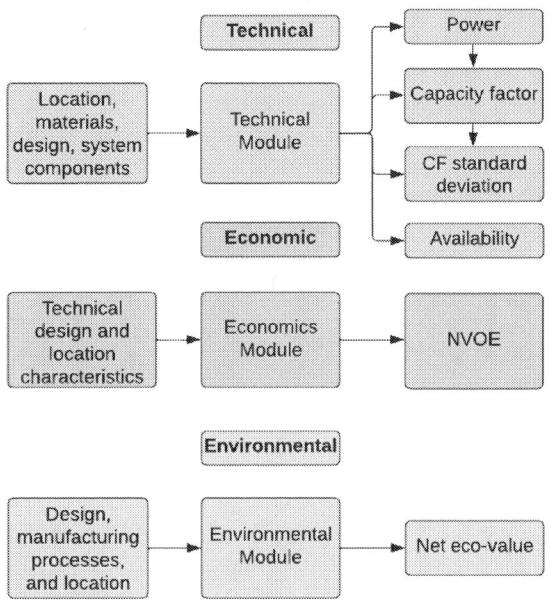

FIGURE 3. Flowchart of parameters and metrics

3.1 Technical Metrics

In addition to the environmental implications, it is necessary to select metrics that capture the technical performance of the WEC. These technical metrics reflect key attributes such as power and energy production that connect the economic and environmental aspects with the technical design. The metrics chosen are availability, capacity factor, and the standard deviation of the capacity factor, as shown in Figure 4.

WEC availability (A) is the proportion of active energy production time to the total lifetime of the WEC (T). To determine this proportion, we rely on three probabilities that encapsulate the causes of inactivity: insufficient wave activity, maintenance, and catastrophic failure, denoted p_w, p_m, and p_f respectively. These probabilities reflect technical parameters and design decisions such as materials and location. Using these probabilities, A is computed using the following equation:

$$A = (1 - p_w)(1 - p_m)(1 - p_f) \quad (1)$$

The capacity factor CF and its standard deviation σ_{CF} are important metrics that quantify the power variation of the WEC. Capacity factor represents the ratio of average power to maximum rated power. Thus, technologies with more consistent power output have higher capacity factors as they are able to pro-

duce at or near the maximum rated power for much of the time. Meanwhile, σ_{CF} represents the ratio of the standard deviation of power to maximum rated power. CF and σ_{CF} are defined with the following two equations.

$$CF = A \frac{P_{avg}}{P_{max}} \quad (2)$$

$$\sigma_{CF} = \frac{\sqrt{\frac{1}{n} \sum_{i=1}^{n} (P_i - P_{avg})^2}}{P_{max}} \quad (3)$$

The average power is assumed to be strictly positive, reflecting only the average wave activity of the region, and does not account for any periods of device downtime. The availability factor considers circumstances in which the device is inoperable or down due to maintenance or repairs.

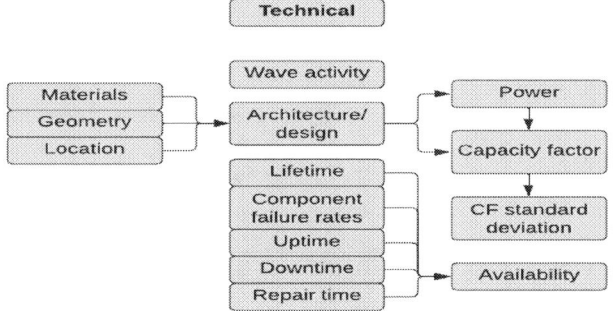

FIGURE 4. Flowchart of technical parameters and metrics

3.2 Economic Metrics

The primary economic metric chosen for the analysis is the net value of energy (NVOE). NVOE is the difference between the levelized value of energy (LVOE) and the levelized cost of energy (LCOE): $NVOE = LVOE - LCOE$. NVOE was selected because it is a comprehensive indicator of economic competitiveness, taking both costs and benefits into account [20]. LCOE is the discounted proportion of total capital expenditures (CapEx) and operational expenditures (OpEx) to the device's total average annual energy production over the course of the device's economic lifetime:

Copyright © 2023 by ASME

$$LCOE = \frac{CapEx + \sum\limits_{t=1}^{T} \frac{OpEx(t)}{(1+r)^t}}{\sum\limits_{t=1}^{T} \frac{AEP(t)}{(1+r)^t}} \qquad (4)$$

In the above equation, T is the project lifetime in years, r is the discount rate, and AEP is the annual energy production.

The LCOE captures the performance of a WEC with respect to costs, but does not reflect the actual benefits of the WEC. LVOE includes the financial value of the revenue generated by the energy generation of the WEC as well as external benefits of the WEC including avoided costs at the system level. Avoided costs include additional capacity that otherwise would have needed to be built without wave; for instance, without wave energy and its winter power peak, excess solar capacity or battery capacity would be required to meet winter energy demands. To compute the revenue, we sum the product of the instantaneous power and instantaneous price. The instantaneous power is derived from by putting hourly wave resource data through a simulation to determine the instantaneous WEC power $P(t)$ at any given time t. Instantaneous price of energy $R(t)$ can be obtained from hindcast and forecast data, or as the dual value to the power demand constraint in a CEM. Revenue value is then scaled by the availability factor A to account for disruptions to energy production.

$$LVOE = A\frac{\sum\limits_{t=1}^{T} P(t)R(t)}{\sum\limits_{t=1}^{T} \frac{AEP(t)}{(1+r)^t}} + \delta V \qquad (5)$$

LVOE is levelized at a discount rate r and reflects value per kilowatt hour. T represents the project lifetime and δV represents the additional non-monetizable avoided costs of the system. All of these metrics are encapsulated in Figure 5

3.3 Environmental Metrics

In this work's eco-cost system, eco-costs quantify the negative environmental impacts of a system [14]. Eco-costs are divided into two categories: local costs and global costs, LC and GC. Global pressures include the overall impact of a WEC on the environment due to the materials that comprise it and the processes used to manufacture, implement, and deploy it. Local pressures refer to the WEC's impact on a certain region, as determined by how the system interacts with the ecosystem near it. Examples of local pressures are the potential for a system to leak hydraulic fluid, collide with local wildlife, or increase the temperature and noise level of the surrounding water. On the

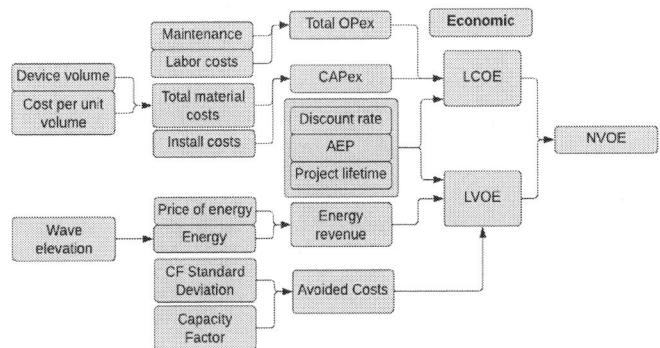

FIGURE 5. Flowchart of economic parameters and metrics

other hand, avoided eco-cost, or eco-value, refers to the positive environmental impacts associated with using a renewable energy source in place of a non-renewable source. Eco-values are divided into global and local values (GV and LV). Global benefits are calculated as the eco-cost per unit of energy of a fossil fuel multiplied by the amount of fossil fuel displaced by renewable sources. Local benefits include the positive impacts of a WEC near its deployment site, such as providing new breeding grounds for fish or protecting coastlines by damping the wave climate. Just as NVOE is the difference between LVOE and LCOE, net eco-value is defined as the difference between eco-value and eco-cost divided by the amount of energy produced.

Figure 6 summarizes the eco-costs and eco-values framework.

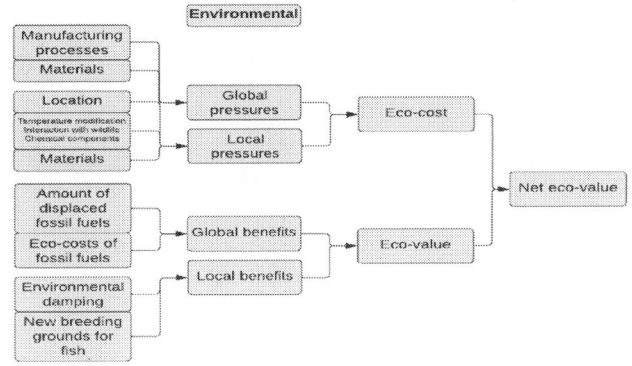

FIGURE 6. Flowchart of environmental parameters and metrics

4 METRIC WEIGHTINGS

Two weighting systems, PBE weighting and grid weighting, were chosen to capture the process of optimization along specific

Copyright © 2023 by ASME

systems and the process of optimization across multiple disciplines and sets of variables and parameters. The weightings of metrics that compose a subset assigns values to different metrics. This process helps ensure that variables will be treated differently when optimizing a particular set of metrics, with more weighted metrics receiving more attention, balancing the optimization process based on these weights.

By clearly noting the system of weights assigned to metrics and subsets, this helps ensure that the optimization process will treat variables and parameters according to their relative importance.

4.1 Weights for Powering the Blue Economy Markets

Multiple applications exist for WECs within PBE markets, such as saltwater desalination and marine aquaculture. Each application produces a different end product that is subject to distinct technical, economic and environmental constraints. To optimize wave energy to suit these different constraints, the optimization process must consider how the unique goals of each application alter the weights placed on metrics and types of metrics. For example, offshore aquaculture has proven to be a sustainable and relatively environmentally-friendly means of food production [16]. To optimize WECs for offshore aquaculture, more energy and resources should be devoted to optimizing the technical performance and economic viability of the device to meet the high power demands of aquaculture and avoid raising costs for an already expensive industry.

The markets within the PBE sector can be listed in a set. Let K represent the set of all PBE applications and let $k_i \in K$ be any singular application.

Every market k_i will be optimized according to two metrics: NVOE and the net ecovalue. There exists a Pareto front consisting of NVOE and net ecovalue. In accordance with Pareto efficiency, optimizing one metric will be at the expense of the other. Every application exists at a different point on the Pareto front, as the nature of the application determines whether more weight is placed on maximizing net value or net ecovalue. Let J_{nvoe} denote NVOE and J_{ev} denote net ecovalue.

The metrics J_j are composed of n inputs, each with a value m_j, and the these values m_j are assigned a weight value v_j. The value of J_j sum of the product of weights and metric values.

$$J_j = \sum_{j=1}^{n} m_j v_j \qquad (6)$$

4.2 Weights for the Grid Market

Section 2 discussed the need for a surrogate economic dispatch (ED) and capacity expansion model (CEM), the details of which are discussed here. To provide initial insights, existing wave energy CEM/ED results are used to create a surrogate. Un-

NVOE	NVOE	v_j	EV	EV	v_j
LCOE		0.5	Eco-Cost	LC	0.3
LVOE	σ_{CF}	0.06		GC	0.7
	Payback Period	0.11	Eco-Value	LV	0.5
	CF	0.11		GV	0.5
	A	0.11			
	δV	0.11			

TABLE 1. Table of weights for individual metrics

fortunately, very few such studies exist. Only four relevant papers were identified, three of which were published within the last year. The authors of [21] create a simplified CEM with monthly resolution to determine the optimal grid mix of a 100% renewable California 2045 grid, focusing on the sensitivity to wave energy $LCOE$ and CF. The EU-funded Evolve project considers an ED model with hourly resolution in 2030, 2040, and 2050 in the UK, Ireland, and Portugal [22]. The authors of [23] create a capacity expansion model with 3 hour resolution for North Carolina to find the pareto-efficient tradeoff between $LCOE$ and σ_{CF} for portfolios containing various capacities of wave and tidal energy. Finally, [24] runs a CEM for the entire US at various levels of wave energy $LCOE$. Despite its comprehensiveness, the study is over a decade old and assumes low renewables penetration, so its results have limited utility here.

The following section combines results of the first three of these studies to present sensitivities to three metrics: $LCOE$, σ_{CF}, and CF. Due to widely different assumptions and scopes of the existing studies, the weights presented here should be regarded as approximate, serving primarily to demonstrate the feasibility and intuition behind such a process. Future work includes running capacity expansion models with consistent assumptions and comprehensive scope to conclusively assess sensitivities. MATLAB scripts created for this section are accessible at https://github.com/symbiotic-engineering/wec-systems.

The total added value δV ($) of the WEC can be expressed as the sum of the avoided investment, δC_{INV}, and operating costs, δC_{OPR}. These avoided costs can be approximated by their variation due to the three metrics of interest.

$$\delta V = \delta C_{INV} + \delta C_{OPR}$$
$$\approx \frac{\partial C_{INV}}{\partial LCOE} \delta LCOE + \frac{\partial C_{INV}}{\partial CF} \delta CF + \frac{\partial C_{INV}}{\partial \sigma_{CF}} \delta \sigma_{CF} \qquad (7)$$
$$+ \frac{\partial C_{OPR}}{\partial LCOE} \delta LCOE + \frac{\partial C_{OPR}}{\partial CF} \delta CF + \frac{\partial C_{OPR}}{\partial \sigma_{CF}} \delta \sigma_{CF}$$

Copyright © 2023 by ASME

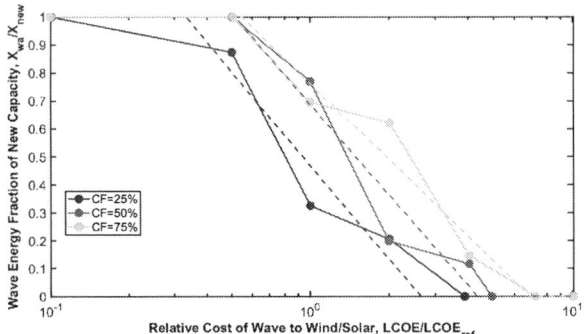

FIGURE 7. Optimal wave energy capacity as a function of LCOE results from [21], with a linear semilog fit.

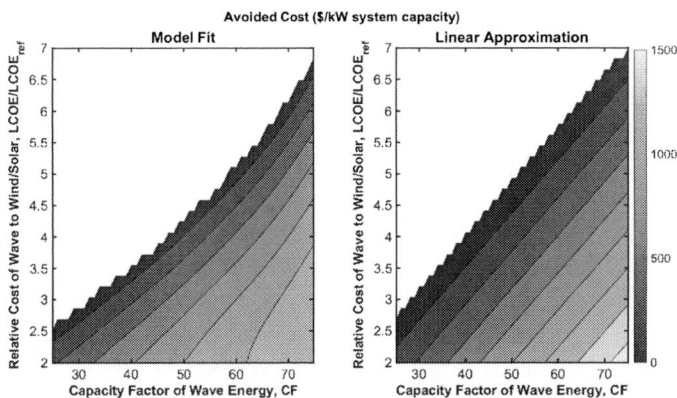

FIGURE 8. Avoided cost as a function of $LCOE$ and CF, nonlinear and linearized.

In the absence of a comprehensive CEM where the partial derivatives could be obtained by varying each of the metrics independently, we will reformulate the derivatives to make use of the information available from [21, 22, 23]. The results of [21] can be used to get $\frac{\partial C_{INV}}{\partial LCOE}$ and $\frac{\partial C_{INV}}{\partial CF}$, and the results of [23] can be used to get the sum $\frac{\partial C_{INV}}{\partial \sigma_{CF}} + \frac{\partial C_{OPR}}{\partial \sigma_{CF}}$. Finally, the results of [22] can be combined with those of [21] to get $\frac{\partial C_{OPR}}{\partial LCOE}$ and $\frac{\partial C_{OPR}}{\partial CF}$.

Upon observation of an approximately linear relationship between the relative wave cost $LCOE/LCOE_{ref}$ and optimal wave new capacity fraction X_{wa}/X_{new} on a semilog scale, as in Figure 1 of [21], the following curve fit was developed:

$$\frac{X_{wa}}{X_{new}} = \max(\min(-m\log\left(\frac{LCOE}{LCOE_{ref}}\right)+b,1),0) \qquad (8)$$

The fit parameters m and b represent the negative slope and y-intercept respectively, and the max and min serve to limit the result to $[0,1]$. This fit is shown in Figure 7.

The original study only included three discrete capacity factors, 25%, 50%, and 75%. In order to find the sensitivity to CF, the fit parameters b and m were assumed to vary quadratically with CF, allowing smooth interpolation with no loss of accuracy because three points determine a quadratic. This is shown in Equation 9.

$$b = -1.15 \cdot 10^{-4}\,CF^2 + 1.74 \cdot 10^{-2}\,CF + 0.105$$
$$m = -4.83 \cdot 10^{-5}\,CF^2 + 3.08 \cdot 10^{-3}\,CF + 0.435 \qquad (9)$$

Using these fit parameters, we can now compute $\frac{X_{wa}}{X_{tot}}$ as a function of CF and $LCOE/LCOE_{ref}$. To go from capacity to avoided cost, we use the fact that a CEM assumes equilibrium, so that at the optimum, the cost equals the value. Thus, we have:

$$\delta C_{INV}(LCOE,CF) = LCOE\,X_{wa}(LCOE,CF) \qquad (10)$$

where $X_{wa}(LCOE,CF)$ is computed from Equation 8.

Moving onto the EVOLVE report [22], they provide avoided operating costs and avoided carbon as a function of the fraction of total wave capacity X_{wa}/X_{tot}, for different years and locations. Here the 2050 results for Great Britain are used because this is the only year and location in the study for which the grid is 100% renewable, matching the assumptions in [21]. After converting 1 £ = 1.2 USD, the March 2023 exchange rate, and normalizing by the 290 GW capacity, we obtain the following sensitivity:

$$\frac{\partial C_{OPR}/X_{tot}}{\partial X_{wa}/X_{tot}} = 114\;\$/kW \qquad (11)$$

This can be combined with Equation 8 and the CF-dependent ratio X_{new}/X_{tot} to yield $\delta C_{OP}(LCOE,CF)$. Together with Equation 10, we have now fully quantified the dependence of δV on $LCOE$ and CF, and this is shown in Figure 8. Notably, the avoided operating costs make up less than 4% of the total avoided costs. This suggests a possible simplification that the operating costs could be neglected when obtaining future CEM sensitivities, so we must solve only the capacity expansion model and not the economic dispatch model.

Finally, the nonlinear relation can be linearized, shown on the right side of Figure 8. The linearization slightly overpredicts the avoided cost for high capacity factors and low LCOEs but is reasonable given the uncertainty of the fits and results used to generate the model. Using $LCOE_{ref} = \$0.15/kWh$, the linearized equation is

$$\delta V(LCOE,CF) = 28.5\,CF - 2220\,LCOE \qquad (12)$$

where CF is on a 0-100 scale and LCOE is in $/kWh.

The final step is to get the dependence of δV on the standard deviation of capacity factor, σ_{CF}. This is done by plotting the breakeven LCOE against the corresponding σ_{CF}, where the data is found in Table 1 and Figure 6 respectively of [23]. This

Copyright © 2023 by ASME

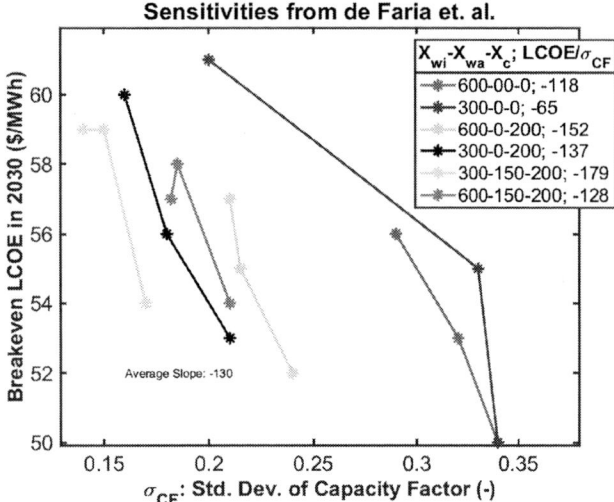

FIGURE 9. CEM results from [23] showing the tradeoff between LCOE and σ_{CV}. X_{wi}, X_{wa}, and X_c are the capacities in MW of wind, wave, and ocean current energy. The slope from a least-squares linear fit is shown for each portfolio, as well as the average slope $LCOE/\sigma_{CF} = -130$ \$/MWh per %.

relationship was not examined in [23] and is shown here in Figure 9. Taking linear regressions for each portfolio and computing the average slope yields \$130/MWh = \$0.13/kWh, which tells us the how the *LCOE* sensitivity compares to the σ_{CV} sensitivity.

The resulting sensitivity is:

$$
\begin{aligned}
\frac{\partial C_{INV+OP}}{\partial \sigma_{CF}} &= 0.13 \frac{\partial C_{INV+OP}}{\partial LCOE} \\
&= 0.13(-2220) \\
&= -289
\end{aligned}
\tag{13}
$$

With the weights now established, they can be summarized as follows:

$$
\left[\frac{\partial V}{\partial LCOE}, \frac{\partial V}{\partial CF}, \frac{\partial V}{\partial \sigma_{CF}} \right] = [-2220, 28.5, -289]
\tag{14}
$$

5 OPTIMIZATION

Section 2 introduced the Multidisciplinary Design Optimization (MDO) and Control Co-Design (CCD) framework used in this paper. A practical discussion of the techno-economic part of the formulation and its results as applied to a two-body point absorber can be found in [7]. This section focuses on the conceptual aspects of extending [7] to consider environmental impacts in addition to techno-economic objectives, and it proposes methods to include multiple architectures, markets, locations. The framework articulated here will be implemented in future work.

5.1 Design Optimization Formulation

The MDO problem is set up as a standard optimization problem, shown in equation 15.

$$
\begin{aligned}
\min \quad & \mathbf{J}(\mathbf{x}, \mathbf{p}) \\
\text{subject to} \quad & \mathbf{g}(\mathbf{x}, \mathbf{p}) \leq 0 \\
& \mathbf{h}(\mathbf{x}, \mathbf{p}) = 0 \\
& x_{i,LB} \leq x_i \leq x_{i,UB} \\
\text{where} \quad & \mathbf{x} = [x_1 \ldots x_N]^T,
\end{aligned}
\tag{15}
$$

where $\mathbf{x}$ and $\mathbf{p}$ are the problem design variables and parameters, respectively. The design variables are the model inputs that the optimization algorithm iterates until a feasible, optimal solution is found. They are bounded by lower and upper bounds, $x_{i,LB}$ and $x_{i,UB}$. The parameters are the inputs that are held constant. The objectives, $\mathbf{J}$, are the outputs to be maximized or minimized. The inequality, $\mathbf{g}$, and equality, $\mathbf{h}$, constraints bound the problem to ensure the optimizer returns feasible and physically realizable solutions.

For the WEC system-level MDO-CCD problem in particular, the objectives J are taken to be the metrics developed in the previous sections, indexed by j:

$$
j \in J = \{\text{NVOE, net ecovalue}\}
\tag{16}
$$

Additionally, there are several categorical design variables and parameters that make matters more complicated because standard gradient-based optimizers can only handle continuous variables. The design variables are

$$
\mathbf{x} = \{a, s, g, x_{continuous}\}
\tag{17}
$$

where a is the hydrodynamic architecture, s is the structural material, g is the type of generator or powertrain system, and $x_{continuous}$ is a vector consisting of geometric dimensions, structural thicknesses, controller maximum force and power, and powertrain sizing variables. The architecture, material, and generator can take on the following 7, 5, and 2 values, respectively:

$$
\begin{aligned}
a \in A = \{ & \text{ point absorber, attenuator, terminator, oscillating} \\
& \text{surge, oscillating water column, overtopping}\} \\
s \in S = \{ & \text{ steel, aluminum, concrete, plastic, composite}\} \\
g \in G = \{ & \text{hydraulic, electric}\}
\end{aligned}
\tag{18}
$$

The parameters $\mathbf{p}$ can likewise be described:

$$
\mathbf{p} = \{l, k, p_{continuous}\}
\tag{19}
$$

where l is the location, k is the market, and $p_{continuous}$ is a vector with physical parameters like the density of seawater and density of steel, economic parameters like the interest rate and the cost of certain components, and design parameters like powertrain ef-

Copyright © 2023 by ASME

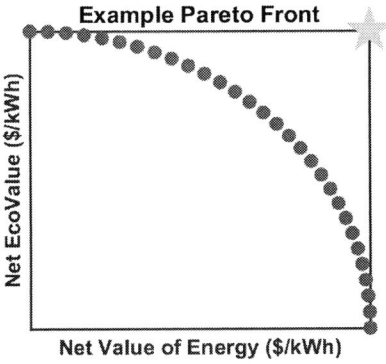

FIGURE 10. Sample Pareto front

ficiency and required factor of safety. The location and market can take on the following 9 and 3 discrete values, respectively:

$$l \in L = \{\text{ME, MA, NC, FL, OR, N.CA, S.CA, HI, AK}\} \tag{20}$$
$$k \in K = \{\text{grid, PBE coastal communities, PBE at sea}\}$$

Reviewing the categorical design variables and parameters, there is a total of $7 \cdot 5 \cdot 2 \cdot 9 \cdot 3 = 1890$ unique combinations. Running the optimization separately for all 1890 combinations is computationally prohibitive—if each optimization takes 5 minutes (a lower bound estimate that corresponds to 20 iterations, 15 seconds per iteration), this would be around 6.5 days of continuous computation. This points to the need for an optimization algorithm that can work with categorical design variables and a global sensitivity methodology that does not involve trying every combination of locations and markets.

Heuristic optimization schemes like genetic algorithms, particle swarm optimization, and simulated annealing can handle discrete variables but do not utilize gradient information, making them slower and with no guarantee of optimality. Subproblem optimization schemes where a heuristic optimizer drives an internal gradient-based optimizer or hybrid mixed-integer gradient-based optimization schemes like AMIEGO [9] are therefore attractive for this problem. Meanwhile, efficient sampling strategies inspired by design of experiments are available from toolboxes like DAKOTA and SALib, which can decrease the number of parameter combinations necessary. Future work involves implementing these algorithms with design optimization to evaluate the extent to which the computational complexity can be reduced.

Finally, the constraints **g** and **h** include hydrostatic equilibrium and stability, structural factors of safety, satisfying dynamic assumptions like linear hydrodynamic theory, and constraints to prevent edge conditions like negative power or undefined forces.

A sample Pareto front from running the multi-objective optimization a single time is shown in Figure 10.

5.2 Sensitivities and Technology Roadmapping

In this framework, local sensitivities are found as total derivatives through OpenMDAO [25] and global sensitivities would be found via sampling and reoptimization, as described above. $\frac{\partial x*}{\partial p}$ can be used to consider robustness to parameters, including discrete parameters like location and market. Designs that are robust to highly uncertain parameters are preferred. Sensitivities can also be used to inform R&D priorities and industry roadmaps, using technology roadmapping. $\frac{\partial J}{\partial p}$ tells the influence of parameters on performance. For parameters which are not necessarily uncertain but that represent the performance of a particular component, such as generator cost per kW or structural factor of safety, a large value of $\frac{\partial J}{\partial p}$ indicates that more effort should be spent developing that technology, for example with experimental campaigns to increase confidence in loadcases.

Roadmapping can also help decide between different pathways that could achieve that optimal solution, such as deciding whether to try to make a real-time causal controller get nearly as much power as the optimal controller, or finding a way to cheaply manufacture a unique geometry that is hydrodynamically optimal, by assessing the R&D required for each one. Roadmapping can take the form of a separate optimization problem over a few R&D scenarios, by quantifying how much a certain R&D focus is expected to improve performance, with the overall objective being net present value for a corporation.

6 CONCLUSION

Overcoming the techno-economic viability bottleneck is necessary to achieve widespread WEC adoption and enhance global decarbonization. Meanwhile, considering environmental sustainability in the design process can ensure that the resulting technology aligns with its designers' intent and maximizes positive impacts. Determining the ideal WEC design with the optimization process outlined here is an important step in helping the industry to converge to a design, thereby providing clarity, unlocking economies of scale, and improving investors' confidence. If successful, it could result in a society in which the ocean acts as a source of renewable, affordable, and economically-viable energy.

In future work, the authors intend to implement the process articulated here. Specifically, the work will integrate OpenM-DAO, a leading MDO tool developed by NASA [25], and We-cOptTool, a wave energy control co-design tool developed by Sandia National Labs [26], with MDOcean, a codebase developed to perform basic MDO for WECs [7]. This combination of tools will form a unified open-source WEC MDO-CCD optimization framework which, when used in conjunction with the industry-level impact-oriented objective functions provided here, will advance wave energy design towards full decarbonization of the electricity sector. Other work in the future should investigate value metrics for PBE markets in more detail.

Copyright © 2023 by ASME

ACKNOWLEDGEMENTS

This material is based on work supported by the National Science Foundation (NSF) Graduate Research Fellowship under Grant DGE–2139899. Any opinion, findings, and conclusions or recommendations expressed in this material are those of the authors and do not necessarily reflect the views of the NSF.

REFERENCES

[1] Bhatnagar, D., Bhattacharya, S., Preziuso, D., Hanif, S., O'Neil, R., Alam, M., Chalishazar, V., Newman, S., Lessick, J., Medina, G. G., Douville, T., Robertson, B., Busch, J., Kilcher, L., and Yu, Y., 2021. Grid Value Proposition of Marine Energy. Tech. Rep. PNNL-31123, Pacific Northwest National Lab, Richland, WA (United States).

[2] Kilcher, L., Fogarty, M., and Lawson, M., 2021. Marine Energy in the United States: An Overview of Opportunities. Tech. Rep. NREL/TP-5700-78773, NREL, Feb.

[3] Sobieszczanski-Sobieski, J., and Haftka, R., 1997. "Multidisciplinary aerospace design optimization: survey of recent developments". *Structural optimization, 14*(1), Aug., pp. 1–23.

[4] Garcia-Sanz, M., 2019. "Control Co-Design: An engineering game changer". *Advanced Control for Applications, 1*(1), p. e18.

[5] Herber, D. R., 2014. "Dynamic system design optimization of wave energy converters utilizing direct transcription".

[6] Giannini, G., Rosa-Santos, P., Ramos, V., and Taveira-Pinto, F., 2022. "Wave energy converters design combining hydrodynamic performance and structural assessment". *Energy, 249*, June, p. 123641.

[7] McCabe, R., Murphy, O., and Haji, M., 2022. "Multidisciplinary Optimization to Reduce Cost and Power Variation of a Wave Energy Converter". American Society of Mechanical Engineers Digital Collection.

[8] Ricks, W., Norbeck, J., and Jenkins, J., 2022. "The value of in-reservoir energy storage for flexible dispatch of geothermal power". *Applied Energy, 313*, May, p. 118807.

[9] Roy, S., Crossley, W., Stanford, B., Moore, K., and Gray, J. "A Mixed Integer Efficient Global Optimization Algorithm with Multiple Infill Strategy - Applied to a Wing Topology Optimization Problem". In *AIAA Scitech 2019 Forum*. American Institute of Aeronautics and Astronautics.

[10] Sterman, J., 2000. *Business Dynamics: Systems Thinking and Modeling for a Complex World*. Irwin/McGraw-Hill.

[11] Triton | PNNL.

[12] Galparsoro, I., Korta, M., Subirana, I., Borja, , Menchaca, I., Solaun, O., Muxika, I., Iglesias, G., and Bald, J., 2021. "A new framework and tool for ecological risk assessment of wave energy converters projects". *Renewable and Sustainable Energy Reviews, 151*, Nov., p. 111539.

[13] Araignous, E., and Safi, G., 2020. Environmental and So-cial Acceptance Tools - alpha version. Deliverable D6.5, France Energies Marines, Feb.

[14] Vogtlander, J. G., Baetens, B., Bijma, A., Brandjes, E., Lindeijer, E., Segers, M., Witte, F., Brezet, J., and Hendriks, C., 2010. *LCA-based assessment of sustainability: the Eco-costs/Value Ratio (EVR)*. Sustainable Design Series of Delft University of Technology. Sustainability Impact Metrics, Oegetgeest, The Netherlands.

[15] Driscoll, F. R., Weber, J. W., Jenne, D. S., Thresher, R. W., Fingersh, L. J., Bull, D., Dallman, A., Gunawan, B., Ruehl, K., Newborn, D., Quintero, M., LaBonte, A., Karwat, D., and Beatty, S., 2018. Methodology to Calculate the ACE and HPQ Metrics Used in the Wave Energy Prize. Tech. Rep. NREL/TP-5000-70592, National Renewable Energy Lab. (NREL), Golden, CO (United States), Mar.

[16] Jenne, S., 2021. Powering the Blue Economy: Economics of Marine Renewable Energy Systems. Tech. Rep. NREL/PR-5700-78328, National Renewable Energy Lab. (NREL), Golden, CO (United States), July.

[17] Caio, A., Davey, T., and McNatt, C., 2019. "Tackling the Wave Energy Paradox - Stepping Towards Commercial Deployment". OnePetro.

[18] McCabe, A. P., 2013. "Constrained optimization of the shape of a wave energy collector by genetic algorithm". *Renewable Energy, 51*, Mar., pp. 274–284.

[19] Weber, J., 2012. "WEC Technology Readiness and Performance Matrix – finding the best research technology development trajectory".

[20] Mai, T., Mowers, M., and Eurek, K., 2021. Competitiveness Metrics for Electricity System Technologies. Tech. Rep. NREL/TP-6A20-72549, National Renewable Energy Lab. (NREL), Golden, CO (United States), Feb.

[21] Coe, R., Lavidas, G., Bacelli, G., Kobos, P., and Neary, V., 2022. "Minimizing Cost in a 100% Renewable Electricity Grid: A Case Study of Wave Energy in California". American Society of Mechanical Engineers Digital Collection.

[22] , 2023. The system benefits of ocean energy to European power systems. Tech. rep., EVOLVE Consortium, Jan.

[23] de Faria, V. A. D., de Queiroz, A. R., and DeCarolis, J. F., 2022. "Optimizing offshore renewable portfolios under resource variability". *Applied Energy, 326*, Nov., p. 120012.

[24] Previsic, M., Eppler, J., Hand, M., Heimiller, D., Short, W., and Eurek, K., 2012. The Future Potential of Wave Power in the United States. Tech. rep., Aug.

[25] Gray, J., Hwang, J., Martins, J., Moore, K., and Naylor, B., 2019. "OpenMDAO: an open-source framework for multidisciplinary design, analysis, and optimization". *Structural and Multidisciplinary Optimization, 59*(4), pp. 1075–1104.

[26] Coe, R. G., Bacelli, G., Olson, S., Neary, V. S., and Topper, M. B. R., 2020. "Initial conceptual demonstration of control co-design for WEC optimization". *Journal of Ocean Engineering and Marine Energy, 6*(4), Nov., pp. 441–449.

Proceedings of the ASME 2023
International Design Engineering Technical Conferences and
Computers and Information in Engineering Conference
IDETC-CIE2023
August 20-23, 2023, Boston, Massachusetts

DETC2023-114762

A MULTI-FIDELITY GAUSSIAN PROCESS REGRESSION METHOD FOR PROBABILISTIC WIND FARM POWER CURVE ESTIMATION

Honglin Li[*]
The University of Texas at Dallas
Richardson, TX 75080
Email: honglin.li@utdallas.edu

Cong Feng[†]
National Renewable Energy Laboratory
Golden, CO 80401
Email: cong.feng@nrel.gov

Jie Zhang[‡]
The University of Texas at Dallas
Richardson, TX 75080
Email: jiezhang@utdallas.edu

ABSTRACT

Accurate estimation of the power curve for wind turbines or wind farms is crucial to ensure their efficient operation and management. However, conventional methods for power curve estimation rely either on expensive and infrequent measurements or on low-quality numerical simulations. Moreover, the majority of previous studies on power curve estimation for wind turbines or wind farms focused on deterministic estimation, which provides a point estimate of the relationship between wind speed and power generation. Nevertheless, the deterministic approach fails to consider the inherent uncertainty associated with wind energy production resulting from varying turbine characteristics. This can lead to inaccurate power generation estimation and suboptimal decisions regarding energy management. In this paper, a kernel density estimation (KDE) based Multi-Fidelity Gaussian Process Regression (MFGPR) model is proposed to fuse theoretical power curve data and the ground true measurements to create a mapping of wind speed and wind power. By conducting a case study on an actual wind farm in China, the efficacy of the proposed MFGPR model was demonstrated in characterizing the variability of wind power. The probabilistic MFGPR

model was also able to generate confidence intervals that encompassed the measured power, thereby improving the accuracy and confidence in wind power estimation or wind resource assessment. Overall, the proposed MFGPR model offers a reliable approach to integrate high-fidelity ground measurements and theoretical power curve data, resulting in precise wind resource assessment and power estimation.

Keywords: Probabilistic wind power curve estimation, Multi-fidelity modeling, Gaussian process regression, Kernel density estimation.

1 Introduction

Wind energy has emerged as a promising source of renewable energy, contributing significantly to the global energy mix. However, the intermittent nature of wind poses significant challenges for grid integration and energy management. Accurately estimating the power curve of wind turbines or wind farms is essential for both wind farm operations and long-term power grid planning. This is because precise power curve estimation can decrease the uncertainty in short-term and long-term wind power forecasting [1, 2].

Most of existing research on wind turbine/farm power curve estimation has focused on deterministic estimation [3, 4, 5], which provides a point estimate of the relationship between wind speed and power generation. However, the deterministic approach does not account for the inherent uncertainty of wind energy production due to varying

---

[*]Ph.D. Student, Department of Mechanical Engineering and the Center for Wind Energy.

[†]Director's Postdoctoral Fellow.

[‡]Associate Professor, Department of Mechanical Engineering, Department of Electrical and Computer Engineering (Affiliated), and the Center for Wind Energy (Affiliated), ASME Professional Member. Address all correspondence to this author.

Copyright © 2023 by ASME

turbine characteristics, resulting in inaccurate power generation estimation and suboptimal energy management decisions. Probabilistic wind farm power curve estimation [6], on the other hand, characterizes the stochastic nature of the wind power generation by providing confidence intervals in wind power estimation.

Accurate wind farm power curve estimation is crucial for the efficient and reliable operations of wind farms. Probabilistic power curve estimation has the potential to help improve the economic viability and competitiveness of wind energy by reducing the uncertainty and risk associated with its operation and grid integration.

1.1 Literature Review

Accurate models of wind turbine power curves are important for understanding turbine performance and aiding operational decision-making. Lydia et al. [7] have discussed the significance of wind turbine power curve in understanding the performance of wind turbines, monitoring and controlling turbines, estimating wind energy potential, and forecasting wind power. Li et al. [8] have compared different machine learning models for estimating wind turbine power curves.

Multi-fidelity modeling (MFM) is a technique that aims to leverage information from various sources of data with different levels of fidelity to improve the accuracy of predictions [9, 10]. In the context of energy prediction, MFM has been shown to be effective in reducing the computational cost of predicting wind energy output while maintaining high accuracy. Luo et al. [11] proposed a cascaded multi-fidelity deep learning (CMF-DL) framework that utilizes high-fidelity numerical weather observation data and low-fidelity numerical weather prediction (NWP) data to improve the accuracy of photovoltaic power forecasting. An MFM based Super-Fidelity Network (SFNet) deep learning method for wind farm wake modeling was proposed in [12]. The SFNet combines advantages of analytical models and numerical models to predict high-fidelity flow fields using low-fidelity data. The model has a mean absolute error of 1.9% when trained on only 45 samples and can predict flow fields of a wind farm consisting of 100 turbines within several seconds on a standard desktop.

Gaussian Process Regression (GPR) is a popular method for modeling the output of complex systems, such as probabilistic wind power curve estimation, due to its ability to handle non-linear relationships and uncertainties. For example, Rogers et al. [13] have proposed a heteroscedastic Gaussian process model that automatically quantifies variance in predictions, to generate accurate mean predictions and higher likelihoods than a homoscedastic model on operational wind turbine data. Manobel et al. [14] proposed

a wind turbine power curve estimation method using Artificial Neural Networks (ANN) with Gaussian Processes prefiltering, showing a 25% improvement in root mean square error for the predicted power compared to benchmark parametric and non-parametric methods.

Kernel density estimation (KDE) is another technique that has been explored to help quantify the uncertainty in wind energy. For example, Juban et al. [15] proposed a KDE-based method for wind power forecasting. Zhang et al. [16, 17] proposed a KDE-based method to accurately characterize and predict the annual variation of wind conditions.

1.2 Research Objective

This paper aims to develop an enhanced probabilistic wind farm power curve estimation model using a KDE-based Multi-Fidelity Gaussian Process Regression (MFGPR) method, by combining intermittent high-fidelity measurements with continuous low-fidelity numerical simulations from models such as the theoretical power curve. By considering the varying characteristics in wind turbines and uncertain wind conditions, the KDE-based MFGPR model could predict the mean value and confidence intervals of the wind power output. The main contributions of this paper include: (1) Develop a kernel density estimation based multi-fidelity Gaussian Process Regression model for wind turbine/farm power curve estimation; and (2) Explore the impact of linear and non-linear combination methods of Gaussian kernels in multi-fidelity models on wind turbine/farm power curve estimation.

The remainder of the paper is organized as follows. The developed KDE-based MFGPR method is described in Section 2. Section 3 describes the experimental setup for the case study, and Section 4 discusses the results in probabilistic wind power curve estimation. Concluding remarks and future work are discussed in Section 5.

2 Methodology

In machine learning or deep learning modeling, it is a common practice in the data pre-processing phase to clean the data, filter out most of the noisy data, and keep the data concentrated in the theoretical range of values. This can greatly improve the model's ability to fit the theoretical values. However, adopting such an approach may result in the loss of actual data characteristics, especially when estimating models in uncertain conditions, such as power curve estimation for wind turbines, which is considerably influenced by mechanical malfunctions as well as the variability of wind speed. Simply cleaning out these discrete values would severely reduce the ability of the model to fit the real

Copyright © 2023 by ASME

situation. If the amount of data is simply reduced, it is challenging for the GPR model to accurately capture the characteristics within the data. On the other hand, if the full dataset is used to model the GPR model, there could exist over-fitting issues with the highly noisy data.

This paper utilizes the KDE algorithm to sample the entire dataset, benefiting from its various advantages. First, the amount of data is significantly reduced while retaining the original data distribution characteristics. Second, the sampling process only follows the density of the data distribution without assuming any underlying distribution of the data. Third, KDE estimates the probability density function (PDF) at each data point using a kernel function that assigns more weight to nearby points. This local estimation can be advantageous when the underlying distribution varies across the data space as seen in the wind energy data. The weakening or disappearance of the underlying signal due to data sampling can be compensated for by low-fidelity data.

2.1 Kernel Density Estimation (KDE)

KDE [15, 16, 18] is a non-parametric way to estimate the probability density function of a random variable. The basic idea of KDE is to estimate the density function of a random variable by placing a kernel (usually a Gaussian function) on each data point and summing the resulting kernel functions. The resulting density estimate is then normalized so that it integrates to one. KDE is a flexible method that can be used to estimate the PDF of data with any distribution, including multimodal distributions. This makes it useful for a wide range of applications where the underlying distribution is unknown or complex. Unlike parametric methods, KDE does not make any assumptions about the underlying distribution of the data. This can be an advantage when the data is highly skewed, has outliers, or comes from a distribution that is difficult to model parametrically. KDE preserves all the information in the data, unlike methods that involve binning the data or fitting a parametric distribution, which can be an advantage when analyzing data with complex distributions or small sample sizes. The density is estimated by:

$$\hat{f}_h(x) = \frac{1}{nh} \sum_{i=1}^{n} K_h \left(\frac{x - x_i}{h} \right) \tag{1}$$

where $\hat{f}_h(x)$ is the estimated density at x, $K_h(x)$ is the kernel function with bandwidth h, and x_i is the ith data point. The kernel function $K_h(x)$ is defined as:

$$K_h(x) = \frac{1}{h} \phi \left(\frac{x}{h} \right) \tag{2}$$

where $\phi(\cdot)$ is the standard normal probability density function. The bandwidth h is a smoothing parameter that controls the width of the kernel function. The choice of h has a significant impact on the smoothness of the estimated density. The standard normal probability density function $\phi(\cdot)$ is defined as:

$$\phi(x) = \frac{e^{-x^2/2}}{\sqrt{2\pi}} \tag{3}$$

2.2 Gaussian Process Regression (GPR)

GPR is a flexible supervised machine learning technique used for regression analysis that models the distribution of a function using a Gaussian process (GP), which can provide accurate predictions and uncertainty estimates for complex functions [19]. The Gaussian process is defined as a collection of random variables, and any finite number of variables has a joint Gaussian distribution.

The key components of the GPR model include the mean function $m(\mathbf{x})$ and the covariance function $c(\mathbf{x}, \mathbf{x}')$, which together fully specify the Gaussian process. The mean function represents the expected value of the function at each input point, while the covariance function specifies the degree of similarity between the function values at different input points. The covariance function is an appropriate kernel function parameterized by a vector of hyperparameters θ that results in a symmetric positive-definite covariance matrix $C_{ij} = c(\mathbf{x}_i, \mathbf{x}_j; \theta), C \in \mathbb{R}^{n \times n}$ [20].

The goal of GPR is to estimate the underlying function $f(\mathbf{x})$ given a set of input-output pairs $(\mathbf{x}_i, y_i)$, where $\mathbf{x}_i \in \mathbb{R}^D$ and $y_i \in \mathbb{R}$. The GPR model assumes that the output y_i is a noisy observation of the function value $f(\mathbf{x}_i)$, and that the noise follows a Gaussian distribution with a zero mean and variance σ_n^2, i.e. $f \sim \mathcal{GP}\left(f \mid \mathbf{0}, \sigma_n^2\right)$, where $\sigma_n^2 = c(\mathbf{x}_i, \mathbf{x}_j; \theta)$

The predicted function value $\hat{f}(\mathbf{x})$ at a new input point $\mathbf{x}$ is given by the posterior mean of the Gaussian process, which is a weighted sum of the observed function values:

$$\hat{f}(\mathbf{x}) = \mathbf{c}_*^\top \left(\mathbf{C} + \sigma_n^2 \mathbf{I} \right)^{-1} \mathbf{y} \tag{4}$$

where $\mathbf{c}_*$ is a vector of covariance between the new input point $\mathbf{x}$ and the training input points, $\mathbf{C}$ is the covariance matrix of the training input points, and $\mathbf{y}$ is a vector of the observed function values.

The uncertainty in the predicted function value is characterized by the posterior variance of the Gaussian process, which measures the degree of uncertainty in the predicted function value at a new input point:

$$\text{Var}[\hat{f}(\mathbf{x})] = c(\mathbf{x}, \mathbf{x}) - \mathbf{c}_*^\top \left(\mathbf{C} + \sigma_n^2 \mathbf{I} \right)^{-1} \mathbf{c}_* \tag{5}$$

Copyright © 2023 by ASME

where $c(\mathbf{x}, \mathbf{x})$ is the covariance between the new input point $\mathbf{x}$ and itself.

A systematic extension of the GPR framework can be employed to create probabilistic models by recursion, allowing for the amalgamation of data from sources with variable fidelity [20, 21, 22].

$$g_t(\mathbf{x}) = \rho g_{t-1}(\mathbf{x}) + \delta_t(\mathbf{x}) \tag{6}$$

Equation 6 defines a framework in which Gaussian process g_{t-1} and g_t model data at different levels of fidelity. The scaling constant ρ quantifies the correlation between the model outputs y_t and y_{t-1}, while $\delta_t(x_t)$ is a GP with a mean of $\mu \delta_t$ and a covariance function of c_t. In other words, μ_t follows a Gaussian process with the mean $\mu \delta_t$ and covariance function $c_t(\mathbf{x}_t, \mathbf{x}'_t; \theta_t)$, denoted as $\mu_t \sim \mathcal{GP}(\delta_t | \mu \delta_t, c_t(\mathbf{x}_t, \mathbf{x}'_t; \theta_t))$.

2.3 Multi-fidelity GPR Modeling

Multi-fidelity modeling is a technique used to improve the efficiency and accuracy of computer simulations by incorporating information from simulations of varying levels of fidelity. In many engineering applications, simulations of high-fidelity (i.e., detailed and accurate) models can be computationally expensive and time-consuming, while simulations of low-fidelity (i.e., simplified and approximate) models are much faster but may be less accurate. Multi-fidelity modeling seeks to combine the advantages of both high- and low-fidelity models to produce accurate predictions in a computationally efficient manner.

One common approach to multi-fidelity modeling is to use a set of low-fidelity models, denoted by $f_1(\mathbf{x}), f_2(\mathbf{x}), \ldots, f_M(\mathbf{x})$, along with a high-fidelity model, denoted by $f_{M+1}(\mathbf{x})$. The low-fidelity models are typically less expensive to simulate, while the high-fidelity model provides the most accurate predictions but at a higher computational cost.

The goal of multi-fidelity modeling is to construct a surrogate model, denoted by $\hat{f}(\mathbf{x})$, which accurately approximates the high-fidelity model using information from the low-fidelity models. One typical method is to combine the low-fidelity models in a weighted manner with an adjustment term to account for the difference between the low-fidelity and high-fidelity models [9, 23]:

$$\hat{f}(\mathbf{x}) = \sum_{m=1}^{M} w_m(\mathbf{x}) f_m(\mathbf{x}) + \epsilon(\mathbf{x}) \tag{7}$$

where $w_m(\mathbf{x})$ are the weights assigned to each low-fidelity model, and $\epsilon(\mathbf{x})$ is the correction term that accounts for the discrepancy between the low- and high-fidelity models.

The weights $w_m(\mathbf{x})$ are typically chosen to be a function of the low-fidelity model predictions and some measure of their accuracy or fidelity, such as the discrepancy between the low- and high-fidelity models:

$$w_m(\mathbf{x}) = \frac{\omega_m(\mathbf{x})}{\sum_{i=1}^{M} \omega_i(\mathbf{x})} \tag{8}$$

where $\omega_m(\mathbf{x})$ is a fidelity measure that depends on the discrepancy between the low-fidelity model $f_m(\mathbf{x})$ and the high-fidelity model $f_{M+1}(\mathbf{x})$. One common fidelity measure is the normalized square error:

$$\omega_m(\mathbf{x}) = \frac{\|f_m(\mathbf{x}) - f_{M+1}(\mathbf{x})\|^2}{\sigma_{f_{M+1}}^2} \tag{9}$$

where $\sigma_{f_{M+1}}^2$ is the variance of the high-fidelity model predictions.

The weighted sum method is a technique that involves a linear combination of different fidelity-level models, and it is solely suitable when the models are linearly correlated. For non-linear models of different levels of fidelity, we can use Eq. 6 to fuse them together recursively by replacing the GP prior f_{t-1} with the GP posterior $f_{*t-1}(\mathbf{x})$ from the previous inference level [24].

$$f_t(\mathbf{x}) = g_t(\mathbf{x}, f_{*t-1}(\mathbf{x})) \tag{10}$$

where $g_t \sim \mathcal{GP}\left(c_t \mid \mathbf{0}, c_t\left((\mathbf{x}, f_{*t-1}(\mathbf{x})), (\mathbf{x}', f_{*t-1}(\mathbf{x}')); \theta_t\right)\right)$. Since the inputs $f_{*t-1}(\mathbf{x})$ and $\mathbf{x}$ belong to inherently different spaces, sometimes the chosen structure for the covariance function of g_t may not be appropriate. To address this challenge, Perdikaris et al. [20] proposed an extension to the methodology by introducing a more structured prior for g_t, which better reflects the autoregressive nature of Eq. 10. Specifically, the covariance kernel can be decomposed as:

$$\begin{aligned} c_{t_g} = c_{t_\rho}\left(\mathbf{x}, \mathbf{x}'; \theta_{t_\rho}\right) \cdot c_{t_f}\left(f_{*t-1}(\mathbf{x}), f_{*t-1}(\mathbf{x}'); \theta_{t_f}\right) \\ + c_{t_\delta}\left(\mathbf{x}, \mathbf{x}'; \theta_{t_\delta}\right) \end{aligned} \tag{11}$$

where θ_{t_ρ}, θ_{t_f}, and θ_{t_δ} represent the hyperparameters of valid covariance functions, namely c_{t_ρ}, c_{t_f} and c_{t_δ}, respectively. All the kernel functions mentioned above are selected to have the form of the squared exponential, which is also known as a radial basis function (RBF) kernel [19].

$$c_t \left(\mathbf{x}, \mathbf{x}'; \theta_t \right) = \sigma_t^2 \exp \left(-\frac{1}{2} \sum_{i=1}^{d} w_{i,t} \left(\mathbf{x}_i - \mathbf{x}_i' \right)^2 \right) \quad (12)$$

where σ_t^2 denotes the variance parameter and $(w_{i,t})_{i=1}^{d}$ represents automatic relevance determination weight corresponding to the fidelity level t.

3 Case Study
3.1 Data Summary

We have applied the proposed MFGPR method to a Spatial Dynamic Wind Power Forecasting (SDWPF) dataset, which is obtained from a real wind farm owned by the Longyuan Power Group Corp. Ltd. The dataset consists of the wind power data of 134 1.5 MW wind turbines, wind farm layout and weather conditions [25]. Table 1 summarizes the statistics of the SDWPF dataset.

TABLE 1: Statistics of the SDWPF dataset

Parameter	Value
Data length (Days)	245
Resolution	10 minutes
No. of Turbines	134
No. of Features	13
No. of Records	4,727,520

Figure 1 depicts 200 high-fidelity real data sampled from the 4,727,520 data points using the KDE algorithm, and 200 low-fidelity theoretical data points obtained from the theoretical power curve, which is a 1.5 MW wind turbine theoretical power curve data obtained from the Energy Systems Integration Group (ESIG)[26]. The cut-in, rated, and cut-out wind speeds are 3 m/s, 14 m/s, and 25 m/s, respectively. It is seen from the high-fidelity data that the actual power generation does not exactly follow the theoretical power curve, due to uncertainties, turbine conditions, and other factors.

3.2 Deterministic Evaluation Metrics

Standard metrics such as root mean square error (RMSE), mean absolute error (MAE), mean absolute per-

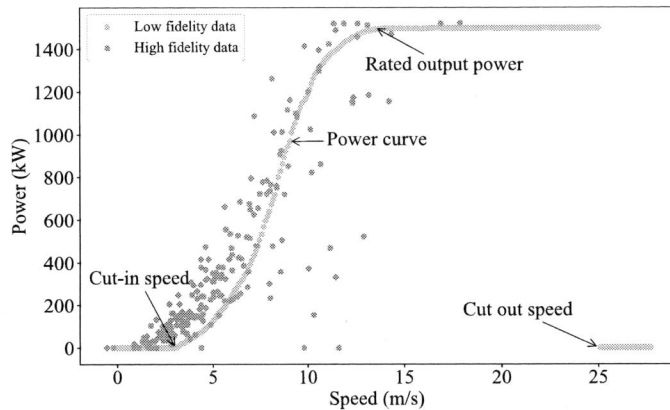

FIGURE 1: High- and low-fidelity training data used in this study

centage error (MAPE), normalized RMSE (NRMSE), and normalized MAE (NMAE) are used to assess the accuracy of deterministic power curve estimation in this paper. These metrics provide a quantitative measure of the deviation between predicted and actual power values. RMSE and MAE are based on the difference between the predicted and actual values, while MAPE measures the percentage deviation. NRMSE and NMAE are normalized versions of RMSE and MAE, respectively, which account for the scale of the data being analyzed. These metrics are defined as:

$$RMSE = \sqrt{\frac{\sum_{i=1}^{n} \left(\hat{x}_i - x_i \right)^2}{n}} \quad (13)$$

$$NRMSE = \frac{1}{x_{\max}} \sqrt{\frac{\sum_{i=1}^{n} \left(\hat{x}_i - x_i \right)^2}{n}} \quad (14)$$

$$MAE = \frac{1}{n} \sum_{i=1}^{n} |\hat{x}_i - x_i| \quad (15)$$

$$NMAE = \frac{1}{n} \sum_{i=1}^{n} \left| \frac{\hat{x}_i - x_i}{x_{\max}} \right| \quad (16)$$

$$MAPE = \frac{100\%}{n} \sum_{i=1}^{n} \left| \frac{\hat{x}_i - x_i}{x_i} \right| \quad (17)$$

Copyright © 2023 by ASME

where $\hat{x}$ represents the predicted value, x is the actual value, and n is the number of samples.

3.3 Probabilistic Evaluation Metrics

The pinball loss is a measure of the deviation between the predicted and actual values, which is commonly used in quantile regression to evaluate the accuracy of predictions [27]. The pinball loss function is defined as:

$$L_\tau(y, \hat{y}) = \begin{cases} (y - \hat{y})\tau, & \text{if } y \geq \hat{y} \\ (\hat{y} - y)(1 - \tau), & \text{if } y < \hat{y} \end{cases} \quad (18)$$

where y is the actual value, $\hat{y}$ is the predicted value, and τ is the quantile of interest.

In probabilistic estimation, sharpness refers to the degree of concentration or spread of the forecast distribution around its mean. A sharp forecast means that the predicted distribution is concentrated around the mean, while a flat forecast means that the distribution is spread out. The sharpness of a probabilistic forecast can be measured using various metrics, such as the variance, standard deviation, or entropy of the forecast distribution. The smaller the value of these metrics, the sharper the forecast [28]. Mathematically, sharpness can be defined as follows:

$$S = \frac{1}{N} \sum_{i=1}^{N} \sum_{j=1}^{M} (\hat{p}_{ij} - \bar{p}_j)^2 \quad (19)$$

where S is the sharpness, N is the number of forecast samples, M is the number of forecast probabilities in each sample, $\hat{p}_{ij}$ is the forecast probability for the j-th category in the i-th sample, and $\bar{p}_j$ is the average probability for the j-th category across all forecast samples.

Reliability of probabilistic estimation refers to the degree to which the forecast probabilities match the observed frequencies [29]. It measures how well the forecast probabilities reflect the actual outcomes. A well-calibrated probabilistic forecast is considered reliable. The most commonly used measure of reliability is the probability integral transform (PIT) histogram, which is the cumulative distribution function (CDF) evaluated at the observation. If the forecast is well-calibrated, the PIT values should be uniformly distributed between 0 and 1. Mathematically, the reliability can be expressed as:

$$Re = \frac{1}{N} \sum_{i=1}^{N} \int_{-\infty}^{\infty} [\hat{F}_i(z) - I(Y_i \leq z)]^2 dP(z) \quad (20)$$

where N is the number of observations, Y_i is the actual observation, $\hat{F}_i(z)$ is the forecast CDF at z, and I is the indicator function.

4 Results and Discussion

In the case study, we have compared the performance of three power curve estimation models, including a nonlinear kernel based MFGPR, a linear kernel based MFGPR, and a standard GPR method. All three models were tested on 9 wind turbines that were selected randomly.

The deterministic power curve estimation results are summarised in Table 2, and a smaller metric value indicates better performance. The results show that, on average, the nonlinear kernel fusion method utilizing the MFGPR approach performed better than the standard GPR method for the first eight wind turbines, demonstrating a decrease of 3.9% in RMSE and 3.5% in MAE.

Figure 2 also displays a comparison of the three models' performance in terms of deterministic power curve estimation for the 9 wind turbines across 50 consecutive time intervals. The figure shows that the non-linear kernel fusion approach consistently outperforms the other two models, namely the standard GPR and the linear kernel fusion approach. The non-linear kernel fusion approach produces the most accurate estimation for the power output of the wind turbines in the vast majority of the time intervals. In contrast, the standard GPR and the linear kernel fusion approaches show less consistent performance, with both models producing higher errors in certain time intervals. Overall the non-linear kernel fusion approach is more robust and reliable in predicting the power output of wind turbines.

Figures 3 and 4 provide further insights into the performance of the three models for the probabilistic estimation of wind turbine power output. Figure 3 shows that the non-linear kernel fusion method produces more accurate probabilistic estimation results, especially for extreme value points. This suggests that the non-linear kernel fusion approach is better able to capture the underlying patterns and uncertainties in the data, resulting in more accurate probabilistic estimation.

Figure 4 compares the performance of the three models across three probabilistic evaluation criteria: sharpness, reliability, and pinball loss. The results show that the standard GPR achieves the best results for sharpness across all nominal proportions, indicating that it produces more tightly concentrated predictive distributions. For reliability, the non-linear kernel fusion method performs worse than the linear fusion method for smaller nominal proportions, but performs better for larger nominal proportions, with both models outperforming the standard GPR model. This suggests that the non-linear approach is better able to pro-

Copyright © 2023 by ASME

TABLE 2: Deterministic power estimation results using nonlinear kernel, linear kernel, and standard GPR

		T1	T2	T3	T4	T5	T6	T7	T8	T9
Nonlinear kernel MFGPR	RMSE (kW)	188.75	225.88	225.97	210.56	201.77	232.98	181.60	210.72	250.11
	NRMSE (%)	0.12	0.15	0.15	0.14	0.13	0.15	0.12	0.14	0.16
	MAE (kW)	112.27	117.05	143.92	115.39	122.34	134.12	105.77	127.14	155.75
	NMAE (%)	0.07	0.08	0.09	0.08	0.08	0.09	0.07	0.08	0.10
Linear kernel MFGPR	RMSE (kW)	212.61	239.48	238.97	222.44	212.92	244.84	203.28	212.45	262.50
	NRMSE (%)	0.14	0.16	0.16	0.15	0.14	0.16	0.13	0.14	0.17
	MAE (kW)	150.48	162.02	178.65	150.01	158.50	171.81	145.24	162.13	198.50
	NMAE (%)	0.10	0.11	0.12	0.10	0.10	0.11	0.10	0.11	0.13
Standard GPR	RMSE (kW)	201.07	226.36	241.07	216.55	209.82	238.48	198.96	217.74	248.67
	NRMSE (%)	0.13	0.15	0.16	0.14	0.14	0.16	0.13	0.14	0.16
	MAE (kW)	113.00	126.19	146.09	118.69	126.31	139.77	106.74	132.85	166.71
	NMAE (%)	0.07	0.08	0.10	0.08	0.08	0.09	0.07	0.09	0.11

duce probabilistic estimations that are consistent with the true outcomes. For pinball loss that measures the accuracy of probabilistic estimation, the non-linear kernel fusion method achieves better results than the other two models in extreme cases.

Figure 5 depicts the NRMSE results of the non-linear MFGPR model for all turbines in the wind farm. With the exception of a few turbines for which relatively large values are obtained, the non-linear MFGPR model has accurately estimated the power curve for the majority of wind turbines.

5 Conclusion

In this study we proposed and investigated the the multi-fidelity Gaussian process regression (GPR) approach for wind farm power curve estimation. By comparing the results of three different GPR models on nine randomly selected wind turbines, the study found that the non-linear kernel fusion method MFGPR consistently outperformed the standard GPR model, achieving lower errors across multiple metrics such as RMSE, reliability, and pinball loss. Moreover, the non-linear kernel fusion method MFGPR demonstrated superior performance for both deterministic and probabilistic power curve estimation.

Overall, these findings highlight the potential of the multi-fidelity GPR approach, and in particular, the non-linear kernel fusion method, for improving the accuracy and reliability of wind energy estimation. This has significant implications for the efficient operation of wind farms and the maximization of wind energy utilization. Therefore, this study provides a valuable contribution to the field of wind energy forecasting and emphasizes the potential of the proposed method for future applications.

Potential future research directions include (i) examining the applicability of the MFGPR model in forecasting wind power uncertainty based on existing wind farms, and (ii) exploring ways to enhance the accuracy of estimation while reducing computational costs through the integration of additional techniques.

ACKNOWLEDGMENT

This paper was developed based upon funding from the Alliance for Sustainable Energy, LLC, Managing and Operating Contractor for the National Renewable Energy Laboratory for the U.S. Department of Energy.

Copyright © 2023 by ASME

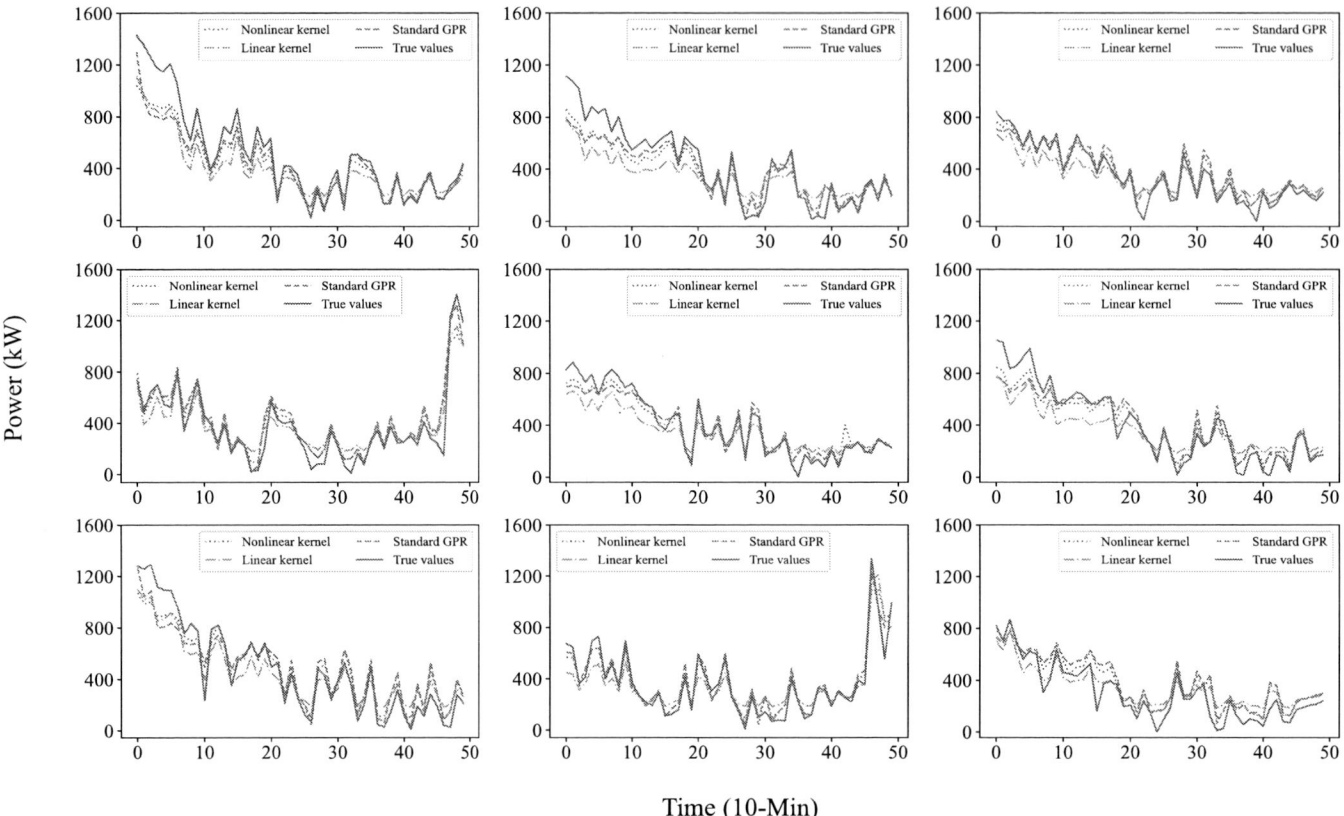

FIGURE 2: Deterministic single turbine power curve and power generation estimation for nine randomly selected wind turbines. Row 1: T1 – T3; Row 2: T4 – T6; Row 3: T7 – T9

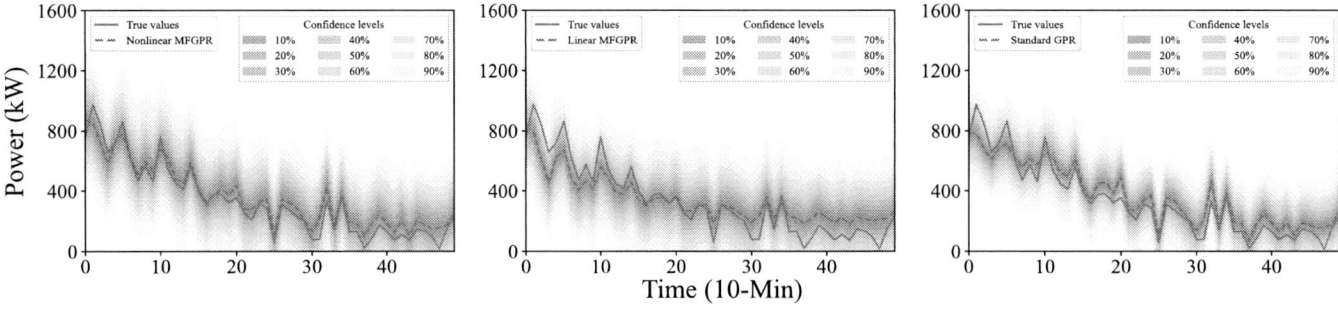

FIGURE 3: Probabilistic power curve and power generation estimation for one selected wind turbine. From left to right, the three subplots show the results of probabilistic estimates of the nonlinear kernel MFGPR, linear kernel MFGPR, and standard GPR models on the same turbine.

REFERENCES

[1] Arwade, S. R., Lackner, M. A., and Grigoriu, M. D., 2011. "Probabilistic models for wind turbine and wind farm performance". *Journal of Solar Energy Engineering*, **133**(4).

[2] Olaofe, Z., and Folly, K., 2012. "Wind power esti-

mation using recurrent neural network technique". In IEEE Power and energy society conference and exposition in africa: intelligent grid integration of renewable energy resources (PowerAfrica), IEEE, pp. 1–7.

[3] Chang, T.-P., Liu, F.-J., Ko, H.-H., Cheng, S.-P., Sun, L.-C., and Kuo, S.-C., 2014. "Comparative analysis

Copyright © 2023 by ASME

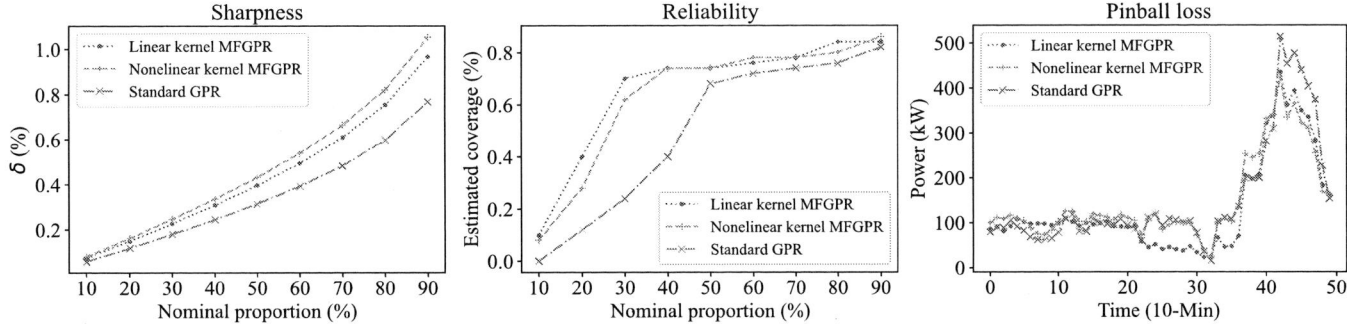

FIGURE 4: The sharpness, reliability, and pinball loss metrics for evaluating the performance of probabilistic power estimation

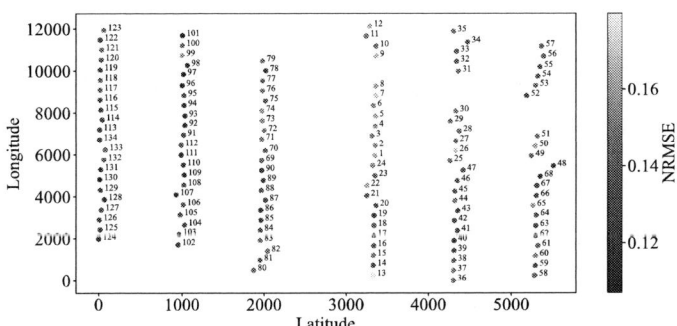

FIGURE 5: The NRMSE results of all turbines in the wind farm based on the nonlinear MFGPR power curve estimation (the units of latitude and longitude are meter)

on power curve models of wind turbine generator in estimating capacity factor". *Energy, **73**,* pp. 88–95.

[4] Lydia, M., Selvakumar, A. I., Kumar, S. S., and Kumar, G. E. P., 2013. "Advanced algorithms for wind turbine power curve modeling". *IEEE Transactions on sustainable energy, **4**(3),* pp. 827–835.

[5] Pei, S., and Li, Y., 2019. "Wind turbine power curve modeling with a hybrid machine learning technique". *Applied Sciences, **9**(22),* p. 4930.

[6] Yun, E., and Hur, J., 2021. "Probabilistic estimation model of power curve to enhance power output forecasting of wind generating resources". *Energy, **223**,* p. 120000.

[7] Lydia, M., Kumar, S. S., Selvakumar, A. I., and Kumar, G. E. P., 2014. "A comprehensive review on wind turbine power curve modeling techniques". *Renewable and Sustainable Energy Reviews, **30**,* pp. 452–460.

[8] Li, S., Wunsch, D. C., O'Hair, E., and Giesselmann, M. G., 2001. "Comparative analysis of regression and artificial neural network models for wind turbine power curve estimation". *J. Sol. Energy Eng., **123**(4),* pp. 327–332.

[9] Song, X., Lv, L., Sun, W., and Zhang, J., 2019. "A radial basis function-based multi-fidelity surrogate model: exploring correlation between high-fidelity and low-fidelity models". *Structural and Multidisciplinary Optimization, **60**,* pp. 965–981.

[10] Wang, X., Liu, Y., Sun, W., Song, X., and Zhang, J., 2018. "Multidisciplinary and multifidelity design optimization of electric vehicle battery thermal management system". *Journal of Mechanical Design, **140**(9),* p. 094501.

[11] Luo, X., and Zhang, D., 2023. "A cascaded deep learning framework for photovoltaic power forecasting with multi-fidelity inputs". *Energy,* p. 126636.

[12] Li, R., Zhang, J., and Zhao, X., 2022. "Multi-fidelity modeling of wind farm wakes based on a novel super-fidelity network". *Energy Conversion and Management, **270**,* p. 116185.

[13] Rogers, T., Gardner, P., Dervilis, N., Worden, K., Maguire, A., Papatheou, E., and Cross, E., 2020. "Probabilistic modelling of wind turbine power curves with application of heteroscedastic gaussian process regression". *Renewable Energy, **148**,* pp. 1124–1136.

[14] Manobel, B., Sehnke, F., Lazzús, J. A., Salfate, I., Felder, M., and Montecinos, S., 2018. "Wind turbine power curve modeling based on gaussian processes and artificial neural networks". *Renewable Energy, **125**,* pp. 1015–1020.

[15] Juban, J., Siebert, N., and Kariniotakis, G. N., 2007. "Probabilistic short-term wind power forecasting for the optimal management of wind generation". In 2007 IEEE Lausanne Power Tech, IEEE, pp. 683–688.

[16] Zhang, J., Chowdhury, S., Messac, A., and Castillo, L. "Multivariate and multimodal wind distribution model based on kernel density estimation". In ASME 2011 5th International Conference on Energy Sustainability, American Society of Mechanical Engineers Digital Collection, pp. 2125–2135.

[17] Zhang, J., Chowdhury, S., Messac, A., and Castillo, L.,

Copyright © 2023 by ASME

2013. "A multivariate and multimodal wind distribution model". *Renewable Energy,* **51**, pp. 436–447.

[18] Parzen, E., 1962. "On estimation of a probability density function and mode". *The annals of mathematical statistics,* **33**(3), pp. 1065–1076.

[19] Seeger, M., 2004. "Gaussian processes for machine learning". *International journal of neural systems,* **14**(02), pp. 69–106.

[20] Perdikaris, P., Raissi, M., Damianou, A., Lawrence, N. D., and Karniadakis, G. E., 2017. "Nonlinear information fusion algorithms for data-efficient multi-fidelity modelling". *Proceedings of the Royal Society A: Mathematical, Physical and Engineering Sciences,* **473**(2198), p. 20160751.

[21] Kennedy, M. C., and O'Hagan, A., 2000. "Predicting the output from a complex computer code when fast approximations are available". *Biometrika,* **87**(1), pp. 1–13.

[22] Le Gratiet, L., and Garnier, J., 2014. "Recursive cokriging model for design of computer experiments with multiple levels of fidelity". *International Journal for Uncertainty Quantification,* **4**(5).

[23] Fernández-Godino, M. G., Park, C., Kim, N.-H., and Haftka, R. T., 2016. "Review of multi-fidelity models". *arXiv preprint arXiv:1609.07196.*

[24] Cutajar, K., Pullin, M., Damianou, A., Lawrence, N., and González, J., 2019. "Deep gaussian processes for multi-fidelity modeling". *arXiv preprint arXiv:1903.07320.*

[25] Zhou, J., Lu, X., Xiao, Y., Su, J., Lyu, J., Ma, Y., and Dou, D., 2022. "Sdwpf: A dataset for spatial dynamic wind power forecasting challenge at kdd cup 2022". *arXiv preprint arXiv:2208.04360.*

[26] ESIG. 1.5 mw wind turbine theoretical power curve. [Online] Available at: `https://www.esig.energy/wiki-main-page/general-electric-1-5-mw-series/`. [Accessed: 01 Mar. 2023].

[27] Sun, M., Feng, C., Chartan, E. K., Hodge, B.-M., and Zhang, J., 2019. "A two-step short-term probabilistic wind forecasting methodology based on predictive distribution optimization". *Applied energy,* **238**, pp. 1497–1505.

[28] Feng, C., Cui, M., Hodge, B.-M., and Zhang, J., 2017. "A data-driven multi-model methodology with deep feature selection for short-term wind forecasting". *Applied Energy,* **190**, pp. 1245–1257.

[29] Sun, M., Feng, C., and Zhang, J., 2019. "Conditional aggregated probabilistic wind power forecasting based on spatio-temporal correlation". *Applied Energy,* **256**, p. 113842.

Proceedings of the ASME 2023
International Design Engineering Technical Conferences and
Computers and Information in Engineering Conference
IDETC-CIE2023
August 20-23, 2023, Boston, Massachusetts

DETC2023-115046

EXPLORATION OF BUILDING CLUSTERING POTENTIAL WITH ENERGY STORAGE IN NEW YORK CITY

Gregory Kaminski[1]   Philip Odonkor[1,*]

[1] School of Systems and Enterprises, Stevens Institute of Technology, Hoboken, NJ 07030

ABSTRACT

The adoption of distributed energy resources has increased in recent years due to decreasing implementation costs and increasing regulatory incentives to reduce energy consumption. This trend has been driven by a surge in energy consciousness and the availability of energy-saving products and technologies. Many energy consumers are also electing to use battery systems to store energy locally, further reducing reliance on the electrical grid and maximizing the benefits of distributed energy systems. For owners of multiple buildings, or multiple owners willing to share the operational cost, building clusters may be formed to more effectively take advantage of these distributed resources and storage systems. However, determining what makes a "good" building cluster when implementing these systems in existing buildings is a challenge. Additional challenges are presented when geographical proximity of these buildings is considered. Using metrics derived from comparison of unique five-building clusters from a population of sixteen stock buildings with a given distributed energy resource (in this case, a solar photovoltaic panel array) and energy storage system, we apply performance ranking information to a population of actual building data from New York City. In doing so, the authors aim to build upon the concept of logical building clusters with consideration of physical building location to develop an understanding of which geographic areas present the greatest opportunity for clustering.

Keywords: Building clusters, energy storage, distributed energy resources, building proximity

1. INTRODUCTION

In recent years, the rapid increase in the adoption of distributed energy resources (DER) has been fueled by several factors, including the decreasing cost of implementation, growing energy consciousness, and increased desire for energy resilience. In addition, regulatory incentives and limitations have made these resources more appealing. For instance, a recently-enacted law in New York City imposes penalties on building owners who exceed baseline carbon emissions (which are based on building size and use) by relying on non-renewable energy sources [1]. However, there are still challenges to implementing DERs. For example, distributed sources like solar and wind power can be intermittent and require reliance on the local electrical grid when they are unavailable or cannot satisfy the building's demand [2]. Specifically, solar arrays rely on consistent sunlight to generate power reliably and wind farms rely on steady wind speeds. When sunlight and wind are unreliable, generation becomes intermittent.

A way to potentially address this issue is through the utilization of energy storage systems. These systems can store energy produced on site and can aid in decreasing net power usage during periods of high demand. Various companies such as Tesla [3], Johnson Controls [4], GE [5], and SunPower [6] provide energy storage solutions for both residential and commercial purposes. Typically, these solutions consist of inverters and battery cabinets that are modular, thus enabling flexibility in instant power delivery and overall energy storage capacity. The size and control methods of these systems can differ significantly and, in some cases, may lead to an increase in energy consumption [7]. As a result, significant effort has been

---

* Corresponding author: podonkor@stevens.edu

Copyright © 2023 by ASME

directed towards developing optimal sizing and control algorithms suited for specific buildings.

After deployment of these distributed sources and energy storage systems, significant variation in building energy demand may result in either insufficient energy production to meet building needs or excess energy that is not utilized by the building. This leads to the need for grid power consumption or the possibility of energy sale back to the grid. Two emerging trends in the energy sector aim to tackle these imbalances in energy supply and demand - the concept of net-zero energy buildings (NZEBs) and the Smart Grid. The goal of a NZEB is generally to achieve a balance between energy supply and demand on an annual basis, utilizing mechanisms for communication and control between a building's energy-consuming systems and sources [8]. However, this does not eliminate the need for a building to rely on the grid during periods when energy generated and/or stored is inadequate to satisfy demand. The Smart Grid links networks of decentralized energy sources and consumers via bi-directional communication and control algorithms to optimize the utilization of these sources [9].

Multiple buildings may be used to form clusters which share distributed energy resources as well as energy storage systems. A great deal of effort has been placed on sizing and control strategy optimization for these clusters. Xie et al. [10] proposed a mixed-integer linear programming process, and Odonkor et al. [11,12] used reinforcement learning to develop a control strategy for DERs. Current literature [13–15] also addresses DER sizing and siting concerns with mixed integer linear programming and game theory-based approaches.

There has been considerably less attention paid to the role of composition and energy consumption as a factor in building cluster performance. There is a wealth of research focused on sorting building load profiles using clustering techniques [16]. However, few examine the implications of cluster composition on the overall performance of the cluster, and, by extension, the utility of these energy generation and storage systems. This represents a crucial step towards the employment of these techniques, as detailed load profile data, particularly over extended periods, is often not obtainable for existing buildings. Despite the growing use of energy metering as a result of increased energy consciousness and regulations, owners of buildings lacking such monitoring technologies may only have access to general building characteristics (such as size and use) and information on an energy bill. Consequently, building owners seeking to integrate alternative energy sources and/or energy storage systems may need to rely on this limited information to determine appropriate system sizes and capacities.

Emphasis has been more heavily placed on the 'logical' building cluster, which considers building characteristics in a vacuum – independent of geographical concerns. This work aims to build upon the idea of the logical cluster by incorporating considerations for geographic proximity. To this end, we build upon the work completed in [17] by expanding our building cluster data to include all available building types and increasing

the cluster size. We then calculate cluster scores based on composition and use this to assign a clustering potential score to each building in New York City. This information is then visualized to observe emerging characteristics. Leveraging insights from this information, we aim to generate new and generalizable knowledge to support optimal building cluster formation. While there are several other physical and infrastructural considerations needed for optimal clustering, this work serves as the first step in addressing the knowledge gap present when composing building clusters.

The remainder of the paper details how this was achieved and is structured as follows. Section 2 provides an overview of the methodology, focusing on the process by which building clusters were scored and the assumptions associated with each cluster component. Section 3 presents the results, including key metrics observed for the buildings in the dataset. Sections 4 and 5 close the paper with a discussion of our results and concluding remarks, respectively.

2. METHODOLOGY

Before each building could be assigned a score, all possible clusters needed to be assigned a score based on composition. To accomplish this, all unique five-building clusters were simulated using the stock building data and battery discharge algorithm described in sections 2.1 through 2.4. Once these clusters were simulated, several performance metrics were calculated, including average load, peak load, peak-to-average ratio, energy use intensity, and grid independence. Then a composite score for each cluster was calculated based on the equally weighted sum of the normalized value for each metric. These metrics and the related composite score calculation are further detailed in section 2.5.

2.1 Cluster Ranking Model Overview

Each building cluster consists of a solar photovoltaic (PV) system, an energy storage system (battery), and five buildings, each with unique electrical demand load profiles. It is important to note that this configuration is designed as a preliminary framework for performance evaluation, rather than an optimized solution. The total load profile of the cluster is taken as the sum of the demand loads of each building. The PV system is first used to satisfy the demand load of the building cluster, then charge the battery system, or both if enough energy is produced. The models were created using the Python 3.9 scripting language in the Spyder 5.1.5 development environment. Figure 1 provides a diagrammatic representation of the system configuration. Arrows in the figure indicate the potential directions of power flow in the system. Each part of the system, including the functions, data, and modeling methods, are explained in the remainder of section 2.

2.2 Stock Building Demand Load Profiles

Building load profile data for this work was sourced from the Department of Energy's (DOE) Office of Energy Efficiency & Renewable Energy (EERE) [18]. The dataset contains one year of hourly load profile data for 16 commercial building types

Copyright © 2023 by ASME

in each of the 16 climate zones within the United States. These load profiles are not taken from specific existing buildings. Rather, they were developed to be representative of a 'standard' building of each type. For this model, data was used that is reflective of buildings within New York City. It is worth noting that the energy models used to generate this dataset are also used as a baseline for determining compliance with energy codes and regulations in building construction projects by governing agencies. Table 1 outlines some physical characteristics for each of the 16 building types included in the dataset. Assumptions surrounding the building, occupant, and equipment characteristics are outlined in [19] and [20].

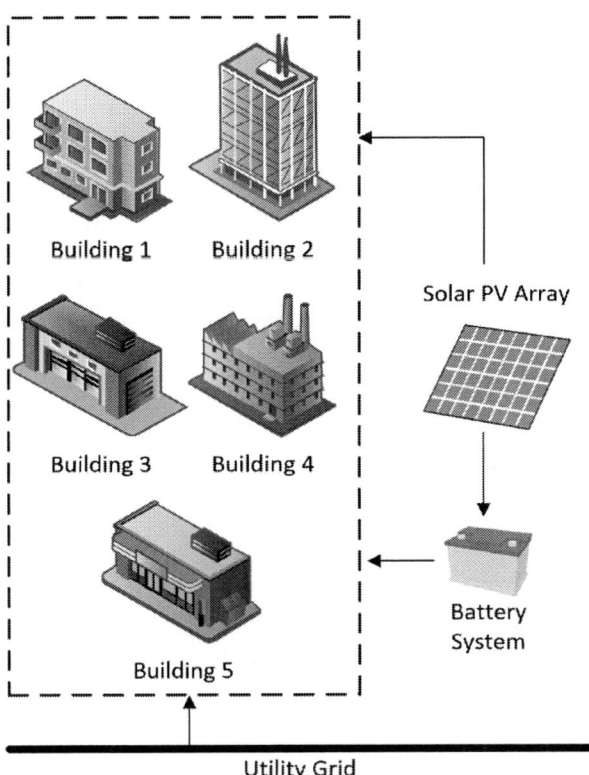

FIGURE 1: SYSTEM MODEL CONFIGURATION

To build upon the previous work in [17], all sixteen building load profiles were used in this analysis and the cluster size (i.e., the number of buildings in each cluster) was increased from three to five. The electrical use characteristics for each building were greatly varied. For this reason, constant PV array and battery system sizes used in previous work could not be used for all clusters and a different method for sizing these systems was required. As such, an external tool was used to determine sizes of these systems in each cluster. Further information regarding this tool is found in section 2.3.

Using the 16 building types available, 15,504 unique clusters were formed, representing all possible combinations of five buildings, and allowing for multiple instances of the same building type within a cluster. These combinations were not intended to be reflective of buildings that may exist within close proximity to one another. The overall load profile of the cluster is used as the baseline for comparison later in this work.

For calculations related to building size presented later in this work, it was necessary to consider the floor areas of each building. These areas are referenced from [19] and [20]. Natural gas consumption data is also used to calculate certain metrics. This information is also present in the dataset.

TABLE 1: REFERENCE BUILDING CHARACTERISTCS

Building Type	Total Floor Area (ft²)	No. of Floors
Small Office	5,500	1
Medium Office	53,628	3
Large Office	498,588	12*
Primary School	73,960	1
Secondary School	210,887	2
Stand-Alone Retail	24,962	1
Strip Mall	22,500	1
Supermarket	45,000	1
Quick-Service Restaurant	2,500	1
Full-Service Restaurant	5,500	1
Small Hotel	43,200	4
Large Hotel	122,120	6
Hospital	241,351	5*
Outpatient Healthcare	40,946	3
Warehouse	52,045	1
Midrise Apartment	33,740	4

* Plus basement (not included in table number)

2.3 PV System

It should be noted that the building load profiles outlined in the previous section do not have solar PV system parameters or generation data associated with them. Consequently, we sourced supplementary data using the National Renewable Energy Laboratory's (NREL) "REopt" tool [21]. This is similar to the approach used in [22]. However, outage scenarios were not considered in this analysis. REopt is used to determine the viability of renewable energy sources for a given energy use case or to recommend optimal sizing of sources and energy storage systems based on user-defined input parameters and goals. It allows for robust definition of system constraints, such as detailed load profiles, building system properties, detailed rate structures, and power failure resilience options. It also provides

Copyright © 2023 by ASME

hourly generation data for the recommended PV system. For this work, REopt was used to determine appropriate sizing of the solar PV and battery systems in each cluster. The associated hourly PV generation data provided by REopt was also used. Due to the extensive number of clusters considered in this work, REopt's application program interface (API) was used via Python scripting.

The sizing of the PV system for each cluster was recommended based on default economic parameters (e.g., installation cost per kilowatt, operating cost per kilowatt, etc.), physical parameters (e.g., watts per square foot of space available for solar panel installation), and available space for solar panel installation. For the purposes of this analysis, the aggregate roof area of all buildings in the cluster was considered to be the available space for solar panel installation. Also, since REopt performs economic analysis to recommend system sizing, a New York City commercial electrical rate structure was used.

As with the building load profiles and rate structure, the PV system energy production was simulated using New York City location data. The 'DC System Size' was also selected based on the recommendation from REopt. Figure 2 shows a sample of solar PV output data for four different days during the year of hourly data used. Having developed the PV generation model and energy storage system size, the next section provides an overview of the battery model we developed to allow for energy storage.

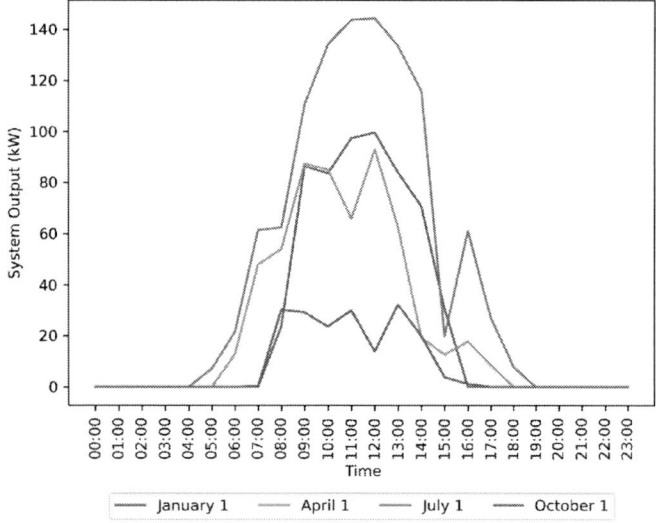

FIGURE 2: PV SYSTEM SAMPLE OUTPUT

2.4 Energy Storage Discharge Model

The energy storage system in this work is modeled as a battery system, with parameters based on those recommended by the REopt tool as noted in the previous section. The parameters used in the model to define the battery system are as follows: P_{rated} is the output power for which the system is rated (measured in kilowatts, kW). E_{rated} is the energy capacity of the system

(measured in kilowatt-hours, kWh). η_{rt} is the round-trip efficiency of the system.

The modular and scalable nature of these products allows system designers to adapt system configurations to the specific scenario in which the system is being deployed. For the purposes of this analysis, the battery storage system power and capacity values used were the direct outputs determined by the REopt tool. The round-trip efficiency was modeled after that of the Tesla 'Powerpack' [3] at $\eta_{rt} = 88\%$.

For this work, the equations for the battery storage system model outlined below are derived from the 'target zero' operational approach described in the literature [7]. This logic aims to minimize the net grid demand to the extent possible, with no consideration for future building demand or solar power availability. Any available stored energy from the system will be discharged if the power available from the PV system cannot satisfy the demand.

$$P_{targ}(C,t) = P_{use}(C,t) - P_{gen}(C,t) \qquad (1)$$

The target function (Eqn. 1) is the first step taken based on the given building load, P_{use} and power available from the PV system, P_{gen}. It defines the energy that the battery needs to provide in order to arrive at a net-zero power use for a given cluster, C, and given time step, t. Once this is calculated, the target value, P_{targ} and current state of the battery, E_{bat}, are used to decide whether the battery will be charging, discharging, or neither (Eqn. 2).

If the value of P_{targ} is positive (meaning P_{use} is greater than P_{gen}), the cluster is a net energy consumer, and the battery will be discharged if possible to satisfy the demand. Otherwise, the cluster is a net energy producer, and the battery will be charged. The energy supplied by the battery, P_{bat}, is set to zero if the battery will be neither charged nor discharged.

$$P_{bat}(C,t) =$$
$$\begin{cases} P_{bat,dichg} & P_{targ}(C,t) > 0 \text{ and } 0 \leq E_{bat} \\ P_{bat,chg} & P_{targ}(C,t) < 0 \text{ and } 0 < E_{bat} \\ 0 & \text{otherwise} \end{cases} \qquad (2)$$

The rated power of the battery system, P_{rated}, is the maximum charge or discharge rate. If the battery is determined to be charging or discharging, the model then confirms that the constraints of the battery system will not be violated (i.e., the energy discharge exceeds the available energy, or the state of the battery charge exceeds the maximum rated capacity). The power output supplied by the battery is then set accordingly.

$$P_{bat,chg} =$$
$$\begin{cases} \max(P_{targ}, -P_{rated}) & E_{bat} - P_{targ}\Delta t\kappa < E_{rated} \\ -\frac{E_{rated} - E_{bat}}{\Delta t} & E_{bat} - P_{targ}\Delta t\kappa \geq E_{rated} \end{cases} \qquad (3)$$

Copyright © 2023 by ASME

$$P_{bat,dischg} =$$
$$\begin{cases} \frac{E_{bat}}{\Delta t} & E_{bat} - P_{targ}\Delta t\kappa \leq 0 \\ \min(P_{targ}, P_{rated}) & E_{bat} - P_{targ}\Delta t\kappa > 0 \end{cases} \quad (4)$$

The constants Δt and κ represent the time step duration used in the dataset (in data points per hour) and an efficiency constant, respectively. For this dataset, Δt is set equal to 1 since the dataset contains hourly demand loads. The constant κ is defined in Equation. 5.

$$\kappa = \begin{cases} 1/\sqrt{\eta_{rt}} & P_{bat}(C,t) > 0 \quad \text{(discharging)} \\ \sqrt{\eta_{rt}} & P_{bat}(C,t) < 0 \quad \text{(charging)} \end{cases} \quad (5)$$

The logic shown is utilized by iterating through the dataset, determining the battery power (P_{bat}), calculating the change in battery charge level (ΔE_{bat}), and determining the next state of the battery (E_{bat}). The values of ΔE_{bat} and E_{bat} for a given cluster and time are represented in Equations 6 and 7, respectively. $E_{bat,i}$ is the initial state of battery charge, which set to zero in this model for the first day of the year, representing the start of the year with an empty battery. After this, stored energy is permitted to carry from day to day if unused.

$$\Delta E_{bat}(C,t) = -P_{bat}(C,t)\kappa\Delta t \quad (6)$$

$$E_{bat}(C,t) = E_{bat,i}(C) + \sum_{x=1}^{t} \Delta E_{bat}(C,x) \quad (7)$$

Degradation of battery capacity and performance over time is typically considered when modeling energy storage systems. This has not been considered for this work, however, as the relatively short time period being modeled would not significantly impact battery capacity and performance. It is also worth noting that for the purpose of this analysis, energy cost is not considered (except by the REopt tool), as these costs will vary based on customer type, location, energy rates, peak demand rates, and incentives.

2.5 Cluster Composite Score Determination

Once all unique clusters were simulated, five metrics were calculated and used to ultimately determine the cluster's composite score. These metrics include the reduction in average consumption (AR), reduction in peak demand load (PR), change in peak-to-average consumption ratio, (PAR) grid independence/reliance (GI, measured in time during which the cluster was completely self-sufficient, with the current power consumption met or exceeded by the PV system and/or battery system), and energy use intensity reduction (EUI, measured in annual energy usage per unit of building floor area).

$$S_n = \frac{AR_n - AR_{min}}{AR_{max} - AR_{min}} + \frac{PR_n - PR_{min}}{PR_{max} - PR_{min}} + \frac{EUI_n - EUI_{min}}{EUI_{max} - EUI_{min}} +$$

$$\frac{GI_n - GI_{min}}{GI_{max} - GI_{min}} + \left(1 - \frac{PAR_n - PAR_{min}}{PAR_{max} - PAR_{min}}\right) \quad (8)$$

Each individual cluster's change in performance was normalized, weighted, and summed to arrive at a composite score between zero and five (five being the highest performance improvement). This is shown in equation 8. It should be noted that while each parameter was weighted equally, this process can be generalized to more heavily weigh parameters that are more important for a given use case. Of the 15,504 possible unique clusters, approximately 95% yielded recommendations from REopt for PV and battery systems. This is because REopt only recommends a PV system size only if the life cycle cost of energy is reduced [23]. The remainder were assigned a score of zero, as they did not show performance improvement. Analysis of this cluster performance is the focal point of separate ongoing research by the authors. Composite scores for each cluster were used to evaluate each building in the New York City buildings dataset.

2.6 Building Score Determination

The building geographic data for this work was sourced from the New York City Department of Planning's Primary Land Use Tax Lot Output (PLUTO) dataset [24]. The dataset contains numerous parameters for each building in New York city, including location data, building type, size, owner information, and several others. For this work, building class and location were the primary parameters used. ArcGIS Pro 3.0.4 was used to filter data, prepare the relevant building data for export to Python, and visualize the subsequent outputs from Python.

FIGURE 3: VISUAL REPRESENTATION OF BUILDING DATA

To determine clustering potential scores for each building, it was first necessary to limit the scope to the buildings in the dataset for which stock building data is available (i.e. filtering out buildings that do not fall into one of the sixteen building

FIGURE 4: OVERALL NYC BUILDING HEATMAP

types contained within the DOE dataset noted in section 2.1). New York City is particularly unique in this regard, as there are 214 different building classes defined within the dataset. Certain building types were clearly matched to those in the DOE stock building dataset, but others were grouped together into the most similar category. As a result, only 42 building classes were mapped to their respective category, reducing the building population used for this work from 858,208 buildings to only 58,503 buildings. This reduction, however, can mostly be attributed to the removal of single-family residential buildings, and small multi-family residential buildings, which together account for almost 700,000 buildings alone.

Once the unnecessary buildings were filtered out of the dataset, ArcGIS was employed to derive an inventory of all buildings situated within a 500-foot perimeter surrounding any given building of interest. This was used as a basis for determining the potential clusters any given building could form. This data was exported from ArcGIS and imported to Spyder for analysis via Python.

To provide a basis for further insight, data surrounding the connections of these building 'nodes' was viewed in the context of the surrounding streets. Connecting lines were defined between each building and its surrounding buildings within 500 feet. Intersections between these lines and streets were tabulated and compiled. A visual representation of the building data points, connecting lines, public streets, and intersection points is illustrated in Figure 3. Buildings are shown as colored nodes, while roads are shown as dark grey lines. Red lines visually represent the potential physical links between buildings that are within 500 feet of one another. 'X' symbols show where these connecting lines intercept roads.

To determine the score for a given building, the building, and all other buildings within 500 feet, were compiled into a "population." All possible five-building clusters were then formed from this population. Each of these potential clusters were then assigned a composite score based on their composition (values defined in section 2.5). The average of all these

Copyright © 2023 by ASME

individual cluster scores is taken as the building's "clustering potential" score.

In short, the EERE dataset was used to determine scores of unique cluster compositions, and the PLUTO dataset was used to apply these scores to actual buildings present in the population.

3. RESULTS

After the methodology described in the previous section was applied to each of the buildings in the dataset, the scores were loaded back in ArcGIS for analysis and visualization. The complete heatmap of buildings with color-coded data points is shown in Figure 4. Buildings marked in light blue show the lowest clustering potential, with scores improving as points change to darker blue and purple. The highest scores are shown in bright pink. The calculated scores range from 0 (least potential) to a maximum of approximately 4.371 (most potential).

Unsurprisingly, areas of higher relative potential tend to occur along more main roads compared to areas further from these roads. This is expected, as this work is focused on commercial buildings, which are more common on main roads. Since residential buildings are removed from the dataset for this analysis, the potential number of clusters that can be formed by buildings in areas with predominantly residential building populations is reduced. Also, higher building scores tend to have a higher concentration in areas with diverse building types – mainly neighborhood commercial districts.

Surprisingly, the borough of Manhattan has the lowest average of building scores. This is visible on the overall map, as the buildings in this borough are generally colored in lighter blue – especially in upper Manhattan. The exceptions are two areas - one in the northern Manhattan neighborhood of East Harlem, and one in the Lower East Side. Both areas have small gatherings of buildings scored relatively high in comparison to the rest of the borough. The generally low potential of buildings in the borough comes as a surprise, particularly because it is the most dense borough in this analysis, with buildings having an average of 22.7 other nearby buildings. The likely reason for this is that although building density is high, building diversity is low. This leads to mostly homogenous areas of buildings, providing little opportunity for different cluster compositions.

The borough of Staten Island, which has the highest average building score, but the least buildings, also shows the trend of higher scores along main roads. Buildings in this borough are generally more spread out – more than 20% of the buildings do not have another building within 500 feet, and buildings have an average of only 7.3 nearby buildings.

Building score trends in the Bronx exhibit many similar characteristics to others, with buildings on main roads achieving generally higher scores. There is a particularly dense area of highly-scored buildings in the St. George neighborhood, which contains a relatively diverse group of commercial buildings in close proximity to residential buildings. However, average scores in the borough are surprisingly low, given buildings have

an average of 14.3 other buildings nearby - greater than that of Queens, and only slightly below that of Brooklyn.

Queens and Brooklyn show the most varied results. Average building scores are high in both boroughs, with both boroughs having large areas with low scores and few areas of dense building populations with high scores. In Queens, the Sunnyside and Long Island City neighborhoods show particularly high density of highly-scored buildings. Both neighborhoods have particularly high concentrations of diverse commercial buildings. In Brooklyn, the two neighborhoods of Williamsburg and Borough Park stand out as the most densely populated high-scoring buildings on the map. Both areas are also unique in that the highly-scored buildings do not appear to exist along any one particular street. Instead, they appear scattered throughout the neighborhood. We will focus more on these areas in the next section.

TABLE 2: AVERAGE BUILDING SCORE

Borough	Building Count	Average Raw Building Score	Average Adjusted Building Score
Bronx	9,721	0.896	0.985
Brooklyn	20,149	1.039	1.153
Manhattan	13,435	0.790	0.933
Queens	12,843	1.093	1.214
Staten Island	2,355	1.103	1.192
Overall	**58,503**	**0.972**	**1.089**

Table 2 outlines the building counts and average building score by borough. While the building scores only account for the potential of the building for clustering based on the averages of all possible clusters, this measure is highly sensitive to the number of nearby buildings. In some cases, system designers may place a higher emphasis on flexibility and connectedness. For this reason, additional parameters were incorporated into additional "adjusted" building scores. Rewards to account for the quantity of nearby buildings and penalties for streets crossed were applied to the original raw building scores. These adjustments are intended to account for logistical and physical implications of forming clusters. A higher quantity of nearby buildings is seen as a benefit, as it allows greater flexibility when forming clusters in the context of the larger population. Road crossings represent potential physical barriers to connections between buildings and, therefore, impose a penalty on the score. For the purposes of this exercise, the reward and penalty weights were chosen to be +1% for every 10 nearby buildings and -1% for every unique street crossed, respectively. While these values were chosen somewhat arbitrarily to demonstrate the concept, different weights can be assigned to these values depending on a system designer's goals and preferences. Table 2 also includes each borough's average adjusted score for comparison.

Copyright © 2023 by ASME

Table 3 outlines average score, average number of nearby buildings, and average number of unique street crossings by building type.

TABLE 3: SCORES BY BULIDING TYPE

Building Type	Average Score	Avg. No. of Nearby Buildings	Avg. No. of Unique Crossings
Small Office	1.230	9.8	2.4
Medium Office	1.498	16.6	3.2
Large Office	1.117	25.0	3.9
Primary School	1.537	12.4	2.4
Secondary School	2.001	7.9	1.9
Stand-Alone Retail	1.287	14.8	3.1
Strip Mall	0.775	4.6	1.1
Full-Service Restaurant	1.144	9.4	2.5
Small Hotel	1.013	16.4	3.1
Large Hotel	0.687	22.2	3.6
Hospital	0.623	8.2	1.9
Outpatient Healthcare	1.196	11.4	2.7
Warehouse	1.798	11.6	2.5
Midrise Apartment	0.573	17.1	3.1

4. DISCUSSION

As noted in the previous section, buildings with higher clustering potential scores tended to be concentrated along major local roadways and in groups where several diverse commercial building types exist. This allows for the greatest possible amount of possible cluster combinations and, consequently, the greatest ability to build effective clusters. As mentioned in Section 3, the Brooklyn neighborhood of Borough Park is an excellent example of this.

Figure 5 shows an enlarged view of this neighborhood with buildings color-coded by score. This is a particularly noteworthy location, as the most highly-scored buildings do not all fall on one main street, but they are scattered throughout the neighborhood. To gain further insight, Figure 6 shows the same area, but the buildings are color-coded by type instead. This gives a sense for how varied the buildings within the neighborhood are. Nearly all building types are present within this area.

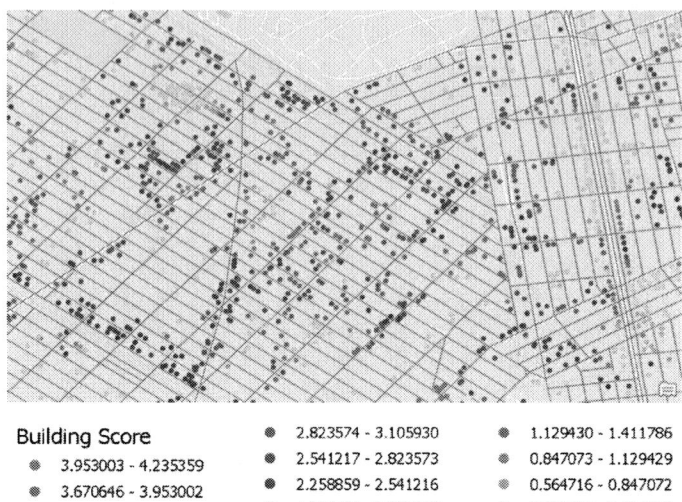

FIGURE 5: BOROUGH PARK BUILDING MAP COLOR CODED BY SCORE

Building scores in Manhattan also demonstrate the preference toward areas with heterogenous groups of buildings. This is visible quantitatively when observing the color differences of nodes in Figure 7. In northern and central Manhattan, where midrise apartments are overwhelmingly prevalent, scores tend to be lower. Scores raise in East Harlem, and in the Financial district, where a greater variety of building types are present. Building scores in midtown Manhattan, where large offices become much more common, raise slightly, but do not match those in more diverse areas.

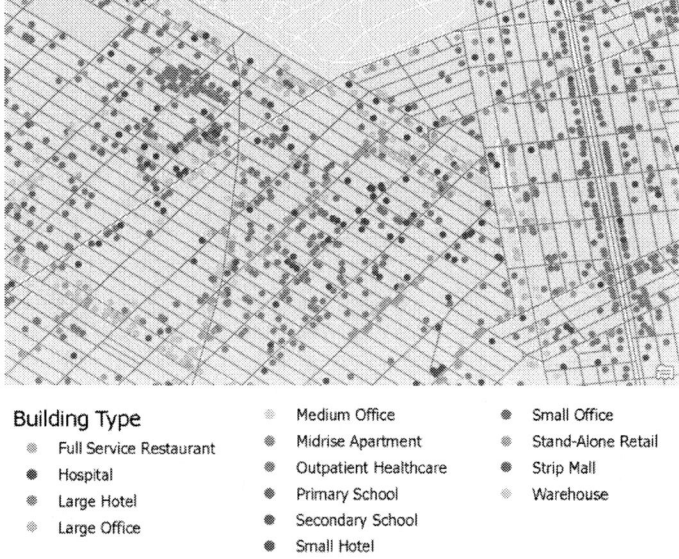

FIGURE 6: BOROUGH PARK BUILDING MAP COLOR CODED BY TYPE

Copyright © 2023 by ASME

In addition, certain building types appear to score generally higher than others. Examples include secondary schools, primary schools, warehouses, and medium offices, as shown in Table 3. For the school buildings, one possible explanation for this is that these buildings can act as 'bridges' between predominantly commercial and predominantly residential zones. While these buildings are constructed to serve residential areas, they also benefit from being in areas with some commercial presence. Medium offices and warehouses may also be constructed to serve particular residential areas, leading to areas with greater building diversity.

FIGURE 7: COMPARISON OF MANHATTAN BY BUILDING SCORE AND BUILDING TYPE

While this work isn't intended to identify universally optimal clusters in any situation, it does provide and demonstrate a process for identifying areas within a given population of buildings that may show a greater potential for clustering. For designers, this can serve as an accessible starting point when evaluating candidates for distributed generation and energy storage systems. Several additional factors can greatly impact the viability of distributed resources in a building cluster, such as available site/roof area for solar panels, interior space for battery systems, shadows present from nearby buildings, physical constraints associated with electrical connections between buildings, and the economics of implementation for specific buildings. A noteworthy contribution for future work would be the consideration of residential buildings in clustering potential. Since the focal point of this work was on commercial buildings, the dataset was truncated to exclude residential buildings.

However, inclusion of these buildings in this analysis could greatly impact the potential of buildings and lead to more effective use of resources. Moreover, while the authors made a concerted effort to accurately map stock building types to building classes found in the dataset, value and accuracy could be added by incorporating other building characteristics into this categorization, such as number of stories, lot area, and building area. The dataset includes extensive characteristics of buildings that could aid in better mapping. This is a particular challenge in urban environments, where multi-use buildings are common and generalized data to represent these buildings is difficult to find. Addition of these factors to better categorize buildings is also an avenue for future investigation.

5. CONCLUSION

The need for distributed energy resources and energy storage systems has been driven by regulatory incentives, increased energy awareness, desire to reduce dependence on public energy grids, and desire to reduce energy costs. This has created motivation to cluster buildings in order to maximize the utility of these systems. However, optimizing the composition of building clusters to make the most efficient use of these resources and storage systems poses a challenge. In this paper, buildings in New York city were scored based on their potential for clustering. By analyzing the trends found in these scores, we are able to make meaningful connections between cluster composition and potential performance benefits. We are also able to observe tendencies of an existing population to determine which circumstances may be conducive to clustering and which may not. These connections can help system designers and building owners make informed decisions about cluster formation based on composition of the buildings involved, depending on their specific goals.

ACKNOWLEDGEMENTS

The authors would like to thank the Stevens School of Systems and Enterprises for support of this work.

REFERENCES

[1] New York City Council, 2019, *Local Law No. 97 of 2019*, New York City Department of Buildings.

[2] Castillo, A., and Gayme, D. F., 2014, "Grid-Scale Energy Storage Applications in Renewable Energy Integration: A Survey," Energy Convers Manag, **87**, pp. 885–894.

[3] "Tesla Powerpack" [Online]. Available: http://bit.ly/3c9ohPg. [Accessed: 07-Feb-2021].

[4] "Johnson Controls L1000 Modular Distributed Energy Storage System" [Online]. Available: http://on.jci.com/2OxmpHW. [Accessed: 02-Jan-2023].

[5] "GE Reservoir Solutions" [Online]. Available: http://invent.ge/3qv0wpS. [Accessed: 02-Jan-2023].

[6] "SunPower Helix Storage" [Online]. Available: https://bit.ly/2PAJudd. [Accessed: 02-Jan-2023].

Copyright © 2023 by ASME

[7] Fares, R. L., and Webber, M. E., 2017, "The Impacts of Storing Solar Energy in the Home to Reduce Reliance on the Utility," Nat Energy, **2**(2), p. 17001.

[8] Odonkor, P., Lewis, K., Wen, J., and Wu, T., 2016, "Adaptive Energy Optimization Toward Net-Zero Energy Building Clusters," Journal of Mechanical Design, Transactions of the ASME, **138**(6), pp. 1–12.

[9] U.S. Department of Energy, 2011, "The Smart Grid: An Introduction," pp. 1–45 [Online]. Available: http://bit.ly/3eiwGmh. [Accessed: 07-Feb-2021].

[10] Xie, C., Wang, D., Lai, C. S., Wu, R., Wu, X., and Lai, L. L., 2021, "Optimal Sizing of Battery Energy Storage System in Smart Microgrid Considering Virtual Energy Storage System and High Photovoltaic Penetration," J Clean Prod, **281**, p. 125308.

[11] Odonkor, P., and Lewis, K., 2019, "Data-Driven Design of Control Strategies for Distributed Energy Systems," Journal of Mechanical Design, **141**(11), pp. 1–10.

[12] Odonkor, P., and Lewis, K., 2019, "Automated Design of Energy Efficient Control Strategies for Building Clusters Using Reinforcement Learning," Journal of Mechanical Design, Transactions of the ASME, **141**(2), pp. 1–9.

[13] Mashayekh, S., Stadler, M., Cardoso, G., and Heleno, M., 2017, "A Mixed Integer Linear Programming Approach for Optimal DER Portfolio, Sizing, and Placement in Multi-Energy Microgrids," Appl Energy, **187**, pp. 154–168.

[14] Novoa, L., Flores, R., and Brouwer, J., 2019, "Optimal Renewable Generation and Battery Storage Sizing and Siting Considering Local Transformer Limits," Appl Energy, **256**.

[15] Gautam, M., Bhusal, N., and Benidris, M., 2021, "A Cooperative Game Theory-Based Approach to Sizing and Siting of Distributed Energy Resources," *2021 North American Power Symposium, NAPS 2021*, Institute of Electrical and Electronics Engineers Inc.

[16] Rajabi, A., Eskandari, M., Ghadi, M. J., Li, L., Zhang, J., and Siano, P., 2020, "A Comparative Study of Clustering Techniques for Electrical Load Pattern Segmentation," Renewable and Sustainable Energy Reviews, **120**, p. 109628.

[17] Kaminski, G., and Odonkor, P., 2021, "Evaluation of Commercial Building Clusters With Energy Storage to Reduce Reliance on Electrical Utility Grids," *Volume 3A: 47th Design Automation Conference (DAC)*, American Society of Mechanical Engineers, pp. 1–10.

[18] Office of Energy Efficiency & Renewable Energy (EERE), "Commercial and Residential Hourly Load Profiles for All TMY3 Locations in the United States (Dataset)" [Online]. Available: http://bit.ly/3enuZ78. [Accessed: 06-Feb-2021].

[19] Field, K., Deru, M., and Studer, D., 2010, "Using DOE Commercial Reference Buildings for Simulation Studies," *SimBuild 2010*, pp. 85–93.

[20] Deru, M., Field, K., Studer, D., Benne, K., Griffith, B., Torcellini, P., Liu, B., Halverson, M., Winiarski, D., Rosenberg, M., Yazdanian, M., Huang, J., and Crawley, D., 2011, "U.S. Department of Energy Commercial Reference Building Models of the National Building Stock," pp. 1–118.

[21] "REopt Web Tool" [Online]. Available: https://reopt.nrel.gov/tool. [Accessed: 30-Apr-2022].

[22] Thompson, J., and Krarti, M., 2021, "Cost-Effectiveness and Resiliency Evaluation of Net-Zero Energy U.S. Residential Communities," ASME Journal of Engineering for Sustainable Buildings and Cities, **2**(3), pp. 1–13.

[23] Anderson, K., Olis, D., Becker, B., Parkhill, L., Laws, N., Li, X., Mishra, S., Kwasnik, T., Jeffery, A., Elgqvist, E., Krah, K., Cutler, D., Zolan, A., Muerdter, N., Eger, R., Walker, A., Hampel, C., Tomberlin, G., and Farthing, A., *The REopt Web Tool User Manual*.

[24] New York City Department of City Planning, "Primary Land Use Tax Lot Output (PLUTO) (Dataset)" [Online]. Available: https://data.cityofnewyork.us/City-Government/Primary-Land-Use-Tax-Lot-Output-PLUTO-/64uk-42ks. [Accessed: 10-Mar-2023].

Copyright © 2023 by ASME

Proceedings of the ASME 2023
International Design Engineering Technical Conferences and
Computers and Information in Engineering Conference
IDETC-CIE2023
August 20-23, 2023, Boston, Massachusetts

DETC2023-116914

MULTI-FIDELITY MODELING FOR DYNAMIC POWER CONTROL AND OPTIMIZATION OF NUCLEAR-RENEWABLE HYBRID ENERGY SYSTEMS

In-Bum Chung
Department of Industrial and Enterprise
Systems Engineering,
University of Illinois,
Urbana, IL 61801, United States.

Pingfeng Wang*
Department of Industrial and Enterprise
Systems Engineering,
University of Illinois,
Urbana, IL 61801, United States.

ABSTRACT

The attempt to utilize renewable energy that are not as stable as carbon-based power has led to hybrid energy systems (HES), where multiple generation sources are combined to supply the target sector. Nuclear-renewable hybrid energy system (N-R HES) is a promising technology that couples nuclear power plant to the renewable energy sources to send the generated power to the grid. Due to the fluctuations in the generation as well as the demand, an industrial process is typically connected to the system to utilize the produced surplus or byproducts. However, it is important to build a control strategy that can manage the power distribution between the grid and the industrial process. This study focuses on creating a multi-fidelity model to predict the appropriate control state for satisfying each demand on the given timescale of the nuclear renewable HES. A high-fidelity model was constructed using Simulink and a forward calculation was used as a low-fidelity model, where the data generated from the high-fidelity model are a nested set of the low-fidelity model's input domain. The multi-fidelity surrogate model was trained using co-kriging method and the surrogate was tested on a synthesized example problem to validate its practicality in controlling a dynamic energy system.

Keywords: Hybrid Energy Systems, Multi-fidelity Modeling, Dynamic Control

NOMENCLATURE

Variables
PV_{Gen}^t Solar PV electricity generation at time step t [MW_e]
E_{Demand}^t Power grid electrical demand at time t [MW_e]
t Time step of the considered operation cycle

Parameters
PHG_{Gen} Nuclear reactor thermal power generation [MW_{th}]

ρ_{TEC} Thermal to electrical power conversion efficiency

Responses
$HTSE_{Th}^t$ HTSE thermal energy usage time step t [MW_{th}]
$HTSE_E^t$ HTSE electrical energy usage at time step t [MW_e]
$HTSE_{state}^t$ HTSE plant state at time step t
$E_{violations}^t$ Power grid demand violation at time t [MW_e]

Performance Measures
$\Sigma HTSE_{state}$ Average sum of HTSE plant control state
$\Sigma E_{violation}$ Average sum of grid demand violations [MW_e]
$N_{violation}$ Number of violated grid demand instances

1. INTRODUCTION

The primary source of energy consumed today heavily relies on fossil fuels such as oil, coal, and natural gas resources for heating homes and fueling vehicles as well as supplying power to industrial facilities. Along with the concerns for global climate change and the depletion of fossil fuel reserves, the advancements in exploiting renewable energy sources has started a transformation in the energy sector. As the dependence on carbon-based energy has led to an unsustainable environment, the attempt to utilize renewable energy sources has increased and extensive studies were conducted to develop environment-friendly production systems with higher efficiency [1, 2]. Typical examples of clean-energy are the wind and solar energy generated using wind turbines and photovoltaic (PV) panels, respectively. However, due to its nature of variability, such renewable energy sources are less reliable when used alone. The effort to accommodate renewables to a practical application resulted in an integrated energy systems (IES) or hybrid energy systems (HES) [3], where each subsection was traditionally regarded independent. Such hybrid energy system is able to provide economic advantage along with environmental benefits by flexibly manipulating different sources of energy, exploiting the strengths of each subsystem [4], and improving system resilience [5–9]. One of the promising

*Corresponding author: pingfeng@illinois.edu

Copyright © 2023 by ASME

technology for the integration of various energy sources is the nuclear-renewable hybrid energy system (N-R HES), which integrates nuclear power plants that can provide clean but also stable generation of power added to a renewable energy source [10]. In addition, instead of connecting the source and destination as a pair, the energy can be distributed to multiple components for flexibility and efficiency. Such hybrid system has proven to be effective at mitigating the fluctuations and uncertainties caused by the renewable energy sources and enhancing the economic value of nuclear energy [11–14].

An N-R HES consists of a nuclear reactor as a primary heat generator (PHG) that outputs steam, which is then fed to a turbine-generator to produce electrical energy. This energy is added to electricity generated by renewable power sources such as wind turbines and solar panels, where an electricity distribution center supplies the demand made by the electric grid. As mentioned before, the renewable sources are intrinsically variable depending on the environmental conditions throughout the production time. On the other hand, nuclear generators needs to be operated at a constant rate, which can not be turned on or off depending on its needs. Hence, it is a common practice to incorporate an appropriate industrial process to the integrated energy system to utilize the excess plant capacity for the periods when a surplus is generated compared to the normal state, e.g., when the electrical demand has decreased or renewable generation has increased. One example of an industrial process connected to N-R HES is a hydrogen production plant using the thermal and electrical energy generated through multiple sources of power. A high temperature steam electrolysis (HTSE) plant can utilize the steam generated from the PHG, where the overall excess electricity is used to split water into hydrogen and oxygen [15, 16]. Kim et al. [17] has given a thorough investigation on N-R HES and HTSE plant, building a high-fidelity model for the hydrogen production considering PID control over the system components to test the resilience under a profile of variable electric load. However, its focus was on the high temperature steam electrolysis instead of the overall power system and energy distribution.

A system of hybrid energy sources can be an effective solution for economic and environmental pressure. However, adding several components to a system is prone to result in a set of complex configurations, which causes difficulties that may rise in terms of design and control. It becomes important to devise an intricate control strategy considering the inter-dependencies between the system components to build a robust design and operation schedule. Wu and Wang investigated various control strategies for dynamic system control [5], comparing traditional proportional-integral-derivative (PID) control, model predictive control (MPC), and adaptive MPC for system resilience. Such controllers were applied to several engineering applications and interdependent infrastructures to show its benefit[18–20]. The difficulty of using such control method lies in the optimization. The controllers either require an offline optimization of its parameters for tuning or an online optimization during the actual execution of the process. Hence, a much cheaper alternative can be of great benefit.

Surrogate modeling offers a cost-efficient alternative of predicting a system behavior. It is employed in circumstances where real or computational simulations are excessively expensive. They replace the high-cost real responses to mathematical models that are easily computed [21]. However, even building a surrogate model requires a number of computationally expensive or time consuming sample points. Such simulations become increasingly demanding as its accuracy of the behavior gains precision. Hence, there were extensive attempts to build an accurate surrogate model using as less evaluations as possible by adaptively adding new sample points in an iterative manner [22–24]. Another approach is building multi-fidelity models that incorporate relatively less expensive but less accurate low-fidelity data to aid the expensive but accurate high-fidelity data [25]. While high-fidelity models that reproduce the actual behavior of the engineering system are obtained through expensive computations, low-fidelity models are acquired by various methods, e.g., linearization, dimensionality reduction, or using simpler physics models, etc. By using less data from a high-fidelity model and many from a low-fidelity model, a multi-fidelity model may generate an approximate of high accuracy with relatively lower cost.

In this paper, a multi-fidelity surrogate modeling technique is utilized for controlling the electric distribution of N-R HES. An empirically tuned PID controller is added to a simulation model that contains the components of N-R HES, which includes the HTSE plant as an industrial process. The objective of this study is to control the generated electrical power and distribute to supply the power grid as well as the HTSE plant in order to maximize the production of hydrogen, while satisfying the grid demand throughout the simulation time. This simulation model was used as a high-fidelity model, while some assumptions were made to remove ancillary components to utilize as a low-fidelity model. The control state was sampled from each model during simulation time step to generate a multi-fidelity surrogate. This surrogate model was then applied back to the given simulation conditions to validate its feasibility and to prove the practicality of incorporating low-fidelity data for dynamic control.

The rest of the paper is organized as follows: section 2 introduces the components that are considered for the N-R HES, section 3 describes the multi-fidelity model in more detail, where the model is validated in section 4 using a synthesized case study. Finally, a conclusion is drawn in section 5 along with plans for the application in future work.

2. COMPONENTS OF N-R HES

There are five primary components considered for the N-R HES. The nuclear reactor serves as a primary heat generator that produces steam which is fed to the turbine generator to generate electricity. The electricity generated through the PV system is combined with the electricity produced through the turbine generator, which is delivered to the power grid. For both thermal and electrical energy, a fraction is distributed to the HTSE plant to produce hydrogen. Figure 1 is a diagram of the configuration that illustrates the described N-R HES. The arrows depict the flow of the energy, where the thermal and electrical flow is indicated by continuous lines and broken lines, respectively.

Copyright © 2023 by ASME

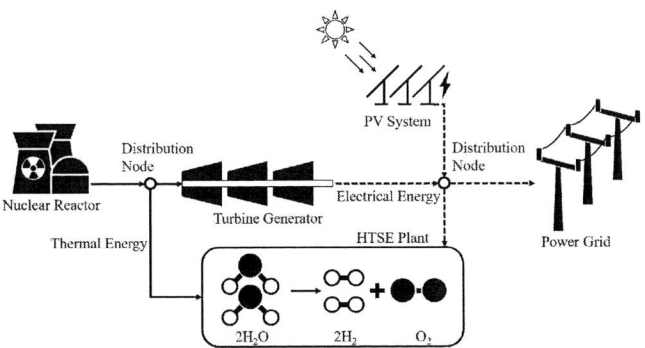

FIGURE 1: CONFIGURATION OF A N-R HES SYSTEM

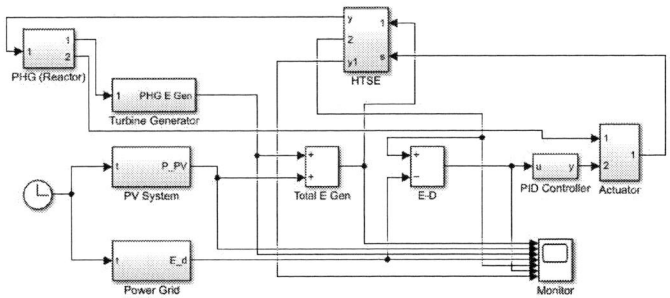

FIGURE 2: SIMULINK MODEL OF THE N-R HES FOR HTSE CONTROL SIMULATION

2.1 Power Generation and Consumption

The nuclear reactor is the primary source of energy of the N-R HES considered for this paper. Hence, most of the thermal and electrical energy used for the turbine generator and the steam electrolysis is produced from the reactor. The reactor is assumed to generate a constant amount of thermal energy, where the turbine converts the input power to electrical energy with an efficiency of 31.8%. The thermal distribution node represents an energy manifold for thermal energy distribution, which controls the amount of thermal energy sent to the HTSE plant and the turbine.

The solar power generation is a fluctuating source of energy, which gradually increases during the day time and decrease after reaching its peak. The given PV system can be replaced by any other renewable source or may be supplemented with another such as wind turbine. For the purposes of this study, a single source of solar PV is used as a representative renewable power generation. Additional power systems can simply add up linearly to the total to the generated renewable energy. The electricity generated through the PV system is then merged with the electricity generated through the turbine at the electrical distribution node, which represents a switch yard for distributing electrical energy to the power grid and the HTSE plant.

A primary destination for the generated energy is the power grid. The electrical demand of the power grid is a variable about the operation time. Coupled with the variability of the power generation, this creates the need for controlling the two distribution nodes in order to meet the electrical demand from the power grid, while sending as much power as possible to the HTSE plant to produce maximum amount of hydrogen.

The main focus of this study is to create a surrogate-based control model for the energy distribution. Hence, the components are simplified to a degree that neglects its inner mechanisms. For example, it is assumed that the power generation is directly proportional to the solar irradiance, the loss between the component connections are ignored, and overheating of components are not considered. The operation ranges of each component and the approach for obtaining low-fidelity data is specified in the following section.

2.2 Simulation Model

The simulation model that operates the N-R HES control process was built using Simulink [26]. Figure 2 shows the Simulink model that consists of the components described in the previous section. The inputs for the model are the time-dependent elements, the power grid electrical demand and the solar PV power generation. These blocks consists of MATLAB functions that approximate the temporal energy profiles, which are explained in more detail in section 4. The reactor generates a constant thermal energy, which is then sent to the turbine to be combined with the PV output. Before entering the turbine, a fraction of the thermal energy is used by the HTSE plant. After the electrical energy from the two sources are combined, a fraction of it is used by the HTSE plant. The rest of the electrical energy (E) is subtracted by the energy demand of the power grid (D), which goes into the PID controller. The PID controller outputs a degree of energy that needs to be increased or decreased according to the output of difference between E and D, which is converted as a status for the HTSE plant operation. For instance, if the demand exceeds the generation, the controller receives a negative error, which requires the reduction in consumption. As the grid demand is input variable, which is not subjected to control, this results in reducing the HTSE consumption by decreasing the HTSE state variable. Here, the status is assigned within the range of zero and one, which corresponds to the minimum and maximum of HTSE plant operation, respectively. The output of the model is time-dependent HTSE status variable that is used for building the multi-fidelity model, which is aimed to replace the PID controller.

2.3 Mathematical Relation between the Components

The interconnections between the five major components, nuclear reactor, turbine generator, solar PV system, power grid, and HTSE plant can be represented by a set of mathematical functions. The total electrical energy generated at time step t can be calculated by adding the electricity generated at the turbine and the PV system. Although the thermal energy generated from the nuclear reactor is assumed to be constant, a fraction of this energy is diverted to the industrial process. Hence, the electrical energy generated by the turbine generator is dependent of the HTSE state, which in turn, is dependent of time. Note that the HTSE plant state is normalized to a range within [0, 1], zero being minimum and one being the maximum hydrogen production, respectively.

$$HTSE_{Th}^{t} = HTSE_{Th}^{t}(HTSE_{state}^{t}(t)) \qquad (1)$$

$$HTSE_{E}^{t} = HTSE_{E}^{t}(HTSE_{state}^{t}(t)) \qquad (2)$$

Copyright © 2023 by ASME

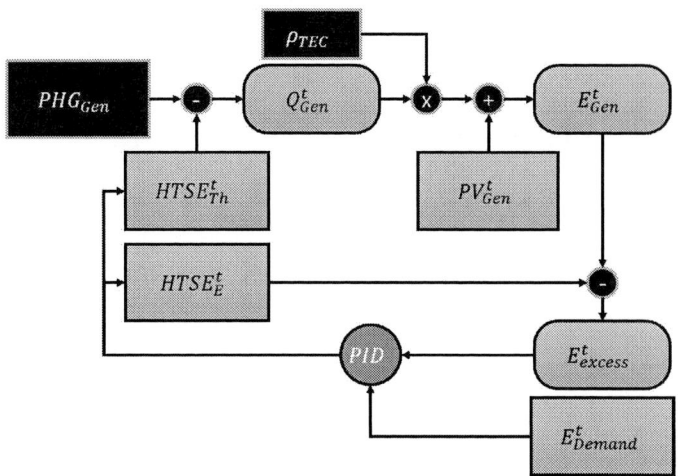

FIGURE 3: DIAGRAM OF ENERGY FLOW CALCULATION

Equations (1) and (2) indicates that the thermal and electrical energy consumed at the HTSE plant are functions of the time-dependent variable $HTSE_{state}^t$.

$$Q_{Gen}^t = PHG_{Gen} - HTSE_{Th}^t \qquad (3)$$

$$E_{Gen}^t = \rho_{TEC} Q_{Gen}^t + PV_{Gen}^t \qquad (4)$$

$$E_{excess}^t = E_{Gen}^t - HTSE_E^t \qquad (5)$$

The aforementioned electrical energy generation is calculated using Eq. (3) and (4). The final destination of the electrical energy is the power grid, where the amount of excess energy left for the grid is calculated using the Eq. (5). The given set of equations are used to calculate a performance measure that is presented in the following section 4.

$$HTSE_{state}^t = PID(E_{excess}^t, E_{Demand}^t) \qquad (6)$$

After the excess electrical energy and the power grid demand is fed to the PID controller, it determines the HTSE plant state for the next time step. As the final stage of the equations affect the initial input of the system, it creates a control loop with the objective of meeting the power grid electrical demand.

Figure 3 is a diagram of the elaborated energy distribution and flow calculation. The black rectangles are fixed parameters of the system, the gray rectangles are the time-dependent variables, and the gray rounded rectangles are the byproducts of the equations used for calculating the next step.

3. MULTI-FIDELITY MODELING FOR ENERGY FLOW CONTROL

The built simulation model is used to generate high-fidelity data for the purposes of this study, while a simple linear calculation of excess energy serves as the low-fidelity model. The data generated from the two models are combined using co-kriging, a conventional method of data fusion.

3.1 Co-Kriging based Multi-fidelity Modeling

Co-kriging is an extension of a popular surrogate modeling technique, kriging. Hence, the basics of the original kriging method is explained briefly before diving deeper in to its use for multi-fidelity modeling. The kriging surrogate model consists of two terms, the global model and the correlation model as shown in Eq. (7), each representing the predicted mean and standard deviation at a point within a design domain [27].

$$\hat{y}(x_{n+1}) = \hat{\mu} + \sum_{i=1}^{n} b_i \Psi_i(x_{n+1}, x_i) \qquad (7)$$

The x depicts a data point, where Ψ is the basis function for calculating the correlations between a new prediction point x_{n+1} and x_i sample points used for building the kriging model. Typically, a Gaussian function is used for the correlation model (basis function), Eq. (8), and the parameters of the kriging model are determined using maximum likelihood estimate (MLE).

$$\Psi_i = corr[Y(x_{n+1}), Y(x_i)] = exp\left(-\sum_{j=1}^{k} \hat{\theta} \|x_{n+1}^j - x_i^j\|^{\hat{p}^j}\right) \qquad (8)$$

The co-kriging model follows the same concept but with using multiple datasets. Here, the number of dataset is limited to two, an expensive high-fidelity data and a cheap low-fidelity data, for simplicity. The idea is to build a model with cheap data, which is multiplied by a scaling factor and add it to a model that estimates the difference between the cheap model and the expensive model as shown in Eq. (9) [28].

$$Z_e(x) = \rho Z_c(x) + Z_d(x) \qquad (9)$$

Each Z is a Gaussian process and ρ is the scaling factor, where subscripts c, e, and d denote cheap, expensive, and difference, respectively. For kriging, the Gaussian process is represented using a correlation function and the parameters are once again determined using MLE, assuming that the cheap and expensive data are independent. Similar to the ordinary kriging, co-kriging also adopts the correlation functions using a covariance matrix to model the relationship between the cheap and expensive sample points as shown in Eq. (10). It can be observed that each entry is a function of the correlation function between the cheap sample points, between the cheap data and expensive data, and between expensive sample points.

$$C = \begin{pmatrix} \sigma_c^2 \Psi_c(X_c, X_c) & \rho\sigma_c^2 \Psi_c(X_c, X_e) \\ \rho\sigma_c^2 \Psi_c(X_e, X_c) & \rho^2\sigma_c^2 \Psi_c(X_e, X_e) + \sigma_d^2\Psi_d(X_e, X_e) \end{pmatrix} \qquad (10)$$

Equation (12) is used for predicting a new high-fidelity sample point through a co-kriging model, where the correlation between the low and high-fidelity data and the new sample point is defined using Eq. (11). Note that the parameters with hat symbol denotes that they are estimates using MLE.

$$c = \begin{pmatrix} \hat{\rho}\hat{\sigma_c^2}\Psi_c(X_c, x_{n+1}) \\ \hat{\rho}\hat{\sigma_c^2}\Psi_c(X_e, x_{n+1}) + \hat{\sigma_d^2}\Psi_d(X_e, x_{n+1}) \end{pmatrix} \qquad (11)$$

$$\hat{y}_e(x_{n_e+1}) = \hat{\mu} + c^T C^{-1}(y - 1\hat{\mu}) \qquad (12)$$

In this paper, kriging and co-kriging methods are implemented based on the MATLAB kriging toolbox [29, 30].

Copyright © 2023 by ASME

TABLE 1: OPERATION RANGE OF EACH COMPONENT

Component	Minimum	Maximum
PV_{Gen}^t	$0 \, [MW_e]$	$28 \, [MW_e]$
E_{Demand}^t	$125 \, [MW_e]$	$160 \, [MW_e]$
$HTSE_{Th}^t$	$7 \, [MW_{th}]$	$15 \, [MW_{th}]$
$HTSE_e^t$	$24 \, [MW_e]$	$54 \, [MW_e]$

3.2 Low-fidelity Model using Component Relations

A low-fidelity model for generating cheap data was obtained through simple calculations from the N-R HES model. The basic assumption was that all excess electrical power after distributing to the power grid, is used for the HTSE plant. Hence, the electrical power is the only factor driving the HTSE state. The steady nuclear power generation is added to the solar PV generation, then the electrical demand is subtracted at each time-step. The resulting output is the excess electrical power. This is scaled to a range of [0, 1] by the minimum and maximum operation range of the HTSE plant to obtain the required HTSE state. For time-steps where the state exceeds the maximum range, it is set to 1. The calculation procedure for the described process is explained in section 2.3 in detail. For the low-fidelity model, all of the terms for HTSE process is excluded, resulting the difference between the electricity generation and grid demand as shown in Eq. (13).

$$E_{residue}^t = PHG_{Gen} * \rho_{TEC} + PV_{Gen}^t - E_{Demand}^t \qquad (13)$$

It should be noted that the thermal energy consumed by the HTSE plant is not being considered for the low-fidelity model. This results in neglecting its affect to the thermal energy going into the turbine generator, which in reality should reduce the electricity being generated at the turbine. Hence, it is shown in the following section that this results in a heavy violation to the power grid electrical demand. However, coupled with the high-fidelity data, as it provides crude but valuable information on the critical regions, it results in a model with higher accuracy.

4. CASE STUDY

In order to test the efficacy of the proposed method, a time-dependent problem is synthesized using an electrical demand and PV generation profile from another HES study [31]. The profiles of the data were scaled to match the size and operation range of each component based on the paper on N-R HES by Kim et al. [17]. This information is organized in Table 1. The nuclear reactor is assumed to generate a constant $600MW_{th}$ thermal energy. Applying the conversion efficiency of 31.8%, this is equivalent to $190.8MW_e$.

For the two time-dependent variables, grid demand and PV generation, 24 time-steps representing an hour of a single day is used to build a 1-D surrogate model, in order to execute a continuous simulation. The data and surrogates are illustrated in Fig. 4. For the electrical demand, a radial basis function network was used to approximate the profile as it contained irregular nonlinearities [32]. This approximation allows more data to be generated on top of the given time-steps when needed. As for the solar PV generation profile, due to its near symmetry and both

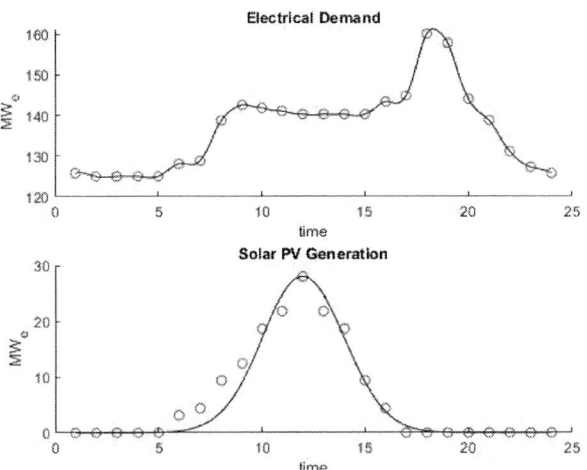

FIGURE 4: POWER GRID ELECTRICAL DEMAND AND SOLAR PV GENERATION PROFILES

ends converging to zero, a simple Gaussian function was used for the approximation.

There are 24 time-steps, each time step representing an hour of a single day. It can be observed that the maximum PV generation occurs around time-step 12 after a gradual increase and falls back to zero at night. The grid electrical demand also decreases gradually as it approaches night time and starts to increase in the morning around time steps 5 to 7. It also contains a critical region where the demand greatly increases at time step 18. The non-linearity of the two functions should be a representative example for a typical day cycle of an energy system. In this paper, as the detailed internal mechanisms of the steam electrolysis is ignored, it is assumed that the thermal and electrical consumption are linearly proportional to the hydrogen generation, which in extent proportional to the HTSE state.

4.1 Performance Measures

For the purpose of this study, two performance measures were used to capture the quantitative behavior of the multi-fidelity model. As the power system have its physical objective for control, instead of using a surrogate accuracy measure such as root-mean-square-error (RMSE), performance measures were selected to incorporate its operational purpose. Under the given generation and demand scenario, it is preferable to control the HTSE plant so that it would generate as much hydrogen as possible. Hence, one measure is the sum of HTSE operation state, which is assumed to be proportional to the production magnitude, which in extent proportional to the energy consumption at the HTSE plant. However, it should only produce to a point where the electrical demand is not violated as this is the primary destination of the generated electricity. Thus, the second measure is the violation of the electrical demand calculated using Eq. (14) given below.

$$E_{violation}^t = min(0, (PHG_{Gen} - HTSE_{Th}^t) * \rho_{TEC} + \\ PV_{Gen}^t - HTSE_E^t - E_{Demand}^t) \qquad (14)$$

Copyright © 2023 by ASME

The superscript denotes the value of each term at time-step i. It should be noted that these measures, sum of states and sum of violation, are not in the same units. Therefore, they should not be added together but only compared individually between different control strategies. Should there be a case where the trade-off is too great to determine which of the control strategy is better, the violation of the electrical demand should be considered a higher priority.

4.2 Multi-fidelity Model Performance Results

The objective of this study is to show the practicality of using a low-fidelity model for a time-dependent control signal to enhance the accuracy of a high-fidelity approximation. Hence, the multi-fidelity model is compared to a surrogate that was built using only the high-fidelity data. For the purposes of this simulation, 24 time-steps of low-fidelity data and 5 randomly selected time-steps of high-fidelity data were used. To avoid a case where the randomly selected high-fidelity samples produce a better multi-fidelity model by chance, an extensive repetition was performed using 5,000 combinations of selected high-fidelity data, and the multi-fidelity model and high-fidelity model were built using the same selection of high-fidelity data for each case. Note that each of the 5,000 combinations were ensured to have a unique subset of high-fidelity data. To prevent confusion of the terminology, multi-fidelity model refers to a co-kriging surrogate model built using both the low-fidelity data and selected high-fidelity data, high-fidelity model refers to a kriging surrogate model built using only the selected high-fidelity data, and the high-fidelity simulation model refers to the Simulink model, which is set as the target response. After the co-kriging and kriging models were built, they were used to predict every 0.05 time-steps between 1 to 24, resulting 461 discrete time-steps, enough to generate a near continuous curve.

An illustrative example on one of the case results is given in Fig. 5. The red circle and blue square points depict the low-fidelity data and high-fidelity data, respectively. The bold blue line is the actual control response from the high-fidelity simulation model, the black continuous line is the multi-fidelity co-kriging model prediction, and the red broken line is the high-fidelity kriging model prediction. It can be observed that even when the important peaks of the simulation response was not selected from the high-fidelity set, the low-fidelity data was able to inform the multi-fidelity model to avoid operating the HTSE plant to an extent where grid demand violations would occur, while the kriging model had no information about the minimum peak at the 18th time step causing a high violation. For the purposes of demonstration, the figure was plotted using only the time domain as a variable. The rest of the results are from surrogates that considers the electrical demand and solar PV generation along with time as variables.

The statistical results are organized in Table 2. The second column and the last column are the results of high-fidelity simulation model and low-fidelity model. As they are calculated using the same data, they are not subjected to any variations through the repetitive evaluation process. Before going over the comparison, it should be noted that although the low-fidelity model has the highest sum of HTSE state, this is at the expense of a

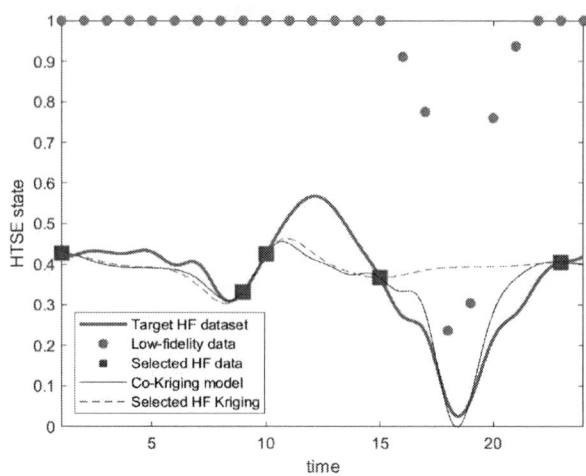

FIGURE 5: SINGLE CASE OF HTSE STATES ABOUT TIME FOR THE ACTUAL RESPONSE AND EACH SURROGATE MODEL

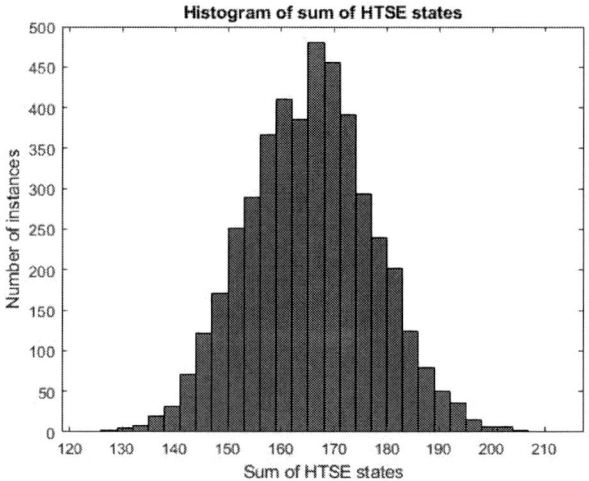

FIGURE 6: HISTOGRAM OF CO-KRIGING SUM OF STATES FOR 4520 CASES WITHOUT VIOLATION

enormous amount of grid demand violation. The rest of the results hold a similar result in terms of the HTSE plant state, where the kriging has slightly higher hydrogen production. However, it suffers greatly in terms of grid demand violation compared to co-kriging. The average violation is approximately seven times higher and more than half of the repetition consists of violations, while the co-kriging model has violation in less than 10% of cases.

Figure 6 shows a histogram of the co-kriging prediction for cases that had no grid demand violations. It can be observed that the majority of results is distributed around the target sum of states of the high-fidelity simulation. The reason for some deviation is that the PID controller was tuned empirically rather than optimized using an algorithm. Further plans on improving the simulation can be found in section 5.

The degree and instances of grid demand violation becomes more clear by observing the two histograms given in Fig. 7 and

Copyright © 2023 by ASME

TABLE 2: COMPARISON BETWEEN SIMULATION MODELS, CO-KRIGING MODEL, AND KRIGING MODEL

Measure	High-fidelity Simulation	Co-Kriging	Kriging	Low-fidelity Model
$\Sigma HTSE_{state}$	167.29	167.92	170.63	420.27
$\Sigma E_{violation}$	0	18.31	127.50	560.17
$N_{violation}$	-	480 / 5000	3249 / 5000	-

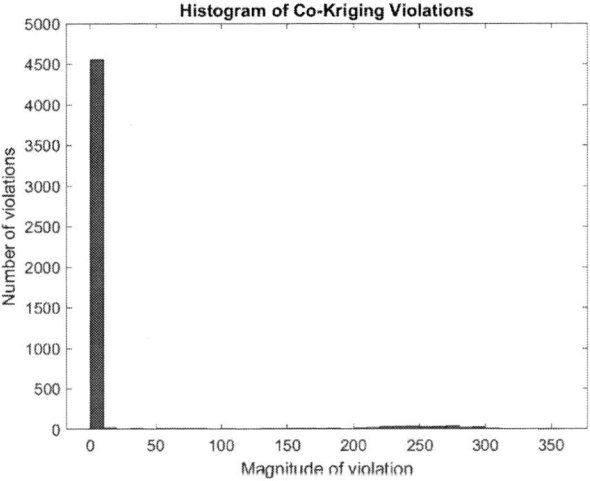

FIGURE 7: HISTOGRAM OF VIOLATIONS FOR THE 5000 CASES USING CO-KRIGING MODEL

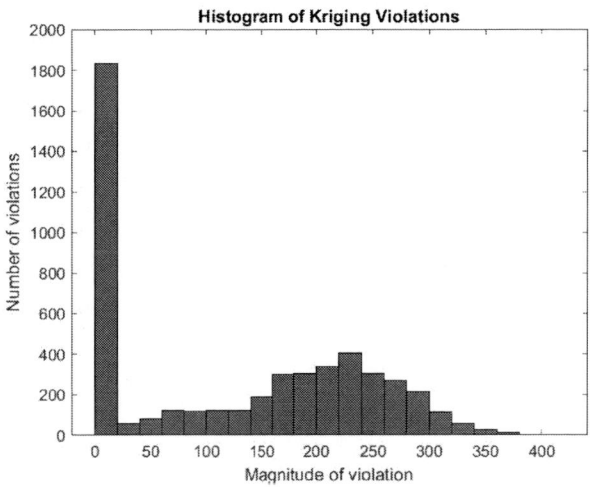

FIGURE 8: HISTOGRAM OF VIOLATIONS FOR THE 5000 CASES USING KRIGING MODEL

Fig. 8. The histogram shows the number of violations in the y-axis against a range of the violation magnitude in the x-axis. It is apparent that the kriging model using only the high-fidelity subset of data performs poorly compared to the co-kriging multi-fidelity model. For the co-kriging multi-fidelity model, more than 90% of the cases satisfied all grid demands, which is why almost all data are concentrated on the left. For the kriging model prediction, more than half of the cases are violated, where most of the violation degree are within the range 200 and 250 MW_e. This is due to the electrical demand surging at time-step 18, where the kriging surrogate was not able to predict the appropriate control state at this critical region when the selected high-fidelity data was not at its vicinity, whereas the co-kriging model was able to obtain this information through the low-fidelity data even with the absense of the high-fidelity data points.

For the co-kriging, although only in a small fraction of cases, it seemed that there were a instances that when the grid demand is violated, it does to a high degree. Some of the instances were investigated to find the cause of this anomaly. It was shown that for such cases, the random high-fidelity data selection were clustered into a narrow range of design domain, causing the overall model to be skewed to a point where the low-fidelity data could not induce a positive influence. Among the 5000 cases, the worst case was extracted and presented in Fig. 9. As it can be observed, although a couple of the selected high-fidelity data are positioned in the far ends to give a appropriate range estimate, the rest of the points are grouped in a narrow range of time that is farther away from the lower peaks. Although the low-fidelity model has

useful information in the vicinity of the minimum peak, because the high-fidelity sample points are especially sparse around this region, the correlation function between the high and low fidelity data was low (calculated using Eq. (10)), and consequently the low-fidelity information was merged only by a insufficient degree, resulting an inaccurate model similar to the kriging model at the critical region.

Based on the case study results, the co-kriging multi-fidelity model for control state estimation has proved its usefulness by incorporating low-fidelity data to enhance its accuracy, avoiding grid demand fulfillment failure by a large difference compared to the kriging model.

4.3 Future-work for Performance Enhancement

From the analysis for the violated cases, the importance of the locations of the high-fidelity data was confirmed. Hence, a method for acquiring an appropriate location within the design domain, in this case the time-step within the operation period, needs to be established in order to increase the efficiency and the reliability of the multi-fidelity model. One of the most popular method of adaptive sampling is efficient global optimization (EGO) using the expected improvement (EI) criterion that can be calculated utilizing the uncertainty prediction of kriging models [33]. Co-kriging, as an extension from kriging, also has the intrinsic feature of predicting the magnitude of uncertainty in the variable domain. Utilizing this information allows sampling additional points in an adaptive manner. It is a common practice to execute such technique, i.e., sequential sampling or adaptive

Copyright © 2023 by ASME

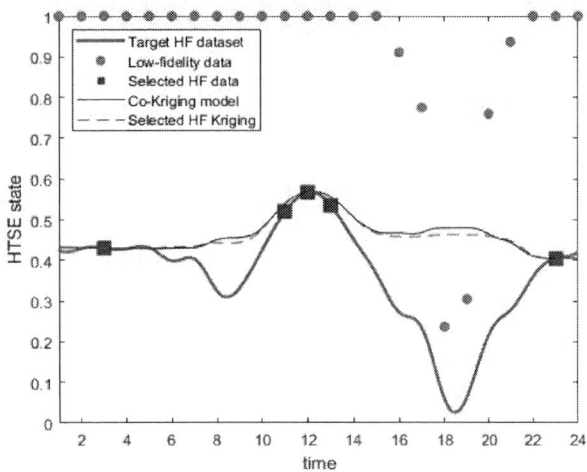

FIGURE 9: CASE OF HTSE STATES ABOUT TIME WITH THE WORST SUM OF GRID DEMAND VIOLATIONS

sampling, for gaining information using as few resources as possible.

The scope of this paper was to use a surrogate to approximate a control strategy for a dynamic power system using a multi-fidelity model. Other than the apparent time variable, component operation profiles were considered as design variables as well. Taking such time-dependent profiles into consideration lets the model to predict the control state considering more information. However, for a single case study with fixed nuclear reactor size, generation profile, and demand profile, using only time as the variable performs slightly better for the prediction. This is due to the dimensionality, where given the same number of sample points, lower dimensional problems are easier to approximate to higher accuracy. In addition, the current surrogate is validated to the same application to approximate the tuned controller. Although this can be an efficient alternative for contorl strategies with online optimizations such as MPC, it does not provide additional benefits for an already fine tuned offline control strategy. Hence, the next step is to gather more realistic cases to test the expanded practicality of a multi-fidelity control surrogate. With enough time and operation profile data coupled with various reactor sizes, the goal is to predict the control state of a new power system with a similar configuration based on the energy profiles without additional controller tuning.

Finally, to test the efficacy of a multi-fidelity model, this paper used a Simulink simulation model as the high-fidelity model. This method is useful for attempting various approaches as the assumed high-fidelity model is still a fast-computing model. However, in practical engineering applications, a typical high-fidelity models are real experiments or time consuming computational simulations. Hence, the final objective of this research is to build an experiment setup that can generate real-time high-fidelity data. To create such a test environment for an application as large as a power system, hardware-in-the-loop (HIL) technique, which utilize not only expensive real-time computational simulation but also a hardware connected to the interface is required. HIL setup allows hardware and controllers to be linked with real-time simulators to conduct experiments with reduced costs compared to fully hardware-based experiments [34]. Implementing a real-time simulator coupled with hardware set up that uses an actuator valve that works as a thermal energy distributor, and a heat exchanger that represents the industrial process, should allow generating a high-fidelity data with much higher accuracy. For such cases, the current simulation model can act as the low-fidelity model for building a multi-fidelity model. Along with an accurate simulation model, a more sophisticated control strategy needs to be implemented as the objective of the study is to approximate and replace the control strategy more computational efficiency. Such examples may include PID controllers with optimized gains or MPC controllers.

5. CONCLUSION

In this paper, replacing a conventional control strategy using a co-kriging multi-fidelity model for a N-R HES was attempted. The hybrid energy system consisted of a nuclear reactor that generates a constant thermal energy, a Rankine-cycle turbine process, a solar photovoltaic system that generates fluctuating electrical energy, a power grid with a time-dependent electrical demand, and a HTSE plant connected to the system as an industrial process. The objective was to control the flow of the energy in order to meet the grid demand while operating the industrial process as much as possible. The high-fidelity data was generated using a Simulink simulation model, while the low-fidelity data was generated using a simple forward calculation that neglected any feedback operation. Using co-kriging method, the time-dependent operations were used as variables along with time-steps, and the response was set as the state of HTSE at the given time-step. The proposed method of control was applied to a synthesized problem of hybrid energy system to validate its practicality.

The results showed that the multi-fidelity model could approximate the control strategy to a comparable degree, where the use of low-fidelity model aided greatly regardless of its simplicity. Based on this study, a future research is planned to include adaptive sampling scheme for better efficiency as well as performing actual real-time experiment that would function as an expensive model, while the simulation model serve as the generator for cheap data, in order to approximate actual hardware control for an N-R HES experiment setup.

ACKNOWLEDGMENTS

This research is partially supported by the National Science Foundation through the Engineering Research Center for Power Optimization of Electro-Thermal Systems (POETS) with cooperative agreement EEC-1449548, and by the U.S. Department of Energy's Office of Nuclear Energy under Award No. DE-NE0008899.

REFERENCES

[1] Twidell, John. *Renewable energy resources*. Routledge (2021).

[2] Vakulchuk, Roman, Overland, Indra and Scholten, Daniel. "Renewable energy and geopolitics: A review." *Renew-*

Copyright © 2023 by ASME

able and Sustainable Energy Reviews Vol. 122 (2020): p. 109547.

[3] Garcia, Humberto E, Chen, Jun, Kim, Jong Suk, McKellar, Michael George, Deason, Wesley R, Bragg-Sitton, Shannon M and Boardman, Richard D. "Nuclear hybrid energy systems regional studies: West texas & northeastern arizona." Technical report no. Idaho National Lab.(INL), Idaho Falls, ID (United States). 2015.

[4] Forsberg, Charles and Aumeier, Steven. "Nuclear-Renewable Hybrid System Economic Basis for Electricity, Fuel, and Hydrogen." Technical report no. Idaho National Lab.(INL), Idaho Falls, ID (United States). 2014.

[5] Wu, Jiaxin and Wang, Pingfeng. "A comparison of control strategies for disruption management in engineering design for resilience." *ASCE-ASME Journal of Risk and Uncertainty in Engineering Systems, Part B: Mechanical Engineering* Vol. 5 No. 2 (2019): p. 020902.

[6] Yodo, Nita, Wang, Pingfeng and Rafi, Melvin. "Enabling resilience of complex engineered systems using control theory." *IEEE Transactions on Reliability* Vol. 67 No. 1 (2017): pp. 53–65.

[7] Wu, Jiaxin and Wang, Pingfeng. "Post-disruption performance recovery to enhance resilience of interconnected network systems." *Sustainable and Resilient Infrastructure* Vol. 6 No. 1-2 (2021): pp. 107–123.

[8] Wu, Jiaxin, Chen, Xin, Badakhshan, Sobhan, Zhang, Jie and Wang, Pingfeng. "Spectral graph clustering for intentional islanding operations in resilient hybrid energy systems." *IEEE Transactions on Industrial Informatics* (2022).

[9] Wu, Jiaxin and Wang, Pingfeng. "Generative Design for Resilience of Interdependent Network Systems." *Journal of Mechanical Design* Vol. 145 No. 3 (2023): p. 031705.

[10] Adler, David B, Jha, Akshaya and Severnini, Edson. "Considering the nuclear option: Hidden benefits and social costs of nuclear power in the US since 1970." *Resource and Energy Economics* Vol. 59 (2020): p. 101127.

[11] Redfoot, Emma K., Verner, Kelley M. and Borrelli, R.A. "Applying analytic hierarchy process to industrial process design in a Nuclear Renewable Hybrid Energy System." *Progress in Nuclear Energy* Vol. 145 (2022): p. 104083.

[12] Arefin, Md Arman, Islam, Mohammad Towhidul, Rashid, Fazlur, Mostakim, Khodadad, Masuk, Nahid Imtiaz and Islam, Md Hasan Ibna. "A Comprehensive Review of Nuclear-Renewable Hybrid Energy Systems: Status, Operation, Configuration, Benefit, and Feasibility." *Frontiers in Sustainable Cities* Vol. 3 (2021): p. 723910.

[13] Suman, Siddharth. "Hybrid nuclear-renewable energy systems: A review." *Journal of Cleaner Production* Vol. 181 (2018): pp. 166–177.

[14] Bragg-Sitton, Shannon M, Boardman, Richard, Rabiti, Cristian and O'Brien, James. "Reimagining future energy systems: Overview of the US program to maximize energy utilization via integrated nuclear-renewable energy systems." *International Journal of Energy Research* Vol. 44 No. 10 (2020): pp. 8156–8169.

[15] Bragg-Sitton, Shannon M, Boardman, Richard D, Cherry, Robert S, Deason, Wesley R and McKellar, Michael G. "An analysis of methanol and hydrogen production via high-temperature electrolysis using the sodium cooled advanced fast reactor." Technical report no. Idaho National Lab.(INL), Idaho Falls, ID (United States). 2014.

[16] O'brien, JE, McKellar, MG, Harvego, EA and Stoots, CM. "High-temperature electrolysis for large-scale hydrogen and syngas production from nuclear energy–summary of system simulation and economic analyses." *International journal of hydrogen energy* Vol. 35 No. 10 (2010): pp. 4808–4819.

[17] Kim, Jong Suk, Boardman, Richard D and Bragg-Sitton, Shannon M. "Dynamic performance analysis of a high-temperature steam electrolysis plant integrated within nuclear-renewable hybrid energy systems." *Applied energy* Vol. 228 (2018): pp. 2090–2110.

[18] Yodo, Nita, Wang, Pingfeng and Rafi, Melvin. "Enabling resilience of complex engineered systems using control theory." *IEEE Transactions on Reliability* Vol. 67 No. 1 (2017): pp. 53–65.

[19] Liu, Xing, Ferrario, Elisa and Zio, Enrico. "Resilience analysis framework for interconnected critical infrastructures." *ASCE-ASME J Risk and Uncert in Engrg Sys Part B Mech Engrg* Vol. 3 No. 2 (2017): p. 021001.

[20] Kozák, Štefan. "From PID to MPC: Control engineering methods development and applications." *2016 cybernetics & informatics (K&I)*: pp. 1–7. 2016. IEEE.

[21] Queipo, Nestor V, Haftka, Raphael T, Shyy, Wei, Goel, Tushar, Vaidyanathan, Rajkumar and Tucker, P Kevin. "Surrogate-based analysis and optimization." *Progress in aerospace sciences* Vol. 41 No. 1 (2005): pp. 1–28.

[22] Bhosekar, Atharv and Ierapetritou, Marianthi. "Advances in surrogate based modeling, feasibility analysis, and optimization: A review." *Computers & Chemical Engineering* Vol. 108 (2018): pp. 250–267.

[23] Chung, In-Bum, Park, Dohyun and Choi, Dong-Hoon. "Surrogate-based global optimization using an adaptive switching infill sampling criterion for expensive black-box functions." *Structural and Multidisciplinary Optimization* Vol. 57 (2018): pp. 1443–1459.

[24] Zhang, Yiming, Kim, Nam H. and Haftka, Raphael T. "General-Surrogate Adaptive Sampling Using Interquartile Range for Design Space Exploration." *Journal of Mechanical Design* Vol. 142 No. 5 (2019): p. 051402.

[25] Fernández-Godino, M. Giselle, Park, Chanyoung, Kim, Nam H. and Haftka, Raphael T. "Issues in Deciding Whether to Use Multifidelity Surrogates." *AIAA Journal* Vol. 57 No. 5 (2019): pp. 2039–2054.

[26] Documentation, Simulink. "Simulation and Model-Based Design." (2020). URL https://www.mathworks.com/products/simulink.html.

[27] Forrester, Alexander IJ, Keane, Andy J and Bressloff, Neil W. "Design and analysis of" Noisy" computer experiments." *AIAA journal* Vol. 44 No. 10 (2006): pp. 2331–2339.

[28] Forrester, Alexander IJ, Sóbester, András and Keane, Andy J. "Multi-fidelity optimization via surrogate modelling." *Proceedings of the royal society a: mathemati-*

Copyright © 2023 by ASME

cal, physical and engineering sciences Vol. 463 No. 2088 (2007): pp. 3251–3269.

[29] Lophaven, Søren N, Nielsen, Hans Bruun and Søndergaard, Jacob. "AMatlab Kriging Toolbox." Technical report no. Technical University of Denmark, Kongens Lyngby, Technical Report No. IMM-TR-2002-12. 2002.

[30] Kitson, Ryan and Cesnik, Carlos E. "High Speed Vehicle Fluid-Structure-Jet Interaction Analysis and Modeling." *58th AIAA/ASCE/AHS/ASC Structures, Structural Dynamics, and Materials Conference*: p. 0405. 2017.

[31] Li, Dongze, Wu, Jiaxin, Zhang, Jie and Wang, Pingfeng. "Co-design optimization of a combined heat and power hybrid energy system." *International Design Engineering Technical Conferences and Computers and Information in Engineering Conference*, Vol. 85383: p. V03AT03A028. 2021. American Society of Mechanical Engineers.

[32] Orr, Mark JL et al. "Introduction to radial basis function networks." (1996).

[33] Jones, Donald R, Schonlau, Matthias and Welch, William J. "Efficient global optimization of expensive black-box functions." *Journal of Global optimization* Vol. 13 No. 4 (1998): p. 455.

[34] Gabbar, Hossam A and Esteves, Otavio Lopes Alves. "Real-Time Simulation of a Small Modular Reactor in-the-Loop within Nuclear-Renewable Hybrid Energy Systems." *Energies* Vol. 15 No. 18 (2022): p. 6588.

[35] Wu, Jiaxin and Wang, Pingfeng. "Risk-averse optimization for resilience enhancement of complex engineering systems under uncertainties." *Reliability Engineering & System Safety* Vol. 215 (2021): p. 107836.

[36] Yodo, Nita and Wang, Pingfeng. "A control-guided failure restoration framework for the design of resilient engineering systems." *Reliability Engineering & System Safety* Vol. 178 (2018): pp. 179–190.

Copyright © 2023 by ASME

**Proceedings of the ASME 2023
International Design Engineering Technical Conferences and
Computers and Information in Engineering Conference
IDETC-CIE2023
August 20-23, 2023, Boston, Massachusetts**

DETC2023-117113

MULTIPHYSICS-INFORMED MACHINE LEARNING FOR BATTERY DESIGN AND HEALTH MONITORING

Parth Bansal[1], Yumeng Li[1,*]

[1]University of Illinois at Urbana-Champaign, Urbana, IL

ABSTRACT

Current Lithium-ion battery (LIBs) designs are nearing the end of their performance capabilities. As the application and demand on these LIBs are growing continuously, there is also a need for continuous innovation both in the area of battery design and the area of battery state of health (SoH) monitoring. Using a silicon (Si) anode instead of a traditionally used graphite electrode allows an increase in the performance of existing LIBs. However, this increased performance comes at the cost of large stresses that develop in the anode as a result of large volumetric changes due to the unique alloying mechanism of the Lithium (Li) into the Si during the battery cycling. These volumetric stresses cause the development of cracks within the Si anode and delamination of the anode from the substrate, which leads to the capacity loss in the battery along with the growth of solid-electrolyte interface (SEI) on the exposed surfaces of the anode. In this study, we develop a physics-informed machine learning (PIML) model to monitor the SoH of the battery based on the output of these coupled failure modes of Si. 3D finite element (FE) models are built to explore how the large volumetric changes cause cracking and delamination in Si and how the SEI growth on the resultant exposed interfaces leads to the degradation of the battery. The outputs of these FE simulations will finally be used to train multiple Gaussian process regression (GPR) models, which combine together to estimate the SoH of the battery under the interactive effects of the studied failure modes. From the results of the study, it is evident that the developed PIML model which uses the output of LIB FE models, which include coupled failure modes, can be used to efficiently (using only 11.4% of charging time) estimate the SoH of the battery with reasonable accuracy (3.04% error).

Keywords: Li Ion Battery, Si Anode, Battery State of Health, Surrogate Model, SEI Growth

---

*Corresponding author: yumengl@illinois.edu

NOMENCLATURE

f	Stress vector [N m^{-2}]
d	Damage variable [-]
k	Adhesive stiffness vector [N m^{-3}]
u	Displacement jump vector [m]
u_m	Norm of u [m]
F^{-1}	Damage evolution function [-]
J	Transportation flux [mol m^{-2} s^{-1}]
D	Diffusion coefficient [cm^2 s^{-1}]
∇c	Concentration gradient [mol m-1]
t	Time [s]
I_n	Electric current density [A m^{-2}]
F_c	Faraday constant [C mol^{-1}]
I_0	Exchange current density [A m^{-2}]
α_a	Anodic charge transfer coefficient [-]
α_c	Cathodic charge transfer coefficient [-]
z	Number of electrons involved in electrode reaction [-]
η	Overpotential [V]
T	Temperature [K]
R	Universal gas constant [J K^{-1} mol^{-1}]
I_{SEI}	Parasitic current due to SEI [A m^{-2}]
Q_{SEI}	SEI related capacity loss [mAH]
$G_k(x)$	Prediction result of performance function at point x [-]
$T(x)$	Trend function [-]
$S(x)$	Gaussian stochastic process [-]
σ^2	Variance [-]
β	Regression coefficient [-]
f	Regression basis function [-]
x_i, x_j	Input points [-]
$R(x_i, x_j)$	Correlation function matrix [-]
a_p, b_p	Kriging parameters [-]
X	Input data set [-]
Y	Input performance [-]
F	Regression basis function [-]
x'	New sample point [-]
r	Correlation vector between x' and X [-]
e	Mean square error [-]

Copyright © 2023 by ASME

1. INTRODUCTION

Rechargeable Lithium-ion batteries are widely used in a variety of applications for energy storage such as mobile phones, portable PCs, electric drones, electric cars etc. These Li-ion batteries are preferred over other available battery technologies due to their high power densities and good life cycle performances [1, 2]. However, there is a constant need for innovation in the LIB technology since the traditionally made LIBs, which are made up of graphite anodes and lithium oxide/phosphate cathodes, are nearing their performance limits. The need for higher performance, better battery life, higher energy density and better safety is driving this need for innovation [3, 4].

Several different innovations have been proposed towards meeting these goals. One of such efforts, that focuses on improving the lithium storage capacity of battery is to use materials such as silicon as the anode material. Si anode material helps in increasing the specific capacity of the battery due to the difference in the intercalation mechanism of the Si for Li-ion storage, i.e. Li-ion diffusion into the interstitial sites of the host lattice . In this mechanism Si atoms react with lithium, which leads to the bonds between the Si atoms giving way for the formation of Li_xSi alloy [5]. Since there are no constraints from the atomic framework of the host material, Si is able to store much more Li as compared to other electrodes [6] leading to dramatically increased specific capacities [7]. However, this substitution of the anode material comes with its own set of issues. The lithium storage mechanism for Si anode can lead to the formation of large internal stresses within the anode as a result of the substantial increase in the volume of Si during the lithiation/delithiation cycles [6]. All of this leads to mechanical-induced failure modes in the battery such like cracking and delamination during the charging/discharging cycles , which cause the overall capacity degradation of the battery through the coupled mechanical-electromechanical process.

The growth of the solid-electrolyte interface on the surface of the anode, that is exposed to the electrolyte, is another major capacity fade mechanism in LIBs. This SEI growth causes capacity fade in the battery by removing the active material from the cycling process along with increasing the resistance of the cell [8–11]. These side reactions mainly occur during the cycling phase when the SEI grows on the interface between the anode and the electrolyte. While this layer starts as a protective barrier between the two phases, by allowing the transfer of Li while maintaining a physical barrier, it ultimately leads to lower power and energy capacity of the battery because its growth results in the increased resistance of the battery and the removal of active Li from the cycling environment [12–15]. Therefore, there is a need for the study of the coupled failure modes to understand the degradation mechanism of the batteries, and further diagnose the SoH of the batteries and the remaining battery capacity as well as useful life.

To this effect, some numerical simulation work has been done on understanding the failure mechanisms in silicon anode. Finite element models have been utilized to understand the lithiation/delithiation process and fundamental failure modes [16–19]. From the standpoint of volumetric changes induced cracking, it is seen that the thickness of anode greatly effects the cracking pattern, wherein thinner anodes are plagued by wiggly cracks while thicker anodes experience straight cracks [20]. Zheng et. al. [21] used a multiphysics based finite element model to investigate the impact of Si layer delamination and resulted capacity degradation in Si anodes. The issue with most of these studies is that they consider relatively simple anode structures which are far away from the actual cases. Also, these studies are lacking in the consideration of the coupled mechanics and electrochemistry on the capacity degradation mechanism of the battery.

Different model-based approaches have also been utilized to tackle the LIBs. Machine learning techniques are widely utilized for development and failure/reliability analysis of systems [22–30]. Applications in the area of LIB such as the battery fault diagnosis, prognostics and health monitoring are also studied [31–34]. A novel methodology for the SoH estimation using ensemble learning models, which integrates multiscale logic regression (LR) and GPR, was proposed by [35]. This design scheme captures the time-varying degradation behavior and reduces the effects of local regeneration phenomenon in LIBs. Although, the ML techniques are easier to implement and are less complex as compared to physics-based models, their accuracy depends on the amount of historically available data and the complexity of the system itself. In addition, the performance of these pure data driven methods is difficult to improve as they are not designed to reveal the underlying physics [31].

Hence, in this study, finite element modeling will be used along with Gaussian process regression models to develop a physics-informed machine learning model. Through the inclusion of the coupled failure modes in the FE study, the proposed model can be used to perform a more robust battery SoH monitoring. First, two FE models are developed to conduct the multiphysics simulations for investigating the major capacity degradation mechanisms: one to study the development of cracks and delamination due to volumetric changes (solid mechanics model) and the other to simulate the cycling of the battery along with the growth of the SEI on the exposed anode surfaces (electrochemical model). Various operating conditions will be utilized for the multiphysics simulation. A synthetic microstructure is also used for the solid mechanics model to find the cracking pattern inside the Si anode. The use of these models allows for the estimation of the battery performance and the battery degradation. The data output from the FE models will be used to develop physics-informed machine learning model, using multiple GPRs, over designated operation conditions by generating SEI layer thickness based charging voltage curves. The developed PIML approach is validated by using a partial charging voltage segment along with its charging temperature through a similarity analysis. The proposed method is a much more efficient and less time consuming method for the SoH estimation of Si anode based LIBs.

2. METHODOLOGY

In this section, the physics-informed machine learning method, using FE models, for battery SoH monitoring is introduced. Section 2.1 details the the multiphysics FE models and their implementation in COMSOL, while section 2.2 describes the physics-informed machine learning technique that is used in this paper.

Copyright © 2023 by ASME

2.1 Multi-Physics Model for Battery Degradation

3D FE modes are developed to simulate the effects of cracking, delamination and the growth of SEI layer on the capacity degradation of the LIB. These three major failure modes are fully coupled during the lithiation and delithiation processes of the battery. However, to reduce the computational costs of physics-based simulation, the cracking and delamination are decoupled with the SEI growth in our models. The first FE model is a solid mechanics model which is used to simulate the cracking and delamination of the Si anode, that has been placed on the Ni current collector substrate, as it undergoes the massive volumetric changes due to the lithiation and delithiation during the charging and the discharging cycles of the battery. The second FE model, takes the input from this first model to calculate the capacity degradation of the LIB due to the seepage of the electrolyte in the cracks and delaminated interfaces seen in the first model. The rest of this sub-section introduces the governing equations for both the FE models which is followed by the description of the implementation of the FE model in COMSOL multi-physics software (Version 6.0).

To investigate the cracking and delamination, a polycrystal FE model is developed to consider the microstructure of Si. Cohesive zone model (CZM) and theory are used to simulate the adhesive/decohesive interactions along the Si grain boundaries and the Si/Ni interface. As previously mentioned, there is a large volumetric change that is seen in the Si anode as a result of the charging/discharging cycles. This model simulates this volumetric change through the use of the state of charge variable (SoC). This variable has a value of 0 when the battery is fully discharges and a value of 1 when the battery is fully charged. The initial strain at each charging/discharging step is proportional to this SoC variable.

Equation 1 gives the stress vector, f, for the CZM that is used to simulate the cracking and delamination:

$$f = (1 - d)ku \qquad (1)$$

where, d, k and u are the damage variable, adhesive stiffness and the displacement jump vector respectively. The interface contact is observed to be broken (complete loss of contact) when the damage variable has a value of 1. The damage variable results in the softening behavior of the interface due to the growth of d during the crack opening. The damage, d, is defined in terms of the norm of the displacement jump vector u_m (Equation 2), since a displacement-based damage model is used for this simulation:

$$d = \begin{cases} 0, & \text{if } u_{m,max} < u_{0m}. \\ min[F^{-1}(u_{m,max}), 1], & \text{else.} \end{cases} \qquad (2)$$

where, F^{-1} is the damage evolution function, $u_{m,max}$ is the max value of u_m over the loading history and is used to keep track of the current state of damage and finally u_{0m} denotes the onset of damage.

The Si island size and shape are determined through the mechanics model, which is used as the input in the electrochemistry model.. The growth of solid electrolyte interface (SEI) on the surfaces of anode, that are exposed to electrolyte, in

the charging-discharging cycles is simulated through the electrochemistry model using the equations that are described below. The transportation flux of Li-ions within the electrode material is calculated through Fick's law (Equation 3):

$$J = -D\nabla c \qquad (3)$$

where, J, D and ∇c is the transportation flux, diffusion coefficient, and the concentration gradient of Li ions respectively. The combination of this flux equation with mass conservation equation (Equation 4) results in Equation 5:

$$\partial c/\partial t + \nabla J = 0 \qquad (4)$$

$$\partial c/\partial t - D\nabla(\nabla c) = 0 \qquad (5)$$

The boundary condition of Equation 5 at the electrode-electrolyte interface is controlled by the electric current density I_n:

$$J.n = I_n/F_c \qquad (6)$$

where, n is the normal vector to the electrode–electrolyte surface and F_c is the Faraday's constant. Butler-Volmer kinetics can be used to determine the electric current density I:

$$I = I_0[exp(\alpha_a z F_c \eta/RT) - exp(\alpha_c z F_c \eta/RT)] \qquad (7)$$

where, I_0 is exchange current density, α_a and α_c are the anodic and cathodic charge transfer coefficient respectively, z is the number of electrons involved in the electrode reaction, η is the overpotential, T is temperature and R is the universal gas constant.

The parasitic current on the surface of the anode as a result of the SEI formation, I_{SEI} reaction is obtained using Equation 8:

$$I_{SEI} = -(1 + HK_{crd})\frac{J.I}{exp(\frac{\alpha\eta_{SEI}F_c}{RT}) + \frac{Q_{SEI}F_cJ}{I}} \qquad (8)$$

where, H, J, α, and f are the operating temperature dependent fitting parameters related to the volumetric changes in the anode particles [36, 37]. The formation of the SEI layer results in the capacity loss Q_{SEI} in the LIB which is related to the parasitic SEI current through Equation 9:

$$\frac{dQ_{SEI}}{dt} = -I_{SEI} \qquad (9)$$

These multi-physics models are solved in 3D continuum space using COMSOL Multiphysics software. The mechanics model is used to simulate the cracking and delamination of the Si anode using the solid mechanics module with embedded cohesive zone elements. Fig. 1 shows this FE model with the Si anode (blue) on top and the Ni current collector substrate (yellow) on the bottom. The Si anode is made up of a synthetic microstructure that is artificially generated. This anode comprises of the different grains and the grain boundaries. By using this approach the cracking pattern between the Si anode grains, as a result of

Copyright © 2023 by ASME

TABLE 1: SOLID MECHANICS MODEL PARAMETERS

Parameter	Value
Thickness of Si, t_{Si}	100 nm
Young's Modulus of Ni, E_{Ni}	200 GPa
Poisson's Ratio of Ni, ν_{Ni}	0.3
Young's Modulus of Si at full discharge, $E_{Si,0}$	80 GPa
Young's Modulus of Si at full charge, $E_{Si,1}$	15 GPa
Poisson's Ratio of Si at full discharge, $\nu_{Si,0}$	0.3
Poisson's Ratio of Si at full charge, $\nu_{Si,1}$	0.22
Yield Strength of Ni, σ_{Ni}	600 MPa
Yield Strength of Si, σ_{Si}	170 MPa
Cracking Toughness, G_c	25 J/m^2
Delamination Toughness, G_d	15 J/m^2

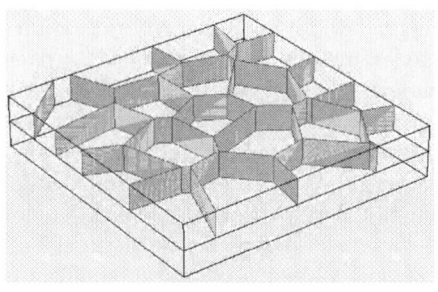

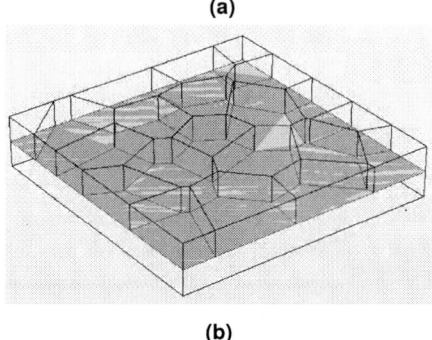

FIGURE 2: CZM INTERFACE SETUP FOR THE SIMULATION OF (A) CRACKING IN THE SI ANODE AT GRAIN BOUNDARIES (B) DELAMINATION OF THE SI ANODE FROM THE NI METAL SUBSTRATE

the volumetric changes during charging/discharging, can be observed as the separation is preferred along the interface between the grains i.e. the grain boundaries. Hence, the cohesive zones are placed along the grain boundaries as shown in Fig. 2a for a 100 nm thick Si anode. Fig. 2b shows the placement of the cohesive zones between the Si/Ni interface to simulate the volume change induced delamination. By using a different grain size and orientation for the Si microstructure, different cracking and delamination patterns will be observed. A roller boundary condition (symmetry) is setup on the edges of the model and the bottom of the metal substrate is fixed in place. The parameters used in this model are given in Table 1

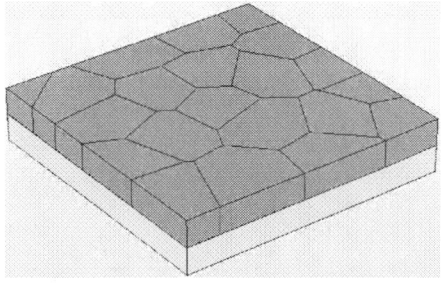

FIGURE 1: FE MODEL FOR CRACKING OF SI ANODE (TOP, BLUE) ON NI CURRENT COLLECTOR (BOTTOM, YELLOW)

The electrochemistry FE model uses the Li-Ion Battery module and the Transport of Diluted Species module to simulate the electrochemical reactions in the battery, including the SEI growth on the exposed anode surfaces. An accelerated battery degradation is simulated in this model so as to see a larger capacity loss and SEI growth for less battery cycles. This is done so that less computational costs are needed to show the merits of this approach. The setup of this model in COMSOL is shown in Fig. 3 and is consistent with other Li ion battery models that have been studied using COMSOL [37, 38]. The top part (blue) is the cathode, which is made up of a nickel, manganese and cobalt (NMC), (1:1:1) active material, a 1M Lithium Hexafluorophosphate (LiPF6) in 1:1 Ethylene Carbonate: Diethyl Carbonate (EC: DEC) is selected as the electrolyte (grey) and the Si anode is highlighted in red. The red surfaces shown in Fig. 3. are the interfaces that are available for the growth of SEI in the electrochemical FE

model.

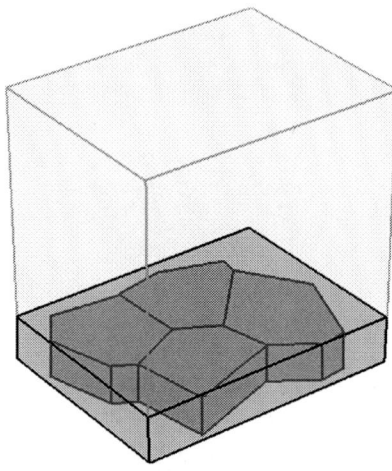

FIGURE 3: IMPLEMENTATION OF THE 3D FE ELECTROCHEMICAL MODEL WITH THE CATHODE (BLUE), ELECTROLYTE (GREY) AND ANODE (RED)

The following boundary conditions are used for the model: the bottom of the anode is grounded and provides the electronic path for the redox reaction and the top of the cathode experiences the electrode current for simulating the charging/discharging. The charging protocol that is employed in the model is the constant current constant voltage (CCCV) protocol. Three different charging/discharging rates (C-rate) are used for the model to study the effect of the charging rates on the battery degradation, where a 1C rate means that it takes 1 hour to fully charge the battery, a 0.5C rate implies that it takes 2 hours and a 2C rate indicates a

Copyright © 2023 by ASME

0.5 hour time to fully charge the battery. The outputs from this model, including the charging voltage curves, capacity loss and SEI thickness, are used as inputs for the PIML model which uses multiple GPR models. The parameter used in this FE model are summarized in Table 2.

TABLE 2: ELECTROCHEMICAL MODEL PARAMETERS

Parameter	Value
Li diffusion coefficient in Si, D_{Li}	$1 \times 10^{-16} \ m^2/s$
Exchange Current Density, I_0	$1 \ mA/cm^2$
Max Capacity of Si, $c_{max,Si}$	$2.78 \times 10^5 \ mol/m^3$
$\alpha_{45°C}$	0.67
$\alpha_{25°C}$	0.69
$J_{45°C}$	8.4×10^{-4}
$J_{25°C}$	1.9×10^{-4}
$f_{45°C}$	200 1/s
$f_{25°C}$	1100 1/s
$H_{45°C}$	6.7
$H_{25°C}$	11

2.2 PIML Model

Multiple GPR models are developed to link the charging voltage curves with the SEI thickness, based on a set of training points, to estimate the SoH of the battery. Gaussian process is chosen as the ML method in this paper as it is the most powerful and accurate ML method for the given application and has been the preferred ML method in battery analysis studies [39, 40]. By using the inputs of the operating conditions such as the temperature and the C-rates, the SEI thickness-wise voltage curves at unobserved points can be outputted. A general Kriging model can be expressed as:

$$G_K(x) = T(x) + S(x) \tag{10}$$

where, $G_K(x)$ denotes the prediction result of the performance function at point x, T(x) is the trend function which is a polynomial term of x and S(x) is the Gaussian stochastic process with zero mean and variance σ^2. T(x) is used to interpolate the input sample points with the polynomial term f(x) which is the product of two vectors, the regression coefficient β and the regression basis function f:

$$T(x) = f^T(x)\beta \tag{11}$$

Equation 12 can be used to define the covariance function between any two input points x_i and x_j:

$$Cov[S(x_i), S(x_j)] = \sigma^2 R(x_i, x_j) \tag{12}$$

where, $R(x_i, x_j)$ defines the correlation function matrix which can be given by Equation 13 using the Kriging parameters a_p and b_p:

$$R(x_i, x_j) = Corr(x_i, x_j) = exp\left[-\sum_{p=1}^{N} a_p |x_i^p - x_j^p|^{b_p}\right] \tag{13}$$

The log-likelihood function of the Kriging model with n observations O = [X,Y], where X and Y are the input data set and performance respectively, can be given as:

$$\ln L = \frac{1}{2}[n \ln 2\pi + n \ln \sigma^2 + \ln |R|$$
$$+ \frac{1}{2\sigma^2}(Y - F^T\beta)^T R^{-1}(Y - F^T\beta)] \tag{14}$$

where, F is the regression basis function at X and $F_{ij} = f_i(x_i)$. Maximizing the likelihood functions can help in solving for σ^2 and β as shown in Equations 15 and 16 respectively:

$$\sigma^2 = \frac{(Y - F^T\beta)^T R^{-1}(Y - F^T\beta)}{n} \tag{15}$$

$$\beta = \left[F^T R^{-1} F\right]^{-1} F^T R^{-1} Y \tag{16}$$

Hence, the response for any given new sample point x' can be estimated using the Kriging model as:

$$G_k(x') = f^T\beta + r^T R^{-1}(Y - F^T\beta) \tag{17}$$

where, r is the correlation vector between x' and the sampled point X. The mean square error $e(x')$ can be found using:

$$e(x') = \sigma^2\left[1 - r^T R^{-1} r + \frac{(1 - F^T R^{-1} r)^2}{F^T R^{-1} F}\right] \tag{18}$$

Therefore, the prediction of the response value at the new sample point x' using the Kriging model can be considered as a random variable that follows a normal distribution.

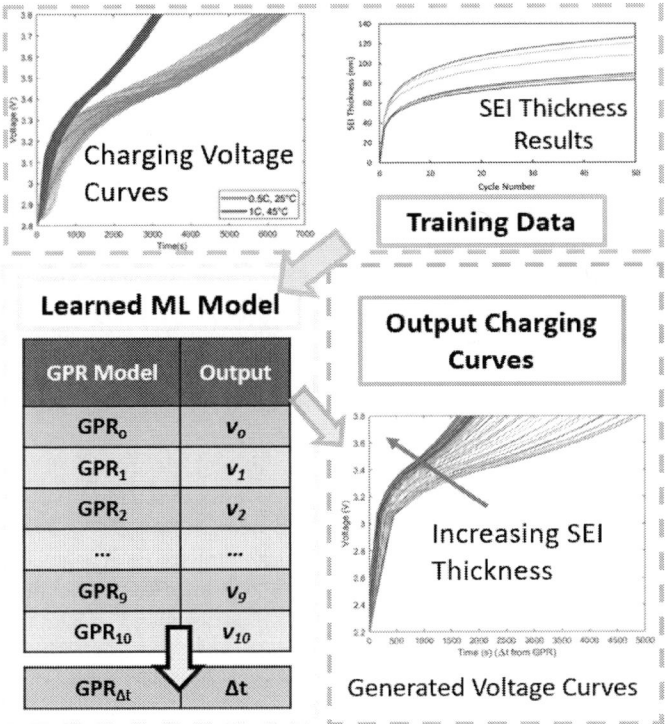

FIGURE 4: PIML MODEL FRAMEWORK

Copyright © 2023 by ASME

Now, the SEI thickness is directly related to the capacity loss and is therefore used as the indicator of the SoH of the LIBs. As the battery ages, the time taken to fully charge the battery decreases because the overall capacity of the battery decreases. This results in a changed shape and a shortening of the charging time curve for this degraded battery as compared to non-degraded battery. Furthermore, a voltage curve with a specific shape and time can be generalized to specific SEI thickness levels. Therefore, based on the overall SEI thickness ranges observed through the FE models, the SEI levels for the PIML model can be defined for the same range with a step size of 1.

Three different charging rates (0.5C, 1C and 2C) along with two different temperatures (25°C and 45°C) are used to generate the charging voltage curves for 40 cycles of charging-discharging. The principle idea of the proposed method is to capture the shortening of the charging curve, as a result of capacity degradation, as an indicator of the SEI level of the battery. PIML method will therefore output the complete SEI thickness-wise charging curves based on the input data from the FE models for each of the defined SEI levels. Using this method, segments of the testing voltage curves can be analyzed using a similarity analysis between the test segment and the PIML generated curves to identify the voltage curve that best resembles the test SEI level. This will allow us to quickly find out the SoH of the battery using only a partial charging segment.

The PIML model uses 11 GPR models to train a singular voltage point each (v_1 v_2 ... v_{11}) within a charging curve and 1 GPR model for the prediction of time (Δt) within these curves. This combination allows for the generation of a full voltage curve for each individual SEI level from the training data. Charging rate (current), temperature and SEI thickness are the inputs while the charging voltage points are the outputs for this model. Also, the obtained voltage points are used to train the GPR model that predicts Δt. Hence, the whole SEI dependent voltage curve can be estimated using the combination of the voltage GPRs and the time GPR models. The framework for this PIML is shown in Fig.4. To test this PIML method, a random partial charging voltage curve, whose SEI thickness is known, is used to test and validate the model by performing a similarity analysis between this curve and the PIML output. This PIML method does take some computational costs and time to setup and be developed, however, once the model is setup and validated it becomes efficient and less time consuming than using the existing methods to study battery SoH.

3. RESULTS AND DISCUSSION

In this section, the results from the FE models and the PIML models will be discussed. Section 3.1 shows the determination of cracking and delamination in the Si anode through the solid mechanics FE model, Section 3.2 talks about the SEI growth and the resultant capacity loss in the electrochemical FE model and finally the prognosis of the SoH of LIB is discussed in Section 3.3.

3.1 Cracking and Delamination

Through the use of the CZM and the interfaces shown in Fig. 2, the cracking pattern and the area of delaminated Si/Ni interface

are determined. The isometric and top view for the cracked and delaminated Si anode blocks, along with the Von-Mises stresses, are shown in Fig. 5a and 5b respectively. The damage parameter plot for the cohesive zones are shown in Fig. 5c. Damage parameter reaching a value of 1 results in the cracking and delamination and this happening is highlighted by the color dark red in this plot. For the cracking failure model, the resultant island with the maximum size is measured and used for electrochemical model, which is highlighted by the black boundary in Fig. 5b. While delamination is included as a failure mode in the solid mechanics model, for the selected Si anode island with the maximum size, little to no delamination was observed. Hence, there is no delaminated Si/Ni interface in the electrochemical model for the current iteration of the FE simulations. The reason for choosing the highlighted Si island is that this block of anode material provides the maximum surface area, i.e. the exposed upper area and cracked area, for the growth of SEI layer in the electrochemical model. The existing mechanics model of the thin anode layer do not consider the influence of SEI growth on cracking, along the thickness direction of the anode and the delamination of the Si/Ni interface. This is adopted in the current work to simplify the anode model in FE analysis to enable the simulation of the capacity degradation of anodes under certain operation conditions.

3.2 Li-Ion Battery Electrochemical Study

The results from Fig. 5 are used to setup the electrochemical study as shown in Fig. 3. The battery is cycled 50 times and the total SEI thickness change for the 0.5C, 45°C case is shown in Fig. 6. From the pattern of the SEI growth it can be seen that there is a uniform increase in SEI thickness on the anode with slightly less growth seen on the sharp edges of the Si anode microstructure. A possible explanation behind this is that there is less penetration of electrolyte in these corners and hence there is less SEI growth.

Fig. 7a and Fig. 7b show the relative capacity loss and the average SEI growth per cycle for the 6 combinations of temperature and C-rates that are studied. From these figures, it can be observed that both the SEI growth and relative capacity loss are higher at the start and that both of these trends slow down as the cycles progress. By comparing the results between the different temperatures (25°C and 45°C) that are simulated, it can be seen that higher operating temperatures lead to higher SEI growth rates and higher relative capacity loss as well. This is due to the fact that an increase in temperature leads to an application of a higher applied parasitic current (T_{SEI}) as described by Equation 8. The study on different charging rates (2C, 1C and 0.5C) reveals that a faster charging rate leads to both higher capacity degradation and higher SEI growth. A faster charging rate implies that a larger current is applied to the battery which leads to a larger applied SEI layer parasitic current.

Another way to visualize the degradation of the battery is to look at the CCCV protocol charging voltage curves as shown in Fig. 8 for two of the studied cases. It can be observed that as the battery degrades, the capacity decreases and hence the time required to reach full charge also decreases. Therefore, the charging curves move towards the left as the number of cycles increase indicating this decrease in time reach full charge.

Copyright © 2023 by ASME

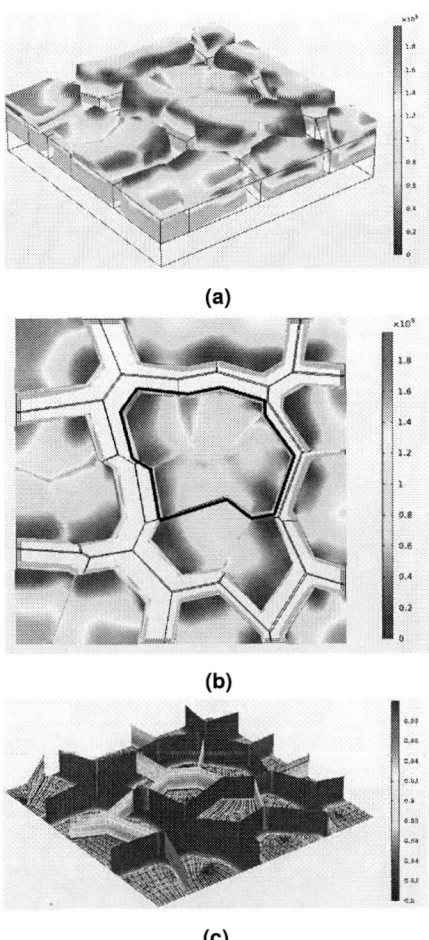

(a)

(b)

(c)

FIGURE 5: (A) ISOMETRIC VIEW AND (B) TOP VIEW SHOWING THE VON-MISES STRESSES AND THE CRACKED AND DELAMINATED SI ANODE AS A RESULT OF VOLUMTERIC EXPANSION/CONTRACTION ALONG WITH THE (C) DAMAGE PARAMETER PLOT

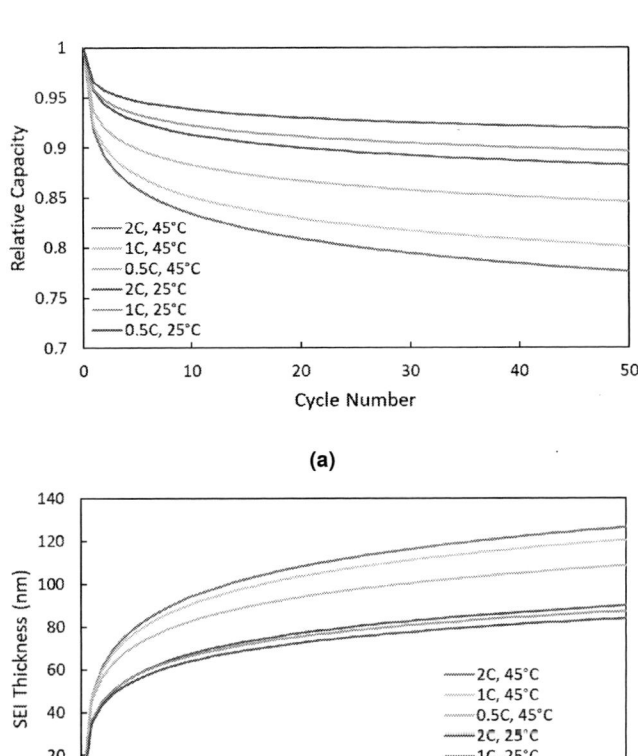

(a)

(b)

FIGURE 7: CYCLE-WISE (A) RELATIVE CAPACITY LOSS AND (B) SEI THICKNESS GROWTH

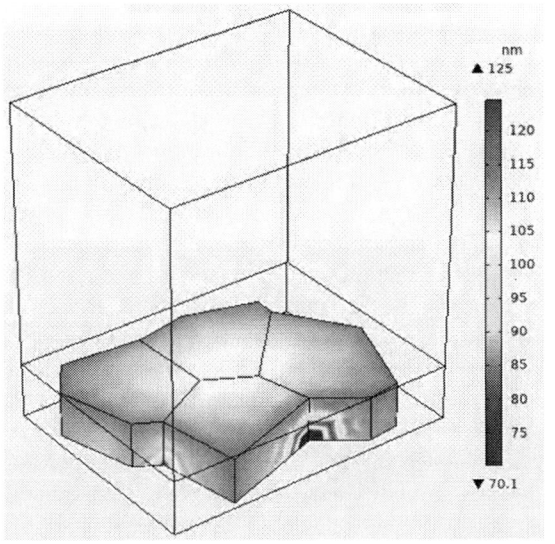

FIGURE 6: SEI GROWTH PATTERN AFTER 50 BATTERY CYCLES FOR 0.5C, 45°C CASE

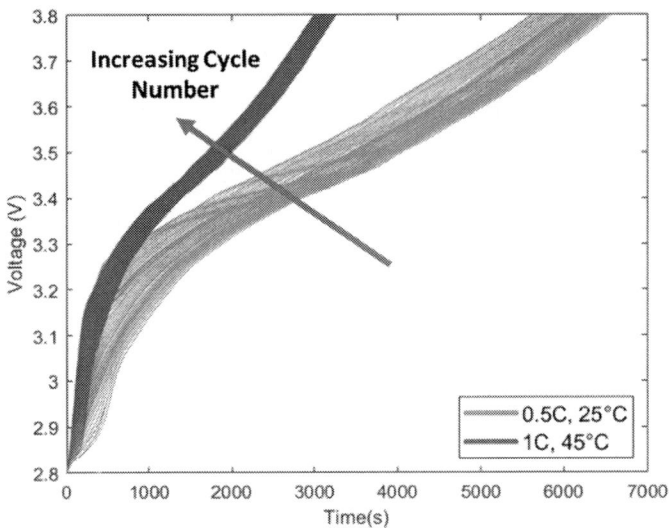

FIGURE 8: VOLTAGE CURVES FOR 50 CHARGING CYCLES

Copyright © 2023 by ASME

3.3 PIML Based SoH Estimation

The SoH of the battery is estimated using a partial charging voltage segment by performing a similarity analysis on the PIML generated SEI thickness dependent voltage charging curves. The developed PIML model uses the outputs derived from the electrochemical model for the curve generation. The inputs to this PIML model are the current, SEI thickness and temperature. As seen through Fig. 7b, the SEI thickness varied from 0-126 nm, therefore 127 SEI thickness dependent voltage curves are generated and shown in Fig. 9. Again, as the SEI grows and battery degrades, the generated curves become shorter as the time required for complete charge decreases.

The partial voltage segment that is used to test the PIML model is shown in Fig. 10 along with the framework for the similarity analysis. The testing data is from the 2C, 45°C case and is not included in the training data. The start and end point voltages for this segment are 3.537V and 3.621V respectively, while the Δt is 200s. This extracted segment data is then used for a similarity analysis on the PIML curves generated in Fig. 9 and the closest curve which matches the test V_1, V_2 and Δt gives the SEI thickness for this partial test segment. The simulated value of SEI thickness for this segment is 113.54nm whereas the similarity analysis estimated thickness is 117nm. This result shows that the proposed method is an efficient and time saving method as by using only 11.4% of the total charging segment and time, the SoH can be estimated with an error of just 3.04%. Since the error in the simulated and estimated SEI thickness is relatively low, this method can be used for the SEI thickness based estimation of the SoH of LIBs. Finally, comparing the developed model with other SoH estimation methods for LIBs, it can be observed that the accuracy and efficiency of the developed model is on par with these some of these models [41, 42] and is better than some other ones [43, 44].

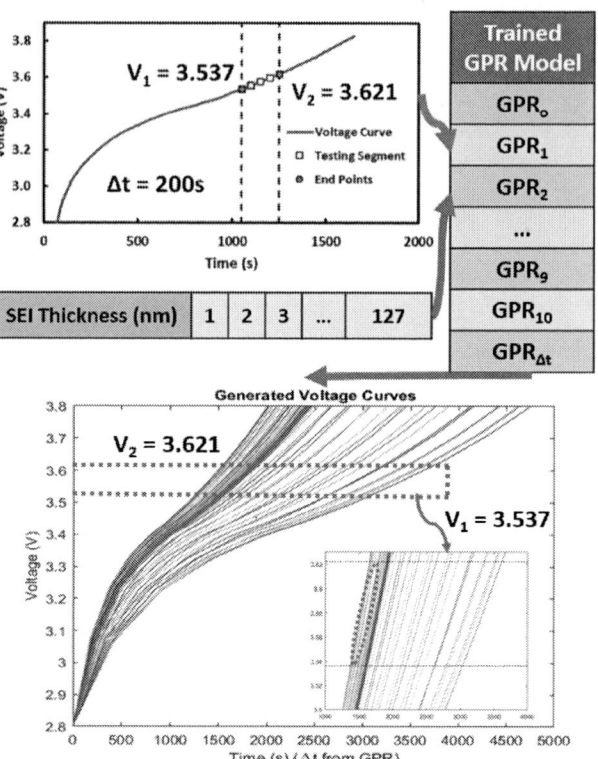

FIGURE 10: IMPLEMENTATION OF SEI ESTIMATION USING PARTIAL CHARGING SEGMENT

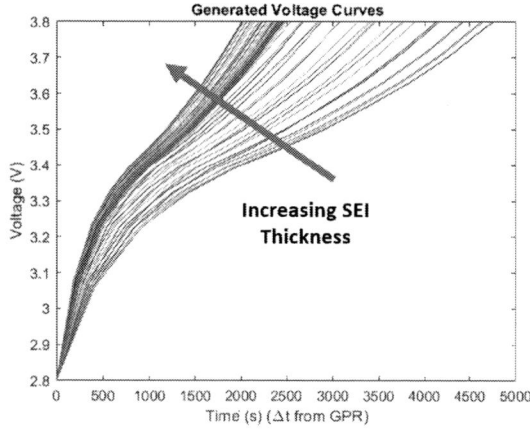

FIGURE 9: SEI THICKNESS-WISE VOLTAGE CHARGING CURVES

4. SUMMARY AND CONCLUSION

In this paper the SoH of Si anode based Li-ion battery is estimated using GPR based PIML method. A multiphysics-informed machine learning model is developed through integrating physics-based battery simulations and nested GPR models. Three different failure models are considered in FE simulations, i.e. cracking, delamination and SEI growth. The developed nested surrogate models are demonstrated to predict the SEI thickness of the anode under certain current and the temperature.

The first part of the illustrates the two multiphysics FE models (a solid mechanics and an electrochemical model) to simulate the different failure modes such as cracking/delamination in the Si anode due to volumetric changes and the growth of SEI layer on the exposed anode surfaces as a result of the charge/discharge cycling of the battery. The Si island size and shape gathered from the solid mechanics model acts as the input for the electrochemical model for the battery cycling and SEI growth induced capacity degradation. The multiphysics FE models are demonstrated to investigate the battery degradation in terms of SEI thickness and capacity loss under the influence of three failure modes in anode materials. The voltage curves of the battery during charging and discharge cycles can be predicted under certain operation conditions. It indicates that elevated temperature and risen charging rate can accelerate the degradation of Li-ion batteries.

The SEI thickness growth, temperature and charging rate obtained from these models are further used to train the GPR models which in turn output the SEI thickness dependent charging voltage curves. The SEI thickness of a partial test segment of the charging voltage curve is determined using the PIML model output through similarity analysis. The simulated value for SEI thickness of this partial segment (11.4% of total charging curve segment) is known and when this value is compared with the PIML result a good agreement is seen between the values (a 3.04% error), which also partially validates the developed model.

Copyright © 2023 by ASME

The multiphysics-informed machine learning model based on GPR can predict the degradation history of the battery in terms of voltage curves under various operation conditions. Through integrating multiphysics simulations with data driven model, it enables the health monitoring of the battery with minimal partial charging segment.

REFERENCES

[1] Nitta, Naoki, Wu, Feixiang, Lee, Jung Tae and Yushin, Gleb. "Li-ion battery materials: present and future." *Materials today* Vol. 18 No. 5 (2015): pp. 252–264.

[2] Tarascon, J-M and Armand, Michel. "Issues and challenges facing rechargeable lithium batteries." *nature* Vol. 414 No. 6861 (2001): pp. 359–367.

[3] Koohi-Fayegh, Seama and Rosen, Marc A. "A review of energy storage types, applications and recent developments." *Journal of Energy Storage* Vol. 27 (2020): p. 101047.

[4] Armand, Michel and Tarascon, J-M. "Building better batteries." *nature* Vol. 451 No. 7179 (2008): pp. 652–657.

[5] Limthongkul, Pimpa, Jang, Young-Il, Dudney, Nancy J and Chiang, Yet-Ming. "Electrochemically-driven solid-state amorphization in lithium-silicon alloys and implications for lithium storage." *Acta Materialia* Vol. 51 No. 4 (2003): pp. 1103–1113.

[6] Kasavajjula, Uday, Wang, Chunsheng and Appleby, A John. "Nano-and bulk-silicon-based insertion anodes for lithium-ion secondary cells." *Journal of power sources* Vol. 163 No. 2 (2007): pp. 1003–1039.

[7] Boukamp, BA, Lesh, GC and Huggins, RA. "All-solid lithium electrodes with mixed-conductor matrix." *Journal of the Electrochemical Society* Vol. 128 No. 4 (1981): p. 725.

[8] Hamon, Yohann, Brousse, Thierry, Jousse, Franck, Topart, Patrice, Buvat, Pierrick and Schleich, Donald M. "Aluminum negative electrode in lithium ion batteries." *Journal of Power Sources* Vol. 97 (2001): pp. 185–187.

[9] Kim, Il-seok, Blomgren, GE and Kumta, PN. "Nanostructured Si/TiB2 composite anodes for Li-ion batteries." *Electrochemical and Solid-State Letters* Vol. 6 No. 8 (2003): p. A157.

[10] Arora, Pankaj, White, Ralph E and Doyle, Marc. "Capacity fade mechanisms and side reactions in lithium-ion batteries." *Journal of the Electrochemical Society* Vol. 145 No. 10 (1998): p. 3647.

[11] Ramadesigan, Venkatasailanathan, Chen, Kejia, Burns, Nancy A, Boovaragavan, Vijayasekaran, Braatz, Richard D and Subramanian, Venkat R. "Parameter estimation and capacity fade analysis of lithium-ion batteries using reformulated models." *Journal of the electrochemical society* Vol. 158 No. 9 (2011): p. A1048.

[12] Etacheri, Vinodkumar, Marom, Rotem, Elazari, Ran, Salitra, Gregory and Aurbach, Doron. "Challenges in the development of advanced Li-ion batteries: a review." *Energy & Environmental Science* Vol. 4 No. 9 (2011): pp. 3243–3262.

[13] Spotnitz, Robert. "Simulation of capacity fade in lithium-ion batteries." *Journal of power sources* Vol. 113 No. 1 (2003): pp. 72–80.

[14] Bashash, Saeid, Moura, Scott J, Forman, Joel C and Fathy, Hosam K. "Plug-in hybrid electric vehicle charge pattern optimization for energy cost and battery longevity." *Journal of power sources* Vol. 196 No. 1 (2011): pp. 541–549.

[15] Randall, Alfred V, Perkins, Roger D, Zhang, Xiangchun and Plett, Gregory L. "Controls oriented reduced order modeling of solid-electrolyte interphase layer growth." *Journal of Power Sources* Vol. 209 (2012): pp. 282–288.

[16] Chew, Huck Beng, Hou, Binyue, Wang, Xueju and Xia, Shuman. "Cracking mechanisms in lithiated silicon thin film electrodes." *International Journal of Solids and Structures* Vol. 51 No. 23-24 (2014): pp. 4176–4187.

[17] Shi, Feifei, Song, Zhichao, Ross, Philip N, Somorjai, Gabor A, Ritchie, Robert O and Komvopoulos, Kyriakos. "Failure mechanisms of single-crystal silicon electrodes in lithium-ion batteries." *Nature communications* Vol. 7 No. 1 (2016): pp. 1–8.

[18] Yang, Hui, Fan, Feifei, Liang, Wentao, Guo, Xu, Zhu, Ting and Zhang, Sulin. "A chemo-mechanical model of lithiation in silicon." *Journal of the Mechanics and Physics of Solids* Vol. 70 (2014): pp. 349–361.

[19] Liu, Xiao Hua, Zheng, He, Zhong, Li, Huang, Shan, Karki, Khim, Zhang, Li Qiang, Liu, Yang, Kushima, Akihiro, Liang, Wen Tao, Wang, Jiang Wei et al. "Anisotropic swelling and fracture of silicon nanowires during lithiation." *Nano letters* Vol. 11 No. 8 (2011): pp. 3312–3318.

[20] Li, Juchuan, Dozier, Alan K, Li, Yunchao, Yang, Fuqian and Cheng, Yang-Tse. "Crack pattern formation in thin film lithium-ion battery electrodes." *Journal of The Electrochemical Society* Vol. 158 No. 6 (2011): p. A689.

[21] Zheng, Zhuoyuan, Liu, Zheng, Wang, Pingfeng and Li, Yumeng. "Numerical modeling on the delamination-induced capacity degradation of silicon anode." *Journal of Energy Storage* Vol. 43 (2021): p. 103190.

[22] Bansal, Parth, Zheng, Zhuoyuan, Shao, Chenhui, Li, Jingjing, Banu, Mihaela, Carlson, Blair E and Li, Yumeng. "Physics-informed machine learning assisted uncertainty quantification for the corrosion of dissimilar material joints." *Reliability Engineering & System Safety* Vol. 227 (2022): p. 108711.

[23] Bansal, Parth, Zheng, Zhuoyuan and Li, Yumeng. "Uncertainty Quantification for Dissimilar Material Joints Under Corrosion Environment." *International Design Engineering Technical Conferences and Computers and Information in Engineering Conference*, Vol. 86229: p. V03AT03A039. 2022. American Society of Mechanical Engineers.

[24] Wu, Hao, Zhu, Zhifu and Du, Xiaoping. "System reliability analysis with autocorrelated kriging predictions." *Journal of Mechanical Design* Vol. 142 No. 10 (2020).

[25] Singh, Akash and Li, Yumeng. "Uncertainty Management and Reduction of Machine Learning Potential." *AIAA Scitech 2021 Forum*: p. 1962. 2021.

[26] Singh, Akash and Li, Yumeng. "Machine Learning Potentials for Graphene." *ASME International Mechanical Engineering Congress and Exposition*, Vol. 86656: p. V003T03A036. 2022. American Society of Mechanical Engineers.

Copyright © 2023 by ASME

[27] Wu, Hao, Hu, Zhangli and Du, Xiaoping. "Time-dependent system reliability analysis with second-order reliability method." *Journal of Mechanical Design* Vol. 143 No. 3 (2021).

[28] Wu, Hao and Du, Xiaoping. "Envelope Method for Time- and Space-Dependent Reliability Prediction." *ASCE-ASME Journal of Risk and Uncertainty in Engineering Systems, Part B: Mechanical Engineering* Vol. 8 No. 4 (2022): p. 041201.

[29] Singh, Akash, Chen, Xin, Li, Yumeng, Koric, Seid and Guleryuz, Erman. "Development of Artificial Neural Network Potential for Graphene." *AIAA Scitech 2020 Forum*: p. 1861. 2020.

[30] Bansal, Parth, Zheng, Zhuoyuan and Li, Yumeng. "Uncertainty Quantification on Galvanic Corrosion Based on Adaptive Surrogate Modeling." *ASME International Mechanical Engineering Congress and Exposition*, Vol. 86717: p. V009T12A005. 2022. American Society of Mechanical Engineers.

[31] Wu, Lifeng, Fu, Xiaohui and Guan, Yong. "Review of the remaining useful life prognostics of vehicle lithium-ion batteries using data-driven methodologies." *Applied Sciences* Vol. 6 No. 6 (2016): p. 166.

[32] Bansal, Parth, Zhang, Zhuoyuan, Wang, Pingfeng and Li, Yumeng. "Multiphysics Modeling on the Capacity Degradation of Silicon Anode." *AIAA SCITECH 2023 Forum*: p. 0772. 2023.

[33] Cao, Wen, Wang, Shun-Li, Fernandez, Carlos, Zou, Chuan-Yun, Yu, Chun-Mei and Li, Xiao-Xia. "A novel adaptive state of charge estimation method of full life cycling lithium-ion batteries based on the multiple parameter optimization." *Energy Science & Engineering* Vol. 7 No. 5 (2019): pp. 1544–1556.

[34] Kim, Minho and Han, Soohee. "Novel data-efficient mechanism-agnostic capacity fade model for Li-ion batteries." *IEEE Transactions on Industrial Electronics* Vol. 68 No. 7 (2020): pp. 6267–6275.

[35] Yu, Jianbo. "State of health prediction of lithium-ion batteries: Multiscale logic regression and Gaussian process regression ensemble." *Reliability Engineering & System Safety* Vol. 174 (2018): pp. 82–95.

[36] Ekström, Henrik and Lindbergh, Göran. "A model for predicting capacity fade due to SEI formation in a commercial graphite/LiFePO4 cell." *Journal of The Electrochemical Society* Vol. 162 No. 6 (2015): p. A1003.

[37] Dasari, Harika and Eisenbraun, Eric. "Predicting capacity fade in silicon anode-based Li-ion batteries." *Energies* Vol. 14 No. 5 (2021): p. 1448.

[38] Ning, Gang, White, Ralph E and Popov, Branko N. "A generalized cycle life model of rechargeable Li-ion batteries." *Electrochimica acta* Vol. 51 No. 10 (2006): pp. 2012–2022.

[39] Wang, Jiwei, Deng, Zhongwei, Yu, Tao, Yoshida, Akihiro, Xu, Lijun, Guan, Guoqing and Abudula, Abuliti. "State of health estimation based on modified Gaussian process regression for lithium-ion batteries." *Journal of Energy Storage* Vol. 51 (2022): p. 104512.

[40] Lyu, Zhiqiang and Gao, Renjing. "Li-ion battery state of health estimation through Gaussian process regression with Thevenin model." *International Journal of Energy Research* Vol. 44 No. 13 (2020): pp. 10262–10281.

[41] Roman, Darius, Saxena, Saurabh, Robu, Valentin, Pecht, Michael and Flynn, David. "Machine learning pipeline for battery state-of-health estimation." *Nature Machine Intelligence* Vol. 3 No. 5 (2021): pp. 447–456.

[42] Feng, Xuning, Weng, Caihao, He, Xiangming, Han, Xuebing, Lu, Languang, Ren, Dongsheng and Ouyang, Minggao. "Online state-of-health estimation for Li-ion battery using partial charging segment based on support vector machine." *IEEE Transactions on Vehicular Technology* Vol. 68 No. 9 (2019): pp. 8583–8592.

[43] Deng, Zhongwei, Hu, Xiaosong, Xie, Yi, Xu, Le, Li, Penghua, Lin, Xianke and Bian, Xiaolei. "Battery health evaluation using a short random segment of constant current charging." *Iscience* Vol. 25 No. 5 (2022): p. 104260.

[44] Ng, Man-Fai, Zhao, Jin, Yan, Qingyu, Conduit, Gareth J and Seh, Zhi Wei. "Predicting the state of charge and health of batteries using data-driven machine learning." *Nature Machine Intelligence* Vol. 2 No. 3 (2020): pp. 161–170.

Copyright © 2023 by ASME

Proceedings of the ASME 2023
International Design Engineering Technical Conferences and
Computers and Information in Engineering Conference
IDETC-CIE2023
August 20-23, 2023, Boston, Massachusetts

DETC2023-117236

MODELING A CONCENTRATED SOLAR COLLECTOR (CSC) - CONVECTION-ENHANCED EVAPORATION (CEE) SYSTEM FOR SMALL-SCALE BRINE MANAGEMENT

Nallely Guillen Rodriguez[1,*], Mustafa F. Kaddoura [1], Natasha C. Wright[1,*]

[1]University of Minnesota - Twin Cities, Minneapolis, MN

ABSTRACT

Desalination is the process of removing salts from a saline water source to obtain fresh water. All desalination processes produce a salt-rich brine. If not disposed of properly, brine can lead to negative environmental consequences. Solar evaporation ponds facilitate its adequate disposal; however, this method requires a large footprint. As an alternative for small-scale desalination plants, a convection-enhanced evaporation (CEE) system has been proposed.

A CEE unit consists of several packed surfaces, distributed by uniform spacing. Brine is injected along the width of the surfaces, forming a liquid thin film. As ambient temperature and relative humidity vary on an hourly and daily basis, a controller is necessary to select the optimal brine injection rate, brine temperature, and air speed that maintains a constant evaporation rate. The optimized operating conditions results in a thermal energy load required to preheat the brine before it enters the evaporation system. This thermal load oscillates in magnitude, as optimal brine inlet temperature varies from less than 1°C up to 90°C, depending on time of the day and year.

In this paper, a model is developed to evaluate the use of a concentrated solar collector (CSC) system for the supply of the CEE system's thermal energy requirements. The model predicts the performance of the coupled CSC-CEE system in terms of specific energy consumption, solar fraction, and evaporation fraction (relative to the target).

Findings establish that due to the high thermal energy demand of the system, the solar fraction can be relatively low, with values of less than 20% for small collector-tank configurations. However, it was also observed that low solar fractions do not necessarily translate to low evaporation fractions, which can be attributed to the high variability in the magnitude of the hourly heat load, and the fact that a high percentage of the evaporating hours depend exclusively on forced air convection through the CEE system without preheating the brine. Further work is

needed to understand the tradeoff in footprint required for the CSC system, CEE system, and any non-evaporated brine storage.

Keywords: brine management, concentrated solar, desalination, solar thermal heating, thermal energy storage, immersed heat exchanger

1. INTRODUCTION

Lack of access to an improved water source is a reality for more than 733 million people around the world, a problem exacerbated by the physical water scarcity caused by over extraction of ground water, contamination of water supplies, and climate change [1]. Desalination is the process of removing salts and other contaminants from a water source to obtain potable water. Desalination is of high interest as more than 40% of the world's population lives within 100 kilometers of seas or oceans [2]. Furthermore, it is estimated that around 1.1 billion people worldwide live in areas affected by saline groundwater, particularly prevalent in arid and semi-arid areas [3].

As important as desalination will likely be in securing long-term safe water supply, it also comes with concerns associated with its implementation: high costs, high energy consumption, and, the focus of this work, management of its highly saline waste brine. Brine, the by-product of desalination, is a residual amount of water with a high concentration of salts, as well as pretreatment chemicals, heavy metals, and microbial contaminants [4]. There are several methods of brine disposal available to large-volume brine producers. Some of these methods include discharge to surface water, discharge to sewers, deep-well injection, evaporation ponds, and land application, all of them having different associated potential for environmental impact, such as degradation of marine life habitat and health and contamination of underground aquifers [4, 5]. For small desalination plants ($<100m^3$ product water / day) these solutions are challenging to implement. Land application permits are costly and time intensive, brine volumes are too small to warrant a deep-well injection site, and sewer networks may not be available. Most brine reduction strategies are thermally based and cost-intensive to implement at

---

*Corresponding author: guill139@umn.edu, natasha@umn.edu
Documentation for `asmeconf.cls`: Version 1.34, May 16, 2023.

Copyright © 2023 by ASME

small-scale, as thermal efficiency decreases with system size [5]. For small-volume, decentralized desalination plants to be scalable, new methods of brine management are needed which are environmentally sustainable, energy efficient, and cost-effective.

1.1 Convection-Enhanced Evaporation

Convection-enhanced evaporation (CEE) is an alternative technology for small-volume brine producers that has been proposed by the Wright Lab at the University of Minnesota in collaboration with QuadSun Solar [6]. The technology is described and modeled in depth in the work of Kaddoura et al. (Figure 1) [5]. One can think of a CEE system as a "folded and stacked" version of an evaporation pond. A CEE unit consists of several packed horizontal or vertical surfaces, distributed by uniform spacing. Brine is injected by orifices on the pipes located along the width of the evaporation surfaces, forming a liquid thin film. A fan drives airflow parallel to the surfaces in a co-flow (horizontal surfaces) or cross-flow (vertical surfaces) configuration [5].

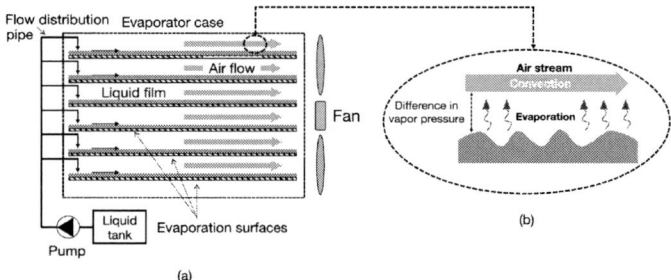

FIGURE 1: SCHEMATIC OF THE CONVECTION-ENHANCED EVAPORATION (CEE) SYSTEM (A). EVAPORATION IS DRIVEN BY A DIFFERENCE IN VAPOR PRESSURE BETWEEN THE TOP LAYER OF THE LIQUID AND AIR (B) [5].

The rate of rate of water evaporation from the brine film is driven by a difference in vapor pressure between the brine and the air flowing above it. The moisture content at the interface between the brine film and the air depends on the saturation pressure, which in turn depends on the temperature of the liquid [7]. Therefore, an increase in brine temperature will promote a higher evaporation rate. Higher airspeed is also associated with an increase in the evaporation rate, though at smaller effect versus that of preheating brine [5].

Increasing the brine temperature and the air speed are associated with an increase in thermal and electrical power consumption, respectively, and therefore an increase in operation costs (OC). In exchange, the evaporation rate per surface will increase, which means that for a specific evaporation target, fewer surfaces will be required, and in consequence the capital cost (CC) of the system will decrease. So, for any specific application of interest, it is necessary to determine the optimal operating conditions to achieve the best tradeoff between capital costs and operation costs, within the context of the ambient conditions that also influence the evaporation rate (air temperature and relative humidity).

1.2 Optimal Operation of CEE Systems

Kaddoura et al. conducted a multi-objective design optimization (MDO) to analyze the trade-off between capital and operational costs [8]. As the cost of thermal and electric energy and materials can depend on location, the approach was to develop and use cost ratios rather than absolute costs: the material cost ratio (r_1) and energy cost ratio (r_2) are given by:

$$r_1 = \frac{K_m}{K_{el}} \qquad (1)$$

$$r_2 = \frac{K_{th}}{K_{el}} \qquad (2)$$

where $K_{el}(\$/kWh_{el})$ and $K_{th}(\$/kWh_{th})$ are the costs of an electric and thermal energy unit, respectively, and where $K_m(\$/m^2)$ is the material cost per evaporation surface.

An MDO having as objective functions the minimization of the normalized capital cost and the normalized operation cost was performed [8]. The optimization assumes fixed design parameters including the spacing between evaporation surfaces, the dimensions of the surfaces, and a target hourly evaporation rate, which depends on the evaporation rate per surface and the number of surfaces in the system. Optimization variables included brine injection rate, brine temperature, air speed, and number of surfaces. For any given combination of ambient temperature and relative humidity (RH), the result is a set of non-dominated optimal design points that can be represented by a Pareto front showing the compromise between the operating cost and the capital cost. An example of a Pareto front for an ambient temperature of 20 °C and a RH of 60% is shown in Figure 2. An increase in operating cost (caused by an increase in brine heating and/or air speed), is associated with a decrease in footprint and capital cost (caused by a decrease in the required number of surfaces).

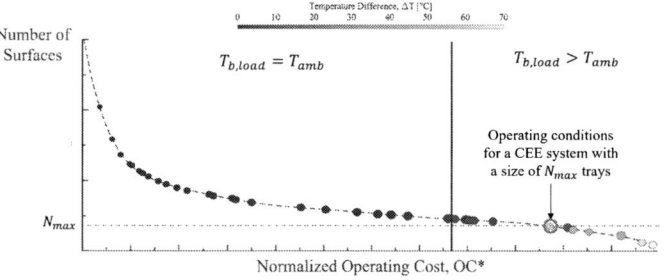

FIGURE 2: EXAMPLE OF A PARETO FRONT DEMONSTRATING THE TRADEOFF BETWEEN NORMALIZED CAPITAL COST (NUMBER OF SURFACES) AND OPERATING COST. FIGURE FOR $T_{amb} = 20°C$, $RH = 60\%$, $r_1 = 55\,kWh/surface$, AND $r_2 = 0.25$ $[kWh_{el}/kWh_{th}]$ [8].

These Pareto fronts offer insight into the best operation conditions depending on the intended size (number of surfaces) in the system, where one would always want to operate with the maximum number of surfaces in the field to minimize operating cost. Note that once a system of a specific size is installed, the operating conditions will vary as ambient conditions, and the resulting Pareto front, vary. The hourly ambient temperature and RH for a CEE system operating in Alamogordo, NM are seen in Figures 3a and 3b. As a result of the changing ambient conditions, the optimized hourly thermal load oscillates substantially throughout

Copyright © 2023 by ASME

the year (Figure 3d), even through the hourly evaporation target is remains the same.

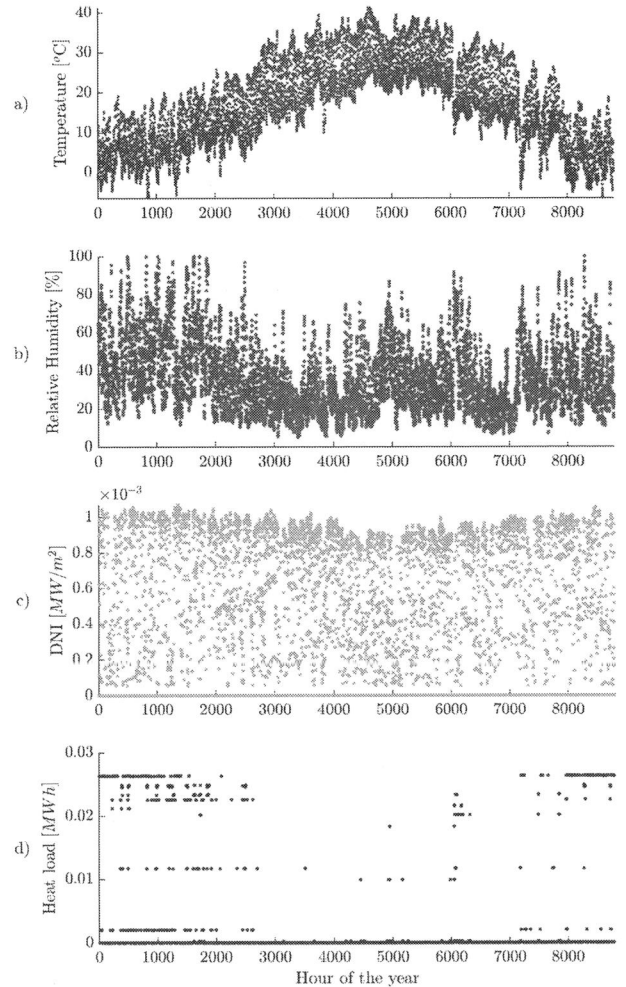

FIGURE 3: HOURLY A) AMBIENT TEMPERATURE, B) RELATIVE HUMIDITY, C) DIRECT INCIDENT RADIATION AND D) THERMAL LOAD FOR A YEAR IN ALAMOGORDO NM.

1.3 Solar Thermal Heating for Brine Management

The goal of CEE is to offer a brine management alternative for small scale desalination and industrial waste. The systems may be located in remote areas or otherwise be off-grid. The objective of this work is to model the annual performance of a concentrated solar collector (CSC) field for the supply of a CEE system's optimized heat load, in order to determine the feasibility of using this sustainable source of energy as a solution for the high thermal energy load. The goal it to provide an alternative to the issues related to the use of fossil fuels in remote areas, such as high transportation costs, price fluctuations, shortages, and environmental pollution [9].

Methods of coupling desalination systems with solar energy can be divided into two broad categories: (1) technologies where the solar energy is directly absorbed by the brine, such as solar stills and solar ponds [10, 11], and (2) technologies where solar collectors or concentrators are used to heat a working fluid; the thermal energy later transferred to the saline brine or used for the steam generation required for the desalination processes (such as in multi-effect distillation or multi-stage flash) [12, 13]. The system designs common in the latter group offer a more adequate reference to follow for the configurations in this work, consisting of a technology for evaporation (the CEE system), which has energy supplied to it via a working fluid that will not be in direct contact with the fluid to be treated (brine).

As is inherent in solar technologies, the thermal output from a CSC system is not stable through the day nor the year. The CEE thermal load also varies on an hourly and annual basis because of its dependence on local ambient conditions (Figure 3) [5]. A model is developed in order to understand the coupled behavior of this highly variable supply and load.

2. MATERIALS AND METHODS

The thermal heating system consists of the CSCs, a thermal storage tank, and a brine storage tank. In this section, a transient model of the thermal heating system is developed. It takes in time-variant ambient conditions and desired loads, and predicts the use of thermal energy by the CEE system. The equations required to understand the energy balance in each module are described here, with full details in [14].

2.1 CSC Model General Description

The thermal heating system configuration used in this model (Figure 4) is based on existing solar thermal desalination system configurations, as the literature offers a rich variety of examples [11, 15]. Incident radiation strikes the CSC and is reflected to the receiver, where the working fluid absorbs the energy raising its temperature. Then, the fluid is pumped to one of the immersed coil heat exchangers (IHX) in the storage tank, releasing heat to the tank, and returning to the CSC. At the same time, brine at an initial temperature equal to the ambient temperature is pumped through the second IHX. Stored thermal energy in the tank acts to heat the brine to the desired target temperature.

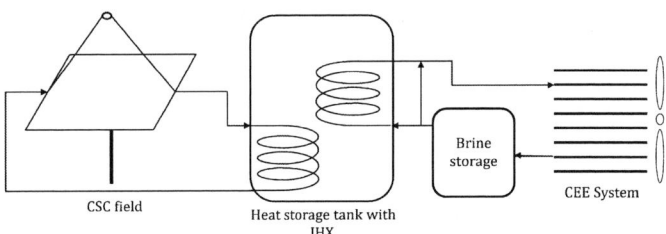

FIGURE 4: CSC MODEL SYSTEM CONFIGURATION

2.2 Solar Collector

Incident solar radiation depends on factors such as the geographic location, date, time of day, and atmospheric conditions. The annual direct normal irradiance (DNI) for Alamogordo, NM is shown in Figure 3c, in comparison to the variable thermal load (Figure 3d) required to obtain a constant target evaporation rate in the CEE system.

This model assumes that operating conditions are constant over the course of each hour. The useful energy $\dot{Q}_{col}$ absorbed by

Copyright © 2023 by ASME

a collector is defined by Equation 3, derived from the definition of instantaneous efficiency (η_{col}) described by Duffie and Beckman [11]:

$$\dot{Q}_{col} = N_{col}\eta_{col}A_c DNI \qquad (3)$$

DNI is acquired from meteorological databases. The aperture area of the collectr A_c is obtained from the equipment data sheet. The modeled CSC is the CST H3500 equipment from QuadSun Solar Solutions. The efficiency function for this collector η_{col} was found by performing a linear fit to a set of its experimental performance data points:

$$\eta_{col} = 0.823 - 2.041 \times \frac{(T_{i,col} - T_{amb})}{DNI} \qquad (4)$$

Here $T_{i,col}$ is the temperature of the working fluid at the inlet of the collector. Notice that for CSC collectors installed in series, the efficiency of each subsequent collector decreases due to the increase in $T_{i,col}$ and resulting increase in thermal losses. To decrease the computational cost, the present model assumes the entire CSC array operates at the efficiency of the first collector. The hourly CSC working fluid outlet temperature $T_{o,col}$ is calculated by:

$$T_{o,col} = T_{i,col} + \frac{\dot{Q}_{col}}{\dot{m}_{col} C_{p,b}}, \qquad (5)$$

where $\dot{m}_{col}$ and $C_{p,b}$ are the mass flow rate and specific heat capacity of the collector working fluid.

To calculate how much energy the working fluid delivers to the storage tank, $\dot{Q}_{C-T}$, and its temperature after leaving the IHX, $T_{o,IHX}$, we define:

$$\dot{Q}_{C-T} = \eta_{IHX,C-T} \dot{m}_{col} C_{p,col} (T_{i,IHX} - T_{tank}) \qquad (6)$$

and,

$$T_{o,IHX} = T_{i,IHX} + \frac{\dot{Q}_{C-T}}{\dot{m}_{col} C_{p,col}}. \qquad (7)$$

The outlet temperature of the working fluid, $T_{o,IHX}$, after one hour is the inlet temperature of the working fluid that returns to the solar field in the subsequent hour. Pipe losses from the tank to the collector are neglected.

Several control parameters where established to decide whether or not the collector pump will run in a given hour. The pump will not run: (i) overnight, when $DNI = 0$, (ii) during sunlight hours when low values of DNI result in negative values of $\dot{Q}_{col}$ (Eq. 4), and (iii) when the collector, IHX, or storage tank reaches temperatures that exceed their design specification. Whenever any of those temperature thresholds are reached the pump turns off and, in a real installation, the collector moves such that the aperture points down, allowing the system to cool down through losses to the ambient or and energy transfer to the CEE system brine.

2.3 Thermal Energy Storage Tank

Unlike the collector model, in which a specific piece of equipment was selected for analysis, the thermal storage tank needed to vary in size in order to see the relevant trends. A theoretical tank model was developed by compiling tank specifications for different models of thermal storage tanks [16].

Storage tanks with two immersed heat exchangers (IHX) were selected. This type of heat exchanger configuration is highly compatible for a system such as the CEE, as salt scaling and subsequent clogging is less prevalent than that linked to other types of heat exchangers. The addition of thermal energy to the tank is carried out by circulating the CSC working fluid through one of the IHXs in the storage tank. In the other IHX, energy is transferred to the brine. Thermal losses from the tank to the ambient are also considered.

The energy balance associated with the storage tank [11], is calculated for an increment of time Δt equal to 1 hour, and results in a final tank temperature T_{tank}^+, defined as:

$$T_{tank}^+ = T_{tank} + \frac{\Delta t}{m_{tank} C_{p,tank}} (Q_{C-T} - Q_{delivered} - Q_{loss}) \qquad (8)$$

Care should be taken to track units as Equation 6 is in units of power (typically MW) whereas Equation 8 is in units of energy (typically MJ) such that,

$$Q_{C-T} = \dot{Q}_{C-T} \times 3600. \qquad (9)$$

$Q_{delivered}$ in Equation 8 is a control-oriented term. It is set equal to Q_{load} when there is enough energy available in the storage tank to supply the thermal load to the brine, and to zero if not. Q_{load} is calculated as,

$$Q_{load} = C_{p,b} \dot{m}_b (T_{b,load} - T_{i,b}) \times 3600, \qquad (10)$$

and Q_{stored} is calculated as,

$$Q_{stored} = m_{tank} C_{p,tank} (T_{tank} - T_{t,ref}), \qquad (11)$$

where $\dot{m}_b$ and $C_{p,b}$ are the mass flow rate and specific heat capacity of the brine, respectively. Q_{stored} is the thermal energy stored in the tank and available for transfer to the brine, calculated using a reference temperature, $T_{t,ref}$, dependent on the heat exchanger effectiveness. For an immersed coil heat exchanger, this effectiveness is defined by Klett et al. [17] via Equation 12. 100% percent efficiency is reached when the coil fluid exits at a temperature equal to the tank temperature at the end of the time period.

$$\eta_{IHX} = \frac{(T_{coil,out} - T_{coil,in})}{(T_T - T_{coil,in})} \qquad (12)$$

The term T_T of Eq. 12 is the average temperature in the vicinity of the coil. To decrease the computational cost for this control-oriented model, a fully-mixed tank and one average tank temperature is assumed. This leads to a conservative estimate of thermal performance in comparison to a stratified tank model [18–20].

Given the fully-mixed tank assumption, the terms in Eq. 12 are solved for the variables representing the inlet and outlet

Copyright © 2023 by ASME

temperatures of the brine, the effectiveness of the immersed heat exchanger and the temperature of the tank. The value of $T_{t,ref}$ is defined as:

$$T_{t,ref} = T_{b,in} + \frac{(T_{b,load} - T_{b,in})}{\eta_{IHX,T-Ev}} \quad (13)$$

If there is enough thermal energy available to deliver the entire demanded load (Q_{load}) to the brine, and the tank temperature is high enough to drive heat transfer for the entire time step (one hour), then the load will be satisfied for the hour. If those conditions are not satisfied, then no thermal energy will be delivered to the brine.

Finally, the tank energy loss to the ambient Q_{loss} is calculated with Equation 14, where (UA_s) is the thermal loss coefficient-area product, with units of $[MW/°C]$.

$$Q_{loss} = (UA_s) \times 3600 \times (T_{tank} - T_{amb}) \quad (14)$$

The coefficient (UA_s) depends on the area and the volume of the storage tank. To depict how this coefficient (and therefore, the energy loss) varies as the tank size changes, data for existing commercial tanks [16] were used. A function was established for the relationship between the tank's surface area to volume ratio and the volume of the tank V_{tank} (Eq. 15); and for the heat loss coefficient dependence on surface area (Eq. 16). By obtaining the thermal loss coefficient U from Eq. 16, the value of (UA_s) can be calculated for any desired tank size. It is important to note however, that this model over estimates the heat loss coefficient for tanks with a volume smaller than 750 L (surface area = 5.4m^2), so it is not recommended for such sizes.

$$\frac{A_{tank}}{V_{tank}} = 0.17 \times V_{tank}^{-0.459} \quad (15)$$

$$U = 0.004 \times A_{tank}^2 - 0.112 \times A_{tank} + 1.246 \quad (16)$$

After the performing a given hour's energy balance, T_{tank}^+ is assigned as the new value of T_{tank}. A temperature constraint is applied to the tank, as per the tank data sheet.

2.4 Model outputs

2.4.1 Solar fraction. The *solar fraction* is considered to assess the value of the CSC implementation. Solar fraction relates the thermal energy required by the application (the load) to that met by the solar thermal system. One can then readily calculate the portion of the load to be met through an auxiliary source of energy. For j hours, the solar fraction f_s is defined as:

$$f_s = \frac{\sum_{i=1}^{j} Q_{delivered,i}}{\sum_{i=1}^{j} Q_{load,i}}. \quad (17)$$

The solar fraction can be calculated on a hourly, daily, monthly, and annual basis, allowing the user to make decisions based on seasonal trends.

2.4.2 Evaporation fraction. In order to couple the solar model with the CEE system model, it is necessary to determine the relationship between the solar heating performance and the evaporation performance. A $f_s < 1$ indicates that time intervals are present in which the thermal load cannot be met. If the full thermal load cannot be supplied for a given hour, no thermal energy is delivered. The CEE system will not run, and therefore the evaporation target cannot be met. The concept of the *evaporation fraction*, f_{ev} is introduced, and is defined as,

$$f_{ev} = \frac{h_{achieved}}{h_{target}} \quad (18)$$

where $h_{achieved}$ is the number of hours in which evaporation was achieved, and h_{target} is the number of hours in which the CEE system was intended to run to reach the evaporation target.

Note that depending on the system location and number of included evaporation surfaces, there may be hours in which no brine heating is required [8]; in these cases, the target evaporation for the hour is still achieved. All the hours in which evaporation is achieved (both those requiring heating, and those that do not) are counted towards the value of $h_{achieved}$.

2.4.3 Specific energy consumption. A common metric used to compare the performance of different desalination and brine minimization technologies is the *specific energy consumption*, SEC, given in units of energy per unit of treated water volume (kWh/m^3) [21]. The thermal and electric SEC for a given time period and target evaporation hours $h_{target} = j$, and for an hourly evaporation target V_{ev}, are given by:

$$SEC_{th} = \frac{f_s \sum_{i=1}^{j} Q_{load,i}}{f_{ev} h_{target} V_{ev}} \quad (19)$$

$$SEC_{el} = \frac{\sum_{i=1}^{j} L_{el,i}}{f_{ev} h_{target} V_{ev}} \quad (20)$$

where $L_{el,i}$ is the electric load of the system for a given hour.

3. RESULTS AND DISCUSSION

A case study was performed for Alamogordo, New Mexico, USA for the year 2020, as this was the site and time frame of the optimized operation schedule used in the CEE system performance evaluation by Kaddoura et al. [22].

3.1 Simulation Parameters

The variables associated with the CEE system are listed in Table 1 and match those originally used by Kaddoura et al. to generate the Pareto fronts [22]. For each hour, the actual ambient temperature and relative humidity were rounded up or down to increments of 5 °C and in increments of 10%, respectively. The discretized ambient temperature and RH levels were used to select the relevant CEE system Pareto front.

The operating specifications of the CST H3500 are summarized in Table 2. The meteorological data was originally downloaded from the National Aeronautics and Space Administration's (NASA) POWER Data access viewer [23]. The hourly TMY irradiance data was obtained from the National Solar Radiation

Copyright © 2023 by ASME

TABLE 1: DESIGN PARAMETERS FOR THE CEE [8]

Element	Value
Material cost ratio r_1	$55\ [kWh_{el}/surface]$
Energy cost ratio r_2	0.25
Target evaporation per day	$1 m^3$
Liquid flow rate	$1-6\ [L/hr \cdot surface]$
Liquid temperature	$T_{amb}\ up\ to\ 90°C$
Salinity of liquid	$0\ [g/kg]$
Evaporation surface dimensions	$0.6\ m \times 0.6\ m$
Surfaces per evaporation unit	100
Evaporation surfaces spacing	$2.4\ [cm]$

TABLE 2: COLLECTOR OPERATING DATA

Element	Value
Aperture Area	$4.4\ [m^2]$
Receiver area	$0.0064\ [m^2]$
Area concentration ratio	687.5
Working fluid	Glycol/Water Mix 50/50
Flow rate	0.0667 [L/s] or 0.0719 [kg/s]
Specific heat	$3.473\ [kJ/kg \cdot C]$
Maximum work fluid temperature	180 [°C]

Database (NSRDB) provided by the National Renewable Energy Laboratory (NREL) [24].

A series of parametric studies were carried out in order to evaluate the performance of the coupled CSC-CEE system for the optimized load. The list of input and output variables used can be found in Table 3. The range for collectors number N_{col} was chosen to see the trends in the behavior of the CSC-CEE system, although it would be unlikely that such number would be used in practice, due to the footprint associated with the CSC system of such dimensions (1050 m^2 at least for $N_{col} = 200$, 5.25 m^2 per collector [6]) against the footprint of a CEE system (7.65 m^2 at least for a system of 5 units with 100 trays each). The number of trays was selected following ranges for the analysis performed by Kaddoura et al. [8], for the CEE system. The sizes of the storage tank were based on the commercially available data sheets [16]. In addition, the analysis was stopped at 30 m^3 due the asymptotic behavior found in the performance with bigger tank sizes for the chosen maximum number of collectors (Figures 5 7, 9, 10).

3.2 Solar Fraction and Storage Tank Size

Appropriate selection of the solar field and tank size is important to maximize the energy delivery, and at the same time maintain system temperature and pressure at safe levels. First, solar fractions was plotted against the number of collectors for different tank sizes and a CEE system of 500 surfaces (5).

The first behavior to notice is that regardless of size, the increase in the solar fraction is not "smooth" as the number of collectors increases. This can be explained by the dynamics of the heat load required by the CEE system. The controller will request brine temperatures varying from 0 to 90 °C during different hours of the year, which translates to a highly variant heat load, as can be observed in Figure 6. When the number of collectors increases

TABLE 3: INPUT AND OUTPUT VARIABLES USED IN THE CSC-CEE MODEL PARAMETRIC STUDY

Input variables	Range	Output variables
N_{col}	0-200	f_s
N_{max}	100,200,500	f_{ev}
m_{tank}	$1\text{-}30 \times 10^3\ [kg]$	SEC_{th}
Max T_{IHX}/T_{tank}	110 / 95 [°C]	SEC_{el}
	125 / 110 [°C]	

by one, the stored energy increases, but not always by enough to cover the next "high" heat load hour. The overall energy stored in the tank needs to accumulate with the increase in collectors until more high-heat loads can be supplied. Additionally, the immediate decrease in solar fraction after a peak might occur because the system achieved the heat supply for an hour with a high heat load, without being able to supply heat to the following hours with low heat loads, decreasing the solar fraction.

The next trend to note in Figure 5 is the presence of a sharp increase in solar fraction with only 1-2 collectors. This is due to the high number of hours in which only a very small heat load is required (Figure 6). This initial increase in solar fraction is followed by a plateau before it starts increasing again. Larger tanks have a longer plateau. This can be attributed to the fact that it will require more collectors to elevate the temperature of a bigger tank to a point where heat transfer is possible for high temperature hours (higher hourly heat loads).

Even when there are no collectors (that is, when there is only a heat storage tank coupled to the CEE system) there is a non-zero solar fraction value ($f_s > 0$). This behavior can be explained by the definition of the tank losses to the ambient (Q_{loss}, Equation 14). The rate of heat transfer from the tank depends on the thermal loss coefficient-area product UA_s. As a result, the tank will not change temperature at the same rate as the ambient (and therefore the incoming brine). When there are low-heat load hours there are times when the energy accumulated by the tank during warmer hours of the day (Q_{stored}, Equation 11) can get transferred to the brine before it gets lost to the ambient.

Whenever the efficiency of the heat exchanger is $\eta_{IHX,2} < 1$, the temperature of the liquid in the storage tank needs to be higher than the target temperature of the brine to drive heat transfer. In cases where the brine needs to be heated to 90 °C, the tank temperature will need to be $> 90°C$. However, the water tank has a maximum temperature specification that is usually around 95 °C. By increasing the tank maximum specified holding temperature, more of the high-heat loads can be supplied, and the solar fraction increases in comparison to the original characteristics of the tank (Figure 5b). In addition to an overall increase in solar fraction, higher allowable tank temperatures also decrease the length of the plateaus because the tank offers a higher thermal capacity, allowing the system to store energy and start supplying higher heat loads with a fewer number of solar collectors.

Finally, it is observed that regardless of the maximum temperature of the tank, small tank sizes offer better performance, reaching the point of increasing solar fraction as soon as with 4

Copyright © 2023 by ASME

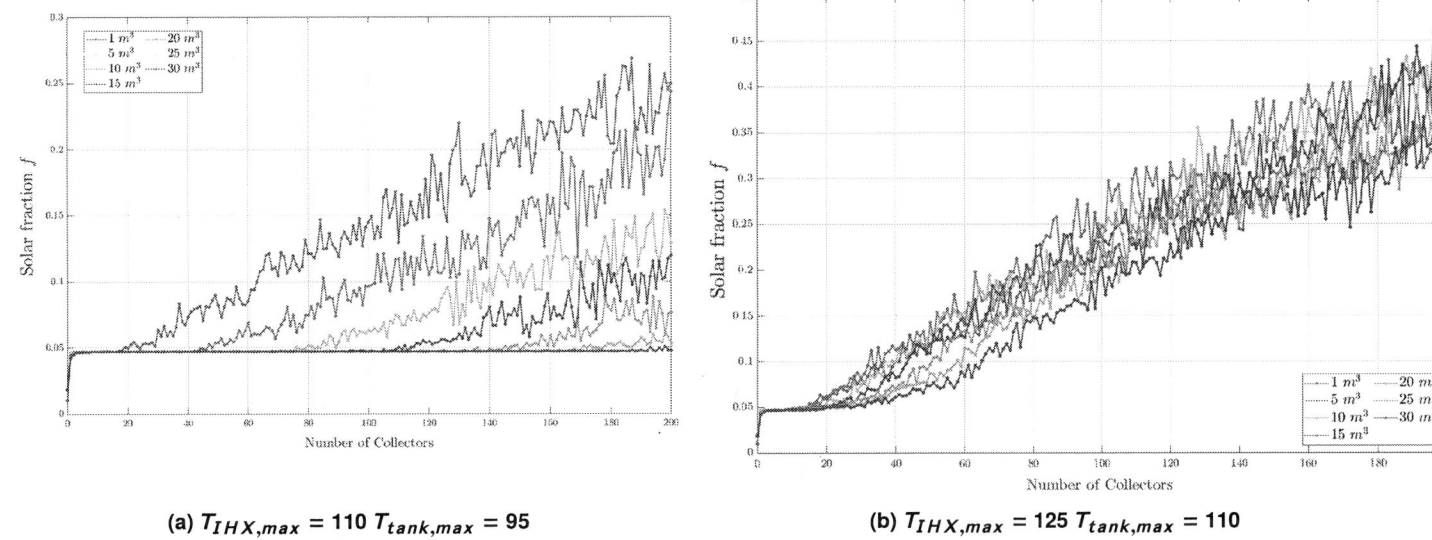

(a) $T_{IHX,max} = 110$ $T_{tank,max} = 95$

(b) $T_{IHX,max} = 125$ $T_{tank,max} = 110$

FIGURE 5: SOLAR FRACTION VS. NUMBER OF COLLECTORS FOR A CEE SYSTEM OF 500 TRAYS AND DIFFERENT MAXIMUM IHX AND MAXIMUM TANK TEMPERATURES.

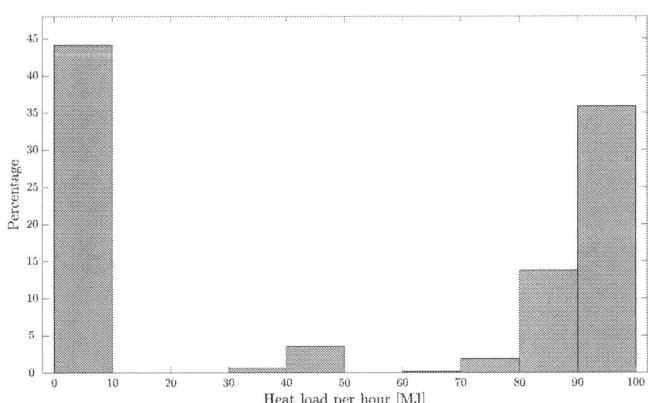

FIGURE 6: DISTRIBUTION OF THE ANNUAL HOURLY HEAT LOADS FOR A CEE SYSTEM OF 500 TRAYS.

collectors in the case of the 5 m^3 tank. Therefore, at least for the scale of the heat load required by the CEE evaporator, small tanks are recommended for coupling the solar heating system. It is also worth noting that while there is an overall increase in the solar fraction with a higher number of collectors, after the initial sharp increase in solar fraction, it can require up to 10 collectors before the solar fraction starts increasing again (the plateau is long). This is inconvenient for the project's goal of reducing the overall foot print of brine management relative to alternative options, such as evaporation ponds. More collectors also translates to higher costs.

3.3 Solar Fraction and Evaporation Fraction

From what has been observed in the previous section, it is noted that a reasonably sized solar array (<10 collectors) can achieve a solar fraction of no more than 5%; this means that the solar thermal system cannot reasonably supply more than 5% of

the originally targeted thermal load. This behavior raises the question of whether solar heating is useful at all for the system. So, the next step is to look at how these solar fraction values translate to evaporation performance.

As shown in Figure 7 the variations in evaporation fraction have a behavior that mimics those of the solar fraction; however, in terms of performance, the results look better: it is seen that by reaching just 5% solar fraction one can account for almost 87% of the total evaporation fraction. Furthermore, it is also worth noting that while the addition of the tank only covered 1% of the annual target heat requirements, in terms of evaporation performance it can represent up to 9.5% of the total (up to 84% of total evaporation minus the 74.5 % reached by using only electrical energy).

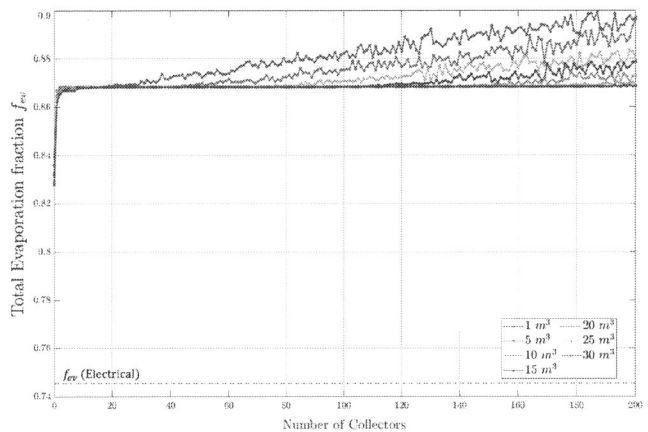

FIGURE 7: EVAPORATION FRACTION VS. NUMBER OF COLLECTORS FOR A CEE SYSTEM OF 500 TRAYS.

Increasing the number of collectors up to 200 collectors (an unrealistic number for the considered use scenarios) for the

Copyright © 2023 by ASME

case of the 1 and 5 m^3 storage tank, increases the solar fraction by about 20%, however, the evaporation fraction only increases by approximately 3%. It is therefore not recommended to add additional solar collectors to try to increase the performance after the first plateau in the solar and evaporation fraction value is reached.

Such behavior is explained, again, by how the hourly heat loads are distributed in magnitude and number since, as shown in Figure 6, the number of hours with the highest heat loads represent less than 10% of the total number of evaporation hours for the year, therefore, a low solar fraction does not translate to a low evaporation fraction, as long as the rest of the hours with low heat loads can be supplied.

3.4 Solar Heating Performance vs. CEE system size

Looking at the performance of the solar heating system for an initial CEE system size of 500 trays, it is important to note that out of the 0.87 total evaporation fraction, around 0.75 is achieved with electrical evaporation only. As explained in Kaddoura et al.'s optimization work [8], a higher number of evaporation trays tends to allow the system to reach target evaporation rates by adjusting the fan speed (electrical energy) only; brine heating is less frequently required.

To test the hypothesis that solar fraction and the role of the heating system would increase if the CEE system size were decreased, the next step was to redo the previously described simulations over differing smaller system sizes (fewer surfaces), while maintaining the same daily target evaporation. As indicated in Table 3, two new sets of CEE system sizes were selected, N_{max} of 100 and 200 trays. The increase in the annual heat load is shown in Figure 8. Additionally, there is an associated increase in electric load, caused by the increase in pressure drop through the evaporation trays associated with the increase in brine temperature.

The evaporation fraction and the solar fraction were calculated for different numbers of collectors and different tank sizes, to see how the optimal solar heating system configuration (number of collectors and size of the tank) varies depending on the size of the evaporation system. The results show that better evaporation performances are obtained with bigger systems, even if smaller systems present better solar fraction performances (Figures 9, 10).

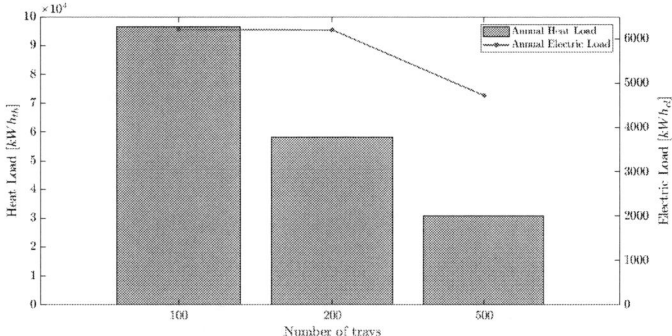

FIGURE 8: ELECTRIC AND HEAT LOADS FOR THREE DIFFERENT CEE SYSTEM SIZES

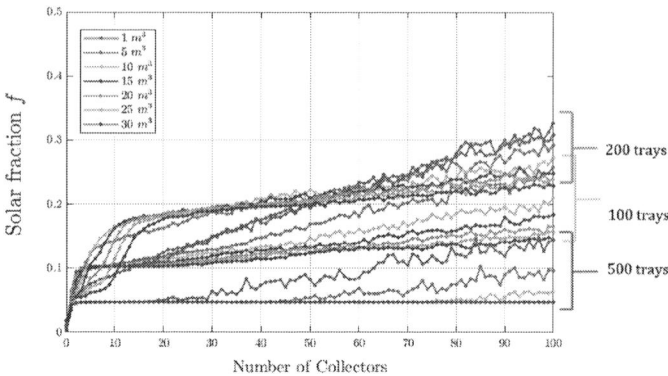

FIGURE 9: SOLAR FRACTIONS VS. NUMBER OF COLLECTORS FOR THREE DIFFERENT CEE SYSTEM SIZES.

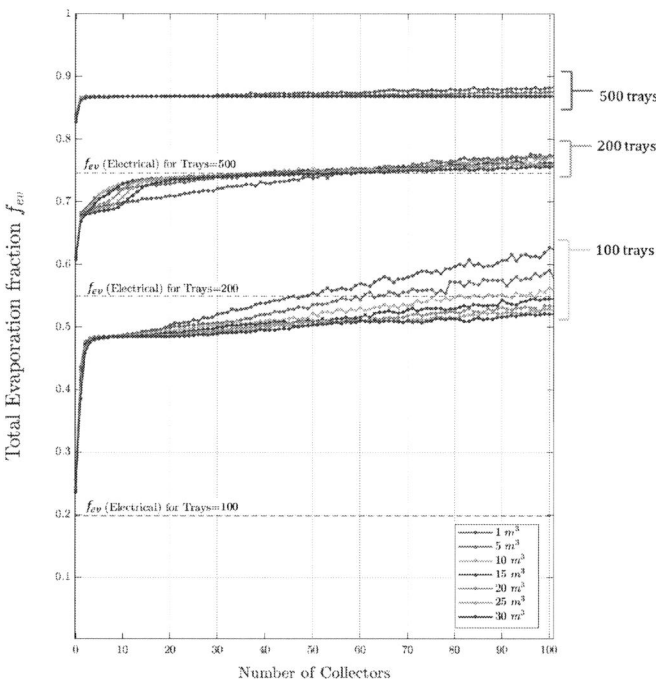

FIGURE 10: EVAPORATION FRACTIONS VS. NUMBER OF COLLECTORS FOR THREE DIFFERENT CEE SYSTEM SIZES.

3.5 Specific Energy Consumption (SEC)

After the parametric study performed in the previous sections, a final set of conditions were selected for further analysis. A solar collector and tank sizing was selected for evaluation and comparison across CEE system sizes. The selected number of collectors was 6 and the size of the tank was 5 m^3 (Figure 11).

If is found that the estimated SEC for the CSC-CEE system is similar to those of other commercial options, which cover a range of 50-500 $[kWh/m^3]$ for SEC_{th} and 2-50 $[kWh/m^3]$ for SEC_{el} [8]. It is important to note that these values are associated with partial evaporation fractions, and therefore there exists a tradeoff between the SEC and reliability of the system. Further analysis is needed to compare the footprint of such technologies against the CSC-CEE system together with the storage of the non-evaporated brine.

Copyright © 2023 by ASME

4. NON-EVAPORATED BRINE

A total evaporation fraction <1 indicates that there is a volume of brine that will not get evaporated, either due to lack of adequate thermal energy or due to insufficient surface area (in the case of the solar-to-electric operation mode). Calculating this volume is important for overall system sizing and budgeting when designing a CEE system for a specific location. For a constant hourly evaporation target, V_{ev}:

$$Non - evaporated\ brine = h_{target}\ V_{ev}\ (1 - f_{ev}) \quad (21)$$

Annual non-evaporated brine volumes are depicted in Figure 12. Future work should include analysis of the cost and footprint implications of this non-evaporated brine. The brine could be disposed through a different method, stored and evaporated during hours with conditions of low energy consumption, or transported (trucked) to a disposal location. Further cost analysis on a case-by-case basis will be needed to determine the trade-offs between costs of transportation, extra energy consumption, and footprint.

5. CONCLUSION

The project presents a computational model that has the goal of evaluating the use of a concentrated solar collector (CSC) system for the heat load supply of a convection-enhanced evaporation system (CEE), which is a novel proposed method to evaporate waste brine for small-scale plants (<100 m³/day). The model predicts the performance of the coupled CSC-CEE system in terms of specific energy consumption, solar fraction, and evaporation fraction (relative to the target).

The configuration of the model includes a CSC field; the thermal energy is supplied and stored using a fully mixed storage tank with two immersed-coil heat exchangers (IHX). A series of control-oriented parameters were included in the model to ensure precise compliance with CEE system operating conditions.

Findings establish that due to the high thermal energy demand of the system, the solar fraction can be relatively low (<20% for small collector-tank configurations). However, this does not necessarily translate to low evaporation performance (up to 86% of the target volume). The specific energy and evaporation performance of the CSC-CEE system are similar to larger,

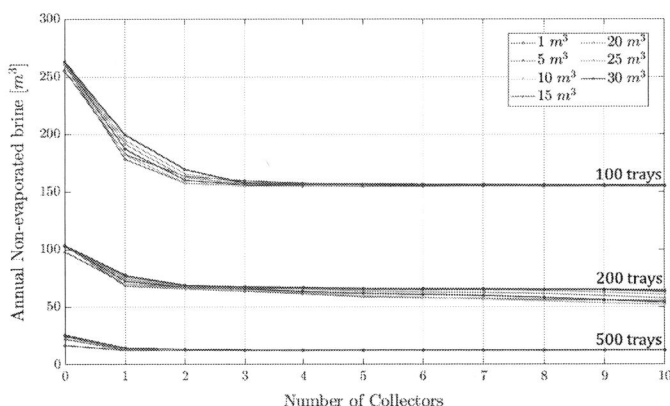

FIGURE 12: ANNUAL NON-EVAPORATED BRINE VOLUMES FOR DIFFERENT SIZES OF CEE SYSTEMS, DEPENDING OF THE NUMBER OF COLLECTORS AND SIZE OF THE HEAT STORAGE TANK.

commercially available systems. Further work is needed to understand the tradeoff in footprint required for the CSC system, the CEE system, and any non-evaporated brine storage.

NOMENCLATURE

Symbols

A	Area [m^2]
C_p	Specific heat [kJ kg^{-1} °C^{-1}]
DNI	Direct Normal Irradiance [MW]
$\dot{m}$	Mass flow rate [kg s^{-1}]
N	Number, count [-]
Q	Energy [MJ]
$\dot{Q}$	Energy per unit time [MW]
$Q_{delivered}$	Energy delivered to the brine [MJ]
Q_{stored}	Energy stored on the storage tank [MJ]
Q_{loss}	Energy losses from the tank to the ambient [MJ]
T	Temperature [°C]
$T_{t,ref}$	Minimum temperature of the tank to drive the target heat transfer [°C]
$(UA)_s$	Tank loss coefficient-area product [MW/°C]

Greek letters

η	Efficiency [-]

Superscripts and subscripts

a	aperture
amb	ambient
av	available
b	brine
col	Concentrated Solar Collector (CSC)
C-T	CSC to Tank
ev	evaporation
IHX	Immersed Coil Heat Exchanger
i	inlet
load	
o	outlet
r	receiver
T-Ev	tank to evaporation system
t,tank	tank

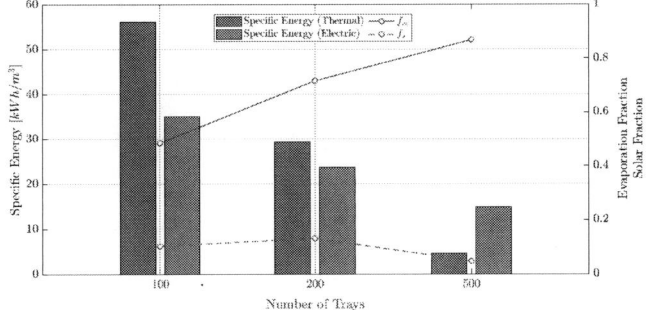

FIGURE 11: TOTAL SPECIFIC ENERGY CONSUMPTION FOR DIFFERENT SYSTEM SIZES ALONG WITH THEIR EVAPORATION AND SOLAR HEATING PERFORMANCE.

Copyright © 2023 by ASME

ACKNOWLEDGMENTS

The authors would like to acknowledge QuadSun Solar for providing the actual certification test data for the concentrated solar collector modeled in this study.

REFERENCES

[1] Nations, United. "Goal 6 | Department of Economic and Social Affairs." Accessed 2023-01-10, URL https://sdgs.un.org/goals/goal6.

[2] Vizzuality. "Populations in Coastal Zones." Accessed 2023-02-10, URL https://resourcewatch.org/data/explore/Populations-in-Coastal-Zones.

[3] Li, Chengcheng, Gao, Xubo, Li, Siqi and Bundschuh, Jochen. "A review of the distribution, sources, genesis, and environmental concerns of salinity in groundwater." *Environmental Science and Pollution Research* Vol. 27 No. 33 (2020): pp. 41157–41174. DOI 10.1007/s11356-020-10354-6. Accessed 2023-02-27, URL https://doi.org/10.1007/s11356-020-10354-6.

[4] Panagopoulos, Argyris, Haralambous, Katherine-Joanne and Loizidou, Maria. "Desalination brine disposal methods and treatment technologies - A review." *Science of The Total Environment* Vol. 693 (2019): p. 133545. DOI 10.1016/j.scitotenv.2019.07.351. Accessed 2023-01-23, URL https://www.sciencedirect.com/science/article/pii/S0048969719334655.

[5] Kaddoura, Mustafa F., Chosa, Matthew, Bhalekar, Prakash and Wright, Natasha C. "Mathematical modeling of a modular convection-enhanced evaporation system." *Desalination* Vol. 510 (2021): p. 115057. DOI 10.1016/j.desal.2021.115057. Accessed 2022-07-21, URL https://linkinghub.elsevier.com/retrieve/pii/S0011916421001284.

[6] Solar, QuadSun. "QuandSun Solar Official Website." Accessed 2023-01-11, URL http://quadsunsolar.com/.

[7] ASHRAE. *ASHRAE Handbook of fundamentals* (1997).

[8] Kaddoura, Mustafa F. and Wright, Natasha C. "Optimization of convection-enhanced evaporation (CEE) using generalized cost ratios." *Water Research* Vol. 219 (2022): p. 118491. DOI 10.1016/j.watres.2022.118491. Accessed 2023-01-24, URL https://www.sciencedirect.com/science/article/pii/S0043135422004456.

[9] Liu, Yanfeng, Yu, Sisi, Zhu, Ying, Wang, Dengjia and Liu, Jiaping. "Modeling, planning, application and management of energy systems for isolated areas: A review." *Renewable and Sustainable Energy Reviews* Vol. 82 (2018): pp. 460–470. DOI 10.1016/j.rser.2017.09.063. Accessed 2023-05-14, URL https://www.sciencedirect.com/science/article/pii/S136403211731314X.

[10] Sharon, H. and Reddy, K. S. "A review of solar energy driven desalination technologies." *Renewable and Sustainable Energy Reviews* Vol. 41 (2015): pp. 1080–1118. DOI 10.1016/j.rser.2014.09.002. Accessed 2023-01-22, URL https://www.sciencedirect.com/science/article/pii/S1364032114007758.

[11] Duffie, John A., Beckman, William A. and Blair, Nathan. *Solar Engineering of Thermal Processes, Photovoltaics and Wind*, 5th ed. John Wiley & Sons (2020). Google-Books-ID: 4vXPDwAAQBAJ.

[12] Curto, Domenico, Franzitta, Vincenzo and Guercio, Andrea. "A Review of the Water Desalination Technologies." *Applied Sciences* Vol. 11 No. 2 (2021): p. 670. DOI 10.3390/app11020670. Accessed 2023-01-22, URL https://www.mdpi.com/2076-3417/11/2/670. Number: 2 Publisher: Multidisciplinary Digital Publishing Institute.

[13] Palenzuela, Patricia and Alarcón-Padilla, Diego C. "Concentrating Solar Power and Desalination Plants." Polo, Jesús, Martín-Pomares, Luis and Sanfilippo, Antonio (eds.). *Solar Resources Mapping: Fundamentals and Applications*. Green Energy and Technology. Springer International Publishing, Cham (2019): pp. 327–340. DOI 10.1007/978-3-319-97484-2_14. Accessed 2022-08-16, URL https://doi.org/10.1007/978-3-319-97484-2_14.

[14] Guillen Rodriguez, Nallely. "Modeling of a Solar Thermal Energy Input for an Optimized Convection-Enhanced Evaporation System." Ph.D. Thesis, University of Minnesota. 2023.

[15] Ben Bacha, H., Dammak, T., Ben Abdalah, A.A., Maalej, A.Y. and Ben Dhia, H. "Desalination unit coupled with solar collectors and a storage tank: modelling and simulation." *Desalination* Vol. 206 No. 1-3 (2007): pp. 341–352. DOI 10.1016/j.desal.2006.05.018. Accessed 2022-07-15, URL https://linkinghub.elsevier.com/retrieve/pii/S0011916406014329.

[16] Winkelmann, Reflex. "Hot Water Storage Tanks & Heat Exchangers." Accessed 2023-02-08, URL https://www.reflex-winkelmann.com/en/products/reflex_products/warmwasserspeicher/products/reflex_products/warmwasserspeicher/.

[17] Klett, D. E., Goswami, D. Y. and Saad, M. T. "Thermal Performance of Submerged Coil Heat Exchangers Used in Solar Energy Storage Tanks." *Journal of Solar Energy Engineering* Vol. 106 No. 3 (1984): pp. 373–375. DOI 10.1115/1.3267612. Accessed 2023-01-15, URL https://asmedigitalcollection.asme.org/solarenergyengineering/article/106/3/373/440601/Thermal-Performance-of-Submerged-Coil-Heat.

[18] Nash, Austin L., Badithela, Apurva and Jain, Neera. "Dynamic modeling of a sensible thermal energy storage tank with an immersed coil heat exchanger under three operation modes." *Applied Energy* Vol. 195 (2017): pp. 877–889. DOI 10.1016/j.apenergy.2017.03.092. Accessed 2022-07-21, URL https://www.sciencedirect.com/science/article/pii/S0306261917303343.

[19] Ghaddar, N. K. "Stratified storage tank influence on performance of solar water heating system tested in Beirut." *Renewable Energy* Vol. 4 No. 8 (1994): pp. 911–925. DOI 10.1016/0960-1481(94)90225-9. Accessed 2023-01-15, URL https://www.sciencedirect.com/science/article/pii/0960148194902259.

[20] Jordan, Ulrike and Furbo, Simon. "Thermal stratification in small solar domestic storage tanks caused by draw-offs." *Solar Energy* Vol. 78 No. 2 (2005): pp. 291–300. DOI 10.1016/j.solener.2004.09.011. Accessed 2023-

Copyright © 2023 by ASME

01-15, URL https://www.sciencedirect.com/science/article/pii/S0038092X04002841.

[21] Guo, Penghua, Li, Tiantian, Wang, Yi and Li, Jingyin. "Energy and exergy analysis of a spray-evaporation multi-effect distillation desalination system." *Desalination* Vol. 500 (2021): p. 114890. DOI 10.1016/j.desal.2020.114890. Accessed 2023-02-01, URL https://www.sciencedirect.com/science/article/pii/S001191642031568X.

[22] Kaddoura, Mustafa F. and Wright, Natasha C. "Novel data-driven optimal control methods for Convection-Enhanced Evaporation (CEE) systems." (2022).

[23] National Aeronautics and Space Administration. "POWER | Data Access Viewer." (2021). Accessed 2022-11-26, URL https://power.larc.nasa.gov/data-access-viewer/.

[24] National Solar Radiation Database. "Typical Meteorological Year (TMY)." Accessed 2022-11-23, URL https://nsrdb.nrel.gov/.

Proceedings of the ASME 2023
International Design Engineering Technical Conferences and
Computers and Information in Engineering Conference
IDETC-CIE2023
August 20-23, 2023, Boston, Massachusetts

DETC2023-111278

EMPIRICALLY TUNED MECHANICAL SIMULATION MODEL OF 3D-PRINTED BIAXIAL WEAVES

Marc Wirth
Engineering Design and Computing Laboratory
Dept. of Mechanical and Process Engineering
ETH Zurich, Switzerland

Kristina Shea
Engineering Design and Computing Laboratory
Dept. of Mechanical and Process Engineering
ETH Zurich, Switzerland

ABSTRACT

Additive manufacturing facilitates the realization of previously infeasible designs. Amongst them are 3D-printed textiles and specifically 3D-printed weaves. Textile designs can be spatially tuned in their mechanical properties by changing the pattern and material locally, when printed on a multi-material 3D printer. To inversely design weaves with desired mechanical properties, an efficient simulation of the mechanical behavior of textiles is essential. State-of-the-art textile models are predominantly set up for nonrecurring simulations. They are accurate but slow. Textile simulations designed for efficiency are seldom capable of capturing nonlinear yarn interaction effects. In this work, we propose a textile model consisting of 1D truss elements with strain and shear dependent stiffness. The truss member properties are defined individually based on the local weave design. The properties are tuned for each textile design by matching load-extension curves and sample deformations of virtual tensile tests to test data of real tensile tests. The simulation approach is validated by tuning the algorithm for two configurations and simulating three designs not previously used for tuning. The simulation can estimate the load-extension curve of all validation samples with high accuracy and shows runtimes of under 40s on an Intel Core i7-10750H for structures with more than 3500 elements. These results pave the way towards a simulation-based approach to synthesize and optimize architected 3D-printed textiles.

Keywords: additive manufacturing, textile model, discrete simulation, mechanical testing, parameter tuning

1. INTRODUCTION

Additive manufacturing, or commonly called 3D printing, enables the design and manufacturing of previously infeasible structures, such as lattice structures [1,2], topology optimized designs [3], metamaterials [4], and active structures that react to external stimuli [5]. Recently, interest has grown in the field of 3D-printed textiles [6,7]. The design freedom inherited from the 3D-printing process is exploited to design visually appealing apparel [8,9], to personalize wearables [10], and to tune textile properties [11,12].

Properties in 3D-printed textiles can be tuned by varying the yarn geometry and yarn material and by varying the local structure. This is fundamentally different from traditional textiles, which are made up from a continuous yarn woven, knitted, or braided into a structure and tuned mainly by determining the yarn material [13] and adjusting the interlacing structure [14–16]. Previous research has shown that varying individual design parameters, such as the yarn diameter, yarn spacing in the in-plane and out-of-plane direction, and yarn pattern, influences the overall mechanical properties of textiles [11]. See FIGURE 1 for an illustration of possible design parameters to tune 3D-printed textiles and an example of a 3D-printed weave. These results illustrate the current capability to spatially tune the mechanical properties of textiles by adjusting the textile parameters locally. Tuning mechanical properties locally is beneficial in applications that benefit from personalization, like functional clothes and rehabilitation devices, for example orthoses [17].

To reach the potential of 3D-printing textiles with varying geometric and material properties, inverse design is required to tune them. Inverse design approaches typically consist of an algorithm that alters a design in a certain direction and an estimator to provide the behavior of a given design that is used to evaluate the performance given certain metrics [18,19]. In the case of mechanical behavior, the estimator often is a finite element (FE) simulation. Textiles consist of multiple small-scale bodies interacting with each other through non-linear contact

Copyright © 2023 by ASME

effects. Due to this hierarchical nature, setting up a textile FE simulation is especially challenging.

a) Design Parameters

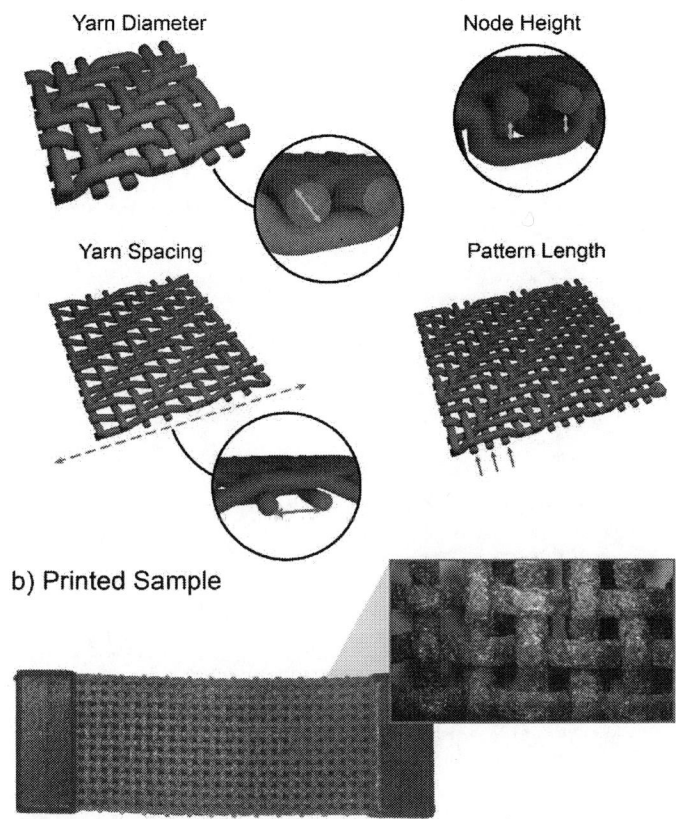

b) Printed Sample

FIGURE 1: a) Examples of parameters for the design of customized 3D-printed weaves b) Example of a 3D-printed weave. Images adapted from [11].

Current textile models can be classified into micro-, meso- and macroscale models [20]. Microscale models simulate the individual fibers or fiber bundles constituting the yarns. Mesoscale models treat the yarns as the smallest geometric unit and model them directly [21]. Macroscale models consider the textile structure as a continuous structure and model the mechanical behavior by a continuum mechanics approach or by discretizing the structure into smaller connected elements like truss members [22,23], beams [24], or membranes [25]. Continuum mechanics based models assume the textile to be a homogenous material. It is thus challenging to model yarn interaction effects [20] and to simulate textiles with spatially varying properties. Discrete truss models are a time-efficient way of modeling textiles [26] but lack accuracy for the behavior observed at high strains [27].

In this work, we address the challenge of simulating the in-plane mechanical behavior of textiles with spatially varying properties in a time-efficient and accurate way. We propose a truss model with strain-dependent truss members and strain-dependent shear elements only active under compression. The model is set up to correspond to the physical structure of a 3D-printed biaxial weave, such that the truss members correspond to the textile yarns. We further propose a shear-dependent yarn softening factor to account for the stiffness overestimation in high strain regions. The truss member properties are defined on a meso-scale by assigning each pin-joint distinct properties depending on the local weave design. We tune the truss member properties using an optimization loop comparing simulation results to corresponding measurement data. This tunes the model with respect to empirical measurements so that the model reflects locally defined properties. The aim of this approach is to balance runtime efficiency with accuracy for the in-plane mechanical properties of the modeled weaves. This will be a significant step towards creating a simulation that can be used in an inverse design approach to help designers quickly iterate alternative designs and precisely tune textiles to achieve desired properties.

2. BACKGROUND

Current research in simulation of textile mechanics categorizes textile models either according to the model scale [20] or the type of approximating elements [28]. In terms of scale, the model types are differentiated by micro-, meso- and macroscale models. Macroscale models approximate the textile sheet as a continuous structure. Meso- and microscale models approximate the yarns and fibers making up the yarns with discrete elements, respectively. In terms of element type, the models are differentiated by continuous, discrete, semi-discrete and hybrid models. Continuous models regard textiles as structures with homogenized properties. Discrete models approximate the textiles as an assembly of smaller structures. Semi-discrete models discretize the textile into smaller elements, which do not necessarily correspond to the true textile structure. Hybrid models are a combination of several approximation types.

One of the early textile models is the pin-joint-net approximation (PJN) [29]. This macroscale semi-discrete model approximates a textile sheet as inextensible bars connected by frictionless hinges. Such a model is extremely fast, but cannot model mechanical behavior [28]. The PJN-model is altered in consecutive work by exchanging the inextensible bars with spring-like truss members and by adding additional elements to model observed effects more accurately. Sharma *et al.* added shear elements connecting nodes in a rectangular unit cell diagonally to introduce shear stiffness [22]. Further research changed the formulation and configuration of the shear element stiffness to include time dependency [23] and to generalize the model [27]. It has been shown that PJN-based approaches are computationally very efficient, but only partially capture yarn defects and fail to accurately depict shear [26].

Other approaches in semi-discrete macroscale simulation proposes modeling the textiles as an assembly of pantographic beams [30], as a mass-spring structure [31] and as hybrid models combining trusses, pantographic beams, and membranes [32]. Due to the model-imposed element connectivity, these models are often not designed to depict yarn interaction effects but to approximate the overall mechanical behavior.

Copyright © 2023 by ASME

To capture interaction effects like yarn sliding, friction, and yarn jamming, discrete meso- and microscale models are beneficial since they incorporate these effects by design. The yarns and fibers are often approximated by chains of elements such as trusses [33] and beams [34,35]. Such models can be very accurate, but are challenging to set up since the parameters determining friction, tension, compression, shear, and torsion have to be set individually [21] and contact effects are difficult to model [36]. Further, these simulations typically take from hours to days to converge [21,26].

A more efficient and therefore widely researched model of simulating textiles is the continuous macroscale model. Models in this category assume continuity in the textile sheet. The textile is viewed as a continuum with homogenous properties and modeled as a sheet body. Common approaches use either non-orthogonal elastic models [37], hypoelastic models [38], hyperelastic models [39], dissipative models [40], or second gradient models [41]. While all of these models are very sophisticated, the homogenization approach requires a continuum assumption. Hence, spatial and material variation of textile properties, as possible in 3D-printed textiles, cannot be modeled.

3. METHODS

The textile model proposed in this work is a pin-jointed truss network enhanced by diagonal shear elements. All elements are assigned a non-linear, piecewise defined strain- and shear-dependent stiffness depending on the physical weave design. The stiffness curves are tuned by matching data of tensile tests of samples with a uniform pattern to their virtual counterpart.

3.1 Simulation Model

The simulated geometry is a network of 1D truss elements connected by pin-joint nodes. The geometry is fixed in the out-of-plane direction. The nodes and truss elements of the model are depicted in FIGURE 2. The truss network is set up to mimic the physical 3D-printed weave. The nodes are placed at the yarn crossings and the truss members are connecting the nodes along the yarn direction and diagonally. The truss elements connecting the nodes along the yarn direction (thick lines in FIGURE 2) are called the main elements. The truss elements connecting the nodes diagonally (thin lines in FIGURE 2) are called the shear elements. They account for shear stiffness. The angles, $\varphi_{1,1}$ to $\varphi_{1,4,}$ between the main elements are used to determine the shear in the network at each node. The shear value τ_i at a node p_i is defined to be the average absolute change in the corresponding n_i angles according to Equation (1).

$$\tau_i = \frac{1}{n_i} \sum_{n_i} |\Delta\varphi_{i,n}| \qquad (1)$$

To represent nonlinearities typically occurring in textile deformation, such as material nonlinearities, friction, yarn sliding and yarn jamming, the truss members have a non-linear stiffness formulation. The stiffness $k_{i,j}$ of an element $b_{i,j}$ is strain- and shear dependent according to Equation (2).

$$k_{i,j} = f(\varepsilon_{i,j}) \cdot ys(\tau_{avg}) \qquad (2)$$

Where τ_{avg} is the average shear in the nodes p_i and p_j adjacent to the element $b_{i,j}$ according to Equation (3).

$$\tau_{avg} = 0.5 \cdot (\tau_1 + \tau_2) \qquad (3)$$

The truss member stiffness $k_{i,j}$ is therefore a function of the truss member strain $\varepsilon_{i,j}$ and modulated by the yarn softening factor ys depending on the average shear τ_{avg}. For shear elements, ys is inactive and set to one.

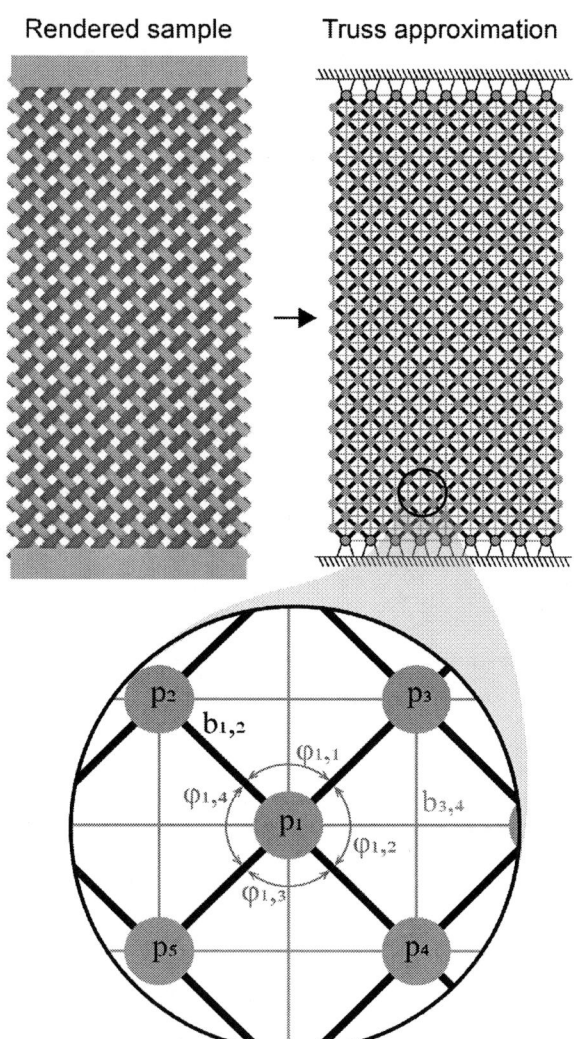

FIGURE 2: Truss representation of a weave. The nodes p_1 to p_5 are connected by the main elements (thick lines) mimicking the physical weave geometry. Shear elements (thin lines) connect the nodes diagonally. The angles between the main members are used to evaluate the local shear value.

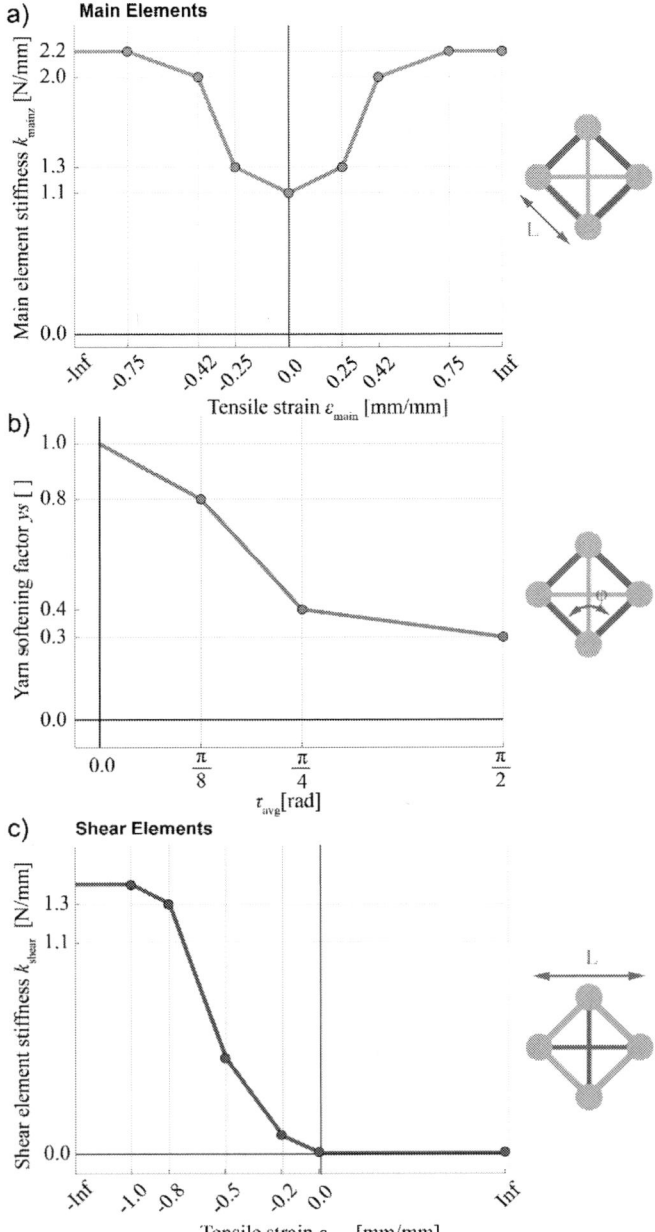

FIGURE 3: Examples of piecewise linear defined stiffness functions. a) Example of the dependency between the main element stiffness k_{main} and the truss member strain. b) The yarn softening factor, ys, depending on the average shear value. c) Example of the strain-dependent shear element stiffness k_{shear}.

Both functions, $f(\varepsilon)$ and $ys(\tau_{avg})$, are defined as being piecewise linear. The functions are uniquely defined by sets of control points. The number and location of the control points can be set arbitrarily, which enables the creation of a large variety of piecewise defined functions. Examples of these piecewise linear functions are depicted in FIGURE 3. In this work, the $f(\varepsilon)$ of the main elements (FIGURE 3a) is called the main element stiffness k_{main}. It is symmetric since the truss members will

never undergo compression due to the designated axial tension load case. The yarn softening factor $ys(\tau_{avg})$ (FIGURE 3b) for the main elements is strictly decreasing and set to one for $\tau_{avg}=0$. The function $f(\varepsilon)$ of the shear elements (FIGURE 3c) is called k_{shear}. The function is constantly increasing with increasing compressive strain and set to 0.01 N/mm for tensile strains. This is to prevent numerical issues in the FE simulation while being close to zero in accordance with previous research [27].

The stiffness functions $f(\varepsilon)$ for the main and shear elements and the yarn softening factor function $ys(\tau_{avg})$ are defined as being properties of the networks nodes. Each node can be assigned a unique set of functions. This effectively enables defining the network locally and thus enables the simulation of textiles with local variations in both design and properties. In this work, the node properties are defined based on the local configuration of the physical textile sample. A textile configuration is defined as the values of the design parameters evaluated at the target location. If a textile changes in its design throughout the sample, for example by gradually varying the yarn diameter, the configuration is varying as well. Since in a pin-jointed network, the stiffness values have to be assigned to the truss members and not to the nodes, each truss member inherits the average properties of its adjacent nodes.

3.2 Finite Element Algorithm

The FE algorithm is shown schematically in FIGURE 4. The user inputs consist of the geometry (nodes p and bars b), the material properties of each textile configuration (stiffness and yarn softening factor), the loads and boundary conditions and the solver options. The input is assembled, and each bar in the model is assigned its stiffness properties depending on the adjacent nodes. The simulation developed in this work is written as a custom FE framework in MATLAB.

The algorithm solves the problem in two intertwined loops. The outer loop increases the load on the structure by an incremental amount and checks the convergence of the inner loop. The inner loop performs displacement update steps on the structure with a Newton-Raphson scheme and a defined update step length. In each run, the inner loop assembles the tangential stiffness matrix based on the current truss member stiffness, estimates the update step u_{target}, scales the update step such that the maximum updated position is equal or below the maximum step size, and updates the node positions accordingly. The convergence value is defined as the maximum entry of the target displacement u_{target}. Whenever this value falls below a defined convergence threshold, the load step is converged. If the convergence value is increasing, the maximum step size is decreased by a step size reduction factor. If the simulation does not converge within a certain number of iterations, the algorithm stops and flags that the solution is not converged.

Copyright © 2023 by ASME

Inputs

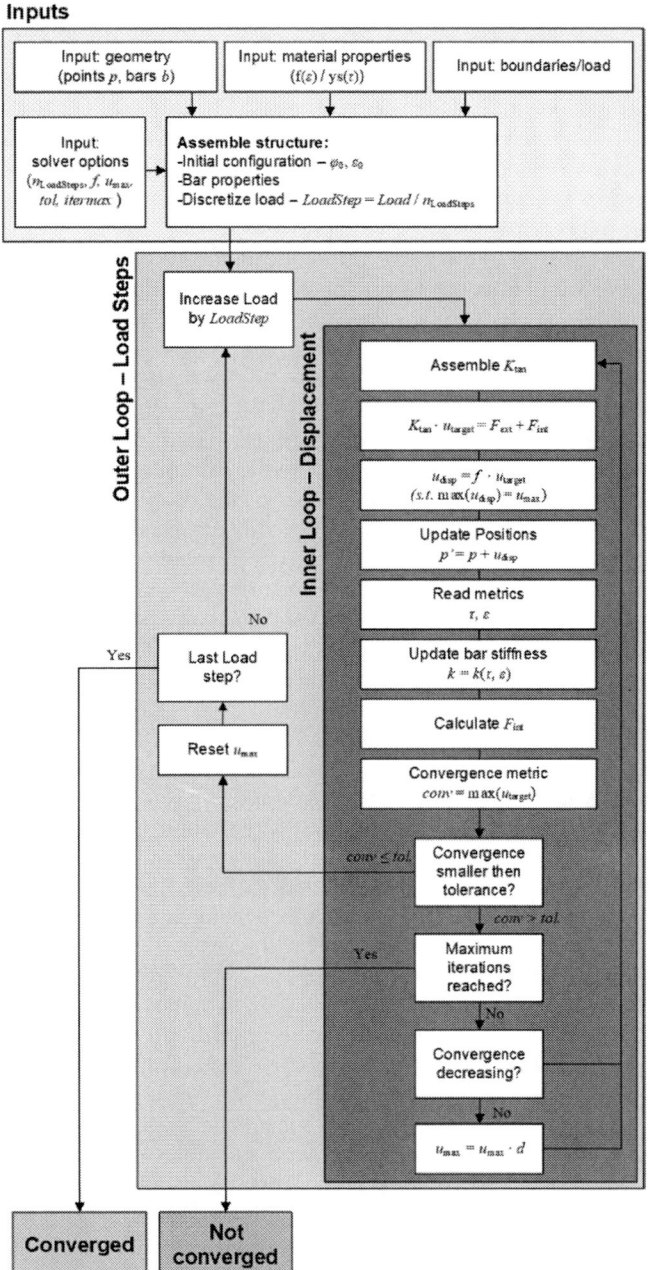

FIGURE 4: Flowchart of the FE simulation algorithm. The input is given by the user. The outer loop increases the load step gradually when the previous load step is converged. The inner loop performs update steps via a Newton-Raphson scheme with a defined step length until convergence.

3.3 Simulation Tuning

The simulation can be tuned by defining the stiffnesses of the main and shear elements and the yarn softening factor ys. The tuning is performed by shifting the control points defining the shape of the stiffness and yarn softening curves such that the result of the FE simulation mimics the measurement data of the corresponding 3D-printed weave with the same load and

boundary conditions. In this case, we carry out tensile tests on 3D-printed weaves and compare the measurements to virtual tensile tests. The simulation is tuned for each textile configuration individually. This means that the 3D-printed samples used for tuning have design with one configuration only. These samples have a uniform pattern, yarn diameter and yarn spacing. This restricts the FE simulation to only have nodes with equal properties and restricts the number of unknown variables. With this approach, the functions for the stiffnesses and yarn softening factor can be evaluated for each configuration individually and a set of different configurations can be built up consecutively.

The tuning for each configuration is performed in two steps using two different 3D-printed samples. First, the structure is simulated and tested in the yarn direction (rotation $\alpha=0°$), such that only the main elements in one direction are under tension (see FIGURE 5a). This step only tunes the stiffness of the main elements. The main element stiffness is set to be defined by three parameters; the initial stiffness k_{init}, the transition strain ε_{tr} and the final stiffness k_{∞}. The yarn softening factor and the shear stiffness are not tuned in this step and set to one and zero, respectively. To prevent numerical issues in the FE simulation due to zero shear stiffness, rotational springs are added between each pair of truss members. These rotational springs act as a balancing force on the nodes if the angle between two truss members deteriorates from $\pi/2$. They are exclusively added in the first tuning step due to the shear elements having no designated properties yet. In the second tuning step, a sample is simulated and tested in the diagonal direction (rotation $\alpha=45°$, see FIGURE 5b). Here, the shear element stiffness and the yarn softening factor are tuned simultaneously. The shear element stiffness is set to be defined by five control points varying in their stiffness value. The strain values of the control points are fixed and equally spaced between -1.25 and 0. Since the shear element stiffness is strictly increasing for increasing compressive strain, the design variables in the optimization scheme are defined being the stiffness differences $\Delta k_{shear,i}$ between the six control points. The yarn softening factor is strictly decreasing with increasing shear value and is thus tuned analogous to the shear element stiffness by adjusting the difference Δys_i between five adjacent control points. The control points shear values are predetermined such that the points are equally spread between 0 and $\pi/2$. The total number of design variables for the optimization in step two is nine; five for k_{shear} and four for ys. The main element stiffness in this step is set to the stiffness determined in tuning step one.

The tuning is performed with a MATLAB implementation of the NOMAD optimization algorithm [42]. The design variables for the optimization are the tuning parameters defined previously. The objective function runs the FE simulation with the curves determined by the current design variables and calculates an objective function by comparing the simulated load-extension curve and the simulated sample deformation at maximum extension to the load-extension curve and sample deformation measured during experiments.

Copyright © 2023 by ASME

a) Step 1 - Main Element

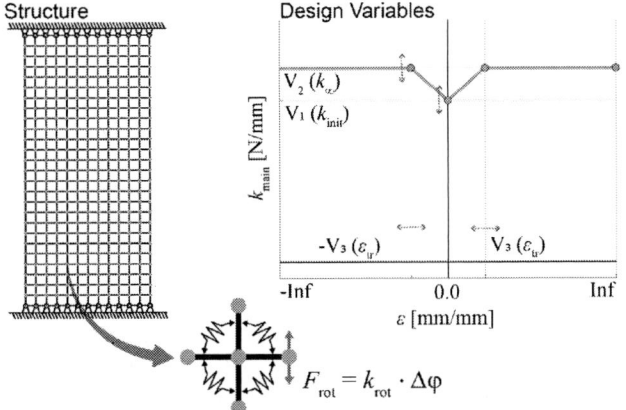

b) Step 2 - Yarn Softening / Shear Element Stiffness

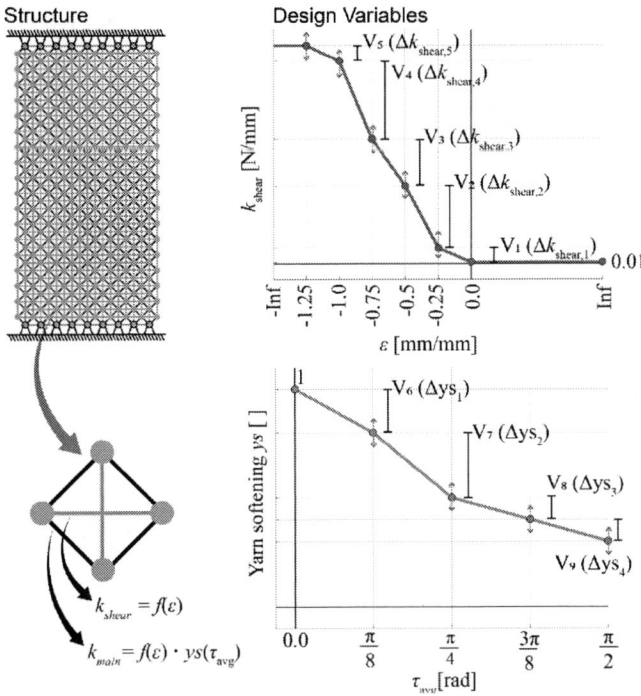

FIGURE 5: Procedure for tuning the properties of one configuration. a) Step 1 tunes the main element stiffness by comparing the FE simulation enhanced by rotational springs to samples tested in the yarn direction (rotation $\alpha=0°$). b) Step 2 tunes the shear element stiffness and the yarn softening factor by comparing the FE results to samples tested in diagonal direction (rotation $\alpha=45°$).

The objective function, OV, is calculated according to Equation (4). It is a weighted multi-objective function combining the deviation of the displacement field and the load-extension between the model and measurement results. Δpos is the distance between the simulated and measured deformation at maximum extension evaluated by the point cloud representing the recorded image data, scaled by the length of the main truss members. ΔF is the difference between the simulated and measured load scaled by the maximum measured load. The

weights W_1 and W_2 are set to 1 and 3, respectively, to promote the fit of the load-extension curve.

$$OV = W_1 \cdot rms(\Delta pos) + W_2 \cdot rms(\Delta F) \qquad (4)$$

The objective function is set to $1x10^7$ whenever the simulation does not converge. This effectively prevents these solutions from being valid. The NOMAD solver is set to stop after the objective function falls below 0.2 or if the maximum runtime exceeds the time of one and two hours for tuning step 1 and 2, respectively.

3.4 Experimental Procedure

All physical samples are generated with a custom MATLAB framework [11]. The target sample size is set to 25x50mm, which is considered a tradeoff between the minimum width stated in ASTM D5035-95 [43], creating a pure shear zone in the sample middle [44] and the testing machine's crosshead extension limits. Each sample consists of three weave layers of the same configuration to improve structural integrity. All layers are attached to solid blocks to clamp them into the tensile testing machine.

The samples are 3D-printed on a Stratasys J850 Prime (Stratasys Ltd., Eden Prairie, MN, USA) with the digital material FLX 9885. The print process is chosen in accordance with previous research that shows its capability to 3D print textiles with variable dimensions and weave patterns [11], as required by this investigation. The samples are post-processed in cycles of dissolving the support material in an alkaline solution and rinsing them with water. After all support material is dissolved, the samples are neutralized in a glycerol-water solution (15/85%) and cured for at least 24h.

All samples are tested on an Instron ElectroPuls E3000 (all Instron devices by Instron, Norwood, MA, USA) with a crosshead speed of 20mm/min. The samples are spray painted with acrylic paint on one side to create a speckle pattern. The speckle pattern is recorded during the tensile test by the video extensometer Instron AVE2 for digital image correlation. Further measurements are the crosshead position and the load on the sample recorded by Instron Dynacell 2527. The tests are stopped when the first yarn breaking is detected by eye.

The measurement data is post-processed in MATLAB. The data is cut off manually at the strain of the first load drop due to a yarn breaking. The load is divided by the number of layers. The image data is processed with Instron DIC replay. Since the software cannot output the correlated image data directly, the data is exported as a video file and a point cloud is fitted to the individual images with a custom MATLAB script. The post-processed data thus contains the load-extension curves and sample deformations throughout the test.

Whenever multiple samples with the same overall design are tested, the data is averaged. The averaging is performed depending on how much the samples maximum extension differ. If the maximum extensions of the shortest and longest curve do not differ by more than a factor of 1.5, the load-extension curves and sample deformation of all samples are averaged normally. If they differ by a factor larger than 1.5, the load-extension curve

Copyright © 2023 by ASME

of the sample with the highest extension is scaled to match the average load-extension curve of all samples in the low strain region. The average sample deformation is set to be the deformation of the sample with the maximum extension.

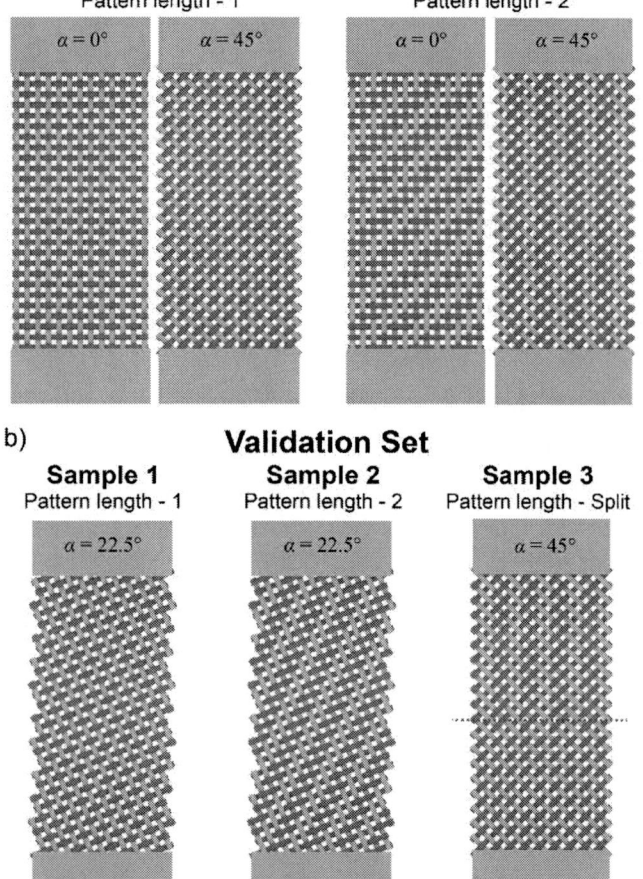

FIGURE 6: a) Two configurations are tested to tune the model. Each configuration is tested for two rotations (0° and 45°). b) Three samples are used to validate the model. They either vary in the rotation or change the pattern length throughout the sample.

4. RESULTS
4.1 Sample Configuration

The simulation is validated by tuning it for two configurations and testing the tuned simulation with three validation samples. The two configurations for the tuning set both have a yarn diameter of 1mm and the spacing between the yarns set to 2.096mm, which is the minimum distance given by the weave generation algorithm for the chosen diameter. The configurations vary in the length of the weave pattern. Configuration one has a weave pattern length of one, while configuration two has a weave pattern length of two. See FIGURE 1 for the definition of the weave pattern length. Configuration one corresponds to a plain weave while

configuration two corresponds to a 2/2 twill weave. Each configuration is tested with a rotation of 0° and 45°. Two samples per configuration and rotation are tested and the measurements are averaged.

The validation set consists of three samples. Samples one and two have the same configuration as the tuning samples one and two respectively but with a rotation of 22.5°. Sample three has a rotation of 45° and changes the pattern length throughout the sample. The top half has a pattern length of two while the bottom half has a pattern length of one. All configuration and validation samples are shown in FIGURE 6.

4.2 Simulation Accuracy

All results are generated with the simulation set to have a maximum step size of 0.2mm, a convergence threshold of 0.2mm, a step size reduction factor of 50% and a maximum iteration count of 1000. The curves for the stiffness and yarn softening for both tuning configurations are shown in FIGURE 7. Configuration one is shown in bold lines while configuration two is shown in dashed lines.

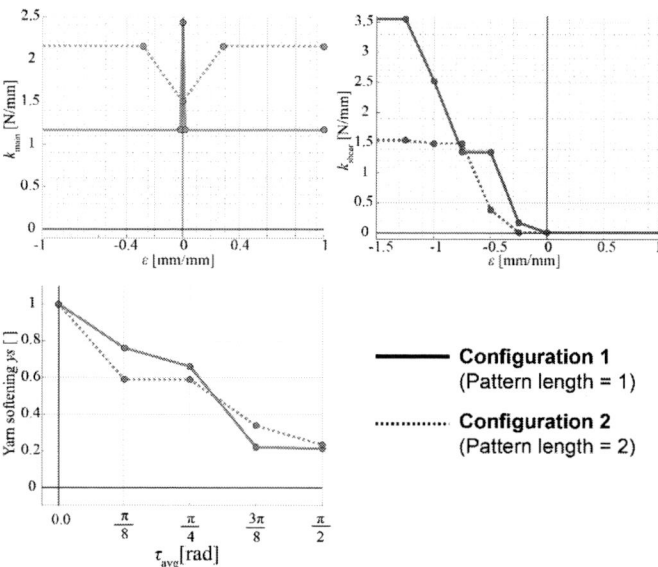

FIGURE 7: Tuned stiffness curves (top left: main elements, top right: shear elements) and yarn softening factor (bottom left) for two configurations varying in the pattern length.

The tuning accuracy is qualitatively depicted in FIGURE 8. For both configurations, the simulation results and the measured data is overlaid. The tuning accuracy can be quantitatively expressed by the objective function value of the tuning process. Since the objective function value is a multi-objective value, the accuracy of the single objectives, namely the match of the load-extension curve and the match of the sample deformation, provides more insight.

TABLE 1 lists the objective function value and further accuracy metrics for the individual objectives for each tuning configuration (C_1 and C_2) and tuning step ($\alpha_{0°}$ and $\alpha_{45°}$). The

Copyright © 2023 by ASME

accuracy of the load-extension data is expressed by two metrics; the average deviation of the simulated load from the measured load ΔF_{mean} and the r^2 value as a correlation value between the simulated and measured load-extension curve. The accuracy of the sample deformation is expressed by two metrics as well. Both metrics only assess the match of the vertical sample edges, since they are not constrained by boundary conditions. $\Delta p_{x,mean}$ is the average horizontal distance of the vertical side edges. The r^2 value is the correlation value when comparing the simulated to the measured edge position.

TABLE 1: Accuracy of the simulation tuning for each configuration and sample rotation.

Sample	OV	Load-extension		Deformation	
		ΔF_{mean}	r^2	$\Delta p_{x,mean}$	r^2
$C_1, \alpha_{0°}$	0.340	0.053N	0.998	0.146mm	0.999
$C_1, \alpha_{45°}$	0.427	0.023N	0.998	0.706mm	0.994
$C_2, \alpha_{0°}$	0.524	0.034N	0.999	0.491mm	0.999
$C_2, \alpha_{45°}$	0.270	0.025N	0.998	0.531mm	0.998

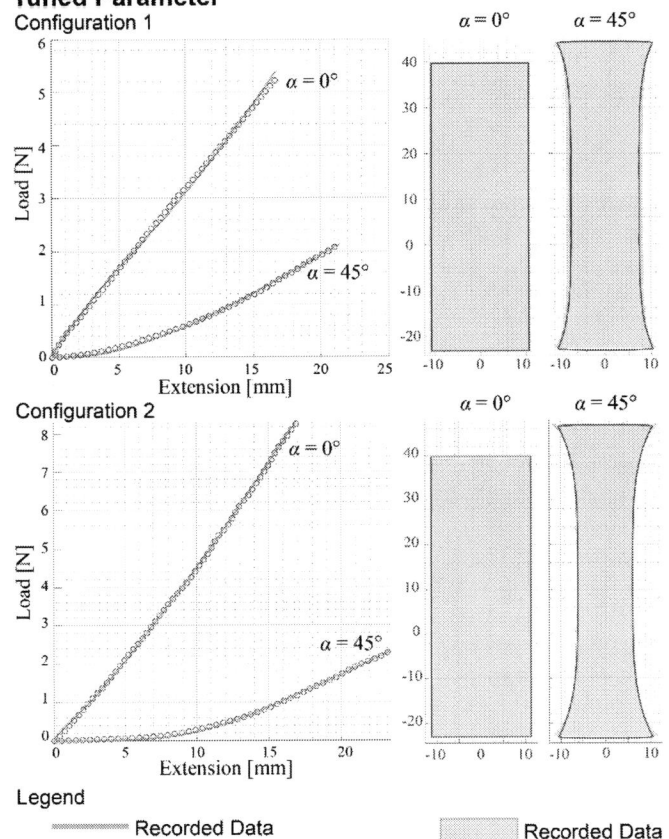

FIGURE 8: Accuracy of the tuning procedure. For each configuration and sample rotation, the load-extension curves and the sample deformation at maximum extension is shown. The dotted lines (left) and the bold borders (right) show the simulated results. The bold curves (left) and the filled patch (right) show the experimental data.

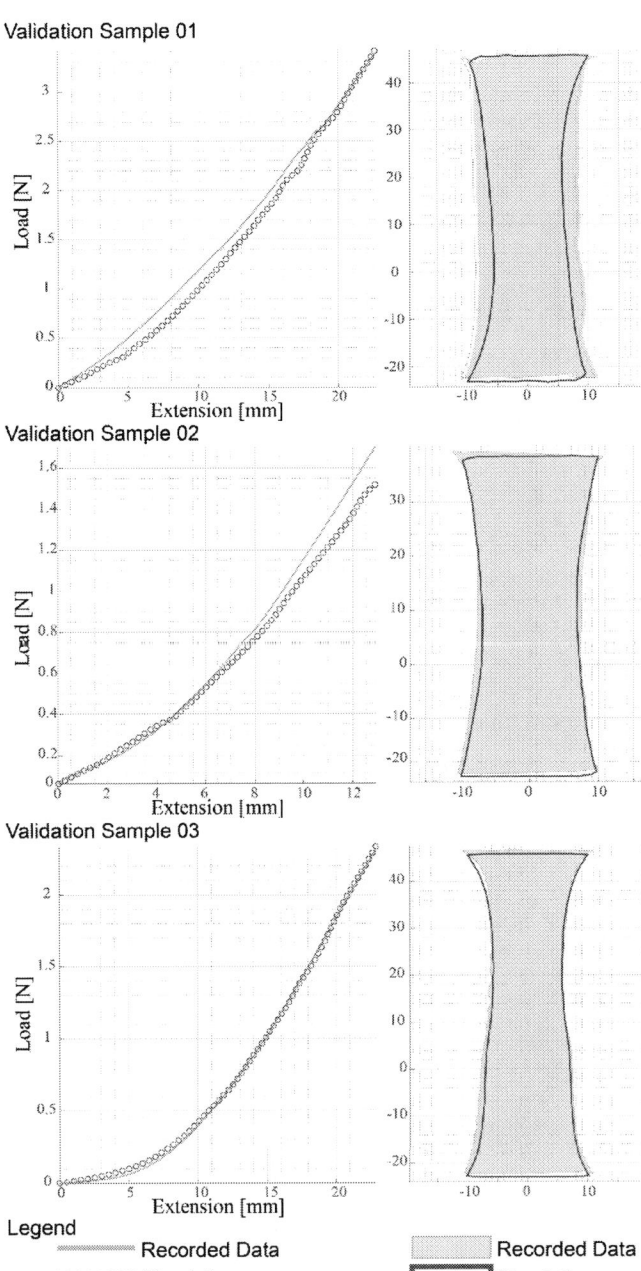

FIGURE 9: Accuracy of the simulation shown qualitatively for each sample of the validation set. The left side shows the load-extension curve for the simulation and the tested sample in dotted and bold, respectively. The right side shows the sample deformation under maximum extension. The bold outline shows the simulation results while the filled patch shows the measured data.

The accuracy of the validation samples is evaluated analogous to the accuracy of the tuning set. It is given by the average deviation of the load ΔF_{mean}, the average distance of the vertical edges $\Delta p_{x,mean}$, and the r^2 values when the simulated data is assumed to be a model of the experimental data. All accuracy metrics are provided in TABLE 2. A qualitative depiction of the

Copyright © 2023 by ASME

accuracy is shown in FIGURE 9. The load-extension curve and sample deformation of the simulation is shown in dotted lines and bold borders respectively. The measured data is shown in a bold line and in a filled patch for the load-extension curve and the deformed sample, respectively.

TABLE 2: Accuracy of the simulation for each validation sample.

Sample	Load-extension		Deformation	
	ΔF_{mean}	r^2	$\Delta p_{x,mean}$	r^2
V_1	0.114N	0.983	2.042mm	0.949
V_2	0.052N	0.978	0.829mm	0.996
V_3	0.018N	0.999	0.998mm	0.995

4.3 Simulation Runtime

The runtime of the simulation is evaluated by running simulations with a varying number of elements. All simulations are virtual tensile tests analogous to the samples used for tuning. The bottom nodes of the sample are thus constrained to have zero displacement while the top nodes are extension-controlled depending on the load step. The simulations are performed in 25 load steps. The maximum deformation is set to 25mm. All samples only have nodes set to configuration one. Since the runtimes vary significantly when changing the sample rotation from 0° to 45°, the results for the two rotation angles are shown separately. The simulations are run on an Intel Core i7-10750H processor using one core. In FIGURE 10, the results of this preliminary runtime investigation are shown.

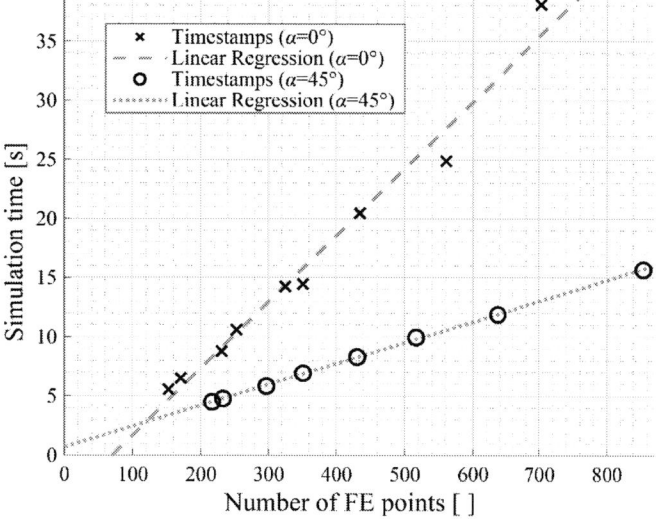

FIGURE 10: Runtime for simulations with configuration one varying in the number of elements. The crosses show the runtime for simulations with load in yarn direction ($\alpha=0°$). The circles show the runtime for simulations with load in diagonal direction ($\alpha=45°$). The dashed and dotted lines show the linear regression for both simulation types.

5. DISCUSSION
5.1 Interpretation of Results

The tuned curves shown in FIGURE 7 show the differences between the configurations. Configuration one has a significantly lower stiffness of the main elements while having stiffer shear elements in high strain regions. The yarn softening is similar for both configurations. The stiffness increase of the main elements for an increase in pattern length as well as the effective decrease in shear stiffness seen in the shear element stiffness is in accordance with the observations of previous work and can be attributed to the amount of crimp in the yarns [11]. The initial stiffness peak for configuration one has not been observed in previous investigations. However, the transition strain of configuration one is small, and the initial stiffness has a negligible impact on the overall simulation.

The accuracy of the tuning samples with a minimum similarity value of $r^2=0.994$ is high. This justifies stopping the parameter tuning although the objective value did not reach the set threshold. The high accuracy can also be regarded as a justification for the proposed approach of modulating the main element stiffness with the yarn softening factor. The yarn softening factor merges multiple effects happening due to yarn interactions and yarn bending, which could not be represented in previous pin-joint truss simulations. Lumping these effects in a modulating parameter delivers the expected results in an effective and fast manner.

The accuracy of the validation samples shows almost always an r^2 value above 0.95 for both, the load-extension curves and the sample deformation. An exception is the accuracy of the sample deformation of validation sample one. The simulation is overestimating the width decrease by 2.024mm on average. An explanation for this can be found in the physical tensile test. Since the sample has a rotation of $\alpha=22.5°$, there are yarns directly connecting the two clamps. This results in stress concentrating in these yarns and therefore stress concentrating in the two corresponding sample corners. In the experiment, the sample in these corners deform almost immediately after force application. It can be assumed that the yarns in the corners lose their load bearing capacity at low strains due to detachment from the clamps. This means that the sample does not need to deform as drastically to achieve an equilibrium. The sample width decrease could therefore not be seen in the experiment.

The efficiency of the nonlinear simulation is shown with simulations of samples having between 700 and 800 nodes, and thus, between 2800 and 3600 truss members, having a runtime below 40s on a modern mobile processor. This could be further improved by less conservative update rules and larger load steps, effectively decreasing the number of iterations of the inner simulation loop. Previous research has presented FE truss simulations with a faster convergence by only using linear truss elements [26]. The approach presented in this work needs to read stiffness values and forces in the truss member for each iteration, which increases the runtime compared to a linear simulation. However, the approach allows properties for each element to change and thus allows simulating inhomogeneous textiles in a

Copyright © 2023 by ASME

semi-discrete approach with a high accuracy. To our knowledge, this is a significant contribution to textile simulation.

5.2 Possibility for Generalization

The simulation is currently tuned for two configurations. This limits the design space to samples varying in only the pattern length. In theory, the presented approach does not restrict the number of configurations. The configuration set can thus be upgraded by consecutively tuning the property curves of new samples and adding them to the set. Examples for further design parameters include the yarn diameter, the spacing between the yarns, the yarn material, and the vertical distance between the yarns at the yarn crossings. Each node can be set to a different configuration, which enables a gradual design change throughout the weave. Examples of such designs are depicted in FIGURE 11. However, a systematic increase of the design space by providing more tuning samples becomes more expensive, the more design parameters are added. Each design parameter must be tested in combination with each existing parameter combination and the tuning effort increases factorially. A possibility to reduce the tuning effort would be to use design of experiments by only testing and tuning a reduced sample set and statistically estimating the properties of the untested samples.

5.3 Limitations

The simulation presented in this work is created to estimate the in-plane behavior of biaxial weaves. The out-of-plane behavior is not depicted due to the set displacement constraints. However, the simulation setup would be capable of simulating 3D truss structures. Since the weaves are currently approximated as pin-jointed 2D trusses, they have no resistance against bending. Simulating 3D deformation would need a different way of approximating the behavior by either adding a bending resistance via rotational springs at the nodes or by adding additional truss layers parallel to the main layer, effectively stiffening the structure in bending deformation. The bending properties of all configurations must be evaluated empirically and translated to the corresponding FE representation. However, for many applications, an in-plane simulation is thought to be suitable to estimate the overall behavior of the design. Especially when the textiles are constrained by an underlying object, like human limbs for wearables or machine elements in soft robotics, it is likely that the overall behavior is dominated by the in-plane properties and the out-of-plane behavior can be neglected.

Another currently neglected phenomenon is the compressive behavior of yarns. Yarns cannot withstand substantial amounts of compression before buckling is observed experimentally. This induces the formation of wrinkles and folds, which is per definition an out-of-plane phenomenon. A possibility to account for yarns under compression would be the deactivation of yarns whenever they reach a compression threshold. This does not directly model wrinkling, but could improve the simulation accuracy for textiles under compression.

a) Gradual change of yarn diameter ($\alpha = 0°$)

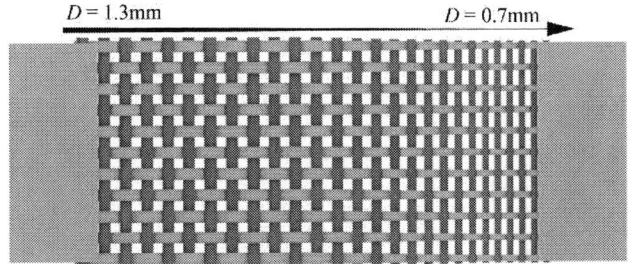

b) Gradual change of yarn diameter ($\alpha = 45°$)

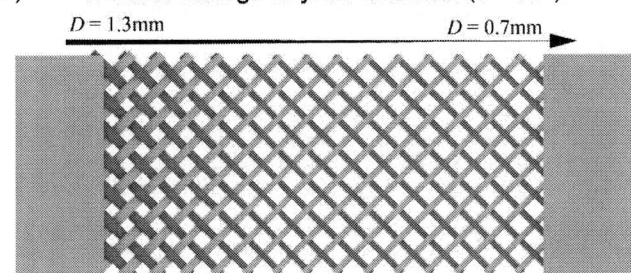

c) Yarn diameter, weave pattern, yarn spacing and rotation arbitrary

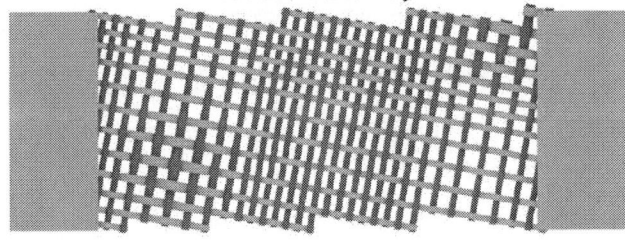

FIGURE 11: Examples of weave designs within a larger design space. (a) Diameter changes from 1.3mm to 0.7mm gradually along a yarn direction (b) Diameter changes from 1.3mm to 0.7mm gradually in diagonal direction (c) Yarn diameter, weave pattern, and yarn spacing vary at arbitrary locations

6. CONCLUSION

We propose a semi-discrete textile model based on 1D truss members with strain-dependent and shear-dependent stiffness to simulate 3D-printed biaxial weaves. The stiffness properties are defined for each node depending on the weave design. We show an approach to tune the properties by matching a virtual tensile test to the physical counterpart using optimization. The tuned simulation is capable of estimating load-extension curves and the mechanical deformation of samples with varying dimensions and weave pattern across the textile patch. This shows the model's capability to simulate locally defined structures within the tuned design space. Since the presented tuning approach is generally applicable for any weave configuration, the design space can be expanded progressively, for example through design of experiments. This enables the estimation of the mechanical properties of various weave designs and will ultimately enable rapid iteration to inversely design biaxial weaves.

Copyright © 2023 by ASME

REFERENCES

[1] Mueller J, Shea K. Stepwise graded struts for maximizing energy absorption in lattices. Extreme Mech Lett 2018;25:7–15.

[2] Lumpe TS, Shea K. Computational design of 3D-printed active lattice structures for reversible shape morphing. J Mater Res 2021;36:3642–55. https://doi.org/10.1557/s43578-021-00225-2.

[3] ZHU J, ZHOU H, WANG C, ZHOU L, YUAN S, ZHANG W. A review of topology optimization for additive manufacturing: Status and challenges. Chinese Journal of Aeronautics 2021;34:91–110. https://doi.org/https://doi.org/10.1016/j.cja.2020.09.020.

[4] Ion A, Kovacs R, Schneider OS, Lopes P, Baudisch P. Metamaterial Textures 2018:1–12–1–12.

[5] Momeni F, Hassani.N SMM, Liu X, Ni J. A review of 4D printing. Mater Des 2017;122:42–79.

[6] Sitotaw DB, Ahrendt D, Kyosev Y, Kabish AK. Additive Manufacturing and Textiles—State-of-the-Art. Applied Sciences 2020;10.

[7] Keefe EM, Thomas JA, Buller GA, Banks CE. Textile additive manufacturing: An overview. Cogent Eng 2022;9:2048439. https://doi.org/10.1080/23311916.2022.2048439.

[8] Danit Peleg - 3D Printed Fashion n.d. https://danitpeleg.com/ (accessed March 9, 2022).

[9] Rosenkrantz J, Louis-Rosenberg J. Dress/Code Democratising Design Through Computation and Digital Fabrication. Architectural Design 2017;87:48–57. https://doi.org/https://doi.org/10.1002/ad.2237.

[10] Perry A. 3D-printed apparel and 3D-printer: exploring advantages, concerns, and purchases. Null n.d.;11:95–103–95–103.

[11] Wirth M, Shea K, Chen T. 3D-printing textiles: multi-stage mechanical characterization of additively manufactured biaxial weaves. Mater Des 2023;225:111449. https://doi.org/https://doi.org/10.1016/j.matdes.2022.111449.

[12] Chan KP, He F, Atwah AA, Khan M. Experimental investigation of self-cleaning behaviour of 3D-printed textile fabrics with various printing parameters. Polym Test 2023;119:107941. https://doi.org/https://doi.org/10.1016/j.polymertesting.2023.107941.

[13] Rossi RM, Fortunato G, Nedjari S, Morel A, Heim F, Osselin J-F, et al. 9 - Mechanical properties of medical textiles. In: Schwartz PBT-S and M of TFA (Second E, editor. The Textile Institute Book Series, Woodhead Publishing; 2019, p. 301–40. https://doi.org/https://doi.org/10.1016/B978-0-08-102619-9.00009-2.

[14] Luo Y, Wu K, Spielberg A, Foshey M, Rus D, Palacios T, et al. Digital Fabrication of Pneumatic Actuators with Integrated Sensing by Machine Knitting. CHI Conference on Human Factors in Computing Systems, New York, NY, USA: Association for Computing Machinery; 2022. https://doi.org/10.1145/3491102.3517577.

[15] Mueller KMA, Mulderrig S, Najafian S, Hurvitz SB, Sodhani D, Mela P, et al. Mesh manipulation for local structural property tailoring of medical warp-knitted textiles. J Mech Behav Biomed Mater 2022;128:105117. https://doi.org/https://doi.org/10.1016/j.jmbbm.2022.105117.

[16] Meiklejohn E, Devlin F, Dunnigan J, Johnson P, Zhang JX, Marschner S, et al. Woven Behavior and Ornamentation: Simulation-Assisted Design and Application of Self-Shaping Woven Textiles. Proc ACM Comput Graph Interact Tech 2022;5. https://doi.org/10.1145/3533682.

[17] Cheng T, Tahouni Y, Wood D, Stolz B, Mülhaupt R, Menges A. Multifunctional Mesostructures: Design and Material Programming for 4D-printing. Symposium on Computational Fabrication, New York, NY, USA: Association for Computing Machinery; 2020. https://doi.org/10.1145/3424630.3425418.

[18] Deng B, Zareei A, Ding X, Weaver JC, Rycroft CH, Bertoldi K. Inverse Design of Mechanical Metamaterials with Target Nonlinear Response via a Neural Accelerated Evolution Strategy. Advanced Materials 2022;n/a:2206238. https://doi.org/https://doi.org/10.1002/adma.202206238.

[19] Bastek J-H, Kumar S, Telgen B, Glaesener RN, Kochmann DM. Inverting the structure–property map of truss metamaterials by deep learning. Proceedings of the National Academy of Sciences 2022;119:e2111505119. https://doi.org/10.1073/pnas.2111505119.

[20] LIANG B, BOISSE P. A review of numerical analyses and experimental characterization methods for forming of textile reinforcements. Chinese Journal of Aeronautics 2021;34:143–163–143–163.

[21] Cirio G, Lopez-Moreno J, Miraut D, Otaduy MA. Yarn-Level Simulation of Woven Cloth. ACM Trans Graph 2014;33. https://doi.org/10.1145/2661229.2661279.

Copyright © 2023 by ASME

[22] Sharma SB, Sutcliffe MPF. A simplified finite element model for draping of woven material. Compos Part A Appl Sci Manuf 2004;35:637–643–637–643.

[23] Skordos AA, Monroy Aceves C, Sutcliffe MPF. A simplified rate dependent model of forming and wrinkling of pre-impregnated woven composites. Compos Part A Appl Sci Manuf 2007;38:1318–1330–1318–1330.

[24] Turco E, dell'Isola F, Cazzani A, Rizzi NL. Hencky-type discrete model for pantographic structures: numerical comparison with second gradient continuum models. Zeitschrift Für Angewandte Mathematik Und Physik 2016;67:85. https://doi.org/10.1007/s00033-016-0681-8.

[25] Giorgio I, Harrison P, dell'Isola F, Alsayednoor J, Turco E. Wrinkling in engineering fabrics: a comparison between two different comprehensive modelling approaches. Proceedings of the Royal Society A: Mathematical, Physical and Engineering Sciences 2018;474:20180063. https://doi.org/10.1098/rspa.2018.0063.

[26] Sun X, Belnoue JP-H, Thompson A, Said B el, Hallett SR. Dry Textile Forming Simulations: A Benchmarking Exercise. Front Mater 2022;9:831820.

[27] Beex LAA, Verberne CW, Peerlings RHJ. Experimental identification of a lattice model for woven fabrics: Application to electronic textile. Compos Part A Appl Sci Manuf 2013;48:82–92. https://doi.org/https://doi.org/10.1016/j.composi tesa.2012.12.014.

[28] Xie J, Guo Z, Shao M, Zhu W, Jiao W, Yang Z, et al. Mechanics of textiles used as composite preforms: A review. Compos Struct 2023;304:116401. https://doi.org/https://doi.org/10.1016/j.compstr uct.2022.116401.

[29] van der Weeën F. Algorithms for draping fabrics on doubly-curved surfaces. Int J Numer Methods Eng 1991;31:1415–26. https://doi.org/https://doi.org/10.1002/nme.162 0310712.

[30] d'Agostino M v, Giorgio I, Greco L, Madeo A, Boisse P. Continuum and discrete models for structures including (quasi-) inextensible elasticae with a view to the design and modeling of composite reinforcements. Int J Solids Struct 2015;59:1–17. https://doi.org/https://doi.org/10.1016/j.ijsolstr. 2014.12.014.

[31] Boubaker B ben, Haussy B, Ganghoffer J-F. Discrete models of fabric accounting for yarn interactions. Revue Européenne Des Éléments Finis 2005;14:653–75. https://doi.org/10.3166/reef.14.653-675.

[32] Harrison P. Modelling the forming mechanics of engineering fabrics using a mutually constrained pantographic beam and membrane mesh. Compos Part A Appl Sci Manuf 2016;81:145–57. https://doi.org/https://doi.org/10.1016/j.composi tesa.2015.11.005.

[33] Daelemans L, Faes J, Allaoui S, Hivet G, Dierick M, van Hoorebeke L, et al. Finite element simulation of the woven geometry and mechanical behaviour of a 3D woven dry fabric under tensile and shear loading using the digital element method. Compos Sci Technol 2016;137:177–87. https://doi.org/https://doi.org/10.1016/j.compsci tech.2016.11.003.

[34] Durville D. Simulation of the mechanical behaviour of woven fabrics at the scale of fibers. International Journal of Material Forming 2010;3:1241–51. https://doi.org/10.1007/s12289-009-0674-7.

[35] Wang J, Zhou H, Ouyang Z, Peng X, Zhou H. A mesoscale tensile model for woven fabrics based on Timoshenko beam theory. Textile Research Journal 2022:00405175221117616. https://doi.org/10.1177/00405175221117616.

[36] Kyosev YK. 6 - The finite element method (FEM) and its application to textile technology. In: Veit D, editor. Simulation in Textile Technology, Woodhead Publishing; 2012, p. 172–222e.

[37] Peng XQ, Cao J. A continuum mechanics-based non-orthogonal constitutive model for woven composite fabrics. Compos Part A Appl Sci Manuf 2005;36:859–74. https://doi.org/https://doi.org/10.1016/j.composi tesa.2004.08.008.

[38] Chen B, Colmars J, Naouar N, Boisse P. A hypoelastic stress resultant shell approach for simulations of textile composite reinforcement forming. Compos Part A Appl Sci Manuf 2021;149:106558. https://doi.org/https://doi.org/10.1016/j.composi tesa.2021.106558.

[39] Aimène Y, Vidal-Sallé E, Hagège B, Sidoroff F, Boisse P. A Hyperelastic Approach for Composite Reinforcement Large Deformation Analysis. J Compos Mater 2009;44:5–26. https://doi.org/10.1177/0021998309345348.

[40] Denis Y, Guzman-Maldonado E, Hamila N, Colmars J, Morestin F. A dissipative constitutive model for woven composite fabric under large strain. Compos Part A Appl Sci Manuf 2018;105:165–79.

Copyright © 2023 by ASME

https://doi.org/https://doi.org/10.1016/j.composi tesa.2017.11.018.

[41] Ferretti M, Madeo A, dell'Isola F, Boisse P. Modeling the onset of shear boundary layers in fibrous composite reinforcements by second-gradient theory. Zeitschrift Für Angewandte Mathematik Und Physik 2014;65:587–612. https://doi.org/10.1007/s00033-013-0347-8.

[42] le Digabel S. Algorithm 909: NOMAD: Nonlinear Optimization with the MADS Algorithm. ACM Trans Math Softw 2011;37. https://doi.org/10.1145/1916461.1916468.

[43] Standard Test Method for Breaking Force and Elongation of Textile Fabrics (Strip Method) 1 n.d.

[44] Lee W, Padvoiskis J, Cao J, de Luycker E, Boisse P, Morestin F, et al. Bias-extension of woven composite fabrics. International Journal of Material Forming 2008;1:895–898–895–898.

Proceedings of the ASME 2023
International Design Engineering Technical Conferences and
Computers and Information in Engineering Conference
IDETC-CIE2023
August 20-23, 2023, Boston, Massachusetts

DETC2023-111726

Build Orientation Optimization for Five-Axis 3D Printing

Ghazi Alonayni, Matthew I. Campbell[1]
School of Mechanical, Industrial, and Manufacturing Engineering
Oregon State University
Corvallis, OR, USA

ABSTRACT

It is well understood that printing orientation plays a critical role in the quality of a final printed product and the total print time. In five-axis 3D printers, the added rotations can further reduce time and material, but initial orientation is still a significant factor. The build orientation is a critical factor in the printed part's surface quality and the need for additional material as support structures. In this research, we develop build orientation sampling techniques and evaluate them based on their computational cost and accuracy. The evaluation of a build orientation is based on three criteria: the volume of the added support structure, the surface quality, and the stability of the part while rotating and tilting on the printing platform. The creation of the support structure uses Boolean polygon operations to compute the amount of material required as support. The surface quality and stability of a part at a given orientation are also calculated. Our proposed optimization scheme was tested on seven 3D parts and was found to only require a small number of evaluations to achieve a Pareto-optimal set of build orientations. This method can be utilized as a first step in planning the fabrication of parts printed on five-axis printers.

1   INTRODUCTION

The orientation of a printed part is crucial to the surface quality, structural integrity, and need for support structures. Five-axis 3D printers can produce intricate and custom parts that are often impossible to create with traditional manufacturing methods. However, the added complexity of rotational axes presents significant challenges in pre-processing, particularly in determining the best print orientation. Poor print orientation can result in visible stair-stepping on the surface, decreased structural integrity, and increased use of support structures, which can lead to costly reprints and post-processing efforts [1,2].

This paper presents an optimization approach based on a combination of deterministic techniques. In an effort to find the optimal orientation efficiently, we seek a method that requires the fewest number of objective function evaluations. Function evaluations are especially time-consuming for creating support structures since many polygon operations are required to define these. The specific results and trade-offs of our optimization approach are presented in detail in this paper. Section 2 summarizes recent multi-axis 3D printing and build orientation optimizations in the literature. Section 3 describes the general methodology of generating build orientations. In Section 4, we describe how we evaluate the build orientations. We test different approaches and discuss the performance and outcomes in Section 5. Section 6 draws conclusions based on results and presents potential barriers and future work.

2   BACKGROUND

Due to the potential advantages of multi-axis 3D printing, numerous studies in recent years have been conducted on process planning [3–11]. The printer's kinematic configuration varies in these studies, and Jetton et al. [12] presented an analysis of possible configurations and concluded that one promising configuration that has complete control throughout the workspace consists of three prismatic joints followed by two revolute joints. Some of the proposed approaches in recent studies [4,7–10] used a segmentation approach, splitting a 3D part into regions to print along their unique orientations. Additionally, printing on non-planar surfaces is another consideration in some research [4,13]. These types of studies often tried to generate support-free parts, while the surface quality of the parts might decline.

Finding the optimum build orientation in five-axis 3D printing is a key factor that can significantly impact the printed part's surface quality, material utilization, and stability. Several research papers specifically focus on five-axis 3D printing and address the build orientation problem. One study by Liu et al.

---

[1] Contact author: Matt.Campbell@oregonstate.edu

Copyright © 2023 by ASME

[14] develops a custom method that finds the build orientation based on an exhaustive search of the initial base planes, which are the largest faces on a part. However, the build orientation algorithm only considers large faces, potentially missing an optimum build orientation on an edge or smaller face.

Numerous studies have focused on finding the optimal build orientation, but most were limited to the conventional 3-axis 3D printers [15–19]. One study by Meifa et al. [18] developed a multi-objective optimization that generates build orientations from facet clustering and evaluates them based on a weighted sum model. Jaiswal et al. [19] implemented a surrogate model to solve the build orientation optimization, focusing on minimizing the surface quality due to the staircase effect and support structure. However, using voxels as shape representation is time-consuming for high-fidelity, and the build orientations were generated using a spherical unit . One noteworthy method by Li et al. [20] controlled the infill pattern and density to achieve self-supporting parts, but this study uses stereolithography printing. Additionally, hollow features were found difficult to deal with, as in one case where the part was split and glued together after being printed separately.

All of these studies demonstrate the complexity of finding the optimal build orientation and this is especially true when dealing with five-axis 3D printing. Furthermore, they highlight the importance of carefully considering the objectives and trade-offs when optimizing the build orientation. In some cases, an optimization method might find a single optimum orientation for all objectives. In other cases, the build orientation strikes a good balance between some objectives but compromises on others. A research gap in five-axis 3D printing can be observed based on the mentioned related works. In this paper, we present a novel optimization scheme that aims to find the best orientation specific to five-axis printing while considering three factors: the volume of the support structure, surface quality, and stability of the part.

3 BUILD ORIENTATION SAMPLING TECHNIQUES

In this section, we present the methodology of generating candidate build orientations. We have implemented three techniques and evaluated them based on computational cost and accuracy. Two orientation generation techniques use a unit sphere, a common and simple method to generate orientations. The third technique uses the convex hull of a given 3D part to generate building orientations. This technique considers orientations based on the faces and edges of the convex hull.

As detailed in the next section, three independent objective functions are defined: the support structure volume, surface quality, and stability. In order to avoid issues of simplifications to a single objective function, the methods exhaustively define a set of candidates from which a Pareto optimal set is extracted [21]. Initially, traditional optimization was considered. A multi-start Nelder-Mead [22] method was implemented to find the optimal unit sphere variables θ and ϕ for a weighted sum of the three objectives. Initial samples were placed evenly about the unit sphere (based on the twenty faces of an icosahedron) for the multi-start aspect of the search. But, in the end, this method tended to vary significantly in the number of function evaluations that were needed and the weights between the objective functions were not easy to define universally.

The geometric-intensive nature of this problem demands numerous operations related to tessellated parts and polygon operations. These were performed using Tessellation and Voxelization Geometry Library (TVGL) [23]. The build search candidate sampling techniques are described in the remainder of this section. Each candidate direction is then evaluated on the three objectives presented later (Section 4). The mathematical condition of pareto optimality is used to reduce the large number to the subset of optimally nondominated solutions.

3.1 Latitude-longitude Lattice Spherical Sampling

The latitude-longitude lattice generates the build orientations via two angles, polar angle, θ, and azimuthal angle, ϕ, as shown in Figure 1. The user must determine an angle step value to populate orientation candidates. For instance, angle step value of 1 degree would generate 64,442 unique orientations over a unit sphere, where θ is from 0 to 180 and ϕ is from 0 to 359. Although this method can have considerably high number of search candidates, the build orientation can be very sensitive to an angle even less than 1 degree. Additionally, evaluating a build orientation can be computationally expensive, so 65 thousand evaluations could take an impractically-long amount of time. The angle step for a high-fidelity spherical sampling can go as low as 1 degree for both θ and ϕ, where the total number of generated unique orientations is 64,442, the low-fidelity spherical sampling is 9°, with a total of 762 unique build orientations. The approximated computational time for a single evaluation is 1.24 seconds for this sampling technique based upon the tested parts in Section 5. This means that the high-fidelity sampling would take 22.19 hours to complete. In comparison, the low-fidelity case takes 0.26 hours to evaluate 762 orientations. Moreover, this sampling technique creates a

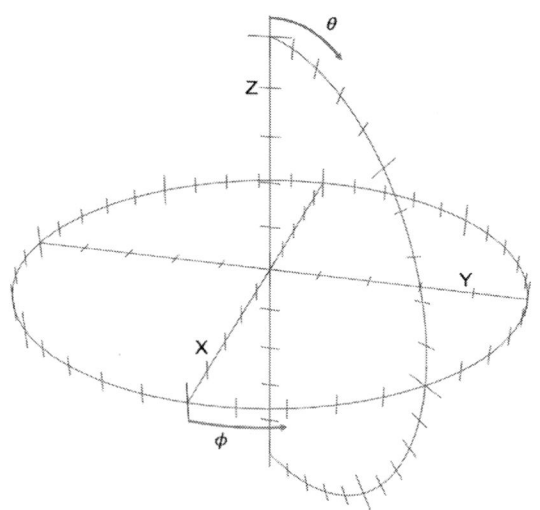

FIGURE 1: SPHERICAL SAMPLING USING θ AND ϕ ANGLES.

Copyright © 2023 by ASME

	Inputs: Tessellated Part, TS; Convex Hull, CH; Minimum Angle Between Two Orientations, $\alpha = 5°$
	Output: List of Candidate Orientations C
1	Sort faces of the CH by area from largest to smallest → Sorted Faces
2	Sort edges of CH by length from longest to shortest → Sorted Edges
3	for each face in Sorted Faces
4	If min angle between face's orientation and candidate orientation list, C, is larger than α
5	Add face's orientation to C
6	End If
7	End for each
8	for each edge in Sorted Edges
9	Edge's orientation is computed from adjacent two faces
10	If min angle between edge's orientation and candidate orientation list, C, is larger than α
11	Add edge's orientation to C
12	End If
13	End for each
14	Switch the direction of all candidate orientations C by 180°

FIGURE 2: PSEUDOCODE FOR FINDING CANDIDATE ORIENTATIONS GENERATION VIA CONVEX HULL

high concentration in the polar regions because they are not evenly distributed over a sphere.

3.2 Near-Equidistant Spherical Sampling

This method is more sophisticated than using a simple spherical sampling mentioned in the previous section. Using a Fibonacci lattice [24] generates the spherical directions that are more evenly spaced. However, it still relies on the user to determine the number of populated orientation candidates. Furthermore, the number of orientations required to achieve a high-fidelity search is high.

3.3 Convex Hull

The convex hull of a part is a mathematical construct that represents the smallest volume that encloses the part [25] without producing any concavities. Figure 3 shows an example of the convex hull of a 3D part. Any face of a convex hull of 3D part also represents possible coincident faces with the ground plane. Although, such face orientations are not guaranteed to be statically stable. Additionally, if only the edges or vertices of the convex hull coincide with a ground plane, then supports will be needed to make it stable. By using information from the convex hull, we can generate fewer search candidates that are meaningfully determined from the shape of the part. This method has the potential to improve the accuracy and efficiency of the optimization process. To generate search candidates using the convex hull, we developed a set of conditions to select the normal orientation of a face or edge as a valid search candidate.

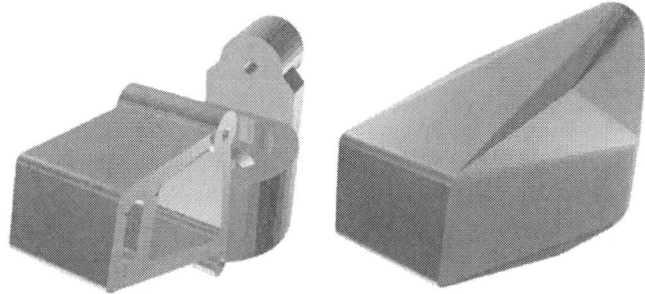

FIGURE 3. THE CONVEX HULL OF A PART.

The pseudocode of the candidate selection via convex hull is illustrated in Figure 2. We first sort the faces by decreasing area and edges by decreasing length. The sorting step prioritizes the largest faces and longest edges. Then, we consider a face's normal direction as a valid search candidate if the smallest angle between it and the chosen candidates is greater than a pre-defined value α. This pre-defined value helps to limit the number of candidates generated by numerous small triangles in the convex hull. The selection of normal orientations for edges follows a similar process, with the normal of an edge being considered the average normal of its adjacent faces. The calculated orientation of faces and edges are the opposite direction which points to the printing platform. Thus, the final step is to switch the direction of all selected orientations by 180°. The output of this algorithm is a list of candidates where the smallest angle between them is larger than α.

4 OBJECTIVE FUNCTIONS

There are many factors that are considered to determine the printing orientation such as the volume of support structure, surface quality, strength, processing time, cost, and volumetric error. Some of the objectives are conflicting such as the surface quality and processing time where an improvement in the surface quality could increase the processing time [1,26]. The three objectives developed in this work are discussed as subsections below.

4.1 Support Structure

One of the main goals of our optimization scheme is to find a print orientation that minimizes or eliminates support structure. This is important because adding support to a region can degrade the surface quality, increase printing time, and increase the cost of the post-processing steps. The orientation of the print can significantly influence the reachable regions, as certain orientations may cause the toolhead to collide with the printing platform or result in exceeding overhang limits. The added material while printing a part to supporting its structure is degrading its surface [27]. Perhaps a water-soluble support structure could be used to reduce the impact on the surface's quality, but this requires using a dual extruder system, which would increase the cost and add additional complexity to the system.

This paper introduces a two-step algorithm that performs two separate passes through the additive layers' polygons. The first pass computes the reachable regions that require no support. The second pass creates the support structure where needed. In the support creation process, the printing layers are simplified as planar layers along the build orientations. Also, the performed operations on layers are polygon offsetting, Boolean intersection, and Boolean subtraction operations to simulate the increase of maximum overhang angle. The concept of this method is illustrated in Figure 4. The process starts with inflating the previous layer by a specific value, t_L . Then, we compute the reachable region by Boolean intersection operation between current and inflated layers. By the end of this process, we have the reachable regions that require no support, shown in green in Figure 4. The second pass goes from the last to the first layer to create support by performing a Boolean subtraction operation between the results of the first step and the original layers. When the subtraction operation detects a non-empty shape, the resulting shape is connected to previous layers until fully connected to a layer or ground, as shown in the red regions in Figure 4.

Some assumptions were made to simplify the problem. For instance, we assumed that the maximum overhang angle is $90°$ from the first printing direction. Recall that as a five-axis additive process, the print platform can generally be rotated by $90°$. However, due to potential collisions with the platform, this angle is not achievable until the part is built a certain distance above the platform. We set this safe distance h at 50 mm (125th layer for 0.4 mm printing thickness). In the region below this limit, the overhang angle starts from $45°$ at the first layer and linearly increases until 50 mm is reached (see Figure 5). However, this is a conservative generalization, and these values could be adjusted for every deposition material and a given printer's geometry. Equation 1 is the polygon inflate value t_L,

which starts with the value of the printing thickness t, the equivalent of $45°$. As the layer number, i, increases, the value of polygon inflation increases.

The objective function of the support structure is based on the volume of the added support. The objective value is zero when no support is needed. However, when the algorithm detects the need for support, the volume is the summation of the area of the added support layers multiplied by printing thickness.

$$t_L = tan\left(\frac{\pi}{4}\left(1 + \frac{(i-2)}{h*t}\right)\right)*t \qquad (1)$$

4.2   Surface Quality

One of the major sources of surface degradation is the staircase effect. The staircase effect is a common issue in 3D printing that is caused by the difference in the print direction and the surface normal of the part. Figure 6 shows the staircase effect on a triangle with normal direction n and build orientation d. To measure the staircase effect, we can use the angle between the two orientations [28]. When the orientations are parallel or perpendicular, this angle is 0 or 90. The cosine of this angle, which is the dot product of the two orientation unit vectors, captures the perpendicular case. The parallel case can be captured using the sine of the angle. Thus, the function value is the area of the triangle multiplied by sine and cosine, as in Equation 2.

$$f_2 = Area * cos(\theta) * sin(\theta) \qquad (2)$$

By squaring the previous equation, we remove negative values and obtain a smooth curve ranging from 0 to 0.25.

$$f_2 = Area * (cos(\theta) * sin(\theta))^2 \qquad (3)$$

For speed, note that a common trigonometric identify, $sin(\theta)^2 + cos^2(\theta) = 1$, allows us to rewrite the equation as:

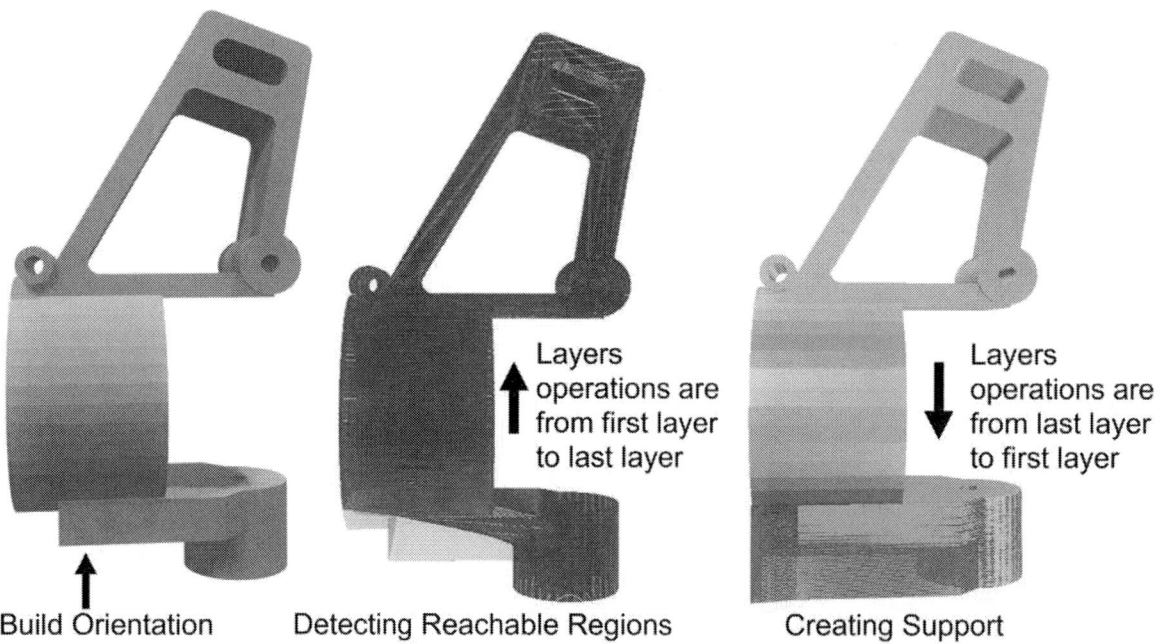

FIGURE 4. THE PROCESS OF CREATING SUPPORT

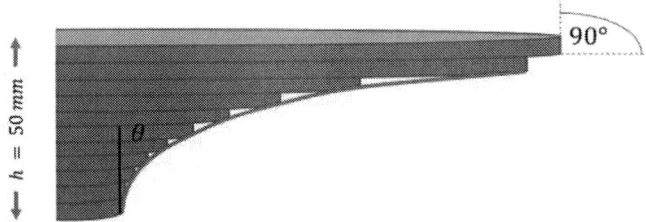

FIGURE 5. OVERHANG ANGLE INCREASES UP TO 90 DEGREES.

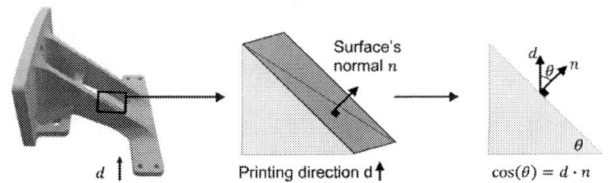

FIGURE 6. STAIRCASE EFFECT FROM THE ORIENTATION OF PRINTING SURFACE AND BUILD ORIENTATION.

$$f_2 = Area * \left(cos(0) * (1 \quad cos^2(\theta))\right)^2 \qquad (4)$$

Since the dot-product is quick and simple to find, this reduces further to:

$$f_2 = Area * ((d \cdot n)^2 - (d \cdot n)^4) \qquad (5)$$

Equation 5 represents the cosine of angle as the dot product of two vectors shown in Figure 6, which are printing orientation d and face normal n. By minimizing Equation 5, we could reduce the staircase effect and improve the surface quality of the printed part. Equation 5 measures the surface quality over a single triangle which creates a smooth curve as shown in Figure 7. The overall value of the surface quality, f_2, is a summation over all triangles as in Equation 6.

$$f_2 = \sum_{i=1}^{Triangles} Area_i * ((d \cdot n_i)^2 - (d \cdot n_i)^4) \qquad (6)$$

4.3 Print Stability

Stability is the third factor we considered when determining the optimal build orientation for a 3D part. Since the printing platform is rotating and tilting while extruding materials, the contact area between the first layer and the printing platform must be high enough not to cause a printing failure.

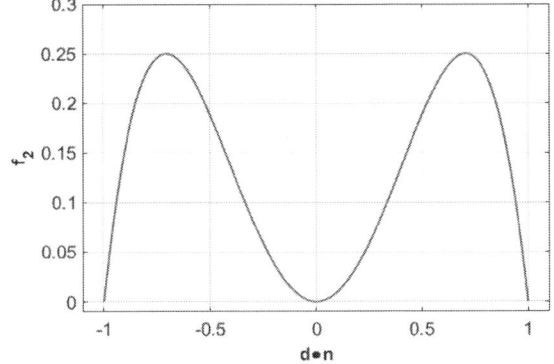

FIGURE 7. THE STAIRCASE EFFECT FUNCTION.

Figure 8 shows first-layer cases where the difference in build orientation changes the contacting layer with the printing platform. Generally, the desirable cases for the stability of the print are printing on large faces. Starting with a large face provides a large contacting surface with the printing platform. However, placing a part on an edge as a contacting layer would be a fair compromise if this significantly improves the other objectives. Printing on only vertices is the least desirable case that needs to be avoided. Positioning a part on vertices increases the probability of printing failure, and it reduces the surface quality. Thus, the candidate orientations are generated based on only placing the part on its faces and edges. However, printing on the largest face does not always mean it is the optimum build orientation. One way to evaluate the stability of a print for a build orientation is by calculating the area of the first printing layer. Then, the area is multiplied by -1 to convert the objective to a minimization problem, as shown in Equation 7.

$$f_3 = -Area_{Layer\ 1} \qquad (7)$$

It is worth mentioning that a better way to analyze the stability of the print is to include a part's center of mass [20]. However, we assume that we are using a heated platform, which has not been the case in recent studies for five-axis 3D printers. The heated platform will enhance the stability of the part while printing as long as the contact area is large enough [29].

5 RESULTS AND DISCUSSION

We implemented the proposed approach in C# on a Windows laptop computer with an Intel i7-11800H @2.30 GHz processor and 64 GB RAM. We applied our proposed optimization scheme to seven 3D parts with varying shapes and sizes, shown in Figure 9 and with numerical data in Table 1. While the parts have varying degrees of complexity and curvature changes, the build orientations are always quite sensitive to small changes. Since many features on these test parts line up with the default coordinate axes, we test the robustness of the optimization techniques by testing if a method can find the optimum build orientation both in the nominal orientation and after slightly rotating the model.

Part-1 was designed for this project to have one obvious solution along the positive z-axis. Coincidentally, Part-2 would

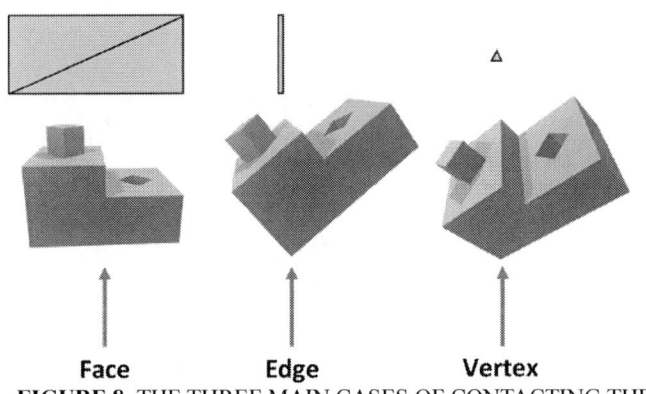

FIGURE 8. THE THREE MAIN CASES OF CONTACTING THE FIRST LAYER.

Copyright © 2023 by ASME

TABLE 1. GEOMETRIC INFORMATION OF TESTED PARTS

Part	Number of triangles		Length, Width, Height [mm]
	Original	Convex hull faces	
Part-1	56	24	150.0, 100.0, 110.0
Part-2	5,370	1,724	331, 131.0, 140
Part-3	3,062	176	190.5, 214.3, 260.4
Part-4	2,662	278	100, 121.7, 233.4
Stanford Bunny	69,664	3004	77.8, 60.3, 77.2
Utah Teapot	9,438	2,496	175.8, 109.3, 85.7
3D Benchy	225,154	42,484	60.0, 31.0, 48.0

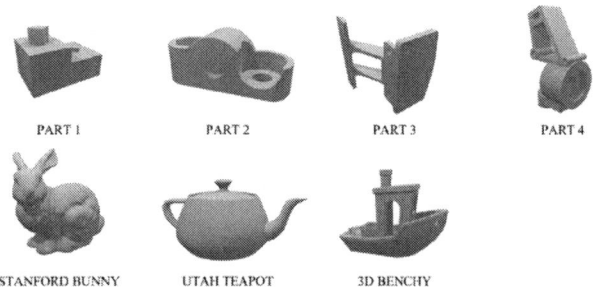

FIGURE 9. TESTED PARTS.

also seem to exhibit this trait. When using any other orientation in parts 1 and 2, the surface quality would suffer due to the staircase effect or require additional support for regions beyond the reachable overhang angle.

The results of the three objective functions for Part-2 are shown in Figure 10, Figure 11, and Figure 12. Note that the optimal at $\theta = 0$ and $\phi = 0$ is the lowest for all three objective functions. These plots are misleading in the sense that θ and ϕ are not truly orthogonal variables. When θ is 0 or π, the entire row of points (all values of ϕ) are technically the same point as these correspond to the north and south poles on the unit sphere. This demonstrates why the latitude-longitude sampling technique is inefficient. Another thing to glean from these plots is the multi-modal nature of these objective functions. Since Part-2 is a fairly simple part, one can observe more multi-modality as the parts become geometrically complex. While these spaces are smooth, this multi-modality, along with the multi-objective nature of the space, makes choosing a robust optimization method difficult. Hence, our approaches are focused on explicit enumeration of the space.

The rest of the tested parts are more complex in shape and are computationally expensive to evaluate, especially in computing the exact value of the support structure volume. A dynamic programming trick was employed to improve the efficiency while evaluating the first objective function. Objective functions 2 and 3 (described in Sections 4.2 and 4.3, respectively) are first calculated. Since we are only interested in keeping the Pareto (or non-dominated) solutions, we can

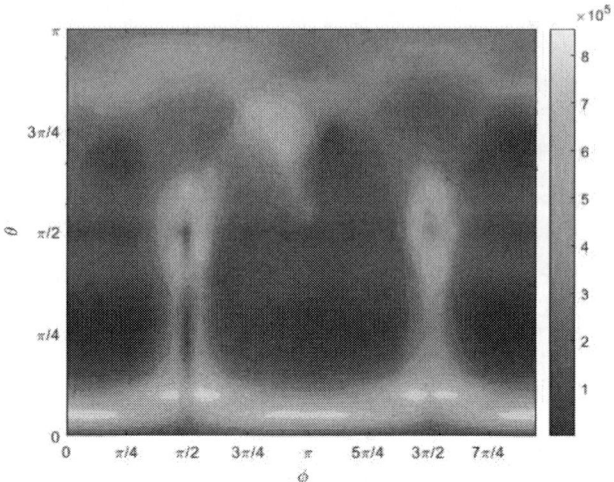

FIGURE 10. THE VOLUME OF THE SUPPORT STRUCTURE FOR PART 2

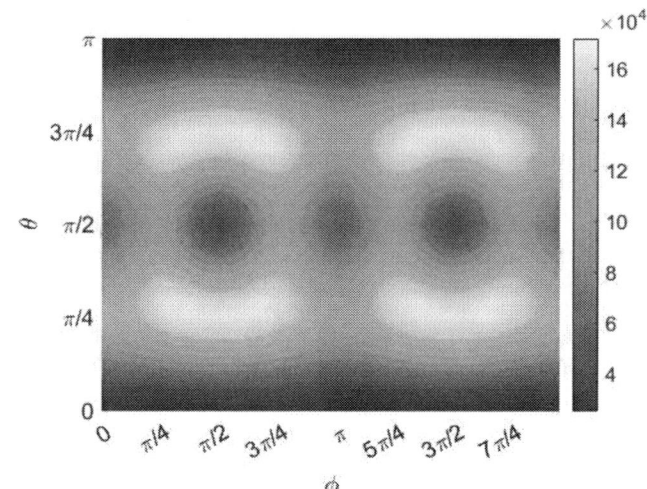

FIGURE 11. THE SURFACE QUALITY FOR PART 2.

determine a value of f_1 that would make a current solution dominated (since f_2 and f_3 have already been calculated). As the

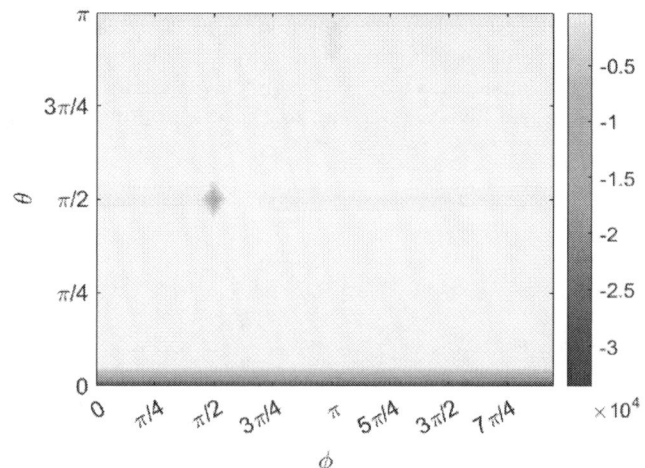

FIGURE 12. THE PRINTING STABILITY FOR PART 2.

Copyright © 2023 by ASME

TABLE 2. COMPARISON BETWEEN ORIENTATION SAMPLING TECHNIQUES

Part	Number of nondominated candidates/ overall searched candidates		
	Latitude-longitude lattice	Fibonacci lattice	Convex hull
Part-1	1 / 762	1 / 100	1 / 15
Part-2	1 / 762	1 / 100	1 / 264
Part-3	3 / 762	1 / 100	3 / 55
Part-4	29 / 762	2 / 100	4 / 99
Bunny	15 / 762	3 / 100	1 / 277
Utah Teapot	45 / 762	1 / 100	2 / 479
3DBenchy	8 / 762	1 / 100	2 / 134

TABLE 3. RESULTS AFTER APPLYING 5° ROTATION AROUND Z, Y, AND X-AXIS.

Part	Number of pareto front candidates/ overall searched candidates		
	Latitude-longitude lattice	Fibonacci lattice	Convex hull
Part 1	14 / 762	5 / 100	1 / 15
Part 2	50 / 762	3 / 100	1 / 264
Part 3	10 / 762	3 / 100	3 / 55
Part 4	67 / 762	4 / 100	4 / 99
Bunny	10 / 762	3 / 100	1 / 277
Utah Teapot	44 / 762	2 / 100	2 / 479
3DBenchy	22 / 762	9 / 100	2 / 134

simulation for creating the support structure, f_1 ensues by defining supports layer by layer, the algorithm tracks its increasing value. If the f_1 exceeds the value necessary to ensure that it is non-dominated, then the candidate is aborted since it will provably no longer be on the Pareto front solutions. Applying the Pareto front in this way reduces the computation time significantly from hours to minutes in all tested cases.

Table 2 and Table 3 show the number of Pareto front – or nondominated – candidates out of the overall searched candidates. For instance, Part-3 in Table 2 has 3 nondominated candidates out of 762 searched candidates for the latitude-longitude lattice, 1 out of 100 for the Fibonacci lattice, and 3 out of 55 for the convex hull method. As the part's complexity increases, the number of nondominated solutions using the convex hull approach is still low, as shown in Table 2 and Table 3. In the nominal or original orientation, the Pareto front solutions were similar for all approaches except for tested Part-4. In the slightly rotated case, the results of the three approaches are not similar. In both the nominal orientation and slightly rotated cases, the convex hull approach found the exact same nondominated orientations with respect to the part orientation as expected. In contrast, the Latitude-longitude and Fibonacci lattice could not perform as well and produced relatively more solutions than their nominal orientations. This is because the discretization in these approaches skips over the dominating solution found in Table 2. The convex hull approach's computation time ranges from 0.8 to 481.8 seconds over these seven test parts. While the computational time for the Latitude-longitude lattice was ranging 31.2 to 1834.55 seconds, the Fibonacci lattice was from 2.0 to 141.0 seconds. The results showed that using a convex hull to generate orientations was highly effective in finding the optimal build orientation with the fewest evaluations, especially for support-free parts. The Fibonacci lattice outperformed the convex hull approach in time, but the slightly rotated case was significantly longer for the Fibonacci lattice, ranging from 42.9 to 1021.4 seconds. The latitude-longitude and Fibonacci lattice in the slightly rotated case has several Pareto front candidates ranging from 10 to 67 and 2 to 9, respectively. In contrast, the convex hull produces 1

to 4 candidates – all of which appear to be legitimate options in choosing an orientation.

For parts where the support structure was not needed, the convex hull approach captured the optimal build orientation for both the nominal orientation and, after slightly rotating the model. The other build orientation generation approaches captured all optimal cases that perfectly align along the positive direction of the z-axis. However, neither the Latitude-longitude nor the Fibonacci lattice captured the optimal build orientation for the slightly rotated cases.

The support creation algorithm needs further optimization for parts that require support structures. For instance, all approaches found build orientation requiring support for Part-4 (see Figure 13). A possible approach for parts that require a support structure is to build a small support that elevates the part above a certain height above the printing platform. However, the stability of the printing would be worse since the support structure might be hard to remove in post-processing or it may not be strong enough while printing.

6 CONCLUSION

This paper presented a novel optimization scheme for finding the optimal print orientation in five-axis 3D printing. Our approach is based on three criteria: the volume of the support structure, the surface quality caused by the staircase effect, and the stability of the part while rotating and tilting the printing platform. Through testing on seven 3D parts, we demonstrated the effectiveness of our new method in finding the Pareto optimal set of print orientations within a convenient amount of time.

Our scheme outperformed traditional methods for defining print orientation since neither the Latitude-longitude lattice approach nor the Fibonacci lattice approach captured the optimal build orientation for the slightly rotated cases. The proposed optimization scheme offers a practical solution for addressing the added complexity introduced by rotational axes in five-axis 3D printing and finding the optimal print orientation for high-quality printed parts.

The optimization scheme could improve by incorporating additional criteria or constraints into the optimization process, like printing time or cost. One potential limitation of the current

Copyright © 2023 by ASME

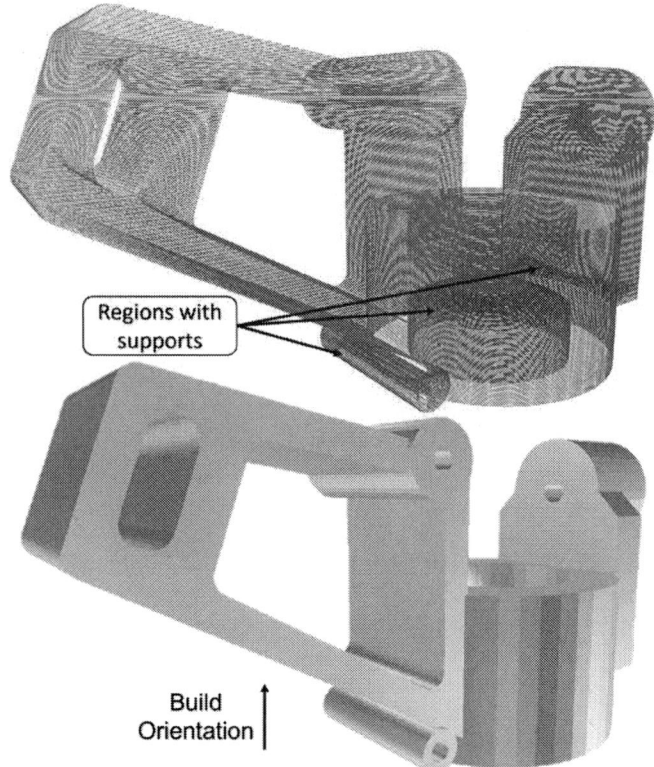

FIGURE 13. PART-4 IS SHOWN IN ITS OPTIMAL BUILD ORIENTATION. NOTE THE RED AREAS WHERE THE SIMULATION FOR THE FIRST OBJECTIVE FUNCTION FINDS THE MINIMAL SUPPORT STURCTURE.

optimization scheme is that it is based on a fixed set of criteria, which may not be suitable for all 3D parts or printing applications. For example, the optimization scheme currently only considers orientation-related factors and does not consider other aspects of the printing process, such as material properties or tool head characteristics. Additionally, it may be valuable to consider integrating other factors, such as material properties or tool head characteristics, into the optimization scheme to fully capture the five-axis 3D printing capability. By addressing these limitations, it may be further possible to improve the optimization scheme's accuracy and efficiency and make it more applicable to a broader range of 3D printing scenarios. We will test printing orientations on a five-axis 3D printer in future work.

ACKNOWLEDGEMENTS
The authors would like to acknowledge the financial support from Saudi Arabia Cultural Mission.

REFERENCES
[1] Alexander, P., Allen, S., and Dutta, D., 1998, "Part Orientation and Build Cost Determination in Layered Manufacturing," Computer-Aided Design, **30**(5), pp. 343–358.

[2] Gibson, I., Rosen, D., Stucker, B., and Khorasani, M., 2021, *Additive Manufacturing Technologies*, Springer Cham.

[3] Li, Y., Tang, K., He, D., and Wang, X., 2021, "Multi-Axis Support-Free Printing of Freeform Parts with Lattice Infill Structures," CAD Computer Aided Design, **133**.

[4] Xie, F., Jing, X., Zhang, C., Chen, S., Bi, D., Li, Z., He, D., and Tang, K., 2022, "Volume Decomposition for Multi-Axis Support-Free and Gouging-Free Printing Based on Ellipsoidal Slicing," Computer-Aided Design, **143**, p. 103135.

[5] Isa, M. A., and Lazoglu, I., 2019, "Five-Axis Additive Manufacturing of Freeform Models Through Buildup of Transition Layers," J Manuf Syst, **50**, pp. 69–80.

[6] Kapil, S., Negi, S., Joshi, P., Sonwane, J., Sharma, A., Bhagchandani, R., and Karunakaran, K. P., 2017, "5-Axis Slicing Methods for Additive Manufacturing Process," *Solid Freeform Fabrication Symposium*, University of Texas at Austin, pp. 1886–1896.

[7] Liu, H., Liu, L., Li, D., Huang, R., and Dai, N., 2020, "An Approach to Partition Workpiece CAD Model Towards 5-Axis Support-Free 3D Printing," International Journal of Advanced Manufacturing Technology, **106**(1–2), pp. 683–699.

[8] Schuh, G., Bergweiler, G., Lukas, G., Hohenstein, S., and Schenk, J., 2020, "Feature-Based Print Method for Multi-Axis Material Extrusion in Additive Manufacturing," *Procedia CIRP*, Elsevier B.V., pp. 885–890.

[9] Hu, Q., Feng, D., Zhang, H., Yao, Y., Aburaia, M., and Lammer, H., 2020, "Oriented to Multi-Branched Structure Unsupported 3D Printing Method Research," Materials, **13**(9).

[10] Xiao, X., and Joshi, S., 2020, "Process Planning for Five-Axis Support Free Additive Manufacturing," Addit Manuf, **36**.

[11] Dai, F., Zhang, H., and Li, R., 2020, "Process Planning Based on Cylindrical or Conical Surfaces for Five-Axis Wire and Arc Additive Manufacturing," Rapid Prototyp J, **26**(8), pp. 1405–1420.

[12] Jetton, C., Rudd, L., and Campbell, M. I., 2022, "Systematic Generation of 5-Axis Manufacturing Machines," Proceedings of the ASME Design Engineering Technical Conference, **7**, pp. 1–10.

[13] Rodriguez-Padilla, C., Cuan-Urquizo, E., Roman-Flores, A., Gordillo, J. L., and Vázquez-Hurtado, C., 2021, "Algorithm for the Conformal 3d Printing on Non-Planar Tessellated Surfaces: Applicability in Patterns and Lattices," Applied Sciences (Switzerland), **11**(16).

[14] Liu, H., Liu, L., Li, D., Huang, R., and Dai, N., 2020, "An Approach to Partition Workpiece CAD Model Towards 5-Axis Support-Free 3D Printing," International Journal of Advanced Manufacturing Technology, **106**(1–2), pp. 683–699.

Copyright © 2023 by ASME

[15] Di Angelo, L., Di Stefano, P., and Guardiani, E., 2020, "Search for the Optimal Build Direction in Additive Manufacturing Technologies: A Review," Journal of Manufacturing and Materials Processing, 4(3).

[16] Qin, Y., Qi, Q., Scott, P. J., and Jiang, X., 2019, "Determination of Optimal Build Orientation for Additive Manufacturing Using Muirhead Mean and Prioritised Average Operators," J Intell Manuf, 30(8), pp. 3015–3034.

[17] Alomari, Y., Birosz, M. T., and Andó, M., 2023, "Part Orientation Optimization for Wire and Arc Additive Manufacturing Process for Convex and Non-Convex Shapes," Sci Rep, 13(1), p. 2203.

[18] Meifa, H., Zheng, N., Qin, Y., Tang, Z., Zhang, H., Fan, B., and Qin, L., 2022, "Description Logic Ontology-Supported Part Orientation for Fused Deposition Modelling," pp. 1–20.

[19] Jaiswal, P., Patel, J., and Rai, R., 2018, "Build Orientation Optimization for Additive Manufacturing of Functionally Graded Material Objects," International Journal of Advanced Manufacturing Technology, 96(1–4), pp. 223–235.

[20] Li, D., Dai, N., Zhou, X., Huang, R., and Liao, W., 2018, "Self-Supporting Interior Structures Modeling for Buoyancy Optimization of Computational Fabrication," International Journal of Advanced Manufacturing Technology, 95(1–4), pp. 825–834.

[21] Luc, D. T., 2008, "Pareto Optimality," Springer Optimization and Its Applications, 17, pp. 481–515.

[22] Luersen, M. A., and Le Riche, R., 2004, "Globalized Nelder–Mead Method for Engineering Optimization," Comput Struct, 82(23–26), pp. 2251–2260.

[23] Campbell, M. I., "TVGL: Tessellation and Voxelization Geometry Library" [Online]. Available: https://github.com/DesignEngrLab/TVGL. [Accessed: 31-Jan-2019].

[24] González, Á., 2010, "Measurement of Areas on a Sphere Using Fibonacci and Latitude-Longitude Lattices," Math Geosci, 42(1), pp. 49–64.

[25] Edelsbrunner, H., 1984, *Algorithms in Combinatorial Geometry*, Springer Berlin, Heidelberg.

[26] Xiao, X., Roh, B. M., and Zhu, F., 2021, "Strength Enhancement in Fused Filament Fabrication via the Isotropy Toolpath," Applied Sciences (Switzerland), 11(13).

[27] Zhang, H., Liu, D., Huang, T., Hu, Q., and Lammer, H., 2020, "Three-Dimensional Printing of Continuous Flax Fiber-Reinforced Thermoplastic Composites by Five-Axis Machine," Materials, 13(7).

[28] Leirmo, T. L., and Semeniuta, O., 2022, "Minimizing Form Errors in Additive Manufacturing with Part Build Orientation: An Optimization Method for Continuous Solution Spaces," Open Engineering, 12(1), pp. 227–244.

[29] Böğrekci, İ., Demircioğlu, P., Sucuoğlu, H. S., and Turhanlar, O., 2019, "The Effect of The Infill Type and Density on the Hardness of 3D Printed Parts," International Journal of 3D Printing Technologies And Digital Industry, 3, pp. 212–219.

Copyright © 2023 by ASME

Proceedings of the ASME 2023
International Design Engineering Technical Conferences and
Computers and Information in Engineering Conference
IDETC-CIE2023
August 20-23, 2023, Boston, Massachusetts

DETC2023-111826

PROCESS-AWARE PREDICTION OF GEOMETRIC ACCURACY FOR ADDITIVE MANUFACTURING VIA TRANSFER LEARNING

Daphne Lin
daphne15lin@utexas.edu
University of Texas at Austin
Austin, TX

Carolyn Seepersad
ccseepersad@mail.utexas.edu
University of Texas at Austin
Austin, TX

ABSTRACT

While additive manufacturing offers unprecedented design freedom to engineers, challenges associated with part accuracy and quality prevent additive manufacturing from becoming as widespread as traditional manufacturing techniques. Statistical quality control is typically applied to mass-produced parts in which large numbers of identical (or similar) parts are fabricated in a production line context, but for customized, additively manufactured parts, part qualification must be performed with a much more limited number of samples— ideally, just one customized part—or the benefit of customized production is lost to the expense of repeated production and refinement. Building upon previous work that established a statistical database of geometric design allowables for a laser sintering additive manufacturing process via extensive geometric metrology, this research establishes transfer learning techniques for transferring those databases to new materials, machines, or parts. The goal of this research is to utilize transfer learning to maximize the accuracy of design allowables for new materials, machines, or parts, while limiting the required number of expensive new metrology studies. Transfer learning is implemented via a two-stage Gaussian process regression method. In the first stage, a trend model is built using pre-existing metrology data to capture general geometric characteristics of the process. In the second stage, a deviation model is trained on the difference between the pre-existing data and a small amount of data gathered from the new material/machine/process. The trend and deviation models are then aggregated to create a surrogate model referred to as the transfer learning model. The proposed method is evaluated using a multi-material polymer selective laser sintering metrology dataset. The performance of the transfer learning model is compared to that of simpler alternative models. Results show that transfer learning models are generally more accurate in
predicting geometric dimensions and less sensitive to changes in training data points.

Keywords: Additive Manufacturing, Predictive metrology, Polymer selective laser sintering, Gaussian process regression, transfer learning

1. INTRODUCTION

Engineering design may be the largest barrier to broader adoption of additive manufacturing (AM) for the fabrication of customized parts [1]. On the one hand, AM is pushing engineering design to be more opportunistic and fully leverage all of the complexities and freedoms afforded by AM: hierarchical structures, customized geometry, multiple materials, and many others [2]. The ability to fully leverage all of these opportunities exceeds the capabilities of modern design tools and methods. On the other hand, engineering design is pushing AM to be more predictable and reliable in terms of (1) material properties, (2) geometric accuracy and resolution, and (3) its ability to reduce the presence of defects and variability that have catastrophic effects on both material properties and geometry. It is essential to be able to build parts within acceptable tolerances in terms of geometry and mechanical properties on the initial attempt. Otherwise, it is cost-prohibitive to build customized parts because each one must be fabricated repeatedly to achieve acceptable results. Additionally, the refinement process must be repeated when the design is fabricated on another machine or using a different material. As such, there is a need for AM design predictability and quality control before engineers can fully take advantage of the design freedom AM offers [3].

One method to increase design predictability is through process-specific design-for-manufacturing guidelines. Traditional manufacturing techniques such as injection molding and casting have extensive guidelines [4,5], but for AM, a

Copyright © 2023 by ASME

relatively new manufacturing method, design guidelines are still being established through research efforts. Examples of general efforts include the creation of a Design for AM repository [6] and the proposition of a modular method to standardize interpretation of design rules across AM [7]. Metrology test parts and studies have also been prevalent in AM research and aim to characterize various AM processes. Moylan et al. designed a general AM metrology part that contains various geometric features such as holes, pins, and staircases [8]. Process-specific metrology test parts have also been designed to capture properties and characteristics that are not shared between AM methods, allowing designers to better understand and leverage the specific technology in their designs. For example, Techniek et al. and Kranz et al. designed metrology parts for direct metal laser sintering to study complex geometries and lightweighting, respectively [9,10]. Additional process-specific metrology parts include designs for material jetting [11], fused deposition modeling [12], and polymer selective laser sintering [13,14].

While these studies provide designers with general design guidelines, they can be tedious and require substantial time, test parts, and measurements. The predictive capabilities of design guidelines are also limited and cannot fully inform material-specific or machine-specific geometries [15]. Figure 1 below highlights the difference in as-built gap sizes between two different nylons used in polymer selective laser sintering (SLS). Despite having the same as-designed dimension, the two materials result in different as-built dimensions.

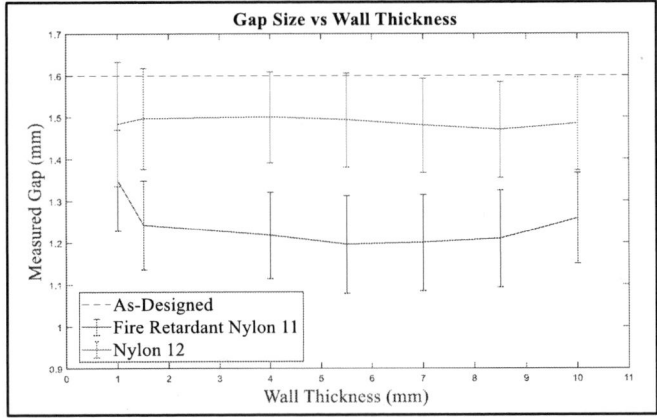

FIGURE 1. Average gap measurements with standard deviations as a function of wall thickness for fire retardant nylon (blue) and nylon 12 (red). The as-designed gap dimension is shown as a dashed, black line at 1.6mm.

Thus, when printing with a new material or machine, additional tests must be conducted to tune the designs and process. However, with the advancement of computational tools such as transfer learning, designers can efficiently transfer previous metrology knowledge to new machines and materials with minimal effort and additional measurements.

Transfer learning is a powerful tool in computational fields and machine learning, as it substantially decreases the need to gather new, large datasets [16,17]. Instead, existing similar datasets along with a predictive function can be substituted as training data for a new model. While transfer learning emerged

from machine learning, its applications have become widespread, ranging from medical [18,19] to transportation [20,21]. Within design and manufacturing, transfer learning or surrogate models have proven their usefulness by increasing prediction accuracy while decreasing the need for physical data [22,23]. Tercan et al. applied transfer learning in the context of injection molding to create a neural network predictor trained on simulation data and transferred to experimental data [24]. Transfer learning has also been used to decrease the computational requirements of complex design simulations by building surrogate models trained primarily on simpler simulations and corrected by fewer complex simulation results [25].

This paper proposes a transfer learning method composed of two Gaussian process models similar to that of Qian et al. [25]. The transfer learning method is then demonstrated in the case of predicting material-specific geometric characterization in polymer selective laser sintering (SLS). The method assumes an abundance of baseline metrology data that can be used to build the first Gaussian process model which is referred to as the trend model. The second Gaussian process model relies on much fewer data points – called target data points – to correct the first model for a new material or machine. This surrogate model, referred to as the transfer learning model, is then able to characterize geometries in the new material or machine much more accurately than if the model were built purely on the baseline data or the new target data alone. The proposed method can be applied across various AM processes to reduce the number of metrology prints required when tuning for a new material or machine while still accurately informing the designer of the geometric characterization of the specific process.

2. BACKGROUND
2.1 Gaussian Process Regression

The predictor follows a two-stage Gaussian process modeling approach originally proposed for combining data from low- and high-fidelity simulations [25]. The Gaussian process regression (GPR) model assumes the following structure shown in Equation 1 where $f(x) = [f(x_1), ..., f(x_m)]^t$ is a set of predefined functions of m input parameters (regression model), $\beta = (\beta_1, ..., \beta_m)^t$ is a set of unknown coefficients, and $\varepsilon(x)$ is a realization of a stationary, typically Gaussian, process for a data set of n vectors of input variable values (correlation model).

$$y(x_i) = \beta^t f(x_i) + \varepsilon(x_i), i = 1, ..., n \qquad (1)$$

For the proposed application in predictive additive manufacturing metrology, the baseline metrology data is assumed to be abundant and is used to build a Gaussian process model that captures the trends but may produce biased results for the new machine/material. A baseline model of each dependent parameter, y_B, is built as a function of d input parameters where the first two terms are the linear part and the final term is the residual part, which is assumed to be a stationary Gaussian process as shown in Equation 2.

$$y_B(x) = \beta_{a0} + \sum_{h=1}^{d} \beta_{ah} x_h + \varepsilon_a(x) \qquad (2)$$

Copyright © 2023 by ASME

With a large quantity of baseline metrology data available, y_B can be approximated accurately. Then, the fitted model is adjusted to an updated model, y_U, by incorporating small amounts of metrology data from the new machine/material. As shown in Figure 1, there may be a nonlinear discrepancy between the baseline and material/machine-specific data, which leads to the updated model shown in Equation 3 where $\rho(x_i)$ is a linear regression function with a scaling term and a shifting term and $\delta(x_i)$ is another stationary Gaussian process that can be used to accommodate nonlinear residual errors between the baseline and updated data. An optimization process is used to fit all of the unknown parameters, as described in Qian et al. [25].

$$y_U(x_i) = \rho(x_i)y_B(x_i) + \delta(x_i) \qquad (3)$$

GPR modeling has several properties that make it an attractive candidate for applications in predictive metrology. First, in regards to the tradeoff between implementation complexity and prediction accuracy, GPR modeling offers relatively high prediction accuracy with relatively simple implementation compared to other metamodeling techniques [26]. Compared to polynomial regression, GPR models are typically more accurate, as they interpolate between training points and are not limited in predictions by a specific function [25,26]. Although polynomial regression models are simpler to create, generating accurate predictions is of higher priority. Alternatively, support vector regression (SVR) offers higher prediction accuracy than GPR models; however, SVR has a higher computational cost in optimizing the unknown parameters [29]. Second, GPR models tend to perform well even in cases with limited data or small datasets. Unlike other metamodeling methods, GPR models are robust against noise because they do not require fixed basis functions [30]. This property of GPR is especially important for the application of interest, as it allows designers to generate accurate geometric characterizations without the need for large volumes of metrology data. Third, GPR models are able to provide error bounds for each prediction as shown in Figure 2. The benefit to having error bounds is that it allows for "Intelligent Sampling." The model is able to improve iteratively by using the error bounds to select the next training point [28].

A drawback to GPR models is that the model traditionally must have unique responses for each input data point. One simple way to achieve uniqueness is to take the mean or median of the data. Weighting functions is another method to achieve uniqueness and combine multiple outputs or responses into a single value. Another slightly more involved method is to add slight offsets in the input variable values such that there is only one response for each input. Another solution is to modify the covariance functions such that multiple responses can be related back to a single input, creating a multi-response GPR [29,30]. For the given application, the desired prediction is simply the geometric dimension for a given input variable value. As such, the use of a multi-response GPR is unnecessary, and the simpler approach of taking the mean or median of the responses at a single input variable value is applied.

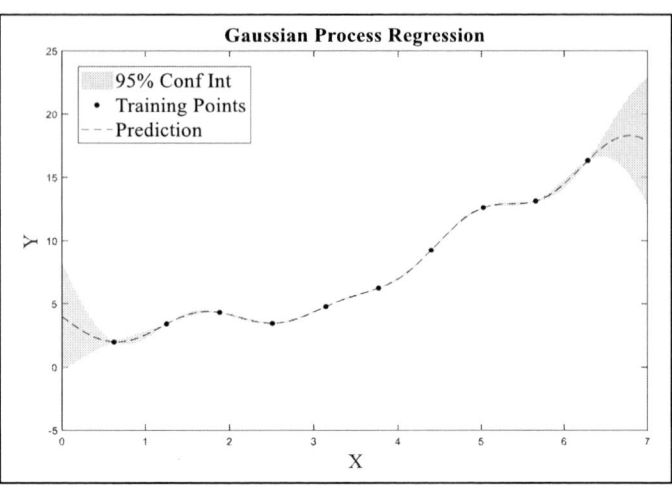

FIGURE 2. Example of Gaussian process regression with confidence interval. Training points are shown as black points. Predictions are shown by a dashed line. A 95% confidence interval is shown in red.

The work in this paper uses an open-source implementation of GPR in Matlab called DACE [33]. The models generated in this paper all use a zero-order regression model and a Gaussian correlation model.

2.2 Polymer SLS Metrology Dataset

To characterize a new process (i.e., new material on an existing machine, known material on a new machine, etc.), metrology test parts are printed and measured to determine as-printed geometries and material properties [8,13,15,32,33,34,35]. These studies generate large quantities of measurement data; however, the data only applies to the specific process through which the test parts were printed. The methodology proposed in this paper leverages existing data to predict geometric characteristics of new processes. For the purposes of this paper, a metrology study performed by Allison et al. [15] served as the dataset to test the proposed transfer learning method. The study collected metrology data on three different materials: nylon 12 (PA 12), glass-filled nylon 12 (GF Nylon 12), and fire-retardant nylon 11 (FR PA 11). As such, the study served as an effective dataset to test the predictive performance of the transfer learning method in the context of geometric characterization of a new material.

Allison et al.'s metrology study was conducted using a test part specifically designed for polymer SLS [15,35]. Because polymer SLS parts can be printed without support structures, the test part, shown below in Figure 3, can fit a large number of geometric features within a 5cm x 5cm x 5cm volume. The test part yielded both quantitative and qualitative data, as summarized in Table 1.

Copyright © 2023 by ASME

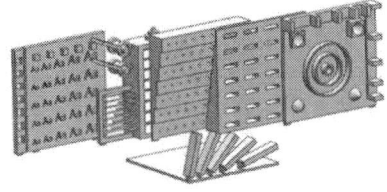

FIGURE 3. Polymer SLS test part designed to print as a cube (left) and to disassemble for measurements (right) **[37]**.

TABLE 1. Summary of geometric features on polymer SLS test part. Feature ranges reported from Allison et al. (2017) **[37]**. Measurement type reported from Allison et al. (2019) [15].

Feature	Feature Ranges	Measurement Type
Holes	0.8 - 2.60 mm diameter 1.0 - 10.0 mm wall thickness	Quantitative
Thin Rods	0.3 - 0.9 mm diameter	Qualitative
Thin Walls	0.2 - 0.8 mm	Quantitative
Gaps	1.4 - 2.0 mm gaps 1.0 - 10.0 mm wall thickness	Quantitative
Cylinders	2.0 - 8.0 mm diameter	-
Hollow Cylinders	5.0 - 25 mm diameter	-
Domes	6.0 mm diameter	-
Cones	6.0 mm diameter 5.2 mm height	-
Linear Accuracy	5.0 - 12.5 mm 2 build directions	Quantitative
Surface Roughness	0 - 90°	Quantitative
Hinges	0.6 mm and 1.0 mm shaft clearances	Qualitative
Lettering	10 - 18 pt font 0.5 - 2.5 mm embedded/raised depth	Qualitative
Snap Fits	-0.5 to 0.25 mm offsets	Qualitative

Figure 4 shows the design of hole and gap features from the polymer SLS test part. The dataset used in this paper focused primarily on the following features as functions of wall thickness: 2.4mm diameter holes, 2.6mm diameter holes, 1.4mm gaps, and 1.6mm gaps. The hole diameters and gap widths are expected to change with wall thickness because thicker walls lead to increased local temperatures and oversintering in the powder bed. The specific hole diameters were selected to maximize the size of the dataset, as smaller hole diameters did not always resolve and were not measurable. All studied gap sizes resolved, so the gap sizes used to test the transfer learning method were arbitrarily chosen.

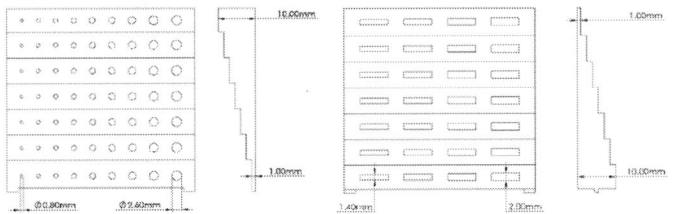

FIGURE 4. Holes and gaps as functions of wall thickness [15].

Measurements were collected at wall thicknesses of 1.0mm, 1.5mm, 4.0mm, 5.5mm, 7.0mm, 8.5mm, 10.0mm. The final datasets for each feature consisted of 48 measurements at each wall thickness for each material.

3. METHODOLOGY

The transfer learning predictor is a linear combination of two models: a trend model and a deviation model. Figure 5 below outlines the general steps to create the transfer learning (TL) model proposed in this paper.

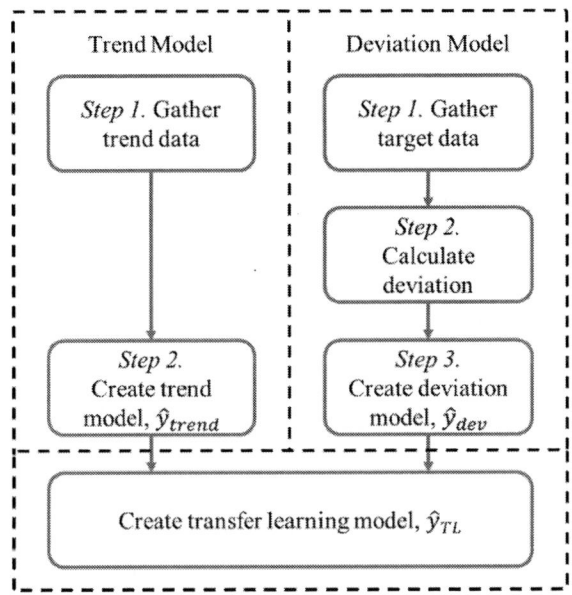

FIGURE 5. Generalized flow to create the trend and deviation models which are linearly combined to create the transfer learning model.

Allison's metrology data was used to demonstrate the predictive capabilities of the TL model. In this case study, a designer is interested in printing a nominal 1.6mm gap or slot in various wall thicknesses using a new, target material (Fire Retardant Nylon - FR PA 11). For the purposes of this example, it is assumed that while the designer does not have metrology data on final slot geometry in FR PA 11, they do have data from previous builds on the polymer SLS machine in Glass-Filled Nylon (GF PA 12) and Nylon 12 (PA 12). The hypothesis is that by leveraging a TL model, the designer is able to better predict the final geometry of their designed, nominal 1.6mm gap in the new material versus traditional curve fitting.

Trend Model Step 1. To create the trend model, the trend data is first gathered and preprocessed to be unique (i.e., each input variable value should have a single value). As discussed in the Background section, this can be done in a number of ways. In this example, each wall thickness (input variable) has 48 measured gap values which are summarized in the box plots shown in Figure 6. As shown in Figure 7, the median of the dataset is used to make the trend data unique while minimize bias from outlying data points.

Trend Model Step 2. The GPR trend model, $\hat{y}_{trend}$, with a zero-order polynomial regression model and a Gaussian correlation model is created by training the model using all available trend data. The trend model for the case study is shown in Figure 8.

Copyright © 2023 by ASME

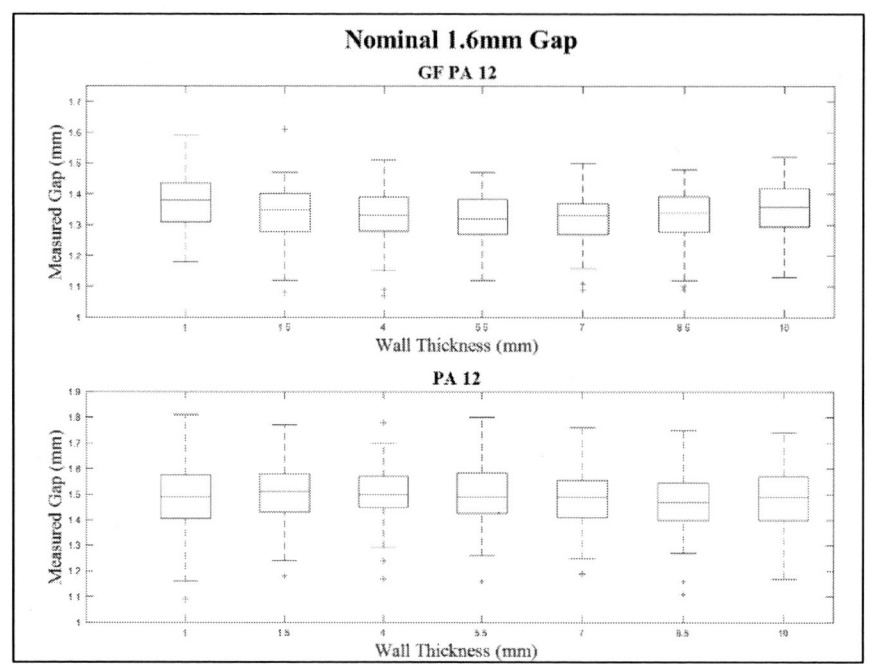

FIGURE 6. Measured gap sizes in varying wall thickness printed in GF PA 12 and PA 12. The nominal, as-designed gap size was 1.6mm.

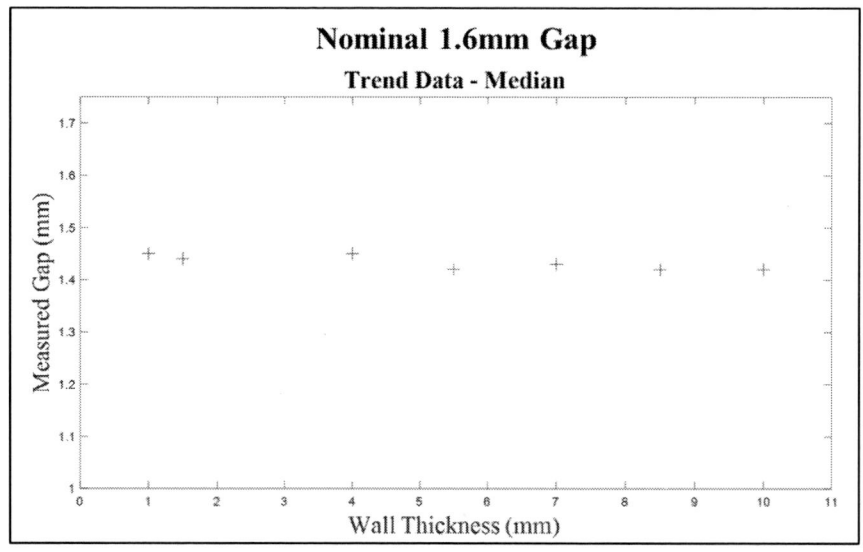

FIGURE 7. Median of measured gap sizes in varying wall thicknesses printed in GF PA 12 and PA 12. The nominal, as-designed gap size was 1.6mm.

Deviation Model Step 1. The deviation model is created from both trend and target data. When collecting training target points, data should ideally be collected at input variable values where trend data is available, so the deviation can be directly calculated. For example, the trend data shown above in Figure 7 does not exist for a wall thickness of 5mm, so target data should not be taken at a 5mm wall thickness. While it is still possible to collect training target data at other locations, the model would have to rely on the trend model to determine the deviation,

adding uncertainty which would propagate to the geometric predictions. In terms of the quantity of training target points to collect, a benefit of the transfer learning methodology presented in this paper is that few training target points are required to generate a reasonably accurate predictor. In the case study, the designer takes three measurements in FR PA 11 as shown in Figure 8; this could be due to limitations in time or cost which prevent the designer from conducting a full metrology study to characterize the new material.

Copyright © 2023 by ASME

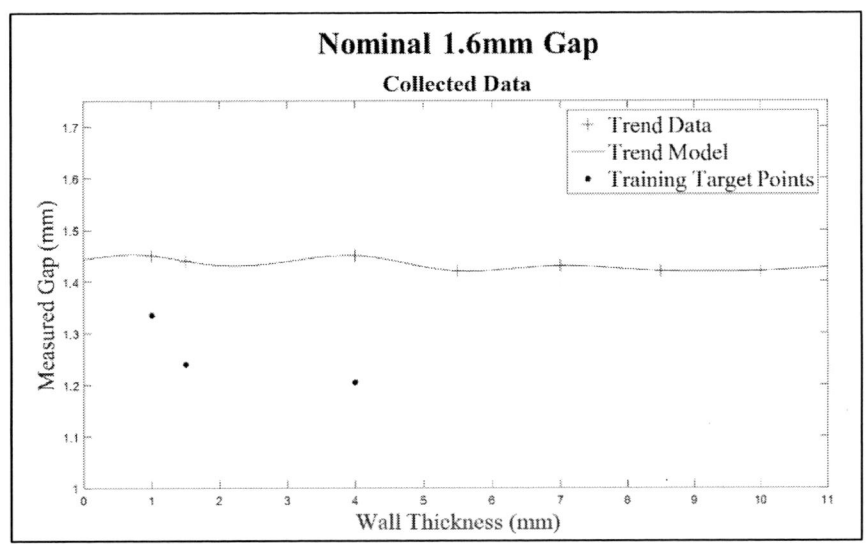

FIGURE 8. Trend and training target data shown for a nominal, as-designed 1.6mm gap with varying wall thicknesses.

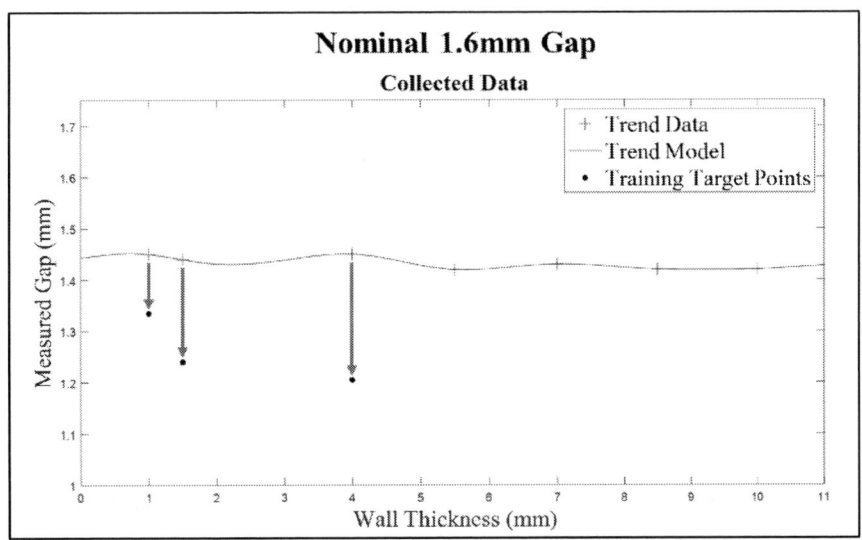

FIGURE 9. Deviation between trend and target data shown by red arrows for a nominal, as-designed 1.6mm gap with varying wall thicknesses.

Deviation Model Step 2. The deviation is calculated by taking the difference between the trend and training target data at each training input variable value as shown in Equation 4 and graphically shown in Figure 9. Similar to *Trend Model Step 1,* the deviations must be unique for each input variable value.

$$deviation_i = trend_i - target_i \qquad (4)$$

Deviation Model Step 3. Once the deviation has been calculated, the deviation model, $\hat{y}_{dev}$, is created by training another GPR model on the deviations calculated in the previous step. The deviation model can be created with the same or different parameters as the trend model. For the case study, both

GPR models used zero-order polynomial regression models and Gaussian correlation models.

Transfer Learning Model. The TL model is created by simply combining the trend and deviation models as shown in Equation 5.

$$\hat{y}_{TL} = \hat{y}_{trend} - \hat{y}_{dev} \qquad (5)$$

Figure 10 shows the transfer learning model predictions (black, dashed line) from the case study along with the FR PA 11 target data measurements (black points for training, red points for test). As seen above, the TL model is capable of making predictions at wall thicknesses where no trend or training target data exists.

Copyright © 2023 by ASME

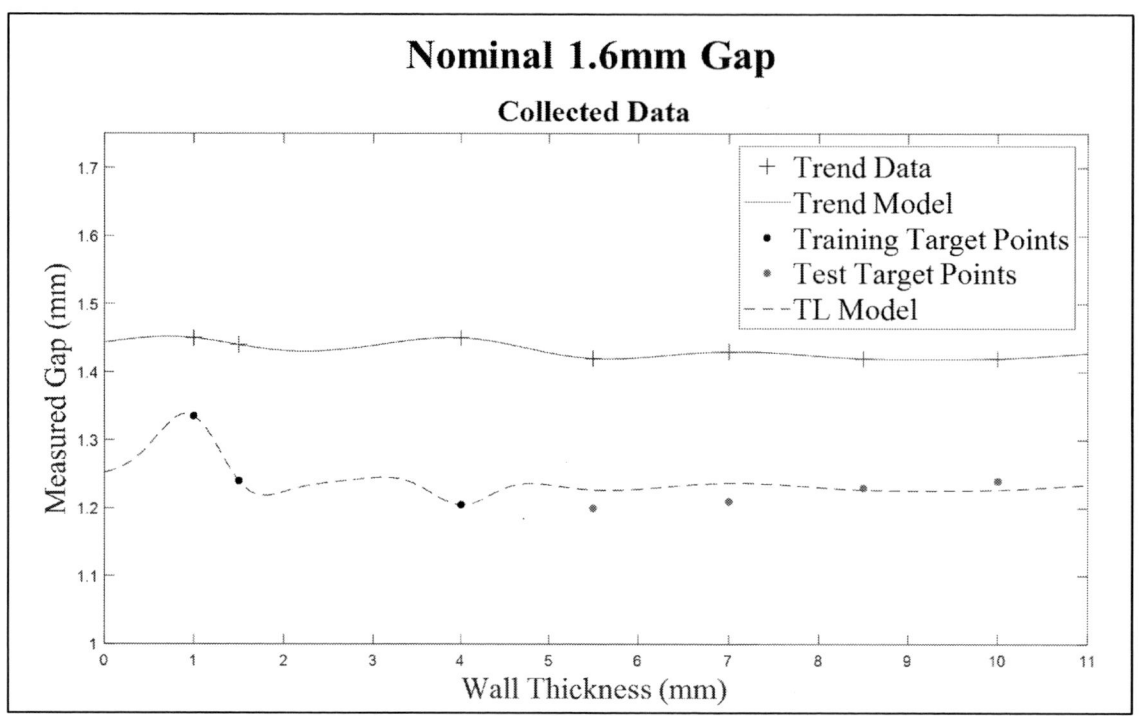

FIGURE 10. Transfer learning model predictions for FR PA 11 nominal 1.6mm gaps shown as a black, dashed line. Target data is shown as points (black for training, red for test). Trend data and model are shown as plus signs and a solid blue line, respectively.

4. RESULTS AND DISCUSSION

To gauge the predictive powers of the proposed TL model, three additional common fit methods were tested for comparison: linear fit, quadratic fit, and linear shift (LS). Both the linear and quadratic fits were created by directly fitting the training target points. The LS predictor was created by first generating the trend model detailed in the Methodology section. Then, the trend model was shifted by the average deviation between trend points and training target points. Figure 11 below shows all predictors in the context of the case study performed above. In the case with three training target points, both the linear and quadratic fits were unable to accurately predict the remaining four target points (test points). The LS and TL models captured the general trend of the target points much more accurately. However, one key difference between these two models, which can be seen in Figure 11, is that the TL model always forced the predictor through the training target points used to create the model (i.e., the error at the training target points will always be zero). The LS could not guarantee zero error at the training target points even though these points were known measurements. The linear and quadratic fits also could not guarantee zero error. In this regard, the TL model was more accurate than the other three methods.

All predictors depended on the target points used in training. Another example is shown in Figure 12 where an additional two training points were used to train each predictor. Again, only the TL model prediction interpolated the training target points used to create the model. However, compared to the example in Figure 11, the linear and quadratic fits performed much better, likely due to the increased number of fitted target points.

In Figures 11 and 12, it is difficult to distinguish the LS model and TL model predictions past wall thicknesses of 4mm. This can be attributed to the deviations between the trend and target data which remained relatively constant. As such, the LS model characterized the target data reasonably well. However, there exist cases where deviations between the trend and target data vary at different *x*-locations; in these cases, the TL model is expected to perform better than the LS model, as it is able to capture the changes in deviation. Figure 13 shows a case for a 2.4mm diameter hole. The trend points are shown as plus signs while the generated trend model is shown with a blue line. Again, GF PA 12 and PA 12 made up the trend data while FR PA 11 was the target material. The hole diameter trend remained relatively flat as wall thickness changed. On the other hand, the FR PA 11 hole diameters appeared to decrease as wall thickness increased. While the TL model captured this decrease, the LS model did not. As such, the TL model is generally a less risky option than using a LS model, as it is able to predict a wider variety of potential target behaviors with little to no additional costs.

Copyright © 2023 by ASME

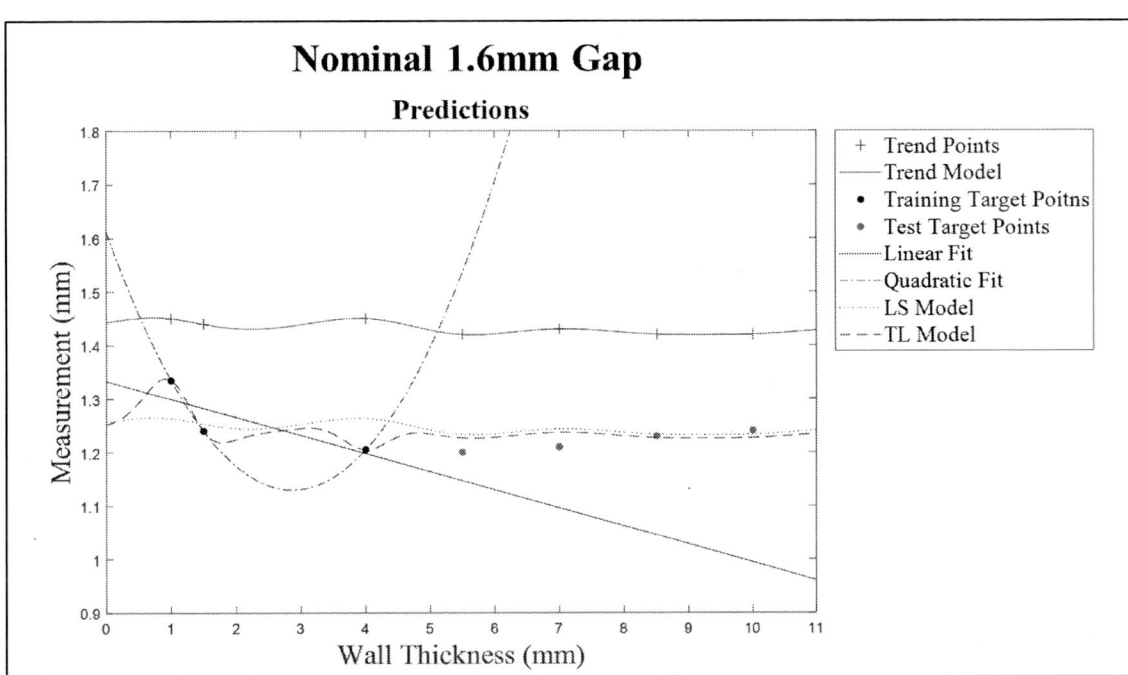

FIGURE 11. Linear Fit, Quadratic Fit, Linear Shift model, and Transfer Learning model gap predictions using three training target points. Training target points are shown in black; test target points are shown in red. the linear and quadratic fits do not predict the target points. both the linear shift and transfer learning models are able to approximate the test target points. however, only the transfer learning model is able to interpolate through the training target points.

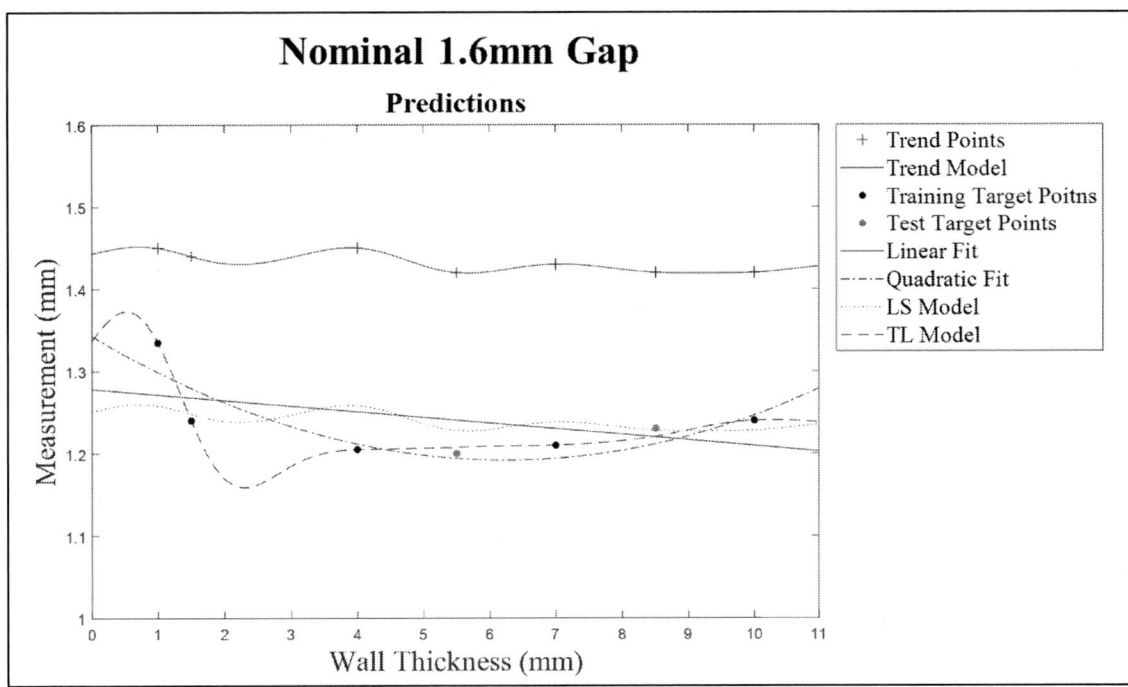

FIGURE 12. Linear Fit, Quadratic Fit, Linear Shift model, and Transfer Learning model gap predictions using five training target points. Training target points are shown in black; test target points are shown in red. the linear fit, quadratic fit, and linear shift model do not fully capture the target points. the transfer learning model is able to capture the target points' behavior for small wall thicknesses through large wall thicknesses.

Copyright © 2023 by ASME

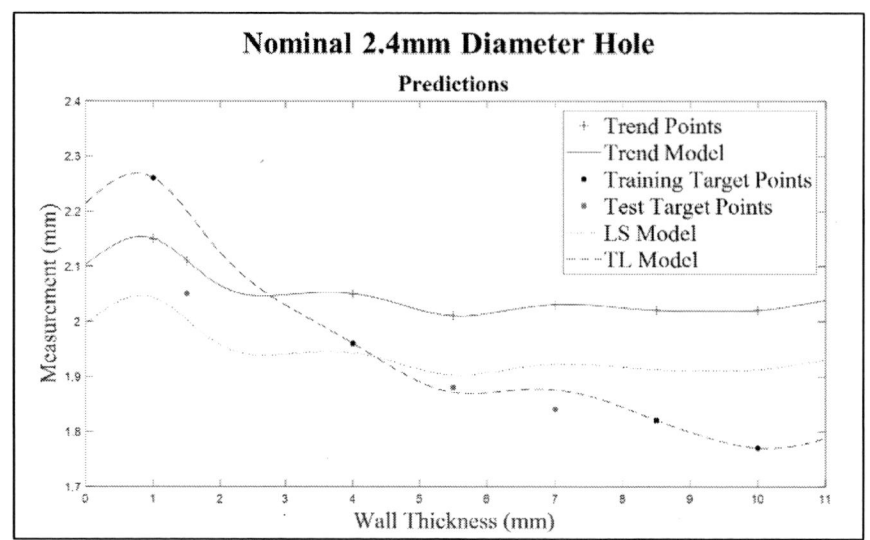

FIGURE 13. Linear Shift model and Transfer Learning model hole diameter predictions using four training target points. Trend points shown as black crosses with the generated trend model plotted in blue. Training target points are shown in black; test target points are shown in red. the transfer learning model is able to interpolate through training target points and better predict hole diameters than the linear shift model.

As discussed previously, the target points chosen to create the models had the potential to influence predictions heavily, particularly with the linear and quadratic fits. To quantify the effect of training data selection on prediction errors, maximum absolute error (MAE) and average absolute error (AAE) were calculated for each combination of n target points using Equations 6 and 7, where $\hat{y}_i$ is the model-based prediction and y_i is the value of the median measurement at target point i. To preserve the physical meaning behind the errors (i.e., error in measurement prediction), MAE and AAE were not normalized by standard deviation. For example, in the case of $n=4$ training target points, every combination of four training points out of seven potential training points (35 combinations) was used to train each model. Each model's MAE and AAE were then evaluated.

$$MAE = max|y_i - \hat{y}_i| \tag{6}$$

$$AAE = \frac{y_i - \hat{y}_i}{\sum i} \tag{7}$$

Figures 14 and 15 show the respective MAE and AAE error ranges from the case study discussed previously. The markers show the average error while the tops and bottoms of the error bars show maximum and minimum errors. Tables 2, 3, 4, and 5 in Appendix A summarize the MAE and AAE errors for gaps of nominal sizes 1.4mm and 1.6mm as well as holes with nominal diameters of 2.4mm and 2.6mm when using GF PA 12 and PA 12 as trend data and FR PA 11 as target data.

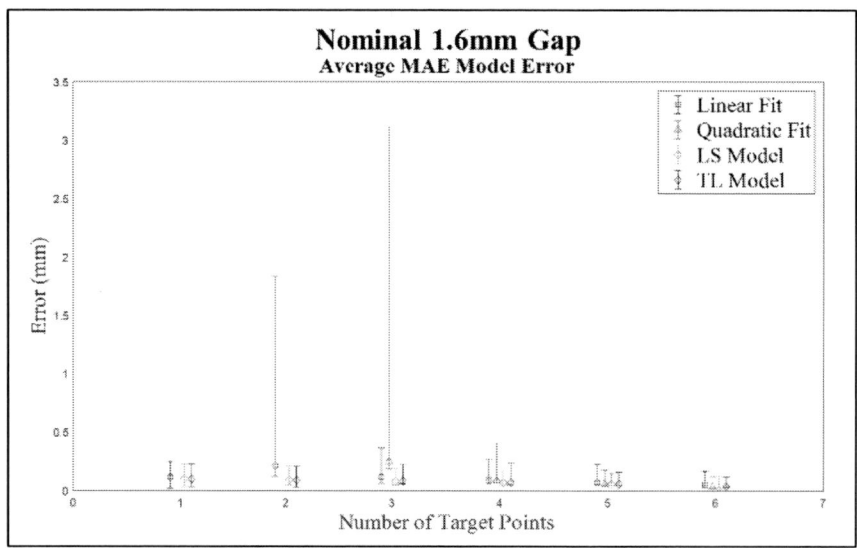

FIGURE 14. Average MAE errors with maximum and minimum error bars for varying numbers of target points. Nominal gap size of 1.6mm using GF PA 12 and PA 12 as trend data and FR PA 11 as target data.

Copyright © 2023 by ASME

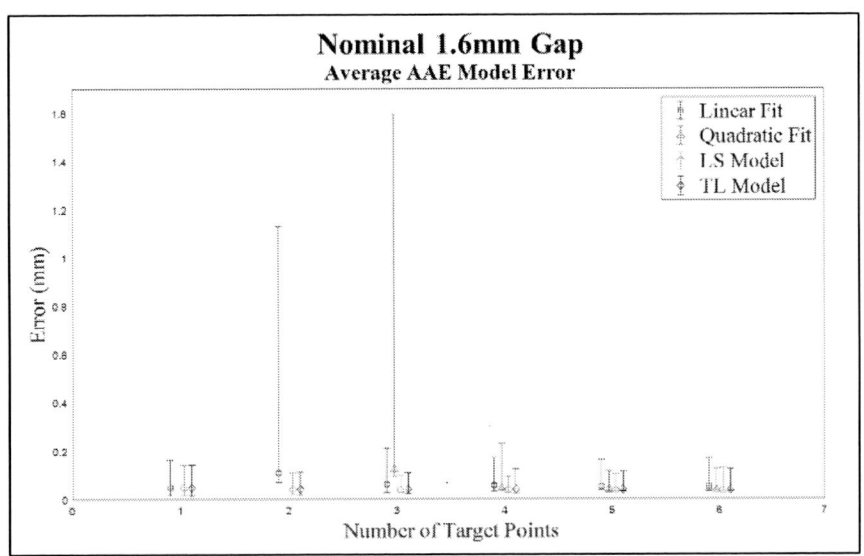

FIGURE 15. Average AAE errors with maximum and minimum error bars for varying numbers of target points. Nominal gap size of 1.6mm using GF PA 12 and PA 12 as trend data and FR PA 11 as target data.

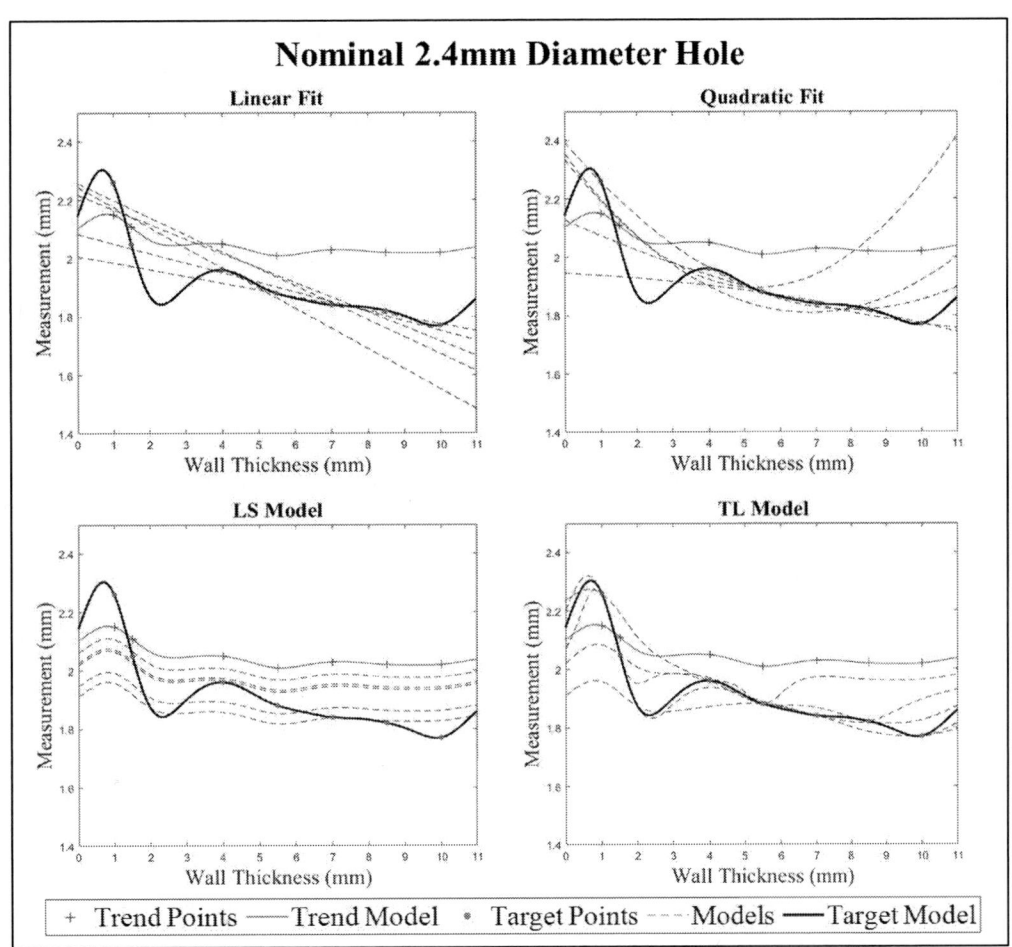

FIGURE 16. Models, shown as dashed black lines, trained on various sets of four target points for nominal 2.4mm diameter holes. The trend points and trend model are shown in each subplot as crosses and a solid blue line, respectively. Target points are shown as red points. The target model, shown as a solid black line, was created by training a GPR on all seven target points.

Copyright © 2023 by ASME

While there may have been combinations of target points which resulted in more accurate predictions using the linear and quadratic fits, the TL and LS models typically resulted in equivalent or lower minimum and maximum errors. This result has two major implications. First, the worst-case predictions (maximum errors) from the linear and quadratic fits were typically worse than those from the TL and LS models, particularly at small numbers of training target points. This result suggests that using TL or LS models to inform design geometry is a safer option, as the worst-case prediction errors will likely be less than traditional fitting techniques.

The second implication is that the spread of errors for the TL and LS models is smaller than that of the linear and quadratic fits for almost any number of training target points. This result suggests that the TL and LS models are less sensitive to the target points chosen to train the models. In application, this is a great benefit, as there may not be obvious choices regarding which target points to collect initially for training. To demonstrate the sensitivity of each model to the choice of training target points, models trained on different sets of four target points were plotted alongside a target model (which is a GPR model trained on all available target data points) as shown in Figure 16. The linear and quadratic fits vary significantly with the sampled training target points. Additionally, they are both constrained by explicit equations such that if the target point behaviors differ from a first or second order fit, both fits are unable to compensate and lose accuracy. The limitations of the LS models are more visible, as the models are heavily influenced by the trend model and cannot capture target data behaviors that vary from trend data behaviors. While the overall shape of the TL models varies with the choice of training target points, the TL models have the benefit of interpolating the training target points and therefore begin to capture the underlying behavior of the target data unlike the LS models.

Figure 16 also highlights the potential for "intelligent sampling" to refine the TL model. GPR models are typically used to support sequential sampling by sampling in locations that maximize the expected improvement in the function or reduce the uncertainty in its predictions. In the TL model in this example, there is a lack of target points between a wall thickness of 1.5mm and 4mm. A designer could choose to run a small metrology study to fill in target points in this region and retrain the TL model to increase prediction confidence

Given the small dataset, it was difficult to compare computational costs of each model, as each model built within milliseconds. To quantify the computational tradeoffs of the TL model, a larger metrology dataset with more design sites would be necessary. Typically, Gaussian process regression models train on the order of $\mathcal{O}(n^3)$ where n is the number of training points [38,39]. Polynomial regression computational costs are on the order of $\mathcal{O}((nm)^3)$ where n is the number of points fitted and $m-1$ is the order of the polynomial [39]. As such, there would likely not be a noticeable computational tradeoff between the TL model and linear or quadratic fit for small datasets. With larger datasets, the TL model may be more costly because of the two-stage surrogate model.

The research presented in this paper explores the use of transfer learning to minimize the cost of polymer SLS metrology studies by predicting geometric features using existing data and

significantly fewer new data points. Although the three materials (PA 12, GF PA 12, and FR PA 11) have different material properties and even classifications (as GF PA 12 is a composite whereas PA 12 is a pure material) [40], the underlying physics behind polymer SLS remains the same, and the materials undergo the same laser-based melting process. Thus, the transfer learning model is able to take implicit information from one or more material datasets and apply it to a new material dataset. In this same vein, it is hypothesized that the transfer learning model presented in this research could extend to direct metal laser sintering (DMLS), as the metal powders experience a similar melting process as the polymer powders; although, significant differences between the two processes (temperatures, conductivities, etc.) may require significantly more training data. The transfer learning method would not perform well in cases where datasets do not share similar processes. For example, using a trend dataset comprised of polymer SLS measurements to predict a target in stereolithography (SLA) would perform poorly, as SLS and SLA do not share common underlying physics. However, for different materials or machines within SLA, the transfer learning model is expected to perform well in predicting geometric features.

5. CONCLUSION AND FUTURE WORK

The proposed transfer learning model offers a cost-effective alternative to process-specific metrology. Instead of printing large quantities of test parts when switching to a new material or machine, designers can leverage existing metrology data from known materials or machines to predict the geometric characterization in the new process. The model consists of two Gaussian process regression (GPR) models: the trend model and the deviation model. The trend model is trained solely on a large dataset of existing metrology data from similar processes. The designer then collects a small number of data points in the new, target process. The deviation model is trained on the difference between the existing metrology data and target data. The trend and deviation models are then combined to create the transfer learning model which predicts as-printed geometries in the target process beyond the few sites where data was collected.

Based on error comparisons across four models, the greatest benefit from choosing a transfer learning model is that the model performs consistently well regardless of the training points used to create the deviation model. In practice, designers may not have control over which points to sample. With the other three models, the predictions can vastly vary based on the collected data points, adding uncertainty to whether or not the predictions are accurate and trustworthy. However, with the transfer learning model, the predictions are both accurate and insensitive to the chosen training points.

The natural next step in this research is to validate the polymer SLS predictions with physical prints at wall thicknesses where hole and gap data were not previously collected. This would better evaluate the accuracy of the predictions. Additionally, it would increase the dataset size which was limited in this paper. A larger dataset size would likely result in a more accurate trend model which could affect the overall performance of the transfer learning model. Another next step is to validate this model in other applications such as switching to a new machine. A metrology study, similar to the study

Copyright © 2023 by ASME

conducted by Allison et al. [15], could supply a dataset where various polymer SLS machines print metrology parts in the same material. The transfer learning model then predicts the geometric characterization of a new machine rather than a new material. The challenge would include accessing multiple polymer SLS machines. Alternatively, the method could be validated in a new AM process such as direct metal laser sintering where metrology data in various metals could serve as the dataset.

Future work on the model includes incorporating the error bounds inherent to GPR models to inform which initial target data points to collect. The case studied in this paper assumed the designer collected specific target data points to use as training data. However, in practice, a designer could initially collect a small number of target data points. Then, by evaluating where the model error bounds are largest, collect additional specific target data points to refine the transfer learning model and decrease model error.

ACKNOWLEDGEMENTS

The authors are grateful for the financial support from the Virginia & Ernest Cockrell Jr. Fellowship. The authors would also like to thank Dr. Jared Allison for access to the multi-material polymer SLS metrology data.

REFERENCES

[1] Thompson, M. K., Moroni, G., Vaneker, T., Fadel, G., Campbell, R. I., Gibson, I., Bernard, A., Schulz, J., Graf, P., Ahuja, B., and Martina, F., 2016, "Design for Additive Manufacturing: Trends, Opportunities, Considerations, and Constraints," CIRP Annals - Manufacturing Technology, p. 24.

[2] Gibson, I., Rosen, D., Stucker, B., and Khorasani, A., 2021, "Design for Additive Manufacturing," pp. 555–607.

[3] Colosimo, B., Huang, Q., Dasgupta, T., and Tsung, F., 2018, "Opportunities and Challenges of Quality Engineering for Additive Manufacturing," Journal of Quality Technology, 50, pp. 233–252.

[4] Boothroyd, G., Dewhurst, P., and Knight, W., 2011, *Product Design for Manufacture and Assembly*, CRC Press.

[5] Swift, K. G., and Booker, J. D., 2013, *Manufacturing Process Selection Handbook*, Butterworth-Heinemann.

[6] Dinar, M., and Rosen, D. W., 2017, "A Design for Additive Manufacturing Ontology," Journal of Computing and Information Science in Engineering, 17(2).

[7] Jee, H., and Witherell, P., 2017, "A Method for Modularity in Design Rules for Additive Manufacturing," Rapid Prototyping Journal, 23, pp. 00–00.

[8] Moylan, S., Slotwinski, J., Cooke, A., Jurrens, K., and Donmez, M. A., 2014, "An Additive Manufacturing Test Artifact," J. RES. NATL. INST. STAN., 119, p. 429.

[9] Techniek, T. I. en, and Castillo, L., 2005, "Study about the Rapid Manufacturing of Complex Parts of Stainless Steel and Titanium."

[10] Kranz, J., Herzog, D., and Emmelmann, C., 2015, "Design Guidelines for Laser Additive Manufacturing of Lightweight Structures in TiAl6V4," Journal of Laser Applications, 27(S1), p. S14001.

[11] Meisel, N., and Williams, C., 2015, "An Investigation of Key Design for Additive Manufacturing Constraints in Multimaterial Three-Dimensional Printing," Journal of Mechanical Design, 137(11).

[12] Vicente, M. F., Canyada, M., and Conejero, A., 2015, "Identifying Limitations for Design for Manufacturing with Desktop FFF 3D Printers," International Journal of Rapid Manufacturing, 5(1), p. 116.

[13] Seepersad, C. C., Govett, T., Kim, K., Lundin, M., and Pinero, D., 2012, "A Designer's Guide for Dimensioning and Tolerancing SLS Parts," University of Texas at Austin.

[14] Sippel, D., 2008, "Investigation of Detail Resolution on Basic Shapes and Development of Design Rules."

[15] Allison, J., Sharpe, C., and Seepersad, C. C., 2019, "Powder Bed Fusion Metrology for Additive Manufacturing Design Guidance," Additive Manufacturing, 25, pp. 239–251.

[16] Weiss, K., Khoshgoftaar, T. M., and Wang, D., 2016, "A Survey of Transfer Learning," Journal of Big Data, 3(1), p. 9.

[17] Zhuang, F., Qi, Z., Duan, K., Xi, D., Zhu, Y., Zhu, H., Xiong, H., and He, Q., 2021, "A Comprehensive Survey on Transfer Learning," Proceedings of the IEEE, 109(1), pp. 43–76.

[18] Byra, M., Wu, M., Zhang, X., Jang, H., Ma, Y.-J., Chang, E. Y., Shah, S., and Du, J., 2020, "Knee Menisci Segmentation and Relaxometry of 3D Ultrashort Echo Time Cones MR Imaging Using Attention U-Net with Transfer Learning," Magn Reson Med, 83(3), pp. 1109–1122.

[19] Zeng, M., Li, M., Fei, Z., Yu, Y., Pan, Y., and Wang, J., 2018, "Automatic ICD-9 Coding via Deep Transfer Learning," Neurocomputing, 324.

[20] Lu, C., Hu, F., Cao, D., Gong, J., Xing, Y., and Li, Z., 2019, "Transfer Learning for Driver Model Adaptation in Lane-Changing Scenarios Using Manifold Alignment," IEEE transactions on intelligent transportation systems, 21(8), pp. 3281–3293.

[21] Angkititrakul, P., Miyajima, C., and Takeda, K., 2011, "Modeling and Adaptation of Stochastic Driver-Behavior Model with Application to Car Following," pp. 814–819.

[22] Francis, J., Sabbaghi, A., Ravi Shankar, M., Ghasri-Khouzani, M., and Bian, L., 2020, "Efficient Distortion Prediction of Additively Manufactured Parts Using Bayesian Model Transfer Between Material Systems," Journal of Manufacturing Science and Engineering, 142(051001).

[23] Sabbaghi, A., and Huang, Q., 2018, "Model Transfer Across Additive Manufacturing Processes via Mean Effect Equivalence of Lurking Variables," The Annals of Applied Statistics, 12(4).

[24] Tercan, H., Guajardo, A., Heinisch, J., Thiele, T., Hopmann, C., and Meisen, T., 2018, "Transfer-Learning: Bridging the Gap between Real and Simulation Data for Machine Learning in Injection Molding," Procedia CIRP, 72, pp. 185–190.

[25] Qian, Z., Seepersad, C., Joseph, V. R., Allen, J., and Wu, C.-F., 2006, "Building Surrogate Models Based on

Detailed and Approximate Simulations," Journal of Mechanical Design - J MECH DESIGN, **128**.

[26] Clarke, S. M., Griebsch, J. H., and Simpson, T. W., 2004, "Analysis of Support Vector Regression for Approximation of Complex Engineering Analyses," Journal of Mechanical Design, **127**(6), pp. 1077–1087.

[27] Jin, R., Chen, W., and Simpson, T. W., 2001, "Comparative Studies of Metamodelling Techniques under Multiple Modelling Criteria," Struct Multidisc Optim, **23**(1), pp. 1–13.

[28] Wang, G. G., and Shan, S., 2006, "Review of Metamodeling Techniques in Support of Engineering Design Optimization," Journal of Mechanical Design, **129**(4), pp. 370–380.

[29] Bhosekar, A., and Ierapetritou, M., 2018, "Advances in Surrogate Based Modeling, Feasibility Analysis, and Optimization: A Review," Computers & Chemical Engineering, **108**, pp. 250–267.

[30] Zhu, Q.-X., Chen, Z.-S., Zhang, X.-H., Rajabifard, A., Xu, Y., and Chen, Y.-Q., 2020, "Dealing with Small Sample Size Problems in Process Industry Using Virtual Sample Generation: A Kriging-Based Approach," Soft Comput, **24**(9), pp. 6889–6902.

[31] Wang, B., and Chen, T., 2015, "Gaussian Process Regression with Multiple Response Variables," Chemometrics and Intelligent Laboratory Systems, **142**, pp. 159–165.

[32] Kleijnen, J. P. C., and Mehdad, E., 2014, "Multivariate versus Univariate Kriging Metamodels for Multi-Response Simulation Models," European Journal of Operational Research, **236**(2), pp. 573–582.

[33] Nielsen, H. B., Lophaven, S. N., and Sondergaard, J., 2002, "DACE - A Matlab Kriging Toolbox."

[34] Mahesh, M., Wong, Y. S., Fuh, J. Y. H., and Loh, H. T., 2004, "Benchmarking for Comparative Evaluation of RP Systems and Processes," Rapid Prototyping Journal, **10**(2), pp. 123–135.

[35] Gong, X., Cheng, B., Price, S., and Chou, K., 2013, *Powder-Bed Electron-Beam-Melting Additive Manufacturing: Powder Characterization, Process Simulation and Metrology*.

[36] Teeter, M. G., Kopacz, A. J., Nikolov, H. N., and Holdsworth, D. W., 2015, "Metrology Test Object for Dimensional Verification in Additive Manufacturing of Metals for Biomedical Applications," Proc Inst Mech Eng H, **229**(1), pp. 20–27.

[37] Allison, J., Sharpe, C., and Seepersad, C. C., 2017, "A Test Part for Evaluating the Accuracy and Resolution of a Polymer Powder Bed Fusion Process," Journal of Mechanical Design, **139**(10), p. 100902.

[38] Keerthi, S., and Chu, W., 2005, "A Matching Pursuit Approach to Sparse Gaussian Process Regression," *Advances in Neural Information Processing Systems*, MIT Press.

[39] Zhou, Z., Ong, Y. S., Nguyen, M. H., and Lim, D., 2005, "A Study on Polynomial Regression and Gaussian Process Global Surrogate Model in Hierarchical Surrogate-Assisted Evolutionary Algorithm," *2005 IEEE Congress on Evolutionary Computation*, pp. 2832-2839 Vol. 3.

[40] Kruth, J. P., Levy, G., Craeghs, T., and Yasa, E., 2008, "Consolidation of Polymer Powders by Selective Laser Sintering," Proceedings of the 3rd international conference on polymers and moulds innovations, pp. 15–30.

Copyright © 2023 by ASME

APPENDIX A

TABLE 2. Summary of MAE and AAE errors for a nominal gap size of 1.6mm using GF PA 12 and PA 12 as trend data and FR PA 11 as target data.

	1.6mm Gap							
	MAE				AAE			
	Average	Std Dev	Minimum	Maximum	Average	Std Dev	Minimum	Maximum
1 Target Point								
Transfer Learning	0.0993	0.0256	0.0650	0.1300	0.0476	0.0221	0.0333	0.0942
Linear Fit	0.1171	0.0182	0.0950	0.1350	0.0481	0.0293	0.0333	0.1142
Quadratic Fit	-	-	-	-	-	-	-	-
Linear Shift	0.0993	0.0256	0.0650	0.1300	0.0476	0.0221	0.0333	0.0942
2 Target Points								
Transfer Learning	0.0901	0.0144	0.0573	0.1175	0.0415	0.0109	0.0267	0.0699
Linear Fit	0.2128	0.3253	0.0880	1.6150	0.1075	0.2111	0.0389	1.0220
Quadratic Fit	-	-	-	-	-	-	-	-
Linear Shift	0.0898	0.0173	0.0400	0.1175	0.0409	0.0108	0.0220	0.0675
3 Target Points								
Transfer Learning	0.0869	0.0235	0.0268	0.1379	0.0402	0.0107	0.0175	0.0682
Linear Fit	0.1188	0.0459	0.0540	0.2453	0.0596	0.0233	0.0349	0.1491
Quadratic Fit	0.2506	0.5227	0.0652	2.8730	0.1227	0.2673	0.0267	1.4725
Linear Shift	0.0898	0.0173	0.0400	0.1175	0.0376	0.0109	0.0125	0.0596
4 Target Points								
Transfer Learning	0.0737	0.0271	0.0232	0.1618	0.0388	0.0138	0.0155	0.0853
Linear Fit	0.0939	0.0396	0.0366	0.1780	0.0547	0.0198	0.0232	0.1141
Quadratic Fit	0.0874	0.0533	0.0133	0.3248	0.0432	0.0295	0.0084	0.1845
Linear Shift	0.0898	0.0173	0.0400	0.1175	0.0358	0.0143	0.0100	0.0596
5 Target Points								
Transfer Learning	0.0574	0.0280	0.0094	0.1021	0.0390	0.0186	0.0084	0.0752
Linear Fit	0.0723	0.0370	0.0166	0.1580	0.0515	0.0227	0.0150	0.1088
Quadratic Fit	0.0603	0.0326	0.0108	0.1180	0.0380	0.0194	0.0108	0.0768
Linear Shift	0.0898	0.0173	0.0400	0.1175	0.0347	0.0197	0.0050	0.0675
6 Target Points								
Transfer Learning	0.0383	0.0305	0.0085	0.0841	0.0383	0.0305	0.0085	0.0841
Linear Fit	0.0491	0.0342	0.0185	0.1189	0.0491	0.0342	0.0185	0.1189
Quadratic Fit	0.0363	0.0324	0.0017	0.0883	0.0363	0.0324	0.0017	0.0883
Linear Shift	0.0898	0.0173	0.0400	0.1175	0.0340	0.0318	0.0050	0.0942

Copyright © 2023 by ASME

TABLE 3. Summary of MAE and AAE errors for a nominal gap size of 1.4mm using GF PA 12 and PA 12 as trend data and FR PA 11 as target data.

1.4mm Gap								
	MAE				AAE			
	Average	Std Dev	Minimum	Maximum	Average	Std Dev	Minimum	Maximum
1 Target Point								
Transfer Learning	0.1221	0.0198	0.1050	0.1500	0.0467	0.0329	0.0292	0.1175
Linear Fit	0.1300	0.0200	0.1000	0.1500	0.0533	0.0326	0.0367	0.1267
Quadratic Fit	-	-	-	-	-	-	-	-
Linear Shift	0.1221	0.0198	0.1050	0.1500	0.0467	0.0329	0.0292	0.1175
2 Target Points								
Transfer Learning	0.1229	0.0244	0.0450	0.1485	0.0443	0.0140	0.0325	0.0713
Linear Fit	0.2579	0.4950	0.0867	2.4100	0.1382	0.3256	0.0434	1.5560
Quadratic Fit	-	-	-	-	-	-	-	-
Linear Shift	0.1086	0.0196	0.0450	0.1350	0.0430	0.0124	0.0325	0.0675
3 Target Points								
Transfer Learning	0.1179	0.0415	0.0156	0.1525	0.0444	0.0136	0.0106	0.0726
Linear Fit	0.1255	0.0352	0.0389	0.2267	0.0589	0.0186	0.0283	0.1136
Quadratic Fit	0.4148	0.9191	0.0835	5.1380	0.2119	0.4784	0.0371	2.6800
Linear Shift	0.0934	0.0319	0.0250	0.1283	0.0401	0.0080	0.0188	0.0537
4 Target Points								
Transfer Learning	0.1031	0.0563	0.0075	0.1553	0.0435	0.0212	0.0045	0.0703
Linear Fit	0.1049	0.0371	0.0355	0.1840	0.0557	0.0200	0.0150	0.1123
Quadratic Fit	0.1113	0.0430	0.0157	0.1883	0.0562	0.0260	0.0105	0.1348
Linear Shift	0.0764	0.0398	0.0137	0.1237	0.0377	0.0137	0.0104	0.0537
5 Target Points								
Transfer Learning	0.0790	0.0667	0.0012	0.1507	0.0447	0.0365	0.0010	0.0912
Linear Fit	0.0825	0.0421	0.0142	0.1501	0.0537	0.0267	0.0073	0.0952
Quadratic Fit	0.0782	0.0518	0.0109	0.1359	0.0474	0.0288	0.0067	0.0920
Linear Shift	0.0569	0.0433	0.0060	0.1200	0.0356	0.0230	0.0060	0.0750
6 Target Points								
Transfer Learning	0.0455	0.0675	0.0006	0.1522	0.0455	0.0675	0.0006	0.1522
Linear Fit	0.524	0.0462	0.0009	0.1345	0.0524	0.0462	0.0009	0.1345
Quadratic Fit	0.0467	0.0528	0.0032	0.1277	0.0467	0.0528	0.0032	0.1277
Linear Shift	0.0336	0.0412	0.0050	0.1175	0.0336	0.0412	0.0050	0.1175

Copyright © 2023 by ASME

TABLE 4. Summary of MAE and AAE errors for a nominal hole diameter of 2.4mm using GF PA 12 and PA 12 as trend data and FR PA 11 as target data.

	2.4mm Hole							
	MAE				AAE			
	Average	Std Dev	Minimum	Maximum	Average	Std Dev	Minimum	Maximum
1 Target Point								
Transfer Learning	0.1657	0.0479	0.1200	0.2300	0.0857	0.0346	0.0583	0.1400
Linear Fit	0.3314	0.0745	0.2300	0.4100	0.1695	0.0619	0.1250	0.3033
Quadratic Fit	-	-	-	-	-	-	-	-
Linear Shift	0.1657	0.0479	0.1200	0.2300	0.0857	0.0346	0.0583	0.1400
2 Target Points								
Transfer Learning	0.1435	0.0305	0.1023	0.2050	0.0719	0.0214	0.0268	0.1190
Linear Fit	0.2780	0.5075	0.1000	2.4700	0.1475	0.3323	0.0430	1.5880
Quadratic Fit	-	-	-	-	-	-	-	-
Linear Shift	0.1481	0.0312	0.0650	0.2050	0.0726	0.0202	0.0250	0.1190
3 Target Points								
Transfer Learning	0.1216	0.0212	0.0790	0.1783	0.0577	0.0194	0.0303	0.1120
Linear Fit	0.1435	0.0396	0.0646	0.2800	0.0676	0.236	0.0416	0.1367
Quadratic Fit	0.4330	0.9045	0.0670	4.9760	0.2315	0.4651	0.0446	2.5775
Linear Shift	0.1332	0.0332	0.0467	0.1800	0.0667	0.216	0.0267	0.1050
4 Target Points								
Transfer Learning	0.1089	0.0267	0.0297	0.1506	0.0518	0.0203	0.0164	0.1063
Linear Fit	0.1170	0.0408	0.0309	0.2000	0.0639	0.0197	0.0280	0.1297
Quadratic Fit	0.1148	0.0651	0.0212	0.4104	0.0617	0.0407	0.0165	0.2497
Linear Shift	0.1160	0.0410	0.0250	0.1625	0.0632	0.0262	0.0150	0.1050
5 Target Points								
Transfer Learning	0.0875	0.0354	0.0167	0.1384	0.0520	0.0215	0.0110	0.1128
Linear Fit	0.0905	0.0457	0.0286	0.1600	0.0632	0.0249	0.0260	0.1058
Quadratic Fit	0.0825	0.0445	0.0234	0.1433	0.0528	0.0251	0.0137	0.0945
Linear Shift	0.0929	0.0499	0.0100	0.1500	0.0610	0.0343	0.0100	0.1190
6 Target Points								
Transfer Learning	0.0603	0.0470	0.0006	0.1193	0.0603	0.0470	0.0006	0.1193
Linear Fit	0.0625	0.0435	0.0283	0.1506	0.0625	0.0435	0.0283	0.1506
Quadratic Fit	0.0550	0.0488	0.0102	0.1333	0.0550	0.0488	0.0102	0.1333
Linear Shift	0.0600	0.0545	0.0117	0.1400	0.0600	0.0545	0.0117	0.1400

Copyright © 2023 by ASME

TABLE 5. Summary of MAE and AAE errors for a nominal hole diameter of 2.6mm using GF PA 12 and PA 12 as trend data and FR PA 11 as target data.

	2.6mm Hole							
	MAE				AAE			
	Average	Std Dev	Minimum	Maximum	Average	Std Dev	Minimum	Maximum
1 Target Point								
Transfer Learning	0.1079	0.0260	0.0850	0.1450	0.0548	0.0209	0.0383	0.0883
Linear Fit	0.3179	0.0737	0.2150	0.3950	0.1710	0.0535	0.1250	0.2808
Quadratic Fit	-	-	-	-	-	-	-	-
Linear Shift	0.1079	0.0260	0.0850	0.1450	0.0548	0.0209	0.0383	0.0883
2 Target Points								
Transfer Learning	0.0886	0.0222	0.0488	0.1225	0.0466	0.0126	0.0238	0.0745
Linear Fit	0.2205	0.3876	0.0750	1.8700	0.1181	0.2502	0.0373	1.1880
Quadratic Fit	-	-	-	-	-	-	-	-
Linear Shift	0.0926	0.0220	0.0275	0.1225	0.0463	0.0125	0.0205	0.0745
3 Target Points								
Transfer Learning	0.0749	0.0207	0.0396	0.1164	0.0429	0.0125	0.0194	0.0730
Linear Fit	0.1110	0.0308	0.0671	0.2046	0.0537	0.0170	0.0291	0.1130
Quadratic Fit	0.3386	0.6605	0.0760	3.5870	0.1840	0.3411	0.0394	1.8587
Linear Shift	0.0832	0.0227	0.0317	0.1117	0.0429	0.0127	0.0187	0.0713
4 Target Points								
Transfer Learning	0.0675	0.0229	0.0166	0.1146	0.0424	0.0163	0.0117	0.0811
Linear Fit	0.0944	0.0252	0.0402	0.1343	0.0511	0.0123	0.0266	0.0816
Quadratic Fit	0.1043	0.0739	0.0364	0.4472	0.0634	0.0505	0.0179	0.2866
Linear Shift	0.0719	0.0270	0.0175	0.1063	0.0414	0.0150	0.0158	0.0712
5 Target Points								
Transfer Learning	0.0512	0.0216	0.0140	0.0921	0.0372	0.0166	0.0076	0.0701
Linear Fit	0.0766	0.0322	0.0129	0.1240	0.0506	0.0199	0.0126	0.0881
Quadratic Fit	0.0692	0.0281	0.0168	0.1149	0.0486	0.0215	0.0086	0.1005
Linear Shift	0.0585	0.0305	0.0160	0.0970	0.0407	0.0195	0.0150	0.0745
6 Target Points								
Transfer Learning	0.0358	0.0249	0.0048	0.0736	0.0358	0.0249	0.0048	0.0736
Linear Fit	0.0504	0.0384	0.0100	0.1144	0.0504	0.0384	0.0100	0.1144
Quadratic Fit	0.0468	0.0341	0.0063	0.0989	0.0468	0.0341	0.0063	0.0989
Linear Shift	0.0407	0.0309	0.0167	0.0883	0.0407	0.0309	0.0167	0.0883

Copyright © 2023 by ASME

Proceedings of the ASME 2023
International Design Engineering Technical Conferences and
Computers and Information in Engineering Conference
IDETC-CIE2023
August 20-23, 2023, Boston, Massachusetts

DETC2023-112499

TOPOLOGY OPTIMIZATION IN CONSIDERATION OF OVERHANG CONSTRAINT FOR ADDITIVE MANUFACTURING BASED ON COUPLED FICTITIOUS PHYSICAL MODEL IN THERMAL DESIGN PROBLEM

Mikihiro Tajima[1], Takayuki Yamada[1,*]

[1]The University of Tokyo, Tokyo, Japan

ABSTRACT

Additive manufacturing can be used to create complex optimal structures through topology optimization. However, the overhang constraint, which requires costly support materials, is a problem when part of the structure is unsupported. Previously, the fictitious physical model was proposed to consider geometric manufacturing requirements, e.g., the overhang constraint. However, converging to a single optimal structure is challenging in the fictitious physical model because the model acts noninteractively as the mechanical model describing physical phenomena. To solve the convergence problem, we propose a coupled fictitious physical model, where two models interact by changing the material constants of the mechanical model in regions where geometric constraints defined in the fictitious physical model are violated such that the objective function worsens. In this paper, the proposed model is formulated based on the steady-state heat conduction problem with overhang constraint. Then, the optimization problem and sensitivity analysis are discussed. Finally, the proposed model is verified using an example two-dimensional heat conduction problem.

Keywords: Additive manufacturing, Overhang constraint, Topology optimization, Level set method, Fictitious physical model

1. INTRODUCTION

Topology optimization [1, 2] is a method to find an optimal structure while allowing for changes in topology, e.g., the number of holes. Generally, the optimal structures obtained by topology optimization are overly complex from a geometric perspective. Thus, the additive manufacturing (AM) method can be used to manufacture the optimal structures because it offers a high degree of freedom in terms of shapes that can be fabricated.

However, the AM method has several requirements, e.g., overhang constraint, which requires support structures if there is

an overhang in the void domain. However, the support structures incur high time and financial burdens, and they are difficult to remove when the optimal structure has closed regions. Thus, a self supporting optimal structure is desirable.

Various topology optimization methods that consider the overhang constraint have been proposed in previous studies [3–8]. For example, several studies [4–6] have proposed a method to restrict the design space such that a structure can only be added within the allowable overhang angles. However, obtaining an optimal structure with high performance is difficult because this method does not temporarily tolerate regions that violate the overhang constraint during the optimization process.

In addition, the fictitious physical model [9–12] was proposed to consider geometric manufacturing requirements. The fictitious physical model describes the fictitious governing equation to consider geometric constraints, and the advantage of this model is that it conforms to topological optimization methods and temporarily allows a structure that does not satisfy geometric constraints, thereby making the solution less localized.

However, converging to a single optimal shape is challenging because the fictitious physical model acts noninteractively with a mechanical model that describes physical phenomena.

In this paper, we propose a coupled fictitious physical model to solve this problem. In the proposed coupled fictitious physical model, the material constants of the mechanical model are changed in regions that violate the geometric constraints introduced by the fictitious physical model such that the objective function worsens. As a result, two coupled models can improve convergence.

The remainder of this paper is organized as follows. In Section 2, we briefly discuss the level set-based topology optimization method. Then, a steady-state heat conduction problem is adopted as the mechanical model and formalized in Section 3. In Section 4, the formalization of the fictitious physical model that extracts geometric features is then considered, and the overhang constraint is formalized. Next, in Section 5, the fictitious

---

*Corresponding author: t.yamada@mech.t.u-tokyo.ac.jp
Documentation for `asmeconf.cls`: Version 1.34, May 10, 2023.

Copyright © 2023 by ASME

thermal conductivity coefficient reflecting the concept of the coupled fictitious physical model is formalized. In addition, numerical examples demonstrate that the fictitious thermal conductivity worsens the objective function. Then, formalization of the optimization problem and sensitivity design is performed in Section 6. In Section 7, the validity of the proposed method is verified by considering a two-dimensional steady-state heat conduction problem. Finally, Section 8 presents the conclusion of this study.

2. LEVEL SET-BASED TOPOLOGY OPTIMIZATION

2.1 Basic concept

This study refers to the level set-based topology optimization method proposed by Yamada et al [13]. Here, the domain that contains the design target is defined as the fixed design domain D, and the domain occupied by the structure is the material domain Ω. Then, a level set function is introduced as follows:

$$
\begin{cases}
0 < \phi(\boldsymbol{x}) \le 1 & \text{if} \quad \boldsymbol{x} \in \Omega \\
\phi(\boldsymbol{x}) = 0 & \text{if} \quad \boldsymbol{x} \in \partial\Omega \\
-1 \le \phi(\boldsymbol{x}) < 0 & \text{if} \quad \boldsymbol{x} \in D\backslash\Omega
\end{cases}
\tag{1}
$$

In addition, a characteristic function that represents the material is defined as follows:

$$
\chi(\phi(\boldsymbol{x})) = \begin{cases}
1 & \text{if} \quad \phi \ge 0 \\
0 & \text{if} \quad \phi < 0.
\end{cases}
\tag{2}
$$

The basic concept of topology optimization is to formulate a structural optimization problem to find a structure that minimizes (or maximizes) the objective function J with the material distribution as the design variable. The structural optimization problem is defined as follows:

$$
\min_{\chi} \qquad J[u, \chi(\phi)] = \int_D j(u)\chi(\phi)d\Omega
\tag{3}
$$

$$
\text{subject to:} \qquad \text{governing equation}
\tag{4}
$$

where u represents a state variable of the governing equation, and $j(u)$ represents the desired objective function.

In the following, we consider how to update the level set function. Here, we assume that the level set function ϕ changes in proportion to the design sensitivity J' as fictitious time t elapses. Then, the reaction diffusion equation to update the level set function is defined as follows:

$$
\frac{\partial\phi(\boldsymbol{x},t)}{\partial t} = -K(J' - \tau\nabla^2\phi(\boldsymbol{x},t))
\tag{5}
$$

where $K \in \mathbb{R}^+$ represents a proportional constant. In Eq. (5), the geometric complexity of the material domain can be controlled by adjusting the regularization parameter $\tau \in \mathbb{R}^+$.

3. FORMULATION OF MECHANICAL MODEL

A mechanical model is a system of governing equations representing physical phenomena, e.g., elasticity or heat diffusion equations. Typically, topology optimization is formulated based on the mechanical model.

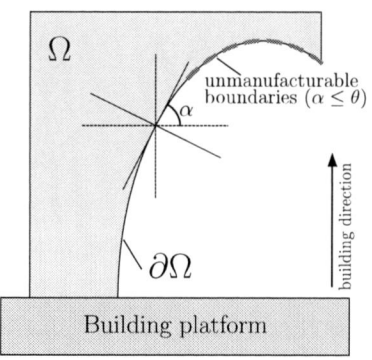

FIGURE 1: THE DEFINITION OF OVERHANG CONSTRAINT

3.1 Thermal conductivity problem

In this study, we consider a steady-state heat conduction problem as an example of the mechanical model. Here, a heat source Q is given in the fixed design domain D, and the temperature field $u_t \in H^1(\Omega)$ is fixed to 0 at the boundary Γ_t. In this case, the governing equation is defined as follows:

$$
\begin{cases}
-\operatorname{div}(K_t\nabla u_t) = Q & \text{in } \Omega \\
u_t = 0 & \text{on } \Gamma_t \\
\nabla \cdot u_t = 0 & \text{on } \partial\Omega\backslash\partial\Gamma_t
\end{cases}
\tag{6}
$$

where K_t represents thermal conductivity. Then, the topology optimization problem is defined as follows:

$$
\inf_{\chi} \qquad J(u_t) = \int_{\Gamma_t} Qu_t d\Gamma
\tag{7}
$$

where $J(u_t)$ represents the objective function referred to as thermal compliance [14].

4. FORMULATION OF FICTITIOUS PHYSICAL MODEL

A fictitious physical model is a system of fictitious governing equations to consider manufacturing requirements, e.g., manufacturing and production processes.

4.1 Definition of overhang constraint

In this study, the overhang constraint is taken as an example of the manufacturing requirement. Figure 1 shows the definition of the overhang constraint. Generally, in the AM method, support materials are required when an inclination angle α exceeds threshold value θ. The inclination angle α and threshold value θ are referred to as the overhang angle and the minimum overhang angle, respectively [8]. However, producing support materials incurs significant costs, and they cannot be removed when the optimal structure includes closed regions. Thus, support-free structures are desirable. The constraint that ensures that $\alpha > \theta$ is satisfied at the structural boundary is an overhang constraint.

4.2 Formulation to detect overhang constraint

Here, we discuss how to detect the overhang constraint using the fictitious physical model and how the model interacts with

Copyright © 2023 by ASME

482

the thermal diffusion problem. Here, the fixed design domain is denoted D.

First, to detect boundaries that violate the overhang constraint, we utilize a fictitious physical model that approximates the normal vector of the boundary of the structural domain [9]. The fictitious physical model used to extract the geometric features is defined as follows:

$$
\begin{cases}
-\operatorname{div}(a_s L^2 \nabla s_i - e_i \chi) + (1 - \chi)s_i = 0 & \text{in } D \\
s_i = 0 & \text{on } \partial D
\end{cases}
\tag{8}
$$

where $s_i \in H^1(D)$ represents a state variable, $a_s \in \mathbb{R}^+$ represents a diffusion coefficient, e_i represents an orthonormal basis vector in $\mathbb{R}_n$, and L represents a characteristic length.

Then, the overhang constraint is detected with the state variable s on the structural boundaries as follows [8, 12]:

$$
g(s)(1 - \chi)^2 = g_1(s)g_2(s)(1 - \chi)^2
\tag{9}
$$

$$
g_1(s) = \frac{a_s}{(1 - \cos\theta)^2}\left\{\mathscr{R}(s \cdot d - \sqrt{\|s\|^2 + \epsilon_s}\cos\theta)\right\}^2
\tag{10}
$$

$$
g_2(s) = \mathscr{R}\left(1 + \beta \nabla \cdot \frac{s}{\sqrt{\epsilon_c + \|s\|^2}}\right)
\tag{11}
$$

where c_s and c_{sr} represent tiny parameters, and $\mathscr{R}(x)$ represents a approximate ramp function defined as follows:

$$
\mathscr{R}(x) = \frac{1}{2}\left(x + \sqrt{x^2 + \epsilon_r}\right)
\tag{12}
$$

where ϵ_r represents a parameter to smooth the shape of the function near $x = 0$.

To introduce regions that violate the overhang constraint from the boundaries defined in Eq. (9), the following fictitious heat diffusion equation is defined as follows:

$$
\begin{cases}
-\operatorname{div}(a_p L^2 \nabla p) + p = g(s)(1 - \chi)^2 & \text{in } D \\
\nabla \cdot p = 0 & \text{on } \partial D
\end{cases}
\tag{13}
$$

where a_p represents a diffusion coefficient, and p represents a fictitious heat. Here, we define the regions that violate the overhang constraint as the product set of the fictitious heat and the material domain $p\chi$.

5. FORMULATION OF COUPLED FICTITIOUS PHYSICAL MODEL

With the fictitious physical model, converging to a single optimal structure is difficult because the fictitious physical model acts independently and noninteractively on the structure as the thermal conductivity problem, i.e., the mechanical model. To solve this problem, we propose a coupled fictitious physical model, in which the material constants of the mechanical model are changed in regions that violate the geometric constraints introduced by the fictitious physical model such that the objective function worsens. As a result, convergence is improved.

To adapt the proposed model to the thermal conductivity problem with the overhang constraint, the thermal conductivity is reduced in the intersections of the fictitious heat and the material domain to increase the thermal compliance (to worsen

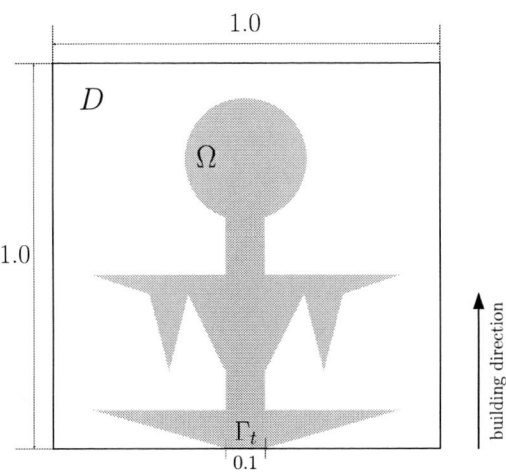

FIGURE 2: A DISTRIBUTION OF THE MATERIAL DOMAIN AND BOUNDARY CONDITION

the objective function). Here, the fictitious thermal conductivity $K_f(p, \chi; \gamma)$ is defined as follows:

$$
K_f(p, \chi; \gamma) = K_t\left(1 - \frac{2}{\pi}\arctan(\gamma p \chi)\right)
\tag{14}
$$

which K_t represents the original thermal conductivity, and γ represents the degree of reduction of the material constant.

5.1 Numerical examples

In the following, numerical examples are discussed to verify whether the fictitious thermal conductivity increases the thermal compliance in regions that violate the overhang constraint. In the following examples, the material domain and boundary condition in Eq.(7) are defined as shown in Fig. 2. In addition, the characteristic length is set to $L = 1$, the minimum overhang angle θ is set to 45°, and the parameter in the approximate ramp function is set to $\epsilon_r = 1.0 \times 10^{-9}$.

First, the distribution of the state variable s in the fictitious physical model is as shown in Fig. 3a. In addition, the diffusion coefficient in Eq. (8) is set to $a_s = 3.0 \times 10^{-4}$.

Second, the distribution of the boundaries that violate the overhang constraint is as shown in Fig. 3b. In Eq. (10), $\epsilon_s = 1.0 \times 10^{-3}$. In Eq. (11), $\beta = 5.0 \times 10^{-2}$, $n_c = 1$ and $\epsilon_c = 1.0 \times 10^{-3}$.

Third, the distribution of regions that violate the overhang constraint is as shown in Fig. 3c. Here, the diffusion coefficient in Eq. (8) is set to $a_p = 1.0 \times 10^{-3}$.

Finally, the distribution of the fictitious thermal conductivity is as shown in Fig. 4 when the thermal conductivity $K_t = 1.0$ and $\gamma = 50$ in Eq. (14). Figure 5 shows the relationship between the degree γ of reduction in thermal conductivity and the objective function $J(u_t)$. As can be seen, the reduction in thermal conductivity in regions that violate the overhang constraint increases the thermal compliance.

Copyright © 2023 by ASME

(a) The state variable s

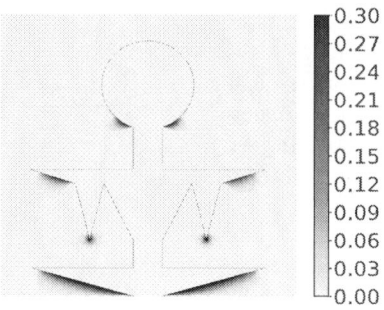

(b) The boundaries that violate overhang constraint $g(s)(1-\chi)^2$

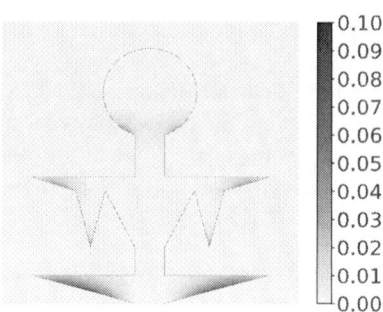

(c) The regions that violate overhang constraint $p\chi$

FIGURE 3: THE RESPECTIVE DISTRIBUTIONS FOR THE PROBLEM SETTING AS SHOWN IN FIG. 2.

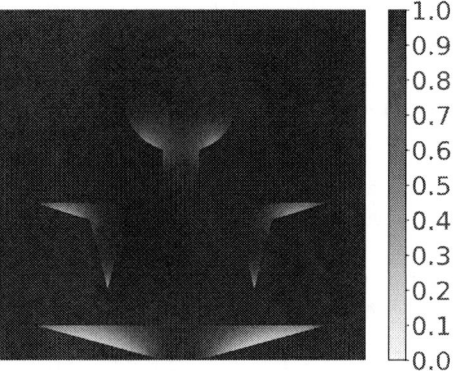

FIGURE 4: A DISTRIBUTION OF THE FICTITIOUS THERMAL CONDUCTIVITY $K_f(p,\chi;\gamma)$

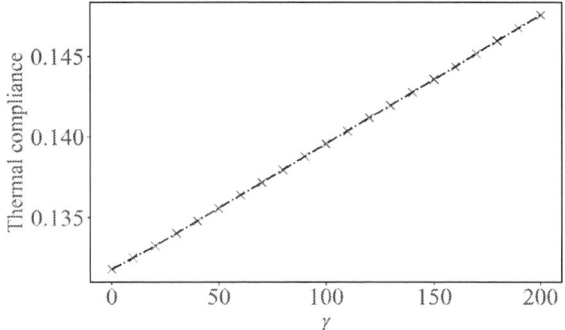

FIGURE 5: RELATIONSHIP BETWEEN THE DEGREE OF REDUCTION IN THERMAL CONDUCTIVITY AND THE OBJECTIVE FUNCTION

6. FORMULATION OF OPTIMIZATION PROBLEM AND SENSITIVITY ANALYSIS

6.1 Formulation of optimization problem

Here, the topology optimization problem is formulated based on the concept of the coupled fictitious physical model for the steady-state heat conduction problem with the overhang constraint. The objective function is the same as Eq. (7). In addition, the fixed design domain is denoted D. The heat source Q is added in the fixed design domain D, and the temperature field p is fixed to 0 on the boundary Γ_t. Under these boundary conditions, a volume constraint G_v is imposed such that the overall volume does not exceed the given threshold $V_{\max}$. Then, the optimization problem is defined as follows:

$$
\begin{cases}
G_v = \displaystyle\int_D \chi(\phi)\mathrm{d}\Omega - V_{\max} \le 0 & \\
-\,\mathrm{div}(a_s L^2 \nabla s_i - e_i \chi) + (1-\chi)s_i = 0 & \text{in } D \\
s_i = 0 & \text{on } \partial D \\
-\,\mathrm{div}(a_p L^2 \nabla p) + p = g(s)(1-\chi)^2 & \text{in } D \\
\nabla \cdot p = 0 & \text{on } \partial D \\
-\,\mathrm{div}(K_f(p,\phi)\nabla u_t) = Q & \text{in } \Omega \\
u_t = 0 & \text{on } \Gamma_t \\
\nabla \cdot u_t = 0 & \text{on } \partial\Omega\backslash\partial\Gamma_t
\end{cases}
\tag{15}
$$

6.2 Numerical scheme for the governing equation

Note that the material domain Ω continues to change during the optimization process; thus redefining the finite element mesh is required in each step of the optimization process. However, redefining the mesh is computationally expensive. Thus, we assume that the void domain $D\backslash\Omega$ has a relatively tiny thermal conductivity compared to the material domain [15]. In addition, the boundary between the material and the void domain is smoothed using a Heaviside function. Then, the fictitious thermal conductivity is redefined as follows:

$$
\tilde{K}_f(p,\phi) = \{(1-c)H_\phi(\phi;\epsilon_u)+c\}K_f(p,\phi)
\tag{16}
$$

Copyright © 2023 by ASME

484

where $H_\phi(\phi; \epsilon_u)$ is defined as follows:

$$H_\phi(\phi; \epsilon_u) = \begin{cases} 0 & \text{if} \quad \phi < -\epsilon_u \\ \dfrac{1}{2} + \left(\dfrac{15}{16} \dfrac{\phi}{\epsilon_u} - \dfrac{5}{8} \left(\dfrac{\phi}{\epsilon_u} \right)^3 + \dfrac{3}{16} \left(\dfrac{\phi}{\epsilon_u} \right)^5 \right) & \text{if} \quad \phi \in [-\epsilon_u, \epsilon_u] \\ 1 & \text{if} \quad \phi > \epsilon_u \end{cases}$$

(17)

where ϵ_u is the width of the transition, and c represents the ratio of thermal conductivity between the material and the void domain. In this study, ϵ_u is set to 0.5, and c is set to 1.0×10^{-2}.

6.3 Sensitivity analysis

Here, the design sensitivity is introduced based on the adjoint variable method. The heat conduction problem with the thermal compliance as the objective function is known to be a self-adjoint problem. Thus, the adjoint variable v_t of variable u_t coincides with u_t. In addition, the adjoint variables of p and s are denoted λ_p and λ_s, respectively. Then, the adjoint equations to calculate the two adjoint variables are defined as follows:

$$\begin{cases} -u_p \nabla^2 \lambda_p + \lambda_p - \dfrac{\partial \tilde{K}_f(p, \phi)}{\partial p} \nabla u_t \cdot \nabla v_t & \text{in } D \\ \nabla \cdot \lambda_p = 0 & \text{on } \partial D \\ -a_s \nabla^2 \lambda_{si} + (1 - \chi) \lambda_{si} = \lambda_p \dfrac{\partial g(s)}{\partial s_i} (1 - \chi)^2 & \text{in } D \\ \lambda_{si} = 0 & \text{on } \partial D \end{cases}$$

(18)

Finally, the design sensitivity is derived using the adjoint variables as follows:

$$J' = \left(\tilde{K}_f(p, \phi) \nabla u_t \cdot \nabla v_t + \dfrac{\partial \tilde{K}_f(p, \phi)}{\partial \phi} \nabla u_t \cdot \nabla v_t \right.$$
$$\left. - e_i \nabla \lambda_{si} - s_i \lambda_{si} + 2 \left(1 - H_\phi(\phi) \right) \lambda_p g(s) \right) H_\phi(\phi) \quad (19)$$

7. OPTIMIZATION EXAMPLES

7.1 Problem setting

To verify the validity of the proposed coupled fictitious physical model, a two-dimensional steady-state heat conduction problem is considered as an optimization example. Here, the design domain D is set as shown in Fig. 6. $V_{\max}$ is set to 25% of the initial volume $V_{\text{init}} = 2.0$. In addition, the thermal conductivity K_t is set to 26.0, the heat source Q is set to 1.0, and the characteristic length L is set to 1.0. The domain for Eq. (8), Eq. (13), and Eq. (18) is defined as a rectangular domain with widths of 0.2 on the left, right and upper edges, extended to encompass the fixed design domain as shown in Fig. 6 because the geometric feature is not detected near the boundary ∂D of the fixed design domain if the boundary condition is imposed on ∂D. In Eq. (8), the diffusion coefficient a_s is set to 3.0×10^{-4}. In Eq. (9), $\epsilon_s = 1.0 \times 10^{-6}$, $\epsilon_c = 1.0 \times 10^{-6}$, $\beta = 1.0 \times 10^{-2}$, and $n_c = 4$. In the approximate ramp function in Eq. (10) and Eq. (11), parameter ϵ_r is set to 1.0×10^{-9} and 0.75, respectively. Finally, in Eq. (13), the diffusion coefficient a_p is set to 1.0×10^{-1}.

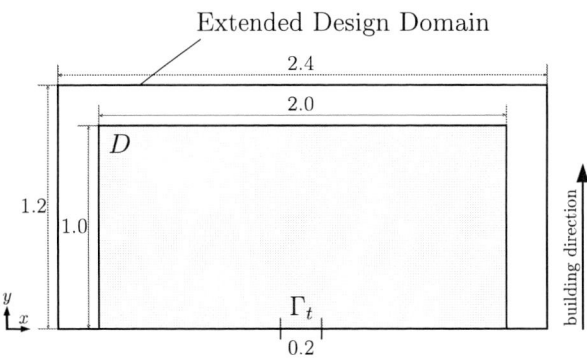

FIGURE 6: PROBLEM SETTING

7.2 Result without overhang constraint

First, a simulation was performed without considering the overhang constraint by setting the degree of reduction γ in thermal conductivity to zero. Figure 7a shows the optimization result obtained after 500 steps. As can be seen, some parts of the optimal structure are unsupported.

7.3 Result with overhang constraint

Next, a simulation was performed with considering the overhang constraint by setting the degree of reduction γ in thermal conductivity to 200. Figures 7b and 7c show the optimization results obtained after 500 steps (each minimum overhang angle is set to 45° and 60°). As shown in Figs. 7b and 7c, the optimal structures satisfy the overhang constraint because none of the overhang angles on the structural boundary exceed the minimum overhang angle.

7.4 Comparison of convergence with and without overhang constraint

Finally, Figs. 8 and 9 illustrate the change in the thermal compliance and ratio of volume with and without the overhang constraint at each step, respectively. As shown in Fig. 8, the thermal compliance with the overhang constraint is greater than without the constraint, but not significantly worse. In addition, from Figs. 8 and 9, the step-by-step changes in the thermal compliance and volume exhibit no significant difference in terms of convergence with and without the overhang constraint. Thus, the validity of the proposed coupled fictitious physical model is verified.

8. CONCLUSION

In this paper, to overcome the convergence problem in the fictitious physical model, a coupled fictitious physical model was proposed to interact the mechanical model with the fictitious physical model by changing the material constants in regions where the geometric constraint is violated such that the objective function worsens. The concept of the coupled fictitious physical model was adapted to the steady-state heat conduction problem with the overhang constraint and formulated. In addition, the formulation was verified by numerical examples to confirm if the concept of the proposed model was satisfied. Then, the formulation of the optimization problem was discussed, and sensitivity analysis

Copyright © 2023 by ASME

(a) Without overhang constraint

(b) With overhang constraint: the minimum overhang angle θ is 45°

(c) With overhang constraint: the minimum overhang angle θ is 60°

FIGURE 7: OPTIMAL STRUCTURES OBTAINED AFTER 500 STEPS WITH AND WITHOUT OVERHANG CONSTRAINT

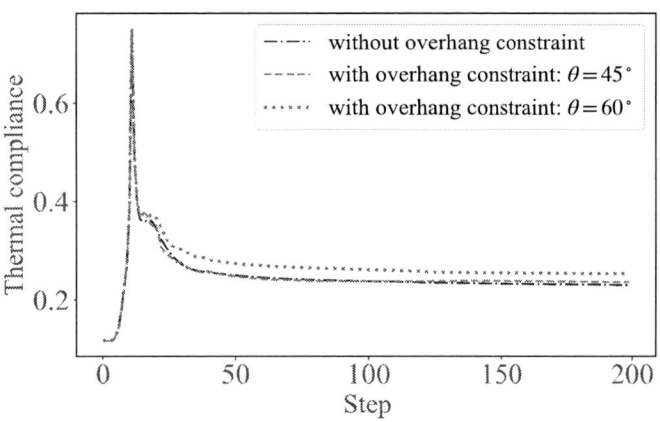

FIGURE 8: THE RELATIONSHIP BETWEEN STEPS AND THE THERMAL COMPLIANCE WITH AND WITHOUT OVERHANG CONSTRAINT

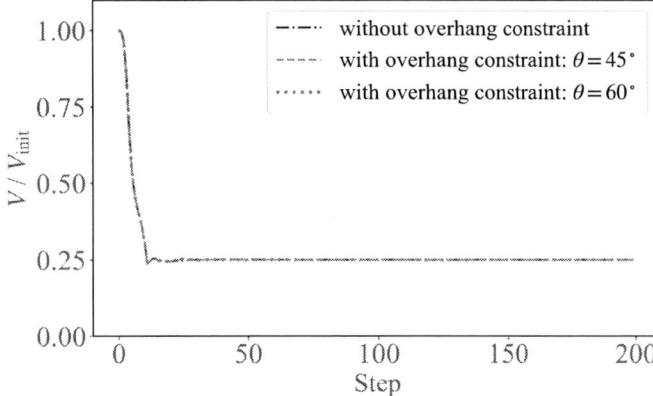

FIGURE 9: THE RELATIONSHIP BETWEEN STEPS AND THE RATIO OF VOLUME WITH AND WITHOUT OVERHANG CONSTRAINT

was performed based on the coupled fictitious physical model. Here, using a two-dimensional heat conduction problem as an example, we verified that the coupled fictitious physical model made no difference in terms of convergence with and without the overhang constraint.

ACKNOWLEDGMENTS

This work was supported in part by a project subsidized by the New Energy and Industrial Technology Development Organization (NEDO). The authors would like to thank Enago (www.enago.jp) for the English language review.

REFERENCES

[1] Bendsøe, Martin Philip and Kikuchi, Noboru. "Generating optimal topologies in structural design using a homogenization method." *Computer methods in applied mechanics and engineering* Vol. 71 No. 2 (1988): pp. 197–224.

[2] Suzuki, Katsuyuki and Kikuchi, Noboru. "A homogenization method for shape and topology optimization." *Computer methods in applied mechanics and engineering* Vol. 93 No. 3 (1991): pp. 291–318.

[3] Leary, Martin, Merli, Luigi, Torti, Federico, Mazur, Maciej and Brandt, Milan. "Optimal topology for additive manufacture: A method for enabling additive manufacture of support-free optimal structures." *Materials & Design* Vol. 63 (2014): pp. 678–690.

[4] Langelaar, Matthijs. "Topology optimization of 3D self-supporting structures for additive manufacturing." *Additive manufacturing* Vol. 12 (2016): pp. 60–70.

[5] Gaynor, Andrew T and Guest, James K. "Topology optimization considering overhang constraints: Eliminating sacrificial support material in additive manufacturing through design." *Structural and Multidisciplinary Optimization* Vol. 54 No. 5 (2016): pp. 1157–1172.

[6] Bi, Minghao, Tran, Phuong and Xie, Yi Min. "Topology optimization of 3D continuum structures under geometric self-supporting constraint." *Additive Manufacturing* Vol. 36 (2020): p. 101422.

[7] Allaire, Grégoire, Dapogny, Charles, Estevez, Rafael, Faure, Alexis and Michailidis, George. "Structural optimization under overhang constraints imposed by additive manufacturing technologies." *Journal of Computational Physics* Vol. 351 (2017): pp. 295–328.

Copyright © 2023 by ASME

[8] Wang, Yaguang, Gao, Jincheng and Kang, Zhan. "Level set-based topology optimization with overhang constraint: Towards support-free additive manufacturing." *Computer Methods in Applied Mechanics and Engineering* Vol. 339 (2018): pp. 591–614.

[9] Yamada, Takayuki. "Geometric shape features extraction using a steady state partial differential equation system." *Journal of Computational Design and Engineering* Vol. 6 No. 4 (2019): pp. 647–656.

[10] Sato, Yuki, Yamada, Takayuki, Izui, Kazuhiro and Nishiwaki, Shinji. "Manufacturability evaluation for molded parts using fictitious physical models, and its application in topology optimization." *The International Journal of Advanced Manufacturing Technology* Vol. 92 (2017): pp. 1391–1409.

[11] Yamada, Takayuki and Noguchi, Yuki. "Topology optimization with a closed cavity exclusion constraint for additive manufacturing based on the fictitious physical model approach." *Additive Manufacturing* Vol. 52 (2022): p. 102630.

[12] Yamada, Takayuki, Jun, Masamune, Teramoto, Hiroshi, Hasebe, Takahiro and Kuroda, Hirotoshi. "Topology optimization with geometrical feature constraints based on the partial differential equation system for geometrical features." *Transactions of the JSME* Vol. 85 No. 877 (2019): pp. 19–00129.

[13] Yamada, Takayuki, Izui, Kazuhiro, Nishiwaki, Shinji and Takezawa, Akihiro. "A topology optimization method based on the level set method incorporating a fictitious interface energy." *Computer Methods in Applied Mechanics and Engineering* Vol. 199 No. 45-48 (2010): pp. 2876–2891.

[14] Bendsoe, Martin Philip and Sigmund, Ole. *Topology optimization: theory, methods, and applications.* Springer Science & Business Media (2003).

[15] Allaire, Grégoire, Jouve, François and Toader, Anca-Maria. "Structural optimization using sensitivity analysis and a level-set method." *Journal of computational physics* Vol. 194 No. 1 (2004): pp. 363–393.

Copyright © 2023 by ASME

Proceedings of the ASME 2023
International Design Engineering Technical Conferences and
Computers and Information in Engineering Conference
IDETC-CIE2023
August 20-23, 2023, Boston, Massachusetts

DETC2023-116644

APPLICATION OF A CONTINUOUS VARIABLE DENSITY INFILL: MANIPULATING CENTER OF GRAVITY

Patrick N. Murphy[1,2], Bashir Khoda[1]

[1] University of Maine, Department of Mechanical Engineering, Orono, Maine
[2] Advanced Structures and Composites Center, Orono, Maine

ABSTRACT

In mechanical design mass properties, such as center of gravity and total mass, are an important consideration. As such the distribution of mass, i.e., the effective density distribution of a part, is a key consideration. When producing parts with additive manufacturing techniques like Fused Deposition Modeling, parts are typically hollowed to reduce manufacture time and replaced with a uniform infill structure. This process will affect the mass properties and provides only adjustment to the total mass. However, manipulating the center of gravity of a part can beneficial as it can change the total mass, moment of inertia, and vibration response of a part. Methods have been proposed in the past like the works of Chiu et. al., Yamanaka et. al., and Grigolato et. al. with each seeking to overcome certain limitations. In this paper, a method is proposed for the manipulation of the center of gravity of an additively manufactured part. The method first creates a density distribution with the desired mass properties then a novel process-aware continuous variable density infill algorithm is utilized to create the infill structure. With this algorithm it is ensured that a manufacturable functionally graded infill structure can be generated. The algorithm proposed for realizing this density distribution achieves this in two steps: variation of extrusion line spacing and width of the standard rectilinear infill pattern. This paper focuses on a 2D case, as implementing variation in the z axis was shown to be trivial. The methodology has been implemented within MATLAB to first generate the density distribution, then apply the continuous variable density infill generation algorithm, and finally generate the g-code. All sample prints were made with an extrusion-based, fused deposition modeling (FDM) printer. A preliminary test showed inconsistent behavior from the variable extrusion width. As such, each test case was printed with and without the variable extrusion width to further evaluate. While relatively high, the total error from input density gradient to print for the samples with only VLS showed very good consistency, both being ~10%. This is favorable as a consistent error can be corrected for with pre/post processing. The samples with variable extrusion width did improve the results, however the degree of the improvement was

not consistent, varying from ~3% to ~8%. This behavior was predicted as the range of extrusion width variation is rather limited at $\pm 50\%$. The feasibility of the proposed method has been demonstrated and could be implemented for 3D arbitrary shapes in future work.

1. INTRODUCTION

Following its mass availability in the past decade, polymer based Additive Manufacturing (AM) has become common place in both research and industry [1]–[4]. With this, has come a new set of design and process considerations when optimizing parts. In particular, variable density lattice structures have emerged in design for additive manufacture as one of the most popular strategies for functional optimization, allowing for lightweight structures [5]–[15]. However, there are many processes that fall under AM, each with their own design considerations.

Stereolithography (SLA) is an AM process that utilizes UV light to selectively cure a polymer resin, through either a UV laser or a Liquid Crystal Display. The result is superior 'resolution' and smaller layers heights than other AM processes. However, SLA can be limited in material choice, as it requires UV polymerizing resins. This typically means expensive proprietary resins, that need added material hazard precautions, are required. Another very important limitation of SLA is the inability to create fully enclosed voids within a part, as resin needs to be able to drain. Otherwise, the void will be filled with uncured resin [1], [2].

Another AM process that utilizes laser light is Selective Laser Sintering (SLS). Known as a power bed process, SLS utilizes a laser to sinter a powered polymer layer by layer. This process is like SLA and carries the same limitations in terms of parts with internal voids. While SLS typically offers a greater range of materials and mechanical performance than SLA, these materials can once again be expensive. In addition, SLS machines are typically the most expensive of the common AM processes [1], [2], [4].

The most prevalent of these AM methods is Fused Filament Deposition (FDM). It offers relative simplicity and low cost, when compared to SLM and SLA [2]. This extends to both the

Approved for Public Release – DEVCOM SC PR2023_14292

Copyright © 2023 by ASME

material and the machine itself. FDM forms a 3D part by the building up of planner layers through the extrusion of thermoset polymer resins. This process allows of the rapid design and manufacture of complex geometries, including internal voids. FDM is the AM process that was selected as the focus of this work.

When designing any part, there are always particular requirements that must be met. A very common and important consideration is the mass properties of a part, i.e., total mass, center of gravity, and moment of inertia. Undesired mass properties can lead to poor part performance, due to excessive fatigue and vibration. However, it's not always possible to change the size or shape of a part to achieve the desired mass properties. As such variable density design methods are often favorable for achieving desired mass properties when shape and size cannot change.

An algorithm for controlling the center of gravity of a hallowed part was developed by Chiu et. al. with a focus on additive manufacturing. The hallowing portion of their algorithm was built on the dexel model, as developed by van Hook [16]. Control of the center of gravity is achieved by adding the condition of accounting for the change in center of gravity when removing voxels in the hallowing process. In this way, the algorithm can maintain the external geometry of the part while affecting the mass properties as desired [17]. This can be seen as a subtractive approach to center of gravity control. However, this algorithm doesn't account for the AM process limitations and could lead to many unsupported internal overhangs.

Another method developed by Yamanaka et. al. sought to control the mass properties, namely center of gravity, through variable density of an internal lattice structure. In this method an optimum density distribution is formulated and realized as a truss-based lattice structure, with the objective being the minimization of error between target density and the truss density. While this method does create support for internal overhangs, it suffers from a deferent set of limitations. The lattice structure must be fully defined in the CAD model. This results in a very complex STL to be sliced, meaning long slicing computation times. This geometry also has no consideration of the AM process limitations, like overhangs and small features [8].

The method proposed by Grigolato et. al. sought to overcome some of the issues in geometric modeling of complex variable density lattice structures. The proposed method utilizes the infill patterns within the CAM software used to slice or generate the G-code, to define a graded density lattice structure rather than the shell-based lattice approach. The method consisted of two steps: first slicing the part as normal and generating a homogenous infill, and then post processing the g-code by modifying the extrusion width of the infill. This served to create variable density within the infill structure based of a given density gradient [9]. The limitation of this method is that the algorithm is inherently unaware of process and machine limitations when generating or applying a variable density gradient.

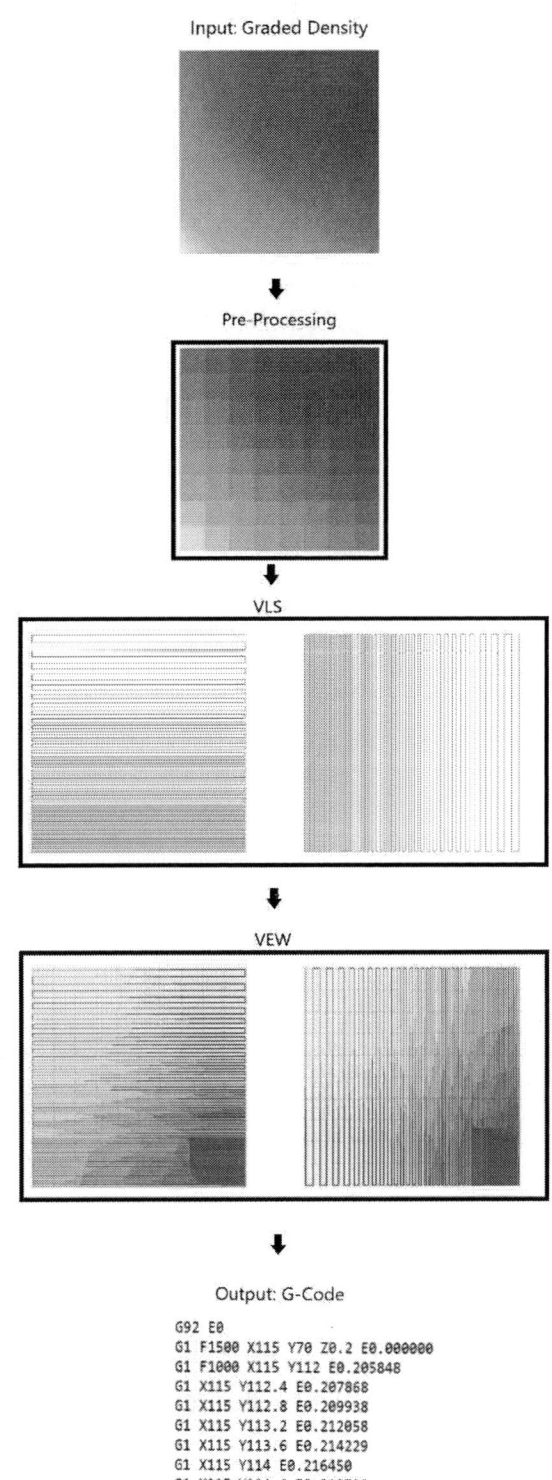

Figure 1:Method Overview

In this paper, a method is established for manipulating the center of gravity of the infill structure of an AM part. This could be utilized to maintain the CG of a part when hallowed by a slicer or change the CG to achieve desired mass properties. This

Copyright © 2023 by ASME

method consists of two phases: the first is generating a function to define the density distribution that results in the desired CG change, and the second is a novel continuous variable density infill. The first step of the method is to determine the desired center of gravity and total mass. This can then be used to calculate the coefficients of a functionally defined density gradient with the desired properties. Finally, an algorithm is proposed for realizing this density distribution through variable density infill achieved in two parts: variation of extrusion line spacing and width of the standard rectilinear infill pattern.

As proposed by Grigolato et. al. this algorithm seeks to avoid the limitations of modeling a functionally graded infill structure in CAD by defining and generating the structure at the G-code level, as would be done by a commercial slicer to create uniform infill [9]. This algorithm differs from those proposed by Chiu and Yamanaka, as it aims to account for process limitations and reduce time, both process and computational [8], [17].

2. MATERIALS AND METHODS

2.1 Functionally Defining Density

The center of gravity, $\bar{x}$ and $\bar{y}$, of an object in which density is a function of position is proportional to the integral of that density function over the total mass, M.

$$\bar{x} = \frac{\int_0^y \int_0^x y * \rho(x,y)\, dx\, dy}{M} \quad (1)$$

$$\bar{y} = \frac{\int_0^y \int_0^x x * \rho(x,y)\, dx\, dy}{M} \quad (2)$$

$$M = \int_0^y \int_0^x \rho(x,y)\, dx\, dy \quad (3)$$

The values of $\bar{x}$ and $\bar{y}$ are given in this method as user defined parameters. Therefore, it is advantageous to define the density as a smooth and regular function across the domain i.e., the bounds of the object. The definition of this function can be simplified when not considering adjustment of the center of gravity in the z axis. As this method is to be applied on a per layer basis for additively manufactured objects this simplification can be made as variation within a layer in the z axis is only possible through controlling the total mass of each layer. However, this does present a third user defined parameter, average density,

$$\rho_{Avg} = \frac{M}{X * Y} \quad (4)$$

which can be used to find total mass of a layer. This is comparable to the infill precent parameter for homogenous infill generated by a standard slicer.

With the requirement for a smooth and regular function of x and y for the density, the function can easily be defined as any polynomial representing a smooth and regular surface. To that end, a 2d quadric offers a good basis for defining density as a function of position, i.e.

$$\rho = F(x,y) = Ax^2 + By^2 + Cxy + Dx + Ey \quad (5)$$

As this is the relative density the range should from 0 to 1, therefore it'd be desirable to normalize the function. Furthermore, this method is simplified to work with only a square domain, so the equation is simplified and rewritten as:

$$\rho(x,y) = \frac{Ax^2 + By^2 + Cxy + Dx + Ey}{S^2 + S} \quad (6)$$

where S is the side length of the square object, note this equation has 5 unknown coefficients. If given the desired $\bar{x}$, $\bar{y}$, and average relative density, there is only 3 equations with which to solve for 3 unknowns. Thus, to reduce the number of unknowns and further simplify its assumed D and E, the linear terms, are equal to 0. These terms have the smallest effect on the density distribution.

$$\rho(x,y) = \frac{Ax^2 + By^2 + Cxy}{S^2} \quad (7)$$

Finally, given the parameters of S, $\bar{x}$, $\bar{y}$, and ρ_{Avg}, the unknown coefficients can be solved for using *eq. 1*, *eq. 2*, and *eq. 3*, creating a functionally graded density with the desired mass properties.

Expanding this methodology to manipulate the center of gravity in 3 dimensions i.e., adding control of $\bar{z}$ would be trivial. The 2.5D nurture of AM through FDM, control of $\bar{z}$ becomes a matter of defining the weight for a given layer. Hence the desire to not only have control of $\bar{x}$ and $\bar{y}$, but also the average relative density for each layer. This method allows for the creation of a wide range of physically meaningful density gradients but does have limitations.

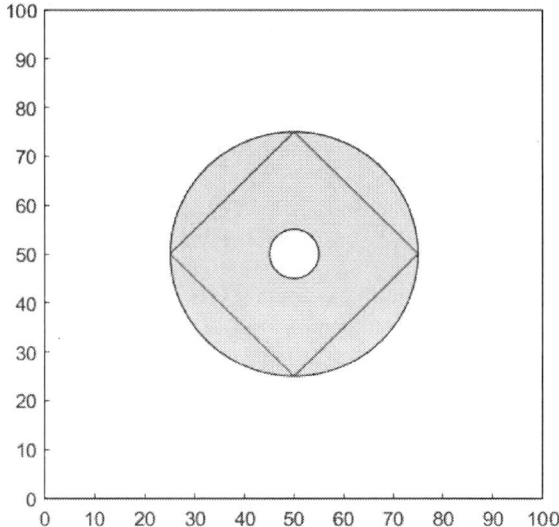

Figure 2: Viable CG Manipulation Region

The viable range of the center of gravity for this method is shown in (*fig. 2*), the green region. A key behavior of the this is that the maximum average density that can be achieved decreases as the as the CG moves away from the red line shown in (*fig. 2*). This line represents all possible locations of the center of gravity if the coefficient C = 0. Also note that with this method that achieving a center of gravity that is within 5% of the geometric center is not viable.

2.2 Variable Density infill: VLS+VEW

There are many different infill patterns utilized in additive manufacturing. For the algorithm proposed, the rectilinear infill pattern was selected as the focus. While it could be applied to similar patterns, rectilinear is one of the most basic patterns, offering relative ease in implementing variable line spacing and g-code generation. In contrast a parametric pattern like gyroid is too complex to simply change the spacing between extrusion lines as proposed.

This functionally graded density is discretized to match the element size of variable density infill algorithm parameters. Since the rectilinear infill creates a grid in only one axis per layer, an additional pre-processing step can be performed, in which the discretized graded density can transformed into two vectors: one being a vector representing the mean of the rows, and the other representing the mean of the columns. Only the vector corresponding to the working axis of the layer would be needed to determine line spacing.

$$L_{spacing} = \frac{Extrution\ Width}{\rho} \qquad (8)$$

To further reduce computation time, another transform can be applied during preprocessing which converts the vector of densities into the corresponding line spacing that would result in that density.

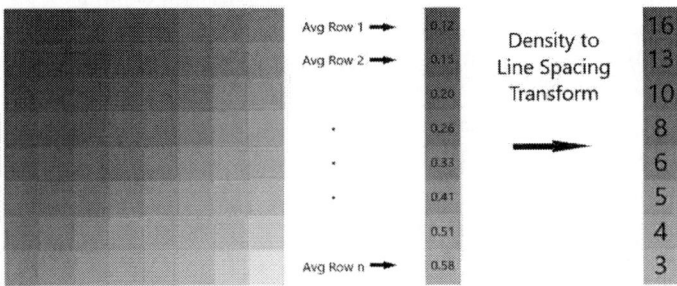

Figure 3: VLS Pre-processing for a single layer, working in the y axis

This creates a vector of the desired spacing for each discretized element. This preprocessing eliminates the need to apply this function at each iteration across the whole discretized density distribution in the process of determining the overall line spacing. This allows for decreased processing time.

Finding the line spacing at a given point is best defined as an optimization problem, where the objective is to minimize error between average enclosed element value and the given line spacing. This can be written as

$$min:\ e_i(u) = u_{avg} - L_{spacing} \qquad (5)$$
$$s.t.$$
$$L_{spacing} > Extruction\ width$$
$$L_{spacing} < L_{Max}$$

where u_{avg} is the average enclosed element value and $L_{spacing}$ is the line spacing. A simple bisection method is employed to solve the minimization for each line spacing in series, as the problem is one-dimensional.

Algorithm 1: Variable Line Spacing

$x \leftarrow 1, i \leftarrow 1;$
while $Break == False$ **do**
  $a = L_{Min};$
  $b = L_{Max};$
  while $b - a > 1$ **do**
    $c = (a + b)/2;$
    $L_c =;$
    $E_c =;$
    if $sign(E_c) == sign(E_a)$ **then**
      $a = c;$
      $E_a = E_c;$
    else
      $b = c;$
    end
  end
  $L(i) = c;$
  $x = x + L(i);$
  if $x + L_{Min} > L_{Max}$ **then**
    $dx = x - L_{Max};$
    $L(i) = L(i) - dx;$
    $x = L_{Max};$
    $Break = true;$
  end
  $i = i + 1;$
end

The output of this gives the number of extrusion lines in the working axis and their spacing relative to each other across the domain of that layer. From this a toolpath, a matrix representing the relative density can be constructed. The matrix is used along with the discretized input density matrix to calculate the error in the relative density.

$$[E] = [\rho_i] - [\rho_{VLS}] \qquad (6)$$

Matrix E represents the discretized error in density across the domain, which is used to generate a matrix of extrusion width modulation. The purpose is to use the error matrix in determining the extrusion width variation that will reduce error.

2.3 G-Code Generation

A custom g-code generation script was written to handle slicing. The script is able to take in the extrusion line spacing and width data as vectors generated by the VLS+VEW method. The vectors are preprocessed and binned by extrusion width, creating the longest possible line segments per g-code line to help reduce the total number of lines. Without this, the number of lines would quickly grow proportionally with the step size of the extortion width variation.

Algorithm 2: G-Code Generation

```
for j := 1 to Number of lines do   comment
    Bearkup extrution line by simalar E value;
    i = 1;
    p1 = 1;
    p2 = 2;
    while P_2 do
        while ExtModX_{P_1,j} do
            P_2 = P_2 + 1 ;
        end
        subline(i,:) =;
        P_1 = P_2;
        i = i + 1;
    end
    Generate x,y, and E data;
end
```

3. RESULTS AND DISCUSSION

Initial test pieces showed a poor correlation between the design and printed parts in terms of center of gravity. It was hypothesized that the variable extrusion width wasn't preforming as expected, as originally the extrusion width was given an arbitrary limit of +200% and -75% the nominal width of 0.4 mm. Following further review, it was decided that the percent change in extrusion width should be lowered to a more achievable 50%. It's unknown why exactly, therefore 4 new tests are to be printed to investigate. Two prints have variable extrusion width effectively turned off. This is more in line with the literature regarding the maximum difference between extrusion width and nozzle size for reliable printing on common FDM printers [2], [18].

Table 1: Density Gradient Parameters

Sample	A	B	C
I	1	1	1
II	1	1	1
III	3	0	0
IV	3	0	0

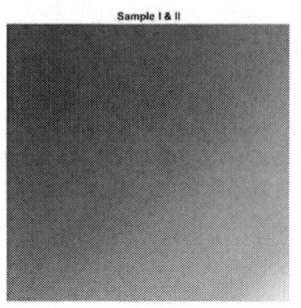 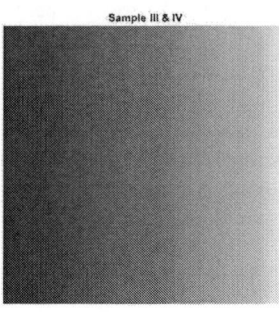

Figure 4: Input Functionally Graded Density

All manufacturing process and algorithm parameters remained the same across the four samples except for the variable extrusion width, which was excluded from two of the samples.

Table 2: Process & Algorithm Parameters

Sample	Side length	Step Size	E width	E variation
I	100 mm	0.20 mm	0.4 mm	N/A
II	100 mm	0.20 mm	0.4 mm	50%
III	100 mm	0.20 mm	0.4 mm	N/A
IV	100 mm	0.20 mm	0.4 mm	50%

The results show that the location of the center of gravity can be controlled for a given geometry. However, those print samples contain higher error than just the infill spacing design. This is due to the lack of fine control in extrusion width with the G-code. The samples without variation in extrusion width showed a consistent 10% error from density distribution to parted sample.

Table 3: Center of Gravity

Sample	Input	Design	Printed
I	63.65mm, 63.65mm	57.52mm, 57.52mm	56.84mm, 56.65mm
II		62.30mm, 62.30mm	61.87mm, 61.93mm
III	75.02mm, 50.00mm	61.42mm, 50.00mm	61.57mm, 49.83mm
IV		70.36mm, 50.01mm	64.82mm, 50.33mm

Copyright © 2023 by ASME

While all the printed samples with variation in extrusion width performed better in each case, they once again showed inconsistent behavior with error ranging from 3% to 8%. Some of this behavior is believed to be the result of the limited variation in extrusion width of 50%. This value was determined from the literature as the upper limit without effecting quality or speed, within reason.

Table 4: CG Error

Sample	Error in Design	Error Design to print	Error total
I	-9.63%	-1.35%	-10.98%
II	-2.12%	-0.64%	-2.76%
III	-10.88%	-0.02%	-10.90%
IV	-3.72%	-4.34%	-8.06%

Believed to be the result of a lack of retraction and z-hop, a lot of un-wanted extrusion and 'stringing' during travel moves can be observed on the sample prints.

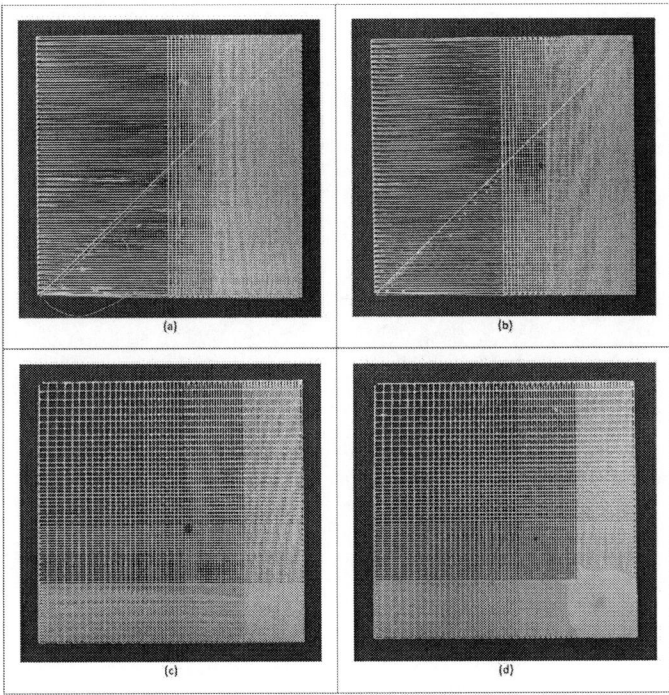

Figure 5: Printed Samples, (a) Sample I, (b) Sample II, (c) Sample III, and (d) Sample IV

This is particularly noticeable in Sample I and II, in which an unintended extrusion line can be seen running diagonally across the part. These types of defects can be mitigated or even eliminated with more sophisticated g-code generation. As proposed, this would be accomplished by integrating the current extrusion width variation g-code generation into an existing commercial slicer. This would also allow for the inclusion of the outer shell structure of the part. Such implementation was deemed outside the scope of this research.

4. CONCLUSION

In this paper, a method has been proposed for manipulating the center of gravity of an additively manufactured part, utilizing a novel variable density infill algorithm. The feasibility and effectiveness of the proposed method has been demonstrated for the 2D case of a square design domain. The Method described herein could be improved if utilized alongside a 'hallowing' method for controlling center of gravity, as this method only adjusts the infill of an AM part. Therefore, without a method to control the center of gravity of the shell of a part, results of the infill method could be negated. Future work would include implementing the method for arbitrary shapes and implementing manipulation of $\bar{z}$ as proposed in this work, as well as further investigation of fine extrusion width control.

ACKNOWLEDGEMENTS

This material is based upon work supported by the U.S. Army Combat Capabilities and Development Command – Soldier Center (DEVCOM SC) under Contract Nos. W911QY-18-C-0101 and W911QY-20-C-0053. Any opinions, findings and conclusions or recommendations expressed in this material are those of the author(s) and do not necessarily reflect the views of the DEVCOM SC.

REFERENCES

[1] A. Bacciaglia, A. Ceruti, and A. Liverani, "Additive Manufacturing Challenges and Future Developments in the Next Ten Years," in *Design Tools and Methods in Industrial Engineering*, C. Rizzi, A. O. Andrisano, F. Leali, F. Gherardini, F. Pini, and A. Vergnano, Eds. Cham: Springer International Publishing, 2020, pp. 891–902. doi: 10.1007/978-3-030-31154-4_76.

[2] I. Gibson, D. Rosen, B. Stucker, and M. Khorasani, *Additive Manufacturing Technologies*. Cham: Springer International Publishing, 2021. doi: 10.1007/978-3-030-56127-7.

[3] W. Oropallo and L. A. Piegl, "Ten challenges in 3D printing," *Engineering with Computers*, vol. 32, no. 1, pp. 135–148, Jan. 2016, doi: 10.1007/s00366-015-0407-0.

[4] E. O. Olakanmi, R. F. Cochrane, and K. W. Dalgarno, "A review on selective laser sintering/melting (SLS/SLM) of aluminium alloy powders: Processing, microstructure, and properties," *Progress in Materials Science*, vol. 74, pp. 401–477, Oct. 2015, doi: 10.1016/j.pmatsci.2015.03.002.

[5] A. O. Aremu *et al.*, "A voxel-based method of constructing and skinning conformal and functionally graded lattice structures suitable for additive manufacturing," *Additive Manufacturing*, vol. 13, pp. 1–13, Jan. 2017, doi: 10.1016/j.addma.2016.10.006.

Copyright © 2023 by ASME

[6] D. J. Brackett, I. A. Ashcroft, R. D. Wildman, and R. J. M. Hague, "An error diffusion based method to generate functionally graded cellular structures," *Computers & Structures*, vol. 138, pp. 102–111, Jul. 2014, doi: 10.1016/j.compstruc.2014.03.004.

[7] D. Li, N. Dai, X. Jiang, Z. Shen, and X. Chen, "Density Aware Internal Supporting Structure Modeling of 3D Printed Objects," in *2015 International Conference on Virtual Reality and Visualization (ICVRV)*, Xiamen, China, Oct. 2015, pp. 209–215. doi: 10.1109/ICVRV.2015.22.

[8] D. Yamanaka, H. Suzuki, and Y. Ohtake, "Density aware shape modeling to control mass properties of 3D printed objects," in *SIGGRAPH Asia 2014 Technical Briefs*, Shenzhen China, Nov. 2014, pp. 1–4. doi: 10.1145/2669024.2669040.

[9] L. Grigolato, S. Rosso, R. Meneghello, G. Concheri, and G. Savio, "Design and manufacturing of graded density components by material extrusion technologies," *Additive Manufacturing*, vol. 57, p. 102950, Sep. 2022, doi: 10.1016/j.addma.2022.102950.

[10] R. Xie, C. Ulven, and B. Khoda, "Design and Manufacturing of Variable Stiffness Mattress," *Procedia Manufacturing*, vol. 26, pp. 132–139, 2018, doi: 10.1016/j.promfg.2018.07.016.

[11] I. Maskery, A. O. Aremu, L. Parry, R. D. Wildman, C. J. Tuck, and I. A. Ashcroft, "Effective design and simulation of surface-based lattice structures featuring volume fraction and cell type grading," *Materials & Design*, vol. 155, pp. 220–232, Oct. 2018, doi: 10.1016/j.matdes.2018.05.058.

[12] E. Cuan-Urquizo *et al.*, "Effective Stiffness of Fused Deposition Modeling Infill Lattice Patterns Made of PLA-Wood Material," *Polymers*, vol. 14, no. 2, Art. no. 2, Jan. 2022, doi: 10.3390/polym14020337.

[13] A. Pasko, O. Fryazinov, T. Vilbrandt, P.-A. Fayolle, and V. Adzhiev, "Procedural function-based modelling of volumetric microstructures," *Graphical Models*, vol. 73, no. 5, Art. no. 5, Sep. 2011, doi: 10.1016/j.gmod.2011.03.001.

[14] A. Panesar, M. Abdi, D. Hickman, and I. Ashcroft, "Strategies for functionally graded lattice structures derived using topology optimisation for Additive Manufacturing," *Additive Manufacturing*, vol. 19, pp. 81–94, Jan. 2018, doi: 10.1016/j.addma.2017.11.008.

[15] F. Tamburrino, S. Graziosi, and M. Bordegoni, "The Design Process of Additively Manufactured Mesoscale Lattice Structures: A Review," *Journal of Computing and Information Science in Engineering*, vol. 18, no. 4, Art. no. 4, Dec. 2018, doi: 10.1115/1.4040131.

[16] T. V. Hook, "Real-time shaded NC milling display," vol. 20, no. 4, 1986.

[17] W. K. Chiu and S. T. Tan, "An automatic method for controlling the centre of gravity of a model," *Computer-Aided Design*, vol. 34, no. 13, Art. no. 13, Nov. 2002, doi: 10.1016/S0010-4485(01)00158-0.

[18] A. Alafaghani, A. Qattawi, B. Alrawi, and A. Guzman, "Experimental Optimization of Fused Deposition Modelling Processing Parameters: A Design-for-Manufacturing Approach," *Procedia Manufacturing*, vol. 10, pp. 791–803, 2017, doi: 10.1016/j.promfg.2017.07.079.

Copyright © 2023 by ASME

Proceedings of the ASME 2023
International Design Engineering Technical Conferences and
Computers and Information in Engineering Conference
IDETC-CIE2023
August 20-23, 2023, Boston, Massachusetts

DETC2023-116865

A FRAMEWORK ESTABLISHING THE BOUNDS OF SMALL ANGLE ASSUMPTIONS IN MULTI-MATERIAL ADDITIVELY MANUFACTURED COMPLIANT MECHANISMS

Evelyn Thomas, Nicholas Meisel, Jared Butler*

School of Engineering Design and Innovation
The Pennsylvania State University
213 Hammond Building
University Park, PA 16802
Email: evelynthomas@psu.edu; nam20@psu.edu; butler@psu.edu

ABSTRACT

Additively-manufactured multi-material compliant mechanisms are an emergent class of mechanisms; these mechanisms have a higher reliance on material properties to drive deflection when compared to other compliant mechanisms. As a result, material-specific properties such as Young's modulus, strain response, and Poisson's ratio must be analyzed to properly predict the performance of multi-material additively manufactured compliant mechanisms. This paper creates a framework for additive manufacturing users to characterize acceptable force-deflection relationship predictors using strain data associated with a generalized multi-material cantilever beam. This allows designers to have more confidence in the performance of their designs prior to production. When applied, the framework allows the user to identify the applied deflections where Hooke's Law and small angle assumptions are, or are not, appropriate. This process will be useful for multi-material additive manufacturing processes capable of producing multi-material prints: directed energy deposition (metals) and material extrusion (thermoplastics). This process will serve as the basis for encouraging the widespread adoption of additively manufactured multi-material compliant mechanisms.

Keywords: Additive Manufacturing, Compliant Mechanisms, Strain, Material Extrusion, Directed Energy Deposition

1. INTRODUCTION

Compliant mechanisms are mechanical devices that obtain their motion through the deflection of flexible members [1]. Using compliant mechanisms instead of rigid mechanisms often leads to a reduction in parts, cost, and weight without compromising performance[1]. These mechanisms have been used in several industries such as aerospace, medical, energy, and consumer

products. In aerospace, Butler, et al. created a highly compressible bellows that could contain debris in hostile environments[2]. Compliant mechanisms-based spinal[3], finger[4], and wrist implants [4] have also been developed to improve patient outcomes. In the energy sector, researchers Zirbel, et al. developed an origami-based deployable solar panel array [5] and Harb, et al. created a proof of concept for a hybrid micropower supply [6]. Compliant consumer products such as bicycle parts [7], centrifugal clutches[8], and switches [8] have been used as well.

Traditionally, these mechanisms have been produced using subtractive manufacturing [8]. Currently, subtractive manufacturing is used more than additive manufacturing in compliant mechanisms manufacturing, which can cause an over-reliance on single-material parts. Although subtractive manufacturing will remain a common method of production, additive manufacturing is rapidly becoming a viable process due to its ability to produce parts that subtractive manufacturing cannot. Given that compliant mechanisms achieve their motion based on boundary conditions, geometry, and material properties,[1] the unique geometric, functional, and hierarchical capabilities[9] of additive manufacturing will be groundbreaking for compliant mechanisms. The ability to produce multi-material parts using material jetting (MJ)[10], directed energy deposition (DED)[11], and material extrusion (MEX)[12] is unrivaled by other traditional manufacturing methods.

Material jetting uses up to three photopolymer resins to deposit material droplets. These droplets can create a gradient of material properties by varying the volumetric percent of these resins. Each resin can have distinct colors, rigidities, yield characteristics, and applications (bio-compatible, tissue-like, etc.)[13]. By contrast, directed energy deposition is a metal-based additive manufacturing practice that can produce parts of an ever-expanding list of wire- or powder-based metal alloys[14]. This process can also lead to the creation of metal functionally graded

---

*Corresponding author: butler@psu.edu

Copyright © 2023 by ASME

parts[11]. Lastly, material extrusion is one of the most popular additive manufacturing processes used in homes, schools, libraries, and laboratories across the world. Filaments used in material extrusion include thermoplastic polyurethane (TPU), high-impact polystyrene (HIP), nylon, acrylonitrile butadiene styrene (ABS), polylactic acid (PLA), carbon fiber, wood PLA, and more[15].

While it is possible to additively manufacture multi-material compliant mechanisms with these materials, their usefulness lies in their performance. To make multi-material compliant mechanisms useful, it is essential to be able to predict the force-displacement relationships for these parts. The easiest method for doing this is using small angle assumptions. With small angle assumptions, it is assumed any deformation perpendicular to the applied force is negligible. These equations can also be rearranged to solve for force, given a vertical deflection. While these assumptions are helpful and quick, they can be very inaccurate as deflections increase [16]. When these models yield inaccurate results, large deflection models must be used. Models that account for large deflections can become limited and/or computationally expensive. As a result, knowing the limit of small angle assumptions allows users to know if other approaches are necessary.

This paper creates a framework for identifying when small angle assumptions are no longer valid for multi-material compliant mechanisms. By determining the limitations of small angle assumptions within their compliant mechanisms, the user is better able to understand the error associated with their design's performance prior to using additive manufacturing.

2. BACKGROUND

2.1 Multi-material Additively Manufactured Compliant Mechanisms

The material complexity of Additive Manufacturing has allowed for multi-material and functionally-graded parts [10][17][18]. The processes capable of multi-material additive manufacturing allow for parts that use rigid and comparatively flexible materials side-by-side. Multi-material additive manufacturing specifically enables material and geometric opportunities for compliant mechanisms that traditionally manufactured, single-material compliant mechanisms cannot achieve.

Multi-material compliant mechanisms across the 3 most common multi-material processes are experiencing an increase in adoption due to their potential. An example from each of these processes follows. DED has been used to produce a functionally-graded, shape memory compliant mechanism using Nitinol[11]. In this paper, Jovanova et. al. provided an algorithm for optimizing the geometry, mechanical properties, and performance of an initially-flat shape memory alloy structure. As a proof of concept, alternating segments of linear elastic and super elastic Nitinol were additively manufactured to create an origami-inspired foldable part. The change in elasticity between the linear elastic and super-elastic regions were critical to establishing the form as well as the flexibility desired for compliant mechanisms design[11]. Talken et al. created compliant mechanisms with a TPU strain joint using MEX[12]. Using TPU and PLA to create a multi-material version of three joints enabled the authors to determine an ideal joint with respect to tensile stiffness, torsional stiffness,

pivot deviation, and range of motion are of particular importance. The performance of compliant mechanisms is subject to the application of the mechanisms. As a result, it is important to keep these four criteria in mind when designing compliant mechanisms. Meisel et al. created a topologically-optimized compliant inverter using MJ[10]. The authors used a solid isotropic material with penalization (SIMP) material interpolation to determine where the different materials should be placed and to decide the geometry of the inverter. By using TangoBlack+, a rubberlike elastomeric photopolymer resin, and VeroWhite, a rigid photopolymer resin, the authors created discrete regions of each material[10]. Additionally, the materials were mixed to create other potential stiffness ratios. As expected, the greater the stiffness ratio between the more rigid and flexible material, the greater the deflection the inverter achieved[10]. Unfortunately, most of these sources use FEA to predict displacement due to the lack of an analytical model for predicting deflection. This can be computationally expensive for large structures or highly complex structures that require several nodes. Therefore, it may be helpful to start with small-angle assumptions and examine their performance.

2.2 Deflection of Non-Constant Members

Non-constant members include beams with material and/or geometric properties that change throughout the beam. Examples may include regions with a stepped cross-sectional area change, a continuous cross-sectional area change, a discrete material change, and a continuous material change.

Parkinson, et al. created a model for examining deflection in a tapered beam using the equation for the path of an ellipse [19]. In this method, the tapered beam is a cantilever with a vertical force applied to the end. The tapered beam has a larger cross-sectional area near the beam and decreases linearly along the beam's length. This beam has uniform material properties.

Howell's Pseudo Rigid Body Model (PRBM) is a widely used method for characterizing compliant mechanisms because it produces useful solutions and is accurate within 0.5% for end angles less than or equal to 77°[1]; this is ideal for predicting force-displacement relationships where small angle assumptions are inappropriate. Despite the PRBM's easy-of-use and accuracy, this method is only useful for beams with constant cross-sections and material properties[1]. Although it is quite accurate for beams with uniform material properties and cross-sections, further iterations were needed to increase the complexity of the models. Sargent, et al.'s Mixed Body Model (MBM) enhanced the PRBM to account for a non-constant cross-section [20], but there still needs to be a method that accounts for varied material properties.

The inability to account for material property differences that occur in multi-material additively manufactured compliant mechanisms can also lead to a higher error when using small angle assumptions. This is because large deflections are already associated with geometric nonlinearities; however, certain materials have special, material-based nonlinearities when subjected to large deflection[21]. Examples of these included hyperelasticity in rubber-like materials and elastomeric foams[21]. These nonlinearities can compound to create a larger error when using small

Copyright © 2023 by ASME

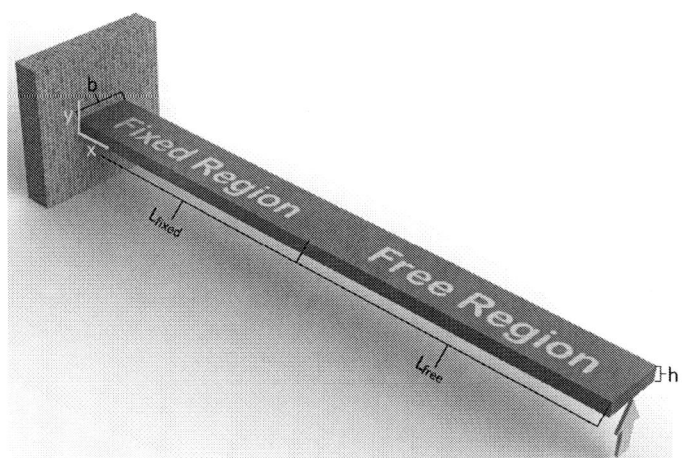

FIGURE 1: ANNOTATED DUAL-MATERIAL CANTILEVER BEAM

angle assumptions. To advance the use of multi-material compliant mechanisms, it is essential to start by examining the limits of acceptability when using small-angle assumptions. Evaluating a cantilever beam with a vertical force on the end, as shown in Figure 1, creates a baseline for understanding the small angle assumptions.

Small angle assumptions (SAA) are appropriate at small deflections. Small angle assumptions (SAA) declare the horizontal deflection to be negligible, $\tan\theta \approx \theta$, and $\sin\theta \approx \theta$, where θ is the angle of the beam as shown in Figure 2. Figure 2 also exhibits the differences in small deflections and large deflections (LD) compared to an undeflected beam when the more flexible material is fixed (Figure 2a) and when the more rigid material is fixed (Figure 2b).

Small angle assumptions are found to follow the linear-elastic region of Hooke's Law and the error is considered small; however, as more error is introduced with greater deflections [16]. The disparities between the large displacement a part is undergoing and small angle assumptions utilized can be a sign that the model is nearing the end of the linearly-elastic region. As a result, examining strain will allow the user to properly gauge where they are on the stress-strain curve with respect to the deforming material.

2.2.1 Force-Deflection Relationships. The relation of force to deflection is vital to understanding compliant mechanisms design. This work uses two approaches to describe the deflection of cantilevered beams. Although the approaches to solving for a force or deflection using small-angle assumptions are different, both methods produce the same result. Two approaches were taken because they are two of the most common methods for determining force-deflection relationships. They also examine the problem in different ways. For example, Castigliano's Theorem involves working from the free end, while the Bernoulli-Euler approach involves working from the fixed end of the beam. Using Castigliano's Theorem approach enables the user to quickly pivot to looking at internal strain energy, while the Bernoulli-Euler approach originates from curvature relationships.

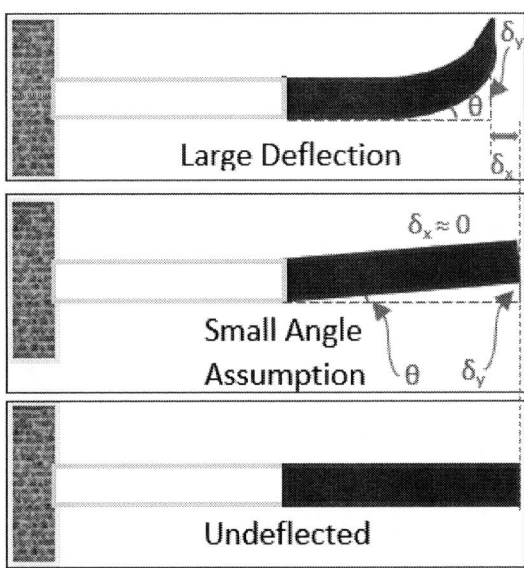

(a) Beam with the more flexible material fixed

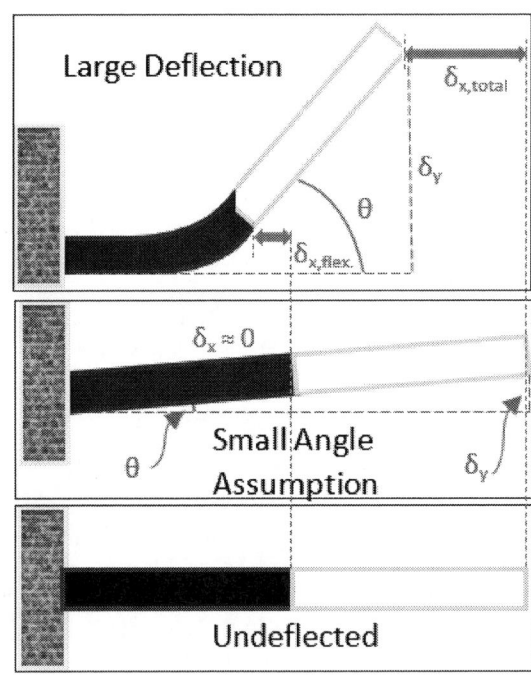

(b) Beam with the more flexible material free

FIGURE 2: BEAM BEHAVIOR COMPARISONS

Castigliano's Theorem for bending is one such method. The internal strain energy of the beam, U, is,

$$U = \int \frac{M^2 dp}{2EI} \tag{1}$$

where M is the moment, E is Young's Modulus, I is the second moment of the area, and p is the distance from the end of the beam[22]. Equation 1 can be plugged into Equation 2 to define

Copyright © 2023 by ASME

the deflection at the end of the beam as δ_i,

$$\delta_i = \frac{\partial U}{\partial F_i} \qquad (2)$$

which is dependent on U and a non-distributed force, F[22].

Another example of predicting force-displacement relationships is Bernoulli-Euler Theory, which states the moment is proportional to the curvature [1]. Through integration, a relationship between force and vertical deflection is established.

2.3 Strain Relationships

There are generally two major categories of strain. The first category is small strain (or infinitesimal strain)[23]. This strain estimation is performed on the initial geometry and is not updated to account for deformation that occurs[24]. Large strain, also known as finite strain or Green-Lagrange Strain, has an extra term that accounts for geometric nonlinearities that may occur during the loading process. The difference between the two is that the large strain equation accounts for geometric nonlinearities that can arise [25][26]. Examples of geometric nonlinearities include contact between boundaries, a "moving mesh, large strain plasticity," and "hyperelastic materials" [26]. These cases are extremely important for multi-material compliant mechanisms. These mechanisms will involve boundary conditions with different material properties that must be preserved. The large deflections may cause plasticity. Additionally, some rubbers and rubber-like elastomers near a Poisson's ratio of nearly 0.5[27] such as TangoBlack+'s can become hyperelastic[28] and incompressible [29] due to the bulk modulus approaching infinity[30]. Both small and large strains must be used together to evaluate the appropriateness of small angle assumptions. These strains can be evaluated to determine relations of stress, loads, and/or Poisson's ratio in the linear elastic region.

Hooke's Law is a proportional relationship between stress and strain

$$\epsilon = \frac{(1+\nu)\sigma - \nu I tr(\sigma)}{E} \qquad (3)$$

where ϵ is this matrix containing strain in three dimensions, σ is a matrix containing the stress in three dimensions, ν is the Poisson's ratio, and I is the identity matrix[30]. When analyzing if a material fits within Hooke's Law, it is important to have a stress-strain curve at hand.

For structures built from additively manufactured polymers, such as TPU 95A, several variables can impact the stress-strain curve for a material. These factors include strain rates[31], printing temperature, testing temperature, and infill [32]. Polymers' mechanical properties are more dependent on strain rate than metals due to stress relaxation[33]. Estimates for the strain at the proportional limit are approximately 0.02 for a range of strain rates and has been shown at $0.006s^{-1}$, $1s^{-1}$, $2s^{-1}$, and $3s^{-1}$[34]. Because metals do not experience as much stress relaxation as polymers do, it is possible to get more uniform stress-strain curves; the strain at the proportional limit for Aluminum, specifically 6061-T, is approximately 0.0023[35].

Evaluating the acceptable strain range in additively-manufactured multi-material parts is essential because it can confirm if the region experiencing the deformation is still within the

bounds of Hooke's Law and if small angle assumptions are appropriate. This is critical to understanding the performance of the parts and allows for a more robust iterative design prior to constructing the part. As a result, we have created a framework to help users understand the performance of their additively manufactured multi-material compliant mechanism.

3. METHODS

This section introduces displacement field, small and large strains with small angle assumptions, small and large strains without small angle assumptions, and Hooke's Law.

3.1 Forces and Deflections in Multi-material Beams

Bernoulli-Euler Beam Deflection Theory. For a multi-material cantilever beam with a purely vertical force at the end, the moment is defined as

$$M = F(L_{fixed} + L_{free} - x), \qquad (4)$$

where M is the moment, F is the force, L_{fixed} is the length of the fixed end of the beam, L_{free} is the length of the free end of the beam, and x is the position of along beam. The moment is at its maximum value, $F(L_{fixed} + L_{free})$ when $x = 0$. The Bernoulli-Euler equation states the moment is proportional to the curvature[1]. The curvature is defined by

$$\frac{d\theta}{ds} = \frac{\frac{d^2y}{dx^2}}{[1 + \frac{dy}{dx}^{\frac{2}{3}}]} \qquad (5)$$

for large and small deflections [1]. When deflections are very small, $\frac{dy}{dx}$ Equation 5 reduces to

$$\theta \approx \frac{dy}{dx}. \qquad (6)$$

This forms the basis of small angle assumptions. which is depicted in Figure 2. A piecewise form of the Bernoulli-Euler Equation is

$$y'' = \theta = \begin{cases} \frac{M}{(EI)_{fixed}}, & \text{if } 0 \leq x \leq L_{fixed} \\ \frac{M}{(EI)_{free}}, & \text{if } L_{fixed} \leq x \leq L_{fixed} + L_{free} \end{cases} \qquad (7)$$

where y'' and θ represent the angle of the beam, M represents the moment, and $(EI)_{fixed}$ and $(EI)_{free}$ represent the rigidity of the fixed and free region, respectively. Within the rigidity term, E is the Young's Modulus and I is the second moment of the area.

Although L_{fixed} and L_{free} are equal as depicted in the beam in Figure 1, the equations outlined in this paper are valid for beams where L_{fixed} and L_{free} are not equal.

Integrating Equation 7 twice and applying boundary conditions, yields

$$\delta_y =$$

Copyright © 2023 by ASME

$$
\left\{
\begin{array}{l}
\dfrac{Fx^2(3(L_{free}+L_{fixed})-x)}{6(EI)_{fixed}}, \text{if } 0 \leq x \leq L_{fixed} \\[2em]
\dfrac{F}{6}\Big(\dfrac{L_{fixed}(3x(2L_{free}+L_{fixed}) - L_{fixed}(3L_{free}+L_{fixed}))}{(EI)_{fixed}} + \\[1.5em]
\qquad \dfrac{(L_{fixed}-x)^2(3L_{free}+L_{fixed}-x)}{(EI)_{free}} \Big), \\[1.5em]
\qquad \text{if } L_{fixed} \leq x \leq L_{fixed} + L_{free}
\end{array}
\right\}
\tag{8}
$$

which represents the end deflection of a dual-material beam experiencing a vertical load. In Equations 4 and 7, M is the moment caused by the vertical force, F.

Castigliano's Theorem. Another method for calculating force-displacement relationships using small angle assumptions is Castigliano's Theorem. A piece-wise approach is used to modify Castigliano's Theorem for a multi-material beam. Equations 2 - 4 can be combined to establish a relationship for the vertical deflection,

$$
\delta_y = \int_0^{L_{free}} \frac{M}{(EI)_{free}} \frac{\partial M}{\partial F} dp + \int_{L_{free}}^{L_{free}+L_{fixed}} \frac{M}{(EI)_{fixed}} \frac{\partial M}{\partial F} dp
\tag{9}
$$

where δ_y is the vertical deflection, M is the moment, EI represents the rigidity of the corresponding section, F is the force, and p represents the distance from the end of the beam as shown in Figure 1. This approach is unique because the analysis of the beam is undertaken from the free end of the beam to the fixed end of the beam, where $p = 0$. As previously stated, this method for calculating the vertical deflection will yield the same result as the Bernoulli-Euler beam theory method; however, it may be more productive to work from the free end of the beam ($p = 0$) or fixed end ($x = 0$) depending on the user's preference. The force can be determined from the vertical deflection if Equation 8 is used or if Equation 4 is substituted into Equation 9.

3.2 Evaluating Deflections

In a multi-material vertically-loaded cantilever beam, deflection will predominately occur within the region with the lowest rigidity, EI, if the rigidity is vastly different between the segments of different materials. As previously mentioned in Figure 2, the location of the flexible region has a large impact on the deflection of the flexible region. The lower the rigidity, the greater the curvature and strain; therefore, this region is where the strain is most likely to exceed the strain at the proportional limit. Therefore, it is essential to determine the four critical points that define the corners of the flexible region in Figure 3. To identify the critical points, start with the bounding points. The bounding points define the start and end of each segment of different rigidities. As a result, the number of bounding points, bp, is

$$
bp = 2n + 4
\tag{10}
$$

where n is the number of segments. This paper is focused on dual-material beams, where the number of bounding points is six, as seen in Table 1. For materials with regions greater than

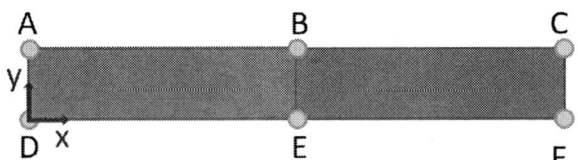

FIGURE 3: CRITICAL POINTS ON MULTI-MATERIAL CANTILEVER BEAM.

two, there will be more than six boundary points, according to Equation 3.2. The critical points are non-"N/A" bounding point values. These critical points surround the region with the lowest rigidity (EI). There can only be four critical points, regardless of the number of segments.

These critical points are the points of interest for the fixed and free positions in Table 1. These points are x and y data points that will be used to determine the horizontal component of deflection, u, where

$$
u = a_0 + a_1 x + a_2 y + a_3 xy
\tag{11}
$$

and the vertical component of deflection, v,

$$
v = b_0 + b_1 x + b_2 y + b_3 xy [36].
\tag{12}
$$

Coefficients are needed to solve Equations 11 and 12. These coefficients will be obtained through solving a system of equations with conditions from Tables 2 or 3. If the more flexible material is in the fixed region, use Table 2; if the more rigid material is in the fixed region, use Table 3. The conditions in each table are divided based on the size of the deflection (SAA vs. LD) and the component of the deflection (u vs. v).

To find u_{SAA} and v_{SAA}, use the Equations 11 and 12 respectively along with the x and y values associated with the points from Table 1. To find u_{LD} and v_{LD}, use the Equations 11 and 12 respectively along with the x and y values associated with the points from Table 1.

Within Table 2 and Table 3, the large deflection data must be supplemented using Finite Element Analysis (FEA) since there is no prior model for obtaining predicted deflections for multi-material cantilever beams. These cells are labeled with "FEA" in Table 2 and Table 3. Note that horizontal deflection occurs, and is characterized by, δ_x at that node's deformed location in your FEA software of choice. In this paper, ABAQUS is used for the hybrid formulation option, which must be used for materials nearing incompressibility [21]. Geometric nonlinearities should also be used for large deflection analysis.

As previously stated, the location of the flexible region has an impact on the displacement characteristics of the beam. When the more flexible material is in the fixed region, Points A and D from Figure 1 remain unchanged in Table 2 because these fixed points do not deflect. The applied vertical deflection (δ_y) accounts for the deflection at the free end of the beam, while $\delta_{y,flex.}$ is the net vertical deflection of the flexible region. Obtaining this value for the large deflection case where the flexible region is in the free region requires using FEA and subtracting any deflection caused by the rigid region. The small-angle case is slightly more

Copyright © 2023 by ASME

complicated. It is possible to obtain a force using the end-of-beam conditions for Equation 8 and δ_y. Since the force in a cantilever is unchanged, this force can be plugged into Equation 8 using the 0 to L_{fixed} conditions to obtain $\delta_{y,flex.}$. Since these equations are built upon small angle assumptions, the resultant displacement is free from nonlinearities associated with large displacement data.

Similar to $\delta_{y,flex.}$, the net horizontal deflection of the flexible region, $\delta_{x,flex.}$, changes values depending on the location of the flexible material. It is important to note that $\delta_{x,flex.}$ must be subtracted from the large deflection δ_x values obtained from FEA; this eliminates any horizontal deflections attributed to the rigid portion of the beam.

After obtaining the coefficients, Equations 11 and 12 will be used to calculate large and small strains for both deflection cases.

3.3 Large and Small Strain for Small vs. Large Deflection Analysis

Strain is defined as the change in length divided by the original length. As a result, the displacement vectors greatly inform calculations for small strain and large strain. The horizontal displacement component, u, and the vertical displacement component, v, from Equations 11 and 12 can be used to find the displacement vector, ∇u, as shown in Equation 13.

Next,

$$\nabla u = \begin{bmatrix} \dfrac{\partial u}{\partial x} & \dfrac{\partial u}{\partial y} \\ \dfrac{\partial v}{\partial x} & \dfrac{\partial v}{\partial y} \end{bmatrix} \tag{13}$$

and its transform,

$$\nabla u^T = \begin{bmatrix} \dfrac{\partial u}{\partial x} & \dfrac{\partial v}{\partial x} \\ \dfrac{\partial u}{\partial y} & \dfrac{\partial v}{\partial y} \end{bmatrix} \tag{14}$$

are needed to calculate

$$\epsilon_{small} = \frac{(\nabla u + \nabla u^T)}{2} \tag{15}$$

and

$$\epsilon_{large} = \frac{(\nabla u + \nabla u^T + \nabla u \nabla u^T)}{2}, \tag{16}$$

where ϵ_{small} is small strain and ϵ_{large} is large strain [23].

After examining the displacement field of the beam section experiencing the greatest curvature and obtaining the strains, these values can be analyzed to determine if small angle assumptions are appropriate.

3.4 Examining the Appropriateness of Small Angle Assumptions

Figure 4 provides a framework for determining the importance of large vs. small strain and the applicability of Hooke's Law within additively manufactured multi-material compliant mechanism. First, the strain values obtained from Section 3.3 will be evaluated to determine whether large or small strain should be used for small angle assumptions. Next, the chosen strain value will be compared to the large and small strains associated with the large deflection case. This will determine if large deflection approaches must be used when calculating deflections. If so, using Hooke's Law and small angle assumptions is automatically inappropriate. If not, the chosen strain must be compared to the strain at the proportional limit. If this chosen strain exceeds the proportional limit strain, Hooke's Law cannot be used. Through this process, the appropriate use of small angle assumptions and Hooke's Law is determined for multi-material compliant mechanisms.

Absolute Difference Between Small and Large Strain Values. Determining if there are geometric nonlinearities that matter lies in the difference between the large and small strain values. If there is a significant difference between the two values, the large strain must be used since it is likely more accurate[37]. If the strain values are close together, it is possible to simplify the values and, thus, we can conclude there are no geometric nonlinearities that are impacting our part. It is typically unlikely that these values will be close together unless the deflection is incredibly small. We recommend using 0.01% highest acceptable strain difference that would make using small strain permissible. This value was obtained by adding the uncertainty of the first strain measurement (0.005%) and the second strain measurement (0.005%). The uncertainty of the strain was obtained through

TABLE 1: THE CRITICAL POINTS ON THE UNDEFLECTED BEAM ARE REPRESENTED BY THESE (X,Y) COORDINATES

Point	Initial Coordinate (x, y)
A	$(0, h)$
B	(L_{fixed}, h)
C	$(L_{fixed}+L_{free}, h)$
D	$(0, 0)$
E	$(L_{fixed}, 0)$
F	$(L_{fixed}+L_{free}, 0)$

TABLE 2: WHEN THE MORE FLEXIBLE MATERIAL IS IN THE FIXED REGION, THESE CONDITIONS CAN BE USED TO SOLVE FOR COEFFICIENTS IN DISPLACEMENT VECTORS.

Points	u_{SAA}	v_{SAA}	u_{LD}	v_{LD}
A & D	0	0	0	0
B & E	0	$\delta_{y,flex.}$	FEA: $\delta_{x,flex.}$	FEA : $\delta_{y,flex.}$
C & F	N/A	N/A	N/A	N/A

TABLE 3: WHEN THE MORE FLEXIBLE MATERIAL IS IN THE FREE REGION, THESE CONDITIONS CAN BE USED TO SOLVE FOR COEFFICIENTS IN DISPLACEMENT VECTORS.

Points	u_{SAA}	v_{SAA}	u_{LD}	v_{LD}
A & D	N/A	N/A	N/A	N/A
B & E	0	0	FEA: $\delta_{x,flex.}$	FEA: $\delta_{y,flex.}$
C & F	0	δ_y	FEA: $\delta_{x,flex.}$	FEA: $\delta_{y,flex.}$

Copyright © 2023 by ASME

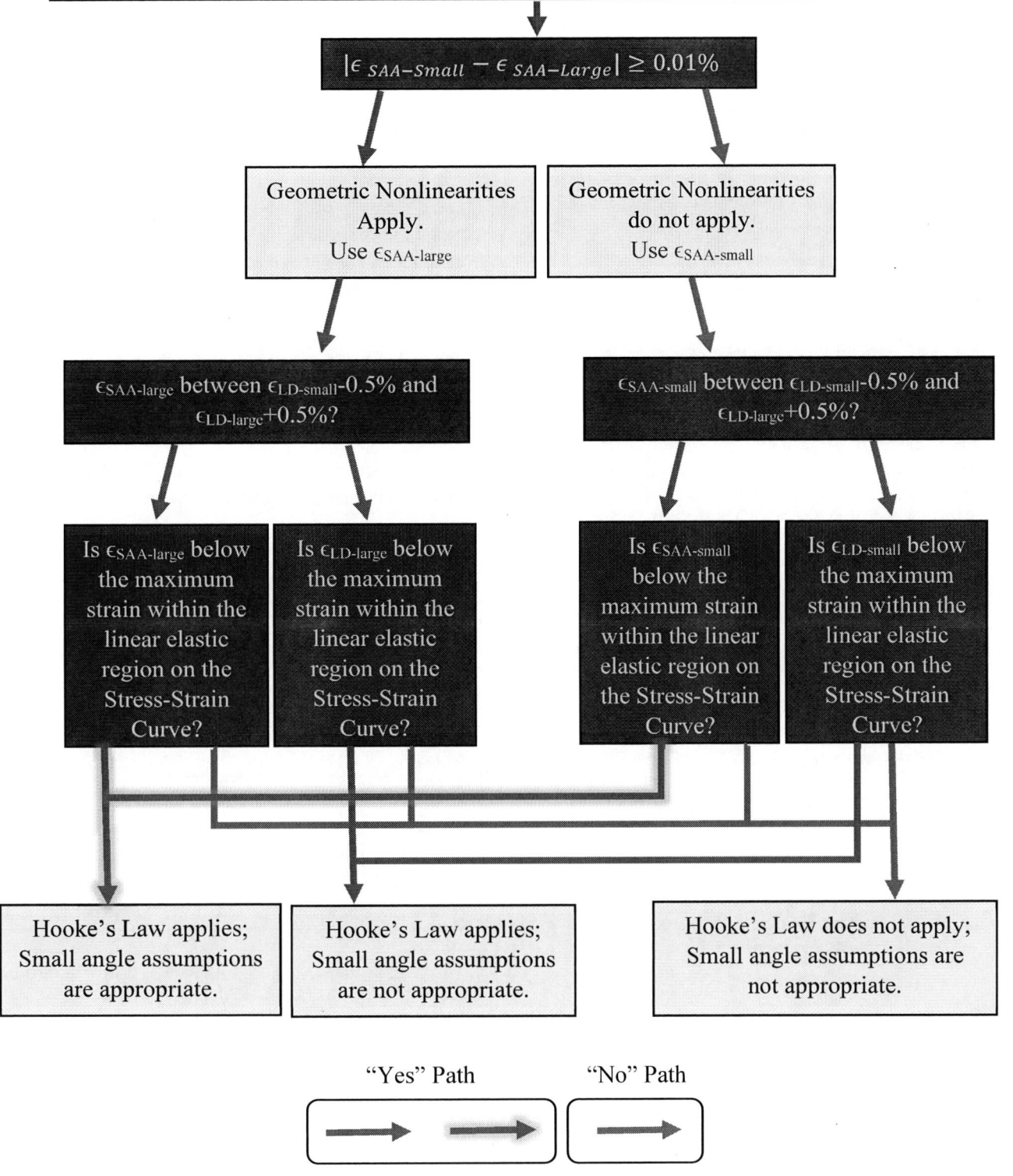

FIGURE 4: THIS FRAMEWORK IDENTIFIES THE BOUNDARIES AND OPPORTUNITIES OF SMALL ANGLE ASSUMPTION (SAA).

Copyright © 2023 by ASME

the resolution of the smallest strain. If the difference between the strains exceeded 0.01%, the geometric nonlinearities are too large to ignore and $\epsilon_{SAA-Large}$ should be used in subsequent steps; if not, $\epsilon_{SAA-Small}$ should be used.

Comparing the Small Angle Assumption Strain Value to Large Deflections. Next, it is important to evaluate the sensibility of the strain chosen in the previous step when compared to the large deflection strains. If it falls between the two strain values obtained for the large deflection case ±0.50%, we can conclude that using small angle assumptions is tentatively appropriate, assuming Hooke's Law is valid. If these assumptions are proven appropriate, Castigliano's Theorem and/or the Bernoulli-Euler Beam Theory equations would be appropriate to use. The ±0.50% is taken from the PRBM's margin of error[1]; however, if the application supports a greater margin of error, the user may decide to deviate from the recommended margin of error. The recommended margin of error was chosen because the strain does not deviate too much from the accepted values of large deflections.

If the small angle assumption strain was not within the large deflection strain range, automatically, small angle assumptions are not appropriate. The large deflection version of the strain must be used instead. For example, if $\epsilon_{SAA-Large}$ was used in the last step and it does not lie within the large deflection strain range, then $\epsilon_{LD-Large}$ must be used for the remainder of the framework.

If the small angle assumption is within the large deflection strain range, then the strain used in the previous step must be evaluated against the maximum strain in the proportional limit.

Hooke's Law. Strain values above the proportional limit cannot be accurately predicted by Hooke's Law and small angle assumptions automatically will not apply [38] This is crucial around the error conditions; it can be possible for small angle assumptions to seem appropriate but lie just outside of the proportional limit. If the maximum strain is exceeded, small angle assumptions are not appropriate. It is possible for Hooke's Law to apply while small angle assumptions are not appropriate, but the inverse will not be true. This framework specifically highlights locations where 1) Hooke's Law applies and small angle assumptions are inappropriate, 2) Hooke's Law applies but small angle assumptions are not appropriate, and 3) Hooke's Law does not apply and small angle assumptions are not appropriate.

4. RESULTS

As a proof of concept for this framework, a 1-D beam with a height of 5 mm, a width of 20 mm, a fixed length of 100 mm, and a free length of 100 mm was created in ABAQUS. These dimensions were chosen because the total length, 200 mm, is under 8" and, therefore, is within the range of typical material extrusion printers. The chosen width was intentionally larger than the chosen height to maximize the efficiency of the bending. Within ABAQUS, hybrid formulation was used in the flexible region due to TPU's Poisson's ratio. ABAQUS automatically uses a strain rate of 0.9[39]. The mesh size was 200. Timoshenko beam elements were used over Euler-Bernoulli beam elements because

Timoshenko beam elements are applicable for stout beams as well as slender beams[21].

For material extrusion, the flexible material was TPU 95A and the rigid material was PLA. These materials were used because they are common material extrusion materials and are low-cost. These materials also have vastly-different Young's Moduli ($E_{PLA} = 2.3465$ GPa [27] and $E_{TPU95A} = 26$ MPa [27]). This allows users to intuitively understand where the maximum curvature will occur, to focus more on the deflection occurring in the flexible region, and, potentially, to approximate the rigid material as a "fixed" region.

For these materials, the differences between the endpoints when using SAA versus LD at increasing deflections are located in Figure 5. Vertical deflections, δ_y, ranging from 0.25 mm to 100 mm were applied to the free end of the beam. The first term, such as "TPU Free" in Figure 5a or "Aluminum Free" in Figure 5b refers to whether the flexible region lies in the free region or the fixed region. The next term is "End" or Middle". "End" refers to the end of the beam and "Middle" refers to the portion of the beam where the two different materials meet. The term "Middle" is appropriate here because the lengths of the two regions are the same. The TPU Free - Middle was not graphed because the

TABLE 4: FRAMEWORK RESULTS WITH FOR THE MEX CASE

Applied Deflection	Flexible Region	ϵ_{SAA} in LD Range?	Hooke's Law?	SAA Acceptable?
0.00025 m	Free	Yes	Yes	Yes
0.0005 m	Free	Yes	Yes	Yes
0.001 m	Free	Yes	Yes	Yes
0.0025 m	Free	Yes	Yes	Yes
0.004 m	Free	Yes	No	No
0.005 m	Free	No	No	No
0.00025 m	Fixed	Yes	Yes	Yes
0.0005 m	Fixed	Yes	Yes	Yes
0.001 m	Fixed	Yes	Yes	Yes
0.0025 m	Fixed	Yes	Yes	Yes
0.005 m	Fixed	Yes	Yes	Yes
0.01 m	Fixed	No	Yes	No
0.015 m	Fixed	No	No	No

TABLE 5: FRAMEWORK RESULTS WITH FOR THE DED CASE

Applied Deflection	Flexible Region	ϵ_{SAA} in LD Range?	Hooke's Law?	SAA Acceptable?
0.00025 m	Free	Yes	Yes	Yes
0.0005 m	Free	Yes	No	No
0.00025 m	Fixed	Yes	Yes	Yes
0.0005 m	Fixed	Yes	Yes	Yes
0.001 m	Fixed	Yes	Yes	Yes
0.0025 m	Fixed	Yes	No	No
0.005 m	Fixed	Yes	No	No
0.01 m	Fixed	No	No	No

Copyright © 2023 by ASME

fixed region here is not of interest; the small angle assumption and large deflection data were even more visually identical than the TPU Free - End - SAA and TPU Fixed - End - SAA data; this is because the rigid region experienced insignificant deflection.

For the first set of trials, the TPU 95A was in the free region while the PLA was in the fixed region. The first data point where Hooke's Law was no longer applicable was $\delta_y = 4$ mm. While the SAA was still in the LD range, SAA is not appropriate due to the Hooke's Law condition found in Figure 4. As a result, the last test where SAA was appropriate was $\delta_y = 2.5$ mm.

For the second set of trials, the PLA was in the free region while the TPU 95A was in the fixed region. Hooke's Law was longer applicable up to $\delta_y = 10$ mm; however, the small angle assumption strain was outside the acceptable range. As a result, the highest deflection where small angle assumptions were valid was $\delta_y = 5$ mm. For small angle assumptions to be valid, all conditions must be "Yes" as shown in Table 4.

For the metals, aluminum ($E = 68.9$ GPa[40]) and steel ($E=$ GPa[41]) were selected due to their compatibility with DED [42]. Additionally, here is a large amount of information concerning their performance and material properties due to their centuries-long use in manufacturing. Initially, deflections of 0.25 mm to 25 mm were applied to the beam and recorded in Table 5. When the aluminum was fixed, the beam was not able to deflect much while still observing Hooke's Law. The last acceptable point when the aluminum was on the free end of the beam was 0.25 mm and the last acceptable point when the aluminum was fixed was 1 mm. Hooke's Law was the limiting criterion for both cases. The strains and displacement regions were smaller because the Young's Moduli were much higher than the polymers' Young's Moduli.

5. DISCUSSION

The number of data points that fit within Hooke's Law was low when the flexible materials were in the free region. This is because the flexible material had to travel further to obtain the applied deflection when compared to cases where the flexible materials were in the fixed region. The added travel increased the strain, which quickly approached the proportional limit's strain value. This was quite apparent in the DED conditions due to the small proportional limit strain within metals, relative to polymers. Increased data points would help the users to evaluate the last acceptable point with greater accuracy.

Despite this, the framework is beneficial because it provides more granularity by accounting for geometric and material non-linearities present in multi-material compliant mechanisms. This framework was tested for material extrusion and directed energy deposition; stress-strain curves for material jetting materials could not be obtained at the time of authorship.

Through Equations 8 and 9, this paper also provided small angle assumption equations created for the vertical deflection in non-constant cantilever beams based on the equations used for constant cantilever beams. These equations can be used to evaluate additively manufactured multi-material compliant mechanisms designs quicker than FEA. If these equations were used outside the recommended cases provided by the framework, error would increase. This is shown by the widening gap between small

Material Extrusion Materials

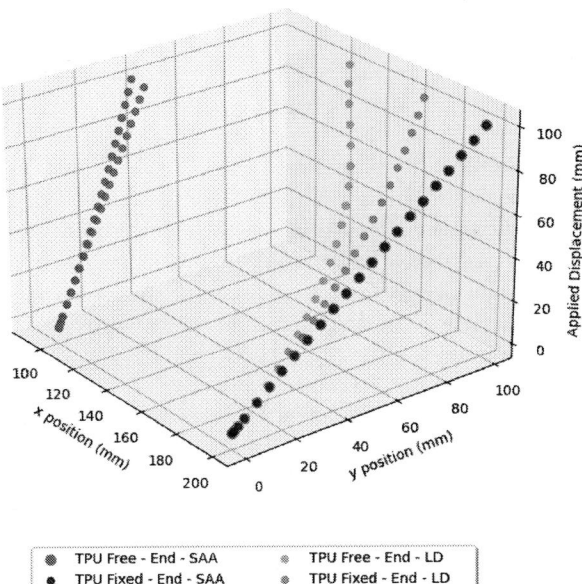

(a) SAA vs. LD data for the MEX materials

Directed Energy Deposition Materials

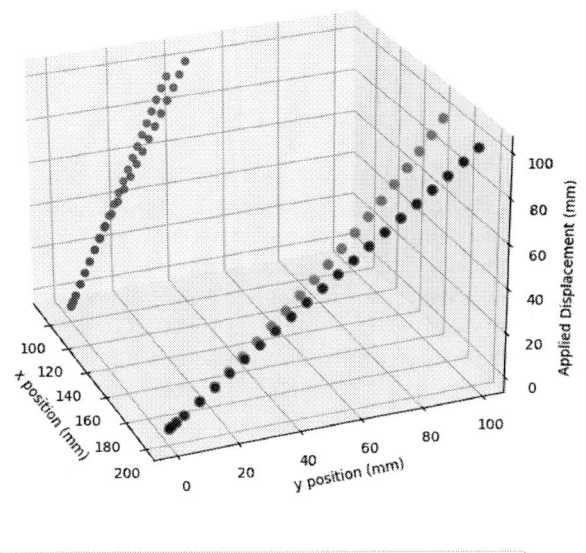

(b) SAA vs. LD data for the DED materials

FIGURE 5: THE GRAPHS DETAIL THE DEFLECTION BEHAVIOR AT KEY POINTS ON THE BEAM.

Copyright © 2023 by ASME

angle assumptions and large deflection in Figure 5. This framework also creates a method for understanding the limitations of Hooke's Law. If someone inappropriately assumed that Hooke's Law ended exactly where small angle assumptions began, they would not be able to use Equation 3 for the material extrusion case where a deflection of 10 mm was applied when TPU was fixed. This is still within the linear elastic region, though the error is slightly beyond the appropriate margin of error for assuming small angle assumptions can be used.

6. CONCLUSIONS

It is important to know where Hooke's Law ends because the law quickly allows stress to be obtained from strain. As a result, the user is able to understand the relationship between stress, strain, and displacement with ease. Using small angle assumptions as much as possible is ideal; however, with compliant mechanisms, minimizing the gap between theoretical and actual performance is more important than simplicity.

This paper creates a guideline to assist with understanding where small angle assumptions for multi-material compliant cantilever beams. It is beneficial to use the material properties and geometry associated with the beams to understand the strain within the beams. The strain elucidates areas where nonlinearities are too large to ignore. This gives the user more confidence in the performance of their multi-material compliant mechanism designs prior to additive manufacturing.

ACKNOWLEDGEMENTS

The authors express their appreciation to The Pennsylvania State University for the Robert W. Graham Endowed Graduate Fellowship, which partially funded this work.

REFERENCES

[1] Howell, L.L. *Compliant Mechanisms*. A Wiley-Interscience publication, Wiley (2001).

[2] Butler, Jared, Morgan, Jessica, Pehrson, Nathan, Tolman, Kyler, Bateman, Terri, Magleby, Spencer P and Howell, Larry L. "Highly compressible origami bellows for harsh environments." *International Design Engineering Technical Conferences and Computers and Information in Engineering Conference*, Vol. 50169: p. V05BT07A001. 2016. American Society of Mechanical Engineers.

[3] Howell, Larry L. "Compliant Mechanisms." McCarthy, J. Michael (ed.). *21st Century Kinematics*: pp. 189–216. 2013. Springer London, London.

[4] Huxman, Connor and Butler, Jared. "A Systematic Review of Compliant Mechanisms as Orthopedic Implants." *Journal of Medical Devices* Vol. 15 No. 4 (2021). DOI 10.1115/1.4052011. URL https://asmedigitalcollection.asme.org/medicaldevices/article-pdf/15/4/040802/6755217/med_015_04_040802.pdf, URL https://doi.org/10.1115/1.4052011. 040802.

[5] Zirbel, Shannon A., Lang, Robert J., Thomson, Mark W., Sigel, Deborah A., Walkemeyer, Phillip E., Trease, Brian P., Magleby, Spencer P. and Howell, Larry L. "Accommodating Thickness in Origami-Based Deployable Arrays1." *Journal of Mechanical Design* Vol.

135 No. 11 (2013). DOI 10.1115/1.4025372. URL https://asmedigitalcollection.asme.org/mechanicaldesign/article-pdf/135/11/111005/6223673/md_135_11_111005.pdf, URL https://doi.org/10.1115/1.4025372. 111005.

[6] Harb, John N., LaFollette, Rodney M., Selfridge, Richard H. and Howell, Larry L. "Microbatteries for self-sustained hybrid micropower supplies." *Journal of Power Sources* Vol. 104 No. 1 (2002): pp. 46–51. DOI https://doi.org/10.1016/S0378-7753(01)00904-1. URL https://www.sciencedirect.com/science/article/pii/S0378775301009041.

[7] Mattson, Christopher A, Howell, Larry L and Magleby, Spencer P. "Development of commercially viable compliant mechanisms using the pseudo-rigid-body model: case studies of parallel mechanisms." *Journal of intelligent material systems and structures* Vol. 15 No. 3 (2004): pp. 195–202.

[8] Perai, Seberang. "Methodology of compliant mechanisms and its current developments in applications: a review." *American Journal of Applied Sciences* Vol. 4 No. 3 (2007): pp. 160–167.

[9] Gibson, Ian, Rosen, David and Stucker, Brent. *Additive Manufacturing Technologies*. Springer New York (2015). DOI 10.1007/978-1-4939-2113-3. URL https://doi.org/10.1007/978-1-4939-2113-3.

[10] Gaynor, Andrew T., Meisel, Nicholas A., Williams, Christopher B. and Guest, James K. "Multiple-Material Topology Optimization of Compliant Mechanisms Created Via PolyJet Three-Dimensional Printing." *Journal of Manufacturing Science and Engineering* Vol. 136 No. 6 (2014). DOI 10.1115/1.4028439. URL https://asmedigitalcollection.asme.org/manufacturingscience/article-pdf/136/6/061015/6267652/manu_136_06_061015.pdf, URL https://doi.org/10.1115/1.4028439. 061015.

[11] Jovanova, Jovana, Frecker, Mary, Hamilton, Reginald F and Palmer, Todd A. "Target shape optimization of functionally graded shape memory alloy compliant mechanisms." *Journal of Intelligent Material Systems and Structures* Vol. 30 No. 9 (2019): pp. 1385–1396. DOI 10.1177/1045389X17733057. URL https://doi.org/10.1177/1045389X17733057, URL https://doi.org/10.1177/1045389X17733057.

[12] Talken, Independence, Liang, Zijuan and Plecnik, Mark. "Approximating Hinges With Multimaterial Compliant Joints." Vol. Volume 8B: 45th Mechanisms and Robotics Conference (MR). 2021. DOI 10.1115/DETC2021-67865. URL https://asmedigitalcollection.asme.org/IDETC-CIE/proceedings-pdf/IDETC-CIE2021/85451/V08BT08A015/6801157/v08bt08a015-detc2021-67865.pdf, URL https://doi.org/10.1115/DETC2021-67865. V08BT08A015.

[13] "List of Stratasys Industrial 3D Printing Materials." URL https://www.stratasys.com/en/materials/materials-catalog/.

[14] Ahn, Dong-Gyu. "Directed energy deposition (DED) process: State of the art." *International Journal of Precision Engineering and Manufacturing-Green Technology* Vol. 8 No. 2 (2021): p. 703–742. DOI 10.1007/s40684-020-00302-7.

Copyright © 2023 by ASME

[15] "Ultimaker Marketplace Materials." URL https://marketplace.ultimaker.com/app/cura/materials.

[16] Li, Xue, He, Xiao-Ting, Ai, Jie-Chuan and Sun, Jun-Yi. "Large Deformation Problem of Bimodular Functionally-Graded Thin Circular Plates Subjected to Transversely Uniformly-Distributed Load: Perturbation Solution without Small-Rotation-Angle Assumption." *Mathematics* Vol. 9 No. 18 (2021). DOI 10.3390/math9182317. URL https://www.mdpi.com/2227-7390/9/18/2317.

[17] Stoner, Brant, Bartolai, Joseph, Kaweesa, Dorcas V., Meisel, Nicholas A. and Simpson, Timothy W. "Achieving functionally graded material composition through bicontinuous mesostructural geometry in material extrusion additive manufacturing." *JOM* Vol. 70 No. 3 (2017): p. 413–418. DOI 10.1007/s11837-017-2669-z.

[18] Zhang, Chi, Chen, Fei, Huang, Zhifeng, Jia, Mingyong, Chen, Guiyi, Ye, Yongqiang, Lin, Yaojun, Liu, Wei, Chen, Bingqing, Shen, Qiang, Zhang, Lianmeng and Lavernia, Enrique J. "Additive manufacturing of functionally graded materials: A review." *Materials Science and Engineering: A* Vol. 764 (2019): p. 138209. DOI https://doi.org/10.1016/j.msea.2019.138209. URL https://www.sciencedirect.com/science/article/pii/S0921509319309955.

[19] Parkinson, Matthew B., Roach, Gregory M. and Howell, Larry L. "Predicting the large deflection path of end-loaded tapered cantilever beams." *American Society of Mechanical Engineers, Pressure Vessels and Piping Division (Publication) PVP* Vol. 415 (2000): pp. 195–200.

[20] Sargent, Brandon S., Ynchausti, Collin R., Nelson, Todd G. and Howell, Larry L. "The Mixed-Body Model: A Method for Predicting Large Deflections in Stepped Cantilever Beams." Vol. Volume 8A: 45th Mechanisms and Robotics Conference (MR). 2021. DOI 10.1115/DETC2021-71332. URL https://asmedigitalcollection.asme.org/IDETC-CIE/proceedings-pdf/IDETC-CIE2021/85444/V08AT08A022/6801895/v08at08a022-detc2021-71332.pdf, URL https://doi.org/10.1115/DETC2021-71332. V08AT08A022.

[21] Smith, Michael. *ABAQUS/Standard User's Manual, Version 6.6*. Dassault Systèmes Simulia Corp, United States (2006).

[22] Gere, J.M. and Goodno, B.J. *Mechanics of Materials*. Cengage Learning (2012).

[23] Bower, Allan. "Kinematics." (2012). URL https://www.brown.edu/Departments/Engineering/Courses/En221/Notes/Kinematics/Kinematics.htm.

[24] Chambon, René. "Small strain vs large strain formulation in computational mechanics: Numerical Modelling in Geomechanics." *Revue française de génie civil* Vol. 6 No. 6 (2002): pp. 1037–1049.

[25] COMSOL. *COMSOL Multiphysics: Structural mechanics module. User's guide.* v 6.1, Comsol (2022).

[26] COMSOL. *COMSOL Multiphysics: Structural mechanics module. User's guide.* v 5.4, Comsol (2018).

[27] Juwita, Adinda Bunga. "Strength and Fatigue Analysis of FDM-fabricated Non-assembly Bi-material Compliant Mechanisms." (2019).

[28] Li, Dong, Dong, Liang and Lakes, Roderic S. "A unit cell structure with tunable Poisson's ratio from positive to negative." *Materials Letters* Vol. 164 (2016): pp. 456–459. DOI https://doi.org/10.1016/j.matlet.2015.11.037. URL https://www.sciencedirect.com/science/article/pii/S0167577X15308661.

[29] Javanmardi, Yousef, Colin-York, Huw, Szita, Nicolas, Fritzsche, Marco and Moeendarbary, Emad. "Quantifying cell-generated forces: Poisson's ratio matters." *Communications Physics* Vol. 4 No. 1 (2021). DOI 10.1038/s42005-021-00740-y.

[30] Bower, Allan. "Hooke's Law." (2012). URL https://www.continuummechanics.org/hookeslaw.html.

[31] Mueller, Jochen, Matlack, Kathryn H., Shea, Kristina and Daraio, Chiara. "Energy absorption properties of periodic and stochastic 3D lattice materials." *Advanced Theory and Simulations* Vol. 2 No. 10 (2019): p. 1900081. DOI 10.1002/adts.201900081.

[32] Wang, Jun, Yang, Bin, Lin, Xiang, Gao, Lei, Liu, Tao, Lu, Yonglai and Wang, Runguo. "Research of TPU materials for 3D printing aiming at non-pneumatic tires by FDM Method." *Polymers* Vol. 12 No. 11 (2020): p. 2492. DOI 10.3390/polym12112492.

[33] Richeton, J., Ahzi, S., Vecchio, K.S., Jiang, F.C. and Adharapurapu, R.R. "Influence of temperature and strain rate on the mechanical behavior of three amorphous polymers: Characterization and modeling of the compressive yield stress." *International Journal of Solids and Structures* Vol. 43 No. 7 (2006): pp. 2318–2335. DOI https://doi.org/10.1016/j.ijsolstr.2005.06.040. URL https://www.sciencedirect.com/science/article/pii/S0020768305003677.

[34] Haid, Daniel Matthias, Duncan, Olly, Hart, John and Foster, Leon. "Characterisation of thermoplastic polyurethane (TPU) for additive manufacturing." .

[35] Yun, Xiang, Wang, Zhongxing and Gardner, Leroy. "Full-Range Stress–Strain Curves for Aluminum Alloys." *Journal of Structural Engineering* (2021).

[36] Ray, Sujit K., Utku, Senol and Wada, Ben K. "Generalization of rectangular element stiffness matrix and thermal load vector associated with a0 + a1x + a2y + a3xy type interpolation rule." *Computers Structures* Vol. 24 No. 6 (1986): pp. 949–951. DOI https://doi.org/10.1016/0045-7949(86)90303-2. URL https://www.sciencedirect.com/science/article/pii/0045794986903032.

[37] FLAC3D. *FLAC3D 7.0 documentation*. Itasca Consulting Group, Inc. (2022).

[38] Langer, Patrick, Jelich, Christopher, Guist, Christian, Peplow, Andrew and Marburg, Steffen. "Simplification of Complex Structural Dynamic Models: A Case Study Related to a Cantilever Beam and a Large Mass Attachment." *Applied Sciences* Vol. 11 No. 12 (2021): p. 5428. DOI 10.3390/app11125428. URL http://dx.doi.org/10.3390/app11125428.

[39] Manual, Abaqus Scripting User's. "Abaqus 6.11." *http://130.149* Vol. 89 No. 2080 (2012): p. v6.

Copyright © 2023 by ASME

[40] Riahi, Mohammad and Nazari, Hamidreza. "Analysis of transient temperature and residual thermal stresses in friction stir welding of aluminum alloy 6061-T6 via numerical simulation." *International Journal of Advanced Manufacturing Technology* Vol. 55 (2011).

[41] Aloraier, Abdulkareem, Al-fadhalah, Khaled, Paradowska, Anna M. and Alfaraj, Emad. "Effect of welding polarity on bead geometry, microstructure, microhardness, and residual stresses of 1020 Steel." *Journal of Engineering Research* Vol. 2 No. 4 (2014). DOI 10.7603/s40632-014-0029-5.

[42] Palmero, Ester M. and Bollero, Alberto. "3D and 4D Printing of Functional and Smart Composite Materials." Brabazon, Dermot (ed.). *Encyclopedia of Materials: Composites.* Elsevier, Oxford (2021): pp. 402–419. DOI https://doi.org/10.1016/B978-0-12-819724-0.00008-2. URL https://www.sciencedirect.com/science/article/pii/B9780128197240000082.

Proceedings of the ASME 2023
International Design Engineering Technical Conferences and
Computers and Information in Engineering Conference
IDETC-CIE2023
August 20-23, 2023, Boston, Massachusetts

DETC2023-115114

EVOLUTIONARY CO-MENTION NETWORK ANALYSIS VIA SOCIAL MEDIA MINING

Phillip A. O. Gavino
Walker Dept. of Mechanical
Engineering
The University of Texas at Austin
Austin, TX 78712-1591

Yinshuang Xiao
Walker Dept. of Mechanical
Engineering
The University of Texas at Austin
Austin, TX 78712-1591

Yaxin Cui
Dept. of Mechanical Engineering
Northwestern University
Evanston, IL 60208

Wei Chen
Dept. of Mechanical Engineering
Northwestern University
Evanston, IL 60208

Zhenghui Sha*
Walker Dept. of Mechanical Engineering
The University of Texas at Austin
Austin, Texas 78712-1591

ABSTRACT

The immense volume of user-generated content on social media provides a rich data source for big data research. Co-mentioned entities in social media content offer valuable information that can support a broad range of studies, from product market competition to dynamic social network mining and modeling. This paper introduces a new approach that combines named entity recognition (NER) and network modeling to extract and analyze co-mention relationships among entities in the same domain from unstructured social media data. This approach contributes to design for market systems literature because little research has investigated product competition via co-mention networks using large-scale unstructured social media data. In particular, the proposed approach provides designers with a new way to gain insight into market trends and aggregated customer preferences when customer choice data is insufficient. More-over, our approach can easily support the evolution analysis of co-mention relationships beyond cross-sectional analysis of co-mention networks in a single year due to the abundance of social media data in multiple years. To demonstrate the approach to

supporting multi-year product competition analysis, we perform a case study on mining co-mention networks of car models with Twitter data. The result shows that our approach can success-fully extract the co-mention relationships of car models in mul-tiple years from 2016 to 2019 from massive Twitter content; and enables us to conduct evolutionary co-mention network analysis with temporal network modeling and descriptive network analy-sis. The analysis confirmed that the co-mention network is capa-ble of identifying frequently discussed entities and topics, such as car model pairs that often involve in competition and emerging vehicle technologies such as electric vehicles (EV). Furthermore, conducting evolutionary co-mention network analysis provides designers with an efficient way to monitor shifts in customer preferences for car features and to track trends in public dis-cussions such as environmental issues associated with EVs over time. Our approach can be generally applied to other studies on co-mention relationships between entities, such as emerging technologies, cellphones, and political figures.

Keywords: Social media content mining; Named entity recognition; co-mention network; Network evolution.

---

*Corresponding author: zsha@austin.utexas.edu

Copyright © 2023 by ASME

1 INTRODUCTION

The growing popularity of social media platforms has led to the accumulation of a vast amount of user-generated content, making it an attractive data source for big data research [1, 2, 3]. The data generated by social media platforms contains a wealth of information, ranging from social connections [4] and public opinions [5] to behavior patterns [6] and engineering design [7, 8, 9], which can be extracted and analyzed for a variety of research purposes. In engineering design, researchers are interested in extracting product feature-related information shared by customers to assist designers in creating products that meet customer preferences. For example, Lim and Tucker developed a Bayesian sampling algorithm to determine optimal search keywords for the accurate extraction of product features from Twitter [8]. In some other domains such as social science, scientists have proposed various approaches to analyze social relationships on social media by modeling followed & following connections or direct communications (e.g., a user posts a direct update to a specific person and engaging in conversation with him/her) [10, 11, 12]. For example, in the study by Huberman et al. (2008) [10], the authors found that the number of friends, had a stronger correlation with user activity than the number of followers. This suggests that when attempting to promote an idea or trend through word of mouth on Twitter, focusing on the number of friends should be the primary strategy.

Besides structured data sources such as the count of followers, post metrics, and comments, there exists additional valuable information embedded within the textual content of social media that remains largely unexplored. One such example is entities co-mentioned in the text, and its theoretical support can trace back to the co-word analysis literature [13, 14]. Researchers revealed that co-word analysis was a powerful tool for discovering associations among research areas in science [13]. In addition, Popovic et al. (2014) extracted the country co-occurrence networks from a large set of financial news and validated its significant overlap with the network built upon the correlation between Credit Default Swaps (CDS) of countries [15]. All of these works demonstrate the value of co-mentioned entity information.

In social media, the entities that are commonly mentioned include people, places, products, organizations/brands, social events, artworks (e.g., music, movies), and terminology [16]. Accordingly, the possible relationships between these entities could be: 1) *association*, denoting that two entities are connected via shared attributes. For example, *My Heart Will Go On* is the theme song for the movie *Titanic*. 2) *Causation.* An example is that the long-term inhalation of specific chemicals is a cause of certain cancers. 3) *Comparison.* For example, two products are co-considered and compared by customers. 4) *Random co-occurrence*, which captures all other undefined relationships. As an example, dealers often announce the arrival of new car models, such as the Toyota Camry, Honda Accord, and Mazda CX-3, and their in-stock status.

Previous research has investigated the potential of co-mention data from different social media platforms, such as car forum posts and Amazon reviews for marketing research [2, 17, 18] and engineering design [19, 20, 21]. The rationale behind these studies is that consumers frequently compare products [22] and their opinions drive content generation and attract visitors [23]. Netzer et al. (2012) work also demonstrated that the co-mention measure from text mining exhibited external validity comparable to that derived from a consumer survey [2]. These findings suggest that co-mention information in user-generated data can provide valuable and reliable insights into product competition for marketing researchers. Meanwhile, knowledge of the structure and evolution of market competition can further benefit the engineering design community [24, 25], as it helps identify customer-desired features through their frequently co-mentioned products.

However, exploring co-mention relationships between entities presents several challenges. First, existing research is based on data collected from specialized platforms, targeting specific groups of users. In contrast, social media platforms (e.g., Twitter) have a more diverse user base and a broader range of product information, but are not fully explored. Second, social media data is often unstructured, making it difficult to accurately extract entity information from short, informal posts that contain emojis, URLs, grammatical errors, and misspellings [26, 27]. The third challenge pertains to characterizing and modeling co-mention relationships and their evolutionary dynamics. Addressing this challenge requires appropriate models to quantitatively and effectively represent the co-mention relationship. To address these challenges, we develop a new approach that integrates named entity recognition (NER) and network modeling to extract and analyze co-mention relationships among entities within the same domain from unstructured social media data. We demonstrate our approach in a case study on co-mention networks of car models from mining multi-year Twitter data.

Compared to existing work, such as the study on co-occurrence networks of car models [2], our study contributes to the literature in two aspects. First, the data source for the existing work is from discussion forums, which are organized by specific topics with more structured and centralized product information. Social media data, on the other hand, could cover discussions beyond pure product comparison, making the extracted co-mention networks more susceptible to noise information. Our approach solved this problem using NER techniques from Natural Language Processing (NLP). Second, this study collects multi-year co-mention network data, enabling the analysis of dynamic competition relations. In particular, we conducted descriptive network analysis on four co-mention networks of car models from 2016 to 2019, thus laying the foundation for our future study on temporal network modeling of product competition in support of design configurations.

The remainder of the paper is organized as follows. In Sec-

Copyright © 2023 by ASME

tion 2, we introduce our method for identifying product data from social media content using NER and the methods for constructing and analyzing product co-mention networks. Then, we present a case study on car models co-mentioned on Twitter in Section 3. In Section 4, the benefits and limitations of our work as well as its implementation for engineering design are discussed. Finally, in Section 5, we summarize the key findings and suggest future research directions.

2 Research Approach

Figure 1 shows an overview of the proposed approach for co-mention network analysis using social media mining. The approach consists of five steps that are detailed in the following sections.

2.1 Step 1: Social Media Data Collection

Step 1 starts with collecting text data from social media. There are two common tools for data collection: 1) social media application programming interfaces (APIs). Most mainstream social media platforms, including Twitter, Facebook, and Instagram, provide APIs that offer developers and researchers access to various features and data, such as posting updates, accessing user data, reading user profiles, and more. 2) Third-party tools/databases. There are several third-party tools and databases that can help collect data from social media and even provide readily available datasets. One such example is snscrape[1], a Python library that facilitates the scraping of data from multiple platforms.

Due to the massive volume of data on social media, it is impractical to collect all available data. Therefore, it is essential to adopt a data collection strategy, which may vary according to research objectives. For the relationship analysis between co-mentioned entities, one collection strategy is a reference-based keyword search, which involves developing a reference list that encompasses all the named entities that are of interest. The subsequent step entails using these entity names as keywords to filter and extract relevant data from a specific time frame.

2.2 Step 2: Text Data Preprocessing

Step 2 involves pre-processing the obtained textual data, which is crucial for the subsequent analysis. As shown in Figure 1, a typical text data processing pipeline includes six sub-steps [28, 29], each of which is described below.

1) **Text cleaning.** This involves removing any noisy characters such as punctuation marks, digits, and emojis. This can be achieved by using regular expressions [30].
2) **Tokenization.** This step is to break up the text into tokens where the notion of the token can be individual words,

phrases, or other essential units depending on the specific objectives of the task at hand [31].
3) **Stop word removal.** Stop words denote common words that are used for the purpose of joining sentences and do not carry much meaning. Examples include "the," "and," "a," and "an." Removing stop words can reduce the size of the text corpus and herein improve the efficiency of downstream tasks [32].
4) **Stemming / Lemmatization.** Stemming and lemmatization are to normalize words to their base form. Stemming refers to a process that reduces words to their root form by crudely removing ends and derivational affixes. Lemmatization is more advanced and properly reduces words to their base form by removing only inflectional endings based on a vocabulary and morphological analysis [33].
5) **Spell-checking and correction.** Spell-checking and correction is the process of identifying misspelled words in a text and replacing them with their correct spelling [34]. This step is essential for improving the accuracy and readability of the text corpus and the data quality for the downstream tasks.
6) **Removing duplicates.** Removing duplicates involves identifying and removing repeated instances of the same text data within a given dataset or document. This process helps to streamline and simplify text analysis tasks by eliminating redundant information [35].

Although this processing pipeline is general, the inclusion and order of steps may differ depending on different task requirements [29]. For instance, stemming or lemmatization might not be relevant or suitable for performing named entity recognition (NER) on Twitter data, given that such data often comprises informal language and slang that do not conform to conventional grammar rules.

2.3 Step 3: Named Entity Recognition (NER)

In Step 3, the objective is to recognize the named entities of interest that appear together in each text sample, that is, a single post (e.g., one tweet) in the case study. This task, commonly known as Named Entity Recognition (NER) in the field of NLP, can be achieved through two methods. The first method involves importing pre-trained NER models from mainstream NLP libraries such as NLTK [2] and spaCy [3]. However, as these models are typically trained on general data, they may not perform optimally on complex social media data. Therefore, the second method is proposed to train a custom model that can be fine-tuned to better identify entities in specific domains [36].

To create a custom NER model using spaCy, three primary steps must be taken. The first step is to prepare the training and testing data, which includes labeling the training and testing data

---

[1] Snscrape: https://github.com/JustAnotherArchivist/snscrape

[2] NLTK: https://www.nltk.org/#
[3] spaCy: https://spacy.io/

Copyright © 2023 by ASME

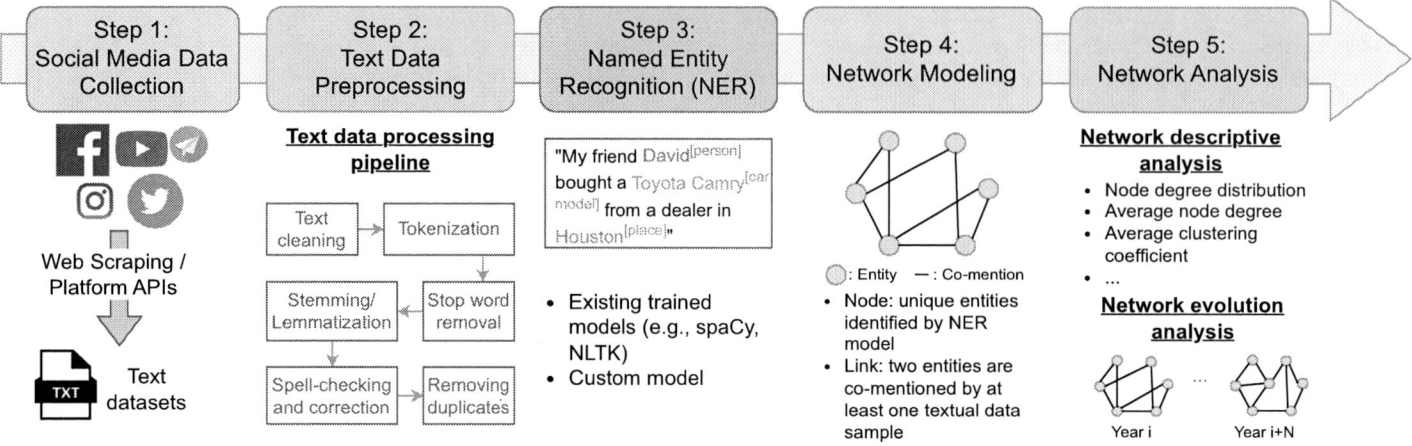

FIGURE 1: Framework of content-based co-mention network mining from social media

and transforming them into the format expected by spaCy. The second step involves modifying the architecture of the spaCy model based on the tasks and specifying the training data and hyperparameters such as batch size and learning rate. In the third step, the model is evaluated using testing data. Evaluation metrics include commonly used machine-learning accuracy scores such as F1-Score, precision, and recall [37].

2.4 Step 4: Network Modeling

In Step 4, the co-mention relationships identified in Step 3 are modeled using a complex network. Complex networks have proven to be a powerful domain-independent representation of complex interactions between entities. For example, social networks have been consistently recognized as an effective means of studying human relationships over the past few decades [38, 39]. Meanwhile, researchers in the engineering domain have also demonstrated the effectiveness of complex networks in system engineering [40] and product design research [41].

Network modeling is a process that encompasses the identification of whether the network is directed or undirected, weighted or unweighted, and the determination of nodes, links, and their corresponding link weights in the case of weighted networks. In the context of co-mention entities, we define nodes as the singular entities identified by the NER model, links as instances where two entities are co-mentioned in at least one text sample, and link weights as the number of times two entities are co-mentioned. Existing work has also used the *lift* value to define link weights [2, 42]. When a causal relationship between entities is unclear, it is reasonable to define the network as undirected.

2.5 Step 5: Network Analysis

In Step 5, following the development of the network model, two types of network analysis are proposed to provide quantitative insight into the co-mention relationships. The first approach involves the application of common network metrics, such as unweighted/weighted degree, network density, global/local average clustering coefficient, and betweenness centrality, to conduct descriptive network analysis [43]. During this process, it is important to connect each metric to the application context for a meaningful interpretation of observed phenomena. For instance, in a co-mention social network, where the co-mentioned entities are individuals' names, individuals with high betweenness centrality can be identified as key connectors between different social clusters.

The second approach is to conduct the network evolution analysis. This is often conducted by performing a time series analysis of the network metrics collected over time and by employing statistical network models to predict network evolution. For example, a temporal exponential random graph model (TERGM) can be used to describe how the probability of edge formation changes over time as a function of various network structures that can incorporate either nodal attributes or edge attributes [24]. A deeper understanding of network evolution obtained from these two types of analysis can support downstream tasks, such as the development of network science-informed deep learning models (e.g., graph neural networks (GNNs)) to predict future co-mention relationships.

3 CASE STUDY

In this section, we present a case study to demonstrate the capability of the proposed approach in co-mention networks of car models using Twitter data in support of the design for vehicle

Copyright © 2023 by ASME

market systems.

3.1 Step 1: Twitter Data Collection

In this study, we collected data from Twitter based on a reference-based keyword search strategy. To begin, we compiled a list of 949 unique car models from 2010 to 2022 by scraping mainstream model names in English from Cars.com. Then, the third-party tool, snscrape, in conjunction with Twitter internal query search function was utilized to collect tweets from 2016 through 2019. Car models from the reference list were the objects searched for in Twitter's database. To allow for consistent samplings across different time periods, a limit of 20 tweets was collected monthly for each car model. This summed up to 240 tweets per car model, with up to 227,760 in total per year. Due to the lack of tweets on specific car models in some months, the number of tweets was less than the maximum number of tweets that are possibly collected. From 2016 to 2019, the number of tweets collected was 86,962; 90,670; 93,861; and 94,302; respectively. The number of tweets increased over the years, influencing the size of the networks generated during this case study.

3.2 Step 2: Twitter Data Preprocessing

The pipeline used for Twitter text data preprocessing in this study is shown in Figure 2. The first step was removing the URLs from the data frames. This was performed at the outset to simplify the subsequent removal of punctuations. If URLs remained in the data frames, they would be fragmented by the punctuation removal step, rendering their deletion challenging. Then, all punctuation marks were removed. Tokenization was conducted in the third step to split each tweet into individual words. Then, the NLTK library was employed to convert all words to lowercase and remove stopwords, such as "a," "the," and "this". Finally, duplicated tweets were removed from the datasets. These duplicates were deemed to have a high probability of being tweeted by bots whose content was meaningless [44]. After deleting duplicates, the total number of tweets kept was 34,278; 36,940; 43,347; and 49,895 from 2016 to 2019, respectively.

3.3 Step 3: Named Entity Recognition (NER) for Twitter Data

To start this process, we first generated training and testing data using the NER Annotator[4], an online annotation tool, to manually mark the car models in tweets that were processed through Step 2 with the label "CAR." This marking method identifies the beginning and ending indices of each entity in a tweet and subsequently converts this information, along with the tweet content, into the format that is expected by spaCy. We used 2,003

---

[4]NER Annotator: https://tecoholic.github.io/ner-annotator/

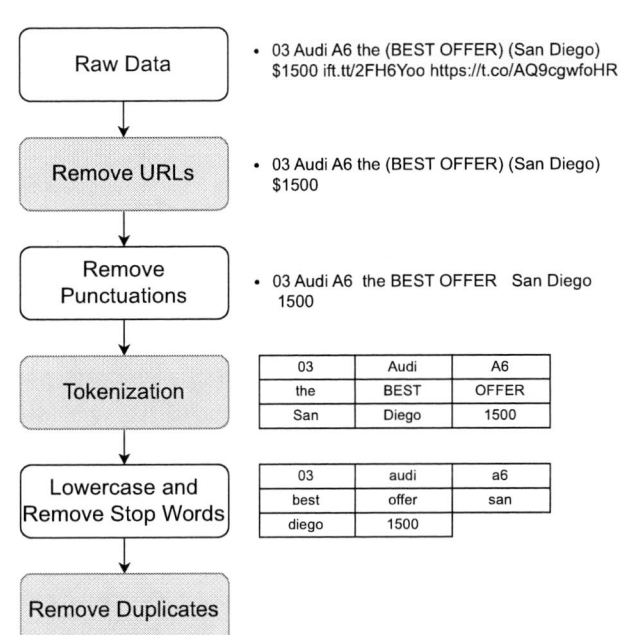

FIGURE 2: Flowchart of the preprocessing method

marked tweets from 2018 as training data and 189, 195, 193, and 193 tweets from 2016 to 2019 as testing data, respectively.

The NER model was then trained using the annotated training data. Upon completion of the training phase, model performance was evaluated using independent test data sets spanning 2016 to 2019. The results of these evaluations were tabulated in Table 1. The results demonstrate that the model achieved F1-scores greater than 69% over the four years, despite being trained solely on data from 2018. Additionally, all precision values were found to be higher than 74%, indicating that more than 74% of the car models recognized by the NER model were correct identification. Furthermore, all recall values exceeded 66%, signifying that the NER model successfully extracted more than 66% of the ground-truth car models (all the manually marked car models within the testing data) from tweets. Finally, we refer to the results of the Twitter Named Entity Recognition shared task associated with the second Workshop on Noisy User-generated Text (W-NUT 2016) to gain general insights into the overall performance of our model. Their results were generated by ten teams. The average F1-Score of these ten NER models was 38.19%, and the highest value was 52.41% [45]. Our model achieved an F1-Score that is approximately 20% higher than the highest F1-Score of the Twitter Named Entity Recognition shared task, justifying the reliability of our trained model.

3.4 Step 4: Twitter Co-Mention Network Modeling

In the process of co-mention network modeling, the cleaned four-year tweet data were processed through the trained NER

Copyright © 2023 by ASME

TABLE 1: The testing results of NER model by year

Year	F1-Score	Precision	Recall
2016	73.25%	80.42%	67.26%
2017	71.50%	77.67%	66.23%
2018	74.83%	74.03%	75.67%
2019	69.96%	74.37%	66.04%

model to identify car model names in each tweet. Subsequently, only tweets containing more than one car model were retained. The resulting count of retained tweets was 4,747; 6,040; 8,408; 11,220 for the years 2016 to 2019, indicating that the percentages of tweets collected that co-mentioned at least two car models were 13.85%, 16.35%, 19.40%, and 22.49%, which is below 50%. This suggests that the co-mention information of cars is dispersed throughout Twitter. Next, given that there existed multiple variant names for some car models in the extracted model name sets, e.g., Ford F 150 being called "Ford F150", "F 150 Ford", and "FordF150", etc., we only generate co-mention connections between identified car models with names consistent with our reference list from Cars.com, resulting in partial entity information loss. For example, we identified three models from one tweet, including "F150 Ford", "Toyota Highlander", and "Subaru Crosstrek". But given that "F150 Ford" differed from the name "Ford F 150" that was recorded in our reference list, we thereby only generated a co-mention connection between Toyota Highlander and Subaru Crosstrek based on this tweet. [5] Our decision to prioritize an accurate network model over one with more information but greater noise reflects a trade-off. In future work, our aim is to develop a more robust similarity algorithm to address this limitation.

Figure 3 illustrates a co-mention network model based on three annotated tweets. The nodes represent unique car models, links signify two car models being co-mentioned in at least one tweet, and link weights denote the total number of tweets that co-mentioned any two models. Note that no sentiment analysis is conducted in this study; thus,the possible relationship between these comorbid car models could be all four possible relationships stated in Section 1, i.e., association, causation, comparison, and random co-occurrence.

3.5 Step 5: Twitter Co-Mention Network Analysis

Figure 4 visualizes the co-mention networks from 2016 to 2019, and the corresponding unweighted degree distributions are

---

[5]We utilized text cosine similarity algorithm [46] to detect car models and their variants. However, this algorithm is not robust against some models, such as Nissan Z which was difficult to distinguish from other Nissan models due to its name in the short letter "Z".

- Tweet 1: "lexus lc 500[CAR] save get porsche 911 gts[CAR]"
- Tweet 2: "say goodbye my old toyota rav4[CAR] thinking buy new ford f150[CAR] chevy silverado 1500[CAR]"
- Tweet 3: "my friends have toyota rav4[CAR] lexus lc 500[CAR] want subaru wrx[CAR] bad raelene first need learn drive"

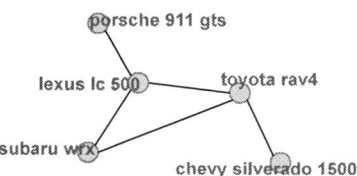

FIGURE 3: An example of co-mention network modeling. These tweets displayed here have undergone processing following Step 2. Nodes are unique car models that were mentioned by all three tweets, and links denote co-mention relationships. For example, a link is built between lexus lc 500 and porsche 911 gts because they were co-mentioned by Tweet 1. We did not include "ford f150" in the network modeling because it is inconsistent with the recalled name listed as "ford f 150" in the reference list.

provided in Figure 5. According to Figure 4, we can observe that the dimensions of the network, including the number of nodes and links, have experienced a marked expansion from 2016 to 2019. This augmentation in the count of nodes signifies an escalation in the number of car models that were co-mentioned in tweets, which can be explained by the number of existing unique car models being cumulative with time. The increasing number of links indicates that more car models are connected through co-mentions. This rise in connections may be the result of a growing interest in car-related topics, such as new automotive technologies (e.g., the surging popularity of electric vehicles), leading to increased discussion.

Furthermore, Figure 5 shows that the degree distributions for both 2016 and 2017 exhibit skewness with a long tail, implying a power law distribution characterized by the fitting curves with R^2 values of 0.862 and 0.812. This suggests the presence of network hubs and the fact that most nodes have relatively few connections. In other words, in 2016 and 2017, a small number of car models were frequently co-mentioned, while the majority of car models (more than 30%) were only co-mentioned once. In contrast, the degree distributions of the 2018 and 2019 networks are spread out (i.e., less skewed than those in 2016 and 2017), indicating that more number of car models were co-mentioned. One possible explanation for this trend is the increase in the overall number of popular car models and heated discussions fueled by the boom of electric vehicles.

By examining the top five hubs, i.e., the top five car models

Copyright © 2023 by ASME

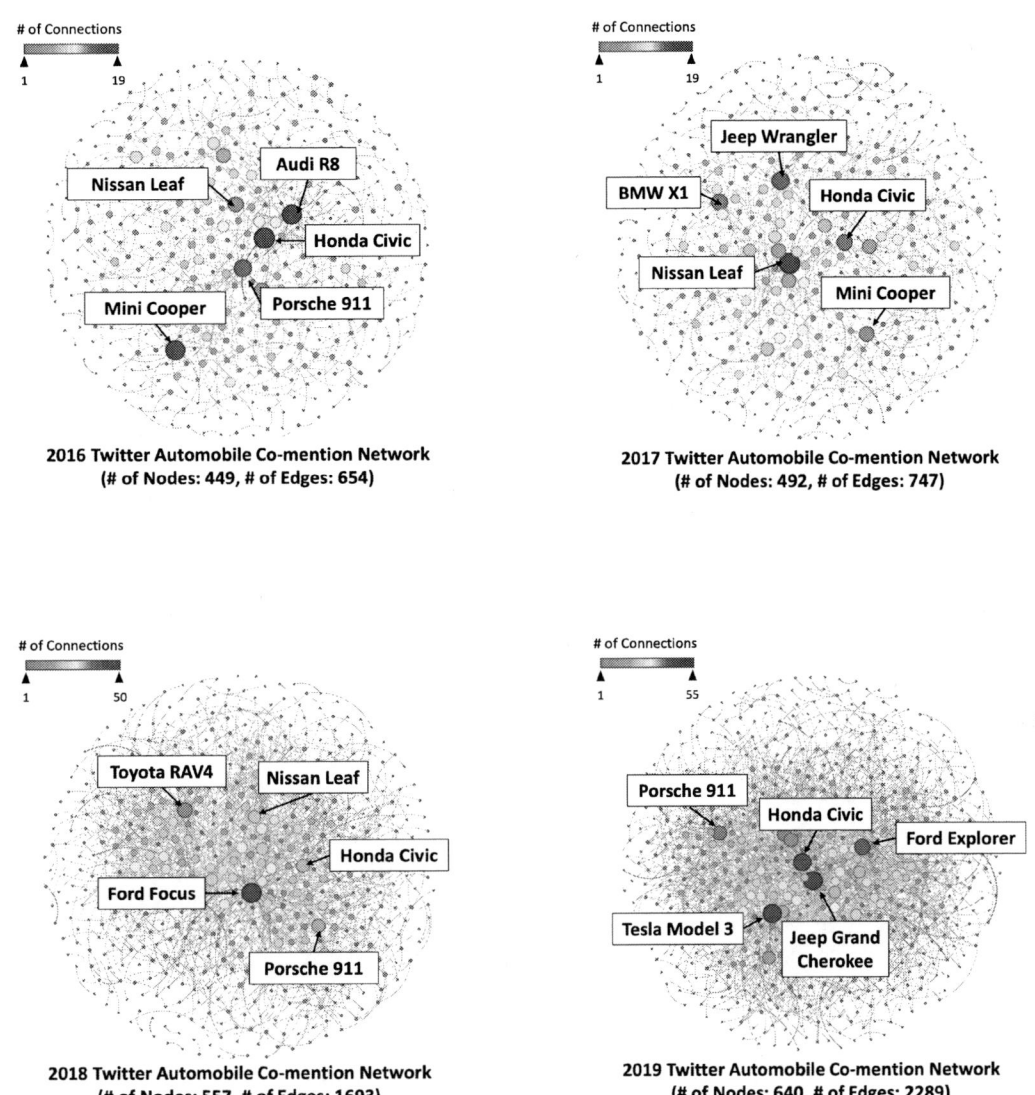

FIGURE 4: Visualizations of co-mention networks from 2016 to 2019. The labeled nodes represent the top-five car models with the most connections in each network, and the corresponding rank information is presented in Table 2.

with the highest unweighted degree, from 2016 and 2019, we observe the evolution of the most frequently discussed car models over time. As presented in Table 2, it is observed that the top five car models vary over time, with Honda Civic being the only model that remains on the list throughout the four years. We also checked the weighted versions of the top five and discovered that the top-one models remain unchanged. For example,

Tesla Model 3 holds the top position in both the unweighted and weighted lists in 2019, with an unweighted degree of 55 and a weighted degree of 79. This suggests that Tesla Model 3 was mentioned alongside 55 distinct models and had a total of 79 comentions.

Furthermore, we compared the car models that were comentioned on social media and their performance in the actual

513

Copyright © 2023 by ASME

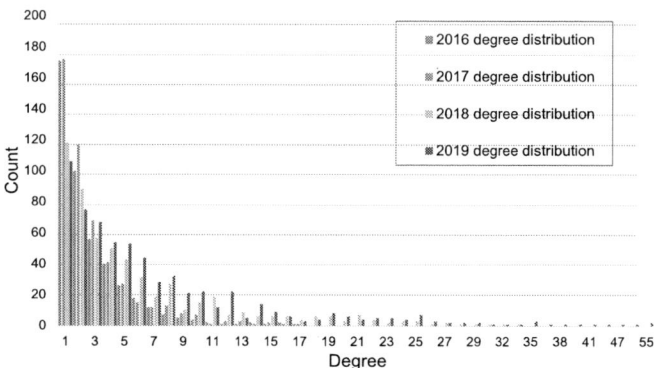

FIGURE 5: Unweighted degree distribution of co-mention networks from 2016 to 2019

sales market, using data of the top five cars with the best sales in 2016 and 2017 from Cars.com [47, 48]. It was observed that the top five best-selling cars were entirely different from the popular cars that were identified in our co-mention networks, indicating whether or not a car model is widely discussed on social media does not necessarily correlate with its sale performance. One possible reason could be that human behavior on social networks can be irrational while purchasing behavior in the real world tends to be more rational and informed by rigorous comparisons. This discrepancy in economics is often described as "talk is cheap." [49]. Another possible reason could be that social media platforms have a diverse user base beyond actual buyers. Hence, while a product or a brand may generate a lot of topics on social media, only a small percentage may come from potential buyers. This further underscores the importance of performing sentiment analysis in our future work, which will allow us to have a better estimate of the type of co-mention relations.

Finally, we identified the car pairs most frequently mentioned in each year, as shown in Table 3, and computed the network metrics associated with each network in Table 4. The results in Table 3 indicate that our approach can successfully capture the car models that were frequently co-mentioned in particular market segments. For example, we found the co-mention of luxury midsize SUVs, BMW X5 vs. Volvo XC90 in 2016, the co-mention of electric vehicles, Tesla Model 3 vs. Chevrolet Bolt EV in 2017, and the co-mention of Jeeps Wrangler and Wrangler Unlimited in 2019. In addition, the trend of the market can also be inferred from the data. For example, the frequent association between Tesla Model 3 and Chevrolet Bolt EV in 2017 serves as evidence of the increasing popularity of electric vehicles.

Regarding the results of the network metrics, the density of a network refers to the proportion of actual connections in a network compared to the total number of possible connections. The reported densities of the co-mention networks are very low (less than 1% on average), indicating sparse car co-mention relations

in Twitter data. The weighted and unweighted average degrees measure the average number of co-mentioning occurrences and the average number of co-mentioned car models of a car. The results show that both measures have increased from 2016 to 2019. The local clustering coefficient is computed for each node and indicates how likely the neighbors of a node are also connected. It measures a network's local link density: The more densely interconnected the neighborhood of a node, the higher its local clustering coefficient. The average local clustering coefficients capture the degree of clustering of a whole network by taking the average of local clustering coefficients. The increased local clustering coefficient observed in 2018 and 2019 suggests that there was a greater tendency for a group of car models to be discussed together, compared to 2016 and 2017. This trend may be attributed to the segmentation of car-related discussions on Twitter, where certain car models from the same segment are discussed more frequently. For instance, the co-occurrence of SUV vs. SUV, rather than SUV vs. Sedan, could reflect the clustering of discussions around SUVs during this period.

4 DISCUSSION

In this section, we discuss two benefits of evolutionary co-mention network analysis for product designers and marketing stakeholders, two limitations of this study that motivate our future research, as well as the potential of the proposed approach in support of product design.

Benefits There are two implications of the evolutionary co-mention network analysis. First, the analysis can help designers track changes in customer preferences for car features and emerging technology. In particular, our analysis indicates an increasing co-mention of electric vehicles from 2016 to 2019. This could be due to the rapid development of EV technologies in the last decade which has attracted a lot of attention on social networks. Second, co-mention relations between brands can provide insights into consumer perceptions and potential market competition structures [2]. Although our current results could not directly imply competition relations, the data generated from this preliminary work provide a foundation for our future work on dynamic competition analysis.

Limitations The current study has some limitations that need to be addressed in our future work. As highlighted in Section 3.4, the primary drawback of the proposed approach is its inability to standardize the names of the vehicle models extracted. This has led to the presence of multiple variant names for certain car models, which could introduce noise into our network. To mitigate this issue, we opted to incorporate only models that have names included in our reference list. While this decision allows us to maintain greater consistency and accuracy in our analysis, it sacrifices some co-mention relationships, making them left un-

Copyright © 2023 by ASME

TABLE 2: Top five car models with the largest unweighted degree (UWD)

2016		2017		2018		2019	
Model	UWD	Model	UWD	Model	UWD	Model	UWD
Honda Civic	19	Nissan Leaf	19	Ford Focus	50	Jeep Grand Cherokee	55
Mini Cooper	18	Jeep Wrangler	16	Toyota RAV4	37	Tesla Model 3	55
Audi R8	18	Honda Civic	15	Porsche 911	33	Honda Civic	51
Porsche 911	16	BMW X1	15	Honda Civic	32	Ford Explorer	47
Nissan Leaf	14	Mini Cooper	14	Nissan Leaf	31	Porsche 911	42

TABLE 3: The most frequently co-mentioned car pairs by year

Year	Linked Car Models	# of Co-mentions
2016	BMW X5 vs. Volvo XC90	6
2017	Tesla Model 3 vs. Chevrolet Bolt EV	8
2018	Buick Envision vs. Cadillac CT6	13
2019	Jeep Wrangler vs. Jeep Wrangler Unlimited	11

TABLE 4: Twitter co-mention network metrics by year

Year	2016	2017	2018	2019
Density	0.007	0.006	0.011	0.012
Unweighted Avg. Deg.	2.913	3.037	6.079	7.466
Weighted Avg. Deg.	3.523	3.793	8.047	9.934
Avg. Local Cluster Coeff.	0.125	0.128	0.229	0.249

mined. In the future, a robust text similarity algorithm is required to address this issue. A second limitation of this study is that we did not classify the co-mention relationships into more granular categories or perform spam detection. As a result, our generated networks may contain instances of *random co-occurrence* relationships or advertising information. To achieve a more impartial understanding of the co-mention relationships among car models, additional noise removal techniques may be required to enhance the precision of our findings.

Implementation for engineering design As illustrated in Figure 6, the mined co-mention network from social media data enables us to uncover potential competition relationships between products. By conducting sentiment analysis, we can categorize the co-mention relationships and extract a sub-network dedicated to product competition. In our previous studies, similar co-consideration networks generated by customer choice sets have proven to be effective for demand analysis, which can inform better design decisions [24, 41, 42]. For instance, in a case study on the vehicle market, an exponential random graph model (ERGM) was developed by incorporating the competition network data and the node (car) attributes data such as engine power to estimate the effects of different attributes on the formulation and evolution of competitive relationships. The estimated ERGM can be further integrated into the decision-based design (DBD) framework [50, 51] to use optimization to determine the preferred design alternative. Therefore, the proposed co-mention network mining approach from social media data, i.e., the starting point of this competition network-based design framework, plays a crucial role in providing the competition network input for enterprise-driven design decisions.

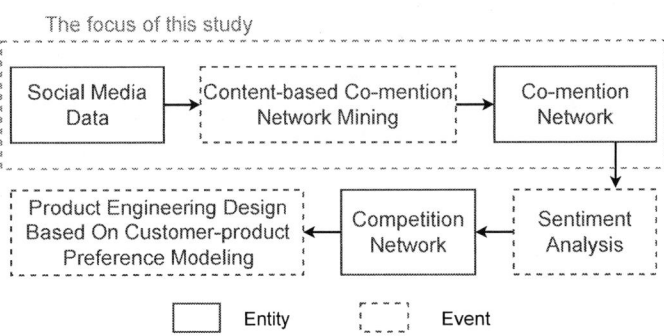

FIGURE 6: The role of this work in competition network-based design framework

Copyright © 2023 by ASME

5 CONCLUSION

This paper proposes an approach that combines NER and network modeling to extract and analyze co-mention relations among entities embedded in social media data. The proposed method is demonstrated using a case study on a car model co-mention network. Despite the challenges posed by the unstructured and noisy nature of Twitter data, our NER algorithm achieved an F1 score of 70%, a performance that is better than the state-of-the-art. The evolving network models showed that the co-mention network is capable of identifying the competing car models that are frequently discussed on social networks and trending vehicle technologies, such as electric vehicles. Furthermore, the results indicate that the popularity of a product on social media may not necessarily indicate its sales performance. This study offers a unique data mining approach for marketing researchers and product designers in the study of the evolution of marketing structures (competition networks).

In future work, we plan to improve our approach by developing a robust text similarity algorithm that can effectively deal with variant names for specific car models. Also, we plan to carry out a sentiment analysis on social media data to detect spam and classify co-mention relationships into more detailed categories. Lastly, we intend to employ temporal network modeling techniques such as TERGM to identify key factors that affect the formation of co-mention relationships and predict products' future co-mention networks in support of product design and development.

ACKNOWLEDGMENT

The authors acknowledge collaborators Noshir S Contractor, Johan Koskinen, Neelam Modi, and Jonathan Haris Januar for their inputs provided during research meetings. We also greatly acknowledge the funding support from NSF CMMI #2005661 and #2203080.

REFERENCES

[1] Zafarani, R., Abbasi, M. A., and Liu, H., 2014. *Social media mining: an introduction.* Cambridge University Press.

[2] Netzer, O., Feldman, R., Goldenberg, J., and Fresko, M., 2012. "Mine your own business: Market-structure surveillance through text mining". *Marketing Science, 31*(3), pp. 521–543.

[3] Aichner, T., Grünfelder, M., Maurer, O., and Jegeni, D., 2021. "Twenty-five years of social media: a review of social media applications and definitions from 1994 to 2019". *Cyberpsychology, behavior, and social networking, 24*(4), pp. 215–222.

[4] del Fresno García, M., Daly, A. J., and Segado Sanchez-Cabezudo, S., 2016. "Identifying the new influences in the internet era: Social media and social network analysis.". *Revista Española de Investigaciones Sociológicas*(153).

[5] Jungherr, A., 2016. "Twitter use in election campaigns: A systematic literature review". *Journal of information technology & politics, 13*(1), pp. 72–91.

[6] De Choudhury, M., Gamon, M., Counts, S., and Horvitz, E., 2021. "Predicting Depression via Social Media". *Proceedings of the International AAAI Conference on Web and Social Media, 7*(1), Aug., pp. 128–137.

[7] Stone, T., and Choi, S.-K., 2013. "Extracting consumer preference from user-generated content sources using classification". In International Design Engineering Technical Conferences and Computers and Information in Engineering Conference, Vol. 55881, American Society of Mechanical Engineers, p. V03AT03A031.

[8] Lim, S., and Tucker, C. S., 2016. "A bayesian sampling method for product feature extraction from large-scale textual data". *Journal of Mechanical Design, 138*(6).

[9] Singh, A. S., and Tucker, C. S., 2015. "Investigating the heterogeneity of product feature preferences mined using online product data streams". In International Design Engineering Technical Conferences and Computers and Information in Engineering Conference, Vol. 57083, American Society of Mechanical Engineers, p. V02BT03A020.

[10] Huberman, B. A., Romero, D. M., and Wu, F., 2008. "Social networks that matter: Twitter under the microscope". *arXiv preprint arXiv:0812.1045.*

[11] Tang, L., and Liu, H., 2011. "Leveraging social media networks for classification". *Data Mining and Knowledge Discovery, 23*, pp. 447–478.

[12] Yoo, E., Rabinovich, E., and Gu, B., 2020. "The growth of follower networks on social media platforms for humanitarian operations". *Production and Operations Management, 29*(12), pp. 2696–2715.

[13] He, Q., 1999. "Knowledge discovery through co-word analysis". *Libr. Trends.*

[14] Chen, X., Chen, J., Wu, D., Xie, Y., and Li, J., 2016. "Mapping the research trends by co-word analysis based on keywords from funded project". *Procedia computer science, 91*, pp. 547–555.

[15] Popović, M., Štefančić, H., Sluban, B., Kralj Novak, P., Grčar, M., Mozetič, I., Puliga, M., and Zlatić, V., 2014. "Extraction of temporal networks from term co-occurrences in online textual sources". *PloS one, 9*(12), p. e99515.

[16] Room, C., 2020. "Named entity recognition". *Algorithms, 8*(3), p. 48.

[17] Won, E. J. S., Oh, Y. K., and Choeh, J. Y., 2018. "Perceptual mapping based on web search queries and consumer forum comments". *International Journal of Market Research, 60*(4), pp. 394–407.

[18] Won, E. J., Oh, Y. K., and Choeh, J. Y., 2023. "Analyzing

Copyright © 2023 by ASME

competitive market structures based on online consumer-generated content and sales data". *Asia Pacific Journal of Marketing and Logistics, 35*(2), pp. 307–322.

[19] Jin, J., Ji, P., and Gu, R., 2016. "Identifying comparative customer requirements from product online reviews for competitor analysis". *Engineering Applications of Artificial Intelligence, 49*, pp. 61–73.

[20] Suryadi, D., and Kim, H. M., 2019. "A data-driven methodology to construct customer choice sets using online data and customer reviews". *Journal of Mechanical Design, 141*(11).

[21] Joung, J., and Kim, H. M., 2021. "Approach for importance–performance analysis of product attributes from online reviews". *Journal of Mechanical Design, 143*(8).

[22] Pang, B., Lee, L., et al., 2008. "Opinion mining and sentiment analysis". *Foundations and Trends® in Information Retrieval, 2*(1–2), pp. 1–135.

[23] Schindler, R., and Bickart, B., 2005. *Published word of mouth: Referable, consumer-generated information on the internet.* Lawrence Erlbaum Associates, Jan., pp. 32–57.

[24] Xie, J., Bi, Y., Sha, Z., Wang, M., Fu, Y., Contractor, N., Gong, L., and Chen, W., 2020. "Data-driven dynamic network modeling for analyzing the evolution of product competitions". *Journal of Mechanical Design, 142*(3), p. 031112.

[25] Wang, M., Sha, Z., Huang, Y., Contractor, N., Fu, Y., and Chen, W., 2018. "Predicting product co-consideration and market competitions for technology-driven product design: a network-based approach". *Design Science, 4*, p. e9.

[26] Li, C., Weng, J., He, Q., Yao, Y., Datta, A., Sun, A., and Lee, B.-S., 2012. "Twiner: named entity recognition in targeted twitter stream". In Proceedings of the 35th international ACM SIGIR conference on Research and development in information retrieval, pp. 721–730.

[27] Suman, C., Reddy, S. M., Saha, S., and Bhattacharyya, P., 2021. "Why pay more? a simple and efficient named entity recognition system for tweets". *Expert Systems with Applications, 167*, p. 114101.

[28] Pradha, S., Halgamuge, M. N., and Vinh, N. T. Q., 2019. "Effective text data preprocessing technique for sentiment analysis in social media data". In 2019 11th international conference on knowledge and systems engineering (KSE), IEEE, pp. 1–8.

[29] Kathuria, A., Gupta, A., and Singla, R., 2021. "A review of tools and techniques for preprocessing of textual data". *Computational Methods and Data Engineering: Proceedings of ICMDE 2020, Volume 1*, pp. 407–422.

[30] Goyvaerts, J., 2017. "Regular expression tutorial-learn how to use regular expressions". *Available online: www. regular-expressions. info (accessed on 31 October 2021).*

[31] Webster, J. J., and Kit, C., 1992. "Tokenization as the ini-

tial phase in nlp". In COLING 1992 volume 4: The 14th international conference on computational linguistics.

[32] Nothman, J., Qin, H., and Yurchak, R., 2018. "Stop word lists in free open-source software packages". In Proceedings of Workshop for NLP Open Source Software (NLP-OSS), pp. 7–12.

[33] Schütze, H., Manning, C. D., and Raghavan, P., 2008. *Introduction to information retrieval*, Vol. 39. Cambridge University Press Cambridge.

[34] Hládek, D., Staš, J., and Pleva, M., 2020. "Survey of automatic spelling correction". *Electronics, 9*(10), p. 1670.

[35] Ramya, R., and Venugopal, K., 2016. "Feature extraction and duplicate detection for text mining: A survey". *Global Journal of Computer Science and Technology, 16*(C5), pp. 1–20.

[36] Shelar, H., Kaur, G., Heda, N., and Agrawal, P., 2020. "Named entity recognition approaches and their comparison for custom ner model". *Science & Technology Libraries, 39*(3), pp. 324–337.

[37] Brownlee, J., 2018. *Better deep learning: train faster, reduce overfitting, and make better predictions.* Machine Learning Mastery.

[38] Wasserman, S., and Faust, K., 1994. *Social network analysis: Methods and applications*, Vol. 8. Cambridge university press.

[39] Freeman, L., 2004. "The development of social network analysis". *A Study in the Sociology of Science, 1*(687), pp. 159–167.

[40] Xiao, Y., and Sha, Z., 2022. "Robust design of complex socio-technical systems against seasonal effects: a network motif-based approach". *Design Science, 8*, p. e2.

[41] Cui, Y., Ahmed, F., Sha, Z., Wang, L., Fu, Y., Contractor, N., and Chen, W., 2022. "A weighted statistical network modeling approach to product competition analysis". *Complexity, 2022*, pp. 1–16.

[42] Sha, Z., Huang, Y., Fu, J. S., Wang, M., Fu, Y., Contractor, N., and Chen, W., 2018. "A network-based approach to modeling and predicting product coconsideration relations". *Complexity, 2018*, pp. 1–14.

[43] Wang, M., Sha, Z., Huang, Y., Contractor, N., Fu, Y., and Chen, W., 2016. "Forecasting technological impacts on customers' co-consideration behaviors: a data-driven network analysis approach". In International Design Engineering Technical Conferences and Computers and Information in Engineering Conference, Vol. 50107, American Society of Mechanical Engineers, p. V02AT03A040.

[44] Tao, K., Abel, F., Hauff, C., Houben, G.-J., and Gadiraju, U., 2013. "Groundhog day: near-duplicate detection on twitter". In Proceedings of the 22nd international conference on World Wide Web, pp. 1273–1284.

[45] Strauss, B., Toma, B., Ritter, A., De Marneffe, M.-C., and Xu, W., 2016. "Results of the wnut16 named entity recog-

Copyright © 2023 by ASME

nition shared task". In Proceedings of the 2nd Workshop on Noisy User-generated Text (WNUT), pp. 138–144.

[46] Rahutomo, F., Kitasuka, T., and Aritsugi, M., 2012. "Semantic cosine similarity". In The 7th international student conference on advanced science and technology ICAST, Vol. 4, p. 1.

[47] Cars.com, 2016. What were the best-selling cars in 2016? https://www.cars.com/articles/what-were-the-best-selling-cars-in-2016-1420692870639/. Accessed on March 12, 2023.

[48] Cars.com, 2017. What were the best-selling cars in 2017? https://www.cars.com/articles/what-were-the-best-selling-cars-in-2017-1420698582900/. Accessed on March 12, 2023.

[49] Farrell, J., and Rabin, M., 1996. "Cheap talk". *Journal of Economic perspectives,* *10*(3), pp. 103–118.

[50] Chen, W., Hoyle, C., and Wassenaar, H. J., 2013. *Decision-based design: Integrating consumer preferences into engineering design.* Springer.

[51] Sha, Z., Cui, Y., Xiao, Y., Stathopoulos, A., Contractor, N., Fu, Y., and Chen, W., 2023. "A network-based discrete choice model for decision-based design". *Design Science,* *9*, p. e7.

Copyright © 2023 by ASME

Proceedings of the ASME 2023
International Design Engineering Technical Conferences and
Computers and Information in Engineering Conference
IDETC-CIE2023
August 20-23, 2023, Boston, Massachusetts

DETC2023-117051

EXPLORING HOW THE DESIGN HIERARCHY OF NEEDS EXPLAINS CORRELATIONS BETWEEN DESIGN DECISIONS AND ONLINE CUSTOMER REVIEW RATINGS

Lisa Retzlaff
Graduate Research Assistant
North Carolina State University
Raleigh, NC 27695, USA
ljretzla@ncsu.edu

Scott Ferguson
Associate Professor
North Carolina State University
Raleigh, NC 27695, USA
scott_ferguson@ncsu.edu

ABSTRACT

Understanding and meeting customer needs plays a significant role in the success of released products. Consumer shopping behavior has also moved toward an online experience, where customers learn of, read about, and make selections based on what they see online. Customer reviews offer a wide range of commentary about the shopping, delivery, and product use experience. The comments specific to product use can provide significant insights into where a product is failing to meet (or exceeds) expectations. In this paper, we establish a frame that the impact of design decisions is often reflected in these comments, and that the star rating a product receives can be correlated to how well a product fulfills the Design Hierarchy of Needs. The Design Hierarchy of Needs is an adaptation of Maslow's Hierarchy of Needs, where needs are associated with the levels of functionality, reliability, usability, proficiency, and creativity. The tenet of the Design Hierarchy of Needs is that a design must first serve lower-level needs (functionality) before addressing higher-level needs (creativity). In this paper, we explore five hypotheses that are posed with the goal of exploring how the Design Hierarchy of Needs can be used for explaining online review star ratings by identifying where products fail to address (or exceed) needs. Reviews from five products across multiple star ratings are coded, and these codes are analyzed to determine the frequency of categories where negative comments about needs are being addressed. We show that the number of negative comments addressing functionality and reliability decrease as the star ratings increase, and that 1- and 2-star reviews are dominated by negative comments about these need levels. We also find that as the star ratings increase, negative comments are focused on the hierarchical levels of proficiency and usability, reflecting the

tradeoffs that customers perceive themselves making while using the product. Finally, we also show that positive comments about proficiency and creativity primarily exists only in 4- and 5-star reviews. We believe that these findings can help drive conversations around design decisions when considering the extent to which needs should be addressed, particularly when tied to product cost, and the tradeoffs that are made when addressing mid-tier levels of the hierarchy. This work further establishes that design decisions must be made in the context of a market system, where need fulfilment, cost and performance, and the online shopping experience must be simultaneously considered.

Keywords: Design Hierarchy of Needs, customer needs, online customer reviews

1. INTRODUCTION

Meeting customer needs and defining a customer's experience with a product are central aspects of engineering design. Building brand loyalty, maintaining customer retention and a reputation of satisfaction, and ultimately promoting innovation requires that companies and designers understand the needs of their customers. However, when customer needs are not met, these aspects of company success will deteriorate over time [1].

We know that it is challenging to meet all customer needs, especially because they may be in conflict. Customers have become more reliant on online shopping, and it is necessary to consider the effects of product reviews and star ratings. We seek to understand how customer needs relate to reviews so we can design products that perform better in online environments. The

Copyright © 2023 by ASME

motivation for this work is based on Maslow's Hierarchy of Needs, shown in Figure 1. Maslow asserted that humans have needs that can be categorized into five hierarchical levels of psychological (food, water), safety (security), belonging (friends), self-esteem (feelings of accomplishment), and self-actualization (achieving one's full potential) [2]. Inspired by Maslow's theory, the Design Hierarchy of Needs states that for the design of a product to be successful, the design must satisfy basic needs before it can meet higher level needs. Maslow's Hierarchy of Needs was re-envisioned to the space of product design where the updated hierarchical levels are functionality, reliability, usability, proficiency, and creativity, as shown in Figure 2 [3]. The importance of this hierarchy is emphasized in how it can show that a failure to meet basic needs can ultimately undermine the product even if exceptional aspects exist. This paper explores how the Design Hierarchy of Needs can be used for explaining online review star ratings by identifying where products fail to address (or exceed) needs. We use five hypotheses to explore the potential impact of applying the Design Hierarchy of Needs to understanding online customer review ratings and the potential ramifications that these insights give us about design decisions.

Hypothesis 1: The number of negative functionality, reliability, and usability comments will decrease as the star ratings of reviews increase.

Hypothesis 2: A defining factor of 1- and 2-star reviews are negative comments addressing functionality and reliability since they fail to lay an effective foundation within the Design Hierarchy of Needs.

Hypothesis 3: Customers who leave 4- and 5-star reviews focus more on negative proficiency aspects because they are able to consider ways to optimize their interaction with the product.

Hypothesis 4: Negative comments about usability appear more often in reviews that do not contain negative comments about functionality and reliability.

Hypothesis 5: There will be a significant number of positive comments addressing proficiency and/or creativity in 4- and 5-star reviews compared to 1 through 3 star reviews.

2. BACKGROUND

Maslow's Hierarchy of Needs, as shown in Figure 1, is a theory within the domain of psychology that describes human needs using 5 hierarchical levels [2]. Maslow establishes that before humans can begin satisfying needs at a higher level of the hierarchy, needs in the lower portion of the hierarchy must be satisfied first. Researchers have explored how the Hierarchy of Needs can be used for explaining consumer purchase motivation and behavior [4] and for guiding product design [5–7]. Yalch and Brunel, for example, applied Maslow's theory and studied if consumers would assess plain products as being comparable to more aesthetic products when considering if basic needs were

met [5]. They examined the usefulness of need hierarchies within the domain of product design through survey responses. However, participants in the study were not able to physically interact with the selected products, they only viewed images of them. Furthermore, Khalid and Helander explore how customer needs can be thought of as 3 dimensions: holistic design, product styling, and functionality [6]. They verified this by distributing a survey to collect customer preference data while also discovering that products the customer is more familiar with will project different design attributes rather than novel devices.

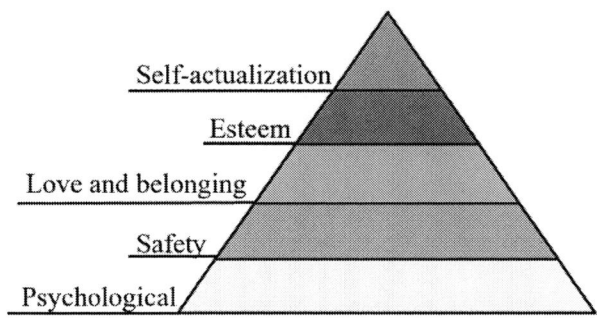

FIGURE 1: MASLOW'S HIERARCHY OF NEEDS

However, these prior works do not develop a hierarchy for the context of product design; they relate to Maslow's criteria instead. Since Maslow's hierarchy is structured for human behavior, it is difficult to apply to products. Maslow's Hierarchy of Needs has been translated to product design, as shown in Figure 2, in the form of the Design Hierarchy of Needs [3,8]. This hierarchy establishes that a design must first satisfy all functional needs before other needs can be considered. After functionality is met, reliability can be targeted where the design should yield stable and consistent performance. Then, usability can be assessed, such as how easy the user can interact with the design. Proficiency assists the user do more, and creativity describes the aesthetic beauty of a design and whether it promotes innovative interactions with the user. This hierarchy inspired by Maslow's work has not yet been applied in the domain of engineering design research.

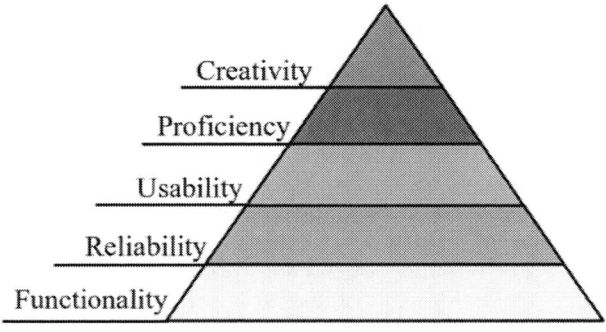

FIGURE 2: DESIGN HIERARCHY OF NEEDS

Copyright © 2023 by ASME

While the concept of the Design Hierarchy of Needs has not been applied to studies involving customer reviews, online reviews have been used in other ways to extract insights about product attributes and customer interaction and needs [9,10]. There has been increased popularity in using user-generated content (UGC) in machine-learning applications to communicate customer needs to product designers due to the speed at which algorithms can parse and analyze large sizes of online reviews. Companies rely on customer input data to assist in developing future products and this automatic mining process to extract words and phrase of interest can greatly assist [9,11–16]. Some studies have taken the results from passing online reviews through a learning model and translated them to engineering characteristics (ECs) to be used for quality function deployment (QFD) [17] which assists in understanding customer needs and can reveal where some design flaws rest [13]. Other studies have implemented the Kano model to assist in coupling different types of needs and customer emotions [16,18–20].

However, without applying the context of the Design Hierarchy of Needs, designers may struggle to identify the correct feature of the design that needs to be improved. For example, if a reliability characteristic is identified but functionality issues are the source of the failure, designers may not be able to make substantial progress in enhancing the product. Since the success of a product is dependent on the level of the customer's satisfaction [21], it is of utmost importance that designers understand the needs of the customer fully. Since Maslow's theory is thought to be ingrained in human nature, the act of applying the Design Hierarchy of Needs may assist in helping the expectations of customer and designer align if they are thinking about product needs using the same scale. Therefore, we focus on how to gain insight into the product itself rather than how to assess product reviews in this work.

3. RESEARCH METHODOLOGY

The research method implemented follows the process presented in Figure 3. In this section, we provide insight into the products and reviews selected, the development and application of codes to reviews, the categorization of codes, and the analysis procedure used in this exploratory study.

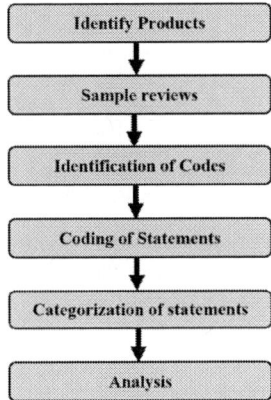

FIGURE 3: RESEARCH METHODOLOGY FRAMEWORK

The data collected for this study comes from five products identified in three different markets. We focus on electromechanical products by selecting three gas-powered handheld leaf blowers from different brands at homedepot.com [22–24], the Dyson v8 Absolute vacuum cleaner from dyson.com [25], and the Vicks SpeedRead Digital Thermometer from amazon.com [26].

We selected products where online reviews were readily available. From the entire body of customer reviews for each product, 100 customer reviews were randomly selected after a filtering and elimination technique was applied. The filtering and elimination technique involved skimming reviews to ensure that content about how the product was used or a developed description about what the reviewer did not like about the product appeared in the review. Reviews that were filtered out of the candidate set only listed non-product related attributes (customer service, shipping time, delivery, etc.) or were broad statements ("I did not like this product"). The final set of 100 reviews were evenly distributed across each star rating (20 reviews for each). The Design Hierarchy of Needs provided the foundation for our deductive coding process. The initial set of codes were functionality, reliability, usability, proficiency, and creativity. Before the coding process began, we also created definitions and explanations for each level of the hierarchy – as shown in Table 1. A description of how Table 1 was developed is reported in the Appendix.

It is important to note that in the context of the Design Hierarchy of Needs, creativity refers to 'innovative interactions.' 'Innovative interactions' has been expanded upon to explain how the customer interacts with the product in a way that has not been done before. When deciding whether a product can be considered to have innovative characteristics, this definition must be considered due to the broad category that surrounds the concept and measure of creativity [26].

When applying the Design Hierarchy of Needs to the selected reviews, we coded only negative statements. The negative statements are believed to be "information rich" [28] in demonstrating where products fail to meet customer needs within the Design Hierarchy of Needs. After applying the coding process to the reviews, each coded statement was categorized in a respective sub-code that described the failure. Sub codes were generated inductively as reviews were being coded and were meant to succinctly describe the failure. Different sub-codes were developed for each product type (the same sub-codes were used for all three leaf blowers, while the Dyson v8 and Vicks Thermometer had their own, unique sub-codes).

We acknowledge that our coding of online reviews is a subjective process and biases can exist. The deductive coding strategy was reviewed by a faculty member and an additional graduate student to reduce such biases. Each person was given reviews and Table 1. We assessed whether each person identified the same codes out of the reviews. We then discussed any differences that were found to establish a uniform coding procedure.

Copyright © 2023 by ASME

TABLE 1: REVISED DESIGN HIERARCHY OF NEEDS

Creativity - Aesthetic beauty, innovative interactions
• Customer interacts with product in a way that has not been done before
• Acknowledgment of design, uniqueness, or new feature
Proficiency – Empower people to do more and better
• Time and efficiency
• Wishes or hopes for future products
• Accessories
• Quality of life
Usability – Design is forgiving, easy to use
• Interaction is easy or difficult
• Convenience
• Comfortability and ergonomics
• Operability
• Safety
• Instruction manual and/or directions
• Set-up and/or installation
Reliability – Stable and consistent performance
• Ability to maintain functionality over time
• Lifespan of product
• Dependability
• Longevity of features
Functionality – Design works, meets basic functional needs
• Accomplishes what it was made to do
• Performance

4. CODING PRODUCT REVIEWS

Products selected from different domains with different uses were chosen for broadness and variety. Three different gas-powered handheld leaf blowers were selected. The leaf blowers are referenced by their respective brand names: Echo, Troy-Bilt, and Homelite. Additionally, the Dyson v8 Absolute vacuum cleaner and Vicks SpeedRead Digital Thermometer were selected. A full breakdown of the statistics associated with the reviews for each product is provided in Table 2.

TABLE 2: SUMMARY OF AVERAGE RATING, TOTAL REVIEWS, CUSTOMER RECOMMENDATION VALUES, AND NUMBER OF STARS FOR SELECTED PRODUCTS

Product	Echo	Troy-Bilt	Homelite	Dyson v8	Vicks
Average Rating	4.5/5	3.8/5	3.6/5	4.6/5	4.6/5
Customer rec.	86%	59%	57%	90%	N/A
Total Reviews	2,222	334	1,027	3,401	1,178
5 Star	1,541	164	405	2,685	704
4 Star	409	61	238	438	67
3 Star	96	31	98	104	54
2 Star	67	24	68	80	50
1 Star	109	54	218	94	303

4.1. Leaf Blowers

When selecting the different leaf blowers, it was desired to select products with similar capabilities such as blowing speed, airflow volume, and engine types. With speed ranges from 150-200mph, airflow ranges from 400-453 CFM, and approximately a 25 cc, 2 cycle gas engine, the three leaf blowers selected are comparable in their capabilities.

4.1.1. Echo Leaf Blower

A sample of customer reviews and coded statements for the Echo leaf blower are presented in Table 3. After all reviews were coded, the number of negative codes pertaining to each hierarchical level was organized into Figure 4.

TABLE 3: EXAMPLES OF CODED STATEMENTS IDENTIFYING FUNCTIONALITY, RELIABILITY, USABILITY, AND PROFICIENCY

Codes	Coded Statement
2 proficiency statements	"Pros: Decent hand blower with enough power to handle most cleanups. If wind challenged then best go with back pack blower but for most standard jobs the blowing power is good. Relatively light weight and easy to handle. Also upgraded metal rod attaches to carb. Not a cable that's prone to wear out. Cons: General hard plastic housing gets in the way of chaining it with other equipment. Chain gets caught between plastic and carb area. Fuel tank too small."
1 usability, 1 functionality statement	"Extremely difficult to start, does not blow very hard."
1 usability, 1 reliability statement	"Started the first time but a week later I hurt my arm trying to start it. It never started again but store did take it back with no hassle."

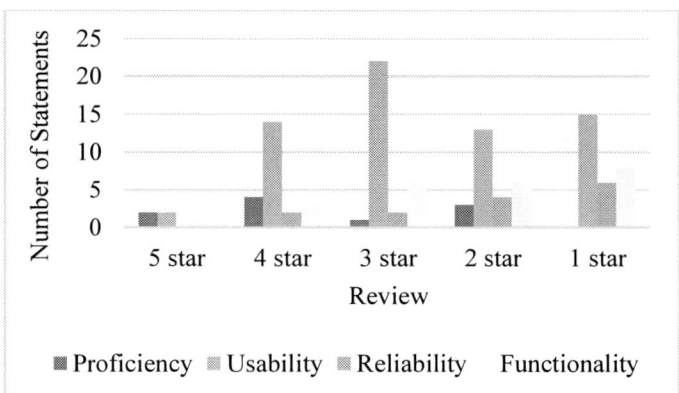

FIGURE 4: ECHO – NUMBER OF CODED STATEMENTS WITH A NEGATIVE PRODUCT ELEMENT WITHIN THE DESIGN HIERARCHY OF NEEDS

Additionally, each coded statement was categorized into a respective sub-code with the intent of demonstrating where product needs are not satisfied. Some example sub-codes that were identified include: 'Attachment issues,' 'Hard to start,' 'Ergonomics,' and 'Safety', as listed in Table 4. The frequency of each identified sub-code is listed in Table 5 and either directly corresponds to attributes within the revised Design Hierarchy of needs or are statements that address one of the attributes.

TABLE 4 EXAMPLE OF SUB-CODES FOR A CODED USABILITY STATEMENT

Sub-code	Coded Usability Statement
Attachment Issues	"Very light, and easy to use. Used it for an hour and was not fatigued holding it. Only downfall was the tube would fly off every so often. Need to place a screw in the tube to hold it for future use."
Hard to start	"Overall very nice… only used a couple of times so far so can't rate its durability but its lightweight enough so not to strain my arms… has plenty of power for my needs… only negative I can say is seems a little hard to get started… at least 6-7 pulls… Overall good product…"
Ergonomics and safety	"My only problem with the ECHO vs the Stihl is the blower intake. The Stihl is on the right side making it ideal for a right-handed operator. The ECHO's is on the left side making it ideal for a left-handed operator, there is a second guard, but the blower still sucks up loose-fitting pants. The Blower gas excellent air volume."

TABLE 5: ECHO – CODED STATEMENT SUB-CODES

Sub-code	5 star	4 star	3 star	2 star	1 star
Efficiency	0	0	0	1	0
Accessories	0	2	1	1	0
Broad use cases	1	2	0	0	0
Wish for larger fuel tank	1	0	0	0	0
Quality of life	0	0	0	1	0
Attachment issues	1	4	7	8	7
Hard to start	0	3	7	2	3
Ergonomics	0	1	2	0	0
Instructions	0	1	2	0	0
Safety	0	2	1	0	1
Noise	0	1	2	0	0
Unforgiving design	0	1	0	1	2
Heavy	1	1	0	1	0
Vibrations	0	0	1	1	2
Does not start after initial use	0	0	0	2	5
Product stops working	0	0	0	0	1
Power and warm up issues	0	2	2	2	0

Does not run properly	1	1	0	1	4
Inadequate power	0	2	6	5	2
Does not start	0	0	0	0	2

4.1.2. Troy-Bilt leaf blower

While we normally sampled 20 reviews from each star rating, only 19 reviews were obtained for the 2-star category due to the limited sample size. All relevant reviews with text were selected, however some reviews only gave a star rating and no explanation thus leading to only 19 reviews being usable. Figure 5 depicts the total number of coded statements for each star rating. Furthermore, the occurrence of each sub-code within the coded statements is presented in Table 6.

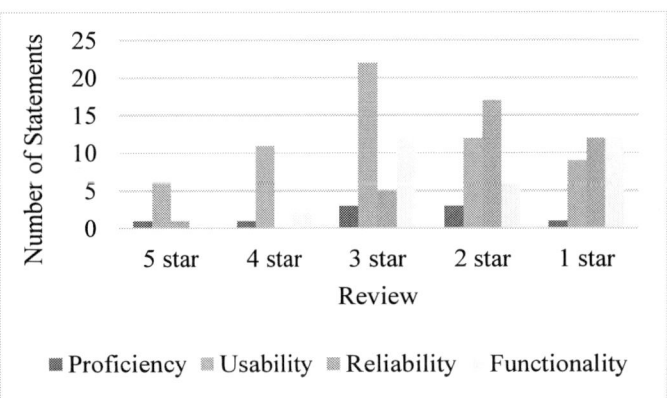

FIGURE 5: TROY-BILT – NUMBER OF CODED STATEMENTS WITH A NEGATIVE PRODUCT ELEMENT WITHIN THE DESIGN HIERARCHY OF NEEDS

TABLE 6: TROY-BILT – CODED STATEMENT SUB-CODES

Sub-code	5 star	4 star	3 star	2 star	1 star
Efficiency	0	0	1	0	1
Wish for intake modification	1	0	1	1	0
Wish for shredder/bagger	0	0	1	0	0
Wish for electric start	0	0	0	1	0
Quality of life	0	0	0	1	0
Accessories	0	1	0	0	0
Difficult to use	1	4	9	2	1
Hard to start	2	1	7	5	4
Heavy	2	4	1	0	1
Ergonomics	1	0	1	3	3
Safety	0	2	0	1	0
Noise	0	0	4	0	0
Attachment issues	0	0	0	1	0
Product stops working	0	0	1	3	5

Copyright © 2023 by ASME

Components breaking	0	0	0	0	1
Does not start after initial use	0	0	1	3	4
Inconsistent performance	1	0	3	9	2
Longevity	0	0	0	2	0
Does not run properly	0	1	1	2	2
Inadequate power	0	1	11	4	6
Would not start	0	0	0	0	1
Would not turn off	0	0	0	0	1
Components do not work	0	0	0	0	2

Safety	0	1	3	1	2
Instructions	1	0	0	0	1
Vibration	0	1	0	0	1
Noise	0	1	0	0	0
Inconsistent performance	0	0	2	1	3
Components breaking	0	0	0	2	3
Product stops working	0	0	0	2	0
Product does not start	0	0	0	2	0
Inadequate vacuum	1	2	2	2	5
Inadequate blower	1	3	6	7	3
Blower would not start	0	0	0	0	3

4.1.3. Homelite leaf blower

This Homelite tool was selected because it is not only a leaf blower, but also has the option to be used as a leaf vacuum and mulcher. The summary of the coded statements can be seen in Figure 6 and Table 7 displays the occurrences of the sub-codes of the statements.

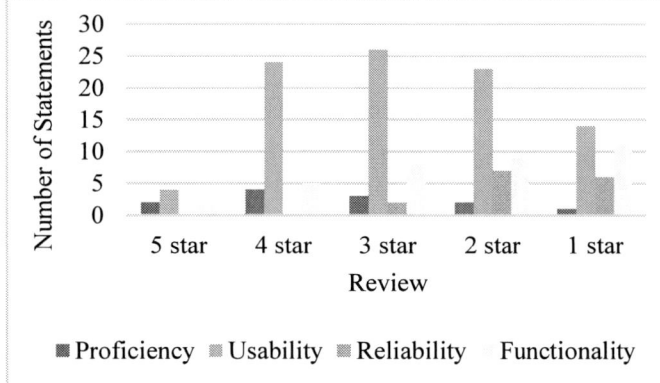

FIGURE 6: HOMELITE – NUMBER OF CODED STATEMENTS WITH A NEGATIVE PRODUCT ELEMENT WITHIN THE DESIGN HIERARCHY OF NEEDS

TABLE 7: HOMELITE – CODED STATEMENT SUB-CODES

Sub-code	5 star	4 star	3 star	2 star	1 star
Efficiency	0	2	0	0	0
Mention of additional tools	1	1	1	1	0
Emptying bag frequency	1	0	0	1	0
Accessories	0	1	2	0	1
Difficult to use	1	8	7	7	6
Hard to swap functions	1	5	7	9	0
Hard to start	1	3	3	3	0
Clogging	0	1	2	2	2
Heavy	0	4	4	1	2

4.2. Dyson v8 Absolute Vacuum

Since the first 3 products analyzed were yard equipment and belong to their own individual market, a product marketed to a different group of people was also examined. The number of coded statements for each star rating can be seen below in Figure 7 and the frequency of the sub-codes for each coded statement is shown in Table 8.

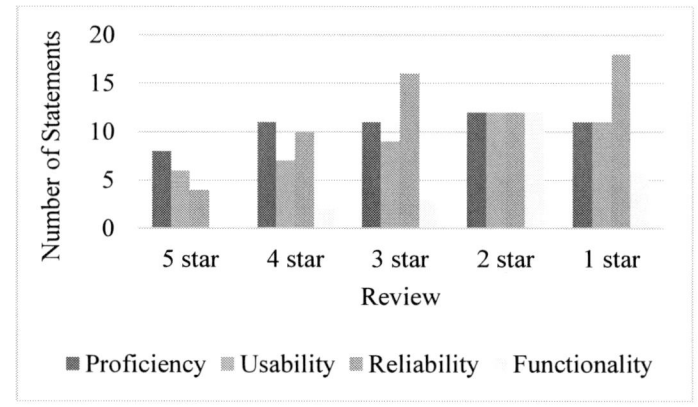

FIGURE 7: DYSON V8 – NUMBER OF CODED STATEMENTS WITH A NEGATIVE PRODUCT ELEMENT WITHIN THE DESIGN HIERARCHY OF NEEDS

TABLE 8: DYSON V8 – CODED STATEMENT SUB-CODES

Sub-code	5 star	4 star	3 star	2 star	1 star
Efficiency	1	1	2	8	3
Wish for better or more batteries	1	1	4	0	1
Wish for hold switch	2	3	0	0	0
Wish for adjustable length	1	0	0	0	0
Wish for light	1	2	0	0	1

Wish for better canister	0	1	1	1	3
Wish for corded stick vacuum	0	0	0	1	1
Enhancement of wall mount	2	2	0	0	0
Quality of life	0	0	1	0	1
Accessories	0	0	1	1	1
Use of additional tools	0	1	2	1	0
Ergonomics	4	5	5	3	4
Clogging	0	0	1	4	1
Difficult to clean	0	0	1	0	4
Operability	1	0	1	0	0
Difficult to use canister	1	1	1	2	2
Heavy	0	1	0	0	0
Accessories difficult to use	0	0	0	1	0
Noise	0	0	0	1	0
Safety	0	0	0	1	0
Battery issues	4	9	11	8	13
Product stops working	0	1	2	3	3
Inconsistent performance	0	0	0	0	1
Longevity	0	0	2	0	1
Product behaves unexpectedly	0	0	0	1	0
Broken features	0	0	1	0	0
Inadequate suction	0	1	2	4	3
Does not work on all surfaces	0	1	1	2	1
Does not collect/hold debris	0	0	0	6	2

4.3. Vicks SpeedRead Digital Thermometer

The final product that was reviewed also belongs to a different market than the previously mentioned products. The Vicks SpeedRead digital thermometer from amazon.com, shown in Figure 8, was selected for its innovative fever insight feature [29]. After the user takes a temperature, the thermometer will display the temperature numerically in °F or °C in addition to a correlating light. If the thermometer screen is green, the user has a normal temperature. If the screen is yellow, the user has a mild fever, and if it turns red, the user has a high fever. This product was characterized as an award-winning product and shown to "modify information flow, modify cognitive demands, and modify physical demands" [29]. The breakdown of coded statements for each star rating for this unique product is summarized in Figure 9 and the determined sub-codes for each star are presented in Table 9.

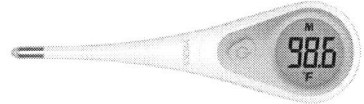

FIGURE 8: VICKS SPEEDREAD DIGITAL THERMOMETER

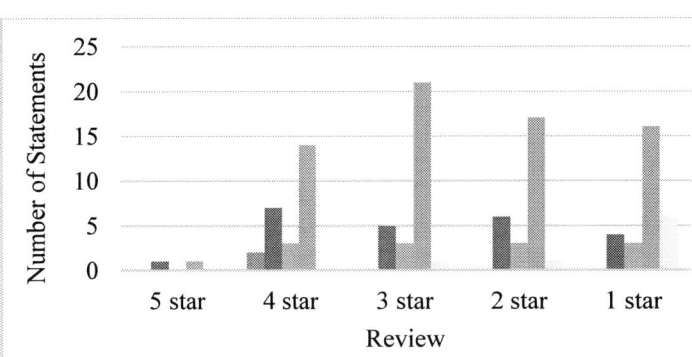

FIGURE 9: VICKS THERMOMETER – NUMBER OF CODED STATEMENTS WITH A NEGATIVE PRODUCT ELEMENT WITHIN THE DESIGN HIERARCHY OF NEEDS

TABLE 9: VICKS THERMOMETER – CODED STATEMENT SUB-CODES

Sub-code	5 star	4 star	3 star	2 star	1 star
Dislike of colors	0	2	0	0	0
Efficiency and 8 second reading	1	2	3	2	2
Battery replacement issues	0	2	1	0	2
Lack of extra sleeves	0	1	1	1	0
Cannot turn sound off	0	1	0	0	0
Wish for low battery warning	0	1	0	0	0
Does not change to °C	0	0	0	1	0
Chemical taste	0	0	0	1	0
Take battery out after use	0	0	0	1	0
Button difficult to use	0	1	1	1	0
Comfortability	0	1	0	0	0
User friendly	0	0	0	0	1
Noise	0	1	0	1	0
Colors too bright	0	0	0	1	0
Battery case difficult to remove	0	0	0	0	2
Difficult to read	0	0	1	0	0
No battery pull tab	0	0	1	0	0
Inaccurate reading	1	4	10	7	9
Battery issues	0	7	4	7	3

Copyright © 2023 by ASME

Light/beep issues	0	2	4	1	2
Stopped working	0	1	1	1	2
Product broke	0	0	2	1	0
Battery dead on arrival	0	0	1	0	3
Would not take temperature	0	0	0	1	1
Would not turn on	0	0	0	0	2

5. RESVISITING HYPOTHESES 1-4

Having collected product reviews and coding statements based on the Design Hierarchy of Needs, we revisit the hypotheses posed in Section 1. We aggregate data across all products with the goal of testing the original hypotheses. The number of negatively coded statements for the five hierarchical levels was summed for each product and is reported in Table 10.

TABLE 10: TOTAL NUMBER OF NEGATIVE CODED STATEMENTS ACROSS ALL PRODUCTS

	5 star	4 star	3 star	2 star	1 star
Creativity	0	2	0	0	0
Proficiency	14	27	23	26	17
Usability	18	59	82	63	53
Reliability	6	26	46	57	57
Functionality	3	12	30	34	43

Hypothesis 1: The number of negative functionality, reliability, and usability statements will decrease as the star ratings of reviews increase.

Hypothesis 1 is partially satisfied. After summing the statements for functionality, reliability, and usability for all star ratings, we show that the total number of comments remains relatively consistent for the 1–3-star ratings from 153-158 statements, and then decreases at the 4- and 5-star mark to 97 and 27 statements, respectively, in Table 11.

TABLE 11: SUMMED NUMBER OF NEGATIVE CODED STATEMENTS FOR FUNCTIONALITY, RELIABILITY, AND USABILITY LEVELS ACROSS ALL PRODUCTS

	5 star	4 star	3 star	2 star	1 star
Usability	18	59	82	63	53
Reliability	6	26	46	57	57
Functionality	3	12	30	34	43
Total	27	97	158	154	153

Although Hypothesis 1 is partially satisfied, we do show that the number of functionality comments decreases after 1-star, and the number of reliability comments also decrease after the 2-star mark. However, a sharp increase exists in the number of usability statements for the 3-star reviews.

Hypothesis 2: A defining factor of 1- and 2-star reviews are negative statements addressing functionality and reliability since they fail to lay an effective foundation within the Design Hierarchy of Needs.

We show that Hypothesis 2 is satisfied due to the increased number of statements for the sum of the functionality and reliability categories when compared to the higher hierarchical levels shown in Table 12.

TABLE 12: TOTAL NUMBER OF NEGATIVE CODED STATEMENTS FOR FUNCTIONALITY AND RELIABILITY LEVELS COMPARED TO NUMBER OF USABILITY, PROFICIENCY, AND CREATIVITY STATEMENTS ACROSS ALL PRODUCTS

	2 star	1 star
Creativity	0	0
Proficiency	26	17
Usability	63	53
Reliability	57	57
Functionality	34	43
Sum of Functionality and Reliability	91	100

Upon examining the sub-codes identified for selected products, there is an increased number of statements addressing how all the products fail at the functional level such as not functioning as expected, being inadequate to perform the task, or not turning on. The correlation between these categories and the number of negative comments in the 1-star reviews suggests that the first level of the Design Hierarchy of Needs has not been satisfied, further implicating where additional changes to the design of the product may need to occur. Furthermore, the highest number of negative comments addressing functionality and reliability can be found within the 1- and 2-star reviews. When summing the number of functionality and reliability statements for each star column, we see that the number of statements for the remainder of hierarchical levels is less than this number. This implies that when there is an increased number of functionality and reliability statements, the foundational elements of the product are weak and therefore the other levels will not be acknowledged by the customer.

Hypothesis 3: Customers who leave 4- and 5-star reviews focus more on negative proficiency aspects because they are able to consider ways to optimize their interaction with the product.

A transition to examining the 4- and 5-star reviews allows us to gain further insight into the ways in which the customer interacts with the product. We stated in Hypothesis 3 that customers who give higher star reviews will be able to address more areas where the design is able to be improved because they are not preoccupied with making sure the product will work and will work for a long time. Table 13 shows Hypothesis 3 is partially verified because there are more comments about proficiency than the total number of statements for the reliability and usability levels in the 5-star reviews, but less proficiency comments for the 4-star reviews.

Copyright © 2023 by ASME

TABLE 13: TOTAL NUMBER OF NEGATIVE CODED STATEMENTS FOR FUNCTIONALITY AND RELIABILITY LEVELS COMPARED TO NUMBER OF PROFICIENCY STATEMENTS ACROSS ALL PRODUCTS

	5 star	4 star
Proficiency	14	27
Reliability	6	26
Functionality	3	12
Sum of Functionality and Reliability	9	38

Even though this hypothesis is only partially verified, we believe that because customers are not as concerned with the product meeting the first two levels of the hierarchy, they are able to list ways in which the product could be improved that would ultimately allow them to interact with it in an enhanced manner. For example, throughout all products sampled, customers stated how they wished the product would help them accomplish the task in less time so they could be more efficient. They also list how accessories could be improved or included with the product, additional features that could be added, and use cases where the product could be optimized to help them better complete the task. It is not unexpected that a total of only 14 proficiency comments were identified in the 5-star category whereas the 4-star category had 27 since the 5-star reviews are believed to better correspond to products that meet the customer's overall needs.

Hypothesis 4: Negative statements about usability appear more often in reviews that do not contain negative comments about functionality and reliability.

We show that Hypothesis 4 is verified when comparing the total number of functionality and reliability statements to the number of usability statements in Table 14. While there are only 9 total comments for functionality and reliability in the 5-star reviews, there are 18 comments addressing usability. A similar trend occurs for the 4-star reviews, with 38 functionality and reliability statements and 59 usability statements. This trend is consistent until the 1- and 2-star categories, which is expected.

TABLE 14: SUMMED NUMBER OF NEGATIVE CODED STATEMENTS FOR FUNCTIONALITY, RELIABILITY, AND USABILITY LEVELS ACROSS ALL PRODUCTS

	5 star	4 star	3 star	2 star	1 star
Usability	18	59	82	63	53
Reliability	6	26	46	57	57
Functionality	3	12	30	34	43
Total	9	38	76	91	100

The results surrounding the usability level yield further insights about the way that consumers interact with products. We consider usability as one of the most subjective levels on the Design Hierarchy of Needs due to the personalized nature that is associated with the level. A product that is easy for one person to use may not be equally as easy for another. This can be identified in many of the sub-codes statements were binned in, such as weight and ergonomics. Furthermore, sub-codes such as how easy the instruction manual is to use, or how forgiving the design is, will be described differently for customers with different experience levels and therefore lead to a wide range of statements. For example, many customers described some leaf blowers to be hard to start, but some commented about how they did not have any issues starting it because they have been starting and using leaf blowers for many years. Customers who are more experienced in using a product are able to focus on higher levels of the hierarchy whereas inexperienced customers tend to have an increased focus on lower levels [4]. Additionally, it should be noted that the Homelite leaf blower was a product with two functions, blowing and vacuuming. We believe that the reason usability comments for that product surpassed those of the other surveyed products was due to the presence of these two functions. The blowing feature also received more comments than the vacuuming feature and we suspect this is because products with two separate functions will have one function that is focused on less by the customer because the other fails to satisfy needs.

6. HYPOTHESIS 5 – POSITIVE PROFICIENCY AND CREATIVITY

We also applied the Design Hierarchy of Needs to code positive statements addressing proficiency and creativity. The coding process was applied to the same reviews but with the intent of identifying statements describing how the product had useful accessories, helped the user be more efficient, improved their quality of life, or features they wished the product had. Additionally, any comments describing an innovative interaction the customer had with the product were also coded. Each coded statement was categorized in a sub-code that explained how the product was perceived to be proficient or creative. The same sub-codes were used for all three leaf blowers, while the Dyson v8 and Vicks Thermometer had their own sub-codes.

We now revisit Hypothesis 5 and address the positively coded statements for proficiency and creativity in their respective subsections. We expect to observe an increased number of positive statements acknowledging proficiency and creativity in 4- and 5-star reviews when compared to 1–3-star reviews. Comments addressing an innovative feature are only applicable to the Vicks Thermometer.

Hypothesis 5: There will be a significant number of positive statements addressing proficiency and/or creativity in 4- and 5-star reviews compared to 1- through 3-star reviews.

6.1. PROFICIENCY

The number of positively coded statements for the hierarchical level of proficiency were summed across all products and summarized in Table 15. We see that Hypothesis 5 is verified due to the large increase in comments acknowledging proficiency within the 5-star category. While the 4-star category is not as large as the 5- star, an increase still exists when compared to the 1–3-star reviews.

Copyright © 2023 by ASME

TABLE 15: TOTAL NUMBER OF POSITIVE CODED STATEMENTS WITHIN LEVEL OF PROFICIENCY ACROSS ALL PRODUCTS

	5 star	4 star	3 star	2 star	1 star
Proficiency	72	28	19	3	3

The coded proficiency statements yield insight into attributes that empowered customers while they interacted with the product. A consistent theme that customers identified across all products was how efficiency and versatility impacted their experiences using the products. All sub-codes generated are displayed in Table 16. Customers appreciated being able to complete the task quicker as well as the flexibility to use the product in different situations. Similar to how customers are able to consider how a product might be improved when lower hierarchical levels are satisfied, we believe they are also more likely to acknowledge features they enjoyed interacting with and ways in which the product empowered them in completing the task when functionality, reliability, and usability are satisfied.

Additionally, it is noted that the language customers used in the 4- and 5-star reviews when addressing proficiency was noticeably different than the lower star reviews. Customers seemed excited and wanted others to know how much they loved the product. For example, users of the Dyson v8 expressed how they took pleasure in vacuuming and how using the different accessories ultimately made it more enjoyable. People who used the Vicks thermometer explained how they could better take care of their families because of the features it offered. It was evident that users experienced empowerment and desired to express this to others.

Table 16: POSITIVE CODED STATEMENT SUB-CODES FOR ALL PRODUCTS

Leaf blowers	Dyson v8	Vicks Thermometer
Efficiency	Efficiency	Efficiency
Versatility	Versatility	Versatility
Quality of life	Quality of life	Quality of life
Better results	Cordless	Size
Cordless	Wall mount	Battery
	Attachments	Case/covers

6.2. CREATIVITY

The creativity portion of Hypothesis 5 was also satisfied since it was found that the 5-star reviews had overwhelmingly more positive comments about the innovation present in the product as opposed to the 1-star reviews. The hierarchical level of creativity was only observed in the Vicks Thermometer and the total number of statements addressing creativity can be seen in Table 17.

TABLE 17: NUMBER OF CODED STATEMENTS WITH A POSITIVE CREATIVITY ELEMENT WITHIN THE DESIGN HIERARCHY OF NEEDS

	5 star	4 star	3 star	2 star	1 star
Creativity	13	7	3	2	1

The 4- and 5-star reviews had the greatest number of statements addressing the innovative feature whereas the 1–3-star reviews had significantly fewer comments. It should also be noted that the 5-star reviews acknowledged the innovative feature differently than the lower star reviews. Customers in the 5-star category were excited and explained how the feature made their lives better when taking a temperature. They began to repeat what the different colors meant and how it helped them. One customer said, "Advantages of this thermometer is that . . . the red yellow green display is instant feedback" and another remarked how they were "able to grab the thermometer from [their] nightstand in the middle of the night, without turning lights on, and still see if [they] have a fever or not simply based on the color. Green means that [the] fever is normal, yellow means [they] have a slight fever, and red means [they] have a fever." This innovative feature ultimately contributed positively to proficiency since the user's quality of life was improved due to the reduced demands in deciding if a fever is present or not. It appeared that the customers liked the feature so much they had the desire to explain how it worked to others through their review. In contrast, the lower star reviews made simple comments such as "[it] makes colors" and "so far the color has worked with temperature" but they did not discuss the innovative characteristics to the same extent as the 5-star reviewers. This is a critical insight since it demonstrates how innovation must inherently be designed into a product and how customers will be more receptive to the additions only when their basic needs are satisfied. Innovative features become lost and considered unimportant to the customer when the functionality, reliability, and usability levels of the hierarchy are compromised.

7. RAMIFICATIONS AND CONCLUSIONS

An important aspect of engineering design is meeting customer needs to maintain customer retention, brand loyalty, and satisfaction. These factors influence the ability of a company to develop innovative solutions but can also only be achieved when customer needs are understood. The purchase decisions customers make when shopping online are influenced by the comments and the star ratings of a review.

In this paper, we propose that customer needs can be classified based on the Design Hierarchy of Needs and thus can reveal where needs are not being met or at what level products are failing. We sampled 100 customer reviews for five different products and used a deductive coding process to code negative statements according to the revised Design Hierarchy of Needs. These statements were then categorized into sub-codes that revealed the frequency at which certain attributes of each level failed. This method was used to test five hypotheses and we were successful in verifying three of the hypotheses and partially verifying the other two hypotheses. In Hypothesis 1, we see that the number of negative functionality, reliability, and usability statements is relatively consistent for 1–3-star ratings and then decreases in the 4- and 5-star mark whereas we expected to see a decrease in the lower ratings as well. In Hypothesis 2, we establish that negative comments addressing functionality and reliability are a defining factor in 1- and 2-star reviews and

Copyright © 2023 by ASME

therefore fail to construct an effective foundation within the Design Hierarchy of Needs. Additionally, we partially verify that customers who leave 4- and 5-star reviews focus more on negative proficiency aspects of the product. While this was only true for the 5-star category, we believe that customers are able to better acknowledge proficient elements of the product because the lower levels of the Design Hierarchy of Needs are satisfied. Finally, we verify Hypothesis 4 in demonstrating that negative comments about usability appear more often in higher star reviews. A unique discussion was also centered around the hierarchical level of usability because it is inherently very subjective, and customers will interpret it differently based on their experience level with the product.

We additionally hypothesized that positive statements addressing the proficiency and creativity levels of the hierarchy would be greater in the 4- and 5-star categories. Ultimately, we do satisfy this hypothesis and provide insight into the product attributes that empowered the customer when they used the product. Furthermore, we establish that the frequency of creativity comments increases as star ratings progress upwards and can be found in reviews for products that have an innovative feature designed into them, such as the Vicks Thermometer.

Our study, based on an initial study of five products, suggests that 1- and 2-star reviews are primarily a function of negative comments around functionality and reliability. Given this insight, companies must ensure functionality and navigate the costs of ensuring reliability if they want to minimize the number of low star reviews. To avoid violating the functionality and reliability levels of the hierarchy, engineers may be prone to overdesign which could impart costly consequences on the design. In the instance of reliability, it is known that some products may be defective due to the cost associated with testing a component of the design, such as battery life. Engineers must consider what tradeoffs they are willing to make and determine if there are acceptable functionality and reliability sub-codes that simply come at the cost of designing and distributing a product.

As companies manage higher star reviews, the tradeoffs being made are focused on usability and proficiency. This suggests the need for properly identifying relevant market segments [30], understanding human variability [31], and translating needs into technical specifications.

We acknowledge that exclusively defining the levels of the Design Hierarchy of Needs is challenging due to the lack of orthogonality that exists within levels. Coding of the reviews and assigning subcodes relied heavily on the context in which the customer wrote the review and we attempted to mitigate possible discrepancies by conversing with a faculty member and graduate student during the development and assignment of the codes. The existence of this challenge provides further insight into the levels of the Design Hierarchy of Needs and how they can appear to be coupled at times. For example, when both functionality and reliability issues are prevalently addressed, the review receives a lower star rating. Since products can have shortcomings existing across multiple levels, it may be necessary to consider a reconstruction of the Design Hierarchy of Needs to better

accommodate for the inherent blending of levels that can occur when considering what level a failure should be categorized as.

There are several opportunities for expanding this study to include a more focused set of products, an increase in the number of reviews sampled, and the applications to design for market systems. Additionally, we believe it would be of interest to introduce the variable of price and examine how it interacts with the Design Hierarchy of Needs. Ultimately, we believe the Design Hierarchy of Needs can assist engineers in making design decisions because it provides a framework that is psychologically intuitive that may be easier for engineers to apply than other frameworks.

REFERENCES

[1] Riaz, A., Hanif, M., and Hafeez, S., 2010, *Factors Affecting Customer Satisfaction.*

[2] McLeod, S., 2018, "Maslow's Hierarchy of Needs," SimplyPsychology.

[3] Steven Bradley, 2018, "Designing for a Hierarchy of Needs," Smashing Magazine.

[4] Cui, L., Wang, Y., Chen, W., Wen, W., and Han, M. S., 2021, "Predicting Determinants of Consumers' Purchase Motivation for Electric Vehicles: An Application of Maslow's Hierarchy of Needs Model," Energy Policy, **151**.

[5] Yalch, R., and Brunei, F., 1996, *Need Hierarchies in Consumer Judgments of Product Designs: Is It Time to Reconsider Maslow's Theory?*

[6] Khalid, H. M., and Helander, M. G., 2004, "A Framework for Affective Customer Needs in Product Design," Theor Issues Ergon Sci, **5**(1), pp. 27–42.

[7] Gershenson, J. A., Khadilkar, D. V., and Stauffer, L. A., 1994, "Organizing and Managing Customer Requirements During the Product Definition Phase of Design," *6th International Conference on Design Theory and Methodology*, American Society of Mechanical Engineers, pp. 99–107.

[8] Lidwell, W., Holden, K., and Butler, J., 2010, *Universal Principles of Design*, Rockport Publishers.

[9] Rai, R., 2012, "Identifying Key Product Attributes and Their Importance Levels From Online Customer Reviews," *Volume 3: 38th Design Automation Conference, Parts A and B*, American Society of Mechanical Engineers, pp. 533–540.

[10] Association for Computing Machinery., and ACM Digital Library., 2007, *ICEC 2007 : Proceedings of the Ninth International Conference on Electronic Commerce, August 19-22, 2007, Minneapolis, Minnesota, USA.*, Association for Computing Machinery.

[11] Timoshenko, A., and Hauser, J. R., 2019, "Identifying Customer Needs from User-Generated Content," Marketing Science, **38**(1), pp. 1–20.

[12] Htay, S. S., and Lynn, K. T., 2013, "Extracting Product Features and Opinion Words Using Pattern Knowledge

Copyright © 2023 by ASME

in Customer Reviews," The Scientific World Journal, **2013**.

[13] Ireland, R., and Liu, A., 2018, "Application of Data Analytics for Product Design: Sentiment Analysis of Online Product Reviews," CIRP J Manuf Sci Technol, **23**, pp. 128–144.

[14] Saidani, M., Kim, H., Ayadhi, N., and Yannou, B., 2021, "Can Online Customer Reviews Help Design More Sustainable Products? A Preliminary Study on Amazon Climate Pledge Friendly Products," *Volume 6: 33rd International Conference on Design Theory and Methodology (DTM)*, American Society of Mechanical Engineers.

[15] El Dehaibi, N., Goodman, N. D., and MacDonald, E. F., 2019, "Extracting Customer Perceptions of Product Sustainability from Online Reviews," Journal of Mechanical Design, Transactions of the ASME, **141**(12).

[16] Ayoub, J., Zhou, F., Xu, Q., and Yang, J., 2019, "Analyzing Customer Needs of Product Ecosystems Using Online Product Reviews," *Volume 2A: 45th Design Automation Conference*, American Society of Mechanical Engineers.

[17] Jin, J., Ji, P., Liu, Y., and Johnson Lim, S. C., 2015, "Translating Online Customer Opinions into Engineering Characteristics in QFD: A Probabilistic Language Analysis Approach," Eng Appl Artif Intell, **41**, pp. 115–127.

[18] Kano, N., Seraku, N., Takahaashi, F., and Tsuji, S., 1984, "Attractive Quality and Must-Be Quality," The Journal of the Japanese Society for Quality Control, **14**(2), pp. 39–48.

[19] Jin, J., Jia, D., and Chen, K., 2022, "Mining Online Reviews with a Kansei-Integrated Kano Model for Innovative Product Design," Int J Prod Res, **60**(22), pp. 6708–6727.

[20] Bi, J. W., Liu, Y., Fan, Z. P., and Cambria, E., 2019, "Modelling Customer Satisfaction from Online Reviews Using Ensemble Neural Network and Effect-Based Kano Model," Int J Prod Res, **57**(22), pp. 7068–7088.

[21] Liu, Y., Jin, J., Ji, P., Harding, J. A., and Fung, R. Y. K., 2013, "Identifying Helpful Online Reviews: A Product Designer's Perspective," *CAD Computer Aided Design*, pp. 180–194.

[22] "Echo Leaf Blower," https://www.homedepot.com/p/ECHO-170-MPH-453-CFM-25-4-cc-Gas-2-Stroke-Cycle-Handheld-Leaf-Blower-PB-2520/303393700.

[23] "Troy-Bilt Leaf Blower," https://www.homedepot.com/p/Troy-Bilt-200-MPH-430-CFM-2-Cycle-25cc-Gas-Handheld-Leaf-Blower-TB430/306716423.

[24] "Homelite Leaf Blower," https://www.homedepot.com/p/Homelite-150-MPH-400-CFM-26cc-Gas-Handheld-Blower-Vacuum-UT26HBV/207000089.

[25] "Dyson v8 Absolute," https://www.dyson.com/vacuum-cleaners/cordless/v8/absolute/silver-nickel.

[26] "Vicks SpeedRead Digital Thermometer." https://www.amazon.com/Vicks-SpeedRead-Digital-Thermometer-V912US/dp/B000GRXHIE/ref=pd_ci_mcx_mh_mcx_views_0?pd_rd_w=tT3sV&content-id=amzn1.sym.1bcf206d-941a-4dd9-9560-bdaa3c824953&pf_rd_p=1bcf206d-941a-4dd9-9560-bdaa3c824953&pf_rd_r=1TA76HFQHCMCT6GDE486&pd_rd_wg=Ojf6V&pd_rd_r=aede5a68-1316-483e-a14a-641f378d1e6b&pd_rd_i=B000GRXHIE

[27] Hazelrigg, G. A., 2012, *Fundamentals of Decision Making for Engineering Design*.

[28] Patton, M. Q., 2005, *Qualitative Research*, John Wiley & Sons, Ltd.

[29] Saunders, M. N., Seepersad, C. C., and Hölttä-Otto, K., 2011, "The Characteristics of Innovative, Mechanical Products," Journal of Mechanical Design, Transactions of the ASME, **133**(2).

[30] Park, S., and Kim, H. M., 2022, "Finding Social Networks Among Online Reviewers for Customer Segmentation," Journal of Mechanical Design, Transactions of the ASME, **144**(12).

[31] Garneau, C. J., and Parkinson, M. B., 2009, "Including Preference in Anthropometry-Driven Models for Design," Journal of Mechanical Design, Transactions of the ASME, **131**(10), pp. 1010061–1010066.

APPENDIX
Development of Table 1:

Table 1 was developed prior to the selection of the reviews used in this study. We read product reviews of various products and extracted the most common themes. When categorizing each statement into a category listed in Table 1, the context in which the statement was written had to be considered. Consider an example involving the usability and proficiency categories. Statements coded under the 'quality of life' sub-code indicated that the product had a more significant impact on the user than simply being "easy to use." The product went beyond fulfilling the customer's basic needs and instead allowed them to use the product in a more efficient way. To assist in determining the context of each statement, Table A1 was constructed that provides questions the coder should ask when coding the statements.

TABLE A1: QUESTIONS TO ASSIST IN CODING STATEMENTS

Creativity
• Does a feature of the product allow the customer to innovatively act with the product?
Proficiency
• Does this product help the user accomplish the task in less time?
• Does the use of this product require more time than normal to complete the task?

- Does the customer indicate wishes for future products?
- Do accessories work poorly?

Usability

- Does the statement describe a way that the user interacts with the product?
- Is the product comfortable to use?
- Are there features of the product that cause the user's safety to be compromised?
- Does the customer indicate that the instruction manual is difficult to interpret?
- Is the set-up challenging for the customer? Do they indicate it took more time than anticipated?

Reliability

- Can the customer depend on the product to work when they need it to?
- Does the product unexpectedly stop working for the customer?
- Do features of the product break?
- Is the behavior of the product anticipated?

Functionality

- Does the product work? Does it turn on?
- Does the product perform the task that it was purchased to do?

Proceedings of the ASME 2023
International Design Engineering Technical Conferences and
Computers and Information in Engineering Conference
IDETC-CIE2023
August 20-23, 2023, Boston, Massachusetts

DETC2023-116586

A RELIABILITY-BASED OPTIMIZATION FRAMEWORK FOR PLANNING OPERATIONAL PROFILES FOR UNMANNED SYSTEMS

Indranil Hazra[1,2], Arko Chatterjee[2,*], Joseph Southgate[2], Matthew J. Weiner[1,2], Katrina M. Groth[1,2], Shapour Azarm[1,2]
[1]Center for Risk and Reliability, [2]Department of Mechanical Engineering
University of Maryland, College Park, MD 20742, USA
*Corresponding author (achatter@umd.edu)

ABSTRACT

Unmanned engineering systems that execute various operations are becoming increasingly complex relying on a large number of components and their interactions. The reliability, maintainability, and performance optimization of these systems are critical due to their intricate nature and inaccessibility during operations. This paper introduces a new reliability-based optimization framework for planning operational profiles for unmanned systems. The proposed method employs deep learning techniques for subsystem health monitoring, dynamic Bayesian networks for system reliability analysis, and Bayesian optimization schemes for optimizing system performance. The proposed framework systematically integrates these schemes to enable their application to a wide range of tasks, including offline reliability-based optimization of system operational profiles. This framework is the first in the literature that incorporates health monitoring of multi-component systems with causal relationships. Using this hybrid scheme on unmanned systems can improve their reliability, extend their lifespan, and enable them to execute more challenging missions. In order to demonstrate the proposed approach, the optimization strategy is implemented and executed using a simulation model for the engine cooling and control system of an unmanned surface vessel.

1 INTRODUCTION

Unmanned systems are gaining popularity for their efficient performance of certain tasks without human intervention. These systems are composed of multiple heterogeneous components and subsystems that interact with each other in complex ways, which can cause inter-dependencies in the event of system failure. Unmanned systems have unique operating conditions and degradation characteristics for each subsystem and component. This degradation leads to differing needs for system reliability and risk assessment. Due to their complexity and possibly inaccessibility during operation, ensuring reliability, maintainability,

and performance optimization of unmanned systems is crucial. Extensive research is necessary to develop reliable and robust maintenance and optimization schemes for these systems.

Many systems are now integrated with advanced sensors and computing technology, resulting in a continuous flow of abundant monitoring data. These real-time condition monitoring data are used to assess the health conditions of components as well as system performance, utilizing the power of prognostics and health management (PHM) technology. Deep learning-based PHM technology is currently on the rise, especially due to its ability to model large data sets and complex nonlinear relationships between different data streams. For instance, several studies have been carried out to develop and validate novel deep learning-based PHM schemes such as for turbofan engines [1], wind turbines [2], and marine diesel engines [3,4]. However, the current PHM technology is mostly concentrated at the component level. The current literature clearly lacks a comprehensive set of applications for the system-level PHM schemes with considerations of PHM at subsystem and component levels. Recent studies by Lewis and Groth [5], Moradi et al. [6], and Guo et al. [7] showed how system-level PHM schemes can be developed and applied to systems with varying levels of complexity, such as nuclear power plants, copper mine stone crushers, and marine electrical propulsion systems, respectively. Moradi and Groth [8] proposed and validated a novel framework that integrates PHM and probabilistic risk assessment tools for health monitoring of complex engineering systems. Additionally, Moradi et al. [9] demonstrated how deep learning-assisted PHM tools integrated with risk assessment techniques can be systematically implemented for health monitoring of a real-world vapor recovery unit at an offshore oil production platform. The authors discussed the advantages of such schemes for health monitoring and failure risk analysis of systems in the context of dynamic risk assessment [10].

Bayesian networks, a type of probabilistic graphical model,

Copyright © 2023 by ASME

are often used for modeling causal relationships in system risk assessment and decision analysis [11]. Numerous applications of Bayesian networks for reliability analysis of multi-component systems can be found in the literature [12, 13]. Bayesian networks with dynamic modeling capabilities have also been extensively utilized for time-dependent reliability analysis of engineering systems. Example applications of dynamic Bayesian networks (DBNs) include reliability and risk analysis of nuclear power plants [5], unmanned surface vessels [14], chemical infrastructures [15], and power distribution systems [16], to name a few. DBNs are essentially system-level models that come with the ability to model time-dependent behavior of complex engineering systems. However, the current literature contains only limited studies that combine a system DBN model with PHM techniques for dynamically updating the model with continuous condition data streams [5].

The operational profiles of unmanned systems can be designed in such a way that they are able to successfully perform certain tasks with high reliability throughout the duration of the mission. In order to enable this, this paper presents a new framework for designing reliability-based optimized operational profiles for unmanned systems in an offline mode for planning. The proposed approach combines deep learning-assisted PHM techniques, Bayesian network-based system reliability monitoring tools, and reliability-based optimization schemes into a single framework. Although numerous applications of reliability-based system performance optimization approaches can be found in the current literature (e.g., tunnel boring machines [17], healthcare systems [18], microelectromechanical systems [19], and structural systems [20]), these approaches do not explore the health monitoring and the causal relationship or dependency aspects of the problem in a systematic manner. To the best of the authors' knowledge, the proposed framework in this paper is the first of its kind that can be utilized in various ways, including: designing (planning) offline operational profiles for unmanned systems prior to a mission; optimizing system operational profiles during mission execution; scheduling maintenance activities, such as determining whether maintenance is necessary immediately or can be postponed until the mission completion; estimating the remaining operational lifespan of systems based on either total mission time or the number of similar missions; and identifying critical components that require spare parts to improve system reliability for extended missions. However, this paper focuses on the problem of offline design of system operational profiles.

In order to demonstrate the proposed approach, a case study is developed using a simulation model for the engine cooling and control system (ECCS) of an unmanned surface vessel (USV). The ECCS controls and cools the marine diesel engine that gives the USV the power it requires for propulsion and to complete its mission. USVs have numerous commercial and military applications as they offer numerous benefits over traditional manned ships. On the commercial side, USVs are (and will

be) used for cargo and supply chain management, offshore oil and gas exploration, marine research and oceanography, fisheries, and environmental monitoring [21]. In military contexts, USVs are employed for intelligence gathering, surveillance and reconnaissance, maritime security and patrol, search and rescue, mine countermeasures, anti-submarine warfare, and fire support [21,22]. They use advanced technology to operate in challenging environments. High operational reliability for USVs allows for longer deployment, enhancing their capability to perform complex missions while reducing the risk to human life. Thus, USVs are versatile platforms for a wide range of tasks, appealing to both civilian and military organizations.

The contributions of this study are summarized as follows:
1. The paper proposes a new and general framework that integrates multiple techniques, such as deep learning for subsystem health monitoring, dynamic Bayesian networks for system reliability analysis, and Bayesian optimization for reliability-based optimization of unmanned systems' operational profiles, which has potential applications in a wide range of fields.
2. The proposed hybrid reliability-based optimization approach is the first in the literature that has an embedded health monitoring of multi-component (or multi-subsystem) systems with their causal (interdependency) relationships.
3. The case study offers a practical demonstration of the proposed method and its effectiveness in optimizing the operational profile of a critical system in a USV.

The remainder of the paper is organized as follows: Section 2 presents the problem aim and definition considered in this study. Section 3 presents the proposed approach and details of the implementation steps. Section 4 presents a case study on the performance optimization of the ECCS of a USV. Finally, Section 5 presents the conclusion of the study.

2 PROBLEM AIM AND DEFINITION

The proposed problem of reliability-based optimization of operational profiles for an unmanned system can be formulated as follows. The mean and variability in the performance over time to complete a task are considered as a good measure of the system performance, and therefore used as the objective function which needs to be optimized, i.e., weighted sum of the negative of the mean and positive of the variability in system performance is minimized. The system is expected to successfully execute any given task with high reliability for the entire duration of its mission. Thus, the system's reliability is assumed to create a limit state, i.e., the system is unable to perform its intended tasks if it crosses a specific limit. The limit state function can be constructed using critical system reliability and can be utilized as a constraint for the optimization problem. The system's reliability is directly dependent on the system's and subsystems' health state. Thus, the amount of time the system takes to complete a given task and the health state of the system create a trade-off by

Copyright © 2023 by ASME

opposing each other. For instance, as the system operates faster to minimize mission time, more stress is placed on the system resulting in a worse health state. Considering the uncertainty, e.g., from the environment, the following optimization problem is constructed:

$$\text{minimize } f(\mathbf{x}, \mathbf{p})$$
$$\text{s.t. } \mathbf{g}(\mathbf{x}, \mathbf{p}) \leq \mathbf{0} \tag{1}$$

where $f(\cdot)$ represents the system performance function, $\mathbf{g}(\cdot)$ represents the limit state function, $\mathbf{x}$ represents the vector of decision variables, e.g., different levels of an operational profile, and $\mathbf{p}$ represents the vector of uncertain parameters for an unmanned system, the performance function can be written in terms of the signal (expected or mean value of the performance) and the noise (standard deviation). For example, the signal may represent the speed at which the system performs a task and the noise can represent the associated variability (standard deviation) in the speed due to uncertainty. As such, the objective function for the system performance can be written in the following form, with the weights (pre-specified by the decision maker) w_1 and w_2 respectively for the weight of the signal and noise:

$$f(\mathbf{x}, \mathbf{p}) = -w_1 \times S + w_2 \times \sigma, \quad 0 \leq (w_1, w_2) \leq 1 \tag{2}$$

where S and σ represent the signal (expected value) and noise (standard deviation) generated from the system performance, respectively. The limit state function can be written as

$$g_j(\mathbf{x}, \mathbf{p}) = R_{j,critical} - R_{j,sys}(\mathbf{x}, \mathbf{p}), \quad j = 1, 2, \cdots, J \tag{3}$$

where $R_{j,sys}(\cdot)$ represents the expected reliability of the system, and $R_{j,critical}$ represents a user-specified critical value, a lower bound on an acceptable system reliability.

3 PROPOSED APPROACH

The framework for the proposed approach consists of six main parts (or Steps 1 to 6): the inputs (Step 1), subsystem and system models (Step 2), subsystem-level health analysis (Step 3), system-level reliability analysis (Step 4), the optimizer (Step 5), and the outputs (Step 6). The framework architecture is presented in Fig. 1, with the step numbers for different parts. Briefly, to diagnose the health of systems, input information is collected from various sources in Step 1, followed by the development of subsystem and system models in Step 2. Using the deep learning-based trained diagnostic models, Step 3 identifies the health states of individual subsystems. Step 4 assigns failure probabilities and evaluates the system reliability profile, while Step 5 optimizes system performance. Finally, in Step 6, the optimized operational profiles and corresponding reliability profiles are collected and stored for future use. A detailed description covering all steps of the framework is presented below in the following subsections.

3.1 Step 1: Input Information

In order to execute the proposed approach, the first task is to collect input information from various sources including system

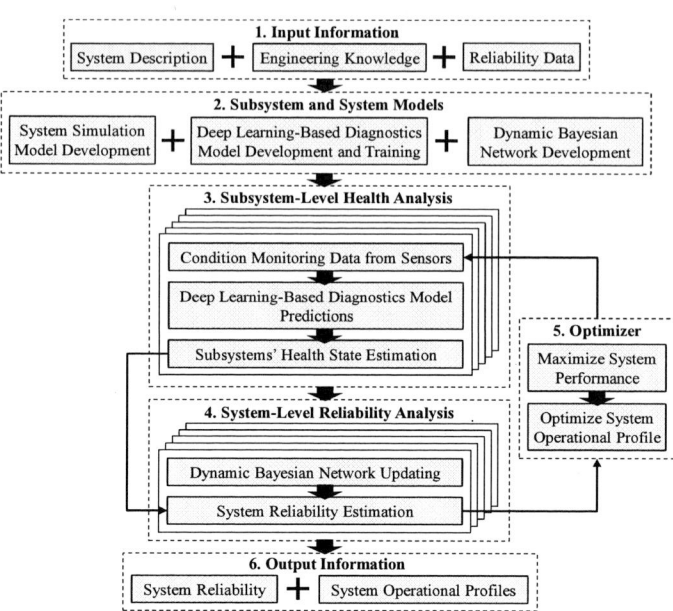

FIGURE 1. A graphical representation of the proposed framework.

description (e.g., list of components and subsystems, their functions and interactions, and system logic), engineering knowledge (e.g., physics-based models), reliability data (e.g., failure modes and associate failure rates), and environmental data. The inputs serve as the basis of developing subsystem- and system-level models.

3.2 Step 2: Subsystem and System Models

The following subsections provide details on how to develop the system simulation model, the deep learning-based diagnostics models, and the DBN model from the input information in order to implement the proposed performance optimization scheme for unmanned systems.

3.2.1 System Simulation Model.

A physics-based system simulation model needs to be developed in order to generate system operational profiles with simulated condition monitoring data from sensors under uncertain environmental conditions. The Simulink environment provided by MathWorks® [23] can be used for this purpose. These operational profiles will be used to identify the reliability-based optimum profile that maximizes the system performance by analyzing subsystems' health data.

3.2.2 Deep Learning Models for Subsystems' Health Diagnosis

Deep learning models can be useful for subsystems' health diagnosis when the underlying physics of a subsystem is either unknown or difficult to derive. An array of sensor measurements can be used as inputs for a deep learning model, and then a model is trained to learn the relationships between data streams from sensors to efficiently classify different health states (e.g., healthy, degraded, and critical) of a component. In the proposed approach, a number of deep learning mod-

Copyright © 2023 by ASME

els needs to be developed for each subsystem. The deep learning models can be trained utilizing real monitoring data and maintenance logs, if available. An illustration of the steps typically involved in data processing and deep learning model training for systems' health diagnosis is presented in Fig. 2. Deep learning uses artificial neural networks to model complex data sets. The process involves the learning of nonlinear relationships between input and output layers. One needs to carefully select and fine-tune its hyperparameters, such as model architecture, activation function, batch size, learning rate, and number of epochs, in order to achieve optimal performance. This allows the deep learning models to achieve higher accuracy and efficiency. Convolutional neural networks (CNNs) are commonly used in the diagnosis of mechanical components. They automatically extract features from layer inputs and control over-fitting using operations like pooling and dropout. The basic building block of a CNN is the convolution operation, which applies a filter to each local region of the input data to produce a feature map. The output of a CNN classifier is a probability distribution on predefined health classes that can be further utilized for system-level analysis [6]. For more details on the deep learning-based PHM approaches, the reader may refer to [24].

3.2.3 Dynamic Bayesian Networks for Uncertainty Propagation.
A DBN is a type of Bayesian network, which models the time dependencies between variable nodes. A Bayesian network shows the conditional probabilities between variables using a directed acyclic graph. It has parent and child nodes that represent discrete or continuous random variables, connected by directed edges. The conditional probability of a parent node is represented using a conditional probability table (CPT). The network uses Bayes' theorem to combine prior beliefs and observed evidence to generate a posterior probability, supporting decision-making and providing valuable insights. A DBN consists of an initial network at $t = 0$ that consists of prior distributions of the node variables, and a two-slice network that defines state transitions through arcs and temporal CPTs between the nodes. Fig. 3 shows an illustration of the two-slice transition network. To calculate the posterior distribution of the network at any time t, the temporal states of the nodes are updated using node evidences and prior information from the time

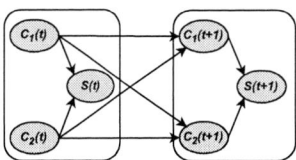

FIGURE 3. Two-slice transition network of a DBN.

slice at $t = 0$. For more details on DBNs, the reader may refer to [11]. DBNs are popular for system reliability and risk analysis as they can combine information from different sources and identify challenging-to-observe system states. They are often constructed from system-level fault trees that are used for failure modeling of multi-component systems.

3.3 Step 3: Subsystems' Health Analysis
Once subsystems' health diagnosis models are developed and trained based on their operational behavior, they can be deployed for online health diagnosis of the components and subsystems. These models need to be updated by retraining them time-to-time. This retraining process addresses changes in subsystem behavior due to changes in the operating environment.

The subsystems' health state classification process using deep learning models is illustrated in Fig. 4 (HS denotes health state). Usually, for components, the health states are classified into three categories: healthy, degraded, and critical. A trained deep learning model uses input condition data from sensors and predicts the current health state of a component or subsystem. These models can also be used to detect early signs of degradation or failure, which can be used to schedule maintenance or repairs before a catastrophic failure occurs. The proposed approach takes the prediction class probabilities into account, and propagates the information to the nodes of a system DBN model for system-level analysis.

3.4 Step 4: System Reliability Analysis
This step uses a system-level DBN created using system failure logic to evaluate the system reliability at specific time steps. The prior failure rates of different components can be obtained from maintenance logs and reliability data handbooks that contain failure rates for different commercial and military equipment. A parameterized DBN model needs to be updated based on

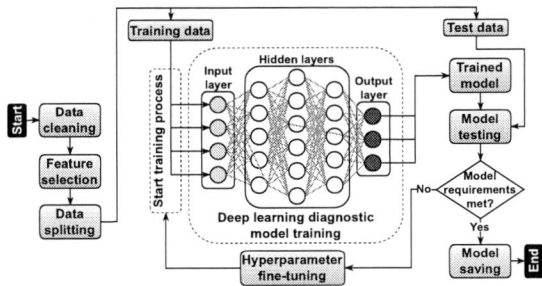

FIGURE 2. Illustration of deep learning model training process.

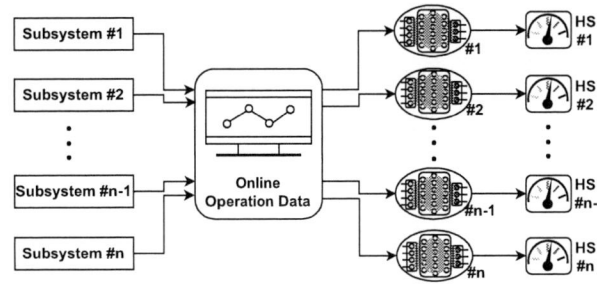

FIGURE 4. An illustration of the subsystems' health analysis process.

Copyright © 2023 by ASME

health information obtained from deep learning-based diagnostic models. In order to connect the subsystem health diagnosis step with the system reliability evaluation step, the identified health states from different subsystems are assigned different failure probabilities to use the health information as temporal evidences for the system DBN. The health state information from different subsystems at different time steps are then used to update the network in real time. The integration process of the health information and the system DBN is illustrated in Fig. 5. The updated reliability profiles at each time step can then be used for monitoring system health and performance optimization.

3.5 Step 5: System Performance Optimization

The system-level reliability analysis step enables us to use the time-dependent system reliability estimates from the DBN for optimization of the system performance. The proposed method seeks to optimize the performance of a system subject to reliability constraints, as opposed to optimizing for reliability whilst still maintaining a particular performance standard. This includes determining an operational profile that maximizes the anticipated performance of a system while satisfying a reliability constraint, such as a maximum allowable failure probability or a minimum system reliability at the end of a task or mission. The optimization scheme entails a predictive model of the design parameters (e.g., a system simulation model), the performance criteria, and a model of the reliability constraint for the system.

3.6 Step 6: Output Information

The proposed optimization scheme provides valuable outputs, such as the optimal system operational profiles and the corresponding system reliability profile for the entire mission. During the mission, these outputs can be deployed in real-time to further enhance the system's performance. In addition, the generated reliability profile may serve as a decision-making tool for scheduling maintenance by determining if the system's reliability at the end of the mission falls below a certain threshold. By utilizing this optimization scheme, mission-critical systems can achieve maximum performance and efficiency while maintaining the highest levels of reliability and safety.

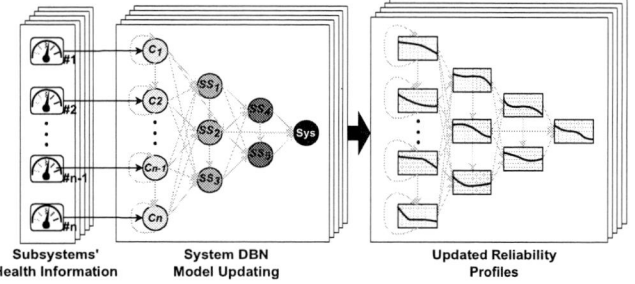

FIGURE 5. An illustration of the DBN model updating process.

Subsystems' Health Information
System DBN Model Updating
Updated Reliability Profiles

4 CASE STUDY

In order to demonstrate the proposed approach, a case study is developed using a simulation model for one of the critical subsystems of a USV. A USV consists of five major subsystems, namely the engine cooling and control system (ECCS), the power system, the navigation system, the data acquisition system, and the communication system [14]. The ECCS is a critical subsystem in the USV that controls and provides cooling for the marine diesel engine. The ECCS consists of the track control devices, the situational awareness devices, and the propulsion subsystem. The propulsion subsystem includes the marine diesel engine, lube oil pump, seawater pump, and freshwater pump, among other components that are monitored by sensors attached to them.

For this case study, the USV is assumed to sail approximately 2,400 nautical miles in the North Pacific Ocean for surveillance, with a sailing duration of around 200 hours. The environmental data with corresponding uncertainty are collected from open source information available on the web [25, 26]. The USV is expected to complete the given task without human intervention and with high reliability for the entire duration of the task. In order to achieve that goal, the problem is divided into several sub-problems that are solved in a sequential fashion as described in Section 3. This case study shows how the proposed approach can optimize the ECCS, a critical system in a USV, in a real-world example that demonstrates its effectiveness.

4.1 System Operational Profile

In order to obtain large quantities of data for training and using the previously discussed diagnosis models, the input data had to be developed for environmental and throttle or loading profile. For each input category, several profiles were created. By executing a given throttle profile through the model while it is subjected to various environmental profiles with uncertainty considerations, sufficient synthetic sensor data was able to be generated for use in the performance-optimizing framework.

The environmental profiles were created incorporating four parameters of air temperature, seawater temperature, air pressure, and seawater current, as shown in Table 1. The "Start" term refers to the environmental conditions that the vessel would experience at the beginning of the vessel's journey and "Finish" term refers to the environmental conditions that the vessel would experience at the end of its journey. Each of the four parameters are assumed to be uniformly distributed. The air and seawater temperature bounds change linearly with time, whereas the air pressure and sea current bounds remain constant. For example, following Table 1, the air temperature is uniformly distributed between 10°C and 22°C at the start of the simulation, and the bounds linearly increase to 20°C and 34°C, respectively, at the end of the mission. These values were based on environmental conditions assuming the vessel is traveling from San Diego to Hawaii during the summer using published data [25, 26]. These

Copyright © 2023 by ASME

four parameters comprise the conditions that a marine diesel engine would be subjected to during operation.

The throttle profiles are the desired loading that the operator is placing on the engine. The throttle was measured using a percentage, from 0 to 100 percent. The throttle profiles were manually generated using multi-level step functions in the MATLAB Simulink environment [23] to assess the performance under varying uncertain environmental conditions. Profiles were for 200 hours of operation which was based on potential mission time and USV fuel capacity.

4.2 System Description

The ECCS of a USV consists of all the components needed to control the vessel's propulsion capabilities. Propulsion is dependent on the marine diesel engine which is dependent on its cooling and control components. As the engine operates it generates a large amount of heat. The heat generation will cause temperatures to rise, potentially becoming detrimental to performance and component health. The cooling support subsystems remove the heat and expel it to the environment. The ECCS contains four cooling subsystems: air, freshwater, lube oil, and seawater. The air, freshwater, and lube oil subsystems all pull heat directly from the engine. The freshwater subsystem also cools the lube oil subsystem. The air and freshwater are then cooled by the seawater where the heat is finally dispelled back to the environment. The flow of these systems can be seen in the simplified block diagram shown in Fig. 6.

During operation, these subsystems are dependent on each other to ensure the proper functionality of the overall system.

TABLE 1. Environment profile uncertainties.

Environment	Upper bound		Lower bound	
	Start	Finish	Start	Finish
Air Temperature (°C)	22	34	10	20
Seawater Temperature (°C)	25	28	19	22
Air pressure (MPa)	0.102	0.102	0.1006	0.1006
Sea current (m/s)	1	1	0.5	0.5

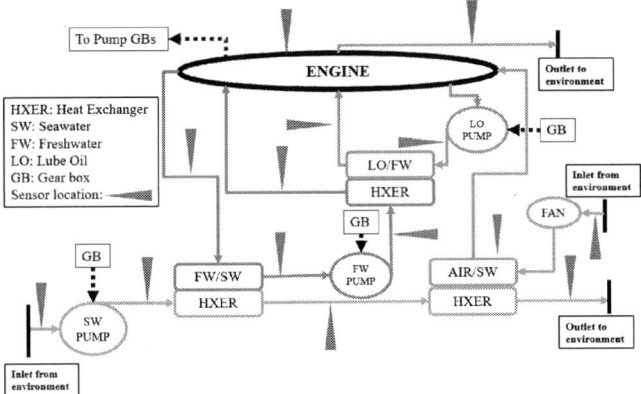

FIGURE 6. Block diagram of ECCS. Note: Some sensor locations contain multiple sensors.

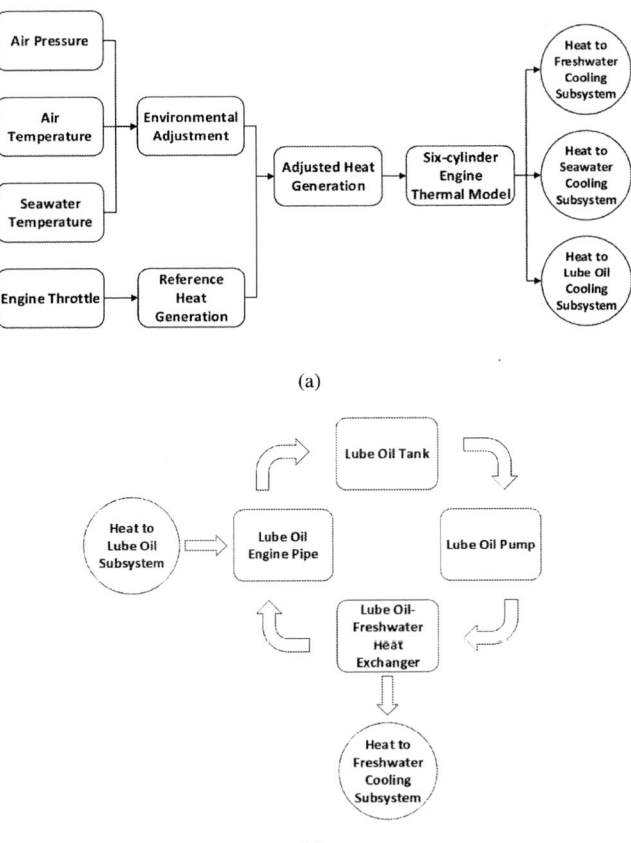

FIGURE 7. Illustrative representations of the Simulink models for (a) engine subsystem, and (b) lube oil subsystem.

Any degradation of one subsystem will result in greater stress on the others and possibly immediate or delayed failure. Sensors and the conditions they monitor are what indicate how well a system is operating. The ECCS model contains 44 sensors monitoring a variety of conditions throughout the system such as temperatures and pressures. If a single or several sensors begin seeing abnormal temperatures based on the current loading and environment conditions, this can be a sign a component has failed or is failing. This is the information passed on to the previously discussed models that are now tasked with determining the new system reliability.

4.3 ECCS Simulation Model and Data Generation

A physics-based simulation model for a USV is required to simulate multiple conditions that the USV could experience. The simulation model was created in MATLAB using MATLAB's Simulink and Simscape Libraries [23]. These proved to be effective tools due to the built-in physics considerations that are programmed into these libraries and the relatively simple approach to running them in parallel. A Python script was used to generate, control, and organize the input data into the model. The use

Copyright © 2023 by ASME

of the Python script (using [27], [28], and [29]) also allowed the integration of the other components of this software framework, such as the DBN structure.

The focus of the simulation model was the ECCS subsystems for the USV. In the ECCS, temperature and pressure sensors are placed throughout to monitor the health and performance of various systems. The engine (Fig. 7(a)) is the primary heat source for the vehicle and is responsible for providing the mechanical power to the other subsystems. This is modeled by considering two aspects: the environmental impact on the engine and the physical model for how the heat is dissipated from the engine through the pistons (based on a 6-cylinder engine design). The cooling subsystems (i.e., lube oil subsystem in Fig. 7(b)) are used to dissipate the heat from the engine to the environment.

The behavior of the engine was simulated by calculating the heat that is expected to be produced by the engine, based on the mechanical power demands of the vessel, the environmental adjustments k (Eq. (4) [30]), the efficiency of the engine, and the expected ratio between heat generated and mechanical power generated (51.5%:48.5%) [31]. So for every 48.5 watts of mechanical power needed, 51.5 watts of heat is generated by the engine before the specific heat adjustment factor β from Eq. (4) is applied.

$$
\begin{aligned}
k &= \left(P_0/P_{ref}\right)^m \left(T_{ref}/T_0\right)^n \left(T_{sw,ref}/T_{sw}\right)^s \\
\alpha &= k - 0.7(1-k)\left((1/\eta_m)-1\right) \\
\beta &= k/\alpha
\end{aligned} \tag{4}
$$

In this equation, k represents an adjustment factor of the engine due to environmental conditions, α represents the mechanical power adjustment factor, β represents the adjustment factor for the specific fuel consumption which is applied as the adjustment factor for the amount of heat that the engine generates, P_0 is the ambient air pressure, P_{ref} is the reference air pressure (1 atm, i.e., the standard air pressure at sea level), T_0 is the air temperature, T_{ref} is the reference air temperature (20°C, i.e., the average air temperature of San Diego during the summer), T_{sw} is the seawater temperature, $T_{sw,ref}$ is the reference seawater temperature (20°C, i.e., the average surface sea water temperature during the summer), m, n, and s are curve fitting importance parameters which would be tuned based on readings for a specific diesel engine, and η_m is the mechanical efficiency of the engine. The reference pressure and temperatures were obtained from [32,33].

In order to simulate the degradation that the different components of the USV would experience, Eq. (5) was used to represent the changing efficiencies of the pumps and engines:

$$
\eta = \frac{p_{in} - p_{leak}}{p_{in}} \tag{5}
$$

where p_{in} represents the input power into the component, p_{leak} represents the amount of power lost by the component, and η represents the efficiency of the component. This Eq.(5) is incorporated into the model (Figs. 7(a) and 7(b)) through the ef-

ficiency performance of the pumps for the freshwater, seawater, and lube oil subsystems. As the component degrades, the leakage or wear on the component is expected to grow. It is assumed that the leakage or wear of the component follows an exponential growth model with an offset that would be caused due to issues related to previous use or manufacturing. In order to perform system-level reliability-based performance optimization of the ECCS, physics-based reliability equations were developed for the components. This allowed the failure rate of components to become functions of information that can be recorded from sensors (such as the inlet temperature of various pumps or engine shaft speed). These equations were separately developed for all subsystems of the ECCS, including the lube oil pump, freshwater pump, seawater pump, and marine diesel engine. Failure rate equations for the pumps were adapted from reliability handbooks [34] and [35]. These handbooks provided the base failure rates as well as factors based on operating conditions that affect the base failure rates. For example, each of the pumps' failure rates ($\lambda_{pump}(\cdot)$) is a function of engine revolution speed (N) in RPM, the temperature of the inlet fluid (T) in °C, and the inlet volumetric flow rate (Q) in m^3/s. The equation for the failure rates of the pumps was composed of the failure rates of the subcomponents, such as λ_{SE} for pump seals, λ_{SH} for pump shaft, λ_{BE} for ball bearings, and λ_{FD} for the fluid driver. Assuming mutually exclusive failure modes, the equations for hourly failure rates of a generic pump, i.e., $\lambda_{Pump} = \lambda_{SE} + \lambda_{SH} + \lambda_{BE} + \lambda_{FD}$, can be written as

$$
\begin{aligned}
\lambda_{SE} &= \left(22.8 \times 10^{-6}\right)\left(0.5^{\frac{250-T}{18}}\right)(0.25C)\left(N/N_{90}\right)v(T) \\
\lambda_{SH} &= \left(1.8145 \times 10^{-10}\right)(460+T) \\
\lambda_{BE} &= \left(0.38 \times 10^{-6}\right)\left(C \times v_{op}/v(T)\right) \\
\lambda_{FD} &= \left(0.4 \times 10^{-6}\right)(1.5C)\left(5 \times N/N_{90}\right)^{1.3}
\end{aligned} \tag{6}
$$

In the above set of equations, C is an impact factor that is different for each type of pump, N_{90} is the shaft speed at 90 percent throttle, $v(T)$ is the liquid viscosity at temperature T, and v_{op} is the liquid viscosity at the expected operating temperature. These factors differ slightly based on the properties of the fluid and the impact factor of that pump. Once these failure rates were determined, an exponential relationship was assumed to find the reliability at any time t as $R(t; T, Q, N) = e^{-\lambda(T,Q,N)t}$. Substituting p_{leak} from Eq. (7) into Eq. (5) results in the Eq. (8), where p_{offset} represents offset in the amount of power loss of the component at the start of a given simulation run. Eq. (8) was used to model the efficiency of the pumps (Fig. 7(b)). The efficiency of the engine was simulated using Eq. (9). This was based on the curve fitting adjustment factor model proposed by [30], where k represents the adjustment factor due to external factors from the engine. This adjustment factor could be tuned to represent the impact of various external factors on an actual engine using physical data.

$$p_{leak} = (p_{in} - p_{offset})\left(1 - e^{-\lambda_{Pump}(T,Q,N)t}\right) + p_{offset} \quad (7)$$

$$\eta = (1 - p_{offset})e^{-\lambda_{Pump}(T,Q,N)t} \quad (8)$$

$$\eta_{eng} = (1 - p_{offset})e^{\frac{-\lambda_{eng}t}{k}} \quad (9)$$

To generate the training data for the diagnosis of the USV, an air temperature, sea temperature, and air pressure profile was used for each simulation. Each simulation was given a starting health state for each component as well as a control profile. The Monte Carlo simulation method was used to generate random samples of environmental profiles, initial health states, and the control profile.

4.4 Deep Learning Model Training for Subsystems' Health Diagnosis

The CNN models were trained for health state classification of four crucial subsystems of the ECCS: the marine diesel engine, the freshwater cooling subsystem, the seawater cooling subsystem, and the lube oil subsystem. The diagnostic models were trained using synthetic sensor data from the system simulation model. In order to minimize the possibility of encountering input data outside the training boundary, CNN models were trained using an exhaustive and balanced synthetic dataset (generated by the ECCS simulator) that includes the most possible conditions that the ECCS may experience due to environmental uncertainties. 20% of the synthetic data were selected randomly for testing, and the rest were used for model training. The data set consists of sensor readings primarily representing temperatures, pressures, and other measurements from different locations of the system in operation (see Fig. 6). To improve the modeling process, the data were scaled between 0 and 1. The data set comes with a total of ten health state classes for different subsystems depending on their efficiencies starting from 100% and gradually decreasing in steps of 10%. However, these classes are also associated with the three main health categories, healthy (efficiency > 80%), degraded, and critical (efficiency < 30%). The classification task was carried out by activating the neurons of the output layer using the softmax function that outputs a probability distribution of classes that can be further utilized as virtual evidence for updating the system DBN [36]. The performances of the CNN-based diagnostic models were measured by simply comparing the accuracy and the confusion matrices in predicting the health state classes. All four CNN models showed high accuracy rate exceeding 97% for both training and testing data sets. These models can be used to output real-time health information, such as determining whether a subsystem is functioning normally, experiencing degradation, or is in a critical state, given the environmental and operational uncertainties.

4.5 Dynamic Bayesian Network for ECCS

In order to create the structure of the DBN and parameterize it for system reliability analysis, we first created a fault tree (FT)

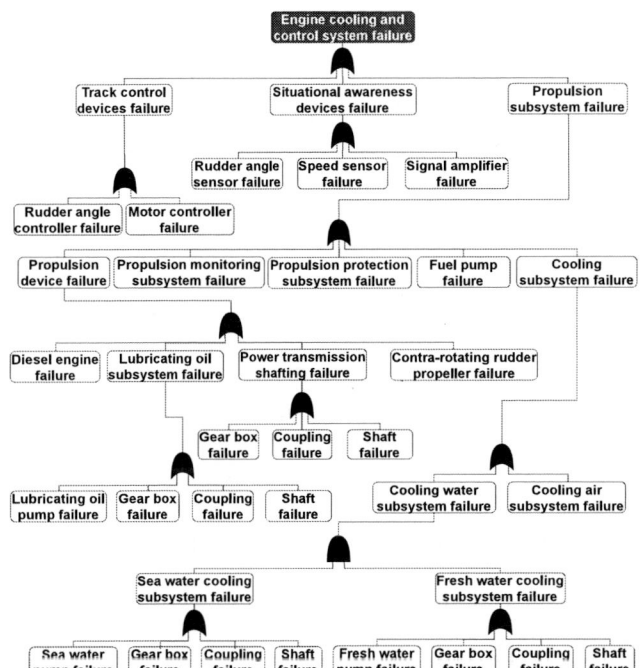

FIGURE 8. The fault tree model for the ECCS (adapted from [14]).

for the ECCS. FTs are developed for reliability modeling and analysis of complex systems that exhibit component degradation over time [37]. Fault trees are useful to understand the combinations of component and subsystem failures (failure paths) that lead to a system-level failure. A fault tree primarily consists of two types of nodes: event nodes and gate nodes. Events represent failures of subsystems and components. The system failure is modeled using the top event of a fault tree. The gates are used to represent dependencies between components. The AND gate represents a parallel system with independent components, and the OR gate represents a series system with dependent components.

The ECCS failure is modeled using a FT shown in Fig. 8; the model is adapted from [14]. The corresponding hourly failure rates of various components are presented in Table 2. The failure rates were obtained from NPRD [38], OREDA [35], and NSWC handbooks [34]. The ECCS is made of track control devices, situational awareness devices, and the propulsion subsystem – all connected through an OR gate. The failure of the propulsion device is modeled using a subtree that has the marine diesel engine and the lube oil pump among other components. Under the engine cooling subsystem, the seawater and freshwater subsystems are connected via an AND gate since failure of both subsystems will result in the failure of the cooling water system.

The FT model shown in Fig. 8 was utilized for developing the structure of the ECCS DBN model. The FT gates were converted to node CPTs in order to parameterize the DBN. The FT events were mapped to the DBN nodes into two states: "work-

Copyright © 2023 by ASME

ing" and "failed". The directed edges between nodes were created by connecting the input events to the output event of a FT gate. The next step was to establish the transition network by adding directed edges from parent nodes to child nodes at different time slices sequentially. A DBN node can also depend on its previous state, thus making it its own parent node in a transition network.

The final step was to create the node CPTs that model the causal relationships between different components and subsystems. The CPTs of the child nodes were determined based on the failure probabilities of their parent nodes and the associated FT gate type. The failure time distribution of individual components were assumed to follow the exponential distribution [14]. Assuming the failure rate of any component to be r, the failure time distribution can be written as $f(t) = re^{-rt}$. Thus, given a component was working at a previous time step, the probability of the same component having a failed state at the current time step is given by $\int_0^{\Delta t} f(t)dt = 1 - e^{-r\Delta t}$. Similarly, if a component has already failed, the failure probability will change to unity. The final DBN structure for the ECCS is shown in Fig. 9, which was developed using the GeNIe Modeler by BayesFusion [36]. The DBN consists of 37 nodes, 62 arcs, and 158 independent parameters. With new health state information, this DBN can be updated to evaluate the reliability profile of the ECCS of a USV. Using prior information obtained from the reliability data handbooks, the DBN model of the ECCS was evaluated for 200 hours with 10-hour time steps. The system reliability after 100 hours of operation was found to be around 0.84, and after 200 hours, it went down to 0.71. The propulsion device showed a prior reliability of 0.8 and the cooling subsystem showed 0.98 after 200 hours, which makes the propulsion device less reliable compared to the cooling subsystem of the ECCS.

TABLE 2. Failure rates for the ECCS components [38], [35], [34].

Failure event	Failure rate (h^{-1})
Air cooling subsystem failure	8.526×10^{-5}
Contra–rotating rudder propeller failure	8.933×10^{-4}
Coupling failure	1.086×10^{-6}
Diesel engine failure	1.537×10^{-4}
Freshwater pump failure	2.388×10^{-5}
Fuel pump failure	1.429×10^{-4}
Gear box failure	1.268×10^{-6}
lube oil pump failure	2.403×10^{-5}
Motor controller failure	4.753×10^{-6}
Propulsion monitoring subsystem failure	1.738×10^{-4}
Propulsion protection subsystem failure	1.738×10^{-4}
Rudder angle controller failure	3.102×10^{-6}
Rudder angle sensor failure	6.239×10^{-6}
Seawater pump failure	1.002×10^{-5}
Shafts failure	1.248×10^{-5}
Signal amplifier failure	1.095×10^{-5}
Speed sensor failure	1.248×10^{-5}

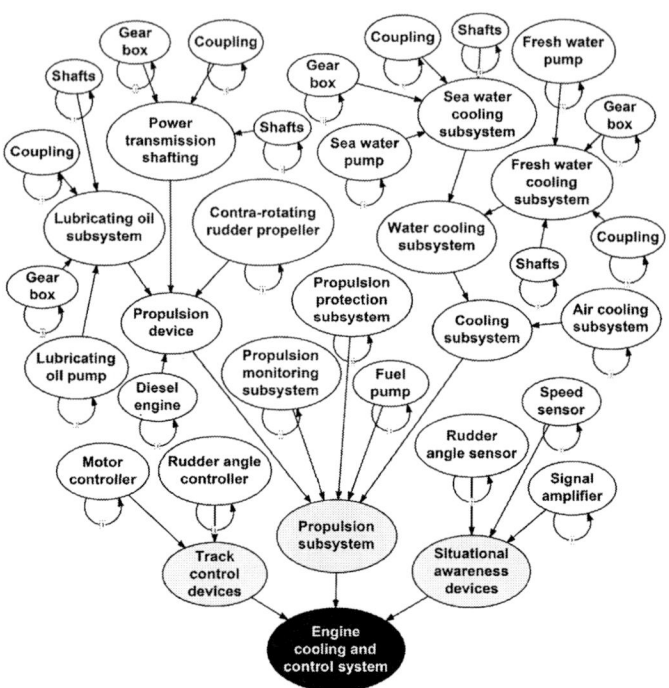

FIGURE 9. The DBN model for the ECCS. (Created using GeNIe Academic [36])

4.6 Optimization of ECCS Operational Profiles

The optimization problem in Section 2 was formulated and solved using a Bayesian optimizer. This problem is computationally expensive, as a single iteration of the optimizer took around 45 minutes to run. The computing platform used is a desktop computer with an Intel® Core™ i9-12900KF processor. As established by Frazier [39], a Bayesian optimizer is an optimization approach designed for optimization problems that take a significant amount of time to evaluate and has less than 20 dimensions. Since the engine throttle profile only had 5 different stages (i.e., design variables), this was an appropriate fit. These stages were the engine throttle levels during different parts of a 200-hour mission, i.e., for five design variables. Each stage or design variable represents a throttle level for 40 hours of the mission. This was implemented using an open-source Python Bayesian optimization library [29]. Eq. (10) was modified with a penalty function when run through the optimizer.

$$\text{minimize} \quad f(\mathbf{x}, \mathbf{p}) = -0.5\mu_{V_{avg}}(\mathbf{x}, \mathbf{p}) + 0.5\sigma_{V_{avg}}(\mathbf{x}, \mathbf{p})$$
$$\text{s.t.} \quad g(\mathbf{x}, \mathbf{p}) : R_{critical} - R_{sys}(\mathbf{x}, \mathbf{p}) \leq 0 \qquad (10)$$

The external penalty approach used can penalize any engine throttle profile that results in the reliability of the system being below the critical system health value, which was set as 0.71 (i.e., $R_{critical} = 0.71$), after some trial and error. The expected value ($\mu_{V_{avg}}$) and standard deviation ($\sigma_{V_{avg}}$) of the speed of the vessel were determined by transmitting the uncertain environmental pa-

Copyright © 2023 by ASME

rameters ($\mathbf{p}$) (from Section 4.1) and the throttle profile through the ECCS simulator.

The way that the Bayesian optimizer worked was by randomly sampling a few throttle profiles and exploring the performance of the different profiles. In this way, by exploring the design space, the optimizer was able to learn about the shape of the objective function and update its uncertainty at a given throttle profile based on this information using the Bayes' theorem. Then the acquisition function that was built into the Bayesian optimizer sampled points to find the throttle profile that resulted in the highest likely increase in value. For more details about the Bayesian optimization, the reader may refer to [39].

4.7 Results and Discussion

At the start of the experiment, the simulated USVs were given an initial health state and various uncertain environmental profiles. The initial health state was established to be a random value within the healthy state for each component. This represents a newly made USV, in which the only damage or degradation would be caused by manufacturer inaccuracies. With a given initial health state, the ECCS model would operate for 200 hours over the provided environment and throttle profiles. As the sensor data was gathered, the optimizer would adjust the throttle. The throttle is the only controllable variable the system can use to prevent excessive component wear. The incorporation of the optimizer seeks to maximize speed, which reduces travel time, while simultaneously reducing component wear. Component wear is guaranteed for any operating system. However, with the addition of the optimizer, the system can reduce component wear at times of more severe operational conditions, such as warmer seawater temperatures. Conversely, the optimizer can take advantage of beneficial conditions, such as low current. Thus, the system is still completing its mission/trip as desired but doing so with higher reliability at completion. As this framework is applied to more missions, the USV can see more time at sea where it belongs and less time in port undergoing repairs. There is also room for improvements in accuracy as the synthetic sensor data becomes replaced with real-time, actual sensor data. It is important to note that, at present, state-of-the-art approaches that combine PHM, system reliability analysis, and performance optimization into a single framework for the planning of operational profiles for unmanned systems do not exist. Thus, individual components of the framework were validated separately. The ECCS model was validated against results obtained from physical equations, while the CNN models were validated by splitting the entire dataset into training (80%) and testing (20%) datasets. Additionally, the DBN and the Bayesian optimizer were validated using results from the literature.

Using the optimizer, an optimized throttle profile was found for the engine to run for the full duration of the mission. This is shown in Fig. 10. During the early iterations of the optimizer, many different throttle profiles were explored. With each itera-

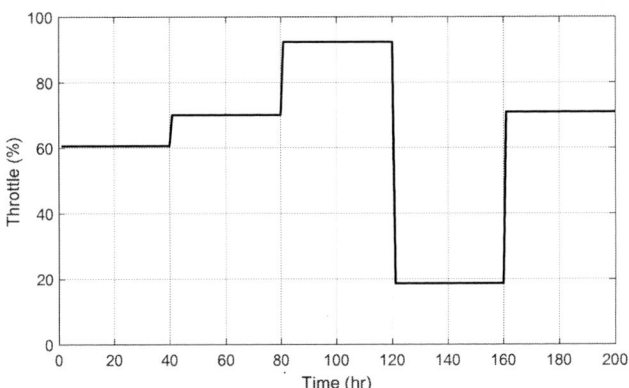

FIGURE 10. Optimized engine throttle profile.

tion, the throttle level increased for each stage of the journey until the optimizer converged to run the vessel at the optimized throttle. The average speed of the vessel was increasing with each iteration. The standard deviation of the vessel speed was found to be relatively low. As explained earlier, as the engine throttle increases, the wear on the components increases. This results in a slight decrease in the system reliability. As the throttle speed increases, expected system reliability decreased and allowed for a larger range of possible expected system reliability results. However, throughout all the iterations of the optimizer, all the system reliability results were above the critical reliability value of 0.71. This means that the throttle profiles tested are all feasible. The optimizer converged to a local optimal throttle profile for the system. The system and subsystem reliability profiles during optimized engine throttle profile load is shown in Fig. 11. The figure shows that the average ECCS reliability decreases to about 0.711 and with a higher rate compared to other critical subsystems. The diesel engine showed only 2.9% decrease in reliability after the 200-hour mission.

These results seem to be reasonable because the system's

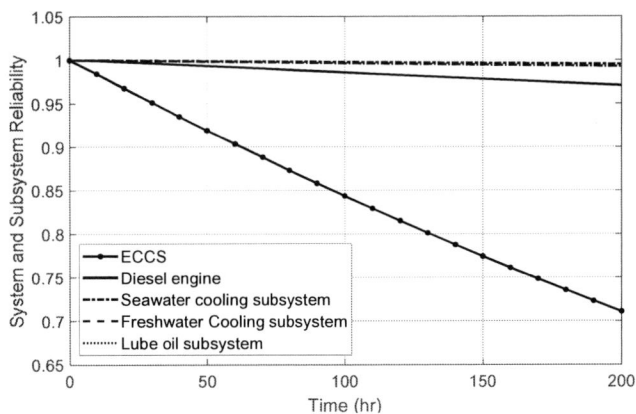

FIGURE 11. System and subsystem reliability during optimized engine throttle profile load.

Copyright © 2023 by ASME

reliability would not decrease to the critical reliability value regardless of what throttle profile was imposed onto it. The goal of the optimizer would be to get the vessel to travel as fast as possible, which is done by running the vessel at the optimized throttle for as long as the system reliability is above the critical value.

5 CONCLUSION

This paper's contributions include a new framework that integrates state-of-the-art techniques for reliability-based optimization of unmanned systems' operational profiles, which has many potential applications. The proposed method is the first to integrate health monitoring of multi-component or multi-subsystem systems with causal relationships.

Unmanned systems have garnered immense popularity due to their potential ability to carry out tasks with high efficiency, all without the need for human intervention. However, their potential ability to achieve such levels of efficiency is heavily reliant on a multitude of components that work synchronously during various operations. Given the complex nature of unmanned systems and the fact that they may be inaccessible during operations, ensuring their reliability and performance is of utmost importance. The proposed reliability-based framework combines state-of-the-art simulation techniques, health diagnosis methods, system reliability analysis, and optimization techniques to enable the operational planning of unmanned system operations. In particular, this approach leverages deep learning techniques for subsystem health monitoring, dynamic Bayesian networks for system reliability analysis, and Bayesian optimization schemes to optimize system performance.

To demonstrate the effectiveness of the proposed approach, the paper implements and executes the framework using a simulation model for the engine cooling and control system of an unmanned surface vessel. The study shows promising results in terms of identifying system operational profiles with high expected reliability under uncertain environmental conditions. Beyond this specific example, the framework's potential for improving the reliability and lifespan of unmanned systems is immense. With its ability to enable longer periods of deployment and the execution of more complex missions, this approach represents a significant step forward in the reliability-based optimization of unmanned systems' operations. However, the proposed concept in this paper is aligned with the understanding that the systems' operational profiles may be altered during a mission by implementing the framework in real-time. Offline operational profile planning can be conducted beforehand to enable the real-time implementation of the proposed system, which can lead to improved performance of unmanned systems with additional knowledge about the system environment. Future work will focus on combining offline and online operational profile planning for such systems.

ACKNOWLEDGMENT

This study was supported by the Office of Naval Research (ONR) under grant number N000142212459. This support does not constitute an endorsement by the funding agency of the opinions expressed in the paper.

REFERENCES

[1] Ruiz-Tagle Palazuelos, A., Droguett, E. L., and Pascual, R., 2020. "A novel deep capsule neural network for remaining useful life estimation". Proceedings of the Institution of Mechanical Engineers, Part O: Journal of Risk and Reliability, **234**(1), pp. 151–167.

[2] Helbing, G., and Ritter, M., 2018. "Deep learning for fault detection in wind turbines". Renewable and Sustainable Energy Reviews, **98**, pp. 189–198.

[3] Ellefsen, A. L., Cheng, X., Holmeset, F. T., Æsøy, V., Zhang, H., and Ushakov, S., 2019. "Automatic fault detection for marine diesel engine degradation in autonomous ferry crossing operation". In 2019 IEEE International Conference on Mechatronics and Automation (ICMA), IEEE, pp. 2195–2200.

[4] Wang, R., Chen, H., and Guan, C., 2021. "A Bayesian inference-based approach for performance prognostics towards uncertainty quantification and its applications on the marine diesel engine". ISA transactions, **118**, pp. 159–173.

[5] Lewis, A. D., and Groth, K. M., 2020. "A dynamic Bayesian network structure for joint diagnostics and prognostics of complex engineering systems". Algorithms, **13**(3), p. 64.

[6] Moradi, R., Ruiz-Tagle Palazuelos, A., Lopez Droguett, E., and Groth, K. M., 2020. "Toward a framework for risk monitoring of complex engineering systems with online operational data: A deep learning-based solution". Proceedings of the Institution of Mechanical Engineers, Part O: Journal of Risk and Reliability, p. 1748006X221079964.

[7] Guo, Y., Wang, H., Guo, Y., Zhong, M., Li, Q., and Gao, C., 2022. "System operational reliability evaluation based on dynamic Bayesian network and xgboost". Reliability Engineering & System Safety, **225**, p. 108622.

[8] Moradi, R., and Groth, K. M., 2020. "Modernizing risk assessment: A systematic integration of pra and phm techniques". Reliability Engineering & System Safety, **204**, p. 107194.

[9] Moradi, R., Cofre-Martel, S., Droguett, E. L., Modarres, M., and Groth, K. M., 2022. "Integration of deep learning and bayesian networks for condition and operation risk monitoring of complex engineering systems". Reliability Engineering & System Safety, **222**, p. 108433.

[10] Villa, V., Paltrinieri, N., Khan, F., and Cozzani, V., 2016. "Towards dynamic risk analysis: A review of the risk as-

Copyright © 2023 by ASME

sessment approach and its limitations in the chemical process industry". Safety science, **89**, pp. 77–93.

[11] Fenton, N., and Neil, M., 2018. Risk assessment and decision analysis with Bayesian networks. CRC Press.

[12] Wilson, A. G., and Huzurbazar, A. V., 2007. "Bayesian networks for multilevel system reliability". Reliability Engineering & System Safety, **92**(10), pp. 1413–1420.

[13] Cai, B., Liu, Y., Liu, Z., Tian, X., Dong, X., and Yu, S., 2012. "Using Bayesian networks in reliability evaluation for subsea blowout preventer control system". Reliability Engineering & System Safety, **108**, pp. 32–41.

[14] Gao, C., Guo, Y., Zhong, M., Liang, X., Wang, H., and Yi, H., 2021. "Reliability analysis based on dynamic Bayesian networks: a case study of an unmanned surface vessel". Ocean Engineering, **240**, p. 109970.

[15] Khakzad, N., 2015. "Application of dynamic Bayesian network to risk analysis of domino effects in chemical infrastructures". Reliability Engineering & System Safety, **138**, pp. 263–272.

[16] Lu, Q., and Zhang, W., 2022. "Integrating dynamic Bayesian network and physics-based modeling for risk analysis of a time-dependent power distribution system during hurricanes". Reliability Engineering & System Safety, **220**, p. 108290.

[17] Wang, L., Sun, W., Long, Y., and Yang, X., 2018. "Reliability-based performance optimization of tunnel boring machine considering geological uncertainties". IEEE Access, **6**, pp. 19086–19098.

[18] Hejazi, T.-H., Badri, H., and Yang, K., 2019. "A reliability-based approach for performance optimization of service industries: An application to healthcare systems". European Journal of Operational Research, **273**(3), pp. 1016–1025.

[19] Kim, C., Wang, S., Hwang, I., and Lee, J., 2007. "Application of reliability-based topology optimization for microelectromechanical systems". AIAA Journal, **45**(12), pp. 2926–2934.

[20] Frangopol, D. M., and Maute, K., 2003. "Life-cycle reliability-based optimization of civil and aerospace structures". Computers & Structures, **81**(7), pp. 397–410.

[21] Manley, J. E., 2008. "Unmanned surface vehicles, 15 years of development". In OCEANS 2008, IEEE, pp. 1–4.

[22] Yan, R.-j., Pang, S., Sun, H.-b., and Pang, Y.-j., 2010. "Development and missions of unmanned surface vehicle". Journal of Marine Science and Application, **9**, pp. 451–457.

[23] The MathWorks Inc., 2022. Simulation and Model-Based Design. MATLAB version: 9.13.0 (R2022b), Natick, Massachusetts, United States. https://www.mathworks.com/products/simulink.html.

[24] Ellefsen, A. L., Æsøy, V., Ushakov, S., and Zhang, H., 2019. "A comprehensive survey of prognostics and health management based on deep learning for autonomous ships". IEEE Transactions on Reliability, **68**(2), pp. 720–740.

[25] Seatemperature Info, 2023. World sea water temperatures: https://seatemperature.info.

[26] Weather Spark, 2023. Climate reports with the weather by month, day, even hour: https://weatherspark.com.

[27] The pandas development team, 2020. pandas-dev/pandas: Pandas, 2.

[28] Harris et al., 2020. "Array programming with NumPy". Nature, **585**(7825), Sept., pp. 357–362.

[29] Nogueira, F., 2023. Bayesian Optimization: Open source (Github) constrained global optimization tool for Python.

[30] Kökkülünk, G., Parlak, A., and Erdem, H. H., 2016. "Determination of performance degradation of a marine diesel engine by using curve based approach". Applied Thermal Engineering, **108**, pp. 1136–1146.

[31] Ouadha, A., and El-Gotni, Y., 2013. "Integration of an ammonia-water absorption refrigeration system with a marine diesel engine: A thermodynamic study". Procedia Computer Science, **19**, pp. 754–761.

[32] Time and Date AS, 2023. Climate & weather averages in San Diego, California, USA: https://www.timeanddate.com/weather/usa/san-diego.

[33] Scripps Institution of Oceanography, 2023. Voyager: How long until ocean temperature goes up a few more degrees?: https://scripps.ucsd.edu/news.

[34] Chesley, J. C., 1998. Handbook of Reliability Prediction Procedures for Mechanical Equipment. Naval Surface Warfare Center (NSWC) Carderock Division, Logistics Engineering Technology Branch.

[35] SINTEF, 2015. Offshore Reliability Data Handbook: 6th Edition. OREDA Participants, Det Norske Veritas (DNV).

[36] BayesFusion,LLC, 2022. "Genie modeler". User Manual. Available online: https://support. bayesfusion. com/docs/GeNIe.pdf.

[37] Ruijters, E., and Stoelinga, M., 2015. "Fault tree analysis: A survey of the state-of-the-art in modeling, analysis and tools". Computer science review, **15**, pp. 29–62.

[38] David, M., William, F., and John, R., 2016. Nonelectronic Parts Reliability Data (NPRD). Reliability Information Analysis Center, Utica, NY.

[39] Frazier, P. I., 2018. "Bayesian optimization". In Recent advances in optimization and modeling of contemporary problems. Informs, pp. 255–278.

Copyright © 2023 by ASME

Proceedings of the ASME 2023
International Design Engineering Technical Conferences and
Computers and Information in Engineering Conference
IDETC-CIE2023
August 20-23, 2023, Boston, Massachusetts

DETC2023-116921

DATA-DRIVEN CONTROL CO-DESIGN FOR INDIRECT LIQUID COOLING PLATE WITH MICROCHANNELS FOR BATTERY THERMAL MANAGEMENT

Zheng Liu
Industrial and Enterprise Systems Engineering, University of Illinois Urbana-Champaign
Urbana, IL

Yanwen Xu
Industrial and Enterprise Systems Engineering, University of Illinois Urbana-Champaign
Urbana, IL

Hao Wu
Industrial and Enterprise Systems Engineering, University of Illinois Urbana-Champaign
Urbana, IL

Pingfeng Wang
Industrial and Enterprise Systems Engineering, University of Illinois Urbana-Champaign
Urbana, IL

Yumeng Li
Industrial and Enterprise Systems Engineering, University of Illinois Urbana-Champaign
Urbana, IL

ABSTRACT

The demand for high-performance electric vehicles keeps increasing with the booming electric vehicles market. Thus, battery cooling is significant in enabling the battery to work under harsh discharge process. Thanks to its high efficiency and low cost, indirect liquid cooling is a widely used cooling method for batteries. Researchers are trying to optimize the plant or control design separately for a better cooling effect. However, they can only produce suboptimal results with low efficiency. Motivated by the imperfections of existing battery cooling systems, we aim to lower the cost of indirect liquid cooling for batteries considering the plant design and control design using data-driven co-design optimization. First, a finite element model of the battery was built to predict the temperature and validate our findings against experimental data. Then, a Gaussian process-based surrogate model combined with Monte Carlo simulation extended the prediction to many scenarios under harsh discharge process. Finally, the surrogate model obtained the optimal plant and control designs. The finite element model validated the optimal design, which lowered the cost by 10%.

Keywords: Battery cooling, Battery thermal management, Lithium-ion batteries, Control co-design, Data-driven method.

NOMENCLATURE

c	concentration of Li$^+$
V	unit of coolant velocity
C	unit C-rate
SOC	state of charge
T	temperature of the battery
T_{set}	target temperature of proportional control
T_{max}	maximum temperature of the battery
t	time
D	diffusivity
∇	gradient operator
J	transportation flux of Li$^+$
n	normal vector of electrode-electrolyte surface
F	Faraday's constant
I_n	electric current density
I_0	exchange current density
T	temperature
R	universal gas constant
α_c	cathodic charge transfer coefficient
α_a	anodic charge transfer coefficient
η	overpotential
Q	heat source
ρ	density

Copyright © 2023 by ASME

C_p	specific heat capacity
k	thermal conductivity
A	cross-section area of the pipe
ρ_l	density of the coolant
$\boldsymbol{u}$	velocity vector of the coolant
P	pouch cell nominal capacity
D_{Li}	Lithium diffusivity in electrode particle
σ	electrical conductivity
r	radius of electrode particle
$\varepsilon_{electrode}$	solid phase fraction in the electrodes
c^{max}	electrode maximum capacity
T_0	inlet coolant temperature
A	pipe cross-section area
V	coolant velocity
V_0	coolant initial velocity
K_p	proportional gain
d	distance between channels
E	cooling cost
Δp	pressure drop of coolant
U	voltage of the battery

1. INTRODUCTION

As the electric vehicles (EVs) market has grown, so that the demand for lithium-ion batteries (LIBs) are increasing [1]. Researchers have spent enormous efforts to explore new materials to enable LIBs to have better performance [2]. At the same time, effective battery thermal management systems (BTMSs) are considered essential to maximize lifetime, enhance power capability, and ensure the safety of LIBs [3]. Effective BTMSs are vital for high-performance EVs, which require LIBs to work under harsh discharge process [4]. Indirect liquid cooling is a practical battery cooling method, which has been adopted by the Chevrolet Volt and Tesla Model S [5]. However, most indirect liquid cooling systems are designed separately regarding physical system and control system [6]. Thus, they can only produce suboptimal results. One way to improve the design of the current BTMSs is applying control co-design.

Control co-design is a branch of multidisciplinary design optimization (MDO), in which the coupled design disciplines include plant and control design [7]. Control co-design has a great improvement from the traditional sequential design method, in which control design is performed after plant design is complete [8]. It uses integrated approach which can fully consider the design coupling between plant and control design, so that it is able to produce a system optimal result [9].

To optimize a design, different methods can be used. Experimental methods have difficulties studying the electrochemical-thermal-mechanical couplings and the capacity fading mechanisms of the LIBs under specific physics of failure [10]. This research adopts numerical methods, including physics-based simulations and a data-driven surrogate model. Compared to the experimental methods, the numerical methods are relatively cheap and time-saving [11]–[14]. Furthermore, data-driven surrogate models can predict the outputs of a large-scale system for any desirable design input through the realization of a regression model and a random process [15]. It

can efficiently capture the general trend of the properties of the system affected by various design variables [16].

Our paper targets the LIB pouch cell with the cold plate indirect cooling system in the Chevrolet Volt battery pack, as shown in Figure 1. Motivated by the imperfections of the existing indirect liquid cooling system, aiming to explore the potential designs of active indirect liquid cooling systems for LIBs to achieve the lowest cooling cost through data-driven design optimization. It uses numerical methods to explore the control co-design (cooling channel design and feedback control design) of the pouch cell cooling system under harsh discharge process. First, based on experimental data, we developed a multiphysics finite element (FE) model based on the pouch cell system to investigate the relationship between cooling channel design, feedback control design, and the thermal behaviors of the system. Then, a Gaussian process-based (GP) surrogate model combined with Monte Carlo simulation (MCS) was constructed, using FE simulation results as training data to scan the design space efficiently. After that, we obtained the optimal design to achieve the lowest cooling cost while ensuring the maximum temperature of LIB below 35°C to ensure performance. Finally, the optimal design was validated by the FE model.

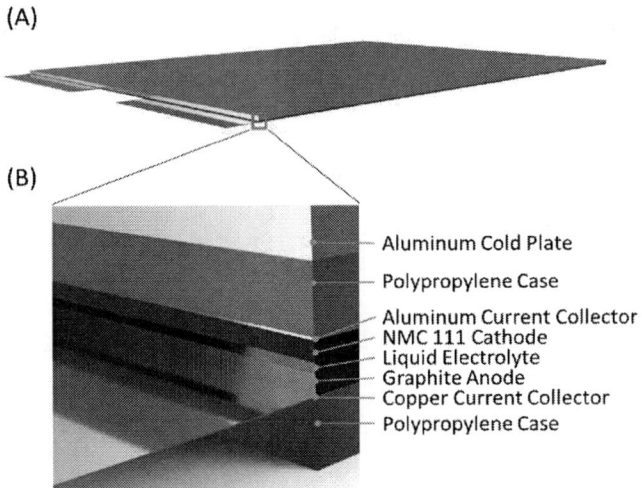

FIGURE 1. (A) The sketch of the pouch cell with the cold plate. **(B)** Layers of the pouch cell with the cold plate.

2. MATERIALS AND METHODS

This section introduces how to optimize the indirect cooling system of the pouch cell using the control co-design method based on multiphysics-based FE model and GP-based surrogate model.

2.1 Multiphysics-Based FE Model

We used COMSOL Multiphysics software to conduct the multiphysics-based FE simulation of the pouch cell with a cold plate, investigating the temperature change during the battery discharging process. Based on Fick's law and the mass

Copyright © 2023 by ASME

conservation equation [17], we can obtain the concentration of Li$^+$ in the electrode particle:

$$\partial c/\partial t - D\nabla(\nabla c) = 0 \qquad (1)$$

The boundary condition of Eqn. (1) is controlled by I_n at the electrode-electrolyte surface:

$$\boldsymbol{J} \cdot \boldsymbol{n} = I_n/F \qquad (2)$$

I_n in Eqn. (2) obeys the Butler-Volmer kinetics [18]:

$$I = I_0\left[exp\left(\frac{\alpha_a zF\eta}{RT}\right) - exp\left(\frac{\alpha_c zF\eta}{RT}\right)\right] \qquad (3)$$

The Joule heat by the charging or discharging process of pouch cell is Q. Then, we can calculate the temperature change of the battery, cold plate, and coolant by heat transfer equation:

$$\rho C_p \frac{\partial T}{\partial t} + \nabla(-k\nabla T) = Q \qquad (4)$$

The thermal conductivity of the pouch cell is anisotropic, with the in-plane thermal conductivity as 25 W/(m·K) and the out-of-plane thermal conductivity as 3.4 W/(m·K). Furthermore, the coolant flow in the cooling channels can be simulated by Navier-Stokes equations [19]:

$$\frac{\partial A\rho_l}{\partial t} + \nabla(A\rho_l\boldsymbol{u}) = 0 \qquad (5)$$

where the velocity vector $\boldsymbol{u}$ is in m/s. Moreover, all the values of parameters used in the FE model are shown in Table 1.

TABLE 1: Values of parameters in FE model

Parameters	Value
P	15 Ah
D_{Li}	1.45×10^{-13} m^2/s
σ	100 S/m
r	2.5 μm
$\varepsilon_{electrode}$	0.384
I_0	0.96 A/m^2
c^{max}	31507 mol/m^3
T_0	25 °C
C_p	1280 J/(kg·K)
A	0.5 mm^2

Fig. 1(A) is the sketch of the LIB pouch cell with cold plate. Besides the aluminum cold plate, which contains ten cooling channels, the LIB pouch cell has 7 layers. As shown in Fig. 1(B), polypropylene case, copper current collector, graphite anode, liquid electrolyte (LiPF$_6$), NMC 111 cathode, aluminum current collector, and polypropylene case make up the pouch cell. Current is assigned through the tab of aluminum current

collector, and the tab of the copper current collector is grounded. The FE model can be easily modified to other LIB chemistries since each battery component in the FE model is designed independently.

2.2 Control Co-Design Method

Co-design connects the control design and physics design [20]. As for the control design of this cooling system, proportional control is adopted to control the coolant velocity. So, the temperature of the pouch cell can be kept within $15-35$ °C to maintain high performance [5]. Since the battery's temperature is controlled within a relatively small range, the influence of the temperature difference of the battery can be ignored. The maximum temperature of the battery is used for proportional control.

$$V = V_0 + K_p(T_{set} - T_{max}) \qquad (6)$$

Due to the power constraints of the pump for coolant, the value of V is between 10^{-6} to 2×10^{-4} kg/s. While K_p [$\times10^{-4}$ kg/(°C·s)] has a limit between -12 to -7. By tuning K_p, we aim to find the best control strategy to save energy. Furthermore, the cooling cost [21] can be calculated as follows:

$$E = \int V\Delta p \, dt \qquad (7)$$

As shown in Fig. 2, the distance between ten channels near current collectors is adjustable. Because the location near current collectors has a higher temperature [22], this phenomenon was also found during our experiments. By adjusting d, overheating can be reduced.

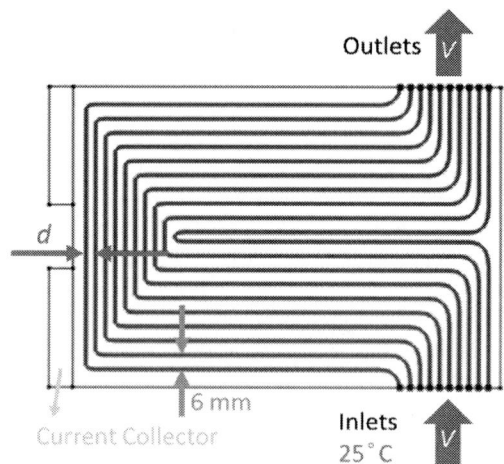

FIGURE 2. The sketch of cooling channels.

So, we have two variables: plant design (d) and control design (K_p). The goal is to lower the cooling cost while ensuring the temperature in the range. The control co-design problem can be formulated as follows:

Copyright © 2023 by ASME

$$\min_{K_p, \ d} E(K_p, \ d)$$

Subject to: $\quad T_{max} - 35°C \leq 0 \qquad (8)$

A simultaneous variation of both sets of variables is required to reach the system's optimal design [7]. So, we apply the GP model to obtain the optimal design to reduce the cooling cost.

2.3 GP-Based Surrogate Model

The surrogate model can reduce the calculation cost by replacing expensive FE simulations with much cheaper calculations [23]. GP-based surrogate model generates an estimated surface from a scattered set of points via a covariance governed GP interpolation method [24]. It can use the training samples to build the model and predict new response sample points [25]. We applied GP-based surrogate model:

$$G_K(\boldsymbol{x}) = T(\boldsymbol{x}) + S(\boldsymbol{x}) \qquad (9)$$

where $G_K(\boldsymbol{x})$ represents the prediction results of the performance function at point $\boldsymbol{x}$ using the GP model, $T(x)$ represents the trend function and is a polynomial term of $\boldsymbol{x}$ that interpolates the input sample points, and $S(x)$ is a Gaussian stochastic process with zero mean and variance σ^2. Furthermore, the polynomial term $T(x)$ is the product of regression basis function f and the regression coefficients $\boldsymbol{\beta}$.

$$T(\boldsymbol{x}) = f^T(\boldsymbol{x})\boldsymbol{\beta} \qquad (10)$$

The co-variance function between two arbitrary input points x_i and x_j can be defined as

$$Cov[S(\boldsymbol{x}_i), S(\boldsymbol{x}_j)] = \sigma^2 R(\boldsymbol{x}_i, \boldsymbol{x}_j) \qquad (11)$$

where $R(x_i, x_j)$ represents the automatic relevance determination-squared exponential correlation function matrix [26], which can be expresses as

$$R(\boldsymbol{x}_i, \boldsymbol{x}_j) = \exp\left[\sum_{p=1}^{n} a_p |x_i^p - x_j^p|^{b_p}\right] \qquad (12)$$

where a_p and b_p are parameters of the model. GP model can also provide the variance of the prediction, which indicates the uncertainty of the prediction. The GP predictor $\hat{g}(\mathbf{x})$ follows a Gaussian distribution at an untried point $\mathbf{x}$.

$$\hat{g}(\mathbf{x}) \sim N\left(\mu_G(\mathbf{x}), \sigma_G^2(\mathbf{x})\right) \qquad (13)$$

where $\mu_G(\mathbf{x})$ is the prediction.

$$\mu_G(\mathbf{x}) = f^T(\boldsymbol{x})\hat{\boldsymbol{\beta}} + r(\mathbf{x})\mathbf{R}^{-1}(\boldsymbol{y} - \mathbf{F}\hat{\boldsymbol{\beta}}) \qquad (14)$$

$\sigma_G^2(\mathbf{x})$ is the variance of $\hat{g}(\mathbf{x})$.

$$\begin{aligned}
\sigma_G^2 &= \hat{\sigma}_G^2\{1 - r(\mathbf{x})^{\mathrm{T}}\mathbf{R}^{-1}r(\mathbf{x}) \\
&+ [\mathbf{F}^{\mathrm{T}}\mathbf{R}^{-1}r(\mathbf{x}) \\
&- f(\mathbf{x})]^{\mathrm{T}}(\mathbf{F}^{\mathrm{T}}\mathbf{R}^{-1}\mathbf{F})^{-1}(\mathbf{F}^{\mathrm{T}}\mathbf{R}^{-1}r(\mathbf{x}) \\
&- f(\mathbf{x}))\}
\end{aligned} \qquad (15)$$

where $\boldsymbol{y}$ is the regression basis function, $\mathbf{F}$ is a $m \times p$ matrix with rows $f^{\mathrm{T}}(\boldsymbol{x})$. $r(\mathbf{x})$ is the correlation vector.

$$r(\mathbf{x}) = [R(\mathbf{x}, \mathbf{x_1}), R(\mathbf{x}, \mathbf{x_2}), \dots, R(\mathbf{x}, \mathbf{x_m})]^{\mathrm{T}} \qquad (16)$$

$\mathbf{R}$ is the correlation matrix, which is composed of correlation functions evaluated.

$$\mathbf{R} = \left[R(\mathbf{x}_i, \mathbf{x}_j)\right]^{\mathrm{T}}, 1 \leq i \leq m, 1 \leq j \leq m \qquad (17)$$

$\hat{\boldsymbol{\beta}}$ is the least-square estimate of $\boldsymbol{\beta}$.

$$\hat{\boldsymbol{\beta}} = (\mathbf{F}^{\mathrm{T}}\mathbf{R}^{-1}\mathbf{F})^{-1}\mathbf{F}^{\mathrm{T}}\mathbf{R}^{-1}\boldsymbol{y} \qquad (18)$$

$\hat{\sigma}_G^2$ can be calculated as shown below.

$$\hat{\sigma}_G^2 = \frac{1}{m}(\boldsymbol{y} - \mathbf{F}\hat{\boldsymbol{\beta}})^{\mathrm{T}}\mathbf{R}^{-1}(\boldsymbol{y} - \mathbf{F}\hat{\boldsymbol{\beta}}) \qquad (19)$$

3. RESULTS AND DISCUSSION

We first validated the outputs of our multiphysics-based FE model with experiment results under the same discharge condition. After that, we got the training set for the GP-based surrogate model from multiphysics-based FE simulation of harsh discharge process. Then, we trained the GP-based surrogate model. Through the plotted space from the surrogate model, we obtained the optimal design with the lowest cooling cost while maintaining the LIB's high performance. Finally, we validated the optimal design through FE simulation.

3.1 Multiphysics-based FE Model Validation

We first conducted simulations under the same condition as the experiments to compare the charging voltage curve to validate the FE model of the pouch cell without cold plate under 1C charging. As shown in Fig. 3(A), the simulation outputs estimate the battery voltage in relatively high fidelity. We also used an IR camera to measure the temperature of the pouch cell to compare with the temperature output from the simulation. As shown in Fig. 3(B), the simulation has the same temperature distribution as the experiment. Moreover, simulation can also accurately predict T_{max} of the pouch cell, which is shown in Fig. 3(C). For our study, the physics-based simulation can save more than 90% of the time compared to the experimental method. Moreover, the experimental method requires manually charging the battery after every discharge process and waiting for the battery to cool down, which consumes much time.

Copyright © 2023 by ASME

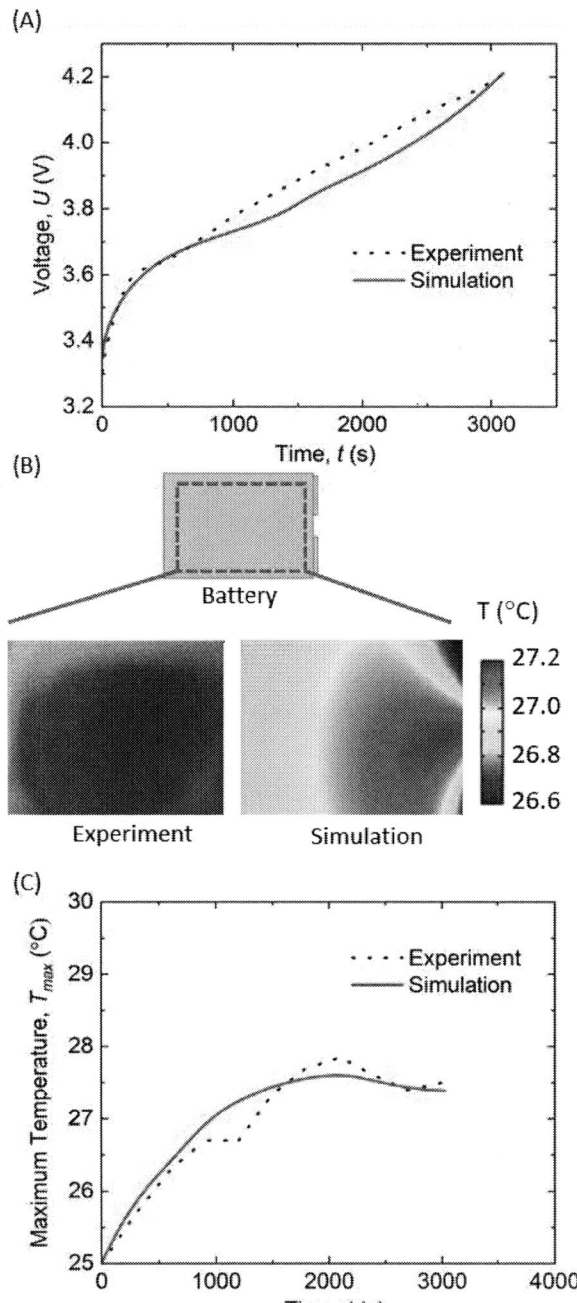

FIGURE 3. (A) Voltage of LIB from experiment and simulation. (B) Temperature distribution of LIB from experiment and simulation. (C) Maximum temperature of LIB from experiment and simulation.

3.2 GP-Based Surrogate Model

After validating the FE model, we moved forward to generate training data for surrogate model. Based on harsh loading condition of 5 C discharging from 100% to 20% of *SOC*. 24 points based on FFD (fractional factorial design) and 24 points based on LHS (Latin hypercube sampling) [27] were generated on the design space (K_p and d). Using the FE model, both cooling cost and maximum temperature can be obtained for these 48 points. Then, we constructed the GP-based surrogate model for both cooling cost and maximum temperature. The generated results are shown in Fig. 4.

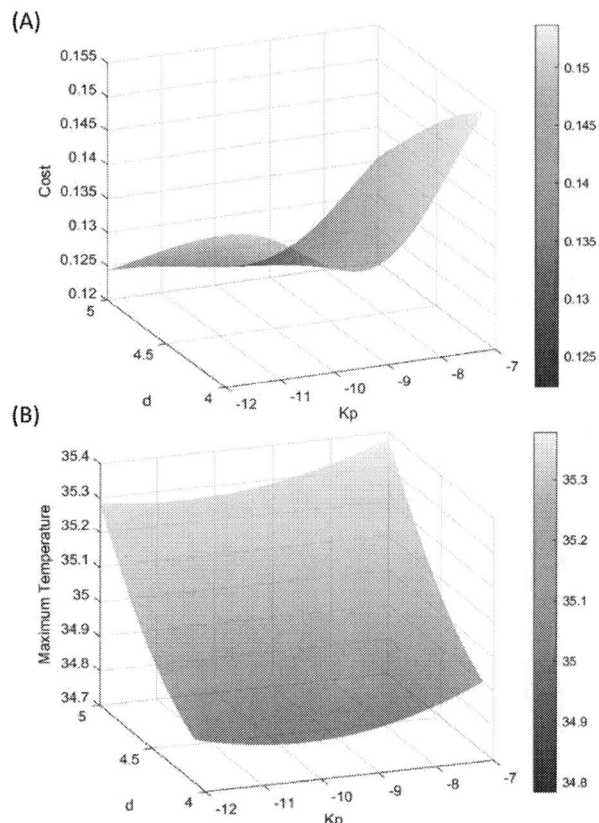

FIGURE 4. (A) Cooling cost response surface over the design space. (B) Maximum temperature difference response surface over the design space.

3.3 Optimal Design and Validation

With the GP-based surrogate model, we used MCS to generate more sampling regarding inputs (K_p and d) and obtained the corresponding outputs (E and T_{max}). The desired response is to lower E while ensuring T_{max} below 35°C. By following the criteria, the optimal design is shown in Table 2. Also, we validated the optimal design through FE simulation. As we can see from the FE simulation results, the data-driven method provides the accurate solution for optimal design with less than 1.6% error of cooling cost. At the same time, the temperature of the LIB is controlled below 35°C.

TABLE 2: The optimal design

K_p	d (mm)	Surrogate		FE	
		E (J)	T_{max} (°C)	E (J)	T_{max} (°C)
-9.40	4.604	0.127	34.99	0.129	34.98

Copyright © 2023 by ASME

Compared to the lowest cooling cost of the feasible designs in the 48 samples initially simulated, the optimal design lowers the cooling cost by more than 1.5%. Besides, the cooling cost of the optimal design is 10% lower than the average of the initial samples. Furthermore, we investigate the temperature of LIB during the discharge process using the optimal design. As shown in Fig. 5, the cooling system effectively controls the temperature of LIB below 35°C.

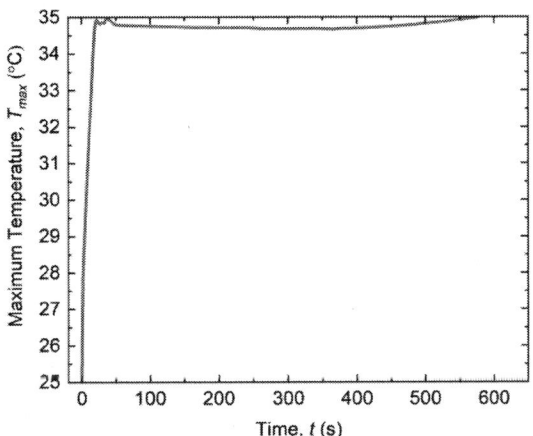

FIGURE 5. The maximum temperature from the FE simulation of the optimal design

4. CONCLUSION

In this study, the LIB's indirect liquid cooling system was explored through data-driven control co-design method. It efficiently achieved the cooling channel design and coolant velocity control with the help of experimental results, the FE model, and the GP surrogate model. Battery cell was tested, and the obtained data was then used to validate the multiphysics-based FE model of battery with indirect cooling system under harsh discharge process. Then, a GP surrogate model was developed to predict the cooling cost and the temperature of the battery. Using the GP surrogate model, the optimal design was obtained. As a result, the cooling cost was lowered by 10%, while the battery performance is ensured. The FE model can be adapted for other LIBs with different chemistry and geometric designs. Moreover, the data-driven control co-design method can be used to design the parameters of other battery thermal management systems.

ACKNOWLEDGEMENTS

The authors thank Long Wu, Yiqi Liu, and Tim Deppen for helpful discussions. This work was supported by the supplement project from the National Science Foundation (NSF) to the Engineering Research Center for Power Optimization of Electro-Thermal Systems (POETS) with cooperative agreement EEC-1449548.

REFERENCES

[1] K. Young, C. Wang, L. Y. Wang, and K. Strunz, "Electric Vehicle Battery Technologies," in *Electric Vehicle Integration into Modern Power Networks*, New York, NY: Springer New York, 2013, pp. 15–56. doi: 10.1007/978-1-4614-0134-6_2.

[2] N. Nitta, F. Wu, J. T. Lee, and G. Yushin, "Li-ion battery materials: present and future," *Mater. Today*, vol. 18, no. 5, pp. 252–264, Jun. 2015, doi: 10.1016/j.mattod.2014.10.040.

[3] C. Roe *et al.*, "Immersion cooling for lithium-ion batteries – A review," *J. Power Sources*, vol. 525, no. February, p. 231094, Mar. 2022, doi: 10.1016/j.jpowsour.2022.231094.

[4] X. Zhou *et al.*, "Strategies towards Low-Cost Dual-Ion Batteries with High Performance," *Angew. Chemie Int. Ed.*, vol. 59, no. 10, pp. 3802–3832, Mar. 2020, doi: 10.1002/anie.201814294.

[5] D. Chen, J. Jiang, G.-H. Kim, C. Yang, and A. Pesaran, "Comparison of different cooling methods for lithium ion battery cells," *Appl. Therm. Eng.*, vol. 94, pp. 846–854, Feb. 2016, doi: 10.1016/j.applthermaleng.2015.10.015.

[6] A. K. Thakur *et al.*, "A state of art review and future viewpoint on advance cooling techniques for Lithium–ion battery system of electric vehicles," *J. Energy Storage*, vol. 32, no. October, p. 101771, Dec. 2020, doi: 10.1016/j.est.2020.101771.

[7] T. Cui and P. Wang, "Reliability-Based Co-Design of Lithium-Ion Batteries for Enhanced Fast Charging and Cycle Life Performances," in *Volume 3A: 47th Design Automation Conference (DAC)*, Aug. 2021, pp. 1–14. doi: 10.1115/DETC2021-71402.

[8] A. L. Mular, D. N. Halbe, and D. J. Barratt, *Mineral processing plant design, practice, and control: proceedings*, vol. 1. SME, 2002.

[9] J. T. Allison and D. R. Herber, "Multidisciplinary design optimization of dynamic engineering systems," *AIAA J.*, vol. 52, no. 4, pp. 691–710, 2014, doi: 10.2514/1.J052182.

[10] Z. Zheng *et al.*, "Lithiation Induced Stress Concentration for 3D Metal Scaffold Structured Silicon Anodes," *J. Electrochem. Soc.*, vol. 166, no. 10, pp. A2083–A2090, 2019, doi: 10.1149/2.1031910jes.

[11] H. Wu, Z. Hu, and X. Du, "Time-Dependent System Reliability Analysis With Second-Order Reliability Method," *J. Mech. Des.*, vol. 143, no. 3, pp. 1–10, Mar. 2021, doi: 10.1115/1.4048732.

[12] Z. Zheng *et al.*, "Electrical and Thermal Active Co-Management for Lithium-ion Batteries," in *2022 IEEE Transportation Electrification Conference & Expo (ITEC)*, Jun. 2022, pp. 1159–1162. doi: 10.1109/ITEC53557.2022.9813807.

[13] Z. Zheng, Z. Liu, P. Wang, and Y. Li, "Numerical modeling on the delamination-induced capacity degradation of silicon anode," *J. Energy Storage*, vol. 43, no. November, p. 103190, Nov. 2021, doi: 10.1016/j.est.2021.103190.

[14] Z. Zheng, Z. Liu, P. Wang, and Y. Li, "Design of Three-Dimensional Bi-Continuous Silicon Based Electrode

Copyright © 2023 by ASME

Materials for High Energy Density Batteries," in *Volume 3A: 48th Design Automation Conference (DAC)*, Aug. 2022, vol. 3-A, pp. 1–8. doi: 10.1115/DETC2022-89652.

[15] H. Wu, Z. Zhu, and X. Du, "System Reliability Analysis With Autocorrelated Kriging Predictions," *J. Mech. Des.*, vol. 142, no. 10, pp. 1–11, Oct. 2020, doi: 10.1115/1.4046648.

[16] D. Beeson and G. Wiggs, "Gaussian process meta-models for efficient probabilistic design in complex engineering design spaces," *Int. Des. Eng. Tech. Conf. Comput. Inf. Eng. Conf.*, vol. 4739, pp. 785–798, 2005.

[17] Y. C. Song, Z. Z. Li, A. K. Soh, and J. Q. Zhang, "Diffusion of lithium ions and diffusion-induced stresses in a phase separating electrode under galvanostatic and potentiostatic operations: Phase field simulations," *Mech. Mater.*, vol. 91, pp. 363–371, Dec. 2015, doi: 10.1016/j.mechmat.2015.04.015.

[18] A. Latz and J. Zausch, "Thermodynamic derivation of a Butler–Volmer model for intercalation in Li-ion batteries," *Electrochim. Acta*, vol. 110, pp. 358–362, Nov. 2013, doi: 10.1016/j.electacta.2013.06.043.

[19] S. Panchal, R. Khasow, I. Dincer, M. Agelin-Chaab, R. Fraser, and M. Fowler, "Thermal design and simulation of mini-channel cold plate for water cooled large sized prismatic lithium-ion battery," *Appl. Therm. Eng.*, vol. 122, pp. 80–90, Jul. 2017, doi: 10.1016/j.applthermaleng.2017.05.010.

[20] T. Cui, Z. Zheng, and P. Wang, "Control Co-Design of Lithium-Ion Batteries for Enhanced Fast-Charging and Cycle Life Performances," *J. Electrochem. Energy Convers. Storage*, vol. 19, no. 3, pp. 1–11, Aug. 2022, doi:

10.1115/1.4053027.

[21] M. Akbarzadeh *et al.*, "A comparative study between air cooling and liquid cooling thermal management systems for a high-energy lithium-ion battery module," *Appl. Therm. Eng.*, vol. 198, no. July, p. 117503, Nov. 2021, doi: 10.1016/j.applthermaleng.2021.117503.

[22] P. Zhu, D. Gastol, J. Marshall, R. Sommerville, V. Goodship, and E. Kendrick, "A review of current collectors for lithium-ion batteries," *J. Power Sources*, vol. 485, no. November 2020, p. 229321, Feb. 2021, doi: 10.1016/j.jpowsour.2020.229321.

[23] Y. Xu and P. Wang, "Adaptive Surrogate Models for Uncertainty Quantification with Partially Observed Information," in *AIAA SCITECH 2022 Forum*, 2022, p. 1439.

[24] Y. Xu and P. Wang, "An Enhanced Squared Exponential Kernel With Manhattan Similarity Measure for High Dimensional Gaussian Process Models," in *Volume 3B: 47th Design Automation Conference (DAC)*, Aug. 2021, vol. 3B-2021, pp. 1–10. doi: 10.1115/DETC2021-71445.

[25] R. Y. Rubinstein and D. P. Kroese, *Simulation and the Monte Carlo method*, vol. 10. John Wiley & Sons, 2016.

[26] S. Fang and H. Chiang, "A High-Accuracy Wind Power Forecasting Model," *IEEE Trans. Power Syst.*, vol. 32, no. 2, pp. 2016–2017, 2017, doi: 10.1109/TPWRS.2016.2574700.

[27] J. C. Helton and F. J. Davis, "Latin hypercube sampling and the propagation of uncertainty in analyses of complex systems," *Reliab. Eng. Syst. Saf.*, vol. 81, no. 1, pp. 23–69, Jul. 2003, doi: 10.1016/S0951-8320(03)00058-9.

Copyright © 2023 by ASME

Proceedings of the ASME 2023
International Design Engineering Technical Conferences and
Computers and Information in Engineering Conference
IDETC-CIE2023
August 20-23, 2023, Boston, Massachusetts

DETC2023-116940

PHYSICS-INFORMED NEURAL NETWORKS FOR DEGRADATION DIAGNOSTICS OF LITHIUM-ION BATTERIES

Sina Navidi[1], Adam Thelen[2], Tingkai Li[1], Chao Hu[1,*]

[1]Department of Mechanical Engineering, University of Connecticut, Storrs, CT 06269, USA
[2]Department of Mechanical Engineering, Iowa State University, Ames, IA 50011, USA

ABSTRACT

Quantifying the extent of degradation in a lithium-ion battery cell can provide valuable insights into the cell's state of health. However, physics-based methods capable of diagnosing component-level degradation typically require long-term aging data and are computationally expensive to deploy. This work investigates combining physics-based modeling and data-driven machine learning to retain high diagnostic accuracy while mitigating the need for long-term degradation data. We develop a physics-informed neural network (PINN) algorithm to diagnose cell health in the late aging stage without needing long-term data. Using a small set of early-life experimental data from an aging experiment, we train a neural network using a physics-informed loss function which penalizes discrepancies between the neural network's and the physics-based model's estimates of four cell health parameters. The trained network can then estimate cell capacity and diagnose three primary degradation modes in the late-aging stage. The proposed method is evaluated and compared with other machine learning algorithms using data from a long-term (3.5 years) cycling experiment of 16 implantable-grade lithium-ion cells. Cross-validation results show that the proposed PINN algorithm can improve the estimation accuracy of the health parameters compared to a purely data-driven approach and has comparable accuracy to the other two physics-informed machine learning methods.

Keywords: physics, machine learning, lithium-ion batteries, degradation diagnostics

1. INTRODUCTION

Lithium-ion batteries have become a popular energy storage technology with wide-ranging industrial applications in electric vehicles (EVs), portable devices, and smart grids [1, 2]. The increasing demand and adoption of lithium-ion batteries require techniques that can proactively manage battery health by diag-

nosing and predicting the degradation patterns of cells operating in the field.

Cycle aging and calendar aging are two main causes of lithium-ion battery degradation driven mainly by parasitic reactions between the charged electrodes and the electrolyte, which lead to capacity and power loss [3]. An example of such reactions is the loss of active materials on the cathode and anode due to metal dissolution [4]. Traditional approaches to battery state of health (SoH) estimation typically only estimate cell capacity and use it as the SoH measure. Compared to these traditional approaches, estimating the SoH of the electrodes and other cell components provides a more complete picture of cell health that can help diagnose more complex degradation modes, such as irreversible lithium plating induced by rapidly shrinking anode capacity and causing an undesirable capacity fade knee point [5]. However, in-situ experimental measurement of these electrode-specific degradation modes is not yet feasible. In turn, non-destructive methods of quantifying component-level degradation modes must be developed [6]. Diagnosing these degradation modes, as a step additional to estimating cell capacity, can more accurately and comprehensively quantify cell SoH. Many existing non-destructive approaches to lithium-ion battery degradation diagnostics estimate the loss of active materials on the positive and negative electrodes (LAM_{PE} and LAM_{NE}) and the remaining lithium inventory (LII) [6–11]. Estimating these metrics along with the cell capacity (Q) provide a more comprehensive measure of cell health. To quantify these degradation metrics, three important degradation parameters (m_{p}, m_{n} which are used to quantify LAM_{PE} and LAM_{NE}, respectively, and LII) need to be identified. The degradation parameters are typically estimated by reconstructing the full-cell open-circuit voltage (OCV) curve as the difference between experimentally obtained positive and negative half-cell OCV curves and matching the reconstructed full-cell OCV curve with the measured curve, following the methods best described by Birkl et al. [6]. The degradation parameters provide a quantitative link between cell's measurable ($dQ/dV(V)$)

---

*Corresponding author: chao.hu@uconn.edu

Copyright © 2023 by ASME

curves and its SoH measured by state of the (LAM_{PE}, LAM_{NE}), (LII) and capacity. However, the method in [6] is time-consuming and often cannot be done automatically, hence the need for new methods that can directly estimate lithium-ion cell degradation modes. Han et al. [7] investigated using a genetic algorithm to quantitatively analyze battery degradation from extensive of-fline aging experiments, and then membership functions could be used for online degradation diagnostics. Similarly, work by Dubbary et al. [8] proposed a degradation diagnostics algorithm that worked by comparing a cell's incremental capacity ($dQ/dV(V)$) and differential voltage ($dV/dQ(Q)$) to an offline degradation path database generated using a physics-based model. While these two degradation diagnostics methods work well, the first method relies on collecting long-term experimental aging data, and both methods use computationally expensive algorithms unsuitable for deployment on a battery management system (BMS). Developing new degradation diagnostics algorithms that leverage a combination of physics and data and remain computationally efficient is of great importance.

Recently, physics-informed machine learning has emerged as an effective method to infuse physical knowledge into machine learning models whose training and inference are otherwise unbounded [12]. Physics-informed machine learning approaches are especially well-known for their high accuracy and improved computational efficiency over traditional physics-based models. More specifically, there has been a growing interest in embedding physical information into neural networks to create a model that trains faster and can better generalize to unseen test data than standard neural networks [13]. In [14], the authors used a reduced-order model to estimate cell discharge voltage as a function of current and used a neural network to learn the non-ideal cell voltage behavior. The trained model could predict a cell's voltage and estimate its maximum discharge capacity. There are also a few studies that enhanced an electrochemical battery model with a neural network to predict terminal voltage [15, 16], state of charge and temperature [17], and electrode-level concentrations and potentials [18]. However, none of these studies target the problem of degradation diagnostics.

A study by Thelen et al. [10] specifically targeted the problem of non-destructive degradation diagnostics by building a lightweight physics-informed machine learning model to estimate cell degradation modes. To enable late-life estimation of cell degradation modes without using late-life aging data, they examined two different methods of training a machine learning model with physics-based simulation data to improve the model's late-life estimation performance. They found that adding simulated late-life aging data to the machine learning model's training dataset decreased estimation error by over 50% compared to the same model trained without the physics-based simulation data. While impressive, the method of infusing physics into a machine learning model by augmenting its training dataset requires a general understanding of the inference space that the model will operate in, which is often not entirely known. Further, having to preemptively define the range of values the model might estimate inadvertently limits its ability to extrapolate to new unseen conditions.

In this work, we develop a physics-informed neural network

(PINN) algorithm to estimate a lithium-ion cell's capacity and the extent of the cell's three primary degradation modes without requiring long-term degradation data. Our PINN models are computationally inexpensive and remain physically compliant (e.g., the physics-informed loss function substantially lowers the chance that models' estimates of the remaining lithium inventory exceed the total cell capacity. Different from the work by Thelen et al. [10] that incorporated the physics of battery degradation by augmenting a training dataset with simulation data, we embed the physics into a PINN model by modifying its loss function. The resulting model is capable of accurately diagnosing cell degradation in the late aging stage without requiring late-life aging data to be collected for training. We evaluate our new PINN and two other physics-informed machine learning methods using data from a long-term (~3.5 years) cycling experiment of 16 implantable-grade lithium-ion cells. Cross-validation results show our developed PINN more accurately estimates the four health parameters than a purely data-driven model and exhibits performance on par with other existing physics-informed models. Most importantly, our PINN trained with physics-based simulation and early-life aging data achieves high accuracy, suggesting that incorporating the physics of degradation can potentially eliminate the need to collect long-term degradation data to diagnose cell health in the late aging stage. Furthermore, this work analyzes the trends of the degradation modes and provides insight into the long-term performance of high-quality implantable-grade lithium-ion cells.

2. DATASET DESCRIPTION

The dataset used in this work comprises 16 implantable-grade lithium cobalt oxide/graphite (LCO) cells tested in a well-controlled laboratory setting for over 3.5 years [10, 11]. The 16 cells are split into 4 groups based on their cycle aging conditions listed in Table 1, and are labeled G1, G2, G3, and G4. All cells are cycled with a fixed charge rate of C/3, where 1C corresponds to a 1-hour charge time. To induce different degrees of degradation, the cells are cycled under two temperatures, 37 °C and 55 °C, and two discharge rates, C/3 and C/24, four unique combinations altogether. During cycling, all cells are charged to an upper cutoff voltage of 4.075 V using a constant current (CC) at C/3, followed by a constant voltage (CV) charging step until the current is below C/50 or the time on the constant voltage step is over 30 minutes. Cells are then discharged to a lower cutoff voltage of 3.4 V at their respective C-rates outlined in Table 1. Approximately every four weeks during regular cycling, a reference performance test (RPT) is performed at 40 °C to measure cell capacity and record capacity-voltage data for degradation diagnostics at a very slow (C/50) to avoid the influence of reaction kinetics on the cell's OCV. Additionally, two cells from each group were removed from the test for destructive analysis, in which the loss on each degradation mode was quantified. The measurements from destructive analysis help us confirm the accuracy of degradation parameters fitted using the half-cell model.

The capacity fade curves of all 16 cells are shown in Fig. 1. The early-life region highlighted in Fig. 1 contains the first 5 data points for all cells, from which the training data is extracted. We observe that the different temperatures and discharge

Copyright © 2023 by ASME

TABLE 1: CELL AGING TEST CONDITIONS

Group	Charge rate	Discharge rate	Temperature	Cells
G1	C/3	C/24	37 °C	4
G2	C/3	C/24	55 °C	4
G3	C/3	C/3	37 °C	4
G4	C/3	C/3	55 °C	4

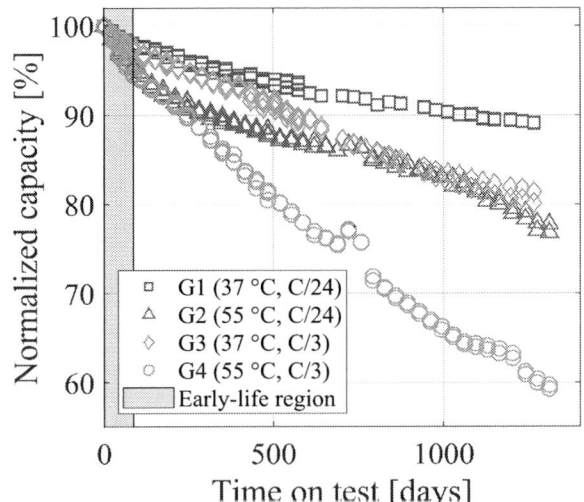

FIGURE 1: CAPACITY FADE CURVES OF THE CELLS TESTED. EACH CELL HAS BETWEEN 17 AND 42 DATA POINTS DEPENDING ON WHETHER OR NOT THE CELL WAS REMOVED FOR DESTRUCTIVE ANALYSIS TO CONFIRM THE EXTENT OF DEGRADATION [19].

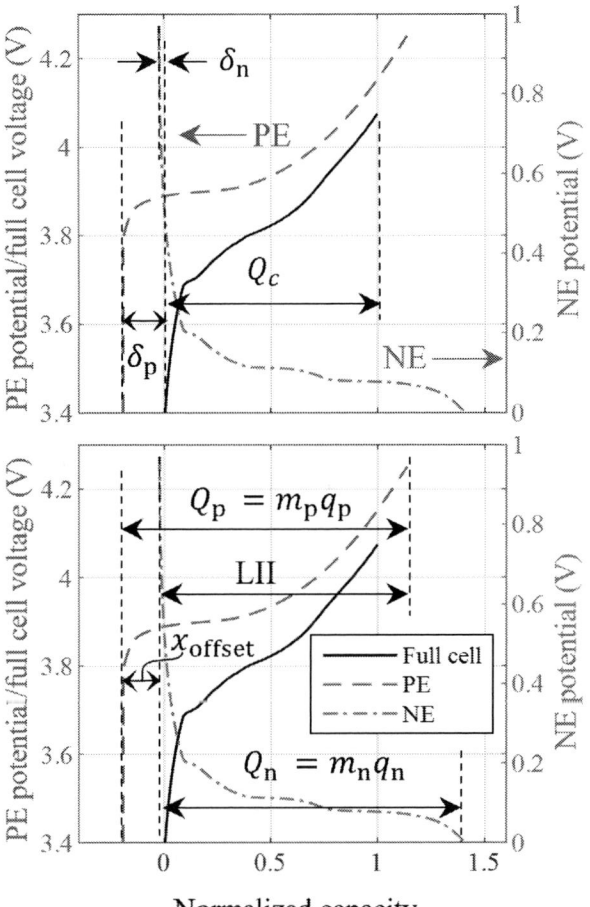

FIGURE 2: VISUAL REPRESENTATION OF THE HALF-CELL MODEL PARAMETERS DEFINED BY THE POSITIVE AND NEGATIVE HALF-CELL CURVES.

C-rates produced unique aging trajectories. Most notably, the higher temperature of 55 °C led to faster capacity fade, although each of the two groups of cells cycled under this elevated temperature exhibited impressively stable and consistent capacity fade performance.

3. HALF-CELL MODEL

To quantify the state of the three degradation modes over the life of a cell, we employ a physics-based half-cell model like those in [6–8, 10]. A half-cell model is a non-destructive degradation diagnostic technique that estimates the states of the three commonly reported degradation modes (LAM_{PE}, LAM_{NE}, and LII) by reconstructing the measured full-cell OCV curve as the sum of the positive and negative half-cell curves. Half-cell OCV data are obtained by constructing and cycling half-cells made using fresh electrode material from a full cell.

With the measured half-cell voltage vs. capacity curves ($V(Q)$) in hand, the state of the three degradation modes can be quantified by translating and scaling the positive and negative half-cell curves such that the generated pseudo-full-cell $V(Q)$ curve matches the experimentally measured curve from an aged cell. This process is visualized in Fig. 2.

Computing the pseudo-full-cell OCV curve can mathematically be written as follows

$$V_c(Q_c)|_{Q=Q_c} = V_p(q_p)|_{q_p = \frac{Q_c - \delta_p}{m_p}} - V_n(q_n)|_{q_n = \frac{Q_c - \delta_n}{m_n}} \quad (1)$$

where $V_c(Q_c)$ is the pseudo-full-cell OCV curve, Q_c refers to the cell capacity, $V(q)$ is a half-cell curve with specific capacity q (mAh/g), m is the active mass (g), and δ is the half-cell curve slippage (mAh). The subscripts p and n correspond to the positive and negative electrodes (PE and NE). The two slippage parameters, δ_p and δ_n quantify the horizontal distance of the left endpoint of the positive and negative half-cell curves with respect to $Q_c = 0$.

These equations mathematically capture the two half-cell curves' location and scale changes, where the magnitudes of the changes to each curve have previously been shown to closely reflect the state of the three degradation modes LAM_{PE}, LAM_{NE}, and LII [6, 10]. The two half-cell model parameters m_p and m_n represent the mass of active materials on each electrode, and estimating them quantifies the state of the degradation modes LAM_{PE} and LAM_{NE}. Further, as a cell degrades, it loses lithium inventory, and the lithium inventory indicator LII reflects the cell's remaining lithium inventory, calculated as $LII = Q_p - (\delta_p - \delta_n)$.

Copyright © 2023 by ASME

4. PHYSICS-INFORMED NEURAL NETWORK

PINNs are specialized neural network architectures that integrate data and mathematical operators to make up for missing physics [13]. A unique property of PINN algorithms is their ability to provide more robust and physically-consistent predictions, especially in the case of extrapolation or generalization tasks. We aim to leverage the physics of the half-cell model described in Sec. 3 to guide the network training and embed knowledge of the degradation parameter relationships into the model so that it can more accurately estimate cell degradation in the late-aging stage.

The architecture of our PINN is shown in Fig. 3. We embed physical knowledge of cell degradation into the model by adding a secondary physics-informed loss term to the standard data-driven loss. We designed the network output in such a way that four of the outputs $(\delta_p, \delta_n, m_p, m_n)$ are constrained by their relationship in the half-cell model from Eqn. 1. Illustrated best in Fig. 2, $\delta_p^{'}$ and δ_n describe the relative position of the positive and negative half-cell curves to one another and are primarily responsible for defining the overall shape of the full-cell pseudo-OCV curve. The active mass parameters, m_p and m_n, directly quantify the remaining active material on the cell electrodes and thus define the capacity of each electrode, which is analogous to the overall length of the half-cell curves shown in Fig. 2. We impose the half-cell model constraints on these four parameters by checking the half-cell generated $dQ/dV(V)$ curve against the experimentally measured $dQ/dV(V)$ curve to ensure they do not differ drastically. In effect, this limits the PINN outputs to take on values that are compliant with the physics of cell degradation, e.g., negative values for m_p and m_n are impossible because that would suggest the electrodes have "negative capacity." At the same time, the outputs of m_p, m_n, Q, and LII which denote as DP in Fig. 3 are compared to the training data in the data-driven loss. The training process consists of several cycles, where the network first generates outputs comprising four half-cell parameters. These outputs are then feed into the half-cell model to generate simulated $dQ/dV(V)$ curves. A comparison is made between these simulated curves and the experimentally measured curves, and the normalized mean squared error between them is computed. This error is incorporated into the loss, which also includes the difference between the predicted degradation parameters by the network and the actual degradation parameters. Subsequently, the network's output is adjusted based on this combined loss during the next training cycle. This iterative process continues until the loss value reaches a sufficiently small threshold, signifying the improved accuracy of the network's predictions over time. As a potential direction for future work, it is worth considering the complete constraint of the output values from the Physics-Informed Neural Network (PINN) model. This can be achieved by outputting only half-cell parameters and passing them through the half-cell model and obtaining the electrode masses as well as capacity (Q) and the lithium inventory (LII).

During model training, we used Adam optimization algorithm [20] to update the network weights. Also, for the activation function, we use the rectified linear unit (ReLU):

$$Relu(z) = max(0, z), \quad (2)$$

The hyper-parameters used for both the PINN and standard NN are outlined in Table 2. As we do not have much early-life experimental data available for training, we opt to use a more shallow network architecture with only 3 layers and 201 trainable parameters.

TABLE 2: NEURAL NETWORK HYPER-PARAMETERS

Hyper-parameter	Value
Number of layers	3
Max epoch	500
Number of hidden neurons	16
Learning rate	0.001
Mini-batch size	8
Trainable parameters	201

5. RESULTS AND DISCUSSION

In this section, we evaluate the performance of a regular neural network, which serves as the baseline model, and a PINN model. These two models have the same hyper-parameters recorded in Table 2 and are cross-validated with the same setup. In particular, we performed a 4-fold cross-validation, similar to the approach employed by Thelen et al. [10], in which the dataset containing 16 battery cells is divided into four mutually exclusive folds. The test data of every single fold consists of one cell from each group. The evaluation criterion is root mean square percentage error (RMSPE), which is a normalized error metric suitable for comparison among outputs with different magnitudes, as follows,

$$RMSPE = \sqrt{\left(\frac{1}{\sum_{k=1}^{4} N_k} \sum_{k=1}^{4} \sum_{i=1}^{N_k} \left(\frac{\hat{y}_i - y_i}{y_i} \times 100\%\right)^2\right)}, \quad (3)$$

where $\hat{y}_i$ and y_i denote the estimated and true values, respectively, and subscript i denotes the test point where $i = 1, 2, \ldots, N_k$. The overall test error of the models is calculated by taking the average of each health parameter's errors across all the 4 folds which is denoted by k.

The respective prediction results for the health parameters m_p, m_n, LII, and Q are shown in Fig. 4 for all cells colored by their groups. As shown in Fig. 4, the estimated values for LII and Q show excellent agreement with the fitted values. This excellent performance may be due to the strong correlations between the simulated $dQ/dV(V)$ curves from the half-cell model and the cell capacity, and also the relatively linear degradation trends for LII and Q with respect to time. However, due to the high dynamics of negative and positive electrodes during discharge which results in more complex degradation trends, estimating the loss of active material on the positive and negative electrodes is more challenging. Yet the prediction of the positive active mass (m_p) is accurate where the mean RMSPE is less than 5%. The performance of the model degrades in estimating remaining negative active mass (m_n) compared with the other three health parameters. However, the performance of the PINN model is still acceptable with low RMSPE for all the groups. To visualize the prediction errors on each health parameter, the normalized predicted values by PINN and baseline neural network are compared to the half-cell fitted

Copyright © 2023 by ASME

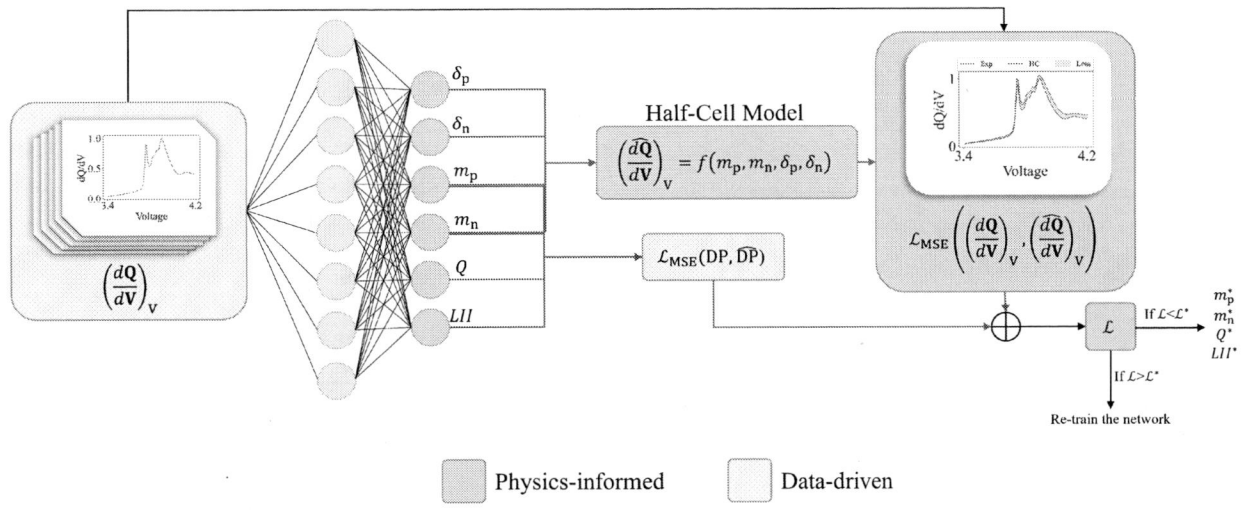

FIGURE 3: PHYSICS-INFORMED NEURAL NETWORK (PINN) ARCHITECTURE.

values in Appendix. A for group G3 cells. The resulting plots highlight the better extrapolation ability of the PINN model compared to the baseline neural network model. Besides, despite the relatively higher prediction errors, the estimated m_n shows good accuracy in following the trend of the true fitted values.

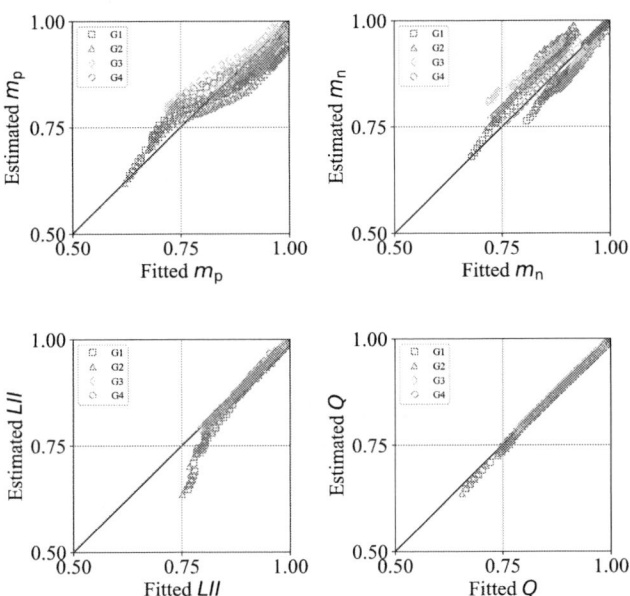

FIGURE 4: NORMALIZED ESTIMATED VS FITTED (TRUE) HEALTH PARAMETERS PREDICTIONS.

Additionally, we compared the prediction accuracy of the PINN and the regular neural network models with the baseline methodologies employed in [10], which are summarized in Table 3. Overall, the mean RMSPE of the PINN model is lower than all the other models considered. For the loss of positive active mass (m_p) and the capacity fade (Q), the testing error of the PINN shows a clear difference compared to that of the lasso, MOGP (multi-

output Gaussian process), ELM (extreme learning machine), and elastic net baseline setups. In predicting the loss of lithium inventory (LII), the PINN model outperformed all the baseline models too, although the difference between the PINN prediction errors and both ELM and elastic net models is not significant. As mentioned before, the estimation error of PINN for the negative active mass (m_n) is lower than the baseline neural network model. However, compared to baseline setups of the lasso, MOGP, ELM, and elastic net models, the test error of the PINN is higher. It is probably because the neural network models could suffer from the limited amount of training data for non-trivial estimation tasks. To highlight the sensitivity of the estimation framework to the neural network architecture while dealing with extrapolation tasks, we change the number of neurons in the single hidden layer of the network. The estimated mean RMSPE results are given in Fig. 5. The estimation errors are the lowest for the number of neurons between 10 and 20. A higher number of neurons in the hidden layer may result in over-fitting considering the limited amount of information contained in the training set is not enough to train all of the neurons.

TABLE 3: ESTIMATION ACCURACY OF THE MODELS

Model	m_p	m_n	LII	Q	Mean
Lasso	8.55	2.75	3.68	4.74	4.93
MOGP	10.3	5.05	4.18	4.58	6.03
ELM	7.41	5.71	2.33	1.71	4.29
Elastic net	8.05	2.44	2.73	3.76	4.25
NN	8.42	9.40	2.20	0.40	5.10
PINN	**4.56**	7.93	**2.17**	**0.38**	**3.76**

6. CONCLUSION

This work aimed to develop a lightweight physics-informed neural network model for diagnosing cell health at a late-life stage by combining the advantages of both physics-based modeling and data-driven machine learning while mitigating the need for long-term data. To this end, a physics-based half-cell model is added

Copyright © 2023 by ASME

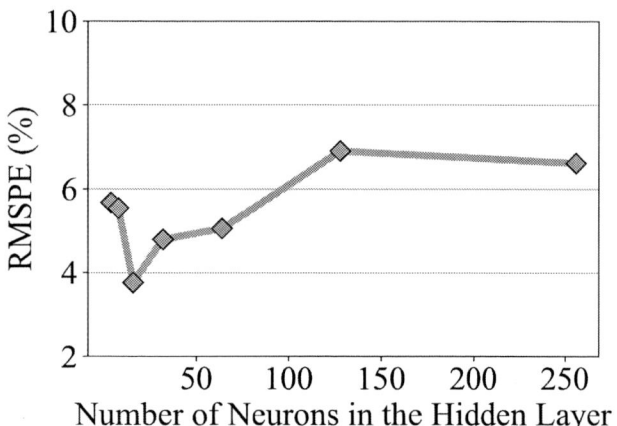

FIGURE 5: SOH ESTIMATION ERROR AVERAGED OVER FOUR HEALTH PARAMETERS TRAINED WITH DIFFERENT NUMBERS OF NEURONS IN THE HIDDEN LAYER OF THE NETWORK.

to a shallow neural network to correct the estimated health parameters along with the data-driven loss. The proposed model can accurately estimate the internal state of health for Lithium-ion batteries merely based on the limited early-life degradation data. The performance of the proposed physics-informed neural network model is evaluated and compared with multiple baseline models, highlighting the PINN model's high extrapolation ability. In future work, fully constraining the PINN model's output through the half-cell model could be considered to enhance the accuracy and reliability of the predictions.

ACKNOWLEDGMENTS

This work was partly supported by Iowa Economic Development Authority under the Iowa Energy Center Grant No. 20-IEC-018 and partly by the US National Science Foundation under Grant No. ECCS-2015710. Any opinions, findings, or conclusions in this paper are those of the authors and do not necessarily reflect the sponsors' views.

REFERENCES

[1] Zhou, Yapeng, Huang, Miaohua, Chen, Yupu and Tao, Ye. "A novel health indicator for on-line lithium-ion batteries remaining useful life prediction." *Journal of Power Sources* Vol. 321 (2016): pp. 1–10.

[2] Zubi, Ghassan, Adhikari, Rajendra S, Sánchez, Nazly E and Acuña-Bravo, Wilber. "Lithium-ion battery-packs for solar home systems: Layout, cost and implementation perspectives." *Journal of Energy Storage* Vol. 32 (2020): p. 101985.

[3] Lin, Cheng, Tang, Aihua, Mu, Hao, Wang, Wenwei and Wang, Chun. "Aging mechanisms of electrode materials in lithium-ion batteries for electric vehicles." *Journal of Chemistry* Vol. 2015 (2015).

[4] Fathi, Reza, Burns, John Chris, Stevens, DA, Ye, Hui, Hu, Chao, Jain, Gaurav, Scott, Erik, Schmidt, Craig and Dahn, Jeff R. "Ultra high-precision studies of degradation mechanisms in aged LiCoO2/graphite Li-ion cells." *Journal of The Electrochemical Society* Vol. 161 No. 10 (2014): p. A1572.

[5] Attia, Peter M, Bills, Alexander, Planella, Ferran Brosa, Dechent, Philipp, Dos Reis, Goncalo, Dubarry, Matthieu, Gasper, Paul, Gilchrist, Richard, Greenbank, Samuel, Howey, David et al. ""Knees" in lithium-ion battery aging trajectories." *Journal of The Electrochemical Society* Vol. 169 No. 6 (2022): p. 060517.

[6] Birkl, Christoph R, Roberts, Matthew R, McTurk, Euan, Bruce, Peter G and Howey, David A. "Degradation diagnostics for lithium ion cells." *Journal of Power Sources* Vol. 341 (2017): pp. 373–386.

[7] Han, Xuebing, Ouyang, Minggao, Lu, Languang, Li, Jianqiu, Zheng, Yuejiu and Li, Zhe. "A comparative study of commercial lithium ion battery cycle life in electrical vehicle: Aging mechanism identification." *Journal of Power Sources* Vol. 251 (2014): pp. 38–54.

[8] Dubarry, Matthieu, Berecibar, Maitane, Devie, A, Anseán, D, Omar, Noshin and Villarreal, Igor. "State of health battery estimator enabling degradation diagnosis: Model and algorithm description." *Journal of Power Sources* Vol. 360 (2017): pp. 59–69.

[9] Tian, Jinpeng, Xiong, Rui, Shen, Weixiang and Sun, Fengchun. "Electrode ageing estimation and open circuit voltage reconstruction for lithium ion batteries." *Energy Storage Materials* Vol. 37 (2021): pp. 283–295.

[10] Thelen, Adam, Lui, Yu Hui, Shen, Sheng, Laflamme, Simon, Hu, Shan, Ye, Hui and Hu, Chao. "Integrating physics-based modeling and machine learning for degradation diagnostics of lithium-ion batteries." *Energy Storage Materials* Vol. 50 (2022): pp. 668–695.

[11] Thelen, Adam, Lui, Yu Hui, Shen, Sheng, Laflamme, Simon, Hu, Shan and Hu, Chao. "Physics-informed machine learning for degradation diagnostics of lithium-ion batteries." *International design engineering technical conferences and computers and information in engineering conference*, Vol. 85383: p. V03AT03A041. 2021. American Society of Mechanical Engineers.

[12] Raissi, Maziar, Perdikaris, Paris and Karniadakis, George E. "Physics-informed neural networks: A deep learning framework for solving forward and inverse problems involving nonlinear partial differential equations." *Journal of Computational physics* Vol. 378 (2019): pp. 686–707.

[13] Karniadakis, George Em, Kevrekidis, Ioannis G, Lu, Lu, Perdikaris, Paris, Wang, Sifan and Yang, Liu. "Physics-informed machine learning." *Nature Reviews Physics* Vol. 3 No. 6 (2021): pp. 422–440.

[14] Nascimento, Renato G, Corbetta, Matteo, Kulkarni, Chetan S and Viana, Felipe AC. "Hybrid physics-informed neural networks for lithium-ion battery modeling and prognosis." *Journal of Power Sources* Vol. 513 (2021): p. 230526.

[15] Park, Saehong, Zhang, Dong and Moura, Scott. "Hybrid electrochemical modeling with recurrent neural networks

Copyright © 2023 by ASME

for li-ion batteries." *2017 American control conference (ACC)*: pp. 3777–3782. 2017. IEEE.

[16] Tu, Hao, Moura, Scott, Wang, Yebin and Fang, Huazhen. "Integrating physics-based modeling with machine learning for lithium-ion batteries." *Applied Energy* Vol. 329 (2023): p. 120289.

[17] Feng, Fei, Teng, Sangli, Liu, Kailong, Xie, Jiale, Xie, Yi, Liu, Bo and Li, Kexin. "Co-estimation of lithium-ion battery state of charge and state of temperature based on a hybrid electrochemical-thermal-neural-network model." *Journal of Power Sources* Vol. 455 (2020): p. 227935.

[18] Li, Weihan, Zhang, Jiawei, Ringbeck, Florian, Jöst, Dominik, Zhang, Lei, Wei, Zhongbao and Sauer, Dirk Uwe. "Physics-informed neural networks for electrode-level state estimation in lithium-ion batteries." *Journal of Power Sources* Vol. 506 (2021): p. 230034.

[19] Lui, Yu Hui, Li, Meng, Downey, Austin, Shen, Sheng, Nemani, Venkat Pavan, Ye, Hui, VanElzen, Collette, Jain, Gaurav, Hu, Shan, Laflamme, Simon et al. "Physics-based prognostics of implantable-grade lithium-ion battery for remaining useful life prediction." *Journal of Power Sources* Vol. 485 (2021): p. 229327.

[20] Kingma, Diederik P and Ba, Jimmy. "Adam: A method for stochastic optimization." *arXiv preprint arXiv:1412.6980* (2014).

[21] Ning, Xiang and Lovell, Mary Rose. "On the Sliding Friction Characteristics of Unidirectional Continuous FRP Deposits." *ASME Journal of Tribology* Vol. 48 No. 5 (2002): pp. 2000–2008. DOI 10.1115/1.4042912.

[22] Takahashi, Kenji and Srinivasan, Venkat. "Examination of graphite particle cracking as a failure mode in lithium-ion batteries: a model-experimental study." *Journal of The Electrochemical Society* Vol. 162 No. 4 (2015): p. A635.

APPENDIX A. VISUALIZATION OF MODEL PREDICTIONS

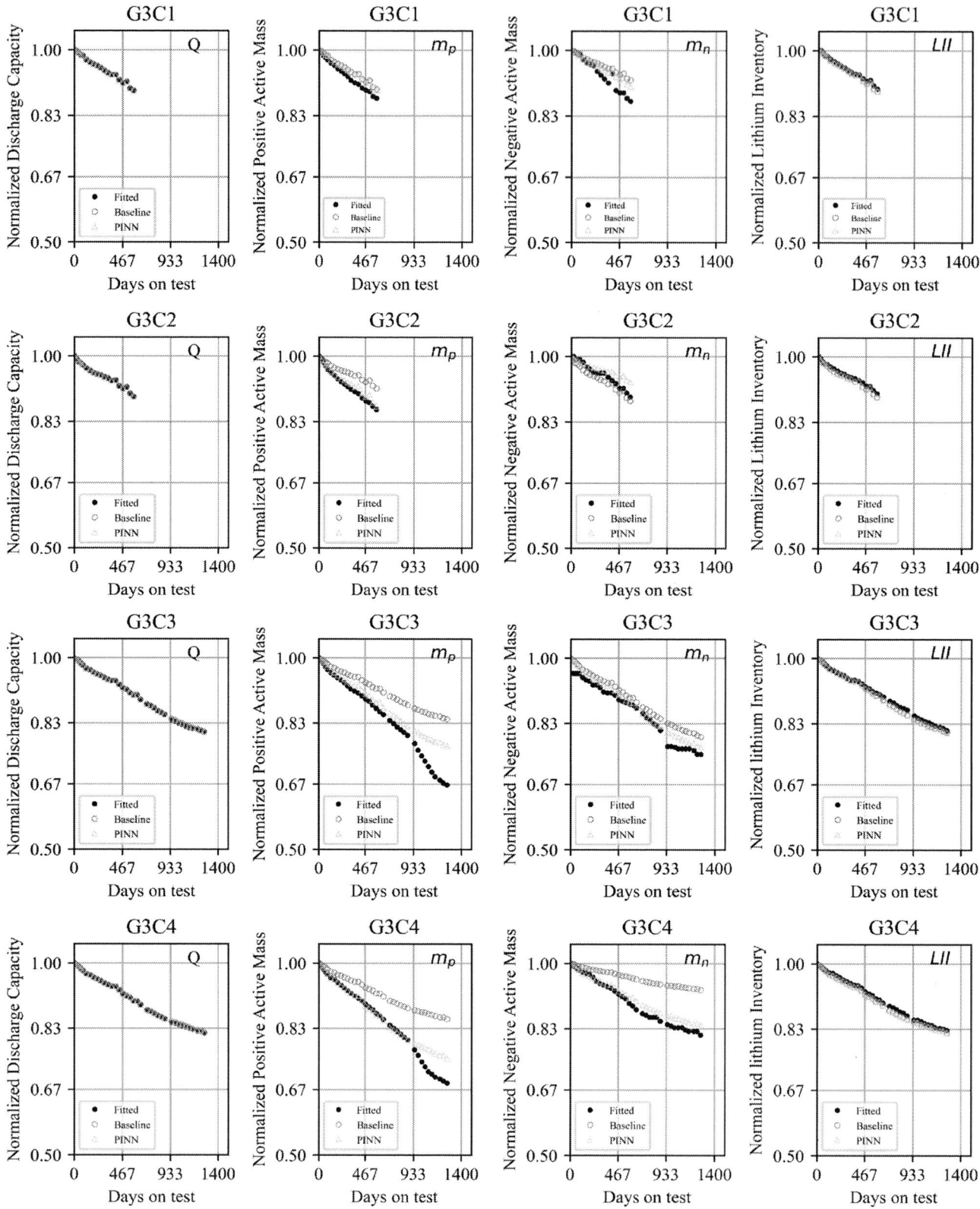

FIGURE 6: COMPARISON OF NORMALIZED PREDICTED AND FITTED (TRUE) HEALTH PARAMETERS OF GROUP G3 FOR THE PINN AND BASELINE MODELS.

Copyright © 2023 by ASME

Proceedings of the ASME 2023
International Design Engineering Technical Conferences and
Computers and Information in Engineering Conference
IDETC-CIE2023
August 20-23, 2023, Boston, Massachusetts

DETC2023-116972

SENSOR NETWORK DESIGN FOR PERMANENT MAGNET SYNCHRONOUS MOTOR FAULT DIAGNOSIS

Sara Kohtz[1], Junhan Zhao[2], Anabel Renteria[1], Anand Lalwani[3], Xiaolong Zhang[2], Kiruba S. Haran[2], Debbie Senesky[3], and Pingfeng Wang[1,*]

[1] Industrial and Enterprise Systems Engineering, University of Illinois at Urbana-Champaign
Urbana, IL 61801, United States

[2] Electrical and Computer Engineering, University of Illinois at Urbana-Champaign
Urbana, IL 61801, United States

[3] Aeronautics and Astronautics, Stanford University, Stanford, CA 94305, United States.

ABSTRACT

Optimal sensor placement is a challenge in many engineering design applications, especially within the field of prognostics and health management. Recently, data-driven approaches have become a staple for solving and addressing these challenges. Machine learning techniques have been applied to solve complex optimization problems in the field of signals processing. However, these methods require a substantial amount of data, which can be difficult to obtain. In addition, the design space may be extremely large, so a deterministic approach may not be possible. Therefore, there is a need for probabilistic frameworks that can simultaneously train a classifier for detection of faults as well as selecting new designs for optimal placement. In this paper, the proposed methodology contains a genetic algorithm embedded with a clustering algorithm to simultaneously train the classifier and determine a sensor network. This novel structure is implemented for detecting short-winding faults of a permanent magnet synchronous motor using magnetic field sensors. The training data is simulated using a finite element model, and the design space is extremely large. Nonetheless, the results of the proposed methodology show accuracy for detection of faults.

Keywords: Optimal sensor placement, permanent magnet synchronous motor, fault detection, genetic algorithm

1. INTRODUCTION

A reliable health monitoring system is crucial for the safety and performance of engineering systems. There have been major advancements in the field of fault diagnosis, prognostics and health management, as well as design for advanced low-cost and long-lasting sensors for the condition monitoring of engineered systems [1-9]. Recently, magnetic field sensing is a field of research that has the potential for satisfying these design constraints. For example, magnetic field sensing can be implemented on numerous wireless sensing applications, as shown in Muhammad et al, which utilizes a surface acoustic wave sensor for detection [10]. Also, fiber optic magnetic field sensors can be utilized under varying conditions and implemented on numerous engineering systems, as shown in Zhou et al [11]. This sensor is especially useful for engineering systems with current sensing Zhou et al implemented fiber optic vector magnetic field sensor under varying temperature and vibration errors [11]. This sensor can be implemented in engineered systems with current sensing, large scale digital networks such as navigation and vehicle detection. For structural health monitoring, hall-effect sensors have shown promise in various fields, including gas turbine speed detection, signals processing, and aerospace engineering [9]. The main benefits of hall-effect sensors include low cost and maintenance, long-lasting, and a reliable response [12].

---

[*] Corresponding Author. Email: pingfeng@illinois.edu.

Copyright © 2023 by ASME

Many complex engineering systems, such as propulsion aircraft and electric vehicles, require high-power motors that are efficient and compact. Permanent magnet synchronous machines (PMSM) are one such motor that has become increasingly popular due to reliability performance and a high energy density to size ratio [13]. However, PMSMs can be a complex structure, and have dynamic interaction between varying temperature, mechanical, and electromagnetic conditions; all of which can lead to failure within the motor itself. In addition, for the application of propulsion aircraft, permanent damage can lead to catastrophic failure [14]. The most common and recurrent faults within the PMSM include magnet loss, phase loss, and short-winding circuit, which can compromise the functionality of the entire system [15]. Consequently, a health monitoring system within the PMSM for fault detection is a necessity. The health states of the motor can be identified by analyzing various signals and measurements from the operating conditions; these include current, voltage, magnetic flux, temperature, vibration, among others. For this paper, the focus is on utilizing the magnetic flux signal as the main identifier for the short-winding fault of the PMSM.

Since the short-winding fault is the most harmful and recurrent within the PMSM, the purpose of this study is to identify these occurrences efficiently and effectively. These faults tend to happen because of insulation degradation within the stator winding [16, 17]. Also, the short circuit may cause overheating within the stator winding, which can result in permanent degradation of winding insulation and demagnetization of magnets [18]. There have been recent studies concentrating on fault detection within the PMSM; most of these make use of the sinusoidal signal from current sensors. Some of these papers have utilized data-driven techniques, such as deep machine learning. For example, Peng et al employed a 1D convolutional neural network to handle the signal from current sensor and predict inter-turn short fault detection [19]. This method had high accuracy for identifying faults, however computational efficiency is an issue for the deep machine learning framework as it is not practical to implement these in real-time. Lee et al also utilized a deep machine learning framework with current sensing, specifically a fully attention-based neural network with longer memory [20]. There have only been a few studies that deviate from using current sensors as the signal for fault detection; and even fewer that implement the magnetic flux as the fault detector. For these scarce studies, simplistic methodologies are utilized, such as mathematical modeling or standard fault trees [21]. None of the methodologies described here consider sensor placement, which if optimized can improve data quality and thus have better accuracy for fault diagnosis. Therefore, sensor placement is an important concept in this paper for fault detection within the PMSM.

In recent literature, optimal sensor placement has been a challenge in many varying applications. There have been varying techniques and hybrid methodologies to tackle this problem. These include optimization techniques, search algorithms, filtering, and machine learning. Often, for real-time prediction, systems are limited to partial information for fault detection. To remedy this, Taher et al incorporates real-time measurements into a Kalman filter for determining optimal sensor placement [22]. The Kalman filter can take many different forms and can be applied to a variety of engineering applications, for instance, Ercan et al utilizes an augmented Kalman Filter to handle vibration measurements [23], where a heuristic methodology with the genetic algorithm was employed for sensor placement optimization while considering different types of sensors. Techniques based upon Information theory have also been extensively implemented to handle noisy signals and determine optimum sensor placement. For example, Zhang et al applied information gain of strain measurements on a beam [24]. Finally, complex optimization approaches have also been implemented, although not as extensively as the previously mentioned methods. For example, particle swarm algorithm can be utilized for discrete optimization; as shown in Kong et al, which applied this method to a hydraulic control system for fault detectability [25]. Also, annealing machine for sensor placement can be implemented, as shown in Inoue et al [26], which is applied to high-dimensional system and can be generalized to any application-specific optimization framework.

Most of these methodologies are application-specific, meaning they cannot be generalized to other engineering systems. Therefore, probabilistic approaches with generic frameworks are a much-needed tool. This paper proposes a methodology that simultaneously trains the classifier for optimal detection of faults, while selecting the best placement of sensor for optimal data quality. Specifically, a clustering algorithm is employed for the classifier, and the optimal solution is found using a genetic algorithm. The remainder of this paper is organized as follows: section II provides a detailed description of the methodology, section III displays the results of the proposed method, and section IV concludes with a discussion and future work.

2. MATERIALS AND METHODS

Figure 1 shows the design space for the placement of sensors for the PMSM. There are a total of 543 possible locations to place a sensor, and the goal is to minimize the number of sensors while obtaining optimal detectability. Due to the abundance of possible coordinates, the number of possible sensor networks is extremely high; and impossible to conduct an exhaustive search. The system of interest in this paper is the permanent magnet synchronous motor, which is utilized in numerous high power engineering applications. The focus is on short-winding faults, specifically inter-turn and inter-phase faults. There are a total of 7 faults that need to be detected, so 8 states need to be classified with a certain accuracy (8 including the healthy state). Essentially, for all faults k, we want the following equation to be satisfied:

$$\forall\, k: D_k(x, y) \geq D_T \qquad (1)$$

where D_T is the detectability threshold, which is the probability of correctly detecting the state; (x, y) are the coordinates of the possible placement of the sensors, which are shown in figure 1. The detectability threshold has a lower threshold of 90%, and an

Copyright © 2023 by ASME

upper threshold of 95%. The focus of this paper is to determine the minimum number of sensors such that the detectability threshold is met. From the 543 possible placements, varying signals will be observed for fault detectability. For visualization, figure 2 shows the signal from 3 given coordinates from figure 1, 1 coordinate from each of the following placements: airgap, inner, and outer. It can be observed that placement in the airgap has a stronger signal; while sensors in the inner have little to no signal, and placement in the outer has a high frequency signal, but not as strong as in the airgap. It is expected that the proposed methodology will select coordinates that are located within the airgap, since it has the strongest signal. The proposed methodology, which is shown in Fig. 3, simultaneously determines the best classifier of faults while selecting the optimal placement of sensors. Similar to the sensor placement optimization formulation as used in the literature [27, 28], however the data for the PMSM can be generated completely offline using a finite element model, so a surrogate model is not needed.

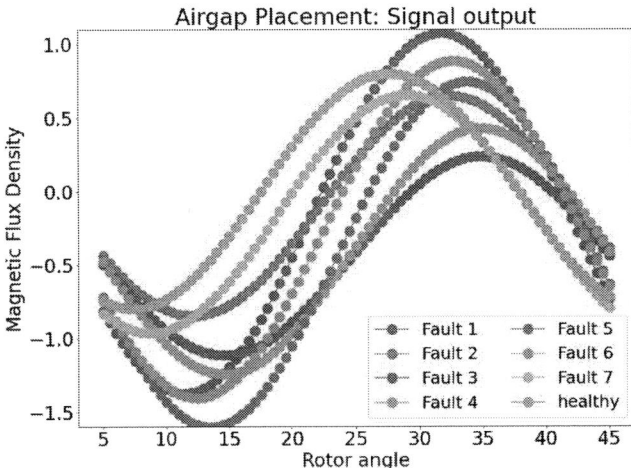

(a) Magnetic Flux in airgap coordinate 1

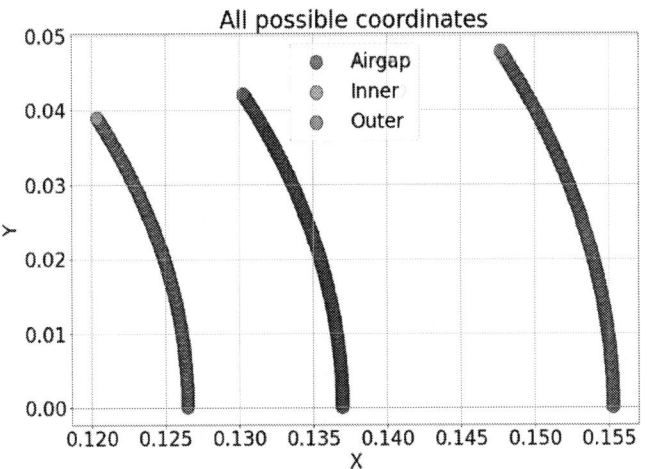

FIGURE 1: THE POSSIBLE 2-D SENSOR PLACEMENT LOCATIONS (COORDINATES) FOR THE PMSM.

The description of the faults is shown in table 1. Each fault corresponds to a kind of short-winding fault. μ defines the scale factor for which the turn has been shorted.

TABLE 1: Fault Descriptions

No.	Fault Name	Description
1	A_1turn	Inter-phase short-winding, $\mu = \frac{1}{6}$
2	A_1turn_1turn	Inter-phase short-winding, $\mu = \frac{1}{3}$
3	AC_1turn_0turn	Phase to phase short-winding, $\mu = \frac{1}{6}$
4	AC_1turn_1turn	Phase to phase short-winding, $\mu = \frac{1}{3}$
5	A_1coil	Inter-phase short-winding, $\mu = \frac{1}{2}$
6	AC_1turn_1coil	Phase to phase short-winding, $\mu = \frac{2}{3}$
7	A_phase	Inter-phase short-winding, $\mu = 1$

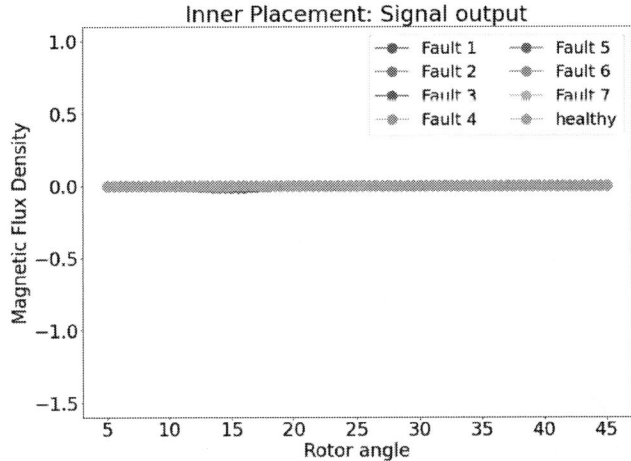

(b) Magnetic Flux in inner coordinate 1

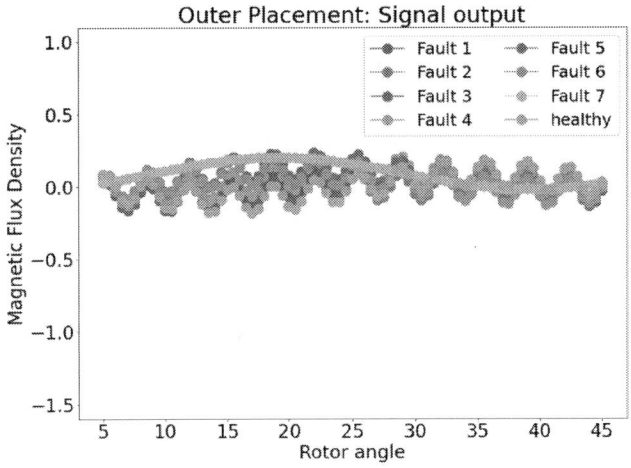

(c) Magnetic Flux in outer coordinate 1

FIGURE 2: SIGNAL OUTPUTS FOR PLACEMENT OF HALL-EFFECT SENSORS

Copyright © 2023 by ASME

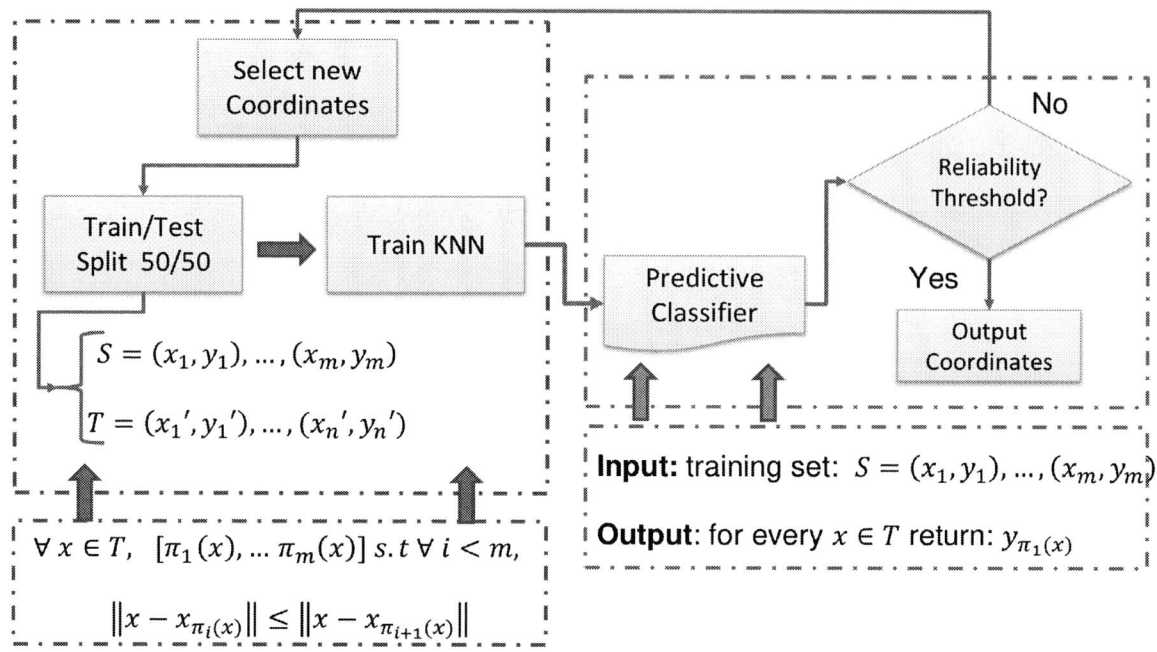

FIGURE 3: THE FLOWCHART OF THE DEVELOPED METHODOLOGY

Figure 3 shows the flowchart of the proposed methodology. The selection of coordinates is determined by a traditional genetic algorithm. Given the total number of sensors (in this case the number of sensors varies from 5-12), the data from the FE model is available to test. This data is split into a training and testing data set, which are labelled S and T, respectively. In this case x is defined as the signal output (which is the magnetic flux density), and y is the categorical state prediction (defined as fault 1-7 or healthy). The classifier for fault detection is determined using a k-nearest neighbor algorithm. From figure 3, $\pi_1(x), \ldots, \pi_m(x)$ is a reordering of testing points based on their distance to x. The label (which is the health state of the motor) associated with $\pi_1(x)$, is the prediction of the classifier; this is shown as $y_{\pi_1(x)}$, in figure 3. This is a distance clustering method, and has the highest computational efficiency in terms of training and testing. Since the classifier will be implemented in real-time, it must be very efficient; in addition, this enables the proposed methodologies to go through numerous combinations of sensor placements and cover more of the design space.

The selection of coordinates using a genetic algorithm is an imperative step within this methodology. The fitness function within the genetic algorithm is defined as the average F1-score across each of the health states. The F1-score, or the harmonic mean, for each fault k is defined in equation 2.

$$F_k = 2 * \frac{p_k r_k}{p_k + r_k} \qquad (2)$$

where p_k is the precision rate, or the proportion of true positives to the total number of positives. r_k is the recall rate, or the proportion of true positives to the total true positives and false negatives.

There is a precision value for each of the k health states, which can be defined as the following:

$$p_k = \frac{TP_k}{TP_k + FP_k} \qquad (3)$$

where the true positive, TP_k, is defined as $\forall x_i' \in T \ s.t. \ y_i' = k$:

$$TP_k = \sum_i \mathbf{1}_{y_{\pi(x_i')}=k} \qquad (4)$$

And the false positive, FP_k, is defined as $\forall x_i' \in T \ s.t. \ y_i' \neq k$:

$$FP_k = \sum_i \mathbf{1}_{y_{\pi(x_i')}=k} \qquad (5)$$

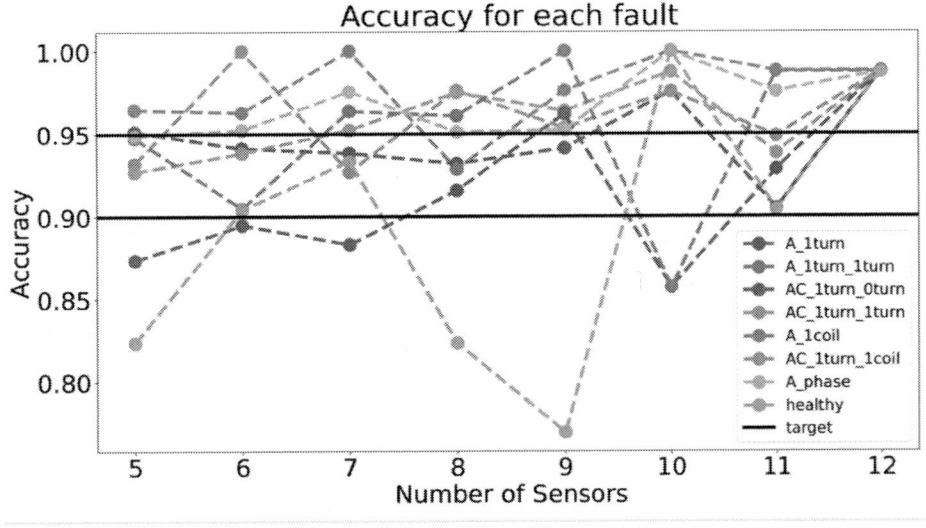

(a) F' results for fault detection

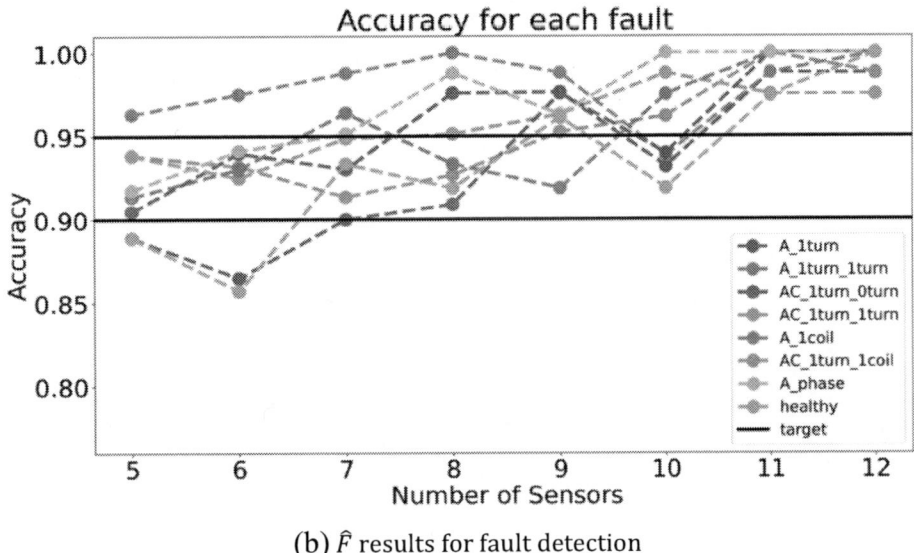

(b) $\hat{F}$ results for fault detection

FIGURE 4: ACCURACY OF FAULT DETECTION: (A) F' FOR EACH FAULT; (B) $\hat{F}$ FOR EACH FAULT

There is also a recall value for each of the k health states, which is defined as the following:

$$r_k = \frac{TP_k}{TP_k + FN_k} \qquad (6)$$

where the false negative, FN_k, is defined as $\forall x_i' \in T \ \ s.t. \ \ y_i' = k$:

$$FN_k = \sum_i \mathbf{1}_{y_{\pi(x_i')} \neq k} \qquad (7)$$

The average F1-score over all states k can be defined as the fitness function for the genetic algorithm.

$$\hat{F} = \frac{1}{K} \sum_k F_k \qquad (8)$$

Where K is the total number of faults. Since it is possible that the algorithm will converge to a solution such that it can predict certain faults with high accuracy, but a single fault can significantly low accuracy (see figure 4 result with 9 sensors).

To remedy this, a fitness function with a penalty term can be implemented:

Copyright © 2023 by ASME

$$F' = \begin{cases} \hat{F} * \min(F_k), & \min(F_k) < 0.90 \\ \hat{F}, & \min(F_k) \geq 0.90 \end{cases} \qquad (9)$$

Essentially, equation 9 asserts a penalty if the detectability

for any of the faults is below 90%. This ensures that the faults have a similar detectability level.

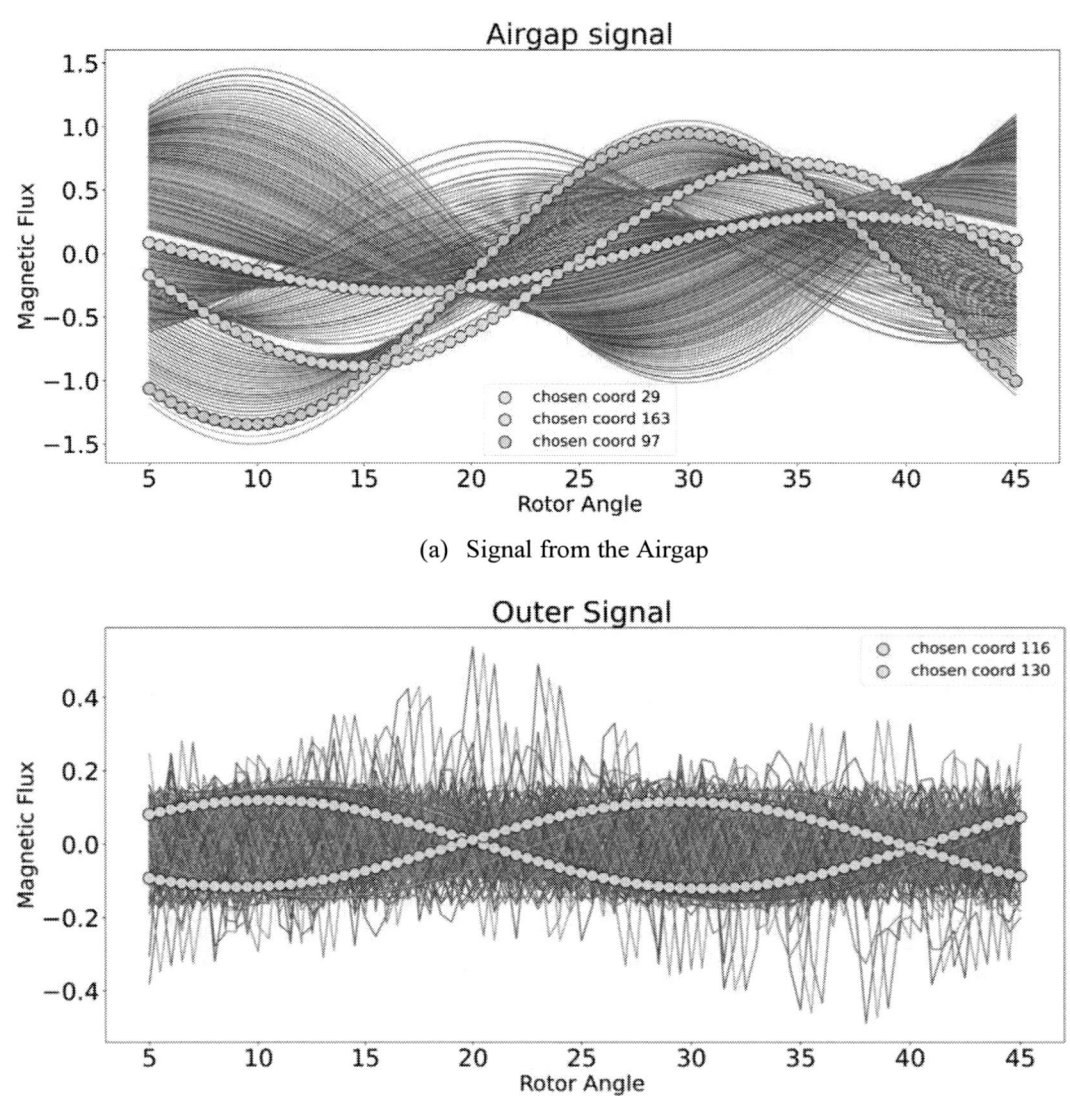

(a) Signal from the Airgap

(b) Signal from the Outer Layer

FIGURE 5: CHOSEN COORDINATES FOR 5 SENSOR RESULTS: (A) CHOSEN COORDINATES FOR 5 SENSORS RESULT WITHIN THE AIRGAP PLACEMENT; (B) CHOSEN COORDINATES FOR 5 SENSOR RESULT WITHIN THE OUTER PLACEMENT

3. RESULTS AND DISCUSSION

The results of the genetic algorithm are shown in figure 4. The top graph shows the accuracy for each fault with regards to the fitness function being defined as $\hat{F}$, while the bottom plot has F' as the fitness function. The bottom axis is the number of sensors, and the y axis is the accuracy for each fault. In general, when the number of sensors are increased, there is a higher overall accuracy; but this increase is not linear. Reasons for this include randomness of the genetic algorithm, and a tradeoff between accuracy for the different faults. When comparing the two plots, it can be observed that the fitness function with penalty term forces the design to have a similar score for each fault. To reach the threshold of 90% for each fault, it seems at least 7 sensors are needed. It is important to maintain this

Copyright © 2023 by ASME

threshold for each fault, as the health monitoring system will be unreliable if the detectability for one of the states falls significantly below the rest (as shown with 9 sensors in the first part of figure 4). For all states to have a F1-score of above 95% accuracy, at least 11 sensors are needed. An overall F1-score of 95% can be obtained with 8 sensors, but not all the faults are detected with 95% accuracy. The explicit values for F' and $\hat{F}$ are shown in table 2. It is clear that the only time the two fitness scores can be the same is when 7 sensors are in place. As shown in Table 2, there are diminishing returns after around 8 sensors.

TABLE 2: Results and F1-Scores

# of sensors	F'	$\hat{F}$
2	0.55	0.8
3	0.65	0.83
4	0.75	0.88
5	0.82	0.92
6	0.79	0.92
7	0.94	0.94
8	0.95	0.95
9	0.96	0.96
10	0.96	0.96
11	0.99	0.99
12	0.99	0.99

In all the results shown in figure 4, the chosen design does not choose any inner placement coordinates. Since there is little to no signal within the inner placement, this is an expected output from the algorithm. Since there is substantial signal in the airgap, the proposed methodology suggests placing the sensors within this location. For the case with 5 sensors, 3 are placed in the airgap and 2 are placed in the outer placement. This is shown in detail in Fig. 5, where the selected signal is highlighted among all possible signals. Ultimately, the proposed solution is the case with 5 sensors, as it is cost-efficient and has a high accuracy. Specifically, it has 93.5% overall accuracy for detecting faults. However, fault 5 is only detected with 89% F1-score, so the F' is lowered to 83%. The overall score is high because this combination of sensors has perfect accuracy in detecting fault 2. To obtain above 90% detectability for all states, 7 sensors are needed.

An algorithm that concurrently trains a classifier and generates new designs is a novel methodology; it can also be generalized any engineering system. In recent literature, the most similar structure is a machine learning technique, namely generative adversarial networks (GAN). This network trains a discriminator (i.e. predictor) and creates new training data as a competition (i.e. generator). This structure can be very useful when the design space if very large and not convex (in this case it is discrete), or when simulation data is limited. In this paper, a PMSM with magnetic field sensing was tested for fault detection, and application-specific future direction can be considered. For this study, only magnetic field sensing was used to determine faults, but other measurements can be incorporated for fault detection, such as voltage or current. Information

fusion techniques can be considered for these new measurements; and combining different sensors and sources of data may improve the accuracy in classifying the faults. Also, the type of error can be considered for the fitness function within the genetic algorithm. For example, a false alarm for a fault is not as serious as a false negative. In the presented study, the classifier would often detect a fault when there was not one (predicting fault 5 when the PMSM was healthy), which is not a fatal error. Nevertheless, this study determines a sensor network for a PMSM with magnetic field sensing, which has not been considered in literature; and the proposed methodology can be generalized to numerous complex engineering systems.

4. CONCLUSION

A generic framework to determine the optimal placement of sensors for the fault diagnosis of permanent magnet synchronous machines is developed in the presented study, where hall effect sensors have been employed. The developed methodology determines sensor placement solutions that provide accurate classification of faults with a minimal number of sensors. While one of the main objectives for sensor placement was to obtain a high detectability for each of the potential faults, it is common for a sensor placement solution to detect certain faults perfectly but fail to detect others. The developed sensor placement methodology, which combines the sensor placement optimization with fault diagnosis, yields high accuracy and computational efficiency for the PMSM application, and could be applied to other sensor network design problems for different engineering applications.

ACKNOWLEDGEMENTS

This research is partially supported by the National Science Foundation (NSF) the Engineering Research Center for Power Optimization of Electro-Thermal Systems (POETS) with cooperative agreement EEC-1449548, and the Alfred P. Sloan Foundation through the Energy and Environmental Sensors program with grant # G-2020-12455.

REFERENCES

[1] Xu, Y., Lalwani, A.V., Arora, K., Zheng, Z., Renteria, A., Senesky, D.G. and Wang, P., 2022. Hall-Effect Sensor Design with Physics-Informed Gaussian Process Modeling. IEEE Sensors Journal, 22(23), pp.22519-22528.

[2] Wang, P., Tamilselvan, P. , and Hu, C., "Health Diagnostics Using Multi-Attribute Classification Fusion," Engineering Applications of Artificial Intelligence, vol.32, pp.192-202, 2014.

[3] Zhang, B., Zhang, L., Xu, J., and Wang, P., "Performance Degradation Assessment of Rolling Element Bearings Based on an Index Combining SVD and Information Exergy," Entropy, 16(10), 5400-5415, 2014.

[4] Xi, Z., Jing, R., Wang, P., and Hu, C., "A Copula-Based Sampling Method for Data-Driven Prognostics and Health Management", Reliability Engineering and System Safety,

Copyright © 2023 by ASME

vol.132, pp. 72-82, 2014.

[5] Tamilselvan, P., Wang, P., Sheng, S., and Twomey, J., "A Two-stage Diagnosis Framework for Wind Turbine Gearbox Condition Monitoring," Int. J. on Prognostics and Health Management, Vol. 4, 010 (11), 2013.

[6] Tamilselvan, P., Wang, P., and Pecht, M., "A Multi-Attribute Classification Fusion System for Insulated Gate Bipolar Transistor Diagnostics," Microelectronics Reliability, Vol. 53, No. 8, pp. 1117-1129, 2013.

[7] Tamilselvan, P., and Wang, P., "Failure Diagnosis Using Deep Belief Learning Based Health State Classification," Reliability Engineering and System Safety, Vol. 115, pp.124-135, 2013.

[8] Bai, G., Wang, P., and Hu, C., "A Self-Cognizant Dynamic System Approach for Prognostics and Health Management", Power Sources, Vol. 278, 15, pp. 163-174, 2015.

[9] Tamilselvan, P., and Wang, P., "A Tri-fold Hybrid Classification Approach for Diagnostics with Unexampled Faulty States", Mechanical Systems & Signal Processing, Vol. 50–51, pp.437–455, 2015.

[10] Muhammad, F., Hong, H., Zhang, P. and Abbas, Q., 2023. Measurement of Magnetic Field Components Using A Single Passive SAW Magnetic Sensor. Sensors and Actuators A: Physical, p.114163.

[11] Zhou, Y., Liu, X., Fan, L., Liu, W., Xing, E., Tang, J. and Liu, J., 2023. Temperature and vibration insensitive fiber optic vector magnetic field sensor. Optics Communications, 530, p.129178.

[12] Popovic, R.E., Randjelovic, Z. and Manic, D., 2001. Integrated Hall-effect magnetic sensors. Sensors and Actuators A: Physical, 91(1-2), pp.46-50.

[13] Morimoto, S., 2007. Trend of permanent magnet synchronous machines. IEEJ Transactions on Electrical and Electronic Engineering, 2(2), pp.101-108.

[14] Choudhary, A., Goyal, D., Shimi, S.L. and Akula, A., 2019. Condition monitoring and fault diagnosis of induction motors: A review. Archives of Computational Methods in Engineering, 26, pp.1221-1238.

[15] Xu, X., Qiao, X., Zhang, N., Feng, J. and Wang, X., 2020. Review of intelligent fault diagnosis for permanent magnet synchronous motors in electric vehicles. Advances in Mechanical Engineering, 12(7), p.1687814020944323.

[16] Yoon, A., Yi, X., Martin, J., Chen, Y. and Haran, K., 2016, February. A high-speed, high-frequency, air-core PM machine for aircraft application. In 2016 IEEE Power and Energy Conference at Illinois (PECI) (pp. 1-4). IEEE.

[17] Hu, C., Wang, P., Youn, B.D., Lee, W.R., and Yoon, J.T., "Copula-Based Statistical Health Grade System against Mechanical Faults of Power Transformers," IEEE Trans. on Power Delivery, 27 (4), pp. 1809-1819, 2012.

[18] Vaseghi, B., Takorabet, N., Nahid-Mobarakeh, B. and Meibody-Tabar, F., 2011. Modelling and study of PM machines with inter-turn fault dynamic model–FEM model. Electric Power Systems Research, 81(8), pp.1715-1722.

[19] Peng, T., Ye, C., Yang, C., Chen, Z., Liang, K. and Fan, X., 2022. A novel fault diagnosis method for early faults of PMSMs under multiple operating conditions. ISA transactions, 130, pp.463-476.

[20] Lee, H., Jeong, H., Koo, G., Ban, J. and Kim, S.W., 2020. Attention recurrent neural network-based severity estimation method for interturn short-circuit fault in permanent magnet synchronous machines. IEEE Transactions on Industrial Electronics, 68(4), pp.3445-3453.

[21] Zeng, C., Huang, S., Lei, J., Wan, Z. and Yang, Y., 2021. Online rotor fault diagnosis of permanent magnet synchronous motors based on stator tooth flux. IEEE Transactions on Industry Applications, 57(3), pp.2366-2377.

[22] Taher, S.A., Li, J. and Fang, H., 2023. Simultaneous seismic input and state estimation with optimal sensor placement for building structures using incomplete acceleration measurements. Mechanical Systems and Signal Processing, 188, p.110047.

[23] Ercan, T., Sedehi, O., Katafygiotis, L.S. and Papadimitriou, C., 2023. Information theoretic-based optimal sensor placement for virtual sensing using augmented Kalman filtering. Mechanical Systems and Signal Processing, 188, p.110031.

[24] Zhang, Z., Peng, C., Wang, G., Ju, Z. and Ma, L., 2023. Optimal sensor placement for strain sensing of a beam of high-speed EMU. Journal of Sound and Vibration, 542, p.117359.

[25] Kong, X., Cai, B., Liu, Y., Zhu, H., Liu, Y., Shao, H., Yang, C., Li, H. and Mo, T., 2022. Optimal sensor placement methodology of hydraulic control system for fault diagnosis. Mechanical Systems and Signal Processing, 174, p.109069.

[26] Inoue, T., Ikami, T., Egami, Y., Nagai, H., Naganuma, Y., Kimura, K. and Matsuda, Y., 2023. Data-driven optimal sensor placement for high-dimensional system using annealing machine. Mechanical Systems and Signal Processing, 188, p.109957.

[27] Eshghi, A. T., Lee, S., Jung, H-J, Wang, P., "Design of structural monitoring sensor network using surrogate modeling of stochastic sensor signal," Mechanical Systems and Signal Processing, vol. 133, 106280, 2019.

[28] Wang, P., Youn, B.D., Hu, C., Ha, J.M. and Jeon, B., 2015. A probabilistic detectability-based sensor network design method for system health monitoring and prognostics. Journal of Intelligent Material Systems and Structures, 26(9), pp.1079-1090.

Copyright © 2023 by ASME

Proceedings of the ASME 2023
International Design Engineering Technical Conferences and
Computers and Information in Engineering Conference
IDETC-CIE2023
August 20-23, 2023, Boston, Massachusetts

DETC2023-117177

MEAN TIME TO FAILURE PREDICTION FOR COMPLEX SYSTEMS WITH ADAPTIVE SURROGATE MODELING

Hao Wu, Yanwen Xu, Zheng Liu and Pingfeng Wang*

Department of Industrial and Enterprise Systems Engineering
University of Illinois Urbana-Champaign
Urbana, IL 61801, United States.

ABSTRACT

The Mean Time to Failure (MTTF) is a critical metric for assessing the reliability of non-repairable systems, and it plays a significant role in incident management. However, accurately estimating MTTF can be challenging due to the limited availability and quality of historical experimental data on a group of identical parts or systems. To address this challenge, we propose an adaptive surrogate modeling method, which approximate the system's performance functions with a more computationally efficient model to predict the MTTF during the design stage. In this paper, we develop an adaptive Gaussian process (GP) approach that combines with Monte Carlo simulation (MCS) to predict MTTF. The proposed method initially trains a GP model for the system's performance function, and then an MTTF surrogate model can be obtained. With a learning function, these models are updated dynamically based on new information as it becomes available. We demonstrate the effectiveness of our method on series system, parallel system, and mixed system with two examples, and the results show that the adaptive surrogate modeling method can accurately predict the MTTF of the system. Our method has the potential to improve the reliability and safety of complex system and reduces the need for costly performance testing.

Keywords: Gaussian process, mean time to failure, complex system, adaptive sampling

1. INTRODUCTION

Mean time to failure (MTTF) is a common metric used in reliability engineering to measure the expected lifespan of a product or system [1–5]. MTTF is an essential parameter used in the design and maintenance of many systems [6–8], such as aircraft engine, power plants, and electronic device. However, estimating MTTF can be challenging due to the difficulty in obtaining accurate data on failure times, which may be rare events and data may be incomplete or unreliable, as well as the presence of multiple factors that can influence failure times, such as uncertainties [9–14], usage patterns, and maintenance practices. There are three types of method that can be used to estimate MTTF of a system, including analytical methods, empirical methods, and simulation-based methods.

Analytical methods involve using mathematical models to estimate MTTF based on the system's failure rate, repair rate, and other relevant parameters. Murphy introduced a diagnosability prediction metric for system modeling of component failure rates and unjustified removals, and the result shows that system improvements based on this prediction technique will increase the quality of a product since increased diagnosability decreases life cycle costs [15]. M. Fitzpatrick and R. Paasch developed an analytical method for predicting life-cycle maintainability labor costs to allow the evaluation of products, based on life-cycle labor costs, early in the design process. These methods are suitable for simple systems with known failure modes and can provide accurate estimates of MTTF [16]. However, applying these analytical methods to complex systems with multiple failure modes can be challenging. For instance, a nuclear power plant may have multiple failure modes that arise from a variety of factors such as equipment malfunction, operator error, and natural disasters. These failure modes can interact in complex ways, making it difficult to accurately model the system's behavior and estimate MTTF.

Empirical methods involve analyzing the failure data of similar systems or components to estimate the MTTF of a

---

* Corresponding Author. Email: pingfeng@illinois.edu.

Copyright © 2023 by ASME

system. These methods are useful when there is limited knowledge of the system's failure modes and can provide reliable estimates of MTTF. Jiang [17] derived a nonparametric estimator of MTTF based on the decomposition of the integral expression of the theoretical MTTF and sample statistical characteristics. The estimated MTTF is then combined with a two-step single parameter maximum likelihood method to estimate the distribution parameters. However, these methods require a significant amount of data, and the budget of the testing is expensive and even unaffordable when the products are expensive.

Simulation-based methods involve modeling the system's behavior using computer simulations and estimating the MTTF based on the simulation results. F.X. Che put forward finite element analysis to investigate solder joint stress strain behavior for solder joint vibration fatigue life prediction [18]. These methods are suitable for complex system with multiple failure modes and can provide detailed insight into the system's performance. However, they can be computationally expensive and require a significant number of computational resources.

In recent years, surrogate modeling methods has emerged as an effective approach to estimate the MTTF of a system. Surrogate models method can approximate the complex relationship between input data and output data, which provide efficient and accurate estimates of MTTF. Surrogate models can be developed using different machine learning and statistical techniques, such as artificial neural networks[19], support vector machines [20, 21], or Gaussian process regression [22–31]. For instance, Wu and Du proposed a new system reliability method that combines Monte Carlo simulation and the Gaussian process method with improved efficiency and accuracy [11]. Li and Wang introduced a feedforward neural network and Gaussian process regression framework for reliability analysis [12]. Hu and Du applied support vector machines (SVM) with first order reliability method to predict system reliability [13]. The use of surrogate modeling methods in estimating MTTF has several advantages. It allows for efficient evaluation of system performance over a wide range of operating conditions and parameters. Additionally, it reduces the reliance on costly and time-consuming simulations or experiments to achieve the failure data to estimate the MTTF.

In this study, we proposed a Gaussian process regression with adaptively modeling method for the estimation of time-dependent system MTTF. Specifically, we adaptively train a Gaussian process (GP) regression model to approximate the limit-state function with respect to the random input variables. A system learning function is proposed to identify the best next training point to refine the GP models of multiple failure modes. When the GP models are accurate, the MTTF of the system is obtained. The contribution of this work is (1) the proposed method utilizes an adaptive surrogate modeling method to predict the system MTTF, and this method can be applicable to series system, parallel system, and mixed system; (2) this method is able to identify both the best next training point and the corresponding component index, which facilities independent training of GP models of multiple failure modes and avoid

unnecessary function calls.

The remainder of this paper is organized as follows. The problem statement is given in Section 2. The proposed method is discussed in Section 3. Two examples are provided in Section 4, followed by conclusions in Section 5.

2. PROBLEM STATEMENT

In this work, the component's limit-state function fails when its repones Y_i is less than zero, which is given by below,

$$Y_i = g_i(\mathbf{X}, t) \tag{1}$$

where $\mathbf{X}$ is a random vector and t is time. Note that the input of g_i might involve with random processes, which can be converted into random variables using the expansion optimal linear estimation method (EOLE) [32, 33]. Therefore, Y_i does not lose generality.

When $\mathbf{x}$ is a realization of $\mathbf{X}$, the response Y_i changes with respect to t. When Y_i becomes negative for the first time, the associated time instance $t = \tau_i(\mathbf{x})$ is named the first time to failure. $\tau_i(\mathbf{x})$ depends on $\mathbf{x}$ and it is determined by Eq. (1) when $Y_i = 0$. For a series system with n_F components, the first time to failure is given by $\tau(\mathbf{x}) = \min(\tau_1(\mathbf{x}), ..., \tau_{n_F}(\mathbf{x}))$. For a parallel system with n_F components, the first time to failure is $\tau(\mathbf{x}) = \max(\tau_1(\mathbf{x}), ..., \tau_{n_F}(\mathbf{x}))$.

Note that the mixed system can be represented by the series system and parallel system configuration. Since $\mathbf{x}$ is a realization of $\mathbf{X}$, the final system MTTF then is expressed as

$$\bar{\tau}(\mathbf{X}) = E(\tau(\mathbf{X})) \tag{2}$$

where $E(\cdot)$ refers to the expectation of τ. However, finding the first root of Eq. (1) is challenging. The purpose of this work is to develop an efficient and accurate method to estimate $\bar{\tau}(\mathbf{X})$.

3. THE PROPOSED METHOD

This section introduces the proposed method, where subsection 3.1 provides an overview of the proposed method, subsection 3.2 discusses the initial sampling, subsection 3.3 presents the estimation of system MTTF, while subsections 3.4, 3.5 and 3.6 details the proposed method.

3.1 Overview of the Proposed Method

The proposed method firstly builds the Gaussian process regression models [26] $\hat{g}_i(\mathbf{X}, t)$ for all components. Then we can obtain each component surrogate model $\hat{\tau}_i(\mathbf{X})$. Since the initial Gaussian process models $\hat{g}_i(\mathbf{X}, t)$ are not accurate, we define the learning function and stop condition to refine models $\hat{g}_i(\mathbf{X}, t)$. Once the algorithm iteratively hits the stop condition, the surrogate model $\hat{\tau}_i(\mathbf{X})$ is accurate and the system MTTF $\bar{\tau}(\mathbf{X})$ can be determined by MCS.

3.2 Initial Sampling

The idea behind designing experiments for constructing a Gaussian process model is to distribute the initial training points uniformly. Latin hypercube sampling [34] and Hammersley sampling [35] are commonly used approaches for this purpose. In this work, the initial samples $\mathbf{x}^s$ and t are generated by

Copyright © 2023 by ASME

Hammersley sampling method to create an initial surrogate model $\hat{g}_i(\mathbf{X}, t)$. Suppose that the dimension of $\mathbf{X}$ is n, and initial sampling points are s. $\mathbf{x}^s$ are generated as follows.

$$\mathbf{x}^s = \begin{bmatrix} x_1^1 & x_2^1 & \dots & x_n^1 \\ x_1^2 & x_2^2 & \dots & x_n^2 \\ \vdots & \vdots & \ddots & \vdots \\ x_1^s & x_2^s & \dots & x_n^s \end{bmatrix} \tag{3}$$

When initial samples t are generated, we then have the following combined initial samples.

$$[\mathbf{x}^s, \boldsymbol{t}^s] = \begin{bmatrix} x_1^1 & x_2^1 & \dots & x_n^1 & t^1 \\ x_1^2 & x_2^2 & \dots & x_n^2 & t^2 \\ \vdots & \vdots & \dots & \vdots & \vdots \\ x_1^s & x_2^s & \dots & x_n^s & t^s \end{bmatrix} \tag{4}$$

We then substitute $[\mathbf{x}^s, \boldsymbol{t}^s]$ into limit-state functions Eq. (1) to obtain $\boldsymbol{y}_i^s$. With the initial training samples $(\mathbf{x}^s, \boldsymbol{t}^s, \boldsymbol{y}_i^s)$, we can build the Gaussian process models $\hat{g}_i(\mathbf{X}, t)$ for all components.

3.3 Estimate System MTTF $\bar{\tau}(\mathbf{X})$

To estimate system MTTF $\bar{\tau}(\mathbf{X})$, we should find the first root $\tau_i(\mathbf{x})$ in Eq. (1). However, it is not practical to find $t = \tau_i(\mathbf{x})$ when the limit-state function is expensive black box function. By referring to [36], instead of finding the accurate first root $\tau_i(\mathbf{x})$, we only need to find the sign $g_i(\mathbf{x}, t)$ of at different time instances after discretizing the time interval. When $Y_i(\tau_i(\mathbf{x})) < 0$ at the first time, then MTTF $\bar{\tau}(\mathbf{X})$ can be estimated. The accuracy of predicting signs of $g_i(\mathbf{x}, t)$ depends on GP model $Y_i = \hat{g}_i(\mathbf{X}, t)$.

Now let's introduce the calculation of system MTTF $\bar{\tau}(\mathbf{X})$. Firstly, the time span [0, T] is evenly discretized into n_t points $t = (t_1, t_2, \dots, t_{n_t})$. Then τ_i for i-th component in series system can be approximated by

$$\hat{\tau}_i(\mathbf{x}) = \min\{t \in \boldsymbol{t} \,|\, \hat{g}_i(\mathbf{x}, t) \leq 0\} \tag{5}$$

$$\tau^S(\mathbf{x}) = \min_{i=1,\dots,n_F} \left(\hat{\tau}_i(\mathbf{x}) \right) \tag{6}$$

where n_F is the number of components in system and $\tau^S(\mathbf{x})$ is first time to failure for a series system when $\mathbf{x}$ is fixed. We then generate n_p samples to estimate system MTTF $\tau^S(\mathbf{x})$, and the series system MTTF is given by

$$\bar{\tau}^S(\mathbf{X}) = \frac{1}{n_p} \sum_{i=1}^{n_p} \tau^S(\mathbf{x}^i) \tag{7}$$

Similarly, the first time to failure for a parallel system is

$$\tau^P(\mathbf{x}) = \max_{i=1,\dots,n_F} \left(\hat{\tau}_i(\mathbf{x}) \right) \tag{8}$$

Then the parallel system MTTF is

$$\bar{\tau}^P(\mathbf{X}) = \frac{1}{n_p} \sum_{i=1}^{n_p} \tau^P(\mathbf{x}^i) \tag{9}$$

Two factors determine the accuracy of $\bar{\tau}^P$ and $\bar{\tau}^S$. The first one is the sample size n_p of $\mathbf{X}$ as $\bar{\tau}^P$ and $\bar{\tau}^S$ are random variables. The other factor is each component's surrogate model

$\hat{g}_i(\mathbf{x}, t)$ as a more accurate model $\hat{g}_i(\mathbf{x}, t)$ will result in a more accurate the model $\hat{\tau}_{g_i}(\mathbf{x})$. In the following, we will introduce how to determine the sample size n_p and the adaptive sampling method to refine model $\hat{g}_i(\mathbf{x}, t)$.

3.4 System Learning Function SYSU

The Gaussian process models' probabilistic information enables the selection of training points to refine the model in an adaptive manner. The most used learning function among the available options is the U learning function [37], which can assess the probability of misjudgment of the sign of limit state functions. If there are n_F components in a system, the U learning function for the ith component is as follows:

$$U_i(\mathbf{x}) = \frac{\left| \mu_{\hat{g}_i}(\mathbf{x}) \right|}{\sigma_{\hat{g}_i}(\mathbf{x})} \tag{10}$$

According to the principles of system reliability theory and U learning function, if there are any components in the series system that have failed, and their highest U learning function is greater than 2, then the system is considered to fail. On the other hand, if all components are functioning properly and their lowest U learning function is greater than 2, then the system is safe. To this end, a new learning function for reliable system, named SYSU learning function, has been developed specifically for series systems, and is expressed as follows:

$$\text{SYSU}^S(\mathbf{x}) = \min_{i=1,\dots,n_F} \frac{\left| \mu_{\hat{g}_i}(\mathbf{x}) \right|}{\sigma_{\hat{g}_i}(\mathbf{x})} \tag{11}$$

The SYSU learning function is designed to assess the likelihood of incorrectly classifying a system as safe or failed. If $\text{SYSU}^S(\mathbf{x}) \geq 2$, it is reasonable to conclude that the current Gaussian process models of all components are capable of precisely determining the system's state.

If the U function involves time, the expected value and standard deviation of the GP model $\hat{g}_i(\mathbf{X}, t)$. are denoted by $\mu_{\hat{g}_i}(\mathbf{x}, t_k)$ and $\sigma_{\hat{g}_i}(\mathbf{x}, t_k)$, respectively. In such cases, the U learning function for a component can be expressed as $U_{t_k}^s(\mathbf{x})$ given below and is used to judge the state of the ith component based on the i-th GP model.

$$U_{t_k}^s(\mathbf{x}) = \min_{k=1,\dots,n_t} \frac{\left| \mu_{\hat{g}_i}(\mathbf{x}, t_k) \right|}{\sigma_{\hat{g}_i}(\mathbf{x}, t_k)} \tag{12}$$

If $SYSU_{t_k}^s(\mathbf{x})$ is greater than or equal to 2, then the state of the i-th component can be accurately determined by the i-th GP model. Then the time index $k_{g_i}^s(\mathbf{x})$ corresponding to $SYSU_{t_k}^s(\mathbf{x})$ is given by

$$k_{g_i}^s(\mathbf{x}) = \operatorname*{argmin}_{k=1,\dots,n_t} \frac{\left| \mu_{\hat{g}_i}(\mathbf{x}, t_k) \right|}{\sigma_{\hat{g}_i}(\mathbf{x}, t_k)} \tag{13}$$

According to Eq.(11), when considering all components in system, SYSU learning function for the above series system is given by

$$SYSU_{n_F}^s(\mathbf{x}) = \min_{i=1,\dots,n_F} \left(\min_{k=1,\dots,n_t} \frac{\left| \mu_{\hat{g}_i}(\mathbf{x}, t_k) \right|}{\sigma_{\hat{g}_i}(\mathbf{x}, t_k)} \right) \tag{14}$$

Copyright © 2023 by ASME

Accordingly, the associated component index corresponding to $SYSU_{n_F}^s(\mathbf{x})$ is identified by:

$$i_{min}^s(\mathbf{x}) = \underset{i=1,\ldots,n_F}{\arg\min} \frac{\left|\mu_{\hat{g}_i}\left(\mathbf{x}, t_{k_{g_i}^s(\mathbf{x})}\right)\right|}{\sigma_{\hat{g}_i}\left(\mathbf{x}, t_{k_{g_i}^s(\mathbf{x})}\right)}$$

$$= \underset{i=1,\ldots,n_F}{\arg\min} \left(\min_{k=1,\ldots,n_t} \frac{\left|\mu_{\hat{g}_i}(\mathbf{x}, t_k)\right|}{\sigma_{\hat{g}_i}(\mathbf{x}, t_k)} \right) \quad (15)$$

It is obvious that $SYSU_{n_F}^s(\mathbf{x})$ depends solely on sample $\mathbf{x}$. Thus, if $SYSU_{n_F}^s(\mathbf{x}) \geq 2$, the MTTF can be accurately predicted by the current GP models. it is suggested that the sample point $\mathbf{x}^*$ that minimizes the value of $SYSU_{n_F}^s(\mathbf{x})$ can be chosen as the most suitable training point shown below.

$$\mathbf{x}^* = \underset{i=1,\ldots,n_p}{\arg\min}(SYSU^S(\mathbf{x}_i)) \quad (16)$$

Once the $\mathbf{x}^*$ is available, the associated component index i^* and time instance index k^* can be obtained by:

$$i^* = i_{min}^s(\mathbf{x}^*) \quad (17)$$

$$k^* = i_{g_{i^*}}^s(\mathbf{x}^*) \quad (18)$$

Once $\mathbf{x}^*$, i^*, and k^* have been achieved, the training set can be updated by the next training point $(\mathbf{x}^*, t_{k^*})$ to update the i^*-th GP model.

Likewise, in the case of a parallel system, if any of the safe components have a highest U learning function value greater than 2, the system is safe. However, if all components fail and their lowest U learning function is greater than 2, then the system is considered to have failed. Therefore, the SYSU learning function for parallel system can be expressed as follows:

$$SYSU^P(\mathbf{x}) = \max_{i=1,\ldots,n_F} \frac{\left|\mu_{\hat{g}_i}(\mathbf{x})\right|}{\sigma_{\hat{g}_i}(\mathbf{x})} \quad (19)$$

The process of selecting the best training sample for a parallel system is comparable to that of a series system, with the only difference being that the parallel system involves in. According to Eq. (14), the SYSU learning function for the above parallel system is given by:

$$SYSU_{n_F}^P(\mathbf{x}) = \max_{i=1,\ldots,n_F} \left(\min_{k=1,\ldots,n_t} \frac{\left|\mu_{\hat{g}_i}(\mathbf{x}, t_k)\right|}{\sigma_{\hat{g}_i}(\mathbf{x}, t_k)} \right) \quad (20)$$

The associated component index corresponding to $SYSU_{n_F}^P(\mathbf{x})$ is identified by:

$$i_{max}^s(\mathbf{x}) = \underset{i=1,\ldots,n_F}{\arg\max} \left(\min_{k=1,\ldots,n_t} \frac{\left|\mu_{\hat{g}_i}(\mathbf{x}, t_k)\right|}{\sigma_{\hat{g}_i}(\mathbf{x}, t_k)} \right) \quad (21)$$

Similarly, the sampling point $\mathbf{x}^*$ with the smallest $SYSU_{t_k}^P(\mathbf{x})$ should be selected as the next training point.

$$\mathbf{x}^* = \underset{i=1,\ldots,n_p}{\arg\min}(SYSU^P(\mathbf{x}_i)) \quad (22)$$

The corresponding model index i^* and time instance index k^* can be obtained by:

$$i^* = i_{max}^s(\mathbf{x}^*) \quad (23)$$

$$k^* = i_{g_{i^*}}^s(\mathbf{x}^*) \quad (24)$$

Finally, the training point $(\mathbf{x}^*, t_{k^*})$ is added into the training set of the i^*-th GP model for refinement.

3.5 Updating Sample Size

To accelerate the convergence of $\bar{\tau}^S$ and $\bar{\tau}^P$, the stopping criterion proposed by Wei [36] is extend here to estimate the system MTTF. Since $\tau(\mathbf{X})$ is random variable, it is desirable to select a good sample size n_s to determine the mean value $\bar{\tau}(\mathbf{X})$ and standard deviation $\sigma_{\bar{\tau}(\mathbf{X})}$ of $\tau(\mathbf{X})$. When $\sigma_{\bar{\tau}(\mathbf{X})}$ is small enough, no more new TPs are needed. Then, the surrogate models $\hat{\tau}_i(\mathbf{X})$ are used to calculate $\bar{\tau}(\mathbf{X})$. With the sample size n_s, the coefficient of variation C_v of $\bar{\tau}$ is given by

$$C_v = \frac{\sigma_{\bar{\tau}}}{\bar{\tau}\sqrt{n_s}} \quad (25)$$

where $\bar{\tau}$ is estimated by Eq. (7) or Eq. (8), and $\sigma_{\bar{\tau}}$ is the standard deviation of $\bar{\tau}$. Eq. (25) shows that the larger n_s is, the smaller the C_v is. A smaller C_v means that $\bar{\tau}$ is more accurate. If the following condition is met, then $\bar{\tau}$ considers to be accurate.

$$C_v \leq \Upsilon \quad (26)$$

where Υ is a threshold and a common value for this small positive number is 0.005.

If the current value of n_s fails to meet the requirements of Eq. (26), the sample size will be increased. By using both Eq. (25) and Eq.(26), we can determine the minimum required sample size, which is as follows:

$$n \geq \left(\frac{\sigma_{\bar{\tau}}}{\bar{\tau}C_v}\right)^2 \quad (27)$$

Assuming ceil represents rounding up to the nearest integer, the value of n_1 can be calculated as $n_1 = \text{ceil}\left(\frac{\sigma_{\bar{\tau}}}{\bar{\tau}\Upsilon}\right)^2$. The value of n_{add} is then determined as follows:

$$n_{add} = n_1 - n_s \quad (28)$$

In situations where $\hat{g}_i(\mathbf{X}, t)$ is not initially accurate during the first few iterations, the accuracy of both $\bar{\tau}$ and $\sigma_{\bar{\tau}}$ may have poor accuracy, which can result in n_{add} being misleading as per Eq. (28). To address this issue, a threshold is set for n_{add}. Consequently, Eq. (28) is revised as follows

$$n_{add} = \begin{cases} \tilde{n}_{add}, & \text{if } n_1 - n_s > \tilde{n}_{add} \\ n_1 - n_s, & \text{otherwise} \end{cases} \quad (29)$$

3.6 Procedure of Proposed Method

In this subsection, the detailed step of the proposed method is given below, and the flowchart is shown in Fig. 1.

Step 1: Discretize the time interval of interest [0, T] into s time instances to obtain time vector $\mathbf{t}^s$. Generate s initial training set $\mathbf{x}^s$ using Hammersley sampling and calculate the corresponding response $\mathbf{y}_i^s$ for each component.

Step 2: Generate n_p random variable samples $\mathbf{X}$ and n_t time instances $\mathbf{t}$ and construct the Gaussian process model $\hat{g}_i$ with respect to $(\mathbf{x}^s, \mathbf{t}^s, \mathbf{y}_i^s)$ for each component.

Step 3: Output the expected values $\mu_{\hat{g}_i}(\mathbf{x}, t_k)$, $\sigma_{\hat{g}_i}(\mathbf{x}, t_k)$ of each GP model $\hat{g}_i$ at all time instances when $\mathbf{x}$ is fixed, $k \in n_t$.

Step 4: Compute the $\hat{\tau}_i(\mathbf{X})$, and find $\tau^S(\mathbf{X})$ using Eq.(6) when the system is series or $\tau^P(\mathbf{X})$ using Eq.(8) when the system is parallel. Then calculate the current $\bar{\tau}^S$ or $\bar{\tau}^P$.

Step 5: Calculate the values of system U function.

For a series system, the equation is given below

$$SYSU_{n_F}^S(\mathbf{X}) = \min_{i=1,\dots,n_F}\left(\min_{k=1,\dots,n_t} \frac{|\mu_{\hat{g}_i}(\mathbf{X}, t_k)|}{\sigma_{\hat{g}_i}(\mathbf{X}, t_k)}\right)$$

Find the best next training point $\mathbf{x}^*$ which satisfied the Eq. (16), then the associate time index k^* and component index i^* are available. The next training point $(\mathbf{x}^*, t_{k^*})$ will be added into training set for updating component i^*-th $\hat{g}_i$.

Likewise, in terms of Eq. (22), the next training point $(\mathbf{x}^*, t_{k^*})$ will be added into training set for updating component i^*-th $\hat{g}_i$ for a parallel system.

Step 6: Check the convergence, once the $SYSU_{n_F}^P(\mathbf{X}) \geq 2$ or $SYSU_{n_F}^S(\mathbf{X}) \geq 2$, calculate the coefficient of variation C_v of $\bar{\tau}$. If $C_v \leq Y$, the process is convergent and return $\bar{\tau}$. If not, add training points n_{add} and go back to step 3 until the process is convergent.

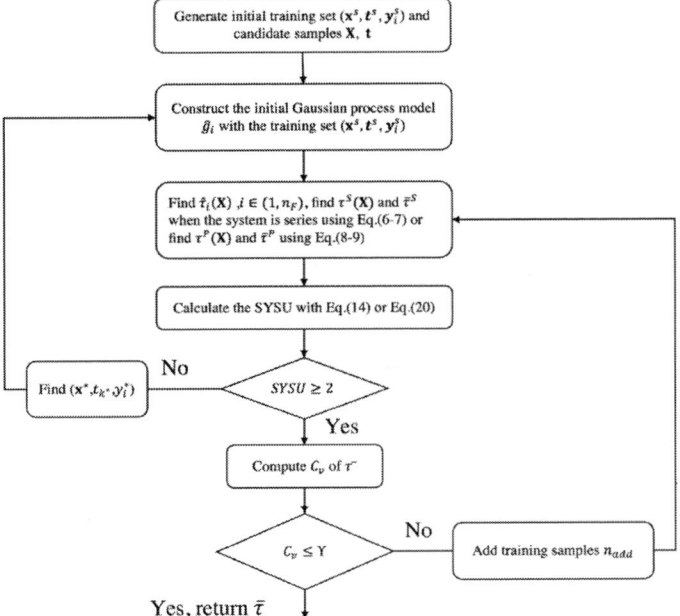

FIGURE 1: FLOWCHART OF PROPOSED METHOD

4. CASE STUDIES

In this section, two case studies have been used to demonstrate our proposed method, denoted as CEFFMTTF. The first example is a mathematic example. It is a mixed system with series system and parallel system configuration. The second example is an engineering example with a random process load. We use number of function calls (Ncall) to show the method's efficiency. The method's error is defined by

$$\text{error} = \left(\frac{\text{MTTF} - \text{MTTF}_{\text{MCS}}}{\text{MTTF}_{\text{MCS}}}\right) \times 100\% \qquad (30)$$

where MTTF is the result obtained by CEFFMTTF. The accuracy and efficiency of our proposed method are calculated compared to Monte Carlo simulation results.

4.1 Case 1: A Math Example

A mixed system as presented in Fig 2. is used to demonstrate the applicability of the proposed method for complex systems. The limit-state functions for components are given by

$$\begin{cases} g_i(\mathbf{X},\mathbf{Y}(t),t) = \pi(x_i - k_i t)^2 y_i - 0.707 P(t), & i = 1,2 \\ g_i(\mathbf{X},\mathbf{Y}(t),t) = \pi(x_i - k_i t)^2 y_i - 0.5 P(t), & i = 3,4 \end{cases} \quad (31)$$

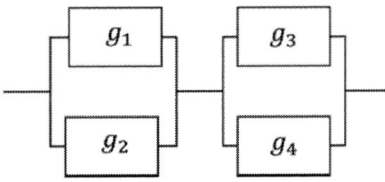

FIGURE 2: A MIXED SYSTEM EXAMPLE WITH FOUR LIMITE STATES

The detailed information of random variables $\mathbf{X}=[x_1, x_2, x_3, x_4, y_1, y_2, y_3, y_4]$ and random process $\mathbf{Y}(t) = [P(t)]$, where $x_1 \sim x_4$ are four normal distribution random variables, and x_1 and x_2 follow $N(2.5, 0.052)$, and x_3 and x_4 follow $N(2.3, 0.052)$. $y_1 \sim y_4$ follow a normal distribution with mean 36×10^3 and standard deviation of 3×10^3, and $P(t)$ is a Gaussian stochastic process with mean of 700×10^3, standard deviation of 70×10^3, and correlation function given by

$$\rho(t_1 - t_2) = \exp\left(-\left(\frac{t_1 - t_2}{5}\right)^2\right) \qquad (32)$$

A failure occurs if the limit-state function $g_i < 0$. Since the limit-state function corporates a random process variable $P(t)$, we use the expansion optimal linear estimation method (EOLE) to expand random process. We use five standard normal variables for the EOLE expansion of $P(t)$. The details implement procedures refer to [32, 33].

In this example, the time interval of interest $[0,10]$ is evenly discretized into 10 different time instances, denoted as $\mathbf{t}^s$. Thus, we initially have 10 training points $\mathbf{X}^s = (\mathbf{x}^s, \mathbf{t}^s)$ with Hammersley sampling method. Substituting the ten training inputs $\mathbf{X}^s$ into Eq. (31), we get the corresponding training outputs $\mathbf{y}^s$. With the training points $(\mathbf{X}^s, \mathbf{y}^s)$ available, the component g_i is initially approximated by the GP model $\hat{g}_i$. Given that the subsystems of $g1$ and $g2$ are configured in parallel and the subsystems of $g3$ and $g4$ are also configured in parallel, these two subsystems are combined to form a series system. As a result, it is easy to estimate the first-time failure of the mixed

Copyright © 2023 by ASME

system at a given point $\mathbf{x}$, which denoted as $\tau_m(\mathbf{x}) = \min\left(\max(\hat{\tau}1(\mathbf{x}), \hat{\tau}2(\mathbf{x})), \max(\hat{\tau}3(\mathbf{x}), \hat{\tau}4(\mathbf{x}))\right)$.

Considering the initial GP model $\hat{g}_i$ is inaccurate, more training pointes are added to update $\hat{g}_i$. By using Eq. (14) and Eq. (20), the sample point $\mathbf{x}^* = (2.4074, 3.6411 \times 10^3)$ with the smallest SYSU($\mathbf{x}$) should be selected as the best training point. The corresponding component index i^* is 2 and time instance t^* is 2.6364. Therefore, $(\mathbf{x}^*, t^*)$ is added into the training set of the i-th GP model for model refinement. The process is iteratively implemented until the system U function SYSU($\mathbf{x}$) >2. When the stop condition is meet by Eq. (26), the algorithm converges and the system MTTF $\bar{\tau}$=7.37. We use "Add 1", "Add 2", "Add 3", and "Add 4" to denote the final training points to update each component, respectively. Results are given in Table 1. The final MTTF of the proposed method for the system is 7.37 with the relative error 1.50% compared to Monte Carlo simulation results.

Table 1 RESULTS OF EXAMPLE 1

Methods	$\overline{T}$	error	Ncall	
CEFFMTTF	7.37	1.50%	Add 1	29
			Add 2	37
			Add 3	54
			Add 4	52
MCS	7.48	N/A	1×10^6	

4.2 Case 2: A Simple Supported Beam

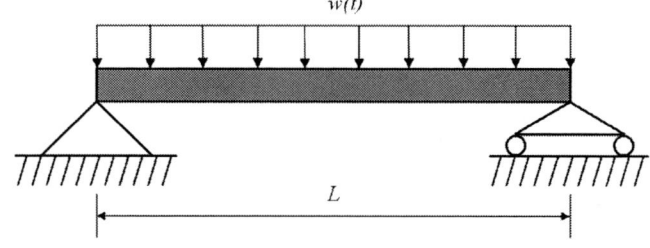

FIGURE 3: A SIMPLE SUPPORTED BEAM

Fig.3 shows a simple supported beam[38]. The length L of the beam is 5 m. The flexural capacity $M(t)$ and shear capacity $V(t)$ of the material follow exponential decay models because of its property degradation.

$$M(t) = M_0 e^{-0.05t} \qquad (33)$$
$$V(t) = V_0 e^{-0.05t} \qquad (34)$$

where M_0 and V_0 are the initial value of $M(t)$ and $V(t)$, respectively. The uniform distributed load $w(t)$ is modeled as a stationary Gaussian process. Table 2 displays the probability distribution characteristics of the random parameters. We consider the beam to have both shear and bending failure modes and treat it as a series system, and the two corresponding performance functions are shown as follows:

$$g_1(t) = M(t) - \frac{1}{8}w(t)L^2 \qquad (35)$$

$$g_2(t) = V(t) - \frac{1}{8}w(t)L \qquad (36)$$

Table 2 VARIABLES OF EXAMPLE 2

Parameter	M_0	V_0	$w(t)$
Distribution	Normal	Normal	Stational Gaussian process
Mean	800 kN	700 kN	50 kN/m
Standard deviation	80 kN	70 kN	5 kN/m
Autocorrelation	N/A	N/A	$\exp\left(-\left(\frac{t_1 - t_2}{5}\right)^2\right)$

Results are given in Table 3. The beam's mean lifetime evaluated by the proposed method is 6.94 years, with a relative error of 0.73 %. Furthermore, the proposed method requires only 29 limit-state function evaluations, making it significantly more cost-effective than MCS.

Table 3 RESULTS OF EXAMPLE 2

Methods	$\overline{T}$	error	Ncall	
CEFFMTTF	6.94	0.73%	Add 1	13
			Add 2	16
MCS	6.89	N/A	1×10^5	

5. CONCLUSION

This study demonstrates the effectiveness of using adaptive surrogate modeling for predicting mean time to failure (MTTF) in engineering systems. The proposed method utilizes Gaussian process-based surrogate models and an adaptive sampling strategy to efficiently estimate the MTTF of complex systems. The results of two case studies indicate that the proposed method is accurate and efficient in predicting MTTF for multiple failure modes. The proposed method is applicable to series system, parallel system, and mixed system. The study shows that the proposed method can identify both the best training sample and component index, which requires a small number of function calls compare to update the training sample for all components. Compared to traditional reliability analysis methods, the adaptive surrogate modeling approach requires significantly fewer evaluations of the limit-state function and thus reduces computational costs. Furthermore, the study shows that the accuracy and efficiency of the proposed method can be enhanced using machine learning algorithms. The learning function designed for the adaptive sampling strategy facilitates the selection of the best training samples and ensures the efficiency of the surrogate model.

Overall, this study contributes to the development of efficient and accurate methods for predicting MTTF, which is essential for assessing the reliability of engineering systems. The proposed adaptive surrogate modeling approach can be a useful tool for design optimization and decision-making in engineering applications, where the accurate prediction of MTTF is crucial.

Copyright © 2023 by ASME

ACKNOWLEDGEMENTS

This research is partially supported by the National Science Foundation (NSF) the Engineering Research Center for Power Optimization of Electro-Thermal Systems (POETS) with cooperative agreement EEC-1449548, and the Alfred P. Sloan Foundation through the Energy and Environmental Sensors program with grant # G-2020-12455.

REFERENCES

[1] X. Liu, Z. Zheng, İ. E. Büyüktahtakın, Z. Zhou, and P. Wang, "Battery asset management with cycle life prognosis," *Reliab Eng Syst Saf*, vol. 216, p. 107948, 2021.

[2] X. Liu and P. Wang, "Maintenance decision making using state dependent markov analysis with failure couplings," in 2020 Asia-Pacific International Symposium on Advanced Reliability and Maintenance Modeling (APARM), IEEE, 2020, pp. 1–6.

[3] P. O'Connor and A. Kleyner, *Practical reliability engineering*. John Wiley & Sons, 2012.

[4] G. Chryssolouris, *Manufacturing systems: theory and practice*. Springer Science & Business Media, 2013.

[5] J. W. McPherson, *Reliability physics and engineering: time-to-failure modeling*. Springer, 2018.

[6] W. Q. Meeker, L. A. Escobar, and F. G. Pascual, *Statistical methods for reliability data*. John Wiley & Sons, 2022.

[7] J. F. Lawless, "Statistical methods in reliability," *Technometrics*, vol. 25, no. 4, pp. 305–316, 1983.

[8] Hao Wu, "PROBABILISTIC DESIGN AND RELIABILITY ANALYSIS WITH KRIGING AND ENVELOPE METHODS," Purdue University Graduate School, 2022. doi: https://doi.org/10.25394/PGS.19653132.v1.

[9] H. Yu *et al.*, "Inlet and Outlet Boundary Conditions and Uncertainty Quantification in Volumetric Lattice Boltzmann Method for Image-Based Computational Hemodynamics," *Fluids*, vol. 7, no. 1, p. 30, 2022.

[10] C. Hu, B. D. Youn, and P. Wang, *Engineering design under uncertainty and health prognostics*. Springer, 2019.

[11] H. Wu and X. Du, "System Reliability Analysis With Second-Order Saddlepoint Approximation," *ASCE-ASME Journal of Risk and Uncertainty in Engineering Systems, Part B: Mechanical Engineering*, vol. 6, no. 4, p. 041001, 2020.

[12] H. Li, J. Yin, and X. Du, "Uncertainty Quantification of Physics-Based Label-Free Deep Learning and Probabilistic Prediction of Extreme Events," in *International Design Engineering Technical Conferences and Computers and Information in Engineering Conference*, American Society of Mechanical Engineers, 2022, p. V03BT03A001.

[13] H. Li and X. Du, "Recovering Missing Component Dependence for System Reliability Prediction via Synergy Between Physics and Data," *Journal of Mechanical Design*, vol. 144, no. 4, 2022.

[14] H. Yu *et al.*, "A new noninvasive and patient-specific hemodynamic index for the severity of renal stenosis and outcome of interventional treatment," *Int J Numer Method Biomed Eng*, vol. 38, no. 7, p. e3611, 2022.

[15] M. D. Murphy and R. Paasch, "Reliability centered prediction technique for diagnostic modeling and improvement," *Res Eng Des*, vol. 9, pp. 35–45, 1997.

[16] M. Fitzpatrick and R. Paasch, "Analytical method for the prediction of reliability and maintainability based life-cycle labor costs," 1999.

[17] R. Jiang, "A novel MTTF estimator and associated parameter estimation method on heavily censoring data," *Qual Reliab Eng Int*, vol. 37, no. 5, pp. 1706–1717, 2021.

[18] F. X. Che and J. H. L. Pang, "Vibration reliability test and finite element analysis for flip chip solder joints," *Microelectronics reliability*, vol. 49, no. 7, pp. 754–760, 2009.

[19] M. Li and Z. Wang, "Deep reliability learning with latent adaptation for design optimization under uncertainty," *Comput Methods Appl Mech Eng*, vol. 397, p. 115130, 2022, doi: https://doi.org/10.1016/j.cma.2022.115130.

[20] Z. Hu and X. Du, "Integration of statistics-and physics-based methods—a feasibility study on accurate system reliability prediction," *Journal of Mechanical Design*, vol. 140, no. 7, 2018.

[21] L. Chengying, W. Hao, W. Liping, and Z. Zhi, "Tool wear state recognition based on LS-SVM with the PSO algorithm," *Journal of Tsinghua University (Science and Technology)*, vol. 57, no. 9, pp. 975–979, 2017.

[22] Y. Xu, S. Kohtz, J. Boakye, P. Gardoni, and P. Wang, "Physics-informed machine learning for reliability and systems safety applications: State of the art and challenges," *Reliab Eng Syst Saf*, vol. 230, p. 108900, 2023, doi: https://doi.org/10.1016/j.ress.2022.108900.

[23] Z. Zheng *et al.*, "A Gaussian Process-Based Crack Pattern Modeling Approach for Battery Anode Materials Design," *Journal of Electrochemical Energy Conversion and Storage*, vol. 18, no. 1, May 2020, doi: 10.1115/1.4046938.

[24] Y. Xu, A. Renteria, and P. Wang, "Adaptive surrogate models with partially observed information," *Reliab Eng Syst Saf*, vol. 225, p. 108566, 2022, doi: https://doi.org/10.1016/j.ress.2022.108566.

[25] H. Wu and X. Du, "Envelope Method for Time-and Space-Dependent Reliability Prediction," *ASCE-ASME Journal of Risk and Uncertainty in Engineering Systems, Part B: Mechanical Engineering*, vol. 8, no. 4, p. 041201, 2022.

[26] H. Wu, Z. Zhu, and X. Du, "System reliability analysis with autocorrelated kriging predictions," *Journal of Mechanical Design*, vol. 142, no. 10, 2020.

[27] H. Wu, Z. Hu, and X. Du, "Time-dependent system reliability analysis with second-order reliability method," *Journal of Mechanical Design*, vol. 143, no. 3, 2021.

[28] Z. Z. Foumani, M. Shishehbor, A. Yousefpour, and R. Bostanabad, "Multi-fidelity cost-aware Bayesian optimization," *Comput Methods Appl Mech Eng*, vol. 407, p. 115937, 2023.

[29] J. Yin and X. Du, "High-dimensional reliability method accounting for important and unimportant input variables,"

Copyright © 2023 by ASME

Journal of Mechanical Design, vol. 144, no. 4, 2022.

[30] J. Yin and X. Du, "Active learning with generalized sliced inverse regression for high-dimensional reliability analysis," *Structural Safety*, vol. 94, p. 102151, 2022.

[31] J. T. Eweis-Labolle, N. Oune, and R. Bostanabad, "Data Fusion With Latent Map Gaussian Processes," *Journal of Mechanical Design*, vol. 144, no. 9, p. 091703, 2022.

[32] H. Wu and X. Du, "Time-and Space-Dependent Reliability-Based Design with Envelope Method," *Journal of Mechanical Design*, pp. 1–34, 2023.

[33] H. Wu and X. Du, "Envelope Method for Time-and Space-Dependent Reliability-Based Design," in *International Design Engineering Technical Conferences and Computers and Information in Engineering Conference*, American Society of Mechanical Engineers, 2022, p. V03BT03A002.

[34] J. C. Helton and F. J. Davis, "Latin hypercube sampling and the propagation of uncertainty in analyses of complex systems," *Reliab Eng Syst Saf*, vol. 81, no. 1, pp. 23–69, 2003.

[35] T.-T. Wong, W.-S. Luk, and P.-A. Heng, "Sampling with Hammersley and Halton points," *Journal of graphics tools*, vol. 2, no. 2, pp. 9–24, 1997.

[36] X. Wei, D. Han, and X. Du, "Physics-Based Gaussian Process Method for Predicting Average Product Lifetime in Design Stage," *J Comput Inf Sci Eng*, vol. 21, no. 4, 2021.

[37] B. Echard, N. Gayton, and M. Lemaire, "AK-MCS: an active learning reliability method combining Kriging and Monte Carlo simulation," *Structural Safety*, vol. 33, no. 2, pp. 145–154, 2011.

[38] C. Jiang, X. P. Wei, Z. L. Huang, and J. Liu, "An outcrossing rate model and its efficient calculation for time-dependent system reliability analysis," *Journal of Mechanical Design*, vol. 139, no. 4, p. 041402, 2017.

Copyright © 2023 by ASME

AUTHOR INDEX

Ahmed, Faez 85, 121, 137, 177, 311, 326
Allen, Janet K. 289, 354
Allison, James T. 15
Alonayni, Ghazi 455
Arechiga, Nikos 326
Azarm, Shapour 532
Baby, Mathew 354
Bagazinski, Noah J. 311
Bansal, Parth 421
Bayat, Saeid 15
Bayrak, Alparslan Emrah 29
Behzadi, Mohammad Mahdi 268
Broussard, Jacob 354
Butler, Jared 495
Cagan, Jonathan 57
Campbell, Matthew I. 455
Chatterjee, Arko 532
Chen, Hongrui 166
Chen, Qiuyi 71
Chen, Wei 238, 507
Chen, Yu-Hsuan 57
Cho, Changik 371
Chung, In-Bum 411
Cui, Yaxin 507
Dietrich, Madison 381
Dolar, Tuba 238
Du, Xianping 38
Feng, Cong 391
Ferguson, Scott 519
Flynn, Nathan 259
Fuge, Mark D. 71
Fuge, Mark 276
Gavino, Phillip A. O. 507
Giannone, Giorgio 137
Groth, Katrina M. 532
Guptan, Akshay 354
Habibi, Milad 276
Haji, Maha 381
Han, Kaixin 198
Haran, Kiruba Sivasubramaniam 559
Hazra, Indranil 532
He, Miao 371
Herber, Daniel R. 108
Hu, Chao 551
Ilies, Horea T. 268
Ji, Juntao 198
Jiang, Mingfei 289
Jin, Yan 153

Joglekar, Aditya 166
Jones, Matthew 121
Kaddoura, Mustafa F. 431
Kaminski, Gregory 401
Kang, Namwoo 98, 190
Kara, Levant Burak 57
Kara, Levent Burak 166
Kelley, Seth 371
Khoda, Bashir 488
Kim, Harrison 218, 250
Kim, Jihoon 190
Kodali, Suhas 121
Kohtz, Sara 559
Lalwani, Anand Vikas 559
Lee, Doksoo 238
Lee, Sumin 190
Li, Chuanhao 289
Li, Honglin 391
Li, Tingkai 551
Li, Yumeng 421, 544
Liang, Jinbin 38
Lin, Daphne 464
Lin, Kangcheng 250
Liu, Alan 381
Liu, Zheng 544, 567
Majowicz, Andrew 302
Malak Jr., Richard J. 1
McCabe, Rebecca 381
Meisel, Nicholas 495
Ming, Zhenjun 289
Mistree, Farrokh 289, 354
Murphy, Patrick N. 488
Nandola, Naresh N. 371
Navidi, Sina 551
Nellippallil, Anand Balu 354
Nobari, Amin Heyrani 121
Obaideh, Yazan Abu 85
Odonkor, Philip 302, 401
Panchal, Jitesh H. 339
Pare, Philip E. 339
Park, Jangseop 98
Park, Seyoung 218
Permenter, Frank 326
Picard, Cyril 177
Qian, Guowei 38
Qian, Xiaoping 208, 259
Qiu, Yunjian 153
Rahn, Christopher D. 371

Regenwetter, Lyle 85
Renteria, Anabel 559
Retzlaff, Lisa..519
Rey, Justin...121
Rodriguez, Nallely Guillen 431
Sadat, Eddieb 29
Saremi, Mostaan Lotfalian 29
Schiffmann, Jurg 177
Seepersad, Carolyn 464
Senesky, Debbie...559
Sha, Zhenghui 507
Shea, Kristina...442
Shi, Shuhui...198
Sirico Jr., Anthony 108
Soh, Gim Song...47
Song, Binyang 326
Song, Jinyu 198
Southgate, Joseph 532
Sridhara, Saketh...228
Sun, Lingyun 198
Suresh, Krishnan...228
Surovi, Nowrin Akter..47
Tajima, Mikihiro 481
Thelen, Adam..551
Thomas, Evelyn 495
Tsai, Ying-Kuan 1
Tu, Ziwei..198
Tylka, Joseph G. 371
Walter, Ian..339
Wang, Jun 276
Wang, Pingfeng............................... 411, 544, 559, 567
Wang, Tianye 208
Weiner, Matthew J. 532
Wirth, Marc..442
Wright, Natasha C..431
Wu, Hao...544, 567
Xiao, Yinshuang..507
Xie, Peng 38
Xu, Hongyi 38
Xu, Yanwen...544, 567
Yamada, Takayuki 481
Yi, Jianbo 38
You, Weitao...198
Yuan, Chenyang 326
Zhang, Jie 391
Zhang, Xiaolong 559
Zhao, Junhan..559

The American Society of Mechanical Engineers (ASME)
Two Park Avenue
New York, NY
10016-5990

ISBN 978-0-7918-8730-1